BASIC BUSINESS STATISTICS

Sixth Edition

BASIC BUSINESS STATISTICS

CONCEPTS AND APPLICATIONS

Mark L. Berenson

David M. Levine

Department of Statistics and Computer Information Systems
Baruch College, City University of New York

Prentice Hall, Upper Saddle River, New Jersey

LIBRARY OF CONGRESS CATALOGING-IN-PUBLICATION DATA

Berenson, Mark L.
Basic business statistics : concepts and applications
Mark L. Berenson, David M. Levine.—6th ed.
p. cm.
Includes bibliographical references and index.
ISBN 0-13-303009-1
1. Commercial statistics. 2. Statistics. I. Levine, David M.
HF1017.B38 1996
519.5—dc 20 94-12551
CIP

Acquisitions Editor: Tom Tucker
Production Editor: Katherine Evancie
Managing Editor: Joyce Turner
Cover Designer: Sue Behnke
Interior Design: Ed Smith
Design Director: Patricia H. Wosczyk
Buyer: Marie McNamara
Assistant Editor: Diane Peirano
Production Assistant: Renée Pelletier
Marketing Manager: Susan McLaughlin
Cover art: Marjory Dressler

Printed in the United States of America

10 9 8 7 6 5 4 3

ISBN 0-13-303009-1

Prentice-Hall International (UK) Limited, *London*
Prentice-Hall of Australia Pty. Limited, *Sydney*
Prentice-Hall Canada Inc., *Toronto*
Prentice-Hall Hispanoamericana, S.A., *Mexico*
Prentice-Hall of India Private Limited, *New Delhi*
Prentice-Hall of Japan, Inc., *Tokyo*
Simon & Schuster Asia Pte. Ltd., *Singapore*
Editora Prentice-Hall do Brasil, Ltda., *Rio de Janeiro*

To our wives,
Rhoda B and Marilyn L
and to our children,
Kathy B, Lori B, and Sharyn L

Brief Contents

Detailed Contents

14 ANOVA and Other *c*-Sample Tests with Numerical Data *525*

15 Hypothesis Testing with Categorical Data *605*

Preface

When planning or revising a textbook, the authors must decide how the text will differ from those already available and what contribution it will make to the field of study. Initially, when we began writing the first edition of *Basic Business Statistics* in 1976, we thought that what was missing from other introductory business statistics texts was a common theme to hold together the various topics and provide a sense of realism for the student. Thus we conceived of a practical, data-analytic approach to the teaching of business statistics through the development and use of a survey (and database) that integrates the various topics, permitting a cohesive study of the subject of business statistics.

In proposing changes for the sixth edition of *Basic Business Statistics,* our major objective is continuous quality improvement of prior editions

- by incorporating trends in pedagogy (for example, active and collaborative learning)
- by demonstrating the increased use of statistical software on personal computers
- by presenting modern statistical developments
- by including trends in business school curriculum (for example, ethics, globalization, and quality)

so that the student will appreciate the value of the subject of statistics in the business school curriculum and find more joy in learning.

As we perceive it, the fundamental strengths of our text are its innovative, data-analytic survey-research approach and its internal pedagogical features.

Main Feature: A Data-Analytic Survey Research Approach

A *scenario* is created in which an employee-benefits consulting firm (the B & L Corporation) is hired to conduct a survey of the full-time employees of an automobile parts manufacturer (Kalosha Industries) in order to develop an employee profile that measures job satisfaction, evaluates longevity and career progress, and assesses attitudes and beliefs. The results of this Employee Satisfaction Survey are intended to assist in the development of an employee-benefits package that would please the workers, strengthen their relationship with management, and demonstrate that Kalosha Industries is taking a proactive role in establishing a TQM environment. The 400 sampled responses to the Employee Satisfaction Survey (that is, the database) obtained in Chapter 2, "Data Collection," are used for examples and student projects throughout the text and serve as a means for integrating such topics as descriptive statistics, probability, statistical inference, and regression analysis.

The use of an actual survey, examined from beginning to end, serves as an integrated case study and provides the students with a cohesive approach to learning the subject of business statistics. In addition, it enables the students to realistically understand the process of data-analytic survey research and aids them in conducting such research in other courses and in occupational settings.

The Employee Satisfaction Survey is developed in Chapter 2 and used as text examples and/or as student project assignments in Chapters 3–6, 8, 10, 12–15, 17, and 18. Section material and student projects dealing with the Survey are highlighted in color and with an icon.

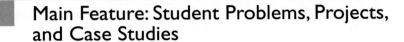

Main Feature: Student Problems, Projects, and Case Studies

Learning results from doing. This text provides the student with the opportunity to select and solve from among the *1,200* problems presented at the end of sections as well as at the end of chapters. Most of these problems apply to realistic situations (using real data whenever possible) in various fields including accounting, economics, finance, health care administration, information systems, management, marketing, and public administration.

- The ***End-of-Section Problems*** give the students the opportunity to reinforce what they have just learned.
- The ***Chapter Review Problems*** included at the end of each chapter are based on the concepts and methods learned throughout the chapter.
- Both the ***End-of-Section*** and ***Chapter Review Problems*** either "stand alone" or refer to other problems within the particular chapter.
- ***Interchapter Problems*** are those that refer to problems in earlier chapters.
- The ***Answers to Selected Problems*** (indicated by the ● symbol) appear at the end of the text.
- A series of ***Collaborative Learning Mini-Case Projects*** are presented throughout the text.
- A set of ***132 Survey/Database Projects*** pertaining to the Employee Satisfaction Survey are presented in the various chapters.
- Detailed ***Case Studies*** are included at the end of each of ten chapters.

These student problems, projects and case studies provide many benefits. The teacher has the opportunity to assign individual "stand alone" problems as well as continuation problems to point out topic connections. For more in-depth assignments, the detailed case studies may be used. Furthermore, the teacher has the opportunity to make assignments from the Survey/Database Projects which serve as an integrated "case study" throughout the text.

Main Feature: Collaborative Learning Mini-Case Projects

Two major pedagagical approaches have begun to filter into the college classroom over the past decade—active and collaborative learning. Interestingly, these two pedagogical approaches are in agreement with the principles expressed in the "management by process" philosophy developed by W. Edwards Deming whose approach to continuous quality improvement is among the major industrial advances of the decade. The adaptation of the TQM philosophy throughout an organization results in a cultural transformation that includes facilitative management, workforce empowerment, and problem solving through the use of cross-functional teams. In a similar vein, if we as teachers employ the principles of active and collaborative learning, we can reduce our lecturing and empower the students to learn more on their own and through teamwork. We must develop the students' critical thinking skills so that we can manage our classrooms more effectively by acting as coaches and facilitators. More importantly, with such developments, future graduates will not only be better prepared to take their place as citizens of their communities but they also will be more prepared to experience a lifetime of self-learning in a dynamic world.

Collaboration enhances learning and collaboration builds teamwork skills needed for participation in business and society. Throughout this text, a series of Collaborative Learning Mini-Case Projects, pertaining to four large data sets (dealing with colleges and universities, cereals, fragrances, and cameras) found in Appendix D, are presented at the end of most chapters. In addition, other Collaborative Learning Projects designed for the classroom are included where appropriate. Thus, a teacher interested in active and collaborative learning now has the opportunity to apply these pedagogical approaches in and out of the classroom by creating student teams and by selecting from the Collaborative Learning Mini-Case Projects and/or the other Colloborative Learning Projects given in the text.

Main Feature: Action (ACTION) and Light Bulb () Problems

Statistics is a living, breathing subject. It is not mere numbers crunching! Emphasis must be placed on understanding and interpretation, and it is essential that students be able to express what they have learned. *Action* ACTION problems enhance literacy by asking the student to write letters, memos, and reports, and to prepare talks. *Light Bulb* problems are particularly thought provoking or have no "exact" answer. Together, *Action* and *light bulb* problems allow the students to think and better enable them to understand the utility of statistical analysis as an aid to the solution of real problems in an organizational setting.

Main Feature: Thought-Provoking Summary Sections Dealing with (Exploratory and Confirmatory) Data Analysis

We feel that observation is the key to understanding. Observation, then, is paramount to developing critical thinking and data-analytical skills. The Berenson–Levine text emphasizes the four components of good data analysis—plotting, observing, computing, and describing—and stresses the importance of meeting assumptions in employing statistical inference techniques. This offers much benefit to the student. Through careful observation of data, the student enhances critical thinking skills and data-analytic skills. Moreover, through careful evaluation of assumptions, the student is likely to select the appropriate statistical inference technique for a given situation.

Main Feature: Thought-Provoking Summary Sections Dealing with Ethical Issues

Ethical issues in business has become a subject of much importance over the years, and AACSB accreditation policy now specifically addresses this within the context of curriculum development. Thus, ethical issues in data analysis are described in all relevant chapters of this text. Through the development of critical thinking skills, the student will be in a good position to understand and appreciate the ramifications of the ethical issues involved in data analysis.

Main Feature: Emphasis on Statistical Software Packages

A major feature of our text is a discussion on the use of such statistical software packages as MINITAB, SAS, SPSS, and STATISTIX. Not only is output from these packages illustrated throughout the text, particularly when describing results from the Employee Satisfaction Survey, but the use of the computer as a tool for assisting in the decision-making process is interwoven in the various chapters. In addition, a data disk in ASCII format that contains almost 200 files pertaining to various problems, projects, and case studies given in the text can be provided to facilitate classroom and homework assignments. The files are identified when they first appear with a disk icon and the file name in the margin. Documentation for these data files is presented in Appendix F.

Demonstrating a variety of statistical software packages is beneficial. Students learn how to interpret output from a variety of packages that they may be using. Moreover, by providing an easily referenced data disk, faculty have the opportunity to assign large-scale, real data problems and projects using statistical software without requiring tedious data entry.

Main Feature: The Deming Philosophy for Quality and Productivity

Over the years, our writing and teaching endeavors have been stimulated by an exchange of ideas at the annual conferences on *Making Statistics More Effective in Schools of Business*. These conferences have dealt with many pedagogical issues, including the importance of real data applications, the use of statistical software, active learning, teamwork, quantitative literacy and statistical thinking. However, the distinguishing thrust of these conferences has been to expound on the impact and importance of the subject of statistics in an organization practicing total quality management (TQM).

The importance of organizational focus on quality is amply demonstrated in Chapter 16 of this text through the presentation of managerial planning (that is, process flow diagrams and Fishbone diagrams) and statistical tools (that is, process control charts) which pinpoint the utility of statistical analysis in an organization practicing total quality management (TQM). Moreover, this text provides detailed coverage of the conceptual basis of total quality management (TQM) with a discussion of the fourteen points of the "management by process" philosophy of W. Edwards Deming. Additional topics on this subject that are presented include operational definitions, statistical thinking, enumerative versus analytic studies, Pareto diagrams, digidot plots, and the parable of the red beads—an experiment intended to demonstrate the concepts of common versus special cause variation.

Main Feature: Modern Statistical Methods

Another important feature of this edition is the inclusion of methodology that, over the past few years, has gained wide-spread usage. As examples, exploratory data analysis (EDA) techniques are presented (Chapters 3 and 4); dot charts, Pareto diagrams, and supertables are discussed (Chapter 5); normal probability plots are described for evaluating the assumption of normality (Chapter 8); bootstrapping estimation methods and prediction intervals for an individual value are developed (Chapter 10); the subject of meta-analysis is introduced (Chapter 11); a p-value approach to hypothesis testing is used (Chapters 11–15); logistic regression is introduced (Chapter 18); residual and influence analysis and model-building in regression are covered (Chapters 17 and 18); and various business forecasting methods are considered (Chapter 19).

A major benefit of the broad coverage of topics in the Berenson-Levine text is flexibility in course development. Our text incorporates relevant, up-to-date methodology in sufficient depth and breadth to be used for either one-semester or two-semester introductory courses at the undergraduate or graduate level, and it also serves as an ample reference for fundamental statistical techniques.

Main Feature: Pedagogical Aids

Our text contains numerous pedagogical aids aimed at enhancing the learning of business statistics.

- **Writing Style** Our basic philosophy is to write for the student, not for the professor. To reduce anxiety, our writing style is as conversational as possible.

- **Real Applications** To provide a sense of realism for the subject matter, we use actual data throughout the text in a variety of examples, problems, projects, and case studies.

- **Reading and Interpreting Statistical Tables** Each of the statistical tables given in Appendix E is examined in depth when it is first presented. We provide detailed explanations and illustrations in order to aid the student in learning how to use the tables. In addition, we present the standard normal distribution on the inside front cover to facilitate its use.

- **Chapter Introductions and Summaries** In the introductory section of each chapter, we provide a list emphasizing what the student is expected to learn. The final section of each chapter reviews what was covered and presents a series of key conceptual questions pertaining to what was learned.

- **Chapter Ending Summary Chart** We end each chapter with a summary chart highlighting significant coverage of material.

- **Key Terms** We provide a listing of key terms with page references at the end of each chapter.

- **Data Diskette and List of Data Files** A data diskette containing almost *200* files is available. The list of data files appears in Appendix F.

It is our hope and anticipation that the pedagogical aids along with the features and unique approaches taken in this textbook will make the study of basic business statistics more meaningful, rewarding, and comprehensible for all readers.

Acknowledgements

We are extremely grateful to the many organizations and companies that generously allowed us to use their actual data for developing problems and examples throughout our text. In particular, we would like to thank the *National Opinion Research Center (NORC)* for providing to the public domain its *General Social Surveys, 1972–1991: Cumulative Codebook*. This source provided the data and enabled the development of the scenario for the Kalosha Industries Employment Satisfaction Survey, which we use as an integrated case study throughout the text.

Moreover, we would like to cite *The New York Times,* CBS Inc. (publisher of *Road & Track*), Consumers Union (publisher of *Consumer Reports*), Moody's Investors Service (publisher of *Moody's Handbook of Common Stocks*), American Hospital Publishing, Inc. (publisher of *Hospitals*), Los Angeles Times Syndicate International (publisher of *New York Newsday*), M.I.T. Center for Advanced Engineering Study, CEEPress Books, Goal/QPC, and Gale Research, Inc.

In addition, we would like to thank the *Biometrika* Trustees, American Cyanamid Company, the Rand Corporation, the Chemical Rubber Company, the Institute of Mathematical Statistics, and the American Society for Testing and Materials for their kind permission to publish various tables in Appendix E and the American Statistical Association for its permission to publish diagrams from *The American Statistician.*

Furthermore, we are indebted to Professor Kristin McDonough, chief librarian at Baruch College, for providing material on library information technology, and we are particularly grateful to Professors George A. Johnson and Joanne Tokle, Idaho State University, and Ed Conn, Mountain States Potato Company, for their kind permission to incorporate parts of their work as our Case Study I, "The Mountain States Potato Company."

A Note of Thanks

We wish to express a note of thanks to some of our colleagues at Baruch College including Ann Brandwein, Stuart Baden, Pasquale DiPillo, Shulamith Gross, Alka Indurkhya, Theodore Joyce, Manus Rabinowitz, and Lawrence Tatum, as well as Terese Bruce, University of Miami; Mark Eakin, University of Texas at Arlington; Rick Edgeman, Colorado State University; Mark Ferris, Saint Louis University; Daniel Gordon, Salem State College; Jacqueline Hoell, Virginia Polytechnic Institute and State University; Donn Johnson, University of Northern Iowa; George Marcoulides, California State University at Fullerton; John McKenzie, Babson College; John Neufeld, University of North Carolina at Greensboro; Alan Olinsky, Bryant College; Barbara Price, Winthrop University; Patricia Ramsey, Fordham University; Ernest Scheuer, California State at Northridge; Michael Sklar, Emory University; and Robert Westerman, California State Polytechnic University at Pomona for their constructive comments during the revision of this textbook. Moreover, we wish to thank Bliss Simon of Baruch College and John Dumo of the Lehman College Educational Computer Center for their assistance in creating the Employee Satisfaction Survey Database as well as Michael Dannenbring, Deryck Fritz, Robert Moran, Hideki Sugiyama, Dessislava Todorova and Richard Whitehead for their assistance in data collection and file development.

A Special Thank You

We would like to close by expressing our thanks to Tom Tucker, Joyce Turner, Richard Wohl, Katherine Evancie, Joanne Jay, Susan McLaughlin, Marie McNamara, Patrice Fraccio, Kate Moore, Diane Peirano, and Renée Pelletier, of the editorial staff and production team at Prentice Hall and to Rachel J. Witty of Letter Perfect, Inc., for their continued encouragement. We would also like to thank Susan L. Reiland, our statistical reader, for her diligence in checking the accuracy of our work. We are indeed fortunate and privileged to work with such a group. Finally, we would like to thank our wives and children for their patience, understanding, love, and assistance in making this book a reality. It is to them that we dedicate this book.

MARK L. BERENSON
DAVID M. LEVINE

chapter 1

Introduction

CHAPTER OBJECTIVE

To present a broad overview of the subject of statistics and its applications, particularly in business.

1.1 What Is Modern Statistics?

A century ago H. G. Wells commented that "statistical thinking will one day be as necessary for efficient citizenship as the ability to read and write." Each day of our lives we are exposed to a wide assortment of numerical information pertaining to phenomena such as stock market activity, market research findings, opinion poll results, unemployment rates, forecasts for the future success of specific industries, and sports data. The subject of **modern statistics** encompasses the collection, presentation, and characterization of information to assist in both data analysis and the decision-making process.

In terms of the functional areas of business, statistics may be applied in

Accounting

- to select samples for auditing purposes
- to understand the cost drivers in cost accounting

Finance

- to track trends in financial measures over time
- to develop ways of forecasting values of these measures at future points of time

Management

- to describe characteristics of employees within an organization
- to improve the quality of the products manufactured or services provided by the organization

Marketing

- to estimate the proportion of customers who prefer one product over another and why they do
- to draw conclusions about what advertising strategy might be most useful in increasing sales of a product

We will begin in this introductory chapter with some important definitions. This will be followed by a discussion of the historical development of the field of statistics and the distinction between different types of statistical studies. We conclude with the role of computer software in statistical analysis.

1.2 The Growth and Development of Modern Statistics

Historically, the growth and development of modern statistics can be traced to two separate phenomena—the needs of government to collect data on its citizenry (see References 6, 7, 10, 14, and 15) and the development of the mathematics of probability theory.

Data have been collected throughout recorded history. During the Egyptian, Greek, and Roman civilizations, data were obtained primarily for the purposes of taxation and military conscription. In the Middle Ages, church institutions often kept records concerning births, deaths, and marriages. In America, various records were kept during colonial times (see Reference 15), and beginning in 1790 the federal Constitution required the taking of a census every ten years. Today these data are used for many purposes, including congressional apportionment and the allocation of federal funds.

1.2.1 Descriptive Statistics

These and other needs for data on a nationwide basis were closely intertwined with the development of descriptive statistics.

> **Descriptive statistics** can be defined as those methods involving the collection, presentation, and characterization of a set of data in order to describe properly the various features of that set of data.

Although descriptive statistical methods are important for presenting and characterizing data (see Chapters 2 through 5), it has been the development of inferential statistical methods as an outgrowth of probability theory that has led to the wide application of statistics in all fields of research today.

1.2.2 Inferential Statistics

The initial impetus for formulation of the mathematics of probability theory came from the investigation of games of chance during the Renaissance. The foundations of the subject of probability can be traced back to the middle of the seventeenth century in the correspondence between the mathematician Pascal and the gambler Chevalier de Mere (see References 9 and 10). These and other developments by such mathematicians as Bernoulli, DeMoivre, and Gauss were the forerunners of the subject of inferential statistics. However, it has only been since the turn of this century that statisticians such as Pearson, Fisher, Gosset, Neyman, Wald, and Tukey pioneered in the development of the methods of inferential statistics that are widely applied in so many fields today.

> **Inferential statistics** can be defined as those methods that make possible the estimation of a characteristic of a population or the making of a decision concerning a population based only on sample results.

To clarify this, a few more definitions are necessary.

> A **population** (or **universe**) is the totality of items or things under consideration.
>
> A **sample** is the portion of the population that is selected for analysis.
>
> A **parameter** is a summary measure that is computed to describe a characteristic of an entire population.
>
> A **statistic** is a summary measure that is computed to describe a characteristic from only a sample of the population.

To relate these definitions to an example, suppose that the president of your college wanted to conduct a survey to learn about student perceptions concerning the quality of life on campus. The population or universe in this instance would be all currently enrolled students, while the sample would consist only of those students who had been selected to participate in the survey. The goal of the survey would be to describe various attitudes or characteristics of the entire population (the parameters). This would be achieved by using the statistics obtained from the sample of students to estimate various attitudes or characteristics of interest in the population. Thus, one major aspect of inferential statistics is the process of using sample statistics to draw conclusions about the population parameters.

The need for inferential statistical methods derives from the need for sampling. As a population becomes large, it is usually too costly, too time consuming, and too cumbersome to obtain our information from the entire population.

Decisions pertaining to the population's characteristics have to be based on the information contained in a sample of that population. Probability theory provides the link by ascertaining the likelihood that the results from the sample reflect the results from the population.

These ideas can also be illustrated by referring to the example of a political poll. If the pollster wishes to estimate the percentage of the votes a candidate will receive in a particular election, he or she will not interview each of the thousands (or even millions) of voters that make up the population. Instead, a sample of voters will be selected. Based on the outcome from the sample, conclusions will be drawn concerning the entire population of voters. Appended to these conclusions will be a probability statement specifying the likelihood or confidence that the results from the sample reflect the voting behavior in the population.

1.3 Statistical Thinking and Modern Management

In the past decade the emergence of a global economy has led to an increasing focus on the quality of products manufactured and services delivered. In fact, more than that of any other individual, it has been the work of a statistician, W. Edwards Deming, that has led to this changed business environment. An integral part of the managerial approach that contains this increased focus on quality (often referred to as **Total Quality Management**) is the application of certain statistical methods and the use of **statistical thinking** on the part of managers throughout a company.

> **Statistical thinking** can be defined as thought processes that focus on ways to understand, manage, and reduce variation.

Statistical thinking includes the recognition that data are inherently variable (no two things or people will be exactly alike in all ways) and that the identification, measurement, control, and reduction of variation provide opportunities for quality improvement. Statistical methods can provide the vehicle for taking advantage of these opportunities. The role of statistical methods in the context of quality improvement can be better understood if we refer to a model of quality improvement as presented in Figure 1.1.

We may observe from Figure 1.1 that the triangle consists of three portions; at the top we have Management Philosophy, and at the two lower corners we have Statistical Methods and Behavioral Tools. Each of these three aspects is indispensable for long-term quality improvement of either the manufactured goods or the services provided by an organization. A management philosophy provides a constant foundation for quality improvement efforts. Among the approaches available are those advocated by W. Edwards Deming (see References 1–3 and Section 16.5) and Joseph Juran (see References 4 and 5).

In order to implement a quality improvement approach in an organization, one needs to use both behavioral tools and statistical methods. Each of these aids in understanding and improving processes. Among the useful behavioral tools are process flow and fishbone diagrams (see Section 16.4), brainstorming, nominal group decision making, and team building. (For further discussion, see Reference 11.) Among the most useful statistical methods for quality improvement are the numerous tables and charts and descriptive statistics discussed in Chapters 3–5 and the control charts developed in Chapter 16.

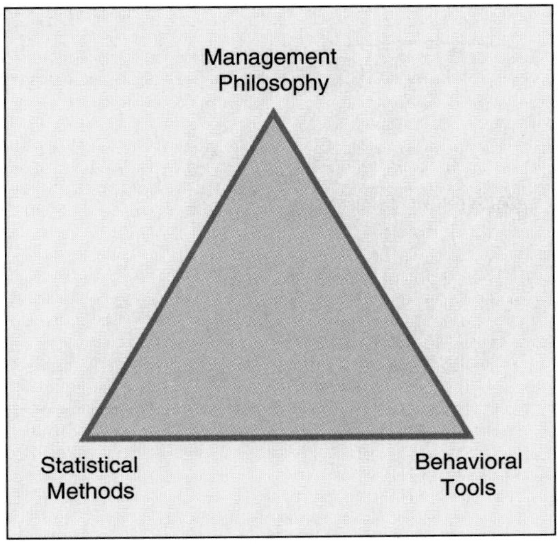

FIGURE 1.1
A model of the quality improvement process.

1.4 Enumerative Versus Analytical Studies

Our discussion of statistical inference in Section 1.2 and the role of statistical methods for quality improvement in Section 1.3 allows us to make an important distinction between two types of statistical studies that are undertaken: **enumerative** studies and **analytical** studies.

Enumerative studies involve decision making regarding a population and/or its characteristics.

The political poll is an example of an enumerative study, since its objectives are to provide estimates of population characteristics and to take some action on that population. The listing of all the units (such as the registered voters) that belong to the population is called the *frame* (see Section 2.7) and provides the basis for the selection of the sample. Thus the focus of the enumerative study is on the counting (or measuring) of outcomes obtained from the frame.

Analytical studies involve taking some action on a process to improve performance in the future.

The investigation of the outcomes of a manufacturing or service process taken over time is an example of an analytical study. The focus of an analytical study is on the prediction of future process behavior and on understanding and improving a process. In an analytical study there is no identifiable universe, as in an enumerative study, and therefore there is also no frame. Perhaps we can highlight the distinction between enumerative and analytical studies by referring to Figures 1.2 and 1.3 on page 6.

In the enumerative study, the bowl represents the population. The questions of interest revolve around the issue of "What's in the bowl?" An example of this might be wanting to know how many of the balls in the bowl are red or what proportion of the balls is red.

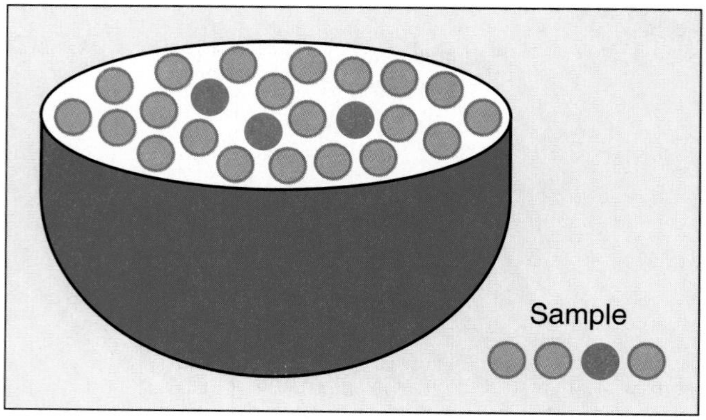

FIGURE 1.2
An enumerative study.

In the analytical study, there are several stages that make up a process. These stages typically include inputs that might involve some combination of people, equipment, material, and information; outputs that are in the form of a product manufactured or a service provided; and the intermediate transformation step that turns the inputs into the desired outputs. A key question of interest revolves around how any data that might be collected as part of the process (often over a period of time) can be used to improve the process in the future. This is indicated in Figure 1.3 by the presence of a feedback loop.

The distinction between enumerative and analytical studies is an important one, since the methods that have been developed primarily for enumerative studies may be misleading or incorrect for analytical studies (see References 1–3). In this textbook we shall develop methods that are appropriate for these two different types of studies. Some of the methods are appropriate for either type of study. Other methods are appropriate primarily for enumerative studies or primarily for analytical studies.

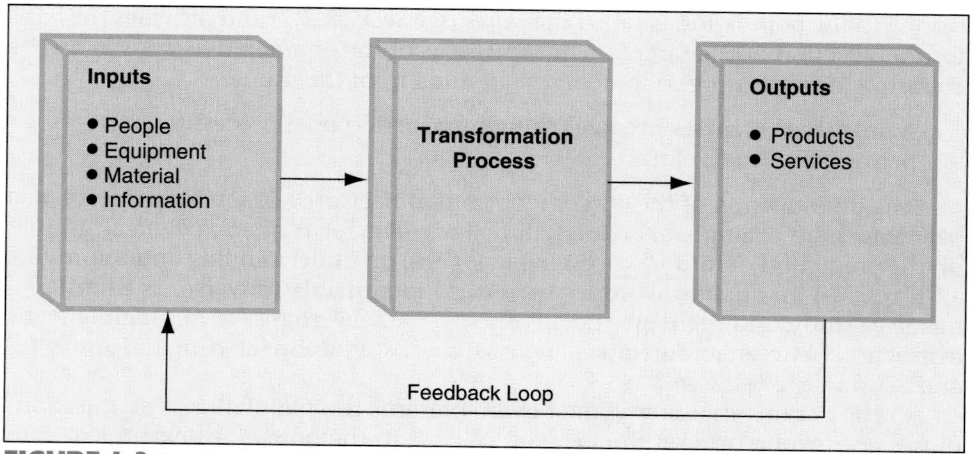

FIGURE 1.3 **An analytical study.**

1.5 The Role of Computer Packages in Statistics

During the past twenty years, the field of statistics has been dramatically changed by the development of computer software specially written for statistical analysis. During the 1980s statistical software experienced a vast technological revolution. Besides the usual improvements manifested in periodic updates, the availability of personal computers led to the development of new packages that used a menu-driven interface. In addition, personal computer versions of existing packages such as SAS, SPSS, and MINITAB (see References 8, 12, and 13) quickly became available. Furthermore, the increasing use of popular spreadsheet packages such as Lotus 1-2-3 and Excel led to the incorporation of statistical features in these packages.

The latter portion of the 1980s and the early 1990s represented a continuing period of technological advances. Whereas the first packages developed for the personal computer were available only for IBM or compatible machines, more recently, packages have been adapted to the environment of the Macintosh personal computer. In addition, rapid advances in computer hardware meant that larger amounts of computer memory were available at lower cost. This enabled package developers to include additional, more sophisticated statistical procedures in each subsequent version of their packages.

Thus, it is easy to understand why the use of these software packages is commonplace throughout the business, academic, and research communities. Therefore, in this text we will take the position that when performing statistical analysis, one is almost certainly going to be accessing some statistical software package or packages (or a spreadsheet package). With this in mind, our focus is geared to the interpretation of the output of several of these packages (primarily MINITAB, SAS, and SPSS), with a more limited emphasis on the steps involved in the computations. This is consistent with the theme of the text, which emphasizes the proper use of statistical methods rather than the mathematical theory underlying the methods.

Although statistical software has made even the most sophisticated analyses feasible, problems arise when statistically unsophisticated users who do not understand the assumptions behind procedures or the limitations of the results obtained are misled by computer-generated statistical output. For pedagogical reasons, we believe it is important that the applications of the methods covered in the text be illustrated through the use of worked-out examples.

1.6 Summary and Overview

As seen in the summary chart for this chapter (page 8), we have presented an introduction to the field of statistics, provided various definitions of terms that will be used throughout the text, and discussed the role of computer software. Chapter 2 will focus on data collection and the selection of samples. These two introductory chapters provide the background for the essential area of descriptive statistics to be discussed in Chapters 3–5.

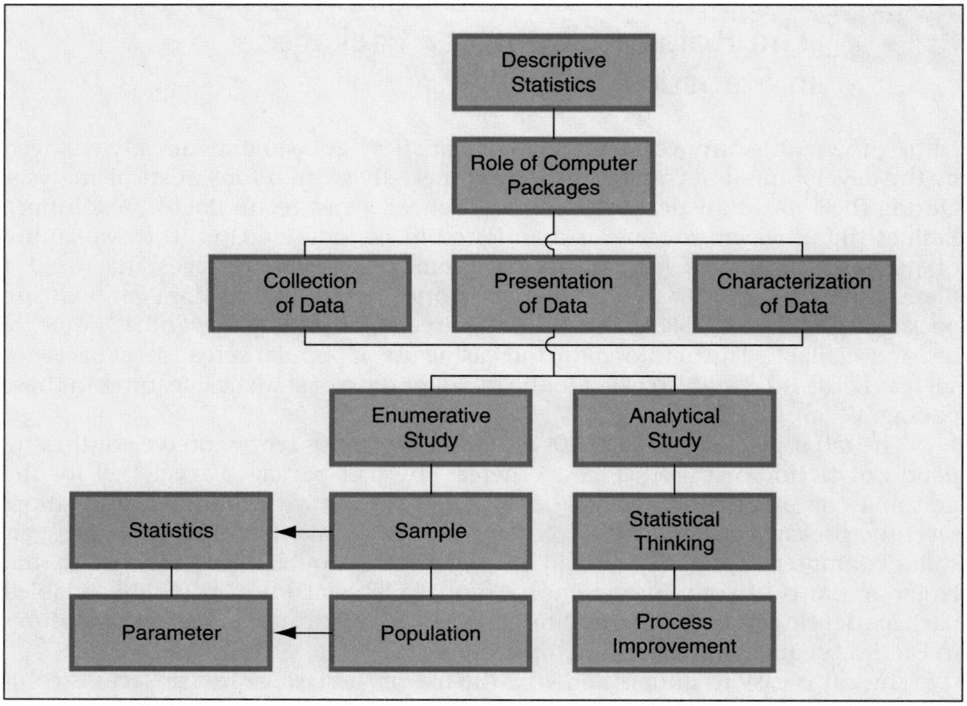

Chapter 1 summary chart.

Getting It All Together

KEY TERMS

analytical studies *5*
descriptive statistics *3*
enumerative studies *5*
inferential statistics *3*
modern statistics *2*
parameter *3*
population *3*
sample *3*
statistic *3*
statistical thinking *4*
Total Quality Management *4*
universe *3*

Chapter Review Problems

To answer the questions below you may wish to go to your library and use the following sources of reference:

Indexes

Business Periodical Index
New York Times Index
Wall Street Journal Index

Business Magazines

Business Week

Forbes

Fortune

General Magazines

Newsweek

Time

U.S. News & World Report

Newspapers

New York Times

U.S.A. Today

Wall Street Journal

Local newspapers

General Information

Statistical Abstract of the United States

For Problems 1.1 to 1.7, specify the general problem to be solved, the specific inference to be made, what the population is, and (if you are describing the results of an actual published study) what the weaknesses of the study might be. Where appropriate, tell what parameters are of primary interest and what statistics are used to arrive at a conclusion.

1.1 Describe an application of statistics to economics or finance.

1.2 Describe an application of statistics to sports.

1.3 Describe an application of statistics to political science or public administration.

1.4 Describe an application of statistics to organizational behavior or operations management.

1.5 Describe an application of statistics to advertising or marketing research.

1.6 Describe an application of statistics to medical research or health care administration.

1.7 Describe an application of statistics to accounting.

1.8 What is the difference between descriptive and inferential statistics? Under what circumstances might each of these areas of statistics be more useful?

1.9 What is the difference between a parameter and a statistic?

● 1.10 For each of the following, indicate whether the study is enumerative or analytical. Discuss.

(a) A college decides to count the total number of students registering for classes that begin before 9 a.m.

(b) A college wishes to determine whether the total number of students registering for classes that begin before 9 a.m. has increased or decreased over different semesters.

(c) A college wishes to determine the reasons for a decrease in the number of applications for undergraduate admission.

Note: Bullet ● indicates that the solutions to these problems are in the *Answers to Selected Problems* section at the back of the book.

1.11 For each of the following, indicate whether the study is enumerative or analytical. Discuss.
(a) A magazine wishes to determine the proportion of its readership that is above fifty years of age.
(b) A magazine would like to determine whether the availability of a discounted price on a five-year subscription renewal will affect the number of subscriptions.
(c) A magazine would like to determine the income level of its readership.
(d) A magazine would like to determine how to reduce the number of errors made in billing subscribers.

References

1. Deming, W. E.,"On Probability as a Basis for Action," *American Statistician*, Vol. 29, 1975, pp. 146–152.
2. Deming, W. E., *Out of the Crisis* (Cambridge, MA: Massachusetts Institute of Technology Center for Advanced Engineering Study, 1986).
3. Deming, W. E., *The New Economics for Industry, Government, Education* (Cambridge, MA: Massachusetts Institute of Technology Center for Advanced Engineering Study, 1993).
4. Juran, J. M., *Juran on Leadership for Quality* (New York: The Free Press, 1989).
5. Juran, J. M., and F. M. Gryna, *Quality Planning and Analysis*, 2d ed. (New York: McGraw-Hill, 1980).
6. Kendall, M. G., and R. L. Plackett, eds., *Studies in the History of Statistics and Probability*, Vol. II (London: Charles W. Griffin, 1977).
7. Kirk, R. E., ed., *Statistical Issues: A Reader for the Behavioral Sciences* (Monterey, CA: Brooks/Cole, 1972).
8. Norusis, M., *SPSS Guide to Data Analysis for SPSS-X: With Additional Instructions for SPSS/PC+* (Chicago, IL: SPSS Inc., 1986).
9. Pearson, E. S., ed., *The History of Statistics in the Seventeenth and Eighteenth Centuries* (New York: Macmillan, 1978).
10. Pearson, E. S., and M. G. Kendall, eds., *Studies in the History of Statistics and Probability* (Darien, CT: Hafner, 1970).
11. Robbins, S. P., *Management,* 4th ed. (Englewood Cliffs, NJ: Prentice-Hall, 1994).
12. Ryan, B. F., and B. L. Joiner, *Minitab Student Handbook,* 3d ed. (North Scituate, MA: Duxbury Press, 1994).
13. *SAS Language and Procedures Usage,* Version 6 (Raleigh, NC: SAS Institute, 1988).
14. Walker, H. M., *Studies in the History of the Statistical Method* (Baltimore, MD: Williams & Wilkins, 1929).
15. Wattenberg, B. E., ed., *Statistical History of the United States: From Colonial Times to the Present* (New York: Basic Books, 1976).

chapter 2

Data Collection

CHAPTER OBJECTIVES

To describe the importance of obtaining good data and to demonstrate how data are collected and prepared for tabular and chart presentation, descriptive summarization, analysis, and interpretation.

2.1 Introduction: The Need for Data

Why do we need to collect data? Four main reasons could be given. Data are needed to:

1. Provide the necessary input to a research study
2. Measure performance in an ongoing service or production process
3. Assist in formulating alternative courses of action in a decision-making process
4. Satisfy our curiosity

As examples:

- The manager wants to monitor a process on a regular basis to find out whether the quality of service being provided or products being manufactured are conforming to company standards.
- The market researcher looks for the characteristics that distinguish a product from its competitors.
- The potential investor wants to determine which firms within which industries are likely to have accelerated growth in a period of economic recovery.
- The pharmaceutical manufacturer needs to determine whether a new drug is more effective than those currently in use.

To the statistician or researcher, the needed information comes from the data. What exactly do we mean by data?

> **Data** can be thought of as the numerical information needed to help us make a more informed decision in a particular situation.

For a statistical analysis to be useful in the decision-making process, the input data must be appropriate. Thus proper data collection is extremely important. If the data are flawed by biases, ambiguities, or other types of errors, even the fanciest and most sophisticated statistical methodologies would not likely be enough to compensate for such deficiencies.

Since the need for useful information is so important to the decision-making process, this chapter is about data collection. In particular, we demonstrate how data are collected and prepared for tabular and chart presentation, descriptive summarization, analysis, and interpretation. To motivate our discussion of data collection, we see from the chapter summary chart on page 47 that data are of two types, the outcomes of numerical random variables measured on interval or ratio scales or the outcomes of categorical variables measured on nominal or ordinal scales. In addition, we note from the chapter summary chart that there are several methods for obtaining data. In this text we focus on survey research through the development of an Employee Satisfaction Survey (Section 2.8) that we highlight as an integrated case throughout, demonstrating how basic research is conducted and used to assist in the decision-making process.

Upon completion of this chapter, you should be able to:

1. Understand why we need data
2. Understand the differences between numerical and categorical data and their levels of measurement
3. Understand the various methods used in obtaining data
4. Develop an appreciation for formulating a research problem and conducting survey research

5. Develop an appreciation for the art of questionnaire design and the importance of objective, meaningful question wording
6. Understand the importance of operational definitions in survey research
7. Understand the importance of obtaining the appropriate population frame
8. Understand how to distinguish between good and bad survey research and the ethical issues involved
9. Use a table of random numbers to select a simple random sample
10. Obtain an appreciation for the problems involved in survey data preparation with respect to editing, coding, and transcribing

2.2 Obtaining Data

There are many methods by which we may obtain needed data. First, we may seek data already published by governmental, industrial, or individual sources. Second, we may design an experiment to obtain the necessary data. Third, we may conduct a survey. Fourth, we may make observations of the behavior, attitudes, or opinions of the individuals in whom we are interested.

2.2.1 Using Published Sources of Data

Owing to some of the most exciting scientific developments in this final decade of the century, we have truly reached an "age of information technology." Bar codes automatically record inventory information as products are purchased from supermarkets, department stores, and other outlets. ATMs enable banking transactions to occur spontaneously with information immediately recorded on account balances. Airline ticketing offices and travel agents have up-to-the-minute information regarding space availability on flights and at hotels. Transactions that took hours, or even days, a decade ago are now accomplished in a matter of seconds. Never before have so much up-to-date data and information been so readily available—and from so many sources.

Use of the library for research has literally taken on a new meaning. No longer does one have to visit the library to access material in printed form—from books, journals, magazines, pamphlets, and newspapers. Although we can still visit the library for these purposes, we can also use its modern multimedia facilities and obtain data electronically through information retrieval systems and on-line databases. On the other hand, our library "visit" can occur electronically through the use of a personal computer with a modem in our home or in our office. CD-ROM has revolutionized information access. As examples, ABI/INFORM Ondisc indexes and abstracts articles from more than 800 journals in the business fields, COMPACT D/SEC has information taken from annual and periodic company reports filed with the Securities and Exchange Commission, and NATIONAL TRADE DATA BANK contains recent trade and international data as well as foreign exchange data.

Regardless of the source used, a distinction is made between the original collector of the data and the organization or individuals who compile the data into tables and charts. The data collector is the **primary source;** the data compiler is the **secondary source.**

The federal government is a major collector and compiler of data for both public and private purposes. The Bureau of Labor Statistics is responsible for collecting data on employment as well as establishing the well-known monthly *Consumer Price Index*. In addition to its constitutional requirement for conducting a decennial census, the Bureau of the Census is concerned with a variety of ongoing surveys regarding population, housing, and manufacturing and from time to time undertakes special studies on topics such as crime, travel, and health care.

In addition to the federal government, various trade publications present data pertaining to specific industrial groups. Investment services such as Moody's display financial data on a company basis. Syndicated services such as A. C. Nielsen provide clients with information enabling the comparison of client products with their competitors. And, of course, daily newspapers are filled with numerical information regarding stock prices, weather conditions, and sports statistics. Throughout this text, various applications will utilize data obtained from such published sources.

2.2.2 Designing an Experiment

A second method for obtaining needed data is through experimentation. In an experiment, strict control is exercised over the treatments given to participants. For example, in a study testing the effectiveness of toothpaste, the researcher would determine which participants in the study would use the new brand and which would not, instead of leaving the choice to the subjects. Proper experimental designs are usually the subject matter of more advanced texts, since they often involve sophisticated statistical procedures. However, in order to develop a feeling for testing and experimentation, the fundamental experimental design concepts will be considered in Chapters 11 through 15.

2.2.3 Conducting a Survey

A third method for obtaining data is to conduct a survey. Here no control is exercised over the behavior of the people being surveyed. They are merely asked questions about their beliefs, attitudes, behaviors, and other characteristics. Their responses are then edited, coded, and tabulated for analysis as in Chapter 3.

2.2.4 Performing an Observational Study

In an observational study, the researcher observes the behavior of interest directly, usually in its natural setting. Most knowledge of animal behavior was originally developed in this way, as was the case in astronomy and geology, where experimentation and surveys are impractical. Observational studies also play a major role in anthropology and sociology because they often provide a richness of description which is absent from more structured methods of data collection such as experiments and surveys.

With respect to the business disciplines, the observational study has a variety of formats, all of which are intended to collect information in a group setting to assist in the decision-making process. As one example, the focus group is a popular marketing research tool that is used for eliciting unstructured responses to open-ended questions. A moderator leads the discussion and all the participants respond to the questions asked. Other, more structured formats involving group dynamics for obtaining information (and consensus building) are various organizational behavior/industrial psychology tools such as brainstorming, the Delphi technique, and the nominal-group method (see Reference 11). These tools have become more popular in recent years owing to the impact of the total quality management (TQM) philosophy on business, because TQM emphasizes the importance of teamwork and employee empowerment in an attempt to improve every product and service.

2.2.5 The Importance of Obtaining Good Data: GIGO

Remember that there are four main reasons for collecting data: (1) to provide input to a research study, (2) to measure performance, (3) to enhance decision making, or (4) to satisfy our curiosity. To emphasize the importance of obtaining good data, researchers have adopted the term **GIGO**—garbage in, garbage out. Regardless of the method utilized for obtaining the data, if a study is to be useful, if performance is to be appropriately monitored, or if the decision-making process is to be enhanced, the data gathered must be *valid*; that is, the "right" responses must be assessed, and in a manner that will elicit meaningful measurements.

In order to design an experiment, conduct a survey, or perform an observational study, one must understand the different types of data and measurement levels. To demonstrate some of the issues involved in obtaining data, we will present them in the context of a survey, although most of the same issues will arise in other types of research.

2.3 Obtaining Data Through Survey Research

A survey statistician will most likely want to develop an instrument that asks several questions and deals with a variety of phenomena or characteristics. These phenomena or characteristics are called **random variables.** The data, which are the observed outcomes of these random variables, may differ from response to response.

2.3.1 Types of Data

As outlined in Figure 2.1 on page 16, there are basically two types of random variables yielding two types of data: **categorical** and **numerical.** Categorical random variables yield categorical responses, while numerical random variables yield numerical responses. For example, the response to the question "Do you currently own United States Government Savings Bonds?" is categorical. The choices are clearly "yes" or "no." On the other hand, responses to questions such as "To how many magazines do you currently subscribe?" or "How tall are you?" are clearly numerical. In the first case the numerical random variable may be considered as *discrete,* while in the second case it can be thought of as *continuous.*

> **Discrete data** are numerical responses that arise from a counting process, while **continuous data** are numerical responses that arise from a measuring process.

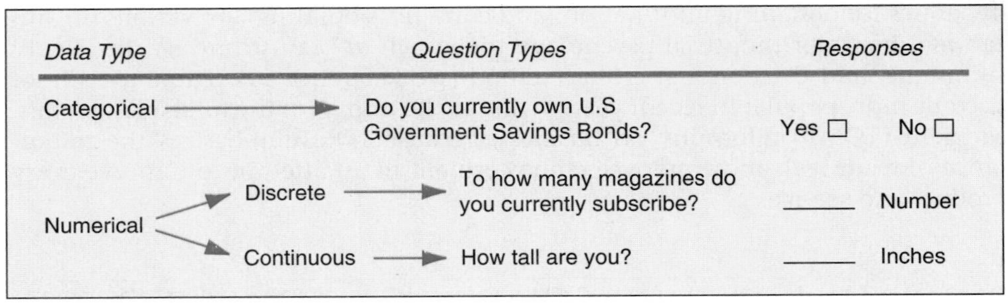

Data Type
Question Types
Responses
Categorical
Do you currently own U.S
Government Savings Bonds?
Yes ☐
No ☐
Numerical
Discrete
To how many magazines do
you currently subscribe?
Number
Continuous
How tall are you?
Inches

Figure 2.1 **Types of data.**

"The number of magazines subscribed to" is an example of a discrete numerical variable, since the response takes on one of a (finite) number of integers. The individual currently subscribes to no magazine, one magazine, two magazines, etc. On the other hand, "the height of an individual" is an example of a continuous numerical variable, since the response can take on any value within a continuum or interval, depending on the precision of the measuring instrument. For example, a person whose height is reported as 67 inches may be measured as $67\frac{1}{4}$ inches, $67\frac{7}{32}$ inches, or $67\frac{58}{250}$ inches if more precise instrumentation is available. Therefore, we can see that height is a continuous phenomenon that can take on any value within an interval.

It is interesting to note that theoretically no two persons could have exactly the same height, since the finer the measuring device used, the greater the likelihood of detecting differences between them. However, most measuring devices are not sophisticated enough to detect small differences, and hence *tied observations* are often found in experimental or survey data even though the random variable is truly continuous.

2.3.2 Levels of Measurement and Types of Measurement Scales

From the above discussion, then, we see that our resulting data may also be described in accordance with the level of measurement attained.

In the broadest sense, all collected data are "measured" in some form. For example, even discrete numerical data can be thought of as arising by a process of *measurement through counting*. The four widely recognized levels of measurement are—from weakest to strongest level of measurement—the nominal, ordinal, interval, and ratio scales.

● **Nominal and Ordinal Scales** Data obtained from a categorical variable are said to have been measured either on a nominal scale or on an ordinal scale. If the observed data are merely classified into various distinct categories in which no ordering is implied, a **nominal** level of measurement is achieved. On the other hand, if the observed data are classified into distinct categories in which ordering is implied, an **ordinal** level of measurement is attained. These distinctions are depicted in Figures 2.2 and 2.3, respectively.

Nominal scaling is the weakest form of measurement because no attempt can be made to account for differences within a particular category or to specify any ordering or direction across the various categories. Ordinal scaling is a somewhat stronger form of measurement, because an observed value classified into one

Categorical Variable		Categories				
Automobile ownership	⟷	Yes	No			
Type of life insurance owned	⟷	Term	Endowment	Straight-Life	Other	None
Political party affiliation	⟷	Democrat	Republican	Independent	Other	

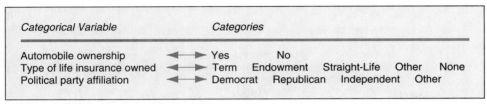

Figure 2.2 Examples of nominal scaling.

Categorical Variable		Ordered Categories
		(Lowest-Highest)
Student class designation	⟷	Freshman Sophomore Junior Senior
Product satisfaction	⟷	Very Unsatisfied Fairly Unsatisfied Neutral Fairly Satisfied Very Satisfied
Movie classification	⟷	G PG PG-13 R X
		(Highest-Lowest)
Faculty rank	⟷	Professor Associate Professor Assistant Professor Instructor
Standard & Poor's bond ratings	⟷	AAA AA A BBB BB B CCC CC C DDD DD D
Restaurant ratings	⟷	***** **** *** ** *
Student grades	⟷	A B C D E F

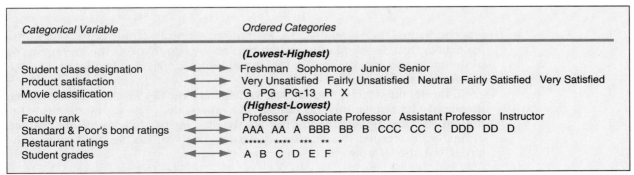

Figure 2.3 Examples of ordinal scaling.

category is said to possess more of a property being scaled than does an observed value classified into another category. Nevertheless, within a particular category no attempt is made to account for differences between the classified values. Moreover, ordinal scaling is still a weak form of measurement, because no meaningful numerical statements can be made about differences between the categories. That is, the ordering implies only *which* category is "greater," "better," or "more preferred"—not *how much* "greater," "better," or "more preferred." For instance, college basketball and college football rankings are other applications of ordinal scaling. The differences in ability between the teams ranked first and second might not be the same as the differences in ability between the teams ranked second and third, or those ranked sixth and seventh, and so on.

● **Interval and Ratio Scales** An **interval scale** is an ordered scale in which the difference between the measurements is a meaningful quantity. For example, a noontime temperature reading of 67 degrees Fahrenheit is 2 degrees Fahrenheit warmer than a noontime temperature reading of 65 degrees Fahrenheit. In addition, the 2 degrees Fahrenheit difference in the noontime temperature readings is the same quantity that would be obtained if the two noontime temperature readings were 76 and 74 degrees Fahrenheit, so the difference has the same meaning anywhere on the scale.

If, in addition to differences being meaningful and equal at all points on a scale, there is a true zero point so that ratios of measurements are sensible to consider, then the scale is a **ratio scale.** A person who is 76 inches tall is twice as tall as someone who is 38 inches tall; in general, then, measurements of length are ratio scales. Temperature is a trickier case: Fahrenheit and centigrade (Celsius) scales are interval but not ratio scales; the "0" demarcation is arbitrary, not real. No one should say that a noontime temperature reading of 76 degrees Fahrenheit is twice as hot as a noontime temperature reading of 38 degrees Fahrenheit. But

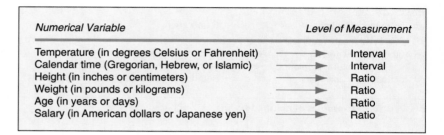

Figure 2.4
Examples of interval and ratio scaling.

Numerical Variable		*Level of Measurement*
Temperature (in degrees Celsius or Fahrenheit)	→	Interval
Calendar time (Gregorian, Hebrew, or Islamic)	→	Interval
Height (in inches or centimeters)	→	Ratio
Weight (in pounds or kilograms)	→	Ratio
Age (in years or days)	→	Ratio
Salary (in American dollars or Japanese yen)	→	Ratio

when measured from absolute zero, as in the Kelvin scale, temperature is on a ratio scale, because a doubling of temperature is really a doubling of the average speed of the molecules making up the substance. Figure 2.4 gives examples of interval- and ratio-scaled variables.

Data obtained from a numerical variable are usually assumed to have been measured either on an interval scale or on a ratio scale. These scales constitute the highest levels of measurement. They are stronger forms of measurement than an ordinal scale because we are able to discern not only which observed value is the largest but also by how much.

● **Caution: The Need for Operational Definitions** Regardless of the level of measurement for our variables, *operational definitions* (see Reference 4) are needed to elicit the appropriate response or attain the appropriate outcome.

> An **operational definition** provides a meaning to a concept or variable that can be communicated to other individuals. It is something that has the same meaning yesterday, today, and tomorrow to all individuals.

As an example, take the word "round." Although the dictionary provides a literal meaning, what is necessary is a meaning that can actually be used in practice. Thus, the issue really is not what is round, but how far something departs from "roundness" before we say that it is not round. This needs to be defined in a way that can be applied consistently from day to day (or, for a production worker, for example, from product to product).

In the context of a survey, consider the question "What is your age?" To avoid problems of ambiguity, we must develop an operational definition for the responses to the question. For example, we must clarify whether age should be reported to the *nearest* birthday or as of the *last* birthday, because if your birthday is next month, you would probably choose the nearest birthday if you were turning 20; but you would be likely to report your present age if you were turning 50!

As a further example of operational definitions, consider the following headline that appeared in a suburban New York county newspaper a few years ago: "Off With a Head Count: Is Suffolk more populous than Nassau? LILCO and the Census Bureau disagree."[1] The article included quotes from the Suffolk county executive (". . . we are confident that Suffolk is No. 1") and the Nassau county executive ("We'll declare it a tie in the spirit of regional cooperation"). Of course, the differences in the two estimates come from the fact that the Census Bureau and the Long Island Lighting Company (LILCO) have different operational definitions used to estimate population in the two counties. The Census Bureau uses birth and death rates, migration patterns as shown on income tax returns, and a demographic formula that estimates that the average number of people per household has been shrinking in the past several years. On the other hand, for its definition, LILCO uses the number of year-round electric and gas meters, building permits, and a factor for the number of people in each house.

Problems for Section 2.3

2.1 Explain the difference between a categorical and a numerical random variable and give an example of each.

2.2 Explain the difference between a discrete and a continuous random variable and give an example of each.

2.3 If two students both score a 90 on the same examination, what arguments could be used to show that the underlying random variable (phenomenon of interest)—test score—is continuous?

● 2.4 Determine whether each of the following random variables is categorical or numerical. If numerical, determine whether the phenomenon of interest is discrete or continuous. In addition, provide the level of measurement and an operational definition for each of the variables.

(a) Number of telephones per household
(b) Type of telephone primarily used
(c) Number of long-distance calls made per month
(d) Length (in minutes) of longest long-distance call made per month
(e) Color of telephone primarily used
(f) Monthly charge (in dollars and cents) for long-distance calls made
(g) Ownership of a cellular phone
(h) Number of local calls made per month
(i) Length (in minutes) of longest local call per month
(j) Whether there is a telephone line connected to a computer modem in the household
(k) Whether there is a FAX machine in the household

2.5 Suppose that the following information is obtained from students upon exiting from the campus bookstore during the first week of classes:

(a) Amount of money spent on books
(b) Number of textbooks purchased
(c) Amount of time spent shopping in the bookstore
(d) Academic major
(e) Gender
(f) Ownership of a personal computer
(g) Ownership of a videocassette recorder
(h) Number of credits registered for in the current semester
(i) Whether or not any clothing items were currently purchased at the bookstore
(j) Method of payment

Classify each of these variables as categorical or numerical. If it is numerical, determine whether the variable is discrete or continuous. In addition, provide the level of measurement and an operational definition for each of the variables.

Note: Bullet ● indicates that the solutions to these problems are in the *Answers to Selected Problems* section at the back of the book.

2.6 Give an example of a numerical variable that is actually discrete but might be considered continuous.

2.7 Give an example in an area of interest to you where data are useful for decision making. What data are useful? How might they be obtained? How might the data be used in the process of making a decision?

2.8 One of the variables most often included in surveys is income. Sometimes the question is phrased "What is your income (in thousands of dollars)?" In other surveys, the respondent is asked to "Place an X in the box corresponding to your income level."

☐ Under $20,000? ☐ $20,000–$39,999 ☐ $40,000 or more

(a) For each of these formats, tell whether the measurement level for the variable is nominal, ordinal, interval, or ratio.
(b) In the first format, explain why income might be considered either discrete or continuous.
(c) Which of these two formats would you prefer to use if you were conducting a survey? Why?
(d) Which of these two formats would likely bring you a greater rate of response? Why?

2.9 Provide an operational definition for each of the following:
(a) An outstanding teacher
(b) A hard worker
(c) A nice day
(d) Fast service
(e) A leader
(f) Commuting time to school or work
(g) A fine quarterback

2.10 Provide an operational definition for each of the following:
(a) A dynamic individual
(b) A boring class
(c) An interesting book
(d) An outstanding performance
(e) A manager
(f) An on-time plane arrival
(g) Study time

Interchapter Problem for Section 2.3

2.11 For each variable in the examples of applications of statistics you mentioned in the answers to Problems 1.1–1.7 on page 9, tell whether the variable is numerical or categorical; if numerical, whether it is discrete or continuous; what level of measurement it has; and, if it is not continuous, whether it could be treated as such.

2.4 Designing the Questionnaire Instrument

Questionnaire development is an art that improves with experience. Remember that the purpose of a questionnaire is to enable us to gather meaningful information that will assist us in some decision-making process. The general procedure for designing a questionnaire will involve:

- Choosing the broad topics that are to reflect the theme of the survey
- Deciding on a mode of response
- Formulating the questions
- Pilot testing and making final revisions

2.4.1 Selection of Broad Topics—Questionnaire Length

The broad topics that are to reflect the theme of the survey must be listed. Before very long a large number of questions will have been created. Unfortunately, however, there is an inverse relationship between the length of a questionnaire and the rate of response to the survey. That is, the longer the questionnaire, the lower will be the rate of response; the shorter the questionnaire, the higher will be the rate of response. It is, therefore, imperative that we carefully evaluate the merits of each question and determine whether the question is really necessary and, if so, how to word it optimally. Questions should be as short as possible. The response categories for categorical questions should be nonoverlapping and complete.

2.4.2 Mode of Response

The particular questionnaire format to be selected and the specific question wording are affected by the intended mode of response. There are essentially three modes through which survey work is accomplished: personal interview, telephone interview, and mail. The personal interview and the telephone interview usually produce a higher response rate than does the mail survey—but at a higher cost.

2.4.3 Formulating the Questions

Because of the inverse relationship between the length of a questionnaire and the rate of response to the survey, each question must be clearly presented in as few words as possible, and each question should be deemed essential to the survey. In addition, questions must be free of ambiguities. Operational definitions are needed to elicit the appropriate response. For example, consider the following two questions:

1. Do you smoke? ___Yes ___No
2. How old are you? ____ (in years)

Question 1 has several possible ambiguities. It is not clear if the desired response pertains to cigarettes, to cigars, to pipes, or to combinations thereof. It is also not clear whether occasional smoking or habitual smoking was the primary concern of the question. If we were interested only in current cigarette consumption, perhaps it would be better to ask

1. About how many cigarettes do you smoke each day? ____

When replying to question 2, as previously pointed out, the respondent may be confused as to whether to base the answer on the *last* birthday or the *nearest* birthday unless the appropriate operational definition is specified. This problem may be avoided, however, if the respondent is merely asked

2. State your date of birth: ______ ______ ______
month day year

2.4.4 Testing the Questionnaire Instrument

Once the pros and cons of each question have been discussed, the instrument is properly organized and made ready for **pilot testing** so that it may be examined for clarity and length. Pilot testing on a small group of subjects is an essential phase in conducting a survey. Not only will this group of individuals be providing an estimate of the time needed for responding to the survey, but they will also be asked to comment on any perceived ambiguities in each question and to recommend additional questions.

Problems for Section 2.4

2.12 Why might you expect a greater response rate from a survey conducted by personal or telephone interview than from one conducted using a mailed questionnaire instrument?

2.13 Suppose that the director of market research at a large department store chain wanted to conduct a survey throughout a metropolitan area to determine the amount of time working women spend shopping for clothing in a typical month.

(a) Describe both the population and sample of interest and indicate the type of data the director primarily wishes to collect.

(b) Develop a first draft of the questionnaire needed in (a) by writing a series of three categorical questions and three numerical questions that you feel would be appropriate for this survey. Provide operational definitions for each question.

2.14 Write a question asking how much education a person has; write three versions of the question that give different levels of detail. Describe situations in which each might be appropriate or inappropriate to use.

2.5 Choosing the Sample Size for the Survey

Rather than taking a complete census, statistical sampling procedures (see Section 2.7) have become the preferred tool in most survey situations. There are three main reasons for drawing a sample. First of all, it is usually too time consuming to perform a complete census. Second, it is too costly to do a complete census. Third, it is just too cumbersome and inefficient to obtain a complete count of the target population.

After the most essential numerical and categorical questions in the survey have been determined, the sample size needed will be based on satisfying the question with the most stringent requirements. Determination of the sample size required for a given survey is a matter that will be examined more appropriately in Chapter 10.

2.6 Selecting the Responding Subjects: Types of Samples

As depicted in Figure 2.5 on page 23, there are basically two kinds of samples: the **nonprobability sample** and the **probability sample.** For many studies only

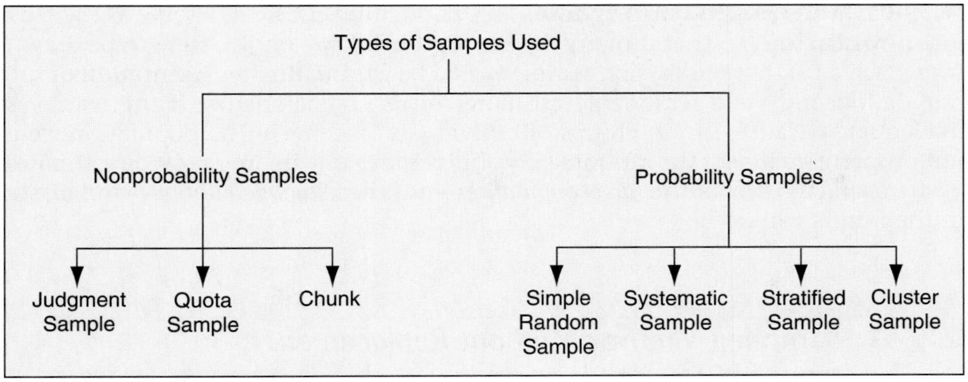

Figure 2.5 **Types of samples.**

a nonprobability sample such as a judgment sample is available. In these instances, the opinion of an expert in the subject matter of a study is crucial to being able to use the results obtained in order to make changes in a process. Some other typical procedures of nonprobability sampling are quota sampling and chunk sampling; these are discussed in detail in specialized books on sampling methods (see References 1, 3, and 8).

In an enumerative study, the only way for us to make correct statistical inferences from a sample to a population is through the use of a probability sample.

> A **probability sample** is one in which the subjects of the sample are chosen on the basis of known probabilities.

The four types of probability samples most commonly used are the simple random sample, the systematic sample, the stratified sample, and the cluster sample.

In a **simple random sample** every individual or item has the same chance of selection as every other, and the selection of a particular individual or item does not affect the chances that any other is chosen. Moreover, a simple random sample may also be construed as one in which each possible sample that is drawn has the same chance of selection as any other sample that could be drawn.

A detailed discussion of systematic sampling, stratified sampling, and cluster sampling procedures can be found in References 1, 3, and 8.

2.7 Drawing the Simple Random Sample

In this section we will be concerned with the process of selecting a simple random sample. Although it is not necessarily the most economical or efficient of the probability sampling procedures, it provides the base from which the more sophisticated procedures have evolved.

The key to proper sample selection is obtaining and maintaining an up-to-date list of all the individuals or items from which the sample will be drawn. Such a list is known as the **population frame.** This population listing will serve as the **target population,** so that if many different probability samples were to be drawn from such a list, hopefully each sample would be a miniature representation of the population and yield reasonable estimates of its characteristics. If the listing is inadequate because certain groups of individuals or items in the population were not properly included, the random probability samples will only provide estimates of the characteristics of the *target population*—not the *actual population*—and biases in the results will occur.

2.7.1 Sampling *With* or *Without Replacement* from Finite Populations

Two basic methods could be used for selecting the sample: The sample could be obtained **with replacement** or **without replacement** from the finite population. The method employed must be clearly stated by the survey statistician, since various formulas subsequently used for purposes of statistical inference are dependent upon the selection method.[2]

Let N represent the population size and n represent the sample size. To draw a simple random sample of size n one could conceivably record the names of the N individual members of the population on separate index cards of the same size, place these index cards in a large fish bowl, thoroughly mix the cards, and then randomly select the n sample subjects from the fish bowl.

When sampling with replacement, the chance that any particular member in the population, say Judy Craven, is selected on the first draw from the fish bowl is $1/N$. Regardless of whoever is actually selected on the first draw, pertinent information is recorded on a master file, and then the particular index card is replaced in the bowl (sampling *with replacement*). The N cards in the bowl are then well shuffled and the second card is drawn. Since the first card had been replaced, the chance for selection of any particular member, including Judy Craven, on the second draw—regardless of whether or not that individual had been previously selected—is still $1/N$. Again the pertinent information is recorded on a master file and the index card is replaced in order to prepare for the third draw. Such a process is repeated until n, the desired sample size, is obtained. Thus, when sampling with replacement, every individual or item on every draw will always have the same 1 out of N chance of being selected.

But should we want to have the same individual or item possibly selected more than once? When sampling from human populations, it is generally felt more appropriate to have a sample of different people than to permit repeated measurements of the same person. Thus we would employ the method of sampling without replacement, whereby once a particular individual is drawn, the same person cannot be selected again. As before, when sampling without replacement, the chance that any particular member in the population, say Judy Craven, is selected on the first draw from the fish bowl is $1/N$. Whoever is selected, the pertinent information is recorded on a master file and then the particular index card is set aside rather than replaced in the bowl (sampling *without replacement*). The remaining $N-1$ cards in the bowl are then well shuffled, and the second card is drawn. The chance that any individual not previously selected will be selected on the second draw is now 1 out of $N-1$. This process of selecting a card, recording the information on a master file, shuffling the remaining cards, and then drawing again continues until the desired sample of size n is obtained.

Regardless of whether we sample with or without replacement, such "fish bowl" methods for sample selection have a major drawback—our ability to thoroughly mix the cards and randomly pull the sample. This became a major issue of contention in the year 1969 when the Selective Service Commission developed a lottery system for choosing males to be drafted into the military because of the Vietnam War.[3]

"Fish bowl" methods of sampling as just described, although easily understandable, are not very useful. Less cumbersome and more scientific methods of selection are desirable to ensure randomization in the selection process. One such method utilizes a **table of random numbers** (see Table E.1 of Appendix E) for obtaining the sample.

2.7.2 Using a Table of Random Numbers

A table of random numbers consists of a series of digits randomly generated and listed in the sequence in which the digits were generated (see Reference 10). Since our numeric system uses 10 digits (0, 1, 2, . . . , 9), the chance of randomly generating any particular digit is equal to the probability of generating any other digit. This probability is 1 out of 10. Hence if a sequence of 500 digits were generated, we would expect about 50 of them to be the digit 0, 50 to be the digit 1, and so on. In fact, researchers who use tables of random numbers usually test out such generated digits for randomness before employing them. Table E.1 has met all such criteria for randomness. Since every digit or sequence thereof in the table is random, we may use the table by reading either horizontally or vertically. The margins of the table designate row numbers and column numbers. The digits themselves are grouped into sequences of five for the sole purpose of facilitating the viewing of the table.

To use such a table in lieu of a fish bowl for selecting the sample, it is first necessary to assign code numbers to the individual members of the population. We then obtain our random sample by reading the table of random numbers and selecting those individuals from the population frame whose assigned code numbers match the digits found in the table. This process will be described in detail in the following section. To better understand the process of survey research from its inception, let us consider the Kalosha Industries Employee Satisfaction Survey.

2.8 Kalosha Industries Employee Satisfaction Survey

As part of its movement to implement a total quality management (TQM) philosophy throughout the company and thereby increase efficiency and productivity, the board of directors of Kalosha Industries, an automobile parts manufacturer with 9,800 employees, wants to study its full-time workforce by developing an employee profile that will measure job satisfaction, evaluate longevity and career progress, and assess the aspirations, attitudes, and beliefs of its workers. Bud Conley, the Vice-President of Human Resources, hires B & L Corporation, an employee-benefits consulting firm, to survey Kalosha employees.

After careful consideration, Conley and the statistician from the B & L Corporation determine that a survey conducted via interoffice mail will be sufficient to obtain an accurate assessment of the desired information on job satisfaction and income and a questionnaire is designed and prepared for pilot testing. Once the vice-president and the statistician have evaluated these results, changes can be made and, if time and budget permit, a second pilot study can be undertaken on a fresh sample of respondents to further improve the questionnaire instrument.

Figure 2.6 on pages 28 and 29 depicts the questionnaire (with 28 questions) that was designed by the vice-president and the statistician in its final form. A major question that had to be addressed by the statistician was how large a sample of full-time employees to draw in order to obtain a worker profile so that B & L Corporation would have the information needed to develop an employee-benefits package that would please the Kalosha Industries workers, strengthen their relationship with management, and demonstrate that the client company was taking a leading role in the industry with respect to establishing a TQM environment—the mandate of the board of directors. Thus it was the goal of B & L Corporation to make inferences about the entire population of Kalosha Industries full-time employees based on the results obtained from the sample.

Determination of the sample size required will be developed in Chapter 10. The required sample size is based on the fact that the vice-president and the statistician have decided that questions 7 and 9 are the most essential numerical and categorical questions, respectively, in the entire survey. As we shall see, the required sample size is 400 full-time workers out of a population of 9,800 full-time employees of Kalosha Industries. However, because not everyone will be willing to respond to the survey, the vice-president must be prepared to have a larger mailing. Based on past experiences with in-house surveys by the consulting firm, nine out of ten full-time workers are expected to respond to such a survey (i.e., a rate of return of 90%); thus a total of 445 such employees must be contacted in order to obtain the desired 400 responses. Therefore the questionnaire instrument was distributed in its final form to 445 full-time employees drawn from the personnel files at Kalosha Industries.

To draw the random sample, the statistician chose to use a table of random numbers. The population frame comprised a listing of the names and company mailbox numbers of all N = 9,800 full-time employees of Kalosha Industries obtained from the personnel files through Bud Conley, the Vice-President of Human Resources. Since the population size (9,800) is a four-digit number, each assigned code number must also be four digits, so that every full-time worker has an equal chance for selection. Thus a code of 0001 is given to the first full-time Kalosha Industries employee in the population listing, a code of 0002 is given to the second full-time employee in the population listing, . . . , a code of 1752 to the one thousand seven hundred fifty-second individual in the population listing, and so on until a code of 9800 is given to the Nth full-time worker in the listing. Since N = 9,800 is the largest possible coded value, all four-digit code sequences greater than N (that is, 9801 through 9999 and 0000) are discarded.

In order to select the random sample, a random starting point for the table of random numbers must be established. One such method is to close one's eyes and strike the table of random numbers with a pencil. Suppose the statistician uses such a procedure and thereby selects row 06, column 05, of Table 2.1 (a replica of Table E.1) as the starting point. Reading from left to right in Table 2.1 in sequences of four digits without skipping, the individual employees are selected for the survey.

Table 2.1 **Using a table of random numbers.**

	Row	Column 00000 12345	00001 67890	11111 12345	11112 67890	22222 12345	22223 67890	33333 12345	33334 67890
	01	49280	88924	35779	00283	81163	07275	89863	02348
	02	61870	41657	07468	08612	98083	97349	20775	45091
	03	43898	65923	25078	86129	78496	97653	91550	08078
	04	62993	93912	30454	84598	56095	20664	12872	64647
	05	33850	58555	51438	85507	71865	79488	76783	31708
Begin Selection (Row 06, Column 05)	06	97340	03364	88472	04334	63919	36394	11095	92470
	07	70543	29776	10087	10072	55980	64688	68239	20461
	08	89382	93809	00796	95945	34101	81277	66090	88872
	09	37818	72142	67140	50785	22380	16703	53362	44940
	10	60430	22834	14130	96593	23298	56203	92671	15925
	11	82975	66158	84731	19436	55790	69229	28661	13675
	12	39087	71938	40355	54324	08401	26299	49420	59208
	13	55700	24586	93247	32596	11865	63397	44251	43189
	14	14756	23997	78643	75912	83832	32768	18928	57070
	15	32166	53251	70654	92827	63491	04233	33825	69662
	16	23236	73751	31888	81718	06546	83246	47651	04877
	17	45794	26926	15130	82455	78305	55058	52551	47182
	18	09893	20505	14225	68514	46427	56788	96297	78822
	19	54382	74598	91499	14523	68479	27686	46162	83554
	20	94750	89923	37089	20048	80336	94598	26940	36858
	21	70297	34135	53140	33340	42050	82341	44104	82949
	22	85157	47954	32979	26575	57600	40881	12250	73742
	23	11100	02340	12860	74697	96644	89439	28707	25815
	24	36871	50775	30592	57143	17381	68856	25853	35041
	25	23913	48357	63308	16090	51690	54607	72407	55538
	⋮	⋮	⋮	⋮	⋮	⋮	⋮	⋮	⋮

Source: Partially extracted from The Rand Corporation, *A Million Random Digits with 100,000 Normal Deviates* (Glencoe, IL: The Free Press, 1955) and displayed in Table E.1 in Appendix E at the back of this text.

The individual with code number 0033 is the first full-time employee in the sample (row 06 and columns 05 through 08). The second individual selected has code number 6488 (row 06 and columns 09 through 12). Individuals with code numbers 4720, 4334, 6391, 9363, 9411, 0959, 2470, and 7054 are selected third through tenth, respectively.

The selection process continues in a similar manner until the needed sample size of 445 full-time employees is obtained. Two pages of a table of random numbers are needed to achieve this. During the selection process, if any four-digit coded sequence repeats, the employee corresponding to that coded sequence is included again as part of the sample if sampling with replacement; however, the repeating coded sequence is merely discarded if sampling without replacement. Note that the coded sequence 4205 appears in row 12, columns 33 through 36, and then again in row 21, columns 21 through 24. Since the statistician from the B & L Corporation is sampling without replacement, the repeating sequences are discarded and a sample of 445 individual full-time employees is obtained.

Figure 2.6 **Questionnaire.**

Employee Satisfaction Survey

Codes	*(In questions 1–28, please* ***insert*** *the value or* ***circle*** *the number as appropriate.)*
___ ___ ___ 1 2 3	*. Identification Number Code _ _ _ (Office use)
___ ___ 5 6	1. How many hours did you work last week, at all jobs? ____
___ 8	2. What is your occupation? 1 Managerial 2 Professional 3 Technical/Sales 4 Admin. Support 5 Service 6 Production 7 Laborer
___ ___ 10 11	3. What is your age (as of last birthday)? ___
___ ___ 13 14	4. How many years of school have you completed? ___
___ 16	5. What is your gender? 1 Male 2 Female
___ 18	6. Among family members living in your household now, how many, including yourself, were employed last year? ___
___ ___ .___ ___ 20 21 22 23	7. What was your "before-taxes" income last year (in thousands of dollars)? ___
___ ___ .___ ___ 25 26 27 28	8. What was your "before-taxes" total family income last year (in thousands of dollars)? ___
___ 30	9. On the whole, how satisfied are you with your job? 1 Very Satisfied 2 Moderately Satisfied 3 A Little Dissatisfied 4 Very Dissatisfied
___ 32	10. If, overnight, you were to become very rich through an inheritance, a gift, or a lottery, would you stop working and retire? 1 Yes 2 No 3 Not Sure
___ 34	11. Which *one* of the following job characteristics is *most important* to you? 1 High Income 2 No Danger of Being Fired 3 Flexible Hours 4 Opportunities for Advancement 5 Enjoying the Work
___ 36	12. Which *one* of the following is the way most people get ahead on the job? 1 Hard Work 2 Hard Work and Luck 3 Luck
___ 38	13. How many traumatic events (death of close relative or friend, divorce/separation, unemployment, disabling illness) did you experience last year? ___
___ 40	14. Are you currently a member of a labor union? 1 Yes 2 No
___ ___ 42 43	15. Since you were 16 years old, about how many years have you worked full-time for pay? ___
___ ___ .___ ___ ___ 45 46 47 48 49	16. How many years altogether have you worked for your present employer? ___

___ 51 17. How many promotions, if any, have you received while working for your present employer? ___

___ 53 18. In the next five years, how likely are you to be promoted?

[1] Very Likely [2] Likely [3] Not Sure
[4] Unlikely [5] Very Unlikely

___ 55 19. Are promotional opportunities better or worse for persons of your gender?

[1] Better [2] Worse [3] Has No Effect

___ 57 20. Since your first full-time job with this organization, how would you describe your "advancement"?

[1] Advanced Rapidly [2] Made Steady Advances
[3] Stayed at about the Same Level [4] Lost Some Ground

___ 59 21. Does your job allow you to take part in making decisions that affect your work?

[1] Always [2] Much of the Time
[3] Sometimes [4] Never

___ 61 22. As part of your job, do you participate in budgetary decisions?

[1] Yes [2] No

___ 63 23. How proud are you to be working for this organization?

[1] Very Proud [2] Somewhat Proud
[3] Indifferent [4] Not At All Proud

___ 65 24. Would you turn down another job for more pay in order to stay with this organization?

[1] Very Likely [2] Likely [3] Not Sure
[4] Unlikely [5] Very Unlikely

___ 67 25. In general, how would you describe relations in your workplace between management and employees?

[1] Very Good [2] Good [3] So So
[4] Bad [5] Very Bad

___ 69 26. In general, how would you describe relations in your workplace between coworkers and colleagues?

[1] Very Good [2] Good [3] So So
[4] Bad [5] Very Bad

___ 71 27. How important was your formal schooling to the job you do now?

[1] Very Important [2] Important
[3] Somewhat Important [4] Not At All Important

___ 73 28. How important was formal on-the-job training to the job you do now?

[1] Very Important [2] Important
[3] Somewhat Important [4] Not At All Important

Problems for Section 2.8

2.15 For each of the 28 questions in the Employee Satisfaction Survey, provide an operational definition that may be necessary to avoid ambiguity.

2.16 When is a sample random? What are some potential problems with using "fish bowl" methods to draw a simple random sample?

2.17 For a study that would involve doing personal interviews with participants (rather than mail or phone surveys), tell why a simple random sample might be less practical than some other methods.

2.18 If you wanted to determine what proportion of movies shown in the United States last year had themes based on sex or violence, how might you get a random sample to answer your question?

2.19 Suppose that I want to select a random sample of size 1 from a population of 3 items (which we can call A, B, and C). My rule for drawing the sample is: Flip a coin; if it is heads, pick item A; if it is tails, flip the coin again; this time, if it is heads choose B, if tails choose C. Tell why this is a random sample, but not a simple random sample.

2.20 Suppose that a population has 4 members (call them A, B, C, and D). I would like to draw a random sample of size 2, which I decide to do in the following way: Flip a coin; if it is heads, my sample will be items A and B; if it is tails, the sample will be items C and D. Although this is a random sample, it is not a simple random sample. Tell why. (If you did Problem 2.19, compare the procedure described there with the procedure described in this problem.)

2.21 For a population list containing $N = 902$ individuals, what code number would you assign for
(a) The first person on the list?
(b) The fortieth person on the list?
(c) The last person on the list?

2.22 For a population of $N = 902$, verify that by starting in row 05 of the table of random numbers (Table E.1) only six rows are needed to draw a sample of size $n = 60$ *without replacement*.

2.9 Obtaining the Responses

Now that the sample of 445 full-time Kalosha Industries employees has been selected and the questionnaires distributed, the responses must be obtained. For this in-house mail survey, sufficient time must be allowed for initial response.

The questionnaire and any set of instructions should have been mailed with a covering letter. The covering letter should be brief and to the point. It should state the goal or purpose of the survey, how the survey will be used, and why it is important that the selected individuals promptly respond. Moreover, it should give any necessary assurance of respondent anonymity and, in some cases (involving regular mail rather than in-house surveys), offer an incentive gift for respondent participation.

Problems for Section 2.9

2.23 Write a draft of the cover letter needed for the Kalosha Industries Employee Satisfaction Survey.

2.24 Write a draft of the cover letter needed for the department store survey developed in Problem 2.13 on page 22.

2.10 Data Preparation: Editing, Coding, and Transcribing

Once the set of data is collected it must be carefully prepared for tabular and chart presentation, analysis, and interpretation. The editing, coding, and transcribing processes are extremely important. Answers to **open-ended questions,** those that require the respondent to articulate a viewpoint, must be properly classified or scored, while the responses to both numerical and categorical questions need to be coded for data entry. All responses are scrutinized for completeness and for errors. If necessary, response validation is obtained by recontacting individuals whose answers appear inconsistent or unusual.

Table 2.2 on page 32 represents the responses of Clark Kent, file code identification number 0033, who was the first employee selected in the sample. To facilitate data entry, each selected individual's four-digit identification number (obtained from the population frame submitted by the company's Office of Human Resources) is replaced by his or her corresponding respondent number, which specifies position in the sample selection process. For example, the first employee selected, Clark Kent (file code number 0033), has coded respondent number 001.

Notice how the questionnaire responses are coded for entry. Categorical questions require a one-digit code such as observed with question 2, "occupation." Clark Kent, a former reporter, is now a publications manager in the Advertising Department at Kalosha Industries and this response is given a code of 1. For numerical questions, however, the number of spaces to allocate for a response must be based on the most extreme answers possible. For example, in question 15 two spaces are required because the total number of years an individual has worked full time for pay cannot exceed two digits. Clark Kent has accumulated 14 years of full-time work experiences since he was 16 years old. Hence a value of 14 is recorded.

The responses from Clark Kent are depicted in Figure 2.7. To enter an individual's responses to the Employee Satisfaction Survey, we observe that a maximum of 73 spaces are required since we have included one blank space between the responses to the different questions. The exact format for the data, however, will depend on the spreadsheet or statistical package utilized, particularly when incomplete responses (i.e., **missing values**) are present.

```
001 50 1 35 20 1 2 78.3 85.3 2 1 4 1 0 2 14 3.00 0 1 1 2 3 1 2 5 3 2 2 1
```

Figure 2.7 **Data entries for the responses of Clark Kent, File Code Identification Number 0033.**

Table 2.3 on pages 33–40 is a printout of the data. This printout corresponds to the responses of the 400 full-time employees who participated in the survey out of the 445 employees asked. We note that Clark Kent's responses appear first, since he was the first Kalosha Industries employee selected in the sample.

Table 2.2 Coding the responses of Clark Kent, File Code Identification Number 0033.

Question	Type of Question	Computer Code	Columns Allocated for Data Entry	Clark Kent's Responses	Coded Response
*	Respondent number	IDNUM	1–3	---	001
1	Hours of work	WORKHRS	5–6	50	50
2	Occupation	OCCUP	8	Managerial	1
3	Age	AGE	10–11	35	35
4	Years of schooling	EDUC	13–14	20	20
5	Gender	SEX	16	Male	1
6	Number of earners	EARNRS	18	2	2
7	Respondent income	RINCOME	20–23	$78,300	78.3
8	Family income	FINCOME	25–28	$85,300	85.3
9	Job satisfaction	SATJOB	30	Moderately satisfied	2
10	Retire if rich	RICHWORK	32	Yes, would retire	1
11	Job characteristics	JOBCHAR	34	Opportunities to advance	4
12	Getting ahead	GETAHEAD	36	Hard work	1
13	Traumatic events	TRAUMA	38	0	0
14	Union membership	MEMUNION	40	Nonunion	2
15	Years worked	WRKYEARS	42–43	14 years	14
16	Years at Kalosha Ind.	EMPYEARS	45–49	3.00 years	3.00
17	Number of promotions	NUMPROMO	51	0	0
18	Future promotion	FUTPROMO	53	Very likely	1
19	Promotional opportunities	SEXPROMO	55	Better	1
20	Advancement progress	ADVANCES	57	Made steady advances	2
21	Decision making	IDECIDE	59	Sometimes decide	3
22	Budgetary decisions	ORGMONEY	61	Yes	1
23	Proud of Kalosha Ind.	PROUDORG	63	Somewhat proud	2
24	Stay in organization	STAYORG	65	Very unlikely	5
25	Union-management	UNMANREL	67	So so	3
26	Coworker relations	COWRKREL	69	Good	2
27	Formal schooling	SCHOOLNG	71	Important	2
28	Formal on-job training	TRAINING	73	Very important	1

Problems for Section 2.10

2.25 Code the following responses for data entry:

(a) Height: 5 feet 2 inches ___inches

(b) Weight at birth: 7 pounds 8 ounces ___pounds

(c) Date of birth: June 27, 1958 ___ years of age

2.26 For each case in Problem 2.25 describe the rules you used for coding. What alternatives could you have considered?

Data file
EMPSAT.DAT

Table 2.3 Computer listing of responses to questionnaire from a sample of full-time employees.

OBS	IDNUM	WORKHRS	OCCUP	AGE	EDUC	SEX	EARNRS	RINCOME	FINCOME	SATJOB	RICHWORK	JOBCHAR	GETAHEAD	TRAUMA	MEMUNION	WRKYEARS	EMPYEARS	NUMPROMO	FUTPROMO	SEXPROMO	ADVANCES	IDECIDE	ORGMONEY	PRODUORG	STAYORG	UNMANREL	COWRKREL	SCHOOLNG	TRAINING
1	1	50	1	35	20	1	2	78.3	85.3	2	1	4	1	0	2	14	3.00	0	1	1	2	3	1	2	5	3	2	2	1
2	2	30	7	64	14	2	2	25.7	81.9	1	1	3	2	0	2	25	11.00	0	5	3	3	2	2	2	2	1	1	2	2
3	3	40	1	33	15	1	2	40.5	85.6	2	1	3	1	0	2	12	9.00	2	5	1	2	2	1	2	2	2	1	1	1
4	4	40	3	23	14	1	1	20.2	20.2	1	1	1	1	0	1	3	1.50	0	1	3	2	1	2	1	5	1	2	2	1
5	5	50	4	33	12	2	1	25.2	25.2	1	1	1	2	0	2	15	4.00	3	1	3	2	1	2	1	3	1	1	3	1
6	6	40	3	60	14	1	1	35.7	35.7	1	1	1	1	0	2	40	20.00	0	3	3	2	1	2	1	1	1	1	1	1
7	7	40	5	37	14	2	1	15.0	15.0	2	2	1	2	0	2	10	1.50	0	2	3	2	3	2	1	4	1	1	1	1
8	8	40	6	25	13	2	3	18.0	49.5	2	1	5	3	1	2	10	1.50	0	5	3	2	3	2	2	5	2	2	1	1
9	9	50	2	39	18	1	1	60.8	60.8	2	1	5	1	0	2	15	5.00	0	4	1	3	3	2	2	4	3	1	3	3
10	10	30	1	35	16	1	1	38.4	38.4	2	2	1	2	0	2	15	3.00	2	3	3	2	1	1	2	4	2	2	4	2
11	11	32	2	35	12	2	1	31.0	33.5	1	1	1	1	0	2	13	1.50	0	2	2	3	1	1	2	4	3	3	1	1
12	12	55	6	49	13	1	2	76.6	93.5	3	2	1	1	0	1	30	5.00	0	5	1	3	1	1	2	5	1	1	3	1
13	13	40	3	34	12	2	2	33.3	55.2	2	1	1	1	0	2	12	6.00	1	1	3	2	2	2	2	4	3	4	2	1
14	14	40	4	50	14	2	1	15.8	17.4	3	2	1	1	0	2	20	10.00	0	1	3	3	3	2	2	4	3	3	2	2
15	15	40	7	49	4	1	1	27.5	27.6	2	2	3	2	1	1	31	22.08	0	4	3	3	4	2	2	2	3	1	4	1
16	16	40	5	39	16	1	2	64.2	67.2	1	1	5	1	0	2	23	0.50	0	1	3	2	1	2	1	4	1	1	1	1
17	17	40	3	61	12	2	1	32.0	34.6	1	1	1	1	0	2	20	0.75	0	5	3	3	1	2	2	4	1	1	4	3
18	18	40	1	59	16	1	1	26.6	35.4	2	1	5	1	1	2	43	1.50	1	2	3	1	4	2	2	4	1	1	1	1
19	19	40	4	25	16	1	1	33.1	33.1	4	1	5	3	0	2	7	1.00	1	1	1	1	1	1	2	4	2	1	3	1
20	20	50	7	20	11	1	2	10.5	14.3	1	1	1	1	0	2	3	1.00	3	1	1	2	3	2	1	4	1	1	3	1
21	21	64	3	37	13	2	1	35.7	35.7	1	1	5	1	1	2	20	3.00	0	4	2	3	1	1	2	2	3	3	1	2
22	22	49	4	24	12	2	2	24.3	27.4	3	1	4	2	0	2	7	5.00	1	1	2	2	3	2	2	5	2	2	3	1
23	23	30	6	33	10	1	2	33.3	42.8	3	1	5	1	0	2	17	5.00	0	2	1	3	1	1	2	2	2	1	2	2
24	24	55	3	30	16	1	1	33.8	33.8	1	1	5	1	0	2	14	5.00	1	2	1	2	2	1	1	2	3	2	2	1
25	25	50	5	43	13	1	4	25.2	91.8	1	1	4	1	0	2	27	12.00	2	1	3	2	2	2	1	4	2	1	2	2
26	26	40	4	56	16	2	1	36.1	36.9	1	2	1	1	0	2	36	30.00	3	1	1	2	2	2	2	4	1	1	1	1
27	27	40	6	35	15	1	2	42.7	51.8	1	1	5	1	0	2	18	0.50	0	2	1	2	1	1	2	2	2	1	1	1
28	28	40	4	35	16	1	2	14.7	46.9	2	1	5	2	3	2	15	0.75	2	1	1	3	3	2	2	4	2	2	2	1
29	29	40	2	42	16	2	3	23.7	51.2	1	1	1	1	1	2	25	5.50	0	5	3	3	2	1	2	4	2	1	1	1
30	30	48	1	34	12	2	2	31.3	67.7	1	1	3	1	0	2	18	3.00	0	4	3	2	1	1	1	4	2	2	3	4
31	31	42	4	50	14	2	3	24.1	69.6	2	1	5	1	0	2	33	3.50	1	2	1	2	1	1	2	4	3	1	1	2
32	32	50	3	34	12	2	4	30.3	34.2	2	2	3	3	1	2	15	8.00	2	2	2	2	2	1	2	2	3	2	1	1
33	33	45	5	41	13	1	2	16.4	18.5	2	2	1	1	0	2	24	2.00	0	5	1	3	2	2	2	4	1	1	1	1
34	34	40	2	44	16	2	3	17.9	34.2	1	2	1	3	0	2	9	5.50	0	5	3	3	1	1	2	5	2	2	1	4
35	35	40	3	27	17	2	2	20.4	36.2	2	2	5	1	0	2	11	7.00	0	2	1	3	3	2	2	4	2	2	2	2
36	36	40	6	40	13	1	3	26.3	29.9	1	1	1	1	0	2	23	10.00	1	1	3	2	3	2	1	1	2	2	1	1
37	37	32	2	33	16	2	2	21.6	46.6	1	2	5	3	0	1	12	2.58	0	4	2	2	4	2	1	2	5	1	1	1
38	38	50	1	38	19	1	2	48.7	68.0	1	1	4	1	0	2	19	15.00	4	1	1	2	2	1	2	2	1	1	1	1
39	39	89	3	41	15	1	1	18.4	19.2	3	1	1	1	0	1	23	1.00	0	4	3	2	4	2	4	5	3	3	2	4
40	40	40	4	32	12	1	1	15.4	15.4	1	1	1	2	0	2	16	0.16	0	2	3	3	4	2	2	4	3	3	4	3
41	41	48	1	58	14	1	2	53.7	56.5	2	1	4	3	0	2	42	36.00	4	5	1	2	1	1	2	5	1	2	1	1
42	42	40	2	28	16	2	1	17.0	20.0	1	1	5	1	0	2	7	12.33	1	1	2	3	1	2	1	4	5	1	3	2
43	43	45	7	18	12	1	5	11.5	55.5	3	2	1	1	0	2	1	1.00	0	2	1	3	4	2	3	4	5	4	4	2
44	44	40	2	33	16	2	1	17.7	19.3	1	1	5	2	0	2	14	3.00	0	5	2	3	2	2	3	4	1	1	3	2
45	45	38	3	34	9	2	3	11.8	43.7	2	2	1	1	0	2	18	1.50	1	2	3	3	3	2	2	2	2	2	3	2
46	46	40	1	48	17	2	1	32.8	33.5	2	1	5	1	0	2	22	6.00	2	4	2	2	4	2	2	4	2	2	1	3
47	47	40	5	21	14	2	2	16.3	60.7	2	1	5	1	0	2	3	0.66	0	3	3	3	2	2	3	2	1	1	4	4
48	48	40	1	26	16	2	2	44.7	65.3	1	1	4	1	0	2	13	3.50	1	3	3	2	1	1	1	2	1	1	2	1
49	49	40	1	39	18	1	1	75.6	91.8	1	1	5	2	1	2	23	3.00	0	5	3	3	1	1	1	1	1	1	1	1
50	50	40	1	29	17	2	1	50.1	50.1	1	1	5	1	0	2	20	16.00	0	1	1	1	1	1	2	2	3	3	1	3

Table 2.3 **(Continued)**

OBS	IDNUM	WORKHRS	OCCUP	AGE	EDUC	SEX	EARNRS	RINCOME	FINCOME	SATJOB	RICHWORK	JOBCHAR	GETAHEAD	TRAUMA	MEMUNION	WRKYEARS	EMPYEARS	NUMPROMO	FUTPROMO	SEXPROMO	ADVANCES	IDECIDE	ORGMONEY	PRODUORG	STAYORG	UNMAREL	COWRKREL	SCHOOLNG	TRAINING
51	51	40	2	31	18	1	1	27.6	29.4	2	2	5	1	0	2	9	4.00	0	2	3	3	2	1	2	2	1	1	1	2
52	52	45	2	33	19	1	1	36.3	37.4	2	1	5	1	0	2	7	4.00	1	1	1	2	2	1	2	2	2	1	2	2
53	53	32	1	36	18	1	2	30.3	71.3	4	1	5	2	0	1	16	4.08	1	4	2	2	1	2	3	4	5	3	1	1
54	54	62	5	28	12	1	2	41.5	47.1	1	1	4	2	0	2	12	5.00	3	1	3	1	1	1	1	1	2	1	4	1
55	55	40	2	40	17	2	1	29.8	44.1	4	1	5	3	0	1	23	4.50	0	5	2	3	4	1	4	5	4	4	1	4
56	56	40	1	25	16	1	2	32.2	43.2	1	1	4	1	0	2	2	2.00	0	1	1	3	1	2	2	4	4	2	1	1
57	57	40	3	39	16	2	1	40.2	45.2	2	2	5	1	0	2	14	2.00	1	1	3	2	3	2	2	4	3	3	2	2
58	58	40	4	27	12	1	3	22.5	32.5	1	1	5	1	0	2	10	5.00	0	4	1	2	2	1	1	2	1	1	1	1
59	59	40	1	40	17	1	2	20.1	35.1	2	1	1	1	0	2	18	11.83	4	1	3	2	1	1	2	4	2	2	1	1
60	60	40	5	27	9	2	3	11.6	48.9	2	2	5	1	0	2	4	4.00	0	5	3	3	1	2	1	2	1	1	2	3
61	61	50	1	53	16	1	3	50.6	75.1	2	1	1	3	0	2	36	10.00	0	2	1	2	2	1	2	4	3	1	2	2
62	62	46	1	39	13	2	2	22.5	50.3	1	1	5	1	0	2	23	1.00	0	5	3	3	1	1	1	1	1	1	1	1
63	63	72	6	47	14	1	1	56.7	59.5	1	1	1	1	0	2	27	5.00	0	5	1	3	3	2	1	5	3	3	4	1
64	64	40	1	27	11	2	2	15.5	30.4	4	2	2	2	1	2	5	2.00	2	5	3	1	2	2	2	4	2	3	1	3
65	65	40	6	40	10	2	2	17.0	19.3	2	1	5	3	0	2	13	4.00	0	5	3	3	4	2	2	5	2	2	4	2
66	66	30	7	27	12	1	1	16.5	17.0	2	1	5	1	2	2	27	0.50	0	1	3	3	4	2	3	5	1	1	4	2
67	67	50	1	35	15	1	1	43.3	43.3	1	1	1	1	1	2	18	10.00	0	5	1	1	2	2	2	4	1	3	1	2
68	68	40	3	25	16	1	3	22.0	52.3	1	1	5	1	0	2	8	2.50	0	5	1	2	1	1	1	2	1	1	1	2
69	69	35	7	53	9	1	1	10.2	15.6	1	1	5	1	1	2	34	5.00	0	1	3	3	2	2	4	5	3	3	2	2
70	70	40	2	35	16	1	1	41.3	41.3	2	1	5	1	0	2	13	6.41	3	1	1	2	2	1	2	2	2	1	2	1
71	71	48	7	43	12	1	2	54.2	67.9	1	2	1	1	0	2	26	11.00	2	1	1	2	1	1	1	4	3	1	1	1
72	72	40	2	33	14	2	3	19.8	62.5	2	2	1	1	1	1	12	2.00	1	2	3	2	3	1	2	5	2	1	1	2
73	73	40	6	48	12	1	3	50.0	54.1	2	1	4	3	2	1	32	21.00	4	5	2	2	4	2	2	5	3	1	4	3
74	74	55	1	29	16	1	1	23.6	34.8	2	1	5	1	0	2	11	1.00	1	1	1	1	2	1	3	4	1	2	1	4
75	75	40	4	26	13	2	1	19.3	19.6	2	2	3	1	0	2	10	4.00	5	4	2	2	1	1	2	4	2	2	3	2
76	76	40	1	23	16	2	1	16.0	16.9	1	1	2	1	0	2	6	0.75	0	1	1	3	1	1	1	1	2	1	1	4
77	77	48	4	30	14	2	2	18.1	41.8	2	1	1	2	1	2	4	1.00	1	1	3	2	1	2	2	4	1	2	1	1
78	78	40	4	53	12	2	1	21.7	35.7	2	1	5	2	0	2	15	12.50	1	1	3	2	3	2	2	4	4	2	2	1
79	79	40	6	53	12	1	2	39.0	43.9	2	2	5	1	0	1	30	0.08	0	2	3	2	2	2	2	3	2	2	2	1
80	80	40	7	43	14	1	1	30.9	30.9	1	1	5	3	0	2	37	2.00	0	4	3	3	3	2	2	3	1	1	4	3
81	81	40	6	30	13	1	2	32.3	39.3	2	1	5	1	0	2	11	10.00	2	2	3	2	2	2	2	3	2	2	3	1
82	82	40	4	27	14	1	2	17.9	46.5	2	2	5	1	2	2	17	0.50	0	2	3	3	3	2	2	3	2	2	2	2
83	83	35	2	59	15	1	2	39.8	69.0	1	1	5	1	0	2	42	6.00	0	5	3	3	1	1	1	2	1	2	1	4
84	84	40	2	36	16	2	1	37.2	38.8	2	1	4	2	0	2	19	12.50	1	4	3	3	3	2	2	2	3	1	1	1
85	85	40	2	52	16	1	2	54.6	56.7	4	2	5	1	0	1	30	21.00	1	4	3	3	2	1	2	3	3	2	2	2
86	86	50	6	40	14	1	2	18.8	57.4	1	1	5	3	0	2	24	1.50	1	5	3	1	1	1	1	1	1	1	1	1
87	87	59	7	29	17	1	2	39.6	46.6	1	1	1	2	0	2	8	2.83	0	5	3	3	1	2	2	2	4	2	3	3
88	88	40	3	40	14	1	2	78.0	92.9	1	1	5	1	0	2	23	4.33	0	1	1	3	1	1	3	4	5	2	1	3
89	89	45	1	36	12	2	1	25.2	25.9	3	2	1	1	0	2	20	10.00	1	2	2	2	3	2	2	3	2	1	4	1
90	90	45	2	38	20	1	1	64.5	65.7	1	1	5	3	0	2	21	4.00	2	1	1	1	1	1	2	2	2	2	1	4
91	91	43	1	64	13	2	1	28.6	35.2	1	2	4	1	0	2	21	7.00	1	5	3	2	1	1	1	1	2	2	2	1
92	92	55	1	43	16	2	2	61.7	78.8	1	1	5	1	0	2	25	18.00	2	5	3	2	2	1	2	2	2	1	2	4
93	93	46	1	28	16	2	1	35.1	38.3	2	2	5	1	0	1	12	0.16	0	2	2	3	2	2	2	5	2	2	1	2
94	94	40	4	52	14	2	1	20.6	47.3	2	1	5	1	0	2	20	3.00	1	2	1	3	1	2	2	4	2	1	2	2
95	95	30	3	48	12	2	2	13.7	36.1	1	2	5	1	0	2	15	1.66	0	5	1	3	1	2	1	2	1	3	1	1
96	96	40	3	52	12	1	2	40.8	96.2	3	1	5	1	1	2	35	0.08	0	5	1	3	1	2	2	4	1	1	3	1
97	97	40	4	43	12	2	2	19.7	57.8	1	2	5	1	0	2	27	9.41	1	4	3	3	3	2	1	2	2	2	1	2
98	98	36	4	39	13	2	3	27.1	31.2	2	2	5	1	0	2	21	16.00	0	5	2	2	2	1	2	4	3	2	2	1
99	99	50	6	29	15	1	1	39.2	63.6	1	2	4	2	0	1	8	1.00	0	4	3	3	2	2	2	4	2	1	2	2
100	100	48	5	34	14	1	2	21.1	33.4	1	1	5	1	0	2	17	10.00	2	1	3	2	1	1	2	4	1	1	1	2

Table 2.3 (Continued)

OBS	IDNUM	WORKHRS	OCCUP	AGE	EDUC	SEX	EARNRS	RINCOME	FINCOME	SATJOB	RICHWORK	JOBCHAR	GETAHEAD	TRAUMA	MEMUNION	WRKYEARS	EMPYEARS	NUMPROMO	FUTPROMO	SEXPROMO	ADVANCES	IDECIDE	ORGMONEY	PRODORG	STAYORG	UNMANREL	COWKREL	SCHOOLNG	TRAINING
101	101	50	5	25	14	1	1	28.4	29.9	2	1	2	1	0	1	8	3.00	0	1	3	3	3	2	4	2	5	4	3	3
102	102	42	6	32	12	1	1	15.0	15.0	3	1	4	3	0	2	14	1.00	0	4	1	3	3	2	3	4	5	2	2	2
103	103	40	7	49	12	2	4	14.6	82.5	2	1	1	1	0	2	20	4.00	0	3	3	2	2	2	2	2	1	1	1	1
104	104	45	4	40	12	2	1	17.8	17.8	1	1	1	1	0	2	20	0.50	0	1	3	3	2	1	1	4	1	1	3	3
105	105	32	2	39	18	1	3	29.2	62.6	1	1	3	2	0	1	18	2.00	0	2	3	3	1	2	3	4	3	2	1	4
106	106	40	3	32	16	1	1	23.0	23.5	2	1	1	1	0	2	15	3.00	3	2	1	1	3	2	2	2	2	2	2	1
107	107	40	6	25	10	1	1	20.0	20.0	3	2	4	3	0	1	8	2.00	3	1	1	2	1	1	2	2	2	1	2	4
108	108	36	4	42	16	2	1	23.1	34.2	4	2	5	2	0	1	25	12.00	1	1	1	3	4	2	3	5	4	1	4	4
109	109	40	4	37	13	1	4	26.8	34.0	3	1	4	1	0	1	18	16.00	0	2	1	3	2	2	2	2	5	3	2	1
110	110	50	4	28	12	2	1	19.3	19.3	2	2	1	1	0	1	8	3.00	0	2	1	3	2	2	2	4	3	3	3	2
111	111	60	3	23	16	1	3	10.3	38.4	2	1	1	3	0	2	6	1.00	1	1	3	2	2	2	3	4	2	1	2	1
112	112	40	4	37	12	2	2	16.7	26.5	2	2	1	1	0	2	20	10.00	0	5	3	3	2	1	2	4	1	2	3	1
113	113	43	4	35	12	2	1	15.9	15.9	1	1	5	1	0	2	9	1.50	1	1	2	2	4	2	2	4	3	1	1	1
114	114	35	4	63	12	2	1	25.8	38.7	1	2	2	3	0	1	45	22.00	1	5	3	2	2	2	1	1	1	1	1	1
115	115	41	4	29	13	1	2	23.9	43.0	2	1	4	3	0	2	10	3.00	2	2	3	2	1	1	2	2	2	2	3	1
116	116	50	7	33	14	1	2	18.9	22.4	2	2	5	2	1	2	16	1.00	0	1	1	3	2	1	2	5	3	5	1	4
117	117	44	6	26	12	1	1	36.3	36.3	2	1	1	1	0	1	10	2.50	0	4	3	3	3	2	2	5	3	3	4	1
118	118	40	2	50	20	2	2	58.0	98.7	2	1	3	3	0	2	34	8.00	0	5	2	3	2	1	2	4	4	3	1	4
119	119	40	5	49	15	2	3	25.7	50.8	2	1	5	1	0	2	25	3.41	0	5	3	3	4	2	2	4	3	2	1	2
120	120	35	2	44	17	1	3	51.8	61.6	2	2	5	3	0	2	27	22.00	0	5	3	3	3	2	4	5	1	3	1	1
121	121	31	5	25	15	1	2	22.1	39.8	1	1	2	1	0	2	7	1.00	1	1	1	3	1	1	1	1	1	1	1	1
122	122	40	1	44	16	2	3	24.8	48.2	1	1	1	1	0	2	14	5.00	3	2	2	2	3	2	2	4	1	1	1	1
123	123	37	4	42	14	2	1	15.6	16.4	4	1	1	2	0	2	24	2.00	0	2	2	3	4	2	3	5	5	3	3	4
124	124	60	1	36	12	2	3	28.5	59.0	4	1	1	1	0	2	17	9.50	1	4	2	3	2	1	3	5	4	4	3	3
125	125	40	5	41	12	1	1	22.5	23.4	1	1	1	1	0	2	24	10.50	1	1	1	2	2	2	2	5	3	2	1	1
126	126	40	5	26	12	1	1	38.2	38.2	1	2	1	3	0	2	8	1.00	0	1	3	2	4	2	3	2	4	1	2	1
127	127	40	4	42	12	2	2	20.0	23.0	1	2	2	3	0	2	15	7.08	0	1	3	2	2	2	2	4	1	1	3	2
128	128	65	7	27	14	1	2	37.3	39.7	2	1	1	1	1	1	10	2.00	0	2	1	2	2	2	2	2	2	1	2	1
129	129	40	7	48	12	2	3	16.7	46.8	2	2	5	1	0	2	20	9.00	0	2	2	2	4	2	2	4	3	2	2	2
130	130	40	5	65	12	1	4	51.8	68.0	1	1	4	1	1	1	49	12.41	2	1	1	2	2	2	1	2	2	3	4	2
131	131	40	3	33	14	2	2	22.7	83.3	2	2	3	3	0	2	15	3.75	0	2	3	2	2	2	2	4	2	2	1	1
132	132	40	6	44	8	1	1	42.5	49.2	2	1	4	3	0	2	27	11.00	1	5	1	2	2	2	2	5	1	2	2	2
133	133	55	2	48	19	1	4	44.4	66.8	2	1	4	3	0	2	28	5.66	0	4	2	2	2	2	2	2	2	2	1	1
134	134	50	4	37	16	1	2	50.5	59.2	2	1	5	1	0	2	20	8.50	2	2	3	2	2	1	2	4	2	2	1	1
135	135	70	3	52	16	1	1	81.7	81.7	1	1	5	1	0	2	36	1.00	2	5	3	1	1	1	2	2	1	1	1	3
136	136	60	3	38	16	1	2	55.3	59.0	1	1	5	1	0	2	20	2.66	0	5	3	3	1	1	1	1	1	1	3	2
137	137	40	4	35	12	2	2	16.2	43.4	2	1	1	1	0	2	10	2.50	3	1	3	1	1	2	2	3	3	1	1	1
138	138	43	1	34	16	1	1	42.6	42.6	1	1	5	3	0	2	12	5.50	1	4	1	2	1	1	1	1	1	2	1	1
139	139	52	6	34	12	1	1	42.3	42.3	2	1	5	1	0	2	16	13.00	2	5	1	2	1	1	2	2	2	2	1	1
140	140	40	4	27	16	2	2	34.6	69.3	1	1	5	1	0	2	10	5.00	4	1	3	1	1	2	2	4	1	1	1	1
141	141	46	7	23	12	1	2	22.5	25.3	2	2	5	1	1	1	6	1.41	0	2	1	2	2	2	2	4	2	2	3	2
142	142	50	1	62	11	1	1	47.1	47.1	1	1	1	2	0	2	45	52.25	4	4	3	2	1	1	2	2	2	1	2	1
143	143	76	5	41	13	1	2	39.1	46.1	1	1	5	1	0	1	24	27.16	4	1	3	1	2	1	1	3	2	2	2	1
144	144	38	7	26	15	1	2	14.9	39.1	1	1	5	1	5	2	9	0.50	1	1	3	1	1	1	1	4	3	3	3	1
145	145	42	3	36	18	2	2	16.0	71.2	4	1	5	1	1	2	19	1.00	0	1	3	3	2	2	2	5	2	2	3	2
146	146	40	3	37	16	1	4	33.1	51.2	2	1	5	1	2	2	19	19.00	4	1	1	2	2	1	3	4	3	2	3	3
147	147	40	1	39	12	2	1	21.2	21.2	2	1	5	1	0	2	20	10.00	0	4	2	2	3	1	1	4	2	2	1	1
148	148	40	6	33	11	1	1	21.7	39.2	2	2	5	1	0	2	15	1.00	2	4	1	1	1	1	2	4	3	3	3	2
149	149	40	3	61	10	2	1	25.1	25.5	1	2	5	2	0	2	30	13.00	3	4	3	2	2	2	1	1	1	2	3	1
150	150	54	3	20	10	1	2	16.1	22.5	2	1	4	2	0	2	3	3.00	0	2	1	3	2	1	3	2	1	1	3	2

Table 2.3 (Continued)

OBS	IDNUM	WORKHRS	OCCUP	AGE	EDUC	SEX	EARNRS	RINCOME	FINCOME	SATJOB	RICHWORK	JOBCHAR	GETAHEAD	TRAUMA	MEMUNION	WRKYEARS	EMPYEARS	NUMPROMO	FUTPROMO	SEXPROMO	ADVANCES	IDECIDE	ORGMONEY	PRODORG	STAYORG	UNMAREL	COWRKREL	SCHOOLNG	TRAINING
151	151	60	6	31	12	1	1	30.4	32.4	2	1	5	2	0	2	14	12.50	3	4	1	2	2	2	2	2	2	2	3	2
152	152	50	2	28	17	1	2	28.6	31.1	2	1	5	1	0	2	6	4.58	2	1	3	2	2	2	2	2	1	1	1	1
153	153	40	3	22	14	1	1	25.2	25.2	2	1	5	2	0	1	9	1.00	0	3	3	3	2	1	2	5	3	2	1	1
154	154	40	3	36	17	1	1	30.6	30.6	1	1	5	1	0	2	15	7.00	3	2	1	2	3	1	2	4	3	1	1	1
155	155	45	3	48	16	2	1	41.5	42.1	1	2	4	2	0	2	26	4.00	0	1	2	3	1	2	1	4	1	1	1	4
156	156	41	2	47	19	2	1	27.0	36.3	4	2	1	1	0	2	15	1.00	0	5	3	4	4	2	3	5	2	3	1	3
157	157	67	6	51	12	1	3	30.9	47.4	1	2	4	3	1	1	33	32.33	0	5	1	3	2	1	4	5	1	2	1	4
158	158	44	7	48	12	1	2	35.1	63.2	2	1	2	3	0	1	31	24.75	2	2	2	2	2	2	3	2	3	2	4	1
159	159	37	7	30	14	2	2	13.9	43.4	2	2	5	1	0	1	10	4.00	0	5	3	2	2	2	2	4	5	4	4	1
160	160	40	5	40	10	2	2	12.1	15.0	1	1	1	1	0	1	20	17.00	0	5	3	3	4	2	2	2	3	3	2	1
161	161	42	4	29	15	2	2	22.7	43.1	2	1	2	3	0	2	13	4.33	0	2	3	3	1	1	2	4	1	1	2	1
162	162	48	7	35	16	1	2	26.7	46.6	2	1	1	1	0	1	19	12.00	1	2	1	3	2	1	2	4	3	2	2	1
163	163	40	4	61	13	2	1	20.0	20.0	1	1	1	2	0	1	25	18.33	0	5	3	3	1	1	2	4	2	1	2	2
164	164	38	5	38	14	1	3	22.7	49.5	2	1	2	1	0	2	21	2.66	2	2	3	1	2	1	2	2	2	2	2	2
165	165	40	5	40	12	1	2	27.1	66.2	1	2	2	1	0	2	23	3.00	1	1	1	2	2	2	1	2	3	2	2	1
166	166	40	4	32	12	2	2	19.7	38.9	2	1	5	1	1	1	15	11.58	3	1	3	1	1	2	2	5	3	2	2	3
167	167	52	7	32	12	1	1	45.3	45.3	1	1	5	1	0	2	14	13.00	3	5	1	3	2	2	2	4	2	2	2	2
168	168	40	7	33	12	1	2	34.0	46.4	3	1	5	1	1	1	12	11.00	0	5	3	3	1	2	4	5	3	3	4	1
169	169	60	2	39	20	1	2	57.8	78.6	1	1	5	1	0	1	22	17.00	2	2	1	3	2	2	2	4	2	2	1	3
170	170	45	1	58	20	2	2	33.3	35.5	1	2	4	2	0	2	20	12.00	0	4	2	3	3	2	2	4	2	2	1	2
171	171	52	7	47	16	1	1	45.3	46.0	2	1	5	3	0	2	30	8.00	0	5	3	2	4	2	2	3	2	1	4	4
172	172	40	3	50	14	1	2	38.1	50.5	3	1	5	2	0	1	32	22.50	2	4	1	3	2	2	2	4	3	2	3	3
173	173	50	1	24	16	1	2	14.7	25.3	2	1	5	2	0	2	7	1.00	0	2	1	2	2	2	2	4	1	1	2	3
174	174	40	2	30	12	1	1	35.5	35.5	1	1	1	1	0	2	13	1.50	0	5	3	3	1	2	1	5	1	1	2	1
175	175	60	2	31	20	1	2	48.6	51.2	1	1	5	1	0	1	15	3.00	1	1	1	2	2	2	2	2	2	2	1	4
176	176	60	6	37	13	1	1	44.8	45.5	2	1	5	2	0	2	18	3.00	0	4	2	2	1	1	1	2	2	2	3	1
177	177	50	3	44	13	2	4	24.7	56.6	1	1	1	1	0	2	27	10.00	0	4	2	1	1	2	1	1	2	1	1	1
178	178	50	2	41	15	1	3	40.1	45.1	2	2	4	1	0	2	25	6.00	2	4	3	2	2	1	2	5	2	3	3	4
179	179	55	2	46	18	1	2	51.4	90.4	1	1	5	2	0	2	29	7.00	4	2	1	1	1	1	2	4	1	2	2	2
180	180	49	2	57	18	1	3	29.9	34.6	2	1	5	1	0	2	38	34.00	0	4	3	3	2	1	1	2	1	2	1	2
181	181	32	7	61	8	1	2	23.7	26.6	2	1	5	1	1	1	40	40.00	0	1	1	3	4	2	2	2	2	2	4	2
182	182	40	2	58	10	1	1	26.4	26.4	1	1	1	2	0	2	40	3.91	0	4	3	3	2	2	2	4	3	1	4	1
183	183	40	6	49	13	1	2	30.8	58.7	4	1	5	2	0	2	25	6.00	0	4	3	3	2	2	2	4	2	2	4	4
184	184	40	3	30	13	2	2	13.6	21.9	4	2	5	2	0	2	5	2.00	0	4	3	3	2	2	2	4	2	2	1	1
185	185	40	4	55	12	2	1	20.5	20.5	2	2	1	2	0	2	20	15.00	2	2	3	2	2	2	2	4	2	2	1	1
186	186	40	3	32	16	1	1	45.7	45.7	2	1	5	1	0	2	15	7.00	1	2	1	3	2	2	2	2	1	1	3	1
187	187	32	4	29	12	2	2	18.9	21.9	1	1	5	1	0	2	12	3.00	1	2	3	2	2	2	2	3	2	2	3	1
188	188	40	2	47	18	2	1	37.9	37.9	3	2	3	1	0	2	30	2.00	0	5	2	3	2	2	2	5	4	4	1	3
189	189	60	1	32	12	1	2	47.7	69.7	2	1	1	1	0	2	15	12.00	5	5	3	2	1	1	1	2	3	3	1	3
190	190	40	3	44	16	1	2	39.0	69.6	1	2	4	1	0	2	27	22.00	4	2	2	2	1	1	2	4	1	1	3	1
191	191	55	1	34	20	1	1	32.4	32.4	1	1	5	1	0	2	11	4.00	1	2	3	2	1	1	2	2	2	1	1	1
192	192	60	7	38	14	1	2	31.2	50.0	2	1	5	1	0	2	21	15.00	2	1	2	2	2	1	1	2	3	3	2	1
193	193	60	1	42	12	2	3	26.7	38.2	2	2	5	1	2	2	26	22.00	3	2	3	2	2	1	2	4	4	2	2	3
194	194	42	7	30	12	1	1	16.4	17.3	3	2	5	2	1	1	13	1.50	0	4	1	3	2	2	2	4	3	2	4	3
195	195	40	7	29	10	1	2	26.1	28.6	2	2	2	3	0	1	13	0.25	0	2	3	3	2	2	2	3	1	2	3	1
196	196	40	4	38	12	2	2	13.2	16.7	2	2	5	1	1	2	22	1.00	0	1	1	2	4	2	2	4	3	1	3	1
197	197	40	4	32	12	2	1	38.7	38.7	2	2	1	1	0	2	8	6.00	0	4	3	3	2	2	1	4	2	2	2	2
198	198	40	7	55	14	2	2	11.9	24.0	1	2	5	1	1	2	35	29.50	1	2	3	3	4	2	1	4	1	3	3	4
199	199	50	5	47	16	1	4	59.3	89.2	1	1	5	1	0	2	30	23.58	3	1	3	2	1	1	2	2	1	1	1	1
200	200	70	1	40	16	1	1	42.3	42.3	1	2	5	1	0	2	18	2.00	0	2	3	1	1	1	1	1	2	2	1	1

Table 2.3 (Continued)

OBS	IDNUM	WORKHRS	OCCUP	AGE	EDUC	SEX	EARNRS	RINCOME	FINCOME	SATJOB	RICHWORK	JOBCHAR	GETAHEAD	TRAUMA	MEMUNION	WRKYEARS	EMPYEARS	NUMPROMO	FUTPROMO	SEXPROMO	ADVANCES	IDECIDE	ORGMONEY	PRODUORG	STAYORG	UNMARREL	COWRKREL	SCHOOLNG	TRAINING
201	201	40	3	37	16	1	1	26.9	26.9	1	2	5	1	0	2	11	5.00	2	1	1	1	1	2	2	4	1	1	2	2
202	202	40	5	43	14	2	2	18.7	21.3	1	1	1	2	0	2	24	22.25	1	4	3	3	3	2	1	4	2	3	1	2
203	203	55	1	32	12	2	2	23.0	32.2	2	1	5	1	0	2	16	11.00	1	4	3	4	1	1	2	5	2	2	4	1
204	204	40	3	31	14	2	2	17.3	27.6	1	1	1	1	0	2	16	10.00	2	1	1	2	2	1	2	2	3	3	1	1
205	205	89	3	44	12	1	2	38.3	46.1	1	1	4	1	0	1	24	0.66	0	5	3	3	1	1	1	5	1	1	2	1
206	206	50	1	39	12	1	2	59.3	74.0	1	1	5	2	0	2	18	11.00	2	4	1	2	1	1	1	2	2	1	3	2
207	207	72	4	41	12	1	1	16.3	16.3	1	1	5	3	1	2	24	18.00	0	2	2	2	1	1	2	4	2	3	2	2
208	208	40	3	30	14	1	2	34.4	38.3	2	1	1	1	0	2	13	11.00	3	5	1	2	2	2	2	4	2	1	2	1
209	209	65	7	58	8	1	3	31.4	42.3	2	1	2	1	0	2	41	3.00	0	4	1	3	4	2	2	2	3	3	3	1
210	210	75	2	55	18	1	2	53.8	74.3	3	1	4	3	4	2	38	29.00	2	5	1	3	1	1	1	2	1	2	1	1
211	211	68	6	60	12	1	1	26.4	35.4	1	1	1	1	0	1	54	10.00	4	2	1	1	1	2	2	4	2	2	1	1
212	212	40	4	29	16	2	3	14.4	28.3	3	1	1	1	0	2	6	3.00	0	4	3	3	4	2	3	4	5	1	1	1
213	213	40	4	42	12	2	1	13.4	29.7	1	1	4	1	0	2	26	1.08	1	1	1	2	4	2	1	4	1	1	1	1
214	214	38	3	41	16	1	1	23.1	26.9	2	1	4	1	0	2	12	12.00	0	3	3	3	2	2	2	4	1	1	1	1
215	215	48	7	35	14	1	2	12.0	15.8	2	2	2	1	0	2	19	7.00	1	2	3	3	3	2	2	4	3	2	4	3
216	216	50	6	24	12	1	2	22.8	34.5	2	1	5	1	0	2	7	3.00	0	2	1	3	1	2	2	2	3	3	2	1
217	217	45	1	39	16	1	1	30.7	34.1	3	1	4	3	0	2	22	9.00	2	4	1	3	4	2	2	5	3	1	2	1
218	218	35	4	38	13	2	1	15.3	15.3	1	2	5	1	1	2	17	4.00	1	1	3	2	1	1	1	4	3	2	1	1
219	219	35	2	37	16	2	1	27.9	35.2	2	2	1	1	0	2	20	14.00	0	2	3	3	2	1	2	4	2	2	1	2
220	220	60	4	47	14	2	3	15.5	68.7	4	2	5	3	0	2	21	1.00	0	1	1	3	3	2	2	5	2	1	3	1
221	221	40	4	38	12	2	4	27.3	72.6	1	1	4	1	0	2	20	10.50	4	4	2	2	3	2	1	1	1	1	4	1
222	222	45	4	41	12	1	1	45.8	49.4	1	2	5	1	0	2	24	8.00	4	1	1	1	3	2	1	1	1	1	2	1
223	223	35	2	41	20	2	2	37.7	61.6	1	1	3	2	0	1	24	17.00	0	5	3	3	1	1	1	1	1	1	1	1
224	224	60	7	38	15	1	2	29.5	49.7	1	1	1	3	0	2	21	4.00	1	5	1	3	1	2	2	5	3	2	1	3
225	225	45	7	29	12	1	2	25.9	33.5	2	2	1	1	0	2	11	3.00	0	2	1	3	1	2	2	5	2	2	4	3
226	226	40	6	38	12	1	2	26.8	34.1	2	2	5	2	0	2	20	7.00	0	4	3	3	3	1	3	4	3	2	3	2
227	227	40	1	33	15	1	1	41.2	41.2	2	1	5	3	0	2	11	10.00	0	5	1	1	1	1	1	4	1	1	4	2
228	228	45	2	26	16	2	1	20.4	53.9	2	2	1	1	0	2	4	4.00	0	5	3	3	1	2	1	1	2	1	1	2
229	229	60	3	25	16	1	1	26.0	39.9	2	1	4	1	0	2	8	3.00	0	5	1	1	1	1	2	5	2	2	1	1
230	230	45	5	62	11	1	2	19.0	34.2	2	1	5	1	0	2	42	23.00	0	5	3	3	2	2	1	1	3	2	2	1
231	231	40	7	42	12	1	2	13.1	25.8	1	2	1	1	0	1	20	12.83	0	4	1	3	1	2	2	2	1	3	1	2
232	232	40	3	31	14	2	2	23.4	45.6	1	2	5	2	0	2	11	5.50	0	5	3	3	2	2	2	4	5	1	1	1
233	233	56	4	33	9	1	1	19.3	23.2	1	1	4	1	0	2	16	4.50	0	3	3	2	2	1	2	4	2	1	1	1
234	234	55	7	21	12	2	4	11.1	35.7	1	2	5	1	0	1	4	2.00	0	1	3	3	3	2	2	2	2	1	2	4
235	235	40	5	50	9	2	2	13.7	24.8	2	2	1	1	0	2	20	10.00	0	1	3	3	4	2	1	5	3	1	3	1
236	236	40	4	36	12	2	1	62.8	63.4	1	1	4	1	0	2	19	2.16	0	5	3	1	1	1	1	1	2	2	2	1
237	237	50	1	41	16	2	1	25.2	26.9	1	1	5	1	0	2	22	2.33	0	2	3	2	2	2	1	4	3	3	1	1
238	238	57	2	27	13	1	2	21.1	28.3	2	1	5	1	0	2	10	0.33	1	2	1	2	3	2	2	4	2	3	2	3
239	239	47	3	42	17	1	1	36.6	39.2	1	2	2	2	2	2	25	5.00	0	4	1	3	1	1	2	5	2	2	1	1
240	240	60	2	28	12	1	2	29.3	44.0	3	1	5	1	0	2	11	5.50	5	2	1	1	1	1	2	5	3	1	1	1
241	241	40	7	46	8	1	4	19.3	24.0	2	1	1	1	1	2	30	2.00	0	1	3	4	2	2	4	5	3	1	1	4
242	242	40	2	32	18	2	1	32.2	32.2	1	1	5	1	0	2	14	5.00	0	5	3	3	2	1	1	2	3	3	1	2
243	243	46	6	40	15	1	1	43.6	51.4	1	1	5	1	0	2	22	18.00	6	2	1	2	3	1	1	2	2	2	1	1
244	244	40	2	39	16	2	2	28.1	68.7	2	1	3	1	1	2	20	5.00	1	2	1	2	2	1	2	2	1	1	1	1
245	245	43	5	30	14	2	2	10.3	28.9	1	2	5	3	3	2	13	0.25	1	1	3	1	1	2	2	1	3	2	3	3
246	246	45	3	33	17	2	1	25.0	26.3	2	2	5	1	0	2	16	1.00	0	4	2	3	3	2	2	5	5	3	1	1
247	247	45	4	27	15	2	1	17.8	17.8	1	1	5	3	0	2	10	2.00	0	2	3	1	1	2	1	1	1	1	1	1
248	248	40	4	37	14	2	1	15.4	15.4	1	1	1	1	1	2	15	15.00	3	1	2	2	2	2	2	2	2	2	4	2
249	249	40	6	56	13	1	1	15.3	15.3	1	1	5	1	0	1	34	1.83	0	2	1	3	1	1	2	2	1	2	1	3
250	250	40	5	30	14	1	2	14.4	32.4	2	1	5	2	0	2	14	6.00	0	5	1	3	2	2	3	5	3	2	2	3

Table 2.3 (Continued)

OBS	IDNUM	WORKHRS	OCCUP	AGE	EDUC	SEX	EARNRS	RINCOME	FINCOME	SATJOB	RICHWORK	JOBCHAR	GETAHEAD	TRAUMA	MEMUNION	WRKYEARS	EMPYEARS	NUMPROMO	FUTPROMO	SEXPROMO	ADVANCES	IDECIDE	ORGMONEY	PRODUORG	STAYORG	UNMANREL	COWORKREL	SCHOOLNG	TRAINING
251	251	65	1	36	14	2	1	25.9	26.6	1	2	5	3	0	2	19	6.00	0	5	3	2	1	1	1	2	1	1	1	4
252	252	40	4	32	12	2	2	15.0	25.4	2	2	3	1	0	2	11	10.00	0	4	3	3	2	2	2	2	1	2	2	4
253	253	50	7	44	13	1	2	30.3	37.7	2	2	1	1	1	1	25	15.00	0	5	1	3	4	2	2	4	5	3	1	1
254	254	80	4	48	14	1	2	52.7	56.1	4	2	1	3	0	2	31	9.00	1	5	3	2	4	2	3	5	5	5	1	1
255	255	40	1	39	16	1	2	45.7	68.5	2	2	4	3	0	2	17	16.00	1	2	3	2	2	2	2	4	3	2	1	1
256	256	45	2	31	20	2	1	28.6	28.6	2	1	5	2	0	2	14	1.00	0	4	2	2	2	1	2	4	4	2	3	4
257	257	60	1	39	16	1	1	16.2	16.8	1	1	1	1	2	2	20	3.25	0	5	3	3	1	1	1	1	1	1	4	1
258	258	40	1	62	15	1	2	44.7	49.4	1	2	2	1	0	2	45	27.00	2	5	1	2	1	1	2	2	1	1	2	2
259	259	72	6	21	12	1	4	18.1	77.9	1	1	1	1	0	2	5	1.08	0	5	3	2	1	2	2	4	2	2	1	1
260	260	40	1	62	14	1	3	55.0	85.4	1	1	1	1	0	2	45	1.91	2	2	2	1	3	2	2	4	4	3	4	1
261	261	45	3	34	12	1	3	27.5	45.2	1	1	4	2	0	2	17	2.08	1	2	3	3	1	1	1	4	2	1	3	3
262	262	40	7	35	11	1	1	25.2	28.4	2	2	1	3	0	1	16	13.00	0	4	3	1	3	2	2	4	2	2	4	2
263	263	45	7	41	12	1	1	21.4	21.4	1	2	1	1	0	1	22	0.25	0	1	3	3	2	2	2	4	2	2	2	4
264	264	50	2	45	16	2	2	25.5	70.4	2	1	1	1	0	2	21	20.00	3	4	1	3	3	1	2	5	2	2	1	2
265	265	40	7	54	10	1	3	16.1	26.4	1	2	2	2	0	2	33	33.00	2	5	3	3	2	2	2	3	2	2	4	2
266	266	40	6	51	16	1	1	31.9	33.5	2	1	5	3	0	2	34	6.00	1	2	1	2	2	2	2	4	3	1	1	1
267	267	40	1	52	14	1	1	50.4	53.7	1	1	2	2	0	2	35	31.91	3	2	1	4	2	1	1	2	2	2	1	3
268	268	60	6	29	13	1	2	16.2	41.2	1	1	1	3	0	2	13	5.00	1	1	1	2	1	2	2	2	3	2	1	2
269	269	40	4	44	18	2	2	40.7	49.2	3	1	1	2	0	1	24	22.00	0	5	2	3	2	2	2	4	3	2	3	1
270	270	50	1	40	16	1	2	54.9	82.7	3	1	5	1	0	2	20	6.00	1	5	1	3	2	2	2	4	3	3	1	2
271	271	40	2	41	18	2	2	18.1	47.1	1	1	5	1	0	2	24	0.25	0	1	3	3	3	2	1	2	1	1	1	2
272	272	45	2	31	12	1	1	50.9	55.1	1	3	1	1	0	2	14	0.66	0	1	1	3	2	1	1	2	3	1	3	1
273	273	37	4	38	17	2	1	14.2	49.2	2	2	5	1	0	2	20	5.00	0	5	3	3	4	2	2	2	2	2	1	3
274	274	46	1	43	12	2	1	27.6	27.6	2	1	5	3	0	2	28	14.00	1	5	3	2	2	1	2	4	2	1	3	2
275	275	40	6	40	14	1	1	26.1	26.1	2	2	5	2	0	2	22	15.00	1	5	3	2	2	1	3	5	3	3	2	4
276	276	40	1	41	19	2	2	23.9	55.3	2	1	5	1	0	1	26	4.00	0	4	3	3	2	1	2	4	3	2	1	1
277	277	48	3	64	12	1	2	58.0	98.6	1	2	5	1	1	2	48	29.00	3	5	3	2	2	2	2	4	2	2	1	1
278	278	51	2	40	13	2	2	27.0	29.7	1	1	5	1	0	2	23	17.00	0	5	2	3	3	2	2	5	2	2	1	1
279	279	40	7	52	8	1	1	33.0	33.2	4	2	3	1	1	2	35	20.00	2	5	1	4	3	2	3	2	5	1	2	2
280	280	40	4	41	13	2	2	17.1	32.4	2	1	5	1	0	2	15	4.00	0	2	3	2	2	2	2	4	2	1	3	3
281	281	40	6	38	12	1	2	26.6	29.6	1	1	1	1	0	1	20	11.00	3	5	1	2	2	2	2	3	3	1	1	1
282	282	29	4	69	12	2	1	16.8	16.8	1	2	5	1	0	2	27	25.00	3	3	1	2	2	2	2	2	1	1	2	4
283	283	40	2	26	18	2	1	26.0	26.0	1	1	5	1	0	1	10	1.00	0	4	3	3	1	1	1	1	2	1	1	2
284	284	40	3	26	14	1	1	20.8	20.8	1	2	4	2	0	2	9	1.50	1	1	1	1	2	2	2	4	2	2	4	2
285	285	40	1	45	12	1	1	21.3	44.1	2	2	5	3	1	2	28	0.58	1	4	1	2	2	1	2	4	4	3	4	2
286	286	40	2	26	16	2	1	23.4	29.0	1	1	5	1	0	2	9	8.00	3	2	3	2	2	2	1	2	2	1	2	2
287	287	55	1	33	14	1	1	41.2	48.8	2	2	1	2	1	2	16	7.00	4	4	3	3	1	1	2	4	2	1	3	1
288	288	32	5	38	12	2	2	20.8	59.4	1	1	5	1	0	2	19	3.50	1	3	3	2	2	2	2	2	2	2	2	1
289	289	40	6	36	14	1	1	39.2	39.8	1	1	2	1	0	2	20	18.00	2	2	1	2	1	2	1	2	3	2	3	1
290	290	60	3	36	16	1	1	51.6	51.6	1	1	3	2	0	2	12	5.00	0	5	3	3	1	2	2	4	1	1	2	1
291	291	36	1	45	13	2	2	28.1	60.3	2	2	5	1	0	2	25	21.00	4	2	2	2	1	1	2	4	3	3	2	1
292	292	50	7	45	12	1	2	27.5	47.6	1	1	1	1	0	1	29	17.00	2	2	1	2	1	2	1	1	2	1	1	1
293	293	40	5	34	14	2	1	19.0	19.5	1	2	5	1	2	2	16	4.00	1	2	3	2	1	2	1	2	2	2	3	1
294	294	40	1	51	13	2	2	39.6	85.9	1	2	5	2	0	2	30	24.00	4	2	1	2	2	2	2	2	1	2	2	1
295	295	40	1	38	16	2	2	30.2	90.5	1	2	1	3	0	2	6	6.00	4	1	3	2	1	2	2	2	3	2	1	1
296	296	32	4	58	13	2	1	23.2	23.2	2	2	5	2	1	2	37	5.00	0	4	1	3	4	2	2	4	1	1	2	4
297	297	50	6	54	14	1	2	51.2	66.7	1	1	5	1	0	2	37	5.00	0	4	1	3	2	1	2	4	2	2	4	1
298	298	32	1	57	12	1	2	76.4	94.2	1	1	1	1	1	2	38	34.00	0	5	3	2	1	1	1	2	2	2	2	1
299	299	54	2	43	18	1	1	19.5	24.6	2	1	2	3	1	2	27	13.00	2	1	1	2	1	2	1	4	4	3	1	1
300	300	45	4	40	13	2	3	17.7	31.4	2	1	5	1	0	2	20	7.00	1	2	1	2	1	2	3	1	3	1	4	1

Table 2.3 **(Continued)**

OBS	IDNUM	WORKHRS	OCCUP	AGE	EDUC	SEX	EARNRS	RINCOME	FINCOME	SATJOB	RICHWORK	JOBCHAR	GETAHEAD	TRAUMA	MEMUNION	WRKYEARS	EMPYEARS	NUMPROMO	FUTPROMO	SEXPROMO	ADVANCES	IDECIDE	ORGMONEY	PRODUORG	STAYORG	UNMANREL	COWRKREL	SCHOOLNG	TRAINING
301	301	40	1	45	16	2	4	29.4	53.3	1	1	5	2	0	1	25	4.00	0	2	1	1	1	1	1	1	2	1	1	1
302	302	50	7	54	14	1	1	42.8	42.8	2	2	5	1	0	2	38	27.00	2	5	3	2	2	2	1	4	2	1	2	1
303	303	40	1	43	13	1	2	14.7	23.3	1	1	5	1	0	2	26	26.00	0	5	3	3	1	1	1	4	2	2	3	2
304	304	35	7	48	16	2	1	20.5	20.9	2	3	5	3	0	2	27	3.00	0	5	2	4	3	2	2	5	3	3	4	2
305	305	40	2	44	16	1	1	20.4	20.4	1	1	5	1	0	2	27	2.00	0	4	3	3	1	2	2	2	2	2	1	1
306	306	40	7	57	11	2	2	18.2	30.1	2	1	1	1	0	2	45	15.00	2	4	3	3	4	2	2	5	2	3	4	2
307	307	40	4	20	12	2	2	15.3	27.1	1	2	4	1	0	2	3	1.50	0	2	1	3	1	2	2	3	1	1	2	1
308	308	28	2	49	19	2	1	43.4	45.3	1	1	5	1	0	2	25	1.00	3	1	2	2	2	2	2	4	3	2	1	1
309	309	40	2	39	14	1	1	38.5	38.5	1	2	5	1	1	2	18	10.00	1	1	1	2	3	2	2	4	3	1	1	1
310	310	40	6	41	14	1	1	36.6	39.4	2	1	5	2	0	2	21	11.00	1	3	1	2	1	1	2	2	2	2	3	1
311	311	40	3	37	15	1	1	31.2	31.2	1	3	5	1	0	2	20	5.00	0	4	3	2	2	1	2	4	1	2	2	2
312	312	40	4	65	13	2	1	18.1	18.2	1	2	5	1	0	2	22	13.00	1	5	3	3	2	2	1	2	2	2	3	2
313	313	40	2	50	18	1	1	30.4	30.4	1	2	3	2	0	1	33	23.00	3	2	2	4	1	1	1	5	2	2	1	1
314	314	34	2	57	16	2	4	33.4	88.0	2	2	3	1	0	2	14	11.00	0	5	3	3	2	1	2	4	1	1	1	2
315	315	40	4	48	12	1	4	46.6	71.0	1	1	5	1	0	1	32	25.00	1	5	2	1	2	1	1	2	3	2	2	2
316	316	50	2	53	19	1	4	67.0	70.1	2	1	5	1	0	2	35	11.00	0	5	1	3	2	1	2	4	5	4	1	3
317	317	60	2	52	16	2	2	33.7	51.4	1	2	4	1	0	2	32	18.00	0	5	3	3	1	2	1	4	1	1	1	1
318	318	52	2	38	14	1	1	39.4	59.9	1	1	5	1	0	2	19	1.50	1	2	1	3	2	1	2	3	2	2	2	1
319	319	40	3	33	16	2	1	31.3	31.3	1	1	1	2	0	2	9	1.00	0	4	2	3	1	2	1	1	2	1	2	4
320	320	45	2	63	16	1	4	91.9	98.3	1	3	5	2	0	2	35	32.00	0	3	3	3	1	2	1	2	1	1	1	2
321	321	40	1	51	16	2	2	33.7	66.1	2	2	5	2	0	2	25	23.00	1	5	3	3	1	1	2	4	1	1	1	1
322	322	38	1	43	15	2	4	16.6	67.6	4	2	5	1	0	2	5	3.00	0	4	3	3	4	2	3	5	4	3	4	4
323	323	40	6	44	12	1	2	43.6	61.2	1	2	3	1	0	1	27	20.00	4	4	3	2	2	1	2	2	2	2	3	2
324	324	50	7	30	12	1	1	22.5	22.5	3	2	5	1	0	2	12	8.00	2	1	1	1	2	1	2	5	3	1	3	1
325	325	40	3	42	12	2	2	17.6	22.7	1	1	5	1	2	2	10	6.00	1	4	3	3	2	2	2	4	2	2	1	1
326	326	40	7	62	8	1	1	31.8	32.0	1	2	5	2	0	2	45	17.00	0	5	3	2	4	2	2	2	3	2	3	3
327	327	60	2	38	18	1	1	27.9	42.7	2	1	5	1	0	1	27	12.00	0	4	3	3	3	2	2	4	4	2	1	1
328	328	40	5	54	12	2	2	14.2	21.3	1	1	5	1	0	2	23	23.00	0	5	3	3	2	1	2	2	2	2	1	1
329	329	40	6	52	12	1	2	25.0	34.6	1	1	5	1	0	2	35	27.00	0	5	1	2	2	2	2	2	3	1	1	1
330	330	48	7	38	12	1	2	36.9	51.7	2	2	4	1	1	1	21	18.00	1	5	3	3	2	2	2	2	3	2	2	1
331	331	44	6	55	12	1	3	42.7	46.9	2	1	5	1	0	1	35	14.00	0	4	1	3	2	1	2	4	3	1	4	4
332	332	60	2	43	15	1	2	37.7	70.8	2	2	1	1	0	2	26	17.00	0	1	2	2	1	1	1	4	3	3	1	1
333	333	73	1	25	14	1	1	52.6	54.4	1	1	5	1	0	2	8	2.50	0	5	3	3	1	1	1	1	1	2	1	1
334	334	48	7	36	11	2	2	25.7	47.5	2	2	5	2	0	2	11	10.00	0	5	3	3	3	2	2	4	3	3	4	2
335	335	40	6	53	12	1	4	32.5	46.5	1	1	5	2	0	2	42	22.00	0	5	3	2	1	1	1	4	1	1	1	1
336	336	60	7	28	16	2	2	21.6	40.0	1	1	5	1	0	2	12	7.00	0	2	3	2	1	1	1	1	1	1	4	1
337	337	40	6	32	14	1	2	47.2	49.3	2	1	5	1	0	1	16	1.00	0	1	3	3	3	1	2	5	5	3	3	3
338	338	60	3	40	12	1	1	36.9	37.5	1	1	5	1	1	2	23	11.25	0	3	3	3	1	1	2	2	2	2	2	2
339	339	42	3	57	12	2	1	17.4	18.7	2	1	5	1	0	2	34	29.25	0	5	2	3	3	2	1	5	2	3	3	3
340	340	32	7	27	12	2	2	14.7	22.0	2	1	4	1	0	2	8	5.00	1	5	3	3	3	2	2	4	1	2	4	4
341	341	40	2	38	16	2	1	22.7	22.7	1	1	4	1	0	1	12	10.00	1	5	2	3	1	1	1	1	1	1	1	1
342	342	75	6	28	14	1	1	13.2	14.5	2	1	4	1	0	2	10	3.00	0	5	3	2	1	1	1	4	1	1	1	1
343	343	50	6	52	11	1	1	22.5	22.5	1	1	5	1	0	2	35	19.16	0	5	1	2	1	1	1	1	1	1	2	1
344	344	50	7	39	12	1	1	23.3	23.3	2	1	5	3	0	2	22	3.41	2	2	3	3	2	2	2	4	2	2	4	3
345	345	60	2	36	18	2	4	25.4	51.2	2	2	5	1	0	2	19	10.00	0	5	3	3	2	1	2	4	1	3	1	1
346	346	60	3	38	16	1	2	17.8	50.6	1	2	5	1	0	2	17	2.00	0	5	3	3	1	1	1	1	1	1	2	2
347	347	40	3	41	12	2	2	38.4	76.4	4	2	5	1	4	1	24	11.00	0	5	2	1	4	2	3	4	5	3	4	1
348	348	50	7	27	14	1	1	31.1	31.1	2	1	5	3	0	2	4	3.00	0	4	3	3	4	2	1	5	2	1	4	4
349	349	80	3	32	10	1	2	16.5	36.3	1	1	2	1	0	2	15	1.50	0	4	3	2	1	2	2	4	2	2	2	1
350	350	40	7	39	14	2	2	18.7	38.3	2	2	1	1	0	2	20	20.00	2	1	3	3	4	2	2	4	3	1	2	1

Table 2.3 (Continued)

OBS	IDNUM	WORKHRS	OCCUP	AGE	EDUC	SEX	EARNRS	RINCOME	FINCOME	SATJOB	RICHWORK	JOBCHAR	GETAHEAD	TRAUMA	MEMUNION	WRKYEARS	EMPYEARS	NUMPROMO	FUTPROMO	SEXPROMO	ADVANCES	IDECIDE	ORGMONEY	PROUDORG	STAYORG	UNMAREL	COWRKREL	SCHOOLNG	TRAINING
351	351	60	6	51	12	1	2	21.8	35.1	3	1	5	1	0	2	44	8.00	0	5	3	3	1	1	1	4	1	1	3	1
352	352	50	6	42	12	1	2	46.9	50.9	1	1	5	2	0	2	25	14.00	0	5	3	3	1	1	1	4	1	1	3	1
353	353	40	7	30	13	1	1	13.7	14.6	2	1	5	3	0	1	10	8.00	0	4	3	3	4	2	2	4	1	1	2	2
354	354	56	2	24	16	2	2	26.6	39.2	2	2	4	1	1	2	9	2.00	0	1	1	1	1	1	2	3	2	2	1	1
355	355	48	7	33	12	2	2	51.9	58.3	2	1	5	3	0	1	16	14.00	1	5	2	3	1	2	2	4	3	2	3	2
356	356	40	4	61	12	2	1	22.5	22.5	1	1	5	3	0	2	25	17.00	1	4	3	3	3	2	2	4	4	3	3	2
357	357	40	7	37	12	2	2	10.8	24.1	3	2	1	2	0	2	10	2.66	0	4	3	3	4	2	3	4	3	3	3	2
358	358	35	3	39	16	1	2	48.4	61.8	1	1	5	2	0	1	18	9.00	0	5	2	2	2	1	1	2	3	2	1	1
359	359	44	4	23	12	2	2	15.0	37.3	3	2	2	1	0	2	6	2.50	0	5	3	2	1	2	3	2	3	3	1	3
360	360	50	4	55	12	2	2	24.2	27.9	2	2	1	2	0	2	30	30.00	3	2	2	3	3	2	2	5	3	2	2	1
361	361	55	1	43	16	1	3	17.5	55.7	2	1	3	1	0	2	27	5.00	2	4	1	2	2	1	2	4	2	1	2	2
362	362	56	7	43	12	1	2	14.9	18.0	2	1	5	1	0	2	27	0.25	0	1	3	2	2	2	2	2	1	1	2	2
363	363	38	1	50	16	2	3	19.9	54.1	1	2	5	1	0	2	14	1.00	0	5	3	3	2	1	2	5	2	4	1	3
364	364	56	2	43	18	1	2	28.0	36.7	3	1	1	1	0	2	26	5.50	0	2	3	2	2	1	2	5	3	2	1	3
365	365	50	3	49	17	1	1	21.9	21.9	2	2	4	1	3	2	32	1.50	0	5	3	4	1	1	1	1	1	1	2	2
366	366	50	2	63	20	1	2	41.1	60.7	1	1	5	2	1	2	45	25.00	2	5	3	2	1	2	2	2	2	2	1	3
367	367	84	6	33	12	1	1	31.2	31.2	1	1	1	2	0	2	16	13.91	3	3	1	2	2	2	2	3	2	2	1	1
368	368	40	3	35	12	2	2	16.4	46.8	2	2	2	2	0	2	16	10.00	0	4	2	3	4	2	2	5	5	1	3	1
369	369	40	6	21	12	1	2	22.0	35.6	2	1	1	2	0	2	4	1.25	3	1	1	2	2	2	3	4	3	2	1	1
370	370	42	5	63	10	1	1	20.2	29.0	2	2	4	1	0	2	46	9.41	7	5	1	1	3	2	2	5	3	2	2	2
371	371	50	6	36	14	1	2	12.6	19.8	2	2	2	2	0	2	18	9.66	5	1	3	2	2	2	1	2	2	1	2	3
372	372	70	6	28	12	1	1	15.2	15.2	1	1	1	2	0	2	10	9.66	2	2	1	2	1	1	1	4	2	1	3	4
373	373	40	6	43	14	1	1	15.8	15.9	2	1	5	3	0	2	27	18.16	7	1	1	3	2	2	2	5	3	3	1	1
374	374	34	6	44	11	1	1	21.3	21.3	2	1	4	2	0	2	24	4.00	0	4	3	2	2	1	3	4	1	1	1	3
375	375	63	7	25	12	1	2	17.6	30.8	1	1	1	1	0	2	8	4.00	1	1	1	2	1	1	2	2	2	2	3	1
376	376	50	2	27	17	2	2	29.7	61.6	1	1	5	1	3	2	7	1.00	0	1	2	2	1	1	2	4	2	2	1	3
377	377	65	1	22	12	1	2	38.5	41.4	1	1	5	1	2	2	6	1.00	1	1	1	1	2	2	2	4	1	1	1	1
378	378	36	1	33	12	2	1	35.7	35.7	1	1	5	1	0	2	17	11.00	0	5	3	2	2	1	2	2	1	1	1	1
379	379	48	6	25	12	2	1	15.3	15.3	2	1	4	1	1	2	4	2.00	0	2	2	3	4	2	3	4	2	2	3	3
380	380	40	6	47	12	1	2	33.9	41.1	2	1	1	1	0	2	28	19.00	1	5	3	2	2	2	2	2	2	2	1	1
381	381	40	2	44	19	2	2	23.9	42.0	1	2	5	1	0	2	28	3.58	0	4	2	3	2	1	1	4	2	3	1	1
382	382	40	7	26	12	2	2	15.4	39.9	2	2	1	1	0	2	6	5.00	1	4	3	3	2	2	2	3	3	2	2	2
383	383	40	7	47	10	1	2	22.6	47.5	1	1	5	1	0	2	32	10.00	1	4	3	2	2	2	2	4	2	2	2	1
384	384	65	7	41	11	1	3	31.4	48.8	2	2	1	3	0	2	24	5.00	1	5	3	2	3	2	1	5	1	1	2	2
385	385	40	4	35	12	1	2	19.3	45.0	2	1	1	1	0	2	18	8.00	0	2	3	3	2	2	2	2	1	1	1	1
386	386	40	7	41	12	2	4	11.8	28.2	2	2	4	1	0	2	25	16.00	0	4	3	4	3	2	2	4	3	2	2	2
387	387	40	2	32	16	2	2	23.6	43.9	1	2	4	1	0	2	16	3.00	0	4	3	3	3	2	2	4	2	2	1	3
388	388	40	7	34	12	1	1	13.3	14.5	2	2	1	2	0	2	16	13.00	0	2	1	3	2	2	2	4	2	2	2	2
389	389	40	7	31	12	2	2	10.1	59.7	2	1	1	1	0	2	13	2.50	0	5	3	3	4	2	2	4	2	2	2	2
390	390	47	4	31	12	2	3	21.9	40.2	1	1	1	1	0	2	21	10.00	0	5	1	3	4	2	2	4	2	2	1	1
391	391	32	4	57	13	2	2	10.3	42.5	2	1	1	2	1	2	20	3.00	0	4	3	2	4	2	2	4	3	1	1	1
392	392	80	7	32	14	1	2	48.4	67.9	1	1	5	1	0	2	15	6.00	2	1	3	2	2	1	1	1	1	1	1	1
393	393	50	5	44	13	1	3	22.5	26.1	1	1	5	2	0	2	24	24.00	0	5	3	3	2	1	2	2	1	1	1	1
394	394	37	3	29	14	2	2	28.9	72.6	1	2	5	1	0	2	10	2.00	0	2	2	2	2	1	1	2	2	2	3	2
395	395	40	4	56	12	2	2	15.5	44.5	2	2	5	1	0	1	34	21.00	0	5	3	3	2	1	2	4	4	3	3	2
396	396	40	2	28	19	2	1	17.9	17.9	1	1	5	1	0	2	10	2.00	1	1	3	2	1	1	1	4	2	3	1	1
397	397	45	6	42	12	1	2	41.1	44.6	2	2	5	3	0	2	26	19.25	0	4	3	3	2	2	2	4	3	2	1	1
398	398	40	1	22	13	1	3	14.5	31.1	2	2	2	1	0	2	2	1.25	1	1	1	2	3	2	2	4	2	2	1	1
399	399	40	7	35	12	1	2	36.3	43.1	2	1	4	3	0	1	16	16.00	4	4	1	2	3	2	2	3	1	1	2	3
400	400	60	6	35	12	1	1	52.8	52.8	3	2	1	3	0	2	15	3.00	2	2	1	2	2	1	2	2	4	2	2	1

2.11 Recognizing and Practicing Good Survey Research and Exploring Ethical Issues

Data collection occurs at an early stage of a statistical analysis. After formulating a problem of interest, we would plan how to obtain the information that will be needed for decision making when attempting to solve the problem. Thus the importance of obtaining good data can never be overstated. Always think of GIGO.

2.11.1 The Sample Survey

Every day, we read about the results of surveys or opinion polls in our newspaper or we hear some interesting or titillating commentary on our radio and television. Clearly, advances in information technology have led to a proliferation of survey research. Not all this research is good, meaningful, or important (Reference 2). It becomes essential that we learn to evaluate critically what we read and hear and discount surveys that lack objectivity or credibility. In particular, we must examine the purpose of the survey, why it was conducted, and for whom. Remember that there are four main reasons for collecting data: (1) to provide input to a research study, (2) to measure performance, (3) to enhance decision making, or (4) to satisfy our curiosity. An opinion poll or survey conducted to satisfy curiosity is mainly for entertainment. Its result is an "end in itself" rather than a "means to an end." We should be more skeptical of such a survey because the result should not be put to further use.

The first step in evaluating the worthiness of a survey is to determine whether it was based on a probability or a nonprobability sample (as discussed in Section 2.6). You may recall that in an enumerative study, the only way for us to make correct statistical inferences from a sample to a population and interpret the results is through the use of a probability sample. Surveys employing nonprobability sampling methods are subject to serious, perhaps unintentional, interview biases that may render its results meaningless. In 1948, for example, each of the major pollsters employed quota sampling and incorrectly predicted the outcome of the presidential election (see Reference 9). As depicted in the photograph on page 42 (Figure 2.8), at least one widely circulated newspaper gambled on the polls' accuracy by printing its early edition based on what was predicted to happen rather than waiting for the ballots to be counted! Embarrassed by the surprising victory of the incumbent, President Harry S. Truman, after they had all predicted the election of Governor Thomas E. Dewey, the polling organizations adopted probability sampling methods for future elections.

Even when surveys employ random probability sampling methods, they are subject to potential errors. There are four types of **survey errors** (Reference 7):

1. Coverage error or selection bias
2. Nonresponse error or nonresponse bias
3. Sampling error
4. Measurement error

A good survey research design attempts to reduce or minimize these various survey errors, often at considerable cost.

Figure 2.8 **President Truman holding *Chicago Tribune* with erroneous headline.**

2.11.2 Coverage Error or Selection Bias

The key to proper sample selection is an adequate population frame or up-to-date list of all the subjects from which the sample will be drawn. **Coverage error** results from the exclusion of certain groups of subjects from this population listing so that they have no chance of being selected in the sample. Coverage error results in a **selection bias**. If the listing is inadequate because certain groups of subjects in the population were not properly included, any random probability sample selected will provide an estimate of the characteristics of the *target* population, not the *actual* population.

The issue of a potential selection bias was raised during the 1992 presidential campaign with a proposal for an "electronic town hall." The idea was to enable viewers to cast an immediate vote by telephone following a televised discussion of an important issue. Unfortunately, one major problem with this otherwise interesting and intriguing proposal was the potential exclusion of millions of registered voters who would be unable to watch the program or respond with an immediate vote. In particular, registered voters without televisions or telephones would be highly under-represented in the resulting responses.

Perhaps the most famous case of selection bias occurred in the 1936 *Literary Digest* poll. In that year, *Literary Digest*, a well-respected magazine, predicted that Governor Alf Landon of Kansas would receive 57% of the votes and overwhelmingly win the presidential election. When the actual votes were counted, Landon received only 38% as President Franklin Delano Roosevelt was easily reelected to a second term in office. The size of the error in the *Literary Digest* poll was considered enormous and unprecedented with respect to any major poll. Having lost its credibility, the magazine went bankrupt.

What went wrong? The prediction from the *Literary Digest* poll was based on the responses from 2.4 million individuals, a huge sample size. A major reason was selection bias. In 1936 the country was still suffering from the Great Depression. However, the *Literary Digest* compiled its population frame from such sources as telephone books, club membership lists, magazine subscriptions, and automobile registrations (Reference 5)—thereby catering to the rich and excluding from its listing the majority of the voting population who, during this period of economic hardship, could not afford such amenities as telephones, club memberships, magazine subscriptions, and automobiles. Thus the 57% estimate for the Landon vote may have been very close to the *target* population, but certainly not the *actual* population.

2.11.3 Nonresponse Error or Nonresponse Bias

Not everyone will be willing to respond to a survey. In fact, research has indicated that individuals in the upper and lower economic classes tend to respond less frequently to surveys than do people in the middle class. **Nonresponse error** results from the failure to collect data on all subjects in the sample. And nonresponse error results in a **nonresponse bias**. Since it cannot be generally assumed that persons who do not respond to surveys are similar to those who do, it is extremely important to follow up on the nonresponses after a specified period of time. Several attempts should be made, either by mail or by telephone, to convince such individuals to change their minds. Based on these results, an effort is made to tie together the estimates obtained from the initial respondents with those obtained from the follow-ups so that we can be reasonably sure that the inferences made from the survey are valid (Reference 1).

As stated in Section 2.4.2, the mode of response affects the rate of response. The personal interview and the telephone interview usually produce a higher response rate than does the mail survey—but at a higher cost.

The 1936 *Literary Digest* poll is also an example of nonresponse bias. A sample of 10 million registered voters were mailed questionnaires and only 2.4 million of them responded. Although a sample of 2.4 million is huge, a response rate of only 24% is far too low to yield accurate estimates of the population parameters without some mechanism to assure that the 7.6 million individuals who didn't respond have similar opinions to those who did. With respect to the *Literary Digest* poll, however, this problem of nonresponse bias was secondary to the problem of selection bias. Even if all 10 million registered voters in the sample had responded, it would not compensate for the fact that the exclusionary target population differed so substantially in composition from the actual voting population.

2.11.4 Sampling Error

There are three main reasons for drawing a sample rather than taking a complete census—it is more expedient, less costly, and more efficient. However, when selecting the subjects using a random probability sample, depending on where one starts in the table of random numbers, chance dictates who in the population frame will or will not be included. Although only one sample is actually selected, if many different samples were to be drawn, hopefully each sample would be a miniature representation of the population and yield reasonable estimates of its characteristics. **Sampling error** reflects the heterogeneity or "chance differences" from sample to sample based on the probability of subjects being selected in the particular samples.

When we read about the results of surveys or polls in newspapers or magazines, there is often a statement regarding margin of error or precision—for example, "the results of this poll are expected to be within ± 4 percentage points of the actual value." Sampling error can be reduced by taking larger sample sizes, although this will increase the cost of conducting the survey.

2.11.5 Measurement Error

In the practice of good survey research, a questionnaire is designed with the intent that it will enable meaningful information to be gathered. The obtained data must be *valid*; that is, the "right" responses must be assessed, and in a manner that will elicit meaningful measurements.

But there is a dilemma here—obtaining meaningful measurements is often easier said than done. Consider the following proverb:

A man with one watch always knows what time it is;
a man with two watches always searches to identify the correct one;
a man with ten watches is always reminded of the difficulty in measuring time.

Unfortunately, the process of obtaining a measurement is often governed by what is convenient, not what is needed. And the measurements obtained are often only a proxy for the ones really desired.

Measurement error refers to inaccuracies in the recorded responses that occur because of a weakness in question wording, an interviewer's effect on the respondent, or the effort made by the respondent.

Much attention has been given to measurement error that occurs because of a weakness in question wording. A question should be clear, not ambiguous. Moreover, it should be objectively presented in a neutral manner; "leading questions" must be avoided.

As an example, in November 1993 the Labor Department reported that the unemployment rate in the United States has been underestimated for more than a decade because of poor questionnaire wording in the Current Population Survey. In particular, the wording led to a significant undercount of women in the labor force. Since unemployment rates are tied to benefit programs such as state unemployment compensation systems, it was imperative that government survey researchers rectify the situation by adjusting the questionnaire wording.

We can demonstrate the impact of question wording on the responses obtained by referring to the following two versions of a question posed by Yankelovich & Partners in national surveys conducted during the 1992 presidential campaign (see Reference 6):

- Do you believe that for every dollar of tax increase there should be $2 in spending cuts with the savings earmarked for deficit and debt reduction?
- Would you favor or oppose a proposal to cut spending by $2 for every dollar in new taxes, with the savings earmarked for deficit reduction, even if that meant cuts in domestic programs like Medicare and education?

The responses to the first version of the question were as follows: 67% said "yes," 18% said "no," and 15% said "don't know." On the other hand, the responses to the alternative version of the question were completely opposite: 33% said "favor," 61% said "oppose," and 6% said "don't know." What has happened here? Why was there such a change from a more positive stance to a more negative position on this issue? Perhaps we can attribute the change to the fact that in the second version of the question an alternative tone was used in its wording and more information regarding potential outcome was provided.

A second source of measurement error occurs in personal interviews and in telephone interviews—a "halo effect" in which the respondent feels obligated to please the interviewer. This type of error can be minimized by proper interviewer training.

A third source of measurement error occurs because of the effort (or lack thereof) on the part of the respondent. Sometimes the measurements obtained are gross exaggerations, either willful or owing to lack of recall on the part of the respondent. In either case, this hampers the utility of the survey—recall GIGO. This type of error can be minimized in two ways: (1) by carefully scrutinizing the data and calling back those individuals whose responses seem unusual and (2) by establishing a program of random callbacks in order to ascertain the reliability of the responses.

2.11.6 Ethical Issues

With respect to the proliferation of survey research (Reference 2), Eric Miller, editor of a newsletter *Research Alert*, stated that "there's been a slow sliding in ethics. The scary part is that people make decisions based on this stuff. It may be an invisible crime, but it's not a victimless one." Not all survey research is good, meaningful, or important and not all survey research is ethical. We must become healthy skeptics and critically evaluate what we read and hear. We must examine the purpose of the survey, why it was conducted, and for whom, and then discard it if it is found to be lacking in objectivity or credibility. In particular, we should be especially wary of the objectivity of polls and surveys commissioned by special-interest groups. Will they be "tooting their own horn?" Moreover, we should be especially wary of nonscientific call-in polls or surveys conducted to satisfy curiosity. Call-in polls such as those in *USA Today* are strictly for fun (see Reference 2). They have no practical value. Their results are heavily biased by a self-selection process. In addition, a particular individual can call in and be tallied more than once.

Ethical considerations arise with respect to the four types of potential errors that may occur when designing surveys that use random probability samples. We must try to distinguish between poor survey design and unethical survey design. The key is *intent*. Coverage error or selection bias becomes an ethical issue only if particular groups or individuals are purposely excluded from the population frame so that the survey results would likely indicate a position more favorable to that of the survey's sponsor. In a similar vein, nonresponse error or nonresponse bias becomes an ethical issue only if particular groups or individuals are less likely to be available to respond to a particular survey format and the sponsor knowingly designs the survey in a manner aimed at excluding such groups or individuals. Sampling error becomes an ethical issue only if the findings are purposely presented without reference to sample size and margin of error so that the sponsor could promote a viewpoint that might otherwise be truly insignificant. Measurement error, however, becomes an ethical issue in any of three ways. A survey sponsor may purposely choose loaded, lead-in questions that would guide the responses in a particular direction. Moreover, an interviewer, through mannerisms and tone, may purposely create a halo effect or otherwise guide the responses in a particular direction. Furthermore, a respondent having a disdain for the survey process may willfully provide false information.

Problems for Section 2.11

2.27 "A survey indicates that Americans overwhelmingly preferred a Chrysler to a Toyota after test-driving both." What information would you want to know before you accept the results of this survey?

2.28 "A survey indicates that the vast majority of college students picked Levi's 501 jeans as the most 'in' clothing." What information would you want to know before you accept the results of this survey?

2.29 "A 900 number call-in survey on rock music indicates that rock group Led Zeppelin's 'Stairway to Heaven' is the most popular song of all time." What information would you want to know before you accept the results of this survey?

2.12 Data Collection: A Review and a Preview

As seen in the chapter summary chart on page 47, this chapter was about data collection. On pages 12–13 of Section 2.1 you were given a list emphasizing the important points to be discussed in the chapter. Check over this list to see whether you feel you have an understanding of these key points. To be sure, you should be able to answer the following conceptual questions:

1. What is the difference between a categorical and a numerical random variable?
2. What is the difference between discrete and continuous data?
3. What are the various levels of measurement?
4. What is an operational definition and why is it so important?
5. What are the main reasons for obtaining data and what methods can be used to accomplish this?
6. What is the difference between probability and nonprobability sampling?
7. Why is the compilation of a complete population frame so important for survey research?
8. What is the difference between sampling with versus without replacement?
9. What are the advantages and disadvantages of personal interviews, telephone surveys, and mail surveys?
10. What is the difference between selection bias and nonresponse bias in surveys?
11. What distinguishes the four potential sources of error when dealing with surveys designed using probability sampling?
12. How do we prepare our collected survey data for tabular and chart presentation and summarization?

Check over the list of questions to see if indeed you know the answers and could (1) explain your answers to someone who did not read this chapter

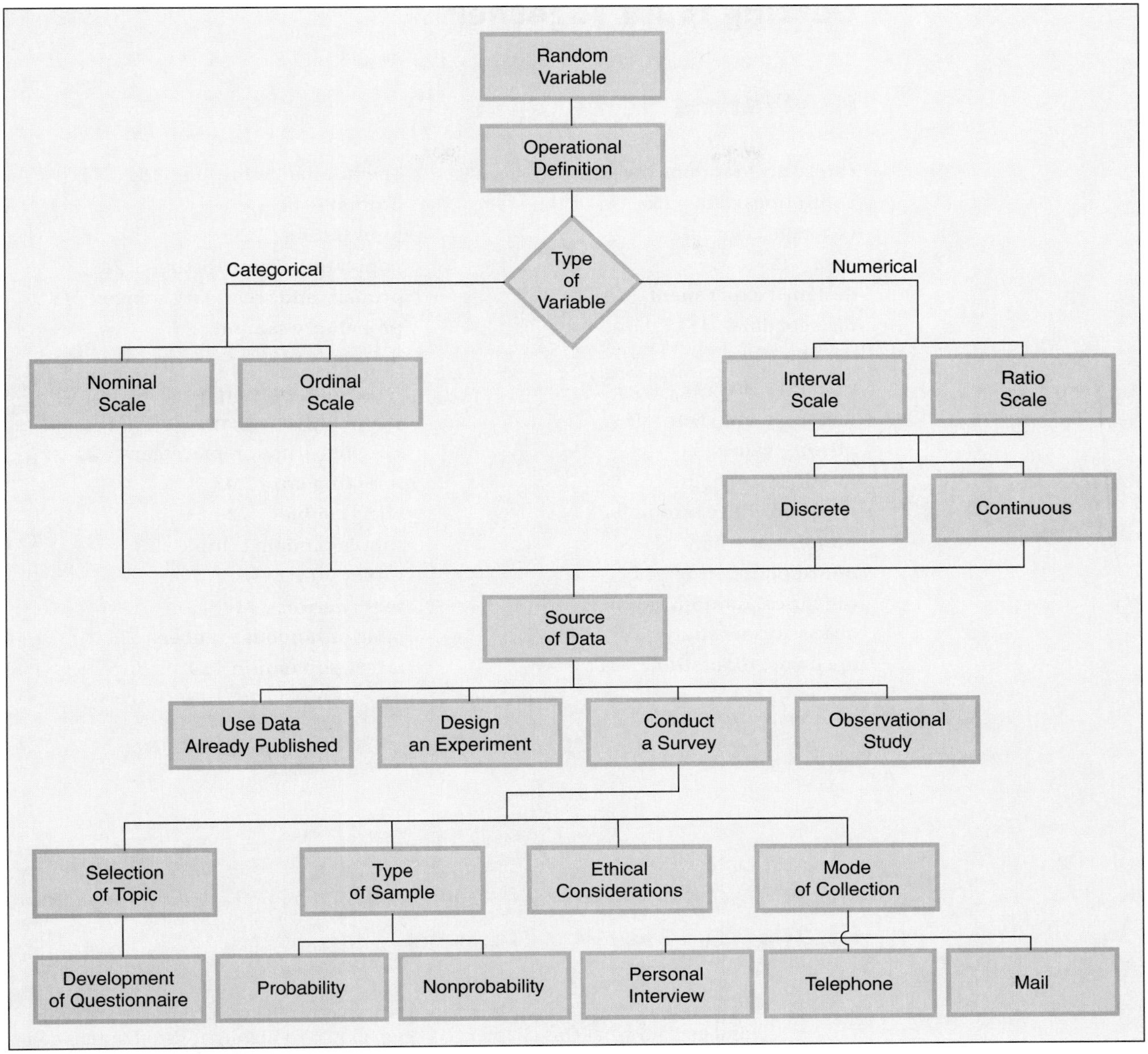

Chapter 2 summary chart.

and (2) give reference to specific readings or examples that support your answer. Also, reread any of the sections that may have seemed fuzzy to see if they make more sense now.

Once data have been collected—be it from a published source, a designed experiment, an observational study, or a survey such as was used at Kalosha Industries—the data must be organized and prepared in order to assist us in making various analyses. In the next three chapters, methods of tabular and chart presentation will be demonstrated, various "exploratory data analysis" techniques will be described, and a variety of descriptive summary measures useful for data analysis and interpretation will be developed.

Getting It All Together

Key Terms

Chapter Review Problems

2.30 ACTION Write a letter to a friend who has not taken a course in statistics and explain what this chapter has been about. To highlight the chapter's content, be sure to incorporate your answers to the 12 review questions on page 46.

2.31 Determine whether each of the following random variables is categorical or numerical. If it is numerical, determine whether the phenomenon of interest is discrete or continuous. In addition, provide the level of measurement and an operational definition for each of the variables.
(a) Brand of personal computer used
(b) Cost of personal computer system
(c) Amount of time the personal computer is used per week
(d) Primary use for the personal computer
(e) Number of persons in the household who use the personal computer
(f) Number of computer magazine subscriptions
(g) Word processing package primarily used

● 2.32 Determine whether each of the following random variables is categorical or numerical. If it is numerical, determine whether the phenomenon of interest is discrete or continuous. In addition, provide the level of measurement and an operational definition for each of the variables.

(a) Amount of money spent on clothing in the last month
(b) Number of women's winter coats owned
(c) Favorite department store
(d) Amount of time spent shopping for clothing in the last month
(e) Most likely time period during which shopping for clothing takes place (weekday, weeknight, or weekend)
(f) Number of pairs of women's gloves owned
(g) Primary type of transportation used when shopping for clothing

● 2.33 Suppose the following information is obtained from Robert Keeler on his application for a home mortgage loan at the Metro County Savings and Loan Association:

(a) Place of Residence: Stony Brook, New York
(b) Type of Residence: Single-family home
(c) Date of Birth: April 9, 1962
(d) Monthly Payments: $1,427
(e) Occupation: Newspaper reporter/author
(f) Employer: Daily newspaper
(g) Number of Years at Job: 4
(h) Number of Jobs in Past Ten Years: 1
(i) Annual Family Salary Income: $66,000
(j) Other Income: $16,000
(k) Marital Status: Married
(l) Number of Children: 2
(m) Mortgage Requested: $120,000
(n) Term of Mortgage: 30 years
(o) Other Loans: Car
(p) Amount of Other Loans: $8,000

Classify each of the responses by type of data and level of measurement.

2.34 Why would you expect a survey conducted by personal or telephone interview to be more costly than one conducted using a mailed questionnaire instrument?

2.35 Suppose that the manager of the Customer Service Division of Xenith was interested primarily in determining whether customers who had purchased a videocassette recorder over the past 12 months were satisfied with their products. Using the warranty cards submitted after the purchase, the manager was planning to survey 1,425 of these customers.

(a) Describe both the population and sample of interest to the manager.
(b) Describe the type of data that the manager primarily wishes to collect.
(c) Develop a first draft of the questionnaire by writing a series of three categorical questions and three numerical questions that you feel would be appropriate for this survey. Provide an operational definition for each variable.
(d) **ACTION** Write a draft of the cover letter needed for this survey.

2.36 In a political poll to try to predict the outcome of an election, what is the population to which we usually want to generalize? How might we get a random sample from that population? From what you know about how such polls are actually conducted, what might be some problems with the sampling in these polls?

● 2.37 Given a population of $N = 93$, draw a sample of size $n = 15$ *without replacement* by starting in row 29 of the table of random numbers (Table E.1). Reading across the row, list the 15 coded sequences obtained.

2.38 Do Problem 2.37 by sampling *with replacement*.

2.39 For a population of $N = 1{,}250$, a *two-stage* usage of the table of random numbers (Table E.1) may be recommended to avoid wasting time and effort. To obtain the sample by a two-stage approach, list the four-digit coded sequences after adjusting the first digit in each sequence as follows: If the first digit is a 0, 2, 4, 6, or 8, change the digit to 0. If the first digit is a 1, 3, 5, 7, or 9, change the digit to 1. Thus, starting in row 07 of the table of random numbers (Table E.1), the sequence 7054 becomes 1054, the sequence 3297 becomes 1297, etc. Verify that only ten rows are needed to draw a sample of size $n = 60$ *without replacement*.

2.40 For a population of $N = 2{,}202$, a *two-stage* usage of the table of random numbers (Table E.1) may be recommended to avoid wasting time and effort. To obtain the sample by a two-stage approach, list the four-digit coded sequences after adjusting the first digit in each sequence as follows: If the first digit is a 0, 3, or 6, change the digit to 0; if the first digit is a 1, 4, or 7, change the digit to 1; if the first digit is a 2, 5, or 8, change the digit to 2; if the first digit is a 9, discard the sequence. Thus, starting in row 07 of the table of random numbers (Table E.1), the sequence 7054 becomes 1054, the sequence 3297 becomes 0297, etc. Verify that only ten rows are needed to draw a sample of size $n = 60$ *without replacement*.

2.41 **ACTION** Write a letter to a friend discussing ethical issues when collecting data through surveys.

2.42 The following computerized output is extracted from a set of data similar to those collected in the Employee Satisfaction Survey (Table 2.3). However, each of the five lines, representing the respective responses of five particular individuals, has one error in it. Use the coded statements shown in Figure 2.7 and Table 2.2 on pages 31 and 32 to determine the particular recording error in each of the five responses.

401	72	6	47	41	1	1	65.7	95.5	1	1	1	1	0	2	27	5.00	0	5	1	3	3	2	1	5	3	3	4	1
402	40	7	48	12	2	3	17.6	48.3	2	4	5	1	0	2	20	9.00	0	2	2	2	4	2	2	4	3	2	2	2
403	42	7	10	12	1	1	20.8	22.3	3	2	5	2	1	1	13	1.50	0	4	1	3	2	2	2	4	3	2	4	3
404	60	2	38	18	1	1	29.7	41.2	2	1	5	1	0	1	17	12.25	0	4	3	3	3	2	2	4	4	2	5	1
405	50	5	44	13	3	3	25.2	28.7	1	1	5	2	0	2	24	24.00	0	5	3	3	2	1	2	2	1	1	1	1

Collaborative Learning Projects

Note: The class is to be divided into groups of three or four students each. One student is initially selected to be project coordinator, another student is project recorder, and a third student is the project timekeeper. In order to give each student experience in developing teamwork and leadership skills, rotation of positions should follow each project. At the beginning of each project, the students should work silently and individually for a short, specified period of time. Once each student has had the opportunity to study the issues and reflect on his/her possible answers, the group is convened and a group discussion follows. If all the members of a group agree with the solutions, the coordinator is responsible for submitting the team's project solution to the instructor with student signatures indicating such agreement. On the other hand, if one or more team members disagree with the solution offered by the majority of the team, a minority opinion may be appended to the submitted project, with signature(s) on it.

CL2.1 Suppose the following information is obtained for Hugh Sain upon his admittance to the Brandwein College infirmary:

(a) Sex: Male
(b) Residence or Dorm: Mogelever Hall
(c) Class: Sophomore
(d) Temperature: 102.2°F (oral)
(e) Pulse: 70 beats per minute
(f) Blood Pressure: 130/80 mg/mm
(g) Blood Type: B Positive
(h) Known Allergies to Medicines: No
(i) Preliminary Diagnosis: Influenza
(j) Estimated Length of Stay: 3 days

Classify each of the ten responses by type of data and level of measurement. Provide an operational definition for each variable. (*Hint*: Be careful with blood pressure; it's tricky.)

CL2.2 Provide an operational definition for each of the following:

(a) A good student
(b) The number of children per household
(c) An excellent movie
(d) A light-tasting beer
(e) A cute dress or outfit
(f) A quality product
(g) A smooth-riding automobile

CL2.3 Suppose that the American Kennel Club (AKC) was planning to survey 1,500 of its club members primarily to determine the percentage of its membership that currently own more than one dog.

(a) Describe both the population and sample of interest to the AKC.
(b) Describe the type of data that the AKC primarily wishes to collect.
(c) Develop a first draft of the questionnaire needed by writing a series of five categorical questions and five numerical questions that you feel would be appropriate for this survey. Provide an operational definition for each variable.
(d) ACTION Write a draft of the cover letter needed for this survey.

Case Study A—Alumni Association Survey

Suppose that the President of the Alumni Association wishes a survey to be taken of its membership from the classes of 1985 and 1986 to determine their past achievements, current activities, and future aspirations. Toward this end, information pertaining to the following areas is desired: sex of the alumnus; major area of study; grade-point index; further educational pursuits (that is, master's degree or doctorate); current employment status; current annual salary; number of full-time positions held since graduation; annual salary anticipated in five years; political party affiliation; and marital status.

As Director of Institutional Research you are asked to write a proposal demonstrating how you plan to conduct the survey. Included in this proposal must be

1. A statement of objectives (that is, what you want to find out and why)

2. A discussion of *how* and *when* the survey will be conducted (that is, how you plan to sample 300 alumni from the list of 3,000 graduates in the two classes)
3. A first draft of the questionnaire instrument (containing an organized sequence of both numerical and categorical questions—including operational definitions for each variable, all category labels, and column allocations for data entry)
4. A first draft of the cover letter to be used with the questionnaire
5. A first draft of any special instructions to respondents to aid them in filling out the questionnaire
6. A discussion of *how* you plan to test the questionnaire for validity and/or ambiguity
7. A demonstration of how the responses will be coded and entered by simulating the data entry for a hypothetical respondent—John Q. Doe, graduate of the class of 1985
8. A statement that you have taken into consideration such things as the costs involved in conducting the survey, personnel needs, and the amount of time required for implementation and completion

Endnotes

1 *Newsday*, April 25, 1988.

2 It is interesting to note that whether we are sampling with replacement from finite populations or sampling without replacement from infinite populations (such as some continuous, ongoing production process), the formulas used are the same.

3 Learning from the experiences of the 1969 draft lottery, the 1970 lottery attempted to correct for the possible mixing and selecting problems. Today, the mixing and selecting process used in televised state lotteries appears to be random—the only human involvement is the announcing of the chosen numbers.

References

1. Cochran, W. G., *Sampling Techniques*, 3d ed. (New York: Wiley, 1977).
2. Crossen, C., "Margin of Error: Studies and Surveys Proliferate, but Poor Methodology Makes Many Unreliable," *The Wall Street Journal*, November 14, 1991, pp. A1 and A9.
3. Deming, W. E., *Sample Design in Business Research* (New York: Wiley, 1960).
4. Deming, W. E., *Out of the Crisis* (Cambridge, MA: Massachusetts Institute of Technology Center for Advanced Engineering Study, 1986).
5. Gallup, G. H., *The Sophisticated Poll-Watcher's Guide* (Princeton, NJ: Princeton Opinion Press, 1972).
6. Goleman, D., "Pollsters Enlist Psychologists in Quest for Unbiased Results," *The New York Times*, September 7, 1993, pp. c1 and c11.
7. Groves, R. M., *Survey Errors and Survey Costs* (New York: Wiley, 1989).
8. Hansen, M. H., W. N. Hurwitz, and W. G. Madow, *Sample Survey Methods and Theory*, Vols. I and II (New York: Wiley, 1953).
9. Mosteller, F., and others, *The Pre-Election Polls of 1948* (New York: Social Science Research Council, 1949).
10. Rand Corporation, *A Million Random Digits with 100,000 Normal Deviates* (New York: Free Press, 1955).
11. Robbins, S. P., *Management*, 4th ed. (Englewood Cliffs, NJ: Prentice Hall, 1994).

chapter 3

Presenting Numerical Data in Tables and Charts

CHAPTER OBJECTIVE

To show how to organize and most effectively present collected numerical data in tables and charts.

3.1 Introduction

In the preceding chapter we learned how to collect data through survey research. As pointed out in Section 2.5, since sampling saves time, money, and labor, we usually deal with sample information rather than data from an entire population. Nevertheless, regardless of whether we are dealing with a sample or a population, as a general rule, whenever a batch of data that we have collected contains about 20 or more observations, the best way to examine such *mass data* is to present it in summary form by constructing appropriate tables and charts. We can then extract the important features of the data from these tables and charts.

Thus, this chapter is about data presentation. In particular, we will demonstrate how large batches of numerical data can be organized and most effectively presented in the form of tables and charts in order to enhance data analysis and interpretation—two key aspects of the decision-making process. To motivate our discussion on the tabular and chart presentation of numerical data, we see from the chapter summary chart on page 94 that the observations in our data batch are of two types, time ordered or independent. Time-ordered observations can be monitored on a digidot plot, while independent observations can be organized into an ordered array or stem-and-leaf display and then presented in tabular form as a frequency distribution or in graphic form as a histogram, polygon, or ogive.

Upon completion of this chapter, you should be able to:

1. Organize a batch of numerical data into an ordered array or stem-and-leaf display
2. Understand how and when to construct and use frequency distributions and percentage distributions
3. Know how and when to construct and use histograms and polygons
4. Understand how and when to construct and use cumulative distributions
5. Know how and when to construct and use ogives (that is, cumulative frequency polygons and cumulative relative frequency polygons)
6. Know how and when to construct and use the digidot plot
7. Appreciate the value of using statistical or spreadsheet packages for presenting numerical data in the form of tables and charts
8. Understand how to distinguish between good and bad numerical data presentation and the ethical issues involved

3.2 Organizing Numerical Data: The Ordered Array and Stem-and-Leaf Display

In order to introduce the relevant ideas for Chapters 3 and 4, let us suppose that a company providing college advisory services to high school students throughout the United States has hired a research analyst to compare the tuition rates charged to out-of-state residents by colleges and universities in different regions of the country. Table 3.1 displays the tuition rates charged to out-of-state residents by each of the 60 colleges and universities in the state of Texas (see Special Data Set 1 of Appendix D, pages D1–D2). When a batch of data such as this one is collected, it is usually in **raw form**; that is, the numerical observations are not arranged in any particular order or sequence. As seen from Table 3.1, as the number of observations gets large, it becomes more and more difficult to focus on the major features in a batch of data and methods are needed to help us organize the observations so that we may better understand what information the data batch is conveying. Two commonly used methods for accomplishing this are the *ordered array* and the *stem-and-leaf display.*

Table 3.1 **Raw data pertaining to tuition rates (in $000) for out-of-state residents at 60 colleges and universities in Texas.**

7.2	4.9	10.7	10.4	6.4	4.8	4.7	4.6	6.0	5.4
4.8	4.7	8.3	3.8	4.8	8.3	6.4	6.6	4.5	8.0
3.6	2.4	8.5	8.8	7.7	4.9	8.6	12.0	4.9	7.0
11.0	4.9	3.9	4.9	4.4	4.9	4.9	8.0	3.6	7.4
7.9	4.9	5.8	3.9	11.6	10.3	3.4	3.9	5.0	3.9
8.0	3.5	4.9	5.8	4.1	3.9	3.5	4.8	5.9	3.6

Data file
C&UTEX.DAT

Source: See Special Data Set 1, Appendix D, pages D1–D2, taken from "America's Best Colleges, 1994 College Guide," *U.S. News & World Report*, extracted from College Counsel 1993 of Natick, Mass. Reprinted by special permission, *U.S. News & World Report*, © 1993 by *U.S. News & World Report* and by College Counsel.

3.2.1 The Ordered Array

If we place the raw data in rank order, from the smallest to the largest observation, the ordered sequence obtained is called an **ordered array.** When the data are arranged into an ordered array, as in Table 3.2, our evaluation of their major features is facilitated. It becomes easier to pick out extremes, typical values, and concentrations of values.

Table 3.2 **Ordered array of tuition rates (in $000) at 60 Texas colleges and universities.**

2.4	3.4	3.5	3.5	3.6	3.6	3.6	3.8	3.9	3.9
3.9	3.9	3.9	4.1	4.4	4.5	4.6	4.7	4.7	4.8
4.8	4.8	4.8	4.9	4.9	4.9	4.9	4.9	4.9	4.9
4.9	4.9	5.0	5.4	5.8	5.8	5.9	6.0	6.4	6.4
6.6	7.0	7.2	7.4	7.7	7.9	8.0	8.0	8.0	8.3
8.3	8.5	8.6	8.8	10.3	10.4	10.7	11.0	11.6	12.0

Source: Table 3.1.

Although it is useful to place the raw data into an ordered array prior to developing summary tables and charts or computing descriptive summary measures (see Chapter 4), the greater the number of observations present in a data batch, the more cumbersome it is to form the ordered array. In such situations it becomes particularly useful to organize the data batch into a stem-and-leaf display in order to study its characteristics (References 1, 13, and 14).

3.2.2 The Stem-and-Leaf Display

A **stem-and-leaf display** separates data entries into "leading digits" or "stems" and "trailing digits" or "leaves." For example, since the tuition rates (in $000) in the Texas data batch all have one- or two-digit integer numbers, either the ones column or the tens column would be the leading digit and the remaining column would be the trailing digit. Thus an entry of 7.2 (corresponding to $7,200) has a leading digit of 7 and a trailing digit of 2.

Figure 3.1 depicts the stem-and-leaf display of the tuition rates for all 60 colleges and universities in Texas. The column of numbers to the left of the vertical line is called the "stem." These numbers correspond to the *leading digits* of the data. In each row the "leaves" branch out to the right of the vertical line, and these entries correspond to *trailing digits*.

2	4
3	869694995956
4	9876878599994999918
5	48089
6	4046
7	27049
8	33058600
9	
10	743
11	06
12	0

N = 60

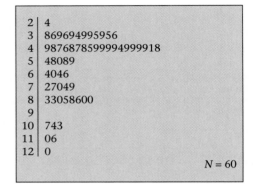

Figure 3.1
Stem-and-leaf display of out-of-state resident tuition rates at 60 Texas colleges and universities.
Source: Table 3.1.

Constructing the Stem-and-Leaf Display Using the data from Table 3.1, the stem-and-leaf display is easily constructed. Note that the first institution, Abilene Christian University, has a tuition rate of 7.2 thousand dollars. Therefore the trailing digit of 2 is listed as the first leaf value next to the stem value of 7 (the leading digit). The second institution, Angelo State University, has a tuition rate of 4.9 thousand dollars. Here the trailing digit of 9 is listed as the first leaf value next to the stem value of 4. Continuing, the third institution, Austin College, has a tuition rate of 10.7 thousand dollars so that the trailing digit of 7 is listed as the first leaf value next to the stem value of 10. The fourth institution, Baylor University, has a tuition rate of 10.4 thousand dollars, so the trailing digit of 4 is listed as the second leaf value next to the stem value of 10.

At this point in its construction, our stem-and-leaf display appears as follows:

2	
3	
4	9
5	
6	
7	2
8	
9	
10	74
11	
12	

Note that two of the four schools have the same stem. As more and more schools are included, those possessing the same stems and, perhaps, even the same leaves within stems (that is, the same tuition rates) will be observed. Such leaf values will be recorded adjacent to the previously recorded leaves, opposite the appropriate stem—resulting in Figure 3.1.

To assist us in further examining the data, we may wish to rearrange the leaves within each of the stems by placing the digits in ascending order, row by row. The **revised stem-and-leaf display** is presented in Figure 3.2.

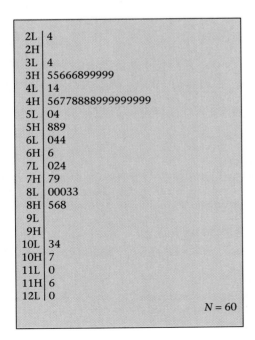

2	4
3	455666899999
4	1456778888999999999
5	04889
6	0446
7	02479
8	00033568
9	
10	347
11	06
12	0
	$N = 60$

Figure 3.2
Revised stem-and-leaf display of out-of-state resident tuition rates at 60 Texas colleges and universities.

Another type of rearrangement is also useful. If we desire to alter the size of the stem-and-leaf display, it is flexible enough for such an adjustment. Suppose, for example, we want to increase the number of stems so that we can attain a lighter concentration of leaves on the remaining stems. This is accomplished in the stem-and-leaf display presented in Figure 3.3.

2L	4
2H	
3L	4
3H	55666899999
4L	14
4H	56778888999999999
5L	04
5H	889
6L	044
6H	6
7L	024
7H	79
8L	00033
8H	568
9L	
9H	
10L	34
10H	7
11L	0
11H	6
12L	0
	$N = 60$

Figure 3.3
Revised stem-and-leaf display of out-of-state tuition rates at 60 Texas colleges and universities using more stems.
Source: Figure 3.2.

Note that each stem from Figure 3.2 has been split into two new stems—one for the *low* unit digits 0, 1, 2, 3, or 4 and the other for the *high* unit digits 5, 6, 7, 8, or 9. These are represented by *L* and *H*, respectively, as indicated in the stem listings of Figure 3.3.

However, some researchers would argue that the data displayed in Figure 3.3 are undersummarized. That is, we are failing to capture how the data are truly clustering within various groupings. Hence, instead of expanding the display, as in Figure 3.3, we might wish to condense the data, as in Figure 3.4.

Figure 3.4
Revised stem-and-leaf display of out-of-state tuition rates at 60 Texas colleges and universities after condensing stems.
Source: Figure 3.2.

2,**3**	**4455666899999**
4,**5**	14567788889999999999**04889**
6,**7**	0446**02479**
8,**9**	00033568
10,**11**	347**06**
12,**13**	0

N = 60

Note that consecutive pairs of stems from Figure 3.2 form the reduced set of stems in Figure 3.4 and the leaves corresponding to the *higher* member of each pair are **boldfaced.**

The (revised) stem-and-leaf display is, perhaps, the most versatile technique in descriptive statistics. It simultaneously organizes the data for further descriptive analyses (as we will see in Chapter 4) and, it prepares the data for both tabular and chart form.

Problems for Section 3.2

● 3.1 Given the following stem-and-leaf display:

9	714
10	82230
11	561776735
12	394282
13	20

(a) Rearrange the leaves and form the revised stem-and-leaf display.
(b) Place the data into an ordered array.
(c) Which of these two devices seems to give more information? Discuss.

Data file MAILORD.DAT

3.2 Upon examining the monthly billing records of a mail-order book company, the auditor takes a sample of 20 of its unpaid accounts. The amounts owed the company were

$4, $18, $11, $7, $7, $10, $5, $33, $9, $12
$3, $11, $10, $6, $26, $37, $15, $18, $10, $21

(a) Develop the ordered array.
(b) Form the stem-and-leaf display.

3.3 The data at the top of page 59 represent the maximum flow rate (in gallons per minute) for a random sample of 34 shower heads tested at 80 pounds per square inch of pressure:

Brand and Model	Max. Flow Rate (at 80 psi)
Sears Energy-Saving Shower Head 20170	2.9
Thermo Saver DynaJet CF01	2.8
Resources Conservation The Incredible Head ES-181	2.0
Zin-Plas Brass Showerhead 14-9601-F	3.6
Zin-Plas Water Pincher 14-9550	2.7
Whedon Saver Shower SS2C	2.5
Great Vibrations Water Saver Massage B28400	2.6
American Standard Shower Head Chrome 10509.0020A	2.9
Teledyne Water Pik Shower Massage 5 SM-2U	2.7
Chatham Solid Brass Shower Head 44-3S	2.8
Teledyne Water Pik Shower Massage 8 SM-4	2.5
Melard Water-saving Adjustable 3610	2.8
Pollenex Dial Massage DM150	2.2
Nova B6402	2.5
Speakman Anystream S2253-AF	2.5
Kohler City Club Z-7351	2.8
NY-Del 550-II	1.8
Ondine Water Saver 28446	2.7
Kohler Trend 11740	2.7
Alsons Somerset 673	4.7
Speakman Cosmopolitan S2270-AF	2.8
Pollenex Dial Massage DM109	2.7
Alsons Alspray Massage Action 690C	3.1
Moen Pulsation 3935	2.9
Sears Personal Hand Shower 20173	3.4
Teledyne Water Pik Shower Massage 5 SM-3U	2.6
Alsons Hand Shower 462PB	2.6
Alsons Massage Action Pulsating 45C	2.7
Moen Pulsation 3981	2.4
Teledyne Water Pik Super Saver SS-3	2.5
Pollenex Dial Massage DM209	5.4
Pollenex Dial Massage/Steamy Mist DM230	4.9
Pryde Splash 2461	2.8
Teledyne Water Pik Shower Massage 8 SM-5	2.5

Source: Copyright 1990 by Consumers Union of United States, Inc., Yonkers, N.Y. 10703. Adapted by permission from *Consumer Reports,* July 1990, pp. 472–473.

Data file SHOWER.DAT

(a) Develop the ordered array.
(b) Form the stem-and-leaf display.

3.4 The following data represent the retail price of a sample of 39 different brands of bathroom scales:

Data file SCALES.DAT

50	50	50	28	65	40	50	22	32	30
79	50	22	20	35	24	25	120	35	35
65	20	14	25	24	48	15	10	17	50
25	22	60	30	12	30	10	12	20	

Source: Copyright 1993 by Consumers Union of United States, Inc., Yonkers, N.Y. 10703. Adapted by permission from *Consumer Reports,* January 1993, pp. 34–35.

(a) Develop the ordered array.
(b) Form the stem-and-leaf display.

Data file STOCK1.DAT

● 3.5 The following data are the book values (i.e., net worth divided by number of outstanding shares) for a random sample of 50 stocks from the New York Stock Exchange:

7	9	8	6	12	6	9	15	9	16
8	5	14	8	7	6	10	8	11	4
10	6	16	5	10	12	7	10	15	7
10	8	8	10	18	8	10	11	7	10
7	8	15	23	13	9	8	9	9	13

(a) Develop the ordered array.
(b) Form the stem-and-leaf display.

3.6 A medical doctor speaking on a late-night television show conjectures that "cancer appears to be more prevalent in states with large urban populations and in states in the eastern part of the United States." The following data represent the incidence rate of cancer (i.e., reported incidence per 100,000 population) in all 50 states during a recent year:

Data file CANCER.DAT

State	Incidence of Cancer per 100,000 Population	State	Incidence of Cancer per 100,000 Population
Alabama	433	Montana	372
Alaska	442	Nebraska	336
Arizona	360	Nevada	422
Arkansas	383	New Hampshire	403
California	366	New Jersey	464
Colorado	282	New Mexico	375
Connecticut	434	New York	329
Delaware	500	North Carolina	355
Florida	367	North Dakota	408
Georgia	406	Ohio	463
Hawaii	371	Oklahoma	326
Idaho	307	Oregon	396
Illinois	402	Pennsylvania	442
Indiana	438	Rhode Island	445
Iowa	377	South Carolina	418
Kansas	345	South Dakota	348
Kentucky	414	Tennessee	408
Louisiana	422	Texas	313
Maine	391	Utah	229
Maryland	491	Vermont	376
Massachusetts	443	Virginia	440
Michigan	454	Washington	364
Minnesota	366	West Virginia	409
Mississippi	438	Wisconsin	398
Missouri	390	Wyoming	238

Source: National Cancer Institute.

(a) Develop the ordered array.
(b) Form the stem-and-leaf display.

3.7 The data at the top of page 61 represent the type (creamy versus chunky), score (0 = poor, 100 = excellent), cost (in cents), and amount of sodium (in mgs) of a sample of 37 brands of peanut butter:

Product	Type	Score	Cost (¢)	Sodium (mgs)
Jif	Creamy	68	22	220
Smucker's Natural	Creamy	65	27	15
Deaf Smith Arrowhead Mills	Creamy	62	32	0
Adams 100% Natural	Creamy	56	26	0
Adams	Creamy	56	26	168
Skippy	Creamy	56	19	225
Laura Scudder's All Natural	Creamy	53	26	165
Kroger	Creamy	50	14	240
Country Pure Brand (Safeway)	Creamy	50	21	225
NuMade (Safeway)	Creamy	45	20	187
Peter Pan	Creamy	44	21	225
Peter Pan	Creamy	41	22	3
A&P	Creamy	40	12	225
Hollywood Natural	Creamy	40	32	15
Food Club	Creamy	39	17	225
Pathmark	Creamy	36	9	255
Lady Lee (Lucky Stores)	Creamy	30	16	225
Albertsons	Creamy	30	17	225
Shur Fine (Shurfine Central Corp.)	Creamy	22	16	225
Smucker's Natural	Chunky	80	27	15
Jif	Chunky	75	23	162
Skippy	Chunky	75	21	211
Adams 100% Natural	Chunky	62	26	0
Deaf Smith Arrowhead Mills	Chunky	62	32	0
Country Pure Brand (Safeway)	Chunky	62	21	195
Laura Scudder's All Natural	Chunky	56	24	165
Smucker's Natural	Chunky	53	26	188
Food Club	Chunky	52	17	195
Kroger	Chunky	50	14	255
A&P	Chunky	47	11	225
Peter Pan	Chunky	47	22	180
NuMade (Safeway)	Chunky	42	21	208
Health Valley 100% Natural	Chunky	42	34	3
Lady Lee (Lucky Stores)	Chunky	40	16	225
Albertsons	Chunky	36	17	225
Pathmark	Chunky	34	9	210
Shur Fine (Shurfine Central Corp.)	Chunky	34	16	195

Data file PEANUT.DAT

Source: Copyright 1990 by Consumers Union of United States, Inc., Yonkers, N.Y. 10703. Adapted by permission of *Consumer Reports*, September 1990, p. 590.

For each of the three variables (score, cost, and sodium)

(a) Develop the ordered array.
(b) Form the stem-and-leaf display.

3.8 The following data represent the amount of time (in seconds) to get from 0 to 60 mph during a road test for a sample of 22 German-made automobile models and a sample of 30 Japanese-made automobile models:

Data file ROAD.DAT

German-made Cars				Japanese-made Cars				
10.0	7.9	7.1	8.6	9.4	7.7	5.7	8.2	9.3
6.4	6.9	8.7	8.3	8.9	9.3	8.3	9.7	8.6
8.5	6.4	7.5	6.7	6.7	9.1	9.5	11.7	10.0
5.5	6.0	5.4	6.9	7.2	6.8	8.0	6.3	8.8
5.1	4.9	8.5	8.8	8.5	7.1	6.5	12.0	9.2
10.9	8.9			9.5	10.5	12.5	6.2	6.6

Source: Data are extracted from *Road & Track,* October 1990, vol. 42, no. 2, p. 47.

(a) Develop the ordered array for each data batch.
(b) Form the stem-and-leaf display for each data batch.

Data file
SHAMPOO.DAT

3.9 The following data are the cost per ounce (in cents) for random samples of 31 conventional shampoos labeled for "normal" hair and 29 shampoos labeled for "fine" hair:

Normal Hair					Fine Hair				
79	63	19	9	37	69	9	23	22	8
49	20	16	55	69	12	32	12	18	74
23	14	9	87	44	19	63	49	37	55
13	16	23	20	64	85	44	87	17	11
28	18	32	81	85	23	50	65	51	35
47	50	8	13	21	14	20	28	8	
9									

Source: Copyright 1992 by Consumers Union of United States, Inc., Yonkers, N.Y. 10703. Adapted by permission from *Consumer Reports*, June 1992, pp. 400-401.

(a) Develop the ordered array for each data batch.
(b) Form the stem-and-leaf display for each data batch.

3.3 Tabulating Numerical Data: The Frequency Distribution

Using either the raw data, ordered array, or revised stem-and-leaf display of out-of-state resident tuition rates for the 60 colleges and universities in Texas (see Tables 3.1 and 3.2 on page 55 and Figure 3.1 on page 56), the research analyst wishes to construct the appropriate tables and charts that will enhance the report she is preparing for the marketing manager of the college advisory service company.

Regardless of whether an ordered array or a stem-and-leaf display is selected for *organizing* the data, as the number of observations gets large it becomes necessary to further condense the data into appropriate summary tables. Thus we may wish to arrange the data into *class groupings* (i.e., *categories*) according to conveniently established divisions of the range of the observations. Such an arrangement of data in tabular form is called a frequency distribution.

> A **frequency distribution** is a summary table in which the data are arranged into conveniently established numerically ordered class groupings or categories.

When the observations are *grouped* or condensed into frequency-distribution tables, the process of data analysis and interpretation is made much more manageable and meaningful. In such summary form the major data characteristics are very easily approximated, thus compensating for the fact that when the data are so grouped the initial information pertaining to individual observations that was previously available is lost through the grouping or condensing process.

In constructing the frequency-distribution table, attention must be given to

1. Selecting the appropriate *number* of class groupings for the table
2. Obtaining a suitable *class interval* or *width* of each class grouping
3. Establishing the *boundaries* of each class grouping to avoid overlapping

3.3.1 Selecting the Number of Classes

The number of class groupings to be used is primarily dependent on the number of observations in the data. That is, larger numbers of observations require a larger number of class groups. In general, however, the frequency distribution should have at least five class groupings, but no more than 15. If there are not enough class groupings or if there are too many, little information would be obtained. As an example, a frequency distribution having but one class grouping that spans the entire range of tuition rates could be formed as follows:

Tuition Rates (in $000)	Number of Schools
2.0-13.0	60
Total	60

From such a summary table, however, no additional information is obtained that was not already known from scanning either the raw data or the ordered array. A table with too much data concentration is not meaningful. The same would be true at the other extreme—if a table had too many class groupings, there would be an underconcentration of data, and very little would be learned.

3.3.2 Obtaining the Class Intervals

When developing the frequency-distribution table it is desirable to have each class grouping of equal width. To determine the width of each class, the *range* of the data is divided by the number of class groupings desired:

$$\text{width of interval} \cong \frac{\text{range}}{\text{number of desired class groupings}} \tag{3.1}$$

Since there are only 60 observations in our tuition rate data, we decided that six class groupings would be sufficient. From the ordered array in Table 3.2 (page 55), the range is computed as 12.0 – 2.4 = 9.6 thousand dollars, and, using Equation (3.1), the width of the class interval is approximated by

$$\text{width of interval} \cong \frac{9.6}{6} = 1.6 \text{ thousand dollars}$$

For convenience and ease of reading, the selected interval or width of each class grouping is rounded up to 2.0 thousand dollars.

3.3.3 Establishing the Boundaries of the Classes

To construct the frequency-distribution table, it is necessary to establish clearly defined class boundaries for each class grouping so that the observations either in raw form or in an ordered array can be properly tallied. Overlapping of classes must be avoided.

Since the width of each class interval for the tuition rate data has been set at 2.0 thousand dollars, the boundaries of the various class groupings must be established so as to include the entire range of observations. Whenever possible, these boundaries should be chosen to facilitate the reading and interpreting of data. Thus the first class interval is established from 2.0 to under 4.0, the second from 4.0 to under 6.0, etc. The data in their raw form (Table 3.1) or from the ordered array (Table 3.2) are then tallied into each class as shown:

Tuition Rates (in $000)	Tallies	Frequency
2.0 but less than 4.0	~~////~~ ~~////~~ ///	13
4.0 but less than 6.0	~~////~~ ~~////~~ ~~////~~ ~~////~~ ////	24
6.0 but less than 8.0	~~////~~ ////	9
8.0 but less than 10.0	~~////~~ ///	8
10.0 but less than 12.0	~~////~~	5
12.0 but less than 14.0	/	1
Total		60

By establishing the boundaries of each class as above, all 60 observations have been tallied into six classes, each having an interval width of 2.0 thousand dollars, without overlapping. From this "worksheet" the frequency distribution is presented in Table 3.3.

Table 3.3 Frequency distribution of tuition rates charged for out-of-state residents at 60 Texas schools.

Tuition Rates (in $000)	Number of Schools
2.0 but less than 4.0	13
4.0 but less than 6.0	24
6.0 but less than 8.0	9
8.0 but less than 10.0	8
10.0 but less than 12.0	5
12.0 but less than 14.0	1
Total	60

Source: Data are taken from Table 3.1 on page 55.

The main advantage of using such a summary table is that the major data characteristics become immediately clear to the reader. For example, we see from Table 3.3 that the *approximate range* of the 60 tuition rates is from 2.0 to 14.0 thousand dollars, with out-of-state tuition at most Texas schools tending to cluster between

4.0 and 6.0 thousand dollars. Other descriptive measures that are obtained from *grouped data* will be presented in Section 4.9.

On the other hand, the major disadvantage of such a summary table is that we cannot know how the individual values are distributed within a particular class interval without access to the original data. Thus for the five schools with out-of-state resident tuition rates between 10.0 and 12.0 thousand dollars, it is not clear from Table 3.3 whether the values are distributed throughout the interval, are close to 10.0 thousand dollars, or are close to 12.0 thousand dollars. The class midpoint, however, is the value used to represent all the data summarized into a particular interval.

The **class midpoint** is the point halfway between the boundaries of each class and is representative of the data within that class.

The class midpoint for the interval "2.0 but less than 4.0" is 3.0 thousand dollars. (The other class midpoints are, respectively, 5.0, 7.0, 9.0, 11.0, and 13.0 thousand dollars.)

3.3.4 Subjectivity in Selecting Class Boundaries

The selection of class boundaries for frequency-distribution tables is highly subjective. Hence for data batches that do not contain many observations, the choice of a particular set of class boundaries over another might yield an entirely different picture to the reader. For example, for the tuition rate data, using a class interval width of 2.5 thousand dollars instead of 2.0 (as was used in Table 3.3) may cause shifts in the way in which the observations distribute among the classes. This is particularly true if the number of observations in the batch is not very large.

However, such shifts in data concentration do not occur only because the width of the class interval is altered. We may keep the interval width at 2.0 thousand dollars but choose different lower and upper class boundaries. Such manipulation may also cause shifts in the way in which the data distribute—especially if the size of the batch is not very large. Fortunately, as the number of observations in a data batch increases, alterations in the selection of class boundaries affect the concentration of data less and less.

Problems for Section 3.3

3.10 A random sample of 50 executive vice-presidents was selected from among the various public relations firms in the United States, and the annual salaries of these company officers were obtained. The salaries ranged from \$52,000 to \$137,000. Set up the class boundaries for a frequency distribution
(a) If 5 class intervals are desired.
(b) If 6 class intervals are desired.
(c) If 7 class intervals are desired.
(d) If 8 class intervals are desired.

3.11 If the asking price of one-bedroom cooperative and condominium apartments in Queens, a borough of New York City, varied from \$103,000 to \$295,000
(a) Indicate the class boundaries of 10 classes into which these values can be grouped.
(b) What class-interval width did you choose?
(c) What are the 10 class midpoints?

3.12 The raw data displayed below are the electric and gas utility charges during the month of July 1993 for a random sample of 50 three-bedroom apartments in Manhattan:

Data file UTILITY.DAT

Raw Data on Utility Charges ($)									
96	171	202	178	147	102	153	197	127	82
157	185	90	116	172	111	148	213	130	165
141	149	206	175	123	128	144	168	109	167
95	163	150	154	130	143	187	166	139	149
108	119	183	151	114	135	191	137	129	158

(a) Form a frequency distribution
(1) having 5 class intervals.
(2) having 6 class intervals.
(3) having 7 class intervals.

[*Hint*: To help you decide how to best set up the class boundaries you should first place the raw data either in a stem-and-leaf display (by letting the leaves be the trailing digits) or in an ordered array.]

(b) Form a frequency distribution having seven class intervals with the following class boundaries: $80 but less than $100, $100 but less than $120, and so on.

3.13 Construct a frequency distribution from the shower heads data in Problem 3.3 on pages 58–59.

3.14 Construct a frequency distribution from the book value data in Problem 3.5 on page 60.

3.15 Construct a frequency distribution from the cancer incidence data in Problem 3.6 on page 60.

3.16 Construct separate frequency distributions for each of the three numerical variables (score, cost, and sodium) from the peanut butter data in Problem 3.7 on pages 60–61.

3.17 Construct separate frequency distributions of the acceleration times for the German-made versus Japanese-made cars in Problem 3.8 on page 61.

3.18 Given the ordered arrays in the accompanying table dealing with the lengths of life (in hours) of a sample of forty 100-watt light bulbs produced by Manufacturer A and a sample of forty 100-watt light bulbs produced by Manufacturer B:

Data file BULBS.DAT

Ordered arrays of length of life of two brands of 100-watt light bulbs (in hours).

Manufacturer A				
684	697	720	773	821
831	835	848	852	852
859	860	868	870	876
893	899	905	909	911
922	924	926	926	938
939	943	946	954	971
972	977	984	1005	1014
1016	1041	1052	1080	1093

Manufacturer B				
819	836	888	897	903
907	912	918	942	943
952	959	962	986	992
994	1004	1005	1007	1015
1016	1018	1020	1022	1034
1038	1072	1077	1077	1082
1096	1100	1113	1113	1116
1153	1154	1174	1188	1230

(a) Form the frequency distribution for each brand. (*Hint*: For purposes of comparison, choose class-interval widths of $100 for each distribution.)

(b) For purposes of answering Problems 3.25, 3.32, and 3.40, form the frequency distribution for each brand according to the following schema [if you have not already done so in part (a) of this problem]:
Manufacturer A: 650 but less than 750, 750 but less than 850, and so on
Manufacturer B: 750 but less than 850, 850 but less than 950, and so on

3.4 Tabulating Numerical Data: The Relative Frequency Distribution and Percentage Distribution

The frequency distribution is a summary table into which the original data are condensed or grouped to facilitate data analysis. To enhance the analysis, however, it is almost always desirable to form either the relative frequency distribution or the percentage distribution, depending on whether we prefer proportions or percentages. These two equivalent distributions are shown as Tables 3.4 and 3.5, respectively.

Table 3.4 Relative frequency distribution of tuition rates charged for out-of-state residents at 60 Texas schools.

Tuition Rates (in $000)	Proportion of Schools
2.0 but less than 4.0	.217
4.0 but less than 6.0	.400
6.0 but less than 8.0	.150
8.0 but less than 10.0	.133
10.0 but less than 12.0	.083
12.0 but less than 14.0	.017
Total	1.000

Source: Data are taken from Table 3.3 on page 64.

Table 3.5 Percentage distribution of tuition rates charged for out-of-state residents at 60 Texas schools.

Tuition Rates (in $000)	Percentage of Schools
2.0 but less than 4.0	21.7
4.0 but less than 6.0	40.0
6.0 but less than 8.0	15.0
8.0 but less than 10.0	13.3
10.0 but less than 12.0	8.3
12.0 but less than 14.0	1.7
Total	100.0

Source: Data are taken from Table 3.3 on page 64.

The **relative frequency distribution** depicted in Table 3.4 on page 67 is formed by dividing the frequencies in each class of the frequency distribution (Table 3.3 on page 64) by the total number of observations. A **percentage distribution** (Table 3.5) may then be formed by multiplying each relative frequency or proportion by 100.0. Thus from Table 3.4 it is clear that the proportion of schools in Texas with out-of-state resident tuition rates from 12.0 to under 14.0 thousand dollars is .017, while from Table 3.5 it is seen that 1.7% of the schools have such tuition rates.

Working with a base of 1 for proportions or 100.0 for percentages is usually more meaningful than using the frequencies themselves. Indeed, the use of the relative frequency distribution or percentage distribution becomes essential whenever one batch of data is being compared with other batches of data, especially if the numbers of observations in each batch differ.

As a case in point, let us suppose that an industrial psychologist wanted to compare daily absenteeism among the clerical workers in two department stores. If, on a given day, 6 clerical workers out of 50 in Store A were absent and 3 clerical workers out of 10 in Store B were absent, what conclusions can be drawn? It is inappropriate to say that *more* absenteeism occurred in Store A. Although we have observed that in Store A there were twice as many absences as there were in Store B, there were also five times as many clerical workers employed in Store A. Hence, in these types of comparisons, we must formulate our conclusions from the *relative rates* of absenteeism, not from the actual counts. Thus it may be stated that the absenteeism rate is two and a half times higher in Store B (30.0%) than it is in Store A (12.0%).

Now suppose, when developing her report for the marketing manager of the college advisory service company, that the research analyst wanted to compare the out-of-state resident tuition rates at the 60 Texas schools with those reported from the 45 higher education institutions in the state of North Carolina. Table 3.6 displays information on the out-of-state resident tuition rate for *each* of the 45 North Carolina colleges and universities (see Special Data Set 1 of Appendix D on page D3).

To compare the tuition rates from the 60 Texas institutions with those from the 45 North Carolina schools, we develop a percentage distribution for the latter group. This new table will then be compared with Table 3.5.

Table 3.6 Raw data pertaining to tuition rates (in $000) for out-of-state residents at 45 colleges and universities in North Carolina.

6.5	4.0	7.1	8.3	5.4	7.6	9.0	15.7	16.7
6.4	5.0	8.5	5.7	7.7	7.2	12.4	7.1	5.5
9.7	4.4	7.0	6.3	8.3	6.9	5.7	7.6	7.9
7.9	6.0	8.2	10.4	9.9	3.9	9.8	8.2	5.6
7.9	6.4	7.4	7.0	13.0	8.7	6.4	6.7	7.4

Data file C&UNC.DAT

Source: See Special Data Set 1, Appendix D, page D3, taken from "America's Best Colleges, 1994 College Guide," *U.S. News & World Report*, extracted from College Counsel 1993 of Natick, Mass. Reprinted by special permission, *U.S. News & World Report*, © 1993 by *U.S. News & World Report* and by College Counsel.

Table 3.7 depicts both the frequency distribution and the percentage distribution of the tuition rates charged to out-of-state residents by the 45 North Carolina schools. This table has been constructed in lieu of two separate tables to save space. Note that the class groupings selected in Table 3.7 match, where possible, those selected in Table 3.3 for the Texas schools. The boundaries of the classes should match or be multiples of each other in order to facilitate comparisons.

Table 3.7 Frequency distribution and percentage distribution of out-of-state resident tuition rates at 45 North Carolina schools.

Tuition Rates (in $000)	Number of Schools	Percentage of Schools
2.0 but less than 4.0	1	2.2
4.0 but less than 6.0	8	17.8
6.0 but less than 8.0	21	46.7
8.0 but less than 10.0	10	22.2
10.0 but less than 12.0	1	2.2
12.0 but less than 14.0	2	4.4
14.0 but less than 16.0	1	2.2
16.0 but less than 18.0	1	2.2
Totals	45	99.9*

* Error due to rounding.
Source: Data are taken from Table 3.6.

Using the percentage distributions of Tables 3.5 and 3.7, it is now meaningful to compare the schools in the two states in terms of the tuition rates charged to out-of-state residents. From the two tables it is apparent that the tuition rates are generally lower in Texas than in North Carolina. For example, in Texas tuition rates are typically clustering between 4.0 and 6.0 thousand dollars (that is, 40.0% of the schools) while in North Carolina the tuition rates are typically clustering between 6.0 and 8.0 thousand dollars (that is, 46.7% of the schools). Moreover, we may note that the *ranges* in tuition rates can be easily approximated from the tables. In North Carolina, the range in tuition rates is approximated to be 16.0 thousand dollars (that is, the difference between 18.0, the upper boundary of the last class, and 2.0, the lower boundary of the first class), while in Texas the range is approximated as 12.0 thousand dollars (that is, 14.0 − 2.0). Other descriptive summary measures that would enhance a comparative analysis of the tuition rates between the two states will be discussed in Chapter 4.

Problems for Section 3.4

Data file UTILITY.DAT

● 3.19 Form the percentage distribution from the frequency distribution developed in Problem 3.12(b) on page 66 regarding utility charges.

Data file SHOWER.DAT

3.20 Form the percentage distribution from the frequency distribution developed in Problem 3.13 on page 66 regarding shower heads.

Data file STOCK1.DAT

● 3.21 Form the percentage distribution from the frequency distribution developed in Problem 3.14 on page 66 regarding book values of companies listed on the NYSE.

Data file CANCER.DAT

3.22 Form the percentage distribution from the frequency distribution developed in Problem 3.15 on page 66 regarding cancer incidence.

Data file PEANUT.DAT

3.23 Form the percentage distributions corresponding to the frequency distributions for each of the three numerical variables (score, cost, and sodium) developed in Problem 3.16 on page 66 regarding peanut butter characteristics.

Data file ROAD.DAT

3.24 Form the percentage distributions from the frequency distributions developed in Problem 3.17 on page 66 concerning the acceleration times for the German-made versus Japanese-made cars.

Data file BULBS.DAT

3.25 Form the percentage distributions from the frequency distributions developed in Problem 3.18 on page 66 concerning the life of light bulbs manufactured by two competing companies, A and B.

3.5 Graphing Numerical Data: The Histogram and Polygon

It is often said that "one picture is worth a thousand words." Indeed, statisticians have employed graphic techniques to more vividly describe batches of data. In particular, histograms and polygons are used to describe numerical data that have been grouped into frequency, relative frequency, or percentage distributions.

3.5.1 Histograms

Histograms are vertical bar charts in which the rectangular bars are constructed at the boundaries of each class.

When plotting histograms, the random variable or phenomenon of interest is displayed along the horizontal axis; the vertical axis represents the number, proportion, or percentage of observations per class interval—depending on whether the particular histogram is, respectively, a frequency histogram, a relative frequency histogram, or a percentage histogram.

Vertical Axis Label	⟷	Type of Chart
Number of observations	⟷	Frequency histogram (or polygon)
Proportion of observations	⟷	Relative frequency histogram (or polygon)
Percentage of observations	⟷	Percentage histogram (or polygon)

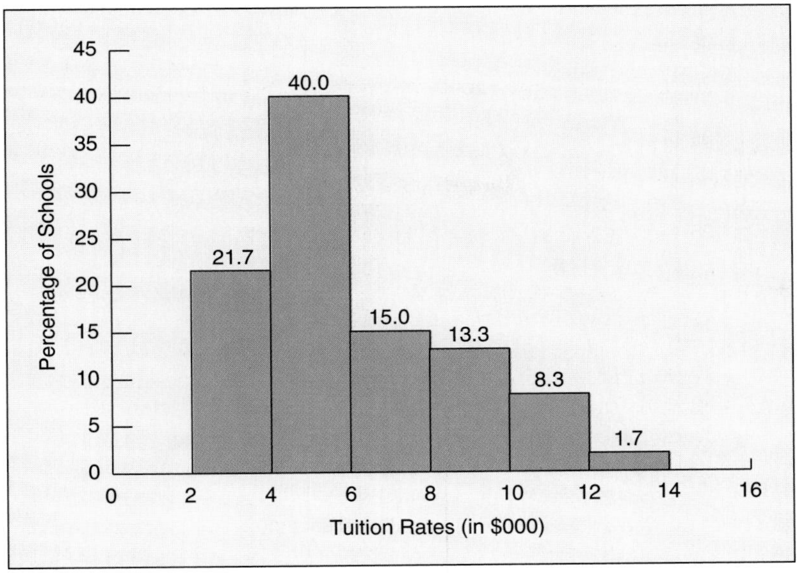

Figure 3.5
Percentage histogram of tuition rates charged for out-of-state residents at 60 Texas schools.
Source: Data are taken from Table 3.5.

A percentage histogram is depicted in Figure 3.5 for the out-of-state resident tuition rates at all 60 colleges and universities in Texas.

It is interesting to note the close visual relationship portrayed by the stem-and-leaf display and the histogram. Look at Figure 3.4 on page 58 and our histogram in Figure 3.5. If we were to rotate the stem-and-leaf display 90° (that is, hold our book sideways), a frequency histogram would be depicted in such a manner that its class groupings would be represented by the stems and its vertical bars would be represented by individual leaves on each stem.[1]

When comparing two or more batches of data, neither stem-and-leaf displays nor histograms can be constructed on the same graph. With respect to the latter, superimposing the vertical bars of one on another would cause difficulty in interpretation. For such cases it is necessary to construct relative frequency or percentage polygons.

3.5.2 Polygons

As with histograms, when plotting polygons the phenomenon of interest is displayed along the horizontal axis and the vertical axis represents the number, proportion, or percentage of observations per class interval.

> The **percentage polygon** is formed by letting the midpoint of each class represent the data in that class and then connecting the sequence of midpoints at their respective class percentages.

Because consecutive midpoints are connected by a series of straight lines, the polygon is sometimes jagged in appearance. However, when dealing with a very large batch of data, if we were to make the boundaries of the classes in its frequency distribution closer together (and thereby increase the number of classes in that distribution), the jagged lines of the polygon would "smooth out."

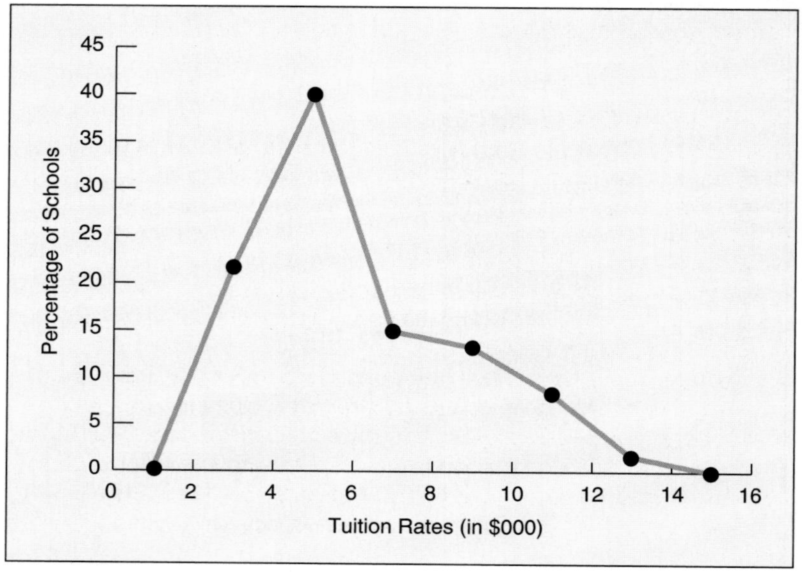

Figure 3.6
Percentage polygon of tuition rates charged for out-of-state residents at 60 Texas schools.
Source: Data are taken from Table 3.5.

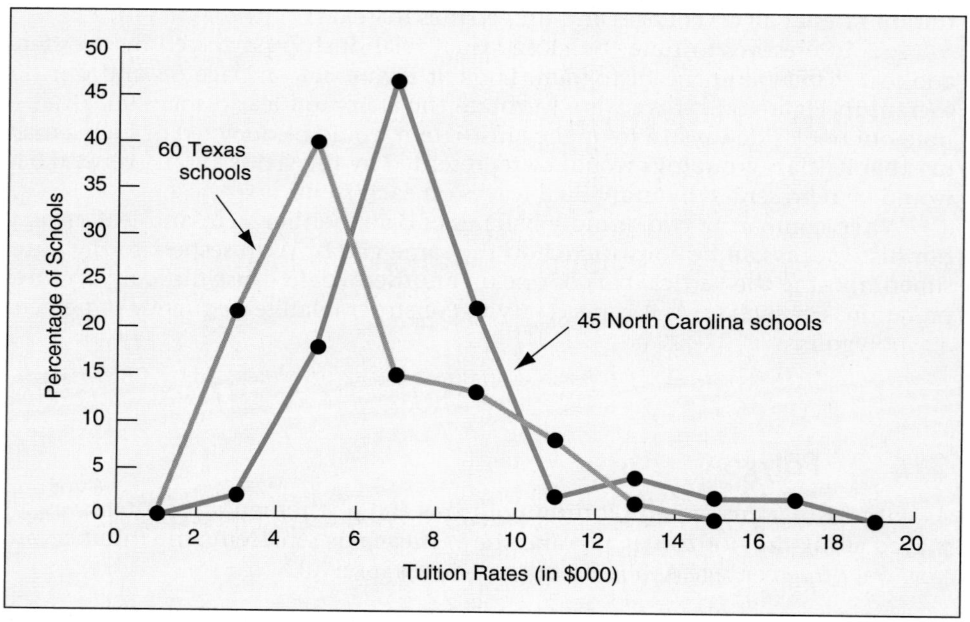

Figure 3.7
Percentage polygons of tuition rates charged for out-of-state residents at 60 Texas schools and 45 North Carolina schools.
Source: Data are taken from Tables 3.5 and 3.7.

Figure 3.6 shows the percentage polygon for the out-of-state resident tuition rates at all 60 Texas schools and Figure 3.7 compares the percentage polygons for the tuition rates at the 60 Texas schools versus the 45 North Carolina schools. The differences in the structure of the two distributions, previously discussed when comparing Tables 3.5 and 3.7, are clearly indicated here.

● **Polygon Construction** Notice that the polygon is a representation of the shape of the particular distribution. Since the area under the percentage distribution (entire curve) must be 100.0%, it is necessary to connect the first and last midpoints with the horizontal axis so as to enclose the area of the observed distribution. In Figure 3.6 this is accomplished by connecting the first observed midpoint with the midpoint of a "fictitious preceding" class (that is, 1.0 thousand dollars) having 0.0% observations and by connecting the last observed midpoint with the midpoint of a "fictitious succeeding" class (that is, 15.0 thousand dollars) having 0.0% observations.

Notice, too, that when polygons (Figure 3.6) or histograms (Figure 3.5) are constructed, the vertical axis must show the true zero or "origin" so as not to distort or otherwise misrepresent the character of the data. The horizontal axis, however, does not need to specify the zero point for the phenomenon of interest. For aesthetic reasons, the range of the random variable should constitute the major portion of the chart, and, when zero is not included, "breaks" in the axis are appropriate.

Problems for Section 3.5

3.26 From the percentage distribution developed in Problem 3.19 on page 70 regarding utility charges
(a) Plot the percentage histogram.
(b) Plot the percentage polygon.

Data file UTILITY.DAT

3.27 From the percentage distribution developed in Problem 3.20 on page 70 regarding shower heads
(a) Plot the percentage histogram.
(b) Plot the percentage polygon.

Data file SHOWER.DAT

3.28 From the percentage distribution developed in Problem 3.21 on page 70 regarding book values of companies listed on the NYSE
(a) Plot the percentage histogram.
(b) Plot the percentage polygon.

Data file STOCK1.DAT

3.29 From the percentage distribution developed in Problem 3.22 on page 70 regarding cancer incidence
(a) Plot the percentage histogram.
(b) Plot the percentage polygon.

Data file CANCER.DAT

3.30 From the percentage distributions developed in Problem 3.23 on page 70 for each of the three numerical variables (score, cost, and sodium) regarding peanut butter characteristics
(a) Plot the respective percentage histogram.
(b) Plot the respective percentage polygon.

Data file PEANUT.DAT

3.31 From the percentage distributions developed in Problem 3.24 on page 70 concerning the acceleration times for the German-made versus Japanese-made cars
(a) Plot the percentage histograms on separate graphs.
(b) Plot the percentage polygons on one graph.

Data file ROAD.DAT

3.32 From the percentage distributions developed in Problem 3.25 on page 70 regarding life of light bulbs
(a) Plot the percentage histograms on separate graphs.
(b) Plot the percentage polygons on one graph.

Data file BULBS.DAT

3.6 Cumulative Distributions and Cumulative Polygons

Two other useful methods of data presentation that facilitate analysis and interpretation are the cumulative distribution tables and the cumulative polygon charts. Both of these may be developed from the frequency distribution table, the relative frequency distribution table, or the percentage distribution table.

3.6.1 The Cumulative Percentage Distribution

Depending on our individual preference for proportions or percentages, when comparing two or more data batches of differing size we select either the relative frequency distribution or the percentage distribution. Since we already have the percentage distributions of out-of-state resident tuition rates at 60 Texas schools and at 45 North Carolina schools in Tables 3.5 and 3.7 (pages 68 and 69), we can use these tables to construct the respective cumulative percentage distributions. See Tables 3.8 and 3.9.

Table 3.8 Cumulative percentage distribution of tuition rates charged for out-of-state residents at 60 Texas schools.

Tuition Rates (in $000)	Percentage of Schools "Less Than" Indicated Value
2.0	0.0
4.0	21.7
6.0	61.7
8.0	76.7
10.0	90.0
12.0	98.3
14.0	100.0

Source: Data are taken from Table 3.5.

Table 3.9 Cumulative percentage distribution of tuition rates charged for out-of-state residents at 45 North Carolina schools.

Tuition Rates (in $000)	Percentage of Schools "Less Than" Indicated Value
2.0	0.0
4.0	2.2
6.0	20.0
8.0	66.7
10.0	88.9
12.0	91.1
14.0	95.6
16.0	97.8
18.0	100.0

Source: Data are taken from Table 3.7.

A **cumulative percentage distribution table** is constructed by first recording the lower boundaries of each class from the percentage distribution and then inserting an extra boundary at the end. We compute the cumulative percentages in the *"less than"* column by determining the percentage of observations less than each of the stated boundary values. Thus from Table 3.5, we see that 0.0% of the out-of-state resident tuition rates in Texas institutions are less than 2.0 thousand dollars; 21.7% of the tuition rates are less than 4.0 thousand dollars; 61.7% of the tuition rates are less than 6.0 thousand dollars; and so on until all (100.0%) of the tuition rates are less than 14.0 thousand dollars. This cumulating process is readily observed in Table 3.10.

Table 3.10 Forming the cumulative percentage distribution.

From Table 3.5		From Table 3.8
Tuition Rates (in $000)	Percentage of Schools in Class Interval	Percentage of Schools "Less Than" Lower Boundary of Class Interval
2.0 but less than 4.0	21.7	0.0
4.0 but less than 6.0	40.0	21.7
6.0 but less than 8.0	15.0	61.7 = 21.7 + 40.0
8.0 but less than 10.0	13.3	76.7 = 21.7 + 40.0 + 15.0
10.0 but less than 12.0	8.3	90.0 = 21.7 + 40.0 + 15.0 + 13.3
12.0 but less than 14.0	1.7	98.3 = 21.7 + 40.0 + 15.0 + 13.3 + 8.3
14.0 but less than 16.0	0.0	100.0 = 21.7 + 40.0 + 15.0 + 13.3 + 8.3 + 1.7

3.6.2 Cumulative Percentage Polygon

To construct a **cumulative percentage polygon** (also known as an **ogive**), we note that the phenomenon of interest—tuition rates—is again plotted on the horizontal axis, while the cumulative percentages (from the *"less than"* column) are plotted on the vertical axis. At each lower boundary, we plot the corresponding (cumulative) percentage value from the listing in the cumulative percentage distribution. We then connect these points with a series of straight-line segments.

Figure 3.8 on page 76 illustrates the cumulative percentage polygon of out-of-state resident tuition rates at the 60 Texas schools. The major advantage of the ogive over other charts is the ease with which we can interpolate between the plotted points.

Approximating the Percentages As one example, the research analyst at the college advisory service company might wish to approximate the percentage of Texas colleges and universities that charge a tuition rate below a specified amount, say 7.0 thousand dollars. To accomplish this, a vertical line is projected upward at 7.0 until it intersects the "less than" curve. The desired percentage is then approximated by reading horizontally from the point of intersection to the percentage indicated on the vertical axis. In this case, approximately 69.2% of the Texas schools have tuition rates under 7.0 thousand dollars. (This, of course, implies that about 30.8% of the schools have tuition rates of at least 7.0 thousand dollars.)

Approximating the Values Even more important, when preparing her report for the marketing manager of the college advisory service company, the

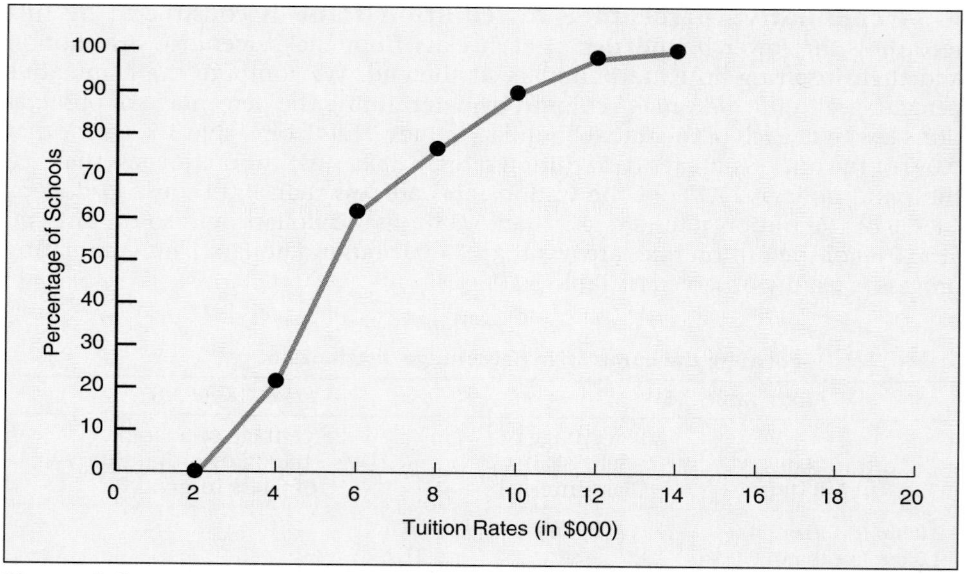

Figure 3.8
Cumulative percentage polygon of tuition rates charged for out-of-state residents at 60 Texas schools.
Source: Data are taken from Table 3.8.

research analyst may also wish to approximate various tuition rates that correspond to particular cumulative percentages. For example, 25.0% of all Texas schools have tuition rates below what amount? To determine this, a horizontal line is drawn from the specified cumulative percentage point (25.0) until it intersects the "less than" curve. The desired tuition rate is then approximated by dropping a perpendicular (a vertical line) at the point of intersection to the horizontal axis. From Figure 3.8 we note that this rate is approximately 4.2 thousand dollars. Other percentage points commonly considered for such analysis (see Chapter 4) are the 50.0% value and the 75.0% value.

● **Comparing Two or More Cumulative Distributions** Such approximations as these are extremely helpful when comparing two or more batches of data. Figure 3.9 on page 77 depicts the cumulative percentage polygons of out-of-state resident tuition rates for both the 60 Texas schools and the 45 North Carolina schools.

From Figure 3.9 we note that in general the Texas ogive is drawn to the left of the North Carolina ogive. For example, in Texas 25% of all tuition rates are below 4.2 thousand dollars, while in North Carolina we see that 25% of all tuition rates are below 6.1 thousand dollars. Moreover, in Texas 50% of all tuition rates are below 5.4 thousand dollars, while in North Carolina 50% of all tuition rates are below 7.2 thousand dollars. Furthermore, in Texas 75% of all tuition rates are below 7.7 thousand dollars, while in North Carolina we see that 75% of all tuition rates are below 8.7 thousand dollars. These comparisons enable us to confirm our earlier impression that tuition rates are lower in Texas than in North Carolina.

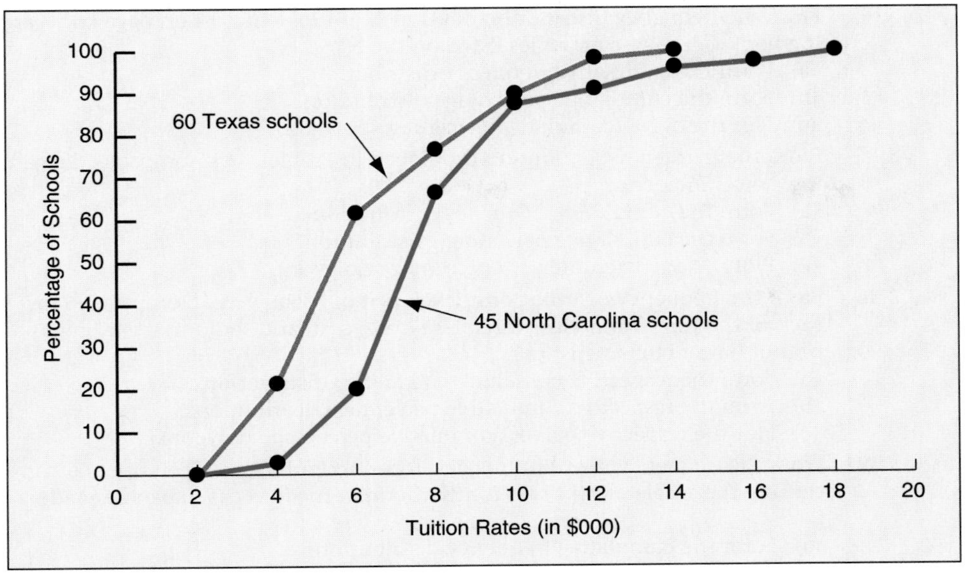

Figure 3.9
Cumulative percentage polygons of tuition rates charged for out-of-state residents at 60 Texas schools and 45 North Carolina schools.
Source: Data are taken from Tables 3.5 and 3.7.

Problems for Section 3.6

Data file C&U.DAT

3.33 Examine Figure 3.9.
(a) 10.0% of the out-of-state resident tuition rates in each state are below what amounts?
(b) 40.0% of the out-of-state resident tuition rates in each state are below what amounts?
(c) 60.0% of the out-of-state resident tuition rates in each state are below what amounts?
(d) 90.0% of the out-of-state resident tuition rates in each state are below what amounts?
(e) What percentage of the out-of-state resident tuition rates in each state are below 5.0 thousand dollars?
(f) What percentage of the out-of-state resident tuition rates in each state are below 11.0 thousand dollars?
(g) Discuss your findings.
(h) How might your information be of assistance to the research analyst at the college advisory service company? Discuss.

3.34 From the frequency distribution developed in Problem 3.12(b) on page 66 regarding utility charges
(a) Form the cumulative frequency distribution.
(b) Form the cumulative percentage distribution.
(c) Plot the ogive (cumulative percentage polygon).

Data file UTILITY.DAT

3.35 From the frequency distribution developed in Problem 3.13 on page 66 regarding shower heads
(a) Form the cumulative frequency distribution.
(b) Form the cumulative percentage distribution.
(c) Plot the ogive (cumulative percentage polygon).

Data file SHOWER.DAT

Data file STOCK1.DAT

● 3.36 From the frequency distribution developed in Problem 3.14 on page 66 regarding book values of companies listed on the NYSE
(a) Form the cumulative frequency distribution.
(b) Form the cumulative percentage distribution.
(c) Plot the ogive (cumulative percentage polygon).

Data file CANCER.DAT

3.37 From the frequency distribution developed in Problem 3.15 on page 66 regarding cancer incidence
(a) Form the cumulative frequency distribution.
(b) Form the cumulative percentage distribution.
(c) Plot the ogive (cumulative percentage polygon).

Data file PEANUT.DAT

3.38 From the frequency distributions developed in Problem 3.16 on page 66 for each of the three numerical variables (score, cost, and sodium) regarding peanut butter characteristics
(a) Form the respective cumulative frequency distributions.
(b) Form the respective cumulative percentage distributions.
(c) Plot the respective ogives (cumulative percentage polygons).

Data file ROAD.DAT

3.39 From the frequency distributions developed in Problem 3.17 on page 66 concerning the acceleration times for the German-made versus Japanese-made cars
(a) Form the cumulative frequency distributions.
(b) Form the cumulative percentage distributions.
(c) Plot the ogives (cumulative percentage polygons) on one graph.

Data file BULBS.DAT

3.40 From the frequency distributions developed in Problem 3.18 on page 66 regarding life of light bulbs from two manufacturers
(a) Form the cumulative frequency distributions.
(b) Form the cumulative percentage distributions.
(c) Plot the ogives (cumulative percentage polygons) on one graph.

3.7 Graphing Numerical Data in Sequence: The Digidot Plot

Thus far in this chapter in our discussion of graphical methods we have not in any way taken into account the sequential order in which the data have been collected. In many situations, particularly in accountancy, economics, and finance, we are interested in studying a set of data collected on a regular daily, weekly, monthly, quarterly, or yearly basis so that it would be natural to plot the outcomes (be they stock price indices, industry wide sales revenues, corporate earnings, etc.) on a graph in which the (horizontal) X axis represents a given period of time. This subject of *time series analysis* is presented in Chapter 19. In other circumstances, particularly in the management of process and product quality, we are also interested in studying the outcomes in a set of data collected in sequential order (be they the number of customers per minute arriving at a branch office of a midtown Manhattan bank during the noon to 1 PM lunch period, the percentage of defective batteries in consecutive samples of 50, the amount of fill in consecutive one-liter apple juice bottles, etc.). The subject of *statistical control of process and product quality* is discussed in Chapter 16 and a variety of control charts are introduced. In this section, as an introduction to these important subjects, we use the processing time (in minutes) spent by a teller handling 24 consecutive customers in a midtown Manhattan bank during the noon to 1 PM lunch period (Figure 3.10). We shall illustrate that plotting the data in sequential order can enhance an analysis.

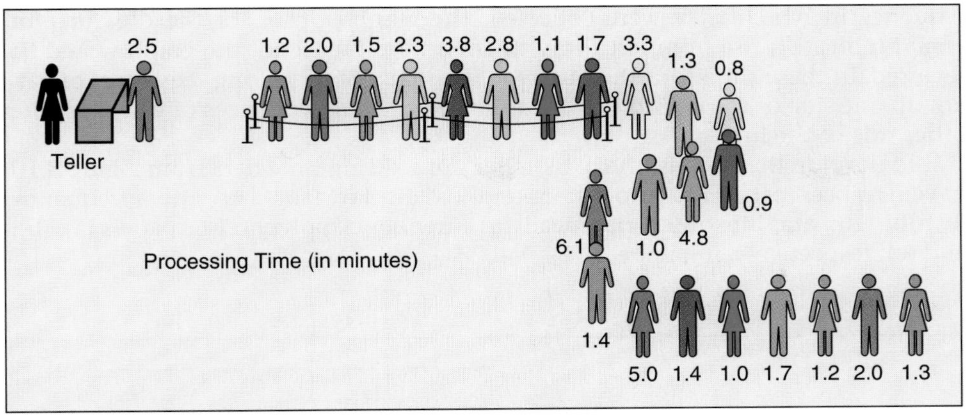

Figure 3.10
Raw data pertaining to teller processing time (in minutes) for 24 consecutive customers in a midtown Manhattan bank.

The data listed in Figure 3.10 appear to be in raw form. Although the data have been recorded chronologically, we should not expect that the processing times (in minutes) pertaining to the 24 consecutive bank customers would follow any observable ordered pattern. (In fact, a major assumption in the inferential procedures we shall be discussing in Chapters 10 through 15 will be that our collected sample observations are randomly and independently drawn.) Here, then, it would be of interest to evaluate graphically whether the data are actually in raw form or whether some unsuspected relationship exists.

3.7.1 Displaying Hunter's Digidot Plot

The **digidot plot** simultaneously presents a stem-and-leaf display and a graph of the observations in the sequential order in which they are obtained. A horizontal line displayed over the sequence usually denotes the *median* or middle value in the ordered array. (The median will be discussed in Section 4.4.2.) This horizontal line permits an easy reference for observing any patterns. For example, as indicated in Figure 3.11, had there been a positive trend in the observations over the ordered

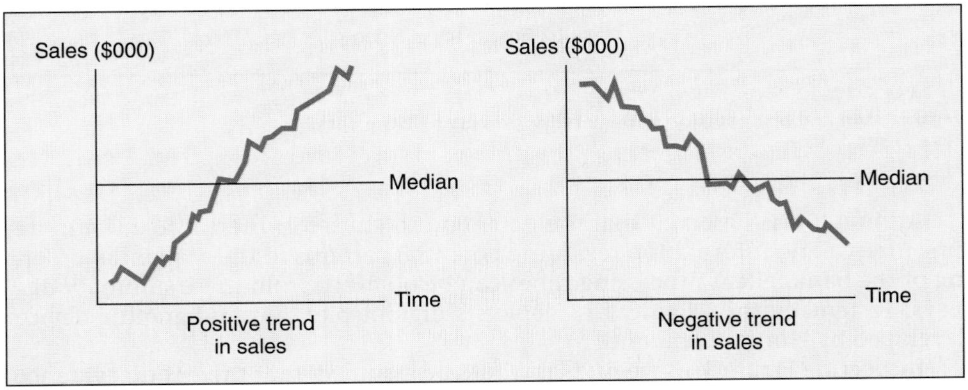

Figure 3.11
Observing trends in data plotted in sequential order.

sequence in which they were collected, the graph portion of the digidot plot would indicate a rise from left to right. For a negative trend, the graph would be reversed. In these situations then we would visually observe long sequences of values on one side of the horizontal line followed by long sequences of values on the other side of the line.

To develop these ideas, the bank teller's processing time data from Figure 3.10 have first been organized into a stem-and-leaf display, tabulated into a frequency distribution, and illustrated graphically as a frequency polygon [see panels (a), (b), and (c), respectively, in Figure 3.12].

0	89
1	251730440723
2	50380
3	83
4	8
5	0
6	1

$n = 24$

(a) Stem-and-leaf display

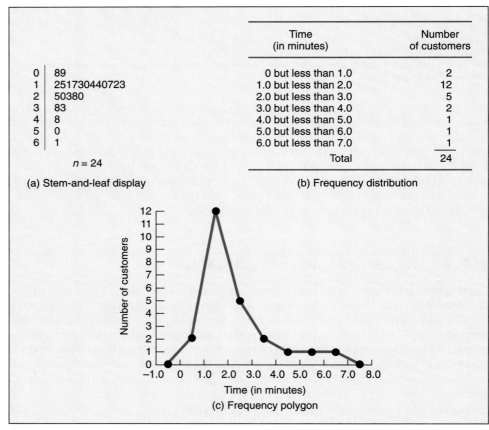

Time (in minutes)	Number of customers
0 but less than 1.0	2
1.0 but less than 2.0	12
2.0 but less than 3.0	5
3.0 but less than 4.0	2
4.0 but less than 5.0	1
5.0 but less than 6.0	1
6.0 but less than 7.0	1
Total	24

(b) Frequency distribution

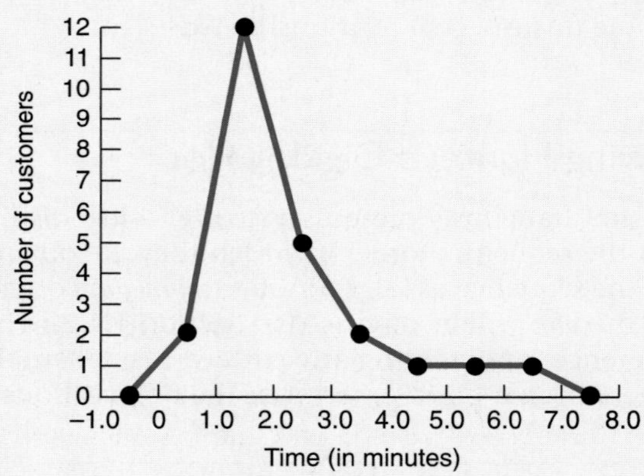

(c) Frequency polygon

Figure 3.12
Organization and presentation of bank teller processing time data.
Source: Figure 3.10.

Although it is observed that the data tend to cluster in the 1.0 to 2.0 minute time interval, no information regarding potential patterns in the sequential ordering of the bank teller's processing times can be obtained from these summary displays. To remedy this, Figure 3.13 depicts a digidot plot, a useful graphic device developed by Hunter (Reference 5).

Inspecting Figure 3.13 we find, as would be assumed, that there is no evidence of any pattern in the graph. There is no relationship (nor should there be) between chronological order and processing time (in minutes). The longest consecutive

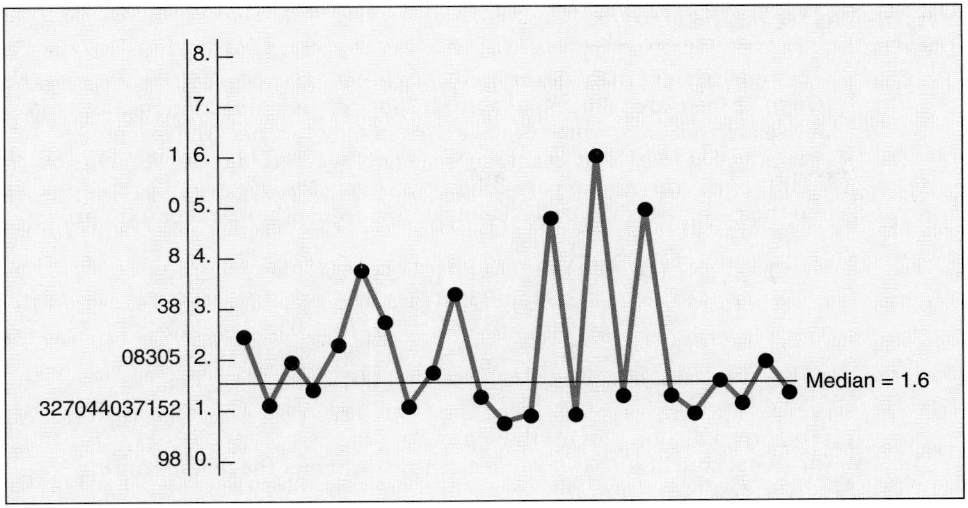

Figure 3.13
Hunter's digidot plot of teller processing time (in minutes) for 24 consecutive customers into a midtown Manhattan bank.

sequence of observations above the center line is 3 (observations 5, 6, and 7) and the longest consecutive sequence below the center line is also 3 (observations 11, 12, and 13). On the other hand, the higher swings of the graph above the center line as compared with the distances below it demonstrate the lack of symmetry in this batch of data.

3.7.2 Constructing Hunter's Digidot Plot

Comparing the stem-and-leaf display in panel (a) of Figure 3.12 with that shown in our digidot plot (Figure 3.13), we note that they would be identical if we were to turn one of them upside down! Thus, in constructing the stem-and-leaf portion of the digidot plot we note that the leaves are branching off to the left of the stems instead of to the right. Moreover, note that the stems are listed high to low from top to bottom instead of low to high as in Figure 3.12. This is done for graphical convenience, since the (vertical) *Y* axis of a graph goes from high to low, top to bottom. To the left of the vertical axis we indicate the stems along with "tick marks" for the processing times (in minutes) on the vertical scale. To the left of the stems we draw another vertical line to permit placement of the leaves. We then simultaneously construct the stem-and-leaf portion and graph the processing times (in minutes) in the order in which they are listed from Figure 3.10. These values are plotted from left to right, with equal distances between. The consecutive dots are then connected and the center line is drawn through the ordered sequence. In Figure 3.13, the center line is plotted from the vertical axis at the value 1.6 minutes. This line represents the processing time (in minutes) for which half of the customer banking transactions are longer and half are shorter. Here the actual center line is plotted because the intent was to check an assumption in a batch of data already obtained. On the other hand, for production or other service processes that are being monitored in progress (that is, the plots are being made interactively), the expected center line (i.e., the target) would be plotted initially so that it would provide for a visual interpretation of patterns over time.

Problems for Section 3.7

Data file
JEANS.DAT

3.41 A manufacturer of men's jeans uses a machine that can be adjusted to vary the length of the material being produced. Suppose the production plan is to produce jeans that are targeted to have a length of 34 inches. The machine is then adjusted to produce jeans whose length is expected to be 34 inches. A sample of 30 consecutive pairs of jeans is selected from the production process and their lengths are recorded below in row sequence (from left to right):

34.02	34.06	34.05	34.01	33.91	33.76
33.89	33.98	33.88	33.96	33.85	33.94
33.91	34.03	34.05	34.00	33.97	33.84
33.74	33.85	33.94	33.99	34.03	34.10
34.02	33.95	33.96	34.01	33.93	33.82

(a) Form a digidot plot for these data.
(b) What conclusions can you reach about whether the manufacturing process is in control?

Data file
QMILE.DAT

3.42 Victor Sternberg was training for a 5K race. As part of his training regimen he ran a quarter-mile interval for speed on the track for 27 consecutive days prior to the race and kept a record of his time trials. The data below are his quarter-mile times (in seconds):

	Sun.	Mon.	Tue.	Wed.	Thu.	Fri.	Sat.
1st Week	90	91	89	88	88	86	84
2nd Week	85	84	83	84	83	82	80
3rd Week	80	81	81	79	79	78	76
4th Week	79	78	75	74	73	72	RACE

(a) Form a digidot plot for these time trials using the center line of 81 seconds for this 27-day period.
(b) What can be concluded from this plot? Discuss.

Data file
DRESS.DAT

3.43 Gross sales receipts (in thousands of dollars) are recorded daily for Ethel's, a dress boutique in New York City, over the 28-day period February 1–February 28, 1993:

	Monday	Tuesday	Wednesday	Thursday	Friday	Saturday	Sunday
Wk 1:	3.3	3.7	3.0	3.5	3.4	5.7	5.0
Wk 2:	3.9	3.8	3.6	3.9	5.6	6.8	3.9
Wk 3:	7.2	4.3	3.8	4.5	3.2	6.6	5.1
Wk 4:	3.1	3.3	3.2	4.2	3.7	6.2	5.4

(Note that federal and state holidays are boxed.)

(a) Analyze the data by constructing a digidot plot with the center line of 3.9 thousand dollars. Describe anything unusual.
(b) Does there appear to be any pattern in gross sales receipts over time?

3.8 Using the Computer for Tables and Charts with Numerical Data: The Kalosha Industries Employee Satisfaction Survey

3.8.1 Introduction and Overview

When dealing with large batches of data, we may use the computer to assist us in our descriptive statistical analysis. In this section we will demonstrate how various statistical software packages can be used for organizing and presenting numerical data in tabular and chart form.

By learning how to access a statistical package, such as MINITAB, SAS, SPSS, or STATISTIX, we will be able to take advantage of recent technological progress and gain an appreciation for the assistance the computer may give us in solving statistical problems—particularly those involving large numbers of variables or large data batches. (See References 7–10.) To accomplish this, let us return to the Kalosha Industries Employee Satisfaction Survey that was developed in Chapter 2.

3.8.2 Kalosha Industries Employee Satisfaction Survey

Data file
EMPSAT.DAT

Bud Conley, the Vice-President of Human Resources, is preparing for a meeting with a representative from B & L Corporation, an employee-benefits consulting firm, to discuss the potential contents of an employee-benefits package that is being developed. Answers to the following two questions would be of particular concern in an initial analysis of the survey data (Table 2.3 on pages 33–40):

1. *General Question A:* What is the distribution of personal income among the full-time employees of Kalosha Industries (see question 7 in the Survey)? That is, how do these data distribute or cluster together?
2. *Specific Question B:* Are there gender differences in the personal incomes of full-time employees of Kalosha Industries (see questions 7 and 5 in the Survey)?

These and other initial questions raised by Bud Conley (see Survey/Database Project at the end of the section) require a detailed descriptive statistical analysis of the 400 responses to the Survey. In practice, a statistician would likely use one or two statistical packages when performing the descriptive statistical analysis. However, computer output from several packages is presented here so that we may demonstrate some of the features of these packages.

3.8.3 Using Statistical Packages for Numerical Data

In response to each of the two questions dealing with personal income, the following would be desired: a stem-and-leaf display, a frequency and percentage distribution, a histogram or polygon, and an ogive. For question A, a characterization of employee personal income, Figure 3.14 depicts the stem-and-leaf display from

SPSS, Figure 3.15 shows the frequency and percentage distributions from STATISTIX, and Figures 3.16 and 3.17 respectively present the histogram and ogive using STATISTIX (pages 85–86).

From the various computer outputs, a response to Bud Conley's first general question can be made. The various displays, tables, and charts indicate that the full-time employee personal income distribution lacks symmetry or balance. Although the annual personal incomes of full-time employees range from 10.1 to 91.9 thousand dollars, the majority (50.25%) of the personal incomes are clustering in the upper teens, low twenties, or high twenties. Furthermore, only 3.25% of the employees have personal incomes of at least 60.0 thousand dollars.

To respond to Bud Conley's specific question B, an evaluation of gender differences in the personal incomes of full-time employees, a sorting of the numerical responses into the two gender categories (male and female) is required. This process can be performed by accessing one of the statistical packages. Once that is achieved, types of output similar to that presented in Figures 3.14 through 3.17 would be needed for each gender grouping. To highlight this, Figure 3.18 (page 87) presents the respective stem-and-leaf displays for the personal incomes of male and female full-time employees using MINITAB.

From Figure 3.18 it appears that the male full-time employees at Kalosha Industries have a higher personal income than do female full-time employees. For males, the incomes range from 10.2 to 91.9 thousand dollars and typically cluster

```
RINCOME:

 Frequency     Stem &  Leaf

    38.00        1 *   000111233333444444
    77.00        1 .   55555555566666666677777777888888899999
    64.00        2 *   00000001111112222222233333344
    60.00        2 .   5555555566666677777788889999
    43.00        3 *   0000011111222333334
    36.00        3 .   55566667788889999
    27.00        4 *   001112223344
    14.00        4 .   556788
    20.00        5 *   000111234
     8.00        5 .   589&
     5.00        6 *   4&
     8.00 Extremes     (67), (76), (76), (78), (82), (92)

 Stem width:       10.0
 Each leaf:        2 case(s)

 & denotes fractional leaves.
```

Figure 3.14

Stem-and-leaf display from SPSS output.

Note: It should be pointed out that in some situations a stem of length ten can be split into five stems based on lowest two digits (*), twos and threes (T), fours and fives (F), sixes and sevens (S), and highest two digits (.) or a stem of length ten can be split into two stems based on low (either L or *) and high (either H or .) digits. As seen in Figure 3.14, SPSS utilizes the * and . symbols for two-way splits of the stems. Moreover, as observed from Figure 3.14, with a sample as large as 400 there is not enough room across the page to print all the leaves (i.e., observations) that are branching off some of the stems. To compensate for this, SPSS designated each leaf to represent two observations and utilized the & symbol to denote a leaf value that repeats an odd number of times.

RINCOME:

LOW	HIGH	FREQ	PERCENT
10	20	115	28.8
20	30	124	31.0
30	40	79	19.8
40	50	41	10.3
50	60	28	7.0
60	70	6	1.5
70	80	5	1.3
80	90	1	0.3
90	100	1	0.3
TOTAL		400	100.0

Figure 3.15
Frequency and percentage distributions from STATISTIX output.

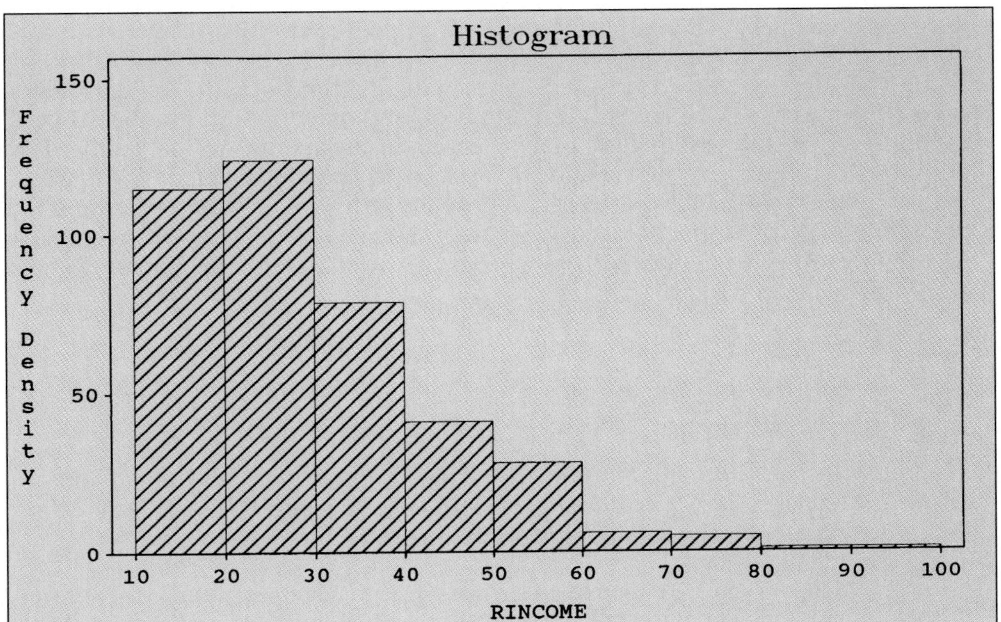

Figure 3.16
Frequency histogram from STATISTIX output.
Note: Figures 3.15, 3.16, and 3.17
As we have discussed in Section 3.3.4 on page 65, there is much subjectivity in selecting class boundaries in frequency distributions. Here we note that the class boundaries in the frequency and percentage distributions obtained by STATISTIX in Figure 3.15 match those for the histogram and ogive obtained by STATISTIX in Figures 3.16 and 3.17. Since each statistical package is programmed differently for establishing the boundaries of the classes in a frequency distribution, other packages might yield different results. However, we could control for this by exercising certain options. We could then set up the lower and upper boundaries of the classes as we desire and our output would be consistent regardless of which package we choose.

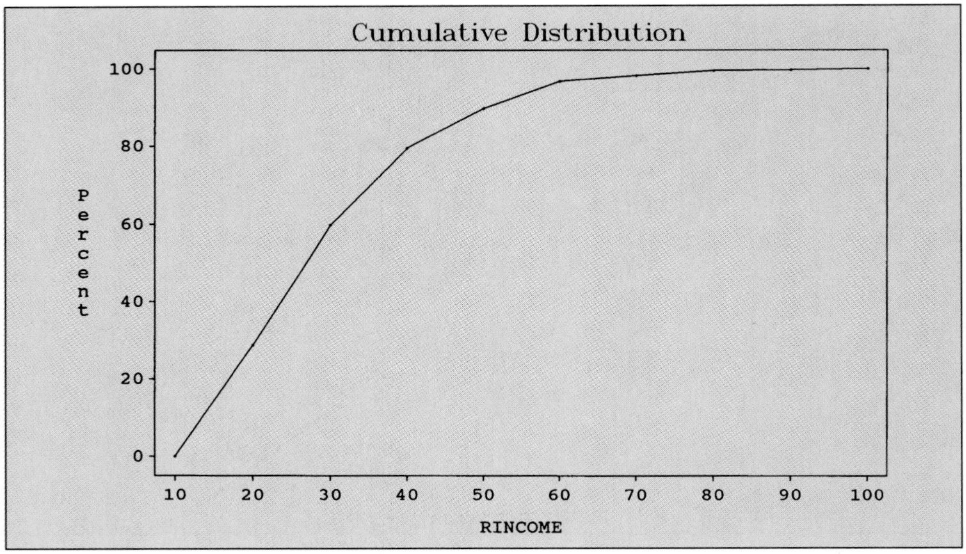

Figure 3.17
Percentage ogive from STATISTIX output.

in the low twenties; for females, the incomes range from 10.1 to 62.8 thousand dollars and overwhelmingly cluster in the high teens. Except for this preponderance of female personal incomes in the high teens, the two distributions are reasonably similar in shape.

Bud Conley was also interested in evaluating other potential gender differences with respect to hours worked, length of employment, and number of promotions. A descriptive statistical analysis based on the answers to these and other questions pertaining to the numerical variables in the Employment Satisfaction Survey (see Survey/Database Project) will provide him with a better understanding of the composition of the Kalosha Industries full-time workforce and assist him in his deliberations with the B&L Corporation toward the development of an employee-benefits package.

Survey/Database Project for Section 3.8

The following problems refer to the sample data obtained from the questionnaire of Figure 2.6 on pages 28–29 and presented in Table 2.3 on pages 33–40. They should be solved with the aid of an available computer package.

Data file EMPSAT.DAT

Suppose you are hired as a research assistant to Bud Conley, the Vice-President of Human Resources at Kalosha Industries. He has given you a list of questions (see Problems 3.44 to 3.59) that he needs answered prior to his meeting with the representatives from the B&L Corporation, the employee-benefits consulting firm that he has retained.

For each of the following problems (3.44 through 3.59) pertaining to the Employee Satisfaction Survey:

(a) Form the stem-and-leaf display.
(b) Form the frequency and percentage distributions.
(c) Plot the histogram.

```
Stem-and-leaf of RINCOME   SEX = 1          N  = 233
Leaf Unit = 1.0

   17     1 00012233334444444
   44     1 555556666666667777888899999
   76     2 00000111111122222222222333333
  106     2 555556666666666667777778889999
  (29)    3 00000000111111112222333333344
   98     3 55566666667788888899999999
   72     4 0001111112222222333444
   50     4 55555667778888
   36     5 000001111122233444
   18     5 5567899
   11     6 044
    8     6 7
    7     7
    7     7 56688
    2     8 1
    1     8
    1     9 1

Stem-and-leaf of RINCOME   SEX = 2          N  = 167
Leaf Unit = 1.0

   21     1 000011111233333344444
   71     1 55555555555555566666666677777777777788888888999999
  (32)    2 00000000111112222233333333344444
   64     2 5555555555556667777777888888999
   34     3 00111222333334
   20     3 5556777889
   10     4 00134
    5     4
    5     5 01
    3     5 8
    2     6 12

MTB > note 'SEX' = 1 is for Male and 'SEX' = 2 is for Female
```

Figure 3.18

MINITAB stem-and-leaf displays of personal income for male and female full-time employees.

Note: In a MINITAB stem-and-leaf display the numbers in the first column are cumulated counts of the observations up to the class containing the median or middle value. In the top panel the (29) means that there are 29 observations in the class containing the middle value. The numbers written below the (29) are the cumulated counts—starting from the largest income back to the class containing the middle value. Moreover, in these MINITAB stem-and-leaf displays, the stems have been split into low (L) and high (H) digits, but these letters do not appear on the printout.

(d) Plot the percentage polygon.
(e) Form the cumulative percentage distribution.
(f) Plot the ogive.
(g) **ACTION** Write a memo to Bud Conley discussing your findings.

3.44 Are there differences in the personal incomes of full-time employees of Kalosha Industries based on an individual's participation in budgetary decisions (see questions 7 and 22)?

3.45 What are the differences in the personal incomes of full-time employees at Kalosha Industries based on occupational grouping (see questions 7 and 2)?

3.46 What are the characteristics of the distribution of the number of hours typically worked per week by all full-time employees of Kalosha Industries (question 1)?

3.47 Are there gender differences in the number of hours typically worked per week by all full-time employees of Kalosha Industries (see questions 1 and 5)?

3.48 Are there differences in the number of hours typically worked per week by all full-time employees of Kalosha Industries based on an individual's participation in budgetary decisions (see questions 1 and 22)?

3.49 Are there differences in the number of hours typically worked per week by all full-time employees at Kalosha Industries based on occupational grouping (see questions 1 and 2)?

3.50 What are the characteristics of the distribution of length of employment (in years) among full-time workers at Kalosha Industries (see question 16)?

3.51 Are there gender differences in the length of employment (in years) among full-time workers at Kalosha Industries (see questions 16 and 5)?

3.52 What are the characteristics of the age distribution (in years) among full-time workers at Kalosha Industries (see question 3)?

3.53 Are there gender differences in the ages of full-time workers at Kalosha Industries (see questions 3 and 5)?

3.54 What are the characteristics of the distribution of attained education (in years of formal schooling) among full-time workers at Kalosha Industries (see question 4)?

3.55 Are there gender differences in attained level of education among full-time workers at Kalosha Industries (see questions 4 and 5)?

3.56 What are the characteristics of the distribution of the number of promotions received while working at Kalosha Industries by all full-time workers (see question 17)?

3.57 Are there gender differences in the number of promotions received at Kalosha Industries by full-time workers (see questions 17 and 5)?

3.58 What are the characteristics of the distribution of total family income among full-time workers at Kalosha Industries (see question 8)?

3.59 What are the characteristics of the distribution of years of full-time employment since age 16 for all full-time workers at Kalosha Industries (see question 15)?

3.9 Recognizing and Practicing Proper Tabular and Chart Presentation and Exploring Ethical Issues

To this point we have studied how a collected batch of numerical data is prepared and then presented in tabular and chart form in order to make the data more manageable and meaningful for purpose of analysis. If our analysis is to be enhanced by a visual display of numerical data, it is essential that the tables and charts be presented clearly and carefully. Tabular frills and "chart junk" must be eliminated so as not to cloud the message given by the data with unnecessary adornments (References 3, 11, 12, and 15). In addition to eliminating chart junk, when displaying charts we must avoid common errors that distort our visual impression (References 2, 4, and 6). Three such errors are:

1. Failing to compare two or more batches of data on a relative basis
2. Compressing the vertical axis
3. Failing to indicate the zero point at the bottom of the vertical axis

3.9.1 Eliminating Chart Junk

As we turn the pages of magazines and newspapers we often find that tables and charts are adorned with various icons and symbols to make them attractive to their readers. Unfortunately, enlivening a table or chart often hides or distorts the intended message being conveyed by the data. For example, some "eye-catching" displays that we typically see in magazines and newspapers erroneously attempt to show icon "areas" representative of numerical information. Can anyone really read and interpret such two-dimensional areas accurately? The answer is no. As seen in Figure 3.19, these graphs may be attractive, but they rarely work!

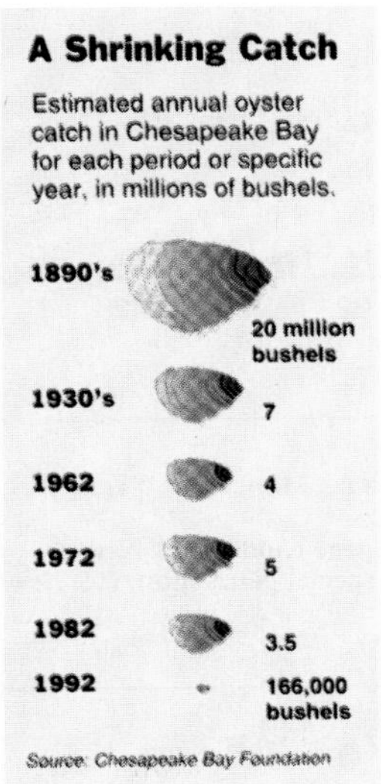

Figure 3.19
"Improper" display of estimated oyster catch (in millions of bushels) in the Chesapeake Bay over various time periods.
Source: *The New York Times*, October 17, 1993, p. 26.

In Figure 3.19, is the icon representing the estimated 20 million bushels of oysters caught in the 1890s really five times the size of the icon representing the estimated 4 million bushels caught in 1962? Such an illustration may catch the eye, but it usually doesn't show anything that could not be presented better in a summary table, a digidot plot, or a plot of the data over time (see Chapter 19).

3.9.2 Failing to Compare Data Batches on a Relative Basis

In Section 3.4 we demonstrated why it is necessary to compare two or more batches of data on a relative basis, and Figures 3.7 (page 72) and 3.9 (page 77), respectively, displayed the proper percentage polygons and percentage ogives comparing the out-of-state resident tuition rates charged by 60 Texas schools and 45 North Carolina schools. Using frequency counts rather than percentages or proportions would be misleading. To show this, Figures 3.20 and 3.21 on page 90 are the respective frequency polygons and frequency ogives "comparing" the tuition rates charged by the 60 Texas schools and the 45 North Carolina schools. In addition, to dramatize the visual distortion, the out-of-state resident tuition rates charged by the 90 colleges and universities in Pennsylvania are included (see Special Data Set 1 of Appendix D on pages D4–D5).

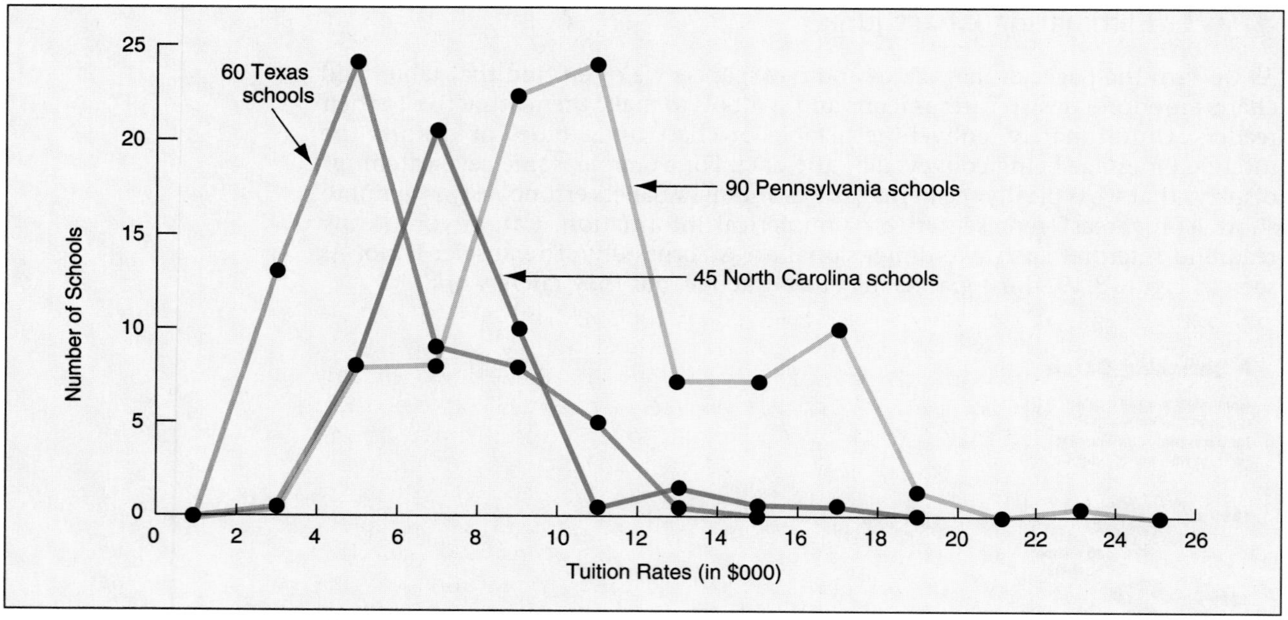

Figure 3.20
"Improper" frequency polygons of tuition rates charged for out-of-state residents at 60 Texas schools, 45 North Carolina schools, and 90 Pennsylvania schools.
Source: Data are taken from Tables 3.3, 3.7, and "America's Best Colleges, 1994 College Guide," *U.S. News & World Report*, extracted from College Counsel 1993 of Natick, Mass. Reprinted by special permission, *U.S. News & World Report*, © 1993 by *U.S. News & World Report* and by College Counsel.

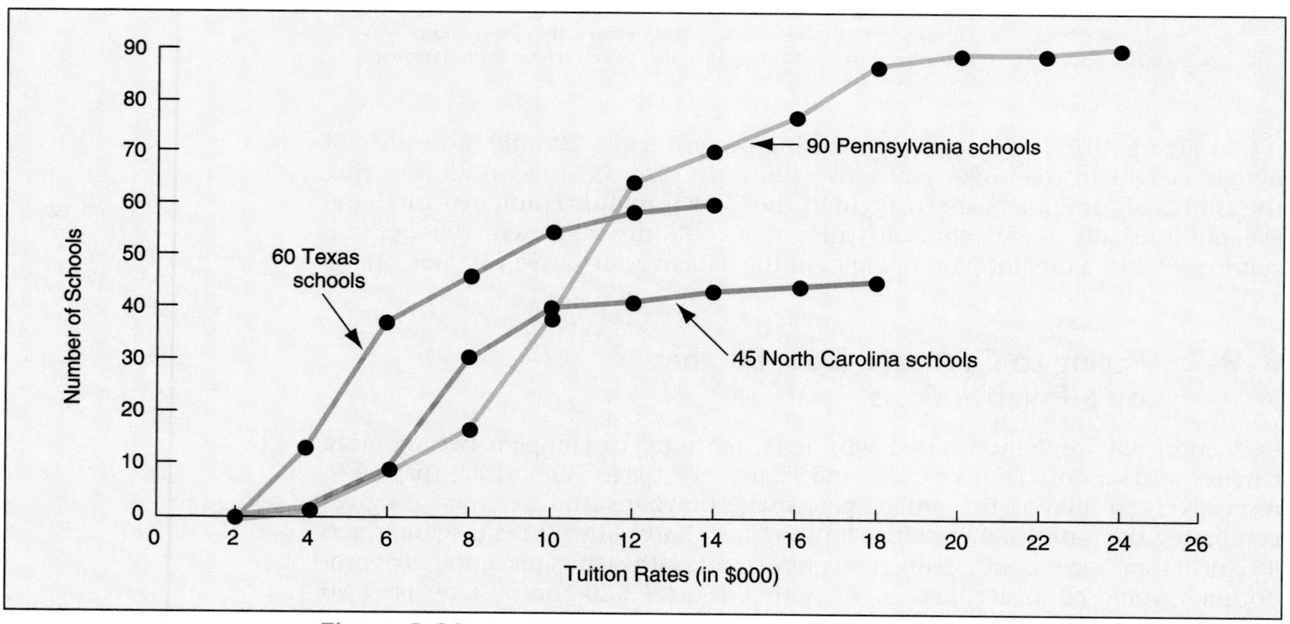

Figure 3.21
"Improper" cumulative frequency polygons of tuition rates charged for out-of-state residents at 60 Texas schools, 45 North Carolina schools, and 90 Pennsylvania schools.
Source: Data are taken from Tables 3.3, 3.7, and "America's Best Colleges, 1994 College Guide," *U.S. News & World Report*, extracted from College Counsel 1993 of Natick, Mass. Reprinted by special permission, *U.S. News & World Report*, © 1993 by *U.S. News & World Report* and by College Counsel.

As can be seen from Figures 3.20 and 3.21, the frequency polygons and frequency ogives for the 60 Texas and 45 North Carolina schools are overwhelmed by those for the 90 Pennsylvania schools and no meaningful comparisons can be made from such distorted charts.

3.9.3 Compressing the Vertical Axis

It is easy to alter a visual impression of a chart by manipulating the scale points on the vertical or horizontal axis. To show this, take a good look at the corresponding polygons—Figure 3.7 on page 72 and Figure 3.20 on page 90. Now take a good look at the corresponding ogives—Figure 3.9 on page 77 and Figure 3.21 on page 90. In our two sets of corresponding charts we kept the scale point dimensions the same on the horizontal axis. For the vertical axis, however, we had to take into account the fact that we added the 90 Pennsylvania schools to Figures 3.20 and 3.21 and yet, for placement in your textbook, we wanted the corresponding charts to take up the same amount of space on the respective pages. Thus, to account for the Pennsylvania schools in Figure 3.20, look at how the "shape" of the curves changes when comparing the polygons representing the Texas and North Carolina schools here and in Figure 3.7. In a similar manner, to account for the Pennsylvania schools in Figure 3.21, look at how the "steepness" or *slope* changes when comparing the ogives representing the Texas and North Carolina schools here and in Figure 3.9.

Although we already knew from Section 3.9.2 that Figures 3.20 and 3.21 were improper representations of their respective polygons and ogives (because the vertical scales in Figures 3.20 and 3.21 did not use percentages or proportions), the important point here is that a compression of the scale on the vertical axis can cause a distortion in the visual information being displayed. For example, had we constructed our percentage histogram (see Figure 3.5 on page 71) by selecting tick marks on the vertical scale from 0 to 100 instead of 0 to 45, our histogram would look much flatter. Moreover, it would appear unaesthetically in the lower half of the boxed frame, leaving unnecessary background space in the top half (see Problem 3.60 on page 93). A good general rule, then, is to construct your charts so that they utilize the entire boxed frame.

3.9.4 Failing to Indicate the Zero Point on the Vertical Axis

The starting point on the vertical axis must be indicated with a zero so as not to distort the visual impression regarding the magnitude of changes occurring in the chart. By taking but a slice of the vertical axis, such changes can be exaggerated. Figure 3.22 on page 92 displays such a visual distortion.

Note that in this chart zero was omitted from the vertical axis. Because of this the reader gets a distorted view of the magnitude of the differences in the daily transactions. For example, over the portrayed time period, the most active trading session occurred on Friday, September 17, while the least active trading session occurred on Monday, October 11 (Columbus Day). However, from the incorrectly drawn graph, the vertical bar representing the most active trading session is three times longer than the vertical bar representing the least active session—leaving the impression that three times as many shares were traded on September 17 than on

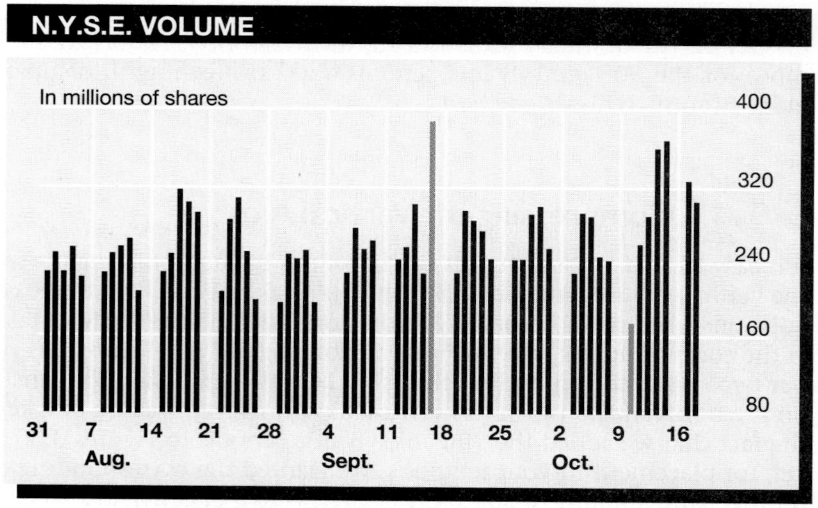

Figure 3.22
"Improper" display of New York Stock Exchange sales volume (in millions of shares traded) over time.
Source: *The New York Times*, October 20, 1993, p. D7.

October 11. Had the zero point been properly displayed on the vertical axis, the graph would have accurately portrayed the fact that only twice as many shares were traded on September 17 than on October 11.

3.9.5 Using Computer Software for Tables and Charts

In Section 3.8 we demonstrated how appropriate computer software can assist us in a descriptive analysis of our data. The computer is an extremely helpful tool that can easily and rapidly store, organize, and process information and provide us with summary results, tables, and charts. Nevertheless, we must bear in mind that the computer is only a tool. We shall see throughout this text, as we both demonstrate and interpret a variety of computer output corresponding to the topics to be studied in the chapters ahead, that it is crucial to use the computer in a manner consistent with correct statistical methodology. Remember GIGO. The computer output we get will depend on four things—the capacity of the hardware used, the quality of the printer chosen, the capability of the statistical software selected, as well as our ability to properly choose and use the software advantageously. And when you are presented with tabular and chart information from the output of some statistical software package, be wary of the extra frills and adornments that might be hiding what the data are intending to convey.

3.9.6 Ethical Issues

Ethical considerations arise when we are deciding what data to present in tabular and chart format and what not to present. It is vitally important when conducting research to document both good and bad results, so that those who continue such research do not have to "reinvent the wheel." Moreover, when making oral presentations and presenting written research reports, it is essential that the results be given in a fair, objective, and neutral manner. Thus we must try to distinguish between poor data presentation and unethical presentation. Again, as in our discussion of ethical considerations in data collection (Section 2.11.6), the key is *intent*. Often, when fancy tables and chart junk are presented or pertinent information is omitted, it is simply done out of ignorance. However, unethical behavior occurs when a researcher willfully hides the facts by distorting a table or chart or by failing to report pertinent findings.

Problems for Section 3.9

3.60 As per the statement in the last paragraph of Section 3.9.3 on page 91, redraw the percentage histogram (Figure 3.5 on page 71) by selecting tick marks on the vertical axis from 0 to 100 and then comment on the aesthetics of your chart.

3.61 **(Student Project)** Bring to class a chart from a newspaper or magazine that you believe to be a poorly drawn representation of some numerical variable. Be prepared to submit the chart to the instructor with comments as to why you feel it is inappropriate. Also, be prepared to present this and comment in class.

3.10 Numerical Data Presentation: A Review and a Preview

As seen in the summary chart on page 94, this chapter was about data presentation. On page 54 of Section 3.1 you were given a list emphasizing the important points to be discussed in the chapter. Check over the list now to see whether you feel you have an understanding of these key points. To be sure, you should be able to answer the following conceptual questions:

1. Why is it necessary to organize a batch of numerical data that we collect?
2. What are the main differences between an ordered array and a stem-and-leaf display?
3. Under what conditions is it most appropriate to construct and use frequency distributions and percentage distributions?
4. How do histograms and polygons differ with respect to their construction and use?
5. When should a percentage ogive (that is, cumulative percentage polygon) be constructed and how should it be used?
6. Why is the percentage ogive such a useful tool?
7. What is the purpose of a digidot plot and how is it constructed?
8. What are some of the ethical issues to be concerned with when presenting numerical data in tabular or chart format?

Check over the list of questions to see if indeed you know the answers and could (1) explain your answers to someone who did not read this chapter and (2) give reference to specific readings or examples that support your answer. Also, reread any of the sections that may have seemed fuzzy to see if they make sense now.

Once the collected numerical data have been presented in tabular and chart format, as was done for Bud Conley at Kalosha Industries, we are ready to make various analyses. In the following chapter, a variety of descriptive summary measures useful for data analysis and interpretation will be developed.

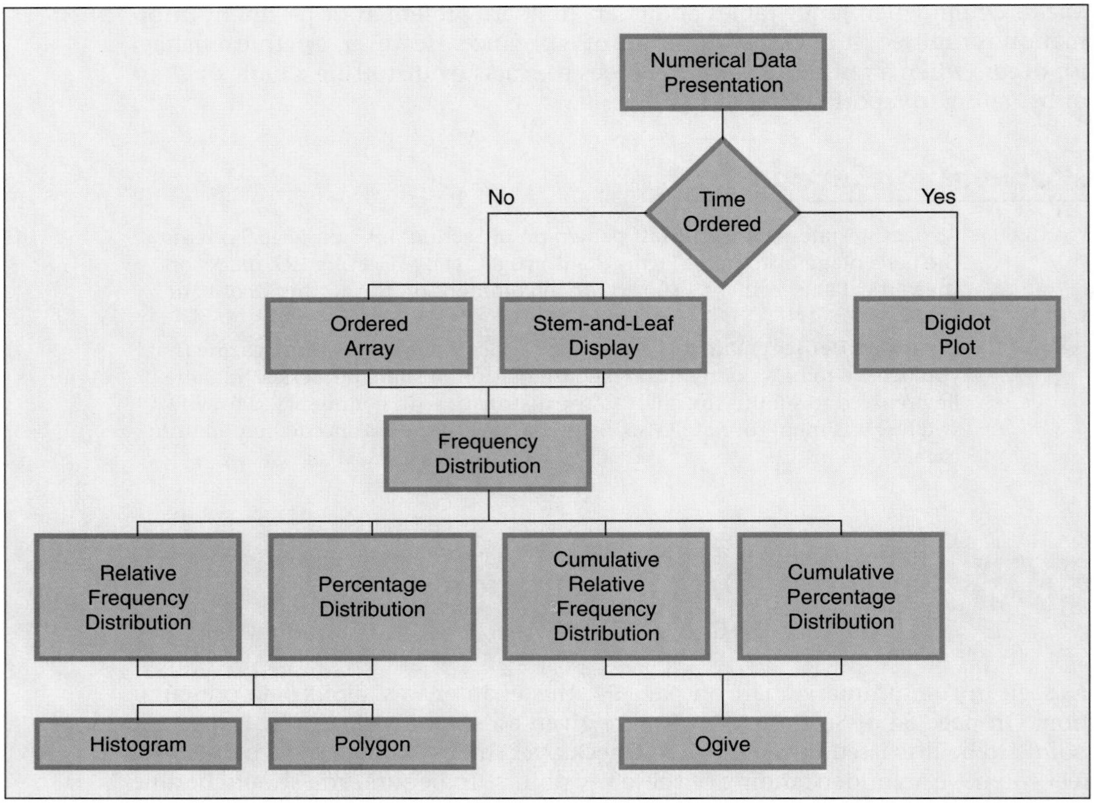

Chapter 3 summary chart.

Getting It All Together

Key Terms

"chart junk" *88*
class boundaries *64*
class groupings *62*
class midpoint *65*
classes *63*
cumulative distribution *74*
digidot plot *79*
frequency distribution *62*
histogram *70*
ogive (cumulative polygon) *75*
ordered array *55*
percentage distribution *68*
polygon *71*
raw form *54*
relative frequency distribution *68*
revised stem-and-leaf display *57*
stem-and-leaf display *55*
width of class interval *63*

Chapter Review Problems

3.62 **ACTION** Write a letter to a friend highlighting what you deem the most interesting or most important features of this chapter.

3.63 In your own words, explain the difference between raw data and an ordered array.

3.64 Why is it advantageous to use a stem-and-leaf display instead of an ordered array?

3.65 Explain the differences between frequency distributions, relative frequency distributions, and percentage distributions.

3.66 When comparing two or more sets of data with different sample sizes, why is it necessary to compare their respective relative frequency or percentage distributions?

3.67 Explain the differences between histograms, polygons, and ogives (cumulative polygons).

3.68 Explain the differences between stem-and-leaf displays and digidot plots.

3.69 The raw data displayed below are the starting salaries for a random sample of 100 computer science or computer systems majors who earned their baccalaureate degrees during the year 1993:

Data file SALARY1.DAT

Starting Salaries ($000)									
24.2	29.9	23.4	23.0	25.5	22.0	33.9	20.4	26.6	24.0
28.9	22.5	18.7	32.6	26.1	26.2	26.7	20.4	22.2	24.7
18.6	18.5	19.6	24.4	24.8	27.8	27.6	27.2	20.8	22.1
19.7	25.3	28.2	34.2	32.5	30.8	26.8	20.6	21.2	20.7
25.2	25.7	32.2	28.8	24.7	18.7	20.5	25.5	19.1	25.5
22.1	27.5	25.8	25.2	25.6	25.2	25.2	27.9	18.9	37.3
29.9	23.2	19.8	20.8	29.5	27.6	21.2	38.7	21.3	24.8
32.3	20.1	26.8	25.4	26.3	21.2	19.5	22.8	21.7	25.3
32.3	28.1	27.5	25.3	19.3	27.4	26.4	20.9	34.5	25.9
31.4	27.4	27.3	20.6	31.8	25.8	25.2	21.9	26.8	26.5

(a) Place the raw data in a stem-and-leaf display. (*Hint*: Let the leaves be the tenths digits.)
(b) Place the raw data in an ordered array.
(c) Form the frequency distribution and percentage distribution.
(d) Plot the percentage histogram.
(e) Plot the percentage polygon.
(f) Form the cumulative percentage distribution.
(g) Plot the ogive (cumulative percentage polygon).
(h) **ACTION** Write a brief report to your dean describing the starting salaries of these recent graduates.

3.70 The following data are the retail prices for a random sample of 30 pencil tire pressure gauge models:

Data file GAUGE.DAT

4.50	6.50	2.00	2.50	4.00	3.50	5.00	3.00	5.00	5.50
1.00	7.50	3.00	2.00	3.00	3.50	3.50	5.00	6.00	4.50
2.00	3.00	3.50	3.50	3.00	3.00	4.00	1.50	1.50	2.50

Source: Copyright 1993 by Consumers Union of United States, Inc., Yonkers, N.Y. 10703. Adapted by permission from *Consumer Reports*, February 1993, pp. 98-99.

(a) Place the raw data in a stem-and-leaf display. (*Hint*: Let the leaves be the tenths digits.)
(b) Place the raw data in an ordered array.
(c) Form the frequency distribution and percentage distribution.
(d) Plot the percentage histogram.
(e) Plot the percentage polygon.
(f) Form the cumulative percentage distribution.
(g) Plot the ogive (cumulative percentage polygon).
(h) ACTION If you were considering purchasing a pencil tire pressure gauge, what else would you want to know? Write a list of questions you would ask in an automobile supply store.

3.71 The data below represent the cost per month of use (in dollars) and the cleaning test score (0 to 100) for a random sample of 39 brands of tubed toothpaste:

Data file TOOTHPST.DAT

Toothpaste	Cost per Month	Score
Ultra brite Original	.58	86
Gleem	.66	79
Caffree Regular	1.02	77
Crest Tartar Control Fresh Mint Gel	.53	75
Colgate Tartar Control Gel	.57	74
Crest Tartar Control Original	.53	72
Ultra brite Gel Cool Mint	.52	72
Colgate Clear Blue Gel	.71	71
Crest Cool Mint Gel	.55	70
Crest Regular	.59	69
Crest Sparkle	.51	64
Close-Up Tartar Control Gel	.67	63
Close-Up Anti-Plaque	.62	62
Colgate Tartar Control Paste	.66	62
Tom's of Maine Cinnamint	1.07	62
Aquafresh Tartar Control	.80	60
Aim Anti-Tartar Gel	.79	58
Aim Extra-Strength Gel	.44	57
Slimer Gel	1.04	57
Arm & Hammer Baking Soda Fresh Mint Gel	1.12	55
Aquafresh	.79	56
Aquafresh Extra Fresh	.81	53
Close-Up Paste	.64	85
Topol Spearment Gel	1.77	82
Topol Spearment	1.32	76
Close-Up Mint Gel	.64	72
Aim Regular-Strength Gel	.55	70
Pepsodent	.39	58
Colgate Baking Soda	1.22	51
Colgate Regular	.74	50
Colgate Junior Gel	.44	39
Colgate Peak	.97	29
Arm & Hammer Baking Soda Fresh Mint	1.26	28
Rembrandt	4.73	53
Sensodyne Original	1.29	80
Sensodyne Gel	1.34	48
Viadent Original Anti-Plaque	1.40	53
Denquel	1.77	37
Butler Protect Gel	1.11	20

Source: Copyright 1992 by Consumers Union of United States, Inc., Yonkers, N.Y. 10703. Adapted by permission from *Consumer Reports,* September 1992, pp. 604–605.

For each of the two numerical variables:
(a) Form the stem-and-leaf display.
(b) Form a combined frequency and percentage distribution table.
(c) Plot the percentage polygon.
(d) Form the cumulative percentage distribution.
(e) Plot the percentage ogive.
(f) ACTION ▶ Write a report for your marketing professor summarizing your findings and characterizing this product.

3.72 Given the batches of data based on closing stock price for random samples of 25 issues traded on the American Exchange and 50 issues traded on the New York Exchange:

American Exchange	New York Exchange	
(25 issues)	(50 issues)	
$ 6.88	$36.50	$26.00
.75	23.50	19.00
3.88	8.25	46.00
4.12	57.50	23.50
11.88	27.12	22.62
15.88	3.75	12.88
16.50	25.00	5.50
8.75	15.50	37.50
9.25	36.12	9.88
7.50	6.00	59.12
5.38	9.12	35.25
14.38	33.38	20.62
2.50	22.50	24.00
4.88	8.75	80.50
6.38	8.62	29.38
33.62	5.75	3.75
4.88	21.88	64.75
9.00	6.12	14.25
2.00	25.00	46.38
20.00	15.88	4.75
14.25	24.00	25.00
4.00	10.88	35.00
15.25	18.75	9.00
2.38	53.88	12.38
49.50	20.38	31.00

Data file NYSEAX.DAT

(a) Using interval widths of $10, form the frequency distribution and percentage distribution for each batch.
(b) Plot the frequency histogram for each batch.
(c) On one graph, plot the percentage polygon for each batch.
(d) Form the cumulative percentage distribution for each batch.
(e) On one graph, plot the ogive (cumulative percentage polygon) for each batch.
(f) ACTION ▶ Write a brief report to your finance professor comparing and contrasting the two batches.

3.73 A wholesale appliance distributing firm wished to study its accounts receivable for two successive months. Two independent samples of 50 accounts were selected for each of the two months. The results are summarized in the table at the top of page 98:

Frequency distributions for accounts receivable.

Amount	March Frequency	April Frequency
\$0 to under \$2,000	6	10
\$2,000 to under \$4,000	13	14
\$4,000 to under \$6,000	17	13
\$6,000 to under \$8,000	10	10
\$8,000 to under \$10,000	4	0
\$10,000 to under \$12,000	0	3
Totals	50	50

(a) Plot the frequency histogram for each month.
(b) On one graph, plot the percentage polygon for each month.
(c) Form the cumulative percentage distribution for each month.
(d) On one graph, plot the ogive (cumulative percentage polygon) for each month.
(e) ACTION Write a brief report to your accounting professor comparing and contrasting the accounts receivable of the two months.

3.74 You are employed as a quality-control engineer at Chrysler Corporation and, in an effort to improve the quality of your company's products, you wish to compare various design features of U.S. and foreign-made automobile models.

The following table contains the cumulative distributions and cumulative percentage distributions of braking distance (in feet) at 80 mph for a sample of 25 U.S.- manufactured automobile models and for a sample of 72 foreign-made automobile models obtained in a recent year.

Cumulative frequency and percentage distributions for the braking distance (in feet) at 80 mph for U.S.-manufactured and foreign-made automobile models.

Braking Distance (in ft.)	U.S.-Made Automobile Models "Less Than" Indicated Values		Foreign-Made Automobile Models "Less Than" Indicated Values	
	Number	Percentage	Number	Percentage
210	0	0.0	0	0.0
220	1	4.0	1	1.4
230	2	8.0	4	5.6
240	3	12.0	19	26.4
250	4	16.0	32	44.4
260	8	32.0	54	75.0
270	11	44.0	61	84.7
280	17	68.0	68	94.4
290	21	84.0	68	94.4
300	23	92.0	70	97.2
310	25	100.0	71	98.6
320	25	100.0	72	100.0

Source: Data are extracted from *Road & Track,* vol. 42, no. 2 (October 1990), p. 47.

Based on these data, answer the following questions:
(a) How many models of U.S.-made automobiles have braking distances of 240 feet or more?
(b) What is the percentage of U.S.-made automobiles with braking distances less than 260 feet?
(c) Which group of car models—U.S. made or foreign made—have the wider *range* in braking distance?

(d) How many foreign-made automobile models have braking distances between 260 feet and 269.9 feet (inclusive)?
(e) Use the cumulative distributions to construct the frequency distributions and percentage distributions for each group of car models.
(f) On one graph, plot the two percentage ogives.
(g) ACTION Write a brief report comparing and contrasting the braking distance information for the two groups of car models.

3.75 You are employed as an analyst for a major building developer who is interested in constructing a shopping mall in either Centerport or Northport—two adjacent communities on the north shore of Long Island in Suffolk County, New York.

The figure below contains the cumulative relative frequency polygons (ogives) of family incomes for two random samples of 200 families each drawn from the two communities.

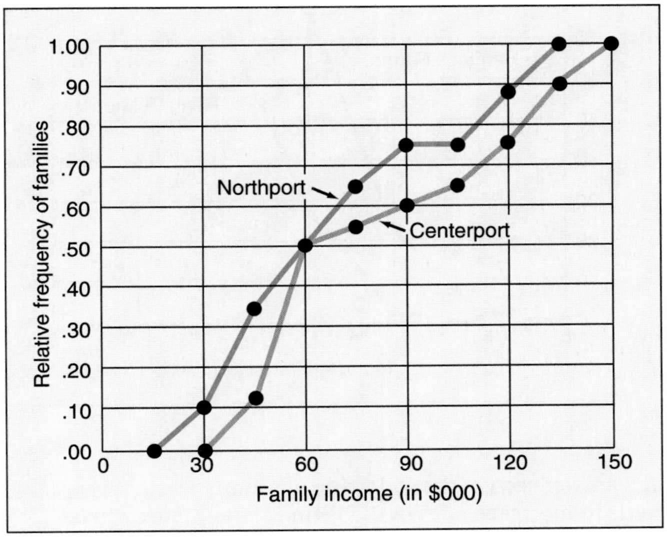

Cumulative relative frequency polygons of family incomes for two communities.

Based on these data, answer each of the following questions:
(a) How many of the families in Centerport have incomes of $120,000 or more?
(b) What is the percentage of families in Centerport with incomes of less than $90,000?
(c) Which sample has a larger *range* of incomes?
(d) How many of the families in Northport have an income of at least $90,000 but less than $105,000?
(e) Does Centerport or Northport have more family incomes of $60,000 or above?
(f) What percentage of Centerport families earn less than $60,000?
(g) What percentage of Centerport families earn $60,000 or more?
(h) Which community has more incomes below $120,000?
(i) Use the ogives to construct the relative frequency distribution and frequency distribution for each community.

(j) On one graph plot the two relative frequency polygons.
(k) ACTION Write a brief report comparing and contrasting the two income distributions.

3.76 You are working for an independent consulting agency hired by a well-known realty company specializing in sales of homes in the Pocono Mountains in northeast Pennsylvania. Your task is to evaluate rates of mortgages held by homeowners in two popular communities.

The figure below contains the percentage ogives of mortgages held by 100 homeowners sampled in Penn Estates and 200 homeowners sampled in Hemlock Farms—two Pocono communities.

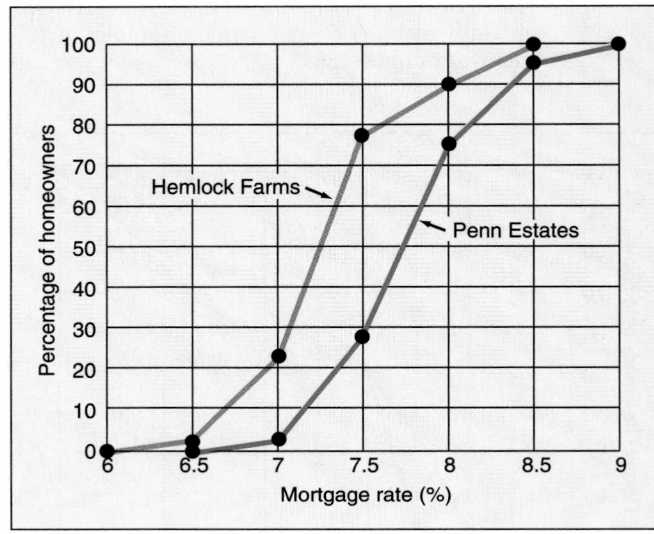

Percentage ogives of mortgage rates for 100 homeowners in Penn Estates and 200 homeowners in Hemlock Farms.

Based on these data, answer each of the following questions:

(a) What is the *range* in mortgage rates of the Penn Estates homeowners?
(b) What is the *range* in mortgage rates of the Hemlock Farms homeowners?
(c) Fifty percent of the Penn Estates homeowners held mortgages with rates of less than what amount?
(d) Fifty percent of the Hemlock Farms homeowners held mortgages with rates of less than what amount?
(e) What percentage of the Penn Estates homeowners held mortgages with rates of at least 7.5% but less than 8%?
(f) What percentage of the Hemlock Farms homeowners held mortgages with rates of less than 8%?
(g) How many of the Penn Estates homeowners held mortgages with rates of 8.5% or more?
(h) Which community contains the largest percentage of homeowners who held mortgages with rates below 7.25%?
(i) Use the ogives to construct the percentage distribution and the frequency distribution for each of the samples.
(j) Plot the two percentage polygons on one graph.
(k) ACTION Write a brief report comparing and contrasting your two distributions. What seems to be apparent about the mortgage rates in these two communities? What reason(s) may be attributed to this? (*Hint*: One of these communities has been steadily growing for 20 years; the other has been rapidly growing for 10 years.)

3.77 **(Student Project)** Choose a stock listed on the NYSE and, starting on a Monday, record its daily closing price for a full four-week (i.e., 20-day) period in which the stock market is open. Also record the changes in closing price from the preceding trading session over this four-week period.
(a) Analyze each batch of data.
(b) Does there seem to be a pattern in the closing prices of the stock over this time period?
(c) Does there seem to be a pattern in the changes in closing prices over time?
(d) ACTION Write a memo to your finance professor based on your findings in (b) and (c).

Collaborative Learning Mini-Case Projects

Note: The class is to be divided into groups of three or four students each. One student is initially selected to be project coordinator, another student is project recorder, and a third student is the project timekeeper. In order to give each student experience in developing teamwork and leadership skills, rotation of positions should follow each project. At the beginning of each project, the students should work silently and individually for a short, specified period of time. Once each student has had the opportunity to study the issues and reflect on his/her possible answers, the group is convened and a group discussion follows. If all the members of a group agree with the solutions, the coordinator is responsible for submitting the team's project solution to the instructor with student signatures indicating such agreement. On the other hand, if one or more team members disagree with the solution offered by the majority of the team, a minority opinion may be appended to the submitted project, with signature(s) on it.

CL 3.1 The research analyst at the college advisory service company has been slightly injured in an automobile accident and requires help to finish her report regarding tuition rates charged to out-of-state residents by colleges and universities in different regions of the country. In order to meet the deadline for a presentation to the board of directors, the marketing manager decides to hire your group, the ________ Corporation, to assist the research analyst in her endeavors. Given Special Data Set 1 of Appendix D on pages D4–D5 regarding the tuition rates charged to out-of-state residents at all 90 colleges and universities in the state of Pennsylvania, the ________ Corporation is ready to:
(a) Outline how the group members will proceed with their tasks
(b) Form the frequency and percentage distribution in the same table
(c) Plot the percentage polygon
(d) Form the cumulative percentage distribution
(e) Plot the percentage ogive
(f) Perform a descriptive analysis comparing the tuition rates in Pennsylvania to those in Texas and North Carolina
(g) Write and submit an executive summary, attaching all tables and charts
(h) Prepare and deliver a ten-minute oral presentation to the marketing manager.

Data file C&UPENN.DAT

Data file C&U.DAT

CL 3.2 A popular family magazine interested in publishing an article on the dietary virtues (or lack thereof) of ready-to-eat cereals hires your group, the ________ Corporation, to study their cost and nutritional characteristics. The theme the article is intending to present is that "ready-to-eat cereals are a quick and efficient way to get the family started each weekday." Armed with Special Data Set 2 of Appendix D on pages D6–D7 displaying useful information on 84 such cereals, the ________ Corporation is ready to:

Data file CEREAL.DAT

4.1 Introduction: What's Ahead

In the preceding chapters we learned how to collect and present numerical data in both tabular and chart format. Now, how do we make sense out of such information? For example, what are the data from the Kalosha Industries Employee Satisfaction Survey in Table 2.3 (pages 33–40) telling us? How can these results ultimately be used by the B&L Corporation, the employee-benefits consulting firm, to develop an employee-benefits package? Although collecting and then presenting data are two essential components of the subject of descriptive statistics, these don't tell the whole story. *Good data analysis* not only involves *presenting* (that is, *plotting*) the collected numerical data and *observing* (that is, *studying*) what the data are trying to convey, it also involves *computing* (that is, *characterizing* or *summarizing*) the key features and *describing* (that is, *analyzing*) the findings. In this chapter we will examine these latter aspects: the summarization, description, and, ultimately, interpretation of the data.

In order to introduce the relevant ideas for the chapter, we see from the chapter summary chart on page 160 that there are three essential characteristics or properties of numerical data: central tendency, variation, and shape. The objective of this chapter is to provide an understanding of these characteristics or properties of numerical data and their corresponding descriptive summary measures as an aid to data analysis and interpretation.

Upon completion of this chapter, you should be able to:

1. Understand the property of central tendency
2. Interpret the differences among the various measures of central tendency such as the mean, median, mode, midrange, and midhinge
3. Understand the difference between central tendency and noncentral tendency
4. Understand the property of variation
5. Interpret the differences among the various measures of variation such as the range, interquartile range, variance, standard deviation, and coefficient of variation
6. Understand the role and use of the Bienaymé-Chebyshev and empirical rules
7. Understand the property of shape
8. Appreciate the value of *exploratory data analysis* techniques: five-number summaries and box-and-whisker plots
9. Know how to approximate descriptive summary measures from a frequency distribution, polygon, or ogive
10. Appreciate the value of statistical software packages for computing the descriptive summary measures
11. Understand how to distinguish between appropriate and inappropriate descriptive summary measures reported in newspapers and magazines as well as the ethical issues involved

4.2 Exploring the Data

In order to introduce the relevant ideas of this chapter, let us return to our research analyst at the college advisory service company who (in Chapter 3) wanted to study tuition rates charged to out-of-state residents by colleges and universities in different regions of the country. In particular, three states have been selected for immediate evaluation—Texas, North Carolina, and Pennsylvania. Suppose, for exploratory purposes, our research analyst was to start by selecting a random sample of six schools from the population listing of 90 colleges and universities in the state of Pennsylvania (see Special Data Set 1 of Appendix D on pages D4–D5).

Pennsylvania School	Tuition (in $000)
University of Pittsburgh	10.3
East Stroudsburg University	4.9
Geneva College	8.9
Drexel University	11.7
California Univ. of Pa.	6.3
Slippery Rock University	7.7

We note that the six schools (recorded in the order in which they were selected) are presented along with their tuition rates (in thousands of dollars) charged to out-of-state residents. What can be learned from such data that will assist our researcher with her evaluation? Based on this sample, we observe the following:

1. The data are in **raw form.** That is, the collected data seem to be in a random sequence with no apparent pattern to the manner in which the individual observations are listed.
2. Each of the tuition rates occurs only once. That is, no one of them is observed more frequently than any other.
3. The spread in the tuition rates ranges from 4.9 to 11.7 thousand dollars.
4. There do not appear to be any unusual or extraordinary tuition rates in this sample. Arranged in numerical order (that is, the *ordered array*), the tuition rates here (in thousands of dollars) are 4.9, 6.3, 7.7, 8.9, 10.3, 11.7. (If the tuition rates, in thousands of dollars, had been 4.9, 6.3, 7.7, 8.9, 10.3, and 28.0, then 28.0 thousand dollars would be considered an extreme observation or **outlier**.)

If our research analyst were to ask us to examine the data and present a short summary of our findings, then comments similar to the four above are basically all that we could be expected to make without more formal statistical training. However, when making such comments, we have both analyzed and interpreted what the data are trying to convey. An analysis is *objective;* we should all agree with these findings. On the other hand, an interpretation is *subjective;* we may form different conclusions when interpreting our analytical findings. From the above, points 2 through 4 are based on analysis, whereas point 1 is an interpretation. With respect to the latter, no formal analytical test (see the *runs test* of Chapter 12) was made—it is simply our conjecture that there is no pattern to the sequence of collected data. Moreover, our conjecture would seem to be appropriate if the sample of six schools was randomly and independently drawn from the population listing using survey methods described in Chapter 2. That was the case here.

Let us now see how we could add to our understanding of what the data are telling us by more formally examining three properties of numerical data.

4.3 Properties of Numerical Data

The three major properties that describe a batch of numerical data are

1. Central tendency
2. Variation
3. Shape

In any analysis and/or interpretation, a variety of descriptive measures representing the properties of central tendency, variation, and shape may be used to extract and summarize the major features of the data batch. If these descriptive summary measures are computed from a sample of data, they are called *statistics;* if they are computed from an entire population of data, they are called *parameters.* Since statisticians usually take samples rather than use entire populations, our primary emphasis in this text is on statistics rather than parameters.

4.4 Measures of Central Tendency

Most batches of data show a distinct tendency to group or cluster about a certain central point. Thus for any particular batch of data, it usually becomes possible to select some typical value or average to describe the entire batch. Such a descriptive typical value is a measure of central tendency or *location.*

Five types of averages often used as measures of central tendency are the arithmetic mean, the median, the mode, the midrange, and the midhinge.

4.4.1 The Arithmetic Mean

The **arithmetic mean** (also called the **mean**) is the most commonly used *average*[1] or measure of central tendency. It is calculated by summing all the observations in a batch of data and then dividing the total by the number of items involved.

● **Introducing Algebraic Notation** Thus, for a sample containing a batch of n observations $X_1, X_2, \ldots, X_n$, the arithmetic mean (given by the symbol $\bar{X}$ —called "X bar") can be written as

$$\bar{X} = \frac{X_1 + X_2 + \cdots + X_n}{n}$$

To simplify the notation, for convenience the term

$$\sum_{i=1}^{n} X_i$$

(meaning the *summation of all the* X_i *values*) is conventionally used whenever we wish to add together a series of observations. That is,

$$\sum_{i=1}^{n} X_i = X_1 + X_2 + \cdots + X_n$$

Rules pertaining to summation notation are presented in Appendix B on pages B1–B5. Using this summation notation, the arithmetic mean of the sample can be more simply expressed as

$$\bar{X} = \frac{\sum_{i=1}^{n} X_i}{n} \tag{4.1}$$

where $\bar{X}$ = sample arithmetic mean

n = sample size

X_i = ith observation of the random variable X

$\sum_{i=1}^{n} X_i$ = summation of all X_i values in the sample (see Appendix B)

For our research analyst's sample, the tuition rates charged to out-of-state residents (in thousands of dollars) are

X_1 = 10.3 at University of Pittsburgh
X_2 = 4.9 at East Stroudsburg University
X_3 = 8.9 at Geneva College
X_4 = 11.7 at Drexel University
X_5 = 6.3 at California Univ. of Pa.
X_6 = 7.7 at Slippery Rock University

The arithmetic mean for this sample is calculated as

$$\bar{X} = \frac{\sum_{i=1}^{n} X_i}{n} = \frac{10.3 + 4.9 + 8.9 + 11.7 + 6.3 + 7.7}{6} = 8.30 \text{ thousand dollars}$$

We observe here that the mean is computed as 8.3 thousand dollars even though not one particular school in the sample actually had that tuition rate. In addition, we see from the **dot scale** of Figure 4.1 on page 108 that for this batch of data three observations are smaller than the mean and three are larger. The mean acts as a *balancing point* so that observations that are smaller balance out those that are larger.

Note that the calculation of the mean is based on all the observations (X_1, X_2, ..., X_n) in the batch of data. No other commonly used measure of central tendency possesses this characteristic. Since its computation is based on every observation, the arithmetic mean is greatly affected by any extreme value or values. In such instances, the arithmetic mean presents a distorted representation of what the data are conveying; hence the mean would not be the best average to use for describing or summarizing such a batch of data.

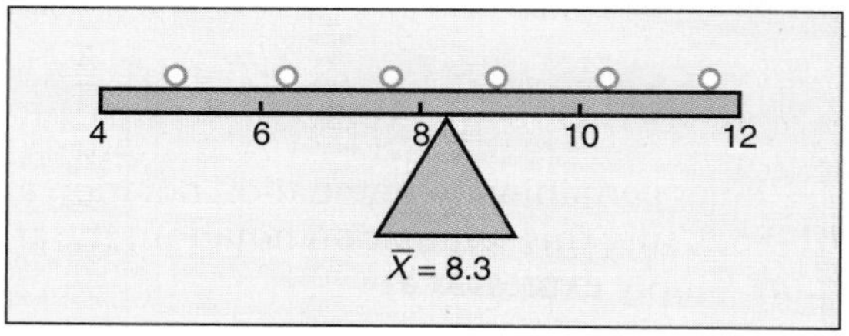

Figure 4.1
Dot scale representing tuition rates (in $000) at six Pennsylvania schools.

To further demonstrate the characteristics of the mean, suppose that our researcher takes a random sample of $n = 6$ schools from the population frame of 60 in the state of Texas and another random sample of $n = 6$ schools from the listing of 45 in the state of North Carolina. The tuition rates charged to out-of-state residents (in thousands of dollars) are reported as follows:

Texas School	Tuition (in $000)
Concordia Lutheran College	6.4
Southern Methodist University	12.0
Texas A & M University	4.9
Lubbock Christian University	6.4
Rice University	8.5
Trinity University	11.6

North Carolina School	Tuition (in $000)
Methodist College	8.3
Warren Wilson College	8.7
Campbell University	7.6
Belmont Abbey College	8.3
Catawba College	9.0
North Carolina State University	7.9

The respective dot scales are displayed in Figures 4.2 and 4.3.

Note that the mean tuition rate for each of these samples is also 8.3 thousand dollars. Nevertheless, as is observed from Figures 4.2 and 4.3, the two samples drawn here have distinctly different features—with respect to each other as well as with respect to the sample of six Pennsylvania schools depicted in Figure 4.1. For example, three of the six Texas schools have tuition rates quite different from those of Trinity University and Southern Methodist University. For this sample, the arithmetic mean is presenting a somewhat distorted representation of what the data are conveying and it may not be the best average to use. On the other hand, for the sample of North Carolina schools and the sample of Pennsylvania schools, the mean is the appropriate descriptive measure for characterizing and summarizing the respective data batches because outliers are not present. In fact, the North Carolina data are quite *homogeneous*. Two of the six schools in this sample have tuition rates equivalent to the mean; furthermore, from Figures 4.1 through 4.3, it is clear that the rates charged by these six North Carolina schools contain the least amount of scatter or variability among the three samples. In addition, it is also observed that the tuition rate data in each of the Pennsylvania and North Carolina samples possess the property of symmetry, while the tuition rate data for the sample in Texas do not. (The properties of variation and shape will be addressed further in Sections 4.5 and 4.6.)

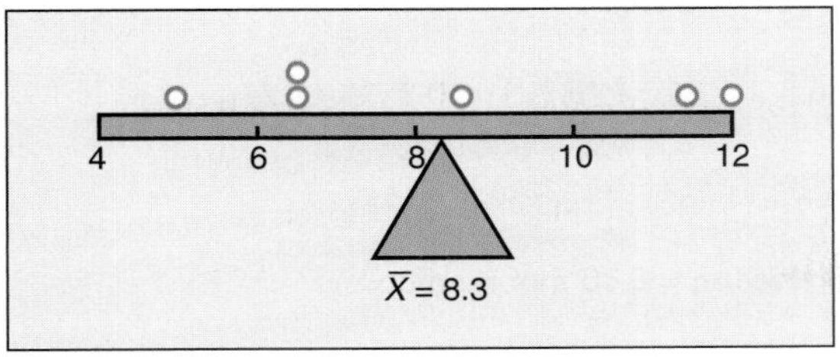

Figure 4.2
Dot scale representing tuition rates (in $000) at six Texas schools.

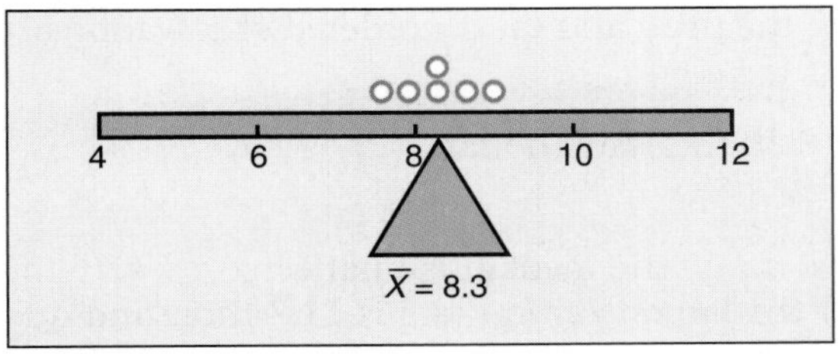

Figure 4.3
Dot scale representing tuition rates (in $000) at six North Carolina schools.

4.4.2 The Median

The **median** is the middle value in an ordered sequence of data. If there are no ties, half of the observations will be smaller and half will be larger. The median is unaffected by any extreme observations in a batch of data. Thus whenever an extreme observation is present, it is appropriate to use the median rather than the mean to describe a batch of data.

To calculate the median from a batch of data collected in its raw form, we must first place the data into an ordered array. We then use the *positioning point formula*

$$\frac{n+1}{2}$$

to find the place in the ordered array that corresponds to the median value. One of two rules is followed:

> **Rule 1** If the size of the sample is an *odd* number, the median is represented by the numerical value corresponding to the positioning point—the $(n + 1)/2$ ordered observation.

> **Rule 2** If the size of the sample is an *even* number, then the positioning point lies between the two middle observations in the ordered array. The median is the *average* of the numerical values corresponding to these two middle observations.

● **Even-Sized Sample** For our researcher's sample of six Pennsylvania schools, the raw data (in thousands of dollars) were

10.3 4.9 8.9 11.7 6.3 7.7

The ordered array becomes

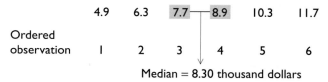

	4.9	6.3	7.7	8.9	10.3	11.7
Ordered observation	1	2	3	4	5	6

Median = 8.30 thousand dollars

For these data, the positioning point is $(n + 1)/2 = (6 + 1)/2 = 3.5$. Therefore the median is obtained by averaging the third and fourth ordered observations:

$$\frac{7.7 + 8.9}{2} = 8.30 \text{ thousand dollars}$$

As can be seen from the ordered array, the median is unaffected by extreme observations. Regardless of whether the largest tuition rate is 11.7 thousand dollars, 21.7 thousand dollars, or 31.7 thousand dollars, the median is still 8.3 thousand dollars.

● **Odd-Sized Sample** Had the sample size been an odd number, the median would merely be represented by the numerical value given to the $(n + 1)/2$ observation in the ordered array. Thus, in the following ordered array for $n = 5$ students' GMAT scores, the median is the value of the third [that is, $(5 + 1)/2$] ordered observation—590:

	500	570	590	600	690
			Median		
Ordered observation	1	2	3	4	5

● **Ties in the Data** When computing the median, we ignore the fact that tied values may be present in the data. Suppose, for example, that the following batch of data represents the starting salaries (in thousands of dollars) for a sample of $n = 7$ marketing majors who have recently graduated from your college:

24.1	22.6	27.0	19.8	21.5	23.7	22.6

The ordered array becomes

	19.8	21.5	22.6	22.6	23.7	24.1	27.0
				Median			
Ordered observation	1	2	3	4	5	6	7

For this odd-sized sample the median positioning point is the $(n + 1)/2 = 4$th ordered observation. Thus the median is 22.6 thousand dollars, the middle value in the ordered sequence, even though the third ordered observation is also 22.6 thousand dollars.

● **Characteristics of the Median** To summarize, the calculation of the median value is affected by the number of observations, not by the magnitude of any extreme(s). Any observation selected at random is just as likely to exceed the median as it is to be exceeded by it.

4.4.3 The Mode

Sometimes, when summarizing or describing a batch of data, the mode is used as a measure of central tendency. The **mode** is the value in a batch of data that appears most frequently. It is easily obtained from an ordered array. Unlike the arithmetic mean, the mode is not affected by the occurrence of any extreme values. However, the mode is not used for more than descriptive purposes because it is more variable from sample to sample than other measures of central tendency.

Using the ordered array for tuition rates charged in a sample of six Pennsylvania schools

4.9	6.3	7.7	8.9	10.3	11.7

we see that there is no mode. None of the tuition rates was "most typical."

Note that there is a difference between *no mode* and a mode of 0, as illustrated in the following ordered array of noontime temperatures (°F) in Duluth during the first week of January:

Ordered array (Duluth, Minnesota)	−4°	−2°	−1°	−1°	0°	0°	0°

Mode = 0°

In addition, a data batch can have more than one mode, as illustrated in the following ordered array of noontime temperatures (°F) in Richmond during the first week of January:

Ordered array (Richmond, Virginia)	21°	28°	28°	35°	41°	43°	43°

In Richmond we see there were two modes—28° and 43°. Such data are described as *bimodal*.

4.4.4 The Midrange

The **midrange** is the average of the *smallest* and *largest* observations in a batch of data. This can be written as

$$\text{Midrange} = \frac{X_{smallest} + X_{largest}}{2} \tag{4.2}$$

Using the ordered array of tuition rates charged from our research analyst's sample of six Pennsylvania schools:

4.9	6.3	7.7	8.9	10.3	11.7

the midrange is computed from Equation (4.2) as

$$\text{Midrange} = \frac{X_{smallest} + X_{largest}}{2}$$

$$= \frac{4.9 + 11.7}{2} = 8.30 \text{ thousand dollars}$$

The midrange is often used as a summary measure both by financial analysts and by weather reporters, since it can provide an adequate, quick, and simple measure to characterize the *entire* data batch—be it a series of daily closing stock prices over a whole year or a series of recorded hourly temperature readings over a whole day.

In dealing with data such as daily closing stock prices or hourly temperature readings, an extreme value is not likely to occur. Nevertheless, in most applications, despite its simplicity, the midrange must be used cautiously. Since it involves only the smallest and largest observations in a data batch, the midrange becomes distorted as a summary measure of central tendency if an outlier is present. In such situations, the midrange is inappropriate; a summary measure somewhat similar in format to the midrange that is always appropriate because it is totally unaffected by outliers is the *midhinge.*

4.4.5 The Midhinge

The **midhinge** is the average of the *first* and *third quartiles* in a batch of data. That is,

$$\text{Midhinge} = \frac{Q_1 + Q_3}{2} \quad \textbf{(4.3)}$$

where Q_1 = first quartile
Q_3 = third quartile

It is a summary measure used to overcome potential problems introduced by extreme values in the data.

● **Quartiles: Measures of "Noncentral" Location** Aside from the measures of central tendency, there also exist some useful measures of "noncentral" location which are employed particularly when summarizing or describing the properties of large batches of numerical data.[2] The most widely used of these measures are the **quartiles.**

Whereas the median is a value that splits the ordered array in half (50.0% of the observations are smaller and 50.0% of the observations are larger), the quartiles are descriptive measures that split the ordered data into four quarters.

The **first quartile, Q_1,** is the value such that 25.0% of the observations are smaller and 75.0% of the observations are larger.

The **second quartile, Q_2,** is the median—50.0% of the observations are smaller and 50.0% are larger.

The **third quartile, Q_3,** is the value such that 75.0% of the observations are smaller and 25.0% of the observations are larger.

To approximate the quartiles, the following *positioning point formulas* are used:

$$Q_1 = \text{value corresponding to the } \frac{n+1}{4} \text{ ordered observation}$$

$$Q_2 = \text{median, the value corresponding to the } \frac{2(n+1)}{4} = \frac{n+1}{2} \text{ ordered observation}$$

$$Q_3 = \text{value corresponding to the } \frac{3(n+1)}{4} \text{ ordered observation}$$

The following rules are used for obtaining the quartile values:

1. If the resulting positioning point is an integer, the particular numerical observation corresponding to that positioning point is chosen for the quartile.
2. If the resulting positioning point is halfway between two integers, the average of their corresponding values is selected.
3. If the resulting positioning point is neither an integer nor a value halfway between two integers, a simple rule used to approximate the particular quartile is to *round off* to the nearest integer positioning point and select the numerical value of the corresponding observation.

To compute our midhinge [see Equation (4.3)], we first need to compute Q_1 and Q_3. For example, from the ordered array pertaining to tuition rates charged (in thousands of dollars) to out-of-state residents at the six Pennsylvania schools we have

$$Q_1 = \frac{n+1}{4} \text{ ordered observation}$$

$$= \frac{6+1}{4} = 1.75\text{th} \cong 2\text{nd ordered observation}$$

Therefore Q_1 can be approximated as 6.30 thousand dollars.

$$Q_3 = \frac{3(n+1)}{4} \text{ ordered observation}$$

$$= \frac{3(6+1)}{4} = 5.25\text{th} \cong 5\text{th ordered observation}$$

Therefore Q_3 can be approximated as 10.30 thousand dollars.

Returning to Equation (4.3), we may now compute the midhinge as

$$\text{Midhinge} = \frac{Q_1 + Q_3}{2}$$

$$= \frac{6.3 + 10.3}{2} = 8.30 \text{ thousand dollars}$$

This result, the average of Q_1 and Q_3 (two measures of noncentral location), cannot be affected by potential outliers, since no observation smaller than Q_1 or larger than Q_3 is considered. Summary measures such as the midhinge and the median, which cannot be affected by outliers, are called **resistant measures.**

Problems for Section 4.4

4.1 Which of the following statements are objective and which are subjective (interpretive)?

(a) The average (mean) price of a home in Bergen County is $184,700.
(b) Housing is expensive in Palo Alto.
(c) The police in New York City are smarter, better educated, and more honest now than they were 30 years ago.
(d) More burglaries per 1,000 homes were reported in Chicago last year than in Des Moines.
(e) The average (mean) score on an IQ test of students at the Wallace School of Science is 145.
(f) Foreign product dumping is crippling our industry.

4.2 Given the following two batches of data—each with samples of size $n = 7$:

Batch 1: 10 2 3 2 4 2 5

Batch 2: 20 12 13 12 14 12 15

(a) For each batch, compute the mean, median, mode, midrange, and midhinge.
(b) Compare your results and summarize your findings.
(c) Compare the first sampled item in each batch, compare the second sampled item in each batch, and so on. Briefly describe your findings here in light of your summary in part (b).

4.3 A track coach must decide on which one of two sprinters to select for the 100-meter dash at an upcoming meet. The coach will base the decision on the results of five races between the two athletes run with 15-minute rest intervals between. The following times (in seconds) were recorded for the five races:

	Race				
Athlete	1	2	3	4	5
Sharyn	12.1	12.0	12.0	16.8	12.1
Tamara	12.3	12.4	12.4	12.5	12.4

(a) Based on these data, which of the two sprinters should the coach select? Why?
(b) Should the selection be different if the coach knew that Sharyn had fallen at the start of the fourth race? Why?
(c) Discuss the differences in the concepts of the mean and the median as measures of central tendency and how this relates to (a) and (b).

4.4 Suppose that, owing to an error, a data batch containing the price-to-earnings (PE) ratios from nine companies traded on the American Stock Exchange was recorded as 13, 15, 14, 17, 13, 16, 15, 16, and 61, where the last value should have been 16 instead of 61. Show how much the mean, median, and midrange are affected by the error (that is, compute these statistics for the "bad" and "good" data sets, and compare the results of using different estimators of central tendency).

4.5 A manufacturer of flashlight batteries took a sample of 13 from a day's production and burned them continuously until they failed. The numbers of hours they burned were

Data file FLASHBAT.DAT

342, 426, 317, 545, 264, 451, 1049,

631, 512, 266, 492, 562, 298

(a) Compute the mean, median, mode, midrange, and midhinge. Looking at the distribution of times, which descriptive measures seem best and which worst? (And why?)

(b) In what ways would this information be useful to the manufacturer? Discuss.

4.6 It is stated in Section 4.5.3 that one important property of the arithmetic mean is

$$\sum_{i=1}^{n}(X_i - \overline{X}) = 0$$

(a) Using the out-of-state resident tuition rates from the sample of six Texas colleges and universities (see page 108), verify that this property holds.

(b) Using the out-of-state resident tuition rates from the sample of six North Carolina colleges and universities (see page 108), verify that this property holds.

4.7 The following data represent the prices (without grass catchers) of a sample of 15 side-bagging lawn mowers having 20-inch swaths:

Data file MOWER.DAT

Brand and Model	Price
Sears Craftsman 38023	$150
Cub Cadet 074R	245
White 072R	220
Sears Craftsman 38033	200
Sears Craftsman 38045	160
Lawn Chief 50-H	130
Mastercut 5020-20	160
Atlas 20 2011 (Winston)	130
Rally A103 CR	119
Murray 20203	112
Sears Companion 38004	114
Sycamore 20-4000	110
Lawn Chief 60-H	160
Wheeler WE20	150
Wheeler WB20	127

Source: Copyright 1990 by Consumers Union of United States, Inc., Yonkers, N.Y. 10703. Adapted by permission from *Consumer Reports*, June 1990, p. 396.

(a) What is the mean price? The median price?
(b) How would this information be of use to a company about to market a new model? Discuss.
(c) If an additional mower were included by mistake in this sample (that is, it was a rear-bagging unit containing a discharge chute and had a total price of $320), what would the mean and median be?
(d) Discuss the reasons for the differences in your responses in (a) and (c).

4.8 The following data are the amount of calories in a 30-gram serving for a random sample of 10 types of fresh-baked chocolate chip cookies:

Data file COOKIES.DAT

Product	Calories
Hillary Rodham Clinton's	153
Original Nestle Toll House	152
Mrs. Fields	146
Stop & Shop	138
Duncan Hines	130
David's	146
David's Chocolate Chunk	149
Great American Cookie Company	138
Pillsbury Oven Lovin'	168
Pillsbury	147

Source: Copyright 1993 by Consumers Union of United States, Inc., Yonkers, N.Y. 10703. Adapted by permission from *Consumer Reports,* October 1993, pp. 646–647.

(a) What is the mean amount of calories? The median amount of calories?
(b) How would this information be of use to a company about to market a new fresh-baked chocolate chip cookie? Discuss.

4.9 The following data are the list prices (in dollars) for a random sample of 17 Type IV (metal) audiotapes:

Data file TAPES.DAT

4.95	18.99	5.29
5.29	3.49	3.99
3.99	5.50	9.95
11.00	5.99	14.99
5.99	3.49	3.50
4.59	2.99	

Source: Copyright 1993 by Consumers Union of United States, Inc., Yonkers, N.Y. 10703. Adapted by permission from *Consumer Reports*, January 1993, pp. 38-39.

(a) What is the mean price? The median price?
(b) ACTION What other information would you want to know before making a purchase of such a tape? Prepare a list of questions that you would ask the salesperson.

4.10 The following data on page 117 represent the amount of funding (in millions $) provided by the Alcohol, Drug Abuse, and Mental Health Administration through grants to a sample of 21 institutions during a recent year:

Name of Institution	Amount Funded (in millions $)
Johns Hopkins University	14.9
University of California at San Francisco	14.1
University of Washington	6.8
Yale University	13.1
Stanford University	7.6
University of California at Los Angeles	13.2
Harvard University	5.1
University of Michigan	11.9
University of Pennsylvania	8.5
Columbia University	7.1
Washington University at St. Louis	5.1
Duke University	5.0
University of Minnesota	6.2
University of California at San Diego	5.5
University of North Carolina	3.8
University of Wisconsin	3.4
University of Rochester	3.5
Yeshiva University	2.8
University of Chicago	4.1
University of Pittsburgh	15.9
Cornell University	5.7

Source: U.S. Department of Health and Human Services, Public Health Services.

Data file ADAMHA.DAT

(a) Organize the data into an ordered array or stem-and-leaf display.
(b) Compute the mean, median, mode, midrange, and midhinge.
(c) Describe the property of central tendency for these data.

4.11 For the last ten days in June, the "Shore Special" train was late arriving at its destination by the following times (in minutes; a negative number means that the train was early by that number of minutes):

Data file TRAIN1.DAT

−3, 6, 4, 10, −4, 124, 2, −1, 4, 1

(a) If you were hired by the railroad as a statistician to show that the railroad is providing good service, what are some of the summary measures you would use to accomplish this?
(b) If you were hired by a TV station that was producing a documentary to show that the railroad is providing bad service, what summary measures would you use?
(c) If you were trying to be objective and unbiased in assessing the railroad's performance, which summary measures would you use? (This is the hardest part, because you cannot answer it without making additional assumptions about the relative costs of being late by various amounts of time.)

4.12 In order to estimate how much water will be needed to supply the community of Falling Rock in the next decade, the town council asked the city manager to find out how much water a sample of families currently uses. The sample of 15 families used the following number of gallons (in thousands) in the past year:

Data file WATER.DAT

11.2, 21.5, 16.4, 19.7, 14.6, 16.9, 32.2, 18.2,
13.1, 23.8, 18.3, 15.5, 18.8, 22.7, 14.0

(a) What is the mean amount of water used per family? The median? The midrange? The midhinge?

(b) Suppose that ten years from now the town council expects that there will be 45,000 families living in Falling Rock. How many gallons of water per year will be needed, if the rate of consumption per family stays the same?
(c) In what ways would the information provided in (a) and (b) be useful to the town council? Discuss.
(d) Why might the town council have used the data from a survey rather than just measuring the total consumption in the town? (Think about what types of users are not yet included in the estimation process.)

Interchapter Problems for Section 4.4

Data file MAILORD.DAT

4.13 Using the monthly billing records of the mail-order book company (Problem 3.2 on page 58):
(a) Compute the mean, median, mode, midrange, and midhinge.
(b) If a total of 350 bills were still outstanding, use the mean to estimate the total amount owed to the company. (*Hint:* Total = $N\overline{X}$.)
(c) ACTION Write a draft of the memo the auditor will want to send to the chief executive officer of the mail-order book company regarding the findings.
(d) In what ways would this information be useful to the chief executive officer? Discuss.

Data file SHOWER.DAT

4.14 Using the data on maximum flow rate of shower heads (Problem 3.3 on page 58):
(a) Compute the mean, median, mode, midrange, and midhinge.
(b) ACTION Describe the property of central tendency for these data.

Data file UTILITY.DAT

● 4.15 Using the data on electric and gas utility charges (Problem 3.12 on page 66):
(a) Compute the mean, median, mode, midrange, and midhinge.
(b) ACTION Describe the property of central tendency for these data.

4.5 Measures of Variation

A second important property that describes a batch of numerical data is variation. **Variation** is the amount of dispersion or "spread" in the data. Two batches of data may differ in both central tendency and variation; or, as shown in Figure 4.1 and 4.3, two batches of data may have the same measures of central tendency but differ greatly in terms of variation. The data batch depicted in Figure 4.3 is much less variable than that depicted in Figure 4.1 (see pages 108 and 109).

Five measures of variation are the range, the interquartile range, the variance, the standard deviation, and the coefficient of variation.

4.5.1 The Range

The **range** is the difference between the largest and smallest observations in a batch of data. That is,

$$\text{Range} = X_{largest} - X_{smallest} \qquad \textbf{(4.4)}$$

Using the ordered array of tuition rates charged (in thousands of dollars) to out-of-state residents from our sample of six Pennsylvania schools:

4.9   6.3   7.7   8.9   10.3   11.7

the range is 11.7 - 4.9 = 6.80 thousand dollars.

The range measures the *total spread* in the batch of data. Although the range is a simple, easily calculated measure of total variation in the data, its distinct weakness is that it fails to take into account *how* the data are actually distributed between the smallest and largest values. This can be observed from Figure 4.4. Thus, as evidenced in scale C, it would be improper to use the range as a measure of variation when either one or both of its components are extreme observations.

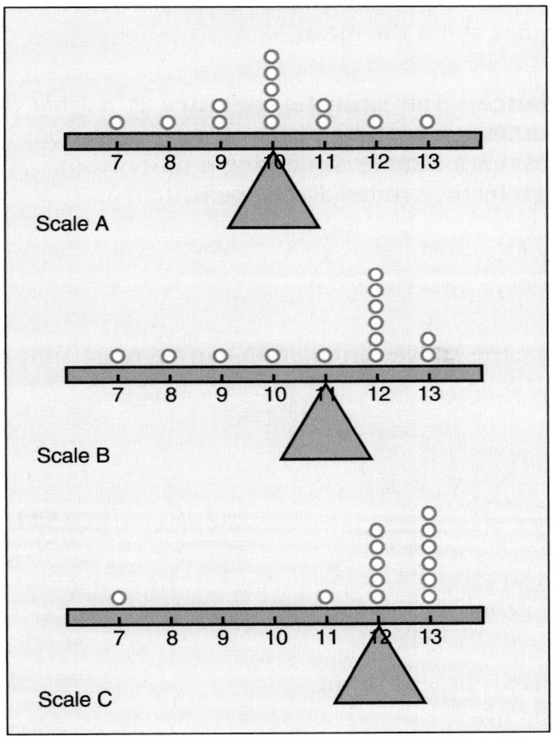

Figure 4.4
Comparing three data sets with the same range.

4.5.2 The Interquartile Range

The **interquartile range** (also called **midspread**) is the difference between the *third* and *first quartiles* in a batch of data. That is,

$$\text{Interquartile range} = Q_3 - Q_1 \quad (4.5)$$

This simple measure considers the spread in the middle 50% of the data and thus is in no way influenced by possibly occurring extreme values.

For the Pennsylvania tuition rate data we have

Interquartile range = $Q_3 - Q_1 = 10.3 - 6.3 = 4.0$ thousand dollars

This is the range in tuition rates for the *middle group* of Pennsylvania schools.

4.5.3 The Variance and the Standard Deviation

Although the range is a measure of the *total spread* and the interquartile range is a measure of the *middle spread,* neither of these measures of variation takes into consideration *how* the observations distribute or cluster. Two commonly used measures of variation that do take into account *how* all the values in the data are distributed are the *variance* and its square root, the *standard deviation.* These measures evaluate how the values fluctuate about the mean.

Defining the Sample Variance The **sample variance** is *roughly* (*or almost*) the average of the squared differences between each of the observations in a batch of data and the mean. Thus, for a sample containing n observations, X_1, X_2, ..., X_n, the sample variance (given by the symbol S^2) can be written as

$$S^2 = \frac{(X_1 - \overline{X})^2 + (X_2 - \overline{X})^2 + \cdots + (X_n - \overline{X})^2}{n - 1}$$

Using our summation notation, the above formulation can be more simply expressed as

$$S^2 = \frac{\sum_{i=1}^{n}(X_i - \overline{X})^2}{n - 1} \tag{4.6}$$

where

$\overline{X}$ = sample arithmetic mean

n = sample size

X_i = ith value of the random variable X

$\sum_{i=1}^{n}(X_i - \overline{X})^2$ = summation of all the squared differences between the X_i values and $\overline{X}$

Had the denominator been n instead of $n - 1$, the average of the squared differences around the mean would have been obtained. However, $n - 1$ is used here because of certain desirable mathematical properties possessed by the statistic S^2 that make it appropriate for statistical inference (see Chapter 9). If the sample size is large, division by n or $n - 1$ doesn't really make much difference.

Defining the Sample Standard Deviation The **sample standard deviation** (given by the symbol S) is simply the square root of the sample variance. That is,

$$S = \sqrt{\frac{\sum_{i=1}^{n}(X_i - \bar{X})^2}{n-1}} \quad (4.7)$$

Computing S^2 and S To compute the variance we

1. Obtain the difference between each observation and the mean
2. Square each difference
3. Add the squared results together
4. Divide the summation by $n - 1$

To compute the standard deviation we merely take the square root of the variance.

For our sample of six Pennsylvania schools, the raw (tuition rate) data (in thousands of dollars) were

10.3 4.9 8.9 11.7 6.3 7.7

and $\bar{X} = 8.30$ thousand dollars.

The sample variance is computed as

$$\begin{aligned} S^2 &= \frac{\sum_{i=1}^{n}(X_i - \bar{X})^2}{n-1} \\ &= \frac{(10.3 - 8.3)^2 + (4.9 - 8.3)^2 + \cdots + (7.7 - 8.3)^2}{6-1} \\ &= \frac{31.84}{5} \\ &= 6.368 \text{ (in squared thousands of dollars)} \end{aligned}$$

and the sample standard deviation is computed as

$$S = \sqrt{S^2} = \sqrt{\frac{\sum_{i=1}^{n}(X_i - \bar{X})^2}{n-1}} = \sqrt{6.368} = 2.52 \text{ thousands of dollars}$$

Obtaining S^2 and S Since in the preceding computations we are squaring the differences, *neither the variance nor the standard deviation can ever be negative.* The only time S^2 and S could be zero would be when there was no variation at all in the data—when each observation in the sample was exactly the same. In such an unusual case the range would also be zero.

But numerical data are inherently variable—not constant. Any random phenomenon of interest that we could think of usually takes on a variety of values. For example, colleges and universities charge different rates of tuition for out-of-state residents just as people have different IQs, incomes, weights, heights, ages, pulse rates, etc. It is because numerical data inherently vary that it becomes so important to study not only measures (of central tendency) that summarize the data but also measures (of variation) that reflect how the numerical data are dispersed.

● **What the Variance and the Standard Deviation Indicate** The variance and the standard deviation measure the "average" scatter around the mean—that is, how larger observations fluctuate above it and how smaller observations distribute below it.

The variance possesses certain useful mathematical properties. However, its computation results in squared units—squared thousands of dollars, squared dollars, squared inches, etc. Thus, for practical work our primary measure of variation will be the standard deviation, whose value is in the original units of the data—thousands of dollars, dollars, inches, etc.

In the Pennsylvania tuition rate sample the standard deviation is 2.52 thousand dollars. This tells us that the *majority* of the tuition rates in this sample are clustering within 2.52 thousand dollars around the mean of 8.30 thousand dollars (that is, between 5.78 and 10.82 thousand dollars).

● **Why We Square the Deviations** The formulas for variance and standard deviation could not merely use

$$\sum_{i=1}^{n}(X_i - \overline{X})$$

as a numerator, because you may recall that the mean acts as a *balancing point* for observations larger and smaller than it. Therefore, the sum of the deviations about the mean is always zero[3]; that is,

$$\sum_{i=1}^{n}(X_i - \overline{X}) = 0$$

To demonstrate this, let us again refer to the Pennsylvania tuition rate data:

10.3 4.9 8.9 11.7 6.3 7.7

Therefore,

$$\begin{aligned}\sum_{i=1}^{n}(X_i - \overline{X}) &= (10.3 - 8.3) + (4.9 - 8.3) + (8.9 - 8.3) \\ &\quad + (11.7 - 8.3) + (6.3 - 8.3) + (7.7 - 8.3) \\ &= 0\end{aligned}$$

This is depicted in the accompanying dot scale diagram displayed in Figure 4.5.

As already noted, three of the observations are smaller than the mean and

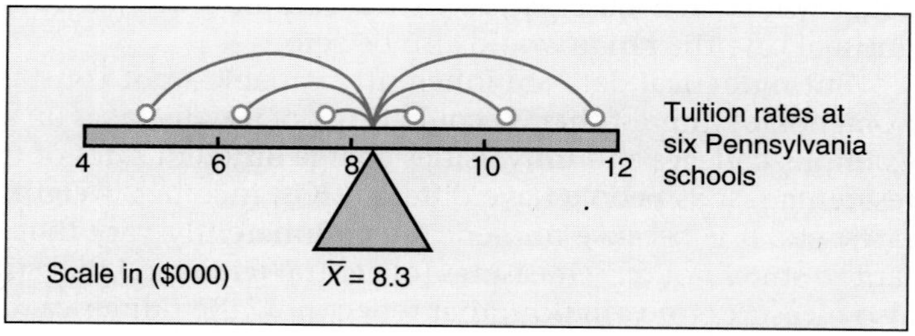

Figure 4.5
The mean as a *balancing point.*

three are larger. Although the sum of the six deviations (2.0, −3.4, 0.6, 3.4, −2.0, and −0.6) is zero, the sum of the squared deviations allows us to study the variation in the data. Hence we use

$$\sum_{i=1}^{n}(X_i - \bar{X})^2$$

when computing the variance and standard deviation. In the squaring process, observations that are farther from the mean get more weight than observations closer to the mean.

The respective squared deviations for the Pennsylvania tuition rate data are

4.00   11.56   0.36   11.56   4.00   0.36

We note that the fourth observation ($X_4 = 11.7$ thousand dollars) is 3.4 thousand dollars higher than the mean, and the second observation ($X_2 = 4.9$ thousand dollars) is 3.4 thousand dollars lower. In the squaring process both these values contribute substantially more to the calculation of S^2 and S than do the other observations in the sample, which are closer to the mean.

Therefore we may generalize as follows:

1. The more spread out or dispersed the data are, the larger will be the range, the interquartile range, the variance, and the standard deviation.
2. The more concentrated or homogeneous the data are, the smaller will be the range, the interquartile range, the variance, and the standard deviation.
3. If the observations are all the same (so that there is no variation in the data), the range, interquartile range, variance, and standard deviation will all be zero.

Computing S^2 and S: "Hand-Held Calculator" Formulas The formulas for variance and standard deviation, Equations (4.6) and (4.7), are *definitional formulas,* but they are often not practical to use—even with a hand-held calculator. For our Pennsylvania tuition rate data the mean, 8.30 thousand dollars, is not an integer. For these more typical situations, where the observations and the mean are unlikely to be integers, the following *"hand-held calculator" formulas* for the variance and the standard deviation are given for practical use:

$$S^2 = \frac{\sum_{i=1}^{n} X_i^2 - n\bar{X}^2}{n-1} \qquad \textbf{(4.8)}$$

$$S = \sqrt{\frac{\sum_{i=1}^{n} X_i^2 - n\bar{X}^2}{n-1}} \qquad \textbf{(4.9)}$$

where $\sum_{i=1}^{n} X_i^2$ = summation of the squares of the individual observations

$n\overline{X}^2$ = sample size times the square of the sample mean

The hand-held calculator formulas, Equations (4.8) and (4.9), are identical to the definitional formulas, Equations (4.6) and (4.7). Since the denominators are the same, it is easy to show through expansion and the use of summation rules (see Appendix B) that

$$\sum_{i=1}^{n} (X_i - \overline{X})^2 = \sum_{i=1}^{n} X_i^2 - n\overline{X}^2$$

Moreover, since S^2 (and S) can never be negative,

$$\sum_{i=1}^{n} X_i^2$$

the summation of squares, must always equal or exceed

$$n\overline{X}^2$$

the sample size times the square of the sample mean.

Returning to the Pennsylvania tuition rate data, the variance and standard deviation are recomputed using Equations (4.8) and (4.9) as follows:

$$\begin{aligned} S^2 &= \frac{\sum_{i=1}^{n} X_i^2 - n\overline{X}^2}{n-1} \\ &= \frac{(10.3^2 + 4.9^2 + \cdots + 7.7^2) - 6(8.3^2)}{6-1} \\ &= \frac{(106.09 + 24.01 + \cdots + 59.29) - 6(68.89)}{5} \\ &= \frac{445.18 - 413.34}{5} \\ &= \frac{31.84}{5} = 6.368 \text{ (in squared thousands of dollars)} \end{aligned}$$

and

$$S = \sqrt{6.368} = 2.52 \text{ thousand dollars}$$

4.5.4 The Coefficient of Variation

Unlike the previous measures we have studied, the **coefficient of variation** is a *relative measure* of variation. It is expressed as a percentage rather than in terms of the units of the particular data.

The coefficient of variation, denoted by the symbol CV, measures the scatter in the data relative to the mean. It may be computed by

$$CV = \left(\frac{S}{\overline{X}}\right)100\% \qquad \textbf{(4.10)}$$

where S = standard deviation in a batch of numerical data
$\overline{X}$ = arithmetic mean in a batch of numerical data

Returning to the tuition rate data obtained from a sample of six Pennsylvania colleges and universities, the coefficient of variation is

$$CV = \left(\frac{S}{\overline{X}}\right)100\% = \left(\frac{2.52}{8.30}\right)100\% = 30.4\%$$

That is, for this sample the relative size of the "average spread around the mean" to the mean is 30.4%.

As a relative measure, the coefficient of variation is particularly useful when comparing the variability of two or more batches of data that are expressed in different units of measurement. As an example from the Kalosha Industries Employee Satisfaction Survey of Chapter 2, suppose that the Vice-President for Human Resources was interested in determining whether the amount of hours worked in a week by a full-time employee (question 1) has greater variability (on a relative basis) than does employee annual personal income (question 7). Since hours worked is a time measurement and annual personal income (in $000) is a monetary amount, it is impossible to compare directly the two standard deviations or the two ranges for these variables. Here, however, the two coefficients of variation can be used to provide the desired answer. (See Problem 4.70 of the Survey/Database Project on page 155.)

The coefficient of variation is also very useful when comparing two or more sets of data that are measured in the same units but differ to such an extent that a direct comparison of the respective standard deviations is not very helpful. As an example, suppose that a potential investor was considering purchasing shares of stock in one of two companies, A or B, which are listed on the American Stock Exchange. If neither company offered dividends to its stockholders and if both companies were rated equally high (by various investment services) in terms of potential growth, the potential investor might want to consider the volatility (variability) of the two stocks to aid in the investment decision. Now suppose that each share of stock in Company A has averaged $50 over the past months with a standard deviation of $10. In addition, suppose that in this same time period the price per share for Company B stock averaged $12 with a standard deviation of $4. In terms of the actual standard deviations the price of Company A shares seems to be more volatile than that of Company B shares. However, since the average prices per share for the two stocks are so different, it would be more appropriate for the potential investor to consider the variability in price relative to the average price in order to examine the volatility/stability of the two stocks. For Company A the coefficient of variation is $CV_A = (\$10/\$50)100\% = 20.0\%$; for Company B the coefficient of variation is $CV_B = (\$4/\$12)100\% = 33.3\%$. Thus relative to the mean, the price of stock B is much more variable than the price of stock A.

Problems for Section 4.5

4.16 Verify that the computation of the standard deviation is identical, regardless of whether the definitional formula (4.7) or the hand-held calculator formula (4.9) is used, for the following:
(a) The out-of-state resident tuition rate data taken from the sample of six colleges and universities in Texas (see page 108).
(b) The out-of-state resident tuition rate data taken from the sample of six colleges and universities in North Carolina (see page 108).

4.17 For each data batch in Problem 4.2 on page 114:
(a) Compute the range, interquartile range, variance, standard deviation, and coefficient of variation.
(b) Compare your results and discuss your findings.
(c) Based on your answers to Problem 4.2 and (a) and (b) above, what can you generalize about the properties of central tendency and variation?

4.18 Using the data from Problem 4.3 on page 114, compute the range, interquartile range, variance, standard deviation, and coefficient of variation for each of the two athletes' race-time trials.

4.19 Using the PE ratio data from Problem 4.4 on page 115, compute the range, interquartile range, variance, standard deviation, and coefficient of variation for the data batch with the error (61) and then recompute these statistics after the PE ratio is corrected to 16.
(a) Discuss the differences in your findings for each measure of dispersion.
(b) Which measure seems to be affected most by the error?

Data file FLASHBAT.DAT

4.20 For the following, refer to the battery life data in Problem 4.5 on page 115:
(a) Calculate the range, variance, and standard deviation.
(b) For many sets of data, the range is about six times the standard deviation. Is this true here? (If not, why do you think it is not?)
(c) Using the information above, what would you advise the manufacturer to do if he wanted to be able to say in advertisements that these batteries "should last 400 hours"? (*Note*: There is no right answer to this question; the point is to consider how to make such a statement precise.)

Data file MOWER.DAT

4.21 Using the lawn mower price data from Problem 4.7 on page 115:
(a) Compute the range, interquartile range, variance, standard deviation, and coefficient of variation in the prices of lawn mowers with 20-inch swaths. (Do not include the price for the rear-bagging unit.)
(b) Discuss the property of variation for these data.

Data file ADAMHA.DAT

4.22 Using the grants data from Problem 4.10 on page 116:
(a) Compute the range, interquartile range, variance, standard deviation, and coefficient of variation for amount funded (in millions of dollars).
(b) Discuss the property of variation for these data.

Data file TRAIN1.DAT

4.23 Using the train time data from Problem 4.11 on page 117:
(a) Compute the range, interquartile range, variance, standard deviation, and coefficient of variation for "lateness" (in minutes).
(b) Discuss the property of variation for these data.

Data file WATER.DAT

4.24 Using the water usage data from Problem 4.12 on page 117:
(a) Compute the range, interquartile range, variance, standard deviation, and coefficient of variation in water consumption.
(b) Discuss the property of variation for these data.

Interchapter Problems for Section 4.5

Data file MAILORD.DAT

4.25 Using the monthly billing records of the mail-order book company (Problem 3.2 on page 58):

(a) Compute the range, interquartile range, variance, standard deviation, and coefficient of variation of the amount owed to the mail-order book company.
(b) Discuss the property of variation for these data.

4.26 Using the data on maximum flow rate of shower heads (Problem 3.3 on page 58): Data file SHOWER.DAT
(a) Compute the range, interquartile range, variance, standard deviation, and coefficient of variation.
(b) Discuss the property of variation for these data.

● 4.27 Using the data on electric and gas utility charges (Problem 3.12 on page 66): Data file UTILITY.DAT
(a) Compute the range, interquartile range, variance, standard deviation, and coefficient of variation.
(b) Discuss the property of variation for these data.

4.6 Shape

A third important property of a batch of data is its **shape**—the manner in which the data are distributed. Either the distribution of the data is **symmetrical** or it is not. If the distribution of data is not symmetrical, it is called asymmetrical or **skewed.**

To describe the shape we need only compare the mean and the median. If these two measures are equal, we may generally consider the data to be symmetrical (or *zero-skewed*). On the other hand, if the mean exceeds the median, the data may generally be described as *positive* or *right-skewed.* If the mean is exceeded by the median, those data can generally be called *negative* or *left-skewed.* That is,

Mean > median: positive or right-skewness
Mean = median: symmetry or zero-skewness
Mean < median: negative or left-skewness

Positive skewness arises when the mean is increased by some unusually high values; negative skewness occurs when the mean is reduced by some extremely low values. Data are symmetrical when there are no really extreme values in a particular direction so that low and high values balance each other out.

Figure 4.6 on page 128 depicts the shapes of three data batches: the data on Scale L are negative or left-skewed (since the distortion to the left is caused by extremely small values); the data on Scale R are positive or right-skewed (since the distortion to the right is caused by extremely large values); and the data on Scale S are symmetrical (the low and high values on the scale balance, and the mean equals the median).

For our sample of six Pennsylvania schools, the tuition rate data were displayed along the dot scale in Figure 4.1 (see page 108). The mean and the median are equal to 8.3 thousand dollars, and the data appear to be symmetrically distributed around these measures of central tendency.

Problems for Section 4.6

● 4.28 For each data batch in Problem 4.2 on page 114:
(a) Describe the shape.
(b) Compare your results and discuss your findings.

4.29 Using the battery life data from Problem 4.5 on page 115, describe the shape. Data file FLASHBAT.DAT

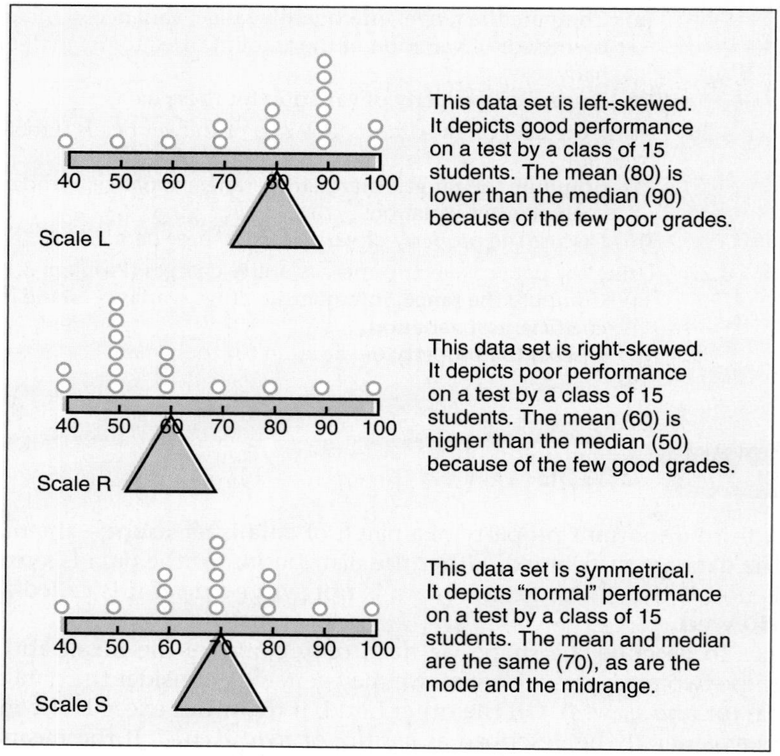

Figure 4.6
A comparison of three data sets differing in shape.

4.30 Using the lawn mower price data from Problem 4.7 on page 115, describe the shape. (Do not include the price for the rear-bagging unit.)

● 4.31 Using the grants data from Problem 4.10 on page 116, describe the shape.

4.32 Using the train "lateness" data from Problem 4.11 on page 117, describe the shape.

4.33 Using the water consumption data from Problem 4.12 on page 117, describe the shape.

Interchapter Problems for Section 4.6

4.34 Using the data on amount owed to the mail-order book company from Problem 3.2 on page 58, describe the shape.

4.35 Using the data on the maximum flow rate of shower heads from Problem 3.3 on page 58, describe the shape.

● 4.36 Using the data on electric and gas utility charges from Problem 3.12 on page 66, describe the shape.

4.7 Five-Number Summary and Box-and-Whisker Plot

Now that we have studied the three major properties of numerical data (central tendency, variation, and shape), it is important that we identify and describe the major features of the data in a summarized format. One approach to this "exploratory data analysis" is to develop a *five-number summary* and to construct a *box-and-whisker plot* (References 10 and 11).

4.7.1 Five-Number Summary

A **five-number summary** consists of

$$X_{smallest} \quad Q_1 \quad \text{median} \quad Q_3 \quad X_{largest}$$

It combines three measures of central tendency (the median, midhinge, and midrange) and two measures of variation (the interquartile range and range) to provide us with a better idea as to the shape of the distribution.

If the data were perfectly symmetrical, the following would be true:

1. The distance from Q_1 to the median would equal the distance from the median to Q_3.
2. The distance from $X_{smallest}$ to Q_1 would equal the distance from Q_3 to $X_{largest}$.
3. The median, the midhinge, and the midrange would all be equal. (These measures would also equal the mean in the data.)

On the other hand, for nonsymmetrical distributions the following would be true:

1. In right-skewed distributions the distance from Q_3 to $X_{largest}$ greatly exceeds the distance from $X_{smallest}$ to Q_1.
2. In right-skewed distributions median < midhinge < midrange.
3. In left-skewed distributions the distance from $X_{smallest}$ to Q_1 greatly exceeds the distance from Q_3 to $X_{largest}$.
4. In left-skewed distributions midrange < midhinge < median.

For our Pennsylvania tuition rate data, the five-number summary is

$$4.9 \quad 6.3 \quad 8.3 \quad 10.3 \quad 11.7$$

We may now use the five-number summary to study the shape of the distribution. From the above rules it is clear that the tuition rate data for our sample of six Pennsylvania schools are perfectly symmetrical.

4.7.2 Box-and-Whisker Plot

In its simplest form, a **box-and-whisker plot** provides a graphical representation of the data through its five-number summary. Such a plot is depicted in Figure 4.7 on page 130 for the tuition rates at the six Pennsylvania schools.

The vertical line drawn within the box represents the location of the median value in the data. Note further that the vertical line at the left side of the box represents the location of Q_1 and the vertical line at the right side of the box represents the location of Q_3. Therefore, we see that the box contains the middle 50% of the observations in the distribution. The lower 25% of the data are represented by a dashed line (that is, a *whisker*) connecting the left side of the box to the location of the smallest value, $X_{smallest}$. Similarly, the upper 25% of the data are represented by a dashed line connecting the right side of the box to $X_{largest}$.

This visual representation of the tuition rates depicted in Figure 4.7 indicates the symmetrical shape of the data. Not only do we observe that the vertical median line is centered in the box, we also see that the whisker lengths are clearly the same.

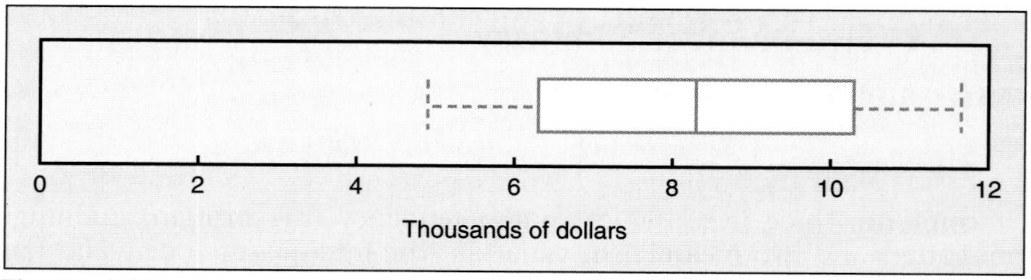

Figure 4.7
Box-and-whisker plot of tuition rates at six Pennsylvania schools.

To summarize what we have learned about graphical representation of our data, Figure 4.8 demonstrates the differences between modern "exploratory data analysis" and traditional displays by depicting five different types of distributions through their box-and-whisker plots and corresponding polygons.

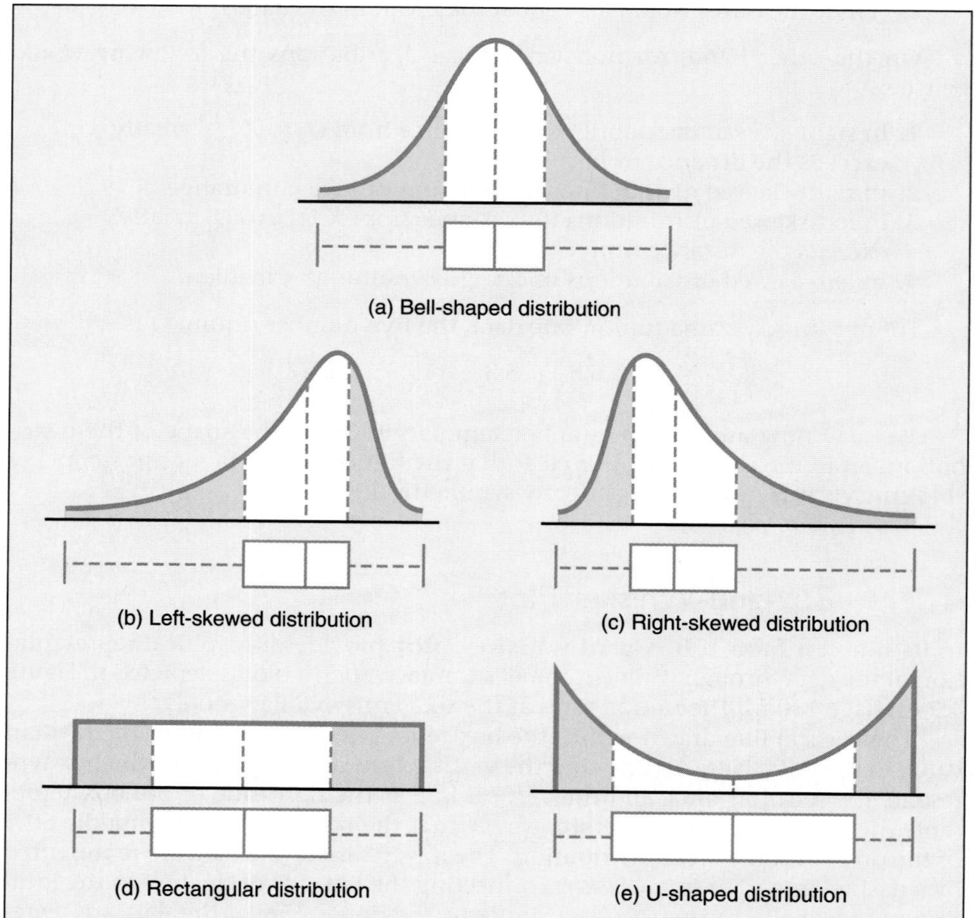

Figure 4.8
Five hypothetical distributions examined through their box-and-whisker plots and their corresponding polygons.
Note: Areas under polygon are split into quartiles corresponding to the five-number summary for the box-and-whisker plots.

When a data batch is perfectly symmetrical, as would be the case in panels (a), (d), and (e), the length of the left whisker will equal the length of the right whisker and the median line will divide the box in half. In practice, it is unlikely that we will observe a data batch that is perfectly symmetrical. However, we should be able to state that our data batch is approximately symmetrical if the lengths of the two whiskers are almost equal and the median line almost divides the box in half.

On the other hand, when our data batch is distinctly left-skewed or right-skewed, as respectively presented in panels (b) and (c) of Figure 4.8, the whisker lengths may vary considerably and the median line will not likely be centered in the box. In panel (b), for example, the skewed (that is, distorted) nature of the data batch indicates that there is a heavy clustering of observations at the high end of the scale (that is, the right side); 75% of all data values are found between the left edge of the box (Q_1) and the end of the right whisker ($X_{largest}$). Therefore the long left whisker contains the distribution of only the smallest 25% of the observations, demonstrating the distortion from symmetry in this data batch.

When observing a right-skewed data batch, as in panel (c) of Figure 4.8, the concentration of data points will be on the low end of the scale (that is, the left side of the box-and-whisker plot). Here, 75% of all data values are found between the beginning of the left whisker ($X_{smallest}$) and the right edge of the box (Q_3), and the remaining 25% of the observations are dispersed along the long right whisker at the upper end of the scale.

Problems for Section 4.7

4.37 Using the battery life data from Problem 4.5 on page 115: Data file FLASHBAT.DAT
(a) List the five-number summary.
(b) Form the box-and-whisker plot and describe the shape.
(c) Compare your answer in (b) with that from Problem 4.29 on page 127. Discuss.

4.38 Using the lawn mower price data (excluding the rear-bagging unit) from Problem 4.7 on page 115: Data file MOWER.DAT
(a) List the five-number summary.
(b) Form the box-and-whisker plot and describe the shape.
(c) Compare your answer in (b) with that from Problem 4.30 on page 128. Discuss.

● 4.39 Using the grants data from Problem 4.10 on page 116: Data file ADAMHA.DAT
(a) List the five-number summary.
(b) Form the box-and-whisker plot and describe the shape.
(c) Compare your answer in (b) with that from Problem 4.31 on page 128. Discuss.

4.40 Using the train "lateness" data from Problem 4.11 on page 117: Data file TRAIN1.DAT
(a) List the five-number summary.
(b) Form the box-and-whisker plot and describe the shape.
(c) Compare your answer in (b) with that from Problem 4.32 on page 128. Discuss.

4.41 Using the water consumption data from Problem 4.12 on page 117: Data file WATER.DAT
(a) List the five-number summary.
(b) Form the box-and-whisker plot and describe the shape.
(c) Compare your answer in (b) with that from Problem 4.33 on page 128. Discuss.

Interchapter Problems for Section 4.7

Data file MAILORD.DAT

4.42 Using the data on amount owed to the mail-order book company from Problem 3.2 on page 58:
(a) List the five-number summary.
(b) Form the box-and-whisker plot and describe the shape.
(c) Compare your answer in (b) with that from Problem 4.34 on page 128. Discuss.

Data file SHOWER.DAT

4.43 Using the data on maximum flow rates for shower heads from Problem 3.3 on page 58:
(a) List the five-number summary.
(b) Form the box-and-whisker plot and describe the shape.
(c) Compare your answer in (b) with that from Problem 4.35 on page 128. Discuss.

Data file UTILITY.DAT

● 4.44 Using the data on electric and gas utility charges from Problem 3.12 on page 66:
(a) List the five-number summary.
(b) Form the box-and-whisker plot and describe the shape.
(c) Compare your answer in (b) with that from Problem 4.36 on page 128. Discuss.

4.8 Calculating Descriptive Summary Measures from a Population

In Sections 4.4 through 4.7 we examined various *statistics* that are utilized to summarize or describe numerical information from a *sample.* In particular, we used these statistics to describe the properties of central tendency, variation, and shape for the tuition rate data obtained from the sample of $n = 6$ Pennsylvania schools. Suppose, however, that our research analyst at the college advisory service company now wishes to conduct a more thorough investigation of the tuition rates charged (in thousands of dollars) to out-of-state residents in each of the 90 colleges and universities in the state of Pennsylvania (that is, the *population*). The resulting measures (that is, the *parameters*) computed from the population of $N = 90$ Pennsylvania schools to summarize and describe the properties of central tendency, variation, and shape could then be used by our research analyst when writing a report to the marketing manager of the college advisory service company comparing and contrasting differences in such tuition charges across regions of the United States.

4.8.1 Population Measures of Central Tendency

● **The Population Mean** The **population mean** is given by the symbol μ_x, the Greek lowercase letter mu subscript X. That is,

$$\mu_x = \frac{\sum_{i=1}^{N} X_i}{N} \tag{4.11}$$

where

N = population size

X_i = ith value of the random variable X

$\sum_{i=1}^{n} X_i$ = summation of all X_i values in the population

● **The Population Median, Mode, Midrange, and Midhinge** The median, mode, midrange, and midhinge for a population of size N are respectively obtained as previously described in Sections 4.4.2 through 4.4.5 for a sample of size n. We merely replace the symbol n with N.

4.8.2 Population Measures of Variation

● **The Population Range and Interquartile Range** The range and interquartile range for a population of size N are respectively obtained as previously described in Sections 4.5.1 and 4.5.2 for a sample of size n.

● **The Population Variance and Standard Deviation** The **population variance** is given by the symbol σ_x^2, the Greek lowercase letter sigma subscript X squared, and the **population standard deviation** is given by the symbol σ_x. That is,

$$\sigma_x^2 = \frac{\sum_{i=1}^{N} (X_i - \mu_x)^2}{N} \tag{4.12}$$

where

N = population size

X_i = ith value of the random variable X

$\sum_{i=1}^{N} X_i$ = summation of all X_i values in the population

$\sum_{i=1}^{N} (X_i - \mu_x)^2$ = summation of all the squared differences between the X_i values and μ_x

and

$$\sigma_x = \sqrt{\frac{\sum_{i=1}^{N} (X_i - \mu_x)^2}{N}} \tag{4.13}$$

We note that the formulas for the population variance and standard deviation differ from those for the sample in that $(n - 1)$ in the denominator of S^2 and S [see Equations (4.6) and (4.7)] is replaced by N in the denominator of σ_x^2 and σ_x.

● **The Population Coefficient of Variation** The **population coefficient of variation**, given by the symbol CV_{pop}, measures the scatter in the data relative to the mean. It may be computed by

$$CV_{pop} = \left(\frac{\sigma_x}{\mu_x}\right)100\% \qquad (4.14)$$

where σ_x = standard deviation in the population

μ_x = arithmetic mean in the population

4.8.3 Results

The raw data of tuition rates charged (in thousands of dollars) at all $N = 90$ colleges and universities in the state of Pennsylvania are presented in Special Data Set 1 of Appendix D on pages D4–D5. From these data, the following revised stem-and-leaf display is obtained (Figure 4.9):

Data file C&UPENN.DAT

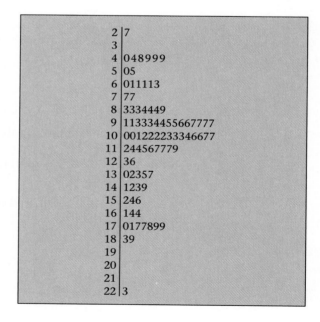

2	7
3	
4	048999
5	05
6	011113
7	77
8	3334449
9	113334455667777
10	001222233346677
11	244567779
12	36
13	02357
14	1239
15	246
16	144
17	0177899
18	39
19	
20	
21	
22	3

Figure 4.9
Revised stem-and-leaf display of out-of-state resident tuition rates at 90 Pennsylvania colleges and universities.
Source: Special Data Set 1 of Appendix D, pages D4–D5.

Using the raw data or the data arranged into the stem-and-leaf display, the following summary measures are obtained:

- **Mean**

$$\mu_x = \frac{\sum_{i=1}^{N} X_i}{N} = \frac{979.8}{90} = 10.89 \text{ thousand dollars}$$

- **Median**

$$\text{Positioning point} = \frac{N+1}{2} \text{ ordered observation}$$

$$= \frac{90+1}{2} = 45.5\text{th ordered observation}$$

To obtain the median we simply count (left to right, row by row) to the 45th and 46th ordered observations and take the average. In our data, these observations are found in the row with a "stem" of 10. The respective "leaves" are 2 and 2—corresponding to tuition rates of 10.2 and 10.2 thousand dollars. Thus the median is (10.2 + 10.2)/2 = 10.20 thousand dollars.

- **Mode** The most frequently observed tuition rates charged to out-of-state residents by colleges and universities in Pennsylvania are 6.1, 9.7, and 10.2 thousand dollars. The data are multimodal.

- **Midrange**

$$\frac{X_{smallest} + X_{largest}}{2} = \frac{2.7 + 22.3}{2} = 12.50 \text{ thousand dollars}$$

- **Q_1**

$$\text{Positioning point} = \frac{N+1}{4} \text{ ordered observation}$$

$$= \frac{90+1}{2} = 22.75\text{th ordered observation}$$

$$\cong 23\text{rd ordered observation}$$

To obtain Q_1 we simply count (left to right, row by row) to the 23rd ordered observation. In our data, the "leaf" is 4, branching off the "stem" of 8. Therefore $Q_1 = 8.40$ thousand dollars.

- **Q_3**

$$\text{Positioning point} = \frac{3(N+1)}{4} \text{ ordered observation}$$

$$= \frac{273}{4} = 68.25\text{th ordered observation}$$

$$\cong 68\text{th ordered observation}$$

To obtain Q_3 we simply count (left to right, row by row) to the 68th ordered observation. In our data, the "leaf" is 3, branching off the "stem" of 13. Therefore $Q_3 =$ 13.30 thousand dollars.

- **Midhinge**

$$\frac{Q_1 + Q_3}{2} = \frac{8.4 + 13.3}{2} = 10.85 \text{ thousand dollars}$$

- **Range**

$$X_{largest} - X_{smallest} = 22.3 - 2.7 = 19.60 \text{ thousand dollars}$$

- **Interquartile Range**

$$Q_3 - Q_1 = 13.3 - 8.4 = 4.90 \text{ thousand dollars}$$

- **Variance**

$$\sigma_x^2 = \frac{\sum_{i=1}^{N}(X_i - \mu_x)^2}{N} = \frac{(14.9 - 10.89)^2 + (16.4 - 10.89)^2 + \cdots + (4.8 - 10.89)^2}{90}$$

$$= 15.594 \text{ (in squared thousands of dollars)}$$

- **Standard Deviation**

$$\sigma_x = \sqrt{\sigma_x^2} = \sqrt{15.594} = 3.95 \text{ thousand dollars}$$

- **Coefficient of Variation**

$$CV_{pop} = \left(\frac{\sigma_x}{\mu_x}\right)100\% = \left(\frac{3.95}{10.89}\right)100\% = 36.3\%$$

4.8.4 Shape

The shape of the population is obtained through a *relative* comparison of the mean and the median, supported by an evaluation of the five-number summary and box-and-whisker plot.

The five-number summary is

$X_{smallest}$	Q_1	median	Q_3	$X_{largest}$
2.70	8.40	10.20	13.30	22.30

and the corresponding box-and-whisker plot is displayed in Figure 4.10 as follows:

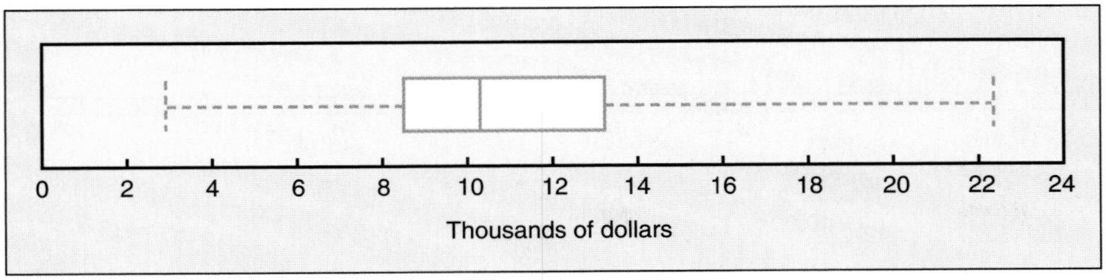

Figure 4.10
Box-and-whisker plot of out-of-state resident tuition charges at 90 Pennsylvania schools.

Among the 90 colleges and universities in the state of Pennsylvania, the population of tuition rates charged to out-of-state residents can be considered as right-skewed in shape because the mean (10.89 thousand dollars) exceeds the median (10.20 thousand dollars). Similar conclusions are drawn from the analysis of the box-and-whisker plot depicted in Figure 4.10.

4.8.5 Summarizing the Findings from the Sample and Population

Table 4.1 summarizes the results of utilizing the various descriptive measures we have investigated in this chapter.

Table 4.1 Using descriptive measures on two data batches.

	Tuition Rates	
Descriptive Measure	Sample ($n = 6$)	Population ($N = 90$)
Mean	8.30	10.89
Median	8.30	10.20
Mode	No mode	Multimodal
$X_{smallest}$	4.90	2.70
$X_{largest}$	11.70	22.30
Midrange	8.30	12.50
Q_1	6.30	8.40
Q_3	10.30	13.30
Midhinge	8.30	10.85
Range	6.80	19.60
Interquartile range	4.00	4.90
Variance	6.368	15.594
Standard deviation	2.52	3.95
Coefficient of variation	30.4%	36.3%
Shape	Symmetrical	Right-skewed

We observe that the various statistics computed from the sample of size six appear to differ from the corresponding characteristics obtained from the population of size 90. Why this has happened, however, is simply a function of chance. When drawing the random sample, our research analyst at the college advisory service company properly used a table of random numbers (Table E.1), as discussed

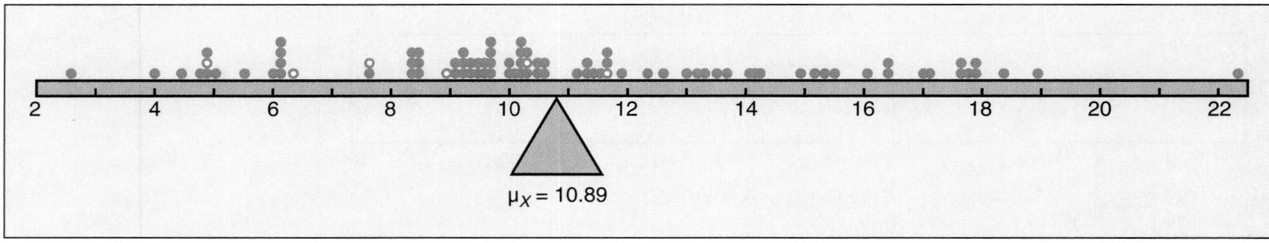

Figure 4.11
Dot scale showing the tuition rates charged (in $000) for 90 Pennsylvania schools.
Source: Figure 4.9.

in Section 2.7. Unfortunately, owing to the small size of the sample and purely by chance, the tuition rates charged by the selected colleges and universities are fairly homogeneous and fail to account for the range in tuition rates that exists in the entire population of 90 schools. This is clearly depicted on the dot-scale diagram in Figure 4.11. The sample data were not right-skewed because not one of the six selected schools had a tuition rate for out-of-state residents (light dots) that was among the highest 30% in the population of schools.

4.8.6 Using the Standard Deviation: The Empirical Rule

In most data batches a large portion of the observations tend to cluster somewhat near the median. In right-skewed data batches this clustering occurs to the left of (that is, below) the median and in left-skewed data batches the observations tend to cluster to the right of (that is, above) the median. In symmetrical data batches, where the median and mean are the same, the observations tend to distribute equally around these measures of central tendency. When extreme skewness is not present and such clustering is observed in a data batch, we can use the so-called *empirical rule* to examine the property of data variability and get a better sense of what the standard deviation is measuring.

> The **empirical rule** states that for most data batches we will find that roughly two out of every three observations (that is, 67%) are contained within a distance of 1 standard deviation around the mean and roughly 90 to 95% of the observations are contained within a distance of 2 standard deviations around the mean.

Hence the standard deviation, as a measure of average variation around the mean, helps us to understand how the observations distribute above and below the mean and helps us to focus on and flag unusual observations (that is, outliers) when analyzing a batch of numerical data.

4.8.7 Using the Standard Deviation: The Bienaymé-Chebyshev Rule

More than a century ago, the mathematicians Bienaymé and Chebyshev (Reference 4) independently examined the property of data variability around the mean.[4] They found that regardless of how a batch of data is distributed, the percentage of observations that are contained within distances of k standard deviations around the mean must be at least

$$\left(1 - \frac{1}{k^2}\right)100\%$$

Therefore, for data with any shape whatsoever

- At least $[1 - (1/2^2)]100\% = 75.0\%$ of the observations must be contained within distances of ±2 standard deviations around the mean.
- At least $[1 - (1/3^2)]100\% = 88.89\%$ of the observations must be contained within distances of ±3 standard deviations around the mean.
- At least $[1 - (1/4^2)]100\% = 93.75\%$ of all the observations must be included within distances of ±4 standard deviations about the mean.

Although the Bienaymé-Chebyshev rule is general in nature and applies to any kind of distribution of data, we will see in Chapter 8 that if the data form the "bell-shaped" normal or Gaussian distribution, 68.26% of all the observations will be contained within distances of ±1 standard deviation around the mean, while 95.44%, 99.73%, and 99.99% of the observations will be included, respectively, within distances of ±2, ±3, and ±4 standard deviations around the mean. These results (among others) are summarized in Table 4.2.

Table 4.2 How data vary around the mean.

	Percentage of Observations Contained Between the Mean and k Standard Deviations Based On		
Number of Standard Deviation Units k	Bienaymé-Chebyshev Rule for Any Distribution	Gaussian Distribution	Pennsylvania School Data
1	Not calculable	Exactly 68.26%	Exactly 64.4%
2	At least 75.00%	Exactly 95.44%	Exactly 96.7%
3	At least 88.89%	Exactly 99.73%	Exactly 100.0%
4	At least 93.75%	Exactly 99.99%	Exactly 100.0%

Specifically, if we knew that a particular random phenomenon followed the pattern of the *bell-shaped* distribution—as many do, at least approximately—we would then know (as will be shown in Chapter 8) *exactly* how likely it is that any particular observation was close to or far from its mean. Generally, however, for any kind of distribution, the **Bienaymé-Chebyshev rule** tells us *at least* how likely it must be that any particular observation falls within a given distance about the mean.

From Table 4.1 we recall that for the population of 90 Pennsylvania colleges and universities, the mean tuition rate charged to out-of-state residents, μ_x, is 10.89 thousand dollars and the standard deviation, σ_x, is 3.95 thousand dollars. From the revised stem-and-leaf display (Figure 4.9 on page 134) we note that 58 out of 90 schools (64.4%) had a tuition rate between $\mu_x - 1\sigma_x$ and $\mu_x + 1\sigma_x$ (that is, between 6.94 and 14.84 thousand dollars). Moreover, we see that 87 out of 90 schools (96.7%) had a tuition rate between $\mu_x - 2\sigma_x$ and $\mu_x + 2\sigma_x$ (that is, between 2.99 and 18.79 thousand dollars). Finally, we note that all 90 schools (100%) had a tuition rate within $\mu_x - 3\sigma_x$ and $\mu_x + 3\sigma_x$ (that is, between 0 and 22.74 thousand dollars).[5] It is interesting to note that even though the tuition rate data are right-skewed in shape, the percentages of colleges and universities with tuition rates falling within one or more standard deviations about the mean are not very different from what would be expected had the data been distributed as a symmetrical, bell-shaped Gaussian distribution.

Problems for Section 4.8

● 4.45 Given the following batch of data for a population of size $N = 10$:

7 5 11 8 3 6 2 1 9 8

(a) Compute the mean, median, mode, midrange, and midhinge.
(b) Compute the range, interquartile range, variance, standard deviation, and coefficient of variation.
(c) Are these data skewed? If so, how?

4.46 Given the following batch of data for a population of size $N = 10$:

7 5 6 6 6 4 8 6 9 3

(a) Compute the mean, median, mode, midrange, and midhinge.
(b) Compute the range, interquartile range, variance, standard deviation, and coefficient of variation.
(c) Are these data skewed? If so, how?
(d) Compare the measures of central tendency to those of Problem 4.45(a). Discuss.
(e) Compare the measures of variation to those of Problem 4.45(b). Discuss.

Data file TAX.DAT

4.47 The following data represent the quarterly sales tax receipts (in $000) submitted to the comptroller of Gmoserville Township for the period ending March 1994 by all 50 business establishments in that locale:

10.3	11.1	9.6	9.0	14.5
13.0	6.7	11.0	8.4	10.3
13.0	11.2	7.3	5.3	12.5
8.0	11.8	8.7	10.6	9.5
11.1	10.2	11.1	9.9	9.8
11.6	15.1	12.5	6.5	7.5
10.0	12.9	9.2	10.0	12.8
12.5	9.3	10.4	12.7	10.5
9.3	11.5	10.7	11.6	7.8
10.5	7.6	10.1	8.9	8.6

(a) Organize the data into an ordered array or stem-and-leaf display.
(b) Compute the mean, median, mode, midrange, and midhinge for this population.
(c) Compute the range, interquartile range, variance, standard deviation, and coefficient of variation for this population.
(d) Form the box-and-whisker plot and describe the shape of these quarterly sales tax receipts data.
(e) What proportion of these businesses have quarterly sales tax receipts
(1) within ±1 standard deviation of the mean?
(2) within ±2 standard deviations of the mean?
(3) within ±3 standard deviations of the mean?
(f) Are you surprised at the results in (e)? (*Hint*: Compare and contrast your findings versus what would be expected based on the empirical rule.)
(g) ACTION Assist the comptroller of this township by writing a draft of the memo that will be sent to the Governor regarding the collected receipts.

(h) How would this information be of use to the Governor? Discuss.

Interchapter Problem for Section 4.8

Data file CANCER.DAT

4.48 Refer to the cancer incidence data from Problem 3.6 on page 60:
(a) Compute the mean, median, mode, midrange, and midhinge for this population.
(b) Compute the range, interquartile range, variance, standard deviation, and coefficient of variation for this population.
(c) Form the box-and-whisker plot and describe the shape of the data.
(d) What proportion of the states have cancer incidence rates
(1) within ±1 standard deviation of the mean?
(2) within ±2 standard deviations of the mean?
(3) within ±3 standard deviations of the mean?
(e) Are you surprised at the results in (d)? (*Hint:* Compare and contrast your findings versus what would be expected based on the empirical rule.)
(f) ACTION Amend your letter to the host of the television show [Problem 3.6(c)] based on your answers to (a)–(e).

4.9 Obtaining Descriptive Summary Measures from Data Grouped into Tables and Charts

It is often necessary to obtain descriptive summary measures from data grouped into frequency distribution tables or presented in histograms, polygons, or ogives. In many cases, we obtain such distributions directly from reports published in magazines, newspapers, or professional journals. In these situations the original (raw) data are just not available. While the descriptive summary measures computed from **ungrouped data**—data in their raw form or in an ordered array or stem-and-leaf display—provide *actual* results, *approximations* for these descriptive measures can be obtained from **grouped data.**

4.9.1 Using Polygons for Comparing Grouped Data Batches

Polygons provide us with a useful visual aid for comparing two or more batches of numerical data in terms of their properties—central tendency, variation, and shape.

Figure 4.12 on page 142 depicts a perfectly symmetrical bell-shaped normal distribution. The mean, median, mode, midrange, and midhinge are theoretically identical.

Figure 4.13 on page 142 displays two identical normal distributions. Polygons A and B are superimposed on one another.

Figure 4.14 on page 142 presents two normal distributions which differ only in central tendency. The mean, median, mode, midrange and midhinge in polygon C exceed (that is, are to the right of) those for polygon A.

Figure 4.15 on page 143 demonstrates two normal distributions that differ only in variation. The range, interquartile range, variance, standard deviation, and coefficient of variation in polygon D are smaller than those for polygon A.

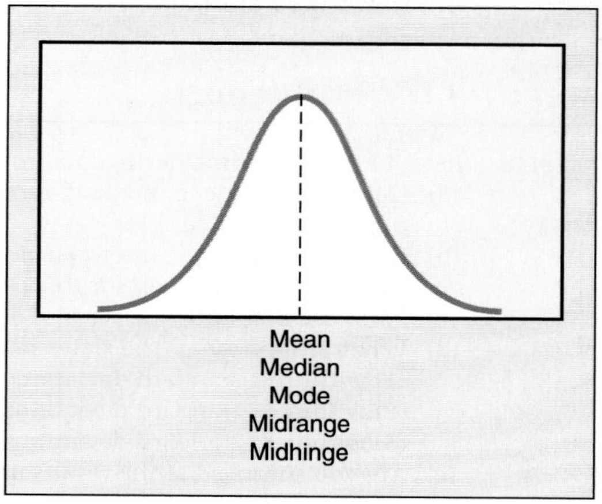

Figure 4.12
Bell-shaped curve.

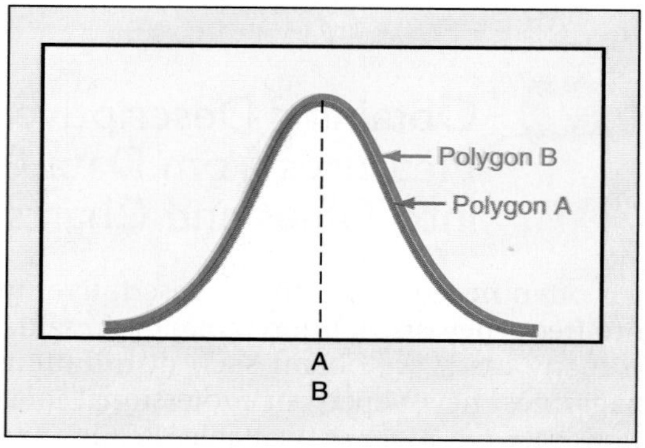

Figure 4.13
Two identical symmetrical bell-shaped normal distributions.

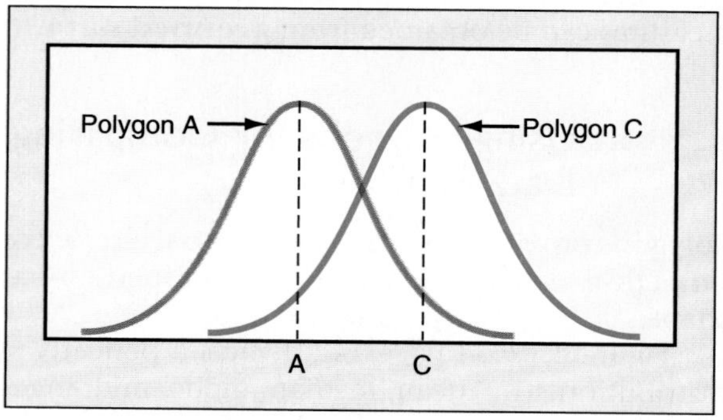

Figure 4.14
Two symmetrical bell-shaped normal distributions differing only in central tendency.

Figure 4.16 depicts three hypothetical polygons: Polygon A is a symmetrical bell-shaped normal distribution; polygon L is negative or left-skewed (since the distortion to the left is caused by extremely small values); and polygon R is positive or right-skewed (since the distortion to the right is caused by extremely large values).

The *relative positions* of the various measures of central tendency (the mean, median, mode, midrange, and midhinge) in skewed distributions can best be examined from Figures 4.17 and 4.18 on pages 143–144.

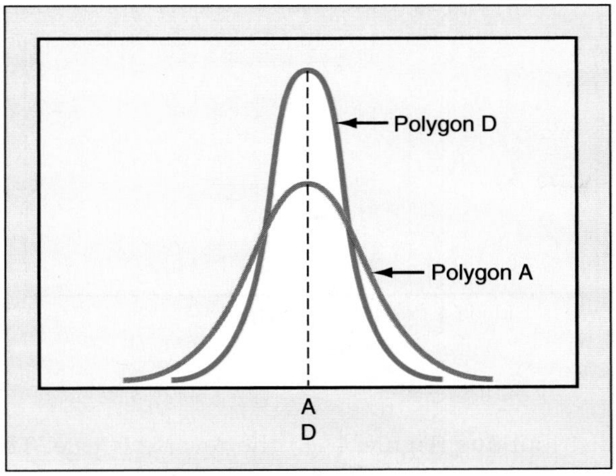

Figure 4.15
Two symmetrical bell-shaped normal distributions differing in variation.

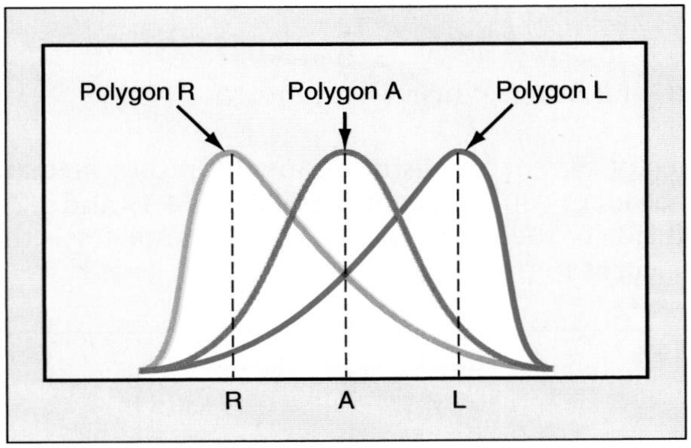

Figure 4.16
Three distributions differing primarily in shape.

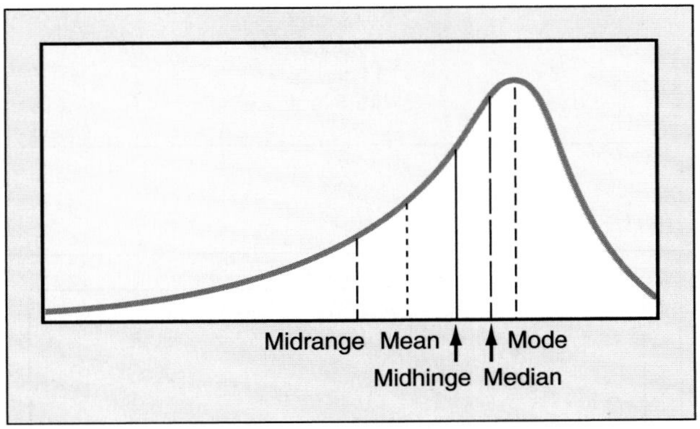

Figure 4.17
Left-skewed distribution.

In left-skewed distributions (Figure 4.17) the few extremely small observations distort the midrange and mean toward the left tail. Hence, we would expect the mode to be the largest value and the midrange to be the smallest. That is,

$$\text{midrange} < \text{mean} < \text{midhinge} < \text{median} < \text{mode}$$

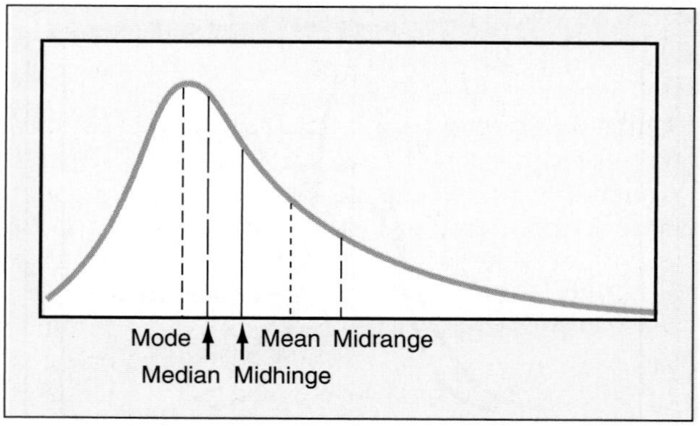

Figure 4.18
Right-skewed distribution.

However, in right-skewed distributions (Figure 4.18) the reverse is true. The few extremely large observations distort the midrange and the mean toward the right tail. Hence, we would expect the midrange to exceed all other measures. That is,

$$\text{mode} < \text{median} < \text{midhinge} < \text{mean} < \text{midrange}$$

On the other hand, in perfectly symmetric distributions the mean, median, midrange, and midhinge will all be identical. As displayed in Figures 4.19 and 4.20, the shape of the curve to the left side of these measures of central tendency is the mirror image of the shape of the curve to their right.

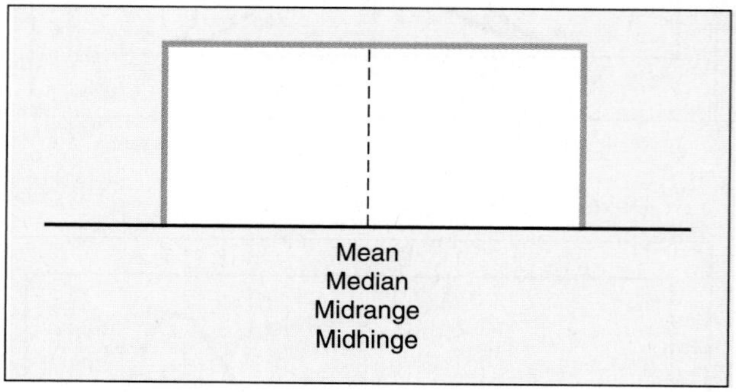

Figure 4.19
Rectangular-shaped curve.
Note: No mode.

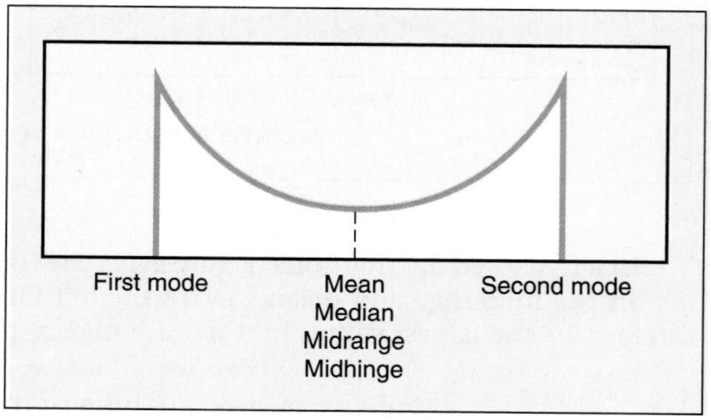

Figure 4.20
U-shaped curve.

4.9.2 Approximating Measures of Central Tendency and Variation

While various formulas exist for approximating the values of the different measures of central tendency and variation when the numerical data have been grouped into a frequency distribution table (Reference 1), it is both simpler and more convenient to use other approaches. Suppose, for example, that our research analyst at the college advisory service company uses Special Data Set 1 in Appendix D (pages D4–D5) to develop the frequency and percentage distributions (Table 4.3) and construct the percentage polygon (Figure 4.21) and percentage ogive (Figure 4.22) depicting tuition rates charged for out-of-state residents at all 90 colleges and universities in the state of Pennsylvania.

Table 4.3 Frequency distribution and percentage distribution of out-of-state resident tuition rates at 90 Pennsylvania schools.

Tuition Rates (in $000)	Number of Schools	Percentage of Schools
2.0 but less than 4.0	1	1.1
4.0 but less than 6.0	8	8.9
6.0 but less than 8.0	8	8.9
8.0 but less than 10.0	22	24.4
10.0 but less than 12.0	24	26.7
12.0 but less than 14.0	7	7.8
14.0 but less than 16.0	7	7.8
16.0 but less than 18.0	10	11.1
18.0 but less than 20.0	2	2.2
20.0 but less than 22.0	0	0.0
22.0 but less than 24.0	1	1.1
Totals	90	100.0

Source: Data are taken from Special Data Set 1 of Appendix D on pages D4–D5.

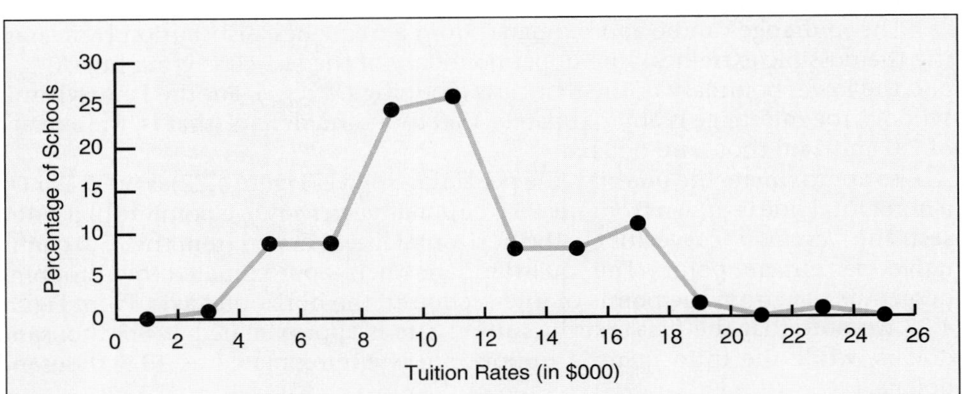

Figure 4.21
Percentage polygon of out-of-state resident tuition rates at 90 Pennsylvania schools.
Source: Data are taken from Table 4.3.

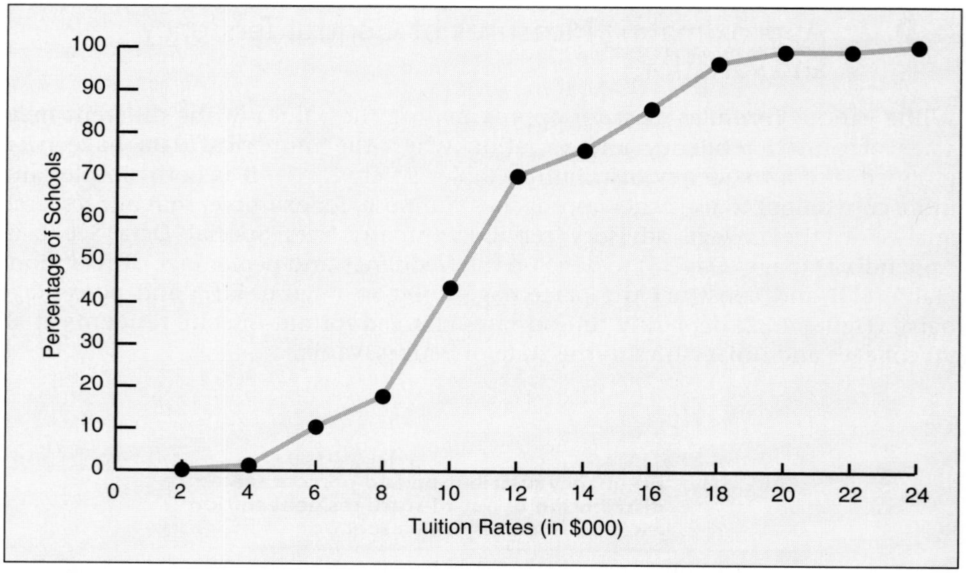

Figure 4.22
Percentage ogive of out-of-state resident tuition rates at 90 Pennsylvania schools.
Source: Data are taken from Table 4.3.

The median can be readily approximated from the percentage ogive depicted in Figure 4.22. That is, 50% of tuition rates are below what amount? To determine this, as shown in Figure 4.23, a horizontal line is drawn from the specified cumulative (50.0) percentage point until it intersects the *"less than"* curve. The median tuition rate is then approximated by dropping a perpendicular (a vertical line) from the point of intersection to the horizontal axis. From Figure 4.23 we note that the median tuition rate is approximated as 10.5 thousand dollars.

The mode can be approximated from a frequency distribution by choosing the midpoint of the class containing the most observations. This class is the most typical or **modal class.** Thus for the 90 Pennsylvania schools (Table 4.3), the modal class contains tuition rates from 10.0 to 12.0 thousand dollars and the mode is approximated as 11.0 thousand dollars.

The midrange can be approximated from a frequency distribution by averaging the possible extremes—the upper boundary of the last class grouping ($X_{largest}$) and the lower boundary of the first class grouping ($X_{smallest}$). For the Pennsylvania schools, the midrange is approximately 13.0 thousand dollars (that is, the average of 2.0 and 24.0 thousand dollars).

To approximate the quartiles the percentage ogive (Figure 4.22) is used. For Q_1, a horizontal line is drawn from the 25.0 cumulative percentage point until it intersects the *"less than"* curve; for Q_3, the horizontal line is drawn from the 75.0 cumulative percentage point. The quartiles are then approximated by dropping perpendiculars from the points of intersection to the horizontal axis. From Figure 4.23 we note that the first quartile tuition rate is approximated as 8.5 thousand dollars, while the third quartile tuition rate is approximated as 13.3 thousand dollars.

The midhinge can be approximated by averaging the quartiles. For the Pennsylvania schools, the midhinge is approximately 10.9 thousand dollars (that is, the average of 8.5 and 13.3 thousand dollars).

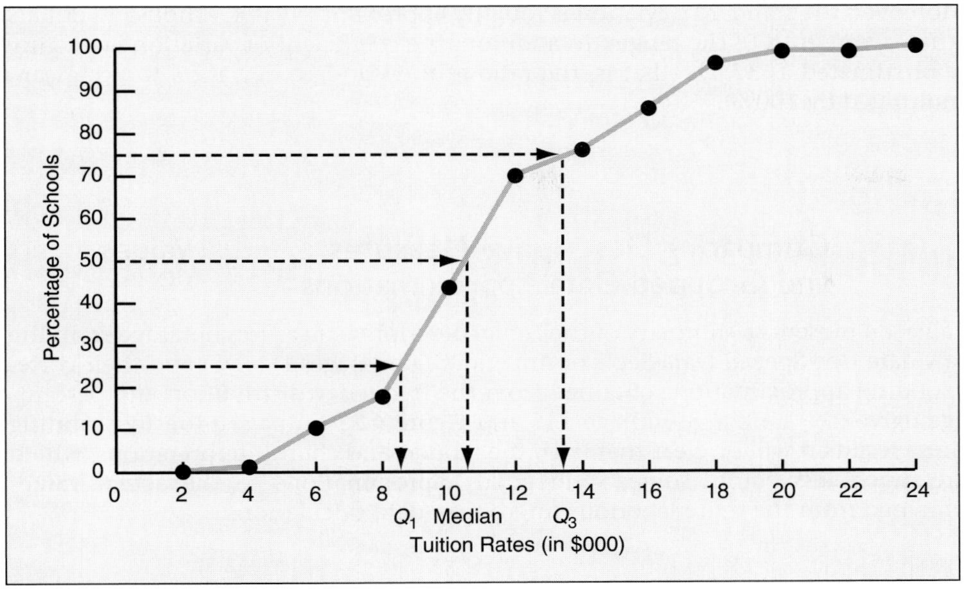

Figure 4.23
Approximating the median and the quartiles from the percentage ogive.
Source: Figure 4.22.

With data grouped into a frequency distribution, the range may be approximated as the difference between the upper boundary of the last class grouping and the lower boundary of the first class grouping. Therefore, from Table 4.3 the range in the tuition rates is approximately 22.0 thousand dollars (that is, 24.0 − 2.0).

The interquartile range can be approximated as the difference between Q_3 and Q_1. For the Pennsylvania schools the interquartile range in tuition rates is approximately 4.8 thousand dollars (that is, 13.3 − 8.5).

Unfortunately, in almost all situations, the mean, variance, standard deviation, and coefficient of variation cannot be accurately approximated once the raw data have been grouped into a frequency distribution without using special formulas (see Reference 1). In some situations, however, useful approximations can be made. For example, if a numerical data batch was perfectly symmetrical, the mean would equal the median, midrange, and midhinge; if the data were approximately symmetrical, the mean could be approximated by the average of these other central tendency measures. In addition, if the histogram or polygon appears to be that of a bell-shaped "normal" distribution, the standard deviation could be approximated as the average between one-sixth of the range and three-fourths of the interquartile range.[6]

On the other hand, if, as in Table 4.3 or Figure 4.21 for the Pennsylvania tuition rates, the frequency distribution or polygon indicates that the data are skewed but have one peak (modal class), a rougher approximation for the mean could be made by averaging the midrange and the median, while a rougher approximation for the standard deviation could be made by taking one-fifth of the range. The coefficient of variation is still defined as a measure of relative scatter around the mean, and it may be approximated from the ratio of the standard deviation to the mean.

Thus, for the grouped Pennsylvania tuition rate data, the mean is roughly approximated as 11.75 thousand dollars (that is, the average between the midrange, 13.0 thousand dollars, and the median, 10.5 thousand dollars).

Moreover, the standard deviation is roughly approximated as 4.4 thousand dollars (that is, one-fifth of the range). In addition, the coefficient of variation is roughly approximated as 37.4% (that is, the ratio of the standard deviation to the mean, multiplied by 100%).

4.9.3 Comparing Descriptive Measures: Actual Values and Grouped Data Approximations

Table 4.4 presents a summary of the actual descriptive measures obtained from the raw data (see Special Data Set 1 of Appendix D on pages D4–D5) and their corresponding approximations obtained from the frequency distribution and percentage ogive (see Table 4.3 on page 145 and Figure 4.22 on page 146). By scanning these results it will be clear that both the tabular and chart interpretations, which are much less cumbersome, yield good approximations to the actual values obtained from the more laborious ungrouped-data calculations.

Table 4.4 A comparison of actual values and grouped data approximations.

	Tuition Rates (in $000) at 90 Pennsylvania Colleges and Universities Obtained from	
Descriptive Measure	Ungrouped (Raw) Data	Grouped Data
Mean	10.89	11.75
Median	10.2	10.5
Mode	Multimodal	11.0
$X_{smallest}$	2.7	2.0
$X_{largest}$	22.3	24.0
Midrange	12.5	13.0
Q_1	8.4	8.5
Q_3	13.3	13.3
Midhinge	10.85	10.9
Range	19.6	22.0
Interquartile range	4.9	4.8
Variance	15.594	19.36
Standard deviation	3.95	4.4
Coefficient of variation	36.3%	37.4%
Shape	Right-skewed	Right-skewed

Problems for Section 4.9

Data file TAX.DAT

4.49 Refer to the quarterly sales tax receipts data in Problem 4.47 on page 140:
(a) Construct a frequency distribution and percentage distribution.
(b) Form the cumulative percentage distribution.
(c) Plot the ogive (cumulative percentage polygon).
(d) Using your tables in (a) and (b) and chart in (c):
(1) Approximate the mean, median, mode, midrange, and midhinge for this population.

(2) Approximate the range, interquartile range, standard deviation, and coefficient of variation for this population.
(3) Describe the shape of the data.

(e) ACTION Compare and contrast your approximations in (d) to that of Problem 4.47(b) and (c). Discuss.

● 4.50 Refer to the grants data in Problem 4.10 on page 116: Data file ADAMHA.DAT

(a) Construct a frequency distribution and percentage distribution.
(b) Form the cumulative percentage distribution.
(c) Plot the ogive (cumulative percentage polygon).
(d) Using your tables in (a) and (b) and chart in (c):
(1) Approximate the mean, median, mode, midrange, and midhinge for this sample.
(2) Approximate the range, interquartile range, standard deviation, and coefficient of variation for this sample.
(3) Describe the shape of the data.
(e) ACTION Compare and contrast your approximations in (d) to that of Problems 4.10 (page 116), 4.22 (page 126), 4.31 (page 128), and 4.39 (page 131). Discuss.

Interchapter Problems for Section 4.9

4.51 Refer to the data regarding the amount owed to a mail-order book company in Problem 3.2 on page 58: Data file MAILORD.DAT

(a) Construct a frequency distribution and percentage distribution.
(b) Form the cumulative percentage distribution.
(c) Plot the ogive (cumulative percentage polygon).
(d) Using your tables in (a) and (b) and chart (c):
(1) Approximate the mean, median, mode, midrange, and midhinge for this sample.
(2) Approximate the range, interquartile range, standard deviation, and coefficient of variation for this sample.
(3) Describe the shape of the data.
(e) ACTION Compare and contrast your approximations in (d) to that of Problems 4.13 (page 118), 4.25 (page 126), 4.34 (page 128), and 4.42 (page 132). Discuss.

4.52 Refer to the data on the maximum flow rate of shower heads in Problem 3.3 on page 58: Data file SHOWER.DAT

(a) Using your tables and charts in Problems 3.13 (page 66), 3.20 (page 70), 3.27 (page 73), and 3.35 (page 77):
(1) Approximate the mean, median, mode, midrange, and midhinge for this sample.
(2) Approximate the range, interquartile range, standard deviation, and coefficient of variation for this sample.
(3) Describe the shape of the data.
(b) ACTION Compare and contrast your approximations in (a) to that of Problems 4.14 (page 118), 4.26 (page 127), 4.35 (page 128), and 4.43 (page 132). Discuss.

● 4.53 Refer to the data on electric and gas utility charges in Problem 3.12 on page 66: Data file UTILITY.DAT

(a) Using your tables and charts in Problems 3.12, 3.19 (page 70), 3.26 (page 73), and 3.34 (page 77):
(1) Approximate the mean, median, mode, midrange, and midhinge for this sample.

(2) Approximate the range, interquartile range, standard deviation, and coefficient of variation for this sample.
(3) Describe the shape of the data.

(b) ACTION Compare and contrast your approximations in (a) to that of Problems 4.15 (page 118), 4.27 (page 127), 4.36 (page 128), and 4.44 (page 132). Discuss.

Data file CANCER.DAT

4.54 Refer to the cancer incidence rate data in Problem 3.6 on page 60:

(a) Using your tables and charts in Problems 3.15 (page 66), 3.22 (page 70), 3.29 (page 73), and 3.37 (page 78):
(1) Approximate the mean, median, mode, midrange, and midhinge for this population.
(2) Approximate the range, interquartile range, standard deviation, and coefficient of variation for this population.
(3) Describe the shape of the data.

(b) ACTION Compare and contrast your approximations in (a) to that of Problem 4.48 (page 141). Discuss.

Data file STOCK1.DAT

● 4.55 Refer to the book values data in Problem 3.5 on page 60:

(a) Using your tables and charts in Problems 3.14 (page 66), 3.21 (page 70), 3.28 (page 73), and 3.36 (page 78):
(1) Approximate the mean, median, mode, midrange, and midhinge for this sample.
(2) Approximate the range, interquartile range, standard deviation, and coefficient of variation for this sample.
(3) Describe the shape of the data.

(b) Using the data from Problem 3.5:
(1) Compute the actual mean, median, mode, midrange, and midhinge for this sample.
(2) Compute the actual range, interquartile range, standard deviation, and coefficient of variation for this sample.
(3) Describe the shape of the data.

(c) ACTION Compare and contrast your approximations in (a) to the actual summary measures in (b). Discuss.

4.10 Using the Computer to Obtain Descriptive Summary Measures: The Kalosha Industries Employee Satisfaction Survey

4.10.1 Introduction and Overview

When dealing with large batches of data, we may use the computer to assist us in our descriptive statistical analysis. In this section we will demonstrate how various statistical software packages can be used for characterizing and summarizing numerical data. We will gain an appreciation for the assistance the computer may give us in solving statistical problems—particularly those involving large numbers of variables or large data batches. (See References 6–9.) To accomplish this, let us return to the Kalosha Industries Employee Satisfaction Survey that was developed in Chapter 2 and for which a descriptive statistical analysis was begun in Section 3.8.2.

Data file
EMPSAT.DAT

4.10.2 Kalosha Industries Employee Satisfaction Survey

We may recall from Section 3.8.2 that Bud Conley, the Vice-President of Human Resources, is preparing for a meeting with a representative from the B & L Corporation to discuss the potential contents of an employee-benefits package that is being developed. Answers to the following two questions would be of particular concern in an initial analysis of the survey data (Table 2.3 on pages 33–40):

1. *General Question A:* What are the characteristics of the distribution of personal income among the full-time employees of Kalosha Industries (see question 7 in the Survey)?
2. *Specific Question B:* Are there gender differences in the personal incomes of full-time employees of Kalosha Industries (see questions 7 and 5 in the Survey)?

These and other initial questions raised by Bud Conley (see Survey/Database Project at the end of the section) require a detailed descriptive statistical analysis of the 400 responses to the Survey. In practice, a statistician would likely use one or two statistical packages when performing the descriptive statistical analysis. However, computer output from several packages is presented here so that we may demonstrate some of the features of these packages.

4.10.3 Using Statistical Packages for Numerical Data

In Section 3.8.3, appropriate tables and charts were presented as part of an initial response to Bud Conley's two questions. To continue with the descriptive analysis, other types of output are needed. For example, in response to question A a set of descriptive summary measures and a box-and-whisker plot would be desired. Figure 4.24 presents computer output displaying the descriptive summary measures for personal income. This output was obtained by accessing MINITAB. In addition, Figure 4.25 on page 152 depicts the box-and-whisker plot for personal income using STATISTIX.

From the computer output here and in Section 3.8.3, a response to Bud Conley's first general question can be made. The various tabular and chart presentations and the summary measures indicate that the distribution of full-time employee personal incomes is right-skewed. The mean income is 29.555 thousand dollars; the median is 26.2 thousand dollars. Moreover, although only 1.75% of the personal incomes exceed 75.0 thousand dollars, the elongation of the whisker

	N	MEAN	MEDIAN	TRMEAN	STDEV	SEMEAN
RINCOME	400	29.555	26.200	28.462	14.106	0.705

	MIN	MAX	Q1	Q3
RINCOME	10.100	91.900	18.725	37.850

Figure 4.24
Summary measures from MINITAB output.
Note: We should be familiar with all the summary measures obtained from the MINITAB output except TRMEAN (which is beyond the scope of this text) and SEMEAN (which will be studied in Chapter 9).

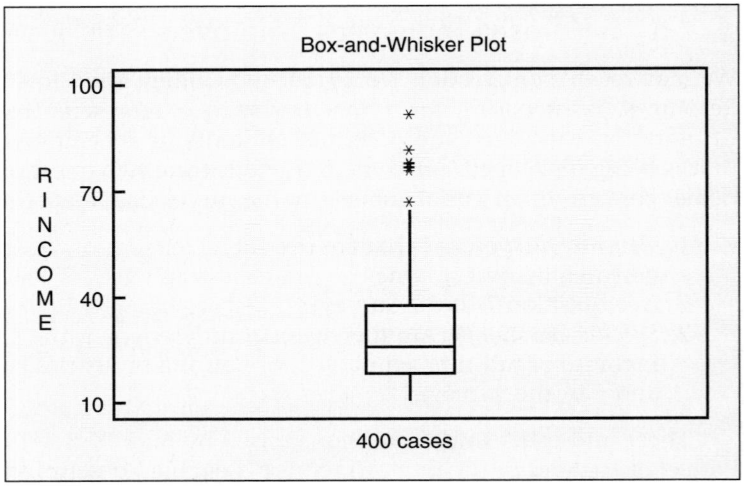

Figure 4.25
Box-and-whisker plot from STATISTIX output.
Note: There is much flexibility among the various statistical software packages with respect to the design and layout of the box-and-whisker plot. As displayed here, a box-and-whisker plot obtained by STATISTIX is printed vertically (with high values at the top of the scale) rather than horizontally (with high values on the right side of the scale). In addition, we observe that extreme values and potential outliers are flagged separately outside the whiskers of the box-and-whisker plot.

at the top part of the box-and-whisker plot (Figure 4.25) indicates that the upper 25% of employee personal incomes are found in the wide range from 37.85 to 91.9 thousand dollars. Nevertheless, a substantial majority of employee personal incomes (72.75%) fall between 15.449 thousand dollars and 43.661 thousand dollars (that is, the interval formed from $\overline{X} \pm S$). In addition (from the five-number summary), although the personal incomes range in value from 10.1 to 91.9 thousand dollars, the "midspread" or interquartile range is from 18.725 to 37.85 thousand dollars.

To respond to Bud Conley's specific question B, an evaluation of gender differences in the personal incomes of full-time employees, a sorting of the numerical responses into the two gender categories (male and female) is required. This process can be performed by accessing one of the statistical packages. Once that is achieved, types of output similar to that presented in Figures 4.24 and 4.25 would be needed for each gender grouping. To highlight this, Figure 4.26 (page 153) presents the set of descriptive summary measures of the personal incomes of full-time male and female employees and Figure 4.27 (page 154) depicts the corresponding box-and-whisker plots. The output displayed in these respective figures was obtained by accessing SAS and SPSS.

From Figures 4.26 and 4.27 here, as well as from the stem-and-leaf displays of Figure 3.18 on page 87, it is observed that while the distributions of employee personal incomes based on gender are each right-skewed in shape, the full-time male employees at Kalosha Industries have substantially higher personal incomes than do the females. The corresponding means, medians, and midhinges each indicate that, on average, the personal incomes of the male employees are 9 to 10 thousand dollars more. In addition, with respect to variation, the personal incomes of the male employees are less homogeneous than those of the female employees. As indicated in the standard deviations, the ranges, and the interquartile ranges obtained from Figure 4.26, there is substantially more variation in the personal incomes of male employees than in those of the females. However, such differences in the personal incomes of the two gender groups dissipate somewhat when a comparison of coefficients of variation is made. For the male employees the relative scatter of personal incomes around the mean is 45.3%; for the female employees, it is 40.9%.

```
------------------------------------------------------ SEX=MALES ----------------
                                                    Univariate Procedure

Variable=RINCOME

                Moments                                       Quantiles(Def=5)

N               233  Sum Wgts        233      100% Max      91.9        99%       78.3
Mean       33.70687  Sum          7853.7       75% Q3       42.7        95%       59.3
Std Dev    15.27374  Variance   233.2872       50% Med      31.2        90%       52.8
Skewness   0.915475  Kurtosis   0.934922       25% Q1         22        10%       16.1
USS        318846.2  CSS        54122.63        0% Min      10.2         5%       14.5
CV         45.31344  Std Mean   1.000616                                 1%       10.5
T:Mean=0   33.68611  Pr>|T|       0.0001      Range         81.7
Num ^= 0        233  Num > 0         233      Q3-Q1         20.7
M(Sign)       116.5  Pr>=|M|      0.0001      Mode          22.5
Sgn Rank    13630.5  Pr>=|S|      0.0001

------------------------------------------------------ SEX=FEMALES --------------
                                                    Univariate Procedure

Variable=RINCOME

                Moments                                       Quantiles(Def=5)

N               167  Sum Wgts        167      100% Max      62.8        99%       61.7
Mean       23.76287  Sum          3968.4       75% Q3       28.1        95%       40.7
Std Dev     9.70901  Variance   94.26488       50% Med      21.9        90%       36.1
Skewness   1.433697  Kurtosis   2.848284       25% Q1       16.7        10%       14.2
USS        109948.6  CSS        15647.97        0% Min      10.1         5%       11.9
CV         40.85789  Std Mean   0.751306                                 1%       10.3
T:Mean=0   31.62877  Pr>|T|       0.0001      Range         52.7
Num ^= 0        167  Num > 0         167      Q3-Q1         11.4
M(Sign)        83.5  Pr>=|M|      0.0001      Mode            15
Sgn Rank       7014  Pr>=|S|      0.0001
```

Figure 4.26
Summary measures from SAS output.
Note: As we see from the output under the headings of Moments and Quantiles, SAS provides an extensive set of summary measures, some of which we have not learned so far and others which we will not cover (see Reference 8). The summary measures of interest to us are highlighted in color.

Bud Conley was also interested in evaluating other potential gender differences with respect to hours worked, length of employment, and number of promotions. A descriptive statistical analysis based on the answers to these and other questions pertaining to the numerical variables in the Employment Satisfaction Survey (see Survey/Database Project) will provide him with a better understanding of the composition of the Kalosha Industries full-time workforce and assist him in his deliberations with the B&L Corporation toward the development of an employee-benefits package.

Survey/Database Project for Section 4.10

The following problems refer to the sample data obtained from the questionnaire of Figure 2.6 on pages 28–29 and presented in Table 2.3 on pages 33–40. They should be solved with the aid of an available computer package.

Data file EMPSAT.DAT

Suppose you are hired as a research assistant to Bud Conley, the Vice-President of Human Resources at Kalosha Industries. He has given you a list of questions (see Problems 4.56 to 4.69) that he needs answered prior to his meeting with a representative from the B&L Corporation, the employee-benefits consulting firm that he has retained.

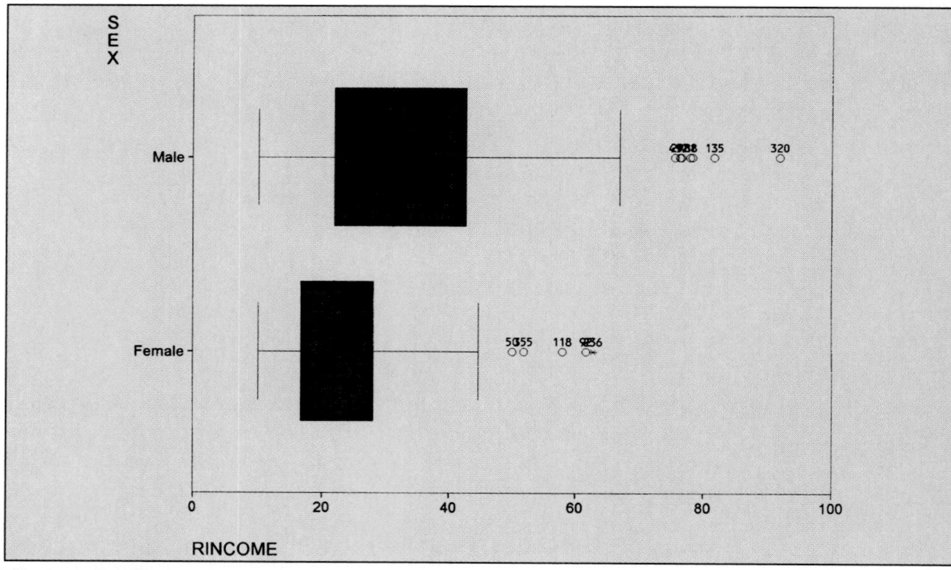

Figure 4.27
Multiple box-and-whisker plots from SPSS.
Note: We observe that the SPSS box-and-whisker plots are displayed here horizontally, with the higher income values on the right side of the scale. In addition, we see that extreme values and potential outliers are flagged separately outside the whiskers of the box-and-whisker plot. SPSS provides the respondent's number next to the flagged outliers. While this is often helpful for locating a particular respondent, when there are several outliers close together as in Figure 4.27, the respondent numbers blur and are of limited use.

From the responses to the questions dealing with numerical variables in the Employee Satisfaction Survey (see pages 33–40), in Problems 4.56 through 4.69 which follow,

(a) Obtain
   (1) the mean
   (2) the median
   (3) the midrange
   (4) the midhinge
   (5) the range
   (6) the interquartile range
   (7) the standard deviation
   (8) the coefficient of variation

(b) List the five-number-summary.

(c) Form the box-and-whisker plot.

(d) ACTION Write a memo to Bud Conley discussing your findings.

4.56 Are there differences in the personal incomes of full-time employees of Kalosha Industries based on an individual's participation in budgetary decisions (see questions 7 and 22)?

4.57 Are there differences in the personal incomes of full-time employees at Kalosha Industries based on occupational grouping (see questions 7 and 2)?

4.58 What are the characteristics of the distribution of the number of hours typically worked per week by all full-time employees of Kalosha Industries (question 1)?

4.59 Are there gender differences in the number of hours typically worked per week by all full-time employees of Kalosha Industries (see questions 1 and 5)?

4.60 Are there differences in the number of hours typically worked per week by all full-time employees of Kalosha Industries based on an individual's participation in budgetary decisions (see questions 1 and 22)?

4.61 Are there differences in the number of hours typically worked per week by all full-time employees at Kalosha Industries based on occupational grouping (see questions 1 and 2)?

4.62 What are the characteristics of the distribution of length of employment (in years) among full-time workers at Kalosha Industries (see question 16)?

4.63 Are there gender differences in the length of employment (in years) among full-time workers at Kalosha Industries (see questions 16 and 5)?

4.64 What are the characteristics of the age distribution (in years) among full-time workers at Kalosha Industries (see question 3)?

4.65 Are there gender differences in the ages of full-time workers at Kalosha Industries (see questions 3 and 5)?

4.66 What are the characteristics of the distribution of attained education (in years of formal schooling) among full-time workers at Kalosha Industries (see question 4)?

4.67 Are there gender differences in attained level of education among full-time workers at Kalosha Industries (see questions 4 and 5)?

4.68 What are the characteristics of the distribution of total family income among full-time workers at Kalosha Industries (see question 8)?

4.69 What are the characteristics of the distribution of years of full-time employment since age 16 for all full-time workers at Kalosha Industries (see question 15)?

4.70 As per the question raised in Section 4.5.4 (page 125) by Bud Conley, the Vice-President for Human Resources, is there greater variability (on a relative basis) in the amount of hours worked in a week by full-time employees (question 1) or in employee annual personal income (question 7)?

4.11 Recognizing and Practicing Proper Descriptive Summarization and Exploring Ethical Issues

In this chapter we have studied how a batch of numerical data is characterized through the computation of various descriptive summary measures regarding the properties of central tendency, variation, and shape. The next step is data analysis and interpretation; the former is *objective*, the latter is *subjective*. How *are* we going to make use of our results and how *should* we make use of our results? Does the drunkard use a lamp post primarily for support or for illumination? In a similar vein, do we use our results primarily to subjectively support a prior position or claim that we have made, or do we use our findings to objectively illuminate what the data are trying to convey?

Since a major role of the statistician is to analyze and interpret results, the computed summary measures should be used primarily to enhance data analysis and interpretation. We must avoid errors that may arise either in the objectivity of what is being analyzed or in the subjectivity of what is being interpreted (References 3 and 5).

4.11.1 Avoiding Errors in Analysis and Interpretation

We may recall that, at the beginning of this chapter (see Section 4.2), prior to learning the descriptive summary measures that characterize the three properties of numerical data (central tendency, variation, and shape), we were asked to examine and describe a set of numerical data pertaining to tuition rates charged to out-of-state residents from a sample of six Pennsylvania colleges and universities.

Thus, without a knowledge of the contents of this chapter, we attempted to analyze and interpret what the data were trying to convey.

Our analysis was *objective;* we should all have agreed with our limited visual findings: there was no typical tuition rate value; the spread in the tuition rates ranged from 4.9 to 11.7 thousand dollars; and there were no outliers present in the data. On the other hand, having now read the chapter and having now gained a knowledge about various descriptive summary measures and their strengths and weaknesses, how could we improve on our previously objective analysis? Since the data are not skewed, shouldn't we report the mean or median? Doesn't the standard deviation provide more information about the property of variation than the range? Shouldn't we describe the data batch as symmetrical in shape?

Objectivity in data analysis is reporting the most appropriate summary measures for a given data batch—those that best meet the assumptions about the given data batch. In our example we properly assumed that the data were in raw form, that is, there was no pattern to the sequence of collected data. Had this assumption been violated, we still would have been able to make objective descriptive comments as indicated here, but we would not have been able to draw inferences on the population of colleges and universities in the state of Pennsylvania; such inferences depend on the assumption that the sampled schools were randomly and independently selected. Thus, only through knowledge and awareness can good, objective data analysis take place.

On the other hand, our data interpretation was *subjective;* we could have formed different conclusions when interpreting our analytical findings. We all see the world from different perspectives. The optimist sees a glass whose volume contains 50% of water as "half full"; the pessimist sees the same glass as "half empty." Some of us will look at the ordered array of tuition rates in thousands of dollars (4.9, 6.3, 7.7, 8.9, 10.3, 11.7) and conclude that out-of-state residents attending schools in Pennsylvania pay too much; others, attending more expensive private institutions, will look at the same data batch and conclude that out-of-state residents pay too little. Thus, since data interpretation is subjective, it must be done in a fair, neutral, and clear manner.

4.11.2 Avoiding Errors in Display: Unnecessary Tabular Frills and Chart Junk

In order to analyze and interpret our data properly, we should first construct appropriate tables and charts as in Chapter 3 and then summarize the results by computing appropriate descriptive measures. All too frequently, as we turn the pages of magazines and newspapers, we find that, in order to avoid dull data presentation, tables and charts are adorned with various icons and symbols to make them attractive to their readers. Unfortunately, "enlivening" a table or chart often hides or distorts the intended message being conveyed by the data.

As examples of good and bad tabular presentation, compare Table 4.5 with Table 4.6 on pages 157 and 158, respectively. In Table 4.5, the adornment (that is, the flags representing the particular countries) actually enhances the information being conveyed.[7] This is not the case in Table 4.6, where the unneeded extra frills dampen the information being conveyed.

Note that in this latter presentation the adornment (that is, the horizontal bars) "attempts" to make a horizontal bar chart out of a table by combining the average verbal and mathematics scores. Why? Let a table be a table and a chart be

Table 4.5 **"Proper" presentation of key global economic indicators.**

World Economies

	U.S.		Japan		Germany		Britain		Canada		Mexico	
	Latest	**Prev.**	**Latest**	**Prev.**	**Latest**	**Prev.**	**Latest**	**Prev.**	**Latest**	**Prev.**	**Latest**	**Prev.**
Industrial production (monthly % change)	**0.2** Sept.	**0.1** Aug.	**1.5** Sept.	**–1.2** Aug.	**–2.0** Sept.	**2.2** Aug.	**0.1** Aug.	**0.9** July	**0.6** Aug.	**–1.1** July	**0.2** June	May
Real G.D.P. (qrt. % chg, annualized)	**2.8** III	**1.9** II	**–2.0** II	**2.2** I	**2.3** II	**–6.4** I	**2.0** II	**2.2** I	**3.4** II	**3.5** I	**0.3** II	**2.4** I
Current Account (billions, local currency)	**–26.9** II	**–22.3** I	**13.3** Sept.	**7.2** Aug.	**–7.4** Aug.	**–8.5** July	**–2.7** II	**–3.0** I	**–6.8** II	**–6.2** I	**–1.7** Aug.	**–1.9** July
Unemployment rate (% of work force)	**6.8** Oct.	**6.7** Sept.	**2.9** Sept.	**2.5** Aug.	**8.8** Oct.	**8.6** Sept.	**10.3** Sept.	**10.4** Aug.	**11.1** Oct.	**11.2** Sept.	**3.9** Aug.	**3.6** July
Consumer inflation (monthly % change)	**0.0** Sept.	**0.3** Aug.	**0.1** Sept.	**0.3** Aug.	**0.6** Oct.	**0.3** Sept.	**0.4** Sept.	**0.4** Aug.	**0.1** Sept.	**0.1** Aug.	**0.7** Sept.	**0.5** Aug.
10-year Government bond (weekly%)	**5.72**	**5.43**	**3.69**	**3.76**	**5.83**	**5.77**	**6.94**	**6.84**	**6.87**	**6.77**	**13.0**	**12.4**
Exchange rate (weekly per $)		**–**	**–**	**108.4**	**1.694**	**1.687**	**0.678**	**0.671**	**1.297**	**1.320**	**3.304**	**3.298**

Data are for the most recent period reported, compared with the previous period. G.D.P. figures for Mexico show growth over 12 months. Current account balances are reported monthly except for the U.S., Britain and Canada, which are reported quarterly; figures for Japan and Mexico are reported in billions of U.S. dollars. The Mexican unemployment rate is for urban areas and may understate the rate nationwide. Bond rates for Japan, Germany and Britain are adjusted to be consistent with U.S. and Canada; Mexican rates are for a 28-day bill. In 1993, Mexico switched to the new peso, which is the old peso divided by 1,000.

Sources: Salomon Brothers; Mexican Government; S.G. Warburg & Company; J.P. Morgan Global Research

Source: *The New York Times*, November 8, 1993, p. D2.

a chart. As we have studied in Chapter 3 (and shall see again in Section 5.3.1), charts such as these require that the "zero point" or "origin" be indicated. This does not happen here and the lengths of the bars are totally meaningless. In fact, the bars cloud the information being presented. SAT scores are measured on an interval scale (see Section 2.3.2) and both the verbal and mathematics components start at 200 points. A reader unfamiliar with such tests would not know this from the display. Thus the reader gets a distorted view of the magnitude of the differences in the average total SAT scores. Actually, some of the bars even appear to be too long! For example, the difference between bar lengths of average total SAT scores for North Dakota (1101) and New Mexico (1003) should be 1.5 times longer than that for Wisconsin (1036) and Wyoming (970). Perhaps a more appropriate and useful display of this data set would be a table having three columns—average total score, average verbal score, and average mathematics score; the rows of the table (that is, the states plus the District of Columbia and the combined national average) could be listed in descending rank order, from highest to lowest, based on average total score (Reference 2).

4.11.3 Using Computer Software for Descriptive Summary Measures

Since the descriptive summary measures are intended to enhance our data analysis and interpretation, we may use the computer to obtain these summary results. Nevertheless, we should not forget that the computer is only a tool. It is crucial that we use the computer in a manner consistent with correct statistical methodology. Remember GIGO. To interact properly with the computer, we must not only be familiar with the particular software package in use but also correctly

Table 4.6 **"Improper" presentation of SAT scores by state.**

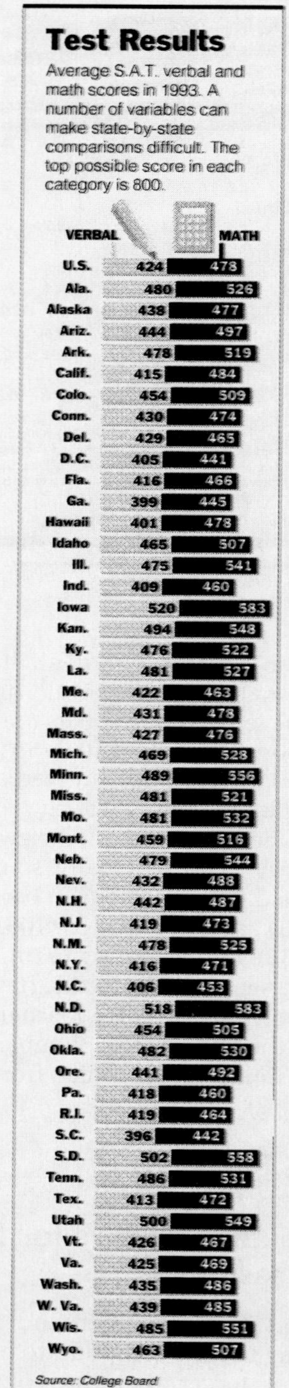

Source: *The New York Times*, August 19, 1993, p. A16.

select statistical procedures that are appropriate for the tasks at hand. For example, you should be wary of the fact that many statistical packages automatically provide, as a default, descriptive summary measures for all the variables in a particular data file—numerical and categorical. However, means, medians, standard deviations, and other descriptive summary measures should be used *only* for numerical variables. It is completely inappropriate to instruct the computer to give such summary measures for categorical variables like occupation or gender. The output would be totally meaningless.

4.11.4 Ethical Issues

Ethical issues are vitally important to all research endeavors. As daily consumers of information, we owe it to ourselves to question what we read about research studies in newspapers and magazines and what we hear on the radio or television. Over time, much skepticism has been expressed about the purpose, the focus, and the objectivity of published studies. Perhaps no comment was ever more biting than a quip attributed[8] to the famous nineteenth-century British statesman Benjamin Disraeli—*"There are three kinds of lies: lies, damned lies, and statistics."*

Again, as was mentioned in Section 3.9.6, ethical considerations arise when we are deciding what results to present in a report and what not to present. It is vitally important when conducting research to document both good and bad results, so that those who continue such research will not have to start from the very beginning. Moreover, when making oral presentations and presenting written research reports, it is essential that the results be given in a fair, objective, and neutral manner. Thus we must try to distinguish between poor presentation of results and unethical presentation. Once more, as in our prior discussions on ethical considerations, the key is *intent*. Often, when pertinent information is omitted, it is simply done out of ignorance. However, unethical behavior occurs when a researcher willfully chooses an inappropriate summary measure (for example, the mean or midrange for a very skewed batch of data) to distort the facts in order to support a particular position. In addition, unethical behavior occurs when a researcher selectively fails to report pertinent findings because it would be detrimental to the support of a particular position.

Problems for Section 4.11

4.71 You receive a telephone call from a friend who is also studying statistics this semester. Your friend has just used a statistical software package to obtain descriptive summary measures for several numerical variables pertaining to a survey concerning student life on campus. He says "I've been asked to write a report and prepare a five-minute classroom presentation on student life on campus. I'm looking at my computer printout—I've got all these descriptive summary measures for each of my seven numerical variables. There's so much information here, I just can't get started. Do you have any suggestions?" You think for a moment, and then reply..........

4.72 An arbitrator was asked to examine a dispute over salaries paid to professional baseball players. The owner of a particular team claimed that the average salary per annum was too high. The agent for the players argued that the average salary for the players on this team was too low. How should the arbitrator evaluate these two conflicting statements?

4.12 Summarizing and Describing Numerical Data: A Review

As observed on the summary chart below, this chapter was about data summarization and description. On page 104 of Section 4.1 you were given a list emphasizing the important points to be discussed in the chapter. Check over the list now to see whether you feel you have an understanding of these key points. To be sure, you should be able to answer the following conceptual questions:

1. What should we be looking for when we attempt to characterize and describe the properties of a batch of numerical data?
2. What do we mean by the property of location or central tendency?
3. What are the differences among the various measures of central tendency such as the mean, median, mode, midrange, and midhinge, and what are the advantages and disadvantages to each?

Chapter 4 summary chart.

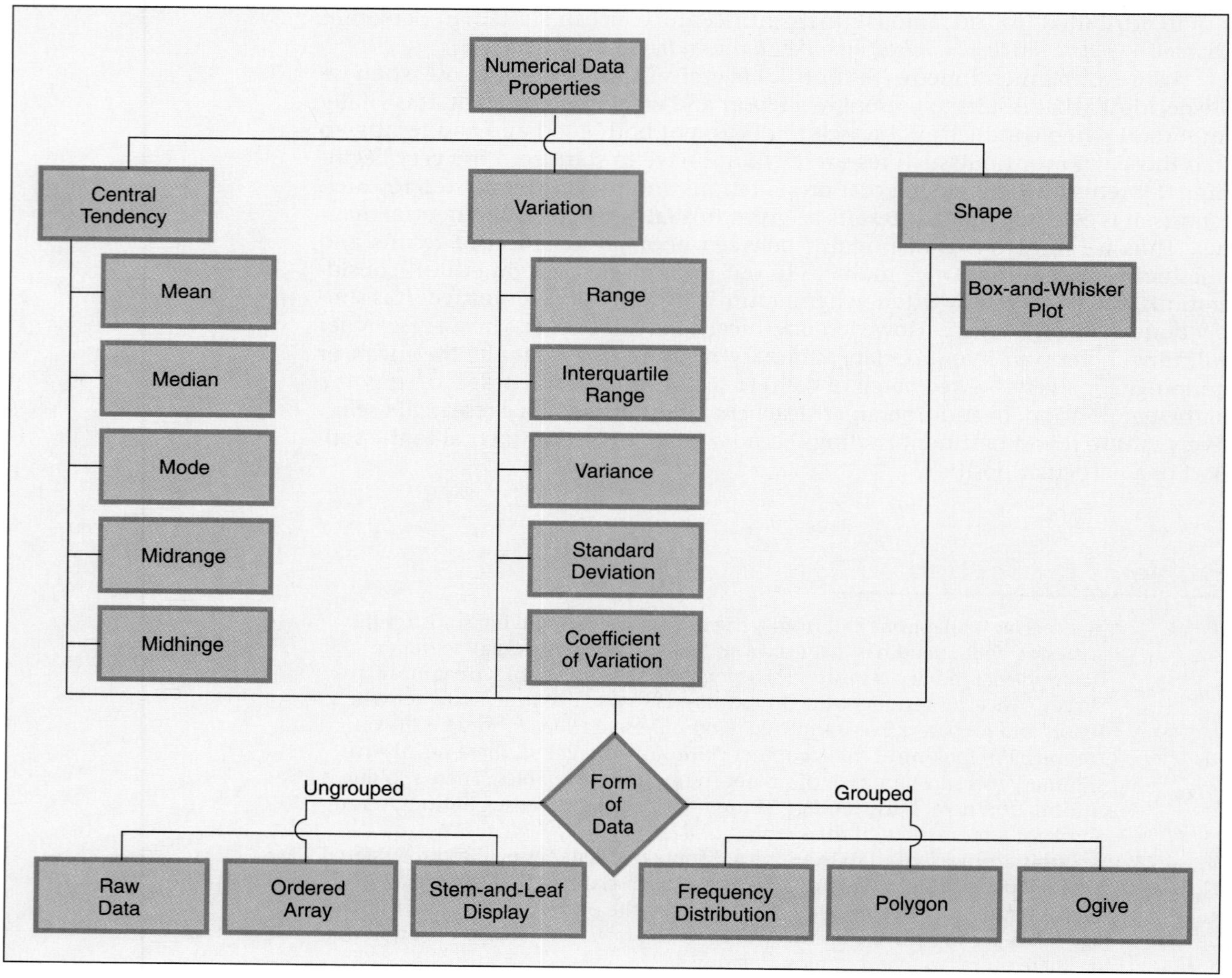

4. What is the difference between measures of central tendency and noncentral tendency?
5. What do we mean by the property of variation?
6. What are the differences among the various measures of variation such as the range, interquartile range, variance, standard deviation, and coefficient of variation, and what are the advantages and disadvantages to each?
7. How do the Bienaymé-Chebyshev and empirical rules help explain the ways in which the observations in a batch of numerical data cluster, congregate, and distribute?
8. What do we mean by the property of shape?
9. Why are such exploratory data analysis techniques as the five-number summary and box-and-whisker plot so useful?
10. How can we approximate descriptive summary measures from a frequency distribution, polygon, or ogive?
11. What are some of the ethical issues to be concerned with when distinguishing between the use of appropriate and inappropriate descriptive summary measures reported in newspapers and magazines?

Check over the list of questions to see if indeed you know the answers and could (1) explain your answers to someone who did not read this chapter and (2) give reference to specific readings or examples that support your answer. Also, reread any of the sections that may have seemed fuzzy to see if they make sense now.

Getting It All Together

Key Terms

arithmetic mean *106*
average *106*
Bienaymé-Chebyshev rule *139*
box-and-whisker plot *129*
central tendency or location *106*
coefficient of variation *124*
data analysis *104*
dot scale *107*
empirical rule *138*
five-number summary *129*
grouped data *141*
interquartile range *119*
left-skew *127*
mean *106*
median *109*
midhinge *112*
midrange *111*
midspread *119*
modal class *146*
mode *111*
outlier or extreme value *105*
population coefficient of variation *134*
population mean *132*
population standard deviation *133*
population variance *133*
properties of numerical data *106*
Q_1: first quartile *113*

Q_2: **second quartile (or median)** *113*
Q_3: **third quartile** *113*
quartiles *112*
range *118*
raw form *105*
resistant measures *114*
right-skew *127*
shape *127*
skewness *127*
standard deviation *120*
symmetry *127*
ungrouped data *141*
variance *120*
variation or dispersion *118*

Chapter Review Problems

4.73 ACTION Write a letter to a friend highlighting what you deem the most interesting or most important features of this chapter.

4.74 In your own words, explain the difference between a statistic and a parameter.

4.75 A batch of numerical data has three major properties. In your own words, define these properties and give examples of each.

Data file CORDPHON.DAT

● 4.76 The following data are the retail prices (in dollars) for a random sample of 32 corded telephone models:

44	35	55	54	78	107	45	63
45	22	36	44	50	50	60	30
39	60	25	25	25	24	46	71
60	40	22	10	20	30	12	10

Source: Copyright 1992 by Consumers Union of United States, Inc., Yonkers, N.Y. 10703. Adapted by permission from *Consumer Reports*, December 1992, pp. 780-781.

(a) Completely analyze the data.
(b) ACTION Write an article for a local newspaper dealing with consumer affairs in order to enlighten its readership on this matter.

Data file STUDIO.DAT

4.77 The following data are the monthly rental prices for a sample of 10 unfurnished studio apartments in Manhattan and a sample of 10 unfurnished studio apartments in Brooklyn Heights:

Manhattan

$955	$1000	$985	$980	$940	$975	$965	$999	$1247	$1119

Brooklyn Heights

$750	$775	$725	$705	$694	$725	$690	$745	$575	$800

(a) For each batch of data compute the mean, median, midhinge, range, interquartile range, standard deviation, and coefficient of variation.
(b) What can be said about unfurnished studio apartments renting in Manhattan versus those renting in Brooklyn Heights?
(c) ACTION How might this information be of use to an individual desiring to move into the New York City area? Write an article about this for the real estate column of your local newspaper.

4.78 Glenn Kramon's article "Coaxing the Stanford Elephant to Dance" (*The New York Times* Sunday Business Section, November 11, 1990) implies that costs at Stanford Medical Center have been driven up higher than at competing institutions because the former is more likely to treat indigent, Medicare, Medicaid, sicker, and more complex patients. To illustrate this, a chart is given that depicts a comparison of average 1989–90 hospital charges for three medical procedures (coronary bypass, simple birth, and hip replacement) at three competing institutions (El Camino, Sequoia, and Stanford).

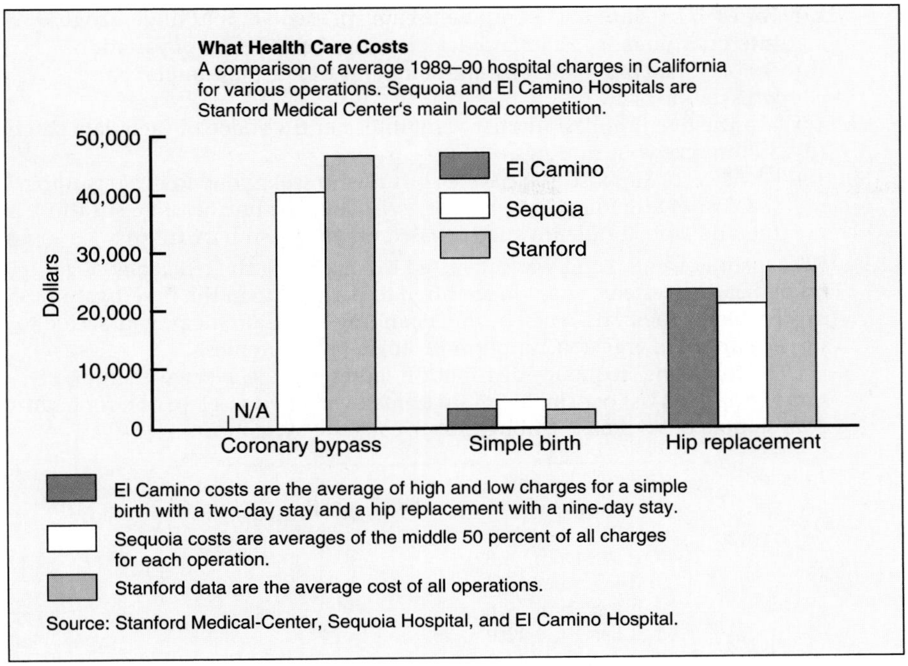

What Health Care Costs
A comparison of average 1989–90 hospital charges in California for various operations. Sequoia and El Camino Hospitals are Stanford Medical Center's main local competition.

El Camino costs are the average of high and low charges for a simple birth with a two-day stay and a hip replacement with a nine-day stay.

Sequoia costs are averages of the middle 50 percent of all charges for each operation.

Stanford data are the average cost of all operations.

Source: Stanford Medical-Center, Sequoia Hospital, and El Camino Hospital.

Your CEO knows you are currently taking a course in statistics and calls you in to discuss the article. She tells you that as she was leaving a meeting of hospital CEOs last night, one of them mentioned that this chart is totally meaningless and asked her opinion. She now requests that you prepare her response. You smile, take a deep breath, and reply..........

4.79 A college was conducting a Phonothon to raise money for the building of an Arts Center. The Provost hoped to obtain one-half million dollars for this purpose. The data below represent the amounts pledged (in $000) by all alumni who were called during the first nine nights of the campaign.

Data file FUNRAISE.DAT

16, 18, 11, 17, 13, 10, 22, 15, 16

(a) Compute the mean, median, and standard deviation.
(b) Describe the shape of this batch of data.
(c) Estimate the total amount that will be pledged (in $000) by all alumni if the campaign is to last 30 nights. (*Hint*: Total = $N\bar{X}$.)
(d) ACTION Write a memo to the Provost summarizing your findings to date and, if necessary, offering any needed recommendations.
(e) How might this information assist the Provost? Discuss.

4.80 The data below represent the tuition charged (in $ thousands) at a sample of 15 preparatory schools in the northeast and at a sample of 15 preparatory schools in the midwest during the academic year 1993–94:

Data file PREP.DAT

Northeast Prep Schools			Midwest Prep Schools		
10.5	8.9	9.6	7.9	10.6	8.4
10.1	9.3	9.1	8.2	10.1	9.2
10.0	9.7	11.2	9.1	8.5	10.7
11.0	10.4	10.5	9.3	7.5	9.5
9.8	10.0	9.9	8.8	9.3	9.8

(a) For each batch of data compute the mean, median, midhinge, range, interquartile range, standard deviation, and coefficient of variation.
(b) For each batch of data form the stem-and-leaf display and box-and-whisker plot.
(c) List the five-number summary and interpret the shape of each data batch.
(d) Summarize your findings.
(e) ACTION Suppose you have a cousin who seeks your advice regarding the cost of attending a preparatory school in the northeast versus those in the midwest. Write him a letter based on your summary in (d).

4.81 Wisconsin Power & Light was interested in improving the efficiency of gas home-heating systems and you are hired to participate in the investigation of this problem. To obtain a better understanding of the problem, you decide to survey current energy consumption in single-family homes.

The following frequency distribution represents the average energy consumption (in BTUs) per single-family home over a two-week period for a random sample of 90 homes throughout the state of Wisconsin:

Energy Consumption (BTUs)	No. of Homes
2.4 but less than 4.8	2
4.8 but less than 7.2	6
7.2 but less than 9.6	25
9.6 but less than 12.0	29
12.0 but less than 14.4	16
14.4 but less than 16.8	8
16.8 but less than 19.2	3
19.2 but less than 21.6	1
Total	90

(a) Form the appropriate tables and charts and completely analyze the data.
(b) ACTION Write a preliminary report for the chief executive officer.

Interchapter Problem

Data file ROAD.DAT

4.82 Refer to the data in Problem 3.8 (page 61) representing the amount of time (in seconds) to get from 0 to 60 mph during a road test for a sample of 22 German-made automobile models and a sample of 30 Japanese-made automobile models:

(a) Using your tables and charts in Problems 3.17 (page 66), 3.24 (page 70), 3.31 (page 73), and 3.39 (page 78):
   (1) Approximate the mean, median, mode, midrange, and midhinge for each sample.
   (2) Approximate the range, interquartile range, standard deviation, and coefficient of variation for each sample.
   (3) Describe the shape of each data batch.
(b) Using your data from Problem 3.8:
   (1) Compute the actual mean, median, mode, midrange, and midhinge for each sample.
   (2) Compute the actual range, interquartile range, standard deviation, and coefficient of variation for each sample.
   (3) Describe the shape of each data batch.
(c) ACTION Compare and contrast your approximations in (a) to the actual summary measures in (b). Discuss.
(d) ACTION Write an article for a magazine dealing with automobiles by summarizing your findings.

Collaborative Learning Mini-Case Projects

Note: The class is to be divided into groups of three or four students each. One student is initially selected to be project coordinator, another student is project recorder, and a third student is the project timekeeper. In order to give each student experience in developing teamwork and leadership skills, rotation of positions should follow each project. At the beginning of each project, the students should work silently and individually for a short, specified period of time. Once each student has had the opportunity to evaluate the issues and reflect on his/her possible answers, the group is convened and a group discussion follows. If all the members of a group agree with the solutions, the coordinator is responsible for submitting the team's project solution to the instructor with student signatures indicating such agreement. On the other hand, if one or more team members disagree with the solution offered by the majority of the team, a minority opinion may be appended to the submitted project, with signature(s) on it.

CL 4.1 Refer to CL 3.1 on page 101. Your group, the ________ Corporation, has been hired to assist the research analyst at the college advisory service company in finishing a report regarding tuition rates charged to out-of-state residents by colleges and universities in different regions of the country. In particular, using Special Data Set 1 of Appendix D on pages D1–D5 regarding the tuition rates charged to out-of-state residents at all 60 colleges and universities in the state of Texas, 45 institutions in North Carolina, and 90 schools in Pennsylvania, the ________ Corporation is ready to:

Data file C&U.DAT

(a) Outline how the group members will proceed with their tasks.
(b) Obtain various descriptive summary measures for each of these populations.
(c) Write and submit an executive summary comparing and contrasting the results across the three states.
(d) Prepare and deliver a ten-minute oral presentation to the marketing manager.

CL 4.2 Refer to CL 3.2 on page 101. Your group, the ________ Corporation, has been hired by the food editor of a popular family magazine to study the cost and nutritional characteristics of ready-to-eat cereals. Having prepared the appropriate tables and charts (see CL 3.2), the ________ Corporation is ready to enhance its preliminary analysis. Armed with Special Data Set 2 of Appendix D on pages D6–D7 displaying useful information on 84 such cereals:

Data file CEREAL.DAT

(a) Outline how the group members will proceed with their tasks.
(b) Obtain various descriptive summary measures on cost, weight, calories, and sugar (in grams per serving)— broken down by type of cereal.
(c) Write and submit an executive summary describing the results.
(d) Prepare and deliver a ten-minute oral presentation to the food editor of the magazine.

CL 4.3 Refer to CL 3.3 on page 102. Your group, the ________ Corporation, has been hired by the marketing director of a manufacturer of well-known men's and women's fragrances to study the characteristics of currently available fragrances. The results of your group's endeavors are to enable the manufacturer to make pricing decisions regarding a new product line that is being planned for distribution in the upcoming holiday season. Having prepared the appropriate tables and charts (see CL 3.3), the ________ Corporation is ready to enhance its preliminary analysis. Armed with Special Data Set 3 of Appendix D on pages D8–D9 displaying useful information on the cost per ounce of 83 such fragrances:

Data file FRAGRAN.DAT

(a) Outline how the group members will proceed with their tasks.
(b) Obtain various descriptive summary measures on the cost per ounce based on:
    (1) Product gender (women's versus men's)
    (2) Type of fragrance (perfume, cologne, or "other")
    (3) Intensity (very strong, strong, medium, or mild)
(c) Write and submit an executive summary describing the results.
(d) Prepare and deliver a ten-minute oral presentation to the marketing director.

Data file CAMERA.DAT

CL 4.4 Refer to CL 3.4 on page 102. Your group, the ________ Corporation, has been hired by the travel editor of a well-known newspaper who is preparing a feature article on compact 35-mm cameras. Having prepared the appropriate tables and charts (see CL 3.4), the ________ Corporation is ready to enhance its preliminary analysis. Armed with Special Data Set 4 of Appendix D on pages D10–D11 displaying useful information on 59 35-mm cameras:
(a) Outline how the group members will proceed with their tasks.
(b) Obtain various descriptive summary measures on such camera features as price, weight, smallest field, range, framing accuracy, and battery life—each broken down by type of 35-mm camera.
(c) Write and submit an executive summary describing the results.
(d) Prepare and deliver a ten-minute oral presentation to the travel editor.

Case Study B—Campus Cafeteria Nutrition Study

Ann Foster, Vice-President for Student Services at a rural liberal arts college, held a meeting with Camille Neller, the newly appointed Director of Food Services, and Dr. Edwina Foxe, Professor of Nutrition, regarding a series of student and parent complaints over the menu offered by the college's cafeteria. Since freshmen were obligated to purchase a meal plan requiring a minimum of two meals per day at the college's cafeteria, there was some concern expressed that the menu did not always offer an inexpensive, quick , and wholesome meal. When asked by Vice-President Foster to respond to these comments, Mrs. Neller stated that she had been on campus only three weeks and had been primarily following the menu provided by her predecessor while experimenting with a gourmet meal selection each day. "Now that these concerns have been called to my attention, I wish to pursue another course," she said. "Considering the fact that the college was situated in a rural area and that, in particular, canned food products needed to be stored for the winter months when deliveries from the nearest town might be delayed, I wish to study the nutritional content of canned soup because this item could easily be made available for all lunch and dinner meals and may even provide the nutrients for the wholesome, inexpensive, quick meal that was requested." Dr. Foxe agreed that such a study would be useful and should provide the necessary information to make a decision regarding implementation. Vice-President Foster asked Dr. Foxe to direct the study and to report her findings to Mrs. Neller in two weeks. The Vice-President offered to support the effort: "Do what it takes," she said, "we must show the students and their parents that we are responsive to their needs." Dr. Foxe requested that a student assistant be provided. "Whomever you want to hire," responded the Vice-President.

Dr. Foxe has hired you to assist her in this study and has provided you with the following data on 47 different canned soup products to investigate their nutritional value.

Nutritional characteristics of 47 different canned soups.

Brand	Product	Type	Cost	Calories	Fat	Calories From Fat	Sodium
Campbell's Homestyle	CN	CC	.35	60	2	30	880
Progresso	CN	CR	.66	75	2	24	730
Campbell's	CN	CC	.18	60	2	30	870
Nissin Cup O'Noodles	CN	DI	.33	170	8	42	970
Progresso Healthy Class.	CN	CR	.77	80	2	23	460
Lipton Soup Mix	CN	DC	.21	80	2	23	700
Campbell's Ramen Noodle	CN	DC	.09	190	8	38	970
Nissin Tip Ramen	CN	DC	.11	200	9	41	960
Campbell's Soup Mix	CN	DC	.26	100	2	18	700
Pathmark	CN	CC	.17	60	2	30	840
ShopRite	CN	CC	.19	60	2	30	840
Maruchan Ramen	CN	DC	.09	190	9	43	780
Lady Lee	CN	CC	.19	60	2	30	840
Weight Watchers	CN	CR	.76	60	1	15	790
Knorr Chicken Flavor	CN	DC	.54	110	2	16	800
Campbell's Home Cookin'	CN	CR	.74	105	3	26	860
Hain	CN	CR	.96	110	4	33	800
Mrs. Grass Noodle Soup	CN	DC	.12	70	2	26	900
Campbell's Cup Instant	CN	DI	.48	105	3	26	1190
Lipton Cup-A-Soup Instant	CN	DI	.36	65	1	14	890
Campbell's Chunky Classic	CN	CR	.74	120	4	30	810
Campbell's Healt. Request	CN	CR	.70	80	2	23	470
Pritikin Chicken Soup	CN	CR	.97	80	1	11	180
Campbell's Low Sodium	CN	CR	.80	125	4	29	65
Healthy Choice	CN	CR	.78	95	2	19	580
Hain Vegetarian	V	CR	.83	125	3	22	670
Campbell's Home Cookin'	V	CR	.53	110	2	16	680
Campbell's Chunky	V	CR	.53	120	3	23	800
Healthy Choice Garden	V	CR	.71	105	1	9	600
Progresso Tomato	V	CR	.46	75	2	24	940
Progresso Vegetable	V	CR	.44	75	1	12	680
Healthy Choice Tomato	V	CR	.73	140	3	19	540
Campbell's Homestyle	V	CC	.34	60	2	30	880
Campbell's Home Cookin'	V	CR	.53	110	1	8	640
Campbell's Made With Beef	V	CC	.23	90	2	20	830
Health Valley Fat-Free	V	CR	.92	55	1	6	280
Campbell's Healt. Request	V	CR	.55	90	1	10	480
Pritikin	V	CR	.94	90	1	10	160
Campbell's	T	CC	.15	90	2	20	670
Campbell's Healt. Request	T	CC	.20	90	2	20	410
ShopRite	T	CC	.13	100	1	9	710
Kroger	T	CC	.14	100	1	9	630
Lady Lee	T	CC	.16	80	0	0	700
Pathmark	T	CC	.15	100	1	9	630
Vons	T	CC	.18	100	1	9	710
Health Valley Org.	T	CR	.87	75	1	12	300
Campbell's Italian	T	CC	.28	90	0	0	740

Notes: For product: CN=chicken noodle, V=vegetable,T=tomato
For type: CC=canned/condensed, CR=canned, ready to serve,
DC=dry/cook-up, DI=dry/instant
Cost in cents per 8-oz serving
Calories per 8-oz serving
Fat in grams per 8-oz serving
Calories as a percentage of fat per 8-oz serving
Sodium level in milligrams per 8-oz serving

Data file
SOUP.DAT

You decide to:

(a) Undertake a complete descriptive evaluation of all numerical variables (cost in cents, calories per 8-ounce serving, fat in grams per 8-ounce serving, calories as a percentage of fat per 8-oz serving, and sodium level in milligrams per 8-oz serving).

(b) Perform a similar evaluation comparing and contrasting each of these numerical variables based on whether the product is a chicken noodle soup or either a vegetable or tomato soup.

(c) Perform a similar evaluation comparing and contrasting each of these numerical variables based on type of soup—canned/condensed, canned and ready to serve, dry/cook-up, or dry/instant.

(d) Make recommendations regarding your findings.

In two weeks, you are scheduled to make a ten-minute oral presentation to Dr. Foxe and Mrs. Neller and to submit a written report, appending all tables and charts. In addition, you have been asked to outline a survey of the student body, with questions pertaining to likes and dislikes with respect to types of soup and various fast foods. Knowing that your findings will be of much value to your fellow students, you set out to accomplish this project.

Endnotes

1. Although the word average refers to any summary measure of central tendency, it is most often used synonymously for the mean.
2. These measures are called **quantiles.** Some of the more widely used quantiles are the **deciles** (which split the ordered data into *tenths*) and the **percentiles** (which split the ordered data into *hundredths*). For further information on these measures, see Reference 1.
3. Using the summation rules from Appendix B, we show the following proof:

$$\sum_{i=1}^{n}\left(X_i - \overline{X}\right) = 0$$

$$\sum_{i=1}^{n} X_i - \sum_{i=1}^{n} \overline{X} = 0$$

$$\sum_{i=1}^{n} X_i - n\overline{X} = 0$$

$$\sum_{i=1}^{n} X_i - \sum_{i=1}^{n} X_i = 0$$

4. The Bienaymé-Chebyshev rule can apply only to distances beyond ±1 standard deviation about the mean.
5. Here $\mu_x \pm 3\sigma_x$ yields the interval −0.96 to 22.74 thousand dollars; however, a negative tuition rate is not meaningful and we record the interval as 0 to 22.74 thousand dollars.
6. We will observe in Section 8.3 that the "practical" range of normally distributed data is six standard deviation distances. Therefore, the standard deviation is approximately one-sixth the range. Moreover, for a data batch that is normally distributed the interquartile range is 1.33 standard deviation distances. Therefore, the standard deviation is approximately three-fourths the interquartile range. With a data batch that is approximately normally distributed, the average of these two approximations would provide a close estimate for the standard deviation.
7. From Table 4.5, if we wished, we could compute, row by row for each of the given economic indicators, various descriptive summary measures across the countries listed.
8. The quip is most often attributed to Benjamin Disraeli (1804–1881), twice Prime Minister of England. However, a recent report [Woerner, D., "Who Really Said It?" *Chance*, vol. 6 (Fall 1993), p. 37] indicates that it might have been first stated by someone else.

References

1. Croxton, F., D. Cowden, and S. Klein, *Applied General Statistics,* 3d ed. (Englewood Cliffs, NJ: Prentice-Hall, 1967).
2. Ehrenberg, A. S. C., "Rudiments of Numeracy," *Journal of the Royal Statistical Society,* Series A, vol. 140 (1977), pp. 277-297.
3. Huff, D., *How to Lie with Statistics* (New York: W.W. Norton, 1954).
4. Kendall, M. G., and A. Stuart, *The Advanced Theory of Statistics,* Vol. I (London: Charles W. Griffin, 1958).
5. Kimble, G. A., *How to Use (and Misuse) Statistics* (Englewood Cliffs, NJ: Prentice-Hall, 1978).
6. *MINITAB Reference Manual Release 8* (State College, PA: MINITAB, Inc., 1992).
7. Norusis, M., *SPSS Guide to Data Analysis for SPSS-X with Additional Instructions for SPSS/PC+* (Chicago, IL: SPSS Inc., 1986).
8. *SAS User's Guide Version 6* (Raleigh, NC: SAS Institute, 1988).
9. *STATISTIX Version 4.0* (Tallahassee, FL: Analytical Software, Inc., 1992).
10. Tukey, J., *Exploratory Data Analysis* (Reading, MA: Addison-Wesley, 1977).
11. Velleman, P. F., and D. C. Hoaglin, *Applications, Basics, and Computing of Exploratory Data Analysis* (Boston, MA: Duxbury Press, 1981).

chapter 5

Presenting Categorical Data in Tables and Charts

CHAPTER OBJECTIVE

To show how to organize and most effectively present collected categorical data in the form of tables and charts.

5.1 Introduction

In Chapter 3 we learned that when collecting a large batch of numerical data, the best way to examine it is first to organize and present it in appropriate tabular and chart format. We can then extract the important features of the data from these tables and charts and use this information along with our computed descriptive summary measures from Chapter 4 to analyze the data and interpret our findings. Often, however, the data batches we collect are categorical, not numerical.

This chapter, like Chapter 3, is about data presentation. In particular, we will demonstrate how batches of categorical data can be organized and most effectively presented in the form of tables and charts in order to enhance data analysis and interpretation—two key aspects of the decision-making process. To motivate our discussion of the tabular and chart presentation of categorical data, we see from the chapter summary chart on page 193 that the type of display that we develop is dependent on the number of categorical variables we are interested in studying. If the observations in our data batch are the outcomes of one categorical variable, we will develop a summary table and a variety of charts. If our interest is in the cross-classification of the outcomes of two categorical variables, we will develop a contingency table. On the other hand, if we wish to examine how several categorical variables relate to a particular categorical variable, we will cross-classify the outcomes into a supertable.

Upon completion of this chapter, you should be able to:

1. Construct and use frequency and percentage summary tables, bar charts, pie charts, dot charts, and Pareto diagrams
2. Cross-classify data based on two categorical variables into contingency tables and interpret the results
3. Cross-classify data based on several categorical variables into supertables and interpret the results
4. Appreciate the value of using statistical software packages for presenting categorical data in tables and charts
5. Understand how to distinguish between good and bad categorical data presentation and ethical issues involved

5.2 Organizing and Tabulating Categorical Data: The Summary Table

In order to introduce the relevant ideas for Chapter 5, let us suppose that our research analyst at the college advisory service wanted to evaluate various features pertaining to colleges and universities in the state of North Carolina. Special Data Set 1 in Appendix D on page D3 displays information on the out-of-state resident tuition rate, type of institution, location of school, academic calendar, and institutional classification for each of the 45 North Carolina colleges and universities. We note that the tuition rate variable is *numerical* while the other variables are all *categorical*. In Chapter 3 we were concerned only with the former; a detailed study of the responses to the categorical variables will be undertaken here.

When dealing with categorical phenomena the observations may be tallied into *summary tables* and then graphically displayed as either *bar charts, pie charts, dot charts,* or *Pareto diagrams.*

To illustrate the development of a **summary table,** let us consider the data obtained by our research analyst on institutional classification. From Special Data Set 1 in Appendix D we see that of the 45 colleges and universities in North Carolina, 2 are classified by the College Counsel as national liberal arts schools (NLA), 16 are regional liberal arts schools (RLA), 4 are classified as national universities (NU), 22 are regional universities (RU), and 1 is a specialty school (SS). This information is presented in the accompanying frequency and percentage summary table, Table 5.1.

Table 5.1 Frequency and percentage summary table pertaining to institutional classification for 45 colleges and universities in North Carolina.

Institutional Classification	Number of Schools	Percentage of Schools
National Liberal Arts Schools (NLA)	2	4.4
Regional Liberal Arts Schools (RLA)	16	35.6
National Universities (NU)	4	8.9
Regional Universities (RU)	22	48.9
Specialty Schools (SS)	1	2.2
Totals	45	100.0

Source: Data are taken from Special Data Set 1, Appendix D, page D3.

Data file C&UNC.DAT

From Table 5.1 we may conclude that the overwhelming majority of schools in North Carolina are classified by the College Counsel as regional universities (48.9%).

5.3 Graphing Categorical Data: Bar, Pie, and Dot Charts

To express the information provided in Table 5.1 graphically, the percentage bar chart (Figure 5.1), percentage pie chart (Figure 5.2), or percentage dot chart (Figure 5.3) can be displayed. These charts are presented on page 172.

5.3.1 The Bar Chart

Figure 5.1 on page 172 depicts a percentage bar chart for the North Carolina institutional classification data presented in Table 5.1. In **bar charts,** each category is depicted by a bar, the length of which represents the frequency or percentage of observations falling into a category.

To construct a bar chart the following suggestions are made:

1. The bars should be constructed horizontally (as in Figure 5.1) when the categorized observations are the outcomes of a categorical variable. The bars should be constructed vertically when the categorized observations are the outcomes of a numerical variable.
2. All bars should have the same width (as in Figure 5.1) so as not to mislead the reader. Only the lengths may differ.

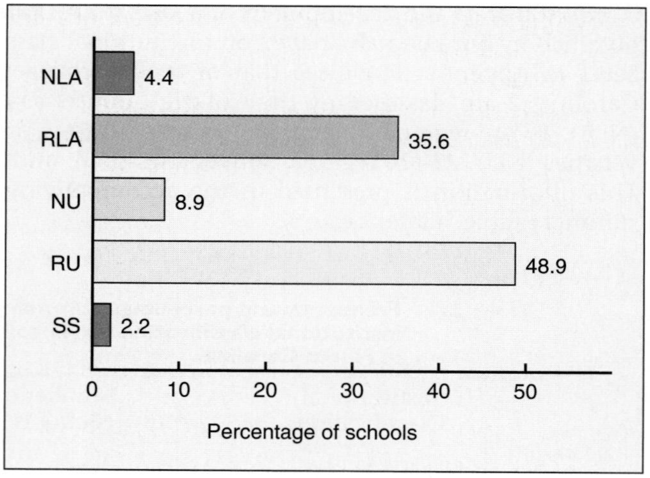

Figure 5.1
Percentage bar chart depicting the institutional classification of 45 colleges and universities in North Carolina.
Source: Data are taken from Table 5.1.

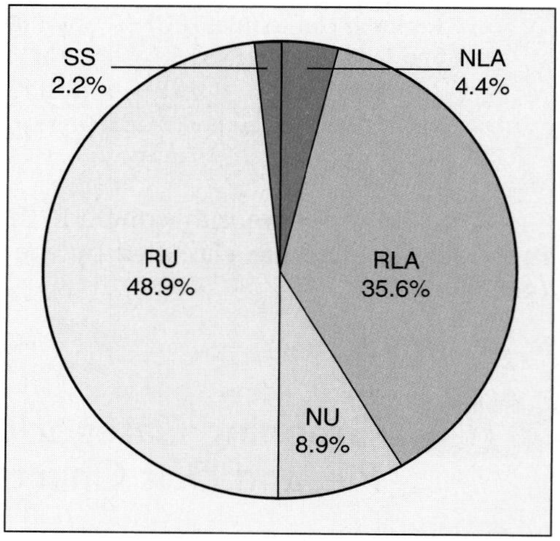

Figure 5.2
Percentage pie chart depicting the institutional classification of 45 colleges and universities in North Carolina.
Source: Data are taken from Table 5.1.

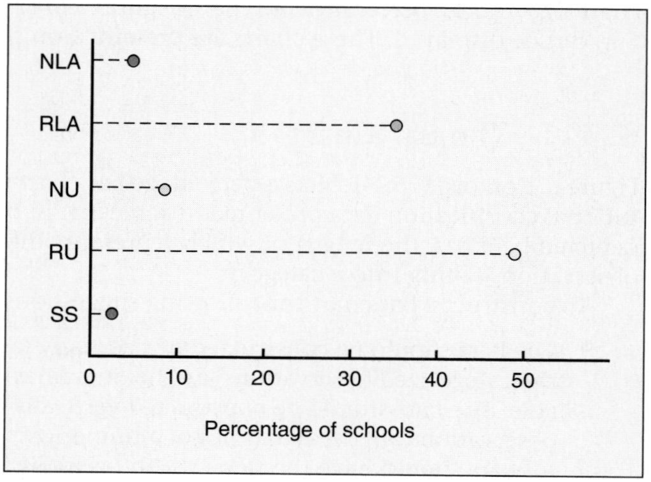

Figure 5.3
Percentage dot chart depicting the institutional classification of 45 colleges and universities in North Carolina.
Source: Data are taken from Table 5.1.

3. Spaces between bars should range from one-half the width of a bar to the width of a bar.
4. Scales and guidelines are useful aids in reading a chart and should be included. The zero point or origin should be indicated.
5. The axes of the chart should be clearly labeled.
6. Any "keys" to interpreting the chart may be included within the body of the chart or below the body of the chart.
7. Footnotes or source notes, when appropriate, are presented after the title of the chart or at the bottom edge of the chart's frame.

5.3.2 The Pie Chart

Figure 5.2 depicts a percentage pie chart for the North Carolina institutional classification data presented in Table 5.1.

To construct a **pie chart** (when appropriate statistical software is not available) we may use both the compass and the protractor—the former to draw the circle, the latter to measure off the appropriate pie sectors. Since the circle has 360°, the protractor may be used to divide up the pie based on the percentage "slices" desired. As an example, in Table 5.1, 8.9% of the institutions in the state of North Carolina are classified by the College Counsel as national universities. Thus, we would multiply 360 by .089, mark off the resulting 32° with the protractor, and then connect the appropriate points to the center of the pie, forming a slice comprising 8.9% of the area of the pie.

Using this procedure with all the categories in Table 5.1 will enable us to construct the entire pie chart displayed in Figure 5.2. However, if a particular summary table contains many categories, construction of the pie chart using a compass and protractor becomes laborious. For this reason, we recommend that statistical software be used for developing a pie chart.

5.3.3 The Dot Chart

Figure 5.3 depicts a percentage dot chart for the North Carolina institutional classification data presented in Table 5.1. In a **dot chart,** each category is depicted by a thin dashed line tipped with a large dot, with the length of the line representing the frequency or percentage of observations falling into a category. Visually, the dot chart takes the form of an unadorned bar chart.

To construct a dot chart the following suggestions are made:

1. The thin dashed lines should be constructed horizontally, as in Figure 5.3, adjacent to the various categories of the variable being studied.
2. Spacing between the thin dashed lines (i.e., between the categories) should be equal.
3. A horizontal scale showing the percentages at the bottom of the frame should be included as in Figure 5.3. The axes should be clearly labeled with the zero point or origin included.
4. Footnotes or source notes, as appropriate, appear after the title of the chart or at the bottom edge of the chart's frame, along with any "keys" to interpreting the chart.

5.3.4 Choosing an Appropriate Chart

The purpose of graphical presentation is to display data accurately and clearly. Figures 5.1, 5.2, and 5.3 all attempt to convey the same information with respect to institutional classification. Whether these charts succeed, however, has been a matter of much concern (see References 1–4, 9, 10). In particular, recent research in the human perception of graphs (Reference 3) concludes that the dot chart portrays the information best and the pie chart presents the weakest display. The dot chart is preferred to the bar chart because simplicity and sparsity of adornment result in greater clarity.[1] However, both of these charts are preferred to the pie chart, because it was observed that the human eye can more accurately judge length comparisons against a fixed scale (as in a dot chart or bar chart) than angular measures (as in a pie chart). Nevertheless, the pie chart has two distinct advantages: (1) it is aesthetically pleasing and (2) it clearly shows that the total for all categories or slices of the pie adds to 100%. Thus, the selection of a particular chart is still highly subjective and often dependent on the aesthetic preferences of the user.

Problems for Section 5.3

5.1 The board of directors of a large housing cooperative wish to investigate the possibility of hiring a supervisor for an outdoor playground. All 616 households in the cooperative were polled, with each household having one vote, regardless of its size. The following data were collected:

Should the co-op hire a supervisor?	
Yes	146
No	91
Not sure	58
No response	321
Total	616

(a) Convert the data to percentages and construct
  (1) a bar chart
  (2) a pie chart
  (3) a dot chart
(b) Which of these charts do you prefer to use here? Why?
(c) Eliminating the "no response" group, convert the 295 responses to percentages and construct
  (1) a bar chart
  (2) a pie chart
  (3) a dot chart
(d) ACTION Based on your findings in (a) and (c), what would you recommend that the board of directors do? Write a letter to the president of the board.

5.2 The following table at the top of page 175 represents the market shares (in percent) held by manufacturers of portable, transportable, and mobile cellular phones sold during 1992:

Manufacturer	Market Share (in %)
Motorola	22
Nokla	14
Mitsubishi	10
NovAtel	9
Toshiba	8
All others	37
Total	100

Source: *The New York Times*, October 31, 1993, Sect. 3, p. 1.

(a) Construct a bar chart.
(b) Construct a pie chart.
(c) Construct a dot chart.
(d) Which of these charts do you prefer to use here? Why?
(e) ACTION Describe these sales results in a brief report and suggest some approaches Mitsubishi could consider for improving its market share position.

5.3 The following data represent the market share (in percent) held by manufacturers of Windows business applications software during 1992:

Manufacturer	Market Share (in %)
Aldus	4.0
Lotus	14.6
Microsoft	60.0
Software Publishing	2.9
Wordperfect	9.6
Others	8.8
Totals	99.9*

* Due to rounding.
Source: *The New York Times*, October 13, 1993, p. D1.

(a) Construct a bar chart.
(b) Construct a pie chart.
(c) Construct a dot chart.
(d) Which of these charts do you prefer to use here? Why?
(e) ACTION Write a report summarizing the above data and offer suggestions as to how Lotus might enhance its market share position.

5.4 Imports to the United States from developing countries accounted for 41.4% of an estimated total of 575.9 billion dollars in the year 1993. On the other hand, exports from the United States to developing countries accounted for 40.7% of an estimated total of 459.6 billion dollars in that year. The following table at the top of page 176 presents a breakdown by country or region (in percent) of United States imports and exports for the year 1993:

Country or Region	Imports into U.S. Percent Share	Exports from U.S. Percent Share
Africa	2.3	1.6
Asia (excluding Japan)	23.5	17.2
Canada	19.2	21.7
European Community	16.6	20.8
Japan	18.4	10.4
Latin America	12.9	16.8
Mideast	2.7	4.7
Other	4.4	6.8
Total	100.0	100.0

Source: *The New York Times*, December 19, 1993, p. F7.

(a) Construct separate bar charts for imports and exports.
(b) Construct separate pie charts for imports and exports.
(c) Construct separate dot charts for imports and exports.
(d) Which of these charts do you prefer to use here? Why?
(e) ACTION Analyze the data and write a memo to your economics professor based on your findings.

5.4 Graphing Categorical Data: The Pareto Diagram

The **Pareto diagram** is a special type of vertical bar chart in which the categorized responses are plotted in the descending rank order of their frequencies and combined with a cumulative polygon on the same scale. The main principle behind this graphical device is its ability to flag the "vital few" from the "trivial many," enabling us to focus on the important responses. Hence the chart achieves its greatest utility when the categorical variable of interest contains many categories. The Pareto diagram is widely used in the statistical control of process and product quality (see Chapter 16).

To illustrate the Pareto diagram, we may observe that in Figure 5.1 on page 172 the bar chart pertaining to institutional classification presents the categories as national liberal arts schools, regional liberal arts schools, national universities, regional universities, and specialty schools. Since regional universities and regional liberal arts schools dominate the College Counsel institutional classifications in the state of North Carolina, a Pareto diagram may be formed by changing the ordering. Such a plot is depicted in Figure 5.4. From the lengths of the vertical bars we observe that almost one out of every two of these schools is classified as a regional university. Moreover, from the cumulative polygon we note that 84.5% of these institutions are classified as either a regional university or a regional liberal arts school.

In constructing the Pareto diagram, the vertical axis contains the percentages (from 100 on top to 0 on bottom) and the horizontal axis contains the categories of interest. The equally spaced bars must also be of equal width and, for visual impact (Reference 9), we suggest that the bars be of the same tint. The point on the cumulative percentage polygon for each category is centered at the midpoint of each respective bar. Hence, when studying a Pareto diagram, we should be focusing on two things—the magnitudes of the differences in bar lengths corresponding to adjacent descending categories and the cumulative percentages of these adjacent categories.

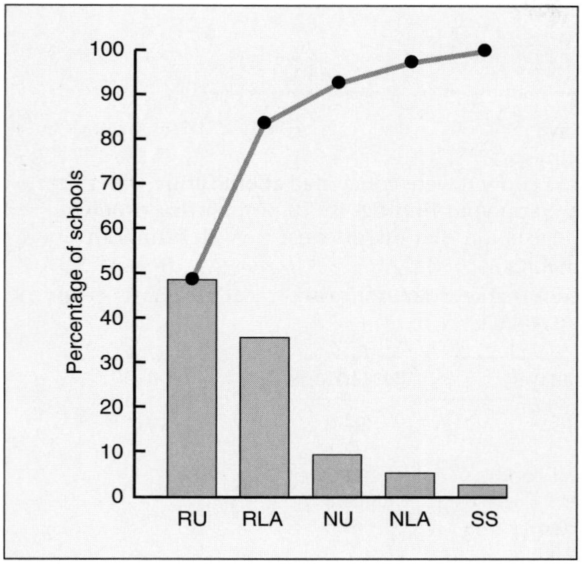

Figure 5.4
Pareto diagram depicting the institutional classification of 45 colleges and universities in North Carolina.
Source: Data are taken from Table 5.1 on page 171.

Problems for Section 5.4

5.5 Refer to the data in Problem 5.3 on page 175 regarding percent market share attained by manufacturers of Windows business applications software.
  (a) Form a Pareto diagram.
  (b) Which of the graphs seems to give the most visual impact—the Pareto diagram here or one of the charts drawn in (a)-(c) of Problem 5.3? Discuss.

5.6 Refer to Problem 5.4 on page 175 regarding imports and exports.
  (a) Set up a table based on estimated balance of trade. That is, for each country or region, calculate the estimated value of import dollars minus export dollars—yielding either a trade deficit or a trade surplus.
  (b) For the countries or regions with which the United States has a trade deficit (i.e., import dollars are more than export dollars), form a Pareto diagram.
  (c) Summarize your findings.
  (d) ACTION Write a report for your economics professor based on your findings in (c). List potential social, political, cultural, and/or economic reasons that may have led to this trade deficit.

5.7 The consulting firm of Holzmacher, McLendon and Murrel reported on daily water consumption per household in the South Farmingdale (New York) Water District during a recent summer. The results of their study indicate the following (see page 178):

Reason for Water Usage	No. Gallons per Day
Bathing and showering	99
Dish washing	13
Drinking and cooking	11
Laundering	33
Lawn watering	150
Toilet	88
Misc.	20
Total	414

(a) Form a Pareto diagram.
(b) Summarize your findings.
(c) **ACTION** If the town council were concerned about future water shortages, write a letter based on your findings in (b) pinpointing problem areas and proposing legislation that might conserve water through changes in personal habits.

5.8 The following data represent market share for ready-to-eat breakfast cereals in 1992:

Company	Percentage
General Mills	24.8
Kellogg	37.8
Kraft General Foods	11.7
Quaker Oats	6.8
Ralston Purina	4.3
RJR Nabisco	2.8
Store brands	8.0
Others	3.8
Totals	100.0

Source: *The New York Times*, November 17, 1992, p. D4.

(a) Form a Pareto diagram.
(b) Summarize your findings.
(c) **ACTION** Write a letter to the food editor of your local newspaper on this matter.

5.9 The following data represent the oil production of OPEC members in December 1992, in millions of barrels per day:

Country	Daily Oil Production (in millions of barrels)
Algeria	0.77
Gabon	0.30
Indonesia	1.35
Iran	3.50
Iraq	0.55
Kuwait	1.30
Libya	1.45
Nigeria	1.90
Qatar	0.42
Saudi Arabia	8.20
United Arab Emirates	2.25
Venezuela	3.50
Total	25.49

Source: *The New York Times*, January 25, 1993, p. D2 .

(a) Form a Pareto diagram.
(b) Summarize your findings.
(c) ACTION▶ Write a letter to the business editor of your local newspaper on this matter.

5.10 A patient-satisfaction survey conducted for a sample of 210 individuals discharged from a large urban hospital during the month of June led to the following list of 384 complaints:

Reason for Complaint	Number
Anger with other patients/visitors	13
Failure to respond to buzzer	71
Inadequate answers to questions	38
Lateness for tests	34
Noise	28
Poor food service	117
Rudeness of staff	62
All others	21
Total	384

(a) Form a Pareto diagram.
(b) Summarize your findings.
(c) ACTION▶ Write a memo to the Chief Executive Officer of the hospital regarding your findings and offer suggestions for improvement.

5.11 The following table presents the number of stockholder meetings held outside the United States at which U.S. pension-fund clients of *Global Proxy Services Corporation* voted in the 1992–93 proxy season:

Country	Number of Meetings Held
Australia	49
Belgium	50
Canada	87
England	374
France	72
Germany	99
Holland	83
Hong Kong	116
Italy	115
Japan	1,249
Switzerland	61
Other	396
Total	2,751

Source: *The New York Times*, July 16, 1993, p. D1.

(a) Form a Pareto diagram.
(b) Summarize your findings.
(c) ACTION▶ Write a letter to the business editor of your local newspaper on this matter.

5.12 In a recent year the U.S. Fire Administration reported on the leading causes of residential fire deaths:

Cause of Death	Percentage
Appliances/Equipment	4.9
Children playing with fire	6.4
Cooking	6.3
Electrical fires	5.0
Heating	12.9
Incendiary	6.7
Open flame/Candles	3.0
Smoking	22.0
Spread from original site	1.1
Unknown origin	31.7
Total	100.0

(a) Form a Pareto diagram.
(b) Summarize your findings.
(c) Reconstruct this table after removing the "unknown origin" category and form a new Pareto diagram.
(d) Summarize your findings.
(e) ACTION Write an article for the "daily living" section of your local newspaper based on your findings in (b) and (d) in order to enlighten the readership on this matter.

5.5 Tabularizing Categorical Data: Contingency Tables and Supertables

5.5.1 The Contingency Table

It is often desirable to examine the responses to two categorical variables simultaneously. For example, our research analyst at the college advisory service company might be interested in examining whether or not there is any pattern or relationship between type of institution (that is, private or public) and the College Counsel's institutional classification. Using Special Data Set 1 in Appendix D on page D3, Table 5.2 depicts this information for all 45 colleges and universities in the state of North Carolina. Such two-way tables of cross-classification are known as **contingency tables.**

To construct Table 5.2, for example, the joint responses for each of the 45 schools with respect to type of institution and institutional classification are tallied into one of the 10 possible "cells" of the table. Thus from Special Data Set 1 in Appendix D on page D3, the first school listed (Appalachian State University) is a public regional university. These joint responses were tallied into the cell composed of the second row and fourth column. The second institution (Barber Scotia

College) is a private regional liberal arts school. These joint responses were tallied into the cell composed of the first row and second column. The remaining 43 joint responses were recorded in a similar manner.

Table 5.2 Contingency table displaying type of institution and institutional classification for 45 colleges and universities in North Carolina.

Type of Institution	Institutional Classification					
	NLA	RLA	NU	RU	SS	Totals
Private	2	16	1	11	0	30
Public	0	0	3	11	1	15
Totals	2	16	4	22	1	45

Source: Data are taken from Special Data Set 1, Appendix D, page D3.

Data file C&UNC.DAT

In order to explore any possible pattern or relationship between type of institution and the College Counsel institutional classification, it is useful first to convert these results into percentages based on

1. The overall total (that is, the 45 colleges and universities in North Carolina)
2. The row totals (that is, private or public)
3. The column totals [that is, national liberal arts school (NLA), regional liberal arts school (RLA), national university (NU), regional university (RU), or specialty school (SS)]

This is accomplished in Tables 5.3, 5.4, and 5.5, respectively.

We will highlight some of the many findings present in these tables for the 45 colleges and universities in the state of North Carolina. From Table 5.3 we note that

1. 66.7% of the institutions in North Carolina are private
2. 8.9% of the institutions in North Carolina are classified as national universities
3. 2.2% of the institutions in North Carolina are private national universities

Table 5.3 Contingency table displaying type of institution and institutional classification for 45 colleges and universities in North Carolina (percentages based on overall total).

Type of Institution	Institutional Classification					
	NLA	RLA	NU	RU	SS	Totals
Private	4.4	35.6	2.2	24.4	0.0	66.7
Public	0.0	0.0	6.7	24.4	2.2	33.3
Totals	4.4	35.6	8.9	48.9	2.2	100.0

Source: Data are taken from Table 5.2.

Table 5.4 **Contingency table displaying type of institution and institutional classification for 45 colleges and universities in North Carolina (percentages based on row totals).**

Type of Institution	Institutional Classification					
	NLA	RLA	NU	RU	SS	Totals
Private	6.7	53.3	3.3	36.7	0.0	100.0
Public	0.0	0.0	20.0	73.3	6.7	100.0
Totals	4.4	35.6	8.9	48.9	2.2	100.0

Source: Data are taken from Table 5.2.

Table 5.5 **Contingency table displaying type of institution and institutional classification for 45 colleges and universities in North Carolina (percentages based on column totals).**

Type of Institution	Institutional Classification					
	NLA	RLA	NU	RU	SS	Totals
Private	100.0	100.0	25.0	50.0	0.0	66.7
Public	0.0	0.0	75.0	50.0	100.0	33.3
Totals	100.0	100.0	100.0	100.0	100.0	100.0

Source: Data are taken from Table 5.2.

From Table 5.4 we note that

1. 53.3% of the private institutions are classified as regional liberal arts schools
2. 73.3% of the public institutions are classified as regional universities

From Table 5.5 we note that

1. 25.0% of the institutions classified as national universities are private
2. 50.0% of the institutions classified as regional universities are public

The tables, therefore, indicate a pattern: North Carolina institutions of higher learning comprise mainly regional universities and regional liberal arts schools—the former are evenly split between public and private schools and the latter are all private schools.

5.5.2 The Supertable

A useful technique for presenting data containing several categorical variables is the supertable (Reference 9). A **supertable** is essentially a collection of contingency tables, each having the same column variable and categories. However, as many row variables as desired are included for comparison against the column variable. The data in each cell of the table are always given as a percentage of its corresponding row total. This permits line-by-line comparisons for the categories *within* a particular row variable as well as for the categories *among* the various row variables.

Table 5.6 is a supertable that investigates possible relationships between a variety of features pertaining to the 45 colleges and universities in North Carolina and the College Counsel's institutional classification. Note the similarity between the top part of Table 5.6 and the contingency table presented as Table 5.4 on page 182.

Table 5.6 A supertable for studying the possible relationships between various features and institutional classification at 45 colleges and universities in North Carolina.

Data file C&UNC.DAT

Variables and Category Percentages	Institutional Classification NLA	RLA	NU	RU	SS
Type of Institution:					
Private (66.7%)	6.7%	53.3%	3.3%	36.7%	0.0%
Public (33.3%)	0.0%	0.0%	20.0%	73.3%	6.7%
Location:					
Rural (22.2%)	0.0%	40.0%	0.0%	60.0%	0.0%
Suburban (24.4%)	18.2%	45.5%	0.0%	36.4%	0.0%
Urban (53.3%)	0.0%	29.2%	16.7%	50.0%	4.2%
Calendar:					
Semester (91.1%)	4.9%	36.6%	9.8%	48.8%	0.0%
Trimester (2.2%)	0.0%	0.0%	0.0%	0.0%	100.0%
4–1–4 (6.7%)	0.0%	33.3%	0.0%	66.7%	0.0%
Out-of-State Resident Tuition:					
$10,000 or more (11.1%)	40.0%	20.0%	20.0%	20.0%	0.0%
Below $10,000 (88.9%)	0.0%	37.5%	7.5%	52.5%	2.5%

Note: For institutional classification: NLA = national liberal arts school, RLA = regional liberal arts school, NU = national university, RU = regional university, SS = school with special focus.
Source: Special Data Set 1 in Appendix D on page D3.

Problems for Section 5.5

5.13 In a recent study, researchers were looking at the relationship between the type of college attended and the level of job that people who graduated in 1975 held at the time of the study. The researchers examined only graduates who went into industry. The cross-tabulation of the data is presented below:

Management Level	Type of College Ivy League	Other Private	Public
High (Sr. V-P or above)	45	62	75
Middle	231	563	962
Low	254	341	732

(a) Construct a table with either row or column percentages, depending on which you think is more informative.
(b) Interpret the results of the study.
(c) What other variable or variables might you want to know before advising someone to attend an Ivy League or other private school if he or she wants to get to the top in business?

5.14 People returning from vacations in different countries were asked how they enjoyed their vacation. Their responses are as follows:

Country	Response to Country: Yuck	So-So	Good	Great
England	5	32	65	45
Italy	3	12	32	43
France	8	23	28	25
Guatemala	9	12	6	2

(a) Construct a table of row percentages.
(b) What would you conclude from this study?
(c) ACTION Write a letter to the travel editor of your local newspaper regarding your findings.

● 5.15 The defeat of the incumbent, George Bush, in the 1992 presidential election was attributed to poor economic conditions and high unemployment. Suppose that a survey of 800 adults taken soon after the election resulted in the following cross-classification of financial condition with education level:

	Education Level			
Financial Conditions	H.S. Degree or Lower	Some College	College Degree or Higher	Totals
Worse off now than before	261	48	38	347
No difference	104	73	41	218
Better off now than before	65	39	131	235
Totals	430	160	210	800

(a) Construct a table of column percentages.
(b) What would you conclude from this study?
(c) ACTION Write a letter to your political science professor regarding your findings.

5.16 Develop (in outline form) a supertable corresponding to stock exchange listing (NYSE, ASE, and OTC) based on such market performance variables as dividend declaration, annual earnings level, change in profits over past year, and level of price to earnings ratio.

5.6 Using the Computer for Tables and Charts with Categorical Data: The Kalosha Industries Employee Satisfaction Survey

5.6.1 Introduction and Overview

When dealing with large batches of data, we may use the computer to assist us in our descriptive statistical analysis. In this section we will demonstrate how various statistical software and spreadsheet packages (see References 5–8) can be used for organizing and presenting categorical data in tabular and chart form. To accomplish this, let us return to the Kalosha Industries Employee Satisfaction Survey that was developed in Chapter 2.

5.6.2 Kalosha Industries Employee Satisfaction Survey

Data file EMPSAT.DAT

We may recall from Sections 3.8.2 and 4.10.2 that Bud Conley, the Vice-President of Human Resources, is preparing for a meeting with a representative from the B & L Corporation to discuss the potential contents of an employee-benefits package that is being developed. Answers to the following three questions (dealing with categorical variables) would be of particular concern in an initial analysis of the survey data (Table 2.3 on pages 33–40):

1. *General Question A:* How do the full-time employees of Kalosha Industries respond to the question on job satisfaction (see question 9 in the Survey)?
2. *Specific Question B:* According to the full-time employees of Kalosha Industries, what would be the most important characteristic to consider when searching for a job (see question 11 in the Survey)?
3. *Specific Question C:* Is there a relationship between gender and chosen occupation among the full-time employees of Kalosha Industries (see questions 5 and 2 in the Survey)?

These and other initial questions raised by Bud Conley (see Survey/Database Project at the end of the section) require a detailed descriptive statistical analysis of the 400 responses to the Survey. In practice, a statistician would likely use one or two statistical or spreadsheet packages when performing the descriptive statistical analysis. However, computer output from several packages is presented here so that we may demonstrate some of the features of these packages.

5.6.3 Using Statistical Software and Spreadsheet Packages for Categorical Data

In response to Bud Conley's first question, a summary table and pie chart are desired. Figure 5.5 on page 186 presents computer output displaying the needed summary table obtained from accessing STATISTIX and Figure 5.6 on page 186 depicts a pie chart from SPSS. From these displays we observe that 46% of the full-time employees at Kalosha Industries are "very satisfied" with their jobs, 43% of the employees are "moderately satisfied," 7% are "a little dissatisfied," and 4% are "very dissatisfied."

In response to Bud Conley's specific question (*B*) regarding important job characteristics, a Pareto diagram is obtained using Microsoft EXCEL for Windows, a spreadsheet package, and displayed in Figure 5.7 on page 186. From this it is observed that three of every four full-time employees at Kalosha Industries believe there are two major features of a job—50.75% of the employees feel that "enjoyment of work" is the most important job characteristic and 25.75% of the employees think that "high income" is the most important job characteristic. Other job features are cited far less frequently—12.75% of the employees claim "opportunities for advancement" is most important, 6.0% state that "job security" is most important, and only 4.75% say that "flexible work hours" is most important.

Figure 5.5
Summary table of job satisfaction from STATISTIX output.
Note: Each of the percentages from the summary table obtained using the STATISTIX package is rounded up and the total adds to 100.2% because of such rounding error.

SUMMARY TABLE OF SATJOB

VALUE	FREQ	PERCENT
Very Satisfied	185	46.3
Moderately Satisfied	171	42.8
A Little Dissatisfied	27	6.8
Very Dissatisfied	17	4.3
TOTAL	400	100.0

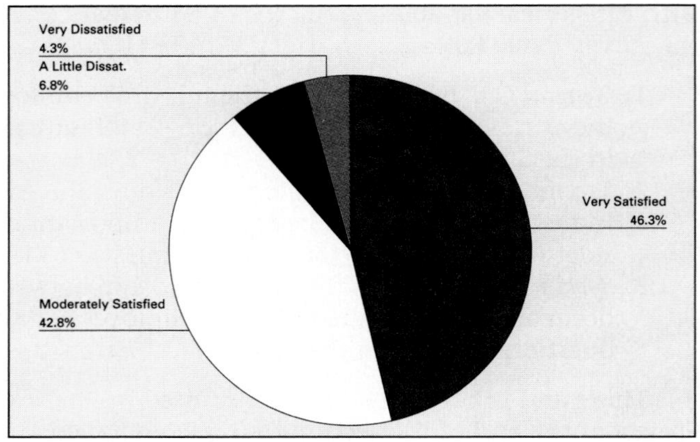

Figure 5.6
Pie chart of job satisfaction from SPSS output.
Note: Each of the percentages from the pie chart obtained using the SPSS package is rounded up and the total adds to 100.2% because of such rounding error.

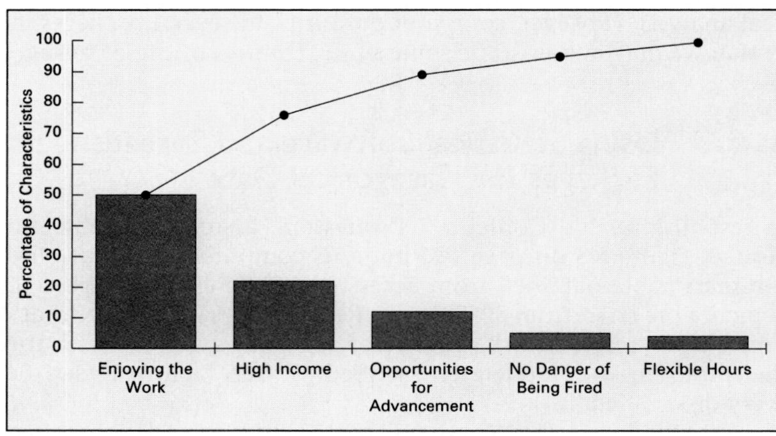

Figure 5.7
Pareto diagram of important job characteristics from Microsoft EXCEL for Windows output.

In response to Bud Conley's specific question (*C*), which deals with a pair of categorical variables, a contingency table is desired. Figure 5.8 displays the needed computer output obtained from accessing SAS. We observe from this 2×7 contingency table that, in general, there appears to be a relationship between gender and occupational grouping at Kalosha Industries. The percentage of males who work in some of the occupational settings differs substantially from that of their female counterparts. In particular, we note from the cells of the table that certain combinations of the two categorical variables stand out. When compared to males, we observe that females are employed in an administrative support capacity with far greater frequency than could be expected if there was no gender-occupation relationship. In addition, when compared to females, we note that males work either in production or as laborers with far greater frequency than could be expected if the two categorical variables were not related.

To prepare for the development of an employee-benefits package, Bud Conley was also interested in evaluating potential indicators of job satisfaction—gender, union membership, perception of getting ahead, promotions, participation in decisions, worker-management relations, and coworker relations (see Survey/Database Project).

TABLE OF SEX BY OCCUP

SEX / OCCUP (Frequency, Percent, Row Pct, Col Pct)	MGL	PROF	TEC/SAL	ADMSPT	SERV	PROD	LABOR	Total
MALES	36	33	34	14	18	51	47	233
	9.00	8.25	8.50	3.50	4.50	12.75	11.75	58.25
	15.45	14.16	14.59	6.01	7.73	21.89	20.17	
	55.38	50.00	59.65	21.54	62.07	94.44	73.44	
FEMALES	29	33	23	51	11	3	17	167
	7.25	8.25	5.75	12.75	2.75	0.75	4.25	41.75
	17.37	19.76	13.77	30.54	6.59	1.80	10.18	
	44.62	50.00	40.35	78.46	37.93	5.56	26.56	
Total	65	66	57	65	29	54	64	400
	16.25	16.50	14.25	16.25	7.25	13.50	16.00	100.00

Figure 5.8
Contingency table of gender and occupation from SAS output.
Note: SAS provides the user with numerous options when developing a contingency table. Invoking such options, as we see from the output in Figure 5.8, each cell contains four pieces of information—the frequency or cell count (highlighted in color), the percent (that is, the cell frequency as a percentage of the grand total), the row percent (that is, the cell frequency as a percentage of the row total), and the column percent (that is, the cell frequency as a percentage of the column total). Note that, by invoking these options, this one table takes the place of four tables. (See Tables 5.2–5.5 on pages 181–182.)

Survey/Database Project for Section 5.6

The following problems refer to the sample data obtained from the questionnaire of Figure 2.6 on pages 28–29 and presented in Table 2.3 on pages 33–40. They should be solved with the aid of an available computer package.

Data file EMPSAT.DAT

Suppose you are hired as a research assistant to Bud Conley, the Vice-President of Human Resources at Kalosha Industries. He has given you a list of questions (see Problems 5.17 to 5.40) that he needs answered prior to his meeting with the representative from the B&L Corporation, the employee-benefits consulting firm that he has retained.

From the responses to the questions dealing with categorical variables in the Employee Satisfaction Survey, in Problems 5.17 through 5.28 that follow,

(a) Form the summary table.
(b) Construct a pie chart or a bar chart.
(c) Construct a Pareto diagram (if appropriate).
(d) **ACTION** Write a memo to Bud Conley discussing your findings.

5.17 Examine occupational grouping (see question 2).

5.18 Examine gender breakdown (see question 5).

5.19 Examine belief on "getting ahead" (see question 12).

5.20 Examine union membership (see question 14).

5.21 Examine perceived likelihood of promotion (see question 18).

5.22 Examine perceived gender-based promotional opportunities (see question 19).

5.23 Examine perceived "advancement" (see question 20).

5.24 Examine perceived participation in decision making (see question 21).

5.25 Examine participation in budgetary decisions (see question 22).

5.26 Examine attitude toward Kalosha Industries (see question 23).

5.27 Examine perception of employee-management relations (see question 25).

5.28 Examine perception of coworker relations (see question 26).

From the responses to the questions dealing with categorical variables in the Employee Satisfaction Survey (see pages 33–40), in Problems 5.29 through 5.40 which follow,

(a) Form contingency table and analyze the data.
(b) **ACTION** Write a memo to Bud Conley discussing your findings.

5.29 Cross-classify gender (question 5) with job satisfaction (question 9).

5.30 Cross-classify job characteristic importance (question 11) with job satisfaction (question 9).

5.31 Cross-classify belief on "getting ahead" (question 12) with job satisfaction (question 9).

5.32 Cross-classify union membership (question 14) with job satisfaction (question 9).

5.33 Cross-classify likelihood of future promotion (question 18) with job satisfaction (question 9).

5.34 Cross-classify perceived gender-based promotional opportunities (question 19) with job satisfaction (question 9).

5.35 Cross-classify perceived "advancement" (question 20) with job satisfaction (question 9).

5.36 Cross-classify perceived participation in decision making (question 21) with job satisfaction (question 9).

5.37 Cross-classify budgetary decision making (question 22) with job satisfaction (question 9).

5.38 Cross-classify attitude toward Kalosha Industries (question 23) with job satisfaction (question 9).

5.39 Cross-classify perception of employee-management relations (question 25) with job satisfaction (question 9).

5.40 Cross-classify perception of coworker relations (question 26) with job satisfaction (question 9).

5.41 **(Class Project)**

(a) As Vice-President of Human Resources at Kalosha Industries, should Bud Conley be happy about the responses to question 9 of the Survey regarding job satisfaction? (While half of the employees are "very satisfied," 11% are disgruntled employees.) Discuss.

(b) What other variables should be considered in a supertable about job satisfaction?

(c) ACTION Suppose you are hired as research assistant to Bud Conley. You are asked to thoroughly analyze the responses in Figures 5.5 and 5.6 (see page 186) and prepare an executive summary that will be submitted to him.

5.7 Recognizing and Practicing Proper Tabular and Chart Presentation and Exploring Ethical Issues

In this chapter we have studied how categorical data are presented in tabular and chart form in order to make the data more manageable and meaningful for purpose of analysis. Again, as in Chapter 3, if our analysis is to be enhanced by a visual display of categorical data, it is essential that the tables and charts be presented clearly and carefully. Tabular frills and "chart junk" must be eliminated so as not to cloud the message given by the data with unnecessary adornments (References 9 and 10). In addition, when displaying charts we must avoid common errors that distort our visual impression.

5.7.1 Tabular Frills, Chart Junk, and Common Errors

Go to your library and compare the tables and charts displayed in a government publication such as the *Survey of Current Business* with those found in a popular weekly magazine or daily newspaper. The government publication is intended for a more sophisticated user and the information is presented in a straightforward manner. Such presentation has often been described in one word—"dull." On the other hand, as we turn the pages of magazines and newspapers we find that tables and charts are adorned with various icons and symbols to make them attractive to their readers. Unfortunately, enlivening a table or chart often hides or distorts the message intended to be conveyed by the data. Overuse of adornments when displaying graphs often results in chart junk. Exaggerated icons and symbols result in a distortion of the visual impact as seen in Figure 5.9 on page 190.

Note that in this chart the magnitude of the 10.8 million transportation, trade, and retail jobs is underrepresented by a truck icon that is smaller than that representing the 5.6 million jobs for farm family members. Also, the can icon representing the 1.3 million food processing jobs is far too large compared to the icons representing the 2.1 million farm worker jobs, the 2.6 million manufacturing jobs, and the 4.1 million jobs in other areas of agriculture and exporting. A simple summary table, or a bar, pie, or dot chart, or a Pareto diagram would have been more effective in accurately portraying the data.

Other types of eye-catching displays that we typically see in magazines and newspapers erroneously attempt to show icon stick figures representative of numerical information. Can anyone really read and interpret such two-dimensional stick figures accurately? The answer is no. As seen in Figure 5.10 on page 190, these graphs may be attractive, but they rarely work!

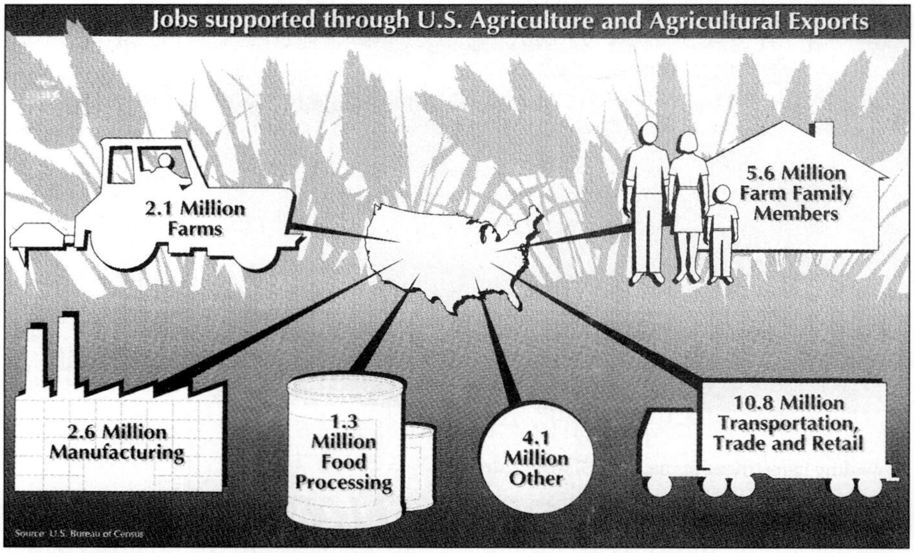

Figure 5.9
"Improper" display of jobs supported through United States agriculture and agricultural exports.
Source: *The New York Times*, October 19, 1993, Advertising Supplement p. D18.

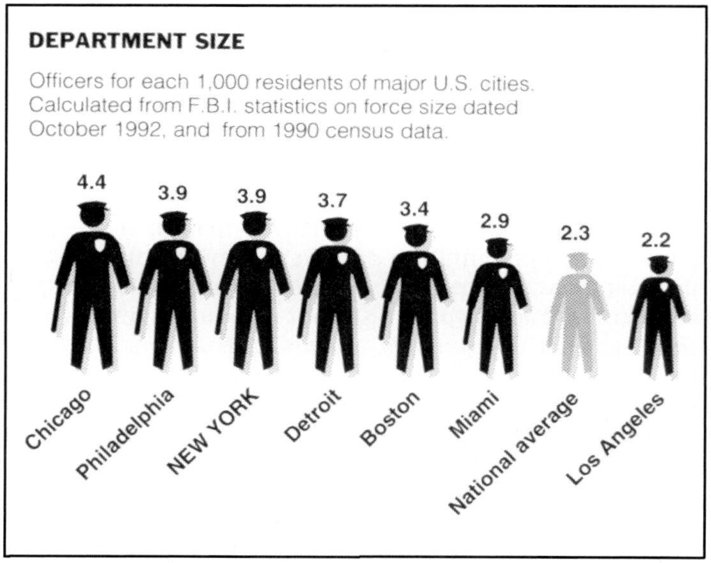

Figure 5.10
"Improper" display of police department size for each 1,000 residents of major cities in the United States.
Source: Extracted from Powell, R., "A Statistical Portrait of the N.Y.P.D.," *The New York Times*, October 10, 1993, p. 35.

From Figure 5.10, is the icon representing police department size per 1,000 residents of Chicago really twice as large as that for Los Angeles? It is supposed to be, but how can this be properly drawn? Such an illustration catches the eye but doesn't really say anything that a summary table, or bar, pie, or dot chart, or Pareto diagram couldn't present better.

In summation, we are active consumers of information that we hear or see daily through the various media. Since much of what we hear or read is junk, we must learn to evaluate critically and discard that which has no real value. We must also keep in mind that sometimes the junk we are provided is based on ignorance; other times, as we shall reiterate in Section 5.7.3, it is planned and malicious. The bottom line—be critical and be skeptical of information provided.

5.7.2 Using Computer Software for Tables and Charts

In Sections 3.8, 4.10, and 5.6 we demonstrated how appropriate computer software can assist us in a descriptive analysis of our data. We have observed that the computer is an extremely helpful tool that can easily and rapidly store, organize, and process information and provide us with summary results, tables, and charts. Nevertheless, we must bear in mind that the computer is only a tool. We shall see throughout this text, as we both demonstrate and interpret a variety of computer output corresponding to the topics to be studied in the chapters ahead, that it is crucial that we use the computer in a manner consistent with correct statistical methodology. Remember GIGO. The computer output we get will depend on four things—the capacity of the hardware used, the quality of the printer chosen, the capability of the statistical software selected, as well as our ability to choose the software properly and use it advantageously. And when you are presented with tabular and chart information from the output of some statistical software package, be wary of the extra frills and adornments that might be hiding what the data are intended to convey.

To interact properly with the computer, we must not only be familiar with a particular software package, but also correctly select the statistical procedures that are appropriate for the tasks at hand. For example, pie charts and contingency tables should be used with data obtained from categorical variables. It is improper to request pie charts or cross-classifications for continuous numerical variables unless they have been first categorized into classes as in a frequency distribution or supertable. On the other hand, descriptive summary measures such as means, medians, and standard deviations should be used *only* for numerical variables. It is completely inappropriate to instruct the computer to give such summary results for categorical variables like occupation or gender. The output would be totally meaningless.

5.7.3 Ethical Issues

Again, as was mentioned in Sections 3.9.6 and 4.11.4, ethical considerations arise when someone is deciding what data to present in tabular and chart format and what not to present. It is vitally important when conducting research to document both good and bad results, so that individuals who continue such research will not have to start from "square one." Moreover, when making oral presentations and presenting written research reports, it is essential that the results be given in a fair, objective, and neutral manner. Thus we must try to distinguish between poor data presentation and unethical presentation. Once more, as in our prior discussions on ethical considerations, the key is *intent*. Often, when fancy tables and chart junk are presented or pertinent information is omitted, it is simply done out of ignorance. However, unethical behavior occurs when an individual willfully hides the facts by distorting a table or chart or by failing to report pertinent findings.

Problems for Section 5.7

5.42 You are planning to study for your statistics examination with a group of classmates, one of whom you particularly want to impress. This individual has volunteered to use a statistical software package to get the needed summary information, tables, and charts for a data set containing several numerical and categorical variables assigned by the instructor for study purposes. This person comes over to you with the printout and exclaims: "I've got it all—the means, the medians, the standard deviations, the stem-and-leafs, the box-and-whisker plots, the pie charts—for all our variables. The problem is, some of the output looks weird—like the stem-and-leafs and the box-and-whiskers for gender and for major, and the pie charts for grade point index and for height. Also, I can't understand why Dr. Hunter said we can't get the descriptive stats for some of our variables—I got it for everything! See, the mean for height is 68.23, the mean for grade point index is 2.76, the mean for gender is 1.50, the mean for major is 4.33." You look at your would-be friend in the eye, take a deep breath, and reply . . . .

5.43 **(Student Project)** Bring to class a chart from a newspaper or magazine that you believe to be a poorly drawn representation of some categorical variable. Be prepared to submit the chart to the instructor with comments as to why you feel it is inappropriate. Also, be prepared to present this and comment in class.

5.8 Categorical Data Presentation: An Overview

As displayed in the summary chart on page 193, this chapter was about categorical data presentation. Data presentation is an essential element of any large-scale statistical investigation and in this chapter we have become familiar with a variety of techniques for tabularizing and charting categorical data. As we have seen in Section 5.6, the rapid developments in computer technology over the past decade have resulted in major advances in computer graphics capabilities, and tables and charts of high quality are now obtainable from a variety of computer software packages.

On page 170 of Section 5.1 you were given a list emphasizing the important points to be discussed in the chapter. Check over the list now to see whether you feel you have an understanding of these key points. To be sure, you should be able to answer the following conceptual questions:

1. Why would you construct a frequency and percentage summary table?
2. How do you construct a frequency and percentage summary table?
3. How do you construct a bar chart, pie chart, dot chart, and Pareto diagram?
4. What are the advantages and/or disadvantages for using a bar chart, a pie chart, a dot chart, or a Pareto diagram?
5. What kinds of percentage breakdowns can assist you in interpreting the results found through the cross-classification of data based on two categorical variables?
6. Why is the formation of a supertable based on the cross-classification of data containing several categorical variables a useful display?

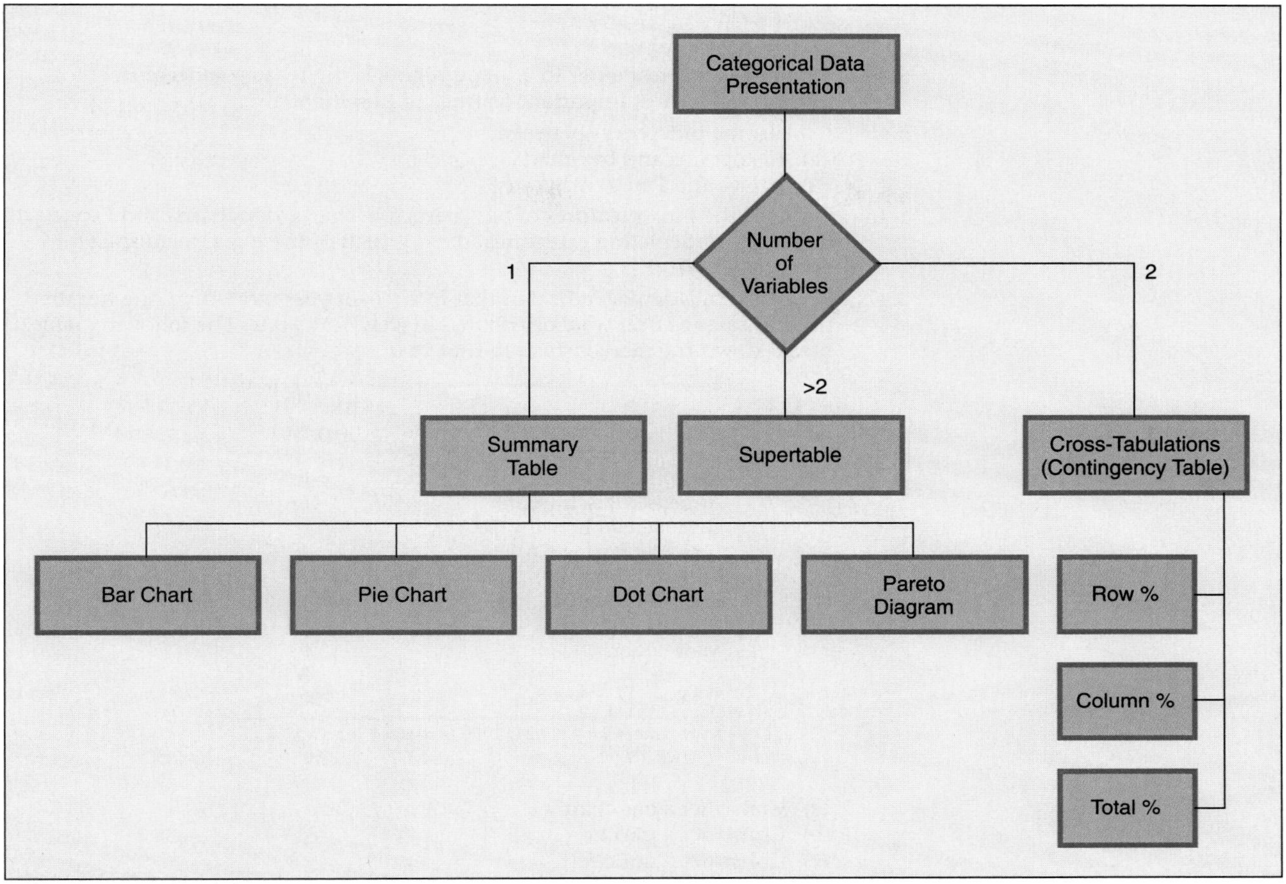

Chapter 5 summary chart.

7. What are some of the ethical issues to be concerned with when presenting categorical data in tabular or chart format?

Check over the list of questions to see if indeed you know the answers and could (1) explain your answers to someone who did not read this chapter and (2) give reference to specific readings or examples that support your answer. Also, reread any of the sections that may have seemed fuzzy to see if they make sense now.

Getting It All Together

Key Terms

bar chart *171*
"chart junk" *189*
contingency table *180*
dot chart *173*
Pareto diagram *176*
pie chart *173*
summary table *171*
supertable *182*

Chapter Review Problems

5.44 **ACTION** Write a letter to a friend highlighting what you deem the most interesting or most important features of this chapter.

5.45 Explain the differences between
(a) histograms and bar charts.
(b) ogives and Pareto diagrams.

5.46 Describe the major features of bar charts, pie charts, dot charts, and Pareto diagrams for depicting categorical data. Which of these graphic displays do you prefer? Why?

5.47 An article in *Newsday* indicated that in a recent year more than one hundred million cases of beer were distributed in New York State. The following table breaks down the market share in that year:

Brewing Company	Market Share (in %)
Adolph Coors Co.	6
Anheuser-Busch Co.	46
G. Heileman Brewing Co.	3
Latrobe Brewing Co.	1
Miller Brewing Co.	20
Stroh Brewing Co.	1
Imported Beers	11
Malt liquor	9
Others	3
Total	100

Source: *Newsday*, May 21, 1990 (extracted from *Beverage World*).

(a) Construct a bar chart.
(b) Construct a pie chart.
(c) Construct a dot chart.
(d) Form a Pareto diagram.
(e) Which of these graphs do you prefer to use here? Why?
(f) **ACTION** Design a survey to be distributed on campus to determine beer preference among students.

5.48 Environmental conservation is a national issue of major importance. It has been said that Americans threw out 227.1 million tons of garbage in a recent year, enough to fill one of the World Trade Center's Twin Towers from bottom to top every day. Typically, disposal of garbage is achieved through landfills (87 percent), incineration (7 percent), and recycling (5 percent).

Suppose that the consulting firm at which you are employed provides the following table showing the percentage breakdown of the sources of waste:

Source	Percentage
Paper and paperboard	37.1
Yard waste	17.9
Glass	9.7
Metals	9.6
Food waste	8.1
Plastics	7.2
Wood	3.8
Rubber and leather	2.5
Textiles	2.1
All other	2.0
Total	100.0

(a) Form the appropriate graph to pinpoint the "vital few" from the "trivial many."
(b) Analyze the data and summarize your findings.
(c) **ACTION▶** Write a letter to the Environmental Protection Agency based on your findings and request government information on the potential for recycling for each of the items.

5.49 Serious cervical injuries to professional football players in the last several years have heightened interest in reducing spinal injuries. The following summary tables respectively display the percentage of spinal injuries categorized by cause and the percentage of sports injuries categorized by particular sport:

Spinal Injury Causes	Percentage
Falls	20.8
Motor Vehicles	47.7
Sports	14.2
Violence	14.6
Other	2.7
Total	100.0

Sports Spinal Injury Causes	Percentage
Diving	66.0
Football	6.1
Gymnastics	2.2
Horseback riding	2.0
Non-skiing winter sports	2.3
Snow skiing	3.8
Surfing	3.1
Trampoline	2.6
Wrestling	2.3
Other	9.6
Total	100.0

Source: The New York Times, November 20, 1991, p. B11.

(a) For the data on overall spinal injury causes, construct:
   (1) a bar chart.
   (2) a pie chart.
   (3) a dot chart.
(b) Which chart do you prefer for purposes of presentation? Why?
(c) For the data on sports spinal injuries, develop the appropriate graph to pinpoint the "vital few" from the "trivial many."
(d) Analyze the data and summarize your findings.
(e) **ACTION▶** Write a letter to the sports editor of your local newspaper explaining your findings.

5.50 The following data represent the market share of all drinks and a stratification of the carbonated soft drink market share based on supermarket sales:

Type of Drink	Market Share (in %)
Beer	12
Carbonated soft drinks	25
Coffee	11
Juice	6
Milk	15
Tap water	19
Other	12
Total	100

Type of Carbonated Soft Drink	Market Share (in %)
Caffeinated cola	48.0
Caffeine-free cola	10.4
Citrus	3.4
Club soda	0.4
Cream	1.4
Dr Pepper	3.9
Ginger ale	3.5
Grape	1.2
Grapefruit	1.0
Lemon-lime	9.8
Mineral water	1.0
Orange	3.7
Root beer	3.7
Sweetened seltzer	0.4
Tonic water	0.7
Unsweetened seltzer	2.2
Other	5.3
Total	100.0

Source: The New York Times, May 2, 1992, p. 19.

(a) For the data on market share of all types of drinks, construct:
    (1) a bar chart.
    (2) a pie chart.
    (3) a dot chart.
(b) Which chart do you prefer for purposes of presentation? Why?
(c) For the data on carbonated soft drink market share, develop the appropriate graph to pinpoint the "vital few" from the "trivial many."
(d) Analyze the data and summarize your findings.
(e) ACTION Write a letter to the food editor of your local newspaper explaining your findings.
(f) **(Class Project)** Let each student in the class respond to the question: "Which type of carbonated soft drink do you most prefer?" so that the teacher may tally the results into a summary table on the blackboard.
    (1) Convert the data to percentages and construct a Pareto diagram.
    (2) Compare and contrast the findings from the class with those obtained nationally based on market shares. What do you conclude? Discuss.

5.51 The following data represent the global market sales of all products manufactured by Motorola, Inc. in 1992 and a stratification of its net sales by business segment:

Region	Market Sales (in %)
Asia-Pacific	15
Europe	21
Japan	7
United States	48
Other	9
Total	100

Business Segment	Net Sales (in %)
Communications	29
General systems	26
Government electronics	5
Information systems	4
Semiconductor	32
Other	4
Total	100

Source: *The New York Times*, October 31, 1993, Sect. 3, p. 6.

(a) For the data on global market sales of all products, construct:
  (1) a bar chart.
  (2) a pie chart.
  (3) a dot chart.
(b) Which chart do you prefer for purposes of presentation? Why?
(c) For the data on net sales by business segment, develop the appropriate graph to pinpoint the "vital few" from the "trivial many."
(d) Analyze the data and summarize your findings.
(e) ACTION Write a letter to your marketing professor explaining your findings.

5.52 The following table provides a percentage breakdown of where personal computers were sold in 1987 and in 1993:

Type	Percentage of Sales	
	1987	1993
Direct response	0	14
Direct sellers	17	4
Mail order	4	3
Mass merchants	3	8
Superstores	0	6
Trade dealers	60	44
Value-added resellers	11	13
Other	5	8
Totals	100	100

Source: The New York Times, May 30, 1993, p. F5.

(a) For each year construct an appropriate graph and analyze the data.
(b) **ACTION** Write a letter to your marketing professor discussing the implications of your analysis of these shifting trends.

5.53 Facility planning has become an important aspect of health care marketing. Hospitals must consider plans for renovation in order to attract and satisfy patients. The following table shows the percentage breakdown within geographical region of the primary reasons hospitals gave for renovation in a recent year. The survey contained 2,770 responses.

Primary Reasons for Hospital Renovation	Percent Response (U.S. region)			
	Midwest	North	South	West
Outdated	25.2	26.0	25.8	28.5
Too small	20.6	21.3	23.0	23.5
New technology	15.6	12.8	15.2	13.6
Consumer demand	12.6	10.6	11.4	12.4
New service	11.2	10.6	11.4	8.4
MD loyalty	7.7	8.9	9.2	7.5
Other	5.8	8.0	3.2	4.9
No answer	1.3	1.8	0.8	1.2
Totals	100.0	100.0	100.0	100.0

Source: Reprinted from *Hospitals*, vol. 64, No. 4, by permission, February 20, 1990.

(a) For each region construct an appropriate graph and analyze the data.
(b) **ACTION** Write a letter to your marketing professor discussing the implications of your analysis.

5.54 **(Class Project)** Let each student in the class be cross-classified based on gender (male, female) and current employment status (yes, no) so that the results are tallied on the blackboard.
(a) Construct a table with either row or column percentages, depending on which you think is more informative.
(b) What would you conclude from this study?
(c) What other variables would you want to know regarding employment in order to enhance your findings?

5.55 Develop (in outline form) a supertable corresponding to job promotion based on gender, race, age group, education level, and occupation level.

5.56 Develop (in outline form) a supertable corresponding to graduate-school intent based on gender, race, age group, employment status, college major, and college grade-point average.

5.57 Develop (in outline form) a supertable corresponding to cigarette smoking based on gender, age group, occupation level, and education level.

Collaborative Learning Mini-Case Projects

Note: The class is to be divided into groups of three or four students each. One student is initially selected to be project coordinator, another student is project recorder, and a third student is the project timekeeper. In order to give each student experience in developing teamwork and leadership skills, rotation of positions should follow each project. At the beginning of each project, the students should work silently and individually for a short, specified period of time. Once each student has had the opportunity to study the issues and reflect on his/her possible answers, the group is convened and a group discussion follows. If all the members of a group agree with the solutions, the coordinator is responsible for submitting the team's project solution to the instructor with student signatures indicating such agreement. On the other hand, if one or more team members disagree with the solution offered by the majority of the team, a minority opinion may be appended to the submitted project, with signature(s) on it.

CL 5.1 Refer to CL 3.1 on page 101 and CL 4.1 on page 165. Your group, the ________ Corporation, has been hired to assist the research analyst at the college advisory service company in finishing a report regarding tuition rates charged to out-of-state residents by colleges and universities in different regions of the country. In particular, using Special Data Set 1 of Appendix D on pages D1–D5 regarding some characteristics of all colleges and universities in the states of Texas, North Carolina, and Pennsylvania, the ________ Corporation is ready to:

(a) Form a separate contingency table (based on row percentages) of type of institution (private or public) and location (rural, suburban, or urban) for each of the three states.
(b) Analyze the data through a comparative analysis of the three states.
(c) Write and submit an executive summary, attaching all tables.
(d) Prepare and deliver a ten-minute oral presentation to the marketing manager.

Data file
C&U.DAT

CL 5.2 Refer to CL 3.2 on page 101 and CL 4.2 on page 165. Your group, the ________ Corporation, has been hired by the food editor of a popular family magazine to study the cost and nutritional characteristics of ready-to-eat cereals. Armed with Special Data Set 2 of Appendix D on pages D6–D7 displaying useful information on 84 such cereals, the ________ Corporation is ready to:

(a) Form a contingency table cross-classifying type of ready-to-eat cereal (high fiber, moderate fiber, low fiber) with level of calories per serving (below 155, at or above 155).
(b) Thoroughly analyze the data.
(c) Write and submit an executive summary, attaching all tables.
(d) Prepare and deliver a ten-minute oral presentation to the food editor of the magazine.

Data file
CEREAL.DAT

CL 5.3 Refer to CL 3.3 on page 102 and CL 4.3 on page 165. Your group, the ________ Corporation, has been hired by the marketing director of a manufacturer of well-known men's and women's fragrances to study the characteristics of currently available fragrances. Armed with Special Data Set 3 of Appendix D on pages D8–D9 displaying useful information on 83 such fragrances, the ________ Corporation is ready to:

(a) Form a contingency table cross-classifying type of fragrance (perfume, cologne, or "other") with intensity of fragrance (very strong, strong, medium, or mild).
(b) Construct a table based on total percentages.
(c) Construct a table based on row percentages.
(d) Construct a table based on column percentages.
(e) Repeat (a) through (d) for women's fragrances only.
(f) Repeat (a) through (d) for men's fragrances only.
(g) Compare and contrast the results in (e) and (f).
(h) Write and submit an executive summary, attaching all tables.
(i) Prepare and deliver a ten-minute oral presentation to the marketing director.

Data file
FRAGRAN.DAT

Data file
CAMERA.DAT

CL 5.4 Refer to CL 3.4 on page 102 and CL 4.4 on page 166. Your group, the ________ Corporation, has been hired by the travel editor of a well-known newspaper who is preparing a feature article on compact 35-mm cameras. Armed with Special Data Set 4 of Appendix D on pages D10–D11 displaying useful information on 59 35-mm cameras, the ________ Corporation is ready to:

(a) Form a contingency table cross-classifying type of 35-mm focal length camera (multiple long, multiple medium, multiple short, automatic single, fixed single) with price level (under $200, $200 or more).
(b) Thoroughly analyze the data.
(c) Write and submit an executive summary, attaching all tables.
(d) Prepare and deliver a ten-minute oral presentation to the travel editor.

Case Study C—Preparing for National Network Television Program

In preparation for the upcoming election year, Karen Miller, the Program Director of a national network television station, wants to present a prime time television special in late October, one week before the election is held. She has just retained the consulting services of Dr. William Gold, an internationally known Professor of Journalism and Politics at a major Ivy League university, and has commissioned him to analyze critically the results of the 1992 presidential election and tie this to current political opinion. "When aired, I want the first half of the program to present the results from two viewpoints—Republican and Democrat. Let's let the audience reflect on the two historical perspectives," said Ms. Miller. She continued: "The second half of the program should present the current political pulse and analyze potential reasons for possibly emerging trends or shifts in current voting behavior. Make sure that key business and economic issues are given proper consideration along with social issues dealing with healthcare, welfare, and crime." Dr. Gold responded: "Karen, I am excited about working with you and your network on this worthwhile project and I am looking forward to this assignment. We really have the opportunity to inform and educate the American people on the current issues and to put this in perspective with respect to the '92 presidential election. I'll have a preliminary report analyzing the '92 presidential election on your desk one month from today. And I'll also include an outline for a survey to measure the current political pulse that you'll want used for the second half of the program."

This semester you have been assigned to be Dr. Gold's research assistant. He has explained to you the essence of the above conversation he had with the network Program Director, Karen Miller, and he has provided you with a supertable (see page 201) describing key aspects of the '92 presidential election. You are to prepare the first draft of the report. You are to meet with Dr. Gold in two weeks, at which time you will submit the first draft and give a fifteen-minute oral presentation. After some thought on this matter, and realizing that the supertable (based on a sample of 15,490 voters) will enable you to provide a thorough analysis, you decide to:

Supertable of the 1992 Presidential Election

% Total Vote		Clinton (%) 43	Bush (%) 38	Perot (%) 19
	Gender:			
46	Men	41	38	21
54	Women	46	37	17
	Race:			
87	Whites	39	41	20
8	Blacks	82	11	7
3	Hispanics	62	25	14
1	Asians	29	55	16
	Marital Status:			
65	Married	40	40	20
35	Unmarried	49	33	18
	Age Group:			
22	18–29 years old	44	34	22
38	30–44 years old	42	38	20
24	45–59 years old	41	40	19
16	60 and older	50	38	12
	Education:			
6	Not a high school graduate	55	28	17
25	High school graduate	43	36	20
29	Some college education	42	37	21
24	College graduate	40	41	19
16	Postgraduate education	49	36	15
	Religion:			
49	White Protestant	33	46	21
27	Catholic	44	36	20
4	Jewish	78	12	10
17	White born-again Christian	23	61	15
	Family Income:			
14	Under $15,000	59	23	18
24	$15,000–$29,999	45	35	20
30	$30,000–$49,999	41	38	21
20	$50,000–$74,999	40	42	18
13	$75,000 and over	36	48	16
	Family's Financial Situation:			
25	Better today	24	62	14
41	Same today	41	41	18
34	Worse today	61	14	25
	Residence:			
24	From the East	47	35	18
27	From the Midwest	42	37	21
30	From the South	42	43	16
20	From the West	44	34	22
	Employment Status:			
68	Employed	42	38	20
5	Full-time student	50	35	15
6	Unemployed	56	24	20
8	Homemaker	36	45	19
13	Retired	51	36	13
	Party Affiliation & View:			
13	Liberal Democrats	85	4	11
20	Moderate Democrats	76	10	14
6	Conservative Democrats	60	24	16
5	Liberal Independents	54	16	30
14	Moderate Independents	42	28	30
7	Conservative Independents	18	54	28
2	Liberal Republicans	17	54	29
15	Moderate Republicans	15	63	21
18	Conservative Republicans	5	82	14

Note: The above data, extracted from *The New York Times*, November 5, 1992, p. B9, were collected by *Voter Research and Surveys* based on a questionnaire completed by 15,490 voters leaving 300 polling places around the nation on election day in 1992.

(a) Develop an outline for the components of the supertable that need to be stressed:
   (1) to present the Democrat viewpoint.
   (2) to present the Republican viewpoint.
(b) Include appropriate charts corresponding to these categorical variables listed in the rows of the supertable.
(c) Include appropriate contingency tables extracted from the supertable.
(d) Prepare the written draft and the oral presentation.

Endnote

1. An interesting question is how to display a category for which no observations are recorded. For example, Table 5.1 on page 171 presents five possible institutional classifications. When constructing such a summary table and tallying the observations into their appropriate institutional classifications, what if one of these categories, say "special schools," contained no observations? It may be argued that a dot chart would be superior to a bar chart in such circumstances because placing a large dot at the origin more accurately portrays a category containing no observations than drawing a vertical line at the origin to represent the width of a bar having no length.

References

1. Cleveland, W. S., "Graphs in Scientific Publications," *The American Statistician,* vol. 38 (November 1984), pp. 261–269.
2. Cleveland, W. S., "Graphical Methods for Data Presentation: Full Scale Breaks, Dot Charts, and Multibased Logging," *The American Statistician,* vol. 38 (November 1984), pp. 270–280.
3. Cleveland, W. S., and R. McGill, "Graphical Perception: Theory, Experimentation, and Application to the Development of Graphical Methods," *Journal of the American Statistical Association,* vol. 79 (September 1984), pp. 531–554.
4. Croxton, F., D. Cowden, and S. Klein, *Applied General Statistics,* 3d ed. (Englewood Cliffs, NJ: Prentice-Hall, 1967).
5. *Microsoft EXCEL for Windows: Step by Step* (Redmond, WA: Microsoft Press, 1993).
6. Norusis, M., *SPSS Guide to Data Analysis for SPSS-X with Additional Instructions for SPSS/PC+* (Chicago, IL: SPSS Inc., 1986).
7. *SAS User's Guide Version 6* (Raleigh, NC: SAS Institute, 1988).
8. *STATISTIX Version 4.0* (Tallahassee, FL: Analytical Software, Inc., 1992).
9. Tufte, E. R., *The Visual Display of Quantitative Information* (Cheshire, CT: Graphics Press, 1983).
10. Tufte E. R., *Envisioning Information* (Cheshire, CT: Graphics Press, 1990).

chapter 6

Basic Probability

CHAPTER OBJECTIVE

To develop an understanding of the basic concepts of probability that are the foundation needed for the study of probability distributions and statistical inference.

6.1 Introduction

In this chapter we will study various rules of basic probability that can be used to evaluate the chance of occurrence of different phenomena. We will begin by discussing three different approaches to determining probabilities that can be used in different situations. We will then learn how to compute a variety of different types of probabilities. We will complete the chapter by discussing rules for counting different types of events, some of which will be revisited when the binomial distribution is discussed in Chapter 7.

Upon completion of this chapter, you should be able to:

1. Understand the different approaches to probability
2. Use either a contingency table or a Venn diagram to find probabilities
3. Understand the rules for finding simple, joint, and conditional probabilities and using the addition rule
4. Distinguish between mutually exclusive, collectively exhaustive, and independent events
5. Use Bayes' theorem to revise probabilities in light of new information
6. Use the various rules for counting the total number of outcomes

6.2 Objective and Subjective Probability

What do we mean by the word probability? **Probability** is the likelihood or chance that a particular event will occur. It could refer to

1. The chance of picking a black card from a deck of cards
2. The chance that an individual selected at random from the Employee Satisfaction Survey is satisfied with his or her job
3. The chance that a new consumer product on the market will be successful

In each of these examples, the probability involved is a proportion or fraction whose values range between 0 and 1 inclusively. We note that an event that has no chance of occurring (i. e., the **null event**) has a probability of zero, while an event that is sure to occur (i.e., the **certain event**) has a probability of one.

Each of the above examples refers to one of three approaches to the subject of probability. The first is often called the ***a priori* classical probability** approach. Here the probability of success is based on prior knowledge of the process involved. In the simplest case, when each outcome is equally likely, this chance of occurrence of the event may be defined as follows:

$$\text{Probability of occurrence} = \frac{X}{T} \qquad (6.1)$$

where

X = number of outcomes for which the event we are looking for occurs
T = total number of possible outcomes

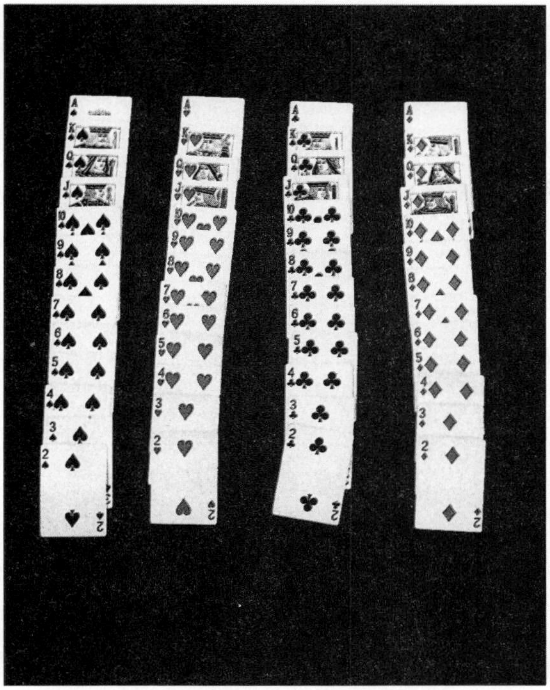

Figure 6.1
Standard deck of 52 playing cards.

A standard deck of cards is presented in Figure 6.1. If we want to find the probability of picking a black card (where we are defining black as a "success") the correct answer would be 26/52 or 1/2, since there are 26 black cards in a standard deck of cards.

What does this probability tell us? If we replace each card after it is drawn, does it mean that one out of the next two cards selected will be black? No, on the contrary, we cannot say for sure what will happen on the next several selections. However, we can say that in the long run, if this selection process is continually repeated, the proportion of black cards selected will approach .50.

In this first example, the number of successes and the number of outcomes are known from the composition of the deck of cards. However, in the second approach to probability, called the **empirical classical probability** approach, although the probability is still defined as the ratio of the number of favorable outcomes to the total number of outcomes, these outcomes are based upon observed data, not upon prior knowledge of a process.

In our second example, from the Employee Satisfaction Survey, the probability that an individual is satisfied with his or her job can be found by selecting a random sample of employees from the entire population. In Chapter 2 such a sample of 400 employees was selected (see Table 2.3 on pages 33–40). Of these 400 employees, 356 were satisfied with their jobs. Therefore, the probability that an employee selected at random is satisfied with his or her job (that is, the probability of occurrence) is 356/400 or .89.

The third approach to probability is called the **subjective probability** approach. Whereas in the previous two approaches the probability of a favorable event was computed *objectively*—either from prior knowledge or from actual data—*subjective* probability refers to the chance of occurrence assigned to an event by a particular individual. This chance may be quite different from the subjective probability assigned by another individual. For example, the inventor of a new toy may assign quite a different probability to the chance of success for the toy than the president of the company that is considering marketing the toy. The assignment of subjective probabilities to various events is usually based upon a combination of an individual's past experience, personal opinion, and analysis of a particular situation. Subjective probability is especially useful in making decisions in situations in which the probability of various events cannot be determined empirically.

Problems for Section 6.2

6.1 For each of the following, indicate whether the type of probability involved is an example of *a priori* classical probability, empirical classical probability, or subjective probability.
(a) That the next toss of a fair coin will land on head
(b) That the New York Mets will win next year's World Series
(c) That the sum of the faces of two dice will be seven
(d) That the train taking a commuter to work will be more than ten minutes late
(e) That a Republican will win the next presidential election in the United States

6.2 Give three examples of a *priori* classical probability.

6.3 Give three examples of empirical classical probability.

6.4 Give three examples of subjective probability.

6.3 Basic Probability Concepts

6.3.1 Sample Spaces and Events

The basic elements of probability theory are the outcomes of the process or phenomenon under study. Each possible type of occurrence is referred to as an event.

A **simple event** can be described by a single characteristic. The collection of all the possible events is called the **sample space.**

We can achieve a better understanding of these terms by referring to two examples. First, let us examine the standard deck of 52 playing cards (see Figure 6.1 on page 205) in which there are four suits (spades, hearts, clubs, and diamonds), each of which has 13 different cards (ace, king, queen, jack, 10, 9, 8, 7, 6, 5, 4, 3, 2).

If we randomly select a card from the deck

1. What is the probability the card is black?
2. What is the probability the card is an ace?
3. What is the probability the card is a black ace?
4. What is the probability the card is black *or* an ace?
5. If we knew that the card selected was black, what is the probability that it is also an ace?

As a second example, let us refer to the data collected in the Employee Satisfaction Survey discussed in Chapter 2. Suppose that from the total sample of 400 employees we pick a single person at random.

1. What is the probability that the employee is satisfied (either very or moderately) with his or her job?
2. What is the probability that the employee has "advanced" in the organization (either rapidly or steadily)?
3. What is the probability that the employee is satisfied *and* has "advanced" in the organization?
4. What is the probability that the employee is satisfied *or* has "advanced" in the organization?
5. Suppose we knew that the employee has "advanced" in the organization. What would be the probability that the employee is satisfied with his or her job?

In the case of the deck of cards, the sample space consists of the entire deck of 52 cards, made up of various events, depending on how they are classified. For example, if the events are classified by suit, there are four events: spade, heart, club, and diamond. If the events are classified by card value, there are 13 events: ace, king, . . . , and 2. On the other hand, from the Employee Satisfaction Survey, the sample space is based on the responses obtained from the 400 employees. The simple events for the two questions of interest here are as follows:

1. For question 9 pertaining to job satisfaction, there are four simple events: very satisfied, moderately satisfied, a little dissatisfied, and very dissatisfied. For our purposes in this chapter, we will collapse these into two simple events: (1) "satisfied," which consists of very satisfied and moderately satisfied, and (2) "dissatisfied," which consists of a little dissatisfied and very dissatisfied.
2. For question 20 pertaining to advancement, there are also four simple events: advanced rapidly, made steady advances, stayed on the same level, and lost some ground. For our purposes in this chapter, we will collapse these into two simple events: (1) "advanced," which consists of advanced rapidly and made steady advances, and (2) "did not advance," which consists of stayed on the same level and lost some ground.

The manner in which the sample space is subdivided depends on the types of probabilities that are to be determined. With this in mind, it is of interest to define both the complement of an event and a joint event as follows:

The **complement of event *A*** includes all events that are not part of event *A*. It is given by the symbol A'.

The complement of the event black would consist of all those cards that were not black (that is, all the red cards). The complement of spade would contain all cards that were not spades (that is, diamonds, hearts, and clubs). The complement of satisfied with the job is not satisfied with the job.

A **joint event** is an event that has two or more characteristics.

The event black ace is a joint event, since the card must be both black *and* ace in order to qualify as a black ace. In a similar manner, the event "the employee is satisfied *and* has advanced in the organization" is a joint event, since the employee must be satisfied with the job *and* also have advanced in the organization.

6.3.2 Contingency Tables and Venn Diagrams

There are several ways in which a particular sample space can be viewed. The first method involves assigning the appropriate events to a **table of cross-classifications.** Such a table is also called a **contingency table** (see Section 5.5).

If the two variables of interest for the card example were "presence of ace" and "color of card," the contingency table would look as shown in Table 6.1.

Table 6.1 Contingency table for face-color variables.

	Red	Black	Totals
Ace	2	2	4
Non-Ace	24	24	48
Totals	26	26	52

The values in each cell of the table were obtained by subdividing the sample space of 52 cards according to the number of aces and the color of the card. It can be noted that if the row and column (margin) totals are known, only one cell entry in this 2 × 2 table is needed in order to obtain the entries in the remaining three cells.

The contingency table for the 400 sampled employees at Kalosha Industries is developed by using a computer package to cross-classify the two variables of interest, employee job satisfaction and advancement in the organization. The table with the categories collapsed into satisfied and not satisfied, and advanced and did not advance, is displayed as Table 6.2.

Table 6.2 Contingency table for job satisfaction and advancement in the organization.

	Advancement		
Job Satisfaction	Yes	No	Total
Yes	194	162	356
No	14	30	44
Totals	208	192	400

As we have seen in Section 5.5, a contingency table provides a clear presentation of the number of possible outcomes of the relevant variables.

The second way to present the sample space is by using a **Venn diagram.** This diagram graphically represents the various events as "unions" and "intersections" of circles.

Figure 6.2 represents a typical Venn diagram for a two-variable situation, with each variable having only two events (A and A', B and B'). The circle on the left (the darker one) represents all events that are part of A. The circle on the right (the lighter one) represents all events that are part of B. The area contained within circle A *and* circle B (center area) is the **intersection** of A and B (written as $A \cap B$), since this area is part of A and is also part of B. The total area of the two circles is the **union** of A and B (written as $A \cup B$) and contains all outcomes that are part

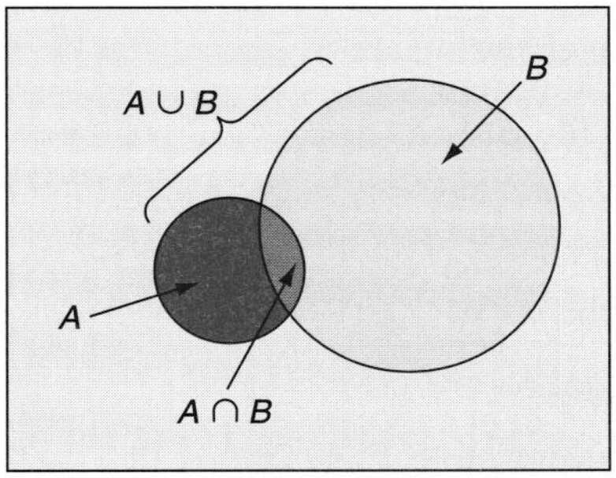

Figure 6.2
Venn diagram for events *A* and *B*.

of event *A*, part of event *B*, or part of both *A* and *B*. The area in the diagram outside $A \cup B$ contains those outcomes that are neither part of *A* nor part of *B*.

In order to develop a Venn diagram, *A* and *B* must be defined. It does not matter which event is defined as *A* or *B*, as long as we are consistent in evaluating the various events.

For the card-playing example, the events can be defined as follows:

$$A = \text{ace} \qquad B = \text{black}$$

$$A' = \text{non-ace} \qquad B' = \text{red}$$

In drawing the Venn diagram (see Figure 6.3), the value of the intersection of *A and B* must be determined so that the sample space can be divided into its parts. $A \cap B$ consists of all black aces in the deck (that is, the two outcomes ace of spades and ace of clubs).

Since there are two black aces, the remainder of event *A* (ace) consists of the red aces (there are two). The remainder of event *B* (black cards) consists of all black cards that are not aces (there are 24). The remaining cards are those that are neither black nor ace (there are also 24).

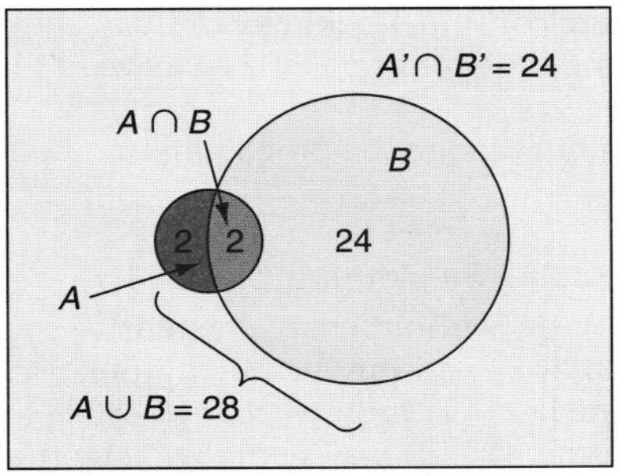

Figure 6.3
Venn diagram for card-deck example.

Problems for Section 6.3

● 6.5 In the past several years, credit card companies have made an aggressive effort to solicit new accounts from college students. Suppose that a sample of 200 students at your college indicated the following information in terms of whether the student possessed a bank credit card and/or a travel and entertainment credit card (see top of page 210):

Bank Credit Card	Travel and Entertainment Credit Card	
	Yes	No
Yes	60	60
No	15	65

(a) Give an example of a simple event.
(b) Give an example of a joint event.
(c) What is the complement of having a bank credit card?
(d) Why is "having a bank credit card *and* having a travel and entertainment credit card" a joint event?
(e) Set up a Venn diagram.

6.6 Numerous intensive studies have been conducted of consumer planning for the purchase of durable goods such as television sets, refrigerators, washing machines, stoves, and automobiles. In one such study, 1,000 individuals in a randomly selected sample were asked whether they were planning to buy a new television in the next 12 months. A year later the same persons were interviewed again to find out whether they actually bought a new television. The response to both interviews is cross-tabulated below:

	Buyers	Nonbuyers	Totals
Planned to buy	200	50	250
Did not plan to buy	100	650	750
Totals	300	700	1,000

(a) Give an example of a simple event.
(b) Give an example of a joint event.
(c) What is the complement of "planned to buy"?
(d) Set up a Venn diagram.

● 6.7 A sample of 500 respondents was selected in a large metropolitan area in order to determine various information concerning consumer behavior. Among the questions asked was "Do you enjoy shopping for clothing?" Of 240 males, 136 answered yes. Of 260 females, 224 answered yes.
(a) Set up a 2 × 2 table or a Venn diagram to evaluate the probabilities.
(b) Give an example of a simple event.
(c) Give an example of a joint event.
(d) What is the complement of "enjoy shopping for clothing"?

6.8 A company has made available to its employees (without charge) extensive health club facilities that may be used before work, during the lunch hour, after work, and on weekends. Records for the last year indicate that of 250 employees, 110 used the facilities at some time. Of 170 males employed by the company, 65 used the facilities.
(a) Set up a 2 × 2 table or a Venn diagram to evaluate the probabilities of using the facilities.
(b) Give an example of a simple event.
(c) Give an example of a joint event.
(d) What is the complement of "used the health club facilities"?

6.9 Each year ratings are compiled concerning the performance of new cars during the first 90 days of use. Suppose that the cars have been categorized according to two attributes, whether or not the car needs warranty-related repair (Yes or No) and the country in which the company manufacturing the car is based

(United States, not United States). Based on the data collected, the probability that the new car needs a warranty repair is .04, the probability that the car is manufactured by an American-based company is .60, and the probability that the new car needs a warranty repair *and* was manufactured by an American-based company is .025.

(a) Set up a 2 × 2 table or a Venn diagram to evaluate the probabilities of a warranty-related repair.
(b) Give an example of a simple event.
(c) Give an example of a joint event.
(d) What is the complement of "manufactured by an American-based company"?

6.4 Simple (Marginal) Probability

Thus far we have focused on the meaning of probability and on defining and illustrating several sample spaces. We shall now begin to answer some of the questions posed in the previous sections by developing rules for obtaining different types of probability.

The most obvious rule for probabilities is that they range in value from 0 to 1. An impossible event has probability 0 of occurring, and a certain event has probability 1 of occurring. **Simple probability** refers to the probability of occurrence of a simple event, *P(A)*, such as

- The probability of selecting a black card
- The probability of selecting an ace
- The probability the employee is satisfied with the job
- The probability that the employee has advanced in the organization

We have already noted that the probability of selecting a black card is 26/52 or 1/2, since there are 26 black cards in the 52-card deck.

How would we find the probability of picking an ace from the deck? We would find the number of aces in the deck by totaling the black aces and the red aces in the deck:

$$P(\text{Ace}) = \frac{\text{number of aces in deck}}{\text{number of cards in deck}} = \frac{\text{number of red aces} + \text{number of black aces}}{\text{total number of cards}} = \frac{2+2}{52} = \frac{4}{52}$$

Simple probability is also called **marginal probability,** since the total number of successes (aces in this case) can be obtained from the appropriate margin of the contingency table (see Table 6.1 on page 208).

The probability of an ace, *P*(*A*), also could be obtained from the Venn diagram (Figure 6.3 on page 209) by looking at the number of outcomes contained in circle *A*. There are four: two contained in $A \cap B$ and two outside $A \cap B$. This, of course, gives us the same result as analyzing the contingency table.

Let us refer to the second example. We want to find the probability that a randomly selected employee is satisfied with his or her job. This probability can be determined by referring to the contingency table (Table 6.2 on page 208):

$$P(\text{satisfied with job}) = \frac{\text{number of employees satisfied with their job}}{\text{total number of employees in the sample}}$$

$$= \frac{356}{400} = .89$$

Problems for Section 6.4

● 6.10 Referring to Problem 6.5 on page 209, if a student is selected at random, what is the probability that
(a) The student has a bank credit card?
(b) The student does not have a bank credit card?
(c) The student has a travel and entertainment credit card?
(d) The student does not have a travel and entertainment credit card?

6.11 Referring to Problem 6.6 on page 210, if an individual is selected at random, what is the probability that in the last year he or she
(a) Has bought a new television?
(b) Planned to buy a new television?
(c) Did not plan to buy a new television?
(d) Has not bought a new television?

● 6.12 Referring to Problem 6.7 on page 210, what is the probability that a respondent chosen at random
(a) Is a male?
(b) Enjoys shopping for clothing?
(c) Is a female?
(d) Does not enjoy shopping for clothing?

6.13 Referring to Problem 6.8 on page 210, what is the probability that an employee chosen at random
(a) Is a male?
(b) Has used the health club facilities?
(c) Is a female?
(d) Has not used the health club facilities?

6.14 Referring to Problem 6.9 on page 210, what is the probability that a new car selected at random
(a) Needs a warranty-related repair?
(b) Is not manufactured by an American-based company?
(c) Does not need a warranty-related repair?
(d) Is manufactured by an American-based company?

6.5 Joint Probability

Whereas marginal probability refers to the occurrence of simple events, **joint probability** refers to phenomena containing two or more events, such as the probability of a black ace, a red queen, or an employee who is satisfied with the job *and* has advanced within the organization.

Recall that a joint event *A and B* means that both event *A and* event *B* must occur simultaneously. Referring to Table 6.1 on page 208, those cards that are black *and* ace consist only of the outcomes in the single cell "black ace." Since there are two black aces, the probability of picking a card that is a black ace is

$$P(\text{black } and \text{ ace}) = \frac{\text{number of black aces}}{\text{number of cards in deck}} = \frac{2}{52}$$

This result can also be obtained by examining the Venn diagram of Figure 6.3 on page 209. The joint event *A and B* (black ace) consists of the *intersection* ($A \cap B$) of events *A* (ace) and *B* (black), which contains two outcomes. Therefore the probability of a black ace is equal to 2/52.

The probability of choosing an employee who is satisfied with his or her job and has advanced within the organization would be obtained from Table 6.2 on page 208 in the following manner:

$$P(\text{satisfied } and \text{ advanced}) = \frac{194}{400} = .485$$

since there are 194 employees who are satisfied with their job *and* have advanced within the organization.

Now that we have discussed the concept of joint probability, the marginal probability of a particular event can be viewed in an alternative manner. We have already shown that the marginal probability of an event consists of a set of joint probabilities. For example, if *B* consists of two events, B_1 and B_2, then we can observe that *P(A)*, the probability of event *A*, consists of the joint probability of event *A* occurring with event B_1 and the joint probability of event *A* occurring with event B_2. Thus, in general,

$$P(A) = P(A \text{ and } B_1) + P(A \text{ and } B_2) + \cdots + P(A \text{ and } B_k) \qquad \textbf{(6.2)}$$

where $B_1, B_2, \ldots, B_k$ are *mutually exclusive* and *collectively exhaustive* events.

Two events are **mutually exclusive** if both events cannot occur at the same time.

Two events are **collectively exhaustive** if one of the events must occur.

For example, being male *and* being female are mutually exclusive and collectively exhaustive events. No one is both (they are mutually exclusive) and everyone is one or the other (they are collectively exhaustive).

Therefore, returning to our first example, the probability of an ace can be expressed as follows:

$$\begin{aligned} P(\text{ace}) &= P(\text{ace } \textit{and} \text{ red}) + P(\text{ace } \textit{and} \text{ black}) \\ &= \frac{2}{52} + \frac{2}{52} \\ &= \frac{4}{52} \end{aligned}$$

This is, of course, the same result that we would obtain if we added up the number of outcomes that made up the simple event "ace."

Problems for Section 6.5

● 6.15 Referring to Problem 6.5 on page 209, what is the probability that if a student is selected at random
(a) The student has a bank credit card *and* a travel and entertainment card?
(b) The student does not have a bank credit card *and* has a travel and entertainment card?
(c) The student has neither a bank credit card *nor* a travel and entertainment card?

6.16 Referring to Problem 6.6 on page 210, what is the probability that in the last year he or she
(a) Planned to buy *and* actually bought a new television?
(b) Planned to buy *and* actually did not buy a new television?
(c) Did not plan to buy *and* actually did not buy a new television?

● 6.17 Referring to Problem 6.7 on page 210, what is the probability that a respondent chosen at random
(a) Is a female *and* enjoys shopping for clothing?
(b) Is a male *and* does not enjoy shopping for clothing?
(c) Is a male *and* enjoys shopping for clothing?

6.18 Referring to Problem 6.8 on page 210, what is the probability that an employee chosen at random
(a) Is a female *and* has used the health club facilities?
(b) Is a male *and* has not used the health club facilities?
(c) Is a female *and* has not used the health club facilities?

6.19 Referring to Problem 6.9 on page 210, what is the probability that a new car chosen at random
(a) Needs a warranty repair *and* is manufactured by a company based in the United States?
(b) Needs a warranty repair *and* is not manufactured by a company based in the United States?
(c) Does not need a warranty repair *and* is not manufactured by a company based in the United States?

6.6 Addition Rule

Having developed a means of finding the probability of event A and the probability of event *"A and B,"* we should like to examine a rule (the **addition rule**) that is used for finding the probability of event *"A or B."* This rule for obtaining the probability of the *union* of A and B considers the occurrence of either event A or event B or both A and B.

The event "black *or* ace" would include all cards that were black, were aces, or were black aces. The event "the employee is satisfied with the job *or* has advanced in the organization" would include all employees who were satisfied with their job, had advanced in the organization, or had both these characteristics.

Suppose we refer to this latter example. Each cell of the contingency table (Table 6.2 on page 208) can be examined to determine whether it is part of the event in question. If we want to study the event "the employee is satisfied with the job *or* has advanced in the organization," from Table 6.2, the cell "is satisfied with the job *and* has not advanced in the organization" is part of the event, since it includes employees who are satisfied with the job. The cell "is not satisfied with the job *and* has advanced in the organization" is included because it contains employees who have advanced in the organization. Finally, the cell "is satisfied with the job *and* has advanced in the organization" has both the characteristics of interest.

Therefore, the probability can be obtained as follows:

$$
\begin{aligned}
P(\text{satisfied } \textit{or} \text{ advanced}) &= P(\text{satisfied } \textit{and} \text{ has not advanced}) \\
&\quad + P(\text{is not satisfied } \textit{and} \text{ has advanced}) \\
&\quad + P(\text{is satisfied } \textit{and} \text{ has advanced}) \\
&= \frac{162}{400} + \frac{14}{400} + \frac{194}{400} = \frac{370}{400} \\
&= .925
\end{aligned}
$$

The computation of $P(A \cup B)$, the probability of the event *A or B,* can be expressed in the following **general addition rule:**

$$P(A \cup B) = P(A \text{ or } B) = P(A) + P(B) - P(A \text{ and } B) \tag{6.3}$$

Applying this addition rule to the preceding example, we obtain the following result:

$$
\begin{aligned}
P(\text{satisfied } \textit{or} \text{ advanced}) &= P(\text{is satisfied with the job}) \\
&\quad + P(\text{has advanced in the organization}) \\
&\quad - P(\text{is satisfied } \textit{and} \text{ has advanced}) \\
&= \frac{356}{400} + \frac{208}{400} - \frac{194}{400} \\
&= \frac{370}{400} = .925
\end{aligned}
$$

The addition rule consists of taking the probability of A and adding it to the probability of B; the intersection of A *and* B must then be subtracted from this total because it has already been included twice in computing the probability of A *and* the probability of B. This can be clearly demonstrated by referring to the contingency table. If the outcomes of the event "is satisfied with the job" are added to those of the event "has advanced in the organization," then the joint event "is satisfied with the job *and* has advanced in the organization" (the intersection) has been included in each of these simple events. Therefore, since this has been "double counted," it must be subtracted to provide the correct result. In fact, in this example, if the joint event is not subtracted our result would be

$$\frac{356}{400} + \frac{208}{400} = \frac{564}{400}$$

which is impossible since no probability can exceed 1.0.

6.6.1 Mutually Exclusive Events

In certain circumstances, however, the joint probability need not be subtracted because it is equal to zero. Such situations occur when no outcomes exist for a particular event. For example, suppose that we wanted to know the probability of picking either a heart *or* a spade if we were selecting only one card from a standard deck of 52 playing cards. Using the addition rule, we have the following:

$$\begin{aligned} P(\text{heart } \textit{or} \text{ spade}) &= P(\text{heart}) + P(\text{spade}) - P(\text{heart } \textit{and} \text{ spade}) \\ &= \frac{13}{52} + \frac{13}{52} - \frac{0}{52} = \frac{26}{52} \end{aligned}$$

We realize that the probability that a card will be both a heart *and* a spade simultaneously is zero, since in a standard deck each card may take on only one particular suit. The intersection in this case is nonexistent (called the **null set**) because it contains no outcomes, since a card cannot be a heart and spade simultaneously.

As mentioned previously, whenever the joint probability does not contain any outcomes, the events involved are considered to be *mutually exclusive*. This refers to the fact that the occurrence of one event (a heart) means that the other event (a spade) cannot occur. Thus, the addition rule for mutually exclusive events reduces to

$$P(A \text{ or } B) = P(A) + P(B) \qquad \textbf{(6.4)}$$

6.6.2 Collectively Exhaustive Events

Now consider what the probability would be of selecting a card that was red *or* black. Since red and black are mutually exclusive events, using Equation (6.4) we would have

$$P(\text{red } or \text{ black}) = P(\text{red}) + P(\text{black})$$
$$= \frac{26}{52} + \frac{26}{52} = \frac{52}{52} = 1.0$$

The probability of red *or* black adds up to 1.0. This means that the card selected must be red or black, since they are the only colors in a standard deck. Since one of these events must occur, they are considered to be *collectively exhaustive* events.

Problems for Section 6.6

6.20 Explain the difference between a collectively exhaustive event and a mutually exclusive event and give an example of each.

6.21 For each of the following, tell whether the events that are created are (i) mutually exclusive, (ii) collectively exhaustive. If they are not, either reword the categories to make them mutually exclusive and collectively exhaustive or tell why this would not be useful.
(a) Registered voters were asked whether they registered as Republican or Democrat.
(b) Respondents were classified on car ownership into the categories American, European, Japanese, none.
(c) People were asked, "Do you currently live in (i) an apartment, (ii) a house?"
(d) A product was classified as defective or not defective.
(e) People were asked, "Do you intend to purchase a color television in the next six months?" (i) Yes, (ii) No

6.22 The probability of each of the following events is zero. For each, tell why. Tell what common characteristic of these events makes their probability zero.
(a) A person who is registered as a Republican and a Democrat.
(b) A product that is defective and not defective.
(c) A house that is heated by oil and by natural gas.

6.23 Referring to Problem 6.5 on page 209, if a student is selected at random, what is the probability that
(a) The student has a bank credit card *or* has a travel and entertainment card?
(b) The student does not have a bank credit card *or* has a travel and entertainment card?
(c) The student has a bank credit card *or* does not have a bank credit card?

6.24 Referring to Problem 6.6 on page 210, if an individual is selected at random, what is the probability that in the last year he or she
(a) Planned to buy a new television *or* actually bought a new television?
(b) Did not plan to buy a new television *or* did not actually buy a new television?
(c) Planned to buy a new television *or* did not plan to buy a new television?

6.25 Referring to Problem 6.7 on page 210, what is the probability that a respondent chosen at random
(a) Is a female *or* enjoys shopping for clothing?
(b) Is a male *or* does not enjoy shopping for clothing?
(c) Is a male *or* a female?

6.26 Referring to Problem 6.8 on page 210, what is the probability that an employee chosen at random
(a) Is a female *or* has used the health club facilities?
(b) Is a male *or* has not used the health club facilities?
(c) Has used the health club facilities *or* has not used the health club facilities?

6.27 Referring to Problem 6.9 on page 210, what is the probability that a new car chosen at random
(a) Needs a warranty repair *or* was manufactured by an American-based company?
(b) Needs a warranty repair *or* was not manufactured by an American-based company?
(c) Needs a warranty repair or does not need a warranty repair?

6.7 Conditional Probability

Each example that we have studied thus far in this chapter has involved the probability of a particular event when sampling from the entire sample space. However, how would we find various probabilities if certain information about the events involved were already known? For example, if we were told that the card was black, what would be the probability that the card was an ace? Or if we were told that the employee had advanced in the organization, what would be the probability he or she was satisfied with the job?

When we are computing the probability of a particular event A, given information about the occurrence of another event B, this probability is referred to as **conditional probability,** $P(A|B)$. The conditional probability $P(A|B)$ can be defined as follows:

$$P(A|B) = \frac{P(A \text{ and } B)}{P(B)} \quad (6.5)$$

where $P(A \text{ and } B)$ = joint probability of A and B
$P(B)$ = marginal probability of B

Rather than using Equation (6.5) for finding conditional probability, we can use the contingency table or Venn diagram. In the first example, we wish to find $P(\text{ace}|\text{black})$. Here the information is given that the card is black. Therefore, the sample space does not consist of all 52 cards in the deck; it consists only of the black cards. Of the 26 black cards, two are aces. Therefore, the probability of an ace, given that we know the card is black, is

$$P(\text{ace}|\text{black}) = \frac{\text{number of black aces}}{\text{number of black cards}} = \frac{2}{26}$$

This result (2/26) can also be obtained by using Equation (6.5) as follows:

If

$$P(A|B) = \frac{P(A \text{ and } B)}{P(B)}$$

event A = ace

event B = black

then

$$P(\text{ace}|\text{black}) = \frac{2/52}{26/52}$$

$$= \frac{2}{26}$$

Let us now examine the second example mentioned, that of determining P(is satisfied with the job|has advanced in organization). Since the information given is that the employee has advanced in the organization, the sample space is reduced to those 208 individuals. Of these 208 employees, from Table 6.2 on page 208 we may observe that 194 are satisfied with the jobs. Therefore, the probability that an employee is satisfied with the job given that he or she has advanced in the organization may be computed as follows:

$$P(\text{is satisfied}|\text{has advanced}) = \frac{\begin{array}{c}\text{number of employees who}\\ \text{are satisfied with the job}\\ \textit{and}\text{ have advanced}\end{array}}{\begin{array}{c}\text{number of employees who}\\ \text{have advanced in the organization}\end{array}}$$

$$= \frac{194}{208}$$

Again, Equation (6.5) would provide the same answer, as follows:

$$P(A|B) = \frac{P(A \text{ and } B)}{P(B)}$$

where

$$\text{event } A = \text{is satisfied with the job}$$

$$\text{event } B = \text{has advanced in the organization}$$

$$P(\text{is satisfied}|\text{has advanced}) = \frac{P(\text{is satisfied } \textit{and} \text{ has advanced})}{P(\text{has advanced})}$$

$$= \frac{194/400}{208/400}$$

$$= \frac{194}{208}$$

Decision Trees In Table 6.2 on page 208, the employees were classified according to being satisfied with their job and also according to whether or not they had advanced in the organization. An alternative way to view the breakdown of the possibilities into four cells is through the use of a **decision tree.** Figure 6.4 on page 220 consists of a decision tree for these data.

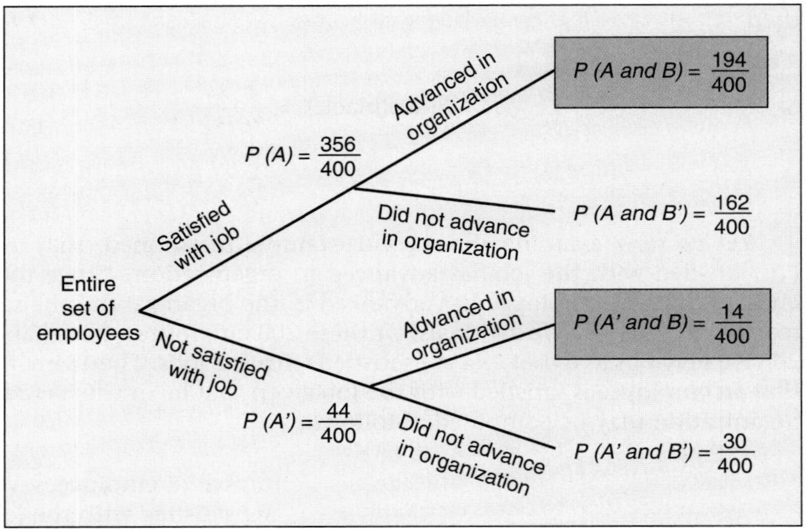

Figure 6.4
Decision tree for the data in Table 6.2.

In Figure 6.4 beginning at the left with the entire set of employees, there are two "branches" according to whether an employee is satisfied or not satisfied with the job. Each of these branches has two subbranches, corresponding to whether an employee has or has not advanced in the organization. The probabilities placed at the end of the initial branches represent the marginal probabilities of A [that is, $P(A)$] and A' [that is, $P(A')$], and the probabilities at the end of each of the four subbranches represent the joint probability for each combination of events A and B. The conditional probability can be obtained by dividing the joint probability of interest by the appropriate marginal probability.

For example, to obtain P(is satisfied with the job|has advanced in organization) we would take P(is satisfied with the job *and* has advanced in organization) and divide it by P(has advanced in organization). From Figure 6.4, we would have

$$P(A|B) = \frac{P(A \text{ and } B)}{P(B)}$$

$$P(\text{is satisfied}|\text{has advanced}) = \frac{194/400}{208/400}$$

$$= \frac{194}{208}$$

Note that the denominator, $P(B)$, is the sum of the probabilities of the two appropriate joint events, $P(A \text{ and } B) + P(A' \text{ and } B)$, the probability of satisfied with the job *and* having advanced in the organization plus the probability of not being satisfied with the job *and* having advanced in the organization.

● **Statistical Independence** In the first example, we observed that the probability that the card picked is an ace, given that we know it is black, is 2/26. We may remember that the probability of picking an ace out of the deck, P(ace), was 4/52, which reduces to 2/26. This result reveals some important information. The prior knowledge that the card was black did not affect the probability that the card was an ace. This characteristic is called **statistical independence** and can be defined as follows:

$$P(A|B) = P(A) \quad (6.6)$$

where $P(A|B)$ = conditional probability of A given B

$P(A)$ = marginal probability of A

Thus we may note that two events A and B are statistically independent if and only if $P(A|B) = P(A)$. In a 2×2 contingency table, once this holds for one combination of A and B it will be true for all others.[1] Here, "color of the card" and "being an ace" are statistically independent events. Knowledge of one event in no way affects the probability of the second event.

We should also like to determine whether being satisfied with the job is independent of having advanced in the organization. The proportion of those employees who are satisfied with the job, given they have advanced in the organization is $194/208 = .933$, and the proportion of all employees who are satisfied with the job is $356/400 = .89$. This result reveals some important information: The prior knowledge of advancement in an organization has slightly affected our prediction of job satisfaction. Thus, from a statistical perspective we can state that these two events can be considered to be somewhat associated, that is, not independent. The proportion of employees satisfied with their job is not precisely the same as the proportion of employees satisfied with their job given that they have advanced in the organization.

Problems for Section 6.7

● 6.28 Referring to Problem 6.5 on page 209
- (a) Assume that we know that the student has a bank credit card. What is the probability then that he or she has a travel and entertainment card?
- (b) Assume we know that the student does not have a travel and entertainment card. What then is the probability that he or she has a bank credit card?
- (c) Are the two events, having a bank credit card and having a travel and entertainment card, statistically independent? Explain.

6.29 Referring to Problem 6.6 on page 210
- (a) If the respondent planned to buy a new television, what is the probability that he or she actually bought one?
- (b) If the respondent did not plan to buy a new television, what is the probability that he or she did not buy a new television?
- (c) Are planning to buy a new television and actually buying one statistically independent? Explain.

● 6.30 Referring to Problem 6.7 on page 210
- (a) Suppose the respondent chosen is a female. What then is the probability that she does not enjoy shopping for clothing?
- (b) Suppose the respondent chosen enjoys shopping for clothing. What then is the probability that the individual is a male?
- (c) Are enjoying shopping for clothing and the gender of the individual statistically independent? Explain.

6.31 Referring to Problem 6.8 on page 210
(a) Suppose that we select a female employee of the company. What then is the probability that she has used the health club facilities?
(b) Suppose that we select a male employee of the company. What then is the probability that he has not used the health club facilities?
(c) Are gender of the individual and use of the health club facilities statistically independent? Explain.

6.32 Referring to Problem 6.9 on page 210
(a) Suppose we know that the car was manufactured by a company that is based in the United States. What then is the probability that the car needs a warranty repair?
(b) Suppose we know that the car was not manufactured by a company that is based in the United States. What then is the probability that the car needs a warranty repair?
(c) Are need for a warranty repair and location of the company manufacturing the car statistically independent?

6.8 Multiplication Rule

The formula for conditional probability can be manipulated algebraically so that the joint probability *P(A and B)* can be determined from the conditional probability of an event. Using Equation (6.5)

$$P(A|B) = \frac{P(A \text{ and } B)}{P(B)}$$

and solving for the joint probability *P(A and B)*, we have the **general multiplication rule:**

$$P(A \text{ and } B) = P(A|B)P(B) \tag{6.7}$$

To demonstrate the use of this multiplication rule we turn to an example. Suppose that 20 marking pens are displayed in a stationery store. Six are red and 14 are blue. We are to select two markers randomly from the set of 20. What is the probability that both markers selected are red? Here the multiplication rule can be used in the following way:

$$P(A \text{ and } B) = P(A|B)P(B)$$

Therefore if

$$A_R = \text{second marker selected is red}$$
$$B_R = \text{first marker selected is red}$$

we have

$$P(A_R \text{ and } B_R) = P(A_R|B_R)\,P(B_R)$$

The probability that the first marker is red is 6/20, since 6 of the 20 markers are red. However, the probability that the second marker is also red depends on the result of the first selection. If the first marker is not returned to the display after its color is determined (sampling *without* replacement), then the number of markers remaining will be 19. If the first marker is red, the probability that the second also is red is 5/19, since 5 red markers remain in the display. Therefore, using Equation (6.7), we have the following:

$$P(A_R \text{ and } B_R) = \left(\frac{5}{19}\right)\left(\frac{6}{20}\right)$$
$$= \frac{30}{380} = .079$$

However, what if the first marker selected is returned to the display after its color is determined? Then the probability of picking a red marker on the second selection is the same as on the first selection (sampling *with* replacement), since there are 6 red markers out of 20 in the display. Therefore, we have the following:

$$P(A_R \text{ and } B_R) = P(A_R|B_R)P(B_R)$$
$$= \left(\frac{6}{20}\right)\left(\frac{6}{20}\right)$$
$$= \frac{36}{400} = .09$$

This example of sampling *with* replacement illustrates that the second selection is independent of the first, since the second probability was not influenced by the first selection. Therefore, the **multiplication rule for independent events** can be expressed as follows [by substituting $P(A)$ for $P(A|B)$]:

$$P(A \text{ and } B) = P(A)P(B) \quad \textbf{(6.8)}$$

If this rule holds for two events, A and B, then A and B are statistically independent. Therefore, there are two ways to determine statistical independence.

1. Events A and B are statistically independent if and only if $P(A|B) = P(A)$.
2. Events A and B are statistically independent if and only if $P(A \text{ and } B) = P(A)P(B)$.

It should be noted that for a 2 × 2 contingency table, if this is true for one joint event, it will be true for all joint events.[1] For example, if the probability of a card being an ace is independent of it being black, then the probability of it being an ace is independent of it being red, the probability of not being an ace is independent of it being black, and the probability of not being an ace is independent of it being red.

Now that we have discussed the multiplication rule, we can write the formula for marginal probability [Equation (6.1)] as follows. If

$$P(A) = P(A \text{ and } B_1) + P(A \text{ and } B_2) + \cdots + P(A \text{ and } B_k)$$

then, using the multiplication rule, we have

$$P(A) = P(A|B_1)P(B_1) + P(A|B_2)P(B_2) + \cdots + P(A|B_k)P(B_k) \quad \textbf{(6.9)}$$

where $B_1, B_2, \ldots, B_k$ are k mutually exclusive and collectively exhaustive events.

We may illustrate this formula by referring to Table 6.1 on page 208. Using Equation (6.9), we may compute the probability of an ace as follows:

$$\begin{aligned} P(A) &= P(A|B_1)P(B_1) + P(A|B_2)P(B_2) \\ &= \left(\frac{2}{26}\right)\left(\frac{26}{52}\right) + \left(\frac{2}{26}\right)\left(\frac{26}{52}\right) \\ &= \frac{2}{52} + \frac{2}{52} \\ &= \frac{4}{52} \end{aligned}$$

Problems for Section 6.8

● 6.33 Referring to Problem 6.5 on page 209, use Equation (6.8) to determine whether having a bank credit card is statistically independent of having a travel and entertainment credit card.

6.34 Referring to Problem 6.6 on page 210, use Equation (6.8) to determine whether planning to buy a new television and actually buying one are statistically independent.

● 6.35 Referring to Problem 6.7 on page 210, use Equation (6.8) to determine whether enjoying shopping for clothing is statistically independent of the gender of the individual.

6.36 Referring to Problem 6.8 on page 210, use Equation (6.8) to determine whether using the health club facilities is statistically independent of the gender of the individual.

6.37 Referring to Problem 6.9 on page 210, use Equation (6.8) to determine whether need for warranty repair is statistically independent of location of the company manufacturing the car.

6.38 Suppose you believe that the probability that you will get an A in Statistics is .6, and the probability that you will get an A in Organizational Behavior is .8. If these events are independent, what is the probability that you will get an A in both Statistics and Organizational Behavior? Give some plausible reasons why these events may not be independent, even though the teachers of these two subjects may not communicate about your work.

6.39 A standard deck of cards is being used to play a game. There are four suits (hearts, diamonds, clubs, and spades), each having 13 cards (ace, 2, 3, 4, 5, 6, 7, 8, 9, 10, jack, queen, and king), making a total of 52 cards. This complete deck is thoroughly mixed, and you will receive the first two cards from the deck *without* replacement.

(a) What is the probability that both cards are queens?

(b) What is the probability that the first card is a 10 *and* the second card is a 5 or 6?

(c) If we were sampling *with* replacement, what would be the answer in (a)?

(d) In the game of blackjack the picture cards (jack, queen, king) count as 10 points and the ace counts as either 1 or 11 points. All other cards are counted at their face value. Blackjack is achieved if your two cards total 21 points. What is the probability of getting blackjack in this problem?

6.40 A box of nine baseball gloves contains two left-handed gloves and seven right-handed gloves.

(a) If two gloves are randomly selected from the box *without* replacement, what is the probability that
(1) both gloves selected will be right-handed?
(2) there will be one right-handed glove *and* one left-handed glove selected?

(b) If three gloves are selected, what is the probability that all three will be left-handed?

(c) If we were sampling *with* replacement, what would be the answers to (a)(1) and (b)?

6.9 Bayes' Theorem

Conditional probability takes into account information about the occurrence of one event to find the probability of another event. This concept can be extended to revise probabilities based on new information and to determine the probability that a particular effect was due to a specific cause. The procedure for revising these probabilities is known as **Bayes' theorem** [since it was originally developed by Rev. Thomas Bayes (1702–1761); see Reference 2].

One interesting application of Bayes' theorem relates to the area of medical diagnosis testing. Suppose that the probability that a person has a certain disease is .03. Medical diagnostic tests are available to determine whether the person actually has the disease. If the disease is actually present, the probability that the medical diagnostic test will give a positive result (indicating that disease is present) is .90. If the disease is not actually present, the probability of a positive test result (indicating that it is present) is .02. Given this information, we would like to know the following:

1. If the medical diagnostic test has given a positive result (indicating that disease is present), what is the probability that the disease is actually present?
2. What proportion of all medical diagnostic tests indicate positive results (that the disease is present)?
3. If the medical diagnostic test has given a negative result (indicating that disease is not present), what is the probability that the disease is not present?

Bayes' theorem can be developed from the definitions of conditional and marginal probability in the following manner:

$$P(A \text{ and } B) = P(A|B)P(B) \quad \textbf{(6.10a)}$$

but also

$$P(A \text{ and } B) = P(B|A)P(A) \quad \textbf{(6.10b)}$$

From Equations (6.10b) and (6.10a) we have

$$P(B|A)P(A) = P(A|B)P(B)$$

so that, by dividing by *P(A)*, we get

$$P(B|A) = \frac{P(A|B)P(B)}{P(A)} \quad \textbf{(6.10c)}$$

From Equation (6.9), however,

$$P(A) = P(A|B_1)P(B_1) + P(A|B_2)P(B_2) + \cdots + P(A|B_k)P(B_k)$$

so that Bayes' theorem is

$$P(B_i|A) = \frac{P(A|B_i)P(B_i)}{P(A|B_1)P(B_1) + P(A|B_2)P(B_2) + \cdots + P(A|B_k)P(B_k)} \quad \textbf{(6.10d)}$$

where B_i is the ith event out of k mutually exclusive events.

Now we can use Bayes' theorem to determine the desired probabilities listed above for the medical diagnosis testing problem. Let

event D = has disease   event T = test is positive

event D' = does not have disease   event T' = test is negative

and

$P(D) = .03$ $P(T|D) = .90$

$P(D') = .97$ $P(T|D') = .02$

Let us answer the first question. If the medical diagnostic test has given a positive result (indicating that disease is present), what is the probability that the disease is actually present, [$P(D|T)$]? Using Equation (6.10d), we would have

$$P(D|T) = \frac{P(T|D)P(D)}{P(T|D)P(D) + P(T|D')P(D')}$$

$$= \frac{(.90)(.03)}{(.90)(.03) + (.02)(.97)}$$

$$= \frac{.0270}{.0270 + .0194} = \frac{.0270}{.0464}$$

$$= .582$$

The computation of the probabilities is summarized in Table 6.3 and displayed in the form of a decision tree in Figure 6.5. The probability that the disease is present given that the test was positive is only .582. This result may seem surprisingly low, given a 90% chance that the test would be positive if the disease is present. However, only 3% of the population have the disease, and there is also a 2% chance that the test will be positive in the 97% of the population that does not have the disease. If the company manufacturing the diagnostic testing equipment wanted to improve the probability that the disease is present given that the test is positive, it would have to increase the chance that the test would be positive if the disease is present and/or decrease the chance that the test will be positive in the 97% of the population that does not have the disease (see Problems 6.41 and 6.42 on page 228, respectively).

Table 6.3 Bayes' theorem calculations for the medical diagnosis problem.

Events D_i	Prior Probability $P(D_i)$	Conditional Probability $P(T\|D_i)$	Joint Probability $P(T\|D_i)P(D_i)$	Revised Probability $P(D_i\|T)$
D = has disease	.03	.90	.0270	.0270/.0464 = .582 = $P(D\|T)$
D' = does not have disease	.97	.02	.0194	.0194/.0464 = .418 = $P(D'\|T)$
			.0464	1.000

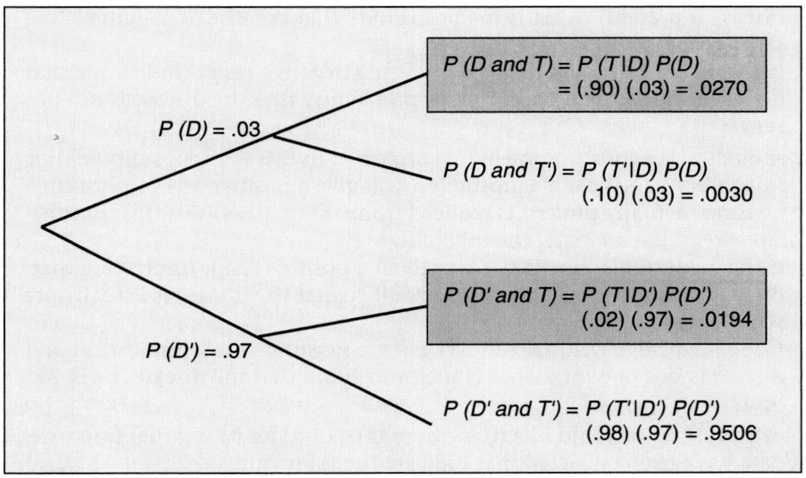

Figure 6.5
Decision tree for the medical diagnosis test problem.

To answer the second question, concerning the proportion of all medical diagnostic tests that indicate positive results (that the disease is present), we examine the denominator of Bayes' theorem. This represents the marginal probability of event T, a positive test result. Therefore the probability of a positive test result is .0464.

Let us now answer the third question. If the medical diagnostic test has given a negative result (indicating that disease is not present), what is the probability that the disease is not present? We would have

$$P(T'|D) = 1 - P(T|D) = 1 - .90 = .10$$
$$P(T'|D') = 1 - P(T|D') = 1 - .02 = .98$$

Using Equation (6.10d), we have

$$\begin{aligned} P(D'|T') &= \frac{P(T'|D')P(D')}{P(T'|D)P(D) + P(T'|D')P(D')} \\ &= \frac{(.98)(.97)}{(.10)(.03) + (.98)(.97)} \\ &= \frac{.9506}{.0030 + .9506} = \frac{.9506}{.9536} \\ &= .997 \end{aligned}$$

Thus, the probability that the disease is not present, given that the test was negative is .997.

Problems for Section 6.9

6.41 In the medical diagnosis problem just discussed in this section, suppose that the probability that the medical diagnostic test will give a positive result if the disease is actually present has been increased from .90 to .95. Given this information, we would like to know the following:
(a) If the medical diagnostic test has given a positive result (indicating that disease is present), what is the probability that the disease is actually present?
(b) If the medical diagnostic test has given a negative result (indicating that disease is not present), what is the probability that the disease is not present?

6.42 In the medical diagnosis problem just discussed in this section, suppose that the probability a medical diagnostic test will give a positive test result if the disease is not actually present is reduced from .02 to .01. Given this information, we would like to know the following:
(a) If the medical diagnostic test has given a positive result (indicating that disease is present), what is the probability that the disease is actually present?
(b) If the medical diagnostic test has given a negative result (indicating that disease is not present), what is the probability that the disease is not present?

6.43 A television station would like to measure the ability of its weather forecaster. Past data have been collected that indicate the following:

1. The probability the forecaster predicted sunshine on sunny days is .80.
2. The probability the forecaster predicted sunshine on rainy days is .40.
3. The probability of a sunny day is .60.

Find the probability that

(a) It will be sunny given that the forecaster has predicted sunshine.
(b) The forecaster will predict sunshine.

6.44 An advertising executive is studying television viewing habits of married men and women during prime-time hours. Based on past viewing records he has determined that during prime time husbands are watching television 60% of the time. It has also been determined that when the husband is watching television, 40% of the time the wife is also watching. When the husband is not watching television, 30% of the time the wife is watching television. Find the probability that

(a) If the wife is watching television, the husband is also watching television.
(b) The wife is watching television during prime time.

6.45 The Olive Construction Co. is determining whether it should submit a bid for the construction of a new shopping center. In the past, Olive's main competitor, Base Construction Co., has submitted bids 70% of the time. If Base Construction Co. does not bid on a job, the probability that the Olive Construction Co. will get the job is .50; if Base Construction Co. does bid on a job, the probability that the Olive Construction Co. will get the job is .25.

(a) If the Olive Construction Co. gets the job, what is the probability that the Base Construction Co. did not bid?
(b) What is the probability that the Olive Construction Co. will get the job?

6.46 A municipal bond rating service has three rating categories (A, B, and C). Suppose that in the past year, of the municipal bonds issued throughout the country, 70% were rated A, 20% were rated B, and 10% were rated C. Of the municipal bonds rated A, 50% were issued by cities, 40% by suburbs, and 10% by rural areas. Of the municipal bonds rated B, 60% were issued by cities, 20% by suburbs, and 20% by rural areas. Of the municipal bonds rated C, 90% were issued by cities, 5% by suburbs, and 5% by rural areas.

(a) If a new municipal bond is to be issued by a city, what is the probability that it will receive an A rating?
(b) What proportion of the municipal bonds are issued by cities?
(c) What proportion of the municipal bonds are issued by suburbs?

6.47 The marketing manager of a toy manufacturing firm is planning to introduce a new toy into the market. In the past, 40% of the toys introduced by the company have been successful and 60% have not been successful. Before the toy is actually marketed, market research is conducted and a report, either favorable or unfavorable, is compiled. In the past, 80% of the successful toys received favorable reports and 30% of the unsuccessful toys also received favorable reports.

(a) Suppose that market research gives a favorable report on a new toy. What is the probability that the new toy will be successful?
(b) What proportion of new toys receive favorable market research reports?

6.10 Counting Rules

Each rule of probability that we have discussed has involved counting the number of favorable outcomes and the total number of outcomes. In many instances, however, because of the large number of possibilities, it is not feasible to list each of the outcomes. In these circumstances, rules for counting have been developed. Five different counting rules will be discussed here.

First of all, suppose a coin was being flipped 10 times. How would we determine the number of different possible outcomes (the sequences of heads and tails)?

Counting Rule 1: If any one of k different mutually exclusive and collectively exhaustive events can occur on each of n trials, the number of possible outcomes is equal to

$$k^n \quad (6.11)$$

If a coin (having two sides) is tossed 10 times, the number of outcomes is $2^{10} = 1,024$. If a die (having six sides) is rolled twice, the number of different outcomes is $6^2 = 36$.

The second counting rule is a more general version of the first. To illustrate this rule, suppose that the number of possible events is different on some of the trials. For example, a state motor vehicle department would like to know how many license plate numbers would be available if the license plate consisted of three letters followed by three digits. The fact that three values are letters (each having 26 possible outcomes) and three positions are digits (each having 10 outcomes) leads to the second rule of counting.

Counting Rule 2: If there are k_1 events on the first trial, k_2 events on the second trial, . . . , and k_n events on the nth trial, then the number of possible outcomes is

$$(k_1)(k_2)\cdots(k_n) \quad (6.12)$$

Thus if a license plate consisted of three letters followed by three digits, the total number of possible outcomes would be $(26)(26)(26)(10)(10)(10) = 17,576,000$. Taking another example, if a restaurant menu had a price-fixed complete dinner that consisted of an appetizer, entree, beverage, and dessert and there was a choice of five appetizers, ten entrees, three beverages, and six desserts, the total number of possible dinners would be $(5)(10)(3)(6) = 900$.

The third counting rule involves the computation of the number of ways that a set of objects can be arranged in order. If a set of six textbooks is to be placed on a shelf, how can we determine the number of ways in which the six books may be arranged? We may begin by realizing that any of the six books could occupy the first position on the shelf. Once the first position is filled, there are five books to choose from in filling the second. This assignment procedure is continued until all the positions are occupied. This situation can be generalized as counting rule 3.

Counting Rule 3: The number of ways that all n objects can be arranged in order is

$$n! = n(n-1)\cdots(1) \quad (6.13)$$

where $n!$ is called *n factorial* and $0!$ is defined as 1.

The number of ways that six books could be arranged is

$$n! = 6! = (6)(5)(4)(3)(2)(1) = 720$$

In many instances we need to know the number of ways in which a subset of the entire group of objects can be arranged in order. Each possible arrangement is called a **permutation.** For example, modifying the preceding problem, if six textbooks are involved, but there is room for only four books on the shelf, how many ways can these books be arranged on the shelf?

Counting Rule 4: *Permutations:* The number of ways of arranging X objects selected from n objects in order is

$$\frac{n!}{(n - X)!} \tag{6.14}$$

Therefore, the number of ordered arrangements of four books selected from six books is equal to

$$\frac{n!}{(n - X)!} = \frac{6!}{(6 - 4)!} = \frac{6!}{2!} = \frac{(6)(5)(4)(3)(2)(1)}{(2)(1)} = 360$$

Finally, in many situations we are not interested in the *order* of the outcomes but only in the number of ways that X objects can be selected out of n objects, *irrespective of order.* This rule is called the rule of **combinations.**

Counting Rule 5: *Combinations:* The number of ways of selecting X objects out of n objects, irrespective of order, is equal to

$$\frac{n!}{X!(n - X)!} \tag{6.15}$$

This expression may be denoted by the symbol $\binom{n}{X}$.

Comparing this rule to the previous one, we see that it differs only in the inclusion of a term $X!$ in the denominator. This is because when we counted permutations, all of the arrangements of the X objects were distinguishable; with combinations, the $X!$ possible arrangements of objects are irrelevant. Therefore, the number of combinations of four books selected from six books is expressed by

$$\frac{n!}{X!(n - X)!} = \frac{6!}{4!(6 - 4)!} = \frac{6!}{4!\,2!} = \frac{(6)(5)(4)(3)(2)(1)}{(4)(3)(2)(1)(2)(1)} = 15$$

Problems for Section 6.10

6.48 If there are ten multiple-choice questions on a exam, each having three possible answers, how many different possibilities are there in terms of the sequence of correct answers?

6.49 A lock on a bank vault consists of three dials, each with 30 positions. In order for the vault to open when closed, each of the three dials must be in the correct position.
(a) How many different possible "dial combinations" are there for this lock?
(b) What is the probability that if you randomly select a position on each dial, you will be able to open the bank vault?
(c) Explain why "dial combinations" are not mathematical combinations expressed by Equation (6.15).

6.50 (a) If a coin is tossed seven times, how many different outcomes are possible?
(b) If a die is tossed seven times, how many different outcomes are possible?
(c) Discuss the differences in your answers to (a) and (b).

6.51 A particular brand of women's jeans can be ordered in seven different sizes, three different colors, and three different styles. How many different jeans would have to be ordered if a store wanted to have one pair of each type?

6.52 If each letter is used once, how many different four-letter "words" can be made from the letters E, L, O, and V?

6.53 There are seven teams in the Atlantic Division of the National Hockey League: Florida, New Jersey, New York Islanders, New York Rangers, Philadelphia, Tampa Bay, and Washington. How many different orders of finish are there for these seven teams? Do you *really* believe that all these orders are equally likely? Discuss.

6.54 Referring to Problem 6.53, how many different orders of finish are possible for the first four positions?

6.55 A gardener has six rows available in his vegetable garden to place tomatoes, eggplant, peppers, cucumbers, beans, and lettuce. Each vegetable will be allowed one and only one row. How many ways are there to position these vegetables in his garden?

6.56 The Big Triple at the local racetrack consists of picking the correct order of finish of the first three horses in the ninth race. If there are 12 horses entered in today's ninth race, how many Big Triple outcomes are there?

6.57 The Quinella at the local racetrack consists of picking the horses that will place first and second in a race *irrespective* of order. If eight horses are entered in a race, how many Quinella combinations are there?

6.58 A student has seven books that she would like to place in an attaché case. However, only four books can fit into the attaché case. Regardless of the arrangement, how many ways are there of placing four books into the attaché case?

6.59 A daily lottery is to be conducted in which two winning numbers are to be selected out of 100 numbers. How many different combinations of winning numbers are possible?

6.60 A reading list of articles for a course contains 20 articles. How many ways are there to choose three articles from this list?

6.11 Ethical Issues and Probability

Ethical issues may arise when any statements relating to probability are being presented for public consumption, particularly when these statements are part of an advertising campaign for a product or service. Unfortunately, a substantial portion of the population is not very comfortable with any type of numerical concept (see

Reference 4) and misinterprets the meaning of the probability. In some instances, the misinterpretation is not intended, but in other cases advertisements may unethically try to mislead potential customers.

One example of a potentially unethical application of probability relates to the sales of tickets to a state lottery in which the customer typically selects a set of numbers (let's say six) from a larger list of numbers (let's say 54 numbers). Although virtually all participants know that they are unlikely to win the lottery, they also have very little idea of how unlikely it is for them to select, for example, all six winning numbers out of the list of 54 numbers. In addition, they have even less idea how likely it is that they can win a consolation prize by selecting either four winning numbers or five winning numbers. Given this background, it seems to us that a recent advertising campaign in which a commercial for a state lottery said "We won't stop until we have made everyone a millionaire" is at the very least deceptive and at the worst unethical. Actually, given the fact that the lottery brings millions of dollars in revenue into the state treasury, the state is never going to stop running it, although in anyone's lifetime no one can be sure of becoming a millionaire by winning the lottery.

A second example of a potentially unethical application of probability relates to an investment newsletter that promises a 20% annual return on investment with a 90% probability. In such a situation, it seems imperative that the investment service needs to (1) explain the basis on which this probability estimate rests, (2) provide the probability statement in another format such as nine chances in ten, and (3) explain what happens to the investment in the 10% of the cases in which a 20% return is not achieved.

Problems for Section 6.11

6.61 **ACTION** Write an advertisement for the state lottery that describes the probability for winning in an ethical fashion.

6.62 **ACTION** Write an advertisement for the investment newsletter that states the probability for a 20% annual return in an ethical fashion.

6.12 Basic Probability: A Review and a Preview

As shown in the chapter summary chart on page 234, this chapter was about basic probability. We examined various rules of probability as well as applications of these rules to a variety of problems. On page 204 of Section 6.1 you were given a list emphasizing the important points to be discussed in the chapter. Check over the list now to see whether you feel you have an understanding of these key points. To be sure, you should be able to answer the following conceptual questions:

1. What are the differences among *a priori* classical probability, empirical classical probability, and subjective probability?
2. What is the difference between a simple event and a joint event?
3. What is the difference between union and intersection?
4. How can the addition rule be used to find the probability of occurrence of event *A or B*?
5. What is the difference between mutually exclusive events and collectively exhaustive events?

6. How does conditional probability relate to the concept of statistical independence?
7. How does the multiplication rule differ for events that are and are not independent?
8. How can Bayes' theorem be used to revise probabilities in the light of available information?
9. Under what situations are the various counting rules used?
10. What is the difference between a permutation and a combination?

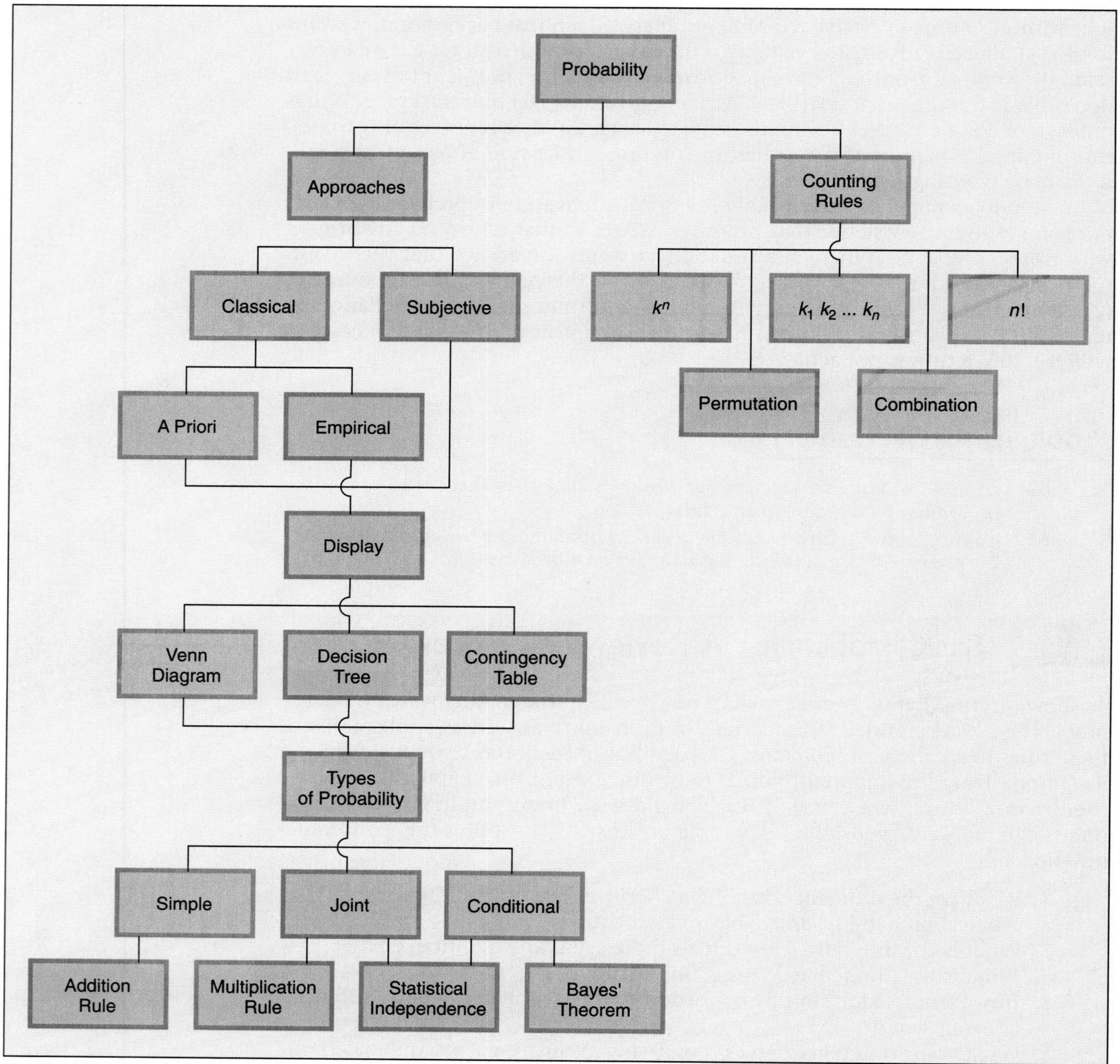

Chapter 6 summary chart.

Probability theory is the foundation of statistical inference. The concepts learned in this chapter will be extended to a variety of situations in subsequent chapters in order to make inferences about populations.

Getting It All Together

Key Terms

Chapter Review Problems

6.63 In evaluating conditional probabilities and using Bayes' theorem, which do you prefer—summary tables such as Table 6.3 or decision trees such as Figure 6.5, both on page 227? Why?

6.64 When rolling a die once, what is the probability that
(a) The face of the die is odd?
(b) The face is even *or* odd?
(c) The face is even *or* a one?
(d) The face is odd *or* a one?
(e) The face is both even *and* a one?
(f) Given the face is odd, it is a one?

● 6.65 The director of a large employment agency wishes to study various characteristics of its job applicants. A sample of 200 applicants has been selected for analysis. Seventy applicants have had their current jobs for at least five years; 80 of the applicants are college graduates; 25 of the college graduates have had their current jobs at least five years.
(a) What is the probability that an applicant chosen at random
(1) Is a college graduate?
(2) Is a college graduate *and* has held the current job less than five years?
(3) Is a college graduate *or* has held the current job at least five years?
(b) Given that a particular employee is a college graduate, what is the probability that he or she has held the current job less than five years?

(c) Determine whether being a college graduate *and* holding the current job for at least five years are statistically independent. (*Hint:* Set up a 2 × 2 table or a Venn diagram or a decision tree to evaluate the probabilities.)

6.66 Suppose that a survey has been undertaken to determine if there is a relationship between place of residence and ownership of a foreign-made automobile. A random sample of 200 car owners from large cities, 150 from suburbs, and 150 from rural areas was selected with the results shown below.

	Type of Area			
Car Ownership	Large City	Suburb	Rural	Totals
Own foreign car	90	60	25	175
Do not own foreign car	110	90	125	325
Totals	200	150	150	500

(a) If a car owner is selected at random, what is the probability that he or she
   (1) Owns a foreign car?
   (2) Lives in a suburb?
   (3) Owns a foreign car *or* lives in a large city?
   (4) Lives in a large city *or* a suburb?
   (5) Lives in a large city *and* owns a foreign car?
   (6) Lives in a rural area *or* does not own a foreign car?

(b) Assume we know that the person selected lives in a suburb. What is the probability that he or she owns a foreign car?

(c) Is area of residence statistically independent of whether the person owns a foreign car? Explain.

6.67 The Finance Society at a college of business at a large state university would like to determine whether there is a relationship between a student's interest in finance and his or her ability in mathematics. A random sample of 200 students is selected and they are asked whether their ability in mathematics and interest in finance is low, average, or high. The results were as follows:

	Ability in Mathematics			
Interest in Finance	Low	Average	High	Totals
Low	60	15	15	90
Average	15	45	10	70
High	5	10	25	40
Totals	80	70	50	200

(a) Give an example of a simple event.

(b) Give an example of a joint event.

(c) Why is "high interest in finance" *and* "high ability in mathematics" a joint event?

(d) If a student is selected at random, what is the probability that he or she
   (1) Has a high ability in mathematics?
   (2) Has an average interest in finance?
   (3) Has a low ability in mathematics?
   (4) Has a high interest in finance?

(5) Has a low ability in mathematics *and* a low interest in finance?
(6) Has a high ability in mathematics *and* an average interest in finance?
(7) Has a high ability in mathematics *and* a high interest in finance?
(8) Has a high interest in finance *or* a high ability in mathematics?
(9) Has an average interest in finance *or* a low ability in mathematics?
(10) Has a low interest in finance *or* an average interest in finance? Are these two events mutually exclusive? Why?
(11) Has a low ability in mathematics *or* an average ability in mathematics *or* a high ability in mathematics? Are these events mutually exclusive? Why? Are they also collectively exhaustive? Why?

(e) Assume we know that the person selected has a high ability in mathematics. What is the probability that this individual has a high interest in finance?
(f) Assume we know that the person selected has an average ability in mathematics. What is the probability that this individual has a low interest in finance?
(g) Are interest in finance and ability in mathematics statistically independent? Explain.

6.68 A soft drink bottling company maintains records concerning the number of unacceptable bottles of soft drink obtained from the filling and capping machines. Based on past data, the probability that a bottle came from machine I *and* was nonconforming is .01 and the probability that a bottle came from machine II *and* was nonconforming is .025. Half the bottles are filled on machine I and the other half are filled on machine II.

(a) Give an example of a simple event.
(b) Give an example of a joint event.
(c) If a filled bottle of soft drink is selected at random, what is the probability that:
(1) it is a nonconforming bottle?
(2) it was filled on machine II?
(3) it is filled on machine I *and* is a conforming bottle?
(4) it is filled on machine II *and* is a conforming bottle?
(5) it is filled on machine I *or* is a conforming bottle?
(d) Suppose we know that the bottle was produced on machine I. What is the probability that it is nonconforming?
(e) Suppose we know that the bottle is nonconforming. What is the probability that it was produced on machine I?
(f) Explain the difference in the answers to (d) and (e).
(*Hint:* Set up a 2×2 table or a Venn diagram to evaluate the probabilities.)

6.69 An art director for a magazine has 12 pictures to choose from for 5 positions in her magazine.

(a) How many different sets of 5 pictures could she choose from the 12 available pictures?
(b) Once she has picked her 5 pictures, in how many ways can she arrange them in the magazine?
(c) How many permutations are there of 12 objects taken 5 at a time? [*Hint:* Show how this answer is related to your answers to (a) and (b).]

Survey Database Project

The following problems refer to the sample data obtained from the questionnaire of Figure 2.6 on pages 28–29 and presented in Table 2.3 on pages 33–40. They should be solved with the aid of an available computer package.

Data file
EMPSAT.DAT

We may recall that as research assistant to Bud Conley, the Vice-President of Human Resources at Kalosha Industries, you were asked to obtain cross-classifications to the questions dealing with categorical variables in the Employee Satisfaction Survey (see Problems 5.29–5.40 on page 188). Suppose that in order to understand the relationships among the categorical variables, we would like to determine the following:

6.70 Referring to gender (question 5) and job satisfaction (question 9)
- (a) What is the probability that a randomly selected employee is:
  - (1) a male?
  - (2) very satisfied with the job?
  - (3) a male *and* very satisfied with the job?
  - (4) a female *or* very satisfied with the job?
- (b) Given that the employee selected is a male, what is the probability that he is very satisfied with his job?
- (c) Given that the employee selected is a female, what is the probability that she is very satisfied with her job?
- (d) Is gender statistically independent of job satisfaction? Explain.

6.71 Referring to the most important job characteristics (question 11) and job satisfaction (question 9)
- (a) What is the probability that a randomly selected employee:
  - (1) has high income as the most important job characteristic?
  - (2) is a little dissatisfied with the job?
  - (3) has high income as the most important job characteristic *and* is a little dissatisfied with the job?
  - (4) has high income as the most important job characteristic *or* is a little dissatisfied with the job?
- (b) Given that the employee selected has high income as the most important job characteristic, what is the probability that he or she is a little dissatisfied with his or her job?
- (c) Is most important job characteristic statistically independent of job satisfaction? Explain.

6.72 Referring to the way people get ahead (question 12) and job satisfaction (question 9)
- (a) What is the probability that a randomly selected employee:
  - (1) feels that the way most people get ahead is hard work?
  - (2) is moderately satisfied with his or her job?
  - (3) feels that the way most people get ahead is hard work *and* is moderately satisfied with the job?
  - (4) feels that the way most people get ahead is hard work *or* is moderately satisfied with the job?
- (b) Given that the employee selected feels that the way most people get ahead is hard work, what is the probability that he or she is moderately satisfied with his or her job?
- (c) Is the way people get ahead statistically independent of job satisfaction? Explain.

6.73 Referring to membership in a labor union (question 14) and job satisfaction (question 9)
- (a) What is the probability that a randomly selected employee:
  - (1) is a member of a labor union?
  - (2) is moderately satisfied with his or her job?
  - (3) is a member of a labor union *and* is moderately satisfied with his or her job?
  - (4) is a member of a labor union *or* is moderately satisfied with his or her job?
- (b) Given that the employee selected is a member of a labor union, what is the probability that he or she is moderately satisfied with his or her job?
- (c) Is membership in a labor union statistically independent of job satisfaction?

6.74 Referring to likelihood of promotion (question 18) and job satisfaction (question 9)
(a) What is the probability that a randomly selected employee:
(1) feels that he or she is likely to be promoted?
(2) is moderately satisfied with his or her job?
(3) feels likely to be promoted *and* is moderately satisfied with his or her job?
(4) feels likely to be promoted *or* is moderately satisfied with his or her job?
(b) Given that the employee selected is likely to be promoted, what is the probability that he or she is moderately satisfied with his or her job?
(c) Is likelihood of promotion statistically independent of job satisfaction? Explain.

6.75 Referring to whether your job allows you to take part in making decisions that affect your work (question 21) and job satisfaction (question 9)
(a) What is the probability that a randomly selected employee:
(1) sometimes takes part in making decisions that affect his or her work?
(2) is very satisfied with his or her job?
(3) sometimes takes part in making decisions that affect his or her work *and* is very satisfied with his or her job?
(4) sometimes takes part in making decisions that affect his or her work *or* is very satisfied with his or her job?
(b) Given that the employee selected sometimes takes part in making decisions that affect his or her work, what is the probability that he or she is very satisfied with his or her job?
(c) Is taking part in making decisions statistically independent of job satisfaction? Explain.

6.76 Referring to whether your job allows you to participate in budgetary decisions (question 22) and job satisfaction (question 9)
(a) What is the probability that a randomly selected employee:
(1) participates in budgetary decisions?
(2) is very satisfied with his or her job?
(3) participates in budgetary decisions *and* is very satisfied with his or her job?
(4) participates in budgetary decisions *or* is very satisfied with his or her job?
(b) Given that the employee selected participates in budgetary decisions, what is the probability that he or she is very satisfied with his or her job?
(c) Is participating in budgetary decisions statistically independent of job satisfaction? Explain.

6.77 Referring to how proud the employee is to work for the organization (question 23) and job satisfaction (question 9)
(a) What is the probability that a randomly selected employee:
(1) is very proud?
(2) is very satisfied with his or her job?
(3) is very proud *and* is very satisfied with his or her job?
(4) is very proud *or* is very satisfied with his or her job?
(b) Given that the employee selected is very proud, what is the probability that he or she is very satisfied with the job?
(c) Is being proud to work in the organization statistically independent of job satisfaction? Explain.

6.78 Referring to relations between management and employees (question 25) and job satisfaction (question 9)
(a) What is the probability that a randomly selected employee:
(1) describes relations as very good?
(2) is very satisfied with his or her job?
(3) describes relations as very good *and* is very satisfied with his or her job?
(4) describes relations as very good *or* is very satisfied with his or her job?

(b) Given that the employee selected describes relations as very good, what is the probability that the employee is very satisfied with his or her job?
(c) Is relations between management and employees statistically independent of job satisfaction? Explain.

6.79 Referring to relations between coworkers and colleagues (question 26) and job satisfaction (question 9)
(a) What is the probability that a randomly selected employee:
(1) describes relations as very good?
(2) is very satisfied with his or her job?
(3) describes relations as very good *and* is very satisfied with his or her job?
(4) describes relations as very good *or* is very satisfied with his or her job?
(b) Given that the employee selected describes relations as very good, what is the probability that he or she is very satisfied with the job?
(c) Is relations between coworkers and colleagues statistically independent of job satisfaction? Explain.

6.80 ACTION Based on the results of Problems 6.70–6.79, write a letter to Bud Conley, the Vice-President of Human Resources at Kalosha Industries, detailing your findings.

Endnote

1. In a contingency table with R rows and C columns, the rule would have to be examined for $(R-1)(C-1)$ separate combinations of A and B.

References

1. Hays, W. L., *Statistics for the Social Sciences,* 3d ed. (New York: Holt, Rinehart and Winston, 1980).
2. Kirk, R. E., ed., *Statistical Issues: A Reader for the Behavioral Sciences* (Belmont, CA: Wadsworth, 1972).
3. Mosteller, F., R. Rourke, and G. Thomas, *Probability with Statistical Applications,* 2d ed. (Reading, MA: Addison-Wesley, 1970).
4. Paulos, J. A. *Innumeracy* (New York: Hill and Wang, 1988).

chapter 7

Some Important Discrete Probability Distributions

CHAPTER OBJECTIVES

To develop an understanding of the concept of mathematical expectation and its applications in decision making and to show how certain types of discrete data may be represented by particular kinds of mathematical models.

7.1 Introduction

In Chapter 6 we established various rules of probability and examined some counting techniques. In this chapter we will utilize this information to develop the concept of mathematical expectation and to develop some probability distribution models that represent discrete phenomena of interest. In particular, we will begin by defining the probability distribution and then discuss the two critical characteristics of any probability distribution, its mean or expected value, $[E(X)]$, and its variance, σ_x^2. We will then develop two important discrete probability distributions, the binomial and the Poisson, and also indicate the circumstances in which the Poisson distribution can be used to approximate the binomial.

Upon completion of this chapter, you should be able to:

1. Compute the expected value and variance of a discrete probability distribution
2. Understand the assumptions of the binomial distribution and know how to find any binomial probability
3. Understand the assumptions of the Poisson distribution and know how to find any Poisson probability
4. Know when and how the Poisson distribution can be used to approximate the binomial distribution

7.2 The Probability Distribution for a Discrete Random Variable

As discussed in Section 2.3, a numerical random variable is some phenomenon of interest whose responses or outcomes may be expressed numerically. Such a random variable may also be classified as discrete or continuous—the former arising from a counting process and the latter from a measuring process. This chapter deals with some probability distributions that represent discrete random variables. As an example from the Employee Satisfaction Survey developed in Chapter 2, responses to the question about (total) number of promotions received pertain to a probability distribution for a discrete random variable.

We may define the probability distribution for a discrete random variable as follows:

> A **probability distribution for a discrete random variable** is a mutually exclusive listing of all possible numerical outcomes for that random variable such that a particular probability of occurrence is associated with each outcome.

Assuming that a fair six-sided die will not stand on edge or roll out of sight (*null events*), Table 7.1 represents the probability distribution for the outcomes of a single roll of the fair die. Since all possible outcomes are included, this listing is complete (or *collectively exhaustive*) and thus the probabilities must sum up to 1. We can then use this table to obtain various probabilities for the rolling of a fair die.

Table 7.1 Theoretical probability distribution of the results of rolling one fair die.

Face of Outcome	Probability
1 ⚀	1/6
2 ⚁	1/6
3 ⚂	1/6
4 ⚃	1/6
5 ⚄	1/6
6 ⚅	1/6
Total	1

The probability of a face ⚂ is

$$P(⚂) = 1/6$$

Using the addition rule for mutually exclusive events, the probability of an odd face is

$$P(\text{odd}) = P(⚀) + P(⚂) + P(⚄)$$
$$= 1/6 + 1/6 + 1/6 = 3/6$$

Moreover, the probability of a face of ⚁ or less is

$$P(⚁ \text{ or less}) = P(⚀) + P(⚁)$$
$$= 1/6 + 1/6 = 2/6$$

And the probability of a face larger than ⚅ is

$$P(> ⚅) = 0$$

7.3 Mathematical Expectation and Expected Monetary Value

In order to summarize a discrete probability distribution we shall compute its major characteristics—the mean and the standard deviation.

7.3.1 Expected Value of a Discrete Random Variable

The mean (μ_x) of a probability distribution is the **expected value** of its random variable.

The **expected value of a discrete random variable** may be considered as its weighted average over all possible outcomes—the weights being the probability associated with each of the outcomes.

This summary measure can be obtained by multiplying each possible outcome X_i by its corresponding probability $P(X_i)$ and then summing up the resulting products. Thus, the expected value of the discrete random variable X, symbolized as $E(X)$, may be expressed as follows:

$$\mu_x = E(X) = \sum_{i=1}^{N} X_i P(X_i) \tag{7.1}$$

where X = discrete random variable of interest

X_i = ith outcome of X

$P(X_i)$ = probability of occurrence of the ith outcome of X

$i = 1, 2, \ldots, N$

For the theoretical probability distribution of the results of rolling one fair die (Table 7.1), the expected value of the roll may be computed as

$$\mu_x = E(X) = \sum_{i=1}^{N} X_i P(X_i) = (1)(1/6) + (2)(1/6) + (3)(1/6) + (4)(1/6) + (5)(1/6) + (6)(1/6)$$

$$= 1/6 + 2/6 + 3/6 + 4/6 + 5/6 + 6/6$$

$$= 21/6 = 3.5$$

Notice that the expected value of the results of rolling a fair die is not "literally meaningful," since we can never obtain a face of 3.5. However, we can expect to observe the six different faces with equal likelihood, so we should have about the same number of ones, twos, . . . , and sixes. In the long run, over many rolls, the average value would be 3.5.

To make this particular situation meaningful, however, we introduce the following carnival game: How much money should we be willing to put up in order to have the opportunity of rolling a fair die if we were to be paid, in dollars, the amount on the face of the die? Since the expected value of a roll of a fair die is 3.5, the expected long-run payoff is \$3.50 per roll. This means that, on any particular roll, our payoff will be \$1.00, \$2.00, . . . , or \$6.00, but over many, many rolls the payoff can be expected to average out to \$3.50 per roll. Now if we want the game to be fair, neither we nor our opponent (the "house") should have an advantage. Thus we should be willing to pay \$3.50 per roll to play. If the house wants to charge us \$4.00 per roll we can expect to lose from such gambling, on the average, \$.50 per roll over time, and unless we derive some intrinsic satisfaction (costing on the average \$.50 per roll), we should refrain from participating in such a game.

Usually, though, in any casino or carnival-type game the expected long-run payoff to the participant is negative, otherwise the house would not be in business (References 5 and 6). Such games as Craps, Under-or-over-seven, Chuck-a-luck, or Roulette (see Reference 5) attract large numbers of participants and, in each case, the expected return over time favors the house. This is the case because something other than expected monetary value is the ultimate criterion used by participants. The concept of the *expected utility of money* is discussed in Reference 2. It is the criterion that rational participants are considering, implicitly or explicitly, when they partake in such games. On the other hand, however, the house uses the expected-monetary-value criterion when it participates in such games.

7.3.2 Variance and Standard Deviation of a Discrete Random Variable

The **variance (σ_x^2) of a discrete random variable** may be defined as the weighted average of the squared differences between each possible outcome and its mean—the weights being the probabilities of each of the respective outcomes.

This summary measure can be obtained by multiplying each possible squared difference $(X_i - \mu_x)^2$ by its corresponding probability $P(X_i)$ and then summing up the resulting products. Hence, the variance of the discrete random variable X may be expressed as follows:

$$\sigma_x^2 = \sum_{i=1}^{N} (X_i - \mu_x)^2 P(X_i) \qquad (7.2)$$

where X = discrete random variable of interest

X_i = ith outcome of X

$P(X_i)$ = probability of occurrence of the ith outcome of X

$i = 1, 2, \ldots, N$

Moreover, the **standard deviation (σ_x) of a discrete random variable** is given by

$$\sigma_x = \sqrt{\sum_{i=1}^{N} (X_i - \mu_x)^2 P(X_i)} \qquad (7.3)$$

For the theoretical probability distribution of the results of rolling one fair die (Table 7.1) the variance and the standard deviation may be computed by

$$\begin{aligned}\sigma_x^2 &= \sum_{i=1}^{N} (X_i - \mu_x)^2 P(X_i) \\ &= (1-3.5)^2(1/6) + (2-3.5)^2(1/6) + (3-3.5)^2(1/6) + (4-3.5)^2(1/6) \\ &\quad + (5-3.5)^2(1/6) + (6-3.5)^2(1/6) \\ &= 2.9166\end{aligned}$$

and

$$\sigma_x = 1.71$$

In terms of our carnival game, the mean payoff per roll is \$3.50 with a standard deviation of \$1.71. According to the Bienaymé-Chebyshev rule (Section 4.8.7), the majority of our payoffs would be expected to be within $\sqrt{2} = 1.414$ standard deviations of the mean (that is, $\mu_x \pm 1.414\sigma_x$). Most likely, then, on a per-roll basis we would expect a payoff between \$2.00 and \$5.00 [that is, the integer outcomes between the values $3.50 \pm (1.414)(1.71)$] and not win very frequently if it is costing us \$4.00 per roll to play.

7.3.3 Expected Monetary Value

As prospective participants in the carnival game, the most important question that we had to address was whether or not it was *profitable* for us to play the game. To answer this question, we had to realize that the random variable of interest from the "gambling viewpoint" was not really X, the *outcome on the face* of the die (as in Table 7.1 on page 243), but rather V, the *dollar value* associated with the resulting outcome from rolling the die. Thus, for participating in the game, the values for V ranged from $-\$3.00$ to $+\$2.00$ since it cost \$4.00 for every roll of the die (see Table 7.2).

Table 7.2 Theoretical probability distribution representing the dollar value for participating in carnival game.

Result (X)	Dollar Value (V)	Probability
1	−3	1/6
2	−2	1/6
3	−1	1/6
4	0	1/6
5	1	1/6
6	2	1/6
		1

For purposes of decision making, the objective is to compare the **expected monetary values** (denoted by *EMV*) among alternative strategies (such as "play the carnival game" versus "don't play"). The expected monetary value indicates the average profit that would be gained if a particular strategy were selected in many decision-making situations (such as "play the game many times"). Hence to play the game

$$EMV(\text{play}) = E(V) = \sum_{i=1}^{N} V_i P(V_i) = (-3)(1/6) + (-2)(1/6) + \cdots + (2)(1/6)$$
$$= -.50$$

while not playing the game

$$EMV \text{ (not play)} = E(V) = 0$$

Thus our long-run expected payoff for participating is negative. On the average, we'd be losing 50 cents each time we decide to roll the die. This means that if we played for an evening in which the die is rolled 100 times, we would expect to collect $350 but pay out $400. Thus, at the end of play we would expect to lose $50 on the 100 rolls—an average of 50 cents per roll (or $12\frac{1}{2}$ cents per dollar wagered).

● **Assigning Probabilities** To compute the expected monetary value for the various strategies, a set of probabilities must be assigned to the mutually exclusive and collectively exhaustive listing of outcomes and events. In many cases, no information is available about the probability of occurrence of the various events and, thus, equal probabilities are assigned. In other instances, the probabilities of the events can be estimated in several ways. First, information may be available from past experience that can be used for estimating the probabilities. Second, a subjective assessment of the likelihood of the various events may be given by managers or other supervisory personnel. Third, the probabilities of the events could follow a particular discrete probability distribution such as the binomial or Poisson distribution. These two probability distributions will be the subject of the remainder of this chapter.

Problems for Section 7.3

● 7.1 Given the following probability distributions:

Distribution A		Distribution B	
X	$P(X)$	X	$P(X)$
0	.50	0	.05
1	.20	1	.10
2	.15	2	.15
3	.10	3	.20
4	.05	4	.50

(a) Compute the mean for each distribution.
(b) Compute the standard deviation for each distribution.
(c) ACTION Compare and contrast the results in (a) and (b). Discuss what you have learned.

7.2 Given the following probability distributions:

Distribution C		Distribution D	
X	$P(X)$	X	$P(X)$
0	.20	0	.10
1	.20	1	.20
2	.20	2	.40
3	.20	3	.20
4	.20	4	.10

(a) Compute the mean for each distribution.
(b) Compute the standard deviation for each distribution.
(c) ACTION Compare and contrast the results in (a) and (b). Discuss what you have learned.

7.3 Using the company records for the past 500 working days, the manager of Torrisi Motors, a suburban automobile dealership, has summarized the number of cars sold per day into the following table:

Number of Cars Sold per Day	Frequency of Occurrence
0	40
1	100
2	142
3	66
4	36
5	30
6	26
7	20
8	16
9	14
10	8
11	2
Total	500

(a) Form the empirical probability distribution (that is, relative frequency distribution) for the discrete random variable *X*, the number of cars sold per day.
(b) Compute the mean or expected number of cars sold per day.
(c) Compute the standard deviation.
(d) What is the probability that on any given day
  (1) fewer than 4 cars will be sold?
  (2) at most 4 cars will be sold?
  (3) at least 4 cars will be sold?
  (4) exactly 4 cars will be sold?
  (5) more than 4 cars will be sold?
(e) ACTION Write a letter to the manager discussing the dealership's performance over the past 500 working days.

7.4 An employee for a vending concession at a baseball stadium must choose between working behind the hot dog counter and receiving a fixed sum of $50 for the evening and walking around the stands selling beer on a commission basis. If the latter is chosen the employee can make $90 on a warm night, $70 on a moderate night, $45 on a cool night, and $15 on a cold night. At this time of year the probabilities of a warm, moderate, cool, or cold night are, respectively, 0.1, 0.3, 0.4, and 0.2.
(a) Determine the mean or expected value to be earned by selling beer that evening.
(b) Compute the standard deviation.
(c) Which product should the employee sell? Why?

7.5 A state lottery is to be conducted in which 10,000 tickets are to be sold for $1 each. Six winning tickets are to be randomly selected: one grand-prize winner of $5,000, one second-prize winner of $2,000, one third-prize winner of $1,000, and three other winners of $500 each.
(a) Compute the expected value of playing this game.
(b) ACTION Would you play this game? Why?

7.6 Let us consider rolling a pair of six-sided dice. The random variable of interest represents the total of the two numbers (that is, faces) that occur when the pair of fair dice are rolled. The probability distribution is given below:

X	$P(X)$
2	1/36
3	2/36
4	3/36
5	4/36
6	5/36
7	6/36
8	5/36
9	4/36
10	3/36
11	2/36
12	1/36
	1

(a) Determine the mean or expected sum from rolling a pair of the fair dice.
(b) Compute the variance and the standard deviation.

The game of Craps deals with the rolling of a pair of fair dice. A *field bet* in the game of Craps is a one-roll bet and is based on the outcome of the pair of dice. For every $1.00 bet you make: you can lose the $1.00 if the sum is 5, 6, 7, or 8; you can win $1.00 if the sum is 3, 4, 9, 10, or 11; or you can win $2.00 if the sum is either 2 or 12.

(c) Form the probability distribution function representing the different outcomes that are possible in a field bet.
(d) Determine the mean of this probability distribution.
(e) What is the player's expected long-run profit (or loss) from a $1.00 field bet? Interpret.
(f) What is the expected long-run profit (or loss) to the house from a $1.00 field bet? Interpret.
(g) **ACTION▶** Would you play this game and make a field bet?

7.7 In the carnival game Under-or-over-seven a pair of fair dice are rolled once and the resulting sum determines whether or not the player wins or loses his or her bet. For example, the player can bet $1.00 that the sum is under 7—that is, 2, 3, 4, 5, or 6. For such a bet the player will lose $1.00 if the outcome equals or exceeds 7 or will win $1.00 if the result is under 7. Similarly, the player can bet $1.00 that the sum is over 7—that is, 8, 9, 10, 11, or 12. Here the player wins $1.00 if the result is over 7 but loses $1.00 if the result is 7 or under. A third method of play is to bet $1.00 on the outcome 7. For this bet the player will win $4.00 if the result of the roll is 7 and lose $1.00 otherwise.

(a) Form the probability distribution function representing the different outcomes that are possible for a $1.00 bet on being under 7.
(b) Form the probability distribution function representing the different outcomes that are possible for a $1.00 bet on being over 7.
(c) Form the probability distribution function representing the different outcomes that are possible for a $1.00 bet on 7.
(d) Prove that the expected long-run profit (or loss) to the player is the same—no matter which method of play is used.
(e) **ACTION▶** Would you prefer to play Under-or-over-seven or make a field bet in Craps (Problem 7.6)? Why?

7.8 Why does the term *expected value* have that name, even though in many cases (such as our carnival game on page 244) you will never see the expected value as the result of any single experiment? (That is, in what sense is a value that never occurs expected?)

7.9 Suppose an author is trying to choose between two publishing companies that are competing for the marketing rights to her new novel. Prentice Hall has offered the author $10,000 plus $2.00 for each book sold. Random House has offered the author $2,000 plus $4.00 for each book sold. The author estimates the distribution of demand for this book as follows:

Number of Books Sold	Probability
1,000	.45
2,000	.20
5,000	.15
10,000	.10
50,000	.10

ACTION Using the expected-monetary-value criterion, determine whether the author should sell the marketing rights to Prentice Hall or to Random House. Discuss.

7.10 The Islander Fishing Co. purchases clams for $1.50 per pound from Peconic Bay fishermen for sale to various New York restaurants for $2.50 per pound. Any clams not sold to the restaurants by the end of the week can be sold to a local soup company for $.50 per pound. The probabilities of various levels of demand are as follows:

Demand (pounds)	Probability
500	.2
1,000	.4
2,000	.4

(*Hint:* The company can purchase 500 pounds, 1,000 pounds, or 2,000 pounds.)

ACTION Using the expected-monetary-value criterion, determine the optimal number of pounds of clams that the company should purchase from the fishermen. Discuss.

7.11 The LeFleur Garden Center chain purchases Christmas trees from a supplier for sale during the holiday season. The trees are purchased for $10.00 each and are sold for $25.00 each. Any trees not sold can be sold for $3.00 each. The probability of various levels of demand is as follows:

Demand (number of trees)	Probability
100	.2
200	.6
500	.2

(*Hint:* The chain can purchase trees in lots of 100, 200, or 500.)

ACTION Using the expected-monetary-value criterion, determine the number of trees that the chain should purchase from the supplier. Discuss.

7.12 An investor has a certain amount of money available to invest now. Three alternative portfolio selections are available. The estimated profits of each portfolio under each economic condition are indicated in the following payoff table:

	Portfolio Selection		
Event	A	B	C
Economy declines	\$500	−\$2,000	−\$7,000
No change	\$1,000	\$2,000	−\$1,000
Economy expands	\$2,000	\$5,000	\$20,000

Based upon his own past experience, the investor assigns the following probabilities to each economic condition:

$$P\text{ (economy declines)} = .30$$
$$P\text{ (no change)} = .50$$
$$P\text{ (economy expands)} = .20$$

ACTION Determine the best portfolio selection for the investor according to the expected-monetary-value criterion. Discuss.

7.4 Discrete Probability Distribution Functions

The probability distribution for a discrete random variable may be

1. A *theoretical* listing of outcomes and probabilities (as in Table 7.1), which can be obtained from a mathematical model representing some phenomenon of interest.
2. An *empirical* listing of outcomes and their observed relative frequencies.
3. A *subjective* listing of outcomes associated with their subjective probabilities representing the degree of conviction of the decision maker as to the likelihood of the possible outcomes (as discussed in Section 6.2).

In this chapter we will be concerned mainly with the first kind of probability distribution—the listing obtained from a mathematical model representing some phenomenon of interest.

A **model** is considered to be a miniature representation of some underlying phenomenon. In particular, a mathematical model is a mathematical expression representing some underlying phenomenon. For discrete random variables, this mathematical expression is known as a **probability distribution function.**

When such mathematical expressions are available, the exact probability of occurrence of any particular outcome of the random variable can be computed. In such cases, then, the entire probability distribution can be obtained and listed.

For example, the probability distribution function represented in Table 7.1 is one in which the discrete random variable of interest is said to follow the **uniform probability distribution.** The essential characteristic of the uniform distribution is that all outcomes of the random variable are equally likely to occur. Thus the probability that the face ⚄ of the fair die turns up is the same as that for any other result—1/6—since there are six possible outcomes.

In addition, other types of mathematical models have been developed to represent various discrete phenomena that occur in the social and natural sciences, in medical research, and in business. The more useful of these represent data characterized by the binomial probability distribution and the Poisson probability distribution. These two distributions will now be developed.

7.5 Binomial Distribution

The **binomial distribution** is a discrete probability distribution function that is extremely useful for describing many phenomena.

The binomial distribution possesses four essential properties:

1. The possible observations may be obtained by two different sampling methods. Either each observation may be considered as having been selected from an *infinite population without replacement* or from a *finite population with replacement.*
2. Each observation may be classified into one of two mutually exclusive and collectively exhaustive categories, usually called *success* and *failure.*
3. The probability of an observation's being classified as success, p, is constant from observation to observation. Thus, the probability of an observation's being classified as failure, $1 - p$, is constant over all observations.
4. The outcome (that is, success or failure) of any observation is independent of the outcome of any other observation.

The discrete random variable or phenomenon of interest that follows the binomial distribution is the number of successes obtained in a sample of n observations. Thus the binomial distribution has enjoyed numerous applications:

- In games of chance:
  What is the probability that red will come up 15 or more times in 19 spins of the roulette wheel?
- In product quality control:
  What is the probability that in a sample of 20 tires of the same type none will be defective if 8% of all such tires produced at a particular plant are defective?
- In education:
  What is the probability that a student can pass a ten-question multiple-choice exam (each question containing four choices) if the student guesses on each question? (Passing is defined as getting 60% of the items correct—that is, getting at least six out of ten items correct.)
- In finance:
  What is the probability that a particular stock will show an increase in its closing price on a daily basis over the next ten (consecutive) trading sessions if stock market price changes really are random?

In each of these examples the four properties of the binomial distribution are clearly satisfied. For the roulette example, a particular set of spins may be construed as the sample taken from an infinite population of spins without replacement. When spinning the roulette wheel, each observation is categorized as red (success) or not red (failure). The probability of spinning red, p, on an American roulette wheel is 18/38 and is assumed to remain stable over all observations. Thus the probability of failure (spinning black or green), $1 - p$, is 20/38 each and every time the roulette wheel spins. Moreover, the roulette wheel has no memory—the outcome of any one spin is independent of preceding or following spins, so that, for example, the probability of obtaining red on the 32nd spin, given that the previous 31 spins were all red, remains equal to p, 18/38, if the roulette wheel is a fair one (see Figure 7.1).

FIGURE 7.1
American roulette wheel.

In the product quality-control example, the sample of tires is also selected without replacement from an ongoing production process, an infinite population of manufactured tires.[1] As each tire in the sample is inspected, it is categorized as defective or nondefective according to the operational definition of conformance to specifications that has been previously developed. Over the entire sample of tires the probability of any particular tire being classified as defective, p, is .08, so that the probability of any tire being categorized as nondefective, p, is .92. (Note that when we are looking for defective tires, the discovery of such an event is deemed a success. This is one of the instances referred to previously where, for statistical purposes, the term "success" may refer to business failures, deaths due to a particular illness, and other phenomena that, in nonstatistical terminology, would be deemed unsuccessful.) The production process is assumed to be stable. In addition, for such a production process, the probability of one tire being classified as defective or nondefective is independent of the classification for any other tire.

Similar statements pertaining to the binomial distribution's characteristics in the education example and in the finance example can be made as well. This is left to the reader. (See Problems 7.13 and 7.14 on page 259.)

The four examples of binomial probability models previously described are distinguished by the parameters n and p. Each time a set of parameters—the number of observations in the sample, n, and the probability of success, p—is specified, a particular binomial probability distribution can be generated.

7.5.1 Development of the Mathematical Model

As another example of a phenomenon that satisfies the conditions of the binomial distribution, and one that is convenient for intuitively deriving an expression for the probabilities that arise in binomial problems, we shall return to the rolling of a fair die which was discussed in Section 7.2. Here, however, we consider success to be the outcome face ⚄ and failure to be any other outcome. Suppose we are now interested in three rolls of this same die in order to determine how frequently the face ⚄ is obtained.[2] What might occur? None of the rolls might land on ⚄; one of the rolls may be a ⚄; two of the rolls may land on ⚄; or all three rolls may land on ⚄. Can the binomial random variable, the number of ⚄ faces occurring on three rolls of a fair die, take on any other value? That would be impossible since if we roll the same die three times and are interested in how often a particular value (face ⚄) occurs, that value cannot exceed the number of rolls n, nor can it be lower than zero. Hence the range of a binomial random variable is from 0 to n.

Suppose then, for example, that we roll a fair die three times and observe the following result:

First Roll	Second Roll	Third Roll
⚄	Not 5	⚄

We now wish to determine the probability of this occurrence; that is, what is the probability of obtaining two successes (face ⚄) in three rolls in the *particular sequence* above? Since it may be assumed that rolling dice is a stable process, the probability that each roll occurs as above is

First Roll	Second Roll	Third Roll
$p = 1/6$	$1 - p = 5/6$	$p = 1/6$

Since each outcome is independent of the others, the probability of obtaining the given sequence is

$$p(1 - p)p = p^2(1 - p)^1 = p^2(1 - p) = (1/6)^2(5/6) = 5/216$$

Thus, out of 216 possible and equally likely outcomes from rolling a fair die three times, five will have the face ⚄ as the first and last roll, with a face other than ⚄ (that is, ⚀, ⚁, ⚂, ⚃, or ⚅) as the middle roll, and the particular sequence above will be obtained.

Now, however, we may ask how many different sequences are there for obtaining two faces of ⚄, out of $n = 3$ rolls of the die? Using the rule of combinations given by Equation (6.15) on page 231, we have

$$\binom{n}{X} = \frac{n!}{X!(n - X)!} = \frac{3!}{2!(3 - 2)!} = 3$$

such sequences. These three possible sequences are

Sequence 1 = ⚄ [Not 5] ⚄ with probability $p(1 - p)\,p = p^2(1 - p)^1 = 5/216$

Sequence 2 = ⚄ ⚄ [Not 5] with probability $pp(1 - p) = p^2(1 - p)^1 = 5/216$

Sequence 3 = [Not 5] ⚄ ⚄ with probability $(1 - p)pp = p^2(1 - p)^1 = 5/216$

Therefore, the probability of obtaining exactly two faces of ⚄ from three rolls of a die is equal to

(number of possible sequences) × (probability of a particular sequence)

$$(3) \times (5/216) = 15/216 = .0694$$

A similar, intuitive derivation can be obtained for the other three possible outcomes of the random variable—no face ⚄, one face ⚄, or all three faces ⚄. However, as *n*, the number of observations, gets large, this type of intuitive approach becomes quite laborious, and a mathematical model is more appropriate. In general, the following mathematical model represents the binomial probability distribution for obtaining the number of successes (X), given a knowledge of the parameters n and p:

$$P(X = x|n, p) = \frac{n!}{x!(n - x)!}\,p^x(1 - p)^{n-x} \qquad \textbf{(7.4)}$$

where $P(X = x|n, p)$ = the probability that $X = x$, given a knowledge of n and p

n = sample size

p = probability of success

$1-p$ = probability of failure

x = number of successes in the sample ($X = 0, 1, 2, \ldots, n$)

We note, however, that the generalized form shown in Equation (7.4) is merely a restatement of what we had intuitively derived.

The binomial random variable X can have any integer value from 0 through n. In Equation (7.4) the product

$$p^x(1 - p)^{n-x}$$

tells us the probability of obtaining exactly x successes out of n observations *in a particular sequence,* while the term

$$\frac{n!}{x!(n - x)!}$$

tells us *how many sequences* or arrangements (i.e., *combinations*—see Section 6.10) of the x successes out of n observations are possible. Hence, given the number of observations n and the probability of success p, we may determine the probability of x successes:

$$\begin{aligned} P(X = x|n, p) &= (\text{number of possible sequences}) \\ &\quad \times (\text{probability of a particular sequence}) \\ &= \frac{n!}{x!(n - x)!}\, p^x (1 - p)^{n-x} \end{aligned}$$

by substituting the desired values for n, p, and x and computing the result.

Thus, as previously shown, the probability of obtaining exactly two faces of  from three rolls of a die is

$$\begin{aligned} P\left(X = 2|n = 3, p = \frac{1}{6}\right) &= \frac{3!}{2!(3 - 2)!}\left(\frac{1}{6}\right)^2\left(1 - \frac{1}{6}\right)^{3-2} \\ &= \frac{3!}{2!1!}\left(\frac{1}{6}\right)^2\left(\frac{5}{6}\right)^1 \\ &= 3\left(\frac{1}{6}\right)\left(\frac{1}{6}\right)\left(\frac{5}{6}\right) = \frac{15}{216} = .0694 \end{aligned}$$

Such computations may become quite tedious, especially as n gets large. However, we may obtain the probabilities directly from Table E.7 of Appendix E or use statistical software and thereby avoid any computational drudgery. Table E.7 provides, for various selected combinations of the parameters n and p, the probabilities that the binomial random variable takes on values of $X = 0, 1, 2, \ldots, n$. However, the reader should be cautioned that the values for p in Table E.7 are taken to only two decimal places; thus, in some circumstances, due to rounding errors, the probabilities will only be approximations to the true result. As a case in point, for our dice-rolling experiment, we first find in Table E.7 the combination $n = 3$ with p rounded to .17. To obtain the approximate probability of exactly two successes, we read the probability corresponding to the row $X = 2$, and the result is .0720 (as demonstrated in Table 7.3).[3] Thus Table E.7 has given us an approximate answer to the true probability, .0694, obtained from Equation (7.4) using the fraction $1/6 = p$ rather than the rounded decimal value .17.

Table 7.3 Obtaining a binomial probability.

		p												
n	x	0.01	0.02	0.03	...	0.10	0.11	0.12	0.13	0.14	0.15	0.16	0.17	0.18
2	0	0.9801	0.9604	0.9409	...	0.8100	0.7921	0.7744	0.7569	0.7396	0.7225	0.7056	0.6889	0.6724
	1	0.0198	0.0392	0.0582	...	0.1800	0.1958	0.2112	0.2262	0.2408	0.2550	0.2688	0.2822	0.2952
	2	0.0001	0.0004	0.0009	...	0.0100	0.0121	0.0144	0.0169	0.0196	0.0225	0.0256	0.0289	0.0324
3	0	0.9703	0.9412	0.9127	...	0.7290	0.7050	0.6815	0.6585	0.6361	0.6141	0.5927	0.5718	0.5514
	1	0.0294	0.0576	0.0847	...	0.2430	0.2614	0.2788	0.2952	0.3106	0.3251	0.3387	0.3513	0.3631
	2	0.0003	0.0012	0.0026	...	0.0270	0.0323	0.0380	0.0441	0.0506	0.0574	0.0645	0.0720	0.0797
	3	0.0000	0.0000	0.0000	...	0.0010	0.0013	0.0017	0.0022	0.0027	0.0034	0.0041	0.0049	0.0058

Source: Extracted from Table E.7.

7.5.2 Characteristics of the Binomial Distribution

Each time a set of parameters—n and p—is specified, a particular binomial probability distribution can be generated. This can be readily seen by examining Table E.7 for various combinations of n and p.

Shape We note that a binomial distribution may be symmetric or skewed. Whenever $p = .5$, the binomial distribution will be symmetric regardless of how large or small the value of n. However, when $p \neq .5$, the distribution will be skewed. The closer p is to .5 and the larger the number of observations, n, the less skewed the distribution will be.

Thus the distribution of the number of occurrences of red in 19 spins of the roulette wheel is only slightly skewed to the right, since $p = 18/38$. On the other hand, with small p the distribution will be highly right-skewed—as is observed for the distribution of the number of defective tires in a sample of 20 where $p = .08$. For very large p, the distribution would be highly left-skewed.

We leave it to the reader to verify the effect of n and p on the shape of the distribution by plotting the histogram in Problem 7.19(c) on page 260. However, to summarize the above characteristics, three binomial distributions are depicted in Figure 7.2 on page 258. Panel A represents the probability of obtaining the face 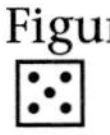 on three rolls of a fair die; panel B represents the probability of obtaining "heads" on three tosses of a fair coin; and panel C represents the probability of obtaining "heads" on four tosses of a fair coin. Thus a comparison of panel A with panel B demonstrates the effect on shape when the sample sizes are the same but the probabilities for success differ. Moreover, a comparison of panel B with C shows the effect on shape when the probabilities for successes are the same but the sample sizes differ.

The Mean The mean of the binomial distribution can be readily obtained as the product of its two parameters, n and p. That is, instead of using Equation (7.1), which holds for all discrete probability distributions, for data that are binomially distributed we simply compute

$$\mu_X = E(X) = np \qquad (7.5)$$

Panel A

Three Rolls of a Fair Die
X = Number of ⚄

$$P\left(X=0 \mid n=3, p=\frac{1}{6}\right)=\frac{3!}{0!3!}=\left(\frac{1}{6}\right)^{0}\left(\frac{5}{6}\right)^{3}=\frac{125}{216}$$

$$P\left(X=1 \mid n=3, p=\frac{1}{6}\right)=\frac{3!}{1!2!}=\left(\frac{1}{6}\right)^{1}\left(\frac{5}{6}\right)^{2}=\frac{75}{216}$$

$$P\left(X=2 \mid n=3, p=\frac{1}{6}\right)=\frac{3!}{2!1!}=\left(\frac{1}{6}\right)^{2}\left(\frac{5}{6}\right)^{1}=\frac{15}{216}$$

$$P\left(X=3 \mid n=3, p=\frac{1}{6}\right)=\frac{3!}{3!0!}=\left(\frac{1}{6}\right)^{3}\left(\frac{5}{6}\right)^{0}=\frac{1}{216}$$

$$\overline{\quad 1 \quad}$$

The probability of getting at least two faces ⚄ is
$\frac{15}{216}+\frac{1}{216}=\frac{16}{216}$

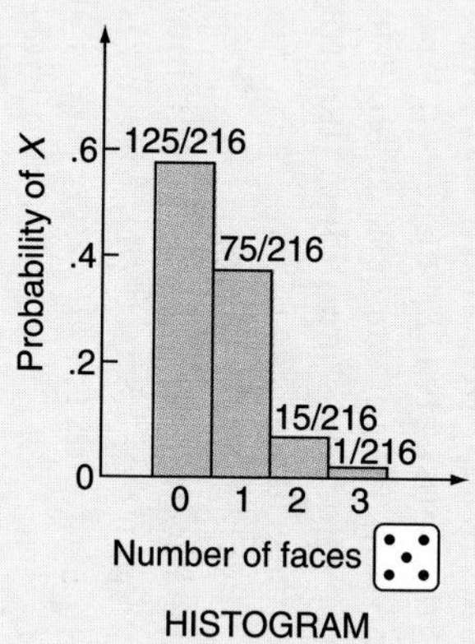

Panel B

Three Tosses of a Fair Coin
X = Number of "Heads"

$$P\left(X=0 \mid n=3, p=\frac{1}{2}\right)=\frac{3!}{0!3!}=\left(\frac{1}{2}\right)^{0}\left(\frac{1}{2}\right)^{3}=\frac{1}{8}$$

$$P\left(X=1 \mid n=3, p=\frac{1}{2}\right)=\frac{3!}{1!2!}=\left(\frac{1}{2}\right)^{1}\left(\frac{1}{2}\right)^{2}=\frac{3}{8}$$

$$P\left(X=2 \mid n=3, p=\frac{1}{2}\right)=\frac{3!}{2!1!}=\left(\frac{1}{2}\right)^{2}\left(\frac{1}{2}\right)^{1}=\frac{3}{8}$$

$$P\left(X=3 \mid n=3, p=\frac{1}{2}\right)=\frac{3!}{3!0!}=\left(\frac{1}{2}\right)^{3}\left(\frac{1}{2}\right)^{0}=\frac{1}{8}$$

$$\overline{\quad 1 \quad}$$

The probability of getting at least two heads is
$\frac{3}{8}+\frac{1}{8}=\frac{4}{8}$

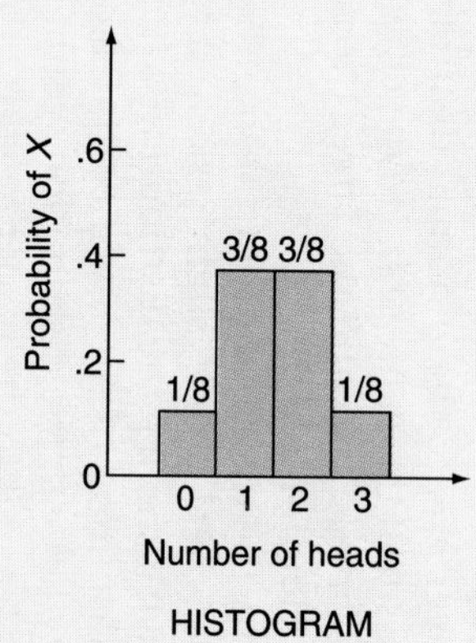

Panel C

Four Tosses of a Fair Coin
X = Number of "Heads"

$$P\left(X=0 \mid n=4, p=\frac{1}{2}\right)=\frac{4!}{0!4!}=\left(\frac{1}{2}\right)^{0}\left(\frac{1}{2}\right)^{4}=\frac{1}{16}$$

$$P\left(X=1 \mid n=4, p=\frac{1}{2}\right)=\frac{4!}{1!3!}=\left(\frac{1}{2}\right)^{1}\left(\frac{1}{2}\right)^{3}=\frac{4}{16}$$

$$P\left(X=2 \mid n=4, p=\frac{1}{2}\right)=\frac{4!}{2!2!}=\left(\frac{1}{2}\right)^{2}\left(\frac{1}{2}\right)^{2}=\frac{6}{16}$$

$$P\left(X=3 \mid n=4, p=\frac{1}{2}\right)=\frac{4!}{3!1!}=\left(\frac{1}{2}\right)^{3}\left(\frac{1}{2}\right)^{1}=\frac{4}{16}$$

$$P\left(X=4 \mid n=4, p=\frac{1}{2}\right)=\frac{4!}{4!0!}=\left(\frac{1}{2}\right)^{4}\left(\frac{1}{2}\right)^{0}=\frac{1}{16}$$

$$\overline{\quad 1 \quad}$$

The probability of getting at least two heads is
$\frac{6}{16}+\frac{4}{16}+\frac{1}{16}=\frac{11}{16}$

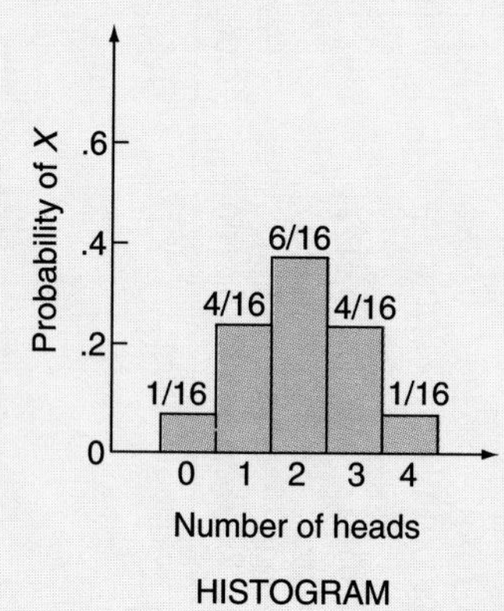

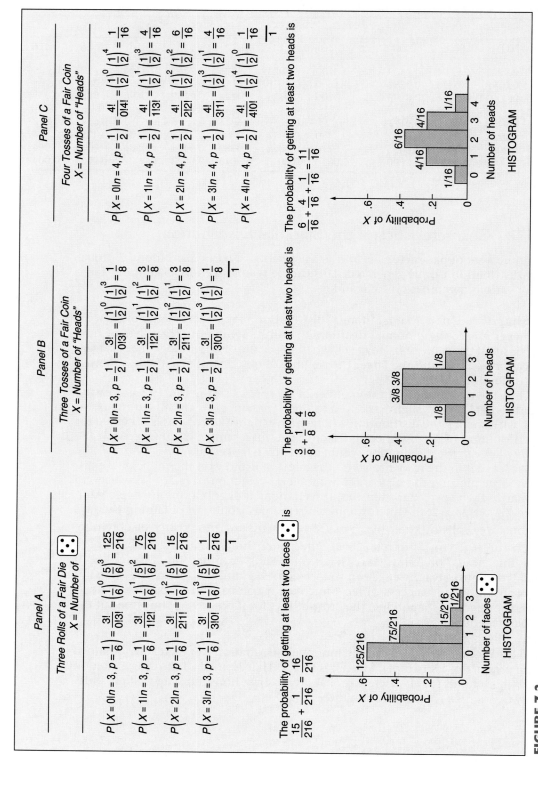

FIGURE 7.2
Comparison of three binomial distributions.

Intuitively, this makes sense. For example, if we spin the roulette wheel 19 times, how frequently should we "expect" the color red to come up? On the average, over the long run, we would theoretically expect

$$\mu_x = E(X) = np = (19)\left(\frac{18}{38}\right) = 9$$

occurrences of red in 19 spins—the same result that would be obtained from the more general expression shown in Equation (7.1).

The Standard Deviation The standard deviation of the binomial distribution is calculated using the formula

$$\sigma_x = \sqrt{np(1 - p)} \qquad \textbf{(7.6)}$$

Referring to our roulette wheel example, we compute

$$\sigma_x = \sqrt{(19)\left(\frac{18}{38}\right)\left(\frac{20}{38}\right)} = \sqrt{4.7368} = 2.18$$

This is the same result that would be obtained from the more general expression shown in Equation (7.3).

Summary In this section we have developed the binomial model as a useful discrete probability distribution in its own right. The binomial distribution, however, plays an even more important role when it is used in statistical inference problems regarding estimating or testing proportions (as will be discussed in Chapters 10 and 15).

Problems for Section 7.5

7.13 Describe how the four properties of the binomial distribution could be satisfied in the education example on page 252.

7.14 Describe how the four properties of the binomial distribution could be satisfied in the finance example on page 252.

7.15 Using Table E.7 determine the following:
- (a) If $n = 4$ and $p = .12$, then $P(X = 0|n = 4, p = .12)$?
- (b) If $n = 10$ and $p = .40$, then $P(X = 9|n = 10, p = .40)$?
- (c) If $n = 10$ and $p = .50$, then $P(X = 8|n = 10, p = .50)$?
- (d) If $n = 6$ and $p = .83$, then $P(X = 5|n = 6, p = .83)$?
- (e) If $n = 10$ and $p = .90$, then $P(X = 9|n = 10, p = .90)$?

7.16 In the carnival game of Chuck-a-luck three fair dice are rolled after the player has placed a bet on the occurrence of a particular face of the dice, say ⚄. For every $1.00 bet that you place you can lose the $1.00 if none of the three dice shows the face ⚄; you can win $1.00 if one die shows the face ⚄;

you can win \$2.00 if two of the dice show the face ⚄; or you can win \$3.00 if all three dice show the face ⚄.

(a) Form the probability distribution function representing the different monetary values (winnings or losses) that are possible (from one roll of the three dice). [*Hint:* Review Section 7.5.1 and see Figure 7.2 (panel A).]
(b) Determine the mean of this probability distribution.
(c) What is the player's expected long-run profit (or loss) from a \$1.00 bet? Interpret.
(d) What is the expected long-run profit (or loss) to the house? Interpret.
(e) **ACTION** Would you play Chuck-a-luck and make a bet? Discuss.

7.17 Suppose that warranty records show that the probability that a new car needs a warranty repair in the first ninety days is .05. If a sample of three new cars is selected,
(a) What is the probability that
  (1) none needs a warranty repair?
  (2) at least one needs a warranty repair?
  (3) more than one needs a warranty repair?
(b) What assumptions are necessary in (a)?
(c) What are the mean and standard deviation of the probability distribution in (a)?

7.18 The probability that a salesperson will sell a magazine subscription to someone who has been randomly selected from the telephone directory is .20. If the salesperson calls 10 individuals this evening, what is the probability that
(a) No subscriptions will be sold?
(b) Exactly two subscriptions will be sold?
(c) At least two subscriptions will be sold?
(d) At most two subscriptions will be sold?

● 7.19 An important part of the customer service responsibilities of a natural gas utility company concerns the speed with which calls relating to no heat in a house can be serviced. Suppose that one service variable of importance refers to whether or not the repair person reaches the home within a two-hour period. Past data indicate that the likelihood is .60 that the repair person reaches the home within a two-hour period. If a sample of five service calls for "no heat" is selected, what is the probability that the repair person will arrive at
(a) All five houses within a two-hour period?
(b) At least three houses within the two-hour period?
(c) Find the probability that the repair person will arrive at zero, one, and two houses and plot the histogram for the probability distribution.
(d) What is the shape of the distribution plotted in (c)? Explain.

7.6 Poisson Distribution

The **Poisson distribution** is another discrete probability distribution function that has many important practical applications. Not only are numerous discrete phenomena represented by a Poisson process, but the Poisson model is also used for providing approximations to the binomial distribution (as will be described in Section 7.6.4).

The following are some examples of Poisson-distributed phenomena:

- *Number of calls per hour* coming into the switchboard of a police station.
- *Number of car arrivals per day* at a toll bridge.
- *Number of major industrial strikes per year* in the United Kingdom.
- *Number of chips per cookie* in a pack of Marilyn's chocolate-chip cookies.
- *Number of blemishes on a square yard* of cloth.
- *Number of defects per batch* in a production process.
- *Number of runs per inning* of a baseball game.

In each of the above cases, the discrete random variable—*number of "successes" per unit* (that is, per interval of time, length, area, etc.)—is representative of a Poisson process.

A **Poisson process** is said to exist if we can observe discrete events in an *area of opportunity*—a continuous interval (of time, length, surface area, etc.)—in such a manner that if we shorten the area of opportunity or interval sufficiently

1. The probability of observing exactly one success in the interval is stable.
2. The probability of observing more than one success in the interval is 0.
3. The occurrence of a success in any one interval is statistically independent of that in any other interval.

To better understand the Poisson process, suppose we examine the number of customers arriving during the 12 noon to 1 PM lunch hour at a bank located in the central business district in a large city. Any arrival of a customer is a discrete event at a particular point over the continuous one-hour interval. Over such an interval of time there might be an average of 180 arrivals. Now if we were to break up the one-hour interval into 3,600 consecutive one-second intervals

1. The expected (or average) number of customers arriving in any one-second interval would be .05.
2. The probability of having more than one customer arriving in any one-second interval is 0.
3. The arrival of one customer in any one-second interval has no effect on (i.e., is statistically independent of) the arrival of any other customer in any other one-second interval.

7.6.1 The Mathematical Model

It is interesting to note that the Poisson distribution has but one parameter, which we will call λ (the Greek lowercase letter lambda). While the Poisson random variable X refers to the *number of successes per unit,* the parameter λ refers to the *average or expected number of successes per unit.* Moreover, we note that in theory the Poisson random variable ranges from 0 to ∞.

The mathematical expression for the Poisson distribution for obtaining X successes, given that λ successes are expected, is

$$P(X = x|\lambda) = \frac{e^{-\lambda}\lambda^x}{x!} \quad (7.7)$$

where $P(X = x|\lambda)$ = the probability that $X = x$ given a knowledge of λ

λ = expected number of successes

e = mathematical constant approximated by 2.71828

x = number of successes per unit

7.6.2 Characteristics

- **Shape** Each time the parameter λ is specified, a particular Poisson probability distribution can be generated. A Poisson distribution will be right-skewed when λ is small and will approach symmetry (with a peak in the center) as λ gets large.

- **The Mean and the Standard Deviation** An interesting property of the Poisson distribution is that the mean μ_x and variance σ_x^2 are each equal to the parameter λ. Thus,

$$\mu_x = E(X) = \lambda \quad (7.8)$$

and

$$\sigma_x = \sqrt{\lambda} \quad (7.9)$$

7.6.3 Applications of the Poisson Model

To demonstrate Poisson model applications, return to the example of customers' arrival at the bank at lunch hour: If, on average, .05 customers arrive per *second,* what is the probability that in a given *minute* exactly two customers will arrive? What is the chance that more than two customers will arrive in a given minute?

To solve these we must convert from seconds into minutes.

Conversions						
Avg. per Second		Avg. per Minute		Avg. per Hour		Avg. per Day
.05	↔	3.0	↔	180.0	↔	4,320.0

λ, the expected number of arrivals per minute, is 3.0. Now, using Equation (7.7), we have, for the first question

$$P(X = 2|\lambda = 3.0) = \frac{e^{-3.0}(3.0)^2}{2!} = \frac{9}{(2.71828)^3(2)} = .2240$$

Fortunately, hand calculations are not necessary here. Referring to Table E.6 in Appendix E, the tables of the Poisson distribution, the result may be obtained. As displayed in Table 7.4, which is a replica of Table E.6, only the values of λ and X are needed. Hence, the probability that exactly two customers arrive, given that 3.0 are expected, is .2240.

TABLE 7.4 Obtaining a Poisson probability.

	λ									
X	2.1	2.2	2.3	2.4	2.5	2.6	2.7	2.8	2.9	3.0
0	.1225	.1108	.1003	.0907	.0821	.0743	.0672	.0608	.0550	.0498
1	.2572	.2438	.2306	.2177	.2052	.1931	.1815	.1703	.1596	.1494
2	.2700	.2681	.2652	.2613	.2565	.2510	.2450	.2384	.2314 →	.2240
3	.1890	.1966	.2033	.2090	.2138	.2176	.2205	.2225	.2237	.2240
4	.0992	.1082	.1169	.1254	.1336	.1414	.1488	.1557	.1662	.1680
5	.0417	.0476	.0538	.0602	.0668	.0735	.0804	.0872	.0940	.1008
6	.0146	.0174	.0206	.0241	.0278	.0319	.0362	.0407	.0455	.0504
7	.0044	.0055	.0068	.0083	.0099	.0118	.0139	.0163	.0188	.0216
8	.0011	.0015	.0019	.0025	.0031	.0038	.0047	.0057	.0068	.0081
9	.0003	.0004	.0005	.0007	.0009	.0011	.0014	.0018	.0022	.0027
10	.0001	.0001	.0001	.0002	.0002	.0003	.0004	.0005	.0006	.0008
11	.0000	.0000	.0000	.0000	.0000	.0001	.0001	.0001	.0002	.0002
12	.0000	.0000	.0000	.0000	.0000	.0000	.0000	.0000	.0000	.0001

Source: Extracted from Table E.6.

To answer the second question—the probability that in any given minute more than two customers will arrive—we have

$$P(X > 2|\lambda = 3.0) = P(X = 3|\lambda = 3.0) + P(X = 4|\lambda = 3.0) + \cdots + P(X = \infty|\lambda = 3.0)$$

Since all the probabilities in a probability distribution must sum to 1, the terms on the right side of the equation can be expressed as

$$1 - P(X \le 2|\lambda = 3.0)$$

Thus

$$P(X > 2|\lambda = 3.0) = 1 - \{P(X = 0|\lambda = 3.0) + P(X = 1|\lambda = 3.0) + P(X = 2|\lambda = 3.0)\}$$

Now, using Equation (7.7), we have

$$P(X > 2|\lambda = 3.0) = 1 - \left\{\frac{e^{-3.0}(3.0)^0}{0!} + \frac{e^{-3.0}(3.0)^1}{1!} + \frac{e^{-3.0}(3.0)^2}{2!}\right\}$$

From Table E.6 (or its replica, Table 7.4) we can readily obtain the probabilities of 0, 1, or 2 successes, given a mean of 3.0 successes. Therefore,

$$P(X > 2|\lambda = 3.0) = 1 - \{.0498 + .1494 + .2240\}$$

$$= 1 - .4232 = .5768$$

Thus we see that there is roughly a 42.3% chance that two or fewer customers will arrive at the bank per minute. Therefore, a 57.7% chance exists that three or more customers will arrive.

7.6.4 Using the Poisson Distribution to Approximate the Binomial Distribution

For those situations in which n is large (≥ 20) and p is very small ($\leq .05$), the Poisson distribution may be used to approximate the binomial distribution. From Equation (7.4) on page 255 it is clearly seen that as n gets large, the computations for the binomial distribution become tedious. However, for situations in which p is also very small, the following mathematical expression for the Poisson model may be used to approximate the true (binomial) result:

$$P(X = x|n, p) \cong \frac{e^{-np}(np)^x}{x!} \tag{7.10}$$

where $P(X = x|n, p)$ = the probability that $X = x$ given a knowledge of n and p

n = sample size

p = true probability of success

e = base of the Napierian (natural) logarithmic system—a mathematical constant approximated by 2.71828

x = number of successes in the sample

It was noted that the Poisson random variable may theoretically range from 0 to ∞. However, when used as an approximation to the binomial distribution, the Poisson random variable—the number of successes out of n observations—clearly cannot exceed the sample size n. Moreover, with large n and small p, Equation (7.10) implies that the probability of observing a large number of successes becomes small and approaches zero quite rapidly. Due to the severe degree of right-skewness in such a probability distribution, no difficulty arises when applying the Poisson approximation to the binomial.

Characteristics As mentioned previously, an interesting characteristic about the Poisson distribution is that the mean μ_x and the variance σ_x^2 are each equal to λ. Thus, when using the Poisson distribution to approximate the binomial distribution, we may compute the mean

$$\mu_x = E(X) = \lambda = np \tag{7.11}$$

and we may approximate the standard deviation

$$\sigma_x = \sqrt{\lambda} = \sqrt{np} \tag{7.12}$$

We note that the standard deviation given by Equation (7.12) agrees with that given for the binomial model [Equation (7.9)] when p is close to zero so that $(1 - p)$ is close to one.

● **Application** To illustrate the use of the Poisson approximation for the binomial, we compute the probability of obtaining exactly one defective tire from a sample of 20 if 8% of the tires manufactured at a particular plant are defective.[4] Thus from Equation (7.10) we have

$$P(X = 1 | n = 20, p = .08) \cong \frac{e^{-(20)(.08)}[(20)(.08)]^1}{1!} = \frac{e^{-1.6}(1.6)^1}{1!}$$

However, rather than having to use the natural logarithmic system to determine this probability, tables of the Poisson distribution (Table E.6) can be employed. Referring to these tables, the only values necessary are the parameter λ and the desired number of successes X. Since in the above example $\lambda = 1.6$ and $X = 1$, we have from Table E.6

$$P(X = 1 | \lambda = 1.6) = .3230$$

This is shown in Table 7.5 (which is a replica of Table E.6).

Had the true distribution, the binomial, been employed instead of the approximation, we would compute

$$P(X = 1 | n = 20, p = .08) = \binom{20}{1}(.08)^1(.92)^{19} = .3282$$

This computation, though, is tedious. Clearly, with Table E.7 available, one could argue that we should look up the binomial probability directly for $n = 20$, $p = .08$, and $X = 1$ and not bother calculating it or using the Poisson approximation. On the other hand, Table E.7 shows binomial probabilities only for particular n from 2 through 20, so that for $n > 20$ the Poisson approximation should certainly be used if p is very small.

To summarize our findings, Figure 7.3 on page 266 compares the binomial distribution (panel A) and its Poisson approximation (panel B) for the number of defective tires in a sample of 20. The similarities of the two results are clearly evident, thus demonstrating the usefulness of the Poisson approximation even when p is as large as .08.

TABLE 7.5 Obtaining a Poisson probability.

	λ									
X	1.1	1.2	1.3	1.4	1.5	1.6	1.7	1.8	1.9	2.0
0	.3329	.3012	.2725	.2466	.2231	.2019	.1827	.1653	.1496	.1353
1	.3662	.3614	.3543	.3452	.3347	.3230	.3106	.2975	.2842	.2707
2	.2014	.2169	.2303	.2417	.2510	.2584	.2640	.2678	.2700	.2707
3	.0738	.0867	.0998	.1128	.1255	.1378	.1496	.1607	.1710	.1804
4	.0203	.0260	.0324	.0395	.0471	.0551	.0636	.0723	.0812	.0902

Source: Extracted from Table E.6.

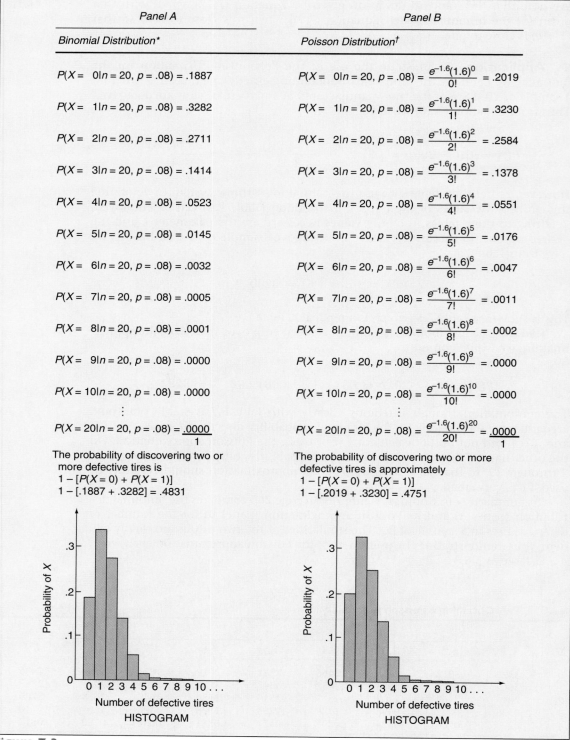

Panel A	*Panel B*
*Binomial Distribution**	*Poisson Distribution†*
$P(X = 0 \mid n = 20, p = .08) = .1887$	$P(X = 0 \mid n = 20, p = .08) = \frac{e^{-1.6}(1.6)^0}{0!} = .2019$
$P(X = 1 \mid n = 20, p = .08) = .3282$	$P(X = 1 \mid n = 20, p = .08) = \frac{e^{-1.6}(1.6)^1}{1!} = .3230$
$P(X = 2 \mid n = 20, p = .08) = .2711$	$P(X = 2 \mid n = 20, p = .08) = \frac{e^{-1.6}(1.6)^2}{2!} = .2584$
$P(X = 3 \mid n = 20, p = .08) = .1414$	$P(X = 3 \mid n = 20, p = .08) = \frac{e^{-1.6}(1.6)^3}{3!} = .1378$
$P(X = 4 \mid n = 20, p = .08) = .0523$	$P(X = 4 \mid n = 20, p = .08) = \frac{e^{-1.6}(1.6)^4}{4!} = .0551$
$P(X = 5 \mid n = 20, p = .08) = .0145$	$P(X = 5 \mid n = 20, p = .08) = \frac{e^{-1.6}(1.6)^5}{5!} = .0176$
$P(X = 6 \mid n = 20, p = .08) = .0032$	$P(X = 6 \mid n = 20, p = .08) = \frac{e^{-1.6}(1.6)^6}{6!} = .0047$
$P(X = 7 \mid n = 20, p = .08) = .0005$	$P(X = 7 \mid n = 20, p = .08) = \frac{e^{-1.6}(1.6)^7}{7!} = .0011$
$P(X = 8 \mid n = 20, p = .08) = .0001$	$P(X = 8 \mid n = 20, p = .08) = \frac{e^{-1.6}(1.6)^8}{8!} = .0002$
$P(X = 9 \mid n = 20, p = .08) = .0000$	$P(X = 9 \mid n = 20, p = .08) = \frac{e^{-1.6}(1.6)^9}{9!} = .0000$
$P(X = 10 \mid n = 20, p = .08) = .0000$	$P(X = 10 \mid n = 20, p = .08) = \frac{e^{-1.6}(1.6)^{10}}{10!} = .0000$
⋮	⋮
$P(X = 20 \mid n = 20, p = .08) = \underline{.0000}$	$P(X = 20 \mid n = 20, p = .08) = \frac{e^{-1.6}(1.6)^{20}}{20!} = \underline{.0000}$
1	1
The probability of discovering two or more defective tires is $1 - [P(X = 0) + P(X = 1)]$ $1 - [.1887 + .3282] = .4831$	The probability of discovering two or more defective tires is approximately $1 - [P(X = 0) + P(X = 1)]$ $1 - [.2019 + .3230] = .4751$

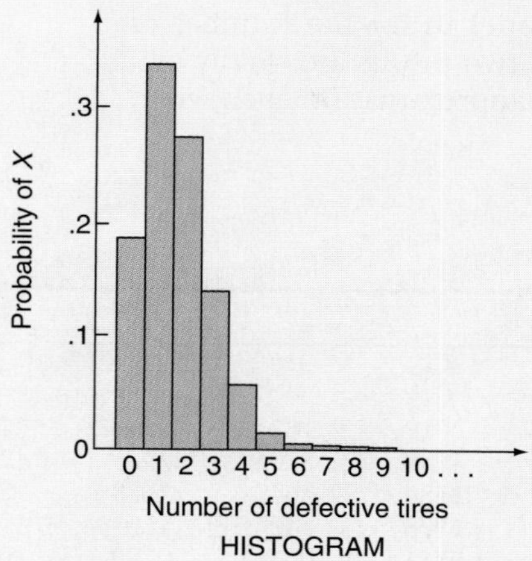

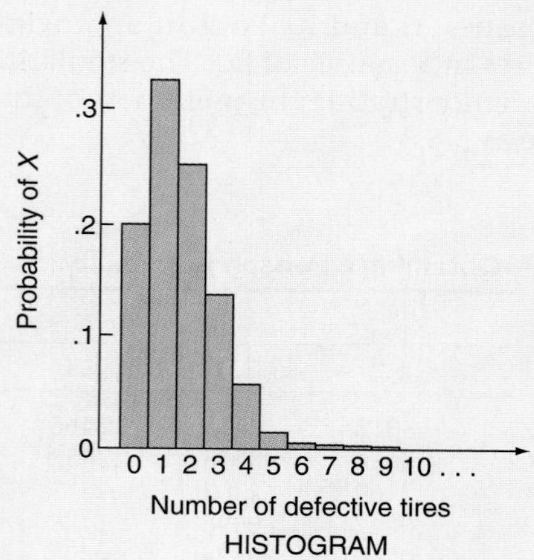

Figure 7.3
Binomial distribution and its Poisson approximation.

* The binomial probabilities are taken from Table E.7.
† The Poisson probabilities are taken from Table E.6.

Problems for Section 7.6

7.20 Using Table E.6, determine the following:
(a) If $\lambda = 2.5$, then $P(X = 2|\lambda = 2.5)$?
(b) If $\lambda = 8.0$, then $P(X = 8|\lambda = 8.0)$?
(c) If $\lambda = 0.5$, then $P(X = 1|\lambda = 0.5)$?
(d) If $\lambda = 3.7$, then $P(X = 0|\lambda = 3.7)$?
(e) If $\lambda = 4.4$, then $P(X = 7|\lambda = 4.4)$?

7.21 The average number of claims per hour made to the Gnecco & Trust Insurance Company for damages or losses incurred in moving is 3.1. What is the probability that in any given hour
(a) Fewer than three claims will be made?
(b) Exactly three claims will be made?
(c) Three or more claims will be made?
(d) More than three claims will be made?

7.22 Based upon past records, the average number of two-car accidents in a New York City police precinct is 3.4 per day. What is the probability that there will be
(a) At least six such accidents in this precinct on any given day?
(b) Not more than two such accidents in this precinct on any given day?
(c) Fewer than two such accidents in this precinct on any given day?
(d) At least two but no more than six such accidents in this precinct on any given day?

● 7.23 The quality control manager of Marilyn's Cookies is inspecting a batch of chocolate-chip cookies that have just been baked. If the production process is in control, the average number of chip parts per cookie is 6.0. What is the probability that in any particular cookie being inspected
(a) Fewer than five chip parts will be found?
(b) Exactly five chip parts will be found?
(c) Five or more chip parts will be found?
(d) Four or five chip parts will be found?

7.24 Refer to Problem 7.23. How many cookies in a batch of 100 being sampled should the manager expect to discard if company policy requires that all chocolate-chip cookies sold must have at least four chocolate-chip parts?

7.25 A natural gas exploration company averages 4 strikes (that is, natural gas is found) per 100 holes drilled. If 20 holes are to be drilled, what is the probability that
(a) Exactly 1 strike will be made?
(b) At least 2 strikes will be made?

Solve this problem using two different probability distributions (the binomial and the Poisson) and briefly compare and explain your results.

7.26 Based upon past experience, 2% of the telephone bills mailed to suburban households are incorrect. If a sample of 20 bills is selected, find the probability that at least one bill will be incorrect. Do this using two probability distributions (the binomial and the Poisson) and briefly compare and explain your results.

7.7 Some Important Discrete Probability Distributions: A Review

As indicated in the summary chart on page 268, this chapter was about some useful discrete probability distributions. On page 242 of Section 7.1 you were given a list emphasizing the important points to be discussed in the chapter. Check over the list now to see whether you feel you have an understanding of these key points.

	Action	
Defective Rate	Do Not Call Mechanic	Call Mechanic
Very low (1%)	$ 20	$100
Low (5%)	$100	$100
Moderate (10%)	$200	$100
High (20%)	$400	$100

Based upon past experience, each defective rate is assumed to be equally likely to occur.

(a) Using the expected-monetary-value criterion, determine whether to call the mechanic.

The manufacturer decides that prior to a final decision a random sample of pens should be studied. Thus, at the end of a particular day's production, a random sample of 15 pens is selected, of which 2 are defective.

(b) The manufacturer wishes to use the *Bayesian decision-making approach* (Reference 7) by revising prior probabilities to take into account the sample information. Use the expected-monetary-value criterion to determine whether to call the mechanic. [*Hint:* Use the binomial distribution and Bayes' theorem (Section 6.9) to determine the conditional probability of this sample outcome, given a particular defective rate.]

7.30 Based upon past experience, 15% of the bills of a large mail-order book company are incorrect. A random sample of three current bills is selected.

(a) What is the probability that
  (1) exactly two bills are incorrect?
  (2) no more than two bills are incorrect?
  (3) at least two bills are incorrect?

(b) What assumptions about the probability distribution are necessary to solve this problem?

7.31 The quality control manager of Ruby's Gambling Equipment Company, which manufactures dice for sale to gambling casinos, must ensure the "fairness" of the dice prior to shipment. Suppose that a particular die is rolled 20 times.

(a) What is the probability that the die lands on an odd face (that is, ⚀, ⚂, or ⚄)
  (1) exactly 17 times?
  (2) at least 17 times?
  (3) at most 17 times?
  (4) more than 17 times?
  (5) fewer than 17 times?

(b) How many times can the die be expected to land on an odd face if the die is truly a fair one?

7.32 Abe Lincoln said that "you can't please all the people all the time." Suppose that you can please each individual nine times out of ten and that there are eight people you want to please.

(a) Calculate
  (1) the probability that you'll please all of them.
  (2) the probability that you'll please at least six of them.
  (3) the probability that you'll please four or fewer.

(b) What is the expected number of people you will please? How likely is it that you will please exactly that number?

(c) What is the standard deviation of the number of people you will please? From this and the expected value, find out approximately how many people you will please at least three-fourths of the time.

7.33 Based on past experience, the main printer in a university computer center is operating properly 90% of the time. If a random sample of ten inspections is made

(a) What is the probability that the main printer is operating properly
  (1) exactly nine times?
  (2) at least nine times?
  (3) at most nine times?
  (4) more than nine times?
  (5) fewer than nine times?

(b) How many times can the main printer be expected to operate properly?

7.34 Records provided by the Vice-President for Human Resources at a large urban hospital indicate that, on any given workday, 10% of the nonclinical workforce (i.e., kitchen, housekeeping and janitorial, electrical and plumbing, security, mailroom, laundry, clerical, and administrative) are absent from work. What is the probability that in a random sample of 10 nonclinical workers

(a) Exactly one will be absent today?
(b) At least two will be absent?

● 7.35 Suppose that on a very long arithmetic test, Donna would get 70% of the items right.

(a) For a ten-item quiz, calculate the probability that Donna will get
  (1) at least seven items right.
  (2) less than six items right (and therefore fail the quiz).
  (3) nine or ten items right (and get an A on the quiz).

(b) Use an appropriate table to check your calculations of the probability distribution.

(c) What is the expected number of items that Donna will get right? What proportion of the time will she get that number right?

(d) What is the standard deviation of the number of items that Donna will get right? Compare the proportion of time that Donna will be within two standard deviations according to the distribution you just calculated with the same probability calculated from the Bienaymé-Chebyshev inequality.

● 7.36 The manufacturer of the disk drives used in one of the well-known brands of microcomputers expects 2% of the disk drives to malfunction during the microcomputer's warranty period.

(a) In a sample of ten disk drives, what is the probability that
  (1) none will malfunction during the warranty period?
  (2) exactly one will malfunction during the warranty period?
  (3) at least two will malfunction during the warranty period?

(b) Solve (a)(1), (a)(2), and (a)(3) using the Poisson distribution as an approximation of the binomial distribution and briefly compare your results.

(c) In a sample of 50 disk drives, what is the *approximate* probability that
  (1) none will malfunction during the warranty period?
  (2) exactly one will malfunction during the warranty period?
  (3) at least two will malfunction during the warranty period?

(d) What assumptions are needed in order to use the two probability distributions in this problem?

7.37 An actuary from the Egan Life Insurance Company has determined that .0001 of the elderly population incur a rare disease each year. A random sample of 10,000 Medicare patient records is to be evaluated. What is the probability that

(a) None of these Medicare patients will have incurred the rare disease?
(b) At least two of the Medicare patients will have incurred the rare disease?
(c) No more than two of the Medicare patients will have incurred the rare disease?

7.38 One out of every 100 light bulbs produced by the Lori Lighting Co. fails before the end of a one-week period when left burning continuously. One bulb is installed on each of the 50 floors of a large apartment building in New York City. What is the *approximate* probability that
(a) One bulb will be burned out at the end of the week?
(b) More than three of the bulbs will be burned out at the end of the week?
(c) Fewer than three of the bulbs will be burned out at the end of the week?
(d) Three of the bulbs will be burned out at the end of the week?

Endnotes

1. As an example of a binomial random variable arising from sampling with replacement from a finite population, consider the probability of obtaining two clubs in five draws from a randomly shuffled deck of cards, where the selected card is replaced and the deck well shuffled after each draw.
2. Three rolls of the same die is equivalent to one roll of each of three dice. See Problem 7.16 on the game of Chuck-a-luck (page 259).
3. Note that for $p > .5$, we read across the bottom and up the right-hand side of the table.
4. Even though $p > .05$, we observe from Figure 7. 3 that the Poisson distribution provides a good approximation to the binomial distribution.

References

1. Derman, C., L. J. Gleser, and I. Olkin, *A Guide to Probability Theory and Application* (New York: Holt, Rinehart and Winston, 1973).
2. Eppen, G. D., F. J. Gould, and C. P. Schmidt, *Introductory Management Science,* 4th ed. (Englewood Cliffs, NJ: Prentice-Hall, 1993).
3. Larsen, R. J., and M. L. Marx, *An Introduction to Mathematical Statistics and Its Applications,* 2d ed. (Englewood Cliffs, NJ: Prentice-Hall, 1986).
4. Miller, I., and J. E. Freund, *Probability and Statistics for Engineers,* 5th ed. (Englewood Cliffs, NJ: Prentice-Hall, 1994).
5. Scarne, J., *Scarne's New Complete Guide to Gambling* (New York: Simon and Schuster, 1974).
6. Thorp, E. O., *Beat the Dealer* (New York: Random House, 1962).
7. Winkler, R. L., *Introduction to Bayesian Inference and Decision* (New York: Holt, Rinehart and Winston, 1972).

chapter 8

The Normal Distribution

CHAPTER OBJECTIVES

To show how the normal probability density function can be used to represent certain types of continuous phenomena and to approximate various models representing discrete phenomena under specific conditions.

8.1 Introduction

In Chapter 7 we developed the concept of a probability distribution for a discrete random variable and, in particular, we studied the binomial and Poisson distributions. In this chapter we will turn our discussion to the most important probability distribution in statistics, the normal distribution. We will begin by discussing the properties of the normal distribution and then develop various applications. We will then study a simple graphical tool, the normal probability plot, that can be used to evaluate whether a set of data appears to be normally distributed. We will conclude the chapter by showing how the normal distribution can be used to approximate the binomial and Poisson distributions under certain circumstances.

Upon completion of this chapter, you should be able to:

1. Understand the properties of the normal distribution
2. Find an area under the normal curve
3. Find the value that corresponds to any percentage point of the normal distribution
4. Develop and interpret a normal probability plot
5. Know when and how the normal distribution can be used to approximate the binomial and Poisson distributions

8.2 Mathematical Models of Continuous Random Variables: The Probability Density Function

Now that we have studied some discrete probability distributions, we turn our attention to **continuous probability density functions**—those that arise due to some measuring process on various phenomena of interest. Continuous models have important applications in engineering and the physical sciences as well as in business and the social sciences. Some examples of continuous random phenomena are: height, weight, time between arrivals (of customers at a bank), and customer servicing times. Moreover, from the Employee Satisfaction Survey developed in Chapter 2, responses to questions regarding hours worked, age, and income also pertain to probability density functions for continuous random variables.

When a mathematical expression is available to represent some underlying continuous phenomenon, the probability that various values of the random variable occur within certain ranges or intervals may be calculated. However, the *exact* probability of a *particular value* from a continuous distribution is zero.

As an example, the probability distribution represented in Table 8.1 is obtained by categorizing a distribution in which the continuous random phenomenon of interest is said to follow the **Gaussian** or "bell-shaped" **normal probability density function.** If the nonoverlapping (mutually exclusive) listing contains all possible class intervals (is collectively exhaustive), the probabilities will again sum to 1. This is demonstrated in Table 8.1. Such a probability distribution may be considered as a relative frequency distribution as described in Section 3.4, where, except for the two open-ended classes, the midpoint of every other class interval represents the data in that interval.

Table 8.1 Thickness of 10,000 brass washers manufactured by a large company.

Thickness (inches)	Relative Frequency or Probability
Under .0180	48/10,000 = .0048
.0180 < .0182	122/10,000 = .0122
.0182 < .0184	325/10,000 = .0325
.0184 < .0186	695/10,000 = .0695
.0186 < .0188	1198/10,000 = .1198
.0188 < .0190	1664/10,000 = .1664
.0190 < .0192	1896/10,000 = .1896
.0192 < .0194	1664/10,000 = .1664
.0194 < .0196	1198/10,000 = .1198
.0196 < .0198	695/10,000 = .0695
.0198 < .0200	325/10,000 = .0325
.0200 < .0202	122/10,000 = .0122
.0202 or above	48/10,000 = .0048
Total	1.0000

Unfortunately, obtaining probabilities or computing expected values and standard deviations for continuous phenomena involves mathematical expressions that require a knowledge of integral calculus and are beyond the scope of this book. Nevertheless, one continuous probability density function that we shall focus upon has been deemed so important for applications that special probability tables (such as Table E.2 of Appendix E) have been devised in order to eliminate the need for what otherwise would require laborious mathematical computations. This particular continuous probability density function is known as the **Gaussian** or **normal distribution.**

8.3 The Normal Distribution

8.3.1 Importance of the Normal Distribution

The normal distribution is vitally important in statistics for three main reasons:

1. Numerous continuous phenomena seem to follow it or can be approximated by it.
2. We can use it to approximate various discrete probability distributions and thereby avoid much computational drudgery (Section 8.6).
3. It provides the basis for *classical statistical inference* because of its relationship to the *central limit theorem* (to be developed in Chapter 9).

8.3.2 Properties of the Normal Distribution

The normal distribution has several important theoretical properties. Among these are

1. It is bell-shaped and symmetrical in its appearance.

2. Its measures of central tendency (mean, median, mode, midrange, and midhinge) are all identical.
3. Its "middle spread" is equal to 1.33 standard deviations. That is, the interquartile range is contained within an interval of two-thirds of a standard deviation below the mean to two-thirds of a standard deviation above the mean.
4. Its associated random variable has an infinite range ($-\infty < X < +\infty$).

In actual practice some of the variables we observe may only approximate these theoretical properties. This occurs for two reasons: (1) the underlying population distribution may be only approximately normal and (2) any actual sample may deviate from the theoretically expected characteristics. For some phenomenon that may be approximated by the normal distribution model:

1. Its polygon may be only approximately bell-shaped and symmetrical in appearance.
2. Its measures of central tendency may differ slightly from each other.
3. The value of its interquartile range may differ slightly from 1.33 standard deviations.
4. Its *practical range* will not be infinite but will generally lie within 3 standard deviations above and below the mean. (That is, range ≈ 6 standard deviations.)

As a case in point, let us refer to Figure 8.1, which depicts the relative frequency histogram and polygon for the distribution of the thickness of 10,000 brass washers presented in Table 8.1 on page 275. For these data, the first three theoretical properties of the normal distribution seem to have been satisfied; however, the fourth does not hold. The random variable of interest, thickness, cannot possibly take on values of zero or below, nor can a washer be so thick that it becomes unusable. From Table 8.1 we note that only 48 out of every 10,000 brass washers manufactured can be expected to have a thickness of .0202 inch or more, while an equal number can be expected to have a thickness under .0180 inch. Thus the chance of randomly obtaining a washer so thin or so thick is .0048 + .0048 = .0096—or almost 1 in 100.

We shall leave it to the reader to verify (see Problem 8.5 on page 291) that 99.04% of these manufactured washers can be expected to have a thickness between .0180 and .0202 inch—that is, 2.59 standard deviations (distances) above and below the mean.

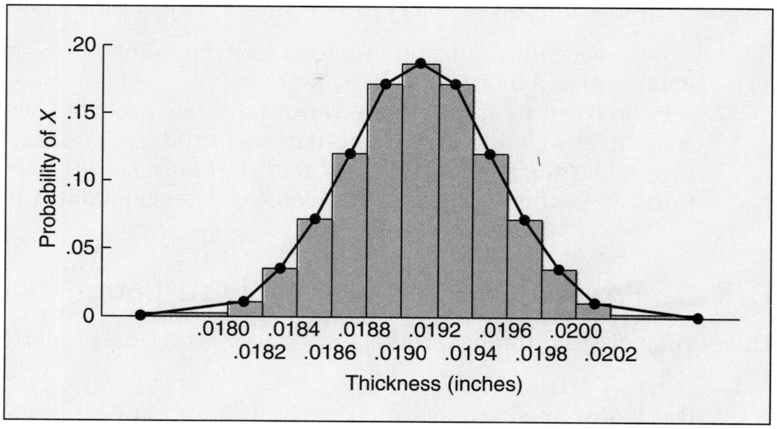

Figure 8.1
Relative frequency histogram and polygon of the thickness of 10,000 brass washers.
Source: Data are taken from Table 8.1.

8.3.3 The Mathematical Model

The mathematical model or expression representing a probability density function is denoted by the symbol $f(X)$. For the normal distribution, the model used to obtain the desired probabilities is

$$f(X) = \frac{1}{\sqrt{2\pi}\,\sigma_x} e^{-(1/2)[(X-\mu_x)/\sigma_x]^2} \tag{8.1}$$

where

e is the mathematical constant approximated by 2.71828
π is the mathematical constant approximated by 3.14159
μ_x is the population mean
σ_x is the population standard deviation
X is any value of the continuous random variable, where $-\infty < X < +\infty$

Let us examine the components of the normal probability density function in Equation (8.1). Since e and π are mathematical constants, the probabilities of the random variable X are dependent only upon the two parameters of the normal distribution—the population mean μ_x and the population standard deviation σ_x. Every time we specify a *particular combination* of μ_x and σ_x, a *different* normal probability distribution will be generated. We illustrate this in Figure 8.2, where three different normal distributions are depicted. Distributions A and B have the same mean (μ_x) but have different standard deviations. On the other hand, distributions A and C have the same standard deviation (σ_x) but have different means. Furthermore, distributions B and C depict two normal probability density functions that differ with respect to both μ_x and σ_x.

Unfortunately, the mathematical expression in Equation (8.1) is computationally tedious. To avoid having to make such computations, it would be useful to have a set of tables that would provide the desired probabilities. However, since an infinite number of combinations of the parameters μ_x and σ_x exists, an infinite number of such tables would be required.

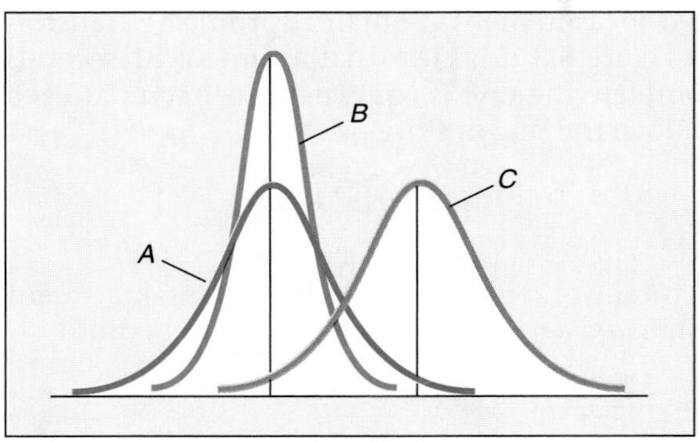

Figure 8.2
Three normal distributions having differing parameters μ_x and σ_x.

8.3.4 Standardizing the Normal Distribution

Fortunately, by *standardizing* the data, we will only need one table. (See Table E.2.) By using the **transformation formula**

$$Z = \frac{X - \mu_x}{\sigma_x} \tag{8.2}$$

any normal random variable X is converted to a standardized normal random variable Z. While the original data for the random variable X had mean μ_x and standard deviation σ_x, the standardized random variable Z will always have mean $\mu_z = 0$ and standard deviation $\sigma_z = 1$.

A **standardized normal distribution** is one whose random variable Z always has a mean $\mu_z = 0$ and a standard deviation $\sigma_z = 1$.

Substituting in Equation (8.1), we see that the probability density function of a standard normal variable Z is

$$f(Z) = \frac{1}{\sqrt{2\pi}} e^{-(1/2)Z^2} \tag{8.1a}$$

Thus, we can always convert any set of normally distributed data to its standardized form and then determine any desired probabilities from a table of the standardized normal distribution.

To see how the transformation formula (8.2) may be applied and how we may then use the results to read probabilities from the table of the standardized normal distribution (Table E.2), let us consider the following problem.

Suppose a consultant was investigating the time it took factory workers in an automobile plant to assemble a particular part after the workers had been trained to perform the task using an individual learning approach. The consultant determined that the time in seconds to assemble the part for workers trained with this method was normally distributed with a mean μ_x of 75 seconds and a standard deviation σ_x of 6 seconds.

● **Transforming the Data** We see from Figure 8.3 that every measurement X has a corresponding standardized measurement Z obtained from the transformation formula (8.2). Hence from Figure 8.3 it is clear that a time of 81 seconds required for a factory worker to complete the task is equivalent to 1 standardized unit (that is, 1 *standard deviation*) above the mean, since

$$Z = \frac{81 - 75}{6} = +1$$

and a time of 57 seconds required for a worker to assemble the part is equivalent to 3 standardized units (that is, 3 *standard deviations*) below the mean because

$$Z = \frac{57 - 75}{6} = -3$$

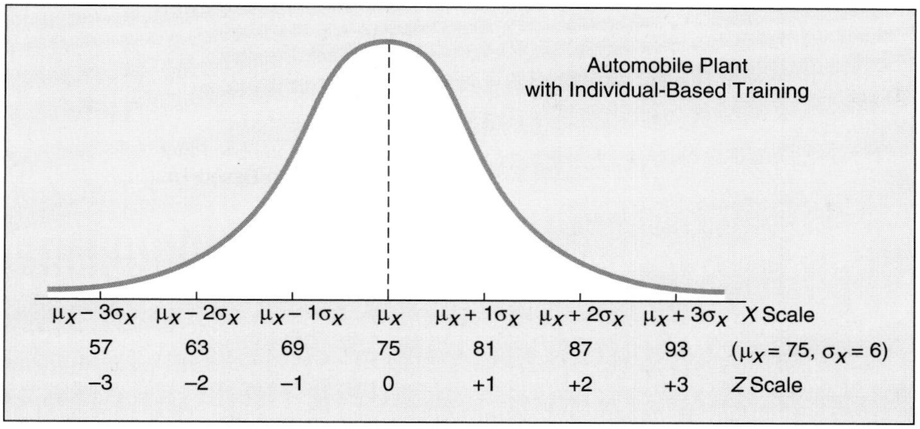

Figure 8.3
Transformation of scales.

Thus, the standard deviation has become the unit of measurement. In other words, a time of 81 seconds is 6 seconds (i.e., 1 standard deviation) higher, or *slower* than the average time of 75 seconds and a time of 57 seconds is 18 seconds (i.e., 3 standard deviations) lower, or *faster* than the average time.

Suppose now that the consultant conducted the same study at another automobile plant, where the workers were trained to assemble the part by using a team-based learning method. Suppose that at this plant the consultant determined that the time to perform the task was normally distributed with mean μ_x of 60 seconds and a standard deviation σ_x of 3 seconds. The data are depicted in Figure 8.4 on page 280. In comparison with the results for workers who had an individual learning method, we note, for example, that at the plant where workers had team-based training, a time of 57 seconds to complete the task is only 1 standard deviation below the mean for the group, since

$$Z = \frac{57 - 60}{3} = -1$$

We may also note that a time of 63 seconds is 1 standard deviation above the mean time for assemblage, since

$$Z = \frac{63 - 60}{3} = +1$$

and a time of 51 seconds is 3 standard deviations below the group mean because

$$Z = \frac{51 - 60}{3} = -3$$

8.3.5 Using the Normal Probability Tables

The two bell-shaped curves in Figures 8.3 and 8.4 depict the relative frequency polygons for the normal distributions representing the time (in seconds) for all factory workers to assemble a part at two automobile plants, one that employed individual-based training and the other that employed team-based training. Since at each plant the times to assemble the part are known for every factory worker, the data represent the entire population at a particular plant, and therefore the

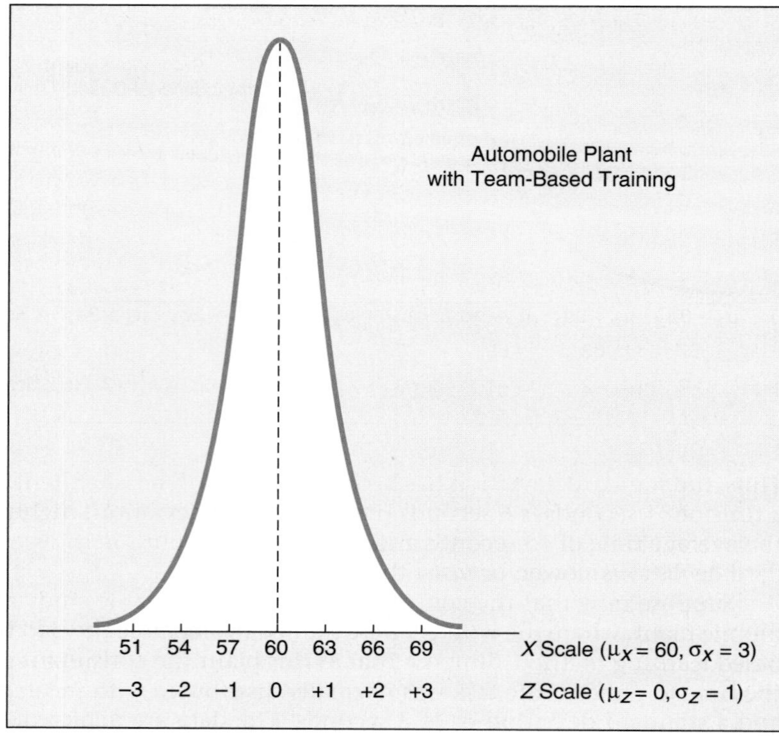

Figure 8.4
A different transformation of scales.

probabilities or proportion of area under the entire curve must add up to 1. Thus, the area under the curve between any two reported time values represents only a portion of the total area possible.

Suppose the consultant wishes to determine the probability that a factory worker selected at random from those who underwent individual-based training should require between 75 and 81 seconds to complete the task. That is, what is the likelihood that the worker's time is between the plant mean and one standard deviation above this mean? This answer is found by using Table E.2.

Table E.2 represents the probabilities or areas under the normal curve calculated from the mean μ_x to the particular values of interest X. Using Equation (8.2), this corresponds to the probabilities or areas under the standardized normal curve from the mean ($\mu_z = 0$) to the transformed values of interest Z. Only positive entries for Z are listed in the table, since for such a symmetrical distribution having a mean of zero, the area from the mean to $+Z$ (that is, Z standard deviations above the mean) must be identical to the area from the mean to $-Z$ (that is, Z standard deviations below the mean).

To use Table E.2 we note that all Z values must first be recorded to two decimal places. Thus our particular Z value of interest is recorded as +1.00. To read the probability or area under the curve from the mean to $Z = +1.00$, we scan down the Z column from Table E.2 until we locate the Z value of interest (in tenths). Hence we stop in the row $Z = 1.0$. Next we read across this row until we intersect the column that contains the hundredths place of the Z value. Therefore, in the body of the table the tabulated probability for $Z = 1.00$ corresponds to the intersection of the row $Z = 1.0$ with the column $Z = .00$ as shown in Table 8.2 (which is a replica of Table E.2). This probability is .3413. As depicted in Figure 8.5, there is a 34.13%

Table 8.2 Obtaining an area under the normal curve.

Z	.00	.01	.02	.03	.04	.05	.06	.07	.08	.09
0.0	.0000	.0040	.0080	.0120	.0160	.0199	.0239	.0279	.0319	.0359
0.1	.0398	.0438	.0478	.0517	.0557	.0596	.0636	.0675	.0714	.0753
0.2	.0793	.0832	.0871	.0910	.0948	.0987	.1026	.1064	.1103	.1141
0.3	.1179	.1217	.1255	.1293	.1331	.1368	.1406	.1443	.1480	.1517
0.4	.1554	.1591	.1628	.1664	.1700	.1736	.1772	.1808	.1844	.1879
0.5	.1915	.1950	.1985	.2019	.2054	.2088	.2123	.2157	.2190	.2224
0.6	.2257	.2291	.2324	.2357	.2389	.2422	.2454	.2486	.2518	.2549
0.7	.2580	.2612	.2642	.2673	.2704	.2734	.2764	.2794	.2823	.2852
0.8	.2881	.2910	.2939	.2967	.2995	.3023	.3051	.3078	.3106	.3133
0.9	.3159	.3186	.3212	.3238	.3264	.3289	.3315	.3340	.3365	.3389
1.0	.3413	.3438	.3461	.3485	.3508	.3531	.3554	.3577	.3599	.3621
1.1	.3643	.3665	.3686	.3708	.3729	.3749	.3770	.3790	.3810	.3830

Source: Extracted from Table E.2.

chance that a factory worker selected at random who has had individual-based training will require between 75 and 81 seconds to assemble the part.

On the other hand, we know from Figure 8.4 that at the automobile plant where the workers received team-based training, a time of 63 seconds is 1 standardized unit above the mean time of 60 seconds. Thus the likelihood that a randomly selected factory worker who received team-based training will complete the assemblage in between 60 and 63 seconds is also .3413.[1] These results are clearly illustrated in Figure 8.6 (page 282), which demonstrates that regardless of the value of the mean μ_x and standard deviation σ_x of a particular set of normally distributed data, a transformation to a standardized scale can always be made from Equation (8.2), and, by using Table E.2, any probability or portion of area under the curve can be obtained. From Figure 8.6 we see that the probability or area under the curve from 60 to 63 seconds for the workers who had team-based training is identical to the probability or area under the curve from 75 to 81 seconds for the workers who had individual-based training.

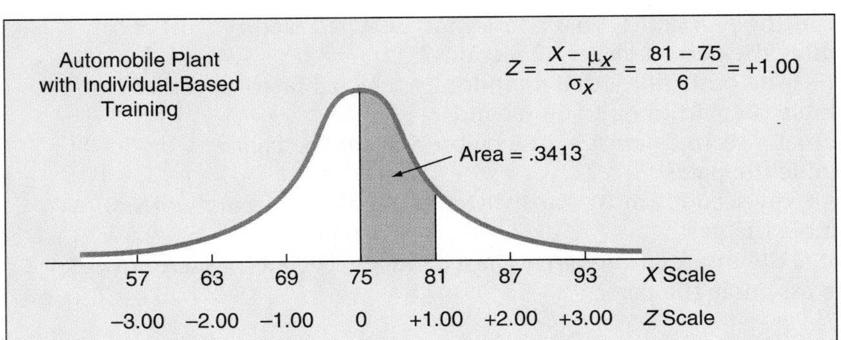

Figure 8.5
Determining the area between the mean and *Z* from a standardized normal distribution.

8.4 Applications

Now that we have learned to use Table E.2 in conjunction with Equation (8.2), many different types of probability questions pertaining to the normal distribution can be resolved. To illustrate this, let us suppose that the consultant raises the

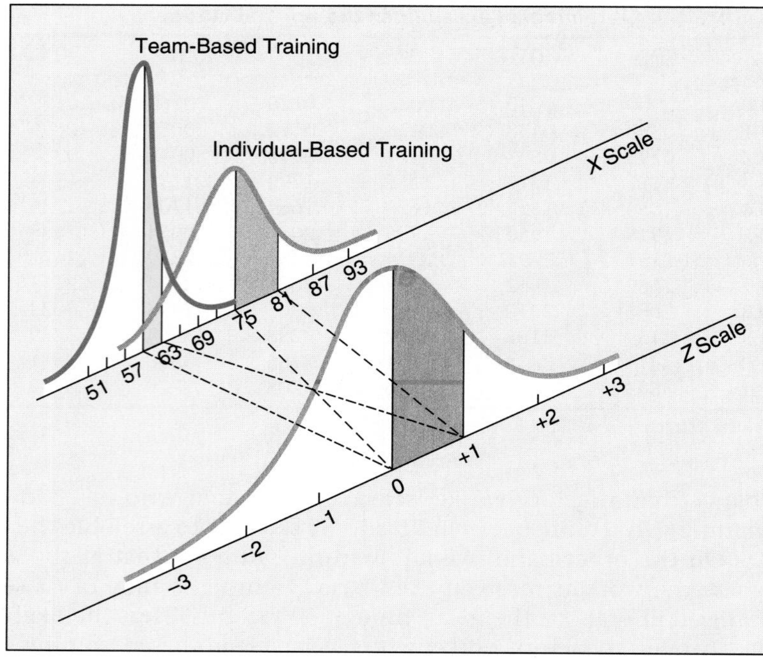

Figure 8.6
Demonstrating a transformation of scales for corresponding portions under two normal curves.

following questions with regard to assembling a particular part by workers who had individual-based training:

1. What is the probability that a randomly selected factory worker can assemble the part in under 75 seconds or in over 81 seconds?
2. What is the probability that a randomly selected factory worker can assemble the part in 69 to 81 seconds?
3. What is the probability that a randomly selected factory worker can assemble the part in under 62 seconds?
4. What is the probability that a randomly selected factory worker can assemble the part in 62 to 69 seconds?
5. How many seconds must elapse before 50% of the factory workers assemble the part?
6. How many seconds must elapse before 10% of the factory workers assemble the part?
7. What is the interquartile range (in seconds) expected for factory workers to assemble the part?

8.4.1 Finding the Probabilities Corresponding to Known Values

We recall from Section 8.3.4 that for workers who had individual-based training the assembly time data are normally distributed with a mean μ_x of 75 seconds and a standard deviation σ_x of 6 seconds. In responding to questions 1 through 4, we shall use this information as we seek to determine the probabilities associated with various measured values.

Question 1: Finding $P(X < 75$ or $X > 81)$ How can we determine the probability that a randomly selected factory worker will perform the task in under 75 seconds or over 81 seconds? Since we have already determined the probability that a randomly selected factory worker will need between 75 and 81 seconds to assemble the part, from Figure 8.5 on page 281 we observe that our desired probability must be its *complement*, that is, $1 - .3413 = .6587$.

Another way to view this problem, however, is to separately obtain both the probability of assembling the part in under 75 seconds and the probability of assembling the part in over 81 seconds and then use the *addition rule for mutually exclusive events* [Equation (6.4)] to obtain the desired result. This is depicted in Figure 8.7.

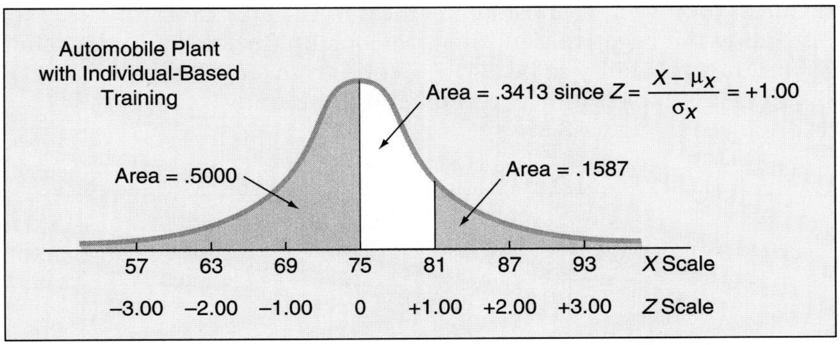

Figure 8.7
Finding $P(X < 75$ or $X > 81)$.

Since the mean and median are theoretically the same for normally distributed data, it follows that 50% of the workers can assemble the part in under 75 seconds.[2] To show this, from Equation (8.2) we have

$$Z = \frac{X - \mu_x}{\sigma_x} = \frac{75 - 75}{6} = 0.00$$

Using Table E.2, we see that the area under the normal curve from the mean to $Z = 0.00$ is .0000. Hence the area under the curve less than $Z = 0.00$ must be $.5000 - .0000 = .5000$ (which happens to be the area for the entire left side of the distribution from the mean to $Z = -\infty$, as shown in Figure 8.7).

Now we wish to obtain the probability of assembling the part in over 81 seconds. But Equation (8.2) only gives the areas under the curve from the mean to Z, not from Z to $+\infty$. Thus we find the probability from the mean to Z and subtract this result from .5000 to obtain the desired answer. Since we know that the area or portion of the curve from the mean to $Z = +1.00$ is .3413, the area from $Z = +1.00$ to $Z = +\infty$ must be $.5000 - .3413 = .1587$. Hence the probability that a randomly selected factory worker will perform the task in under 75 or over 81 seconds, $P(X < 75$ or $X > 81)$, is $.5000 + .1587 = .6587$.

Question 2: Finding $P(69 \leq X \leq 81)$ Suppose that we are now interested in determining the probability that a randomly selected factory worker can complete the part in 69 to 81 seconds, that is, $P(69 \leq X \leq 81)$. We note from Figure 8.8 on page 284 that one of the values of interest is above the mean assembly time of 75 seconds and the other value is below it. Since our transformation formula (8.2) permits us only to find probabilities from a particular value of interest to the mean, we can obtain our desired probability in three steps:

1. Determine the probability from the mean to 81 seconds.
2. Determine the probability from the mean to 69 seconds.
3. Sum up the two mutually exclusive results.

For this example, we already completed step 1; the area under the normal curve from the mean to 81 seconds is .3413. To find the area from the mean to 69 seconds (step 2), we have

$$Z = \frac{X - \mu_x}{\sigma_x} = \frac{69 - 75}{6} = -1.00$$

Table E.2 shows only positive entries for Z. Because of symmetry, it is clear that the area from the mean to $Z = -1.00$ must be identical to the area from the mean to $Z = +1.00$. Discarding the negative sign, then, we look up (in Table E.2) the value $Z = 1.00$ and find the probability to be .3413. Hence, from step 3, the probability that the part can be assembled in between 69 and 81 seconds is .3413 + .3413 = .6826. This is displayed in Figure 8.8.

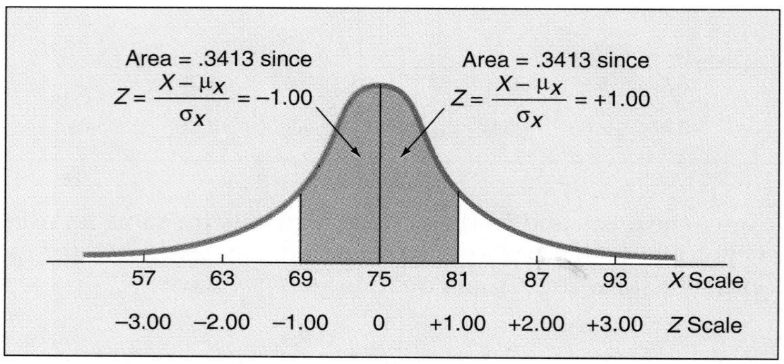

Figure 8.8
Finding $P(69 \le X \le 81)$.

Generalizing from the Standard Normal Distribution The above result is rather important. If we may generalize for a moment, we can see that for any normal distribution there is a .6826 chance that a randomly selected item will fall within ±1 standard deviation above or below the mean. We will leave it to the reader to verify from Table E.2 (see Problem 8.4 on page 291) that there is a .9544 chance that any randomly selected normally distributed observation will fall within ±2 standard deviations above or below the mean and a .9973 chance that the observation will fall between ±3 standard deviations above or below the mean.

For the plant in which the workers received individual-based training, this tells us that slightly more than two out of every three factory workers (68.26%) can be expected to complete the task within ±1 standard deviation from the mean. Moreover, from Figure 8.9, slightly more than 19 out of every 20 factory workers (95.44%) can be expected to complete the assembly within ±2 standard deviations from the mean (that is, between 63 and 87 seconds), and, from Figure 8.10, practically all factory workers (99.73%) can be expected to assemble the part within ±3 standard deviations from the mean (that is, between 57 and 93 seconds).

From Figure 8.10 it is indeed quite unlikely (.0027 or only 27 factory workers in 10,000) that a randomly selected factory worker will be so fast or so slow that he or she could be expected to complete the assembly of the part in under 57 seconds or over 93 seconds. Thus it is clear why $6\sigma_x$ (that is, 3 standard deviations

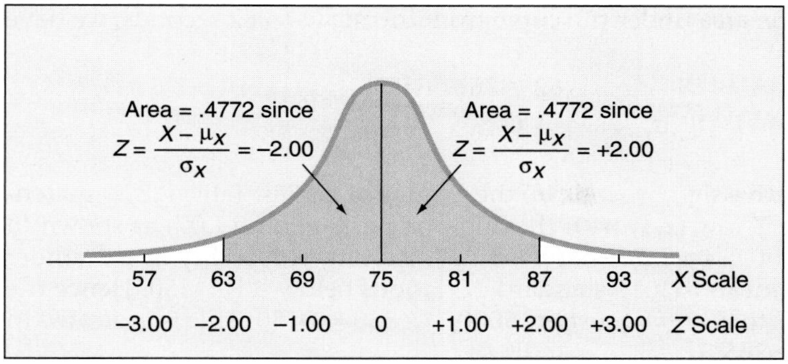

Figure 8.9
Finding $P(63 \le X \le 87)$.

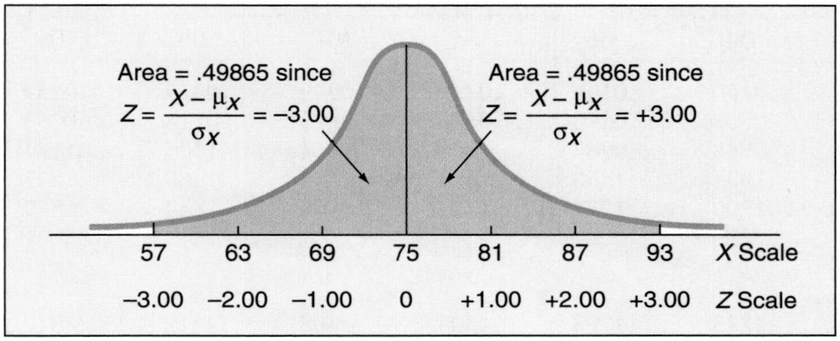

Figure 8.10
Finding $P(57 \le X \le 93)$.

onds or over 93 seconds. Thus it is clear why $6\sigma_x$ (that is, 3 standard deviations above the mean to 3 standard deviations below the mean) is often used as a *practical approximation of the range* for normally distributed data.

● **Question 3: Finding $P(X < 62)$** To obtain the probability that a randomly selected factory worker can assemble the part in under 62 seconds we should examine the shaded lower left-tailed region of Figure 8.11. The transformation formula (8.2) only permits us to find areas under the standardized normal distribution from the mean to Z, not from Z to $-\infty$. Thus, we must find the probability from the mean to Z and subtract this result from .5000 to obtain the desired answer.

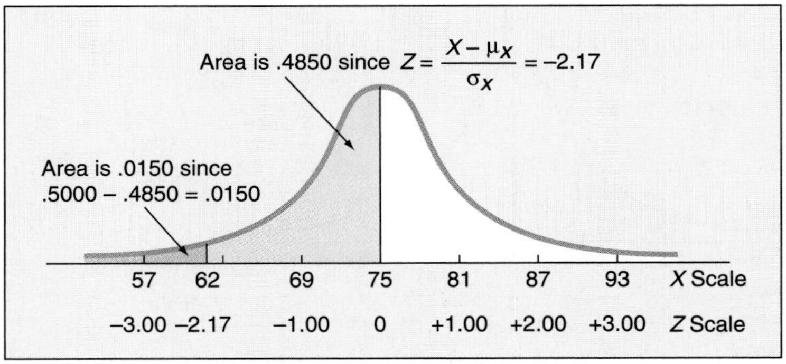

Figure 8.11
Finding $P(X < 62)$.

To determine the area under the curve from the mean to 62 seconds, we have

$$Z = \frac{X - \mu_x}{\sigma_x} = \frac{62 - 75}{6} = \frac{-13}{6} = -2.17$$

Neglecting the negative sign, we look up the Z value of 2.17 in Table E.2 by matching the appropriate Z row (2.1) with the appropriate Z column (.07) as shown in Table 8.3 (a replica of Table E.2). Therefore, the resulting probability or area under the curve from the mean to 2.17 standard deviations below it is .4850. Hence the area from $Z = -2.17$ to $Z = -\infty$ must be $.5000 - .4850 = .0150$. This is indicated in Figure 8.11 on page 285.

Table 8.3 Obtaining an area under the normal curve.

Z	.00	.01	.02	.03	.04	.05	.06	.07	.08	.09
0.0	.0000	.0040	.0080	.0120	.0160	.0199	.0239	.0279	.0319	.0359
0.1	.0398	.0438	.0478	.0517	.0557	.0596	.0636	.0675	.0714	.0753
0.2	.0793	.0832	.0871	.0910	.0948	.0987	.1026	.1064	.1103	.1141
0.3	.1179	.1217	.1255	.1293	.1331	.1368	.1406	.1443	.1480	.1517
0.4	.1554	.1591	.1628	.1664	.1700	.1736	.1772	.1808	.1844	.1879
.	.	.	.	.	.	.	.	.	.	.
.	.	.	.	.	.	.	.	.	.	.
.	.	.	.	.	.	.	.	.	.	.
2.0	.4772	.4778	.4783	.4788	.4793	.4798	.4803	.4808	.4812	.4817
2.1	.4821	.4826	.4830	.4834	.4838	.4842	.4846	.4850	.4854	.4857
2.2	.4861	.4864	.4868	.4871	.4875	.4878	.4881	.4884	.4887	.4890
2.3	.4893	.4896	.4898	.4901	.4904	.4906	.4909	.4911	.4913	.4916
2.4	.4918	.4920	.4922	.4925	.4927	.4929	.4931	.4932	.4934	.4936

Source: Extracted from Table E.2.

Question 4: Finding $P(62 \le X \le 69)$ As a final illustration of determining probabilities from the standardized normal distribution, suppose we wish to find how likely it is that a randomly selected factory worker can complete the task in 62 to 69 seconds. Since both values of interest are below the mean, we see from Figure 8.12 that the desired probability (or area under the curve between the two

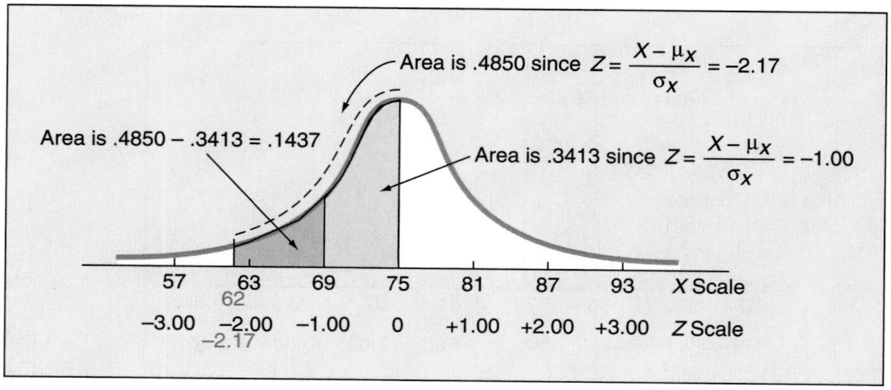

Figure 8.12
Finding $P(62 \le X \le 69)$.

values) is less than .5000. Since our transformation formula (8.2) only permits us to find probabilities from a particular value of interest to the mean, we can obtain our desired probability in three steps:

1. Determine the probability or area under the curve from the mean to 62 seconds.
2. Determine the probability or area under the curve from the mean to 69 seconds.
3. Subtract the smaller area from the larger (to avoid double counting).

For this example, we have already completed steps 1 and 2 in answering questions 3 and 2, respectively. The area from the mean to 62 seconds is .4850, and the area from the mean to 69 seconds is .3413. Hence, from step 3, by subtracting the smaller area from the larger one we determine that there is only a .1437 probability of randomly selecting a factory worker who could be expected to complete the task in between 62 and 69 seconds. That is,

$$P(62 \le X \le 69) = P(62 \le X \le 75) - P(69 \le X \le 75)$$
$$= .4850 - .3413 = .1437$$

8.4.2 Finding the Values Corresponding to Known Probabilities

In our previous applications regarding normally distributed data we have sought to determine the probabilities associated with various measured values. Now, however, suppose we wish to determine particular numerical values of the variables of interest that correspond to known probabilities. As examples, let us respond to questions 5 through 7.

● **Question 5** To determine how many seconds elapse before 50% of the factory workers assemble the part, we should examine Figure 8.13. Since this time value corresponds to the median, and the mean and median are equal in all symmetric distributions, the median must be 75 seconds.

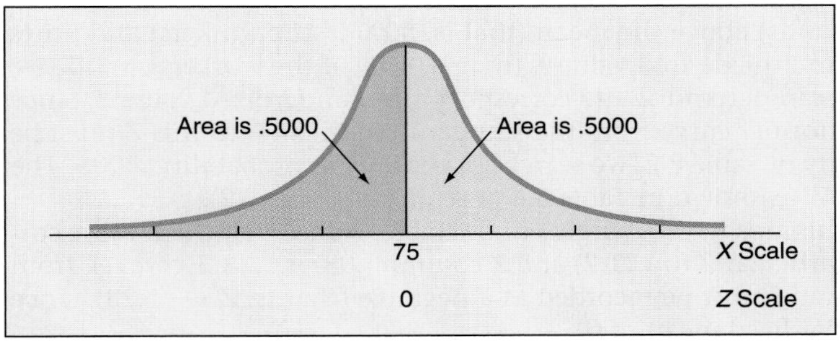

Figure 8.13
Finding *X*.

● **Question 6** To determine how many seconds elapse before 10% of the factory workers assemble the part, we should focus on Figure 8.14 on page 288. Since 10% of the factory workers are expected to complete the task in under X seconds, then 90% of the workers would be expected to require X seconds or more to do the job. From Figure 8.14 we may observe that this 90% can be broken down into two

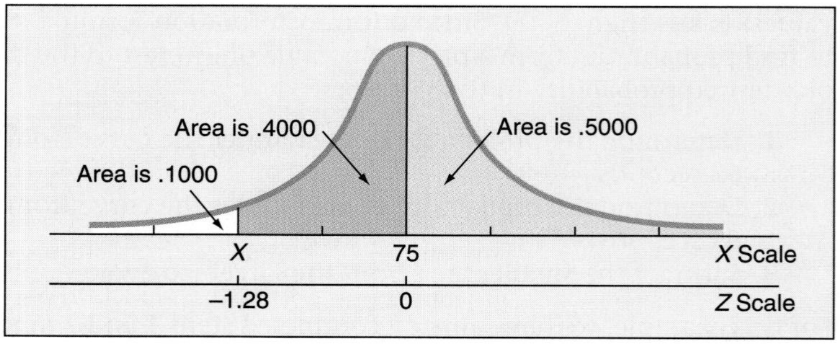

Figure 8.14
Finding *Z* to determine *X*.

Table 8.4 Obtaining a *Z* value corresponding to a particular area under the normal curve.

Z	.00	.01	.02	.03	.04	.05	.06	.07	.08	.09
0.0	.0000	.0040	.0080	.0120	.0160	.0199	.0239	.0279	.0319	.0359
0.1	.0398	.0438	.0478	.0517	.0557	.0596	.0636	.0675	.0714	.0753
0.2	.0793	.0832	.0871	.0910	.0948	.0987	.1026	.1064	.1103	.1141
0.3	.1179	.1217	.1255	.1293	.1331	.1368	.1406	.1443	.1480	.1517
0.4	.1554	.1591	.1628	.1664	.1700	.1736	.1772	.1808	.1844	.1879
.	.	.	.	.	.	.	.	.	.	.
.	.	.	.	.	.	.	.	.	.	.
1.0	.3413	.3438	.3461	.3485	.3508	.3531	.3554	.3577	.3599	.3621
1.1	.3643	.3665	.3686	.3708	.3729	.3749	.3770	.3790	.3810	.3830
1.2	.3849	.3869	.3888	.3907	.3925	.3944	.3962	.3980	.3997	.4015
1.3	.4032	.4049	.4066	.4082	.4099	.4115	.4131	.4147	.4162	.4177
1.4	.4192	.4207	.4222	.4236	.4251	.4265	.4279	.4292	.4306	.4319

Source: Extracted from Table E.2.

parts—times (in seconds) above the mean (that is, 50% of the workers) and times between the mean and the desired value *X* (that is, 40% of the workers). While we do not know *X*, we can determine the corresponding standardized value *Z*, since the area under the normal curve from the standardized mean 0 to this *Z* must be .4000. Using the body of Table E.2, we search for the area or probability .4000. The closest result is .3997, as shown in Table 8.4 (a replica of Table E.2).

Working from this area to the margins of the table, we see that the *Z* value corresponding to the particular *Z* row (1.2) and *Z* column (.08) is 1.28. However, from Figure 8.14, the *Z* value must be recorded as a negative (that is, $Z = -1.28$), since it is below the standardized mean of 0.

Once *Z* is obtained, we can now use the transformation formula (8.2) to determine the value of interest, *X*. Since

$$Z = \frac{X - \mu_x}{\sigma_x}$$

then

$$X = \mu_x + Z\sigma_x \quad \textbf{(8.3)}$$

Substituting, we compute

$$X = 75 + (-1.28)(6) = 67.32 \text{ seconds}$$

Thus we could expect that 10% of the workers will be able to complete the task in less than 67.32 seconds.

As a review, to find a *particular* value associated with a known probability we must take the following steps:

1. Sketch the normal curve and then place the values for the means (μ_x and μ_z) on the respective X and Z scales.
2. Split the appropriate half of the normal curve into two parts—the portion from the desired X to the mean and the portion from the desired X to the tail.
3. Shade the area of interest.
4. Using Table E.2, determine the appropriate Z value corresponding to the area under the normal curve from the desired X to the mean μ_x.
5. Using Equation (8.3), solve for X; that is,

$$X = \mu_x + Z\sigma_x$$

● **Question 7** To obtain the interquartile range we must first find the value for Q_1 and the value for Q_3; then we must subtract the former from the latter.

To find the first quartile value, we must determine the time (in seconds) for which only 25% of the factory workers can be expected to assemble the part faster. This is depicted in Figure 8.15.

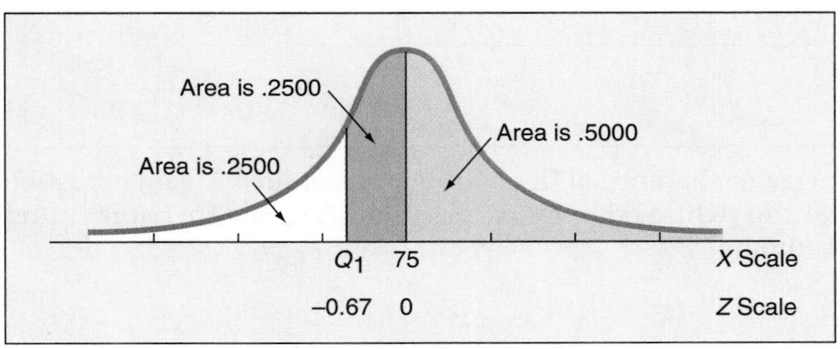

Figure 8.15
Finding Q_1.

Although we do not know Q_1, we can obtain the corresponding standardized value Z, since the area under the normal curve from the standardized mean 0 to this Z must be .2500. Using the body of Table E.2, we search for the area or probability .2500. The closest result is .2486, as shown in Table 8.5 on page 290 (which is a replica of Table E.2).

Working from this area to the margins of the table, we see that the Z value corresponding to the particular Z row (0.6) and Z column (.07) is 0.67. However, from Figure 8.15, the Z value must be recorded as a negative (that is, $Z = -0.67$), since it lies to the left of the standardized mean of 0.

Table 8.5 **Obtaining a *Z* value corresponding to a particular area under the normal curve.**

Z	.00	.01	.02	.03	.04	.05	.06	.07	.08	.09
0.0	.0000	.0040	.0080	.0120	.0160	.0199	.0239	.0279	.0319	.0359
0.1	.0398	.0438	.0478	.0517	.0557	.0596	.0636	.0675	.0714	.0753
0.2	.0793	.0832	.0871	.0910	.0948	.0987	.1026	.1064	.1103	.1141
0.3	.1179	.1217	.1255	.1293	.1331	.1368	.1406	.1443	.1480	.1517
0.4	.1554	.1591	.1628	.1664	.1700	.1736	.1772	.1808	.1844	.1879
0.5	.1915	.1950	.1985	.2019	.2054	.2088	.2123	.2157	.2190	.2224
0.6	.2257	.2291	.2324	.2357	.2389	.2422	.2454	.2486	.2518	.2549
0.7	.2580	.2612	.2642	.2673	.2704	.2734	.2764	.2794	.2823	.2852
0.8	.2881	.2910	.2939	.2967	.2995	.3023	.3051	.3078	.3106	.3133
0.9	.3159	.3186	.3212	.3238	.3264	.3289	.3315	.3340	.3365	.3389
1.0	.3413	.3438	.3461	.3485	.3508	.3531	.3554	.3577	.3599	.3621
1.1	.3643	.3665	.3686	.3708	.3729	.3749	.3770	.3790	.3810	.3830

Source: Extracted from Table E.2.

Once Z is obtained, the final step is to use Equation (8.3). Hence,

$$\begin{aligned} Q_1 &= X = \mu_x + Z\sigma_x \\ &= 75 + (-0.67)(6) \\ &= 75 - 4 \\ &= 71 \text{ seconds} \end{aligned}$$

To find the third quartile, we must determine the time (in seconds) for which 75% of the factory workers can be expected to assemble the part faster (and 25% could complete the task slower). This is displayed in Figure 8.16.

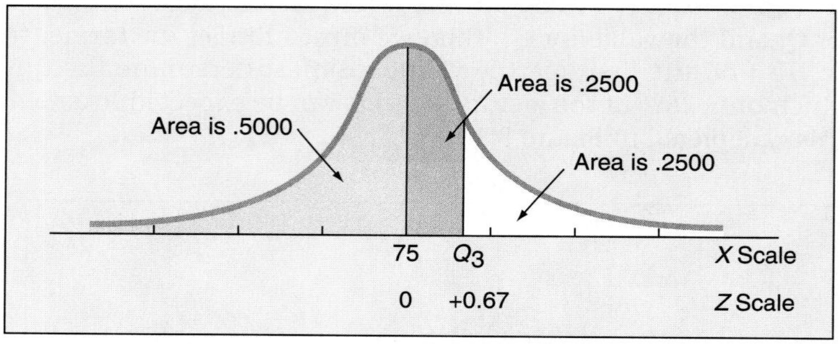

Figure 8.16
Finding Q_3.

From the symmetry of the normal distribution, our desired Z value must be +0.67 (since Z lies to the right of the standardized mean of 0). Therefore, using Equation (8.3), we compute

$$\begin{aligned} Q_3 &= X = \mu_x + Z\sigma_x \\ &= 75 + (+0.67)(6) \\ &= 75 + 4 \\ &= 79 \text{ seconds} \end{aligned}$$

The interquartile range or middle spread of the distribution is

$$\begin{aligned} \text{interquartile range} &= Q_3 - Q_1 \\ &= 79 - 71 \\ &= 8 \text{ seconds} \end{aligned}$$

Problems for Section 8.4

8.1 Given a standardized normal distribution having mean of 0 and standard deviation of 1 (Table E.2)

(a) What is the probability that

(1) Z is less than 1.57?

(2) Z exceeds 1.84?

(3) Z is between 1.57 and 1.84?

(4) Z is less than 1.57 or greater than 1.84?

(5) Z is between -1.57 and 1.84?

(6) Z is less than -1.57 or greater than 1.84?

(b) What is the value of Z if 50.0% of all possible Z values are larger?

(c) What is the value of Z if only 2.5% of all possible Z values are larger?

(d) Between what two values of Z (symmetrically distributed around the mean) will 68.26% of all possible Z values be contained?

8.2 Given a standardized normal distribution (with mean of 0 and standard deviation of 1), determine the following probabilities:

(a) $P(Z > +1.34)$

(b) $P(Z \leq +1.17)$

(c) $P(0 \leq Z \leq +1.17)$

(d) $P(Z < -1.17)$

(e) $P(-1.17 \leq Z \leq +1.34)$

(f) $P(-1.17 \leq Z \leq -0.50)$

● 8.3 Given a standardized normal distribution having a mean of 0 and a standard deviation of 1

(a) What is the probability that

(1) Z is between the mean and +1.08?

(2) Z is less than the mean or greater than +1.08?

(3) Z is between -0.21 and the mean?

(4) Z is less than -0.21 or greater than the mean?

(5) Z is at most +1.08?

(6) Z is at least -0.21?

(7) Z is between -0.21 and +1.08?

(8) Z is less than -0.21 or greater than +1.08?

(b) Determine the following probabilities:

(1) $P(Z > +1.08)$

(2) $P(Z < -0.21)$

(3) $P(-1.96 \leq Z \leq -0.21)$

(4) $P(-1.96 \leq Z \leq +1.08)$

(5) $P(+1.08 \leq Z \leq +1.96)$

(c) What is the value of Z if 50.0% of all possible Z values are smaller?

(d) What is the value of Z if only 15.87% of all possible Z values are smaller?

(e) What is the value of Z if only 15.87% of all possible Z values are larger?

8.4 Verify the following:

(a) The area under the normal curve between the mean and 2 standard deviations above and below it is .9544.

(b) The area under the normal curve between the mean and 3 standard deviations above and below it is .9973.

8.5 The thickness of a batch of 10,000 brass washers of a certain type manufactured by a large company is normally distributed with a mean of .0191 inch and with a standard deviation of .000425 inch. Verify that 99.04% of these washers can be expected to have a thickness between .0180 and .0202 inch.

8.6 Monthly food expenditures for families of four in a large city average \$420 with a standard deviation of \$80. Assuming that the monthly food expenditures are normally distributed:

(a) What percentage of these expenditures are less than \$350?
(b) What percentage of these expenditures are between \$250 and \$350?
(c) What percentage of these expenditures are between \$250 and \$450?
(d) What percentage of these expenditures are less than \$250 or greater than \$450?
(e) Determine Q_1 and Q_3 from the normal curve.

8.7 Toby's Trucking Company determined that on an annual basis the distance traveled per truck is normally distributed with a mean of 50.0 thousand miles and a standard deviation of 12.0 thousand miles.

(a) What proportion of trucks can be expected to travel between 34.0 and 50.0 thousand miles in the year?
(b) What is the probability that a randomly selected truck travels between 34.0 and 38.0 thousand miles in the year?
(c) What percentage of trucks can be expected to travel either below 30.0 or above 60.0 thousand miles in the year? 25.08%
(d) How many of the 1,000 trucks in the fleet are expected to travel between 30.0 and 60.0 thousand miles in the year?
(e) How many miles will be traveled by *at least* 80% of the trucks?

8.8 Plastic bags used for packaging produce are manufactured so that the breaking strength of the bag is normally distributed with a mean of 5 pounds per square inch and a standard deviation of 1.5 pounds per square inch.

(a) What proportion of the bags produced have a breaking strength
  (1) between 5 and 5.5 pounds per square inch?
  (2) between 3.2 and 4.2 pounds per square inch?
  (3) at least 3.6 pounds per square inch?
  (4) less than 3.17 pounds per square inch?
(b) Between what two values symmetrically distributed around the mean will 95% of the breaking strengths fall?

8.9 A set of final examination grades in an introductory statistics course was found to be normally distributed with a mean of 73 and a standard deviation of 8.

(a) What is the probability of getting at most a grade of 91 on this exam?
(b) What percentage of students scored between 65 and 89?
(c) What percentage of students scored between 81 and 89?
(d) What is the final exam grade if only 5% of the students taking the test scored higher?
(e) If the professor "curves" (gives A's to the top 10% of the class regardless of the score), are you better off with a grade of 81 on this exam or a grade of 68 on a different exam where the mean is 62 and the standard deviation is 3? Show statistically and explain.

8.10 At a well-known business school the grade-point indexes of its 1,000 undergraduates are approximately normally distributed with mean $\mu_x = 2.83$ and standard deviation $\sigma_x = .38$.

(a) What is the probability that a randomly selected student has a grade-point index between 2.00 and 3.00?
(b) What percentage of the student body are on probation—that is, have grade-point indexes below 2.00?
(c) How many students at this school are expected to be on the dean's list—that is, have grade-point indexes equal to or exceeding 3.20?
(d) What grade-point index will be exceeded by only 15% of the student body?

8.11 A statistical analysis of 1,000 long-distance telephone calls made from the headquarters of Johnson & Shurgot Corp. indicates that the length of these calls is normally distributed with $\mu_x = 240$ seconds and $\sigma_x = 40$ seconds.

(a) What percentage of these calls lasted less than 180 seconds?
(b) What is the probability that a particular call lasted between 180 and 300 seconds?
(c) How many calls lasted less than 180 seconds or more than 300 seconds?
(d) What percentage of the calls lasted between 110 and 180 seconds?
(e) What is the length of a particular call if only 1% of all calls are shorter?
(f) If we could not assume that the data were normally distributed, what would be the probability that a particular call lasted between 180 and 300 seconds? (*Hint:* Recall the Bienaymé-Chebyshev rule in Section 4.8.7.)
(g) Discuss the differences in your answers to (b) and (f).
(h) ACTION If you were hired as a consultant, write a memo to the vice-president of finance regarding your findings in parts (a)–(e).

8.12 Show that for normally distributed data the interquartile range is approximately equal to 1.33 standard deviations.

8.13 Show that for normally distributed data the standard deviation can be approximated as .75 times the interquartile range.

8.14 A building contractor claims he can renovate a 200-square-foot kitchen and dining room in 40 work hours, plus or minus 5 (i.e., the mean and standard deviation, respectively). The work includes plumbing, electrical installation, cabinets, flooring, painting, and the installation of new appliances. Assuming, from past experience, that times to complete similar projects are normally distributed with mean and standard deviation as estimated above

(a) What is the likelihood the project will be completed in less than 35 hours?
(b) What is the likelihood the project will be completed between 28 hours and 32 hours?
(c) What is the likelihood the project will be completed between 35 hours and 48 hours?
(d) 10 percent of such projects require more than how many hours?
(e) Determine the midhinge for completion time.
(f) Determine the interquartile range for completion time.

8.15 Suppose that the amount of sodium per slice of white bread produced by a particular food processing company is normally distributed with a mean of 110 mg and a standard deviation of 25 mg.

(a) What is the probability that a randomly selected slice will contain between 82 and 100 mg of sodium?
(b) What is the likelihood that a randomly selected slice will contain at least 100 mg of sodium?
(c) What must the amount of sodium (in mg) in a particular slice of bread be if 50.0% of all slices have more sodium?
(d) What must the amount of sodium (in mg) in a particular slice of bread be if 2.5% of all slices have more sodium?
(e) 83% of the slices of bread produced by this food processing company will contain at least how many mg of sodium?

8.16 Suppose that the amount of time it takes the IRS to send refunds to taxpayers is normally distributed with a mean of 12 weeks and a variance of 9.

(a) What proportion of taxpayers should get a refund
  (1) within 6 weeks?
  (2) within 9 weeks?
(b) What proportion of refunds will be sent more than 15 weeks after the IRS receives the tax return?
(c) How long will it take before 90% of taxpayers get their refunds?

8.17 Wages for workers in a particular industry average \$11.90 per hour and the standard deviation is \$.40. If the wages are assumed to be normally distributed:
(a) What percentage of workers receive wages between \$10.90 and \$11.90?
(b) What percentage of workers receive wages between \$10.80 and \$12.40?
(c) What percentage of workers receive wages between \$12.20 and \$13.10?
(d) What percentage of workers receive wages less than \$11.00?
(e) What percentage of workers receive wages more than \$12.95?
(f) What percentage of workers receive wages less than \$11.00 or more than \$12.95?
(g) What must the wage be if only 10 percent of all workers in this industry earn more?
(h) What must the wage be if 25 percent of all workers in this industry earn less?
(i) Determine the midhinge and interquartile range of the wages in this industry.

8.5 Assessing the Normality Assumption: Evaluating Properties and Constructing Probability Plots

Now that we have discussed the importance of the normal distribution and described its properties (Section 8.3) as well as demonstrated how it may be applied (Section 8.4), a very practical question must be considered. That is, we must be able to assess the likelihood that a particular data set can be assumed as coming from an underlying normal distribution or can be adequately approximated by it.

8.5.1 Exploring the Data—The Art of Data Analysis

The reader must be cautioned—*not all continuous random variables are normally distributed!* Often the continuous random phenomenon that we may be interested in studying neither will follow the normal distribution nor can be adequately approximated by it. While some methods for studying such continuous phenomena are outside the scope of this text (see References 1 and 3), *distribution-free* techniques (see Reference 4) which do not depend on the particular form of the underlying random variable will be discussed in Chapters 12 to 15.

Hence, for a descriptive analysis of any particular batch of data, the practical question remains: How can we decide whether our data set seems to follow or at least approximate the normal distribution sufficiently to permit it to be examined using the methodology of this chapter? Two descriptive *exploratory* approaches will be taken here to evaluate the *goodness-of-fit*.

1. A comparison of the data set's characteristics with the properties of an underlying normal distribution.
2. The construction of a normal probability plot.

More formal *confirmatory* approaches to the *goodness-of-fit* of a normal distribution are outside the scope of this text (see References 4 and 5).

8.5.2 Evaluating the Properties

In Section 8.3.2 we noted that the normal distribution has several theoretical properties. We recall that it is bell-shaped and symmetrical in appearance; its measures of central tendency are all identical; its interquartile range is equal to 1.33

standard deviations; and its random variable is continuous in form and has an infinite range.

We also noted that in actual practice some of the continuous random phenomena we observe may only approximate these theoretical properties, either because the underlying population distribution may be only approximately normal or because any obtained sample data set may deviate from the theoretically expected characteristics. In such circumstances, the data may not be perfectly bell-shaped and symmetrical in appearance. Moreover, the measures of central tendency will differ slightly and the interquartile range will not be exactly equal to 1.33 standard deviations. Furthermore, in practice the range of the data will not be infinite—it will be approximately equal to 6 standard deviations.

However, many continuous phenomena are neither normally distributed nor approximately normally distributed. For such phenomena, the descriptive characteristics of the respective data sets will not match well with the above four properties of a normal distribution.

What then should we do in order to investigate the assumption of normality in our data? One approach is to compare and contrast the actual data characteristics against the corresponding properties from an underlying normal distribution. To accomplish this, the following three steps are suggested:

1. Make some tallies and plots and observe their appearance.
   - For small- or moderate-sized data batches construct a stem-and-leaf display and box-and-whisker plot.
   - In addition, for large data batches construct the frequency distribution and plot the histogram and polygon.
2. Compute descriptive summary measures and compare the actual characteristics of the data with the underlying theoretical as well as practical properties of the normal distribution.
   - Obtain the mean, median, mode, midrange, and midhinge and note the similarities or differences in these five measures of central tendency.
   - Obtain the interquartile range and standard deviation. Note how well the interquartile range can be approximated by 1.33 times the standard deviation.
   - Obtain the range and note how well it can be approximated by 6 times the standard deviation.
3. Make some tallies to evaluate how the observations in the data set distribute themselves.
   - Determine if approximately two-thirds of the observations lie between the mean plus or minus 1 standard deviation.
   - Determine if approximately four-fifths of the observations lie between the mean plus and minus 1.28 standard deviations.
   - Determine if approximately 19 out of every 20 observations lie between the mean plus or minus 2 standard deviations.

As good data analysts, these are the kinds of things we should always be thinking of doing: *plotting, observing, computing,* and *describing*. Many of the descriptive statistical techniques we have studied so far come into play here. Nothing is new. Compared with other distributional forms, we know what a normal distribution is supposed to look like [see panel (a) of Figure 4.8 on page 130 comparing the polygon and box-and-whisker plot].

A second approach to evaluating the assumption of normality in our data is through the construction of a normal probability plot.

8.5.3 Constructing the Normal Probability Plot

We may recall that **quantiles** were defined as measures of "noncentral" location that are usually computed for summarizing large batches of numerical data [see Endnote 2 (page 168) pertaining to Section 4.4]. In that section we stressed the median (which splits the ordered observations in half) and the quartiles (which split the ordered observations in fourths) and in that Endnote we mentioned other quantiles such as the deciles (which split the ordered observations in tenths) and the percentiles (which split the ordered observations in hundredths). With this in mind, we may define a normal probability plot as follows:

> A **normal probability plot** is a two-dimensional plot of the observed data values on the *vertical* axis with their corresponding quantile values from a standardized normal distribution on the *horizontal* axis (see References 2 and 7).

If the plotted points seem to lie either on or close to an imaginary straight line rising from the lower left corner of the graph to the upper right corner, we would have evidence to believe that the obtained data batch is (at least approximately) normally distributed. On the other hand, if the plotted points appear to deviate from this imaginary straight line in some patterned fashion, then we would have reason to believe that the obtained data batch is not normally distributed and that the methodology presented in this chapter may not be appropriate.

To construct and use a normal probability plot, the following steps must be taken:

1. Place the values in the data batch into an ordered array.
2. Find the corresponding standard normal quantile values.
3. Plot the corresponding pairs of points using the observed data values on the vertical axis and the associated standard normal quantile values on the horizontal axis.
4. Assess the likelihood that the random variable of interest is (at least approximately) normally distributed by inspecting the plot for evidence of linearity (i.e., a straight line).

These steps will be described in detail.

Obtaining the Ordered Array Since the original data batch is likely to be obtained in raw form, the observations must be rearranged from smallest to largest in order to facilitate a matching with the corresponding standard normal quantile values. Thus, the original data are placed into an ordered array.

Finding the Standard Normal Quantile Values We know that a standard normal distribution is characterized by a mean of 0 and standard deviation of 1. Owing to its symmetry, the median or middle quantile value from a standard normal distribution must also be 0. Therefore, when dealing with a standard normal distribution it should be clear that quantile values below the median will be negative and quantile values above the median will be positive. However, the question we must still answer is, how can we obtain the quantile values from this distribution? The process by which we can accomplish this is known as an **inverse normal scores transformation** (see Reference 4).

The following is noted: Given a data batch containing n observations from a standardized normal distribution, let the symbol q_{z_1} represent its first (and smallest) quantile value, let the symbol q_{z_2} represent its second smallest quantile value, let the symbol q_{z_i} represent the ith smallest quantile value, and let the symbol q_{z_n} represent the largest quantile value. Because of symmetry, the standard normal quantiles q_{z_1} and q_{z_n} will have the same numerical value—except for sign. Of course, q_{z_1} will be negative and q_{z_n} will be positive.

> **The first standard normal quantile, q_{z_1},** is the value on a standard normal distribution below which the proportion $1/(n + 1)$ of the area under the curve is contained.
>
> **The second standard normal quantile, q_{z_2},** is the value on a standard normal distribution below which the proportion $2/(n + 1)$ of the area under the curve is contained.
>
> **The *ith standard normal quantile,* q_{z_i},** is the value on a standard normal distribution below which the proportion $i/(n + 1)$ of the area under the curve is contained.
>
> **The *nth (and largest) standard normal quantile,* q_{z_n},** is the value on a standard normal distribution below which the proportion $n/(n + 1)$ of the area under the curve is contained.

Making the Inverse Normal Scores Transformation As in Section 8.4.2, once we know the probability or area under the curve, we may use the body of Table E.2 to locate the appropriate area and then its corresponding standard normal quantile value in the margins of this table. Thus, in general, to find the ith standard normal quantile value from a data batch containing n observations, we sketch the standard normal distribution and locate the value q_{z_i} such that the proportion $i/(n + 1)$ of the area under the curve is contained below that value. By subtraction, we then compute the area under the curve from q_{z_i} to the mean μ_z of 0. We then find this area in the body of Table E.2 and, working to the margins of that table, we locate the corresponding standard normal quantile value.

To demonstrate this, let us suppose we wish to obtain the set of standard normal quantile values corresponding to a sample of 19 observations. The first standard normal quantile value, q_{z_1}, is that value below which the proportion $\frac{1}{n+1} = \frac{1}{19+1} = \frac{1}{20} = .05$ of the area under the normal curve is contained. From Figure 8.17 on page 298 we see that the area from q_{z_1} to the mean is .45 so that, from the body of Table 8.6 on page 298, q_{z_1} would fall halfway between -1.65 and -1.64. Since the standard normal quantile values are usually reported with two decimal places, the value -1.65 is chosen here.

The second standard normal quantile value, q_{z_2}, is that value below which the proportion $\frac{2}{n+1} = \frac{2}{19+1} = \frac{2}{20} = .10$ of the area under the normal curve is obtained. From Figure 8.18 and Table 8.7 (see pages 298 and 299), q_{z_2} would fall between -1.29 and -1.28 but closer to the latter. Hence, -1.28 is selected here.

Continuing in a similar manner, for example, the tenth standard normal quantile value, $q_{z_{10}}$, is that value below which the proportion $\frac{10}{n+1} = \frac{10}{19+1} = \frac{10}{20} = .50$ of the area under the normal curve is contained. Since we have located the median, this standard normal quantile value must be 0.00. We leave it as an exercise for the reader to show that the second largest standard normal quantile value, $q_{z_{18}}$, is $+1.28$ and the largest standard normal quantile value, $q_{z_{19}}$, is $+1.65$ (Problem 8.18 on page 302).

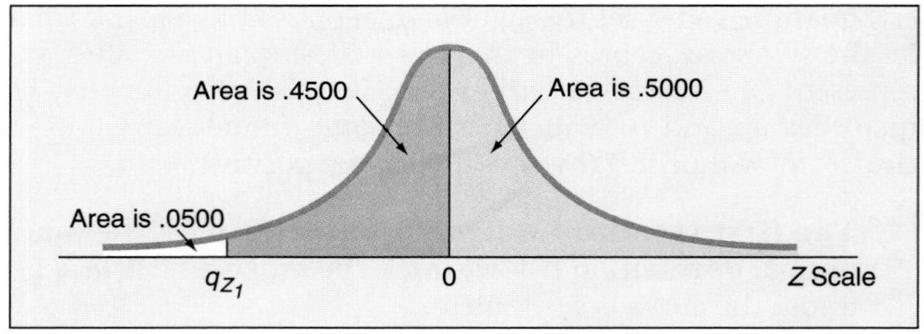

Figure 8.17
Finding the first standard normal quantile value from a data batch with 19 observations.

Table 8.6 Obtaining a standard normal quantile value corresponding to a particular area under the normal curve.

Z	.00	.01	.02	.03	.04	.05	.06	.07	.08	.09
0.0	.0000	.0040	.0080	.0120	.0160	.0199	.0239	.0279	.0319	.0359
0.1	.0398	.0438	.0478	.0517	.0557	.0596	.0636	.0675	.0714	.0753
0.2	.0793	.0832	.0871	.0910	.0948	.0987	.1026	.1064	.1103	.1141
0.3	.1179	.1217	.1255	.1293	.1331	.1368	.1406	.1443	.1480	.1517
0.4	.1554	.1591	.1628	.1664	.1700	.1736	.1772	.1808	.1844	.1879
.	.	.	.	.	.	.	.	.	.	.
.	.	.	.	.	.	.	.	.	.	.
.	.	.	.	.	.	.	.	.	.	.
1.0	.3413	.3438	.3461	.3485	.3508	.3531	.3554	.3577	.3599	.3621
1.1	.3643	.3665	.3686	.3708	.3729	.3749	.3770	.3790	.3810	.3830
1.2	.3849	.3869	.3888	.3907	.3925	.3944	.3962	.3980	.3997	.4015
1.3	.4032	.4049	.4066	.4082	.4099	.4115	.4131	.4147	.4162	.4177
1.4	.4192	.4207	.4222	.4236	.4251	.4265	.4279	.4292	.4306	.4319
1.5	.4332	.4345	.4357	.4370	.4382	.4394	.4406	.4418	.4429	.4441
1.6	.4452	.4463	.4474	.4484	.4495	.4505	.4515	.4525	.4535	.4545
1.7	.4554	.4564	.4573	.4582	.4591	.4599	.4608	.4616	.4625	.4633

Source: Extracted from Table E.2.

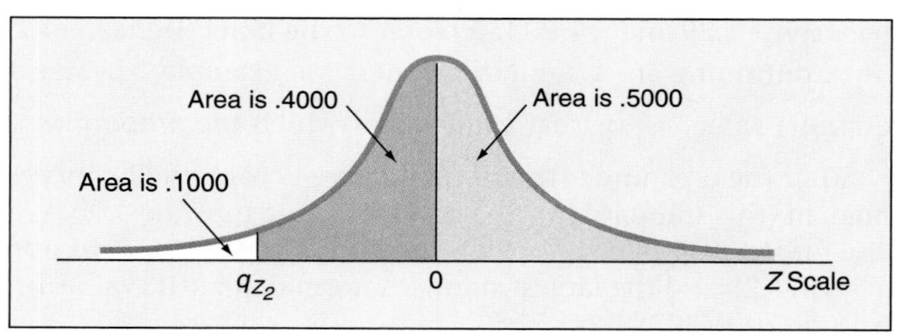

Figure 8.18
Finding the second standard normal quantile value from a data batch with 19 observations.

Table 8.7 Obtaining a standard normal quantile value corresponding to a particular area under the normal curve.

Z	.00	.01	.02	.03	.04	.05	.06	.07	.08	.09
0.0	.0000	.0040	.0080	.0120	.0160	.0199	.0239	.0279	.0319	.0359
0.1	.0398	.0438	.0478	.0517	.0557	.0596	.0636	.0675	.0714	.0753
0.2	.0793	.0832	.0871	.0910	.0948	.0987	.1026	.1064	.1103	.1141
0.3	.1179	.1217	.1255	.1293	.1331	.1368	.1406	.1443	.1480	.1517
0.4	.1554	.1591	.1628	.1664	.1700	.1736	.1772	.1808	.1844	.1879
.	.	.	.	.	.	.	.	.	.	.
.	.	.	.	.	.	.	.	.	.	.
.	.	.	.	.	.	.	.	.	.	.
1.0	.3413	.3438	.3461	.3485	.3508	.3531	.3554	.3577	.3599	.3621
1.1	.3643	.3665	.3686	.3708	.3729	.3749	.3770	.3790	.3810	.3830
1.2	.3849	.3869	.3888	.3907	.3925	.3944	.3962	.3980	.3997	.4015
1.3	.4032	.4049	.4066	.4082	.4099	.4115	.4131	.4147	.4162	.4177
1.4	.4192	.4207	.4222	.4236	.4251	.4265	.4279	.4292	.4306	.4319

Source: Extracted from Table E.2.

8.5.4 Constructing the Normal Probability Plot and Interpreting the Results

Hypothetical Applications Table 8.8 on page 300 presents ordered arrays of hypothetical midterm test scores from 19 students in each of five sections ("A" through "E") of a course in Introductory Calculus. Also shown in Table 8.8 are the corresponding standard normal quantile values obtained from the previously described inverse normal scores transformation. If we were to construct normal probability plots for these five distinct data sets, what would they show us and how can we interpret the plots?

The normal probability plots for the five class sections are depicted in panels (a) through (e) of Figure 8.19. From panel (a) we observe that the points appear to deviate from a straight line in a random manner so we may

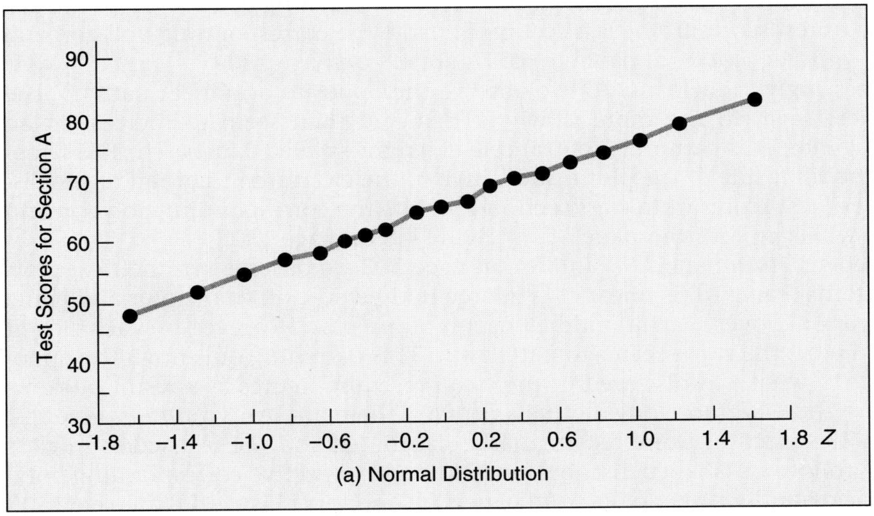

Figure 8.19
Normal probability plots for 5 hypothetical data batches.

Table 8.8 **Ordered arrays of hypothetical midterm test scores obtained from 19 students in each of five sections (A through E) of a course in Introductory Calculus and corresponding standard normal quantile values.**

(A) Bell-Shaped Normal Distribution	(B) Left-Skewed Distribution	(C) Right-Skewed Distribution	(D) Rectangular-Shaped Distribution	(E) U-Shaped Distribution	q_{z_i}
48	47	47	38	41	−1.65
52	54	48	41	42	−1.28
55	58	50	44	43	−1.04
57	61	51	47	45	−0.84
58	64	52	50	47	−0.67
60	66	53	53	49	−0.52
61	68	53	56	52	−0.39
62	71	54	59	55	−0.25
64	73	55	62	59	−0.13
65	74	56	65	65	0.00
66	75	57	68	71	0.13
68	76	59	71	75	0.25
69	77	62	74	78	0.39
70	77	64	77	81	0.52
72	78	66	80	83	0.67
73	79	69	83	85	0.84
75	80	72	86	87	1.04
78	82	76	89	88	1.28
82	83	83	92	89	1.65

conclude that the data batch from class section A is approximately normally distributed. [Note the corresponding polygon and box-and-whisker plot from panel (a) of Figure 4.8 on page 130.]

On the other hand, from panel (b) we observe a nonlinear pattern to the plot. The points seem to rise somewhat more steeply at first and then seem to increase at a decreasing rate. This pattern is an example of a left-skewed data batch. The steepness of the left side of the plot is indicative of the elongated left tail of the distribution of test scores from class section B. [Note the corresponding polygon and box-and-whisker plot from panel (b) of Figure 4.8 on page 130.]

Interestingly, from panel (c) we observe the opposite nonlinear pattern. The points here seem to rise more slowly at first and then seem to increase at an increasing rate. This pattern is an example of a right-skewed data batch. The steepness of the right side of the plot is indicative of the elongated right tail of the distribution of test scores from class section C. [Note the corresponding polygon and box-and-whisker plot from panel (c) of Figure 4.8 on page 130.]

Moreover, from panels (d) and (e) on page 302 we observe symmetrical plots with patterns. Panel (d) is linear over a large middle portion of the plot and panel (e) is linear only over a small middle portion of the plot. However, on each side of these two plots the curve seems to flatten out. This flattening out shows the opposite effect to what was observed in the two preceding figures as a result of skewness. Here there are no elongated tails. In fact, there are really no tails—the test scores in class section D are rectangularly distributed and the test scores in class section E follow a U-shaped distribution. [Note the respective corresponding polygons and box-and-whisker plots from panels (d) and (e) of Figure 4.8 on page 130.]

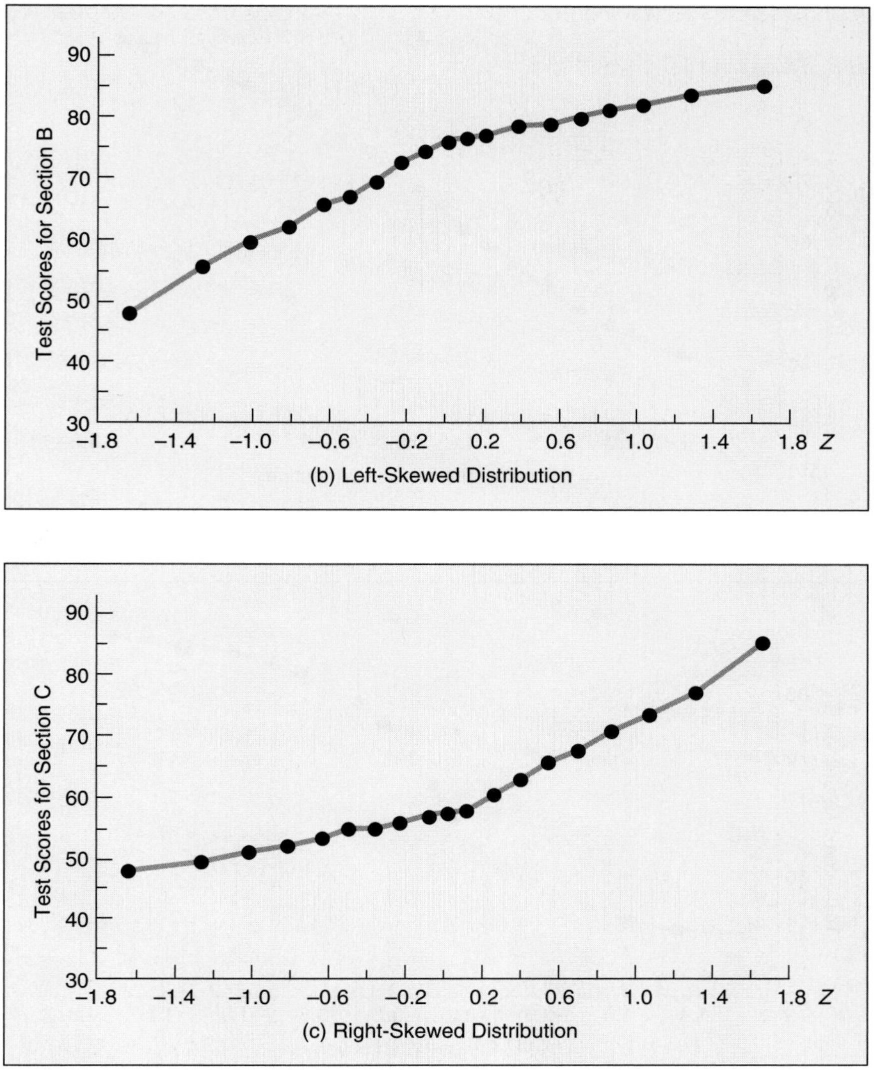

Figure 8.19 **(continued).**

● **Real Application: Comparing Tuition Rates in Two States** Now that we have seen how to interpret normality or lack thereof from a set of hypothetical applications, it is of interest to demonstrate the utility of the normal probability plot using real applications. Using Special Data Set 1 from Appendix D, Figure 8.20 on page 303 depicts the normal probability plots obtained from the MINITAB software package of the annual tuition rates for out-of-state residents at colleges and universities in Texas (see Table 3.1 on page 55) and North Carolina (see Table 3.6 on page 69). (The characteristics of these two data sets have been described in Chapters 3 and 5.) From panel (a) of Figure 8.20 we observe that the out-of-state tuition rates charged in Texas seem to be rising more slowly at first and then seem to be increasing at an increasing rate, confirming our belief that the data batch is right-skewed. A similar phenomenon appears to be occurring in panel (b), representing out-of-state tuition rates in North Carolina.

Figure 8.19
(continued).

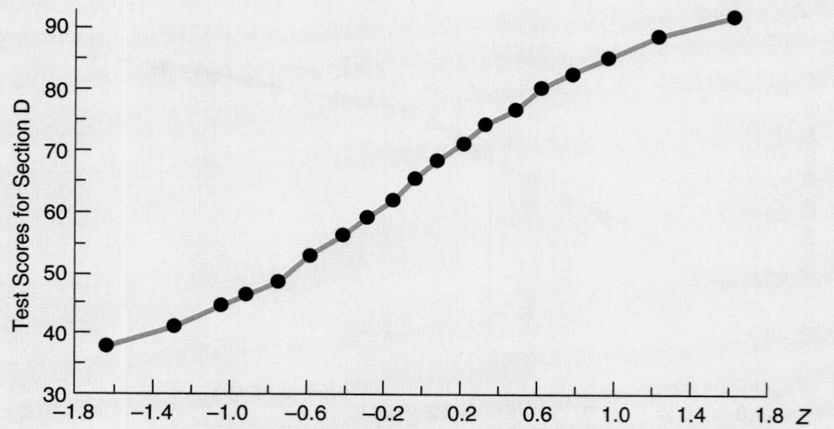

(d) Rectangular-Shaped Distribution

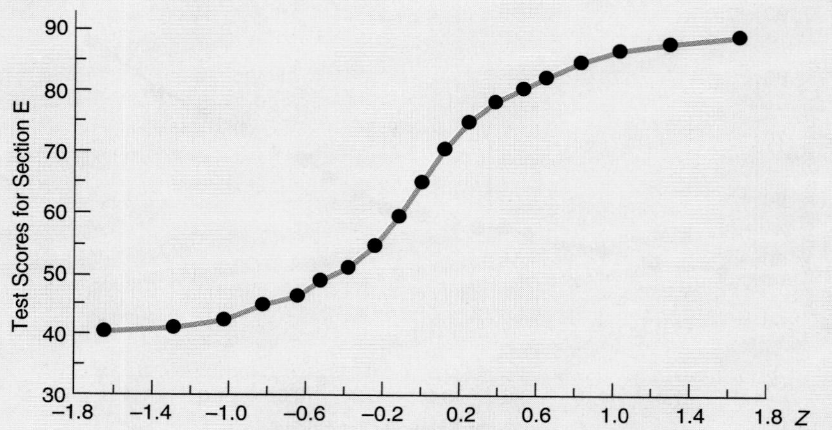

(e) U-Shaped Distribution

Problems for Section 8.5

8.18 Show that for a sample of 19 observations, the 18th smallest (i.e., second largest) standard normal quantile value obtained from the inverse normal scores transformation is +1.28 and the 19th (i.e., largest) standard normal quantile value is +1.65.

8.19 Show that for a sample of 39 observations, the smallest and largest standard normal quantile values obtained from the inverse normal scores transformation are, respectively, −1.96 and +1.96, and the middle (i.e., 20th) standard normal quantile value is 0.00.

● 8.20 Using the inverse normal scores transformation on a sample of 6 observations, list the 6 expected proportions or areas under the standardized normal curve along with their corresponding standard normal quantile values.

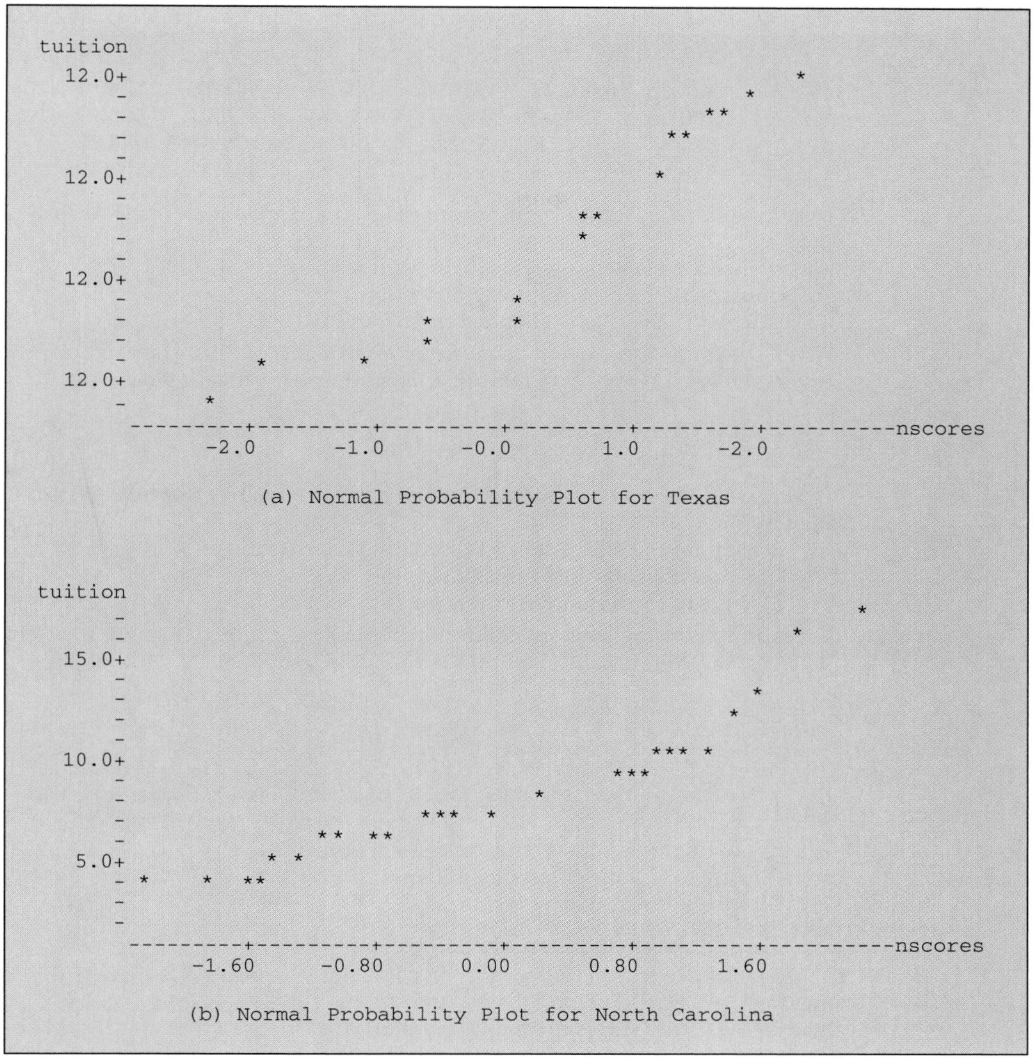

(a) Normal Probability Plot for Texas

(b) Normal Probability Plot for North Carolina

Figure 8.20
Normal probability plots of the tuition rate for out-of-state students in colleges and universities in Texas and North Carolina obtained from MINITAB.
Source: Data are taken from Tables 3.1 and 3.6.

● 8.21 Given the following ordered array (from left to right) of the amount of money withdrawn from a cash machine by 25 customers at a local bank:

Data file ATM1.DAT

$ 40	$ 50	$ 50	$ 70	$ 70	$ 80	$ 80	$ 90	$100	$100
$100	$100	$100	$100	$110	$110	$120	$120	$130	$140
$140	$150	$160	$160	$200					

Decide whether or not the data appear to be approximately normally distributed by
(a) Evaluating the actual versus theoretical properties.
(b) Constructing a normal probability plot.
(c) Discussing the results obtained in (a) and (b).

Data file GROCERY.DAT

8.22 Given the following data on grocery bills paid by a random sample of 28 customers in a local supermarket:

\$44.24	\$35.56	\$45.93	\$49.92	\$38.94	\$41.16	\$44.84
\$27.28	\$50.66	\$50.97	\$45.93	\$46.58	\$28.73	\$25.93
\$24.21	\$23.84	\$54.58	\$52.62	\$47.36	\$30.84	\$48.62
\$31.15	\$38.58	\$34.96	\$45.32	\$53.81	\$40.22	\$37.19

Decide whether or not the data appear to be approximately normally distributed by
(a) Evaluating the actual versus theoretical properties.
(b) Constructing a normal probability plot.
(c) Discussing the results obtained in (a) and (b).

Data file MKTEST.DAT

8.23 Given the following ordered array (from left to right) of final examination results obtained from 19 students in an Introductory Marketing class:

64	66	66	69	70	71	71	73	75	77
78	79	79	81	83	83	88	89	92	

Decide whether or not the data appear to be approximately normally distributed by
(a) Evaluating the actual versus theoretical properties.
(b) Constructing a normal probability plot.
(c) Discussing the results of (a) and (b).

Data file GAS1.DAT

8.24 Given the following data on amount of gasoline (in gallons) filled by an attendant at a highway gasoline station for a random sample of 24 automobiles:

12.78	8.89	10.09	10.64	15.98	13.95	9.48	10.84
10.88	9.93	7.74	5.80	11.84	10.29	10.89	6.68
12.09	8.28	8.83	7.95	7.33	12.56	8.86	9.15

Decide whether or not the data appear to be approximately normally distributed by
(a) Evaluating the actual versus theoretical properties.
(b) Constructing a normal probability plot.
(c) Discussing the results of (a) and (b).

Interchapter Problems for Section 8.5

For Problems 8.25–8.32, decide whether or not the data appear to be approximately normally distributed by
(a) Evaluating the actual versus theoretical properties.
(b) Constructing a normal probability plot.
(c) Discussing the results obtained in (a) and (b).

8.25 Use the data on maximum flow rates of shower heads (Problem 3.3 on page 58).

Data file SHOWER.DAT

8.26 Use the data on cancer incidence rates (Problem 3.6 on page 60).

Data file CANCER.DAT

8.27 Use the data on sodium content of peanut butter brands (Problem 3.7 on page 60).

Data file PEANUT.DAT

8.28 Use the data on electric and gas utility charges (Problem 3.12 on page 66).

Data file UTILITY.DAT

8.29 Use the data on the life of light bulbs for each of two manufacturers (Problem 3.18 on page 66).

Data file BULBS.DAT

8.30 Use the data on cost per month of toothpaste (Problem 3.71 on page 96).

Data file TOOTHPST.DAT

8.31 Use the data on amount of funding through grants (Problem 4.10 on page 116).

Data file ADAMHA.DAT

8.32 Use the data on the cost of corded telephones (Problem 4.76 on page 162).

Data file CORDPHON.DAT

8.6 The Normal Distribution as an Approximation to the Binomial and Poisson Distributions

In the earlier sections of this chapter we demonstrated the importance of the normal probability density function because of the numerous phenomena that seem to follow it or whose distributions can be approximated by it. In this section we shall demonstrate another useful aspect of the normal distribution—how it may be employed to approximate various important discrete probability distributions such as the binomial and the Poisson.

8.6.1 Need for a Correction for Continuity Adjustment

There are two major reasons to employ a **correction for continuity adjustment** here.

First, recall that a discrete random variable can take on only specified values while a continuous random variable can take on any values within a continuum or interval around those specified values. Hence, when using the normal distribution to approximate such discrete distributions as the binomial or the Poisson, more accurate approximations of the probabilities are likely to be obtained if a correction for continuity adjustment is employed.

Second, recall that with a continuous distribution (such as the normal), the probability of obtaining a *particular* value of a random variable is zero. On the other hand, when the normal distribution is used to approximate a discrete distribution, a correction for continuity adjustment can be employed so that we may approximate the probability of a specific value of the discrete distribution.

As a case in point, consider an experiment in which we toss a fair coin 10 times and observe the number of heads. Suppose we want to compute the probability of obtaining *exactly* 4 heads. Whereas a discrete random variable can have only a specified value (such as 4), a continuous random variable used to approximate it could take on any values whatsoever within an interval around that specified value, as demonstrated on the accompanying scale:

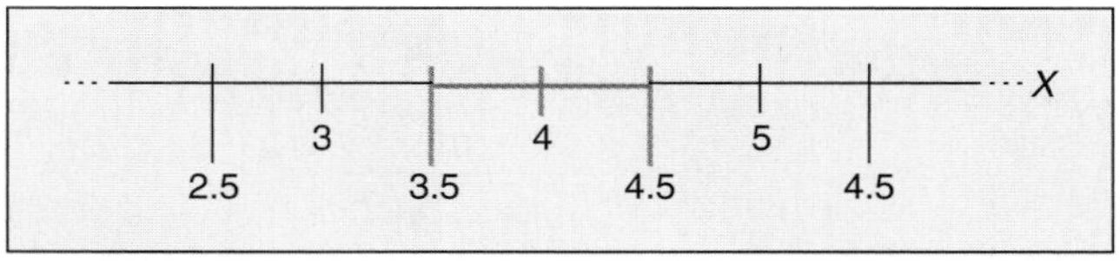

The correction for continuity adjustment requires adding or subtracting 0.5 from the value or values of the discrete random variable X as needed. Hence to use the normal distribution to approximate the probability of obtaining *exactly* 4 heads (i.e., $X = 4$), we would find the area under the normal curve from $X = 3.5$ to $X = 4.5$, the lower and upper boundaries of 4. To determine the approximate probability of observing *at least* 4 heads, we would find the area under the normal curve from $X = 3.5$ and above since, on a continuum, 3.5 is the lower boundary of X. Similarly, to determine the approximate probability of observing *at most* 4 heads, we would find the area under the normal curve from $X = 4.5$ and below since, on a continuum, 4.5 is the upper boundary of X.

When using the normal distribution to approximate discrete probability distributions, we see that semantics becomes important again. To determine the approximate probability of observing *fewer* than 4 heads, we would find the area under the normal curve from $X = 3.5$ and below; to determine the approximate probability of observing *more than* 4 heads, we would find the area under the normal curve from $X = 4.5$ and above; and to determine the approximate probability of observing 4 *through* 7 heads, we would find the area under the normal curve from $X = 3.5$ to $X = 7.5$. The reader will have an opportunity to obtain these results in Problem 8.33 on page 310.

8.6.2 Approximating the Binomial Distribution

In Section 7.5.2 we stated that the binomial distribution will be symmetric (like the normal distribution) whenever $p = .5$. When $p \neq .5$ the binomial distribution will not be symmetric. However, the closer p is to .5 and the larger the number of sample observations n, the more symmetric the distribution becomes.

On the other hand, the larger the number of observations in the sample, the more tedious it is to compute the exact probabilities of success by use of Equation (7.4). Fortunately, though, whenever the sample size is large, the normal distribution can be used to approximate the exact probabilities of success that otherwise would have to be obtained through laborious computations.

As a general rule this normal approximation can be used whenever np and $n(1 - p)$ are at least 5. We recall from Section 7.5.2 that the mean of the binomial distribution is given by

$$\mu_x = np$$

and the standard deviation of the binomial distribution is obtained from

$$\sigma_x = \sqrt{np(1 - p)}$$

Substituting into the transformation formula (8.2)

$$Z = \frac{X - \mu_x}{\sigma_x}$$

we have

$$Z = \frac{X - np}{\sqrt{np(1 - p)}}$$

so that, for large enough n, the random variable Z is approximately normally distributed.

Hence, to find approximate probabilities corresponding to the values of the discrete random variable X we have

$$Z \cong \frac{x_a - np}{\sqrt{np(1 - p)}} \tag{8.4}$$

where

$\mu_x = np$, mean of the binomial distribution

$\sigma_x = \sqrt{np(1-p)}$, standard deviation of the binomial distribution

x_a = adjusted number of successes, x, for the discrete random variable X, such that $x_a = x - .5$ or $x_a = x + .5$ as appropriate

and the approximate probabilities of success are obtained from Table E.2, the table of the standardized normal distribution.

● **Example** To illustrate this, suppose, in the product quality-control example described on page 252, that a sample of n = 1,600 tires of the same type are obtained at random from an ongoing production process in which 8% of all such tires produced are defective. What is the probability that in such a sample *not more than* 150 tires will be defective?

Since both np = 1,600(.08) = 128 and $n(1 - p)$ = 1,600(.92) = 1,472 exceed 5, we may use the normal distribution to approximate the binomial:

$$Z \cong \frac{x_a - np}{\sqrt{np(1-p)}} = \frac{150.5 - 128}{\sqrt{(1,600)(.08)(.92)}} = \frac{22.5}{10.85} = +2.07$$

Here x_a, the adjusted number of successes, is 150.5 and the approximate probability that X does not exceed this value corresponds, on the standardized Z scale, to a value of not more than +2.07. This is depicted in Figure 8.21.

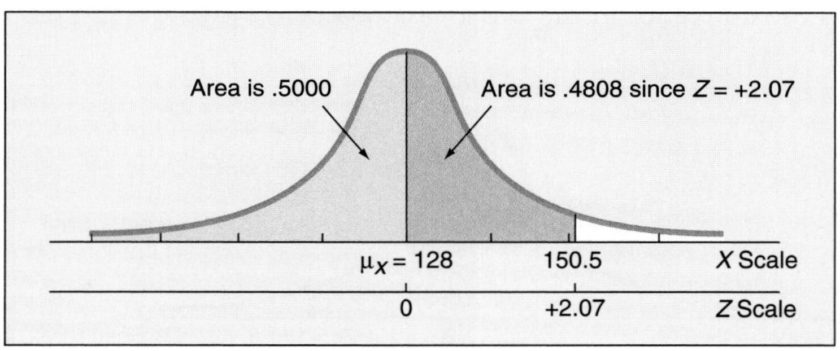

Figure 8.21
Approximating the binomial distribution.

Using Table E.2, the area under the curve between the mean and Z = +2.07 is .4808, so that the approximate probability is given by .5000 + .4808 = .9808.

Under the binomial distribution the probability of obtaining not more than 150 defective tires consists of all events up to and including 150 defectives—that is, $P(X \leq 150) = P(X = 0) + P(X = 1) + \cdots + P(X = 150)$, and the true probability may be laboriously computed from

$$\sum_{X=0}^{150} \binom{1,600}{X} (.08)^X (.92)^{1,600-X}$$

To appreciate the amount of work saved by using the normal approximation to the binomial model in lieu of the exact probability computations, just imagine making the following 151 computations from Equation (7.4) before summing up the results:

$$\binom{1,600}{0}(.08)^0(.92)^{1,600} + \binom{1,600}{1}(.08)^1(.92)^{1,599} + \cdots + \binom{1,600}{150}(.08)^{150}(.92$$

● **Obtaining a Probability Approximation for an Individual Value** Suppose that we now want to approximate the probability of obtaining *exactly* 150 defectives. The correction for continuity defines the integer value of interest to range from one-half unit below it to one-half unit above it. Therefore, the probability of obtaining 150 defective tires would be defined as the area (under the normal curve) between 149.5 and 150.5. Thus by using Equation (8.4), the probability can be approximated as follows:

$$Z \cong \frac{150.5 - 128}{\sqrt{(1,600)(.08)(.92)}} = \frac{22.5}{10.85} = +2.07$$

and

$$Z \cong \frac{149.5 - 128}{\sqrt{(1,600)(.08)(.92)}} = +1.98$$

From Table E.2, we note that the area under the normal curve from the mean to $X = 150.5$ is .4808 and the area under the curve from the mean to $X = 149.5$ is .4761. Thus, as depicted in Figure 8.22, the approximate probability of obtaining 150 defective tires is the difference in the two areas, .0047.

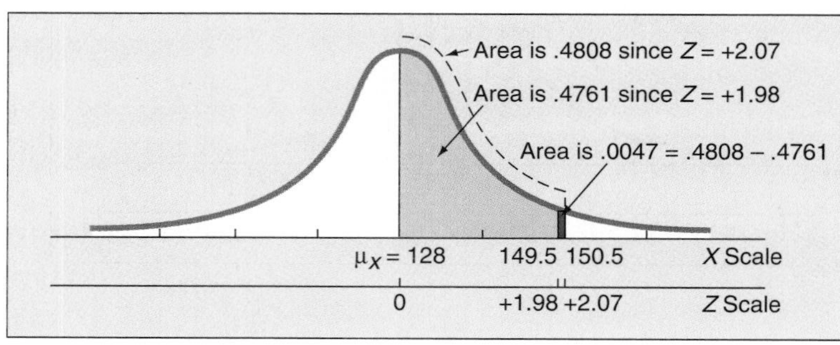

Figure 8.22
Approximating an exact binomial probability.

8.6.3 Approximating the Poisson Distribution

The normal distribution may also be used to approximate the Poisson model whenever the parameter λ, the expected number of successes, equals or exceeds 5. Since the value of the mean and the variance of a Poisson distribution are the same, we have

$$\mu_x = \lambda$$

and

$$\sigma_x = \sqrt{\lambda}$$

and by substituting into the transformation formula (8.2),

$$Z = \frac{X - \mu_x}{\sigma_x}$$

we have

$$Z = \frac{X - \lambda}{\sqrt{\lambda}}$$

so that, for large enough λ, the random variable Z is approximately normally distributed.

Hence, to find approximate probabilities corresponding to the values of the discrete random variable X we have

$$Z \cong \frac{x_a - \lambda}{\sqrt{\lambda}} \tag{8.5}$$

where

λ = expected number of successes or mean of the Poisson distribution

$\sigma_x = \sqrt{\lambda}$, the standard deviation of the Poisson distribution

x_a = adjusted number of successes, x, for the discrete random variable X, such that $x_a = x - .5$ or $x_a = x + .5$ as appropriate

and the approximate probabilities of success are obtained from Table E.2.

● **Example** To illustrate this let us suppose that at a certain automobile plant the average number of work stoppages per day due to equipment problems during the production process is 12.0. What then is the approximate probability of having 15 *or fewer* work stoppages due to equipment problems on any given day? From Equation (8.5) we have

$$Z \cong \frac{x_a - \lambda}{\sqrt{\lambda}} = \frac{15.5 - 12.0}{\sqrt{12.0}} = +1.01$$

Here x_a, the adjusted number of successes, is 15.5. Hence the approximate probability that X does not exceed this value corresponds, on the standardized Z scale, to a value of not more than +1.01. This is depicted in Figure 8.23 on page 310. From Figure 8.23 and Table E.2, we note that the area under the normal curve from the mean to 15.5 is .3438. Hence the area up to 15.5 is .5000 + .3438 = .8438. Therefore, the approximate probability of having 15 or fewer work stoppages due to equipment problems on any given day is .8438. This approximation compares quite favorably to the exact Poisson probability, .8445, obtained from Equation (7.7) on page 261.

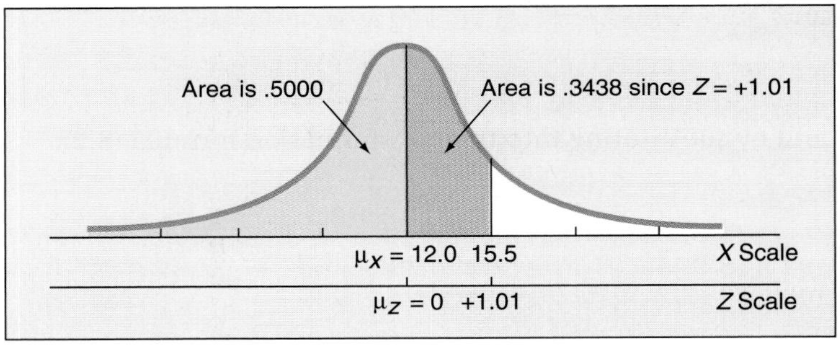

Figure 8.23
Approximating the Poisson distribution.

8.6.4 Large Sample Sizes: Neglecting the Correction for Continuity Adjustment

We have seen from the second example in Section 8.6.2 that if we are interested in obtaining probability approximations for individual values of the random variable, then it is necessary to use the correction for continuity adjustment. On the other hand, for other types of probability approximations, no hard-and-fast rule exists for using the correction for continuity adjustment. Since it is known that the advantages of increased accuracy become minimal with larger sample sizes and since the employment of the correction for continuity adjustment increases the computational complexity of our work, the correction for continuity will not be used in the remainder of this text. In most instances, our sample sizes will be large enough so that differences in the approximations obtained when using or not using the correction for continuity adjustment will be negligible.

Problems for Section 8.6

● 8.33 Consider an experiment in which we toss a fair coin 10 times and observe the number of heads.
- (a) Use Equation (7.4) on page 255 or Table E.7 to determine the probability of observing
  - (1) 4 heads
  - (2) at least 4 heads
  - (3) at most 4 heads
  - (4) fewer than 4 heads
  - (5) more than 4 heads
  - (6) 4 through 7 heads
- (b) Use the normal approximation to the binomial distribution [Equation (8.4)] to approximate the probabilities in (a)(1)–(a)(6).
- (c) Compare and contrast your findings in (a) and (b). Do you think that the normal distribution provides a good approximation to the binomial distribution in (b)?

8.34 For overseas flights, an airline has three different choices on its dessert menu—ice cream, apple pie, and chocolate cake. Based on past experience the airline feels that each dessert is equally likely to be chosen.
- (a) If a random sample of four passengers is selected, what is the probability that at least two will choose ice cream for dessert?
- (b) If a random sample of 21 passengers is selected, what is the *approximate* probability that at least two will choose ice cream for dessert?

8.35 Based upon past experience, 40% of all customers at Miller's Automotive Service Station pay for their purchases with a credit card.
(a) If a random sample of three customers is selected, what is the probability that
(1) none pay with a credit card?
(2) two pay with a credit card?
(3) at least two pay with a credit card?
(4) not more than two pay with a credit card?
(b) If a random sample of 200 customers is selected, what is the *approximate* probability that
(1) at least 75 pay with a credit card?
(2) not more than 70 pay with a credit card?
(3) between 70 and 75 customers, inclusive, pay with a credit card?

8.36 On average, 10.0 persons per minute are waiting for an elevator in the lobby of a large office building between the hours of 8 a.m. and 9 a.m.
(a) What is the probability that in any one-minute period at most four persons are waiting?
(b) What is the *approximate* probability that in any one-minute period at most four persons are waiting?
(c) Compare your results in (a) and (b).

● 8.37 The number of cars arriving per minute at a toll booth on a particular bridge is Poisson distributed with a mean of 2.5.
(a) What is the probability that in any given minute
(1) no cars arrive?
(2) not more than two cars arrive?
(b) If the expected number of cars arriving at the toll booth per ten-minute interval is 25.0, what is the *approximate* probability that in any given ten-minute period
(1) not more than 20 cars arrive?
(2) between 20 and 30 cars arrive?

8.38 Cars arrive at Kenny's Car Wash at a rate of nine per half-hour.
(a) What is the probability that in any given half-hour period at least three cars arrive?
(b) What is the *approximate* probability that in any given half-hour period at least three cars arrive?
(c) Compare your results in (a) and (b).

8.39 Suppose that the number of defective videocassette tapes that are returned to a video rental store has averaged seven per day.
(a) What is the (*exact*) probability that two tapes will be returned today?
(b) What is the (*exact*) probability that at least two tapes will be returned today?
(c) What assumptions were made about the probability distribution selected in (a) and (b)? Discuss.
(d) Obtain *approximate* answers for (a) and (b) using a different probability distribution model. Discuss the differences in your findings.

8.7 The Normal Distribution: A Review

As seen in the summary chart on page 312, in this chapter we have thoroughly examined the normal probability distribution. On page 274 of Section 8.1 you were given a list emphasizing the important points to be discussed in the chapter. Check over the list now to see whether you feel you have an understanding of these key points. To be sure, you should be able to answer the following conceptual questions:

1. Why is it that only one table of the normal distribution is needed in order to find any probability under the normal curve?
2. How would you find the area between two values under the normal curve when both values are on the same side of the mean?
3. How would you find the X value that corresponds to a given percentile of the normal distribution?
4. Why do the individual observations have to be converted to standard normal quantile values in order to develop a normal probability plot?
5. When can the normal distribution be used to approximate the binomial distribution?
6. When can the normal distribution be used to approximate the Poisson distribution?

In this chapter, the normal distribution was shown to be useful in its own right and also useful as an approximation of various discrete models. In the next chapter we shall investigate how the normal distribution provides the basis for classical statistical inference.

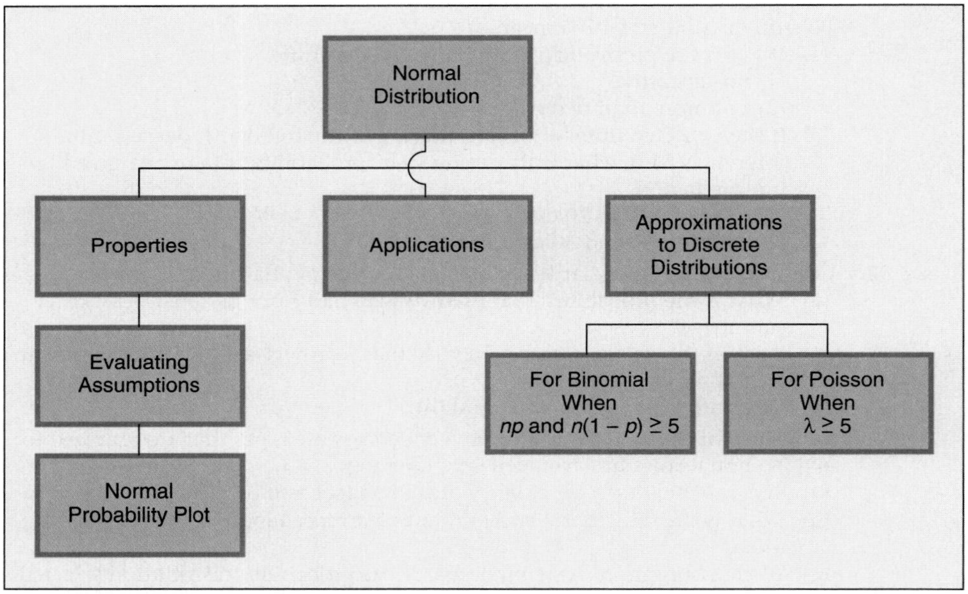

Chapter 8 summary chart.

Getting It All Together

Key Terms

Chapter Review Problems

● 8.40 Suppose the Governor projects that, on a weekly basis, a State Football Lottery Program he has proposed is expected to average 10.0 million dollars in profits (to be turned over to the state for educational programs) with a standard deviation of 2.5 million dollars. Suppose further that the weekly profits data are assumed to be (approximately) normally distributed. The following questions may be raised (or anticipated) at the Governor's next press conference:

(a) What is the probability that on any given week profits will be
  (1) between 10.0 and 12.5 million dollars?
  (2) between 7.5 and 10.0 million dollars?
  (3) between 7.5 and 12.5 million dollars?
  (4) at least 7.5 million dollars?
  (5) under 7.5 million dollars?
  (6) between 12.5 and 14.3 million dollars?

(b) Fifty percent of the time weekly profits (in millions of dollars) are expected to be above what value?

(c) Ninety percent of the time weekly profits (in millions of dollars) are expected to be above what value?

(d) What is the interquartile range in weekly profits expected from the State Football Lottery Program?

(e) ACTION You have been hired as a consultant. Prepare responses for the Governor on these nine anticipated questions and write a report discussing your overall findings.

8.41 An industrial sewing machine uses ball bearings that are targeted to have a diameter of 0.75 in. The specification limits under which the ball bearing can operate are 0.74 in (lower) and 0.76 in. (upper). Past experience has indicated that the actual diameter of the ball bearings is approximately normally distributed with a mean of .753 in. and a standard deviation of .004 in. What is the probability that a ball bearing will be:

(a) Between the target and the actual mean?
(b) Between the lower specification limit and the target?
(c) Above the upper specification limit?
(d) Below the lower specification limit?
(e) Above which value in diameter will 93 percent of the ball bearings be?

8.42 Suppose that the content of bottles of soft drink has been found to be normally distributed with a mean of 2.0 liters and a standard deviation of .05 liter. Bottles that contain less than 95 percent of the listed net content (1.90 liters in this case) can make the manufacturer subject to penalty by the state office of consumer affairs, while bottles that have a net content above 2.10 liters may cause excess spillage upon opening.

(a) What proportion of the bottles will contain:
   (1) Between 1.90 and 2.0 liters?
   (2) Between 1.90 and 2.10 liters?
   (3) Below 1.90 liters?
   (4) Below 1.90 liters or above 2.10 liters?
   (5) Above 2.10 liters?
   (6) Between 2.05 and 2.10 liters?

(b) Ninety-nine percent of the bottles would be expected to contain at least how much soft drink?

(c) Ninety-nine percent of the bottles would be expected to contain an amount that is between which two values (symmetrically distributed)?

(d) Explain the difference in the results in (b) and (c).

(e) Suppose that in an effort to reduce the number of bottles that contain less than 1.90 liters, the bottler sets the filling machine so that the mean is 2.02 liters. Under these circumstances, what would be your answers in (a), (b), and (c)?

8.43 A city agency that processes building renovation permits has a policy that states that the permit is free if it is not ready at the end of five business days from when the application is made. Processing time is measured from when the permit is received (the time is stamped) to when the application has been fully processed.

(a) If the process has a mean of three days and a standard deviation of one day, what proportion of the permits will be free?

(b) If the process has a mean of two days and a standard deviation of 1.5 days, what proportion of the permits will be free?

(c) Which process [(a) or (b)] will result in more free permits? Explain.

(d) For the process described in (a), would it be better to focus on reducing the average to two days, or the standard deviation to 0.75 day? Explain.

8.44 Sally D. is 67 inches tall and weighs 135 pounds. If the heights of women are normally distributed with $\mu_x = 65$ inches and $\sigma_x = 2.5$ inches and if the weights of women are normally distributed with $\mu_x = 125$ pounds and $\sigma_x = 10$ pounds, determine whether Sally's more unusual characteristic is her height or her weight. Discuss.

8.45 The net weight of boxes of packaged cereal follows the normal distribution with mean $\mu_x = 368$ grams. Find the standard deviation σ_x if 98% of the boxes have a net weight below 400 grams.

8.46 The toll charge for telephone calls to Central America follows the normal distribution with mean $\mu_x = \$21.00$. Find the standard deviation σ_x if 80% of the calls have a toll charge above \$17.50.

8.47 It is known that one out of every three people entering Groshen's (a large department store) will make at least one purchase.

(a) If a random sample of $n = 5$ persons is selected, what is the probability that
   (1) two or more of them will make at least one purchase?
   (2) at most four of them will make at least one purchase?

(b) If a random sample of $n = 81$ persons is selected, what is the *approximate* probability that
   (1) 30 or more of them will make at least one purchase?
   (2) at most 40 of them will make at least one purchase?

8.48 The famous parapsychologist Professor Sy Klops decides to investigate whether people can read minds. He makes up five cards, each of which has a different symbol on it. When a subject comes to be tested, Klops picks up a card at random and asks the subject to guess which symbol is on the card. Klops then records whether the subject is correct or not, shuffles the cards again, and repeats the procedure. (Each such cycle on which one of two outcomes can occur is sometimes called a *trial*.)

(a) Suppose each subject is given five trials. Calculate what proportion of people tested will be expected to get none right, one right, and so on.

(b) In the situation described in (a), if the professor tests 1,000 people and says that five "show promise for having telepathic powers" because they got all five trials correct, would you agree or disagree, and why?

(c) Now suppose that the professor uses more trials for each subject, so that everyone is tested on 50 trials. Discuss whether or not you think he can use the normal approximation to the binomial distribution.

(d) Regardless of your answer to (c), use the normal approximation to the binomial to estimate the probability that someone will get at least 15 right by chance.

8.49 Based on past experience the dispatcher of Toby's Trucking Company estimates that on any given day 20% of the trucks will arrive at their destinations more than one hour late.

(a) If a sample of ten trucks is selected, what is the probability that at most one arrives more than one hour late?

(b) If a sample of 100 trucks is selected, what is the *approximate* probability that at most ten arrive more than one hour late?

8.50 (a) What is the (approximate) probability that a student could pass a 100-question true-false examination if the student were to guess on each question? (*Note*: To pass, the student must get at least 60 questions right.)

(b) What is the (*approximate*) probability that a student would get exactly 60 questions right on a 100-question true-false examination if the student were to guess on each question?

8.51 The average number of accidents per day in a tire factory is 4.0.

(a) What is the probability that in any one day
(1) exactly four accidents will occur?
(2) more than four accidents will occur?
(3) at least four accidents will occur?
(4) either three or four accidents will occur?

(b) What is the probability that in any five-day period
(1) exactly 20 accidents will occur?
(2) more than 20 accidents will occur?
(3) at least 20 accidents will occur?
(4) 15 through 20 accidents will occur?

(c) What is the *approximate* probability that in any five-day period
(1) exactly 20 accidents will occur?
(2) more than 20 accidents will occur?
(3) at least 20 accidents will occur?
(4) 15 through 20 accidents will occur?

(d) Compare and contrast your findings in (a), (b), and (c). Discuss.

8.52 The average number of work stoppages per hour in a production process is 0.8.

(a) What is the probability that in any one hour
(1) exactly two stoppages will occur?
(2) at most two stoppages will occur?
(3) fewer than two stoppages will occur?
(4) one or two stoppages will occur?

(b) What is the probability that in any one eight-hour shift
   (1) exactly 16 stoppages will occur?
   (2) at most 16 stoppages will occur?
   (3) fewer than 16 stoppages will occur?
   (4) 8 through 16 stoppages will occur?
(c) What is the *approximate* probability that in any one eight-hour shift
   (1) exactly 16 stoppages will occur?
   (2) at most 16 stoppages will occur?
   (3) fewer than 16 stoppages will occur?
   (4) 8 through 16 stoppages will occur?
(d) Compare and contrast your findings in (a), (b), and (c). Discuss.

8.53 Based upon past experience, 7% of all luncheon expense vouchers are in error. If a random sample of 400 vouchers is selected, what is the *approximate* probability that
(a) exactly 25 are in error?
(b) fewer than 25 are in error?
(c) between 20 and 25 (inclusive) are in error?

Survey Database Project

EMPSAT.DAT

The following problems refer to the sample data obtained from the questionnaire of Figure 2.6 on pages 28–29 and presented in Table 2.3 on pages 33–40. For each problem, decide whether or not the data appear to be approximately normally distributed by
(a) Evaluating the actual versus theoretical properties.
(b) Constructing a normal probability plot.
(c) Discussing your findings.

8.54 Refer to the data on number of hours (question 1).
8.55 Refer to the data on age (question 3).
8.56 Refer to the data on personal income (question 7).
8.57 Refer to the data on total family income (question 8).
8.58 Refer to the data on number of years worked (question 15).

Collaborative Learning Mini-Case Projects

For each of the following, refer to the instructions on page 101.

Data file C&UPENN.DAT

CL8.1 Refer to Special Data Set 1 (see Appendix D) regarding tuition rates for out-of-state students. For the 90 colleges and universities in Pennsylvania, decide whether the tuition rates appear to be approximately normally distributed by
(a) Evaluating the actual versus theoretical properties.
(b) Constructing a normal probability plot.
(c) Stating your conclusions based on (a) and (b).
(d) Compare the conclusions reached in (c) with those for Texas and North Carolina reached in Section 8.5 and in Chapters 3 and 4.

Data file CEREAL.DAT

CL8.2 Refer to Special Data Set 2 (see Appendix D) regarding ready-to-eat cereals and the following variables: cost, weight, and sugars. For each of these numerical variables decide whether or not the data appear to be approximately normally distributed by
(a) Evaluating the actual versus theoretical properties.
(b) Constructing a normal probability plot.
(c) Stating your conclusions based on (a) and (b).

Data file FRAGRAN.DAT

CL8.3 Refer to Special Data Set 3 (see Appendix D) regarding men's and women's fragrances. Decide whether or not the cost per ounce for men's fragrances and women's fragrances each appear to be approximately normally distributed by

(a) Evaluating the actual versus theoretical properties.
(b) Constructing a normal probability plot.
(c) Stating your conclusions based on (a) and (b).
(d) Compare the results for men's and women's fragrances. What conclusions can you reach?

CL8.4 Refer to Special Data Set 4 (see Appendix D) regarding 35mm cameras and the following variables: price, weight, smallest field, range, framing accuracy, and battery life. For each of these numerical variables decide whether or not the data appear to be approximately normally distributed by
(a) Evaluating the actual versus theoretical properties.
(b) Constructing a normal probability plot.
(c) Stating your conclusions based on (a) and (b).

Data file CAMERA.DAT

Case Study D — Playing Roulette

There are nine basic betting strategies in roulette. These are presented at the right along with the corresponding odds of each strategy.

As examples, in strategy 1 a $1 bet on any particular number results in a loss of the dollar or a profit of $35 (that is, a payoff of $36 including the $1 bet) and in strategy 9 a $1 bet on "even–odd" or "red–black" or "1 to 18"–"19 to 36" results in a loss of the dollar or a profit of a dollar (that is, a payoff of $2 including the $1 bet).

Suppose that you were planning a trip to a casino offering the game of roulette with the aforementioned betting strategies and odds. A friend who is a mathematical statistics major has hinted to you that the various betting strategies may not all be equal and that perhaps there is at least one that may give the player a greater advantage or disadvantage. Using the results of mathematical expectation, you decide to check out the various betting strategies based on your intent to make only $1 bets and then write a letter to your friend explaining your findings.

As you are preparing your letter while relaxing at the hotel pool, you overhear a heated conversation on the best method of attempting to maximize profits in 200 plays of the roulette game if one were to continually make $1 bets on each game and not worry about the consequences (that is, the possibility of losing much or all of the $200). The two people were arguing over whether it is a better strategy to bet "one number" (strategy 1) on each play or to continually bet a "three-number street" (strategy 3) on each play or a particular color such as "red" (strategy 9) on each play. Since you just happen to have your table of the normal distribution with you, you decide to analyze these possibilities and impress your friend with your mathematical and statistical abilities by discussing the poolside conversation and describing your findings in your letter.

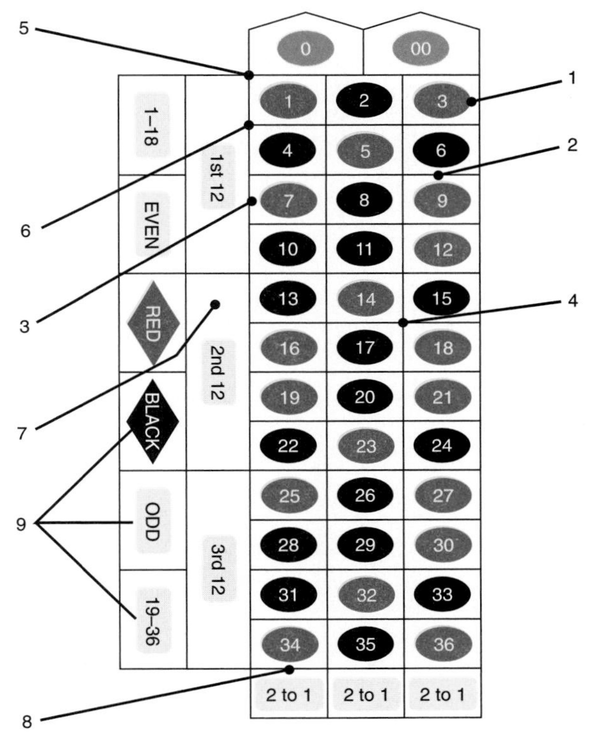

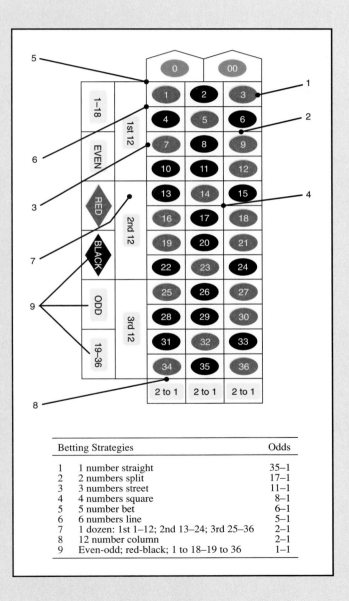

Betting Strategies		Odds
1	1 number straight	35–1
2	2 numbers split	17–1
3	3 numbers street	11–1
4	4 numbers square	8–1
5	5 number bet	6–1
6	6 numbers line	5–1
7	1 dozen: 1st 1–12; 2nd 13–24; 3rd 25–36	2–1
8	12 number column	2–1
9	Even-odd; red-black; 1 to 18–19 to 36	1–1

Endnotes

1. Mathematically this may be expressed as

$$P(60 \le X \le 63) = P(0 \le Z \le 1) = .3413.$$

2. Unlike the case of discrete random variables where the wording of the problem is so essential, we note that for continuous random variables there is much more flexibility in the wording. Hence there are two ways to state our result: We can say that 50% of the workers can assemble the part in *under 75 seconds* or we can say that 50% of the workers can assemble the part in *75 seconds or less*. Semantics are unimportant, because with continuous random variables the probability of assembling the part in exactly 75 seconds (or any other exactly specified time) is 0.

References

1. Derman, C., L. J. Gleser, and I. Olkin, *A Guide to Probability Theory and Application* (New York: Holt, Rinehart and Winston, 1973).
2. Gunter, B., "Q-Q Plots," *Quality Progress* (February 1994), pp. 81–86.
3. Larsen, R. J., and M. L. Marx, *An Introduction to Mathematical Statistics and Its Applications,* 2d ed. (Englewood Cliffs, NJ: Prentice-Hall, 1986).
4. Marascuilo, L. A., and M. McSweeney, *Nonparametric and Distribution-Free Methods for the Social Sciences* (Monterey, CA: Brooks/Cole, 1977).
5. Ramsey, P. P. and P. H. Ramsey, "Simple Tests of Normality in Small Samples," *Journal of Quality Technology*, Vol. 22, 1990, pp. 299–309.
6. Ryan, B. F., and B. L. Joiner, *Minitab Student Handbook,* 3d ed. (North Scituate, MA: Duxbury Press, 1994).
7. Sievers, G. L., "Probability Plotting," in Kotz, S., and N. L. Johnson, Eds., *Encyclopedia of Statistical Sciences,* Vol. 7 (New York: Wiley, 1986), pp. 232–237.

chapter 9

Sampling Distributions

CHAPTER OBJECTIVES

To develop the concept of a sampling distribution for both numerical and categorical variables and to examine the central limit theorem for cases in which a population is either normally or not normally distributed.

9.1 Introduction

A major goal of data analysis is to use statistics such as the sample mean and the sample proportion in order to estimate the corresponding parameters in the respective populations. We should realize that in *enumerative studies,* one is concerned with drawing conclusions about a population, not about a sample. As examples, a political pollster would be interested in the sample results only as a way of estimating the actual proportion of the votes that each candidate will receive from the population of voters. Likewise, an auditor, in selecting a sample of vouchers, is interested only in using the sample mean for estimating the population average amount. Moreover, in our Employee Satisfaction Survey, a statistician would utilize sample information as a way of drawing inferences about the personal income of the population of employees of Kalosha Industries. In each of these situations, the sample is used for drawing conclusions about the population.

In practice, a single sample of a predetermined size is selected at random from the population. The items that are to be included in the sample are determined through the use of a random number generator, such as a table of random numbers (see Section 2.7). Hypothetically, in order to be able to use the sample statistic to estimate the population parameter, we should examine every possible sample that could occur. If this selection of all possible samples actually were to be done, the distribution of the results would be referred to as a **sampling distribution.** The process of generalizing these sample results to the population is referred to as **statistical inference.**

In the preceding three chapters we have examined basic rules of probability and investigated various probability distributions such as the binomial, Poisson, and normal distributions. In this chapter we shall use these rules of probability along with our knowledge of probability distributions to begin focusing on how certain statistics (such as the mean or proportion) can be utilized in making inferences about the true population parameters. We will begin by discussing the properties of sample estimators that are used to estimate population parameters. We will then develop the concept of the sampling distribution and study the central limit theorem. Sampling distributions for the mean and for the proportion will be developed for situations in which sampling occurs with replacement and without replacement.

Upon completion of this chapter, you should be able to:

1. Understand the properties of the arithmetic mean
2. Be familiar with the concept of a sampling distribution
3. Know why the sampling distribution of the mean approaches a normal distribution as the sample size increases
4. Understand the effect on the standard error of sampling from a finite population

9.2 Sampling Distribution of the Mean

9.2.1 Properties of the Arithmetic Mean

In Chapter 4 we discussed several measures of central tendency. Undoubtedly, the most widely used (if not always the best) measure of central tendency is the arithmetic mean. This is particularly the case if the population can be assumed to be normally distributed.

Among several important mathematical properties (see Reference 2) of the arithmetic mean for a normal distribution are

1. Unbiasedness
2. Efficiency
3. Consistency

The first property, **unbiasedness**, involves the fact that the average of all the possible sample means (of a given sample size n) will be equal to the population mean μ_x.

This property can be demonstrated empirically by looking at the following example: Suppose that each of the four typists comprising a population of secretarial support service in a particular department of a company were asked to type the same page of a manuscript. The number of errors made by each typist was

Typist	Number of Errors
A	3
B	2
C	1
D	4

This population distribution is shown in Figure 9.1.

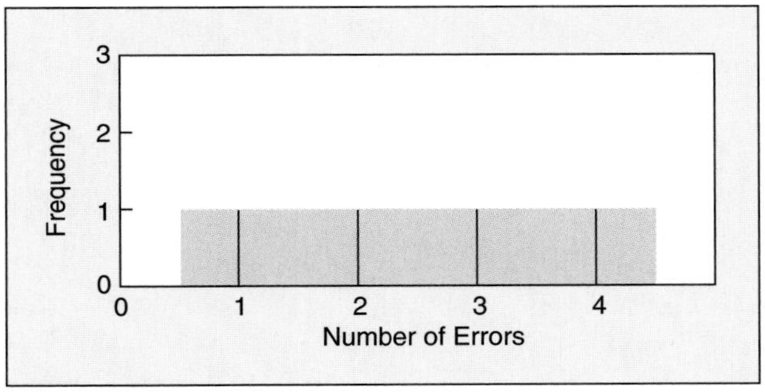

Figure 9.1
Number of errors made by a population of four typists.

We may recall from Section 4.8 that when the data from a population are available, the mean can be computed from

$$\mu_x = \frac{\sum_{i=1}^{N} X_i}{N} \tag{9.1}$$

and the standard deviation can be computed from

$$\sigma_x = \sqrt{\frac{\sum_{i=1}^{N}(X_i - \mu_x)^2}{N}} \tag{9.2}$$

Thus,

$$\mu_x = \frac{3 + 2 + 1 + 4}{4} = 2.5 \text{ errors}$$

and

$$\sigma_x = \sqrt{\frac{(3 - 2.5)^2 + \cdots + (4 - 2.5)^2}{4}} = 1.12 \text{ errors}$$

If samples of two typists are selected *with* replacement from this population, there are 16 possible samples that could be selected ($N^n = 4^2 = 16$). These possible sample outcomes are shown in Table 9.1.

Table 9.1 **All 16 samples of $n = 2$ typists from a population of $N = 4$ typists when sampling *with* replacement.**

Sample	Typists	Sample Outcomes	Sample Mean $\bar{X}_i$
1	A, A	3, 3	$\bar{X}_1 = 3$
2	A, B	3, 2	$\bar{X}_2 = 2.5$
3	A, C	3, 1	$\bar{X}_3 = 2$
4	A, D	3, 4	$\bar{X}_4 = 3.5$
5	B, A	2, 3	$\bar{X}_5 = 2.5$
6	B, B	2, 2	$\bar{X}_6 = 2$
7	B, C	2, 1	$\bar{X}_7 = 1.5$
8	B, D	2, 4	$\bar{X}_8 = 3$
9	C, A	1, 3	$\bar{X}_9 = 2$
10	C, B	1, 2	$\bar{X}_{10} = 1.5$
11	C, C	1, 1	$\bar{X}_{11} = 1$
12	C, D	1, 4	$\bar{X}_{12} = 2.5$
13	D, A	4, 3	$\bar{X}_{13} = 3.5$
14	D, B	4, 2	$\bar{X}_{14} = 3$
15	D, C	4, 1	$\bar{X}_{15} = 2.5$
16	D, D	4, 4	$\bar{X}_{16} = 4$
			$\mu_{\bar{X}} = 2.5$

If all these 16 sample means are averaged, the mean of these values ($\mu_{\bar{X}}$) is equal to 2.5, which is the mean of the population μ_x.

On the other hand, if sampling was being done *without* replacement, there would be six possible samples of two typists:

$$\frac{N!}{n!\,(N-n)!} = \frac{4!}{2!\,2!} = 6$$

These six possible samples are listed in Table 9.2.

Table 9.2 **All six possible samples of $n = 2$ typists from a population of $N = 4$ typists when sampling *without* replacement.**

Sample	Typists	Sample Outcomes	Sample Mean $\overline{X}_i$
1	A, B	3, 2	$\overline{X}_1 = 2.5$
2	A, C	3, 1	$\overline{X}_2 = 2$
3	A, D	3, 4	$\overline{X}_3 = 3.5$
4	B, C	2, 1	$\overline{X}_4 = 1.5$
5	B, D	2, 4	$\overline{X}_5 = 3$
6	C, D	1, 4	$\overline{X}_6 = 2.5$
			$\mu_{\overline{X}} = 2.5$

In this case, also, the average of all sample means ($\mu_{\overline{X}}$) is equal to the population mean, 2.5. Therefore we have shown that the sample arithmetic mean is an unbiased estimator of the population mean. This tells us that although we don't know how close the average of any particular sample selected comes to the population mean, we are at least assured that the average of all the possible sample means that could have been selected will be equal to the population mean.

The second property possessed by the mean, **efficiency,** refers to the precision of the sample statistic as an estimator of the population parameter. For distributions such as the normal, the arithmetic mean is considered to be more stable from sample to sample than are other measures of central tendency. For a sample of size n the sample mean will, on the average, come closer to the population mean than any other unbiased estimator, so that the sample mean is a better estimate of the population mean.

The third property, **consistency,** refers to the effect of the sample size on the usefulness of an estimator. As the sample size increases, the variation of the sample mean from the population mean becomes smaller so that the sample arithmetic mean becomes a better estimate of the population mean.

9.2.2 Standard Error of the Mean

The fluctuation in the average number of typing errors that was obtained from all 16 possible samples when sampling *with* replacement is illustrated in Figure 9.2 on page 324.

In this small example, although we can observe a good deal of fluctuation in the sample mean—depending on which typists were selected—there is not nearly as much fluctuation as in the actual population itself. The fact that the sample means are less variable than the population data follows directly from the **law of large numbers.** A particular sample mean averages together all the values in the sample. A population may consist of individual outcomes that can take on a wide range of values from extremely small to extremely large. However, if an extreme value falls into the sample, although it will have an effect on the mean, the effect will be reduced since it is being averaged in with all the other values in the sample. Moreover, as the sample size increases, the effect of a single extreme value gets even smaller, since it is being averaged with more observations.

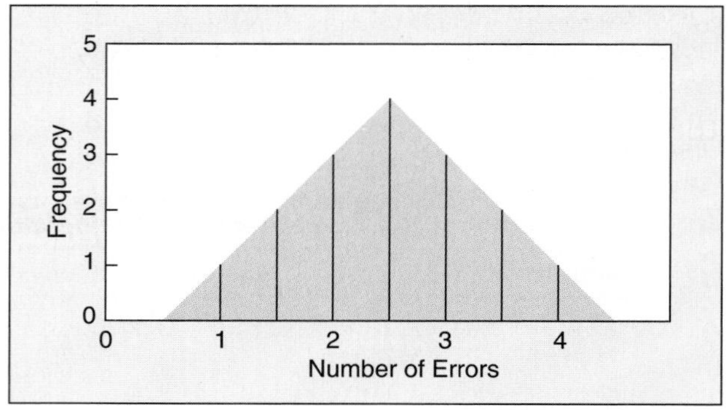

Figure 9.2
Sampling distribution of the average number of errors for samples of two typists.

This phenomenon is expressed statistically in the value of the standard deviation of the sample mean. This is the measure of variability of the mean from sample to sample and is referred to as the **standard error of the mean,** $\sigma_{\overline{X}}$. When sampling *with* replacement, the standard error of the mean is equal to

$$\sigma_{\overline{X}} = \frac{\sigma_x}{\sqrt{n}} \qquad (9.3)$$

the standard deviation of the population divided by the square root of the sample size. Therefore, as the sample size increases, the standard error of the mean will decrease by a factor equal to the square root of the sample size. This relationship between the standard error of the mean and the sample size will be further examined in Chapter 10 when we address the issue of sample size determination.

9.2.3 Sampling from Normal Populations

Now that we have introduced the idea of a sampling distribution and mentioned the standard error of the mean, we need to explore the question of what distribution the sample mean $\overline{X}$ will follow. It can be shown that if we sample *with* replacement from a population that is normally distributed with mean μ_x and standard deviation σ_x, the sampling distribution of the mean will also be normally distributed for *any size n* with mean $\mu_{\overline{X}} = \mu_x$ and have a standard error of the mean $\sigma_{\overline{X}}$.

In the most elementary case, if we draw samples of size $n = 1$, each possible sample mean is a single observation from the population, since

$$\overline{X} = \frac{\sum_{i=1}^{n} X_i}{n} = \frac{X_i}{1} = X_i$$

If we know that the population is normally distributed with mean μ_x and standard deviation σ_x, then the sampling distribution of $\overline{X}$ for samples of $n = 1$ must also follow the normal distribution with mean $\mu_{\overline{X}} = \mu_x$ and standard error of the mean $\sigma_{\overline{X}} = \sigma_x / \sqrt{1} = \sigma_X$. In addition, we note that as the sample size increases, the sampling distribution of the mean still follows a normal distribution with mean $\mu_{\bar{X}} = \mu_x$. However, as the sample size increases, the standard error of the mean decreases, so that a larger proportion of sample means are closer to the population mean. This can be observed by referring to Figure 9.3. In this figure, 500 samples of size 1, 2, 4, 8, 16, and 32 were randomly selected from a normally distributed population. We can see clearly from the polygons in Figure 9.3 that while the sampling distribution of the mean is approximately[1] normal for each sample size, the sample means are distributed more tightly around the population mean as the sample size is increased.

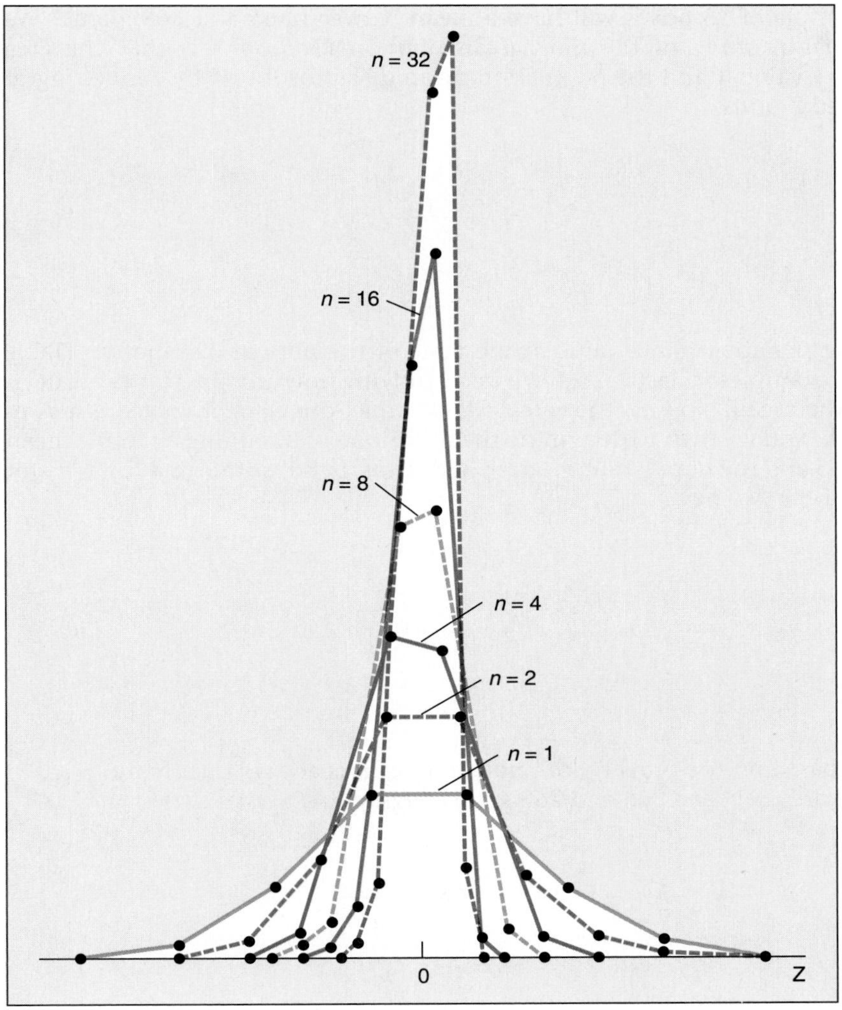

Figure 9.3
Sampling distributions of the mean from 500 samples of size n = 1, 2, 4, 8, 16, and 32 selected from a normal population.

● **Application** We may obtain deeper insight into the concept of the sampling distribution of the mean if we examine the following: Suppose that the packaging equipment in a manufacturing process that is filling 368-gram (13-ounce) boxes of cereal is set so that the amount of cereal in a box is normally distributed with a mean of 368 grams. From past experience, the population standard deviation for this filling process is known to be 15 grams.

If a sample of 25 boxes is randomly selected from the many thousands that are filled in a day, and the average weight is computed for this sample, what type of result could be expected?

For example, do you think that the sample mean would be 368 grams? 200 grams? 365 grams? The sample acts as a miniature representation of the population, so that if the values in the population were normally distributed, the values in the sample should be approximately normally distributed. Moreover, if the population mean is 368 grams, the sample mean has a good chance of being close to 368 grams.

To explore this problem even further, how can we determine the probability that the sample of 25 boxes will have a mean between 365 and 368 grams? We know from our study of the normal distribution (Section 8.3) that the area between any value X and the population mean μ_x can be found by converting to standardized Z units

$$Z = \frac{X - \mu_x}{\sigma_x} \quad \textbf{(9.4)}$$

and finding the appropriate value in the table of the normal distribution (Table E.2). In the examples in Section 8.4, we were studying how any single value X deviates from the mean. Now, in the cereal-fill example, the value involved is a sample mean, $\overline{X}$, and we wish to determine the likelihood of obtaining a sample mean between 365 and the population mean of 368. Thus, by substituting $\overline{X}$ for X, $\mu_{\overline{X}}$ for μ_x , and $\sigma_{\overline{X}}$ for σ_x we have

$$Z = \frac{\overline{X} - \mu_{\overline{X}}}{\sigma_{\overline{X}}} = \frac{\overline{X} - \mu_x}{\frac{\sigma_x}{\sqrt{n}}} \quad \textbf{(9.5)}$$

Note that, based on the property of unbiasedness, it is always true that $\mu_{\overline{X}} = \mu_x$. To find the area between 365 and 368 grams (Figure 9.4) we have

$$Z = \frac{\overline{X} - \mu_x}{\frac{\sigma_x}{\sqrt{n}}} = \frac{365 - 368}{\frac{15}{\sqrt{25}}} = \frac{-3}{3} = -1.00$$

Looking up 1.00 in Table E.2, we find an area of .3413. Therefore, 34.13% of all the possible samples of size 25 would have a sample mean between 365 and 368 grams.

We must realize that this is not the same as saying that a certain percentage of *individual* boxes will have between 365 and 368 grams. In fact, that percentage can be computed from Equation (9.4) as follows:

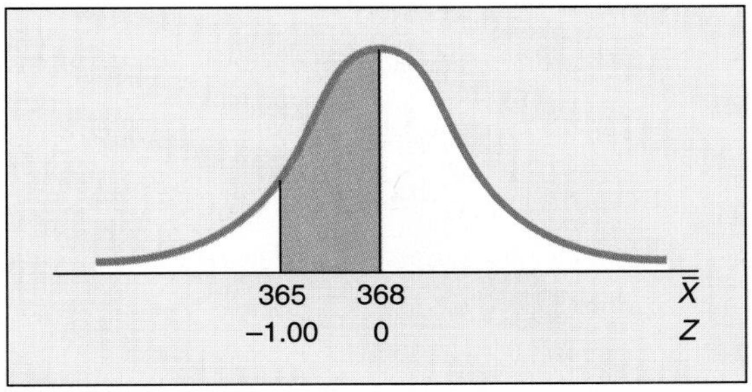

Figure 9.4
Diagram of normal curve needed to find area between 365 and 368 grams.

$$Z = \frac{X - \mu_x}{\sigma_x} = \frac{365 - 368}{15} = \frac{-3}{15} = -0.20$$

The area corresponding to $Z = -0.20$ in Table E.2 is .0793. Therefore, 7.93% of the *individual* boxes are expected to contain between 365 and 368 grams. Comparing these results, we may observe that many more *sample means* than *individual boxes* will be between 365 and 368 grams. This result can be explained by the fact that each sample consists of 25 different values, some small and some large. The averaging process dilutes the importance of any individual value, particularly when the sample size is large. Thus, the chance that the mean of a sample of 25 will be close to the population mean is greater than the chance that a *single individual* value will be.

How would our results be affected by using a different sample size, such as 100 boxes instead of 25? Here we would have the following:

$$Z = \frac{\overline{X} - \mu_x}{\frac{\sigma_x}{\sqrt{n}}} = \frac{365 - 368}{\frac{15}{\sqrt{100}}} = \frac{-3}{1.5} = -2.00$$

From Table E.2, the area under the normal curve from the mean to $Z = -2.00$ is .4772. Therefore 47.72% of the samples of size 100 would be expected to have means between 365 and 368 grams, as compared with only 34.13% for samples of size 25.

Instead of determining the proportion of sample means that are expected to fall within a certain interval, we might be more interested in finding out the interval within which a fixed proportion of the samples (means) would fall. For example, suppose we wanted to find an interval around the population mean that will include 95% of the sample means based on samples of 25 boxes. The 95% could be divided into two equal parts, half below the mean and half above the mean (see Figure 9.5 on page 328). Analogous to Section 8.4, we are determining a distance below and above the population mean containing a specific area of the normal curve. From Equation (9.5) we have

$$Z_L = \frac{\overline{X}_L - \mu_x}{\frac{\sigma_x}{\sqrt{n}}}$$

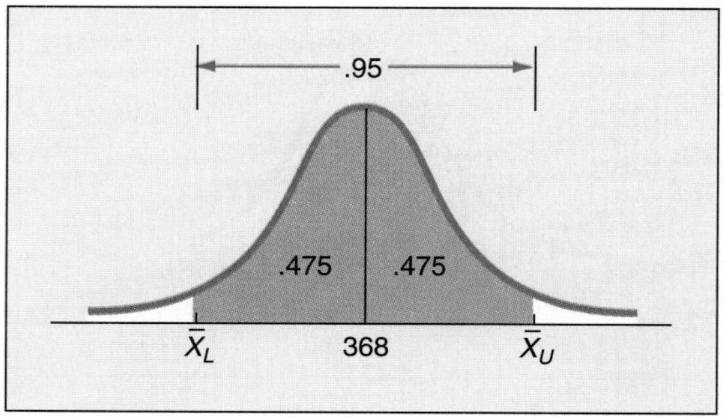

Figure 9.5
Diagram of normal curve needed to find upper and lower limits to include 95% of sample means.

where $Z_L = -Z$

and

$$Z_U = \frac{\bar{X}_U - \mu_x}{\frac{\sigma_x}{\sqrt{n}}}$$

where $Z_U = +Z$.

Therefore, the lower value of $\bar{X}$ is

$$\bar{X}_L = \mu_x - Z\frac{\sigma_x}{\sqrt{n}} \tag{9.6a}$$

and the upper value of $\bar{X}$ is

$$\bar{X}_U = \mu_x + Z\frac{\sigma_x}{\sqrt{n}} \tag{9.6b}$$

Since $\sigma_x = 15$ and $n = 25$ and the value of Z corresponding to an area of .475 from the center of the normal curve is 1.96, the lower and upper values of $\bar{X}$ can be found as follows:

$$\bar{X}_L = 368 - (1.96)\frac{15}{\sqrt{25}} = 368 - 5.88 = 362.12$$

$$\bar{X}_U = 368 + (1.96)\frac{15}{\sqrt{25}} = 368 + 5.88 = 373.88$$

Our conclusion would be that 95% of all sample means based on samples of 25 boxes should fall between 362.12 and 373.88 grams.

9.2.4 Sampling from Nonnormal Populations

In the preceding section we explored the sampling distribution of the mean for the case in which the population itself was normally distributed. However, we should realize that in many instances either we will know that the population is not normally distributed or we may believe that it is unrealistic to assume a normal distribution. Thus, we need to examine the sampling distribution of the mean for populations that are not normally distributed. This issue brings us to an important theorem in statistics, the *central limit theorem.*

> **Central Limit Theorem:** As the sample size (number of observations in each sample) gets *large enough,* the sampling distribution of the mean can be approximated by the normal distribution. This is true regardless of the shape of the distribution of the individual values in the population.

What sample size is large enough? A great deal of statistical research has gone into this issue. As a general rule, statisticians have found that for most population distributions, once the sample size is at least 30, the sampling distribution of the mean will be approximately normal. However, we may be able to apply the central limit theorem for even smaller sample sizes if some knowledge of the population is available (for example, if the distribution is symmetric).

The application of the central limit theorem to different populations can be illustrated by referring to Figures 9.6 to 9.8 on pages 330–332. Each of the depicted sampling distributions has been obtained by using the computer to select 500 different samples from their respective population distributions. These samples were selected for varying sizes ($n = 2, 4, 8, 16, 32$) from three different continuous distributions (normal, uniform, and exponential).

Figure 9.6 on page 330 illustrates the sampling distribution of the mean selected from a normal population. In the preceding section we stated that if the population is normally distributed, the sampling distribution of the mean will be normally distributed regardless of the sample size. An examination of the sampling distributions shown in Figure 9.6 gives empirical evidence for this statement. For each sample size studied, the sampling distribution of the mean is *close* to the normal distribution that has been superimposed.

The second figure, Figure 9.7 on page 331, presents the sampling distribution of the mean based on a population that follows a continuous uniform (rectangular) distribution. As depicted in part (a), for samples of size $n = 1$, each value in the population is equally likely. However, when samples of only two are selected, there is a peaking or *central limiting* effect already working. In this case we can observe more values close to the mean of the population than far out at the extremes. As the sample size increases, the sampling distribution of the mean rapidly approaches a normal distribution. Once there are samples of at least eight observations, the sample mean approximately follows a normal distribution.

Finally, the third figure, Figure 9.8 on page 332, depicts the sampling distribution of the mean obtained from a highly right-skewed population, called the exponential distribution (Reference 2). From Figure 9.8 we note that as the sample size increases, the sampling distribution becomes less skewed. When samples of size 16 are taken, the distribution of the mean is slightly skewed, while for samples of size 32 the sampling distribution of the mean appears to be normally distributed.

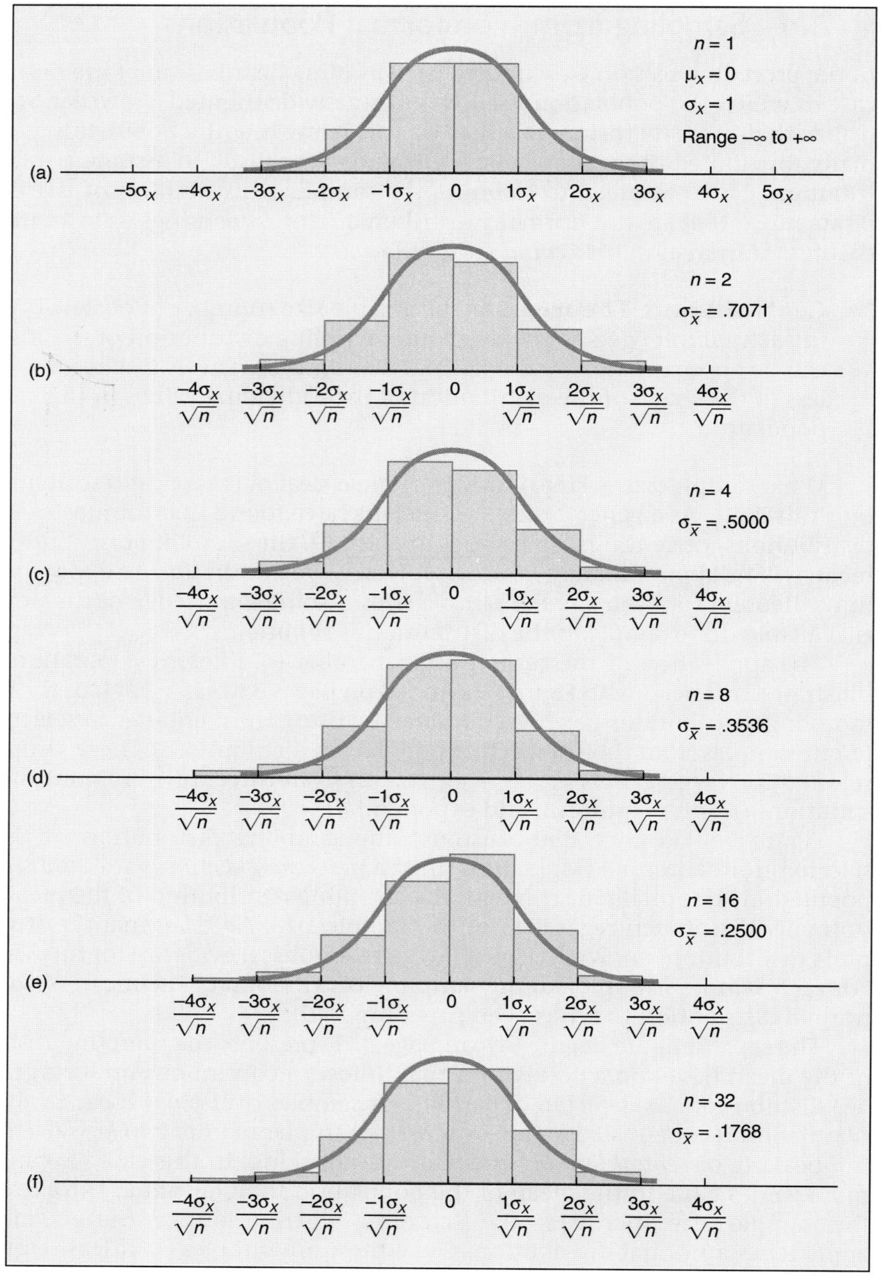

Figure 9.6
Normal distribution and the sampling distribution of the mean from 500 samples of size n = 2, 4, 8, 16, 32.

We may now use the results obtained from our well-known statistical distributions (normal, uniform, exponential) to summarize our conclusions as follows:

1. For most population distributions, regardless of shape, the sampling distribution of the mean will be approximately normally distributed if samples of at least 30 observations are selected.
2. If the population distribution is fairly symmetric, the sampling distribution of the mean will be approximately normal if samples of at least 15 observations are selected.

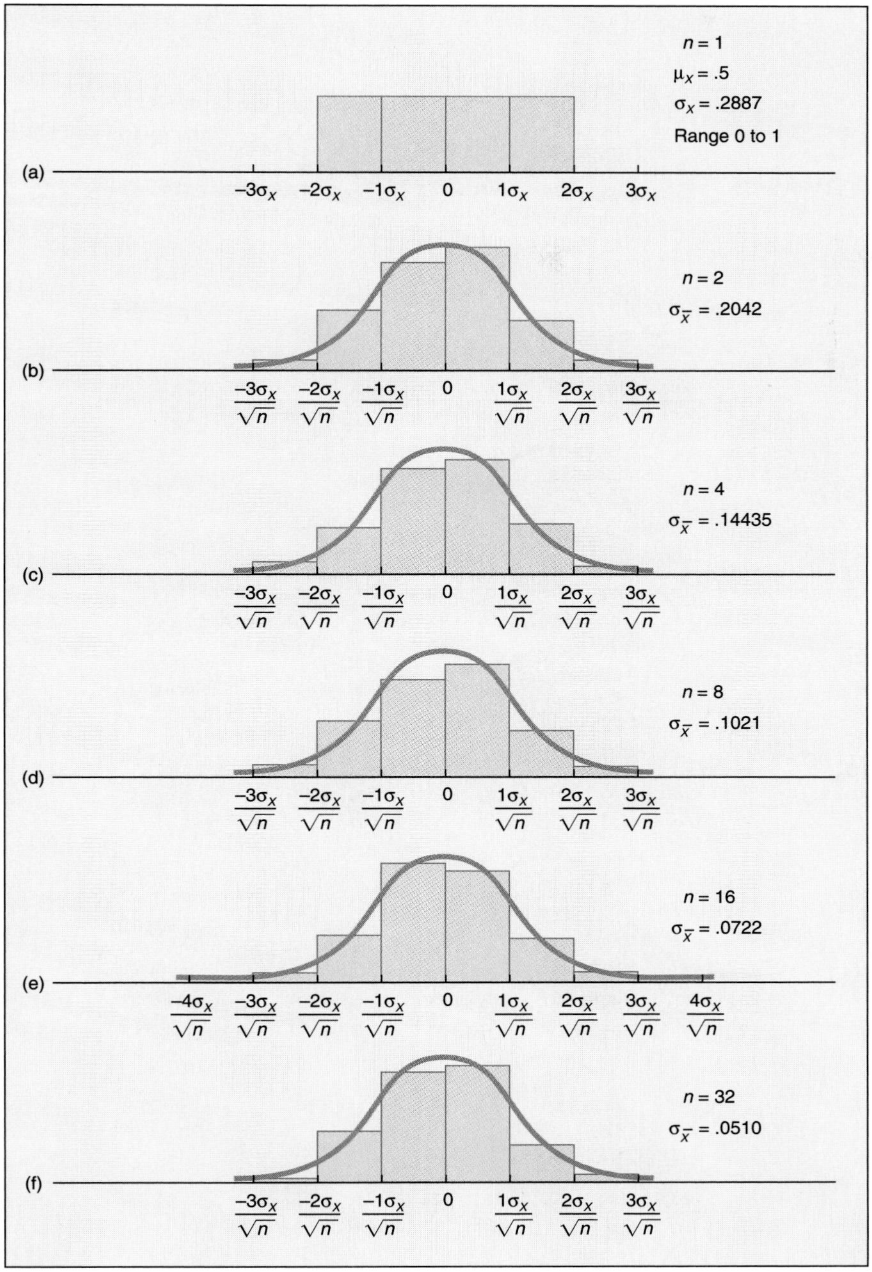

Figure 9.7
Continuous uniform (rectangular) distribution and the sampling distribution of the mean from 500 samples of size n = 2, 4, 8, 16, 32.

3. If the population is normally distributed, the sampling distribution of the mean will be normally distributed regardless of the sample size.

The central limit theorem, then, is of crucial importance in using statistical inference to draw conclusions about a population. It allows us to make inferences about the population mean without having to know the specific shape of the population distribution.

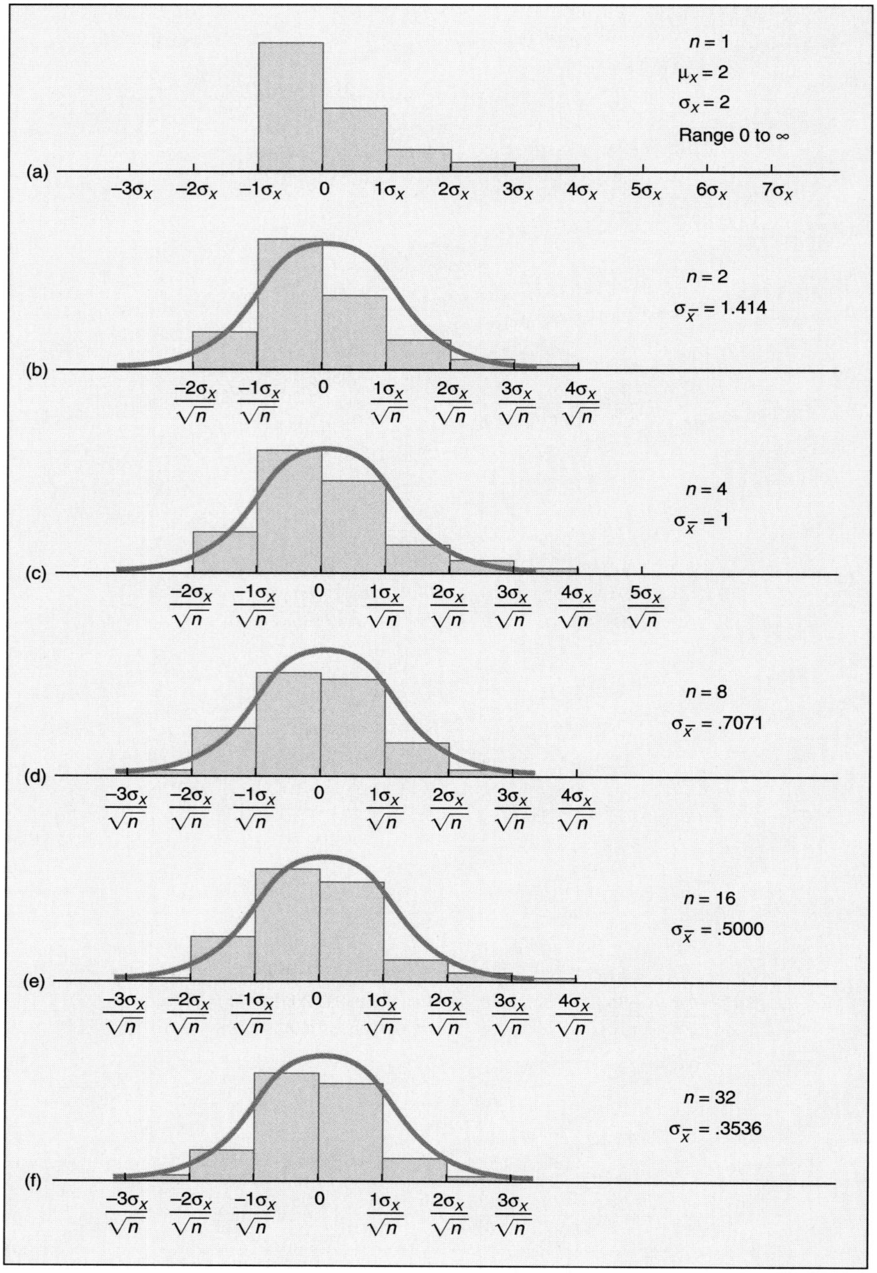

Figure 9.8
Exponential distribution and the sampling distribution of the mean from 500 samples of size n = 2, 4, 8, 16, 32.

Problems for Section 9.2

9.1 Explain why a statistician would be interested in drawing conclusions about a population rather than merely describing the results of a sample.

9.2 Distinguish between a probability distribution and a sampling distribution.

9.3 For each of the following three populations indicate what the sampling distribution for samples of 25 would consist of:

(a) Travel expense vouchers for a university in an academic year.
(b) Absentee records (days absent/year) in 1994 for employees of a large manufacturing company.
(c) Yearly sales (in gallons) of unleaded gasoline at service stations located in a particular county.

9.4 The following data represent the number of days absent per year in a population of six employees of a small company:

1, 3, 6, 7, 7, 12

(a) Assuming that you sample *without* replacement
  (1) Select all possible samples of size 2 and set up the sampling distribution of the mean.
  (2) Compute the mean of all the sample means and also compute the population mean. Are they equal? What property is this called?
  (3) Do parts (1) and (2) for all possible samples of size 3.
  (4) Compare the shape of the sampling distribution of the mean obtained in parts (1) and (3). Which sampling distribution seems to have the least variability? Why?
(b) Assuming that you sample *with* replacement, do parts (1)–(4) of (a) and compare the results. Which sampling distributions seem to have the least variability, those in (a) or (b)? Why?

9.5 Referring to Table 3.6 on page 69 (tuition rates for out-of-state students at 45 colleges and universities in North Carolina), and assuming that you sample *without* replacement

Data file C&UNC.DAT

(a) Select all possible samples of size 2 and set up the sampling distribution of the mean.
(b) Compute the mean of all the sample means and also compute the population mean. Are they equal? What is this property called?

9.6 The diameter of ping-pong balls manufactured at a large factory is expected to be approximately normally distributed with a mean of 1.30 inches and a standard deviation of .04 inch. What is the probability that a randomly selected ping-pong ball will have a diameter
(a) Between 1.28 and 1.30 inches?
(b) Between 1.31 and 1.33 inches?
(c) Between what two values (symmetrically distributed around the mean) will 60% of the ping-pong balls fall (in terms of the diameter)?
(d) If many random samples of 16 ping-pong balls are selected
  (1) What would the mean and standard error of the mean be expected to be?
  (2) What distribution would the sample means follow?
  (3) What proportion of the sample means would be between 1.28 and 1.30 inches?
  (4) What proportion of the sample means would be between 1.31 and 1.33 inches?
  (5) 60% of the sample means will be between what two values?
(e) Compare the answers of (a) with (d)(3) and (b) with (d)(4). Discuss.
(f) Explain the difference in the results of (c) and (d)(5).
(g) Which is more likely to occur—an individual ball above 1.34 in., a sample mean above 1.32 in. in a sample of size 4, or a sample mean above 1.31 in. in a sample of size 16? Explain.

9.7 Long-distance telephone calls are normally distributed with $\mu_x = 8$ minutes and $\sigma_x = 2$ minutes. If random samples of 25 calls were selected
(a) Compute $\sigma_{\bar{X}}$.
(b) What proportion of the sample means would be between 7.8 and 8.2 minutes?

(c) What proportion of the sample means would be between 7.5 and 8 minutes?
(d) If random samples of 100 calls were selected, what proportion of the sample means would be between 7.8 and 8.2 minutes?
(e) Explain the difference in the results of (b) and (d).
(f) Which is more likely to occur—an individual value above 11 minutes, a sample mean above 9 minutes in a sample of 25 calls, or a sample mean above 8.6 minutes in a sample of 100 calls? Explain.

9.8 The amount of time a bank teller spends with each customer has a population mean $\mu_x = 3.10$ minutes and standard deviation $\sigma_x = .40$ minute. If a random sample of 16 customers is selected
(a) What is the probability that the average time spent per customer will be at least 3 minutes?
(b) There is an 85% chance that the sample mean will be below how many minutes?
(c) What assumption must be made in order to solve (a) and (b)?
(d) If a random sample of 64 customers is selected, there is an 85% chance that the sample mean will be below how many minutes?
(e) What assumption must be made in order to solve (d)?
(f) Which is more likely to occur—an individual time below 2 minutes, a sample mean above 3.4 minutes in a sample of 16 customers, or a sample mean below 2.9 minutes in a sample of 64 customers? Explain.

9.3 Sampling Distribution of the Proportion

When dealing with a categorical variable where each individual or item in the population can be classified as either possessing or not possessing a particular characteristic such as male or female, or "satisfied with your job or not satisfied with your job," the two possible outcomes could be assigned scores of 1 or 0 to represent the presence or absence of the characteristic. If only a random sample of n individuals were available, the sample mean for such a categorical variable would be found by summing all the 1 and 0 scores and then dividing by n. For example, if in a sample of five employees three were satisfied with their job and two were not, there would be three ones and two zeros. Summing the three ones and two zeros and dividing by the sample size of five would give us a mean of 0.60, which is also the proportion of individuals in the sample who are satisfied with their job. Therefore, when dealing with categorical data, the sample mean $\overline{X}$ (of the 1 and 0 scores) is the sample proportion p_s having the characteristic of interest. Thus, the sample proportion p_s can be defined as

$$p_s = \frac{X}{n} = \frac{\text{number of successes}}{\text{sample size}} \qquad (9.7)$$

The sample proportion p_s has the special property that it must be between 0 and 1. If all individuals possessed the characteristic, they each would be assigned a score of 1 and p_s would be equal to 1. If half the individuals possessed the characteristic, half would be assigned a score of 1, the other half would be assigned a score of 0, and p_s would be equal to 0.5. If none of the individuals possessed the characteristic, they each would be assigned a score of 0 and p_s would be equal to 0.

While the sample mean $\bar{X}$ is an estimator of the population mean μ_x, the statistic p_s is an estimator of the population proportion p. By analogy to the sampling distribution of the mean, the standard error of the proportion σ_{p_s} would be

$$\sigma_{p_s} = \sqrt{\frac{p(1-p)}{n}} \tag{9.8}$$

The sampling distribution of the proportion would actually follow the binomial distribution discussed in Section 7.5. However, as discussed in Section 8.6, the normal distribution can be used to approximate the binomial distribution when np and $n(1-p)$ are each at least 5. In most cases in which inferences are being made about the proportion, the sample size is substantial enough to meet the conditions for using the normal approximation (see Reference 1). Thus, in many instances, we may use the normal distribution to evaluate the sampling distribution of the proportion. In order to illustrate the sampling distribution of the proportion, let us refer to the following example.

● **Application** The manager of the local branch of a savings bank has determined that 40% of all depositors have multiple accounts at the bank. If a random sample of 200 depositors is selected, what is the probability that the sample proportion of depositors with multiple accounts will be between .40 and .43?

Since $np = 200(.40) = 80$ and $n(1-p) = 200(.60) = 120$, the sampling distribution of the proportion can be assumed to be normally distributed.[2] Thus, we have

$$Z = \frac{\bar{X} - \mu_{\bar{X}}}{\sigma_{\bar{X}}} = \frac{\bar{X} - \mu_x}{\dfrac{\sigma_x}{\sqrt{n}}}$$

and because we are dealing with sample proportions (not sample means), we have

$$p_s = \text{sample proportion}$$

$$p = \text{population proportion}$$

$$\sigma_{p_s} = \sqrt{\frac{p(1-p)}{n}}$$

and, substituting p_s for $\bar{X}$, $\mu_{p_s} = p$ for $\mu_{\bar{X}}$, and $\sigma_{p_s} = \sqrt{p(1-p)/n}$ for $\sigma_{\bar{X}}$, we have

$$Z \cong \frac{p_s - p}{\sqrt{\dfrac{p(1-p)}{n}}} \tag{9.9}$$

Substituting,

$$Z \cong \frac{p_s - p}{\sqrt{\dfrac{p(1-p)}{n}}}$$

$$Z \cong \frac{.43 - .40}{\sqrt{\dfrac{(.40)(.60)}{200}}} = \frac{.03}{\sqrt{\dfrac{.24}{200}}}$$

$$= \frac{.03}{.0346}$$

$$= 0.87$$

Using Table E.2, the area under the normal curve from $Z = 0$ to $Z = 0.87$ is .3078. Therefore the probability of obtaining a sample proportion between .40 and .43 is .3078. This means that if the true proportion of successes in the population were .40, then 30.78% of the samples of size 200 would be expected to have sample proportions between .40 and .43. (See Figure 9.9.)

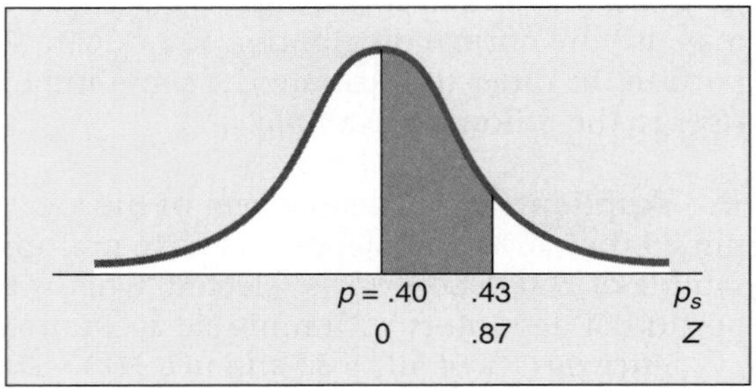

Figure 9.9
Diagram of normal curve needed to find the area between the proportions .40 and .43.

Problems for Section 9.3

9.9 Historically, 10% of a large shipment of machine parts are defective. If random samples of 400 parts are selected, what proportion of the samples will have
(a) Between 9% and 10% defective parts?
(b) Less than 8% defective parts?
(c) If a sample size of only 100 was selected, what would your answers have been in (a) and (b)?
(d) Which is more likely to occur—a percent defective above 13% in a sample of 100 or a percent defective above 10.5% in a sample of 400? Explain.

9.10 A political pollster is conducting an analysis of sample results in order to make predictions on election night. Assuming a two-candidate election, if a specific candidate receives at least 55% of the vote in the sample, then that candidate will be forecast as the winner of the election. If a random sample of 100 voters is selected, what is the probability that a candidate will be forecast as the winner when
(a) The true percentage of his vote is 50.1%?
(b) The true percentage of his vote is 60%?
(c) The true percentage of his vote is 49% (and he will actually lose the election)?
(d) If the sample size was increased to 400, what would be your answer to (a), (b), and (c)? Discuss.

9.11 Based on past data, 30% of the credit card purchases at a large department store are for amounts above $100. If random samples of 100 credit card purchases are selected

(a) What proportion of samples are likely to have between 20% and 30% of the purchases over $100?
(b) Within what symmetrical limits of the population percentage will 95% of the sample percentages fall?

9.12 Suppose that a marketing experiment is to be conducted in which students are to taste two different brands of soft drink. Their task is to correctly identify the brand tasted. If random samples of 200 students are selected and it is assumed that the students have no ability to distinguish between the two brands
(a) What proportion of the samples will have between 50% and 60% of the identifications correct?
(b) Within what symmetrical limits of the population percentage will 90% of the sample percentages fall?
(c) What is the probability of obtaining a sample percentage of correct identifications in excess of 65%?
(d) Which is more likely to occur—more than 60% correct identifications in a sample of 200 or more than 55% correct identifications in a sample of 1,000? Explain.
(*Hint:* If an individual has no ability to distinguish between the two soft drinks, then each one is equally likely to be selected.)

9.13 Historically, 93% of the deliveries of an overnight mail service arrive before 10:30 on the following morning. If random samples of 500 deliveries are selected, what proportion of the samples will have
(a) Between 93% and 95% of the deliveries arriving before 10:30 on the following morning?
(b) More than 95% of the deliveries arriving before 10:30 on the following morning?
(c) If samples of size 1,000 were selected, what would your answers be in (a) and (b)?
(d) Which is more likely to occur—more than 95% of the deliveries in a sample of 500 arriving before 10:30 on the following morning or less than 90% in a sample of 1,000 arriving before 10:30 on the following morning? Explain.

9.4 Sampling from Finite Populations

The central limit theorem and the standard errors of the mean and the proportion were based upon the premise that the samples selected were chosen *with* replacement. However, in virtually all survey research, sampling is conducted *without* replacement from populations that are of a finite size N. In these cases, particularly when the sample size n is not small as compared with the population size N (i.e., more than 5% of the population is sampled), so that $n/N > .05$, a **finite population correction factor (fpc)** should be used in defining both the standard error of the mean and the standard error of the proportion. The finite population correction factor may be expressed as

$$\text{fpc} = \sqrt{\frac{N - n}{N - 1}} \qquad (9.10)$$

where n = sample size
N = population size

Therefore, when dealing with means, we have

$$\sigma_{\bar{X}} = \frac{\sigma_x}{\sqrt{n}}\sqrt{\frac{N-n}{N-1}} \tag{9.11}$$

and when we are referring to proportions, we have

$$\sigma_{p_s} = \sqrt{\frac{p(1-p)}{n}}\sqrt{\frac{N-n}{N-1}} \tag{9.12}$$

Examining the formula for the finite population correction factor [Equation (9.10)], we see that the numerator will always be smaller than the denominator, so the correction factor will be less than 1. Since this finite population correction factor is multiplied by the standard error, the standard error becomes smaller when corrected. That is, we get more accurate estimates because we are sampling a large segment of the population.

● **Application** We may illustrate the application of the finite population correction factor by referring back to the two problems discussed in this chapter. In Section 9.2.3 on page 325, a sample of 25 cereal boxes was selected from a filling process. Suppose that a population of 2,000 boxes were filled on this particular day. Using the finite population correction factor we would have

$$\sigma_x = 15, \quad n = 25, \quad N = 2{,}000$$

$$\begin{aligned}\sigma_{\bar{X}} &= \frac{\sigma_x}{\sqrt{n}}\sqrt{\frac{N-n}{N-1}} \\ &= \frac{15}{\sqrt{25}}\sqrt{\frac{2{,}000-25}{2{,}000-1}} \\ &= 3\sqrt{.988} = 2.982\end{aligned}$$

The probability of obtaining a sample whose mean is between 365 and 368 grams is computed as follows:

$$Z = \frac{\bar{X} - \mu_{\bar{X}}}{\sigma_{\bar{X}}} = \frac{-3}{2.982} = -1.01$$

From Table E.2 the approximate area under the normal curve is .3438.

It is evident in this example that the use of the finite population correction factor had a very small effect on the standard error of the mean and the subsequent area under the normal curve, since the sample was only 1.25% of the population size.

In the example concerning the local savings bank on page 335, suppose that there were a total of 1,000 different depositors at the bank. The previous sample of size 200 out of this finite population results in the following:

$$\sigma_{p_s} = \sqrt{\frac{p(1-p)}{n}}\sqrt{\frac{N-n}{N-1}}$$

$$= \sqrt{\frac{(.4)(.6)}{200}}\sqrt{\frac{1{,}000-200}{1{,}000-1}}$$

$$= \sqrt{\frac{24}{200}}\sqrt{\frac{800}{999}} = \sqrt{\frac{24}{200}}\sqrt{.801}$$

$$= (.0346)(.895) = .031$$

Using $\sigma_{p_s} = .031$ as the standard error of the sample proportion in Equation (9.9), then $Z = .03/.031 = 0.97$, and, from Table E.2, the appropriate area under the normal curve is .3340. In this example, the use of the finite population correction factor had a moderate effect on the standard error of the proportion and on the area under the normal curve, since the sample size is 20% (that is, $n/N = .20$) of the population.

Problems for Section 9.4

- ● 9.14 Referring to Problem 9.6 on page 333, if the population consisted of a box of 200 ping-pong balls, what would be your answer to part (d)(4) of that problem?
- 9.15 Referring to Problem 9.8 on page 334, if there were a population of 500 customers, what would be your answers to (a) and (b) of that problem?
- ● 9.16 Referring to Problem 9.9 on page 336, if the shipment included 5,000 machine parts, what would be your answers to (a) and (b) of that problem?
- 9.17 Referring to Problem 9.13 on page 337, if the population consisted of 10,000 deliveries, what would be your answers to (a) and (b) of that problem?

9.5 Sampling Distributions: A Review

As seen in the summary chart on page 340, in this chapter we have studied the sampling distribution of the sample mean and the sampling distribution of the sample proportion. The importance of the normal distribution in statistics has been further emphasized by examining the central limit theorem. We have seen that knowledge of a population distribution is not always necessary in drawing conclusions from a sampling distribution of the mean or proportion.

On page 320 of Section 9.1 you were given a list emphasizing the important points to be discussed in the chapter. Check over the list now to see whether you feel you have an understanding of these key points. To be sure you should be able to answer the following conceptual questions:

1. Why is the sample arithmetic mean an unbiased estimator of the population arithmetic mean?
2. Why does the standard error of the mean decrease as the sample size n increases?
3. Why does the sampling distribution of the mean follow a normal distribution for a *large enough* sample size even though the population may not be normally distributed?
4. Under what circumstances does the sampling distribution of the proportion approximately follow the normal distribution?

5. What is the effect on the standard error of using the finite population correction factor?

The concepts concerning sampling distributions are central to the development of statistical inference. The main objective of statistical inference is to take information based only on a sample and use this information to draw conclusions and make decisions about various population values. The statistical techniques developed to achieve these objectives are discussed fully in the next six chapters (confidence intervals and tests of hypotheses).

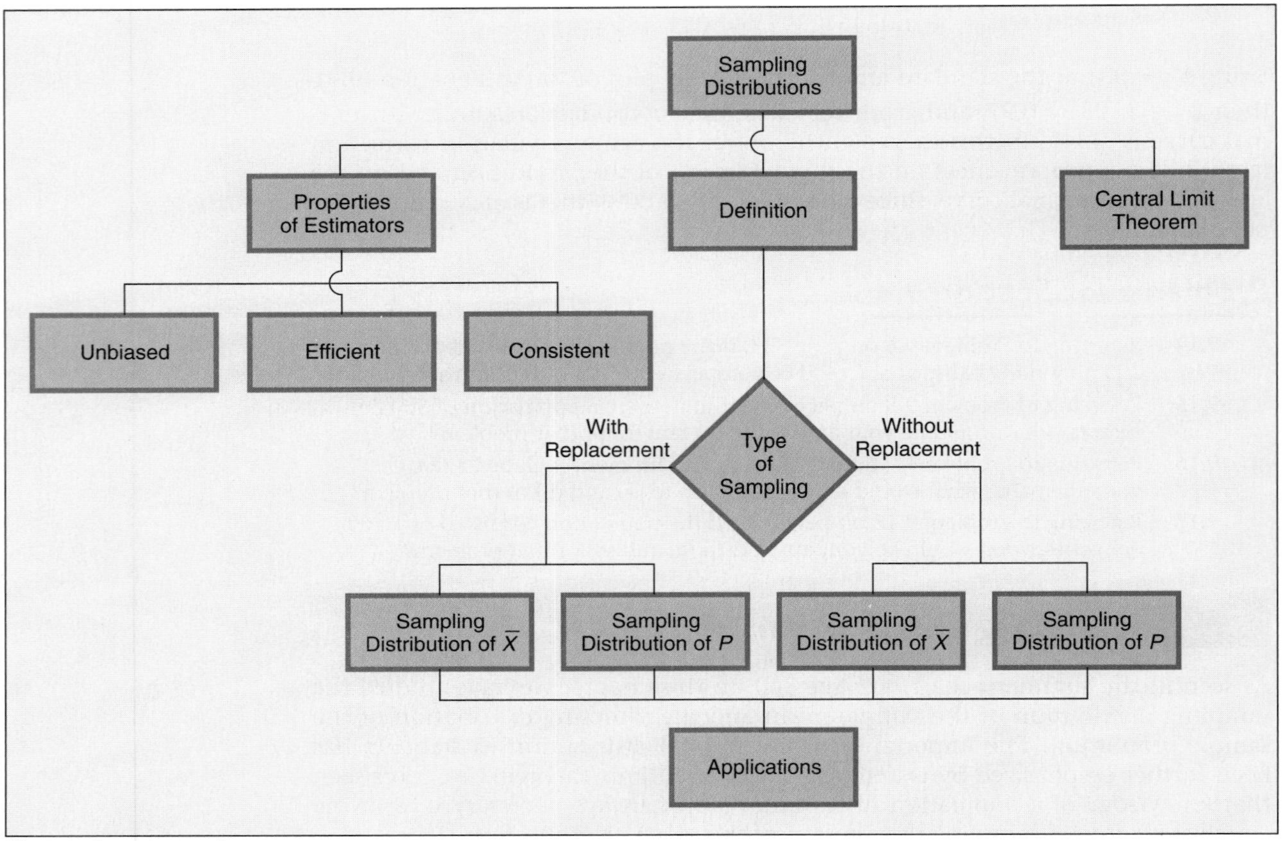

Chapter 9 summary chart.

Getting It All Together

Key Terms

central limit theorem *329*
consistency *323*
efficiency *323*
finite population correction factor *337*
law of large numbers *323*
sampling distribution *320*
sampling distribution of the mean *324*
sampling distribution of the proportion *334*
standard error of the mean *324*
statistical inference *320*
unbiasedness *321*

Chapter Review Problems

9.18 A soft-drink machine is regulated so that the amount dispensed is normally distributed with $\mu_x = 7$ ounces and $\sigma_x = .5$ ounce. If samples of nine cups are taken, what value will be exceeded by 95% of the sample means?

9.19 The life of a type of transistor battery is normally distributed with $\mu_x = 100$ hours and $\sigma_x = 20$ hours.
(a) What proportion of the batteries will last between 100 and 115 hours?
(b) If random samples of 16 batteries are selected
(1) What proportion of the sample means will be between 100 and 115 hours?
(2) What proportion of the sample means will be more than 90 hours?
(3) Within what limits around the population mean will 90% of sample means fall?
(c) Is the central limit theorem necessary to answer (b)(1), (2), and (3)? Explain.

9.20 An orange juice producer buys all his oranges from a large orange orchard. The amount of juice squeezed from each of these oranges is approximately normally distributed with a mean of 4.70 ounces and a standard deviation of .40 ounce.
(a) What is the probability that a randomly selected orange will contain
(1) Between 4.70 and 5.00 ounces?
(2) Between 5.00 and 5.50 ounces?
(b) 77% of the oranges will contain at least how many ounces of juice?
Suppose that a sample of 25 oranges is selected:
(c) What is the probability that the sample mean will be at least 4.60 ounces?
(d) Between what two values symmetrically distributed around the population mean will 70% of the sample means fall?
(e) 77% of the sample means will be above what value?
(f) Are the results of (b) and (e) different? Explain why.

9.21 **(Class Project)** The table of random numbers is an example of a uniform distribution since each digit is equally likely to occur. Starting in the row corresponding to the day of the month in which you were born, use the table of random numbers (Table E.1) to take *one digit* at a time. Select samples of size $n = 2$, $n = 5$, $n = 10$. Compute the sample mean $\overline{X}$ of each sample. For each sample size, each student should select five different samples so that a frequency distribution of the sample means can be developed for the results of the entire class. What can be said about the shape of the sampling distribution for each of these sample sizes?

9.22 **(Class Project)** A coin having one side heads and the other side tails is to be tossed ten times and the number of heads obtained is to be recorded. If each student performs this experiment five times, a frequency distribution of the number of heads can be developed from the results of the entire class. Does this distribution seem to approximate the normal distribution?

9.23 **(Class Project)** The number of cars waiting in line at a car wash is distributed as follows:

Length of Waiting Line (number of cars)	Probability
0	.25
1	.40
2	.20
3	.10
4	.04
5	.01

The table of random numbers can be used to select samples from this distribution by assigning numbers as described as follows:

1. Start in the row corresponding to the day of the month in which you were born.
2. *Two-digit* random numbers are to be selected.
3. If a random number between 00 and 24 is selected, record a length of 0; if between 25 and 64, record a length of 1; if between 65 and 84, record a length of 2; if between 85 and 94, record a length of 3; if between 95 and 98, record a length of 4; if it is 99, record a length of 5.

Select samples of size $n = 2$, $n = 10$, $n = 25$. Compute the sample mean for each sample. For example, if a sample size 2 results in random numbers 18 and 46, these would correspond to lengths of 0 and 1, respectively, producing a sample mean of 0.5. If each student selects five different samples for each sample size, a frequency distribution of the sample means (for each sample size) can be developed from the results of the entire class. What conclusions can you draw about the sampling distribution of the mean as the sample size is increased?

9.24 **(Class Project)** The table of random numbers can be used to simulate the operation of selecting different colored balls from a bowl as follows:

1. Start in the row corresponding to the day of the month in which you were born.
2. *One-digit* numbers are to be selected.
3. If a random digit between 0 and 6 is selected, consider the ball to be white; if a random digit is a 7, 8, or 9, consider the ball to be red.

Select samples of 10, 25, and 50 digits. In each sample, count the number of white balls and compute the proportion of white balls in the sample. If each student in the class selects five different samples for each sample size, a frequency distribution of the proportion of white balls (for each sample size) can be developed from the results of the entire class. What conclusions can be drawn about the sampling distribution of the proportion as the sample size is increased?

9.25 **(Class Project)** Suppose that part 3 of Problem 9.24 uses the following rule: If a random digit between 0 and 8 is selected, consider the ball to be white; if a random digit of 9 is selected, consider the ball to be red. Compare and contrast the results obtained in this problem and in Problem 9.24.

Endnotes

1. We must remember that "only" 500 samples out of an infinite number of samples have been selected, so that the sampling distributions shown are only approximations of the true distributions.
2. When working with the sampling distribution of the proportion for very large samples, the continuity correction factor (see Section 8.6) is usually omitted, since it will have minimal effect on the results.

References

1. Cochran, W. G., *Sampling Techniques,* 3d ed. (New York: Wiley, 1977).
2. Larsen, R. L., and M. L. Marx, *An Introduction to Mathematical Statistics and Its Applications,* 2d ed. (Englewood Cliffs, NJ: Prentice-Hall, 1986).

chapter 10

Estimation

CHAPTER OBJECTIVES

To utilize the sampling distribution to develop a confidence interval estimate for either a mean or a proportion and to determine the sample size necessary to obtain a desired confidence interval.

10.1 Introduction

Statistical inference is the process of using sample results to draw conclusions about the characteristics of a population. In this chapter we shall examine statistical procedures that will enable us to *estimate* either a population mean or a population proportion.

There are two major types of estimates: point estimates and interval estimates. A **point estimate** consists of a single sample statistic that is used to estimate the true value of a population parameter. For example, the sample mean $\bar{X}$ is a point estimate of the population mean μ_x and the sample variance S^2 is a point estimate of the population variance σ_x^2. Recall from Section 9.2.1 that the sample mean $\bar{X}$ possessed the highly desirable properties of unbiasedness and efficiency. Although in practice only one sample is selected, we know that the average value of all possible sample means is μ_x, the true population parameter.[1] Since the sample statistic ($\bar{X}$) varies from sample to sample (i.e., it depends on the elements selected in the sample), we need to take this into consideration in order to provide for a more informative estimate of the population characteristic. To accomplish this, we shall develop an **interval estimate** of the true population mean by taking into account the sampling distribution of the mean. The interval that we construct will have a specified confidence or probability of correctly estimating the true value of the population parameter μ_x. Similar interval estimates will also be developed for the population proportion, *p*. We will then discuss how we can determine the size of the sample to be selected and demonstrate how a finite population can affect the width of the confidence interval developed and the sample size selected.

Upon completion of this chapter, you should be able to:

1. Interpret the meaning of a confidence interval estimate
2. Set up the confidence interval estimate of the mean when σ_x is known or unknown
3. Set up the confidence interval estimate of the proportion
4. Determine the required sample size for either means or proportions
5. Use the finite population correction factor when sampling from a finite population without replacement

10.2 Confidence Interval Estimation of the Mean (σ_x Known)

In Section 9.2 we observed that from either the central limit theorem or knowledge of the population distribution we could determine the percentage of sample means that fell within certain distances of the population mean. For instance, in Section 9.2.3, in the example involving the filling of cereal boxes (in which $\mu_x = 368$, $\sigma_x = 15$, and $n = 25$) we observed that 95% of all sample means would fall between 362.12 and 373.88 grams.

The type of reasoning in this statement (*deductive reasoning*) is exactly opposite to the type of reasoning that is needed here *(inductive reasoning)*. In statistical inference we must take the results of a single sample and draw conclusions about the population, not vice versa. In practice, the population mean is the unknown quantity that is to be estimated. Suppose, for example, in the cereal box packaging process, that the true population mean μ_x was unknown but the true population standard deviation σ_x was known to be 15 grams. Thus, rather than taking $\mu_x \pm$

$(1.96)(\sigma_x/\sqrt{n})$ to find the upper and lower limits around μ_x as in Section 9.2.3, let us determine the consequences of substituting the sample mean $\bar{X}$ for the unknown μ_x and using $\bar{X} \pm (1.96)(\sigma_x/\sqrt{n})$ as an interval within which we estimate the unknown μ_x. Although in practice a single sample of size n is selected and the mean $\bar{X}$ is computed, we need to obtain a hypothetical set of all possible samples, each of size n, in order to understand the full meaning of the interval estimate that will be obtained.

Suppose, for example, that our sample of size $n = 25$ had a mean of 362.3 grams. The interval developed to estimate μ_x would be $362.3 \pm (1.96)(15)/(\sqrt{25})$ or 362.3 ± 5.88. That is, the estimate of μ_x would be

$$356.42 \le \mu_x \le 368.18$$

Since the population mean μ_x (equal to 368) is included *within* the interval, we observe that this sample has led to a correct statement about μ_x (see Figure 10.1).

To continue our hypothetical example, suppose that for a different sample of $n = 25$, the mean was 369.5. The interval developed from this sample would be $369.5 \pm (1.96)(15)/(\sqrt{25})$ or 369.5 ± 5.88. That is, the estimate of μ_x would be

$$363.62 \le \mu_x \le 375.38$$

Since the true population mean μ_x (equal to 368) is also included within this interval, we conclude that this statement about μ_x is correct.

Now, before we begin to think that we will *always* make correct statements about μ_x from the sample $\bar{X}$, suppose we draw a third hypothetical sample of size $n = 25$ in which the sample mean is equal to 360 grams. The interval developed here would be $360 \pm (1.96)(15)/(\sqrt{25})$ or 360 ± 5.88. In this case, the estimate of μ_x is

$$354.12 \le \mu_x \le 365.88$$

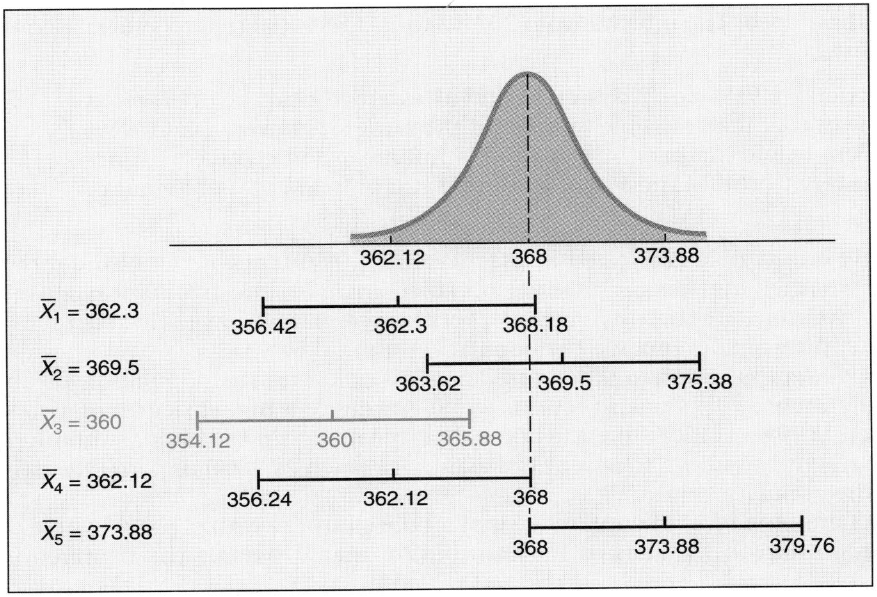

Figure 10.1
Confidence interval estimates from five different samples of size $n = 25$ taken from a population where $\mu_x = 368$ and $\sigma_x = 15$.

Observe that this estimate is *not* a correct statement, since the population mean μ_x is not included in the interval developed from this sample (see Figure 10.1 on page 345). Thus we are faced with a dilemma. For some samples the interval estimate of μ_x will be correct, and for others it will be incorrect. In addition, we must realize that in practice we select only *one* sample, and since we do not know the true population mean, we cannot determine whether our particular statement is correct.

What we can do in order to resolve this dilemma is to determine the proportion of samples producing intervals that result in correct statements about the population mean μ_x. In order to do this, we need to examine two other hypothetical samples: the case in which $\overline{X}$ = 362.12 grams and the case in which $\overline{X}$ = 373.88 grams. If $\overline{X}$ = 362.12, the interval will be 362.12 ± $(1.96)(15)/(\sqrt{25})$ or 362.12 ± 5.88. That is,

$$356.24 \le \mu_x \le 368.00$$

Since the population mean of 368 is at the upper limit of the interval, the statement is a correct one (see Figure 10.1 on page 345).

Finally, if $\overline{X}$ = 373.88, the interval will be 373.88 ± $(1.96)(15)/(\sqrt{25})$ or 373.88 ± 5.88. That is,

$$368.00 \le \mu_x \le 379.76$$

In this case, since the population mean of 368 is included at the lower limit of the interval, the statement is a correct one.

Thus, from these examples (see Figure 10.1) we can determine that if the sample mean based on a sample of $n = 25$ falls anywhere between 362.12 and 373.88 grams, the population mean will be included *somewhere* within the interval. However, we know from our discussion of the sampling distribution in Section 9.2.3 that 95% of the sample means fall between 362.12 and 373.88 grams. Therefore, 95% of all sample means will include the population mean within the interval developed. The interval from 362.12 to 373.88 is referred to as a 95% confidence interval.

> In general, a 95% **confidence interval estimate** can be interpreted to mean that if all possible samples of the same size n were taken, 95% of them would include the true population mean somewhere within the interval around their sample means, and only 5% of them would not.

Since only one sample is selected in practice and μ_x is unknown, we never know for sure whether the specific interval obtained includes the population mean. However, we can state that we have 95% confidence that we have selected a sample whose interval does include the population mean.

In our examples we had 95% confidence of including the population mean within the interval. In some situations we might desire a higher degree of assurance (such as 99%) of including the population mean within the interval. In other cases we might be willing to accept less assurance (such as 90%) of correctly estimating the population mean.

In general, the level of confidence is symbolized by $(1 - \alpha) \times 100\%$, where α is the proportion in the tails of the distribution that is outside the confidence interval. Therefore to obtain the $(1 - \alpha) \times 100\%$ confidence interval estimate of the mean with σ_x known, we have

$$\bar{X} \pm Z\frac{\sigma_x}{\sqrt{n}}$$

or **(10.1)**

$$\bar{X} - Z\frac{\sigma_x}{\sqrt{n}} \le \mu_x \le \bar{X} + Z\frac{\sigma_x}{\sqrt{n}}$$

where Z is the value corresponding to an area of $(1 - \alpha)/2$ from the center of a standardized normal distribution.

To construct a 95% confidence interval estimate of the mean, the Z value corresponding to an area of $.95/2 = .4750$ from the center of the standard normal distribution is 1.96. The value of Z selected for constructing such a confidence interval is called the **critical value** for the distribution.

There is a different critical value for each level of confidence, $1 - \alpha$. A **level of confidence** of 95% led to a Z value of ± 1.96 (see Figure 10.2). If a level of confidence of 99% were desired, the area of .99 would be divided in half, leaving .495 between each limit and μ_x (see Figure 10.3). The Z value corresponding to an area of .495 from the center of the normal curve is approximately 2.58.

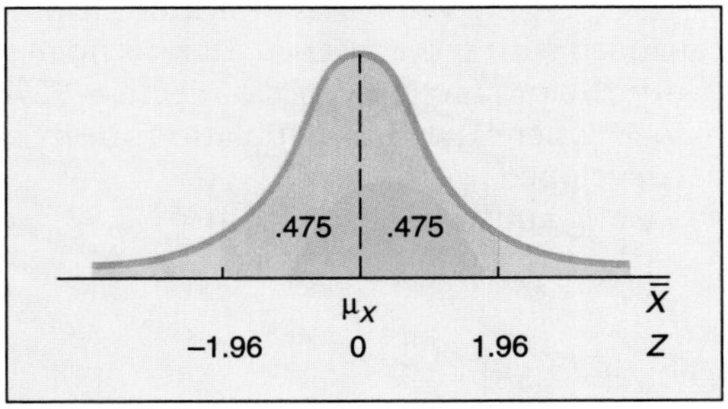

Figure 10.2
Normal curve for determining the Z value needed for 95% confidence.

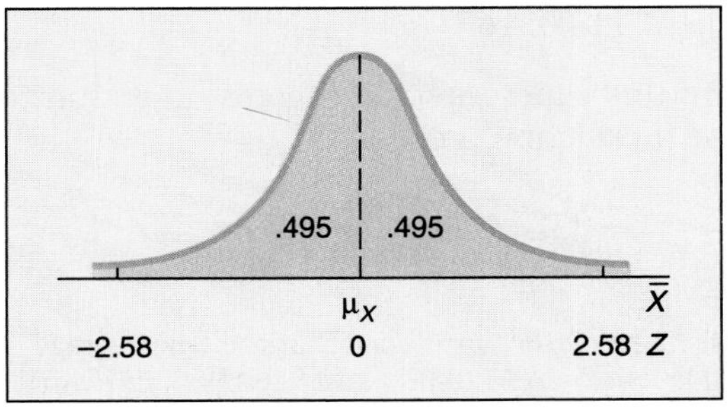

Figure 10.3
Normal curve for determining the Z value needed for 99% confidence.

Now that we have considered various levels of confidence, one might wonder why we wouldn't want to make the confidence level as close to 100% as possible. But any increase in the level of confidence is achieved only by simultaneously widening (and making less precise and less useful) the confidence interval obtained. Thus we would have more confidence that the population mean is within a broader range of values. This tradeoff between the width of the confidence interval and the level of confidence will be discussed in greater depth when we investigate how the sample size n is determined (see Section 10.7).

● **Application** We may illustrate the application of the confidence interval estimate by an example. A manufacturer of computer paper has a production process that operates continuously throughout an entire production shift. The paper is expected to have an average length of 11 inches and the standard deviation is known to be 0.02 inch. At periodic intervals, samples are selected to determine whether the average paper length is still equal to 11 inches or whether something has gone wrong in the production process to change the length of the paper produced. If indeed such a situation has occurred, corrective action must be contemplated. A random sample of 100 sheets has been selected and the average paper length is found to be 10.998 inches. If a 95% confidence interval estimate of the population average paper length were desired, using Equation (10.1), with $Z =$ 1.96 for 95% confidence, we would have

$$\bar{X} \pm Z\frac{\sigma_x}{\sqrt{n}} = 10.998 \pm (1.96)\frac{.02}{\sqrt{100}}$$

$$= 10.998 \pm .00392$$

$$10.99408 \leq \mu_x \leq 11.00192$$

Thus we would estimate, with 95% confidence, that the population mean is between 10.99408 and 11.00192 inches. Since 11, the value that indicates the production process is working properly, is included within the interval, there is no reason to believe that anything is wrong with the production process. There is 95% confidence that the sample selected is one where the true population mean is included somewhere within the interval developed.

If 99% confidence were desired, then using Equation (10.1), with $Z = 2.58$, we would have

$$\bar{X} \pm Z\frac{\sigma_x}{\sqrt{n}} = 10.998 \pm (2.58)\frac{.02}{\sqrt{100}}$$

$$= 10.998 \pm .00516$$

$$10.99284 \leq \mu_x \leq 11.00316$$

Once again, since 11 is included within this wider interval, there is no reason to believe that anything is wrong with the production process.

Problems for Section 10.2

10.1 A market researcher states that she has 95% confidence that the true average monthly sales of a product will be between $170,000 and $200,000. Explain the meaning of this statement.

10.2 Why can't the production manager in the example on page 348 have 100% confidence? Explain.

10.3 Is it true in the example on page 348 pertaining to the production of computer paper that 95% of the sample means will fall between 10.99408 and 11.00192 inches? Explain.

10.4 Is it true in the example on page 348 pertaining to the production of computer paper that we do not know for sure whether the true population mean is between 10.99408 and 11.00192 inches? Explain.

10.5 Suppose that the manager of a paint supply store wanted to estimate the correct amount of paint contained in one-gallon cans purchased from a nationally known manufacturer. It is known from the manufacturer's specifications that the standard deviation of the amount of paint is equal to .02 gallon. A random sample of 50 cans is selected, and the average amount of paint per one-gallon can is 0.995 gallon.

(a) Set up a 99% confidence interval estimate of the true population average amount of paint included in a one-gallon can.

(b) Based on your results, do you think that the store owner has a right to complain to the manufacturer? Why?

(c) Does the population amount of paint per can have to be normally distributed here? Explain.

(d) Tell why an observed value of .98 gallon for an individual can would not be unusual, even though it is outside the confidence interval you calculated.

10.6 The quality control manager at a light bulb factory needs to estimate the average life of a large shipment of light bulbs. The process standard deviation is known to be 100 hours. A random sample of 50 light bulbs indicated a sample average life of 350 hours.

(a) Set up a 95% confidence interval estimate of the true average life of light bulbs in this shipment.

(b) Does the population of light bulb life have to be normally distributed here? Explain.

(c) Tell why an observed value of 320 hours would not be unusual, even though it is outside the confidence interval you calculated.

10.7 The inspection division of the Lee County Weights and Measures Department is interested in estimating the actual amount of soft drink that is placed in 2-liter bottles at the local bottling plant of a large nationally known soft-drink company. The bottling plant has informed the inspection division that the standard deviation for 2-liter bottles is .05 liter. A random sample of 100 2-liter bottles obtained from this bottling plant indicated a sample average of 1.99 liters.

(a) Set up a 95% confidence interval estimate of the true average amount of soft drink in each bottle.

(b) Does the population of soft-drink fill have to be normally distributed here? Explain.

(c) Tell why an observed value of 2.02 liters would not be unusual, even though it is outside the confidence interval you calculated.

10.3 Confidence Interval Estimation of the Mean (σ_x Unknown)

Just as the mean of the population μ_x is usually not known, the actual standard deviation of the population σ_x is also not likely to be known. Therefore, we need to obtain a confidence interval estimate of μ_x by using only the sample statistics of $\bar{X}$ and S. To achieve this we turn to the work of William S. Gosset.

10.3.1 Student's *t* Distribution

At the turn of this century a statistician named William S. Gosset, an employee of Guinness Breweries in Ireland (see Reference 7), was interested in making inferences about the mean when σ_x was unknown. Since Guinness employees were not permitted to publish research work under their own names, Gosset adopted the pseudonym "Student." The distribution that he developed has come to be known as **Student's *t* distribution.** If the random variable X is normally distributed, then the statistic

$$\frac{\bar{X} - \mu_x}{\frac{S}{\sqrt{n}}}$$

has a t distribution with $n - 1$ *degrees of freedom.* Notice that this expression has the same form as Equation (9.5) on page 326, except that S is used to estimate σ_x, which is presumed unknown in this case.

10.3.2 Properties of the *t* Distribution

In appearance the t distribution is very similar to the normal distribution. Both distributions are bell-shaped and symmetric. However, the t distribution has more area in the tails and less in the center than does the normal distribution (see Figure 10.4). This is because σ_x is unknown, and we are using S to estimate it. Since we are uncertain of the value σ_x, the values of t that we observe will be more variable than for Z.

However, as the number of degrees of freedom increases, the t distribution gradually approaches the normal distribution until the two are virtually identical. This happens because, as the sample size gets larger, S becomes a better estimate of σ_x. With a sample size of about 120 or more, S estimates σ_x precisely enough that there is little difference between the t and Z distributions. For this reason, most statisticians will use Z instead of t when the sample size is over 120.

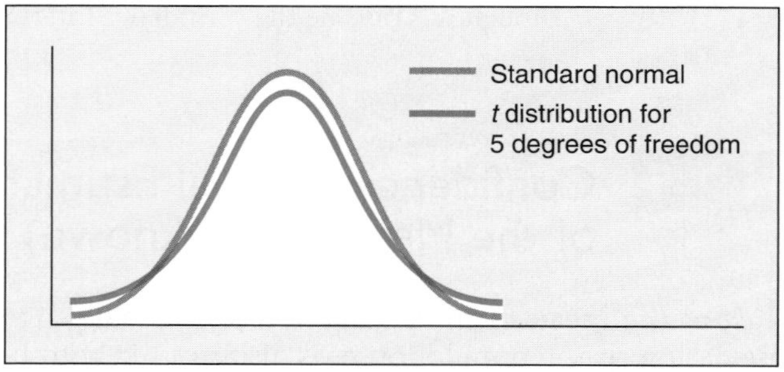

Figure 10.4
Standard normal distribution and t distribution for 5 degrees of freedom.

In practice, as long as the sample size is not too small and the population is not very skewed, the t distribution can be used in estimating the population mean when σ_x is unknown. The critical values of t for the appropriate degrees of freedom can be obtained from the table of the t distribution (see Table E.3). The top of each column of the t table indicates the area in the right tail of the t distribution (since positive entries for t are supplied, the values are for the upper tail); each row represents the particular t value for each specific degree of freedom. For example, with 34 degrees of freedom, if 95% confidence were desired, the appropriate value of t would be found in the following manner (as shown in Table 10.1). The 95% confidence level indicates that there would be an area of .025 in each tail of the distribution. Looking in the column for an upper-tail area of .025 and in the row corresponding to 34 degrees of freedom results in a value of t of 2.0332. Since t is a symmetrical distribution, with a mean of 0, if the upper-tail value is +2.0332, the value for the lower-tail area (lower .025) would be −2.0332. A t value of 2.0332 means that the probability that t would exceed +2.0332 is .025 or 2.5% (see Figure 10.5).

Table 10.1 Determining the critical value from the t table for an area of .025 in each tail with 34 degrees of freedom.

	Upper-Tail Areas					
Degrees of Freedom	.25	.10	.05	.025	.01	.005
1	1.0000	3.0777	6.3138	12.7062	31.8207	63.6574
2	0.8165	1.8856	2.9200	4.3027	6.9646	9.9248
3	0.7649	1.6377	2.3534	3.1824	4.5407	5.8409
4	0.7407	1.5332	2.1318	2.7764	3.7469	4.6041
5	0.7267	1.4759	2.0150	2.5706	3.3649	4.0322
.	.	.	.	.	.	.
.	.	.	.	.	.	.
.	.	.	.	.	.	.
31	0.6825	1.3095	1.6955	2.0395	2.4528	2.7440
32	0.6822	1.3086	1.6939	2.0369	2.4487	2.7385
33	0.6820	1.3077	1.6924	2.0345	2.4448	2.7333
34	0.6818	1.3070	1.6909	2.0322	2.4411	2.7284
35	0.6816	1.3062	1.6896	2.0301	2.4377	2.7238

Source: Extracted from Table E.3.

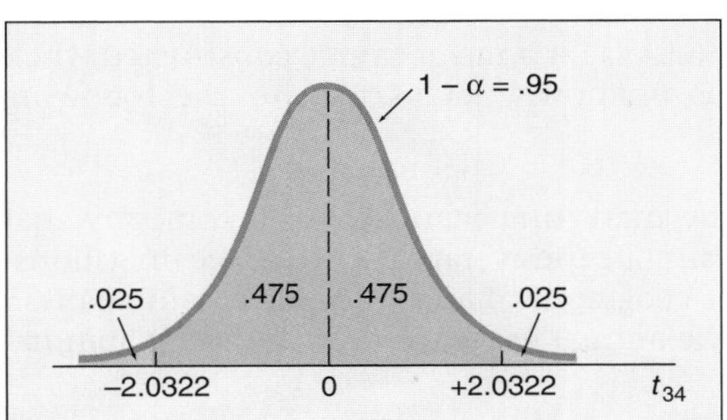

Figure 10.5
t distribution with 34 degrees of freedom.

10.3.3 The Concept of Degrees of Freedom

We may recall from Chapter 3 that the sample variance S^2 requires the computation of

$$\sum_{i=1}^{n}(X_i - \overline{X})^2$$

Thus, in order to compute S^2 we need to first know $\overline{X}$. Therefore we can say that only $n - 1$ of the sample values are free to vary. That is, there are $n - 1$ degrees of freedom.

We may illustrate this concept as follows: Suppose that we had a sample of five values that had a mean of 20. How many distinct values would we need to know before we could obtain the remainder? The fact that $n = 5$ and $\overline{X} = 20$ also tells us that $\sum_{i=1}^{n} X_i = 100$, since $\sum_{i=1}^{n} X_i/n = \overline{X}$. Thus, once we know four of the values, the fifth one will not be *free* to vary, since the sum must add to 100. For example, if four of the values are 18, 24, 19, and 16, the fifth value can only be 23, so that the sum equals 100.

10.3.4 The Confidence Interval Statement

The $(1 - \alpha) \times 100\%$ confidence interval estimate for the mean with σ_x unknown is expressed as follows:

$$\overline{X} \pm t_{n-1}\frac{S}{\sqrt{n}}$$

or **(10.2)**

$$\overline{X} - t_{n-1}\frac{S}{\sqrt{n}} \le \mu_x \le \overline{X} + t_{n-1}\frac{S}{\sqrt{n}}$$

where t_{n-1} is the critical value of the t distribution with $n - 1$ degrees of freedom for an area of $\alpha/2$ in the upper tail.

In order to see how confidence intervals for a mean can be constructed when the population standard deviation is unknown, let us turn to the following application.

● **Application** Suppose that the marketing manager for a company that supplies home heating oil wanted to estimate the average annual usage (in gallons) of single-family homes in a particular geographical area. A random sample of 35 single-family homes is selected and the annual usage for these homes is summarized in Table 10.2.

Table 10.2 Annual amount of heating oil consumed (in gallons) in a sample of 35 single-family houses.

Data file
OILUSE.DAT

1150.25	1352.67	983.45	1365.11	942.71	1577.77	330.00
872.37	1126.57	1184.17	1046.35	1110.50	1050.86	851.60
1459.56	1252.01	373.91	1047.40	1064.46	1018.23	996.92
941.96	767.37	1598.57	1598.66	1343.29	1617.73	1300.76
1013.27	1402.59	1069.32	1108.94	1326.19	1074.86	975.86

For these data we may use a statistical software package to obtain the sample average $\overline{X} = 1{,}122.7$ gallons and the sample standard deviation $S = 295.72$ gallons.

If the marketing manager would like to have 95% confidence that the interval obtained includes the population average amount of heating oil consumed per year, using $\overline{X} = 1{,}122.7$, $S = 295.72$, $n = 35$, and $t_{34} = 2.0322$, we have

$$\overline{X} \pm t_{n-1}\frac{S}{\sqrt{n}} = 1{,}122.7 \pm (2.0322)\frac{295.72}{\sqrt{35}}$$

$$= 1{,}122.7 \pm 101.58$$

$$1{,}021.12 \le \mu_x \le 1{,}224.28$$

We would conclude with 95% confidence that the average amount of heating oil consumed per year is between 1,021.12 and 1,224.28 gallons. The 95% confidence interval states that we are 95% sure that the sample we have selected is one in which the population mean μ_x is located within the interval. This 95% confidence actually means that if all possible samples of size 35 were selected (something that would never be done in practice), 95% of the intervals developed would include the true population mean *somewhere* within the interval.

Problems for Section 10.3

10.8 Determine the critical value of t in each of the following circumstances:
(a) $1 - \alpha = .95$, $n = 10$.
(b) $1 - \alpha = .99$, $n = 10$.
(c) $1 - \alpha = .95$, $n = 32$.
(d) $1 - \alpha = .95$, $n = 65$.
(e) $1 - \alpha = .90$, $n = 16$.

10.9 A new breakfast cereal was test-marketed for one month at stores of a large supermarket chain. The results for a sample of 16 stores indicated average sales of \$1,200 with a sample standard deviation of \$180. Set up a 99% confidence interval estimate of the true average sales of this new breakfast cereal.

10.10 The manager of a branch of a local savings bank wanted to estimate the average amount held in passbook savings accounts by depositors at the bank. A random sample of 30 depositors was selected and the results indicated a sample average of \$4,750 and a sample standard deviation of \$1,200.

(a) Set up a 95% confidence interval estimate of the average amount held in all passbook savings accounts.
(b) If an individual had $4,000 in a passbook savings account, would this be considered unusual? Explain your answer.

10.11 A stationery store would like to estimate the average retail value of greeting cards that it has in its inventory. A random sample of 20 greeting cards indicated an average value of $1.67 and a standard deviation of $0.32. Set up a 95% confidence interval estimate of the average value of all greeting cards that are in its inventory.

Data file DENTAL.DAT

10.12 The personnel department of a large corporation would like to estimate the family dental expenses of its employees in order to determine the feasibility of providing a dental insurance plan. A random sample of 10 employees revealed the following family dental expenses in the preceding year:

$110, 362, 246, 85, 510, 208, 173, 425, 316, 179

(a) Set up a 90% confidence interval estimate of the average family dental expenses for all employees of this corporation.
(b) What assumption about the population distribution must be made in (a)?
(c) Give an example of a family dental expense that would be outside the confidence interval but would not be unusual for an individual family, and tell why this is not a contradiction.
(d) ACTION What should the personnel department tell the president of the corporation concerning family dental expenses? Write a memo.

Data file GASERVE.DAT

10.13 The customer service department of a local gas utility would like to estimate the average length of time between the entry of the service request and the connection of service. A random sample of 15 houses were selected from the records available during the past year. The results recorded in number of days are displayed as follows:

114	78	96	137	78	103	117	
126	86	99	114	72	104	73	86

(a) Set up a 95% confidence interval estimate of the population average waiting time in the past year.
(b) What assumption about the population distribution must be made in (a)?
(c) ACTION Use the results in (a) to provide information to your neighbor, who is thinking about converting to gas heat. Write a letter.

Data file HMO.DAT

10.14 The Director of Quality of a large health maintenance organization wants to evaluate patient waiting time at a local facility. A random sample of 25 patients is selected from the appointment book. The waiting time was defined as the time from when the patient signed in to when he or she was seen by the doctor. The following data represent the waiting time (in minutes):

19.5	30.5	45.6	39.8	29.6
25.4	21.8	28.6	52.0	25.4
26.1	31.1	43.1	4.9	12.7
10.7	12.1	1.9	45.9	42.5
41.3	13.8	17.4	39.0	36.6

(a) Set up a 95% confidence interval estimate of the population average waiting time.
(b) What assumption about the population distribution must be made in (a)?

10.15 Set up a 95% confidence interval estimate for each of the following batches of data:

Batch 1: 1, 1, 1, 1, 8, 8, 8, 8

Batch 2: 1, 2, 3, 4, 5, 6, 7, 8

Tell why they have different confidence intervals even though they have the same mean and range.

10.16 Compute a 95% confidence interval for the numbers 1, 2, 3, 4, 5, 6, 20. Change the number 20 to 7, and recalculate the confidence interval. Using these results, describe the effect of an outlier (or extreme value) on the confidence interval.

Interchapter Problems for Section 10.3

10.17 Referring to the data of Problem 3.3 on page 58 — Data file SHOWER.DAT

(a) Set up a 95% confidence interval estimate of the average flow rate of all fixed-position shower heads.

(b) What assumption about the population distribution must be made in (a)?

(c) Use the results in (a) to provide information to your neighbor, who is thinking about purchasing a fixed-position shower head.

10.18 Referring to the data of Problem 3.12 on page 66 — Data file UTILITY.DAT

(a) Set up a 90% confidence interval estimate of the population average electric and gas utility charges for three-bedroom apartments in Manhattan.

(b) What assumption about the population distribution must be made in (a)?

10.19 Referring to the data of Problem 4.5 on page 115 — Data file FLASHBAT.DAT

(a) Set up a 95% confidence interval estimate of the average life of the flashlight batteries.

(b) What assumption about the population distribution must be made in (a)?

(c) Give an example of a battery life that would be outside the confidence interval but would not be unusual for an individual battery, and tell why this is not a contradiction.

(d) **ACTION** Write a draft of a memo that an advertising agency can use to market the batteries.

10.20 Referring to the data of Problem 4.7 on page 115 — Data file MOWER.DAT

(a) Set up a 99% confidence interval estimate of the average price of the lawn mowers.

(b) What assumption about the population distribution must be made in (a)?

(c) **ACTION** Use the results in (a) to provide information to your neighbor who is thinking about purchasing a side-bagging lawn mower. Write a letter.

10.21 Referring to the data of Problem 4.76 on page 162 — Data file CORDPHON.DAT

(a) Set up a 90% confidence interval estimate of the average retail price of all corded telephones.

(b) What assumption about the population distribution must be made in (a)?

10.4 Estimation through Bootstrapping

Although confidence interval estimates have been widely applied in making inferences about population parameters, the estimating procedure used is based on assumptions that do not always hold. For example, the confidence interval estimate of μ_x discussed in Section 10.3 assumed that the underlying population was normally distributed. Although the confidence interval procedure is fairly insensitive to moderate departures from this assumption, if there is substantial nonnormality in the population, particularly if a small sample size n is used, the confidence interval estimate for the mean may not be precise. In this section we will consider an alternative estimation approach called bootstrapping.

Bootstrapping estimation procedures involve an initial sample and then repeated resampling from the initial sample. These procedures, developed by Efron (see References 2, 3, and 5), are heavily computer intensive. What makes bootstrapping estimation useful is that the procedures are based on the initial sample and make no assumptions regarding the shape of the underlying population. In addition, the procedures do not require knowledge of any population parameters.

The steps in bootstrapping estimation of the mean are as follows:

1. Draw a random sample of size n *without replacement* from a population frame of size N.
2. Resample the initial sample by selecting n observations *with replacement* from the n observations in the initial sample.
3. Compute $\overline{X}$, the statistic of interest, from this *resample*.
4. Repeat steps 2 and 3 m different times (where m is typically selected as a value between 100 and 1,000, depending on the speed of the computer being used).
5. Form the **resampling distribution** of the statistic of interest (i.e., the distribution of the sample mean obtained from the m resamples) using a stem-and-leaf display or an ordered array.
6. To form a $(1 - \alpha) \times 100\%$ bootstrap confidence interval of the population mean μ_x, use the stem-and-leaf display or ordered array for the resampling distribution and find the value that cuts off the smallest $\alpha/2 \times 100\%$ and the value that cuts off the largest $\alpha/2 \times 100\%$ of the statistic. These values provide the lower and upper limits for the bootstrap confidence interval estimate of the unknown parameter.

To demonstrate the bootstrap confidence interval estimation procedure, we return to the example concerning heating oil consumption discussed on page 352. We may recall that the marketing manager wanted to estimate the average annual consumption of heating oil of customers residing in single-family homes. A random sample of 35 customers was selected without replacement and the t distribution was used to develop a 95% confidence interval estimate of μ_x. The estimate obtained was based on the assumption that the underlying population of oil usage consumption was approximately normally distributed, an assumption that is not necessary for the bootstrap procedure.

Following the six-step procedure just described, the sample data of Table 10.2 (page 353) were entered into a computer file and a random resample of 35 observations was selected with replacement using the MINITAB package. The first resample is listed in Table 10.3. The mean of this sample is 1,003.26 gallons.

Table 10.3 First resample of size 35 from the sample of 35 single-family homes (annual amount of heating oil consumed).

1326.19	1150.25	1184.17	1013.27	330.00	1126.57	1343.29
330.00	1013.27	996.92	1047.40	1018.23	1064.46	1598.66
941.96	767.37	1050.86	1459.56	975.86	373.91	1018.23
373.91	1050.86	1074.86	1326.19	1110.50	1402.59	1343.29
1110.50	373.91	1069.32	330.00	1365.11	1110.50	941.96

Notice that in this first resample some values from the original sample such as 330.00 are repeated and others such as 1,352.67 do not occur. If this resampling process is performed $m = 200$ times, the resampling distribution containing 200 resampling means can be developed. Table 10.4 presents the ordered array of the resampling distribution. To form a 95% bootstrap confidence interval estimate of μ_x, the smallest 2.5% and the largest 2.5% of the resample means need to be identified (Step 6).

Table 10.4 Ordered array of 200 resampled means obtained from MINITAB used to form 95% bootstrap confidence interval estimate of μ_x.

988.50	994.49	1003.26	1004.25	1006.53	1014.38	1018.62
1030.21	1030.73	1032.04	1032.77	1034.17	1035.21	1041.45
1042.23	1046.16	1050.01	1050.97	1052.73	1053.00	1054.58
1055.08	1055.39	1055.76	1059.57	1059.88	1060.51	1061.70
1061.79	1062.46	1062.86	1063.31	1066.91	1068.84	1070.53
1071.00	1071.38	1074.99	1077.18	1077.63	1078.62	1079.41
1079.46	1081.45	1083.40	1084.25	1085.21	1085.98	1086.22
1086.83	1089.45	1089.95	1092.37	1092.53	1094.48	1095.25
1095.35	1095.40	1095.88	1096.26	1097.59	1100.06	1100.19
1100.20	1101.59	1103.41	1104.07	1104.92	1105.19	1106.67
1106.84	1111.14	1112.46	1112.47	1112.53	1113.58	1115.09
1115.71	1116.46	1116.51	1117.94	1118.67	1119.17	1119.71
1119.98	1120.31	1121.01	1121.46	1121.68	1121.92	1122.40
1123.34	1123.73	1124.38	1124.98	1125.63	1126.01	1127.00
1127.15	1127.30	1127.55	1127.86	1128.06	1129.03	1129.66
1129.84	1132.30	1132.78	1133.39	1134.03	1134.43	1135.07
1136.67	1136.90	1138.24	1138.52	1139.53	1139.81	1140.60
1142.72	1142.73	1143.03	1143.32	1143.35	1143.70	1144.47
1145.72	1146.21	1149.77	1150.87	1152.88	1153.13	1153.25
1153.35	1156.35	1156.43	1158.49	1158.62	1159.14	1159.45
1159.55	1161.01	1161.16	1161.50	1162.12	1163.15	1163.24
1164.06	1164.39	1165.45	1167.87	1168.50	1170.06	1171.11
1171.21	1172.03	1172.19	1172.90	1173.00	1173.37	1174.25
1176.32	1176.73	1178.39	1178.63	1182.16	1182.37	1183.12
1183.99	1184.27	1184.68	1185.80	1185.97	1186.86	1187.01
1187.27	1188.03	1188.25	1188.53	1189.39	1190.20	1191.10
1191.28	1191.86	1192.90	1193.21	1196.66	1199.30	1206.23
1206.28	1209.61	1216.85	1225.20	1226.46	1226.67	1227.02
1230.11	1233.43	1233.97	1251.88			

When 200 resample means are obtained, the fifth (that is, 200 × .025) smallest value will cut off the lowest 2.5% while the fifth largest value will cut off the highest 2.5%. From Table 10.4, we obtain the values of 1,006.53 gallons as the fifth smallest and 1,227.02 gallons as the fifth largest. Therefore the 95% bootstrap confidence interval for the population average amount of heating oil consumed is 1,006.53 to 1,227.02 gallons. This estimate is fairly close to the traditional confidence interval estimate of 1,021.12 to 1,224.28 obtained in Section 10.3. However, the bootstrap estimate requires less stringent assumptions than does the traditional confidence interval estimate.

Problems for Section 10.4

Data file GASERVE.DAT

10.22 Referring to Problem 10.13 on page 354,
(a) Using a statistical software package, generate 200 resamples each of size 15 and establish a 95% bootstrap confidence interval estimate of the average waiting time for the connection of gas service.
(b) Compare the results of (a) to those of Problem 10.13(a).

Data file HMO.DAT

10.23 Referring to Problem 10.14 on page 354,
(a) Using a statistical software package, generate 200 resamples each of size 25 and establish a 95% bootstrap confidence interval estimate of average waiting time of patients at the HMO.
(b) Compare the results of (a) to those of Problem 10.14(a).

Interchapter Problems for Section 10.4

Data file SHOWER.DAT

10.24 Referring to Problem 3.3 on page 58,
(a) Using a statistical software package, generate 200 resamples each of size 34 and establish a 95% bootstrap confidence interval estimate of the average flow rate of fixed-position shower heads.
(b) Compare the results of (a) to those of Problem 10.17(a) on page 355.

Data file UTILITY.DAT

10.25 Referring to Problem 3.12 on page 66,
(a) Using a statistical software package, generate 200 resamples each of size 50 and establish a 90% bootstrap confidence interval estimate of the population average electric and gas utility charges for three-bedroom apartments in Manhattan.
(b) Compare the results of (a) to those of Problem 10.18(a) on page 355.

Data file FLASHBAT.DAT

10.26 Referring to Problem 4.5 on page 115,
(a) Using a statistical software package, generate 200 resamples each of size 13 and establish a 95% bootstrap confidence interval estimate of average life of the flashlight batteries.
(b) Compare the results of (a) to those of Problem 10.19(a) on page 355.

Data file MOWER.DAT

10.27 Referring to Problem 4.7 on page 115,
(a) Using a statistical software package, generate 200 resamples each of size 15 and establish a 99% bootstrap confidence interval estimate of average price of the lawnmowers.
(b) Compare the results of (a) to those of Problem 10.20(a) on page 355.

Data file CORDPHON.DAT

10.28 Referring to Problem 4.76 on page 162,
(a) Using a statistical software package, generate 200 resamples each of size 32 and establish a 90% bootstrap confidence interval estimate of population average retail price of corded telephones.
(b) Compare the results of (a) to those of Problem 10.21(a) on page 355.

10.5 Prediction Interval for a Future Individual Value

In addition to the need to obtain a confidence interval estimate for the population mean, it is often important to be able to predict the outcome of a future individual value (see References 6 and 9). Although the form of the prediction interval is similar to the confidence interval estimate of Equation (10.2), we must be careful to note that the prediction interval is estimating an *observable future individual value* X_f, *not an unknown* parameter μ_x. Thus the **prediction interval** for a future individual value X_f is provided in Equation (10.3).

$$\bar{X} \pm t_{n-1} S \sqrt{1 + \frac{1}{n}}$$

or (10.3)

$$\bar{X} - t_{n-1} S \sqrt{1 + \frac{1}{n}} \le X_f \le \bar{X} + t_{n-1} S \sqrt{1 + \frac{1}{n}}$$

If we return to our example concerning the consumption of heating oil by single-family homes, suppose that we wanted to obtain a 95% prediction interval estimate of the future amount of heating oil an individual single-family home would use annually. Using Equation (10.3), we would have

$$\bar{X} \pm t_{n-1} S \sqrt{1 + \frac{1}{n}} = 1{,}122.7 \pm (2.0322)(295.72)\sqrt{1 + \frac{1}{35}}$$

$$= 1{,}122.7 \pm 609.45$$

$$513.25 \le X_f \le 1{,}732.15$$

We may observe that this result differs markedly from the one obtained when we were estimating the confidence interval for the population mean. We note again that here we are estimating a *future individual value* X_f, *not an unknown* parameter μ_x.

Problems for Section 10.5

● 10.29 Referring to Problem 10.9 on page 353
- (a) Set up a 99% prediction interval estimate of the monthly sales in a future individual store.
- (b) Explain the difference in the results obtained in (a) and in Problem 10.9.

10.30 Referring to Problem 10.10 on page 353
- (a) Set up a 95% prediction interval estimate of the amount held in a passbook savings account by a future individual depositor.
- (b) Explain the difference in the results obtained in (a) and in Problem 10.10.

10.31 Referring to Problem 10.12 on page 354 — Data file DENTAL.DAT
- (a) Set up a 90% prediction interval estimate of the annual family dental expenses for a future individual employee.
- (b) Explain the difference in the results obtained in (a) and in Problem 10.12.

● 10.32 Referring to Problem 10.13 on page 354 — Data file GASERVE.DAT
- (a) Set up a 95% prediction interval estimate of the waiting time for gas installation for a future individual customer.
- (b) Explain the difference in the results obtained in (a) and in Problem 10.13.

confidence interval estimate of the true proportion of the bank's depositors who are paid weekly.

10.40 An auditor for the state insurance department would like to determine the proportion of claims that are paid by a health insurance company within two months of receipt of the claim. A random sample of 200 claims is selected, and it is determined that 80 were paid out within two months of the receipt of the claim.
(a) Set up a 99% confidence interval estimate of the true proportion of the claims paid within two months.
(b) How can the results of (a) be used in a report to the state insurance department?

10.41 An automobile dealer would like to estimate the proportion of customers who still own the same cars they purchased five years earlier. A random sample of 200 customers selected from the automobile dealer's records indicated that 82 still owned the cars that had been purchased five years earlier. Set up a 95% confidence interval estimate of the true proportion of all customers who still own the same cars five years after they were purchased.

10.42 A stationery supply store receives a shipment of a certain brand of inexpensive ballpoint pens from the manufacturer. The owner of the store wishes to estimate the proportion of pens that are defective. A random sample of 300 pens is tested, and 30 are found to be defective.
(a) Set up a 90% confidence interval estimate of the proportion of defective pens in the shipment.
(b) The shipment can be returned if it is more than 5% defective; based on the sample results, can the owner return this shipment?

10.43 The advertising director for a fast-food chain would like to estimate the proportion of high school students who are familiar with a particular commercial that has been broadcast on radio and television in the last month. A random sample of 400 high school students indicated that 160 were familiar with the commercial. Set up a 95% confidence interval estimate of the true population proportion of high school students who are familiar with the commercial.

10.44 The telephone company would like to estimate the proportion of households that would purchase an additional telephone line if it was made available at a substantially reduced installation cost. A random sample of 500 households was selected. The results indicated that 135 of the households would purchase the additional telephone line at a reduced installation cost. Set up a 99% confidence interval estimate of the true population proportion of households who would purchase the additional telephone line at a reduced installation cost.

10.45 The dean of a graduate school of business would like to estimate the proportion of MBA students enrolled who have access to a personal computer outside the school (either at home or at work). A sample of 150 students revealed that 105 had access to a personal computer outside the school (either at home or at work). Set up a 90% confidence interval estimate of the true population proportion of students who have access to a personal computer outside the school (either at home or at work).

10.7 Sample Size Determination for the Mean

In each of our examples concerning confidence interval estimation, the sample size was arbitrarily determined without regard to the size of the confidence interval. In the business world the determination of the proper sample size is a complicated procedure that is subject to the constraints of budget, time, and ease of selection. For example, if the marketing manager for a company that supplies home heating oil wanted to estimate the average annual usage of single-family homes in a particular geographical area, he would try to determine in advance how good an estimate would be required. This would mean that he would decide how much error he was willing to allow in estimating the population average annual usage. Was accuracy required to be within ±10 gallons, ±25 gallons, ±50 gallons, ±100 gallons, etc.? The marketing manager would also determine in advance how sure (confident) he wanted to be of correctly estimating the true population parameter. In determining the sample size for estimating the mean, these requirements must be kept in mind along with information about the standard deviation.

To develop a formula for determining sample size recall Equation (9.5):

$$Z = \frac{\overline{X} - \mu_x}{\frac{\sigma_x}{\sqrt{n}}}$$

where Z is the critical value corresponding to an area of $(1 - \alpha)/2$ from the center of a standardized normal distribution. Multiplying both sides of Equation (9.5) by $\sigma_x/\sqrt{n}$, we have

$$Z\frac{\sigma_x}{\sqrt{n}} = \overline{X} - \mu_x$$

Thus, the value of Z will be positive or negative, depending on whether $\overline{X}$ is larger or smaller than μ_x. The difference between the sample mean $\overline{X}$ and the population mean μ_x, denoted by e, is called the **sampling error.** The sampling error e can be defined as

$$e = \frac{Z\sigma_x}{\sqrt{n}} \quad \textbf{(10.5a)}$$

Solving this equation for n, we have

$$n = \frac{Z^2\sigma_x^2}{e^2} \quad \textbf{(10.5b)}$$

Therefore, to determine the sample size, three factors must be known:

1. The confidence level desired, which determines the value of Z, the critical value from the normal distribution.[2]
2. The sampling error permitted, e.
3. The standard deviation, σ_x.

In practice, the determination of these three quantities may not be easy. How is one to know what level of confidence to use and what sampling error is desired? Typically these questions can be answered only by the *subject matter expert*, that is, the individual who is familiar with the variables to be analyzed. Although 95% is the most common confidence level used (in which case $Z = 1.96$), if one desires greater confidence, 99% might be more appropriate; if less confidence is deemed acceptable, then 90% might be utilized.

For the sampling error, we should be thinking not of how much sampling error we would like to have (we really don't want any error) but of how much we can live with and still be able to provide adequate conclusions for the data.

Even when the confidence level and the sampling error are specified, an estimate of the standard deviation must be available. Unfortunately, the population standard deviation σ_x is rarely known. In some instances the standard deviation can be estimated from past data. In other situations, one can develop an educated guess by taking into account the range and distribution of the variable. For example, if one assumes a normal distribution, the range is approximately equal to $6\sigma_x$ (that is, $\pm 3\sigma_x$ around the mean), so that σ_x can be estimated as range/6. If σ_x cannot be estimated in this manner, a *pilot* study can be conducted and the standard deviation estimated from the resulting data.

Returning to the earlier example, suppose that the marketing manager would like to estimate the population mean annual usage of home heating oil to within ±50 gallons of the true value and desires to be 95% confident of correctly estimating the true mean. Based on a previous study taken last year, the marketing manager feels that the standard deviation can be estimated as 325 gallons. With this information the sample size can be determined in the following manner for $e = 50$, $\sigma_x = 325$, and 95% confidence ($Z = 1.96$):

$$n = \frac{Z^2\sigma_x^2}{e^2} = \frac{(1.96)^2(325)^2}{(50)^2}$$

$$= \frac{(3.8416)(105{,}625)}{2{,}500} = 162.31$$

Therefore, $n = 163$.

We have chosen a sample of size 163 because the general rule used in determining sample size is to always round up to the nearest integer value in order to slightly oversatisfy the criteria desired.

We may note that, if the marketing manager utilized these criteria, a sample of 163 should have been taken—not a sample of 35. However, the standard deviation that has been used was estimated at 325 based on a previous survey. If the standard deviation obtained in the actual survey is very different from this value, the computed sampling error will be directly affected.

Problems for Section 10.7

10.46 A survey is planned to determine the average annual family medical expenses of employees of a large company. The management of the company wishes to be 95% confident that the sample average is correct to within ±\$50 of the true average family medical expenses. A pilot study indicates that the standard deviation can be estimated as \$400. How large a sample size is necessary?

10.47 If the manager of the paint supply store in Problem 10.5 on page 349 wanted to estimate the average amount in a one-gallon can to within ±.004 gallon with 95% confidence and also assumed that the standard deviation remained at .02 gallon, what sample size would be needed?

10.48 If the quality control manager in Problem 10.6 on page 349 wanted to estimate the average life to within ±20 hours with 95% confidence and also assumed that the process standard deviation remained at 100 hours, what sample size is needed?

10.49 If the inspection division in Problem 10.7 on page 349 wanted to estimate the average amount of soft-drink fill to within ±.01 liter with 95% confidence and also assumed that the standard deviation remained at .05 liter, what sample size would be needed?

10.50 A consumer group would like to estimate the average monthly electric bills for the month of July for single-family homes in a large city. Based upon studies conducted in other cities, the standard deviation is assumed to be \$25. The group would like to estimate the average bill for July to within ±\$5 of the true average with 99% confidence. What sample size is needed?

10.51 A pharmaceutical company is considering a request to pay for the continuing education of its research scientists. It would like to estimate the average amount spent by these scientists for professional memberships. What sample size is required to be 90% confident of being correct to within ±\$10? Based on a pilot study, the standard deviation is estimated to be \$35.

10.52 An advertising agency that serves a major radio station would like to estimate the average amount of time that the station's audience spends listening to radio on a daily basis. What sample size is needed if the agency wants to be 90% confident of being correct to within ±5 minutes? From past studies, the standard deviation is estimated as 45 minutes.

10.53 Suppose that a competitor of the supermarket chain described in Problem 10.9 on page 353 wanted to estimate its population average sales for the breakfast cereal to within ±\$100 with 99% confidence. Since it *does not have access to the sample results of Problem 10.9,* it makes its own independent estimate of the standard deviation, which it believes to be \$200. What sample size is needed?

10.54 Suppose that a gas utility serving a different geographical area than the one in Problem 10.13 on page 354 wishes to estimate its average waiting time to within ±5 days with 95% confidence. Since it *does not have access to the sample results of Problem 10.13,* it makes its own independent estimate of the standard deviation, which it believes to be 20 days. What sample size is needed?

10.8 Sample Size Determination for a Proportion

In Section 10.7 we discussed the determination of sample size needed for the estimation of a population mean. Now suppose that the production manager wishes to determine the sample size necessary for estimating the population proportion of newspapers printed that have a nonconforming attribute, such as excessive ruboff, improper page setup, missing pages, and so on. The methods of sample size determination that are utilized in estimating a population proportion are similar to those employed in estimating a mean.

To develop a formula for determining sample size recall from Equation (9.9) that

$$Z \cong \frac{p_s - p}{\sqrt{\dfrac{p(1-p)}{n}}}$$

where Z is the critical value corresponding to an area of $(1 - \alpha)/2$ from the center of a standardized normal distribution. Multiplying both sides of Equation (9.9) by $\sqrt{\frac{p(1-p)}{n}}$ we have

$$Z\sqrt{\frac{p(1-p)}{n}} = p_s - p$$

The sampling error, e, is equal to $(p_s - p)$, the difference between the sample proportion (p_s) and the parameter to be estimated (p). This sampling error can be defined as

$$e = Z\sqrt{\frac{p(1-p)}{n}} \tag{10.6a}$$

Solving for n, we obtain

$$n = \frac{Z^2 p(1-p)}{e^2} \tag{10.6b}$$

In determining the sample size for estimating a proportion, three unknowns must be defined:

1. The level of confidence desired.
2. The sampling error permitted, e.
3. The true proportion of "success," p.

In practice, the selection of these three quantities is often difficult. Once we determine the desired level of confidence, we will be able to obtain the appropriate Z value from the normal distribution. The sampling error e indicates the amount of error that we are willing to accept or tolerate in estimating the population proportion. The third quantity—the true proportion of success, p—is actually the population parameter that we are trying to find! Thus, how can we state a value for the very thing that we are taking a sample in order to determine?

Here there are two alternatives. First, in many situations past information or relevant experience may be available that enables us to provide an educated estimate of p. Second, if past information or relevant experience is not available, we try to provide a value for p that would never *underestimate* the sample size needed. Referring to Equation (10.6b), we observe that the quantity $p(1 - p)$ appears in the numerator. Thus, we need to determine the value of p that will make $p(1 - p)$ as large as possible. It can be shown that when $p = .5$, then the product $p(1 - p)$ achieves its maximum result. Several values of p along with the accompanying products of $p(1 - p)$ are

$$\begin{aligned}
p &= .5, & p(1-p) &= (.5)(.5) = .25 \\
p &= .4, & p(1-p) &= (.4)(.6) = .24 \\
p &= .3, & p(1-p) &= (.3)(.7) = .21 \\
p &= .1, & p(1-p) &= (.1)(.9) = .09 \\
p &= .01, & p(1-p) &= (.01)(.99) = .0099
\end{aligned}$$

Therefore, when we have no prior knowledge or estimate of the true proportion p, we may want to use $p = .5$ as the most conservative way of determining the sample size. This would produce the largest sample size possible and therefore result in the highest possible cost. However, the use of $p = .5$ may result in an overestimate of the sample size since the actual sample proportion is utilized in the confidence interval. If the actual sample proportion is very different from .5, the width of the confidence interval may be substantially narrower than originally intended.

In our example, suppose that the production manager wanted to have 90% confidence of estimating the proportion of nonconforming newspapers to within ±.04 of its true value. In addition, since the publisher of the newspaper has not previously undertaken such a survey, no information is available from past data. Therefore p will be set equal to .5.

With these criteria in mind, the sample size needed can be determined in the following manner with 90% confidence ($Z = 1.645$), $e = .04$, and $p = .5$:

$$n = \frac{(1.645)^2(.5)(.5)}{(.04)^2} = 422.82$$

Thus, $n = 423$.

Therefore, in order to be 90% confident of estimating the proportion to within ±.04 of its true value, a sample size of 423 would be needed.

Problems for Section 10.8

10.55 A political pollster would like to estimate the proportion of voters who will vote for the Democratic candidate in a presidential campaign. The pollster would like 90% confidence that her prediction is correct to within ±.04 of the true proportion. What sample size is needed?

● 10.56 A cable television company would like to estimate the proportion of its customers who would purchase a cable television program guide. The company would like to have 95% confidence that its estimate is correct to within ±.05 of the true proportion. Past experience in other areas indicates that 30% of the customers will purchase the program guide. What sample size is needed?

● 10.57 A bank manager wants to be 90% confident of being correct to within ±.05 of the true proportion of depositors who have both savings and checking accounts. What sample size is needed?

10.58 An audit test to establish the frequency of occurrence of failures to follow a specific internal control procedure was to be undertaken. The auditor decides that the maximum tolerable error rate that is permissible is 5%. What size sample is required to achieve a sample precision of ±2% with 99% confidence?

10.59 A large shipment of air filters is received by Joe's Auto Supply Co. The air filters are to be sampled in order to estimate the proportion that are unusable. From past experience, the proportion of unusable air filters is estimated to be .10. How large a random sample should be taken to estimate the true proportion of unusable air filters to within ±.07 with 99% confidence?

10.60 Suppose that Milt's Motors, a competitor of the automobile dealer in Problem 10.41 on page 362, also wanted to conduct a survey to determine the proportion of his customers who own their cars five years after purchasing them. Suppose that he wanted to be 95% confident of being correct to within ±.025 of the true proportion. Note that this dealer *does not have access to the sample results of Problem 10.41*. What sample size is needed?

10.9 Estimation and Sample Size Determination for Finite Populations

10.9.1 Estimating the Mean

In Section 9.4 we saw that when sampling without replacement from finite populations, the **finite population correction (fpc) factor** served to reduce the standard error by a value equal to $\sqrt{(N-n)/(N-1)}$. When estimating population parameters from such samples without replacement, the finite population correction factor should be used for developing confidence interval estimates.

Therefore, the $(1-\alpha) \times 100\%$ confidence interval estimate for the mean would become

$$\bar{X} \pm t_{n-1} \frac{S}{\sqrt{n}} \sqrt{\frac{N-n}{N-1}} \qquad \textbf{(10.7)}$$

In the marketing manager's heating oil usage example, a sample of 35 single-family homes was selected. Suppose that there was a population of 500 single-family homes that were served by the company. Using the finite population correction factor, we would have, with $\bar{X} = 1{,}122.7$ gallons, $S = 295.72$, $n = 35$, $N = 500$, and $t_{34} = 2.0322$ (for 95% confidence):

$$\begin{aligned}
\bar{X} \pm t_{n-1} \frac{S}{\sqrt{n}} \sqrt{\frac{N-n}{N-1}} &= 1{,}122.7 \pm (2.0322)\frac{295.72}{\sqrt{35}}\sqrt{\frac{500-35}{500-1}} \\
&= 1{,}122.7 \pm (101.58)(.9653) \\
&= 1{,}122.7 \pm 98.05 \\
1{,}024.65 \le \mu_x &\le 1{,}220.75
\end{aligned}$$

Here, since more than 5% of the population was to be sampled, the finite population correction factor had a moderate effect on the confidence interval estimate.

10.9.2 Estimating the Proportion

When sampling without replacement, the $(1-\alpha) \times 100\%$ confidence interval estimate of the proportion would be

$$p_s \pm Z\sqrt{\frac{p_s(1-p_s)}{n}}\sqrt{\frac{N-n}{N-1}} \qquad \textbf{(10.8)}$$

In the production manager's study of the nonconformance of newspapers, a sample of 200 was selected from a population of 100,000 newspapers that were printed. The 90% confidence interval estimate would be determined in the following manner when sampling without replacement. We have $p_s = 35/200 = .175$, $Z = 1.645$, $n = 200$, and $N = 100{,}000$. Thus

$$p_s \pm Z\sqrt{\frac{p_s(1-p_s)}{n}}\sqrt{\frac{N-n}{N-1}} = .175 \pm (1.645)\sqrt{\frac{(.175)(.825)}{200}}\sqrt{\frac{100{,}000-200}{100{,}000-1}}$$

$$= .175 \pm (1.645)(.0269)\sqrt{.998}$$

$$= .175 \pm .0442(.999)$$

$$= .175 \pm .0442$$

$$.1308 \le p \le .2192$$

In this case, since the sample was a very small fraction of the population, the correction factor had virtually no effect on the confidence interval estimate (as was computed on page 361).

10.9.3 Determining the Sample Size

Just as the correction factor was used in developing confidence-interval estimates, it should also be used in determining sample size when sampling without replacement. For example, when estimating the mean the sampling error would be

$$e = \frac{Z\sigma_x}{\sqrt{n}}\sqrt{\frac{N-n}{N-1}} \qquad (10.9)$$

and when estimating the proportion the sampling error would be

$$e = Z\sqrt{\frac{p(1-p)}{n}}\sqrt{\frac{N-n}{N-1}} \qquad (10.10)$$

In determining the sample size when estimating the mean, we would have, from Equation (10.5b),

$$n_0 = \frac{Z^2\sigma_x^2}{e^2}$$

where n_0 is the sample size without considering the finite population correction factor.

Applying the correction factor to this results in the actual sample size n, computed from

$$n = \frac{n_0 N}{n_0 + (N - 1)} \tag{10.11}$$

In the marketing manager's survey to estimate the annual consumption of heating oil, the sample size needed in order to be 95% confident of being correct to within ±$50 (assuming a standard deviation of 325 gallons) was 163, since n_0 was computed as 162.31. Using the correction factor in Equation (10.11) leads to the following:

$$n = \frac{(162.31)(500)}{162.31 + (500 - 1)} = 122.72$$

Thus $n = 123$.

Here, since more than 30% of the population was to be sampled, the finite population correction factor had a substantial effect on the sample size—reducing it from 163 to 123. However, in general this may not be the case. For example, we may recall that in order to estimate the true proportion of nonconforming newspapers, the production manager needed a sample size of 423 (since n_0 was computed as 422.82). Using the correction factor leads to

$$n = \frac{n_0 N}{n_0 + (N - 1)}$$

$$n = \frac{(422.82)(100{,}000)}{422.82 + (100{,}000 - 1)} = 421.04$$

Thus $n = 422$.

Here the use of the correction factor made virtually no difference in the sample size selected.

● **The Kalosha Industries Employee Satisfaction Survey** We may also recall that in the Kalosha Industries Employee Satisfaction Survey in Chapter 2 we stated that a sample of 400 employees had to be selected. This sample size is based on satisfying the requirements of those questions that are deemed the most important. In this study, Bud Conley, the Vice-President for Human Resources and the statistician for the B & L Corporation have determined that questions 7 and 9 are the most essential numerical and categorical questions, respectively.

Since the random variable personal income, question 7, is numerical, in order to determine the sample size required for estimating the population mean we use Equations (10.5b) and (10.11). Three quantities are needed: the desired confidence level (Z), the sampling error (e), and the standard deviation (σ_x). After considerable thought and consultation, Bud Conley decided that he would like to have 95% confidence that the estimate of the average personal income is correct to within ±1.5 thousands of dollars of the true value. Based on past surveys, the standard deviation in personal income is estimated as 15.62 thousands of dollars. With this information, the sample size can be determined in the following manner, with $e = 1.5$, $\sigma_x = 15.62$ (estimated), and 95% confidence ($Z = 1.96$):

$$n_0 = \frac{Z^2\sigma_x^2}{e} = \frac{(1.96)^2(15.62)^2}{(1.5)^2} = 416.57$$

Thus,

$$n = \frac{n_0 N}{n_0 + (N - 1)} = \frac{(416.57)(9{,}800)}{416.57 + (9{,}800 - 1)} = 399.62$$

Therefore $n = 400$.

However, before deciding upon the sample size needed for the entire survey, we must evaluate the sample size required for question 9, the categorical variable "how satisfied are you with your job?" This can be found by using Equations (10.6b) and (10.11) after determining three quantities—the confidence level desired (Z), the sampling error (e), and an estimate of the true proportion of employees who are satisfied with their job. Once again, as with the numerical variable, considerable thought was given to determining the desired values. Bud Conley concluded that he would like 90% confidence that the estimate of the true proportion of employees who were satisfied with their jobs (question 9, codes 1 and 2) is correct to within ±.045. Based on experience with similar surveys, the population proportion of employees who are satisfied is assumed to be at least .80. With this information, the sample size can be determined in the following manner with $e = .045$, $p = .80$, and 90% confidence ($Z = 1.645$):

$$n_0 = \frac{Z^2 p(1 - p)}{e^2} = \frac{(1.645)^2(.80)(1 - .20)}{(.045)^2} = 213.81$$

Thus,

$$n = \frac{n_0 N}{n_0 + (N - 1)} = \frac{(213.81)(9{,}800)}{213.81 + (9{,}800 - 1)} = 209.27$$

Therefore $n = 210$.

We have seen that a sample of 400 employees is needed to satisfy the requirements for the most important question involving a numerical variable (personal income) and a sample of 210 employees is required to satisfy the requirements for the most important question pertaining to a categorical variable (job satisfaction). However, since we must satisfy both requirements simultaneously with one sample, the larger sample size of 400 must be utilized for the Employee Satisfaction Survey.

Problems for Section 10.9

● 10.61 Refer to Problems 10.6 and 10.48 on pages 349 and 365. If the shipment contains a total of 2,000 light bulbs
(a) Set up a 95% confidence interval estimate of the true average life of light bulbs in this shipment.
(b) Determine the sample size needed to estimate the average life to within ±20 hours with 95% confidence.

10.62 Refer to Problem 10.46 on page 364. What sample size is necessary if the company has 3,000 employees?

10.63 Refer to Problem 10.39 on page 361. If the bank has 1,000 depositors
(a) Set up a 90% confidence interval estimate of the true proportion of depositors who are paid weekly.
(b) Determine the sample size needed to estimate the true proportion to within ±.05 with 90% confidence.

10.64 Refer to Problems 10.41 and 10.60 on pages 362 and 367. Suppose the population consists of 4,000 owners at *each* auto dealer.
(a) From Problem 10.41, set up a 95% confidence interval estimate of the true proportion of customers who still own their cars five years after they purchased them.
(b) From Problem 10.60, determine what sample size is necessary to estimate the true proportion to within ±.025 with 95% confidence.

10.65 Refer to Problems 10.7 and 10.49 on pages 349 and 365. If the population consists of 2,000 bottles
(a) Set up a 95% confidence interval estimate of the true population average amount of soft drink in each bottle.
(b) Determine the sample size that is necessary to estimate the true average amount to within ±.01 liter with 95% confidence.

10.66 Refer to Problem 10.11 on page 354. If the number of greeting cards in its inventory is equal to 300
(a) Set up a 95% confidence interval estimate of the population average value of all greeting cards that are in its inventory.
(b) Compare the results obtained in (a) with those of Problem 10.11.

10.10 The Kalosha Industries Employee Satisfaction Survey Revisited

Data file EMPSAT.DAT

We may recall from Sections 4.10.2 and 5.6.2 that Bud Conley, the Vice-President for Human Resources, is preparing for a meeting with the representative from the B & L Corporation to discuss the potential contents of an employee-benefits package that is being developed. Among the questions that were of particular concern were:

1. The personal income of Kalosha Industries employees (question 7).
2. The proportion of Kalosha Industries employees who were either very satisfied or moderately satisfied with their jobs (question 9, codes 1 and 2).

Now that we have developed the confidence interval estimate approach in this chapter, we may make inferences about the true population characteristics in terms of personal income and job satisfaction.

From Figure 4.24 on page 151, we have determined that $\overline{X}$, the average personal income (in thousands of dollars) in our sample, is 29.555 and the sample standard deviation, S, is 14.106. If a 95% confidence interval estimate were desired, the critical value for t would be approximately 1.96 since the t distribution with $400 - 1 = 399$ degrees of freedom is approximately equivalent to the normal distribution. Using Equation (10.2), we have

$$\bar{X} \pm t_{n-1} \frac{S}{\sqrt{n}} = 29.555 \pm (1.96) \frac{14.106}{\sqrt{400}}$$

$$= 29.555 \pm 1.382$$

$$28.173 \leq \mu_x \leq 30.937$$

We would conclude with 95% confidence that the average personal income of full-time Kalosha Industries employees is between \$28,173 and \$30,937.

Turning to our second question of interest related to job satisfaction, in Table 6.2 on page 208 we indicated that 356 out of the 400 employees in the sample had stated that they were very satisfied or moderately satisfied with their jobs. A 95% confidence interval estimate of the population proportion of the employees who were satisfied with their jobs can be developed as follows:[3]

$$p_s = 356/400 = .89, \text{ and with 95\% confidence } Z = 1.96$$

From Equation (10.4)

$$p_s \pm Z\sqrt{\frac{p_s(1-p_s)}{n}} = .89 \pm (1.96)\sqrt{\frac{(.89)(.11)}{400}}$$

$$= .89 \pm (1.96)(.0156)$$

$$= .89 \pm .031$$

$$.859 \leq p \leq .921$$

Therefore we can conclude with 95% confidence that between 85.9 and 92.1% of the Kalosha Industries employees are satisfied with their jobs.

Survey Database Project for Section 10.10

Data file EMPSAT.DAT

The following problems refer to the sample data obtained from the questionnaire of Figure 2.6 on pages 28–29 and presented in Table 2.3 on pages 33–40. They should be solved with the aid of an available computer package.

Suppose that you have been hired by Bud Conley, the Vice-President for Human Resources at Kalosha Industries. He has given you a list of questions (see Problems 10.67–10.83) for which he wishes to make inferences about the entire population of Kalosha Industries full-time employees. He has decided to use a 95% level of confidence.

ACTION *Write an executive summary to Bud Conley discussing your findings based on:*

10.67 The average number of hours worked last week at all jobs (question 1).

10.68 The average age (question 3).

10.69 The average before taxes total family income (question 8).

10.70 The average number of years of full-time employment (question 15).

10.71 The average number of years employed by Kalosha Industries (question 16).

10.72 The proportion of employees who are classified as managerial or professional (question 2, codes 1 and 2).

10.73 The proportion of employees who would stop working and retire if they became very rich (question 10, code 1).

10.74 The proportion of employees for whom enjoying the work is most important (question 11, code 5).

10.75 The proportion of employees who feel they are very likely or likely to be promoted in the next five years (question 18, codes 1 and 2).

10.76 The proportion of employees who have advanced rapidly or made steady advances (question 20, codes 1 and 2).

10.77 The proportion of employees whose job allows them to take part in making decisions either always or part of the time (question 21, codes 1 and 2).

10.78 The proportion of employees who participate in budgetary decisions (question 22, code 1).

10.79 The proportion of employees who are very proud or somewhat proud to be working for Kalosha Industries (question 23, codes 1 and 2).

10.80 The proportion of employees who would be very likely or likely to turn down a job for more pay to stay at Kalosha Industries (question 24, codes 1 and 2).

10.81 The proportion of employees who describe relations between management and employees as very good or good (question 25, codes 1 and 2).

10.82 The proportion of employees who describe relations among coworkers and colleagues as very good or good (question 26, codes 1 and 2).

10.83 The proportion of employees who state that on the job training was very important or important to the job (question 28, codes 1 and 2).

10.11 Estimation, Sample Size Determination, and Ethical Issues

Ethical issues relating to the selection of samples and the inferences that accompany them from sample surveys can arise in several ways. The major ethical issue relates to whether or not confidence interval estimates are provided along with the point estimates of the sample statistics obtained from a survey. To just indicate a point estimate of a sample statistic without also including the confidence interval limits (typically set at 95%), the sample size used, and an interpretation of the meaning of the confidence interval in terms that a layman can understand can raise ethical issues because of their omission. Failure to include a confidence interval estimate might mislead the user of the survey results into thinking the point estimate obtained from the sample was all that is needed to predict the population characteristics with certainty. Thus, it is important that the interval estimate be indicated in a prominent place in any written communication along with a simple explanation of the meaning of the confidence interval. In addition, the size of the sample should be highlighted so that the reader clearly understands the magnitude of the survey that has been undertaken.

One of the most common areas where ethical issues concerning estimation from sample surveys occurs is in the publication of the results of political polls. All too often the results of the polls are highlighted on page one of the newspaper, and the sampling error involved along with the methodology used is printed on the page where the article is typically continued (often in the middle of the newspaper). During the 1992 Presidential campaign, *The New York Times* (October 7, 1992) presented some of these issues by publishing the results of five different polls that were taken between October 1 and October 4, 1992. Table 10.5 summarizes the results of these five polls.

Table 10.5 Five polls of the 1992 presidential election taken between October 1 and October 4, 1992.

Poll	Candidate		
	Bush	Clinton	Perot
New York Times/CBS News	38%	46%	7%
Washington Post/ABC News	35%	48%	9%
Gallup for CNN/*USA Today*	35%	47%	10%
Harris	36%	53%	9%
Gallup for *Newsweek*	36%	44%	14%

Source: The New York Times, October 7, 1992, p. A.1

Although there are many possible reasons for the differences in the results, including those discussed in Section 2.11, it is also quite possible that most of the differences are merely due to sampling error. If we assume that the sample sizes were sufficiently large to provide a sampling error of ±3.5% with 95% confidence, confidence intervals for each of the five different polls could be obtained. Confidence intervals for the percentage of voters favoring candidate Bill Clinton are presented in Table 10.6.

Table 10.6 Confidence interval estimates based on five polls of the 1992 presidential election taken between October 1 and October 4, 1992.

Poll	Candidate		
	Clinton	Lower Limit	Upper Limit
New York Times/CBS News	46%	42.5	49.5
Washington Post/ABC News	48%	44.5	51.5
Gallup for CNN/*USA Today*	47%	43.5	50.5
Harris	53%	49.5	56.5
Gallup for *Newsweek*	44%	40.5	47.5

We may observe from Table 10.6 that all of the polls result in confidence intervals whose major difference appears to be sampling error. Thus, in summary, in order to ensure an ethical interpretation of statistical results, the confidence levels, sample size, and confidence limits should be made available for any survey that has been conducted.

10.12 Estimation and Statistical Inference: Review and a Preview

As observed in the summary chart on page 376, in this chapter we have developed two approaches for estimating the characteristic of a population, confidence interval estimation and bootstrapping. We have also investigated how we can determine the sample size that is necessary for a survey and have considered the finite population correction factor.

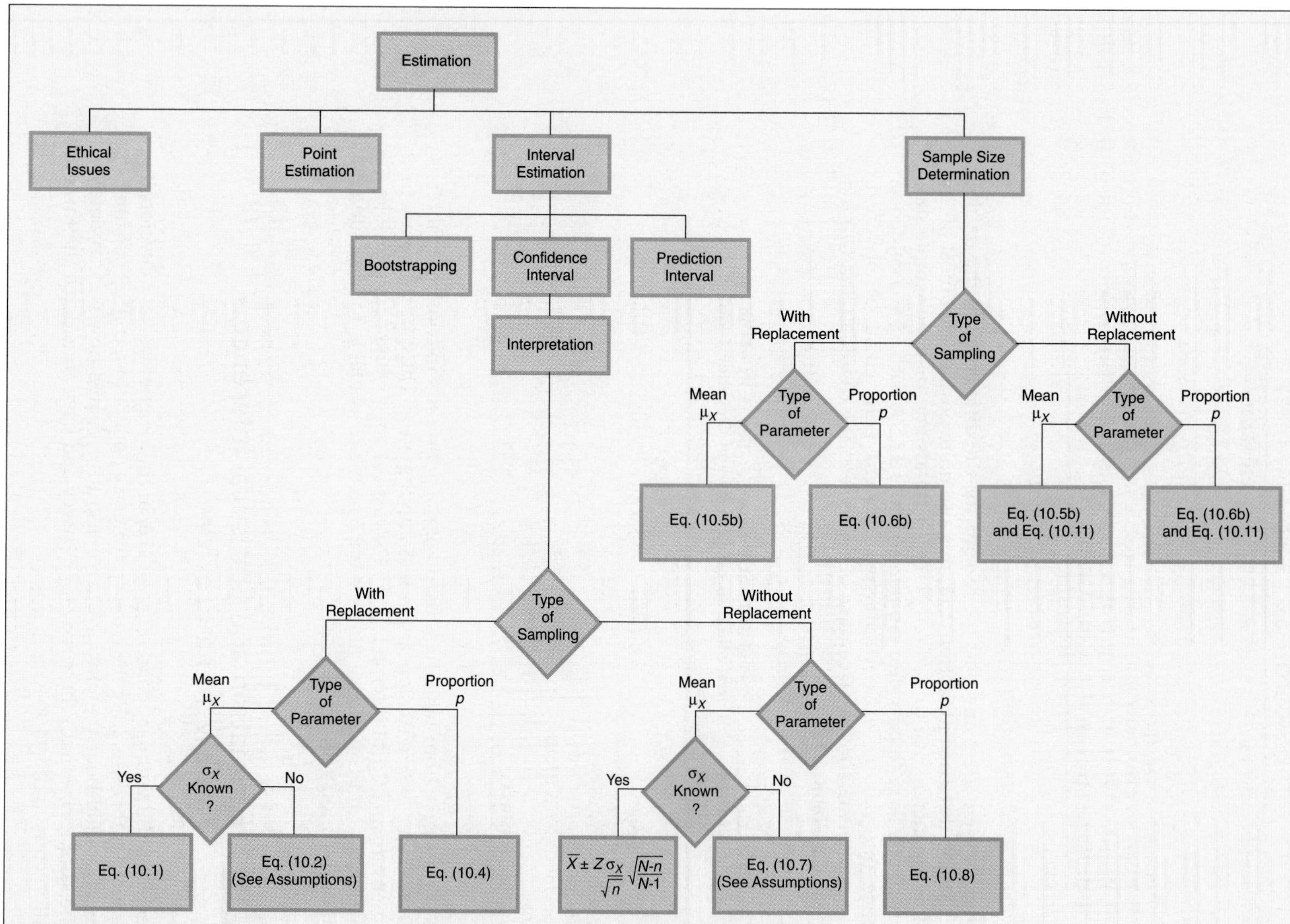

Chapter 10 summary chart.

On page 344 of Section 10.1 you were given a list emphasizing the important points to be discussed in the chapter. Check over the list now to see whether you feel you have an understanding of these key points. To be sure you should be able to answer the following conceptual questions:

1. Why is it that we can never really have 100% confidence of correctly estimating the population characteristic of interest?
2. When is the *t* distribution used in developing the confidence interval estimate for the mean?
3. Under what circumstances might you use the bootstrapping approach rather than the traditional confidence interval estimation?
4. How does the prediction interval differ from the confidence interval?
5. Why is it true that for a given sample size *n*, an increase in confidence is achieved by widening (and making less precise) the confidence interval obtained?
6. How does sampling without replacement from a finite population affect the confidence interval estimate and the sample size necessary?

Now that we have made estimates of population characteristics such as the mean and proportion using confidence intervals, in the next five chapters we turn to a hypothesis-testing approach in which we are making decisions about population parameters.

Getting It All Together

Key Terms

bootstrapping estimation *356*
confidence interval estimate *346*
critical value *347*
degrees of freedom *352*
finite population correction factor *368*
interval estimate *344*
level of confidence *347*
point estimate *344*
prediction interval *359*
resampling distribution *356*
sampling error *363*
Student's *t* distribution *350*

Chapter Review Problems

10.84 Referring to Problem 10.6 (page 349), set up 99% and 90% confidence interval estimates of the true average life of light bulbs in the shipment. Compare and discuss the meaning of the three confidence interval estimates.

● 10.85 A market researcher for a large consumer electronics company would like to study television viewing habits of residents of a particular small city. A random sample of 40 respondents was selected, and each respondent was instructed to keep a detailed record of all television viewing in a particular week. The results were as shown on page 378:

Amount of viewing per week:
$\bar{X} = 15.3$ hours, $S = 3.8$ hours.
27 respondents watched the evening news on at least three weeknights.

(a) Set up a 95% confidence interval estimate for the average amount of television watched per week in this city.
(b) Set up a 95% confidence interval estimate for the proportion of respondents who watch the evening news on at least three nights per week.
(c) Set up a 95% prediction interval estimate of the amount of viewing per week of a future individual respondent.
(d) **ACTION** How can the market researcher use the results of (a) and (c) to determine a plan for television advertising? Write a letter to the vice-president of marketing.

If the market researcher wanted to take another survey in a different city
(e) What sample size is required if he wishes to be 95% confident of being correct to within ±2 hours and assumes the population standard deviation is equal to 5 hours?
(f) What sample size is needed if he wishes to be 95% confident of being within ±.035 of the true proportion who watch the evening news on at least three weeknights *if no previous estimate were available?*

10.86 The real estate assessor for a county government wishes to study various characteristics concerning single-family houses in the county. A random sample of 70 houses revealed the following:

Heated area of the house:
$\bar{X} = 1{,}759$ sq ft, $S = 380$ sq ft.
42 houses had central air conditioning.

(a) Set up a 99% confidence interval estimate of the population average heated area of the house.
(b) Set up a 95% confidence interval estimate of the population proportion of houses that have central air conditioning.
(c) Set up a 99% prediction interval estimate of the heated area of a future individual house.

10.87 The personnel director of a large corporation wished to study absenteeism among clerical workers at the corporation's central office during last year. A random sample of 25 clerical workers revealed the following:

Absenteeism:
$\bar{X} = 9.7$ days, $S = 4.0$ days
12 employees were absent more than 10 days.

(a) Set up a 95% confidence interval estimate of the average number of days absent for clerical workers last year.
(b) Set up a 95% confidence interval estimate of the proportion of clerical workers absent more than 10 days last year.
(c) Set up a 95% prediction interval estimate of the number of days absent for a future individual clerical worker.
(d) How can the personnel director use the results of (a) and (c) to determine how absenteeism can be reduced in the coming year?

If the personnel director also wishes to take a survey in a branch office
(e) What sample size is needed if the director wishes to be 95% confident of being correct to ±1.5 days and the population standard deviation is assumed to be 4.5 days?
(f) What sample size is needed if the director wishes to be 90% confident of being correct to within ±.075 of the true proportion of workers who are absent more than 10 days if no previous estimate is available?

10.88 The market research director for Dotty's department store would like to study women's spending per year on cosmetics. A survey is to be sent to a sample of the store's credit card holders to determine:

- The average yearly amount that women spend on cosmetics.
  (a) If the market researcher wanted to have 99% confidence of estimating the true population average to within ±$5 and the standard deviation is assumed to be $18 (based on previous surveys), what sample size is needed?
- The population proportion of women who purchase their cosmetics primarily from Dotty's department store.
  (b) If the market researcher wishes to have 90% confidence of estimating the true proportion to within ±.045, what sample size is needed?
  (c) **ACTION** Based on the results in (a) and (b), how many of the store's credit card holders should be sampled? Explain.

10.89 The branch manager of an outlet of a large nationwide chain of bookstores wanted to study characteristics of customers of her store, which was located near the campus of a large state university. In particular, she decided to focus on two variables: the amount of money spent by customers and whether the customers would consider purchasing educational videotapes relating to specific courses such as statistics, accounting, or calculus or graduate preparation exams such as GMAT, GRE, or LSAT. The results from a sample of 70 customers were as follows.

Amount of money spent:
$\overline{X} = \$28.52, \; S = \$11.39$

A total of 28 customers stated that they would consider purchasing educational videotapes.

(a) Set up a 95% confidence interval estimate of the population average amount of money spent in the bookstore.
(b) Set up a 90% confidence interval estimate of the proportion of customers who would consider purchasing educational videotapes.
(c) Set up a 95% prediction interval of the amount of money spent by a future individual student.

Suppose that the branch manager of another store from a *different* bookstore chain wishes to conduct a similar survey in his store (which is located near another university).

(d) If he wanted to have 95% confidence of estimating the true population average amount spent in his store to within ±$2 and the standard deviation is assumed to be $10, what sample size is needed?
(e) If he wanted to have 90% confidence of estimating the true proportion of shoppers who would consider the purchase of videotapes to within ±.04, what sample size is needed?
(f) Based on your answers to (d) and (e), what size sample should be taken?

10.90 The branch manager of an outlet (store #1) of a large nationwide chain of pet supply stores wanted to study characteristics of customers of her store. In particular, she decided to focus on two variables: the amount of money spent by customers, and whether the customers owned only one dog, only one cat, or more than one dog and/or one cat. The results from a sample of 70 customers were as follows.

Amount of money spent:
$\overline{X} = \$21.34, \; S = \$9.22$

A total of 37 customers owned only a dog;
26 customers owned only a cat;
7 customers owned at least one dog and at least one cat.

(a) Set up a 95% confidence interval estimate of the population average amount of money spent in the pet supply store.
(b) Set up a 90% confidence interval estimate of the proportion of customers who own *only* a cat.
(c) Set up a 95% prediction interval estimate of the amount of money spent by a future individual customer.

Suppose that the branch manager of another store (outlet #2) wishes to conduct a similar survey in his store (and does not have any access to the information generated by the owner of outlet store #1).

(d) If he wanted to have 95% confidence of estimating the true population average amount spent in his store to within ±$1.50 and the standard deviation is assumed to be $10, what sample size is needed?
(e) If he wanted to have 90% confidence of estimating the true proportion of customers who own only a cat to within ±.045, what sample size is needed?
(f) Based on your answers to (d) and (e), what size sample should be taken?

10.91 The owner of a restaurant serving continental food wanted to study characteristics of customers of her restaurant. In particular, she decided to focus on two variables, the amount of money spent by customers and whether or not customers ordered dessert. The results from a sample of 60 customers were as follows:

Amount spent:
$\bar{X} = \$38.54, \; S = \$7.26$

18 customers purchased dessert.

(a) Set up a 95% confidence interval estimate of the population average amount of money spent per customer in the restaurant.
(b) Set up a 90% confidence interval estimate of the proportion of customers who purchase dessert.
(c) Set up a 95% prediction interval of the money spent by a future individual customer.

Suppose that the owner of a competing restaurant wishes to conduct a similar survey in his restaurant (and does not have access to the information obtained by the owner of the first restaurant).

(d) If he wanted to have 95% confidence of estimating the true population average amount spent in his restaurant to within ±$1.50 and the standard deviation is assumed to be $8, what sample size is needed?
(e) If he wanted to have 90% confidence of estimating the true proportion of customers who purchase dessert to within ±.04, what sample size is needed?
(f) Based on your answers to (d) and (e), what size sample should be taken?

10.92 A representative for a large chain of hardware stores was interested in testing the product claims of a manufacturer of "ice melt" that was reported to melt snow and ice at temperatures as low as 15 degrees Fahrenheit. A shipment of 400 five-pound bags was purchased by the chain for distribution. The representative wanted to know with 95% confidence, within ±.05, what proportion of bags of ice melt would perform the job as claimed by the manufacturer.

(a) How many bags does the representative need to test? What assumption should be made concerning the true proportion in the population? (This is called *destructive testing;* that is, the product being tested is destroyed by the test and is then unavailable to be sold.)

(b) If the representative actually tested 50 bags, out of which 42 did the job as claimed, construct a 95% confidence interval estimate for the population proportion that will do the job as claimed.

(c) How can the representative use the results of (b) to determine whether to sell the "ice melt" product?

Interchapter Problems

10.93 Referring to the data of Problem 3.7 on page 60

Data file PEANUT.DAT

(a) Set up 95% confidence interval estimates of the population average cost and average amount of sodium per serving of peanut butter.

(b) Set up 95% prediction interval estimates of the cost and amount of sodium per serving of a future brand of peanut butter.

(c) **ACTION** Use the results of (a) and (b) as part of a newspaper article about peanut butter.

10.94 Referring to the data of Problem 3.8 on page 61

Data file ROAD.DAT

(a) Set up 90% confidence interval estimates of the average amount of time to go from 0 to 60 mph for German-made and Japanese-made automobile models.

(b) Set up 90% prediction interval estimates of the time for a future individual German-made automobile model and a future Japanese-made automobile model.

(c) **ACTION** Based on your results in (a) and (b), what would you tell your friend who was interested in this feature when making a decision to purchase an automobile? Write her a letter.

10.95 Referring to the data of Problem 3.9 on page 62

Data file SHAMPOO.DAT

(a) Set up 95% confidence interval estimates of the average cost per ounce for shampoos labeled for normal hair and for fine hair.

(b) Set up 95% prediction interval estimates of the cost per ounce for a future shampoo labeled for normal hair and a future shampoo labeled for fine hair.

10.96 Referring to the data of Problem 3.18 on page 66

Data file BULBS.DAT

(a) Set up 95% confidence interval estimates of the average life for light bulbs produced by manufacturer A and manufacturer B.

(b) Set up 95% prediction interval estimates of the life for a future light bulb produced by manufacturer A and a future light bulb produced by manufacturer B.

10.97 Referring to the data of Problem 4.80 on page 163

Data file PREP.DAT

(a) Set up 95% confidence interval estimates of the population average tuition for Northeast prep schools and for Midwest prep schools.

(b) Set up 95% prediction interval estimates of the tuition for future individual prep schools in the Northeast and in the Midwest.

(c) **ACTION** Based on your results in (a) and (b), what would you tell your cousin who is thinking of attending a prep school in the Northeast or Midwest? Write a letter.

Collaborative Learning Mini-Case Projects

Note: The class is to be divided into groups of three or four students each. One student is initially selected to be project coordinator, another student is project recorder, and a third student is the project timekeeper. In order to give each student experience in developing teamwork and leadership skills, rotation of positions should follow each project. At the beginning of each project, the students should work silently and individually for a short, specified period of time. Once each student has had the opportunity to study the issues and reflect on his/her possible answers, the group is convened and a group discussion follows. If all the members of a group agree with the solutions, the coordinator is responsible for submitting the team's project solution to the instructor with student signatures indicating such agreement. On the other hand, if one or more team members disagree with the solution offered by the majority of the team, a minority opinion may be appended to the submitted project, with signature(s) on it.

Data file CEREAL.DAT

CL 10.1 Refer to CL3.2 on page 101. Set up all appropriate estimates of population nutritional characteristics of ready-to-eat cereals. Include these estimates in any written and oral presentation to be made to the food editor of the magazine.

Data file FRAGRAN.DAT

CL10.2 Refer to CL3.3 on page 102. Set up all appropriate estimates of population characteristics of fragrances. Include these estimates in any written and oral presentation to be made to the marketing director.

Data file CAMERA.DAT

CL10.3 Refer to CL3.4 on page 102. Set up all appropriate estimates of population characteristics of cameras. Include these estimates in any written and oral presentation to be made to the travel editor.

Endnotes

1. It is for this reason that the denominator of the sample variance is $n - 1$ instead of n, so that S^2 will be an unbiased estimator of σ_x^2; that is, if

$$S^2 = \frac{\sum_{i=1}^{n}(X_i - \overline{X})^2}{n-1} \quad \text{and} \quad \sigma_x^2 = \frac{\sum_{i=1}^{n}(X_i - \mu_x)^2}{N}$$

then $E(S^2) = \sigma_x^2$, and therefore S^2 is an unbiased estimator of σ_x^2.

2. We use Z instead of t because (1) to determine the critical value of t we would need to know the sample size, which we don't know yet, and (2) for most studies the sample size needed will be large enough that the normal distribution is a good approximation to the t distribution.
3. The finite population correction factor is not used here since the sample size of 400 is less than 5% of the population of 9,800.

References

1. Cochran, W. G., *Sampling Techniques*, 3d ed. (New York: Wiley, 1977).
2. Diaconis, P., and B. Efron, "Computer-Intensive Methods in Statistics," *Scientific American*, 248, 1983, pp. 116–130.
3. Efron, B., *The Jackknife, the Bootstrap, and Other Resampling Plans* (Philadelphia: Society for Industrial and Applied Mathematics, 1982).
4. Fisher, R. A., and F. Yates, *Statistical Tables for Biological, Agricultural and Medical Research*, 5th ed. (Edinburgh: Oliver & Boyd, 1957).
5. Gunter, B. "Bootstrapping: How to Make Something from Almost Nothing and Get Statistically Valid Answers. Part I: Brave New World," *Quality Progress*, 24, December 1991, pp. 97–103.
6. Hahn, G. J., and W. Nelson, "A Survey of Prediction Intervals and Their Applications," *Journal of Quality Technology*, 5, 1973, pp. 178–188.
7. Kirk, R. E., ed., *Statistical Issues: A Reader for the Behavioral Sciences* (Belmont, CA: Wadsworth, 1972).
8. Larsen, R. L., and M. L. Marx, *An Introduction to Mathematical Statistics and Its Applications*, 2d ed. (Englewood Cliffs, NJ: Prentice-Hall, 1986).
9. Scheuer, E. M., "Let's Teach More about Prediction," *Proceedings of the Statistical Education Section of the American Statistical Association*, 1990.
10. Snedecor, G. W., and W. G. Cochran, *Statistical Methods*, 7th ed. (Ames, IA: Iowa State University Press, 1980).

chapter 11

Fundamentals of Hypothesis Testing

CHAPTER OBJECTIVES

To develop hypothesis-testing methodology as a technique for analyzing differences and making decisions; to determine the risks involved in making these decisions based only on sample information; and to study the interrelationship of these risks with the sample size used.

11.1 Introduction

In Chapter 9 we began our discussion of statistical inference by developing the concept of a sampling distribution. In Chapter 10 we considered enumerative studies in which a statistic (such as the sample mean or sample proportion) obtained from a random sample is used to *estimate* its corresponding population parameter.

In this chapter we will begin to focus on another phase of statistical inference that is also based on sample information—hypothesis testing. In particular, we will develop a step-by-step methodology that will enable us to make inferences about a specific value of a population's parameter by *analyzing differences* between the results we actually observe (that is, our sample statistic) and the results we would expect to obtain if some underlying hypothesis was actually true. In addition to developing the **hypothesis-testing methodology** as a technique for analyzing differences and making decisions, we will also be evaluating the risks involved in making these decisions based only on sample information and we will be studying the interrelationship of these risks with the sample size used. Emphasis here is placed on the fundamental and conceptual underpinnings of hypothesis-testing methodology. In the four chapters that follow, numerous hypothesis-testing procedures will be presented that are frequently employed in the analysis of data obtained from studies and experiments designed under a variety of conditions.

Upon completion of this chapter, you should be able to:

1. Distinguish between the null (H_0) and alternative (H_1) hypotheses
2. Distinguish between the risks of committing Type I and Type II errors
3. Understand the concept of the power of a test
4. Distinguish between one-tailed and two-tailed tests
5. Understand the *p* value approach to hypothesis testing
6. Appreciate the connection between confidence intervals and hypothesis tests
7. Understand the interrelationship of α, β, n, and the type of test
8. Apply the step-by-step hypothesis testing methodology

11.2 Hypothesis-Testing Methodology

In order to develop hypothesis-testing methodology, we will focus on some issues based on the cereal box packaging process discussed in Chapters 9 and 10. For example, the production manager is concerned with evaluating whether or not the process is working in a way that ensures that, on average, the proper amount of cereal (i.e., 368 grams) is being filled in each box. He decides to select a random sample of 25 boxes from the packaging process and examine their weights to determine how close each of these boxes comes to the company's specification of 368 grams on average per box. The production manager hopes to find that the process is working properly. However, he might find that the sampled boxes weigh too little or perhaps too much and feel that he should halt production until a maintenance crew can examine and, if necessary, repair or replace a machine part. Therefore, by analyzing the differences between the weights obtained from the sample and the 368 gram expectation obtained from the company's specifications, a decision based on this sample information is going to be made and one of the following two conclusions will be reached:

1. The average fill in the entire cereal box packaging process is 368 grams. No corrective action is needed.
2. The average fill is not 368 grams; either it is less than 368 grams or it is more than 368 grams. Corrective action is needed.

11.2.1 The Null and Alternative Hypotheses

Hypothesis testing begins with some theory, claim, or assertion about a particular parameter of a population. For purposes of statistical analysis, the production manager chooses as his initial hypothesis that the process is in control; that is, the average fill is 368 grams and no corrective action is needed. The hypothesis that the population parameter is equal to the claimed company specification is referred to as the **null hypothesis**.

A null hypothesis is always one of *status quo* or no difference. We commonly identify the null hypothesis by the symbol $\boldsymbol{H_0}$. Our production manager would establish as his null hypothesis that the packaging process is in control and working properly, that the mean amount of cereal per box is the 368 gram company specification. This can be stated as:

$$H_0: \quad \mu_x = 368$$

Note that even though the production manager only has information from the sample, the null hypothesis is written in terms of the population parameter. This is because he is interested in the entire packaging process, that is, (the population of) all cereal boxes being filled. The sample statistics will be used to make inferences about the condition of the entire packaging process. Parallel to the American legal system, in which innocence is presumed until guilt is proven, the theoretical basis of hypothesis testing requires that the null hypothesis be considered true until evidence, such as the results observed from the sample data, indicates that it is false. If the null hypothesis is considered false, something else must be true.

Whenever we specify a null hypothesis we must also specify an **alternative hypothesis**, or one that must be true if the null hypothesis is found to be false. The alternative hypothesis ($\boldsymbol{H_1}$) is the opposite of the null hypothesis (H_0). For the production manager, this can be stated as:

$$H_1: \quad \mu_x \neq 368$$

The alternative hypothesis represents the conclusion that would be reached if there were sufficient evidence from sample information to decide that the null hypothesis is unlikely to be true and therefore we reject it. In our example, if the weights of the sampled boxes were sufficiently above or below the expected 368 gram average specified by the company, the production manager would reject the null hypothesis in favor of the alternative hypothesis that the average amount of fill is different from 368 grams. He would therefore stop production and take whatever action necessary to correct the problem.

Hypothesis-testing methodology is designed so that our rejection of the null hypothesis is based on evidence from the sample that our alternative hypothesis is far more likely to be true. However, failure to reject the null hypothesis is not proof that it is true. We can never prove the null hypothesis is correct because we are basing our decision only on the sample information, not the entire population.

Therefore, if we fail to reject the null hypothesis we can only conclude that there is insufficient evidence to warrant its rejection.

To summarize some key points:

- The null hypothesis (H_0) is the hypothesis that is always tested.
- The alternative hypothesis (H_1) is set up as the opposite of the null hypothesis and represents the conclusion supported if the null hypothesis is rejected.

In what is known as *classical* hypothesis-testing methodology (see References 1 and 2)

- The null hypothesis always refers to a specified value of the *population parameter* (such as μ_x), not a *sample statistic* (such as $\overline{X}$).
- The statement of the null hypothesis *always* contains an equal sign regarding the specified value of the parameter (i.e., H_0: $\mu_x = 368$ grams).
- The statement of the alternative hypothesis *never* contains an equal sign regarding the specified value of the parameter (i.e., H_1: $\mu_x \neq 368$ grams).

11.2.2 The Critical Value of the Test Statistic

We can develop the logic behind the hypothesis-testing methodology by contemplating how we can determine, based only on sample information, the plausibility of the null hypothesis. Our production manager had stated as his null hypothesis that the average amount of cereal per box over the entire packaging process is 368 grams (i.e., the population parameter specified by the company). He then collected a sample of boxes from the packaging process, weighed each box, and computed the sample mean. We recall that a statistic from a sample is an estimate of the corresponding parameter from the population from which the sample was drawn and will likely differ from the actual parameter value because of chance or sampling error. Therefore, even if the null hypothesis were in fact true, the sample statistic would not necessarily be equal to the corresponding population parameter. Nevertheless, under such circumstances, we would expect them to be very similar to each other in value. In such a situation, there would be no evidence to reject the null hypothesis. If, for example, the sample average were 367.6, we would be inclined to conclude that the average has not changed (that is, $\mu_x = 368$), since the sample mean is very close to the hypothesized value of 368. Intuitively, we would be thinking that it is not unlikely that we could obtain a sample mean of 367.6 from a population whose mean was 368. On the other hand, if there was a large discrepancy between the value of the statistic and its corresponding hypothesized parameter, our instinct would be to conclude that the null hypothesis is implausible or unlikely to be true. For example, if the sample average were 320, our instinct would be to conclude that the average is not 368 (that is, $\mu_x \neq 368$), since the sample mean is very far from the hypothesized value of 368. In such a case we would be reasoning that it is very unlikely that the sample mean of 320 could be obtained if the population mean were really 368 and, therefore, it would be more reasonable to conclude that the population mean is not equal to 368. Here we would reject the null hypothesis. In either case, our decision would be reached because of our belief that randomly selected samples are truly representative of the underlying populations from which they were drawn.

Unfortunately, the decision-making process is not always so clear-cut and cannot be left to an individual's subjective judgment as to the meaning of "very close" or "very different." It would be arbitrary for us to determine what is very close and what is very different without using operational definitions. Hypothesis-testing methodology provides operational definitions for evaluating such differences and enables us to quantify the decision-making process so that the probability of obtaining a given sample result if the null hypothesis were true can be found. This is achieved by first determining the sampling distribution for the sample statistic (i.e., the sample mean) and then computing the particular *test statistic* based on the given sample result. Since the sampling distribution for the test statistic often follows a well-known statistical distribution, such as the normal or *t* distribution, we can use these distributions to determine the likelihood of a null hypothesis being true.

11.2.3 Regions of Rejection and Nonrejection

The sampling distribution of the test statistic is divided into two regions, a **region of rejection** (sometimes called the **critical region**) and a **region of nonrejection** (see Figure 11.1). If the test statistic falls into the region of nonrejection, the null hypothesis cannot be rejected. In our example, the production manager would conclude that the average amount filled has not changed. If the test statistic falls into the rejection region, the null hypothesis will be rejected. Here the production manager would conclude that the population mean is not 368.

The region of rejection may be thought of as consisting of the values of the test statistic that are unlikely to occur if the null hypothesis is true. On the other hand, these values are not so unlikely to occur if the null hypothesis is false. Therefore, if we observe a value of the test statistic that falls into this *critical region*, we reject the null hypothesis because that value would be unlikely if the null hypothesis were true.

In order to make a decision concerning the null hypothesis, we must first determine the **critical value** of the test statistic. The critical value divides the nonrejection region from the rejection region. However, the determination of this critical value depends on the size of the rejection region. As we will see in the next section, the size of the rejection region is directly related to the risks involved in using only sample evidence to make decisions about a population parameter.

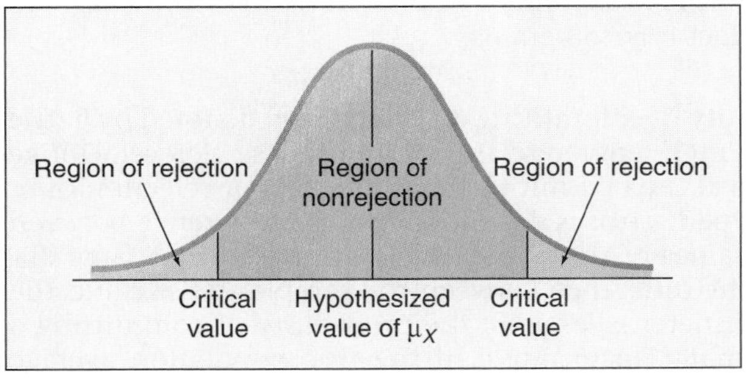

Figure 11.1
Regions of rejection and nonrejection in hypothesis testing.

11.2.4 Risks in Decision Making Using Hypothesis-Testing Methodology

When using a sample statistic to make decisions about a population parameter there is a risk that an incorrect conclusion will be reached. Indeed, two different types of errors can occur when applying hypothesis-testing methodology:

> A **Type I error** occurs if the null hypothesis H_0 is rejected when in fact it is true and should not be rejected.
>
> A **Type II error** occurs if the null hypothesis H_0 is not rejected when in fact it is false and should be rejected.

In our cereal box packaging example, the Type I error would occur if the production manager concluded (based on sample information) that the average population amount filled was *not* 368 when in fact it was 368. On the other hand, the Type II error would occur if he concluded (based on sample information) that the average population amount filled was 368 when in fact it was not 368.

The Level of Significance The probability of committing a Type I error, denoted by α (the lowercase Greek letter alpha), is referred to as the **level of significance** of the statistical test. Traditionally, the statistician controls the Type I error rate by deciding the risk level α he or she is willing to tolerate in terms of rejecting the null hypothesis when it is in fact true. Since the level of significance is specified before the hypothesis test is performed, the risk of committing a Type I error, α, is directly under the control of the individual performing the test. Researchers have traditionally selected α levels of .05 or smaller. The choice of selecting a particular risk level for making a Type I error is dependent on the cost of making a Type I error. Once the value for α is specified, the size of the rejection region is known, since α is the probability of rejection under the null hypothesis. From this fact the critical value or values that divide the rejection and nonrejection regions can be determined.

The Confidence Coefficient The complement $(1 - \alpha)$ of the probability of a Type I error is called the confidence coefficient, which, when multiplied by 100 percent, yields the confidence level that we studied in Section 10.2.

> The **confidence coefficient**, denoted by $1 - \alpha$, is the probability that the null hypothesis H_0 is not rejected when in fact it is true and should not be rejected.

In terms of hypothesis-testing methodology, this coefficient represents the probability of concluding that the specified value of the parameter being tested under the null hypothesis may be plausible. In our cereal box packaging example, the confidence coefficient measures the probability of concluding that the average fill per box is 368 grams when in fact it is 368 grams.

The β Risk The probability of committing a Type II error, denoted by β (the lowercase Greek letter beta), is often referred to as the consumer's risk level. Unlike the Type I error, which statistical tests permit us to control by our selection of α, the probability of making a Type II error is dependent on the difference between the hypothesized and actual values of the population parameter. Since large differences are easier to find, if the difference between the sample statistic and the corresponding population parameter is large, β, the probability of committing a Type II error, will likely be small. For example, if the true population average (which is unknown to us) were 320 grams, there would be a small chance (β) of concluding that the average had not changed from 368. On the other hand, if the difference between the statistic and the corresponding parameter value is small, the probability of committing a Type II error will likely be large. Thus if the true population average were really 367 grams, there would be a high probability of concluding that the population average amount filled had not changed from the specified 368 grams (and we would be making a Type II error).

The Power of a Test The complement ($1 - \beta$) of the probability of a Type II error is called the power of a statistical test.

> The **power** of a statistical test, denoted by $1 - \beta$, is the probability of rejecting the null hypothesis when in fact it is false and should be rejected.

In our cereal box packaging example, the power of the test is the probability of concluding that the average amount of fill is not 368 grams when in fact it is actually not 368. A more detailed discussion of the power of a statistical test will be presented in Section 11.9.

Risks in Decision Making: A Delicate Balance Table 11.1 illustrates the results of the two possible decisions (reject H_0 or do not reject H_0) that can occur in any hypothesis test. Depending on the specific decision, one of two types of errors may occur[1] or one of two types of correct conclusions may be reached.

Table 11.1 **Hypothesis testing and decision making.**

	Actual Situation	
Statistical Decision	H_0 *True*	H_0 *False*
Do not reject H_0	Confidence ($1 - \alpha$)	Type II error (β)
Reject H_0	Type I error (α)	Power ($1 - \beta$)

One way in which we can control the probability of making a Type II error in a study is to increase the size of the sample. Larger sample sizes will generally permit us to detect even very small differences between the sample statistics and the population parameters. For a given level of α, increasing the sample size will decrease β and therefore increase the power of the test to detect that the null hypothesis H_0 is false. Unfortunately, however, there is always a limit to our resources. Thus for a given sample size we must consider the tradeoffs between the two possible types of errors. Since we can directly control our risk of Type I error, we can reduce our risk by selecting a lower level for α (for example, .01 instead of .05). However, when α is decreased, β will be increased so a reduction in risk of Type I error will result in an increased risk of Type II error. If, on the other hand, we wish to reduce β, our risk of Type II error, we could select a larger value for α (for example, .05 instead of .01).

In our cereal box packaging example, the risk of a Type I error involves concluding that the average amount per box has changed from the hypothesized 368 grams when in fact it has not changed. The risk of a Type II error involves concluding that the average amount per box has not changed from the hypothesized 368 grams when in truth it has changed. The choice of reasonable values for α and β depends on the costs inherent in each type of error. For example, if it were very costly to make changes from the *status quo*, then we would want to be very sure that a change would be beneficial, so the risk of a Type I error might be most important and would be kept very low. On the other hand, if we wanted to be very certain of detecting changes from a hypothesized mean, the risk of a Type II error would be most important and we might choose a higher level of α.

Problems for Section 11.2

11.1 Why is it possible for the null hypothesis to be rejected when in fact it is true?

11.2 For a given sample size, if α is reduced from .05 to .01, what will happen to the β?

11.3 Why is it possible that the null hypothesis will not always be rejected when it is false?

11.4 What is the relationship of α to the Type I error?

11.5 What is the relationship of β to the Type II error?

11.6 For H_0: $\mu_x = 100$, H_1: $\mu_x \neq 100$, and for a sample of size n, β will be larger if the actual value of μ_x is 90 than if the actual value of μ_x is 75. Why?

11.7 In the American legal system, a defendant is presumed innocent until proven guilty. Consider a null hypothesis H_0 that the defendant is innocent and an alternative hypothesis H_1 that the defendant is guilty. A jury has two possible decisions: convict the defendant (i.e., reject the null hypothesis) or do not convict the defendant (i.e., do not reject the null hypothesis). Explain the meaning of the risks of committing either a Type I or Type II error in this example.

11.8 Suppose the defendant in Problem 11.7 was presumed guilty until proven innocent. How would the null and alternative hypotheses differ from those in Problem 11.7? What would be the meaning of the risks of committing either a Type I or Type II error here?

11.9 How is power related to the probability of making a Type II error?

11.3 Z Test of Hypothesis for the Mean (σ_x Known)

Now that we have described the hypothesis-testing methodology, let us return to the question of interest to the production manager at the packaging plant. We may recall that he wanted to determine whether or not the cereal box packaging process was in control—that the average fill per box throughout the entire packaging process remains at the specified 368 grams and no corrective action is needed. To study this he planned to take a random sample of 25 boxes, weigh each, and then evaluate the difference between the sample statistic and hypothesized population parameter by comparing the mean weight (in grams) from the sample to the expected mean of 368 grams specified by the company. For this cereal box packaging process, the null and alternative hypotheses were

$$H_0: \quad \mu_x = 368$$

$$H_1: \quad \mu_x \neq 368$$

If we assume that the standard deviation σ_x is known, then based on the central limit theorem the sampling distribution of the mean would follow the normal distribution and the **Z-test statistic** would be

$$Z = \frac{\bar{X} - \mu_x}{\frac{\sigma_x}{\sqrt{n}}} \qquad (11.1)$$

In this formula, the numerator measures how far (in an absolute sense) the observed sample mean $\bar{X}$ is from the hypothesized mean μ_x. The denominator is the standard error of the mean, so Z represents how many standard errors $\bar{X}$ is from μ_x.

If the production manager decided to choose a level of significance of .05, the size of the rejection region would be .05, and the critical values of the normal distribution could be determined. These critical values can be expressed in standard-deviation units. Since the rejection region is divided into the two tails of the distribution (this is called a **two-tailed test**), the .05 is divided into two equal parts of .025 each. A rejection region of .025 in each tail of the normal distribution results in an area of .475 between the hypothesized mean and each critical value. Looking up this area in the normal distribution (Table E.2), we find that the critical values that divide the rejection and nonrejection regions are (in standard-deviation units) +1.96 and −1.96. Figure 11.2 illustrates this case; it shows that if the mean is actually 368 grams, as H_0 claims, then the values of the test statistic Z will have a standard normal distribution centered at $\mu_x = 368$. Observed values of Z greater than 1.96 or less than −1.96 indicate that $\bar{X}$ is so far from the hypothesized $\mu_x = 368$ that it is unlikely that such a value would occur if H_0 were true.

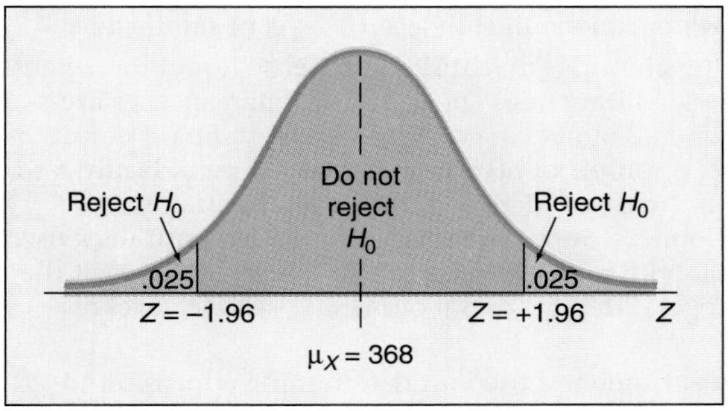

Figure 11.2
Testing a hypothesis about the mean (σ_x known) at the .05 level of significance.

Therefore the decision rule would be

Reject H_0 if $Z > +1.96$

or if $Z < -1.96$;

otherwise do not reject H_0.

Suppose that the sample of 25 cereal boxes indicated a sample mean ($\bar{X}$) of 372.5 grams and the population standard deviation (σ_x) was assumed to remain at

15 grams as specified by the company (see Section 9.2.3). Using Equation (11.1), we have

$$Z = \frac{\overline{X} - \mu_x}{\frac{\sigma_x}{\sqrt{n}}}$$

$$= \frac{372.5 - 368}{\frac{15}{\sqrt{25}}} = +1.50$$

Since $Z = +1.50$, we see that $-1.96 < +1.50 < +1.96$. Thus our decision is not to reject H_0. We would conclude that the average amount filled is 368 grams. Alternatively, to take into account the possibility of a Type II error, we may phrase the conclusion as "there is no evidence that the average fill is different from 368 grams."

Problems for Section 11.3

11.10 Suppose that the director of manufacturing at a clothing factory needed to determine whether a new machine was producing a particular type of cloth according to the manufacturer's specifications, which indicated that the cloth should have a mean breaking strength of 70 pounds and a standard deviation of 3.5 pounds. A sample of 36 pieces revealed a sample mean of 69.7 pounds. Is there evidence that the machine is not meeting the manufacturer's specifications in terms of the average breaking strength? (Use a .05 level of significance.)

11.11 The purchase of a coin-operated laundry is being considered by a potential entrepreneur. The present owner claims that over the past five years the average daily revenue was $675 with a standard deviation of $75. A sample of 30 selected days reveals a daily average revenue of $625. Is there evidence that the claim of the present owner is not valid? (Use a .01 level of significance.)

● 11.12 A manufacturer of salad dressings uses machines to dispense liquid ingredients into bottles that move along a filling line. The machine that dispenses dressings is working properly when 8 ounces are dispensed. The standard deviation of the process is .15 ounce. A sample of 50 bottles is selected periodically, and the filling line is stopped if there is evidence that the average amount dispensed is different from 8 ounces. Suppose that the average amount dispensed in a particular sample of 50 bottles is 7.983 ounces. Is there evidence that the population average amount is different from 8 ounces? (Use a .05 level of significance.)

11.13 Suppose that scores on an aptitude test used for determining admission to graduate study in business are known to be normally distributed with a population mean of 500 and a population standard deviation of 100. If a random sample of 12 applicants from Stephan College have a sample mean of 537, is there any evidence that their mean score is different from the mean expected of all applicants? (Use a .01 level of significance.)

Interchapter Problems for Section 11.3

11.14 Referring to the example concerning the length of computer paper on page 348, is there evidence that the average length is different from 11 inches? (Use a .05 level of significance.)

11.15 Referring to Problem 10.5 on page 349, is there evidence that the average amount is different from 1.0 gallon? (Use a .01 level of significance.)

● 11.16 Referring to Problem 10.6 on page 349, the production process is said to be "in control" (that is, working properly) when the population average life of the light bulbs is 375 hours.

(a) State the null and alternative hypotheses.

(b) Using a .05 level of significance, what should the quality control manager conclude about the process based on sample results?

11.17 Referring to Problem 10.7 on page 349, is there evidence that the average amount in the bottles is not equal to 2.0 liters? (Use a .05 level of significance.)

11.4 Summarizing the Steps of Hypothesis Testing

Now that we have used hypothesis-testing methodology to draw a conclusion about the population mean in situations where the population standard deviation is known, it will be useful to summarize the steps involved.

1. State the null hypothesis, H_0.
2. State the alternative hypothesis, H_1.
3. Choose the level of significance, α.
4. Choose the sample size, n.
5. Determine the appropriate statistical technique and corresponding test statistic to use.
6. Set up the critical values that divide the rejection and nonrejection regions.
7. Collect the data and compute the sample value of the appropriate test statistic.
8. Determine whether the test statistic has fallen into the rejection or the nonrejection region.
9. Make the statistical decision.
10. Express the statistical decision in terms of the problem.

● **Steps 1 and 2.** The null and alternative hypotheses must be stated in statistical terms. In testing whether the average amount filled was 368 grams, the null hypothesis was that μ_x equals 368, and the alternative hypothesis was that μ_x was not equal to 368 grams.

● **Step 3.** The level of significance is specified according to the relative importance of the risks of committing Type I and Type II errors in the problem. We chose $\alpha = .05$. (This, along with the sample size, determines β.)

● **Step 4.** The sample size is determined after taking into account the specified risks of committing Type I and Type II errors (that is, selected levels of α and β) and considering budget constraints in carrying out the study. Here 25 cereal boxes were randomly selected.

● **Step 5.** The statistical technique that will be used to test the null hypothesis must be chosen. Since σ_x was known (that is, specified by the company to be 15 grams), a Z test was selected.

● **Step 6.** Once the null and alternative hypotheses are specified and the level of significance and the sample size are determined, the critical values for the appropriate statistical distribution can be found so that the rejection and nonrejection regions can be indicated. Here the values +1.96 and −1.96 were used to define these regions since the Z test statistic refers to the standard normal distribution.

● **Step 7.** The data are collected and the value of the test statistic is computed. Here, $\bar{X} = 372.5$ grams, so $Z = +1.50$.

● **Step 8.** The computed value of the test statistic is compared with the critical values for the appropriate sampling distribution to determine whether it falls into the rejection or nonrejection region. Here, $Z = +1.50$ is in the region of nonrejection, since $-1.96 < Z = +1.50 < +1.96$.

● **Step 9.** The hypothesis-testing decision is made. If the test statistic falls into the nonrejection region, the null hypothesis H_0 cannot be rejected. If the test statistic falls into the rejection region, the null hypothesis is rejected. Here, H_0 is not rejected.

● **Step 10.** The consequences of the hypothesis-testing decision must be expressed in terms of the actual problem involved. In our cereal box packaging example we concluded that there was no evidence that the average amount of cereal fill was different from 368 grams.

11.5 The *p* Value Approach to Hypothesis Testing: Two-Tailed Tests

In recent years, with the advent of widely available statistical software, an approach to hypothesis testing that has increasingly gained acceptance involves the concept of the ***p* value.**

The ***p* value** is the probability of obtaining a test statistic equal to or more extreme than the result obtained from the sample data, given that the null hypothesis H_0 is really true.

The p value is often referred to as the *observed level of significance,* the smallest level at which H_0 can be rejected for a given set of data.

- If the p value is greater than or equal to α, the null hypothesis is not rejected.
- If the p value is smaller than α, the null hypothesis is rejected.

To understand the p value approach, let us refer to the cereal box packaging example of Section 11.3. In that section we tested whether or not the average amount of cereal fill was equal to 368 grams (page 392). We obtained a Z value of +1.50 and did not reject the null hypothesis since +1.50 was greater than the lower critical value of −1.96 but less than the upper critical value of +1.96.

We may now use the p value approach to find the probability of obtaining a test statistic Z that is *more extreme* than +1.50. When using a *two-tailed test,* this means that we need to compute the probability of obtaining a Z value greater than +1.50 along with the probability of obtaining a Z value less than −1.50. From Table

E.2, the probability of obtaining a Z value above +1.50 is .5000 − .4332 = .0668. Since the standard normal distribution is symmetric, the probability of obtaining a value below −1.50 is also .0668. Thus the p value for this two-tailed test is .0668 + .0668 = .1336 (see Figure 11.3). This result may be interpreted to mean that the probability of obtaining a result equal to or more extreme than the one observed is .1336. Since this is greater than $\alpha = .05$, the null hypothesis is not rejected.

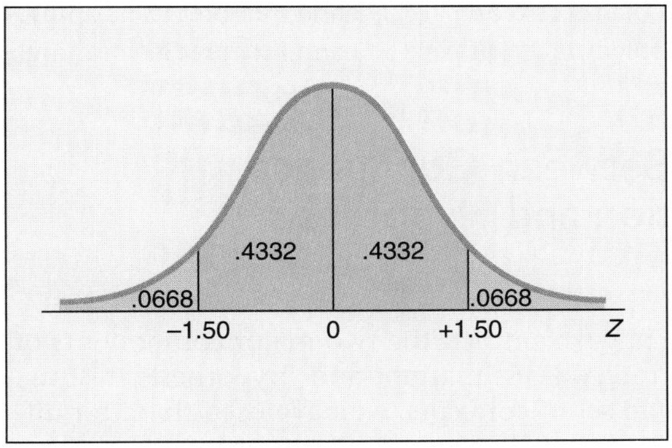

Figure 11.3
Finding the *p* value for a two-tailed test.

Unless we are dealing with a test statistic that follows the normal distribution, the computation of the p value is very difficult. Thus it is fortunate that statistical computer software such as MINITAB, SAS, SPSS, and STATISTIX (see References 6, 7, 9, and 10) routinely present the p value as part of the output for many hypothesis-testing procedures.

Now that we have discussed the p value approach for hypothesis testing, it will be useful to summarize the steps involved.

1. State the null hypothesis, H_0.
2. State the alternative hypothesis, H_1.
3. Choose the level of significance, α.
4. Choose the sample size, n.
5. Determine the appropriate statistical technique and corresponding test statistic to use.
6. Collect the data and compute the sample value of the appropriate test statistic.
7. Calculate the p value based on the test statistic. This involves
   (a) Sketching the distribution under the null hypothesis H_0.
   (b) Placing the test statistic on the horizontal axis.
   (c) Shading in the appropriate area under the curve, based on the alternative hypothesis H_1.
8. Compare the p value to α.
9. Make the statistical decision.
10. Express the statistical decision in terms of the problem.

Problems for Section 11.5

11.18 Compute the p value in Problem 11.10 on page 392 and interpret its meaning.

11.19 Compute the p value in Problem 11.11 on page 392 and interpret its meaning.

- 11.20 Compute the p value in Problem 11.12 on page 392 and interpret its meaning.
- 11.21 Compute the p value in Problem 11.13 on page 392 and interpret its meaning.

Interchapter Problems for Section 11.5

- 11.22 Compute the p value in Problem 11.14 on page 392 and interpret its meaning.
- 11.23 Compute the p value in Problem 11.15 on page 393 and interpret its meaning.
- 11.24 Compute the p value in Problem 11.16 on page 393 and interpret its meaning.
- 11.25 Compute the p value in Problem 11.17 on page 393 and interpret its meaning.

11.6 A Connection Between Confidence Interval Estimation and Hypothesis Testing

In Chapter 10 and this chapter we have examined the two major components of statistical inference—confidence interval estimation and hypothesis testing. Although they are based on the same set of concepts, we have used them for different purposes. In Chapter 10 we used confidence intervals to estimate parameters, and in this chapter we have seen that we can use hypothesis testing for making decisions about specified values of population parameters.

In many situations we can use confidence intervals to do a test of a null hypothesis. This can be illustrated for the test of a hypothesis for a mean. Referring back to the cereal box packaging process, we first attempted to determine whether the population average amount was different from 368 grams. We tested this in Section 11.3 using Equation (11.1)

$$Z = \frac{\overline{X} - \mu_x}{\frac{\sigma_x}{\sqrt{n}}}$$

Instead of testing the null hypothesis that $\mu_x = 368$ grams, we could also solve the problem by obtaining a confidence-interval estimate of μ_x. If the hypothesized value of $\mu_x = 368$ fell into the interval, the null hypothesis would not be rejected. That is, the value 368 would not be considered unusual for the data observed. On the other hand, if the hypothesized value did not fall into the interval, the null hypothesis would be rejected, because 368 grams would then be considered an unusual value. Using Equation (10.1), the confidence-interval estimate could be set up from the following data:

$n = 25$, $\overline{X} = 372.5$ grams, $\sigma_x = 15$ grams (specified by the company)

For a confidence level of 95% (corresponding to a .05 level of significance—that is, $\alpha = .05$) we have

$$\overline{X} \pm Z \frac{\sigma_x}{\sqrt{n}}$$

$$372.5 \pm (1.96) \frac{15}{\sqrt{25}}$$

$$372.5 \pm 5.88$$

so that

$$366.62 \le \mu_x \le 378.38$$

Since the interval includes the hypothesized value of 368 grams, we would not reject the null hypothesis and we would conclude that there is no evidence the mean fill over the entire packaging process is not 368 grams. This is the same decision we reached by using hypothesis-testing methodology.

Interchapter Problems for Section 11.6

11.26 Compare the conclusions obtained from Problems 10.5 and 11.15 on pages 349 and 393. Are the conclusions the same? Why?

11.27 Compare the conclusions obtained from Problems 10.6 and 11.16 on pages 349 and 393. Are the conclusions the same? Why?

11.28 Compare the conclusions obtained from Problems 10.7 and 11.17 on pages 349 and 393. Are the conclusions the same? Why?

11.7 One-Tailed Tests

In Section 11.3 we used hypothesis-testing methodology to examine the question of whether or not the average amount of fill over the entire packaging process (that is, the population) was 368 grams. The alternative hypothesis (H_1: $\mu_x \neq 368$) contained two possibilities: either the average could be less than 368 grams or the average could be more than 368 grams. For this reason the rejection region was divided into the two tails of the sampling distribution of the mean. And, as we have just observed in the previous section, since a confidence interval estimate of the mean contains a lower and upper limit respectively corresponding to the left and right tail critical values from the sampling distribution of the mean, we were able to use the confidence interval to do a test of the null hypothesis that the average amount of fill over the entire packaging process was 368 grams.

In some situations, however, the alternative hypothesis focuses in a particular direction. For example, the chief financial officer (CFO) of the food packaging company would mainly be concerned with excess because, if more than 368 grams of cereal were actually being filled per box but the price charged to the customer was based on the 368 grams labeled on the box, the company would be losing money unnecessarily. Therefore, she would be interested in whether the average amount of fill for the entire packaging process was *above* 368 grams. To her, and strictly from a financial point of view with respect to her responsibility as CFO for the company (the ethics of which will be discussed in Section 11.11), unless the sample mean was significantly above 368 grams, the process would be considered to be working properly. For the CFO, the null and alternative hypotheses would be stated as follows:

H_0: $\mu_x \le 368$ (process is working properly)

H_1: $\mu_x > 368$ (process is not working properly)

The rejection region here would be entirely contained in the upper tail of the sampling distribution of the mean since we want to reject H_0 only when the sample mean is significantly above 368 grams. When such a situation occurs where the entire rejection region is contained in one tail of the sampling distribution of the test statistic, it is called a **one-tailed test.** If we again choose a level of significance α of .05, the critical value on the Z distribution can be determined. As seen from Table 11.2 and Figure 11.4, since the entire rejection region is in the upper tail of the standard normal distribution and contains an area of .05, the area from the mean to the critical value must be .45; thus the critical value of the Z test statistic is +1.645, the average of +1.64 and +1.65. (We should note here that some statisticians would *round off* to two decimal places and select +1.64 as the critical value, whereas others would *round up* to +1.65. We prefer to interpolate between the areas .4495 and .4505 so as to select the critical value with upper tail area as close to .05 as possible. Thus we took the average of +1.64 and +1.65.)

Table 11.2 Obtaining the critical value of the Z test statistic from the standard normal distribution for a one-tailed test with $\alpha = .05$.

Z	.00	.01	.02	.03	.04	.05	.06	.07	.08	.09
0.0	.0000	.0040	.0080	.0120	.0160	.0199	.0239	.0279	.0319	.0359
0.1	.0398	.0438	.0478	.0517	.0557	.0596	.0636	.0675	.0714	.0753
0.2	.0793	.0832	.0871	.0910	.0948	.0987	.1026	.1064	.1103	.1141
0.3	.1179	.1217	.1255	.1293	.1331	.1368	.1406	.1443	.1480	.1517
0.4	.1554	.1591	.1628	.1664	.1700	.1736	.1772	.1808	.1844	.1879
⋮	⋮	⋮	⋮	⋮	⋮	⋮	⋮	⋮	⋮	⋮
1.0	.3413	.3438	.3461	.3485	.3508	.3531	.3554	.3577	.3599	.3621
1.1	.3643	.3665	.3686	.3708	.3729	.3749	.3770	.3790	.3810	.3830
1.2	.3849	.3869	.3888	.3907	.3925	.3944	.3962	.3980	.3997	.4015
1.3	.4032	.4049	.4066	.4082	.4099	.4115	.4131	.4147	.4162	.4177
1.4	.4192	.4207	.4222	.4236	.4251	.4265	.4279	.4292	.4306	.4319
1.5	.4332	.4345	.4357	.4370	.4382	.4394	.4406	.4418	.4429	.4441
1.6	.4452	.4463	.4474	.4484	.4495	.4505	.4515	.4525	.4535	.4545
1.7	.4554	.4564	.4573	.4582	.4591	.4599	.4608	.4616	.4625	.4633

Source: Extracted from Table E.2.

The decision rule would be

$$\text{Reject } H_0 \text{ if } Z > +1.645;$$
$$\text{otherwise do not reject } H_0.$$

Using the Z test given by Equation (11.1) on the information obtained from the sample drawn by the production manager

$n = 25$, $\bar{X} = 372.5$ grams, $\sigma_x = 15$ grams (specified by the company)

we have

$$Z = \frac{\bar{X} - \mu_x}{\frac{\sigma_x}{\sqrt{n}}}$$

$$= \frac{372.5 - 368}{\frac{15}{\sqrt{25}}} = +1.50$$

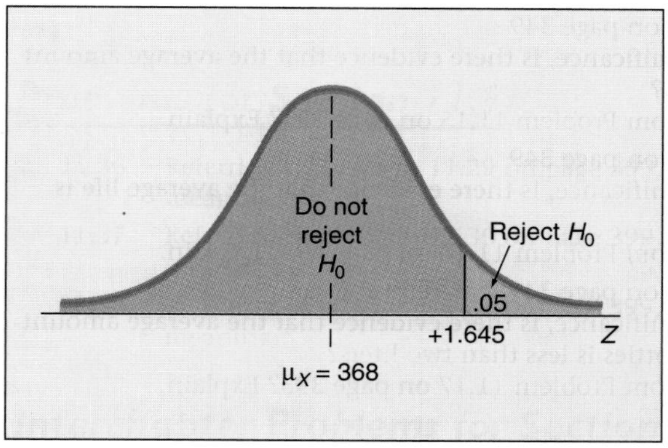

Figure 11.4
One-tailed test of hypothesis for a mean (σ_x known) at the .05 level of significance.

Since $Z = +1.50 < +1.645$, our decision would be not to reject H_0 and we would conclude that there is no evidence that the average amount of cereal fill per box over the entire packaging process is above 368 grams. That is, even though the sample mean $\bar{X}$ exceeded 368 grams, the result from the sample is deemed due to chance or sampling error; it is not statistically significant.

Problems for Section 11.7

● 11.29 The Glen Valley Steel Company manufactures steel bars. If the production process is *working properly*, it turns out steel bars with an average length of at least 2.8 feet with a standard deviation of .20 foot (as determined from engineering specifications on the involved production equipment). Longer steel bars can be used or altered; shorter bars must be scrapped. A sample of 25 bars is selected from the production line. The sample indicates an average length of 2.73 feet. The company wishes to determine whether the production equipment needs any adjustment.
(a) State the null and alternative hypotheses.
(b) If the company wishes to test the hypothesis at the .05 level of significance, what decision would it make?

11.30 Referring to Problem 11.10 on page 392
(a) At the .05 level of significance, is there evidence that the mean breaking strength is less than 70 pounds?
(b) How does (a) differ from Problem 11.10? Explain.

11.31 Referring to Problem 11.12 on page 392
(a) At the .05 level of significance, is there evidence that the average amount dispensed is less than 8 ounces?
(b) How does (a) differ from Problem 11.12? Explain.

Interchapter Problems for Section 11.7

11.32 Referring to the example concerning the length of computer paper on page 348, at the .05 level of significance, is there evidence that the average length is less than 11 inches?

is equal to 15 grams so that the Z test will be appropriate. If a level of significance (α) of .05 is selected and a random sample of 25 boxes is obtained, the value of $\overline{X}$ that will enable us to reject the null hypothesis can be found from Equation (9.6a) as follows:

$$\overline{X}_L = \mu_x - Z\frac{\sigma_x}{\sqrt{n}}$$

Since we have a one-tailed test with a level of significance of .05, the value of Z equal to 1.645 standard deviations *below* the hypothesized mean can be obtained from Table E.2 (see Figure 11.6). Therefore,

$$\overline{X}_L = 368 - (1.645)\frac{15}{\sqrt{25}} = 368 - 4.935 = 363.065$$

The decision rule for this one-tailed test would be

Reject H_0 if $\overline{X} < 363.065$; otherwise do not reject H_0.

The decision rule states that if a random sample of 25 boxes reveals a sample mean of less than 363.065 grams, the null hypothesis will be rejected and the representative will conclude that the process is not working properly. If in fact this is the case, the power of the test measures the probability of concluding that the process is not working properly for differing values of the true population mean.

Suppose, for example, we would like to determine the chance of rejecting the null hypothesis when the population mean is actually 360 grams. Based on our decision rule, we need to determine the probability or area under the normal curve below 363.065 grams. From the central limit theorem and the assumption of normality in the population, we may assume that the sampling distribution of the mean follows a normal distribution. Therefore the area under the normal curve below 363.065 grams can be expressed in standard deviation units, since we are finding the probability of rejecting the null hypothesis when the true mean has shifted to 360 grams. Using Equation (11.1), we have

$$Z = \frac{\overline{X} - \mu_1}{\frac{\sigma_x}{\sqrt{n}}}$$

where μ_1 is the actual population mean. Thus

$$Z = \frac{363.065 - 360}{\frac{15}{\sqrt{25}}} = 1.02$$

From Table E.2, there is a 34.61% chance of observing a Z value between the mean and +1.02 standard deviations. Since we wish to determine the area below 363.065, the area under the curve below the mean (50%) must be added to this value, and the power of the test is found to be 84.61% (see Figure 11.7). β, the probability that the null hypothesis ($\mu_x = 368$) will not be rejected, is $1 - .8461 = .1539$ (or 15.39%). This is the probability of committing a Type II error.

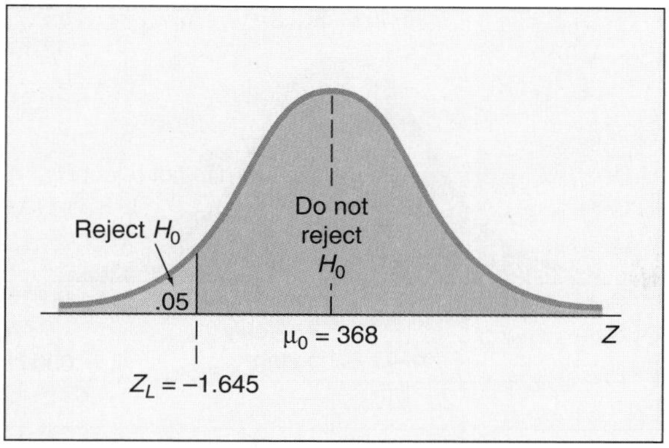

Figure 11.6
Determining the lower critical value for a one-tailed test for a population mean at the .05 level of significance.

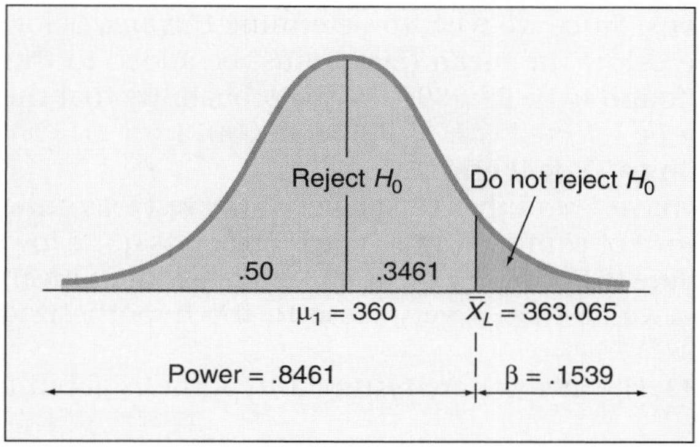

Figure 11.7
Determining the power of the test and the probability of a Type II error when $\mu_1 = 360$ grams.

Now that we have determined the power of the test if the population mean were really equal to 360 we can also calculate the power for any other value that μ_x could attain. For example, what would be the power of the test if the population mean were really equal to 352 grams? Assuming the same standard deviation, sample size, and level of significance, the decision rule would still be

Reject H_0 if $\overline{X} < 363.065$; otherwise do not reject H_0.

Once again, since we are testing a hypothesis for a mean, from Equation (11.1) we have

$$Z = \frac{\overline{X} - \mu_1}{\frac{\sigma_x}{\sqrt{n}}}$$

If the population mean shifts down to 352 grams (see Figure 11.8 on page 404), then

$$Z = \frac{363.065 - 352}{\frac{15}{\sqrt{25}}} = 3.69$$

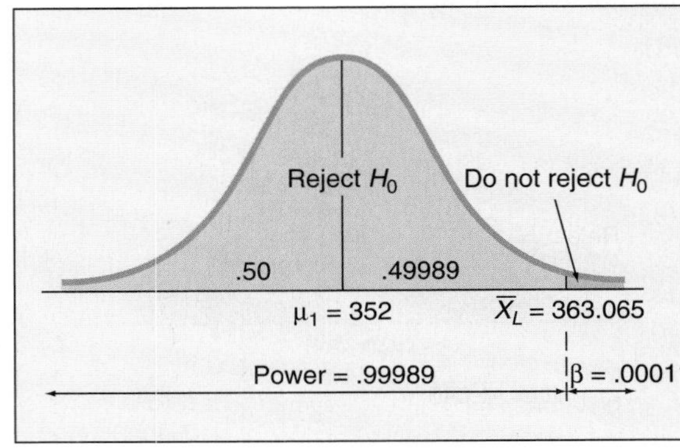

Figure 11.8
Determining the power of the test and the probability of a Type II error when $\mu_1 = 352$ grams.

From Table E.2, there is a 49.989% chance of observing a Z value between the mean and +3.69 standard deviations. Since we wish to determine the area below 363.065, the area under the curve below the mean (50%) must be added to this value, and the power of the test is found to be 99.989%. β, the probability that the null hypothesis ($\mu_x = 368$) will not be rejected is $1 - .99989 = .00011$ (or .011%). This is the probability of committing a Type II error.

In the preceding two cases we have found that the power of the test was quite high, whereas, conversely, the chance of committing a Type II error was quite low. In our next example we shall compute the power of the test if the population mean were really equal to 367 grams—a value that is very close to the hypothesized mean of 368 grams.

Once again, from Equation (11.1), since we are testing a hypothesis about a mean (with σ_x known), we have

$$Z = \frac{\overline{X} - \mu_1}{\frac{\sigma_x}{\sqrt{n}}}$$

If the population mean were really equal to 367 grams (see Figure 11.9), then

$$Z = \frac{363.065 - 367}{\frac{15}{\sqrt{25}}} = -1.31$$

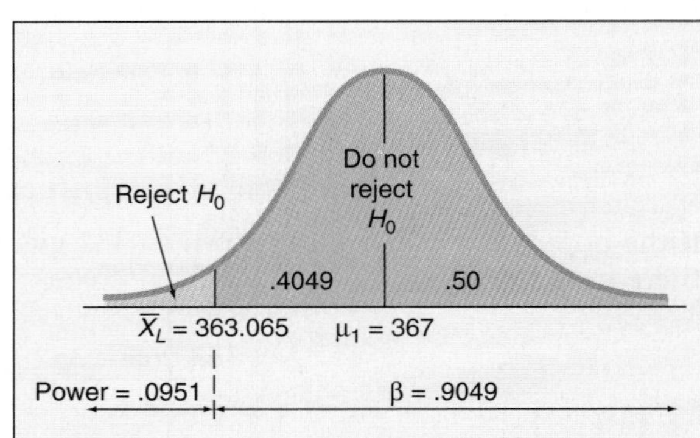

Figure 11.9
Determining the power of the test and the probability of a Type II error when $\mu_1 = 367$ grams.

so that

$$\begin{array}{r} .5000 \\ -.4049 \\ \hline .0951 \end{array} = \text{power} = 1 - \beta$$

From Table E.2, we can observe that the probability (area under the curve) between the mean and -1.31 standard deviation units is .4049 (or 40.49%). Since, in this instance, the rejection region is in the lower tail of the distribution, the power of the test is 9.51%, and the chance of making a Type II error is 90.49%.

Figure 11.10 below illustrates the power of the test for various possible values of μ_1 (including the three cases that we have examined). This is called a **power curve.** The computations for our three cases are summarized in Figure 11.11 on page 406.

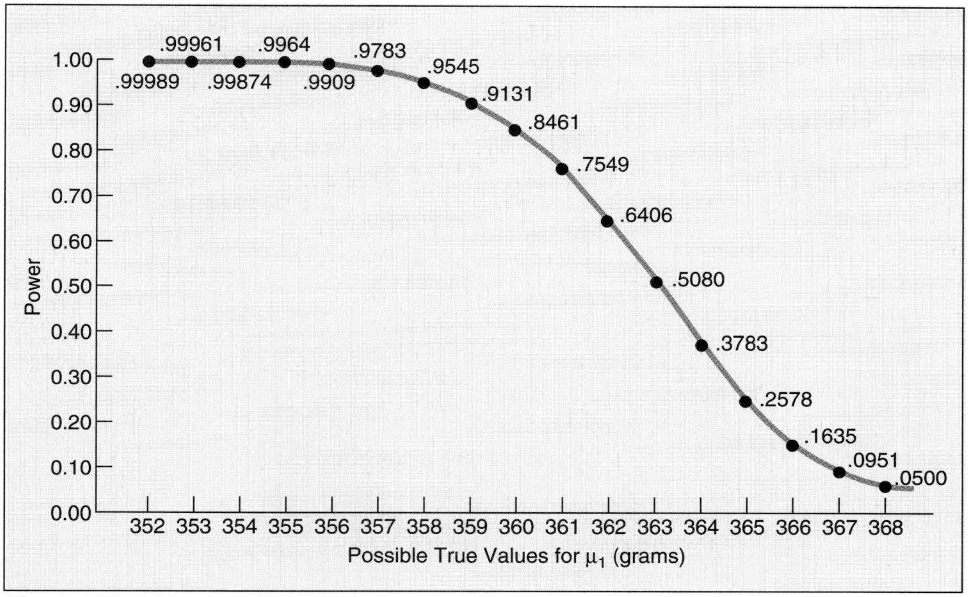

Figure 11.10
Power curve of the cereal box packaging process for the alternative hypothesis H_1: $\mu_x < 368$.

From Figure 11.10 we observe that the power of this one-tailed test increases sharply (and approaches 100%) as the actual population mean takes on values farther below the hypothesized mean of 368 grams. Clearly, for this one-tailed test the smaller the actual mean μ_1 is when compared with the hypothesized mean, the greater will be the power to detect this disparity.[2] On the other hand, for values of μ_1 close to 368 grams the power is rather small, since the test cannot effectively detect small differences between the actual population mean and the hypothesized value of 368 grams. Interestingly, if the population mean were actually 368 grams, the power of the test would be equal to α, the level of significance (which is .05 in this example), since the null hypothesis would actually be true.

The drastic changes in the power of the test for differing values of the actual population means can be observed by reviewing the different panels of Figure 11.11. From panels A and B we can see that when the population mean does not greatly differ from 368 grams, the chance of rejecting the null hypothesis, based on the decision rule involved, is not large. However, once the actual population

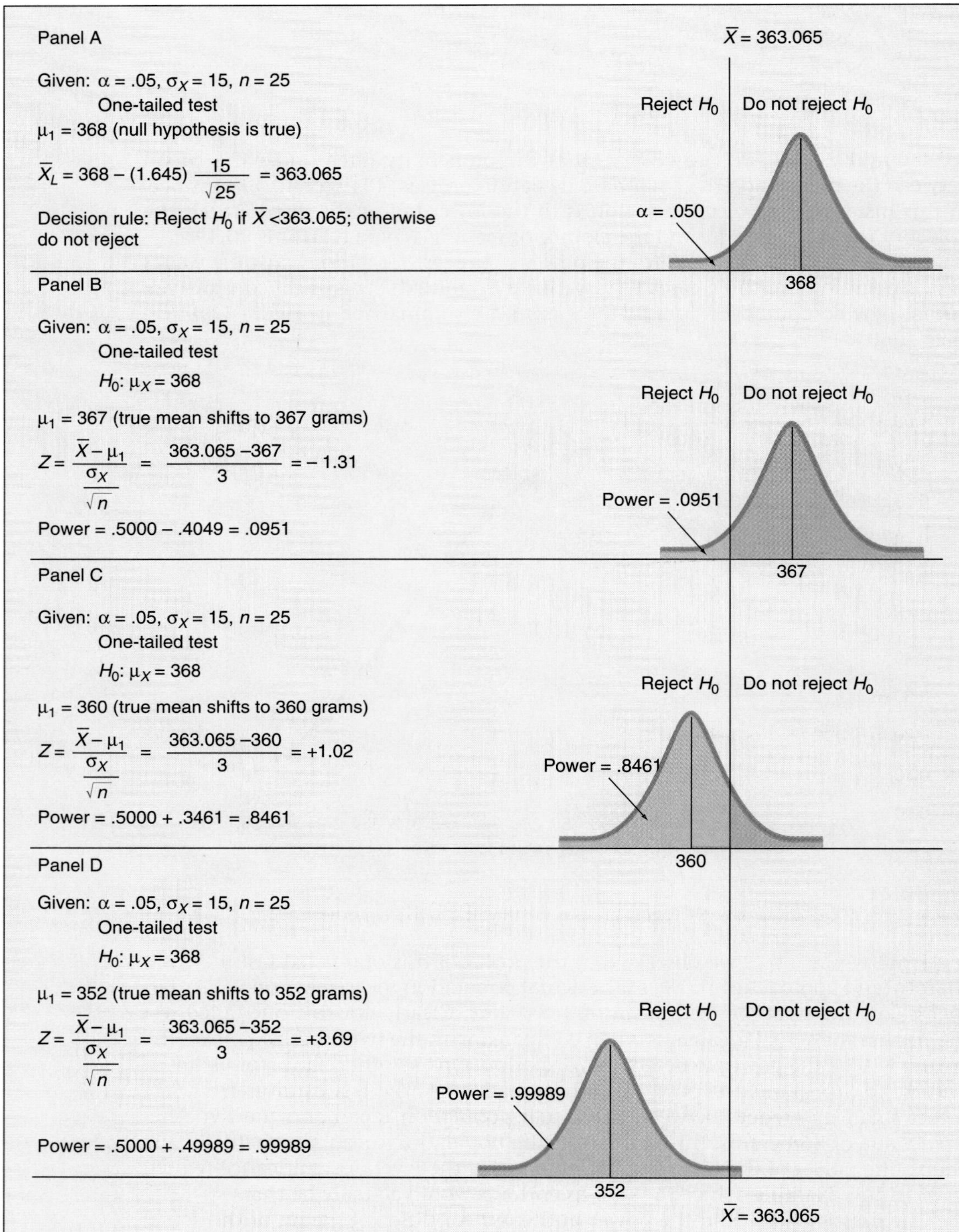

Panel A
Given: $\alpha = .05$, $\sigma_X = 15$, $n = 25$
One-tailed test
$\mu_1 = 368$ (null hypothesis is true)
$\bar{X}_L = 368 - (1.645)\frac{15}{\sqrt{25}} = 363.065$
Decision rule: Reject H_0 if $\bar{X} < 363.065$; otherwise do not reject
$\bar{X} = 363.065$
Reject H_0
Do not reject H_0
$\alpha = .050$
368
Panel B
Given: $\alpha = .05$, $\sigma_X = 15$, $n = 25$
One-tailed test
H_0: $\mu_X = 368$
$\mu_1 = 367$ (true mean shifts to 367 grams)
$Z = \frac{\bar{X} - \mu_1}{\frac{\sigma_X}{\sqrt{n}}} = \frac{363.065 - 367}{3} = -1.31$
Power = .5000 – .4049 = .0951
Reject H_0
Do not reject H_0
Power = .0951
367
Panel C
Given: $\alpha = .05$, $\sigma_X = 15$, $n = 25$
One-tailed test
H_0: $\mu_X = 368$
$\mu_1 = 360$ (true mean shifts to 360 grams)
$Z = \frac{\bar{X} - \mu_1}{\frac{\sigma_X}{\sqrt{n}}} = \frac{363.065 - 360}{3} = +1.02$
Power = .5000 + .3461 = .8461
Reject H_0
Do not reject H_0
Power = .8461
360
Panel D
Given: $\alpha = .05$, $\sigma_X = 15$, $n = 25$
One-tailed test
H_0: $\mu_X = 368$
$\mu_1 = 352$ (true mean shifts to 352 grams)
$Z = \frac{\bar{X} - \mu_1}{\frac{\sigma_X}{\sqrt{n}}} = \frac{363.065 - 352}{3} = +3.69$
Power = .5000 + .49989 = .99989
Reject H_0
Do not reject H_0
Power = .99989
352
$\bar{X} = 363.065$

Figure 11.11
Determining statistical power for varying values of the actual population mean.

mean shifts substantially below the hypothesized 368 grams, the power of the test greatly increases, approaching its maximum value of 1 (or 100%).

In our discussion of the power of a statistical test we have utilized a one-tailed test, a level of significance of .05, and a sample size of 25 boxes. With this in mind we can determine the effect on the power of the test by varying, one at a time

- The type of statistical test—one-tailed versus two-tailed
- The level of significance α
- The sample size n

While we will leave these exercises to the reader (see Problems 11.42–11.48 on pages 407–408 and Problems 11.49 and 11.50 on page 410), the following will be observed:

- A one-tailed test is more powerful than a two-tailed test and should be used whenever it is appropriate to specify the direction of the alternative hypothesis.
- Since the probability of committing a Type I error (α) and the probability of committing a Type II error (β) have an inverse relationship, and the latter is the complement of the power of the test ($1 - \beta$), then α and the power of the test vary directly. An increase in the value of the level of significance (α) chosen would result in an increase in power, and a decrease in α would result in a decrease in power.
- An increase in the size of the sample n chosen would result in an increase in power; a decrease in the size of the sample selected would result in a decrease in power.

Problems for Section 11.9

● 11.42 A coin-operated soft-drink machine was designed to discharge, when it is operating properly, at least 7 ounces of beverage per cup with a standard deviation of 0.2 ounce. If a random sample of 16 cupfuls is selected by a statistician for a consumer testing service, and the statistician is willing to take a risk of $\alpha = .05$ of committing a Type I error, compute the power of the test and the probability of a Type II error (β) if the population average amount dispensed is actually
(a) 6.9 ounces per cup
(b) 6.8 ounces per cup

● 11.43 Refer to Problem 11.42. If the statistician is only willing to take a risk of $\alpha = .01$ of committing a Type I error, compute the power of the test and the probability of a Type II error (β) if the population average amount dispensed is actually
(a) 6.9 ounces
(b) 6.8 ounces
(c) Compare the results in (a) and (b) of this problem and Problem 11.42. What conclusion can you draw here?

● 11.44 Refer to Problem 11.42. If the statistician selects a random sample of 25 cupfuls and is willing to take a risk of $\alpha = .05$ of committing a Type I error, compute the power of the test and the probability of a Type II error (β) if the population average amount dispensed is actually
(a) 6.9 ounces
(b) 6.8 ounces
(c) Compare the results in (a) and (b) of this problem and Problem 11.42. What conclusion can you draw here?

11.45 A tire manufacturer produces tires that last, on average, at least 25,000 miles when the production process is working properly. Based upon past experience, the standard deviation of the tires is assumed to be 3,500 miles. The production manager will stop the production process if there is evidence that the average tire life is below 25,000 miles. If a random sample of 100 tires is selected (to be subjected to destructive testing), and the production manager is willing to take a risk of $\alpha = .05$ of committing a Type I error, compute the power of the test and the probability of a Type II error (β) if the population average life is actually
(a) 24,000 miles
(b) 24,900 miles

11.46 Refer to Problem 11.45. If the production manager is only willing to take a risk of $\alpha = .01$ of committing a Type I error, compute the power of the test and the probability of a Type II error (β) if the population average life is actually
(a) 24,000 miles
(b) 24,900 miles
(c) Compare the results in (a) and (b) of this problem and (a) and (b) in Problem 11.45. What conclusion can you draw here?

11.47 Refer to Problem 11.45. If the production manager selects a random sample of 25 tires and is willing to take a risk of $\alpha = .05$ of committing a Type I error, compute the power of the test and the probability of a Type II error (β) if the population average life is actually
(a) 24,000 miles
(b) 24,900 miles
(c) Compare the results in (a) and (b) of this problem and (a) and (b) in Problem 11.45. What conclusion can you draw here?

11.48 Refer to Problem 11.45. If the production manager will stop the process when there is evidence that the average life is *different* from 25,000 miles (either less than or greater than) and a random sample of 100 tires is selected along with a level of significance α of .05, compute the power of the test and the probability of a Type II error (β) if the population average life is actually
(a) 24,000 miles
(b) 24,900 miles
(c) Compare the results in (a) and (b) of this problem and (a) and (b) in Problem 11.45. What conclusion can you draw here?

11.10 Planning a Study: Determining Sample Size Based on α and β

In planning a statistical study we have already seen in Section 10.7 that the sample size needed can be determined for a specified confidence level and sampling error. In a decision-making procedure such as hypothesis testing, however, assuming a one-tailed test, we may determine the sample size needed for a specified level of significance α and desired power of a test $(1 - \beta)$ as follows:

$$n = \frac{\sigma_x^2 (Z_\alpha - Z_\beta)^2}{(\mu_0 - \mu_1)^2} \qquad (11.2)$$

where σ_x^2 = variance in the population

Z_α = Z value for a given α level of significance

Z_β = Z value for a given probability β of committing a Type II error

μ_0 = value of the population mean under the null hypothesis

μ_1 = value of the population mean under the alternative hypothesis

To demonstrate how we can determine the sample size needed for a specified level of significance α and desired power of a test $(1 - \beta)$, we may refer once again to our cereal box packaging process. Suppose that the representative from the Office of Consumer Affairs wished to have an 80% chance (power) of rejecting the company's claim that the average amount of cereal per box is 368 grams (i.e., the null hypothesis) when the population mean is actually equal to 360 grams and was willing to take a 5% risk of committing a Type I error by rejecting the null hypothesis that the average fill is 368 grams when in fact it really is that amount (i.e., use a level of significance α of .05). How many cereal boxes need to be selected for the sample? Using Equation (11.2), we have

$$n = \frac{\sigma_x^2 (Z_\alpha - Z_\beta)^2}{(\mu_0 - \mu_1)^2}$$

and, for the cereal box packaging process,

$$\sigma_x = 15 \text{ grams}$$

$$\mu_0 = 368 \text{ grams}$$

$$\mu_1 = 360 \text{ grams}$$

Using a level of significance of $\alpha = .05$ for a one-tailed test, the rejection region can be established as follows (see Figure 11.12).

The Z_α value obtained from Table E.2 is equal to -1.645 because the rejection region contains .05 of the area under the normal curve (so that the area between the lower critical value and the null hypothesized mean of 368 grams is .45).

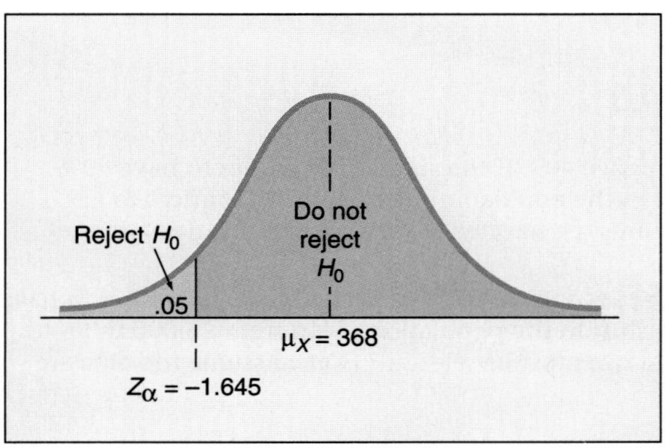

Figure 11.12
Determining the lower critical value in a one-tailed test for the population mean when the sample size is unknown.

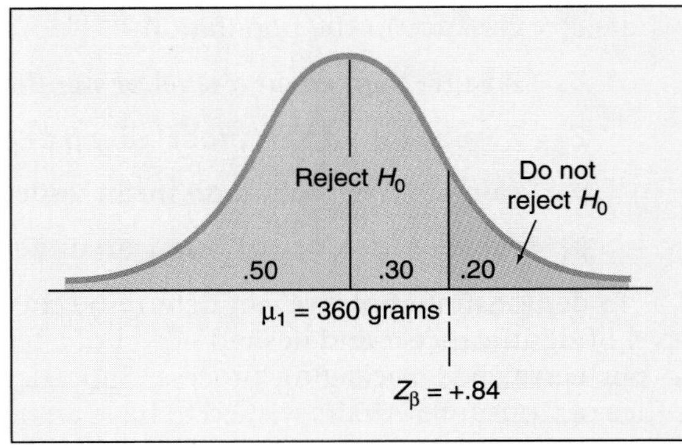

Figure 11.13
Determining the critical value for μ_1 = 360 grams when the sample size is unknown.

If a power of 80% is desired when the actual population mean is 360 grams, the value of Z_β can also be obtained from Table E.2 (see Figure 11.13). Since we wish to have 80% power of rejecting a false null hypothesis (and hence accept a 20% risk of committing a Type II error), we observe that this results in an area of .30 between the actual population mean of 360 grams and the critical value (which corresponds to .84 standard deviation unit above the actual population mean).

Using Equation (11.2), the sample size would be found as follows:

$$n = \frac{(15)^2(-1.645 - .84)^2}{(368 - 360)^2}$$

$$= \frac{(225)(-2.485)^2}{8^2} = 21.71$$

Therefore $n = 22$.

A sample size of 22 boxes would be required if the representative from the Office of Consumer Affairs was willing to take a .05 risk of making a Type I error and desired an 80% chance of rejecting the null hypothesis of 368 grams and detecting that the actual population mean had actually shifted to 360 grams. The fact that the production manager for the food processing company offered the representative an even larger sample (n was 25 boxes) was an indication of "good will" on the part of the company.

Problems for Section 11.10

● 11.49 Refer to Problem 11.42 on page 407. If the statistician wishes to have 99% power of detecting a shift in the population mean from 7.0 ounces to 6.9 ounces, what sample size must be selected? (*Note:* Assume the data are normally distributed.)

11.50 Refer to Problem 11.45 on page 408. If the production manager wishes to have 80% power of detecting a shift in the population mean from 25,000 miles to 24,000 miles, what sample size must be selected? (*Note:* Assume the data are normally distributed.)

11.51 A delivery service company is testing newly developed computer software aimed at increasing efficiency through improved routing schedules. The intent is to decrease expenses in overtime payments to carriers. In order to determine whether the computer software should be adopted by the company, the director of planning wants to test it on a trial basis. How many carriers must be selected to participate in the study if the director of planning wishes to have 95% power of detecting a 30-minute reduction in the average overall daily delivery times and is willing to accept an α risk of .01 (that is, 1%) if the currently used scheduling software indicates the standard deviation in overall daily delivery times to be 45 minutes ? (*Note:* Assume the data are normally distributed.)

● 11.52 Systolic blood pressure represents the pressure in the body's arterial system when the heart is contracting and forcing out blood. Individuals with hypotension or low blood pressure may require their systolic blood pressure to be regulated. Suppose that a pharmaceutical company is planning to test a new drug aimed at raising and stabilizing the systolic blood pressure of hypotensive individuals. How many people having this medical condition must be selected to participate in the study if the research director wishes to have 90% power of detecting a shift in the population mean from 85 mg/mm to 90 mg/mm and is willing to accept an α risk of .05 (that is, 5%) if prior studies indicate the standard deviation to be 7 mg/mm? (*Note:* Assume the data are normally distributed.)

11.11 Potential Hypothesis-Testing Pitfalls and Ethical Issues

To this point we have studied the fundamental concepts of hypothesis-testing methodology. We have learned how it is used for analyzing differences between sample estimates (i.e., statistics) of hypothesized population characteristics (i.e., parameters) in order to make decisions about the underlying characteristics. We have also learned how to evaluate the risks involved in making these decisions.

In particular, when dealing with the numerical outcomes of a random sample taken from some population whose variance σ_x^2 is either known or assumed known, we learned how to test a hypothesis that the population mean μ_x was equal to some specified value. The appropriate statistical procedure for carrying out such a hypothesis test is the Z test given by Equation (11.1), and the sampling distribution of the test statistic Z follows a standard normal distribution. In the following chapter we shall introduce two other statistical test procedures, the t test and the Wilcoxon signed-ranks test, each of which would be more appropriate than our Z test in a given set of circumstances. Part of good data analysis is to understand the assumptions underlying each of the hypothesis test procedures we shall be encountering and to select the one most appropriate for a given set of conditions.

11.11.1 Avoiding Pitfalls

When planning to carry out a test of hypothesis based on some designed experiment or research study under investigation, several questions need to be raised in order to ensure that proper methodology is used:

1. What is the goal of the experiment or research? Can it be translated into a null and alternative hypothesis?
2. Is the hypothesis test going to be two-tailed or one-tailed?
3. Can a random sample be drawn from the underlying population of interest?
4. What kind of *measurements* will be obtained from the sample? Are the sampled outcomes of the random variable going to be numerical or categorical?
5. At what significance level, or risk of committing a Type I error, should the hypothesis test be conducted?
6. What is the desired power to detect a difference of a specific size?
7. Is the intended sample size large enough to achieve the desired power of the test for the level of significance chosen?
8. What statistical test procedure is to be used on the sampled data and why?
9. What kind of conclusions and interpretations can be drawn from the results of the hypothesis test?

Questions like these need to be raised and answered in the planning stage of a designed experiment or research endeavor, so a person with substantial statistical training should be consulted and involved early in the process. All too often such an individual is consulted far too late in the process, after the research has been conducted and the data have been collected. Typically all that could be done in such a late stage is a "mop up." One could choose the statistical test procedure that would be best for the obtained data under the assumption that certain biases that have been built into the study (because of poor planning) are negligible. But this is a large assumption. Good research involves good planning. To avoid biases, adequate controls must be built in from the beginning. Good research is forward looking or prospective in nature, not backward or retrospective. Remember GIGO.

11.11.2 Ethical Issues

We must try to distinguish between what is poor research methodology and what is unethical behavior. Ethical considerations arise when a researcher is manipulative of the hypothesis-testing process. The following are some of the ethical issues that arise when dealing with hypothesis-testing methodology:

- Data collection method—randomization
- Informed consent from human subjects being "treated"
- Type of test —two-tailed or one-tailed
- Choice of level of significance α
- Data snooping
- Cleansing and discarding of data
- Reporting of findings
- Meta-analysis

Data Collection Method—Randomization To eliminate the possibility of potential biases in the results, we must use proper data collection methods. To be able to draw meaningful conclusions, the data we obtain must be the outcomes of a random sample from some underlying population or the outcomes from some experiment in which a randomization process was employed. Potential subjects should not be permitted to self select for a study. In a similar manner, a researcher should not be permitted to pick the subjects for the study. Aside from the potential ethical issues that may be raised, such lacks of randomization can result in serious coverage errors or selection biases and destroy the value of any study.

- **Informed Consent from Human Subjects Being "Treated"** Ethical considerations require that any individual who is to be subjected to some "treatment" in an experiment be apprised of the research endeavor and any potential behavioral or physical side effects and provide informed consent with respect to participation. A researcher is not permitted to dupe or manipulate the subjects in a study.

- **Type of Test—Two-tailed or One-tailed** If we have prior information that leads us to test the null hypothesis against a specifically directed alternative, then a one-tailed test will be more powerful than a two-tailed test. On the other hand, we should realize that if we are interested only in *differences* from the null hypothesis, not in the *direction* of the difference, the two-tailed test is the appropriate procedure to utilize. This is an important point. For example, if previous research and statistical testing has already established the difference in a particular direction or if an established scientific theory states that it is only possible for results to occur in one direction, then a one-tailed or directional test may be employed. However, these conditions are not often satisfied in practice and it is recommended that one-tailed tests be used cautiously. Using arguments based on ethical principles, Fleiss (see Reference 3) and other statisticians have stated that, in the overwhelming majority of research studies, a two-tailed test should be employed, particularly if the intention is to report the results to professional colleagues at meetings or in published journal articles. A major reason for this more conservative approach to testing is to enable us to draw more appropriate conclusions on data that may yield unexpected, counterintuitive results.

- **Choice of Level of Significance α** In a well-designed experiment or study the level of significance α is selected in advance of data collection. In fact, we have seen from Equation (11.2) in Section 11.10 that the desired level of significance along with the desired level of statistical power determines the sample size. One cannot be permitted to alter the level of significance, after the fact, in order to achieve a specific result. This would be "data snooping."

- **Data Snooping** *Data snooping* is never permissible. It would be unethical to perform a hypothesis test on a set of data, look at the results, and then select whether it should be two-tailed or one-tailed and/or choose the level of significance. These steps must be done first, as part of the planned experiment or study, before the data are collected, in order for the conclusions drawn to have meaning. In those situations in which a statistician is consulted by a researcher late in the process, with data already available, it is imperative that the null and alternative hypotheses be established and the level of significance chosen prior to carrying out the hypothesis test.

- **Cleansing and Discarding of Data** Data cleansing is not data snooping. Data cleansing is an important part of an overall analysis—remember GIGO. We may recall from Chapters 2 and 3 that once a set of raw numerical data is collected, it must be prepared, entered, and organized for further analysis. In the data preparation stage of editing, coding, and transcribing, one has an opportunity to review the files or records for any observation whose measurement seems to be extreme or unusual. Once the data set is entered into a computer file it must be proofread against the transcribed list, providing a second opportunity to correct errors. After this has been accomplished, the outcomes of the numerical variables in the data set should be organized into stem-and-leaf displays and box-and-whisker plots in preparation for further data presentation and *confirmatory analysis*. This *exploratory data analysis* stage gives us a third opportunity to cleanse the data set by flagging outlier observations that need to be checked against the original files or records. In addition, the exploratory data analysis enables us to examine the data graphically with respect to the assumptions underlying a particular hypothesis test procedure needed for confirmatory analysis.

The process of data cleansing raises a major ethical question. Should an observation be removed from a study? The answer is a qualified "yes." If it can be determined that a measurement is incomplete or grossly in error owing to some equipment problem or unusual behavioral occurrence unrelated to the study, a decision to discard the observation may be made. Sometimes there is no choice—a mouse in a laboratory experiment inadvertently dies before a final measurement is taken or an individual decides to quit a particular study he has been participating in before a final measurement can be made. The analysis of such **censored data** is not uncommon in the field of biostatistics, where experiments dealing with the possible effectiveness of some drug or treatment used on either laboratory animals or human subjects are regularly conducted (see Reference 5). In a well-designed experiment or study, the researcher would plan, in advance, decision rules regarding the possible discarding of data.

● **Reporting of Findings** When conducting research it is vitally important to document both good and bad results, so that individuals who follow up on such research do not have to "reinvent the wheel." It would be inappropriate to report the results of hypothesis tests that show statistical significance but not those for which there was insufficient evidence in the findings. Reporting all findings on a particular subject is particularly important when a *meta-analysis* is to be conducted.

● **Meta-Analysis** Meta-analysis is a controversial methodology that uses the hypothesis-testing framework as its underpinning.

> **Meta-analysis** is an objective and quantitative methodology used for combining and summarizing previous research endeavors on a particular subject into an overall or global finding.

Proponents of meta-analysis argue that there is really no alternative, objective methodology for synthesizing prior research results (see References 4 and 11). Whenever a researcher is planning an experiment to investigate some theoretical hypothesis, it is essential as part of the background of the project to undertake a literature review that examines previous findings. It has been argued that multiple studies of the same subject or topic should be regarded as a complex data set that requires the same kind of detailed statistical analyses as would any one study that contains numerous observations. Ironically, without meta-analysis, traditional literature searches have been accomplished in an unscientific fashion, subject to the biases and interpretations of the particular researcher. A well-performed meta-analysis overcomes this. And, as leading proponents of the methodology have argued, there is a real need to use meta-analysis to integrate the findings from prior research because it is rare that a single experiment or study would provide sufficiently definitive answers upon which to base government or business policy.

Nevertheless, meta-analysis has its share of skeptics who believe that the methodology contains many pitfalls and raises several ethical issues. Whereas it is intended to provide for an objective compilation or synthesis of several, presumably related studies or experiments in order to arrive at a single, overall conclusion, there are many researchers who believe that meta-analysis methodology is often nothing more than a "statistical fruit salad"—a compilation of studies performed under different conditions that may have utilized different methods, different operational definitions and measuring scales, and different types of subjects (i.e., apples, oranges, pears, and grapes), and that logical conclusions cannot be made from such aggregations. Even a leading proponent of the methodology, Ingram Olkin, has warned that "doing a meta-analysis is easy, doing one well is hard" (Reference 8). Thus we must be cautious about the results from a meta-analytic study that we may read about and we should critically analyze how it was conducted before we accept its conclusions.

Ethical Considerations: A Summary Again, when discussing ethical issues concerning the hypothesis-testing methodology, the key is *intent*. We must distinguish between poor confirmatory data analysis and unethical practice. Unethical behavior occurs when a researcher willfully causes a selection bias in data collection, manipulates the treatment of human subjects without informed consent, uses data snooping to select the type of test (two-tailed or one-tailed) and/or level of significance to his or her advantage, hides the facts by discarding observations that do not support a stated hypothesis, or fails to report pertinent findings.

11.12 Hypothesis-Testing Methodology: A Review and a Preview

A summary of Chapter 11 is depicted in the chart on page 416. This chapter presented the fundamental underpinnings of hypothesis-testing methodology. On page 384 of Section 11.1 you were given a list emphasizing the important points to be discussed in the chapter. Check over the list now to see whether you feel you have an understanding of these key points. To be sure, you should be able to answer the following conceptual questions:

1. What is the difference between a null (H_0) and alternative (H_1) hypothesis?
2. What is the difference between a Type I and Type II error?
3. What is meant by the power of a test?
4. What is the difference between a one-tailed and a two-tailed test?
5. What is meant by a p value?
6. How can a confidence interval estimate for the population mean provide conclusions to the corresponding hypothesis test for the population mean?
7. What are the interrelationships among α, β, n, and the type of test (i.e., one-tailed or two-tailed)?

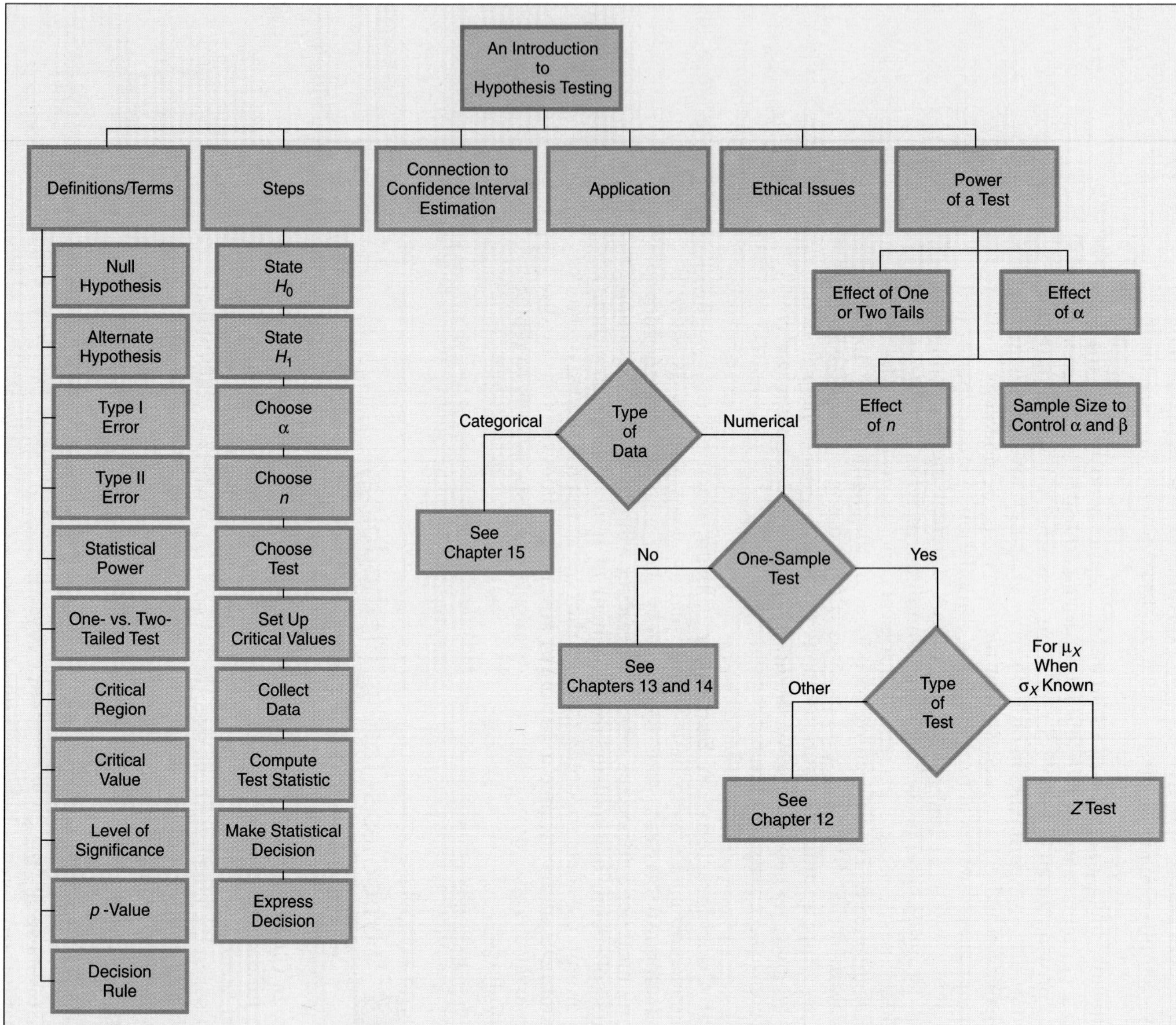

Chapter 11 summary chart.

8. What is the step-by-step hypothesis testing methodology?
9. What are some of the ethical issues to be concerned with when performing a hypothesis test?

Check over the list of questions to see if indeed you know the answers and could (1) explain your answers to someone who did not read this chapter and (2) give reference to specific readings or examples that support your answer. Also, reread any of the sections that may have seemed fuzzy to see if they make sense now.

In the four chapters that follow, we shall be building on the foundations of hypothesis testing that we have discussed here. We will present a set of procedures that may be employed to verify or confirm statistically the results of studies and experiments designed under a variety of conditions.

Getting It All Together

Key Terms

Chapter Review Problems

11.53 ACTION Write a letter to a friend who has not taken a course in statistics and explain what this chapter has been about. To highlight this chapter's content, be sure to incorporate your answers to the nine review questions on page 415 and above.

11.54 When planning to carry out a test of hypothesis based on some designed experiment or research study under investigation, what are some of the questions that need to be raised in order to ensure that proper methodology will be used?

11.55 A businessman was considering the establishment of a Sunday morning bagel and breakfast delivery service in a local suburb and wants to conduct a survey. Based upon the cost of this service and the profits to be made, he has arrived at the following conclusion: If there is evidence that the average order will be more than $14 per household in this suburban area, then the delivery service will be instituted. If no evidence can be demonstrated, the delivery service will not be instituted. Based on past experience with several other suburbs, the standard deviation is estimated to be \$3. The businessman is willing to take a .01 risk of committing a Type I error and institute the service when the actual average order is at most \$14 per household.

(a) If the businessman wishes to have a 97.5% chance of instituting the bagel and breakfast delivery service when the actual population average order is \$17, what sample size should be selected? (*Note:* Assume the data are normally distributed.)

The businessman decides that a random sample of 36 households is to be surveyed.

(b) Compute the probability of instituting the bagel and breakfast delivery service when the average order is actually \$15 per household.

(c) Compute the probability of instituting the bagel and breakfast delivery service when the average order is actually \$17 per household.

(d) Discuss the differences in your results in (a) and (c).

If the businessman was willing to take a .05 risk (rather than a .01 risk) that the service will be instituted when the average order is at most \$14 per household, compute the probability of instituting the bagel and breakfast delivery service when the average order is actually

(e) \$15 per household

(f) \$17 per household

(g) Compare the results in (e) and (f) with those in (b) and (c). What conclusions can you draw here?

If the businessman was willing to select a random sample of 64 households and was also willing to take a risk of $\alpha = .01$ of committing a Type I error, compute the probability of instituting the bagel and breakfast delivery service when the average order is actually

(h) \$15 per household

(i) \$17 per household

(j) Compare the results in (h) and (i) with those in (b) and (c). What conclusion can you draw here?

A sample of 36 households is actually surveyed. From this sample, $\overline{X}$, the average order anticipated, is \$15.66. Assuming the population standard deviation (σ_x) in this suburb is \$3:

(k) Estimate with 99% confidence the population average order anticipated.

(l) Using an α level of .01, determine whether there is evidence that μ_x, the population average order anticipated, exceeds \$14.

(m) Based on your results in (l), what decision should the businessman make regarding the bagel and breakfast delivery service? Why?

11.56 A large chain of discount toy stores would like to determine whether a certain toy should be sold and is considering a survey. Based upon past experience with similar toys, the marketing director of the chain has decided that the toy will be marketed only if there is evidence that monthly gross sales receipts for this toy will average more than \$10,000 throughout the chain of stores. Based on past experience, the standard deviation is estimated to be \$1,000. The marketing director is willing to take a .05 risk of committing a Type I error and market the toy when the average monthly gross sales receipts are actually no more than \$10,000.

(a) If the marketing director wishes to have an 80% chance of marketing the toy when the actual average monthly gross sales receipts are \$10,500, what sample size must be selected? (*Note:* Assume the data are normally distributed.)

A random sample of 25 stores is selected for a test-marketing period of one month.

(b) Compute the probability that the toy will be marketed when the average monthly gross sales receipts are actually \$10,500.
(c) Compute the probability that the toy will be marketed when the average monthly gross sales receipts are actually \$10,800.
(d) What might account for the slight discrepancies in your results in (a) and (b)?

If the marketing director was willing to take a .10 risk (rather than a .05 risk) of selling the toy when the average monthly gross sales receipts are no more than \$10,000, compute the probability that the toy will be marketed when the average monthly gross sales receipts are actually

(e) \$10,500
(f) \$10,800
(g) Compare the results in (e) and (f) with those in (b) and (c). What conclusions can you draw here?

If the marketing director could only select a sample of 16 stores in which to test-market the toy and was willing to take a risk of $\alpha = .05$ of committing a Type I error, compute the probability of marketing that toy when the average monthly gross sales receipts are actually

(h) \$10,500
(i) \$10,800
(j) Compare the results in (h) and (i) with those in (b) and (c). What conclusion can you draw here?

From the sample of 25 stores actually surveyed, $\overline{X}$, the average gross sales receipts for the one-month trial period, is \$10,420. Assuming the population standard deviation (σ_x) based on past experience with similar toys is \$1,000:

(k) Estimate with 95% confidence the population average monthly gross sales receipts.
(l) Using an α level of .05, determine whether there is evidence that μ_x, the population average monthly gross sales receipts, exceeds \$10,000.
(m) Based on your results in (l), what decision should the marketing director make regarding this toy? Why?

Endnotes

1. An easy way to remember which probability goes with which type of error is to note that α is the first letter of the Greek alphabet, and it is used to represent the probability of a Type I error. The letter β is the second letter of the Greek alphabet and is used to represent the probability of a Type II error. (If you have trouble remembering the Greek alphabet, note that the word *alphabet* tells you its first two letters.)
2. For situations involving one-tailed tests in which the actual mean μ_1 really exceeds the hypothesized mean, the converse would be true. The larger the actual mean μ_1 compared with the hypothesized mean, the greater would be the power. On the other hand, for two-tailed tests, the greater the *distance* between the actual mean μ_1 and the hypothesized mean, the greater the power of the test.

References

1. Berenson, M. L., D. M. Levine, and M. Goldstein, *Intermediate Statistical Methods and Applications: A Computer Package Approach* (Englewood Cliffs, NJ: Prentice Hall, 1983).
2. Dixon, W. J., and F. J. Massey, Jr., *Introduction to Statistical Analysis*, 4th ed. (New York: McGraw-Hill, 1983).
3. Fleiss, J. L., *Statistical Methods for Rates and Proportions*, 2nd ed. (New York: Wiley, 1981).
4. Glass, G. V., "Primary, Secondary, and Meta-Analysis of Research," *Educational Researcher,* 1976, Vol. 5, pp. 3–8.
5. Lee, E. T., *Statistical Methods for Survival Data Analysis* (Belmont, CA: Wadsworth, 1980).
6. *MINITAB Reference Manual Release 8* (State College, PA: MINITAB, Inc., 1992).
7. Norusis, M., *SPSS Guide to Data Analysis for SPSS-X with Additional Instructions for SPSS/PC+* (Chicago, IL: SPSS Inc., 1986).
8. Olkin, I., "Meta-Analysis: Current Issues in Research Synthesis," invited presentation at Memorial-Sloan Kettering Cancer Center, New York City, December 5, 1990.
9. *SAS User's Guide Version 6* (Raleigh, NC: SAS Institute, 1988).
10. *STATISTIX Version 4.0* (Tallahassee, FL: Analytical Software, Inc., 1992).
11. Wolf, F. M., *Meta-Analysis: Quantitative Methods for Research Synthesis,* Sage University Paper Series on Quantitative Applications in the Social Sciences (Beverly Hills, CA: Sage, 1986).

chapter 12

One-Sample Tests with Numerical Data

CHAPTER OBJECTIVES

To extend the basic principles of hypothesis-testing methodology to the more commonly used one-sample tests involving numerical data. Tests of hypothesis for the mean, for the median, for the variance or standard deviation, and for randomness are developed and used to demonstrate the differences among parametric procedures, distribution-free procedures, and nonparametric procedures, as well as to indicate their advantages and disadvantages.

12.1 Introduction

In Chapter 11 we introduced the fundamental concepts of hypothesis-testing methodology. When dealing with one sample containing numerical data, we used a Z test to determine whether the population mean μ_x was equal to some specified (i.e., hypothesized) value. The Z test employed is based on the condition that the actual population standard deviation σ_x is known or assumed to take on a specified value. Such hypothesis-testing situations, however, are not common. More typically, hypothesis-testing situations involve making decisions based only on sample information.

In this chapter we extend the basic principles of hypothesis-testing methodology to the more commonly used one-sample tests involving numerical data. In particular, we will describe four useful hypothesis-testing procedures that may be employed. Tests of hypothesis for the mean, for the median, for the variance or standard deviation, and for randomness will be developed. Emphasis is given to the assumptions behind the use of the various tests.

Upon completion of this chapter, you should be able to:

1. Know when and how to use the t test for the population mean μ_x
2. Know when and how to use the Wilcoxon signed-ranks test for the population median M_x
3. Understand the concept of robustness
4. Know when and how to use the χ^2 test for the population variance σ_x^2 or standard deviation σ_x
5. Understand the concept of randomness and the idea of "runs"
6. Know when and how to use the Wald-Wolfowitz one-sample runs test for randomness
7. Distinguish among *classical* parametric tests, distribution-free tests, and nonparametric tests, including their advantages and disadvantages
8. Understand the importance of nominal, ordinal, interval, and ratio scaling in the selection of a statistical hypothesis test procedure

12.2 Choosing the Appropriate Test Procedure

We may recall from Section 11.4 that when summarizing the steps involved in hypothesis-testing methodology (see pages 393–394), a major consideration is the selection of the appropriate statistical technique and its corresponding test statistic. Part of a good data analysis is to understand the assumptions underlying each of the hypothesis-testing procedures we shall be encountering and to select the one most appropriate for a given set of conditions. As we shall see in this chapter, all hypothesis-testing procedures can be broadly described as either *parametric, distribution-free,* or *nonparametric.*

12.2.1 Parametric Procedures

In Chapter 11 we used a parametric procedure, the Z test [given by Equation (11.1)], to test a hypothesis about a population mean. In this chapter we will examine two additional parametric procedures, the *t* test for a population mean and the χ^2 test for a population variance or standard deviation. All parametric procedures have three distinguishing characteristics:

> **Parametric test procedures** may be defined as those that (1) require the level of measurement attained on the collected data to be in the form of an *interval* scale or *ratio* scale (see Section 2.3), (2) involve hypothesis testing of specified parameter values (such as $\mu_x = 368$ grams), and (3) require a stringent set of assumptions.

However, we must decide what kinds of testing procedures to choose if

1. The measurements attained on the data are only categorical (i.e., *nominally* scaled) or in ranks (i.e., *ordinally* scaled).
2. The assumptions underlying the use of the parametric methods cannot be met.
3. The situation requires a study of such features as *randomness, independence, symmetry,* or *goodness of fit* rather than the testing of hypotheses about specific values of particular population parameters.

12.2.2 Distribution-Free and Nonparametric Procedures

When parametric methods of hypothesis testing are not applicable, as in such circumstances as these, distribution-free or nonparametric methods of hypothesis testing can be selected (References 2 and 3).

> **Distribution-free test procedures** may be broadly defined as either (1) those whose test statistic does not depend on the form of the underlying population distribution from which the sample data were drawn, or (2) those for which the data are of insufficient strength (i.e., *nominally* scaled or *ordinally* scaled) to warrant meaningful arithmetic operations.
>
> **Nonparametric test procedures** may be defined as those that are not concerned with the parameters of a population.

In this chapter we will describe a distribution-free procedure, the Wilcoxon signed-ranks test for a hypothesized median, and a nonparametric procedure, the Wald-Wolfowitz one-sample runs test for randomness.

There are five major advantages to using distribution-free or nonparametric procedures:

1. They may be used on all types of data—categorical data (nominal scaling), data in rank form (ordinal scaling), as well as data that have been measured more precisely (interval or ratio scaling).
2. They are generally easy to apply and quick to compute when the sample sizes are small. Sometimes they are as simple as just counting how often some feature appears in the data.
3. They make fewer, less stringent assumptions (which are more easily met) than do the parametric procedures. Hence they enjoy wider applicability and yield a more general, broad-based set of conclusions.
4. Nonparametric methods permit the solution of problems that do not involve the testing of population parameters.
5. Depending on the particular procedure selected, distribution-free methods may be equally (or almost) as powerful as the corresponding parametric procedure when the assumptions of the latter are met, and when they are not met may be quite a bit more powerful.

Although distribution-free or nonparametric methods may be advantageously employed in a variety of situations, they have three major shortcomings:

1. It is disadvantageous to use distribution-free methods when all the assumptions of the parametric procedures can be met.
2. As the sample size gets larger, data manipulations required for distribution-free or nonparametric procedures are sometimes laborious unless appropriate computer software is available.
3. Often special tables of critical values are needed for the test statistics obtained by using distribution-free or nonparametric procedures, and these are not as readily available as are the tables of critical values needed for the test statistics obtained by using parametric procedures (Z, t, and χ^2).

12.2.3 Importance of Assumptions in Test Selection

The sensitivity of parametric procedures to violations in the assumptions has been considered in the statistical literature (References 1 and 2). Some parametric test procedures are said to be **robust** because they are relatively insensitive to slight violations in the assumptions. However, with gross violations in the assumptions both the true level of significance (α) and the power of a test ($1 - \beta$) may differ sharply from what otherwise would be expected. In such cases, a parametric test would be invalid and a distribution-free procedure should be selected instead.

On the other hand, it is disadvantageous to use a distribution-free procedure when all the assumptions of the corresponding parametric test can be met. Unless a parametric procedure is employed in these instances, we would not be taking full advantage of the data. Information is lost when we convert such collected data (from an interval or ratio scale) to either ranks (ordinal scale) or categories (nominal scale). In particular, in such circumstances, some very quick and simple distribution-free tests have much less power than their corresponding parametric procedures and should usually be avoided.

As we investigate a variety of hypothesis-testing procedures in Chapters 12 through 15, we will see how part of a good data analysis is to understand the assumptions underlying each of the hypothesis-testing procedures we shall be encountering and to select the one most appropriate for a given set of conditions.

12.3 t Test of Hypothesis for the Mean (σ_x Unknown)

12.3.1 Introduction

In most hypothesis-testing situations dealing with numerical data, the standard deviation σ_x of the population is unknown. However, the actual standard deviation of the population is estimated by computing S, the standard deviation of the sample. If the population is assumed to be normally distributed, we may recall from Section 10.3 that the sampling distribution of the mean will follow a t distribution with $n - 1$ degrees of freedom. In practice, it has been found that as long as the sample size is not very small and the population is not very skewed, the t distribution gives a good approximation to the sampling distribution of the mean. The test statistic for determining the difference between the sample mean $\overline{X}$ and the population mean μ_x when the sample standard deviation S is used is given by

$$t = \frac{\overline{X} - \mu_x}{\frac{S}{\sqrt{n}}} \tag{12.1}$$

where the test statistic t follows a t distribution having $n - 1$ degrees of freedom.

12.3.2 Application

Data file AMPHRS.DAT

To illustrate the use of the (one-sample) t test, suppose that a manufacturer of batteries claims that the average capacity of a certain type of battery that the company produces is at least 140 ampere-hours. An independent consumer protection agency wishes to test the credibility of the manufacturer's claim and measures the capacity of a random sample of 20 batteries from a recently produced batch. The results, in ampere-hours, are as follows:

137.4	140.0	138.8	139.1	144.4	139.2	141.8	137.3	133.5	138.2
141.1	139.7	136.7	136.3	135.6	138.0	140.9	140.6	136.7	134.1

Since the consumer protection agency is interested in whether or not the manufacturer's claim is being overstated, the test is one-tailed and the following null and alternative hypotheses are established:

$$H_0\text{: } \mu_x \geq 140 \text{ ampere-hours}$$
$$H_1\text{: } \mu_x < 140 \text{ ampere-hours}$$

If a level of significance of $\alpha = .05$ is selected, the critical value of the t distribution with $20 - 1 = 19$ degrees of freedom can be obtained from Table E.3, as illustrated in Figure 12.1 and Table 12.1 on page 426. Since the alternative hypothesis H_1

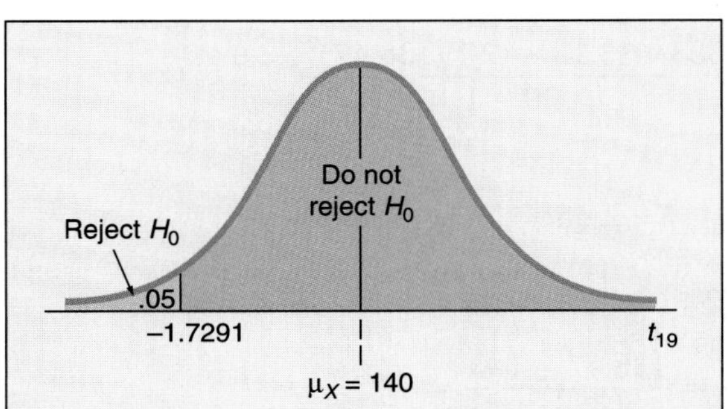

Figure 12.1
Testing a hypothesis about the mean (σ_x unknown) at the .05 level of significance with 19 degrees of freedom.

Table 12.1 Determining the critical value from the t table for an area of .05 in one tail with 19 degrees of freedom.

	Upper-Tail Areas					
Degrees of Freedom	.25	.10	.05	.025	.01	.005
1	1.0000	3.0777	6.3138	12.7062	31.8207	63.6574
2	0.8165	1.8856	2.9200	4.3027	6.9646	9.9248
3	0.7649	1.6377	2.3534	3.1824	4.5407	5.8409
4	0.7407	1.5332	2.1318	2.7764	3.7469	4.6041
5	0.7267	1.4759	2.0150	2.5706	3.3649	4.0322
.	.	.	.	.	.	.
.	.	.	.	.	.	.
.	.	.	.	.	.	.
16	0.6901	1.3368	1.7459	2.1199	2.5835	2.9208
17	0.6892	1.3334	1.7396	2.1098	2.5669	2.8982
18	0.6884	1.3304	1.7341	2.1009	2.5524	2.8784
19	0.6876	1.3277	1.7291	2.0930	2.5395	2.8609
20	0.6870	1.3253	1.7247	2.0860	2.5280	2.8453

Source: Extracted from Table E.3.

that $\mu_x < 140$ ampere-hours is directional, the entire rejection region of .05 is contained in the left tail of the *t* distribution. From the *t* table as given in Table E.3, a replica of which is shown above, the critical value is -1.7291. The decision rule is

$$\text{Reject } H_0 \text{ if } t < t_{19} = -1.7291;$$
$$\text{otherwise do not reject } H_0.$$

For this batch of data

$$\sum_{i=1}^{n} X_i = 2,769.4 \qquad \sum_{i=1}^{n} X_i^2 = 383,613.16 \qquad n = 20$$

Thus,

$$\bar{X} = \frac{\sum_{i=1}^{n} X_i}{n} = \frac{2,769.4}{20} = 138.47$$

and

$$S^2 = \frac{\sum_{i=1}^{n} X_i^2 - n\bar{X}^2}{n-1} = \frac{383,613.16 - (20)(138.47)^2}{20-1} = 7.0706$$

so that

$$S = 2.66$$

Using Equation (12.1), we have

$$t = \frac{\bar{X} - \mu_x}{\frac{S}{\sqrt{n}}} = \frac{138.47 - 140}{\frac{2.66}{\sqrt{20}}} = -2.57$$

Since $t = -2.57 < t_{19} = -1.7291$, the decision is to reject H_0. There is evidence to believe that the manufacturer's claim is overstated, and the consumer protection agency should initiate some corrective measure against the company.

12.3.3 Approximating the *p* Value

Unless we can use a statistical software package, obtaining the *p* value for the *t* distribution is extremely difficult. However, we may use the tables of the *t* distribution (Table E.3) to approximate the *p* value. The computed value for the test statistic with 19 degrees of freedom was −2.57. From Table E.3, we note that owing to the symmetry of the *t* distribution, only upper-tail critical values are shown. Hence, if we neglect the negative sign for purposes of using the table, we observe that the critical value for an upper-tail area of .01 is 2.5395 and the critical value for an upper-tail area of .005 is 2.8609. Since 2.57 is between these values, we also know that the probability of obtaining a *t* value equal to or greater than 2.57 is between .005 and .01. Because of the symmetry of the *t* distribution, the probability of obtaining a *t* value equal to or less than −2.57 is also between .005 and .01. Thus, we can state that the *p* value for this one-sample *t* test is between .005 and .01. Since each of these values is less than .05, the chosen level of significance, the null hypothesis is rejected.

12.3.4 Assumptions of the One-Sample *t* Test

For a given sample size *n*, the test statistic *t* follows a *t* distribution with $n - 1$ degrees of freedom. As we observe in Table E.3, based on available degrees of freedom, each of the rows corresponds to a particular *t* distribution.

The one-sample *t* test is considered a classical parametric procedure. As such, it makes a variety of stringent assumptions which must hold if we are to be assured that the results we obtain from employing the test are valid. In particular, to use the one-sample *t* test it is assumed that the obtained numerical data are independently drawn and represent a random sample from a population that is normally distributed.

As we learned in Section 8.5, the normality assumption can be checked in several ways. A determination of how closely the actual data match the normal distribution's theoretical properties can be made by a descriptive analysis of the obtained statistics along with a graphical analysis to provide a visual interpretation. Thus, by exploring the sample data through a study of its descriptive summary measures along with a graphical analysis (that is, a stem-and-leaf display, a box-and-whisker plot, and a normal probability plot), we may draw our own conclusions as to the likelihood that the underlying population is at least approximately normally distributed. Using the battery capacity ampere-hours data, Figure 12.2 on page 428 depicts MINITAB output displaying the descriptive summary measures, stem-and-leaf, box-and-whisker plot, and a normal probability plot. From these, there is no reason to believe that the assumption of underlying population normality is violated to any great degree and we may conclude that the results obtained by the consumer protection agency are valid.

Although the *t* test is robust if the shape of the population from which the sample is drawn departs somewhat from a normal distribution, particularly when the sample size is large enough to enable the test statistic *t* to be influenced by the central limit theorem (see Section 9.2), erroneous conclusions may be drawn and statistical power can be lost if the *t* test is incorrectly used. Thus, if the sample size

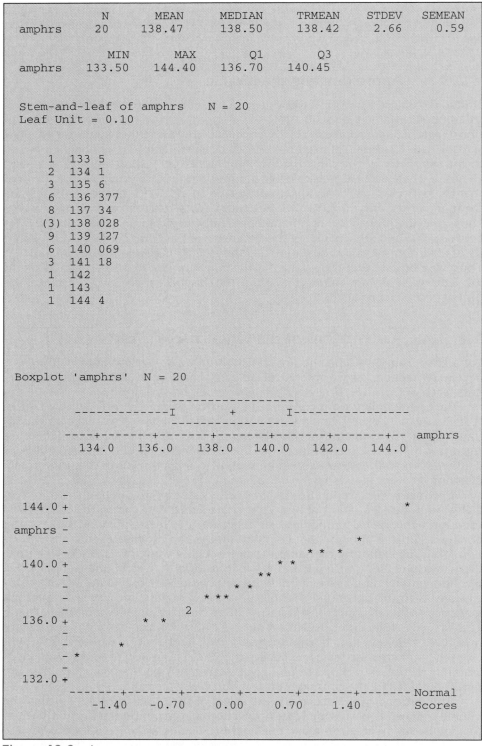

```
             N       MEAN     MEDIAN     TRMEAN     STDEV   SEMEAN
amphrs      20     138.47     138.50     138.42      2.66     0.59

           MIN        MAX         Q1         Q3
amphrs  133.50     144.40     136.70     140.45


Stem-and-leaf of amphrs    N = 20
Leaf Unit = 0.10


    1  133 5
    2  134 1
    3  135 6
    6  136 377
    8  137 34
   (3) 138 028
    9  139 127
    6  140 069
    3  141 18
    1  142
    1  143
    1  144 4


Boxplot 'amphrs'  N = 20

                      -----------------
        -------------I        +       I----------------
                      -----------------
       ----+-------+-------+-------+-------+-------+--  amphrs
         134.0   136.0   138.0   140.0   142.0   144.0


       -
 144.0 +                                              *
       -
amphrs -
       -                                       *
       -                                 * * *
 140.0 +                             * *
       -                          **
       -                       * *
       -                   * **
       -                 2
 136.0 +           * *
       -
       -       *
       - *
       -
 132.0 +
        -------+-------+-------+-------+-------+------- Normal
            -1.40   -0.70    0.00    0.70    1.40       Scores
```

Figure 12.2
MINITAB output for studying the assumptions necessary to employ the *t* test.

n is small (that is, less than 30), and we cannot easily make the assumption that the underlying population from which the sample was drawn is normally distributed, other, distribution-free testing procedures are likely to be more powerful. One such alternative procedure, the one-sample Wilcoxon signed-ranks test, will be described in the next section.

Problems for Section 12.3

● 12.1 A consumers' advocate group would like to evaluate the average energy efficiency rating (that is, EER) of window-mounted, large-capacity (that is, in excess of 7,000 Btu) air-conditioning units. A random sample of 36 such air-conditioning units is selected and tested for a fixed period of time with their EER recorded below:

Data file EER.DAT

8.9	9.1	9.2	9.1	8.4	9.5	9.0	9.6	9.3
9.3	8.9	9.7	8.7	9.4	8.5	8.9	8.4	9.5
9.3	9.3	8.8	9.4	8.9	9.3	9.0	9.2	9.1
9.8	9.6	9.3	9.2	9.1	9.6	9.8	9.5	10.0

(a) Using the .05 level of significance, is there evidence that the average EER is different from 9.0?
(b) What assumptions are being made in order to perform this test?
(c) Find the lower and upper limits for the p value and interpret its meaning.

12.2 A manufacturer of plastics wants to evaluate the durability of rectangularly molded plastic blocks that are to be used in furniture. A random sample of 50 such plastic blocks is examined and the hardness measurements (in Brinell units) are recorded below:

Data file PLASTIC.DAT

283.6	273.3	278.8	238.7	334.9	302.6	239.9	254.6	281.9	270.4
269.1	250.1	301.6	289.2	240.8	267.5	279.3	228.4	265.2	285.9
279.3	252.3	271.7	235.0	313.2	277.8	243.8	295.5	249.3	228.7
255.3	267.2	255.3	281.0	302.1	256.3	233.0	194.4	291.9	263.7
273.6	267.7	283.1	260.9	274.8	277.4	276.9	259.5	262.0	263.5

(a) Using the .05 level of significance, is there evidence that the average hardness of the plastic blocks exceeds 260 (in Brinell units)?
(b) What assumptions are being made in order to perform this test?
(c) Find the lower and upper limits for the p value and interpret its meaning.

● 12.3 The manager of the credit department for an oil company would like to determine whether the average monthly balance of credit card holders is equal to \$75. An auditor selects a random sample of 100 accounts and finds that the average owed is \$83.40 with a sample standard deviation of \$23.65.
(a) Using the .05 level of significance, should the auditor conclude that there is evidence that the average balance is different from \$75?
(b) Find the lower and upper limits for the p value and interpret its meaning.

12.4 A manufacturer of detergent claims that the mean weight of a particular box of detergent is 3.25 pounds. A random sample of 64 boxes revealed a sample average of 3.238 pounds and a sample standard deviation of .117 pound.
(a) Using the .01 level of significance, is there evidence that the average weight of the boxes is different from 3.25 pounds?
(b) Find the lower and upper limits for the p value and interpret its meaning.

12.5 The director of admissions at a large university would like to advise parents of incoming students concerning the cost of textbooks during a typical semester. A sample of 100 students enrolled in the university indicated a sample average cost of \$315.40 with a sample standard deviation of \$43.20.
(a) Using the .10 level of significance, is there evidence that the population average is above \$300?
(b) Find the lower and upper limits for the p value and interpret its meaning.

Interchapter Problems for Section 12.3

Data file MOWER.DAT

12.6 Referring to Problem 4.7 on page 115
(a) Using the .01 level of significance, is there evidence that the average price of the lawn mowers is different from \$175?
(b) What assumptions are being made in order to perform this test?
(c) Find the lower and upper limits for the p value and interpret its meaning.

12.7 Referring to Problem 10.10 on page 353
(a) Is there evidence that the average amount held in passbook savings accounts by depositors is different from \$5,000? (Use the .05 level of significance.)
(b) What assumptions are being made in order to perform this test?
(c) Find the lower and upper limits for the p value and interpret its meaning.

12.8 Referring to Problem 10.11 on page 354
(a) Using the .05 level of significance, is there evidence that the average value of the greeting cards is different from \$1.50?
(b) What assumptions are being made in order to perform this test?
(c) Find the lower and upper limits for the p value and interpret its meaning.

Data file DENTAL.DAT

12.9 Referring to Problem 10.12 on page 354
(a) At the .10 level of significance, is there evidence to enable the personnel manager to conclude that the average family dental expenses of all employees are different from \$320?
(b) What assumptions are being made in order to perform this test?
(c) Find the lower and upper limits for the p value and interpret its meaning.

Data file GASERVE.DAT

● 12.10 Referring to Problem 10.13 on page 354
(a) At the .05 level of significance, is there evidence that the average waiting time between the entry of the service request and the connection of service exceeds 90 days?
(b) What assumptions are being made in order to perform this test?
(c) Find the lower and upper limits for the p value and interpret its meaning.

Data file HMO.DAT

12.11 Referring to Problem 10.14 on page 354
(a) At the .05 level of significance, is there evidence that the average patient waiting time at a local HMO facility is less than 30 minutes?
(b) What assumptions are being made in order to perform this test?
(c) Find the lower and upper limits for the p value and interpret its meaning.

12.4 Wilcoxon Signed-Ranks Test of Hypothesis for the Median

12.4.1 Introduction

The **Wilcoxon signed-ranks test** may be used when we want to test a hypothesis regarding the population median M_x. This distribution-free procedure, which does not make any assumption as to the specific form of the underlying population distribution except that it is approximately symmetrical in shape, may be chosen over its respective parametric counterpart, the t test, when we are able to obtain data measured at a higher level than an ordinal scale but do not believe that the assumptions of the parametric procedure are sufficiently met. When the assumptions of the t test are violated, the Wilcoxon procedure, which makes fewer and less stringent assumptions than does the t test, is likely to be more powerful in detecting the existence of significant differences than its corresponding parametric counterpart. In addition, even under conditions appropriate to the parametric t test, the Wilcoxon signed-ranks test has proven to be almost as powerful. (See References 3 and 4.)

12.4.2 Development

The Wilcoxon signed-ranks test may be used if we are interested in testing a hypothesis regarding a specified population median M_0 based on data from a single sample. The test of the null hypothesis may be one-tailed or two-tailed:

Two-Tailed Test	One-Tailed Test	One-Tailed Test
H_0: Median $= M_0$ H_1: Median $\neq M_0$	H_0: Median $\geq M_0$ H_1: Median $< M_0$	H_0: Median $\leq M_0$ H_1: Median $> M_0$

The assumptions necessary for performing the Wilcoxon signed-ranks test are

1. That the observed data ($X_1, X_2, \ldots, X_n$) constitute a random sample of n independent values from a population with an unknown median.
2. That the underlying random phenomenon of interest is continuous.
3. That the observed data are measured at a higher level than the ordinal scale.
4. That the underlying population is (approximately) symmetrical.

The last assumption represents a major distinction between this distribution-free test procedure and its parametric counterpart, the t test. An assumption of symmetry is not as stringent as an assumption of normality. We should realize from the polygons in Figures 4.19 (rectangular-shaped distribution) and 4.20 (U-shaped distribution) on page 144 that *not all symmetrical distributions are bell-shaped, although all normal distributions are symmetrical and bell-shaped.*

12.4.3 Procedure

To perform the Wilcoxon signed-ranks test the following six-step procedure may be used:

1. We obtain a set of **difference scores D_i** between each of the observed values X_i and the specified value of the hypothesized median M_0—that is, $D_i = X_i - M_0$ where $i = 1, 2, \ldots, n$.
2. We then neglect the "+" and "−" signs and obtain a set of n absolute differences $|D_i|$.
3. We omit from further analysis any absolute difference score of zero, thereby yielding a set of n' nonzero absolute difference scores where $n' \leq n$.

4. We then assign ranks R_i from 1 to n' to each of the $|D_i|$ such that the smallest absolute difference score gets rank 1 and the largest gets rank n'. Owing to a lack of precision in the measuring process, if two or more $|D_i|$ are equal, they are each assigned the average rank of the ranks they would have been assigned individually had ties in the data not occurred.
5. We now reassign the symbol "+" or "−" to each of the n' ranks R_i, depending on whether D_i was originally positive or negative.
6. The Wilcoxon test statistic W is obtained as the sum of the + ranks.

$$W = \sum_{i=1}^{n'} R_i^{(+)} \tag{12.2}$$

Since the sum of the first n' integers $(1, 2, \ldots, n')$ is given by $n'(n'+1)/2$, the Wilcoxon test statistic W may range from a minimum of 0 (where all the observed difference scores are negative) to a maximum of $n'(n'+1)/2$ (where all the observed difference scores are positive). If the null hypothesis were true, we would expect the test statistic W to take on a value close to its mean, $\mu_W = n'(n'+1)/4$. If the null hypothesis were false, we would expect the observed value of the test statistic to be close to one of the extremes.

For samples of $n' \leq 20$, Table E.10 may be used for obtaining the critical values of the test statistic W for both one- and two-tailed tests at various levels of significance. For a two-tailed test and for a particular level of significance, if the observed value of W equals or exceeds the upper critical value or is equal to or less than the lower critical value, the null hypothesis may be rejected. For a one-tailed test in the positive direction the decision rule is to reject the null hypothesis if the observed value of W equals or exceeds the upper critical value. For a one-tailed test in the negative direction the decision rule is to reject the null hypothesis if the observed value of W is less than or equal to the lower critical value.

For samples of $n' > 20$, the test statistic W is approximately normally distributed, and the following large-sample approximation formula may be used for testing the null hypothesis:

$$Z = \frac{W - \mu_W}{\sigma_W} \tag{12.3}$$

where W is the sum of positive ranks; $W = \sum_{i=1}^{n'} R_i^{(+)}$

μ_W is the mean value of W; $\mu_W = \dfrac{n'(n'+1)}{4}$

σ_W is the standard deviation of W; $\sigma_W = \sqrt{\dfrac{n'(n'+1)(2n'+1)}{24}}$

n' is the actual sample size after removing observations with absolute difference scores of zero.

That is,

$$Z = \frac{W - \left(\frac{n'(n'+1)}{4}\right)}{\sqrt{\frac{n'(n'+1)(2n'+1)}{24}}} \quad \textbf{(12.4)}$$

and, based on the level of significance selected, the null hypothesis may be rejected if the computed Z value falls in the appropriate region of rejection, depending on whether a two-tailed test or a one-tailed test is used (see Figure 12.3).

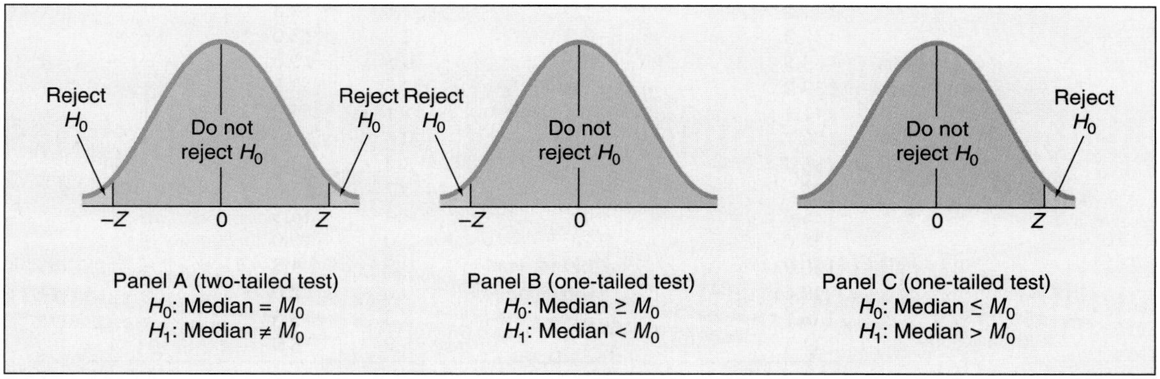

Figure 12.3
Determining the rejection region using the Wilcoxon signed-ranks test.

12.4.4 Application

To illustrate the use of the Wilcoxon signed-ranks test, let us return to the battery capacity data (in ampere-hours) presented in Section 12.3.2 on page 425. Suppose that the independent consumer protection agency wishes to use the random sample of 20 batteries from a recently produced batch to test the credibility of the manufacturer's claim that the average capacity is at least 140 ampere-hours. However, it does not want to make the stringent assumption that the underlying population, that is, the capacity in ampere-hours of all the batteries in the batch, is normally distributed. In such a situation, the distribution-free Wilcoxon signed-ranks test may be employed to test a hypothesis about the population median M_x. Since the consumer protection agency is interested in whether or not the manufacturer's claim is being overstated, the test is one-tailed and the following null and alternative hypotheses are established:

$$H_0: \text{ Median} \geq 140 \text{ ampere-hours}$$

$$H_1: \text{ Median} < 140 \text{ ampere-hours}$$

To perform the one-sample test the first step is to obtain a set of difference scores D_i between each of the observed values X_i and the specified value of the hypothesized median M_0—that is,

$$D_i = X_i - M_0$$

where $i = 1, 2, ..., n$.

The remaining steps of the six-step procedure are developed in Table 12.2.

Table 12.2 Setting up the Wilcoxon (one-sample) signed-ranks test.

Capacity X_i	$D_i = X_i - 140.0$	$\|D_i\|$	R_i	Sign of D_i
137.4	−2.6	2.6	11.0	−
140.0	.0	.0	...	Discard
138.8	−1.2	1.2	7.0	−
139.1	−0.9	0.9	4.5	−
144.4	+4.4	4.4	16.5	+
139.2	−0.8	0.8	3.0	−
141.8	+1.8	1.8	8.5	+
137.3	−2.7	2.7	12.0	−
133.5	−6.5	6.5	19.0	−
138.2	−1.8	1.8	8.5	−
141.1	+1.1	1.1	6.0	+
139.7	−0.3	0.3	1.0	−
136.7	−3.3	3.3	13.5	−
136.3	−3.7	3.7	15.0	−
135.6	−4.4	4.4	16.5	−
138.0	−2.0	2.0	10.0	−
140.9	+0.9	0.9	4.5	+
140.6	+0.6	0.6	2.0	+
136.7	−3.3	3.3	13.5	−
134.1	−5.9	5.9	18.0	−

The test statistic W is obtained as the sum of the positive ranks.

$$W = \sum_{i=1}^{19} R_i^{(+)} = 16.5 + 8.5 + 6 + 4.5 + 2 = 37.5$$

From Table 12.2 we note that after discarding the second observation in the original sample, only 5 of the 19 remaining nonzero absolute-difference scores exceed the hypothesized median of at least 140 ampere-hours. Thus to test for significance we compare the observed value of the test statistic, $W = 37.5$, to the lower-tail critical value presented in Table E.10 for $n' = 19$ and for a level of significance α selected at .05. As shown in Table 12.3 (which is a replica of Table E.10), this critical value is 53. Since $W = 37.5 < W_L = 53$, the null hypothesis may be rejected at the .05 level of significance. There is evidence to believe that the manufacturing claim is overstated, and the protection agency should initiate some corrective measure against the manufacturer.

It is interesting to note that the large-sample approximation formula [Equation (12.4)] for the test statistic yields excellent results for samples as small as 8. With the data of Table 12.2, for a sample of $n' = 19$ (nonzero differences),

$$Z = \frac{W - \left(\frac{n'(n'+1)}{4}\right)}{\sqrt{\frac{n'(n'+1)(2n'+1)}{24}}} = \frac{37.5 - 95}{\sqrt{617.5}} = \frac{-57.5}{24.89} = -2.31$$

Table 12.3 Obtaining lower-tail critical value W for Wilcoxon one-sample signed-ranks test where $n' = 19$ and $\alpha = .05$.

One-Tailed:	$\alpha = .05$	$\alpha = .025$	$\alpha = .01$	$\alpha = .005$
Two-Tailed:	$\alpha = .10$	$\alpha = .05$	$\alpha = .02$	$\alpha = .01$
n'	(Lower, Upper)			
5	0,15	—,—	—,—	—,—
6	2,19	0,21	—,—	—,—
7	3,25	2,26	0,28	—,—
.	.	.	.	.
.	.	.	.	.
.	.	.	.	.
17	41,112	34,119	27,126	23,130
18	47,124	40,131	32,139	27,144
19 →	53,137	46,144	37,153	32,158
20	60,150	52,158	43,167	37,173

Source: Extracted from Table E.10.

Since $Z = -2.31$ is less than the critical Z value of -1.645, the null hypothesis would also be rejected. However, since Table E.10 is available for $n' \leq 20$, it is both simpler and more accurate to just look up the critical value in the table and avoid these computations when possible.

We should note here that for these data the same conclusion has been reached by the independent consumer protection agency, that there is sufficient evidence to reject the manufacturer's claim regarding battery capacity, regardless of whether the parametric t test of Section 12.3 is used or the distribution-free Wilcoxon signed-ranks test is employed. In this situation, the viability of the assumption of underlying normality in the data does not affect the decision reached, regardless of the test procedure used. If, as we suspected from our exploratory data analysis given in Figure 12.2 on page 428, the underlying population is approximately normally distributed, the t test of Section 12.3 would likely be slightly more powerful than the Wilcoxon signed-ranks test in its ability to detect a false null hypothesis. This phenomenon can be observed by comparing the p values of both tests. The more powerful test procedure would result in a smaller p value. Using MINITAB software (see Figure 12.4), the p value associated with the t test is .0093 while the p value associated with the Wilcoxon signed-ranks test is .011.

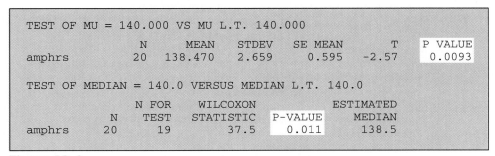

```
TEST OF MU = 140.000 VS MU L.T. 140.000

                  N     MEAN   STDEV   SE MEAN        T   P VALUE
amphrs           20  138.470   2.659     0.595    -2.57    0.0093

TEST OF MEDIAN = 140.0 VERSUS MEDIAN L.T. 140.0

                 N FOR    WILCOXON               ESTIMATED
            N     TEST   STATISTIC   P-VALUE        MEDIAN
amphrs     20       19        37.5     0.011         138.5
```

Figure 12.4
Comparing the p values from the t test and the Wilcoxon signed-ranks test using MINITAB.

Problems for Section 12.4

Data file TAR.DAT

12.12 A cigarette manufacturer claims that the tar content of a new brand of cigarettes is 17 milligrams. A random sample of 24 cigarettes is selected and the tar content measured. The results are shown below in milligrams:

16.9	16.6	17.3	17.5	17.0	17.2	16.1	16.4	17.3	15.9	17.7	18.3
15.6	16.8	17.1	17.2	16.4	18.1	17.4	16.7	16.9	16.0	16.5	17.8

Using a level of significance of $\alpha = .01$, is there evidence that the median tar content of this new brand is different from 17 milligrams?

Data file CLAIM.DAT

12.13 An actuary of a particular insurance company wants to examine the records of larceny claims filed by persons insured under a household goods policy. In the past the median claim was for $125. A random sample of 18 claims is taken and the results are as follows:

$180	$132	$75	$242	$120	$127	$120	$140	$87
$65	$200	$108	$90	$105	$350	$130	$115	$160

Using a level of significance of $\alpha = .05$, is there evidence that the median claim significantly increased?

Data file EER.DAT

12.14 Referring to Problem 12.1 on page 429

(a) Using the .05 level of significance, is there evidence that the median EER of the air conditioners is different from 9.0?

(b) What assumptions are being made in order to perform this test?

(c) Are there any differences between your present results and those obtained using the *t* test in Problem 12.1? Discuss.

Data file PLASTIC.DAT

12.15 Referring to Problem 12.2 on page 429

(a) Using the .05 level of significance, is there evidence that the median hardness (in Brinell units) of the plastic blocks exceeds 260?

(b) What assumptions are being made in order to perform this test?

(c) Are there any differences between your present results and those obtained using the *t* test in Problem 12.2? Discuss.

Interchapter Problems for Section 12.4

Data file MOWER.DAT

12.16 Referring to Problem 4.7 on page 115

(a) Using the .01 level of significance, is there evidence that the median price of the lawn mowers is different from $175?

(b) What assumptions are being made in order to perform this test?

(c) Are there any differences between your present results and those obtained using the *t* test in Problem 12.6 on page 430? Discuss.

12.17 Referring to Problem 10.10 on page 353

(a) What information would you need in order to perform a Wilcoxon signed-ranks test. Discuss fully.

(b) Why might you want to perform a Wilcoxon signed-ranks test here?

(c) What assumptions are being made in order to perform this test?

Data file DENTAL.DAT

12.18 Referring to Problem 10.12 on page 354

(a) Using the .10 level of significance, is there evidence to enable the personnel manager to conclude that the median family dental expenses of all employees are different from $320?

(b) What assumptions are being made in order to perform this test?

(c) Are there any differences between your present results and those obtained using the *t* test in Problem 12.9 on page 430? Discuss.

12.19 Referring to Problem 10.13 on page 354 — Data file GASERVE.DAT

(a) Using the .05 level of significance, is there evidence that the median waiting time between the entry of the service request and the connection of service exceeds 90 days?
(b) What assumptions are being made in order to perform this test?
(c) Are there any differences between your present results and those obtained using the *t* test in Problem 12.10 on page 430? Discuss.

12.20 Referring to Problem 10.14 on page 354 — Data file HMO.DAT

(a) Using the .05 level of significance, is there evidence that the median patient waiting time at a local HMO facility is less than 30 minutes?
(b) What assumptions are being made in order to perform this test?
(c) Are there any differences between your present results and those obtained using the *t* test in Problem 12.11 on page 430? Discuss.

12.5 χ^2 Test of Hypothesis for the Variance (or Standard Deviation)

12.5.1 Introduction

When analyzing numerical data, it is sometimes important to draw conclusions about the variability as well as the average of a characteristic of interest. For example, recall that in our cereal box packaging example (described in Section 11.2), the production manager assumed the company's 15-gram specification for the underlying process standard deviation σ_x was correct and used this parameter value to perform a Z test that the population mean μ_x was 368 grams. Suppose, however, in monitoring whether the equipment used (in the ongoing cereal fill process) is working properly, the production manager is interested in determining whether there is evidence that the standard deviation had changed from the previously specified level of 15 grams. In such a situation, the production manager would be interested in drawing conclusions about the population standard deviation σ_x.

12.5.2 Development

In attempting to draw conclusions about the variability in the population, we first must determine what test statistic can be used to represent the distribution of the variability in the sample data. If the variable (amount of cereal filled in grams) is assumed to be normally distributed, then the test statistic for testing whether or not the population variance is equal to a specified value is

$$\chi^2 = \frac{(n-1)S^2}{\sigma_x^2} \tag{12.5}$$

where n = sample size
S^2 = sample variance
σ_x^2 = hypothesized population variance

and the test statistic χ^2 follows a chi-square distribution with $(n - 1)$ degrees of freedom.

If, as shown in panel A of Figure 12.5, the test of hypothesis is two-tailed, the rejection region is split into both the lower and the upper tails of the chi-square distribution. However, if the test is one-tailed, the rejection region is either in the lower tail (panel B of Figure 12.5) or in the upper tail (panel C of Figure 12.5) of the chi-square distribution, depending on the direction of the alternative hypothesis.

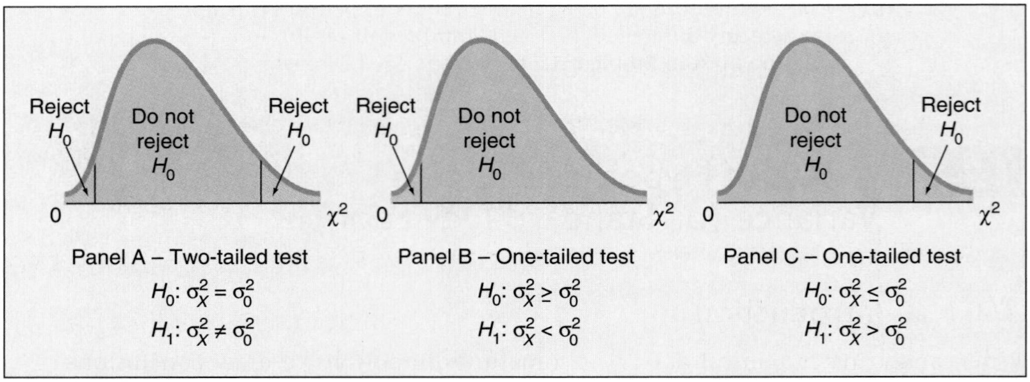

Figure 12.5
Testing a hypothesis about a population variance—one-tailed and two-tailed tests: Panel A, two-tailed test; Panel B, one-tailed test; Panel C, one-tailed test.

For a given sample size n, the test statistic χ^2 follows a chi-square distribution with $n - 1$ degrees of freedom. A **chi-square distribution** is a skewed distribution whose shape depends solely on its number of degrees of freedom. As the number of degrees of freedom increases, a chi-square distribution becomes more symmetrical. Table E.4 contains various upper-tail areas for chi-square distributions pertaining to different degrees of freedom. A portion of this table is displayed as Table 12.4.

Table 12.4 Obtaining the critical values from the chi-square distribution with 1 degree of freedom using a level of significance of $\alpha = .10$.

Degrees of Freedom	Upper-Tail Area									
	.995	.99	.975	.95	.90	.75	.25	.10	.05	.025
1	...	...	0.001	0.004	0.016	0.102	1.323	2.706	3.841	5.024
2	0.010	0.020	0.051	0.103	0.211	0.575	2.773	4.605	5.991	7.378
3	0.072	0.115	0.216	0.352	0.584	1.213	4.108	6.251	7.815	9.348
.	.	.	.	.	.	.	.	.	.	.
.	.	.	.	.	.	.	.	.	.	.
.	.	.	.	.	.	.	.	.	.	.
23	9.260	10.196	11.689	13.091	14.848	18.137	27.141	32.007	35.172	38.076
24	9.886	10.856	12.401	13.848	15.659	19.037	28.241	33.196	36.415	39.364
25	10.520	11.524	13.120	14.611	16.473	19.939	29.339	34.382	37.652	40.646

Source: Extracted from Table E.4.

The value at the top of each column indicates the area in the upper portion (or right side) of a particular chi-square distribution. For example, with 1 degree of freedom the critical value of the χ^2 test statistic corresponding to an upper-tail area of .10 is 2.706 (see Figure 12.6). This means that, for 1 degree of freedom, the probability of exceeding this critical value of 2.706 is .10. Therefore, once we determine the level of significance and the degrees of freedom, the critical value of the χ^2 test statistic can be found from a particular chi-square distribution.

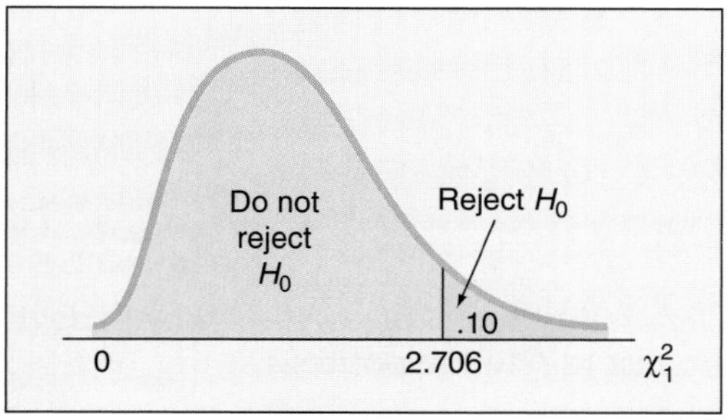

Figure 12.6
Finding the critical value of the χ^2 test statistic from a chi-square distribution with 1 degree of freedom using a .10 level of significance.

12.5.3 Application

To apply the test of hypothesis let us again return to the cereal box packaging example. The production manager is interested in determining whether there is evidence that the standard deviation had changed from the previously specified level of 15 grams. Thus, we have a two-tailed test in which the null and alternative hypotheses can be stated as follows:

$$H_0: \quad \sigma_x = 15 \text{ grams (or } \sigma_x^2 = 225 \text{ “grams squared”)}$$

$$H_1: \quad \sigma_x \neq 15 \text{ grams (or } \sigma_x^2 \neq 225 \text{ “grams squared”)}$$

Since this is a two-tailed test based on a sample of 25 boxes, the null hypothesis would be rejected if the χ^2 test statistic fell into either the lower or upper tail of a chi-square distribution with 24 degrees of freedom as shown in Figure 12.7.

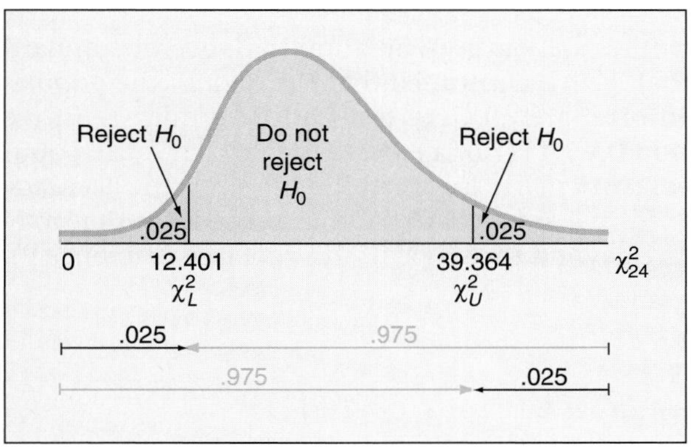

Figure 12.7
Determining the lower and upper critical values of a chi-square distribution with 24 degrees of freedom for a two-tailed test of hypothesis about a population standard deviation using a .05 level of significance.

Since there are 24 degrees of freedom (that is, 25 − 1 = 24), if a level of significance of .05 was selected, the lower (χ^2_L) and upper (χ^2_U) critical values could be obtained from the table of the chi-square distribution (Table E.4). The values at the top of the table indicate the upper-tail areas of a chi-square distribution, that is, the portion under the curve on the right side of the chi-square distribution. Thus we can obtain the lower critical value χ_L^2 of 12.401 from Table E.4 by looking in the column labeled ".975" for 24 degrees of freedom and we can obtain the upper critical value χ_U^2 of 39.364 by looking in the column labeled ".025" for 24 degrees of freedom.

Therefore, the decision rule would be

$$\text{Reject } H_0 \text{ if } \chi^2 > \chi_U^2 = 39.364$$
$$\text{or if } \chi^2 < \chi_L^2 = 12.401;$$
$$\text{otherwise do not reject } H_0.$$

Suppose that from the production manager's sample of 25 boxes the standard deviation (*S*) is computed to be 17.7 grams. To test the null hypothesis at the .05 level of significance using Equation (12.5) we have

$$\chi^2 = \frac{(n-1)S^2}{\sigma_x^2} = \frac{(25-1)(17.7)^2}{15^2} = 33.42$$

Since $\chi_L^2 = 12.401 < \chi^2 = 33.42 < \chi_U^2 = 39.364$, we do not reject H_0. Using the *p* value approach, the probability of obtaining a χ^2 test statistic of 33.42 or larger is slightly less than .10. Since this value is greater than the upper-tail area of .025 (for the two-tailed test), the null hypothesis cannot be rejected. The production manager would conclude that there is no evidence that the actual process (that is, population) standard deviation is different from 15 grams.

When testing a hypothesis about a population variance or standard deviation, it is frequently the case that we are interested in detecting whether the variation in a process has increased. In such circumstances, a one-tailed hypothesis test would be used. (See Reference 5.) The null hypothesis would be rejected at a chosen α level of significance if the computed χ^2 test statistic exceeds the upper-tail critical value (χ_U^2) from a chi-square distribution with $n - 1$ degrees of freedom as in panel C of Figure 12.5 on page 438.

● **Caution** In testing a hypothesis about a population variance or standard deviation, we should be aware that we have assumed that the data in the population are normally distributed. Unfortunately, this χ^2 test statistic is quite sensitive to departures from this assumption (i.e., it is not a *robust* test), so that if the population is not normally distributed, particularly for small sample sizes, the accuracy of the test can be seriously affected (see Reference 1). Other procedures that may be used to address this problem are beyond the scope of this text (see Reference 9).

Problems for Section 12.5

● 12.21 A manufacturer of candy must monitor the temperature at which the candies are baked. Too much variation will cause inconsistency in the taste of the candy. Past records show that the standard deviation of the temperature has been 1.2° F. A random sample of 30 batches of candy is selected and the sample standard deviation of the temperature is 2.1° F.

(a) At the .05 level of significance, is there evidence that the population standard deviation has increased above 1.2° F?
(b) What assumptions are being made in order to perform this test?
(c) Compute the p value in part (a) and interpret its meaning.

12.22 A market researcher for an automobile dealer intends to conduct a nationwide survey concerning car repairs. Among the questions to be included in the survey is the following: "What was the cost of all repairs performed on your car last year?" In order to determine the sample size necessary, he needs to obtain an estimate of the standard deviation. Using his past experience and judgment, he estimates that the standard deviation of the amount of repairs is $200. Suppose that a pilot study of 25 auto owners selected at random indicates a sample standard deviation of $237.52.
(a) At the .05 level of significance, is there evidence that the population standard deviation is different from $200?
(b) What assumptions are being made in order to perform this test?
(c) Compute the p value in part (a) and interpret its meaning.

12.23 The marketing manager of a branch office of a local telephone operating company would like to study characteristics of residential customers served by her office. In particular, she would like to estimate the average monthly cost of calls within the local calling region. In order to determine the sample size necessary, an estimate of the standard deviation must be made. Based on her past experience and judgment, she estimates that the standard deviation is equal to $12. Suppose that a pilot study of 15 residential customers indicates a sample standard deviation of $9.25.
(a) At the .10 level of significance, is there evidence that the population standard deviation is different from $12?
(b) What assumptions are being made in order to perform this test?
(c) Compute the p value in part (a) and interpret its meaning.

12.24 A manufacturer of doorknobs has a production process that is designed to provide a doorknob with a target diameter of 2.5 inches. In the past the standard deviation of the diameter has been .035 inch. In an effort to reduce the variation in the process, various studies have taken place that have resulted in a redesigned process. A sample of 25 doorknobs produced under the new process indicates a sample standard deviation of .025 inch.
(a) At the .05 level of significance, is there evidence that the population standard deviation is less than .035 inch in the new process?
(b) What assumptions are being made in order to perform this test?
(c) Compute the p value in part (a) and interpret its meaning.

12.25 Refer to the battery capacity data on page 425

Data file AMPHRS.DAT

(a) At the .05 level of significance, is there evidence that the population standard deviation exceeds 2.5 ampere-hours?
(b) What assumptions are being made in order to perform this test?
(c) Compute the p value in part (a) and interpret its meaning.

Interchapter Problems for Section 12.5

12.26 In Problem 10.5 on page 349, we noted that the population standard deviation was expected to be equal to .02 gallon. If a random sample of 25 one-gallon cans of paint revealed a sample standard deviation of .025 gallon,
(a) At the .05 level of significance, is there evidence that the population standard deviation has changed?
(b) What assumptions are being made in order to perform this test?
(c) Compute the p value in part (a) and interpret its meaning.

● 12.27 In Problem 10.6 on page 349, we noted that the process standard deviation of the life of electric light bulbs was known to be 100 hours. If a random sample of 20 light bulbs indicates a sample standard deviation of 110 hours,
(a) At the .05 level of significance, is there evidence that the process standard deviation has changed?
(b) What assumptions are being made in order to perform this test?
(c) Compute the p value in part (a) and interpret its meaning.

● 12.28 In Problem 10.7 on page 349, we noted that the standard deviation for 2-liter soda bottles was .05 liter. As part of the quality control process, the bottling company wishes to know whether the standard deviation has increased above .05 liter. A random sample of ten 2-liter bottles indicated a sample standard deviation of .083 liter.
(a) At the .01 level of significance, is there evidence that the process standard deviation has increased?
(b) What assumptions are being made in order to perform this test?
(c) Compute the p value in part (a) and interpret its meaning.

12.6 Wald-Wolfowitz One-Sample Runs Test for Randomness

12.6.1 Introduction

It is usually assumed that the data collected in a study constitute a random sample so that each observation or measurement is randomly and independently drawn from its population. Such an assumption, however, may be tested by the employment of a nonparametric procedure called the **Wald-Wolfowitz one-sample runs test for randomness.** This nonparametric procedure is not concerned with the testing of any particular parameters and as such has no parametric counterpart.

To test for randomness, the null hypothesis would be:

H_0: The process that generates the set of numerical data is random.

The alternative hypothesis would be:

H_1: The process that generates the set of numerical data isn't random.

The null hypothesis of randomness may be tested by observing the *order* or *sequence* in which the items in the sample are obtained. If each item is assigned one of two symbols, such as S and F (for "success" and "failure"), depending on whether its measurement falls above or below a certain value (that is, above or below the median), the randomness of the sequence may be investigated. If the sequence is randomly generated, the value (S or F) of an item will be independent both of its position in the sequence and of the values of the items that precede it and follow it. On the other hand, if the value of an item in the sequence is affected by the values of the items that precede it or succeed it or the probability of its occurrence depends on its position in the sequence, the process generating the sequence is not considered random. In such nonrandom cases either similar items would tend to cluster together (such as when a trend is present in the data) or the similar items would alternatingly mix so that some systematic periodic effect would exist.

To study whether or not an observed sequence is random, we will consider as the test statistic the number of runs present in the data.

> A **run** is defined as a consecutive series of similar items that are bounded by items of a different type or by the beginning or ending of the sequence.

For instance, suppose that the following represents the observed results of an experiment in which a coin was tossed 20 times:

HHHHHHHHHHTTTTTTTTTT

In this sequence there are two runs—a run of 10 heads followed by a run of 10 tails. With similar items tending to cluster, such a sequence would not be considered random even though, as would theoretically be expected when a fair coin is tossed, 10 of the 20 outcomes are heads and 10 are tails.

At the other extreme, suppose that the following sequence is obtained when tossing a coin 20 times:

HTHTHTHTHTHTHTHTHTHT

In this sequence there are 20 runs—10 runs of one head each and 10 runs of one tail each. With such a systematic alternating pattern, this sequence could not be considered random because there are too many runs.

On the other hand, if, as shown below, the sequence of responses to the 20 coin tosses is thoroughly mixed, the number of runs will be neither too few nor too many, and the process may then be considered random:

HHTTHHHHTTTTTHTHTTHH

Therefore, in testing for randomness, what is essential is the *ordering* or *positioning* of the items in the sequence, not just the frequency of items of each type.

12.6.2 Development

To test the null hypothesis of randomness, we may split the total sample size n into two parts, n_1, the number of *successes*, and n_2, the number of *failures*. The test statistic denoted by the symbol U, the total number of runs, is then obtained by counting. For a two-tailed test, if U is either larger or smaller than we would expect in a random series of data, we would reject the null hypothesis of randomness in favor of the alternative that the sequence is not random. If both n_1 and n_2 are less than or equal to 20, Table E.9, Parts 1 and 2, present the critical values for the test statistic U at the $\alpha = .05$ level of significance (two-tailed). If, for a given combination of n_1 and n_2, U is either greater than or equal to the upper critical value or less than or equal to the lower critical value, the null hypothesis of randomness may be rejected at the $\alpha = .05$ level. However, if U lies between these limits, the null hypothesis of randomness cannot be rejected.

On the other hand, tests for randomness are not always two-tailed. If we are interested in testing for randomness against the specific alternative of a **trend effect**—that there is a tendency for like items to cluster together—a one-tailed test is needed. Here we reject the null hypothesis only if too few runs occur—if the observed value of U is less than or equal to the critical value presented in Table E.9, Part 1, at the $\alpha = .025$ level of significance. At the other extreme, if we are interested in testing for randomness against a **systematic** or **periodic effect,** we use a one-tailed test that rejects only if too many runs occur—if the observed value of U is greater than or equal to the critical value given in Table E.9, Part 2, at the $\alpha = .025$ level of significance.

Regardless of whether the test is one-tailed or two-tailed, however, for a sample size n greater than 40 (or when either n_1 or n_2 exceeds 20) the test statistic U is approximately normally distributed. Therefore, the following large-sample approximation formula may be used to test the hypothesis of randomness:

$$Z = \frac{U - \mu_u}{\sigma_u} \tag{12.6}$$

where U = total number of observed runs

μ_u = mean value of U; $\mu_u = \frac{2n_1n_2}{n} + 1$

σ_u = standard deviation of U; $\sigma_u = \sqrt{\frac{2n_1n_2(2n_1n_2 - n)}{n^2(n-1)}}$

n_1 = number of "successes" in sample

n_2 = number of "failures" in sample

n = sample size; $n = n_1 + n_2$

That is,

$$Z = \frac{U - \left(\frac{2n_1n_2}{n} + 1\right)}{\sqrt{\frac{2n_1n_2(2n_1n_2 - n)}{n^2(n-1)}}} \tag{12.7}$$

and, based on the level of significance selected, the null hypothesis may be rejected if the computed Z value falls in the appropriate region of rejection, depending on whether a two-tailed test or a one-tailed test is used (see Figure 12.8).

12.6.3 Application

To illustrate the use of the Wald-Wolfowitz one-sample runs test for randomness, Table 12.5 presents the unemployment rates (per thousand) of clerical workers in the United States from 1960 through 1993.

A distinguishing feature of the Wald-Wolfowitz one-sample runs test for randomness is that it may be used not only on data that constitute a nominal scale,

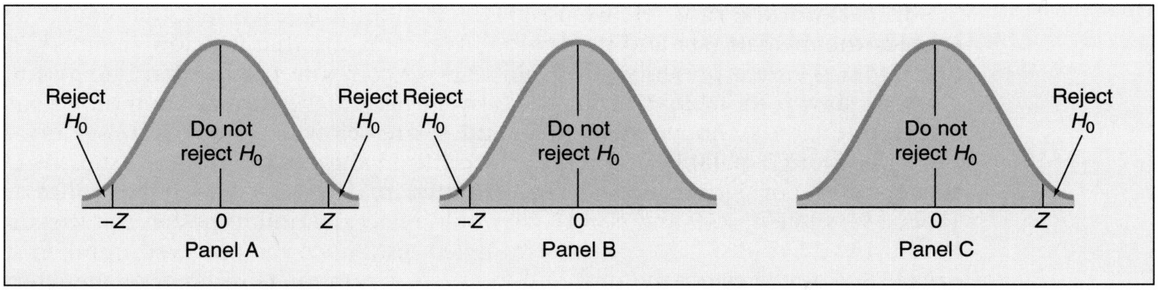

Figure 12.8
Determining the rejection region; Panel A, two-tailed test; Panel B, one-tailed test, trend effect; Panel C, one-tailed test, periodic effect.

Table 12.5 U.S. clerical workers' unemployment rates* (1960–1993).

Year	Unemployment Rates (per thousand)	Relationship to Median Rate of 4.6†	Year	Unemployment Rates (per thousand)	Relationship to Median Rate of 4.6†
1960	3.8	B	1977	5.9	A
1961	4.6	A	1978	4.9	A
1962	4.0	B	1979	4.6	A
1963	4.0	B	1980	5.3	A
1964	3.7	B	1981	5.7	A
1965	3.3	B	1982	7.0	A
1966	2.9	B	1983	6.4	A
1967	3.1	B	1984	5.1	A
1968	3.0	B	1985	4.9	A
1969	3.0	B	1986	4.7	A
1970	4.0	B	1987	4.2	B
1971	4.8	A	1988	3.9	B
1972	4.7	A	1989	3.9	B
1973	4.2	B	1990	4.1	B
1974	4.6	A	1991	5.0	A
1975	6.6	A	1992	5.7	A
1976	6.4	A	1993	5.1	A

* In 1983 the occupational classifications were changed. From that year to the present, clerical workers comprise the major component of administrative support services.

† A, equal or above; B, below.

Sources: Data are extracted from Table 28, *Handbook of Labor Statistics Bulletin* 2175, U.S. Department of Labor, Bureau of Labor Statistics, December 1984, and from Table 10, *Employment & Earnings,* U.S. Department of Labor, Bureau of Labor Statistics, January 1986, 1988, 1990, 1992, 1994.

where each of the items is classified a success or failure, but also on data measured in the strength of an interval or ratio scale. When interval or ratio scaled data are used, each of the items is classified according to its position with respect to the median of the sequence. For example, from Table 12.5 we may wish to test the null hypothesis that the unemployment rates of clerical workers are randomly distributed with respect to the median over time against the alternative that these rates are not randomly distributed with respect to the median over time; that is,

H_0: Unemployment rates of clerical workers are random over time.

H_1: Unemployment rates of clerical workers are not random over time (two-tailed).

To perform the runs test we assign the symbol A to each rate that equals or exceeds the median rate and the symbol B to each rate that is below the median rate. From the data presented in Table 12.5 we compute the median rate as 4.6. Thus, as shown in Table 12.5 for the 34-year period, 19 annual unemployment rates are equal to or above the median and 15 are below it.

Parts 1 and 2 of Table E.9 present the critical values of the runs test statistic U at the .05 level of significance. As demonstrated in Table 12.6 (which is a replica of Table E.9), since $n_1 = 19$ and $n_2 = 15$ we would reject the null hypothesis at the .05 level if $U \leq 11$ or if $U \geq 24$ for this two-tailed test. Since the observed number of runs is 8, we may reject the null hypothesis of randomness in favor of the alternative. There is apparently a pattern over time to the annual rates of unemployment for clerical workers. If the null hypothesis were true, the p value or probability of obtaining such a result as this or one even more extreme would be less than .05.

It is interesting to note that the large-sample approximation formula [Equation (12.7)] for the test statistic U yields excellent results when the sample size n is less than 40. For example, using the data from Table 12.5 on page 445, for a sample of $n = 34$ annual unemployment rates with $n_1 = 19$ at or above the median rate of 4.6 and $n_2 = 15$ below the median rate,

$$
\begin{aligned}
Z &= \frac{U - \left(\dfrac{2n_1n_2}{n} + 1\right)}{\sqrt{\dfrac{2n_1n_2(2n_1n_2 - n)}{n^2(n-1)}}} \\
&= \frac{8 - \left(\dfrac{(2)(19)(15)}{34} + 1\right)}{\sqrt{\dfrac{[(2)(19)(15)][(2)(19)(15) - 34]}{(34^2)(33)}}} \\
&= \frac{-9.765}{\sqrt{8.0088}} \\
&= -3.45
\end{aligned}
$$

Since $Z = -3.45 < -1.96$, the lower-tail critical value from the standard normal distribution using an α level of significance of .05, the null hypothesis of randomness can be rejected. There is a pattern in the annual unemployment rates of clerical workers. If the null hypothesis were true, the p value or probability of obtaining such a result as this (that is, 3.45 standard deviations from μ_u, the expected number of runs) or one even more extreme would be .00056 (that is, the total area in the two tails of the standard normal distribution—the area below $Z = -3.45$ standard deviations and the area above $Z = +3.45$ standard deviations).

Problems for Section 12.6

12.29 Starting at the beginning of the table of random numbers (Table E.1), for the first 100 digits observed, record the sequence of *high digits* (that is, 5, 6, 7, 8, or 9) and *low digits* (that is, 0, 1, 2, 3, or 4) based on whether the digits are above or below the median value of 4.5. Can this resulting sequence of high and low digits be considered random? (Use $\alpha = .05$.)

Table 12.6 Obtaining the lower- and upper-tail critical values *U* for the runs test where $n_1 = 19$, $n_2 = 15$, and $\alpha = .05$.

Part 1. Lower Tail
($\alpha = .025$)

n_1 \ n_2	2	3	4	5	6	7	8	9	10	11	12	13	14	15	16	17	18	19	20
2											2	2	2	2	2	2	2	2	2
3					2	2	2	2	2	2	2	2	2	3	3	3	3	3	3
4				2	2	2	3	3	3	3	3	3	3	3	4	4	4	4	4
5			2	2	3	3	3	3	3	4	4	4	4	4	4	4	5	5	5
6		2	2	3	3	3	3	4	4	4	4	5	5	5	5	5	5	6	6
7		2	2	3	3	3	4	4	5	5	5	5	5	6	6	6	6	6	6
8		2	3	3	3	4	4	5	5	5	6	6	6	6	6	7	7	7	7
9		2	3	3	4	4	5	5	5	6	6	6	7	7	7	7	8	8	8
10		2	3	3	4	5	5	5	6	6	7	7	7	7	8	8	8	8	9
11		2	3	4	4	5	5	6	6	7	7	7	8	8	8	9	9	9	9
12	2	2	3	4	4	5	6	6	7	7	7	8	8	8	9	9	9	10	10
13	2	2	3	4	5	5	6	6	7	7	8	8	9	9	9	10	10	10	10
14	2	2	3	4	5	5	6	7	7	8	8	9	9	9	10	10	10	11	11
15	2	3	3	4	5	6	6	7	7	8	8	9	9	10	10	11	11	11	12
16	2	3	4	4	5	6	6	7	8	8	9	9	10	10	11	11	11	12	12
17	2	3	4	4	5	6	7	7	8	9	9	10	10	11	11	11	12	12	13
18	2	3	4	5	5	6	7	8	8	9	9	10	10	11	11	12	12	13	13
19	2	3	4	5	6	6	7	8	8	9	10	10	11 →	11	12	12	13	13	13
20	2	3	4	5	6	6	7	8	9	9	10	10	11	12	12	13	13	13	14

Part 2. Upper Tail
($\alpha = .025$)

n_1 \ n_2	2	3	4	5	6	7	8	9	10	11	12	13	14	15	16	17	18	19	20
2																			
3																			
4				9	9														
5			9	10	10	11	11												
6			9	10	11	12	12	13	13	13	13								
7				11	12	13	13	14	14	14	14	15	15	15					
8				11	12	13	14	14	15	15	16	16	16	16	17	17	17	17	17
9					13	14	14	15	16	16	16	17	17	18	18	18	18	18	18
10					13	14	15	16	16	17	17	18	18	18	19	19	19	20	20
11					13	14	15	16	17	17	18	19	19	19	20	20	20	21	21
12					13	14	16	16	17	18	19	19	20	20	21	21	21	22	22
13						15	16	17	18	19	19	20	20	21	21	22	22	23	23
14						15	16	17	18	19	20	20	21	22	22	23	23	23	24
15						15	16	18	18	19	20	21	22	22	23	23	24	24	25
16							17	18	19	20	21	21	22	23	23	24	25	25	25
17							17	18	19	20	21	22	23	23	24	25	25	26	26
18							17	18	19	20	21	22	23	24	25	25	26	26	27
19							17	18	20	21	22	23	23 →	24	25	26	26	27	27
20							17	18	20	21	22	23	24	25	25	26	27	27	28

Source: Extracted from Table E.9, Parts 1 and 2.

12.30 Select a page from your local telephone directory and, examining only the last digit of each telephone number, record the sequence of high (5, 6, 7, 8, 9) and low (0, 1, 2, 3, 4) digits. Can the resulting sequence be considered random? (Use $\alpha = .05$.) Present your results in class and discuss.

Data file VET.DAT

12.31 During the period from 1960 to 1993 there was an increase in the federal budget outlays for veterans' benefits and services. During this period, however, total federal outlays for all functions also increased. The data in the accompanying table present the percentage of total federal outlays for veterans' benefits and services during the 34-year period from 1960 to 1993. With respect to fluctuations below versus at or above the median, is there any evidence of trend over the 34-year period? (Use $\alpha = .025$.)

Percentage of total federal outlays to veterans' benefits and services.

Year	Percentage	Year	Percentage
1960	5.9	1977	4.5
1961	5.8	1978	4.2
1962	5.3	1979	4.0
1963	5.0	1980	3.6
1964	4.8	1981	3.4
1965	4.8	1982	3.2
1966	4.4	1983	3.1
1967	4.4	1984	3.0
1968	3.9	1985	2.8
1969	4.1	1986	2.7
1970	4.4	1987	2.6
1971	4.6	1988	2.8
1972	4.6	1989	2.6
1973	4.9	1990	2.3
1974	5.0	1991	2.4
1975	5.1	1992	2.5
1976	5.0	1993	2.4

Source: Data are extracted from Table 512, *Statistical Abstract of the United States,* U.S. Department of Commerce, 1993.

Data file AMPHRS.DAT

12.32 Refer to the battery capacity data on page 425. The battery capacities (in ampere-hours) are listed from left to right in two rows, that is, in the sequence in which the sample of 20 batteries were selected from a recently produced batch. With respect to fluctuations above and below the sample median of 138.5 ampere-hours, is there evidence that this resulting sequence is not random? (Use $\alpha = .05$.)

Interchapter Problems for Section 12.6

Data file TELLER1.DAT

12.33 Figure 3.10 on page 79 lists the length of time (in minutes) it takes a teller to service 24 consecutive customers into a commercial bank during lunch hour. A digidot plot for these data is displayed in Figure 3.13 on page 81. With respect to the fluctuations above and below the median time of 1.60 minutes, can the resulting sequence be considered random? (Use $\alpha = .05$.)

Data file JEANS.DAT

12.34 Use the data in Problem 3.41 on page 82. Is there evidence from the resulting sequence of 30 consecutive pairs of manufactured jeans that the manufacturing process is out of control? (Use $\alpha = .05$.)

Data file QMILE.DAT

12.35 Use the data in Problem 3.42 on page 82. With respect to fluctuations above and below the median, is there evidence of a trend in Victor Sternberg's time trials over the 27-day period? (Use $\alpha = .025$.)

12.7 Using the Computer for Hypothesis Testing: The Kalosha Industries Employee Satisfaction Survey

12.7.1 Introduction and Overview

When dealing with large batches of data, we may use the computer to assist us not only in our descriptive statistical analysis but in our confirmatory analysis as well. In this section we will demonstrate how statistical software packages can be used for performing various tests of hypotheses when we are studying the outcomes of one sample of numerical data. We will gain a further appreciation for the assistance the computer may give us in solving statistical problems—particularly those involving large numbers of variables or large data batches. (See References 6–8 and 10.) To accomplish this, let us return to the Kalosha Industries Employee Satisfaction Survey that was developed in Chapter 2.

12.7.2 Kalosha Industries Employee Satisfaction Survey

Data file EMPSAT.DAT

We may recall from Section 3.8.2 that Bud Conley, the Vice-President of Human Resources, is preparing for a meeting with a representative from the B & L Corporation to discuss the potential contents of an employee-benefits package that is being considered. Even prior to planning for the Employee Satisfaction Survey, Conley had wanted to determine:

1. Whether there was evidence that the average age of all full-time employees at Kalosha Industries is different from 40 years.
2. Whether there was evidence that the median (total) family income of full-time employees at Kalosha Industries exceeds $39,000.

Based on intuition, Bud Conley had claimed that the average age was around 40 and he had believed that the median (total) family income exceeds $39,000. Confirmatory answers to these two questions as well as others that he has raised (see the Survey/Database Project at the end of the section) would be helpful to him now and are required prior to his meeting with the B & L Corporation representative so that he may have some added leverage in the discussions on the employee-benefits package. Answers can be obtained through a confirmatory analysis of the 400 responses to the Survey (see Table 2.3 on pages 33–40).

12.7.3 Using Statistical Packages for Hypothesis Testing on Numerical Data

To test Bud Conley's first assertion, that the average age of all full-time employees at Kalosha Industries is around 40 years, the null and alternative hypotheses would be:

$$H_0: \mu_x = 40 \text{ years}$$

$$H_1: \mu_x \neq 40 \text{ years}$$

Since we are dealing with one sample containing numerical data, a t test will be employed and the test will be performed at the traditional $\alpha = .05$ level of significance. A two-tailed test is used because Bud Conley simply hypothesized that the average age was around 40. The alternative is that the average age is not 40—either it is significantly more or significantly less. Since there are 400 employees in the sample, there are $400 - 1 = 399$ degrees of freedom. Owing to this very large sample size, the critical values at the .05 level of significance would be approximated by ± 1.96, taken from the bottom row of Table E.3, as shown in Figure 12.9. The decision rule would be

$$\text{Reject } H_0 \text{ if } t > +1.96$$
$$\text{or if } t < -1.96;$$
$$\text{otherwise do not reject } H_0.$$

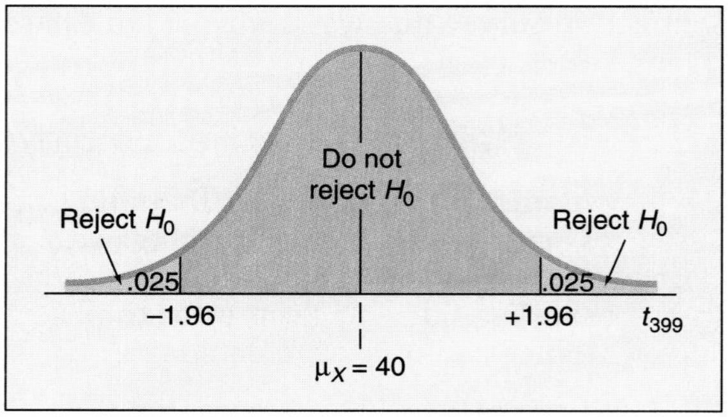

Figure 12.9
A two-tailed test of hypothesis about the mean (σ_x unknown) for a sample of size 400 at the .05 level of significance.

However, we must first determine the appropriateness of the t test by performing a thorough exploratory descriptive analysis. Accessing MINITAB, we obtain various summary statistics, a stem-and-leaf display, a histogram, a box-and-whisker plot, and a normal probability plot regarding the ages of full-time employees. Part of this MINITAB output is illustrated in Figure 12.10.

The data in Figure 12.10 appear to be very slightly right-skewed because the mean slightly exceeds the median, the length of the whisker between Q_3 and X_{largest} in the box-and-whisker plot exceeds the length of the whisker between X_{smallest} and Q_1, and the normal probability plot has some curvature in the tails. However, the t test is robust to such moderate departures from the assumption of normality, particularly with large sample sizes. Here, for random samples as large as 400, the central limit theorem (see Section 9.2) would result in an approximately normal sampling distribution and the t test would seem to be an appropriate procedure to employ.

But is our sample random? The assumption of randomness and independence of observations that comprise the sample is an important one in order to use the t, Wilcoxon, and χ^2 tests developed in this chapter. Accessing MINITAB, we employ the Wald-Wolfowitz one-sample runs test for randomness. The null and alternative hypotheses are:

H_0: The sequence of ages in the sample is random.

H_1: The sequence of ages in the sample is not random (two-tailed).

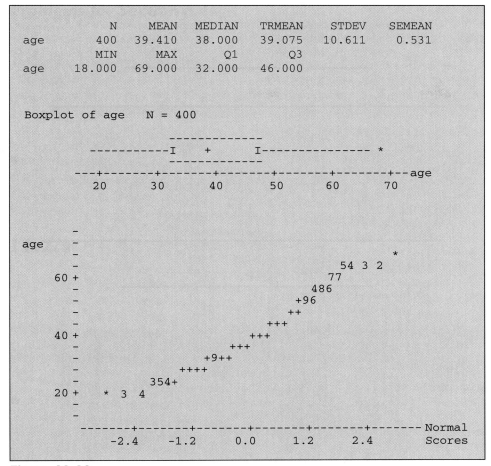

```
             N      MEAN    MEDIAN    TRMEAN     STDEV    SEMEAN
age        400    39.410    38.000    39.075    10.611     0.531
           MIN       MAX        Q1        Q3
age     18.000    69.000    32.000    46.000


Boxplot of age    N = 400

                    -------------
         -----------I    +      I---------------  *
                    -------------
      ---+-------+-------+-------+-------+-------+--age
        20      30      40      50      60      70


      -
age   -
      -                                           *
      -                                    54 3 2
  60 +                                    77
      -                                 486
      -                               +96
      -                              ++
      -                           +++
  40 +                          +++
      -                       +++
      -                    +9++
      -                 ++++
      -             354+
  20 +    *  3  4
      -
      -
      -------+-------+-------+-------+-------+------- Normal
          -2.4    -1.2     0.0     1.2     2.4        Scores
```

Figure 12.10
Part of MINITAB output showing some descriptive information for the age of employees.

From the summary information displayed in the MINITAB output of Figure 12.10, the sample median is 38.0 years. Using a level of significance α of .05, the null hypothesis of randomness is tested as shown in Figure 12.11.

```
age

K = 38.0000

THE OBSERVED NO. OF RUNS = 201
THE EXPECTED NO. OF RUNS = 200.9800
198 OBSERVATIONS ABOVE K 202 BELOW
    THE TEST IS SIGNIFICANT AT 0.9984
    CANNOT REJECT AT ALPHA = 0.05
```

Figure 12.11
MINITAB output displaying the runs test for randomness.

From the MINITAB output, the assumption of randomness (that is, the null hypothesis) cannot be rejected. The p value for the runs test is .9984, far greater than the $\alpha = .05$ chosen level of significance. Thus, we may proceed with the t test.

Using Equation (12.1) on the summary statistics taken from Figure 12.10, we have

$$t = \frac{\overline{X} - \mu_x}{\frac{S}{\sqrt{n}}} = \frac{39.41 - 40}{\frac{10.611}{\sqrt{400}}} = -1.11$$

Since $-1.96 < t = -1.11 < +1.96$, we do not reject H_0. The same conclusion is reached by accessing MINITAB to perform the t test (see Figure 12.12).

```
TEST OF MU = 40.000 V MU N.E. 40.000

           N     MEAN    STDEV  SE MEAN        T  P VALUE
age      400   39.410   10.611    0.531    -1.11     0.27
```

Figure 12.12
MINITAB output for a *t* test.

We may observe from Figure 12.12 that MINITAB displays the null and alternative hypotheses and the level at which the t test is significant (that is, the p value). Here we observe that the p value is .27. Since this is larger than the chosen level of significance, $\alpha = .05$, the null hypothesis cannot be rejected. Therefore, we may conclude that, at the .05 level of significance, there is no evidence to refute Bud Conley's assertion regarding the average age of full-time employees.

To test Bud Conley's second claim, that the median total family income of full-time employees at Kalosha Industries exceeds \$39,000, the null and alternative hypotheses would be:

H_0: Median ≤ 39 thousand dollars

H_1: Median > 39 thousand dollars

Since we are dealing with one sample containing numerical data and are interested in testing a hypothesis about a median value, the Wilcoxon signed-ranks test will be employed and the test will be performed at the traditional $\alpha = .05$ level of significance. A one-tailed test is used because Bud Conley's assertion was directional. He hypothesized that the median total family income exceeds \$39,000. This assertion is the alternative hypothesis. The null hypothesis to be tested (which must include an equal sign) is that the median income is less than or equal to \$39,000. Since there are 400 employees in the sample, the Wilcoxon test statistic W is approximately normally distributed and Equation (12.4) on page 433 may be used to test the null hypothesis. For a one-tailed test at the .05 level of significance, the critical value from the standard normal distribution (Table E.2) is +1.645 (see Figure 12.13). The decision rule would be

Reject H_0 if $Z > +1.645$;

otherwise do not reject H_0.

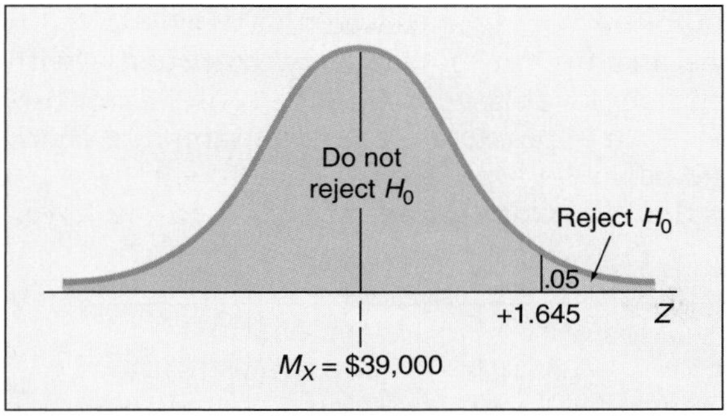

Figure 12.13
A one-tailed test of hypothesis about the median for a sample of size 400 at the .05 level of significance.

To determine the appropriateness of the Wilcoxon test, we perform a thorough exploratory descriptive analysis on the total family income of full-time employees. Part of the output obtained by accessing MINITAB is shown in Figure 12.14.

```
                  N     MEAN   MEDIAN  TRMEAN    STDEV  SEMEAN
fincome         400   42.492   39.200  41.346   19.188   0.959

                MIN      MAX       Q1      Q3
fincome      14.300   98.700   27.600  52.175


Boxplot of 'fincome'   N = 400

                    -------------
          ------I       +       I-------------------*******
                    -------------
       ---+-------+-------+-------+-------+-------+-- fincome
         16      32      48      64      80      96

        -
        -                                       ***** *
     90 +                                     42*
        -                                    *53
fincome -                                    5
        -                                   *86
        -                                 +9
     60 +                                ++
        -                              7+4
        -                            6++8
        -                          +++
        -                        +++7
     30 +                      6++5
        -                  4+++9
        -            3*5454778
        -      *  2*
        -
        ------+-------+-------+-------+-------+------- Normal
            -2.4    -1.2     0.0     1.2     2.4       Scores
```

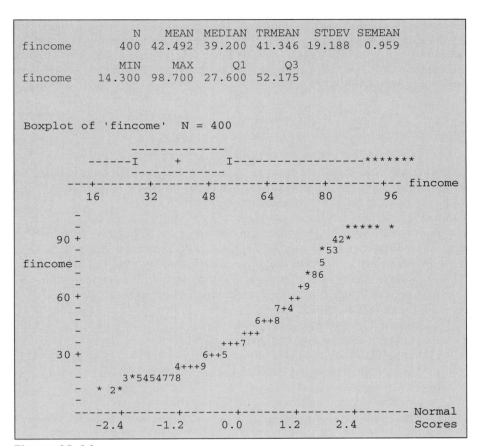

Figure 12.14
Part of MINITAB output showing some descriptive information for the total family incomes of employees.

Although the data in Figure 12.14 appear to be right-skewed, the Wilcoxon test is robust to such departures from the assumption of symmetry, particularly with large sample sizes. Here, for random samples as large as 400, the central limit theorem (see Section 9.2) would result in an approximately normal sampling distribution and the Wilcoxon test would be an appropriate procedure to employ.

Accessing MINITAB to perform the Wilcoxon test (see Figure 12.15) we have:

```
TEST OF MEDIAN = 39.00 VERSUS MEDIAN G.T. 39.00
                          N FOR    WILCOXON              ESTIMATED
                    N     TEST   STATISTIC  P-VALUE         MEDIAN
fincome           400      400     44657.0    0.024          41.00
```

Figure 12.15
MINITAB output for the Wilcoxon one-sample signed-ranks test.

We may observe from Figure 12.15 that MINITAB displays the null and alternative hypotheses and the level at which the test is significant (that is, the *p* value). Here we observe that the test statistic *W* is 44,657, resulting in a *p* value of .024. Since this is less than the selected level of significance, $\alpha = .05$, the null hypothesis is rejected. [If Equation (12.4) had been used, $Z = +1.97 > +1.645$, the upper-tail critical value as shown in Figure 12.13 on page 453, and H_0 would be rejected.] Therefore, we may conclude that, at the .05 level of significance, there is evidence to support Bud Conley's claim that the median total family income of full-time employees exceeds $39,000.

Survey/Database Project for Section 12.7

Data file EMPSAT.DAT

The following problems refer to the sample data obtained from the questionnaire of Figure 2.6 on pages 28–29 and presented in Table 2.3 on pages 33–40. They should be solved with the aid of an available computer package.

Suppose you are hired as a research assistant to Bud Conley, the Vice-President of Human Resources at Kalosha Industries. He has asked you to test two different hypotheses (see Problems 12.36 and 12.37) prior to his meeting with a representative from the B&L Corporation, the employee-benefits consulting firm that he has retained. From the responses to the two questions dealing with numerical variables in the Employee Satisfaction Survey and pertaining to the hypotheses described in Problems 12.36 and 12.37 which follow,

(a) *Obtain a set of descriptive statistics.*
(b) *Develop the stem-and-leaf display.*
(c) *Form the box-and-whisker plot.*
(d) *Develop the normal probability plot.*
(e) *Test the hypothesis of interest.*
(f) ACTION *Write a memo to Bud Conley discussing your findings.*

12.36 At the $\alpha = .05$ level of significance, is there evidence that the average number of hours that full-time employees of Kalosha Industries spend working at all jobs (see question 1) differs from 42?

12.37 At the $\alpha = .05$ level of significance, is there evidence that the median annual personal income (see question 7) of full-time employees of Kalosha Industries exceeds $26,000?

12.8 Potential Hypothesis Testing Pitfalls and Ethical Issues

12.8.1 Potential Pitfalls

In this chapter we introduced four statistical test procedures that may be employed when dealing with a single sample containing numerical data—the t test for a hypothesized mean, the Wilcoxon signed-ranks test for a hypothesized median, the χ^2 test for a hypothesized variance or standard deviation, and the Wald-Wolfowitz one-sample runs test for randomness. Part of a good data analysis is to understand the assumptions underlying each of the hypothesis test procedures and, using this information and other criteria, to select the one most appropriate for a given set of conditions.

The most prominent, distinguishing characteristics of the testing procedures described here versus those that will be developed in the following three chapters are the facts that we have been dealing with situations in which a single random sample of size n is selected and the outcomes are numerical rather than categorical. As observed in the summary chart for Chapter 12 on page 457, if interest is in central tendency, then, depending on certain assumptions, either the t test for a hypothesized mean or the Wilcoxon signed-ranks test for a hypothesized median would be selected. If interest is in studying variation and the sample can be assumed to be drawn from an underlying normal population, the χ^2 test for a hypothesized variance or standard deviation may be employed. If interest is in randomness, the Wald-Wolfowitz one-sample runs test would be used.

12.8.2 Ethical Issues

Ethical considerations arise when a researcher is manipulative of the hypothesis-testing process. As discussed in Section 11.11.2 (pages 412–415), some of the ethical issues that arise when dealing with hypothesis-testing methodology are:

- Data collection method—randomization
- Informed consent from human subjects being "treated"
- Type of test —two-tailed or one-tailed
- Choice of level of significance α
- Data snooping
- Cleansing and discarding of data
- Reporting of findings
- Meta-analysis

Lives can depend on the acceptance and application of faulty research. This becomes more important when ethics are involved. Reread Section 11.11 so it becomes ingrained in your thinking process. Again, when discussing ethical issues concerning the hypothesis-testing methodology, the key is *intent*. We must distinguish between poor confirmatory data analysis and unethical practice. Unethical behavior occurs when a researcher willfully causes a selection bias in data collection, manipulates the treatment of human subjects without informed consent, uses *data snooping* to select the type of test (two-tailed or one-tailed) and/or level of significance to his or her advantage, hides the facts by discarding observations that do not support a stated hypothesis, and fails to report pertinent findings.

12.9 Hypothesis Testing Based on One Sample of Numerical Data: A Review and a Preview

As depicted in the summary chart for Chapter 12, this chapter presented four commonly used hypothesis-testing procedures that involve one sample containing numerical data. On page 422 of Section 12.1 you were given a list emphasizing the important points to be discussed in the chapter. Check over the list now to see whether you feel you have an understanding of these key points. To be sure, you should be able to answer the following conceptual questions:

1. How and when would you use the *t* test for the population mean μ_x?
2. How and when would you use the Wilcoxon signed-ranks test for the population median M_x?
3. What is the meaning of the concept of robustness?
4. How and when would you use the χ^2 test for the population variance σ_x^2 or standard deviation σ_x?
5. What is the meaning of the concept of randomness and the idea of *runs*?
6. How and when would you use the Wald-Wolfowitz one-sample runs test for randomness?
7. What are the distinguishing features among *classical* parametric tests, distribution-free tests, and nonparametric tests?
8. What are the advantages and disadvantages of *classical* parametric tests, distribution-free tests, and nonparametric tests?
9. What are some of the ethical issues to be concerned with when performing a hypothesis test?

Check over the list of questions to see if indeed you know the answers and could (1) explain your answers to someone who did not read this chapter and (2) give reference to specific readings or examples that support your answer. Also, reread any of the sections that may have seemed fuzzy to see if they make sense now.

We shall continue to build on the foundations of hypothesis testing that we have discussed so far. In the following chapter, we will present a set of procedures that may be employed for analyzing the differences between two groups when the data are numerical.

Getting It All Together

Key Terms

χ^2 test for a hypothesized variance or standard deviation *437*
chi-square distribution *438*
difference score D_i *431*
distribution-free test *423*
nonparametric test *423*
one-sample *t* test *425*
(one-sample) Wilcoxon signed-ranks test *430*
parametric or classical test *423*
robust *424*
run *443*
systematic or periodic effect *444*
trend effect *443*
Wald-Wolfowitz one-sample runs test *442*

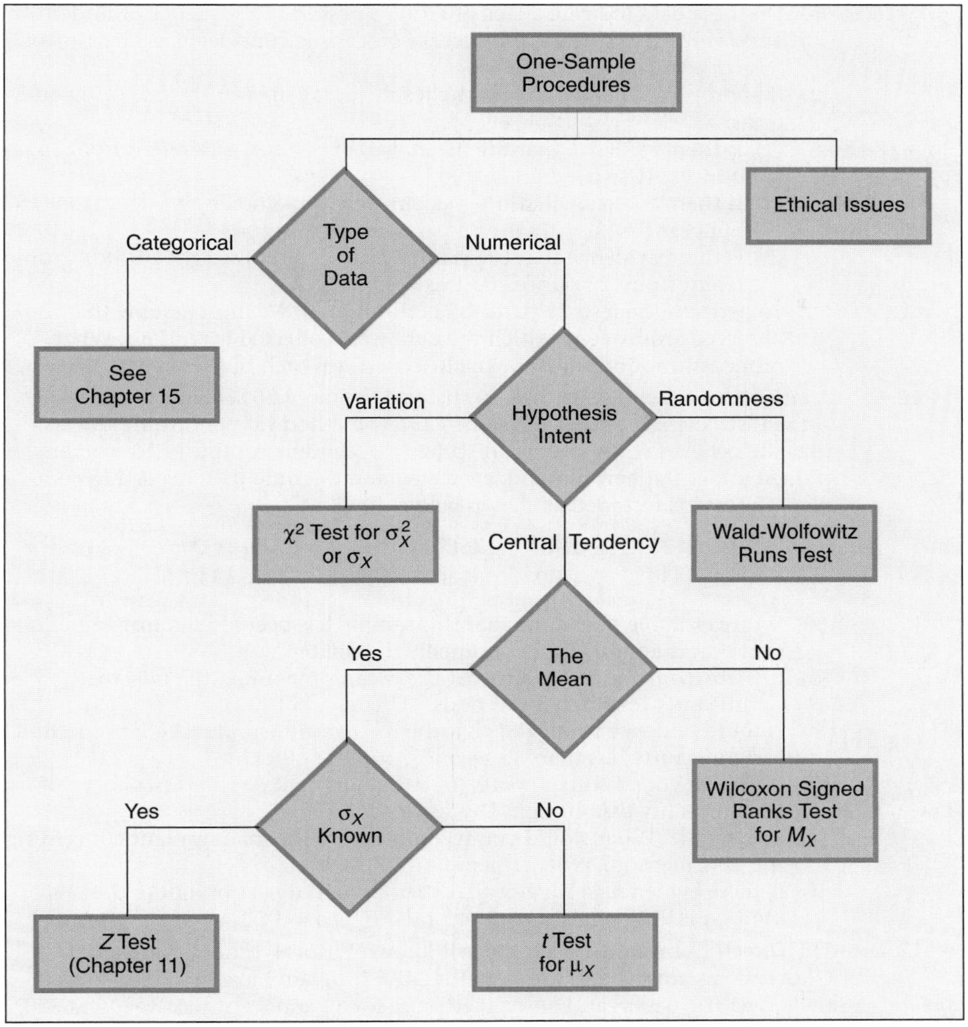

Chapter 12 summary chart.

Chapter Review Problems

12.38 ACTION Write a letter to a friend who has not taken a course in statistics and explain what this chapter has been about. To highlight this chapter's content, be sure to incorporate your answers to the nine review questions on page 456.

12.39 A machine being used for packaging seedless golden raisins has been set so that on the average 15 ounces of raisins will be packaged per box. The quality control engineer wishes to test the machine setting and selects a sample of 30 consecutive raisin packages filled during the production process. Their weights are recorded below in column sequence (from top to bottom, left to right):

Data file RAISINS.DAT

15.2	15.3	15.1	15.7	15.3	15.0	15.1	14.3	14.6	14.5
15.0	15.2	15.4	15.6	15.7	15.4	15.3	14.9	14.8	14.6
14.3	14.4	15.5	15.4	15.2	15.5	15.6	15.1	15.3	15.1

(a) Do these data indicate a lack of randomness in the sequence of underfills and overfills, or can the production process be considered as "in control"? (Use $\alpha = .05$.)

(b) If appropriate, depending on the results in (a), answer the following questions:

(1) Is there evidence that the mean weight per box is different from 15 ounces? (Use $\alpha = .05$.)

(2) Is there evidence that the median weight per box is different from 15 ounces? (Use $\alpha = .05$.)

(3) Is there evidence that the standard deviation of weight per box is different from .25 ounce? (Use $\alpha = .05$.)

(c) To perform the tests in parts (b)(1) through (3) we must assume that the observed sequence in which the data were collected is random. What other assumptions must be made to perform each of these tests? Discuss.

Data file
TASKTIME.DAT

● 12.40 A task on an assembly line has, in the past, required 30 seconds to complete. An industrial engineer has developed a new method for performing the task that she believes will speed up the process. A random sample of 15 workers trained under the new method is selected and the time they needed to complete the task is recorded as shown below:

27.2	31.1	29.0	26.7	28.1	27.3	29.6	30.5
30.0	30.2	25.9	31.3	28.8	27.4	27.0	

(a) Is it reasonable to assume that this sample has been drawn from a population that is approximately normally distributed?

(b) Is there evidence to suggest that the mean time under the new method is significantly less than 30 seconds? (Use $\alpha = .05$.)

(c) Is there evidence to suggest that the median time under the new method is significantly less than 30 seconds? (Use $\alpha = .05$.)

(d) Compare your results in parts (b) and (c) in light of your assessment of the normality assumption. Discuss.

(e) **ACTION** What would you recommend to the management concerning the new method? Write a memo.

(f) Is there evidence to suggest that the standard deviation under the new method is different from 1.2 seconds? (Use $\alpha = .05$.)

Data file
GMAT.DAT

12.41 The Director of Admissions for a well-known business school claims that GMAT scores among applicants to the MBA program have increased substantially over the past year. The average score for all applicants last year was 520. The data below represent the GMAT scores for a random sample of 20 applicants during this year:

560	500	670	460	590	490	540	550	750	620
510	520	380	580	600	550	570	640	490	600

(a) Is there evidence to support the Director's claim? (Use $\alpha = .05$.)

(b) What test did you choose for part (a)? Why?

(c) What assumptions must hold in order to perform the test in part (a)?

(d) Formally evaluate these assumptions through both hypothesis testing (using $\alpha = .05$) and graphical approaches. Discuss.

Data file
CHARRED.DAT

12.42 A consultant from the American Society for Testing and Materials has been asked by a manufacturer of children's clothing to evaluate potential flammability in material being considered for use in toddlers' outfits. The consultant takes a random sample of 20 identical strips of material and subjects each strip to a vertical semirestrained test. The data below indicate the length of charred material (in centimeters) for each of the 20 strips.

6.92	8.39	9.31	9.90	10.63	9.32	9.80	8.98	8.88	8.42
9.47	8.95	8.92	9.44	9.91	8.75	9.24	9.35	11.17	8.01

(a) Is there evidence that the average length (in centimeters) of the charred material is different from 9.25? (Use $\alpha = .05$.)
(b) What test did you choose for part (a)? Why?
(c) What assumptions must hold in order to perform the test in part (a)?
(d) Formally evaluate these assumptions through both hypothesis testing (using $\alpha = .05$) and graphical approaches. Discuss.

12.43 A manufacturer of automobile batteries claims that his product will last on average at least four years (that is, 48 months). A consumers' advocate group would like to evaluate this longevity claim and selects a random sample of 28 such batteries to test. The data below indicate the length of time (in months) that each of these batteries lasted (that is, performed properly before failure).

Data file AUTOBAT.DAT

42.3	39.6	25.0	56.2	37.2	47.4	57.5
39.3	39.2	47.0	47.4	39.7	57.3	51.8
31.6	45.1	40.8	42.4	38.9	42.9	34.1
49.0	41.5	60.1	34.6	50.4	30.7	44.1

(a) Is there evidence that the average battery life is less than 48 months? (Use $\alpha = .05$.)
(b) What test did you choose for part (a)? Why?
(c) What assumptions must hold in order to perform the test in part (a)?
(d) Formally evaluate these assumptions through both hypothesis testing (using $\alpha = .05$) and graphical approaches. Discuss.

12.44 You are having cocktails with a client. Someone has suggested to her that distribution-free or nonparametric techniques might be helpful for her current project. As her statistical confidant, she asks you: "What are distribution-free or nonparametric techniques and when or why may they be helpful?" You sip your drink and you reply. . . . .

Collaborative Learning Mini-Case Projects

For each of the following, refer to the instructions on page 101.

Data file CEREAL.DAT

CL12.1 Refer to CL 3.2 on page 101 and CL 4.2 on page 165. Your group, the ________ Corporation, has been hired by the food editor of a popular family magazine to study the cost and nutritional characteristics of ready-to-eat cereals. Armed with Special Data Set 2 of Appendix D on pages D6–D7, the ________ Corporation is ready to:
(a) Determine if there is evidence that the average cost of all ready-to-eat cereals is different from 30 cents.
(b) Determine if there is evidence that the mean weight of high-fiber cereals exceeds 1.7 ounces.
(c) Determine if there is evidence that the median calories of moderate fiber cereals is greater than 150.
(d) Determine if there is evidence that the standard deviation of sugar (in grams per serving) in low-fiber cereals is different from 0.4 gram.
(e) Write and submit an executive summary describing the results in parts (a)–(d), clearly specifying all hypotheses, selected levels of significance, and the assumptions of the chosen test procedures.
(f) Prepare and deliver a five-minute oral presentation to the food editor of the magazine.

Data file FRAGRAN.DAT

CL12.2 Refer to CL 3.3 on page 102 and CL 4.3 on page 165. Your group, the ________ Corporation, has been hired by the marketing director of a manufacturer of well-known men's and women's fragrances to study the characteristics of currently available fragrances. Armed with Special Data Set 3 of Appendix D on page 000, the ________ Corporation is ready to:

(a) Determine if there is evidence that the mean cost of women's fragrances is different from $120.
(b) Determine if there is evidence that the median cost of perfumes is greater than $200.
(c) Determine if there is evidence that the standard deviation in the cost of mild fragrances exceeds $90.
(d) Write and submit an executive summary describing the results in parts (a)–(c), clearly specifying all hypotheses, selected levels of significance, and the assumptions of the chosen test procedures.
(e) Prepare and deliver a five-minute oral presentation to the marketing director.

Data file CAMERA.DAT

CL12.3 Refer to CL 3.4 on page 102 and CL 4.4 on page 166. Your group, the ________ Corporation, has been hired by the travel editor of a well-known newspaper who is preparing a feature article on compact 35-mm cameras. Armed with Special Data Set 4 of Appendix D on pages D10–D11, the ________ Corporation is ready to:

(a) Determine if there is evidence that the average price of all 35-mm compact cameras is different from $150.
(b) Determine if there is evidence that the mean weight of "multiple long" 35-mm cameras exceeds 17 ounces.
(c) Determine if there is evidence that the median framing accuracy of "automatic single" 35-mm cameras is greater than 80 percent.
(d) Determine if there is evidence that the standard deviation in the battery life of "fixed single" 35-mm cameras is different from 10 exposure rolls.
(e) Write and submit an executive summary describing the results in parts (a)–(d), clearly specifying all hypotheses, selected levels of significance, and the assumptions of the chosen test procedures.
(f) Prepare and deliver a five-minute oral presentation to the travel editor of the newspaper.

References

1. Berenson, M. L., D. M. Levine, and M. Goldstein, *Intermediate Statistical Methods and Applications: A Computer Package Approach* (Englewood Cliffs, NJ: Prentice-Hall, 1983).
2. Bradley, J. V., *Distribution-Free Statistical Tests* (Englewood Cliffs, NJ: Prentice-Hall, 1968).
3. Conover, W. J., *Practical Nonparametric Statistics,* 2d ed. (New York: Wiley, 1980).
4. Daniel. W., *Applied Nonparametric Statistics,* 2d ed. (Boston, MA: Houghton Mifflin, 1990).
5. Dixon, W. J., and F. J. Massey, Jr., *Introduction to Statistical Analysis*, 4th ed. (New York: McGraw-Hill, 1983).
6. *MINITAB Reference Manual Release 8* (State College, PA.: Minitab, Inc., 1992).
7. Norusis, M., *SPSS Guide to Data Analysis for SPSS-X with Additional Instructions for SPSS/PC+* (Chicago, IL: SPSS Inc., 1986).
8. *SAS User's Guide Version 6* (Raleigh, NC: SAS Institute, 1988).
9. Solomon, H., and M. A. Stephens, "Sample variance," in *Encyclopedia of Statistical Sciences,* Vol. 9, Edited by Kotz, S., and N. L. Johnson (New York: Wiley, 1988), pp. 477–480.
10. *STATISTIX Version 4.0* (Tallahassee, FL: Analytical Software, Inc., 1992).

chapter 13

Two-Sample Tests with Numerical Data

CHAPTER OBJECTIVE

To extend the basic principles of hypothesis testing to two-sample tests involving numerical variables. Both independent and related sample procedures are considered.

13.1 Introduction

In the preceding chapter we focused on a variety of commonly used hypothesis-testing procedures that pertain to a single sample of numerical data drawn from a population. In this chapter we shall extend our discussion of hypothesis testing to consider commonly used procedures that enable us to compare statistics computed from two samples of numerical data in order to make inferences about possible differences in the parameters of the two respective populations. In particular, as seen in the summary chart for this chapter on page 519, we will describe several useful hypothesis-testing procedures that may be employed, depending on the situation. Both independent and related sample procedures are considered. Emphasis is given to the assumptions behind the use of the various tests.

Upon completion of this chapter, you should be able to:

1. Distinguish among various criteria used in the selection of a particular hypothesis test procedure
2. Know when and how to use the pooled-variance t test to examine possible differences in the means of two independent populations
3. Know when and how to use the separate-variance t' test to examine possible differences in the means of two independent populations
4. Know when and how to use the Wilcoxon rank sum test to examine possible differences in the medians of two independent populations
5. Know when and how to use the F test to examine possible differences in the variances of two independent populations
6. Know when and how to use the t test for a possible mean difference, μ_D, in two related populations
7. Know when and how to use the Wilcoxon signed-ranks test for a possible median difference, M_D, in two related populations

13.2 Choosing the Appropriate Test Procedure When Comparing Two Independent Samples

Over the years, many statistical test procedures have been developed that enable us to make comparisons and examine differences between two groups based on independent samples containing numerical data. Thus, an important issue faced by anyone involved in hypothesis testing is the criteria to be used for selection of a particular statistical procedure from among the many that are available. Part of a good data analysis is to understand the assumptions underlying each of the hypothesis-testing techniques and to select the one most appropriate for a given set of conditions. Other criteria for test selection involve the simplicity of the procedure, the generalizability of the conclusions to be drawn, the accessibility of tables of critical values for the test statistic, the availability of computer software packages that contain the test procedure, and last, but certainly not least, the statistical power of the procedure.

In each of the following four sections, we shall describe a hypothesis-testing procedure that examines differences between two independent groups based on samples containing numerical data.

13.3 Pooled-Variance t Test for Differences in Two Means

13.3.1 Introduction

Let us first extend the hypothesis-testing concepts developed in Chapters 11 and 12 to situations in which we would like to determine whether there is any difference between the means of two independent populations. Suppose that we consider two independent populations, each having a mean and standard deviation (symbolically represented as follows):

Population 1	Population 2
μ_1, σ_1	μ_2, σ_2

Let us also suppose that a random sample of size n_1 is taken from the first population and a random sample of size n_2 is drawn from the second population and, further, that the data collected in each sample pertain to some numerical random variable of interest.

The test statistic used to determine the difference between the population means is based on the difference between the sample means $(\overline{X}_1 - \overline{X}_2)$. Because of the central limit theorem discussed in Section 9.2, this test statistic will follow the standard normal distribution for large enough sample sizes. The **Z-test** statistic is

$$Z = \frac{(\overline{X}_1 - \overline{X}_2) - (\mu_1 - \mu_2)}{\sqrt{\dfrac{\sigma_1^2}{n_1} + \dfrac{\sigma_2^2}{n_2}}} \quad \textbf{(13.1)}$$

where $\overline{X}_1$ = mean from the sample taken from population 1
μ_1 = mean of population 1
σ_1^2 = variance of population 1
n_1 = size of the sample taken from population 1
$\overline{X}_2$ = mean from the sample taken from population 2
μ_2 = mean of population 2
σ_2^2 = variance of population 2
n_2 = size of the sample taken from population 2

13.3.2 Developing the Pooled-Variance t Test

However, as we mentioned previously, in most cases we do not know the actual standard deviation of either of the two populations. The only information usually obtainable is the sample means ($\overline{X}_1$ and $\overline{X}_2$) and sample standard deviations (S_1

and S_2). If the assumptions are made that the samples are randomly and independently drawn from respective populations that are normally distributed and, further, that the *population variances are equal* (that is, $\sigma_1^2 = \sigma_2^2$), a **pooled-variance *t* test** can be used to determine whether there is a significant difference between the means of the two populations.

The test to be performed can be either two-tailed or one-tailed, depending on whether we are testing if the two population means are merely *different* or if one mean is *greater than* the other mean.

Two-Tailed Test	One-Tailed Test	One-Tailed Test
H_0: $\mu_1 = \mu_2$ or $\mu_1 - \mu_2 = 0$	H_0: $\mu_1 \geq \mu_2$ or $\mu_1 - \mu_2 \geq 0$	H_0: $\mu_1 \leq \mu_2$ or $\mu_1 - \mu_2 \leq 0$
H_1: $\mu_1 \neq \mu_2$ or $\mu_1 - \mu_2 \neq 0$	H_1: $\mu_1 < \mu_2$ or $\mu_1 - \mu_2 < 0$	H_1: $\mu_1 > \mu_2$ or $\mu_1 - \mu_2 > 0$

where μ_1 = mean of population 1
μ_2 = mean of population 2

To test the null hypothesis of no difference in the means of two independent populations

$$H_0: \quad \mu_1 = \mu_2$$

against the alternative that the means are not the same

$$H_1: \quad \mu_1 \neq \mu_2$$

the following **pooled-variance *t*-test** statistic can be computed:

$$t = \frac{(\overline{X}_1 - \overline{X}_2) - (\mu_1 - \mu_2)}{\sqrt{S_p^2\left(\frac{1}{n_1} + \frac{1}{n_2}\right)}} \tag{13.2}$$

where

$$S_p^2 = \frac{(n_1 - 1)S_1^2 + (n_2 - 1)S_2^2}{(n_1 - 1) + (n_2 - 1)}$$

and

S_p^2 = pooled variance
$\overline{X}_1$ = mean of the sample taken from population 1
S_1^2 = variance of the sample taken from population 1
n_1 = size of the sample taken from population 1
$\overline{X}_2$ = mean of the sample taken from population 2
S_2^2 = variance of the sample taken from population 2
n_2 = size of the sample taken from population 2

From Equation (13.2) we may observe that the pooled-variance t test gets its name because the test statistic requires the pooling or combining of the two sample variances S_1^2 and S_2^2 to obtain S_p^2, the best estimate of the variance common to both populations under the assumption that the two population variances are equal.

The pooled-variance t-test statistic follows a t distribution with $n_1 + n_2 - 2$ degrees of freedom. For a given level of significance, α, we may reject the null hypothesis if the computed t-test statistic exceeds the upper-tailed critical value $t_{n_1+n_2-2}$ from the t distribution or if the computed test statistic falls below the lower-tailed critical value $-t_{n_1+n_2-2}$ from the t distribution. That is, the decision rule is

$$\text{Reject } H_0 \text{ if } t > t_{n_1+n_2-2}$$
$$\text{or if } t < -t_{n_1+n_2-2};$$
$$\text{otherwise do not reject } H_0.$$

The decision rule and regions of rejection are displayed in Figure 13.1.

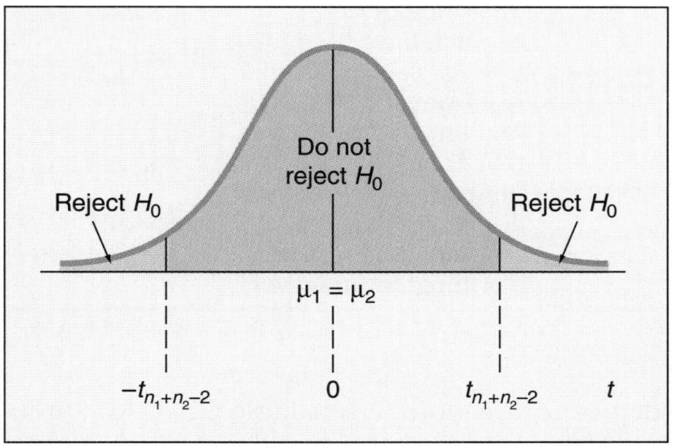

Figure 13.1
Rejection regions for a two-tailed test for the difference between two means.

13.3.3 Application

To demonstrate the use of the pooled-variance t test, suppose that a financial analyst wishes to compare the average dividend yields of stocks traded on the New York Stock Exchange with those traded "over the counter" on the NASDAQ national market listing. Random samples of 21 companies from the New York Stock Exchange and 25 stocks from the NASDAQ national market listing are selected and the results are presented in Table 13.1 on page 466.

If the financial analyst wishes to determine whether there is evidence of a difference in average dividend yield between the two populations of stock listings, the null and alternative hypotheses would be

$$H_0: \mu_1 = \mu_2 \text{ or } \mu_1 - \mu_2 = 0$$
$$H_1: \mu_1 \neq \mu_2 \text{ or } \mu_1 - \mu_2 \neq 0$$

Assuming that the samples are taken from underlying normal populations having equal variances, the pooled-variance t test can be used. If the test were conducted at the $\alpha = .05$ level of significance, the t-test statistic would follow a t

Table 13.1 Comparing dividend yields* of selected issues from the New York Stock Exchange and the NASDAQ national market listing (May 25, 1994).

New York Stock Exchange ($n_1 = 21$)		NASDAQ Listing ($n_2 = 25$)	
Company	Dividend Yield	Company	Dividend Yield
American Express	3.4	Atlantic SE Airlines	1.2
Anheuser-Busch	2.7	Boral Ltd	5.1
Bristol-Myers-Squibb	5.4	Cathay Bancorp	4.3
Dayton-Hudson	2.1	Cit Fed Bancorp	0.8
Dresser Industries	3.0	CPB	3.2
Ford Motor	3.1	First Essex Bancorp	3.0
General Electric	3.0	Goulds Pumps	3.8
General Mills	3.5	Harper Group	1.3
IBM	1.6	Innovex	2.2
Kellogg Co.	2.6	Intel Corp.	0.4
Merck & Co.	3.6	Lindberg Corp.	2.7
NYNEX	6.4	Nature's Sunshine Prod.	1.5
Occidental Petroleum	5.3	Newcor	2.1
Pfizer Inc.	3.0	PCA International	3.3
PPG Inc.	3.0	T Rowe Price Assoc.	1.8
Sara Lee Corp.	2.9	PSB Holdings Corp.	2.4
Texaco Inc.	5.0	Research Inc.	4.6
Texas Instruments	0.9	Seacoast Banking Corp.	2.8
Whirlpool Corp.	2.2	Span-America Med. Sys.	1.8
Winn-Dixie	3.1	Sumitomo Bank of Cal.	3.6
Xerox Corp.	2.9	TCA Cable TV	2.2
		United Fire & Casualty	2.8
		West Coast Bancorp	1.7
		Whitney Holding Corp.	2.6
		Worthington Industries	2.1

*Dividend yield is the ratio of the annual dividend per share to the closing price per share, expressed as a percentage.

distribution with $21 + 25 - 2 = 44$ degrees of freedom. From Table E.3 of Appendix E, the critical values for this two-tailed test are +2.0154 and −2.0154 and, as depicted in Figure 13.2, the decision rule is

$$\text{Reject } H_0 \text{ if } t > t_{44} = +2.0154$$
$$\text{or if } t < -t_{44} = -2.0154;$$
$$\text{otherwise do not reject } H_0.$$

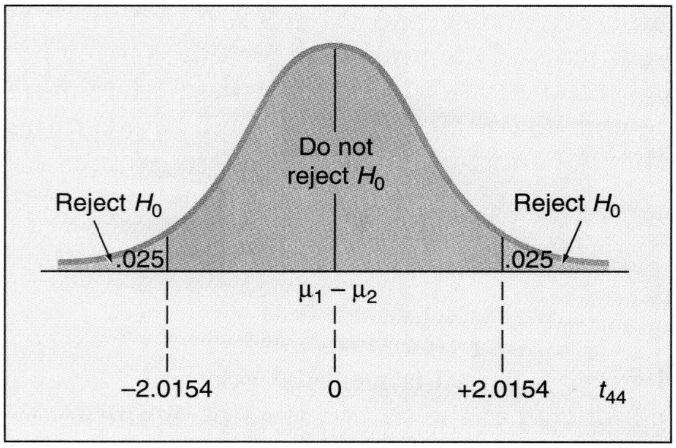

Figure 13.2
Two-tailed test of hypothesis for the difference between the means at the .05 level of significance.

Using the data in Table 13.1, a set of summary statistics are computed and displayed in Table 13.2:

Table 13.2 Some summary statistics on dividend yields

New York Stock Exchange	NASDAQ Listing
$n_1 = 21$	$n_2 = 25$
$\overline{X}_1 = 3.27$	$\overline{X}_2 = 2.53$
$S_1^2 = 1.698$	$S_2^2 = 1.353$
$S_1 = 1.30$	$S_2 = 1.16$
$X_{smallest_1} = 0.9$	$X_{smallest_2} = 0.4$
$Q_{1_1} = 2.65$	$Q_{1_2} = 1.75$
$Median_1 = 3.0$	$Median_2 = 2.4$
$Q_{3_1} = 3.55$	$Q_{3_2} = 3.25$
$X_{largest_1} = 6.4$	$X_{largest_2} = 5.1$

For our data we have

$$t = \frac{(\overline{X}_1 - \overline{X}_2) - (\mu_1 - \mu_2)}{\sqrt{S_p^2\left(\frac{1}{n_1} + \frac{1}{n_2}\right)}}$$

where

$$\begin{aligned} S_p^2 &= \frac{(n_1 - 1)S_1^2 + (n_2 - 1)S_2^2}{(n_1 - 1) + (n_2 - 1)} \\ &= \frac{20(1.30)^2 + 24(1.16)^2}{21 + 25 - 2} \\ &= \frac{66.432}{44} \\ &= 1.510 \end{aligned}$$

and therefore

$$\begin{aligned} t &= \frac{3.27 - 2.53}{\sqrt{1.510\left(\frac{1}{21} + \frac{1}{25}\right)}} \\ &= \frac{0.74}{\sqrt{0.132}} \\ &= \frac{0.74}{0.364} \\ &= 2.03 \end{aligned}$$

Using a .05 level of significance, the null hypothesis (H_0) is rejected because $t = +2.03 > t_{44} = +2.0154$. If the null hypothesis were true, there would be an

$\alpha = .05$ probability of obtaining a *t*-test statistic either larger than +2.0154 standard deviations from the center of the *t* distribution or smaller than −2.0154 standard deviations from the center of the *t* distribution. The *p* value, or probability of obtaining a difference between the two sample means even larger than the 0.74 observed here, which translates to a test statistic *t* with a distance even farther from the center of the *t* distribution than ±2.03 standard deviations, would be slightly less than .05 if the null hypothesis of no difference in the means were true. (Using MINITAB, the *p* value is actually computed as .048.) Since the *p* value is less than α, we have sufficient evidence that the null hypothesis is not true and we reject it.

The null hypothesis is rejected because the test statistic *t* has fallen into the region of rejection. The financial analyst may conclude that there is evidence of a difference in the mean dividend yields for the two groups. Companies traded on the New York Stock Exchange appear to have significantly higher dividend yields than do companies trading over the counter on the NASDAQ national market listing.

We should note that in our financial analyst's study, the two groups had unequal sample sizes. When the two sample sizes are equal (that is, $n_1 = n_2$), the formula for the pooled variance can be simplified to

$$S_p^2 = \frac{S_1^2 + S_2^2}{2}$$

13.3.4 Summary

In testing for the difference between the means, we have assumed that we are sampling from normally distributed populations having equal variances. We must examine the consequences for the pooled-variance *t* test of departures from each of these assumptions. For situations in which we cannot or do not wish to make the assumption that the two populations having equal variances are actually normally distributed, the pooled-variance *t* test is **robust** (that is, not sensitive) to moderate departures from the assumption of normality, provided that the sample sizes are large. In such situations, the pooled-variance *t* test may be used without serious effect on its power. On the other hand, if sample sizes are small and we cannot or do not wish to make the assumption that the data in each group are taken from normally distributed populations, two choices exist. Either some *normalizing transformation* (see Reference 11) on each of the outcomes can be made and the pooled-variance *t* test used, or some distribution-free procedure such as the *Wilcoxon rank sum test* (to be covered in Section 13.5), which does not depend on the assumption of normality for the two populations, can be performed.

For situations in which we cannot or do not wish to make the assumption that the two normally distributed populations from which the sample data are drawn have equal population variances, the *Behrens-Fisher problem* is said to exist (see Reference 9) and a *separate-variance t′* test, developed by Satterthwaite (see Reference 8), is presented in the next section.

Problems for Section 13.3

● 13.1 The quality control manager at a light bulb factory would like to determine if there is any difference in the average life of bulbs manufactured on two different types of machines. The process standard deviation of machine I is 110 hours and of machine II is 125 hours. A random sample of 25 light bulbs obtained from machine I indicated a sample mean of 375 hours, and a similar sample of 25 from machine II indicated a sample mean of 362 hours. Using the .05 level of significance

(a) Is there any evidence of a difference in the average life of bulbs produced by the two types of machines?

(b) Compute the p value in part (a) and interpret its meaning.

13.2 The purchasing director for an industrial parts factory is investigating the possibility of purchasing a new type of milling machine. She has determined that the new machine will be bought if there is evidence that the parts produced have a higher average breaking strength than those from the old machine. The process standard deviation of the breaking strength for the old machine is 10 kilograms and for the new machine is 9 kilograms. A sample of 100 parts taken from the old machine indicated a sample mean of 65 kilograms, whereas a similar sample of 100 from the new machine indicated a sample mean of 72 kilograms. Using the .01 level of significance

(a) Is there evidence that the purchasing director should buy the new machine?

(b) Compute the p value in part (a) and interpret its meaning.

● 13.3 Management of the Sycamore Steel Co. wishes to determine if there is any difference in performance between the day shift of workers and the evening shift of workers. A sample of 100 day-shift workers reveals an average output of 74.3 parts per hour with a sample standard deviation of 16 parts per hour. A sample of 100 evening-shift workers reveals an average output of 69.7 parts per hour with a sample standard deviation of 18 parts per hour. At the .10 level of significance

(a) Is there evidence of a difference in average output between the day shift and the evening shift?

(b) Find the lower and upper limits of the p value in part (a) and interpret its meaning.

13.4 An independent testing agency has been contracted to determine whether there is any difference in gasoline mileage output of two different gasolines on the same model automobile. Gasoline A was tested on 200 cars and produced a sample average of 18.5 miles per gallon with a sample standard deviation of 4.6 miles per gallon. Gasoline B was tested on a sample of 100 cars and produced a sample average of 19.34 miles per gallon with a sample standard deviation of 5.2 miles per gallon. At the .05 level of significance

(a) Is there evidence of a difference in average performance of the two gasolines?

(b) Find the lower and upper limits of the p value in part (a) and interpret its meaning.

13.5 A carpet manufacturer is studying differences between two of its major outlet stores. The company is particularly interested in the time it takes before customers receive carpeting that has been ordered from the plant. Data concerning a sample of delivery times for the most popular type of carpet are summarized as follows:

	Store A	Store B
$\bar{X}$	34.3 days	43.7 days
S	2.4 days	3.1 days
n	41	31

(a) At the .01 level of significance, is there evidence of a difference in the average delivery time for the two outlet stores?
(b) Find the lower and upper limits of the p value in part (a) and interpret its meaning.

13.6 Suppose the manager of a pet supply store wanted to determine whether there was a difference in the amount of money spent by owners of dogs and owners of cats. (Customers who owned both were disregarded in this analysis.) The results for a sample of 37 dog owners and 26 cat owners are summarized as follows:

	Purchases for Dogs	Purchases for Cats
$\bar{X}$	\$26.47	\$19.16
S	\$ 9.45	\$ 8.52
n	37	26

(a) At the .05 level of significance, is there evidence of a difference in the average amount of money spent in the pet supply store between dog owners and cat owners?
(b) What assumptions must be made in order to do part (a) of this problem?
(c) Find the lower and upper limits of the p value in part (a) and interpret its meaning.

13.7 An industrial psychologist wishes to study the effects of motivation on sales in a particular firm. Of 24 new salespersons being trained, 12 are to be paid at an hourly rate and 12 on a commission basis. The 24 individuals were randomly assigned to the two groups. The following data represent the sales volume (in thousands of dollars) achieved during the first month on the job.

Data file SALESVOL.DAT

Hourly Rate		Commission	
256	212	224	261
239	216	254	228
222	236	273	234
207	219	285	225
228	225	237	232
241	230	277	245

(a) Is there evidence that wage incentives (through commission) yield greater average sales volume? (Use $\alpha = .01$.)
(b) What assumptions must be made in order to do part (a) of this problem?
(c) Find the lower and upper limits of the p value in part (a) and interpret its meaning.

13.8 A manufacturer is developing a nickel-metal hydride battery that is to be used in cellular telephones in lieu of nickel-cadmium batteries. The director of quality control decides to evaluate the newly developed battery against the widely used nickel-cadmium battery with respect to performance. A random sample of 25 nickel-cadmium batteries and a random sample of 25 of the newly developed nickel-metal hydride batteries are placed in cellular telephones of the same brand and model. The performance measure of interest is the talking time (in minutes) prior to recharging. The results are as follows:

Data file NICKELBAT.DAT

Nickel-Cadmium Battery		Nickel-Metal Hydride Battery	
54.5	71.0	78.3	103.0
67.0	67.8	79.8	95.4
41.7	56.7	81.3	91.1
64.5	69.7	69.4	46.4
86.8	70.4	82.8	87.3
40.8	74.9	82.3	71.8
72.5	75.4	62.5	83.2
76.9	64.9	77.5	85.0
81.0	104.4	85.3	74.3
83.3	90.4	85.3	85.5
82.0	72.8	86.1	72.1
71.8	58.7	41.1	74.1
68.8		112.3	

(a) Is there evidence of a difference in the two types of batteries with respect to average talking time (in minutes) prior to recharging? (Use $\alpha = .05$.)
(b) What assumptions must be made in order to do part (a) of this problem?
(c) Find the lower and upper limits of the p value in part (a) and interpret its meaning.

13.9 The following data represent the annual effective yields, in percent, on money market accounts from a sample of 10 New York area commercial banks and from a sample of 10 New York area savings banks:

Data file NYBANKS.DAT

Commercial Banks	Yield	Savings Banks	Yield
Banco Popular	2.25	Anchor Savings	2.43
Bank of N.Y.	2.32	Apple Bank Savings	2.53
Chase Manhattan	2.02	Carteret Savings (N.J.)	2.38
Chemical	1.92	Crossland Savings	2.50
Citibank	2.02	Dime Savings Bank	3.00
EAB	1.82	Emigrant Savings	2.50
First Fidelity (N.J.)	2.10	First Fed (Rochester)	2.55
Marine Midland	2.38	Green Point Savings	3.20
Midlantic Bank (N.J.)	2.30	Home Savings Amer	2.50
Republic Nat'l	2.28	People's Bank (Conn.)	2.02

Source: New York Times, May 25, 1994, p. D6.

(a) Is there evidence of a difference in the average effective yields on money market accounts in the two types of banks in the New York area? (Use $\alpha = .05$.)
(b) What assumptions must be made in order to do part (a) of this problem?
(c) Find the lower and upper limits of the p value in part (a) and interpret its meaning.

Interchapter Problems for Section 13.3

13.10 Refer to the monthly rental prices of unfurnished studio apartments in Manhattan and Brooklyn Heights (see Problem 4.77 on page 162).

Data file STUDIO.DAT

(a) Is there evidence that the average rental price is higher in Manhattan than in Brooklyn Heights? (Use $\alpha = .01$.)
(b) What assumptions are needed to do part (a)?
(c) Find the lower and upper limits of the p value in part (a) and interpret its meaning.

Data file PREP.DAT

● 13.11 Refer to Problem 4.80 on page 163.
(a) Is there evidence of a difference in the average tuition at preparatory schools in the Northeast and in the Midwest? (Use $\alpha = .01$.)
(b) Find the lower and upper limits of the p value in part (a) and interpret its meaning.
(c) What would you report to the guidance counselor at your school concerning tuition in these two regions?

Data file SHAMPOO.DAT

13.12 Referring to the data of Problem 3.9 (cost of conventional shampoos) on page 62
(a) Is there evidence of a difference in the average cost between shampoos labeled for normal hair versus those labeled for fine hair? (Use $\alpha = .05$.)
(b) Find the lower and upper limits of the p value in part (a) and interpret its meaning.
(c) If you were scheduled to write an article for a magazine comparing the two types of shampoos, what conclusions would you draw?

13.4 Separate-Variance t' Test for Differences in Two Means

13.4.1 Introduction

In our discussion of testing for the difference between the means of two independent populations in the previous section, we pooled the sample variances together into a common estimate S_p^2 because we assumed that the population variances were equal (that is, $\sigma_1^2 = \sigma_2^2$). This situation is depicted in panel A of Figure 13.3 for the case in which normally distributed population 1 has a higher mean than does normally distributed population 2. However, if, as in panel B of Figure 13.3, we are either unwilling to assume that the two normally distributed populations have equal variances or we have evidence that the variances are not equal, then the **Behrens-Fisher problem** exists (see Reference 9), the pooled-variance t test is inappropriate, and a **separate-variance t' test** developed by Satterthwaite (see Reference 8) may be employed. In the Satterthwaite approximation procedure, the two separate sample variances are included in the computation of the t'-test statistic—hence the name separate-variance t' test.

13.4.2 Development

To test the null hypothesis of no differences in the means of two independent populations

$$H_0: \mu_1 = \mu_2 \text{ or } \mu_1 - \mu_2 = 0$$

against the alternative that the means are not the same

$$H_1: \mu_1 \neq \mu_2 \text{ or } \mu_1 - \mu_2 \neq 0$$

the following separate-variance t'-test statistic can be computed

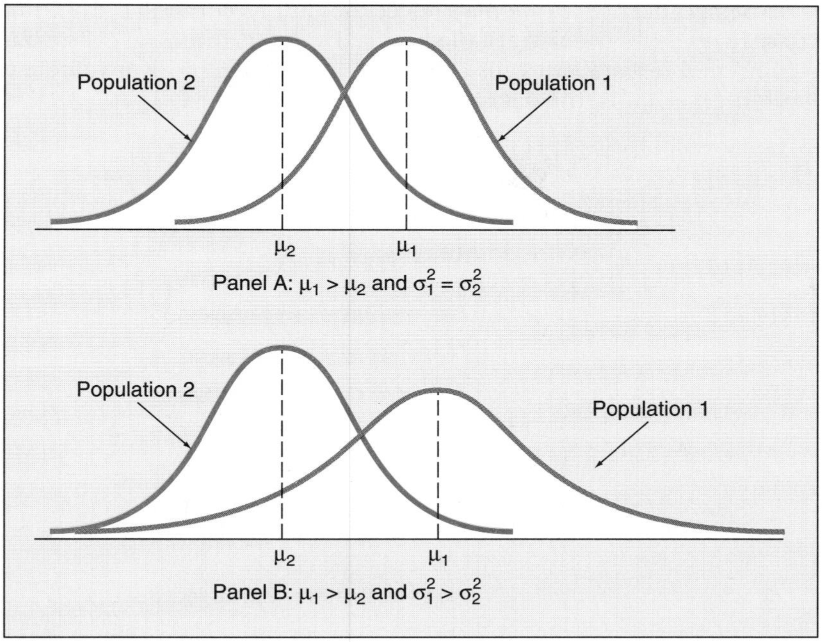

Figure 13.3
Comparing the means from two normally distributed populations.

$$t' = \frac{(\overline{X}_1 - \overline{X}_2) - (\mu_1 - \mu_2)}{\sqrt{\dfrac{S_1^2}{n_1} + \dfrac{S_2^2}{n_2}}} \qquad \textbf{(13.3)}$$

where $\overline{X}_1$ = mean of the sample taken from population 1
S_1^2 = variance of the sample taken from population 1
n_1 = size of the sample taken from population 1
$\overline{X}_2$ = mean of the sample taken from population 2
S_2^2 = variance of the sample taken from population 2
n_2 = size of the sample taken from population 2

The separate-variance t'-test statistic can be approximated by a t distribution with degrees of freedom ν taken to be the integer portion of the computation

$$\nu = \frac{\left(\dfrac{S_1^2}{n_1} + \dfrac{S_2^2}{n_2}\right)^2}{\dfrac{\left(\dfrac{S_1^2}{n_1}\right)^2}{n_1 - 1} + \dfrac{\left(\dfrac{S_2^2}{n_2}\right)^2}{n_2 - 1}} \qquad \textbf{(13.4)}$$

For a given level of significance, α, we may reject the null hypothesis if the computed t' test statistic exceeds the upper-tailed critical value t_ν from the t distribution or if the computed test statistic falls below the lower-tailed critical value $-t_\nu$ from the t distribution. That is, the decision rule is

$$\text{Reject } H_0 \text{ if } t > t_\nu$$
$$\text{or if } t < -t_\nu;$$
$$\text{otherwise do not reject } H_0.$$

The decision rule and regions of rejection are displayed in Figure 13.4.

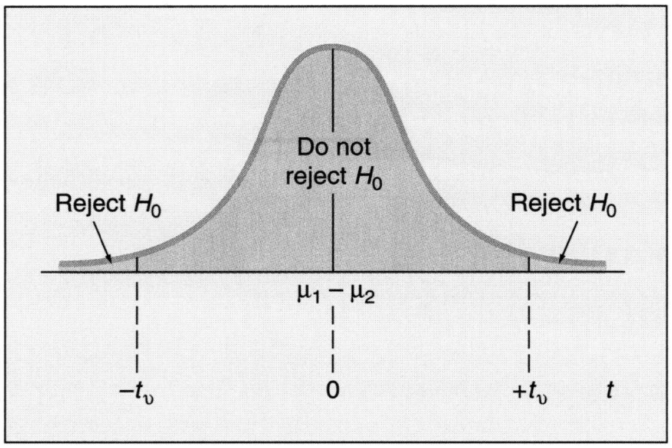

Figure 13.4
Rejection regions for a two-tailed test for the difference between two means.

13.4.3 Application

The separate-variance t' test may be demonstrated by referring to the problem of interest to the financial analyst (see page 465). We recall that the financial analyst wanted to determine whether there was any difference in the mean dividend yield of stocks traded on the New York Stock Exchange versus those traded over the counter on the NASDAQ national market listing. To compare the differences in average dividend yield between the two populations of stock listings, the null and alternative hypotheses would be

$$H_0: \mu_1 = \mu_2 \text{ or } \mu_1 - \mu_2 = 0$$
$$H_1: \mu_1 \neq \mu_2 \text{ or } \mu_1 - \mu_2 \neq 0$$

Dividend yields for random samples of 21 companies from the New York Stock Exchange and 25 stocks from the NASDAQ national market listing are displayed in Table 13.1 (page 466) and summary statistics are presented in Table 13.2 (page 467).

If we can assume that the samples are taken from underlying normal populations but we are unwilling to assume that these populations have equal variances, the separate-variance t' test can be employed. If the test were conducted at the $\alpha = .05$ level of significance, using Equation (13.4) the separate-variance t'-test statistic can be approximated by a t distribution with $\nu = 40$ degrees of freedom, the integer portion of the following computation:

$$
\nu = \frac{\left(\dfrac{S_1^2}{n_1} + \dfrac{S_2^2}{n_2}\right)^2}{\dfrac{\left(\dfrac{S_1^2}{n_1}\right)^2}{n_1 - 1} + \dfrac{\left(\dfrac{S_2^2}{n_2}\right)^2}{n_2 - 1}}
$$

$$
= \frac{\left(\dfrac{1.698}{21} + \dfrac{1.353}{25}\right)^2}{\dfrac{\left(\dfrac{1.698}{21}\right)^2}{20} + \dfrac{\left(\dfrac{1.353}{25}\right)^2}{24}}
$$

$$
= \frac{.018219}{\dfrac{.006538}{20} + \dfrac{.002929}{24}}
$$

$$
= \frac{.018219}{.000449}
$$

$$
= 40.58
$$

From Table E.3 of Appendix E, the upper- and lower-critical values for this two-tailed test are, respectively, +2.0211 and −2.0211, and, as depicted in Figure 13.5, the decision rule is

$$
\text{Reject } H_0 \text{ if } t > t_{40} = +2.0211
$$
$$
\text{or if } t < -t_{40} = -2.0211;
$$
$$
\text{otherwise do not reject } H_0.
$$

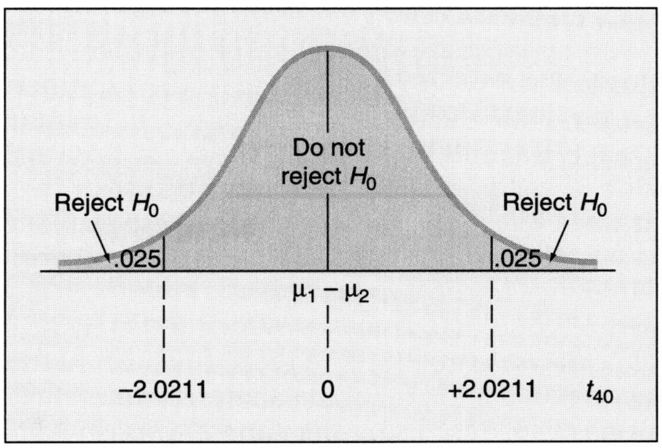

Figure 13.5
Two-tailed test of hypothesis for the difference between the means at the .05 level of significance.

Using the data of Table 13.2 on page 467 we have, from Equation (13.3)

$$
t' = \frac{(\bar{X}_1 - \bar{X}_2) - (\mu_1 - \mu_2)}{\sqrt{\dfrac{S_1^2}{n_1} + \dfrac{S_2^2}{n_2}}}
$$

$$= \frac{3.27 - 2.53}{\sqrt{\frac{1.698}{21} + \frac{1.353}{25}}}$$

$$= \frac{0.74}{0.3674}$$

$$= 2.01$$

Using a .05 level of significance, the null hypothesis (H_0) cannot be rejected here because $t' = 2.01 < t_{40} = 2.0211$. The p value, or probability of obtaining a difference between the two sample means even larger than the 0.74 observed here, which translates to a test statistic t' with a distance even farther from the center of the t distribution than ± 2.01 standard deviations, would be slightly greater than .05 if the null hypothesis of no differences in the means were true. (Using MINITAB, the p value is actually computed as .051.) Since the p value is greater than α of .05, we have no evidence to refute the null hypothesis.

The null hypothesis is not rejected here because the test statistic t' has not fallen into the region of rejection as depicted in Figure 13.5 on page 475. The financial analyst would conclude that there is no evidence of a difference in the mean dividend yields for the two groups.

13.4.4 Dilemma: Conflicting Results

We have an interesting dilemma here. Using the pooled-variance t test of Section 13.3 the financial analyst would have concluded that there is evidence of a difference in the mean dividend yields for the two groups, but using the separate-variance t' test such a conclusion cannot be reached. The major distinction between the t and t' tests is that the former assumes the underlying populations from which the samples were drawn are normally distributed and have equal variances, whereas the latter only assumes the underlying populations are normally distributed (see panels A and B of Figure 13.3 on page 473). As we learned in Section 11.2 (see Table 11.1 on page 389), if the null hypothesis were really true and there is no difference in average dividend yields in the two groups, the use of the t test here would cause us to commit a Type I error while the use of the t' test would result in a correct decision. On the other hand, if the null hypothesis were really false and there is a difference in average dividend yields in the two groups, the use of the t test would result in the correct decision of rejecting a false null hypothesis (i.e., *statistical power*) while the use of the t' test would cause us to commit a Type II error of failing to detect a true average difference in the two groups.

Resolving such a dilemma is part of good data analysis. Which test statistic, t or t', is more credible for the financial analyst's situation? To get to the heart of this matter we should have been conducting an exploratory data analysis and evaluating the plausibility of the assumptions needed for using the t and t' tests. In addition, in Section 13.6 we will develop the F test to determine whether there is evidence of a difference in the two population variances. Based on the results of that test, we may be guided as to which of our previous tests, t or t', is more appropriate for the financial analyst to use.

On the other hand, if our exploratory data analysis reveals that the assumption of underlying normality in the sampled populations is questionable, it might lead us to conclude that neither the t nor the t' test is appropriate. In such a situation, either a *data transformation* (see Reference 11) would be made (and then the assumptions rechecked to determine whether the t or t' test is more appropriate) or a distribution-free procedure which does not make these stringent assumptions would be employed. One such distribution-free procedure, the Wilcoxon rank sum test, will be presented in the next section.

Figure 13.6 depicts MINITAB output presenting the descriptive summary measures, stem-and-leaf displays, box-and-whisker plots, and normal probability plots

```
                N      MEAN    MEDIAN    TRMEAN     STDEV    SEMEAN
NYSE           21     3.271     3.000     3.232     1.303     0.284
NASDAQ         25     2.532     2.400     2.513     1.163     0.233

              MIN       MAX        Q1        Q3
NYSE        0.900     6.400     2.650     3.550
NASDAQ      0.400     5.100     1.750     3.250


Stem-and-leaf of NYSE      N  = 21
Leaf Unit = 0.10

    1    0 9
    1    1
    2    1 6
    4    2 12
    8    2 6799
   (7)   3 0000114
    6    3 56
    4    4
    4    4
    4    5 034
    1    5
    1    6 4

Stem-and-leaf of NASDAQ    N  = 25
Leaf Unit = 0.10

    1    0 4
    2    0 8
    4    1 23
    8    1 5788
   (5)   2 11224
   12    2 6788
    8    3 023
    5    3 68
    3    4 3
    2    4 6
    1    5 1


MTB > boxplot NYSE

                                           ---------
                 *          -----------I   +    I-              *  **         O
                                           ---------
          +---------+---------+---------+---------+---------+---------+------Dyields
         0.0       1.0       2.0       3.0       4.0       5.0       6.0


MTB > boxplot NASDAQ

                                     ---------------
                  -------------I     +       I-----------------
                                     ---------------
          +---------+---------+---------+---------+---------+---------+------Dyields
         0.0       1.0       2.0       3.0       4.0       5.0       6.0
```

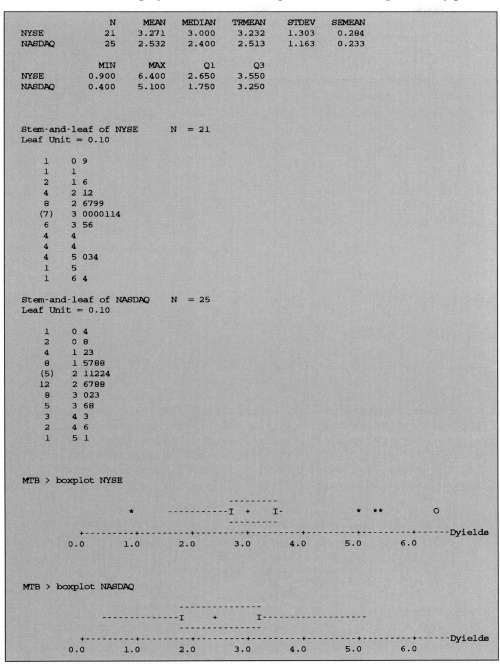

Figure 13.6 **MINITAB output illustrating a descriptive comparison between two sample groups.**

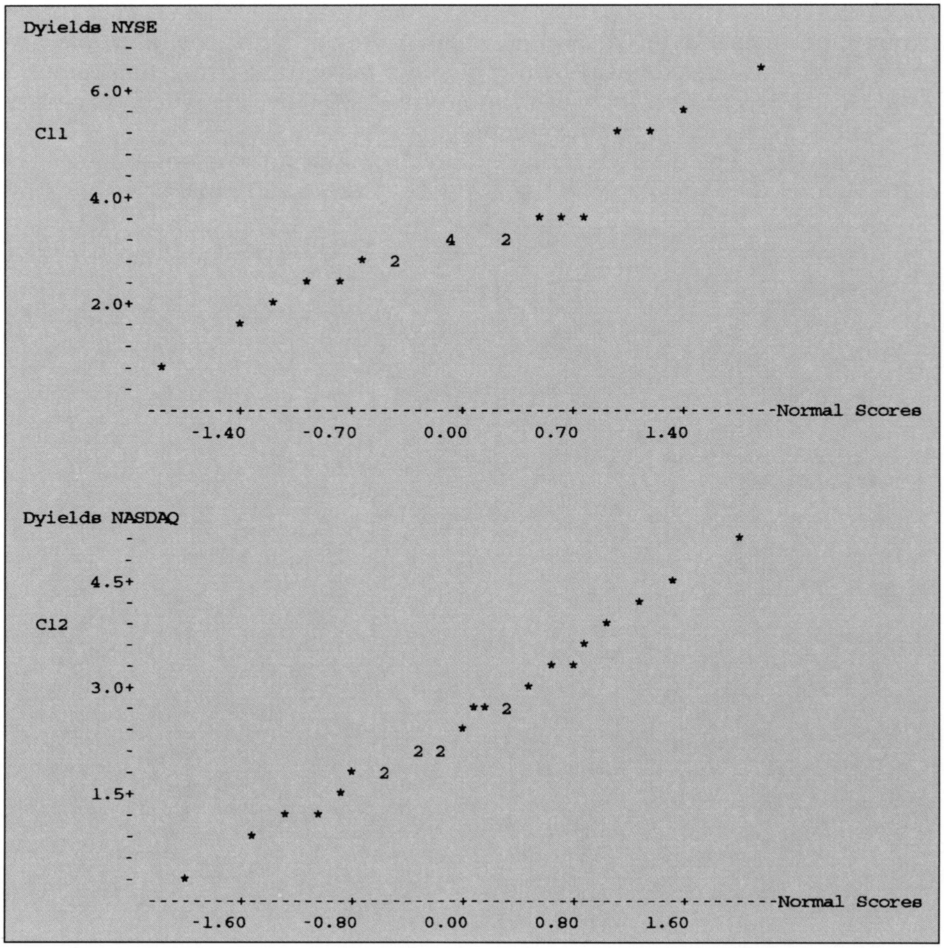

Figure 13.6 ***(Continued)***

for the two sample groups. From this exploratory data analysis, we may question the validity of the normality assumption for the first population—dividend yields of companies listed on the New York Stock Exchange. The sample taken from this group is only of size 21, and this may be too small for us to assume that the central limit theorem (see Section 9.2) has taken effect. If this is the case, the Wilcoxon rank sum test should be used.

Problems for Section 13.4

● 13.13 A real estate agency wanted to compare the appraised values of single-family homes in two Nassau County, New York communities. A sample of 60 listings in Farmingdale and 99 listings in Levittown yielded the following results (in thousands of dollars):

	Farmingdale	Levittown
$\overline{X}$	191.33	172.34
S	32.60	16.92
n	60	99

Assuming that the population variances are not equal, at the .05 level of significance, is there evidence of a difference in the average appraised values for single-family homes in the two Nassau County communities?

13.14 Shipments of meat, meat by-products, and other ingredients are mixed together in several filling lines at a pet food canning factory. Two filling lines in particular are to be compared because, although the average amount filled in the can of pet food is usually the same, the variability of the cans filled in line A is usually much greater than that of line B. The following sample data were obtained (in filling eight-ounce cans):

	Line A	Line B
$\bar{X}$	8.005	7.997
S	0.012	0.005
n	11	16

Assuming that the population variances are not equal, at the .05 level of significance, is there evidence of a difference in the average weight of cans filled on the two lines?

13.15 Refer to Problem 13.5 on page 469. Assume that the population variances are not equal.
(a) At the .01 level of significance, is there evidence of a difference in the average delivery time for the two outlet stores?
(b) What assumptions must be made in order to do part (a) of this problem?
(c) Compare the results obtained in part (a) with those of Problem 13.5.

13.16 Refer to Problem 13.6 on page 470. Assume that the population variances are not equal.
(a) At the .05 level of significance, is there evidence of a difference in the average amount of money spent in the pet supply store between dog owners and cat owners?
(b) What assumptions must be made in order to do part (a) of this problem?
(c) Compare the results obtained in part (a) with those of Problem 13.6.

13.17 Refer to Problem 13.7 on page 470. Assume that the population variances are not equal. Data file SALESVOL.DAT
(a) At the .01 level of significance, is there evidence that wage incentives (through commission) yield greater average sales volume?
(b) What assumptions must be made in order to do part (a) of this problem?
(c) Compare the results obtained in part (a) with those of Problem 13.7.

● 13.18 Refer to Problem 13.8 on page 470. Assume that the variances in the populations are not equal. Data file NICKELBAT.DAT
(a) At the .05 level of significance, is there evidence of a difference in the two types of batteries with respect to average talking time (in minutes) prior to recharging?
(b) What assumptions must be made in order to do part (a) of this problem?
(c) Compare the results obtained in part (a) with those of Problem 13.8.

13.19 Refer to Problem 13.9 on page 471. Assume that the variances in the populations are not equal. Data file NYBANKS.DAT
(a) At the .05 level of significance, is there evidence of a difference in the average effective yields on money market accounts in the two types of banks in the New York area? (Use $\alpha = .05$.)

(b) What assumptions must be made in order to do part (a) of this problem?

(c) Compare the results obtained in part (a) with those of Problem 13.9.

13.20 A public official working on health care reform policy wants to compare occupancy rates (i.e., average annual percentage of beds filled) in urban versus suburban hospitals within his state. A random sample of 16 urban hospitals and a random sample of 16 suburban hospitals are selected within the state and the occupancy rates are recorded as follows:

Data file HOSPITAL.DAT

Urban Hospitals		Suburban Hospitals	
76.5	73.3	71.5	63.0
75.9	77.4	73.4	76.0
79.6	79.0	74.6	75.5
77.5	79.9	74.3	70.7
79.4	70.4	71.2	67.4
78.7	77.7	67.8	62.6
78.6	78.1	76.9	73.0
79.3	75.9	60.0	76.5

(a) Assume that the variances in the populations of hospital types (i.e., urban and suburban) are not equal. Using a .05 level of significance, is there evidence of a difference in the average occupancy rates between urban and suburban hospitals in this state?

(b) What other assumptions must be made in order to do part (a) of this problem?

13.21 The director of training for a company manufacturing electronic equipment is interested in determining whether different training methods have an effect on the productivity of assembly line employees. She randomly assigned 42 recently hired employees into two groups of 21, of which the first received a computer-assisted, individual-based training program and the other received a team-based training program. Upon completion of the training, the employees were evaluated on the time (in seconds) it took to assemble a part. The results are as follows:

Data file TRAINING.DAT

Computer-Assisted, Individual-Based Program		Team-Based Program	
19.4	16.7	22.4	13.8
20.7	19.3	18.7	18.0
21.8	16.8	19.3	20.8
14.1	17.7	15.6	17.1
16.1	19.8	18.0	28.2
16.8	19.3	21.7	20.8
14.7	16.0	30.7	24.7
16.5	17.7	23.7	17.4
16.2	17.4	23.2	20.1
16.4	16.8	12.3	15.2
18.5		16.0	

(a) Assume that the variances in the populations of training methods are not equal. Using a .05 level of significance, is there evidence of a difference in the average assembly times (in seconds) between employees trained in a computer-assisted, individual-based program and those trained in a team-based program?

(b) What other assumptions must be made in order to do part (a) of this problem?

Interchapter Problems for Section 13.4

13.22 Refer to Problem 4.77 on page 162. Assume that the variances in the population are not equal. Data file STUDIO.DAT

(a) Is there evidence that the average rental price in Manhattan is greater than in Brooklyn Heights? (Use the .01 level of significance.)

(b) Compare the results obtained in part (a) with those of Problem 13.10.

● 13.23 Refer to Problem 4.80 on page 163. Assume that the variances in the populations are not equal. Data file PREP.DAT

(a) Is there evidence that the average tuition is higher at preparatory schools in the Northeast than in the Midwest? (Use $\alpha = .01$.)

(b) Compare the results obtained in part (a) with those of Problem 13.11.

13.24 Refer to the data of Problem 3.9 on page 62. Assume that the population variances are not equal. Data file SHAMPOO.DAT

(a) Is there evidence of a difference in the average cost of shampoos labeled for "normal" hair versus those labeled for "fine" hair? (Use the .05 level of significance.)

(b) Compare the results obtained in (a) with those of Problem 13.12.

13.5 Wilcoxon Rank Sum Test for Differences in Two Medians

13.5.1 Introduction

If sample sizes are small and we cannot or do not wish to make the assumption that the data in each group are taken from normally distributed populations, two choices exist. Either the pooled-variance t test or separate-variance t' test may be used, whichever is more appropriate, following some *normalizing transformation* on the data (see Reference 11) or some distribution-free procedure, which does not depend on the assumption of normality for the two populations, can be employed. In this section we introduce the **Wilcoxon rank sum test**, a widely used, very simple, and powerful distribution-free procedure for testing for differences between the medians of two populations. The Wilcoxon rank sum test has proven to be almost as powerful as its parametric counterparts (the t test and the t' test) under conditions appropriate to the latter and is likely to be more powerful when the stringent assumptions of those tests are not met.

In addition, the Wilcoxon rank sum test is an excellent procedure to choose when only ordinal-type data can be obtained, as is often the case when dealing with studies in consumer behavior, marketing research, and experimental psychology. The parametric t and t' tests would not be used in such situations since these procedures require that the obtained data are measured on at least an interval scale.

13.5.2 Procedure

To perform the Wilcoxon rank sum test we must replace the observations in the two samples of size n_1 and n_2 with their combined ranks (unless the obtained data contained the ranks initially). The ranks are assigned in such a manner that rank 1 is given to the smallest of the $n = n_1 + n_2$ combined observations, rank 2 is given to the second smallest, and so on until rank n is given to the largest. If several values are tied, we assign each the average of the ranks that would otherwise have been assigned.

For convenience, whenever the two sample sizes are unequal, we will let n_1 represent the *smaller*-sized sample and n_2 the *larger*-sized sample. The Wilcoxon rank sum test statistic T_1 is merely the sum of the ranks assigned to the n_1 observations in the smaller sample. (For equal-sized samples, either group may be selected for determining T_1.)

For any integer value n, the sum of the first n consecutive integers may easily be calculated as $n(n + 1)/2$. The test statistic T_1 plus the sum of the ranks assigned to the n_2 items in the second sample, T_2, must therefore be equal to this value; that is,

$$T_1 + T_2 = \frac{n(n + 1)}{2} \tag{13.5}$$

so that Equation (13.5) can serve as a check on the ranking procedure.

The test of the null hypothesis can be either two-tailed or one-tailed, depending on whether we are testing if the two population medians are merely *different* or if one median is *greater than* the other median.

Two-Tailed Test	One-Tailed Test	One-Tailed Test
H_0: $M_1 = M_2$	H_0: $M_1 \geq M_2$	H_0: $M_1 \leq M_2$
H_1: $M_1 \neq M_2$	H_1: $M_1 < M_2$	H_1: $M_1 > M_2$

where M_1 = median of population 1 having n_1 sample observations
M_2 = median of population 2 having n_2 sample observations

When the sizes of both samples n_1 and n_2 are ≤ 10, Table E.11 may be used to obtain the critical values of the test statistic T_1 for both one- and two-tailed tests at various levels of significance. For a two-tailed test and for a particular level of significance, α, if the computed value of T_1 equals or exceeds the upper critical value or is less than or equal to the lower critical value, the null hypothesis may be rejected. For one-tailed tests having the alternative H_1: $M_1 < M_2$, the decision rule is to reject the null hypothesis if the observed value of T_1 is less than or equal to the lower critical value. For one-tailed tests having the alternative H_1: $M_1 > M_2$ the decision rule is to reject the null hypothesis if the observed value of T_1 equals or exceeds the upper critical value.

To demonstrate how to use Table E.11 to obtain the critical values for the T_1 test statistic, let us suppose that the sample sizes for our two groups are 8 and 10 and we wish to choose a level of significance α of .05. From Table 13.3, which is a

Table 13.3 Obtaining the lower- and upper-tail critical values T_1 for the Wilcoxon rank sum test where $n_1 = 8$, $n_2 = 10$, and $\alpha = .05$

	α		n_1						
			4	5	6	7	8	9	10
n_2	One-Tailed	Two-Tailed				(Lower, Upper)			
9	.025	.05	14,42	22,53	31,65	40,79	51,93	62,109	
	.01	.02	13,43	20,55	28,68	37,82	47,97	59,112	
	.005	.01	11,45	18,57	26,70	35,84	45,99	56,115	
	.05	.10	17,43	26,54	35,67	45,81	56,96	69,111	82,128
10	.025	.05	15,45	23,57	32,70	42,84	53,99	65,115	78,132
	.01	.02	13,47	21,59	29,73	39,87	49,103	61,119	74,136
	.005	.01	12,48	19,61	27,75	37,89	47,105	58,122	71,139

Source: Extracted from Table E.11.

replica of Table E.11, if $n_1 = 8$, $n_2 = 10$, and $\alpha = .05$, we observe that the lower and upper critical values for a two-tailed test are, respectively, 53 and 99. If the computed value of the test statistic T_1 falls between these critical values, the null hypothesis could not be rejected. If, however, the computed value of the test statistic equals or exceeds 99 or is less than or equal to 53, the null hypothesis would be rejected.

For large sample sizes the test statistic T_1 is approximately normally distributed. The following large-sample approximation formula may be used for testing the null hypothesis when sample sizes are outside the range of Table E.11:

$$Z = \frac{T_1 - \mu_{T_1}}{\sigma_{T_1}} \qquad \textbf{(13.6)}$$

where $n_1 \leq n_2$

T_1 = sum of the ranks assigned to the n_1 observations in sample 1

μ_{T_1} = mean value of T_1

σ_{T_1} = standard deviation of T_1

μ_{T_1}, the mean value of the test statistic T_1, can be computed from

$$\mu_{T_1} = \frac{n_1(n + 1)}{2}$$

and σ_{T_1}, the standard deviation of T_1, is calculated from

$$\sigma_{T_1} = \sqrt{\frac{n_1 n_2 (n + 1)}{12}}$$

so that Equation (13.6) may be rewritten as

$$Z = \frac{T_1 - \frac{n_1(n+1)}{2}}{\sqrt{\frac{n_1 n_2 (n+1)}{12}}} \qquad (13.7)$$

Based on α, the level of significance selected, the null hypothesis may be rejected if the computed Z value falls in the appropriate region of rejection, depending on whether a two-tailed or a one-tailed test is used (see Figure 13.7).

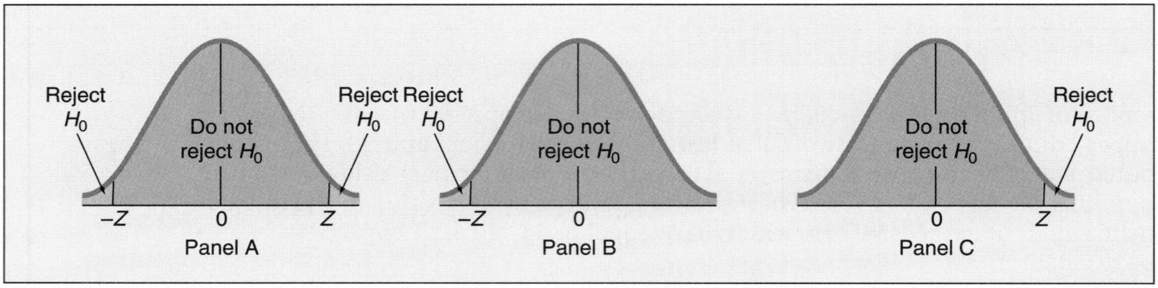

Figure 13.7
Determining the rejection region: Panel A, two-tailed test ($M_1 \neq M_2$); Panel B, one-tailed test ($M_1 < M_2$); Panel C, one-tailed test ($M_1 > M_2$).

13.5.3 Application

To demonstrate the use of the Wilcoxon rank sum test, let us refer back to the problem faced by our financial analyst (see page 465) who wanted to determine whether there was any difference in the average dividend yield of stocks traded on the New York Stock Exchange versus those traded over the counter on the NASDAQ national market listing. Dividend yields for random samples of 21 issues from the New York Stock Exchange and 25 issues from the NASDAQ national market listing are displayed in Table 13.1 (page 466) and summary statistics are presented in Table 13.2 (page 467).

If, as a result of an exploratory data analysis (see Figure 13.6 on pages 477–478), the financial analyst does not wish to make the stringent assumption that the samples were taken from populations that are normally distributed, the Wilcoxon rank sum test may be used for evaluating possible differences in the median dividend yields.[1] Since the financial analyst is not specifying which of the two groups is likely to possess a greater median dividend yield, the test is two-tailed, and the following null and alternative hypotheses are established:

H_0: $M_1 = M_2$ (the median dividend yields are equal)

H_1: $M_1 \neq M_2$ (the median dividend yields are different)

To perform the Wilcoxon rank sum test, we form the combined ranking of the dividend yields obtained from the $n_1 = 21$ companies on the New York Stock Exchange and the $n_2 = 25$ companies on the NASDAQ national market listing. The combined ranking of the dividend yields is displayed in Table 13.4. (We note that a rank of 1 is given to Intel, the company with the lowest dividend yield, and a rank of 46 is given to NYNEX, the company with the highest dividend yield in the combined ranking.)

Table 13.4 Forming the combined ranks

New York Stock Exchange ($n_1 = 21$)		NASDAQ Listing ($n_2 = 25$)	
Company	Dividend Yield Combined Rank	Company	Dividend Yield Combined Rank
American Express	35	Atlantic SE Airlines	4
Anheuser-Busch	20.5	Boral Ltd	43
Bristol-Myers-Squibb	45	Cathay Bancorp	40
Dayton-Hudson	12	Cit Fed Bancorp	2
Dresser Industries	28	CPB	33
Ford Motor	31.5	First Essex Bancorp	28
General Electric	28	Goulds Pumps	39
General Mills	36	Harper Group	5
IBM	7	Innovex	15
Kellogg Co.	18.5	Intel Corp	1
Merck & Co.	37.5	Lindberg Corp	20.5
NYNEX	46	Nature's Sunshine Prod.	6
Occidental Petroleum	44	Newcor	12
Pfizer Inc.	28	PCA International	34
PPG Inc.	28	T Rowe Price Assoc.	9.5
Sara Lee Corp.	24.5	PSB Holdings Corp.	17
Texaco Inc.	42	Research Inc.	41
Texas Instruments	3	Seacoast Banking Corp.	22.5
Whirlpool Corp.	15	Span-America Med. Sys.	9.5
Winn-Dixie	31.5	Sumitomo Bank of Cal.	37.5
Xerox Corp.	24.5	TCA Cable TV	15
		United Fire & Casualty	22.5
		West Coast Bancorp	8
		Whitney Holdings	18.5
		Worthington Industries	12

Source: Data are taken from Table 13.1.

We then obtain the test statistic T_1, the sum of the ranks assigned to the smaller sample:

$$T_1 = 35 + 20.5 + \cdots + 24.5 = 585.5$$

As a check on the ranking procedure we also obtain T_2 and use Equation (13.5) to show that the sum of the first $n = 46$ integers in the combined ranking is equal to $T_1 + T_2$:

$$T_1 + T_2 = \frac{n(n+1)}{2}$$

$$585.5 + 495.5 = \frac{46(47)}{2} = 1{,}081$$

To test the null hypothesis of no difference in the median dividend yields in the two populations, we use the large-sample approximation formula

[Equation (13.7)]. Choosing the .05 level of significance, the critical values from the standard normal distribution (Table E.2) are ±1.96 (see Figure 13.8). The decision rule would be

$$\text{Reject } H_0 \text{ if } Z > +1.96$$
$$\text{or if } Z < -1.96;$$
$$\text{otherwise do not reject } H_0.$$

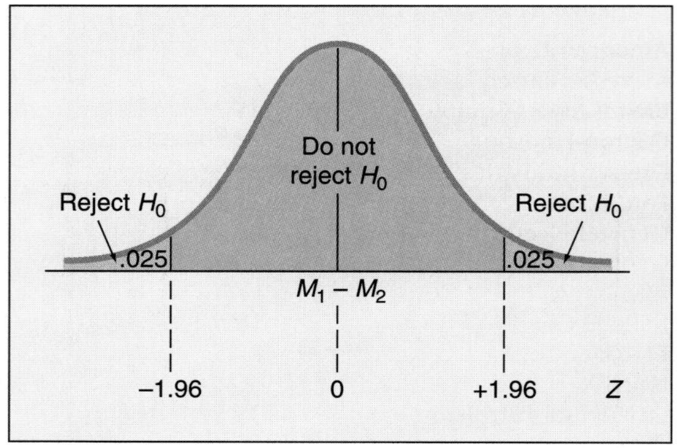

Figure 13.8
A two-tailed test of hypothesis for the difference in medians at the .05 level of significance.

Using Equation (13.7) we have

$$Z = \frac{T_1 - \dfrac{n_1(n+1)}{2}}{\sqrt{\dfrac{n_1 n_2 (n+1)}{12}}}$$

$$= \frac{585.5 - \dfrac{21(47)}{2}}{\sqrt{\dfrac{21(25)(47)}{12}}}$$

$$= \frac{585.5 - 493.5}{45.35}$$

$$= 2.03$$

Since $Z = +2.03 > +1.96$, the decision is to reject H_0. The p value, or probability of obtaining a test statistic W even larger than the 585.5 observed here, which translates to a test statistic Z with a distance even farther from the center of the standard normal distribution than ±2.03 standard deviations, is .0424 if the null hypothesis of no difference in the medians were true. Since the p value is less than $\alpha = .05$, we don't believe that the null hypothesis is true and we reject it.

The null hypothesis is rejected because the test statistic Z has fallen into the region of rejection. Thus, without having to make the stringent assumption of normality in the original populations, the financial analyst may conclude that there is evidence of a difference in the median dividend yields for the two groups. Companies traded on the New York Stock Exchange appear to have significantly higher dividend yields than do companies trading over the counter on the NASDAQ national market listing.

13.5.4 Reflections

This conclusion agrees with the findings from the pooled-variance *t* test, not the separate-variance *t′* test. If we can now demonstrate that there is no evidence of a difference in the variability in the two groups, we would have sufficient reason to conclude that the results from the *t* test and Wilcoxon rank sum tests are plausible. Testing for the difference between two population variances will be the topic of Section 13.6.

Problems for Section 13.5

● 13.25 A statistics professor taught two special sections of a basic course in which the 10 students in each section were considered outstanding. She used a "traditional" method of instruction (T) in one section and an "experimental" method (E) in the other. At the end of the semester she ranked the students on the basis of their performance from 1 (worst) to 20 (best).

Data file TESTRANK.DAT

T	1	2	3	5	9	10	12	13	14	15
E	4	6	7	8	11	16	17	18	19	20

For this instructor, was there any evidence of a difference in performance based on the two methods? (Use $\alpha = .05$.)

13.26 The Director of Human Resources at a 1,200-bed New York City hospital was evaluating candidates for the position of Administrator of the Billings and Payments Department. Among the applicants, 22 were invited for interviews. Following the interviews, the rankings (1 = most preferred) of the candidates (based on interview, academic record, and prior experience) are presented below, broken down by "type" of master's degree obtained—MBA versus MPH.

Data file INTRANK.DAT

MBA Candidates		MPH Candidates	
1	2	3	6
4	5	7	10
8	9	13	14
11	12	16	18
15	17	19	20
		21	22

Is there evidence that the MBA candidates were more preferred than the MPH candidates? (Use $\alpha = .05$.)

13.27 A New York television station decided to do a story comparing two commuter railroads in the area—the Long Island Rail Road (LIRR) and New Jersey Transit (NJT). The researchers at the station sampled the performance of several scheduled train runs of each railroad, 10 for the LIRR and 12 for the NJT. The data on how many minutes early (negative numbers) or late (positive numbers) each train was are presented below:

Data file TRAIN2.DAT

LIRR:	5	−1	39	9	12	21	15	52	18	23		
NJT:	8	4	10	4	12	5	4	9	15	33	14	7

(a) Is there evidence that the railroads differ in their median tendencies to be late? (Use $\alpha = .01$.)
(b) What conclusions about the lateness of the two railroads can be made?

Data file SALESVOL.DAT

13.28 Refer to the data for Problem 13.7 on page 470. Using a .01 level of significance, is there evidence that wage incentives (through commission) yield greater median sales volume?

Data file NICKELBAT.DAT

13.29 Refer to the data for Problem 13.8 on page 470.
(a) Using a .05 level of significance, is there evidence of a difference in the two types of batteries with respect to median talking time (in minutes) prior to recharging?
(b) What assumptions must be made in order to do part (a) of this problem?
(c) Compare the results obtained in part (a) with those of Problem 13.8 on page 470 and Problem 13.18 on page 479. Discuss.

Data file NYBANKS.DAT

13.30 Refer to Problem 13.9 on page 471.
(a) Using a .05 level of significance, is there evidence of a difference in the median effective yields on money market accounts in the two types of banks in the New York area?
(b) What assumptions must be made in order to do part (a) of this problem?
(c) Compare the results obtained in part (a) with those of Problem 13.9 on page 471 and Problem 13.19 on page 479. Discuss.

Data file HOSPITAL.DAT

13.31 Refer to Problem 13.20 on page 480.
(a) Using a .05 level of significance, is there evidence of a difference in the median occupancy rates between urban and suburban hospitals in this state?
(b) What other assumptions must be made in order to do part (a) of this problem?
(c) Compare the results obtained in part (a) with those of Problem 13.20. Discuss.

Data file TRAINING.DAT

13.32 Refer to Problem 13.21 on page 480.
(a) Using a .05 level of significance, is there evidence of a difference in the median assembly times (in seconds) between employees trained in a computer-assisted, individual-based program and those trained in a team-based program?
(b) What other assumptions must be made in order to do part (a) of this problem?
(c) Compare the results obtained in part (a) with those of Problem 13.21. Discuss.

Interchapter Problems for Section 13.5

Data file STUDIO.DAT

13.33 Refer to the data from Problem 4.77 on page 162.
(a) Test whether there is evidence that the median rent paid for unfurnished studio apartments in Manhattan is higher than in Brooklyn Heights. (Use $\alpha = .01$.)
(b) Compare the results obtained in part (a) with those from Problem 13.10 (page 471) and Problem 13.22 (page 481).

Data file PREP.DAT

13.34 Refer to the data from Problem 4.80 on page 163.
(a) Test whether there is evidence that the median tuition is higher at preparatory schools in the Northeast than in the Midwest. (Use $\alpha = .01$.)
(b) Compare the results obtained in part (a) with those from Problem 13.11 (page 472) and Problem 13.23 (page 481).

13.35 Refer to the data from Problem 3.9 on page 62. Data file SHAMPOO.DAT

(a) Test whether there is evidence of a difference in the median cost of shampoos labeled for "normal" hair versus those labeled for "fine" hair. (Use $\alpha = .05$.)

(b) Compare the results obtained in part (a) with those from Problem 13.12 (page 472) and Problem 13.24 (page 481).

13.6 *F* Test for Differences in Two Variances

13.6.1 Introduction

In the previous three sections we examined some procedures for testing for differences in central tendency (that is, differences in the means or the medians) between two independent populations. In many situations, however, we may also be interested in testing whether two independent populations have the same variability. Either we may be interested in studying the variances of two populations as a "means to an end," that is, testing the assumption of equal variances in order to determine whether the pooled-variance *t* test or the separate-variance *t*′ test is the more appropriate to use when comparing the two population means (Sections 13.3 and 13.4), or we may be truly interested in studying the variances of two populations as an "end in itself."

13.6.2 Development

In order to test for the equality of the variances of two independent populations, a statistical procedure has been devised that is based on the ratio of the two sample variances. If the data from each population are assumed to be normally distributed, then the ratio S_1^2/S_2^2 follows a distribution called the *F* distribution (see Table E.5), which was named after the famous statistician R. A. Fisher. From Table E.5 (a replica of which appears as Table 13.5 on page 491), we can see that the critical values of the *F* distribution depend on *two* sets of degrees of freedom. The degrees of freedom in the numerator of the ratio pertain to the first sample and the degrees of freedom in the denominator of the ratio pertain to the second sample. The ***F*-test** statistic for testing the equality between two variances would be

$$F = \frac{S_1^2}{S_2^2} \tag{13.8}$$

where n_1 = size of sample taken from population 1

n_2 = size of sample taken from population 2

$n_1 - 1$ = degrees of freedom from sample 1 (i.e., the numerator degrees of freedom)

$n_2 - 1$ = degrees of freedom from sample 2 (i.e., the denominator degrees of freedom)

S_1^2 = variance of sample 1
S_2^2 = variance of sample 2

In testing the equality of two variances, either two-tailed or one-tailed tests can be employed, depending on whether we are testing if the two population variances are *different* or if one variance is *greater than* the other variance. These situations are depicted in Figure 13.9.

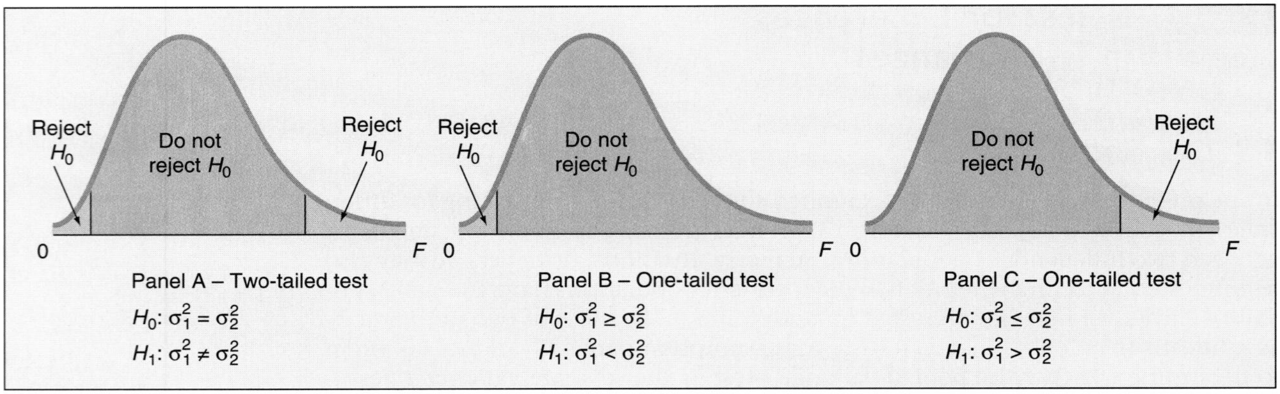

Figure 13.9
Determining the rejection region for testing a hypothesis about the equality of two population variances: Panel A, two-tailed test; Panel B, one-tailed test; Panel C, one-tailed test.

For a given level of significance, α, to test the null hypothesis of equality of variances

$$H_0: \sigma_1^2 = \sigma_2^2$$

against the alternative hypothesis that the two population variances are not equal

$$H_1: \sigma_1^2 \neq \sigma_2^2$$

we may reject the null hypothesis if the computed F test statistic exceeds the upper-tailed critical value $F_{U(n_1-1),(n_2-1)}$ from the F distribution or if the computed test statistic falls below the lower-tailed critical value $F_{L(n_1-1),(n_2-1)}$ from the F distribution. That is, the decision rule is

$$\text{Reject } H_0 \text{ if } F > F_{U(n_1-1),(n_2-1)}$$
$$\text{or if } F < F_{L(n_1-1),(n_2-1)};$$
$$\text{otherwise do not reject } H_0.$$

This decision rule and rejection region are displayed in panel A of Figure 13.9.

13.6.3 Application

In order to demonstrate how we may test for the equality of two variances, we can return to the financial analyst's study of dividend yields from two groups of stocks. The data for this are displayed in Table 13.1 on page 466 and the summary measures for the two samples are presented in Table 13.2 on page 467.

To test for the equality of the two population variances we have the following null and alternative hypotheses:

$$H_0: \sigma_1^2 = \sigma_2^2$$
$$H_1: \sigma_1^2 \neq \sigma_2^2$$

Because this is a two-tailed test, the rejection region is split into the lower and upper tails of the F distribution. If a level of significance of $\alpha = .05$ is selected, each rejection region would contain .025 of the distribution.

The upper-tail critical value of the F distribution with 20 and 24 degrees of freedom can be obtained directly from Table E.5, a replica of which is presented in Table 13.5. Since there are 20 degrees of freedom in the numerator and 24 degrees of freedom in the denominator, the upper-tail critical value can be found by looking in the column labeled "20" and the row labeled "24" which pertain to an upper-tail area of .025. Thus the upper-tail critical value of this F distribution is 2.33.

Table 13.5 Obtaining the critical value of *F* with 20 and 24 degrees of freedom for an upper-tail area of .025.

Denominator df_2	Numerator df_1 1	2	3	. . .	15	20	24	30
1	647.8	799.5	864.2	. . .	984.9	993.1	997.2	1001
2	38.51	39.00	39.17	. . .	39.43	39.45	39.46	39.46
3	17.44	16.04	15.44	. . .	14.25	14.17	14.12	14.08
4	12.22	10.65	9.98	. . .	8.66	8.56	8.51	8.46
⋮	⋮	⋮	⋮	⋮	⋮	⋮	⋮	⋮
15	6.20	4.77	4.15	. . .	2.86	2.76	2.70	2.64
16	6.12	4.69	4.08	. . .	2.79	2.68	2.63	2.57
17	6.04	4.62	4.01	. . .	2.72	2.62	2.56	2.50
18	5.98	4.56	3.95	. . .	2.67	2.56	2.50	2.44
19	5.92	4.51	3.90	. . .	2.62	2.51	2.45	2.39
20	5.87	4.46	3.86	. . .	2.57	2.46	2.41	2.35
21	5.83	4.42	3.82	. . .	2.53	2.42	2.37	2.31
22	5.79	4.38	3.78	. . .	2.50	2.39	2.33	2.27
23	5.75	4.35	3.75	. . .	2.47	2.36	2.30	2.24
24	5.72	4.32	3.72	. . .	2.44	2.33	2.27	2.21

Source: Extracted from Table E.5.

Obtaining Lower-Tail Critical Values Any lower-tail critical value on the F distribution can be obtained from

$$F_{L(n_1-1),(n_2-1)} = \frac{1}{F_{U(n_2-1),(n_1-1)}} \qquad \textbf{(13.9)}$$

where $F_{L(n_1-1),(n_2-1)}$ = lower-tail critical value of the F distribution having n_1-1 and n_2-1 degrees of freedom

$F_{U(n_2-1),(n_1-1)}$ = upper-tail critical value of the F distribution having n_2-1 and n_1-1 degrees of freedom

n_1-1 = degrees of freedom from sample 1

n_2-1 = degrees of freedom from sample 2

Therefore, in this example we have

$$F_{L(20,\ 24)} = \frac{1}{F_{U(24,\ 20)}}$$

To compute our desired lower-tail critical value, we need to obtain the upper .025 value of *F* with 24 degrees of freedom in the numerator and 20 degrees of freedom in the denominator and take its reciprocal. From Table E.5 this upper-tail value is 2.41. Therefore, from Equation (13.9)

$$F_{L(20,\ 24)} = \frac{1}{F_{U(24,\ 20)}} = \frac{1}{2.41} = 0.415$$

As depicted in Figure 13.10, the decision rule is

Reject H_0 if $F > F_{U(20,\ 24)} = 2.33$

or if $F < F_{L(20,\ 24)} = 0.415$;

otherwise do not reject H_0.

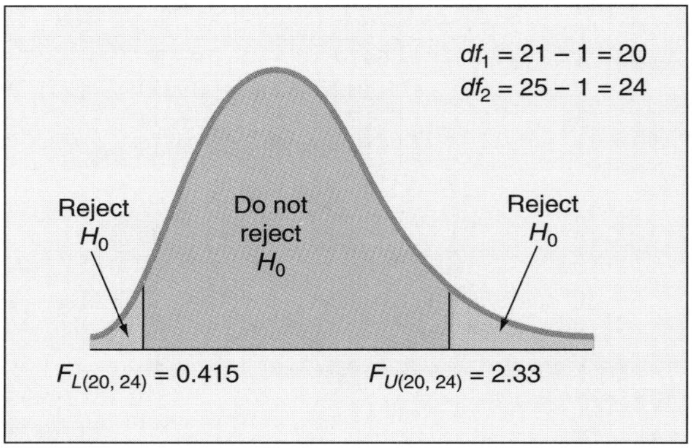

Figure 13.10
Regions of rejection and nonrejection for a two-tailed test for the equality of two variances at the .05 level of significance with 20 and 24 degrees of freedom.

Using Equation (13.8) for the financial analyst's data (see Table 13.2 on page 467), we compute the following *F*-test statistic:

$$F = \frac{S_1^2}{S_2^2}$$

$$= \frac{1.698}{1.353} = 1.25$$

Therefore, since $F_{L(20,\ 24)} = 0.415 < F = 1.25 < F_{U(20,\ 24)} = 2.33$ we do not reject H_0. The financial analyst would conclude that there is no evidence of a difference in the variability of dividend yields for the two populations. Thus, if we could assume that the two populations are normally distributed, the pooled-variance *t* test would be more appropriate than the separate-variance *t*′ test for comparing differences in the average dividend yields because there is no evidence that the population variances are not equal. On the other hand, if we do not feel that the normality assumption is viable, we should use the Wilcoxon rank sum test to determine if there are differences in the median dividend yields in the two populations.

13.6.4 Caution

In testing for the equality of two population variances, we should be aware that the test assumes that each of the two populations is normally distributed. That is, if the assumption of normality for each population is met, the *F*-test statistic follows an *F* distribution with $n_1 - 1$ and $n_2 - 1$ degrees of freedom. Unfortunately, this *F*-test statistic is not *robust* to departures from this assumption (Reference 2), particularly when the sample sizes in the two groups are not equal. Therefore, if the populations are not at least approximately normally distributed, the accuracy of the procedure can be seriously affected (References 2–4 present other procedures for testing the equality of two variances).

Problems for Section 13.6

● 13.36 Suppose that the following information is available for two groups:

$$n_1 = 10 \quad S_1^2 = 13.7 \quad n_2 = 10 \quad S_2^2 = 16.9$$

(a) At the .05 level of significance, is there evidence of a difference between σ_1^2 and σ_2^2?
(b) What is the relationship in (a) between the lower critical value and the upper critical value? Under what conditions will this relationship hold? Explain.
(c) Suppose that we had wanted to perform a one-tailed test. At the .05 level of significance, what is the upper-tailed critical value of the *F*-test statistic to determine if there is evidence that $\sigma_1^2 > \sigma_2^2$?
(d) Suppose that we had wanted to perform a one-tailed test. At the .05 level of significance, what is the lower-tailed critical value of the *F*-test statistic to determine if there is evidence that $\sigma_1^2 > \sigma_2^2$?

13.37 Suppose that the following information is available for two groups:

$$n_1 = 16 \quad S_1^2 = 47.3 \quad n_2 = 13 \quad S_2^2 = 36.4$$

(a) At the .05 level of significance, is there evidence of a difference between σ_1^2 and σ_2^2?
(b) Suppose that we had wanted to perform a one-tailed test. At the .05 level of significance, what is the upper-tailed critical value of the *F*-test statistic to determine if there is evidence that $\sigma_1^2 > \sigma_2^2$?
(c) Suppose that we had wanted to perform a one-tailed test. At the .05 level of significance, what is the lower-tailed critical value of the *F*-test statistic to determine if there is evidence that $\sigma_1^2 > \sigma_2^2$?

13.38 A professor in the Accountancy Department of a business school claims that there is much more variability in the final exam scores of students taking the introductory accounting course as a requirement than for students taking the course as part of their major. Random samples of 13 nonaccounting majors and 10 accounting majors were taken from the professor's class roster in his large lecture and the following results were computed based on the final exam scores:

$$n_{NA} = 13 \qquad S^2_{NA} = 210.2 \qquad n_A = 10 \qquad S^2_A = 36.5$$

(a) At the .05 level of significance, is there evidence to support the professor's claim?
(b) Find upper and lower limits on the p value.

13.39 Refer to Problem 13.5 on page 469.
(a) At the .01 level of significance, is there evidence of a difference in the variances of the shipping time between the two outlets?
(b) Find upper and lower limits on the p value.

13.40 Refer to Problem 13.6 on page 470.
(a) At the .05 level of significance, is there evidence of a difference in the variances of the amount spent between dog and cat owners?
(b) Find upper and lower limits on the p value.

Data file NICKELBAT.DAT

● 13.41 Refer to the data for Problem 13.8 on page 470.
(a) Using a .05 level of significance, is there evidence of a difference in the variances in talking time (in minutes) prior to recharging between the two types of batteries?
(b) Based on the results obtained in part (a), which test should have been chosen, the t test of Problem 13.8, the t' test of Problem 13.18 (page 479), or the Wilcoxon rank sum test of Problem 13.29 (page 488)? Discuss.

Data file NYBANKS.DAT

13.42 Refer to the data for Problem 13.9 on page 471.
(a) Using a .05 level of significance, is there evidence of a difference in the variances in effective yields on money market accounts between the two types of banks in the New York area?
(b) Based on the results obtained in part (a), which test should have been chosen, the t test of Problem 13.9, the t' test of Problem 13.19 (page 479), or the Wilcoxon rank sum test of Problem 13.30 (page 488)? Discuss.

Data file HOSPITAL.DAT

13.43 Refer to the data for Problem 13.20 on page 480.
(a) Using a .05 level of significance, is there evidence of a difference in the variances in occupancy rates between urban and suburban hospitals in this state?
(b) Based on the results obtained in part (a), which test should have been chosen, the t' test of Problem 13.20 (page 480) or the Wilcoxon rank sum test of Problem 13.31 (page 488)? Discuss.

Data file TRAINING.DAT

13.44 Refer to the data for Problem 13.21 on page 480.
(a) Using a .05 level of significance, is there evidence of a difference in the variances in assembly times (in seconds) between employees trained in a computer-assisted, individual-based program and those trained in a team-based program?
(b) Based on the results obtained in part (a), which test should have been chosen, the t' test of Problem 13.21 (page 480) or the Wilcoxon rank sum test of Problem 13.32 (page 488)? Discuss.

Interchapter Problems for Section 13.6

Data file STUDIO.DAT

13.45 Refer to the data of Problem 4.77 on page 162.
(a) Is there evidence that the variance in Manhattan is greater than the variance in Brooklyn Heights? (Use $\alpha = .01$.)
(b) Find upper and lower limits on the p value.

13.46 Refer to the data of Problem 4.80 on page 163.

Data file PREP.DAT

(a) Is there evidence that the variance of tuition differs between preparatory schools in the Northeast and in the Midwest? (Use $\alpha = .01$.)

(b) Find upper and lower limits on the p value.

13.47 Refer to the data of Problem 3.9 on page 62.

Data file SHAMPOO.DAT

(a) Is there evidence of a difference in the variances between the cost of "normal" shampoos and "fine" shampoos? (Use $\alpha = .05$.)

(b) Find upper and lower limits on the p value.

13.7 Using the Computer for Hypothesis Testing with Two Independent Samples: The Kalosha Industries Employee Satisfaction Survey

Bud Conley, the Vice-President of Human Resources at Kalosha Industries, is preparing for another meeting with a representative from the B & L Corporation to discuss the contents of an employee-benefits package that is being developed. Prior to this meeting, an answer to the following question would be of particular concern in a *confirmatory analysis* of the survey data (Table 2.3 on pages 33–40): Is there evidence of a gender difference with respect to average length of time (in years) that full-time workers have been employed at Kalosha Industries (see questions 16 and 5 in the Survey)?

These and other questions raised by Bud Conley (see Survey/Database Project at the end of the section) require a detailed descriptive statistical analysis of the 400 responses to the Survey along with a confirmatory analysis.

13.7.1 Using Statistical Packages for Numerical Data

To determine whether there are significant gender differences with respect to longevity, that is, in the average length of time (in years) that full-time employees have spent working at Kalosha Industries, a choice must be made. Which of the hypothesis-testing procedures—the t test, the t' test, or the Wilcoxon rank sum test—should be chosen here? To assist in the selection of the most appropriate test, a descriptive analysis will be undertaken so that the assumptions of the various test procedures can be evaluated.

Data file EMPSAT.DAT

Figure 13.11 provides computer output presenting the descriptive summary measures regarding longevity at Kalosha Industries for both male and female

	sex	N	MEAN	MEDIAN	TRMEAN	STDEV	SEMEAN
empyears	M	233	9.278	6.000	8.360	9.141	0.599
	F	167	7.812	5.000	7.195	7.117	0.551

	sex	MIN	MAX	Q1	Q3
empyears	M	0.080	52.250	2.040	13.000
	F	0.160	30.000	2.330	11.000

Figure 13.11
MINITAB output of summary measures.
Note: We should be familiar with all the summary measures obtained from the MINITAB output except TRMEAN (which is beyond the scope of this text).

$$H_0: \mu_M = \mu_F$$

against the alternative

$$H_1: \mu_M \neq \mu_F$$

We observe that the t' statistic is equal to +1.80 and, with 395 degrees of freedom, the corresponding two-tailed p value is .0723. For a two-tailed test with a chosen α level of significance of .05, since .0723 > .05 we are unable to reject the null hypothesis. The t'-test statistic of +1.80 falls within the region of nonrejection, between the lower- and upper-tailed critical values of ±1.96 from the t distribution with 395 degrees of freedom. We would conclude that there is no evidence that the average length of employment at Kalosha Industries is different between the full-time male workers and female workers.

Researchers not wishing to perform the t' test owing to the apparent lack of normality in the underlying populations might choose the Wilcoxon rank sum test in its place. Figure 13.14 provides SPSS output illustrating the Wilcoxon rank sum test to investigate differences in median longevity based on gender. To test the null hypothesis

$$H_0: M_M = M_F$$

against the alternative

$$H_1: M_M \neq M_F$$

we note that the test statistic W is 32,498.5 with a p value of .3873. Using a level of significance α of .05, since .3873 > .05 we cannot reject the null hypothesis. There is no evidence that the median length of employment at Kalosha Industries is different between the full-time male workers and female workers.

When the assumption of normality appears to be violated, some researchers would prefer to use the classical parametric t or t' tests after employing a normalizing *data transformation* (see Reference 11) instead of the distribution-free Wilcoxon rank sum test. Figure 13.15 presents STATISTIX output depicting the box-and-whisker plots and the normal probability plots for the two samples following a *natural logarithmic transformation* on the longevity data.[3]

If we carefully observe the sets of plots in Figure 13.15 we would agree that the logarithmic transformation appears to normalize the two samples of data by removing the substantial amount of original skewness. Considering the large sample sizes, it is now reasonable to conclude that there is no longer any serious departure from the normality assumption and hence we may proceed with a classical, parametric test.

Figure 13.16 provides MINITAB output displaying the pooled-variance t test and the separate-variance t' test for the data transformed by natural logarithms.

Applying the F test for the equality of variance assumption to the data in panel A, we observe that

$$F = \frac{S_M^2}{S_F^2} = \frac{(0.9250)^2}{(0.8172)^2} = 1.28$$

Choosing a level of significance, α, of .05, since $F = 1.28 < F_{U(232,166)} = 1.43$ we cannot reject the null hypothesis. The logarithmic transformation has normalized the data and, by reducing the skewness, has stabilized the variability between the two groups.

Under the assumption that the variances of the transformed data in the two populations are similar, the pooled-variance t test to compare the means of the two groups may be employed. Using a .05 level of significance, the test statistic $t = +0.86$ falls between ± 1.96, the lower- and upper-tailed critical values of the t distribution with 398 degrees of freedom, and the null hypothesis of no difference in the (transformed) means cannot be rejected. The p value is .39. Again, there is no evidence of a difference in the means (of the transformed data) between the two groups.

13.7.3 Summary

Regardless of which test procedure is chosen, we can inform Bud Conley that there is no evidence of a difference in the average length of employment between full-time male and female workers at Kalosha Industries. On the other hand, by examining the different p values we can see how important it is to explore the data through a thorough descriptive analysis in order to evaluate the assumptions of the test procedures we are thinking of employing. Many researchers would have chosen the t' test initially. Some would have selected the Wilcoxon rank sum test. Others would have preferred to transform the data following a descriptive analysis and then would have selected the t test. Although the conclusions stated were the same in all the situations, it is somewhat unsettling to note the lack of stability in the t and t' tests even though the sample sizes here (233 and 167) are not small. Observe from Figures 13.13 and 13.16 how varied the respective p values were for tests performed on the original data as opposed to those performed on the transformed data. On the other hand, the Wilcoxon rank sum test, which merely converts measurements to ranks, is totally unaffected by these data transformations.

Survey/Database Project for Section 13.7

Data file EMPSAT.DAT

The following problems refer to the sample data obtained from the questionnaire of Figure 2.6 on pages 28–29 and presented in Table 2.3 on pages 33–40. They should be solved with the aid of an available computer package.

Suppose you are hired as a research assistant to Bud Conley, the Vice-President of Human Resources at Kalosha Industries. He has given you a list of questions (see Problems 13.48 to 13.60) that he needs answered prior to his meeting with a representative from the B&L Corporation, the employee-benefits consulting firm that he has retained. A confirmatory statistical analysis based on the answers to these questions pertaining to the numerical variables in the Employment Satisfaction Survey will provide him with a better understanding of the composition of the Kalosha Industries full-time workforce and assist him in his deliberations with the B&L Corporation toward the development of an employee-benefits package.

From the responses to the questions dealing with numerical variables in the Employee Satisfaction Survey (see pages 33–40), in Problems 13.48 through 13.60 which follow,

(a) Construct the stem-and-leaf display for each of the two samples.
(b) For each of the two samples, obtain
    (1) the mean (5) the range
    (2) the median (6) the interquartile range
    (3) the midrange (7) the standard deviation
    (4) the midhinge (8) the coefficient of variation
(c) List the five-number summary for each of the two samples.
(d) Form the box-and-whisker plot for each of the two samples.
(e) Based on a descriptive analysis of the findings in (a) through (d) with respect to the assumptions of the various hypothesis-testing procedures, select an appropriate procedure and carry out the testing of hypotheses at the $\alpha = .05$ level of significance.
(f) ACTION ▸ Write a memo to Bud Conley discussing your test selection and your findings.

13.48 Is there evidence of a gender difference in the average number of hours typically worked per week by all full-time employees of Kalosha Industries (see questions 1 and 5)?

13.49 Is there evidence of a gender difference with respect to the average age of full-time workers at Kalosha Industries (see questions 3 and 5)?

13.50 Is there evidence of a gender difference with respect to average personal income of full-time workers at Kalosha Industries (see questions 7 and 5)?

13.51 Is there evidence of a difference with respect to the average age of full-time workers at Kalosha Industries based on whether or not they are members of a labor union (see questions 3 and 14)?

13.52 Is there evidence of a difference with respect to average personal income of full-time workers at Kalosha Industries based on whether or not they are members of a labor union (see questions 7 and 14)?

13.53 Is there evidence of a difference in average longevity (that is, length of employment in years) of full-time employees at Kalosha Industries based on whether or not they are members of a labor union (see questions 16 and 14)?

13.54 Is there evidence of a difference in the average number of hours typically worked per week by all full-time employees of Kalosha Industries based on an individual's participation in budgetary decisions (see questions 1 and 22)?

13.55 Is there evidence that the average personal incomes of full-time employees at Kalosha Industries are *higher* if they participate in budgetary decisions than if they don't participate in budgetary decisions (see questions 7 and 22)?

13.56 Is there evidence that the average longevity (that is, length of employment in years) is *greater* for full-time employees at Kalosha Industries who participate in budgetary decisions than for those who don't participate in budgetary decisions (see questions 16 and 22)?

13.57 Is there evidence of a difference in the average number of hours typically worked per week by all full-time employees of Kalosha Industries (question 1) based on whether they are very satisfied (question 9, code 1) or not very satisfied (question 9, codes 2 through 4) with their jobs?

13.58 Is there evidence of a difference with respect to the average age of full-time workers at Kalosha Industries (question 3) based on whether they are very satisfied (question 9, code 1) or not very satisfied (question 9, codes 2 through 4) with their jobs?

13.59 Is there evidence of a difference with respect to average personal income of full-time workers at Kalosha Industries (question 7) based on whether they are very satisfied (question 9, code 1) or not very satisfied (question 9, codes 2 through 4) with their jobs?

13.60 Is there evidence that the average personal income (question 7) is *greater* for full-time employees at Kalosha Industries who feel that formal on-the-job training is important to their work (question 28, code 1) than for those who don't feel that way (question 28, codes 2 through 4)?

13.8 Choosing the Appropriate Test Procedure When Comparing Two Related Samples

In the previous sections of this chapter, we have examined several hypothesis-testing procedures that enable us to make comparisons and examine differences between two independent populations based on samples containing numerical data. In particular, in Sections 13.3 through 13.5 we focused on testing for the difference between the means or medians of two *independent populations*. In the following two sections we shall develop procedures for analyzing the difference between the means or medians of two groups when the sample data are obtained from populations that are **related**—that is, the results of the first group are not independent of the second group. This "dependency" characteristic of the two groups occurs either because the items or individuals are **paired** or **matched** according to some characteristic or because **repeated measurements** are obtained from the same set of items or individuals. In either case, the variable of interest becomes the *difference* between the values of the observations rather than the observations themselves.

In business research it is frequently of interest to examine differences between two *related* groups. For example, in test marketing a product under two different advertising conditions, a sample of test markets can be *matched* (that is, *paired*) on the basis of the test-market population size and/or other socioeconomic and demographic variables. Moreover, when performing a taste-testing experiment, each subject in the sample could be used as his or her own control so that *repeated measurements* on the same individual are obtained.

The first approach to the related-samples problem involves matching of items or individuals according to some characteristic of interest. For example, if the production manager involved in the cereal box packaging process (discussed in Chapters 9 through 11) wanted to study the effect of two different filling machines, one old and one new, on the amount of cereal that is spilled (and thereby wasted), a control for differences in the various types of cereals (which may themselves have different spillage patterns) should be established. In this situation two boxes of each cereal type that is packaged can be tested, with one box assigned to the new machine and the other to the old machine.

The second approach to the related-samples problem involves taking repeated measurements on the same items or individuals. Under the theory that the same items or individuals will behave alike if treated alike, the objective of the analysis is to show that any differences between two measurements of the same items or individuals are due to different treatment conditions. For example, suppose that a software applications manufacturer is developing a new finance package that is intended to be used in education and in business. Since computer processing time is costly, the manufacturer wants the new package to have the same features and capabilities as the current market leader while providing results faster than the current leading package. As a test of the worthiness of the new software package, an experiment is designed where particular financial applications projects must be keyed in for use by the new software package as well as by the current market leader. By using a particular set of financial applications projects on both packages we are in effect using each project as its own control. Therefore, we may simply evaluate differences in the times required to achieve the desired results by comparing the mean (or median) differences in the two time readings rather than comparing the difference in the mean (or median) completion times from two independent samples of financial applications projects, one of which must be keyed in to the new software package while the other is keyed in on the current market leading package. This latter approach of comparing two independent samples was taken in our discussions in Sections 13.3 through 13.6. Here, however, we should note that obtaining the two time readings (one for the new software package and one for the current market leader) for each financial applications project serves to reduce the variability in the time readings compared with what would occur if two independent sets of financial applications projects were used. It also enables us to focus on the differences between the two time readings for each financial applications project in order to measure the effectiveness of the new software package.

Regardless of whether matched (paired) samples or repeated measurements are utilized, the objective is to study the difference between two measurements by reducing the effect of the variability due to the items or individuals themselves. In the two sections that follow, two widely used procedures will be developed—the *t test for the mean difference in related samples* and the *Wilcoxon signed-ranks test for the median difference in related samples*. As discussed in Section 13.2, several criteria can be used for selection of a particular statistical procedure. Part of a good data analysis is to understand the assumptions underlying each of the hypothesis-testing techniques and to select the one most appropriate for a given set of conditions. Other criteria for test selection involve the simplicity of the procedure, the generalizability of the conclusions to be drawn, the accessibility of tables of critical values for the test statistic, the availability of computer software packages that contain the test procedure, and the statistical power of the procedure.

13.9 *t* Test for the Mean Difference

13.9.1 Introduction and Background

In order to determine whether any difference exists between two related groups, the differences in the individual values in each group must be obtained as shown in Table 13.6. To read this table, let $X_{11}, X_{12}, \ldots, X_{1n}$ represent the n observations from a sample. Now let $X_{21}, X_{22}, \ldots, X_{2n}$ represent either the corresponding n matched observations from a second sample or the corresponding n repeated measurements from the initial sample. Furthermore, let $D_1, D_2, \ldots, D_n$ represent the corresponding set of *n difference scores* such that $D_1 = X_{11} - X_{21}$, $D_2 = X_{12} - X_{22}$, ... , and $D_n = X_{1n} - X_{2n}$.

From the central limit theorem, the average difference $\bar{D}$ follows a normal distribution when the population standard deviation of the difference σ_D is known and the sample size is large enough. The **Z-test** statistic is

$$Z = \frac{\bar{D} - \mu_D}{\frac{\sigma_D}{\sqrt{n}}} \quad (13.10)$$

where

$$'' = \frac{\sum_{i=1}^{n} D_i}{n}$$

μ_D = hypothesized mean difference

σ_D = population standard deviation of the difference scores

n = sample size

Table 13.6 Determining the difference between two related groups.

	Group		
Observation	1	2	Difference
1	X_{11}	X_{21}	$D_1 = X_{11} - X_{21}$
2	X_{12}	X_{22}	$D_2 = X_{12} - X_{22}$
.	.	.	.
.	.	.	.
.	.	.	.
i	X_{1i}	X_{2i}	$D_i = X_{1i} - X_{2i}$
.	.	.	.
.	.	.	.
.	.	.	.
n	X_{1n}	X_{2n}	$D_n = X_{1n} - X_{2n}$

13.9.2 Developing the *t* Test for the Mean Difference

However, as mentioned previously, in most cases we do not know the actual standard deviation in a population. The only information usually obtainable is the summary statistics such as the sample mean and sample standard deviation. If the assumptions are made that the sample of difference scores is randomly and independently drawn from a population that is normally distributed, a *t* test can be used to determine whether there is a significant population mean difference. Thus, analogous to the (one-sample) *t* test developed in Section 12.3 [see Equation (12.1)], the *t*-test statistic developed here will follow the *t* distribution with $n - 1$ degrees of freedom. Although the population is assumed to be normally distributed, in practice it has been found that as long as the sample size is not very small and the population is not very skewed, the *t* distribution gives a good approximation to the sampling distribution of the average difference $\bar{D}$. Thus, to test the null hypothesis of no difference in the means of two related populations (i.e., the population mean difference μ_D is 0)

$$H_0: \ \mu_D = 0 \text{ (where } \mu_D = \mu_1 - \mu_2)$$

against the alternative that the means are not the same (i.e., the population mean difference μ_D is not 0)

$$H_1: \ \mu_D \neq 0$$

the following ***t*-test** statistic can be computed

$$t = \frac{\bar{D} - \mu_D}{\frac{S_D}{\sqrt{n}}} \quad (13.11)$$

where

$$\bar{D} = \frac{\sum_{i=1}^{n} D_i}{n}$$

$$S_D = \sqrt{\frac{\sum_{i=1}^{n} D_i^2 - n\bar{D}^2}{n-1}}$$

and

$\sum_{i=1}^{n} D_i^2$ = summation of the squares of each difference score

$n\bar{D}^2$ = sample size times the square of the sample mean difference

and, for a given level of significance, α, we may reject the null hypothesis if the computed *t*-test statistic exceeds the upper-tailed critical value t_{n-1} from the *t* distribution or if the computed test statistic falls below the lower-tailed critical value $-t_{n-1}$ from the *t* distribution. That is, the decision rule is

Reject H_0 if $t > t_{n-1}$

or if $t < -t_{n-1}$;

otherwise do not reject H_0.

However, the test to be performed can be either two-tailed or one-tailed, depending on whether we are testing if the two population means are merely *different* (i.e., the population mean difference $\mu_D \neq 0$) or if one mean is *greater than* the other mean (i.e., the population mean difference $\mu_D > 0$). The three panels of Figure 13.17 indicate the null and alternative hypotheses and rejection regions for the possible two-tailed and one-tailed tests. If, as shown in panel A, the test of hypothesis is two-tailed, the rejection region is split into the lower and upper tails of the *t* distribution. However, if the test is one-tailed, the rejection region is either in the lower tail (panel B of Figure 13.17) or in the upper tail (panel C of Figure 13.17) of the *t* distribution, depending on the direction of the alternative hypothesis.

13.9.3 Application Involving Pairing or Matching

To apply the test for the difference between the means of two related groups, let us refer back to the first example mentioned in Section 13.8. The production manager wanted to determine whether there was evidence that spillage was lower when packages are filled on a new machine than on an old machine. In order to

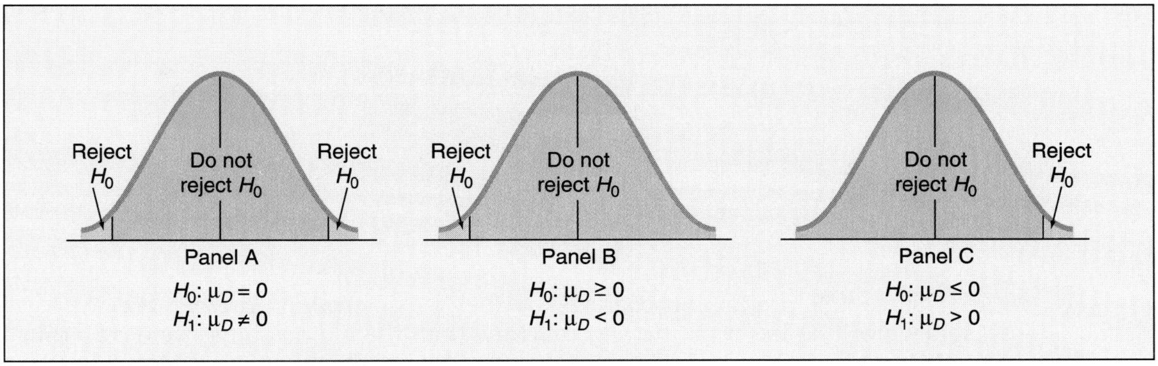

Figure 13.17
Testing for the difference between the means in related samples: Panel A, two-tailed test; Panel B, one-tailed test; Panel C, one-tailed test.

reduce the influence of the variability in type of cereal, a pair of boxes for each of ten different cereal types was randomly selected. One box of each type was filled on the new machine and the other box was filled on the old machine. The assignment of a member of each pair of boxes to a machine (new or old) was done randomly. The results are shown in Table 13.7 below.

Table 13.7 Amount of cereal spilled (in grams) for a random sample of ten types of cereal filled on two machines.

	Machine Type		Difference D_i
Cereal Type	New	Old	$(X_{1i}-X_{2i})$
1	12.73	13.89	−1.16
2	9.75	10.32	−0.57
3	13.78	17.01	−3.23
4	8.37	10.43	−2.06
5	11.71	11.39	+0.32
6	15.47	17.99	−2.52
7	14.56	16.02	−1.46
8	11.74	11.90	−0.16
9	9.76	13.11	−3.35
10	12.47	13.88	−1.41

For these data,

$$\sum_{i=1}^{n} D_i = -15.60, \quad \sum_{i=1}^{n} D_i^2 = 38.1676, \quad n = 10$$

Thus

$$\bar{D} = \frac{\sum_{i=1}^{n} D_i}{n} = \frac{-15.60}{10} = -1.56$$

and

$$S_D^2 = \frac{\sum_{i=1}^{n} D_i^2 - n\bar{D}^2}{n-1} = \frac{38.1676 - 10(-1.56)^2}{9} = 1.537$$

so that

$$S_D = 1.24$$

Since the production manager wishes to determine whether the average spillage will be less with the new machine than the old machine, we have a one-tailed test in which the null and alternative hypotheses can be stated as follows:

$$H_0\colon\ \mu_D \geq 0 \text{ or } \mu_{\text{new}} \geq \mu_{\text{old}}$$

$$H_1\colon\ \mu_D < 0 \text{ or } \mu_{\text{new}} < \mu_{\text{old}}$$

Since samples of ten cereal types have been taken, if a level of significance of .01 is selected, the decision rule is:

Reject H_0 if $t < t_9 = -2.8214$;

otherwise do not reject H_0.

The regions of rejection and nonrejection are illustrated in Figure 13.18.

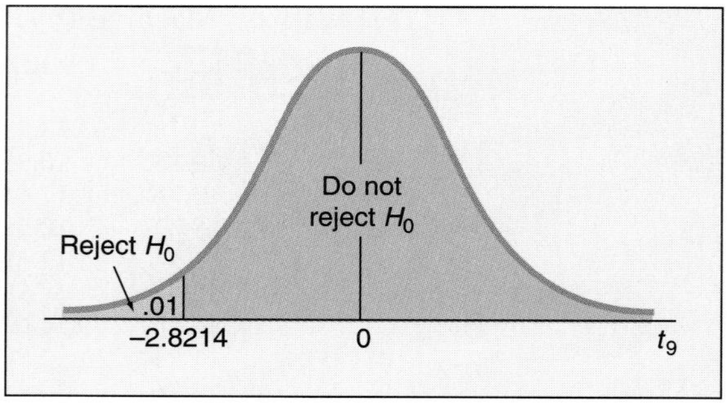

Figure 13.18
One-tailed test for the paired difference at the .01 level of significance with 9 degrees of freedom.

From Equation (13.11) we have

$$t = \frac{\bar{D} - \mu_D}{\frac{S_D}{\sqrt{n}}}$$

so that

$$t = \frac{-1.56 - 0}{\frac{1.24}{\sqrt{10}}} = -3.978$$

Since $t = -3.978 < t_9 = -2.8214$, we reject H_0.

Using the p-value approach, the probability of obtaining a t statistic below -3.978 with 9 degrees of freedom is less than .005. Since this is less than .01, the chosen level of significance α, the null hypothesis is rejected. We would conclude that there is evidence that the average amount of cereal spillage is lower for the new machine than the old machine.

Problems for Section 13.9

13.61 The manager of a nationally known real estate agency has just completed a training session on appraisals for two newly hired agents. To evaluate the effectiveness of his training, the manager wishes to determine whether there is any difference in the appraised values placed on houses by these two different individuals. A sample of 12 houses is selected by the manager and each agent is assigned the task of placing an appraised value (in thousands of dollars) on the 12 houses.

The results are summarized below.

Data file LIRE.DAT

House	Agent 1	Agent 2
1	181.0	182.0
2	179.9	180.0
3	163.0	161.5
4	218.0	215.0
5	213.0	216.5
6	175.0	175.0
7	217.9	219.5
8	151.0	150.0
9	164.9	165.5
10	192.5	195.0
11	225.0	222.7
12	177.5	178.0

(a) At the .05 level of significance, is there evidence of a difference in the average appraised values given by the two agents?
(b) What assumption is necessary to perform this test?
(c) Find the lower and upper limits of the p value in (a) and interpret its meaning.

13.62 Suppose that a shoe company wanted to test material for the soles of shoes. For each pair of shoes the new material was placed on one shoe and the old material was placed on the other shoe. After a given period of time a random sample of ten pairs of shoes was selected and the wear was measured on a ten-point scale (higher is better) with the following results

	Pair Number									
Material	I	II	III	IV	V	VI	VII	VIII	IX	X
New	2	4	5	7	7	5	9	8	8	7
Old	4	5	3	8	9	4	7	8	5	6
Differences	−2	−1	+2	−1	−2	+1	+2	0	+3	+1

Data file SHOESOLE.DAT

(a) At the .05 level of significance, is there evidence of a difference in the average wear for the new material and the old material?

(b) Find the lower and upper limits of the p value in (a) and interpret its meaning.

Data file
GASMILE.DAT

● 13.63 A group of engineering students decide to see whether cars that supposedly do not need high-octane gasoline get more miles per gallon using regular or high-octane gas. They test several cars (under similar road surface, weather, and other driving conditions) using both types of gas in each car at different times. The mileage for each gas type for each car is

	Car									
Gas Type	#1	#2	#3	#4	#5	#6	#7	#8	#9	#10
Regular	15	23	21	35	42	28	19	32	31	24
High-octane	18	21	25	34	47	30	19	27	34	20

(a) Is there any evidence of a difference in the average gasoline mileage between regular and high-octane gas? (Use $\alpha = .05$.)
(b) Find the lower and upper limits of the p value in (a) and interpret its meaning.

13.64 In order to measure the effect of a storewide sales campaign on nonsale items, the research director of a national supermarket chain took a random sample of 13 pairs of stores that were matched according to average weekly sales volume. One store of each pair (the experimental group) was exposed to the sales campaign, and the other member of the pair (the control group) was not. The following data indicate the results over a weekly period:

Data file
SALESCMP.DAT

	Sales ($000) of Nonsale Items	
Store	With Sales Campaign	Without Sales Campaign
1	67.2	65.3
2	59.4	54.7
3	80.1	81.3
4	47.6	39.8
5	97.8	92.5
6	38.4	37.9
7	57.3	52.4
8	75.2	69.9
9	94.7	89.0
10	64.3	58.4
11	31.7	33.0
12	49.3	41.7
13	54.0	53.6

(a) At the .05 level of significance, can the research director conclude that there is evidence that the sales campaign has increased the average sales of nonsale items?
(b) What assumption is necessary to perform this test?
(c) Find the lower and upper limits of the p value in (a) and interpret its meaning.

13.65 A professor in the School of Business wants to investigate the prices of new textbooks in the campus bookstore and the competing off-campus store, which is a branch of a national chain. The professor randomly chooses the required texts for 12 business school courses and compares the prices in the two stores. The results are as follows:

Book	Campus Store	Off-Campus Store
#1	$55.00	$50.95
#2	47.50	45.75
#3	50.50	50.95
#4	38.95	38.50
#5	58.70	56.25
#6	49.90	45.95
#7	39.95	40.25
#8	41.50	39.95
#9	42.25	43.00
#10	44.95	42.25
#11	45.95	44.00
#12	56.95	55.60

Data file
BKPRICE.DAT

(a) At the .01 level of significance, is there any evidence of a difference in the average price of business textbooks between the two stores?
(b) Find the lower and upper limits of the p value in (a) and interpret its meaning.

13.10 Wilcoxon Signed-Ranks Test for the Median Difference

13.10.1 Introduction

For situations involving either matched items or repeated measurements of the same item, the distribution-free **Wilcoxon signed-ranks test for the median difference** may be used when its respective parametric counterpart, the t test for the mean difference that was described in the previous section, is not appropriate. That is, the Wilcoxon signed-ranks test may be chosen over the t test when we are able to obtain data measured at a higher level than an ordinal scale but do not believe that the assumptions of the parametric procedure are sufficiently met. When the assumptions of the t test are violated, the Wilcoxon procedure (which makes fewer and less stringent assumptions than does the t test) is likely to be more powerful in detecting the existence of significant differences than its parametric counterpart. Moreover, even under conditions appropriate to the parametric t test, the Wilcoxon signed-ranks test has proven to be almost as powerful.

13.10.2 Development

The test of the null hypothesis that the population median difference M_D is zero may be two-tailed or one-tailed:

Two-Tailed Test	One-Tailed Test	One-Tailed Test
H_0: $M_D = 0$	H_0: $M_D \geq 0$	H_0: $M_D \leq 0$
H_1: $M_D \neq 0$	H_1: $M_D < 0$	H_1: $M_D > 0$

The assumptions necessary for performing the test are that

1. The observed data constitute a random sample of n independent items or individuals, each with two measurements (X_{11}, X_{21}), (X_{12}, X_{22}), . . ., (X_{1n}, X_{2n}), or the observed data constitute a random sample of n independent pairs of items or individuals so that (X_{1i}, X_{2i}) represents the observed values for each member of the matched pair ($i = 1, 2, \ldots, n$).
2. The underlying variable of interest is continuous.
3. The observed data are measured at a higher level than the ordinal scale—i.e., at the interval or ratio level.
4. The distribution of the population of difference scores between repeated measurements or between matched items or individuals is approximately symmetric.

To perform the Wilcoxon signed-ranks test for the median difference the following six-step procedure may be used:

1. For each item in a sample of n items we obtain a *difference score* D_i (to be described in Section 13.10.3).
2. We then neglect the "+" and "−" signs and obtain a set of n absolute differences $|D_i|$.
3. We omit from further analysis any absolute difference score of zero, thereby yielding a set of n' nonzero absolute difference scores, where $n' \le n$.
4. We then assign ranks R_i from 1 to n' to each of the $|D_i|$ such that the smallest absolute difference score gets rank 1 and the largest gets rank n'. Owing to a lack of precision in the measuring process, if two or more $|D_i|$ are equal, they are each assigned the average rank of the ranks they would have been assigned individually had ties in the data not occurred.
5. We now reassign the symbol "+" or "−" to each of the n' ranks R_i, depending on whether D_i was originally positive or negative.
6. The Wilcoxon test statistic W is obtained as the sum of the positive ranks:

$$W = \sum_{i=1}^{n'} R_i^{(+)} \tag{13.12}$$

Since the sum of the first n' integers $(1, 2, \ldots, n')$ is given by $n'(n' + 1)/2$, the Wilcoxon test statistic W may range from a minimum of 0 (where all the observed difference scores are negative) to a maximum of $n'(n' + 1)/2$ (where all the observed difference scores are positive). If the null hypothesis were true, we would expect the test statistic W to take on a value close to its mean, $\mu_w = n'(n' + 1)/4$. If the null hypothesis were false, we would expect the observed value of the test statistic to be close to one of the extremes.

As with the (one-sample) Wilcoxon signed-ranks test discussed in Section 12.4, Table E.10 may be used for obtaining the critical values of the test statistic W for both one- and two-tailed tests at various levels of significance for samples of $n' \le 20$. For a two-tailed test and for a particular level of significance, if the observed value of W equals or exceeds the upper critical value or is equal to or less than the lower critical value, the null hypothesis may be rejected. For a one-tailed test in the negative direction, the decision rule is to reject the null hypothesis if the observed value of W is less than or equal to the lower critical value. For a one-tailed test in the positive direction, the decision rule is to reject the null hypothesis if the observed value of W equals or exceeds the upper critical value.

For samples of $n' > 20$, the test statistic W is approximately normally distributed, and the following large-sample approximation formula may be used for testing the null hypothesis:

$$Z = \frac{W - \mu_w}{\sigma_w} \quad \textbf{(13.13)}$$

where

W is the sum of positive ranks; $W = \sum_{i=1}^{n'} R_i^{(+)}$

μ_w is the mean value of W; $\mu_w = \frac{n'(n'+1)}{4}$

σ_w is the standard deviation of W; $\sigma_w = \sqrt{\frac{n'(n'+1)(2n'+1)}{24}}$

n' is the actual sample size after removing observations with absolute difference scores of zero

That is,

$$Z = \frac{W - \left(\frac{n'(n'+1)}{4}\right)}{\sqrt{\frac{n'(n'+1)(2n'+1)}{24}}} \quad \textbf{(13.14)}$$

and, based on the level of significance selected, the null hypothesis may be rejected if the computed Z value falls in the appropriate region of rejection, depending on whether a two-tailed test or a one-tailed test is used (see Figure 13.19 on page 514).

13.10.3 Application with Repeated Measurements

Development testing is an important phase for bringing a new product to market. A manufacturer must know what its product's strengths and limitations are so that appropriate promotional strategies can be planned. Therefore, to demonstrate the use of the Wilcoxon signed-ranks test for the median difference, let us refer back to the second example mentioned in Section 13.8.

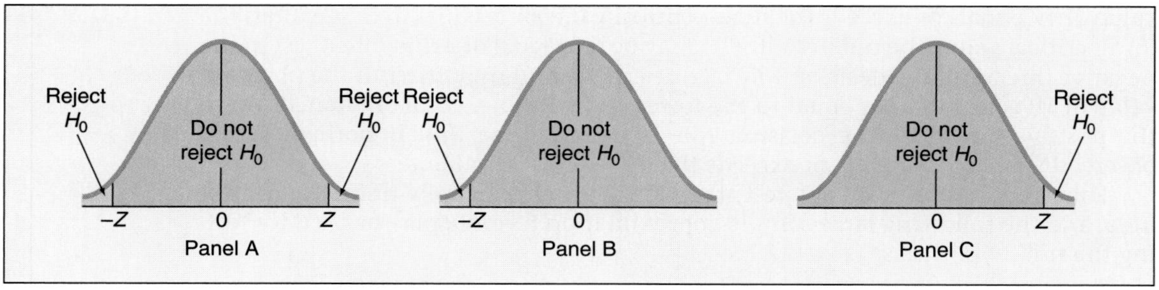

	Panel A	Panel B	Panel C
Paired-difference Procedure:	(two-tailed test) $M_D \neq 0$	(one-tailed test) $M_D < 0$	(one-tailed test) $M_D > 0$

Figure 13.19
Determining the rejection region using the Wilcoxon signed-ranks test.

A software applications manufacturer developing a new finance package intended for educational and business users wants to test the worthiness of the new software package by comparing the differences in computer processing times for particular financial applications projects keyed in for use by the new software package as well as by the current market leader. If the new financial package is effective, it will provide the same results as does the current market leader but it will be quicker. That is, the new software package will require, on average, less computer processing time.

Therefore, we may simply evaluate differences in the times required to achieve the desired results by comparing the mean (or median) differences in the two time readings rather than comparing the difference in the mean (or median) completion times from two independent samples of financial applications projects, one of which must be keyed in to the new software package while the other is keyed in on the current market leading package. This latter approach of comparing two independent samples was taken in our discussions in Sections 13.3 through 13.6. Here, however, we should note that obtaining the two time readings (one for the new software package and one for the current market leader) for each financial applications project serves to reduce the variability in the time readings compared with what would occur if two independent sets of financial applications projects were used. It also enables us to focus on the differences between the two time readings for each financial applications project in order to measure the effectiveness of the new software package.

The results displayed in Table 13.8 are for a sample of $n = 10$ financial applications projects used in the experiment.

The question that must be answered is whether or not this new software package is faster. That is, is there evidence that the average processing time is significantly greater when financial applications projects employ the current market leader rather than the new software package? The following null and alternative hypotheses are established:

$$H_0: \; M_D \leq 0$$

$$H_1: \; M_D > 0$$

and the test is one-tailed.

Table 13.8 Repeated measurements of time in seconds to complete financial applications projects on two competing software packages

Applications Project User	Completion Times (in Seconds) by Current Market Leader	Completion Times (in Seconds) by New Software Package
C.B.	9.98	9.88
T.F.	9.88	9.86
M.H.	9.84	9.75
R.K.	9.99	9.80
M.O.	9.94	9.87
D.S.	9.84	9.84
S.S.	9.86	9.87
C.T.	10.12	9.86
K.T.	9.90	9.83
S.Z.	9.91	9.86

To perform the paired-sample test, the first step of the six-step procedure is to obtain a set of difference scores D_i between each of the n paired observations:

$$D_i = X_{1i} - X_{2i}$$
$$\text{where } i = 1, 2, \ldots, n$$

In our example, we obtain a set of n difference scores from $D_i = X_{current_i} - X_{new_i}$.

If the new software package is effective, the computer processing time is expected to drop, so that the difference scores will tend to be *positive* values (and H_0 will be rejected). On the other hand, if the new software package is not effective, we can expect some D_i values to be positive, others to be negative, and some to show no change (that is, $D_i = 0$). If this is the case, the difference scores will average near zero (that is, $\bar{D} \cong 0$) and H_0 will not be rejected.

The remaining steps of the six-step procedure are developed in Table 13.9 on page 516. Notice that these are exactly the same steps as for the (one-sample) Wilcoxon signed-ranks test described in Section 12.4. From this table we note that project applications user D.S. is discarded from the study (because his difference score is zero) and that eight of the remaining $n' = 9$ difference scores have a positive sign. The test statistic W is obtained as the sum of the positive ranks:

$$W = \sum_{i=1}^{n'} R_i^{(+)} = 7 + 2 + 6 + 8 + 4.5 + 9 + 4.5 + 3 = 44$$

Since $n' = 9$, we use Table E.10 to determine the upper-tail critical value for this one-tailed test with a level of significance, α, selected at .05. The upper-tail critical value is 37. Since $W = 44 > W_U = 37$, the null hypothesis may be rejected. There is evidence to support the contention that the average processing time using the new finance software package is significantly faster than using the current market leader.[4]

Table 13.9 Setting up the Wilcoxon signed-ranks test for the median difference.

Project Applications User	Processing Time (in Seconds) Current Leader X_{1i}	New Package X_{2i}	$D_i = X_{1i} - X_{2i}$	$\lvert D_i \rvert$	R_i	Sign of D_i
C.B.	9.98	9.88	+0.10	0.10	7.0	+
T.F.	9.88	9.86	+0.02	0.02	2.0	+
M.H.	9.84	9.75	+0.09	0.09	6.0	+
R.K.	9.99	9.80	+0.19	0.19	8.0	+
M.O.	9.94	9.87	+0.07	0.07	4.5	+
D.S.	9.84	9.84	0.00	0.00	. . .	Discard
S.S.	9.86	9.87	−0.01	0.01	1.0	−
C.T.	10.12	9.86	+0.26	0.26	9.0	+
K.T.	9.90	9.83	+0.07	0.07	4.5	+
S.Z.	9.91	9.86	+0.05	0.05	3.0	+

Problems for Section 13.10

13.66 A tax preparation company claimed that taxpayers would save money by having its firm prepare their individual tax returns. To evaluate this claim, a consumer protection agency had people who had already prepared their tax forms go to this company's office to get their taxes done by the firm's preparers. The taxes each person would pay if they paid what they calculated and if they paid what the company calculated are presented below:

Data file TAXRET.DAT

Taxpayer	Tax-Return Preparer Firm	Self
Jose	1,459	1,910
Marcia	3,250	2,900
Alexis	1,190	1,200
Harry	8,100	7,650
Jean	13,200	15,390
Marc	9,120	9,100
JR	255,970	33,120
Billy	210	140
Richard	1,290	1,320
Ted	130	0
Bruce	5,190	6,123

(a) Is there evidence that the firm's claim is valid? (Use $\alpha = .05$.)

(b) Discuss the implications of your results.

13.67 The weatherman for a local TV channel reported Wednesday morning, June 1, 1994, that it was expected to be warmer in the continental United States that day than it had been on Tuesday, May 31. To test this claim, a random sample of 22 cities representing a cross section of the continental United States was taken and the following data were recorded:

	High Temperature (°F)			High Temperature (°F)	
City	May 31 (Actual)	June 1 (Predicted)	City	May 31 (Actual)	June 1 (Predicted)
Albany	84	86	Little Rock	84	87
Albuquerque	93	89	Louisville	82	83
Austin	93	95	Miami	85	89
Birmingham	83	81	Nashville	82	83
Boise	79	83	Norfolk	79	79
Boston	84	84	Omaha	93	80
Cleveland	84	85	St. Louis	82	86
Dallas-Ft. Worth	88	92	San Diego	65	70
Denver	91	78	San Jose	80	82
Indianapolis	85	82	Seattle	69	73
Jacksonville	82	85	Tulsa	88	91

Source: New York Times, May 31, 1994, p. D8.

Data file WEATHER.DAT

(a) Is there evidence to support the weatherman's claim that Wednesday is expected to be warmer than Tuesday? (Use $\alpha = .05$.)

(b) Discuss the implications of your results.

13.68 The following data represent the midterm and final examination scores from a sample of 11 students taking Introductory Economics. Both exams were two hours in length and the final covered material learned since the midterm.

Data file ECOTESTS.DAT

	Student										
	N.A.	A.B.	L.B.	M.B.	W.B.	S.D.	T.J.	L.K.	J.M.	H.R.	D.R.
Midterm	80	82	47	75	80	69	83	73	55	70	81
Final Exam	81	85	40	75	83	79	91	72	66	76	79

(a) Is there evidence of an increase in student performance in the second half of the semester? (Use $\alpha = .05$.)

(b) Discuss the implications of your results.

13.69 Refer to Problem 13.63 on page 510. Data file GASMILE.DAT

(a) At the .05 level of significance, is there evidence of a difference in the median gasoline mileage?

(b) Are there any differences in your present results from those obtained using the *t* test? Discuss.

13.70 Refer to Problem 13.64 on page 510. Data file SALESCMP.DAT

(a) At the .05 level of significance, can the research director conclude that there is evidence that the sales campaign has increased the median sales of nonsale items?

(b) Are there any differences in your present results from those obtained using the *t* test? Discuss.

13.11 Potential Hypothesis-Testing Pitfalls and Ethical Issues

13.11.1 Potential Pitfalls

In this chapter we introduced four statistical test procedures that may be employed when analyzing possible differences between the parameters of two independent populations based on samples containing numerical data. In addition, we developed two test procedures that may be used when analyzing possible differences between the parameters of two related populations based on samples containing numerical data. Again, part of a good data analysis is to understand the assumptions underlying each of the hypothesis test procedures and, using this as well as other criteria, select the one most appropriate for a given set of conditions. As observed in the summary chart for this chapter, the major distinction for comparing two groups containing numerical data is based on whether the populations from which the samples were drawn are independent or related. We should not use test procedures designed for independent populations when dealing with paired data, and we should not use test procedures designed for related populations when dealing with two independent samples. After focusing on an appropriate grouping of similar test procedures, we need to look carefully at the assumptions and other criteria prior to selecting a particular procedure.

Appropriate test and/or model selection is of utmost importance to good research and is a very serious matter. In a recent year (see Reference 1), two professors from different universities filed charges of scientific misconduct with the National Institutes of Health against a professor from a third university for his choice of statistical model in a paper he had written on lead poisoning.

13.11.2 Ethical Issues

Ethical considerations arise when a researcher is manipulative of the hypothesis-testing process in ways that enable personal gain. Interestingly, the researcher accused of scientific misconduct was not charged with fraud, plagiarism, faking data, or falsifying results. The charges stemmed from his selection of the statistical modeling and testing procedures as well as the resulting data analysis that presented a divergent viewpoint. For ethics to be an issue, consideration must be given to whether or not this aspect of the research endeavor was willful. For further discussion on ethical issues and hypothesis testing, refer to Section 11.11.2 (pages 412 to 415) and Section 12.8.2 (pages 455 to 456).

13.12 Hypothesis Testing Based on Two Samples of Numerical Data: A Review

This chapter presented several widely used hypothesis-testing procedures that enable us to compare statistics computed from two samples of numerical data in order to make inferences about possible differences in the parameters of the two

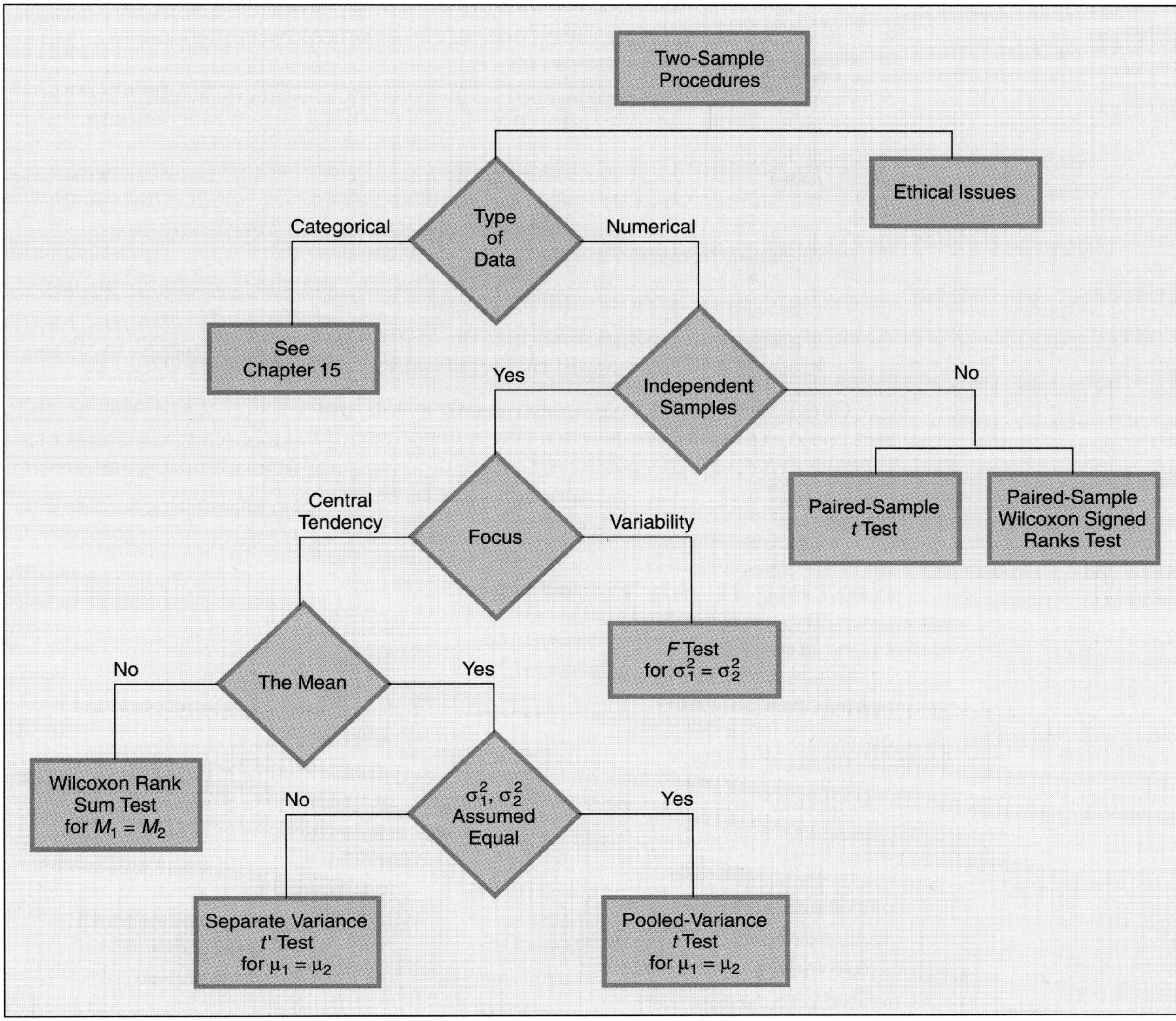

Chapter 13 summary chart.

respective populations. Both independent and related sample procedures were considered and emphasis was given to the assumptions behind the use of the various tests. On page 462 of Section 13.1 you were given a list emphasizing the important points to be discussed in the chapter. Check over the list now to see whether you feel you have an understanding of these key points. To be sure, you should be able to answer the following conceptual questions:

1. What are some of the criteria used in the selection of a particular hypothesis-testing procedure?
2. Under what conditions should the pooled-variance *t* test be selected to examine possible differences in the means of two independent populations?

3. Under what conditions should the separate-variance t' test be selected to examine possible differences in the means of two independent populations?
4. Under what conditions should the Wilcoxon rank sum test be selected to examine possible differences in the medians of two independent populations?
5. Under what conditions should the F test be selected to examine possible differences in the variances of two independent populations?
6. What is the distinction between repeated measurements and matched or paired items?
7. Under what conditions should the t test for the mean difference μ_D in two related populations be selected?
8. Under what conditions should the Wilcoxon signed-ranks test for the median difference M_D in two related populations be selected?

Check over the list of questions to see if indeed you know the answers and could (1) explain your answers to someone who did not read this chapter and (2) give reference to specific readings or examples that support your answer. Also, reread any of the sections that may have seemed fuzzy to see if they make sense now.

Getting It All Together

Key Terms

Behrens-Fisher problem *472*
difference score D_i *504*
***F* test for differences in two variances** *489*
independent populations *463*
mean difference *504*
paired or matched items *503*
pooled-variance *t* test for differences in two means *464*
related populations *503*
related samples *503*
repeated measurements *503*
robust *468*
separate-variance *t′* test for differences in two means *472*
***t* test for the mean difference** *505*
Wilcoxon rank sum test for differences in two medians *481*
Wilcoxon signed-ranks test for the median difference *511*
***Z* test for differences in two means** *463*
***Z* test for the mean difference** *504*

Chapter Review Problems

13.71 ACTION Write a letter to a friend who has not taken a course in statistics and explain what this chapter has been about. To highlight this chapter's content, be sure to incorporate your answers to the eight review questions on pages 519–520.

● 13.72 The R & M department store has two charge plans available for its credit account customers. The management of the store wishes to collect information about each plan and study the differences between the two plans. It is interested in the average monthly balance. A random sample of 25 accounts of plan A and 50 accounts of plan B was selected with the following results:

Plan A	Plan B
$n_A = 25$	$n_B = 50$
$\bar{X}_A = \$75$	$\bar{X}_B = \$110$
$S_A = \$15$	$S_B = \$14.14$

Use statistical inference (confidence intervals or tests of hypothesis) to draw conclusions about *each* of the following:

Note: Use a level of significance of .01 (99% confidence) throughout.

(a) Average monthly balance of all plan B accounts.
(b) Is there evidence that the average monthly balance of plan A accounts is different from \$105?
(c) Is there evidence of a difference in the variances (in the monthly balances) between plan A and plan B?
(d) Is there evidence of a difference in the average monthly balance between plan A and plan B?
(e) Compute lower and upper limits for the *p* values in parts (b)–(d) and interpret their meaning.
(f) Based on the results of parts (a)–(e), what would you tell the management about the two plans?

13.73 A large public utility wishes to compare the consumption of electricity during the summer season for single-family homes in two counties that it services. For each household sampled, the monthly electric bill was recorded with the following results:

	County I	County II
$\bar{X}$	\$115	\$98
S	\$30	\$18
n	25	21

Use statistical inference (confidence intervals or tests of hypothesis) to draw conclusions about *each* of the following:

Note: Use a level of significance of .05 (95% confidence) throughout.

(a) Population average monthly electric bill for County I.
(b) Is there evidence that the average bill in County II is above \$80?
(c) Is there evidence of a difference in the variances between bills in County I and County II?
(d) Is there evidence that the average monthly bill is higher in County I than County II?
(e) Compute lower and upper limits for the *p* values in parts (b)–(d) and interpret their meaning.
(f) Based on the results of parts (a)–(e), what would you tell the utility about the consumption of electricity in the two counties?

13.74 The manager of computer operations of a large company wishes to study computer usage of two departments within the company, the Accounting Department and the Research Department. A random sample of 5 jobs from the Accounting Department in the last week and 6 jobs from the Research Department in the last week were selected, and the processing time (in seconds) for each job was recorded with the results shown on page 522.

Data file ACCRES.DAT

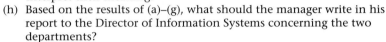

Department	Processing Time (in Seconds)					
Accounting	9	3	8	7	12	
Research	4	13	10	9	9	6

Use statistical inference (confidence intervals or tests of hypothesis) to draw conclusions about *each* of the following:

Note: Use a level of significance of .05 (95% confidence) throughout.

(a) Average processing time for all jobs in the Accounting Department.
(b) Is there evidence that the average processing time in the Research Department is greater than 6 seconds?
(c) Is there evidence of a difference in the variances in the processing time between the two departments?
(d) What assumption must be made in order to do (c)?
(e) Is there evidence of a difference in the mean processing time between the Accounting Department and the Research Department?
(f) What assumption(s) is (are) needed to do (e)?
(g) Compute lower and upper limits for the p values in (b), (c), and (e) and interpret their meaning.
(h) Based on the results of (a)–(g), what should the manager write in his report to the Director of Information Systems concerning the two departments?

Data file
PASCAL.DAT

13.75 A computer professor is interested in studying the amount of time it would take students enrolled in the Introduction to Computers course to write and run a program in PASCAL. The professor hires you to analyze the following results (in minutes) from a random sample of nine students:

10 13 9 15 12 13 11 13 12

(a) At the .05 level of significance, is there evidence that the population average amount is greater than 10 minutes? What would you tell the professor?
(b) Suppose that when checking her results, the computer professor realizes that the fourth student needed 51 minutes rather than the recorded 15 minutes to write and run the PASCAL program. At the .05 level of significance, reanalyze the revised data in part (a). What would you tell the professor now?
(c) The professor is perplexed by these paradoxical results and requests an explanation from you regarding the justification for the difference in your findings in parts (a) and (b). Discuss.
(d) A few days later, the professor calls to tell you that the dilemma is completely resolved. The original number 15 [shown in part (a)] was correct, and therefore your findings in part (a) are being used in the article she is writing for a computer magazine. Now she wants to hire you to compare the results from that group of Introduction to Computers students against those from a sample of 11 computer majors in order to determine whether there is evidence that computer majors can write a PASCAL program (on average) in less time than introductory students. The sample mean for the computer majors is 8.5 minutes and the sample standard deviation is 2.0 minutes. At the .05 level of significance, completely analyze these data. What would you tell the professor?
(e) A few days later the professor calls again to tell you that a reviewer of her article wants her to include the p value for the "correct" result in part (a). In addition, the professor inquires about a "Behrens-Fisher problem," which the reviewer wants her to discuss in her article. In your own words discuss the concept of p value and describe the Behrens-Fisher problem.

Give the approximate p value in part (a) and discuss whether or not the Behrens-Fisher problem had any meaning in the professor's study.

13.76 A professor of business statistics teaching a large lecture wanted to study scores on the midterm and final exams that are given during the semester. The exams each cover one portion of the semester and are not cumulative. The results for a sample of 33 students were as follows:

	Exam			Exam			Exam	
Student	MT	F	Student	MT	F	Student	MT	F
1	89	80	12	56	71	23	63	43
2	80	68	13	67	55	24	89	80
3	86	76	14	99	95	25	62	23
4	68	77	15	82	45	26	74	91
5	88	95	16	75	71	27	62	57
6	89	66	17	58	44	28	70	51
7	82	83	18	56	50	29	65	78
8	89	86	19	55	14	30	82	53
9	42	58	20	72	59	31	91	90
10	61	54	21	73	80	32	84	83
11	84	84	22	79	68	33	95	88

Data file GRADES.DAT

Use statistical inference (confidence intervals or tests of hypothesis) to draw conclusions about *each* of the following:

Note: Use a level of significance of .05 (95% confidence) throughout.

(a) Average midterm grade of all students.
(b) Is there evidence that the average grade on the midterm exam is higher than on the final exam?
(c) Compute lower and upper limits for the p value in part (b) and interpret their meaning.

Collaborative Learning Mini-Case Projects

Data file CEREAL.DAT

Refer to the instructions on page 101 before beginning the following problems.

CL13.1 Refer to CL 3.2 on page 101 and CL 4.2 on page 165. Your group, the ________ Corporation, has been hired by the food editor of a popular family magazine to study the cost and nutritional characteristics of ready-to-eat cereals. Armed with Special Data Set 2 of Appendix D on pages D6–D7, the ________ Corporation is ready to:

(a) Determine if there is evidence of a difference in the average cost of cereals based on whether or not the level of calories per serving is below 155 or at or above 155.
(b) Determine if there is evidence of a difference in the mean amount of sugar in high-fiber cereals versus that in low- and moderate-fiber cereals combined.
(c) Write and submit an executive summary describing the results in parts (a) and (b), clearly specifying all hypotheses, selected levels of significance, and the assumptions of the chosen test procedures.
(d) Prepare and deliver a five-minute oral presentation to the food editor of the magazine.

CL13.2 Refer to CL 3.3 on page 102 and CL 4.3 on page 165. Your group, the ________ Corporation, has been hired by the marketing director of a manufacturer of well-known men's and women's fragrances to study the characteristics of currently available fragrances. Armed with Special Data Set 3 of Appendix D on pages D8–D9, the ________ Corporation is ready to:

Data file FRAGRAN.DAT

(a) Determine if there is evidence of a difference in the average cost of men's fragrances versus women's fragrances.
(b) Determine if there is evidence of a difference in the mean cost of women's perfumes based on whether the intensity is either very strong or strong versus either medium or mild.
(c) Write and submit an executive summary describing the results in parts (a) and (b), clearly specifying all hypotheses, selected levels of significance, and the assumptions of the chosen test procedures.
(d) Prepare and deliver a five-minute oral presentation to the marketing director.

Data file CAMERA.DAT

CL13.3 Refer to CL 3.4 on page 102 and CL 4.4 on page 166. Your group, the ________ Corporation, has been hired by the travel editor of a well-known newspaper who is preparing a feature article on compact 35-mm cameras. Armed with Special Data Set 4 of Appendix D on pages D10–D11, the ________ Corporation is ready to:
(a) Determine if there is evidence of a difference in the average framing accuracy of cameras with a price level under $200 versus those with a price level of $200 or more.
(b) Determine if there is evidence of a difference in the mean weight of cameras that are classified as either multiple long, medium, or short (combined) versus those that are classified as either automatic single or fixed single (combined).
(c) Write and submit an executive summary describing the results in parts (a) and (b), clearly specifying all hypotheses, selected levels of significance, and the assumptions of the chosen test procedures.
(d) Prepare and deliver a five-minute oral presentation to the travel editor of the newspaper.

Endnotes

1. To test for differences in the median dividend yields, it must be assumed that the distributions of dividend yields in both populations from which the random samples were drawn are identical, except possibly for differences in location (i.e., the medians).
2. Since the 232 and 166 degrees of freedom are not shown in Table E.5, rounding to the nearest tabular values of 120 and 120 degrees of freedom, the upper-tailed critical value, F_U, is 1.43 and the lower-tailed critical value, F_L, is 1/1.43 = 0.699. Since $F = 1.65 > F_U = 1.43$, the null hypothesis is rejected.
3. The original measurements (i.e., length of employment in years) are replaced by the corresponding natural logarithms of the "measurements plus one" so that, for example, a full-time employee with 4 years of service to Kalosha Industries would have a transformed "measurement" of 1.61—the natural logarithm of 5.
4. The large sample approximation formula [Equation (13.14)] would result in a Z of +2.55, which exceeds +1.645, the upper-tail critical value of the standard normal distribution with a level of significance α of .05. Hence the null hypothesis would be rejected .

References

1. Begley, S., "Lead, Lies and Data Tape," *Newsweek*, March 16, 1992, p. 62.
2. Bradley, J. V., *Distribution-Free Statistical Tests* (Englewood Cliffs, NJ: Prentice-Hall, 1968).
3. Conover, W. J., *Practical Nonparametric Statistics,* 2d ed. (New York: Wiley, 1980).
4. Daniel, W., *Applied Nonparametric Statistics,* 2d ed. (Boston, MA: Houghton Mifflin, 1990).
5. *MINITAB Reference Manual Release 8* (State College, PA: MINITAB, Inc., 1992).
6. Norusis, M., *SPSS Guide to Data Analysis for SPSS-X with Additional Instructions for SPSS/PC+* (Chicago,IL: SPSS Inc., 1986).
7. *SAS User's Guide Version 6* (Raleigh, NC: SAS Institute, 1988).
8. Satterthwaite, F. E., "An Approximate Distribution of Estimates of Variance Components," *Biometrics Bulletin,* 1946, Vol. 2, pp.110–114.
9. Snedecor, G. W., and W. G. Cochran, *Statistical Methods,* 7th ed. (Ames, IA: Iowa State University Press, 1980).
10. *STATISTIX Version 4.0* (Tallahassee, FL: Analytical Software, Inc., 1992).
11. Winer, B. J., *Statistical Principles in Experimental Design,* 2nd ed. (New York: McGraw-Hill, 1971).

chapter 14

ANOVA and Other c-Sample Tests with Numerical Data

CHAPTER OBJECTIVES

To introduce the concepts of experimental design through the development of the completely randomized design model and the one-way ANOVA procedure used to test for the differences between the means of c groups and to extend this discussion to include the randomized block and factorial design (with interaction) models.

14.1 Introduction

In Chapter 13 we used hypothesis-testing methodology to draw conclusions about possible differences between the parameters of two groups when dealing with numerical data. Frequently, however, it is necessary to evaluate differences among the parameters of several (c) groups. We might want to compare alternative methods, treatments, or materials according to some predetermined criteria. A consumer organization, for example, may wish to determine which type of tires lasts longest under highway conditions; an agricultural researcher would like to know which variety of green beans provides the highest yield; a medical researcher would like to evaluate the effect of different brands of prescription medicine on the reduction of diastolic blood pressure. In each of these examples several groups are being compared and the data within each group are numerical.

We begin this chapter by examining the *completely randomized design model* having only one *factor* with several groups (such as type of tire, variety of green bean, or brand of drug) and by developing procedures for analyzing the numerical data. We then extend this by describing the *randomized block design model* and the more sophisticated *factorial design model* (where more than one factor at a time is studied in one experiment) and by developing procedures for analyzing the numerical data. Throughout the chapter, emphasis is given to the assumptions behind the use of the various testing procedures.

Upon completion of this chapter, you should be able to:

1. Compare and contrast the distinguishing features of the completely randomized design, randomized block design, and factorial design models
2. Understand the concepts behind partitioning the total variation into its various sources of variation in all three design models
3. Understand the assumptions of ANOVA
4. Know when and how to use Hartley's test for homogeneous variances
5. Know when and how to use the one-way ANOVA F test to examine possible differences in the means of c independent populations
6. Know when and how to use the Kruskal-Wallis rank test to examine possible differences in the medians of c independent populations
7. Know when and how to use multiple comparison procedures for evaluating pairwise combinations of the group means or medians
8. Know when and how to use the randomized block F test to examine possible differences in the means of c related populations
9. Know when and how to use the Friedman rank test to examine possible differences in the medians of c related populations
10. Know when and how to use the two-way ANOVA F test to examine possible differences in the means of each factor in a factorial design
11. Know when and how to use the two-way ANOVA F test to examine possible interaction in the levels of the factors in a factorial design
12. Describe the concept of interaction in a factorial design

14.2 Choosing the Appropriate Test Procedure When Comparing *c* Samples

When preparing to evaluate differences among *c* groups containing numerical data, we must select an appropriate testing procedure. This choice is dependent on several criteria:

- The type of experimental design model developed (for example, completely randomized, randomized block, or factorial).
- The level of measurement attained on the data (i.e., ordinal- versus interval- or ratio-scaled).
- The viability of the assumptions underlying alternative test procedures.
- The generalizability of the conclusions to be drawn.
- The accessibility of tables of critical values for the test statistic.
- The availability of computer software packages that contain the test procedure.
- The statistical power of the test procedure.

In the following sections of this chapter we shall describe the most widely used parametric and distribution-free procedures for evaluating differences in *c* groups.

14.3 The Completely Randomized Model: One-Factor Analysis of Variance

It is frequently of interest to compare differences in results among several groups. Many industrial applications involve experiments in which the groups or levels pertaining to only one **factor** of interest (such as baking temperature or flavor preference) are considered. A factor such as baking temperature may have several *numerical* levels (for example, 300º, 350º, 400º, 450º) or a factor such as flavor preference may have several *categorical* levels (vanilla, chocolate, mocha, strawberry, pistachio). Such designed one-factor experiments in which subjects or experimental units are randomly assigned to groups or levels of a single factor are called **one-way** or **completely randomized design models**.

14.4 One-Way ANOVA *F* Test for Differences in *c* Means

14.4.1 Introduction

When the outcome measurements across the *c* groups are continuous and certain assumptions are met, a methodology known as **analysis of variance** (or **ANOVA**) may be employed to compare the means of the groups. In a sense, the term "analysis of variance" appears to be a misnomer since the objective is to analyze differences among the group means. However, through an analysis of the variation in the data, both among and within the c groups, we will be able to draw conclusions about possible differences in group means. In ANOVA we subdivide the total variation in the outcome measurements into that which is attributable to differences *among* the c groups and that which is due to chance or attributable to inherent variation *within* the c groups (see Figure 14.1). "Within group" variation is considered **experimental error**, while "among group" variation is attributable to treatment effects.

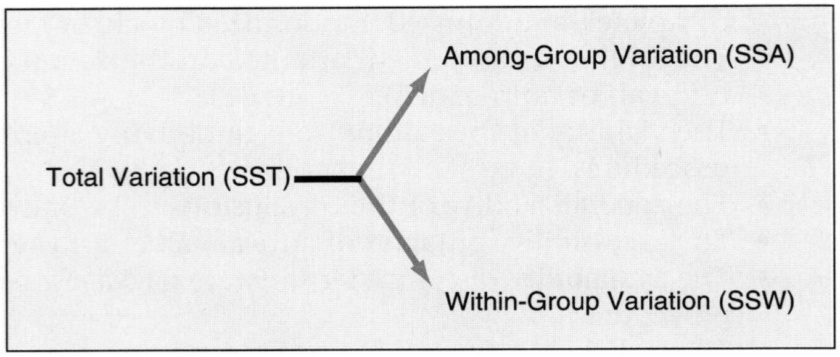

Figure 14.1
Partitioning the total variation in a completely randomized model.

14.4.2 Development

Under the assumptions that the c groups or levels of the factor being studied represent populations whose outcome measurements are randomly and independently drawn, follow a normal distribution, and have equal variances, the null hypothesis of no differences in the population means

$$H_0: \quad \mu_1 = \mu_2 = \cdots = \mu_c$$

may be tested against the alternative that not all the c population means are equal

$$H_1: \quad \text{Not all } \mu_j \text{ are equal (where } j = 1, 2, \ldots, c)$$

Figure 14.2 presents a picture of what a true null hypothesis would look like when five groups are compared and the assumptions of normality and equality of variances hold.

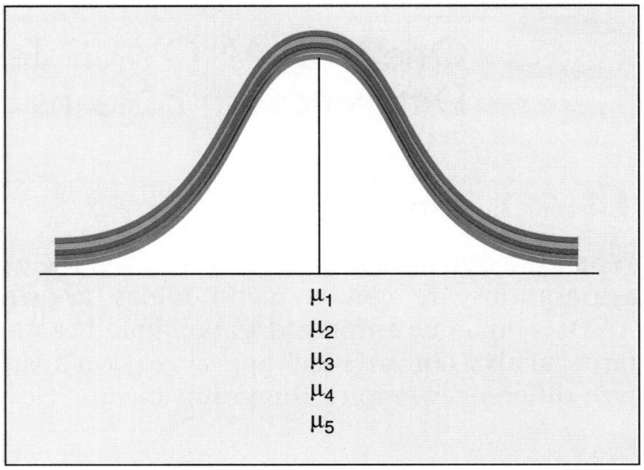

Figure 14.2
All five populations have the same mean: $\mu_1 = \mu_2 = \mu_3 = \mu_4 = \mu_5$.

The five populations representing the different levels of the factor are identical and therefore superimpose on one another. The properties of central tendency, variation, and shape are identical for each.

On the other hand, suppose that the null hypothesis was really false with level 4 having the largest mean, level 1 having the second largest mean, and no differences in the other population means. Figure 14.3 presents a pictorial representation of this.

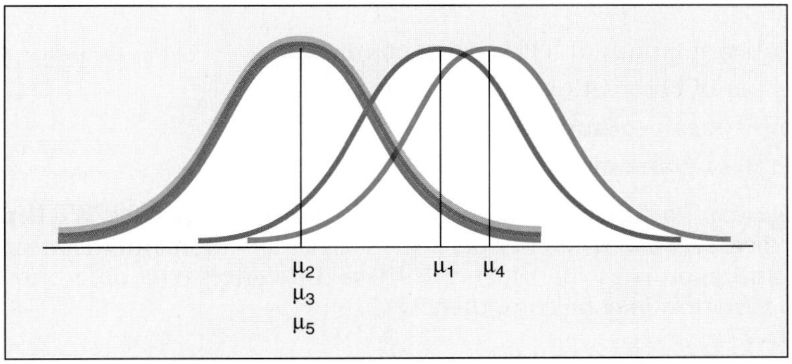

Figure 14.3
A treatment effect is present:
$\mu_4 > \mu_1 > \mu_2 = \mu_3 = \mu_5$.

We note that except for differences in central tendency (that is, $\mu_4 > \mu_1 > \mu_2 = \mu_3 = \mu_5$) the five populations are the same in appearance.

To perform an ANOVA test of equality of population means we subdivide the total variation in the outcome measurements into two parts, that which is attributable to differences among the groups and that which is due to inherent variation within the groups. The **total variation** is usually represented by the **Sum of Squares Total** (or **SST**). Since under the null hypothesis the population means of the c groups are presumed equal, a measure of the total variation among all the observations can be obtained by summing the squared differences between each individual observation and the **overall** or **grand mean** $\overline{\overline{X}}$ that is based on all the observations in all the groups combined. The total variation would be computed as

$$\text{total variation (SST)} = \sum_{j=1}^{c} \sum_{i=1}^{n_j} (X_{ij} - \overline{\overline{X}})^2 \qquad \textbf{(14.1)}$$

where $\overline{\overline{X}} = \dfrac{\sum_{j=1}^{c} \sum_{i=1}^{n_j} X_{ij}}{n}$ is called the overall or grand mean

X_{ij} is the ith observation in group or level j

n_j is the number of observations in group j

n is the total number of observations in all groups combined ($n = n_1 + n_2 + \cdots + n_c$)

c is the number of groups or levels of the factor of interest

The **among-group variation,** usually called the **Sum of Squares Among** groups (or **SSA**), is measured by the sum of the squared differences between the sample mean of each group $\bar{X}_j$ and the overall or grand mean $\bar{\bar{X}}$, weighted by the sample size n_j in each group.[1] The among-group variation is computed from

$$\text{among-group variation (SSA)} = \sum_{j=1}^{c} n_j\left(\bar{X}_j - \bar{\bar{X}}\right)^2 \qquad \textbf{(14.2)}$$

where c is the number of groups or levels being compared

n_j is the number of observations in group or level j

$\bar{X}_j$ is the sample mean of group j

$\bar{\bar{X}}$ is the overall or grand mean

The **within-group variation**, usually called the **Sum of Squares Within** groups (or **SSW**), measures the difference between each observation and the mean of its own group and cumulates the squares of these differences over all groups. The within-group variation may be computed as

$$\text{within-group variation (SSW)} = \sum_{j=1}^{c}\sum_{i=1}^{n_j} (X_{ij} - \bar{X}_j)^2 \qquad \textbf{(14.3)}$$

where X_{ij} is the ith observation in group or level j

$\bar{X}_j$ is the sample mean of group j

Since c levels of the factor are being compared, there are $c - 1$ degrees of freedom associated with the sum of squares among groups. Since each of the c levels contributes $n_j - 1$ degrees of freedom and

$$\sum_{j=1}^{c} (n_j - 1) = n - c$$

there are $n - c$ degrees of freedom associated with the sum of squares within groups. In addition, there are $n - 1$ degrees of freedom associated with the sum of squares total because each observation X_{ij} is being compared to the overall or grand mean $\bar{\bar{X}}$ based on all n observations.

If each of these sums of squares is divided by its associated degrees of freedom, we obtain three *variances* or **mean square** terms—**MSA**, **MSW**, and **MST**:

$$\text{MSA} = \frac{\text{SSA}}{c - 1} \qquad \textbf{(14.4a)}$$

$$\text{MSW} = \frac{\text{SSW}}{n - c} \qquad \textbf{(14.4b)}$$

$$\text{MST} = \frac{\text{SST}}{n - 1} \qquad \textbf{(14.4c)}$$

Since a variance is computed by dividing the sum of squared differences by its appropriate degrees of freedom, the mean square terms are all variances.

Although primary interest is in comparing the means of the c groups or levels of a factor to determine whether a treatment effect exists among the c groups, the ANOVA procedure derives its name from the fact that this is achieved by analyzing variances. If the null hypothesis is true and there are no real differences in the c group means, all three mean square terms—MSA, MSW, and MST—provide estimates of the variance σ^2 inherent in the data. Thus, to test the null hypothesis

$$H_0: \mu_1 = \mu_2 = \cdots = \mu_c$$

against the alternative

$$H_1: \text{ Not all } \mu_j \text{ are equal (where } j = 1, 2, \ldots, c)$$

we compute the test statistic F, the ratio of MSA to MSW, as

$$F = \frac{\text{MSA}}{\text{MSW}} \tag{14.5}$$

The F statistic follows an ***F* distribution** with $c - 1$ and $n - c$ degrees of freedom. For a given level of significance, α, we may reject the null hypothesis if the computed F test statistic exceeds the upper-tailed critical value $F_{U(c-1,n-c)}$ from the F distribution (see Table E.5). That is, as shown in Figure 14.4, our decision rule is

Reject H_0 if $F > F_{U(c-1,n-c)}$;

otherwise don't reject H_0.

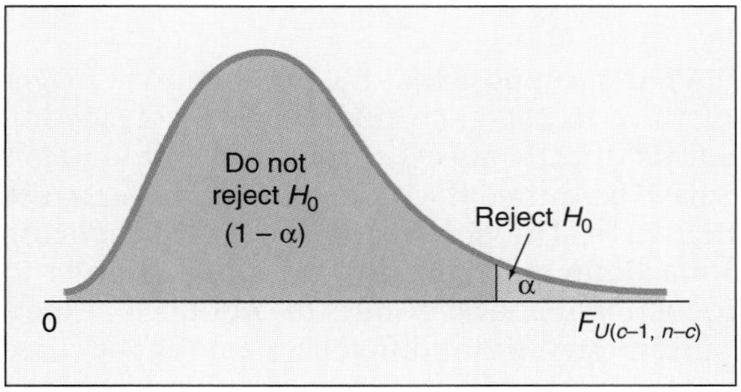

Figure 14.4
Regions of rejection and nonrejection when using ANOVA to test H_0.

If the null hypothesis were true, we should expect the computed F statistic to be approximately equal to 1, since both the numerator and denominator mean square terms are estimating the true variance σ^2 inherent in the data. On the other hand, if H_0 is false (and there are real differences in the means), we should expect the computed F statistic to be substantially larger than 1 because the numerator, MSA, would be estimating the treatment effect or differences among groups in addition to the inherent variability in the data, whereas the denominator, MSW, would be measuring only the inherent variability. Hence the ANOVA procedure yields an F test in which the null hypothesis can be rejected at a selected α level of significance *only* if the computed F statistic is large enough to exceed $F_{U(c-1,n-c)}$, the upper-tail critical value of the F distribution having $c - 1$ and $n - c$ degrees of freedom, as illustrated in Figure 14.4 on page 531.

The results of an analysis of variance procedure are usually displayed in an **ANOVA summary table,** the format for which is presented in Table 14.1. The entries in this table include the sources of variation (i.e., among group, within group, and total), the degrees of freedom, the sums of squares, the mean squares (i.e., the variances), and the calculated F statistic. In addition, the p value (i.e., the probability of obtaining an F statistic as large as or larger than the one obtained, given that the null hypothesis is true) is included in the ANOVA table of most statistical software packages (see Section 14.6). This enables us to make direct conclusions about the null hypothesis without referring to a table of critical values of the F distribution. If the p value is less than the chosen level of significance α, the null hypothesis is rejected.

Table 14.1 Analysis-of-variance summary table.

Source	Degrees of Freedom	Sums of Squares	Mean Square (Variance)	F
Among groups	$c - 1$	$SSA = \sum_{j=1}^{c} n_j (\bar{X}_j - \bar{\bar{X}})^2$	$MSA = \frac{SSA}{c-1}$	$F = \frac{MSA}{MSW}$
Within groups	$n - c$	$SSW = \sum_{j=1}^{c} \sum_{i=1}^{n_j} (X_{ij} - \bar{X}_j)^2$	$MSW = \frac{SSW}{n-c}$	
Total	$n - 1$	$SST = \sum_{j=1}^{c} \sum_{i=1}^{n_j} (X_{ij} - \bar{\bar{X}})^2$		

14.4.3 Application

To illustrate the one-way ANOVA F test, suppose that the production manager at the plant in which cereal is being filled in 368-gram sized boxes is considering the replacement of an old machine that directly affects output in the production process. Three competing suppliers have permitted the production manager to use their particular equipment on a trial basis. The purchasing prices and servicing contracts for these three brands of machines are essentially the same. In order to make a purchasing decision, the production manager decides to conduct an experiment to determine whether there are any significant differences among the three brands of machines in the average time (in seconds) it takes factory workers using them to complete the filling process. Fifteen factory workers of similar experience, ability, and age are randomly assigned to receive training on one of the three brands of machines in such a manner that there are five factory workers for each machine. After an appropriate amount of training and practice, the production manager measures the time (in seconds) it takes the factory workers to complete the filling process using their respective equipment. The results of this experiment are displayed in Table 14.2 along with some summary computations. A *scatter plot* is depicted in Figure 14.5 so that we can visually inspect the data and see how the measurements (in seconds) distribute around their own group means as well as around the overall group mean $\overline{\overline{X}}$. We also get a sense for how each group mean compares to the overall mean. Thus, by examining Figure 14.5, we have the opportunity to observe possible trends or relationships across the groups as well as patterns within groups and, most importantly, this enables us to consider any potential violations in assumptions required by a particular testing procedure. Had the sample sizes in each group been larger, stem-and-leaf displays and a box-and-whisker plot, which provide additional visual information, would also have been obtained.[2]

Table 14.2 Time (in seconds) to complete a filling process using three different machines.

	Machine		
	I	II	III
	25.40	23.40	20.00
	26.31	21.80	22.20
	24.10	23.50	19.75
	23.74	22.75	20.60
	25.10	21.60	20.40
Mean	$\overline{X}_1 = 24.93$	$\overline{X}_2 = 22.61$	$\overline{X}_3 = 20.59$

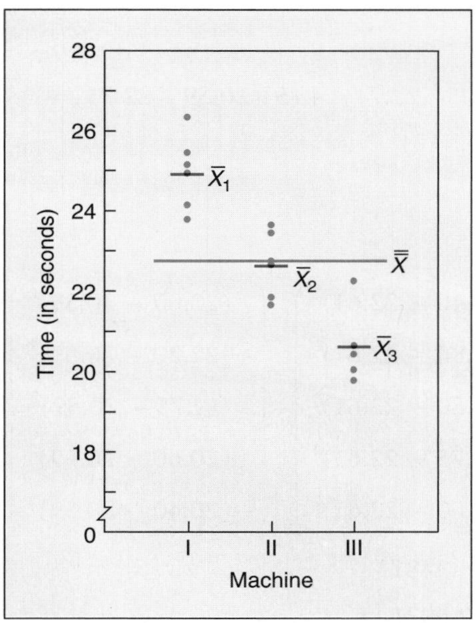

Figure 14.5
Scatter plot of time (in seconds) to complete a task using three different machines.
Source: Table 14.2

We note from Table 14.2 and Figure 14.5 that there are differences in the *sample* means for the three machines. It takes, on average, 24.93 seconds to complete the filling process using machine I, 22.61 seconds using machine II, and 20.59 seconds using machine III. The question that must be answered is whether these sample results are sufficiently different for the production manager to decide that the *population* averages are not all equal.

The null hypothesis states that there is no difference among the groups in mean time to complete the filling process. Thus, substituting 1, 2, 3 for I, II, III we have

$$H_0: \quad \mu_1 = \mu_2 = \mu_3$$

The alternative hypothesis states that there is a *treatment effect*; that is, at least one of the machines differs with respect to the average time required for completing the filling process.

$$H_1: \quad \text{Not all the means are equal}$$

To establish the ANOVA summary table, we first compute the sample means in each group (see Table 14.2 on page 533); then we compute the overall or grand mean

$$\bar{\bar{X}} = \frac{\sum_{j=1}^{c}\sum_{i=1}^{n_j} X_{ij}}{n} = \frac{25.40 + 26.31 + \cdots + 23.40 + \cdots + 20.40}{15}$$

$$= \frac{340.65}{15} = 22.71$$

followed by the sums of squares:

$$\begin{aligned}
\text{SSA} = \sum_{j=1}^{c} n_j\left(\bar{X}_j - \bar{\bar{X}}\right)^2 &= (5)(24.93 - 22.71)^2 + (5)(22.61 - 22.71)^2 + (5)(20.59 - 22.71)^2 \\
&= (5)(2.22)^2 + (5)(-.10)^2 + (5)(-2.12)^2 \\
&= 24.642 + .05 + 22.472 \\
&= 47.164
\end{aligned}$$

$$\text{SSW} = \sum_{j=1}^{c}\sum_{i=1}^{n_j}\left(X_{ij} - \bar{X}_j\right)^2 = \begin{pmatrix} (25.40 - 24.93)^2 \\ +(26.31 - 24.93)^2 \\ +(24.10 - 24.93)^2 \\ +(23.74 - 24.93)^2 \\ +(25.10 - 24.93)^2 \end{pmatrix} + \begin{pmatrix} (23.40 - 22.61)^2 \\ +(21.80 - 22.61)^2 \\ +(23.50 - 22.61)^2 \\ +(22.75 - 22.61)^2 \\ +(21.60 - 22.61)^2 \end{pmatrix} + \begin{pmatrix} (20.00 - 20.59)^2 \\ +(22.20 - 20.59)^2 \\ +(19.75 - 20.59)^2 \\ +(20.60 - 20.59)^2 \\ +(20.40 - 20.59)^2 \end{pmatrix}$$

$$= \begin{pmatrix} .2209 \\ +1.9044 \\ +\ .6889 \\ +1.4161 \\ +\ .0289 \end{pmatrix} + \begin{pmatrix} .6241 \\ +\ .6561 \\ +\ .7921 \\ +\ .0196 \\ +1.0201 \end{pmatrix} + \begin{pmatrix} .3481 \\ +2.5921 \\ +\ .7056 \\ +\ .0001 \\ +\ .0361 \end{pmatrix}$$

$$= 11.0532$$

$$SST = \sum_{j=1}^{c}\sum_{i=1}^{n_j}\left(X_{ij} - \bar{\bar{X}}\right)^2 = \begin{pmatrix} (25.40 - 22.71)^2 \\ +(26.31 - 22.71)^2 \\ +(24.10 - 22.71)^2 \\ +(23.74 - 22.71)^2 \\ +(25.10 - 22.71)^2 \end{pmatrix} + \begin{pmatrix} (23.40 - 22.71)^2 \\ +(21.80 - 22.71)^2 \\ +(23.50 - 22.71)^2 \\ +(22.75 - 22.71)^2 \\ +(21.60 - 22.71)^2 \end{pmatrix} + \begin{pmatrix} (20.00 - 22.71)^2 \\ +(22.20 - 22.71)^2 \\ +(19.75 - 22.71)^2 \\ +(20.60 - 22.71)^2 \\ +(20.40 - 22.71)^2 \end{pmatrix}$$

$$= \begin{pmatrix} 7.2361 \\ +\ 12.9600 \\ +\ 1.9321 \\ +\ 1.0609 \\ +\ 5.7121 \end{pmatrix} + \begin{pmatrix} .4761 \\ +\ .8281 \\ +\ .6241 \\ +\ .0016 \\ +\ 1.2321 \end{pmatrix} + \begin{pmatrix} 7.3441 \\ +\ .2601 \\ +\ 8.7616 \\ +\ 4.4521 \\ +\ 5.3361 \end{pmatrix}$$

$$= 58.2172$$

The respective mean square terms are obtained by dividing these sums of squares by their corresponding degrees of freedom. Since $c = 3$ and $n = 15$, we have

$$MSA = \frac{SSA}{c - 1} = \frac{47.164}{3 - 1} = \frac{47.164}{2} = 23.582$$

$$MSW = \frac{SSW}{n - c} = \frac{11.0532}{15 - 3} = \frac{11.0532}{12} = 0.9211$$

so that using Equation (14.1) to test H_0 we obtain

$$F = \frac{MSA}{MSW} = \frac{23.582}{0.9211} = 25.60$$

If a .05 level of significance is chosen, the critical value of the F statistic would be obtained from Table E.5, a replica of which is presented as Table 14.3 (on page 536). The values in the body of this table refer to selected upper-tailed percentage points of the F distribution. In our productivity study, since there are 2 degrees of freedom in the numerator of the F ratio and 12 degrees of freedom in the

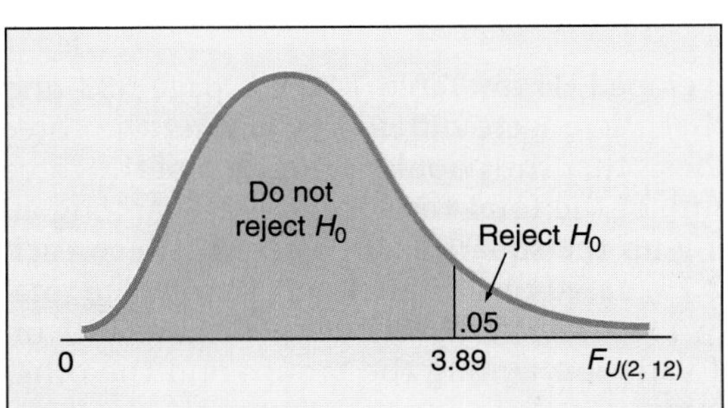

Figure 14.6
Regions of rejection and nonrejection for the analysis of variance at the .05 level of significance with 2 and 12 degrees of freedom.

Table 14.3 Obtaining the critical value of *F* with 2 and 12 degrees of freedom at the .05 level.

Denominator, df_2	Numerator, df_1 1	2	3	4	5	6	7	8	9	10	12
.	.	.	.	.	.	.	.	.	.	.	.
.	.	.	.	.	.	.	.	.	.	.	.
.	.	.	.	.	.	.	.	.	.	.	.
5	6.61	5.79	5.41	5.19	5.05	4.95	4.88	4.82	4.77	4.74	4.68
6	5.99	5.14	4.76	4.53	4.39	4.28	4.21	4.15	4.10	4.06	4.00
7	5.59	4.74	4.35	4.12	3.97	3.87	3.79	3.73	3.68	3.64	3.57
8	5.32	4.46	4.07	3.84	3.69	3.58	3.50	3.44	3.39	3.35	3.28
9	5.12	4.26	3.86	3.63	3.48	3.37	3.29	3.23	3.18	3.14	3.07
10	4.96	4.10	3.71	3.48	3.33	3.22	3.14	3.07	3.02	2.98	2.91
11	4.84	3.98	3.59	3.36	3.20	3.09	3.01	2.95	2.90	2.85	2.79
12	4.75	3.89	3.49	3.26	3.11	3.00	2.91	2.85	2.80	2.75	2.69
13	4.67	3.81	3.41	3.18	3.03	2.92	2.83	2.77	2.71	2.67	2.60
14	4.60	3.74	3.34	3.11	2.96	2.85	2.76	2.70	2.65	2.60	2.53

Source: Extracted from Table E.5.

denominator, the critical value of F at the .05 level of significance is 3.89. Because our computed test statistic $F = 25.60$ exceeds this critical F value, the null hypothesis may be rejected (see Figure 14.6 on page 535). The production manager may conclude that there is a significant difference in the average time required to complete the filling process on the three machines.

The corresponding ANOVA summary table is presented in Table 14.4 and contains the exact p value for the calculated value of F obtained from the MINITAB software package (see Reference 10). Note that the p value or probability of obtaining an F statistic of 25.60 or larger when the null hypothesis is true is 0.000. Since this p value is less than the specified α of .05, the null hypothesis is rejected.

Table 14.4 Analysis-of-variance table for the productivity study.

Source	Degrees of Freedom	Sums of Squares	Mean Square (Variance)	F	p Value
Among groups (machines)	3 − 1 = 2	47.1640	23.5820	25.60	.000
Within groups (machines)	15 − 3 = 12	11.0532	.9211		
Total	15 − 1 = 14	58.2172			

14.4.4 Reflection

Let us review what we have just developed. From Table 14.2 on page 533 and Figure 14.5 on page 533 we noted that there were differences among the three sample means. Under the null hypothesis that the population means of the three groups were presumed equal, a measure of the total variation (or SST) among all the workers was obtained by summing up the squared differences between each observation and the overall mean, 22.71, based on all the observations. The total variation was then subdivided into two separate components (see Figure 14.1 on page 528), one portion consisting of variation among the groups and the other consisting of variation within the groups.

Why is there variation among the values; that is, why are the observations not all the same? One reason is that by treating people differently (in this case, giving them different machines to use) we affect their productivity. This would explain part of the reason why the groups have different means: the bigger the effect of the treatment, the more variation in group means we will find. But there is another reason for variability in the results, which is that people are naturally variable whether we treat them alike or not. So even within a particular group where everyone got the same treatment (i.e., machine) there is variability. Because it occurs within each group, it is called within-group variation (or SSW).

The differences among the group means are called the among-group variation (or SSA). Part of the among-group variation, as we noted above, is due to the effect of being in different groups. But even if there is no real effect of being in different groups (that is, the null hypothesis is true), there will likely be differences among the group means. This is because variability among workers will make the sample means different just because we have different samples. Therefore, if the null hypothesis is true, then the among-group variation will estimate the population variability just as well as the within-group variation. But if the null hypothesis is false, then the among-group variation will be larger. This fact forms the basis for the one-way ANOVA F test of differences in group means.

Now let us review what we have just accomplished. Again, from Table 14.2 on page 533 and Figure 14.5 on page 533 we noted that there were differences among the three sample means. Using the one-way ANOVA F test, the production manager has found sufficient evidence to conclude that there is a significant *treatment effect* across the levels (or groups) of the factor of interest, machine brands. That is, there is evidence that the population means differ with respect to the time required to complete the filling process.

What we don't yet know, however, is which machine or machines differ from the other(s). All we know is that there is sufficient evidence to state that the population means are not all the same; that is, at least one or some combination of them is significantly different. To determine exactly which machine or machines differ, we shall make all possible pairwise comparisons between machines and use a procedure developed by John Tukey (and later modified, independently by Tukey and by C. Y. Kramer, for situations in which the sample sizes differ) to draw our conclusions. (See References 6, 7, and 14.)

14.4.5 Multiple Comparisons: The Tukey-Kramer Procedure

In the productivity study discussed thus far in this chapter, the analysis of variance was used to determine whether there was a difference among several groups in the average time to complete a task. Once differences in the means of the groups are found, it is important that we determine which particular groups are different.

Although many procedures are available (see References 6 and 9), we will focus on the Tukey-Kramer procedure in order to determine which of the c means are significantly different from each other. This method is an example of a *post hoc* (or ***a posteriori***) comparison procedure, since the hypotheses of interest are formulated *after* the data have been inspected.

The Tukey-Kramer procedure enables us to simultaneously examine comparisons between all pairs of groups. The first step involved is to compute the differences $\overline{X}_j - \overline{X}_{j'}$ (where $j \neq j'$) among all $c(c-1)/2$ pairs of means. The **critical range** for the Tukey-Kramer procedure is then obtained from the quantity given in Equation (14.6) on page 538.

$$\text{critical range} = Q_{U(c,n-c)}\sqrt{\frac{\text{MSW}}{2}\left(\frac{1}{n_j}+\frac{1}{n_{j'}}\right)} \quad (14.6)$$

If the sample sizes differ, a critical range would be computed for each pairwise comparison of sample means. The final step is to compare each of the $c(c-1)/2$ pairs of means against its corresponding critical range. A specific pair would be declared significantly different if the absolute difference in the sample means $|\bar{X}_j - \bar{X}_{j'}|$ exceeds the critical range.

To apply the Tukey-Kramer procedure, we return to the productivity study. Using the ANOVA procedure, we concluded that there was a difference in the average time to complete a task by using the three machines. Since there are three groups, there are $(3)(3-1)/2 = 3$ possible pairwise comparisons to be made. From Table 14.2 on page 533, the absolute mean differences are

1. $|\bar{X}_1 - \bar{X}_2| = |24.93 - 22.61| = 2.32$.
2. $|\bar{X}_1 - \bar{X}_3| = |24.93 - 20.59| = 4.34$.
3. $|\bar{X}_2 - \bar{X}_3| = |22.61 - 20.59| = 2.02$.

Only one critical range needs to be obtained here because the three groups had equal-sized samples. To determine the critical range, from Table 14.4 on page 536 we have MSW = .9211 and $n_j = 5$. From Table E.12, for $\alpha = .05$, $c = 3$ and $n - c = 15 - 3 = 12$, the upper-tailed critical value of $Q_{U(3,\,12)}$ is 3.77 (see Table 14.5). From Equation (14.6), we have

$$\text{critical range} = 3.77\sqrt{\left(\frac{.9211}{2}\right)\left(\frac{1}{5}+\frac{1}{5}\right)} = 1.618$$

Since 2.32 > 1.618, 4.34 > 1.618, and 2.02 > 1.618, it would be concluded that there is a significant difference between each pair of means. Hence the production manager would purchase machine III because the average time for completing a task on it is fastest.

Table 14.5 Obtaining the Studentized range Q statistic for α = .05 with 3 and 12 degrees of freedom.

Denominator Degrees of Freedom	Numerator Degrees of Freedom: 2	3	4	5	6	7	8	9	10	11	12	13	14	15	16
1	18.0	27.0	32.8	37.1	40.4	43.1	45.4	47.4	49.1	50.6	52.0	53.2	54.3	55.4	56.3
2	6.09	8.3	9.8	10.9	11.7	12.4	13.0	13.5	14.0	14.4	14.7	15.1	15.4	15.7	15.9
3	4.50	5.91	6.82	7.50	8.04	8.48	8.85	9.18	9.46	9.72	9.95	10.15	10.35	10.52	10.69
.	.	.	.	.	.	.	.	.	.	.	.	.	.	.	.
.	.	.	.	.	.	.	.	.	.	.	.	.	.	.	.
.	.	.	.	.	.	.	.	.	.	.	.	.	.	.	.
11	3.11	3.82	4.26	4.57	4.82	5.03	5.20	5.35	5.49	5.61	5.71	5.81	5.90	5.99	6.06
12	3.08	3.77	4.20	4.51	4.75	4.95	5.12	5.27	5.40	5.51	5.62	5.71	5.80	5.88	5.95
13	3.06	3.73	4.15	4.45	4.69	4.88	5.05	5.19	5.32	5.43	5.53	5.63	5.71	5.79	5.86
14	3.03	3.70	4.11	4.41	4.64	4.83	4.99	5.13	5.25	5.36	5.46	5.55	5.64	5.72	5.79

Source: Extracted from Table E.12.

14.4.6 ANOVA Assumptions

In our productivity study it seems that the analysis is now complete. But is it? Aside from our exploratory investigations in Figure 14.5 on page 533, we have not yet thoroughly evaluated the assumptions underlying the one-way *F* test. How can the production manager know whether the one-way *F* test was an appropriate procedure for analyzing his experimental data?

In Chapters 12 and 13 we mentioned the assumptions made in the application of each hypothesis-testing procedure and the consequences of departures from these assumptions. To employ the one-way ANOVA *F* test, we must also make certain assumptions about the data being investigated. There are three major assumptions in the analysis of variance:

1. Randomness and independence of errors
2. Normality
3. Homogeneity of variance

The first assumption, **randomness and independence of errors,** must be met for all procedures discussed in this chapter, not only those dealing with ANOVA, since the validity of any experiment depends on random sampling and/or the randomization process. To avoid biases in the outcomes, it is essential that either the obtained data be considered as randomly and independently drawn from the *c* populations or that the items or subjects be randomly assigned to the *c* levels of the factor of interest (i.e., the treatment groups). Therefore, the assumption of randomness and independence refers not to haphazard mistakes, but to the difference of each observed value from its own group mean. The assumption is that these differences should be independent for each observed value. That is, the difference (or *error*) for one observation should not be related to the difference (or *error*) for any other observation. This assumption might be violated, for example, in our productivity study if one worker aided another in completing the filling process. The assumption would also be violated if two of the workers in one group were identical twins: their behavior would likely be more similar than would the behavior of any other two individuals in the study. Most often, however, this assumption is violated when data are collected over a period of time, because observations made at adjacent time points may be more alike than those made at very different times. Consider, for example, temperature recorded every day for a month. The temperature on a given day is likely to be near what it was the day before, but less likely to be close to the temperature several weeks later.

Again, departures from this assumption can seriously affect inferences from the analysis of variance. These problems are discussed more thoroughly in References 1 and 6.

The second assumption, **normality,** states that the values in each group are normally distributed. Just as in the case of the *t* test, the one-way ANOVA *F* test is fairly *robust* against departures from the normal distribution; that is, as long as the distributions are not extremely different from a normal distribution, the level of significance of the analysis-of-variance test is usually not greatly affected by lack of normality, particularly for large samples.

When only the normality assumption is seriously violated, distribution-free alternatives to the one-way ANOVA *F* test are available (see Section 14.5).

The third assumption, **homogeneity of variance,** states that the variance within each population should be equal for all populations (that is, $\sigma_1^2 = \sigma_2^2 = \cdots = \sigma_c^2$). This assumption is needed in order to combine or pool the variances within the groups into a single within-group source of variation SSW. If there are equal sample sizes in each group, inferences based on the F distribution may not be seriously affected by unequal variances. If, however, there are unequal sample sizes in different groups, unequal variances from group to group can have serious effects on drawing inferences from the analysis of variance. Thus, from the perspective of computational simplicity, robustness, and power, there should be equal sample sizes in all groups whenever possible.

When only the homogeneity-of-variance assumption is violated, procedures similar to those used in the separate-variance t' test of Section 13.4 are available (see Reference 1). However, if both the normality and homogeneity-of-variance assumptions have been violated, an appropriate *data transformation* may be used that will both normalize the data and reduce the differences in variances (see References 1 and 11) or, alternatively, a more general nonparametric procedure may be employed (see References 2 and 3).

14.4.7 Hartley's F_{max} Test for Homogeneity of Variance

Although the one-way ANOVA F test is relatively robust with respect to the assumption of equal group variances, large departures from this assumption may seriously affect the level and the power of the test. Therefore, various procedures have been developed to more formally test the assumption of homogeneity of variance. Perhaps the simplest and best known is Hartley's F_{max} procedure (see Reference 1). To test for the equality of the c population variances

$$H_0: \sigma_1^2 = \sigma_2^2 = \cdots = \sigma_c^2$$

against the alternative

$$H_1: \text{Not all } \sigma_j^2 \text{ are equal } (j = 1, 2, \ldots, c)$$

we obtain the following F_{max} test statistic from Hartley's F_{max} distribution with c and $(\bar{n} - 1)$ degrees of freedom:

$$F_{max} = \frac{S_{max}^2}{S_{min}^2} \tag{14.7}$$

where S_{max}^2 = largest sample variance

S_{min}^2 = smallest sample variance

$\bar{n} = \dfrac{\sum_{j=1}^{c} n_j}{c} = \dfrac{n}{c}$ (only the integer portion of this value is utilized)

Using a level of significance, α, the null hypothesis of the equality of group variances will be rejected only when the computed F_{max} statistic exceeds the upper-tail critical value of Hartley's F_{max} distribution based upon c and $(\bar{n} - 1)$ degrees of freedom (see Table E.8, a replica of which is presented as Table 14.6). That is,

$$\text{Reject } H_0 \text{ if } F_{max} > F_{max[c,\,(\bar{n}-1)]};$$
$$\text{otherwise don't reject } H_0.$$

In order to illustrate Hartley's F_{max} procedure, let us return to the data for the productivity study, which are presented in Table 14.2 on page 533. Using Equation (4.6) on page 120, we can compute the sample variances of the three groups as follows:

$$S_1^2 = 1.065, \quad S_2^2 = .778, \quad S_3^2 = .921$$

Since each group contains a sample of size 5, $\bar{n} = (5 + 5 + 5)/3 = 5$, and, testing the null hypothesis

$$H_0\text{: } \sigma_1^2 = \sigma_2^2 = \sigma_3^2$$

against the alternative

$$H_1\text{: Not all } \sigma_j^2 \text{ are equal } (j = 1, 2, 3)$$

the F_{max} statistic is computed from Equation (14.7) as

$$F_{max} = \frac{1.065}{.778} = 1.369$$

If the .05 level of significance is selected, the decision rule would be to reject H_0 if $F_{max} > F_{max[3,4]} = 15.5$ (see Table 14.6). In our productivity study, since $F_{max} = 1.369 < F_{max[3,\,4]} = 15.5$, we would not reject H_0 and we would conclude that there is no evidence of a difference in the variances of the three groups.

Although the F_{max} test is simple to utilize, unfortunately it is not robust. It is extremely sensitive to departures from normality in the data. Thus in situations where we are unable to assume normality for each group, other alternative procedures should be applied (see References 2 and 3).

Table 14.6 Obtaining the critical value of F_{max} with 3 and 4 degrees of freedom at the .05 level of significance.

Upper 5% Points ($\alpha = .05$)								
$n - 1 \backslash c$	2	3	4	5	6	7	8	9
2	39.0	87.5	142	202	266	333	403	475
3	15.4	27.8	39.2	50.7	62.0	72.9	83.5	93.9
4	9.60	15.5	20.6	25.2	29.5	33.6	37.5	41.1
5	7.15	10.8	13.7	16.3	18.7	20.8	22.9	24.7
6	5.82	8.38	10.4	12.1	13.7	15.0	16.3	17.5

Source: Extracted from Table E.8.

Problems for Section 14.4

14.1 Explain the difference between the among-groups variance MSA and the within-groups variance MSW.

14.2 How does the one-way ANOVA F test differ from the test for the differences between two variances of Section 13.6?

14.3 Compare and contrast the assumptions of the one-way ANOVA F test and the assumptions of the t test for the difference between the means of two populations. Discuss fully.

14.4 Explain how the graphical methods of Chapters 3 and 4 might be used to evaluate the validity of the assumptions of the analysis of variance.

Data file
INSPGM.DAT

14.5 The personnel manager of a large insurance company wished to evaluate the effectiveness of four different sales-training programs designed for new employees. A group of 32 recently hired college graduates were randomly assigned to the four programs so that there were eight subjects in each program. At the end of the month-long training period a standard exam was administered to the 32 subjects; the scores are given below:

Programs			
A	B	C	D
66	72	61	63
74	51	60	61
82	59	57	76
75	62	60	84
73	74	81	58
97	64	55	65
87	78	70	69
78	63	71	80

(a) Construct an appropriate graph, plot, or chart of the data.
(b) Describe any trends or relationships that might be apparent within or among the groups.
(c) Does the variation within the groups seem to be similar for all groups? Discuss.
(d) At an $\alpha = .05$ level of significance, use Hartley's F_{max} test to test the homogeneity of variance assumption.
(e) Based on your results in (d), can you proceed with a one-way ANOVA F test for differences in the population means or is a data transformation necessary? Discuss.
(f) If conditions are appropriate, at an $\alpha = .05$ level of significance, use the one-way F test to determine whether there is evidence of a difference in the four sales-training programs.
(g) Based on your results in (f), if appropriate, use the Tukey-Kramer procedure to make all pairwise comparisons of the training programs. (Use an overall level of significance of .05.)
(h) ACTION Prepare an executive summary that the personnel manager might send to the vice-president for operations in this company.

14.6 A medical researcher decided to compare several over-the-counter sleep medications. People who had trouble sleeping came to the researcher's sleep laboratory and were randomly assigned to take either a placebo (a pill with no active ingredients), Nighty-night, Snooze-Away, or Mr. Sandman. The number of hours each one slept is given at the top of page 543:

Placebo:	2, 4, 3, 5, 2, 4, 3
Nighty-night:	3, 4, 5, 3, 5, 4, 6, 5
Snooze-Away:	6, 5, 7, 4, 8, 6
Mr. Sandman:	5, 4, 7, 5, 8, 6, 7

Data file SLEEP.DAT

(a) Construct an appropriate graph, plot, or chart of the data.
(b) Describe any trends or relationships that might be apparent within or among the groups.
(c) Does the variation within the groups seem to be similar for all groups? Discuss.
(d) At an $\alpha = .05$ level of significance, use Hartley's F_{max} test to test the homogeneity of variance assumption.
(e) Based on your results in (d), can you proceed with a one-way ANOVA F test for differences in the population means or is a data transformation necessary? Discuss.
(f) If conditions are appropriate, at an $\alpha = .05$ level of significance, use the one-way F test to determine whether there is evidence of a difference in the over-the-counter sleep medications.
(g) Based on your results in (f), if appropriate, use the Tukey-Kramer procedure to make all pairwise comparisons of the medications. (Use an overall level of significance of .05.)
(h) Comment on the precision of the measurements taken in this experiment.
(i) **ACTION** Write a draft of an article that the medical researcher might send to the editor of the health column in your local newspaper.

14.7 Suppose that the assistant superintendent for curriculum affairs wishes to evaluate three alternative sets of mathematics materials so that one set can be selected to be purchased by the entire school district. A third-grade teacher in the district has volunteered to make the comparison. The 24 students in her class are homogeneous with respect to their academic abilities. They are to be divided randomly into three groups, containing 7, 9, and 8 students, respectively. The first group is assigned to set I, the second to set II, and the third to set III.

Data file CURR.DAT

At the end of the year all 24 students are given the same standardized mathematics test. The scores on a scale of 0–100 (low–high) are displayed below:

Material Set		
I	II	III
87	58	81
80	63	62
74	64	70
82	75	64
74	70	70
81	73	72
97	80	92
	62	63
	71	

Based on these results, the assistant superintendent would like to know whether there is any difference among the sets in the scores achieved and, if so, which set(s) are superior to the others.

(a) Completely analyze the data. (Use $\alpha = .05$.)
(b) ACTION Draft a report that the third-grade teacher might send to the assistant superintendent.

14.8 A metallurgist tested five different alloys for tensile strength. He tested several samples of each alloy; the tensile strength of each sample was

Data file TENSILE.DAT

Alloy 1:	12.4, 19.8, 15.2, 14.8, 18.5
Alloy 2:	8.9, 11.6, 10.0, 10.3
Alloy 3:	10.5, 13.8, 12.1, 11.9, 12.6
Alloy 4:	12.8, 14.2, 15.9, 14.1
Alloy 5:	16.4, 15.9, 17.8, 20.3

The metallurgist would like to know if there is evidence of a difference in the tensile strengths of the various alloys and, if so, which ones are significantly stronger than the others.

(a) Completely analyze the data. (Use $\alpha = .05$.)
(b) ACTION Describe the results to the metallurgist in a memo.

14.9 A statistics professor wanted to study four different strategies of playing the game of Blackjack (Twenty-One). The four strategies were

1. Dealer's strategy
2. Five-count strategy
3. Basic ten-count strategy
4. Advanced ten-count strategy

A calculator that could play Blackjack was utilized and data from five sessions of each strategy were collected. The profits (or losses) from each session were:

Data file BLACKJ.DAT

Strategy			
Dealer's	Five Count	Basic Ten Count	Advanced Ten Count
− \$56	− \$26	+ \$16	+ \$60
− \$78	− \$12	+ \$20	+ \$40
− \$20	+ \$18	− \$14	− \$16
− \$46	− \$ 8	+ \$ 6	+ \$12
− \$60	− \$16	− \$25	+ \$ 4

The professor wanted to know whether there is evidence of a difference among the four strategies and, if so, which strategies are superior with respect to potential profitability.

(a) Completely analyze the data. (Use $\alpha = .01$.)
(b) ACTION Write a letter to the professor explaining your findings.

Interchapter Problems for Section 14.4

14.10 In Problem 13.11 on page 472 you used a t test to compare the tuition at preparatory schools in the Northeast and Midwest. (The data are taken from Problem 4.80 on page 163.)

(a) Perform a one-way ANOVA F test on this data set. (Use $\alpha = .01$.)

(b) Square the value of t you computed in Problem 13.11; notice that it is the same (except for rounding error) as the F value. Express in your own words the relationship between t and F.

Data file PREP.DAT

14.11 In Problem 13.12 on page 472 you used a t test to compare the cost of shampoos designated for normal versus fine hair. (The data are taken from Problem 3.9 on page 62.)

(a) Perform a one-way ANOVA F test on this data set. (Use $\alpha = .05$.)

(b) Square the value of t you computed in Problem 13.12; notice that it is the same (except for rounding error) as the F value. Express in your own words the relationship between t and F.

Data file SHAMPOO.DAT

14.5 Kruskal-Wallis Rank Test for Differences in c Medians

14.5.1 Introduction

The **Kruskal-Wallis rank test** for differences in c medians (where $c > 2$) may be considered an extension of the Wilcoxon rank-sum test for two independent samples discussed in Section 13.5. Thus the Kruskal-Wallis test enjoys the same power properties relative to the one-way ANOVA F test as does the Wilcoxon rank-sum test relative to the t test for two independent samples (Section 13.3). That is, the Kruskal-Wallis procedure has proven to be almost as powerful as the F test under conditions appropriate to the latter and even more powerful than the classical procedure when its assumptions (see Section 14.4.6) are violated.

14.5.2 Development

The Kruskal-Wallis rank test is most often used to test whether c independent sample groups have been drawn from populations possessing equal medians. That is, we may test

$$H_0: M_1 = M_2 = \cdots = M_c$$

against the alternative

$$H_1: \text{Not all } M_j \text{ are equal (where } j = 1, 2, \ldots, c).$$

For such situations, it is necessary to assume that

1. The c samples are randomly and independently drawn from their respective populations.
2. The underlying random phenomenon of interest is continuous (to avoid ties).
3. The observed data constitute at least an ordinal scale of measurement, both within and among the c samples.
4. The c populations have the same *variability.*
5. The c populations have the same *shape.*

Interestingly, the Kruskal-Wallis procedure still makes less stringent assumptions than does the F test. To employ the Kruskal-Wallis procedure the measurements need only be ordinal over all sample groups, and the common population distributions need only be continuous—their common shapes are irrelevant. On the other hand, to utilize the classical F test the level of measurement must be more sophisticated, and we must assume that the c samples are coming from underlying normal populations having equal variances.

To perform the Kruskal-Wallis rank test we must first (if necessary) replace the observations in the c samples with their combined ranks such that rank 1 is given to the smallest of the combined observations and rank n to the largest of the combined observations (where $n = n_1 + n_2 + \cdots + n_c$). If any values are tied, they are assigned the average of the ranks they would otherwise have been assigned if ties had not been present in the data.

The Kruskal-Wallis test statistic H may be computed from

$$H = \left[\frac{12}{n(n+1)} \sum_{j=1}^{c} \frac{T_j^2}{n_j}\right] - 3(n+1) \tag{14.8}$$

where n is the total number of observations over the combined samples, i.e., $n = n_1 + n_2 + \ldots + n_c$

n_j is the number of observations in the jth sample; $j = 1, 2, \ldots, c$

T_j is the sum of the ranks assigned to the jth sample

T_j^2 is the square of the sum of the ranks assigned to the jth sample

As the sample sizes in each group get large (greater than 5), the test statistic H may be approximated by the chi-square distribution with $c - 1$ degrees of freedom. Thus for any selected level of significance α, the decision rule would be to reject the null hypothesis if the computed value of H exceeds the upper-tail critical χ^2 value and not to reject the null hypothesis if H is less than or equal to the critical χ^2 value (see Figure 14.7). That is

Reject H_0 if $H > \chi^2_{U(c-1)}$;

otherwise don't reject H_0

The critical χ^2 values are given in Table E.4.

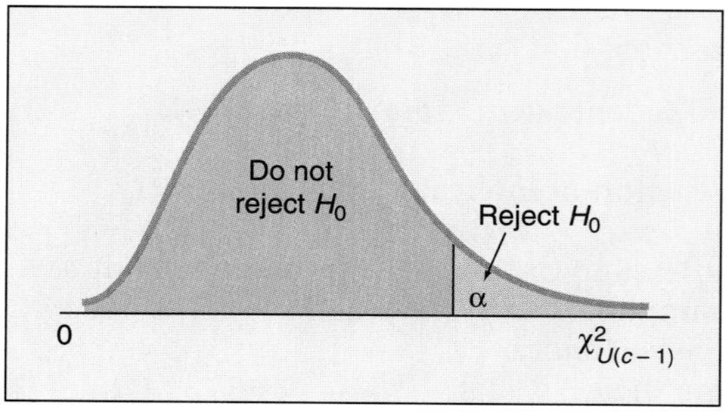

Figure 14.7
Determining the rejection region.

14.5.3 Application

To illustrate the Kruskal-Wallis rank test for differences in c medians, let us return to our productivity study of the previous section. We may recall that the production manager at the plant in which cereal is being filled in 368-gram sized boxes had been considering the replacement of an old machine that directly affects output in the production process and had conducted an experiment to determine whether there are any significant differences among the three brands of machines in the average time (in seconds) it takes factory workers using them to complete the filling process. Fifteen factory workers of similar experience, ability, and age had been randomly assigned to receive training on one of the three brands of machines in such a manner that there are five factory workers for each machine. After an appropriate amount of training and practice, the production manager measured the time (in seconds) it takes the factory workers to complete the filling process using their respective equipment. The results of this experiment were displayed in Table 14.2 on page 533 along with some summary computations and a scatter plot was depicted in Figure 14.5 on page 533 so that a visual, exploratory evaluation of potential trends, relationships, and violations in assumptions of particular testing procedures could be made. If the production manager did not wish to make the assumption that the time measurements (in seconds) were normally distributed across the underlying populations, the distribution-free Kruskal-Wallis rank test for differences in the three population medians could be used.

The null hypothesis to be tested is that the median times to complete the filling process on the three machines are equal; the alternative is that at least one of the machines differs from the others. Thus, substituting 1, 2, 3 for I, II, III we have

$$H_0\text{:}\quad M_1 = M_2 = M_3$$

$$H_1\text{:}\quad \text{Not all the medians are equal}$$

Converting the 15 time measurements in Table 14.2 (page 533) to ranks, we obtain Table 14.7.

Table 14.7 Converting data to ranks.

Machine		
I	II	III
14	9	2
15	6	7
12	10	1
11	8	4
13	5	3

Source: Data are taken from Table 14.2 on page 533.

We note that in the combined ranking, the third employee assigned to machine III completed the filling process fastest and received a rank of 1. The first employee assigned to machine III was second fastest and received a rank of 2. The second employee assigned to machine I completed the filling process in the slowest time and received a rank of 15.

After all the ranks are assigned, we then obtain the sum of the ranks for each group:

$$\text{Rank sums:} \quad T_1 = 65 \quad T_2 = 38 \quad T_3 = 17$$

As a check on the rankings we have

$$T_1 + T_2 + T_3 = \frac{n(n+1)}{2}$$

$$65 + 38 + 17 = \frac{(15)(16)}{2}$$

$$120 = 120$$

Choosing a .05 level of significance, to test the null hypothesis of equal population medians we use Equation (14.8):

$$H = \left[\frac{12}{n(n+1)} \sum_{j=1}^{c} \frac{T_j^2}{n_j}\right] - 3(n+1)$$

$$= \left\{\frac{12}{(15)(16)}\left[\frac{(65)^2}{5} + \frac{(38)^2}{5} + \frac{(17)^2}{5}\right]\right\} - 3(16)$$

$$= \left(\frac{12}{240}\right)[1,191.6] - 48$$

$$= 59.58 - 48 = 11.58$$

Using Table E.4, the upper-tail critical χ^2 value with $c - 1 = 2$ degrees of freedom and corresponding to a .05 level of significance is 5.991 (see Table 14.8, which is a replica of Table E.4). Since the computed value of the test statistic H exceeds the critical value, we may reject the null hypothesis and conclude that not all the machines were the same with respect to median time required for a worker to complete the filling process. (That is, if the null hypothesis were really true, the probability of obtaining such a result or one even more extreme is less than .05.)

We may note that these are the same results that were obtained using the one-way ANOVA F test in Section 14.4.

Table 14.8 Obtaining the approximate χ^2 critical value for the Kruskal-Wallis test at the .05 level of significance with 2 degrees of freedom.

Degrees of Freedom	Upper Tail Area									
	.995	.99	.975	.95	.90	.75	.25	.10	.05	.025
1	—	—	0.001	0.004	0.016	0.102	1.323	2.706	3.841	5.024
2	0.010	0.020	0.051	0.103	0.211	0.575	2.773	4.605	5.991	7.378
3	0.072	0.115	0.216	0.352	0.584	1.213	4.108	6.251	7.815	9.348
4	0.207	0.297	0.484	0.711	1.064	1.923	5.385	7.779	9.488	11.143
5	0.412	0.554	0.831	1.145	1.610	2.675	6.626	9.236	11.071	12.833

Source: Extracted from Table E.4.

14.5.4 Multiple Comparisons: Dunn's Procedure

Since we rejected the null hypothesis and concluded that there was evidence of a significant difference among the machines with respect to the median time it takes a worker to complete the filling process, the next step would be a simultaneous comparison of all possible pairs of machines to determine which one or ones differ from the others. As a follow up to the Kruskal-Wallis rank test, a *post hoc* or *a posteriori* multiple comparison procedure proposed by O. J. Dunn (see References 3 and 4) will be presented.

In general, with c groups or levels of a factor of interest, there are $c(c-1)/2$ possible pairwise comparisons to be made. The first step is to obtain the average rank $\bar{R}_j$ for each of the j groups (where $j = 1, \ldots, c$). We may recall from Equation (14.8) that when computing the test statistic H we obtained the total rank T_j for each group. Thus to compute the average rank $\bar{R}_j$ for the jth group we have

$$\bar{R}_j = \frac{T_j}{n_j} \quad \text{(where } j = 1, \ldots, c\text{)}$$

We then compute the differences $\bar{R}_j - \bar{R}_{j'}$ (where $j \neq j'$) among all $c(c-1)/2$ pairs of average ranks. The critical range for Dunn's procedure is obtained from

$$\text{critical range} = Z_U \sqrt{\frac{n(n+1)}{12}\left(\frac{1}{n_j} + \frac{1}{n_{j'}}\right)} \tag{14.9}$$

where n is the total number of observations in all groups combined, n_j and $n_{j'}$ are, respectively, the number of observations in groups j and j', and, for a selected α overall level of significance, Z_U is the critical value from a standardized normal distribution (Table E.2) containing an area of $\alpha/[c(c-1)]$ in the upper tail.[3]

If the sample sizes differ, the critical range must be computed for each pairwise comparison.

The final step is to compare the difference in each of the $c(c-1)/2$ pairs of average ranks against the corresponding critical range obtained from Equation (14.9). A specific pair of groups would be declared significantly different if the absolute differences in their corresponding average ranks exceeds the critical range.

To apply the Dunn procedure in our productivity study, we first obtain the average ranks over the three groups from the corresponding rank totals obtained from Table 14.7 on page 547:

$$\bar{R}_1 = 13.0 \quad \bar{R}_2 = 7.6 \quad \bar{R}_3 = 3.4$$

There are $(3)(3-1)/2 = 3$ pairwise comparisons to be made because there are three machines. The absolute differences in the average ranks are

1. $|\bar{R}_1 - \bar{R}_2| = |13.0 - 7.6| = 5.4$
2. $|\bar{R}_1 - \bar{R}_3| = |13.0 - 3.4| = 9.6$
3. $|\bar{R}_2 - \bar{R}_3| = |7.6 - 3.4| = 4.2$

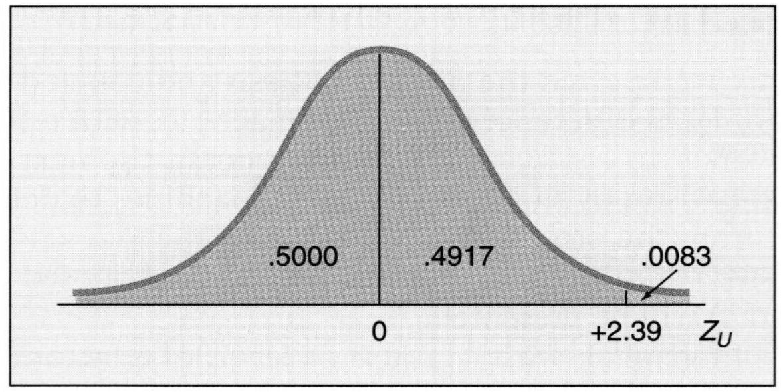

Figure 14.8
Obtaining critical values Z_U for upper-tail area α of .0083 used in Dunn's procedure to establish the critical range.

Since the three groups each have the same sample size, there is only one critical range that would be used in all possible comparisons. Choosing an overall level of significance of .05, we determine the critical range from Equation (14.9) by first obtaining $Z_U = +2.39$ (since the upper-tail area under the curve is .05/6 or .0083 as in Figure 14.8) so that

$$\begin{aligned}\text{critical range} &= Z_U\sqrt{\frac{n(n+1)}{12}\left(\frac{1}{n_j}+\frac{1}{n_{j'}}\right)}\\ &= 2.39\sqrt{\frac{(15)(16)}{12}\left(\frac{1}{5}+\frac{1}{5}\right)}\\ &= 6.76\end{aligned}$$

We note that only the second pairwise comparison (that is, $|\bar{R}_1 - \bar{R}_3| = |13.0 - 3.4| = 9.6$ exceeds the critical range of 6.76, so the production manager may conclude that machine III is significantly faster than machine I but there is no evidence of a significant difference between machines I and II as well as between II and III. We may recall that had the production manager utilized the parametric Tukey-Kramer procedure described in Section 14.4.5, he would have concluded that there were significant differences between all pairs of machines and thus machine III would be purchased because the average time for completing the filling process on it is shortest.

Owing to our conflicting results, the Tukey-Kramer procedure should be used after performing a one-way ANOVA F test if the assumption of normality in the three underlying populations is viable. If not, the Dunn procedure should be employed as a follow-up to the Kruskal-Wallis rank test.

Problems for Section 14.5

● 14.12 An industrial psychologist desires to test whether the reaction times of assembly-line workers are equivalent under three different learning methods. From a group of 25 new employees, nine are randomly assigned to method A, eight to method B, and eight to method C. The data at the top of page 551 present the rankings from 1 (fastest) to 25 (slowest) of the reaction times to complete a task given by the industrial psychologist after the learning period.

Method		
A	B	C
2	1	5
3	6	7
4	8	11
9	15	12
10	16	13
14	17	18
19	21	24
20	22	25
23		

Data file
INDPSYCH.DAT

Is there evidence of a difference in the reaction times for these learning methods? (Use $\alpha = .01$.)

14.13 A quality engineer in a company manufacturing electronic audio equipment was inspecting a new type of battery that was being considered for use. A batch of 20 batteries were randomly assigned to four groups (so that there were five batteries per group). Each group of batteries was then subjected to a particular pressure level—low, normal, high, very high. The batteries were simultaneously tested under these pressure levels and the times to failure (in hours) were recorded:

Data file
BATFAIL.DAT

Pressure			
Low	Normal	High	Very High
8.0	7.6	6.0	5.1
8.1	8.2	6.3	5.6
9.2	9.8	7.1	5.9
9.4	10.9	7.7	6.7
11.7	12.3	8.9	7.8

The quality engineer, by experience, knows such data are coming from populations that are not normally distributed and wants to use a distribution-free procedure for purposes of data analysis.

(a) At the .05 level of significance, completely analyze the data to determine whether there is evidence of a difference in the four pressure levels with respect to median battery life and, if so, which of the pressure levels are worst for battery operation.

(b) ACTION Write a memo to the quality engineer expressing your findings.

(c) Recommend a warranty policy with respect to battery life.

14.14 The quality control engineer in a plant manufacturing stereo equipment wanted to study the effect of temperature on the failure time of a particular electronic component. She designed an experiment wherein 24 of these components, all from the same batch, were randomly assigned to one of three levels of temperature and then simultaneously activated. Recorded at the top of page 552 are the rank order of their times to failure (i.e., a rank of 1 is given to the first component to burn out).

Temperature		
150°F	200°F	250°F
4	2	1
7	8	3
10	11	5
13	12	6
18	17	9
21	19	14
22	20	15
24	23	16

Data file
COMPFAIL.DAT

(a) At the .05 level of significance, is there evidence of a temperature effect on the life of this type of electronic component?

(b) ACTION Draft a report to the quality control engineer based on your findings.

14.15 Use the Kruskal-Wallis rank test and, if appropriate, the Dunn procedure to answer parts (f) and (g) of Problem 14.5 (performance scores) on page 542. (Use $\alpha = .05$.) Are there any differences in your present results from those previously obtained? Discuss.

14.16 Use the Kruskal-Wallis rank test and, if appropriate, the Dunn procedure to answer part (a) of Problem 14.7 (mathematics test scores) on page 543. (Use $\alpha = .05$.) Are there any differences in your present results from those previously obtained? Discuss.

14.17 Use the Kruskal-Wallis rank test and, if appropriate, the Dunn procedure to answer part (a) of Problem 14.9 (profitability) on page 544. (Use $\alpha = .01$.) Are there any differences in your present results from those previously obtained? Discuss.

14.6 Using the Computer for Hypothesis Testing with *c* Independent Samples: The Kalosha Industries Employee Satisfaction Survey

14.6.1 Introduction and Overview

When dealing with large batches of data, we may use the computer to assist us not only in our descriptive statistical analysis but in our confirmatory analysis as well. In this section we will demonstrate how statistical software packages can be used for performing various tests of hypotheses when we are analyzing differences in the outcomes across several groups of numerical data. To accomplish this, let us return to the Kalosha Industries Employee Satisfaction Survey that was developed in Chapter 2.

Data file
EMPSAT.DAT

14.6.2 Kalosha Industries Employee Satisfaction Survey

Bud Conley, the Vice-President of Human Resources, is preparing for yet another meeting with a representative from the B & L Corporation to discuss the potential contents of an employee-benefits package that is being developed. Prior to this meeting, he would like to obtain answers to the following questions through *confirmatory analyses* based on the results of the Employee Satisfaction Survey (see Table 2.3 on pages 33–40):

1. Was there evidence of a difference in average work hours per week (question 1) based on the occupational grouping of all full-time employees at Kalosha Industries (question 2)?
2. Was there evidence of a difference in the median personal incomes of full-time employees at Kalosha Industries (question 7) based on their level of participation in decisions affecting their work (question 21)?

Responses to these two issues as well as others raised by Bud Conley (see Survey/Database Project at the end of the section) require a detailed descriptive analysis of the 400 responses to the Survey along with a confirmatory analysis.

14.6.3 Using Statistical Packages for Evaluating Differences Among *c* Groups

In response to Bud Conley's first question, whether there are significant differences in average work hours per week based on occupational grouping, a descriptive analysis will be performed so that the assumptions of the test procedures can be evaluated.

Figure 14.9 provides Microsoft EXCEL for Windows output (Reference 8) of some of the summary statistics pertaining to a breakdown of work hours by occupational grouping. Although not shown here, stem-and-leaf displays, box-and-whisker plots, and normal probability plots were also obtained and indicated some departure from normality in the underlying populations.

	Occup	N	MEAN	MEDIAN	STDEV	SEMEAN
workhrs	1=M	65	46.00	43.00	9.14	1.13
	2=P	66	45.14	40.00	9.11	1.12
	3=T	57	46.72	40.00	12.51	1.66
	4=A	65	42.23	40.00	8.00	0.99
	5=S	29	43.52	40.00	8.65	1.61
	6=R	54	48.11	41.00	11.84	1.61
	7=L	64	45.88	43.00	9.62	1.20

Figure 14.9
Microsoft EXCEL for Windows output displaying summary statistics.

To test the homogeneity of variance assumption, the Hartley F_{max}-test statistic, representing the ratio of the largest sample variance to the smallest sample variance, was computed from the printout of Figure 14.9. This test must be used cautiously here because it is sensitive to a violation in the assumption of underlying normality. (Some alternative procedures are given in References 1–3.) Using a .05 level of significance, since

$$F_{\max} = \frac{(12.51)^2}{(8.00)^2} = 2.45 > F_{\max[7,57]} \cong 2.26,$$

the upper-tailed critical value obtained by interpolation in Table E.8, the null hypothesis of equal population variances is rejected.

In order to use the one-way ANOVA F test an appropriate transformation must be found that would both normalize the data and stabilize the variances across the groups. Following the use of a natural logarithmic transformation on the original variable (work hours), Figure 14.10 displays MINITAB (Reference 10) output for the one-way ANOVA F test. A check on the homogeneity of variance assumption indicates that the variability in the transformed data has stabilized. That is, using a .05 level of significance, since

$$F_{max} = \frac{(0.2283)^2}{(0.1629)^2} = 1.97 < F_{max[7,57]} \cong 2.26,$$

the null hypothesis cannot be rejected and it may be concluded that there is no evidence of a lack of homogeneity of variances for the transformed data.

Using a .05 level of significance, we may test the null hypothesis regarding the equality of the seven population means

$$H_0: \mu_1 = \mu_2 = \cdots = \mu_7$$

against the alternative

$$H_1: \text{ Not all the } \mu_j \text{ are equal } (j = 1, 2, \ldots, 7)$$

```
MTB > let c36 = loge(c2)
MTB > oneway c36 by c3;
SUBC> tukey.

ANALYSIS OF VARIANCE ON ln(workhrs)
SOURCE      DF        SS         MS          F        p
Occup        6    0.5224     0.0871       2.22    0.041
ERROR      393   15.4160     0.0392
TOTAL      399   15.9383
                                       INDIVIDUAL 95 PCT CI'S FOR MEAN
                                       BASED ON POOLED STDEV
 LEVEL       N      MEAN      STDEV    -------+---------+---------+---------
  1=M       65    3.8105     0.1890                  (-------*-------)
  2=P       66    3.7905     0.1955               (-------*-------)
  3=T       57    3.8158     0.2283                  (--------*--------)
  4=A       65    3.7289     0.1629    (-------*--------)
  5=S       29    3.7572     0.1746     (-----------*-----------)
  6=R       54    3.8472     0.2249                        (--------*--------)
  7=L       64    3.8059     0.1990                 (-------*-------)
                                       -------+---------+---------+---------
POOLED STDEV =    0.1981                    3.720     3.780     3.840

Tukey's pairwise comparisons

    Family error rate = 0.0500
Individual error rate = 0.00338

Critical value = 4.17
```

Figure 14.10
MINITAB output of one-way ANOVA *F* test following a natural logarithmic data transformation.

We observe from the MINITAB output of Figure 14.10 that the one-way F-test statistic, 2.22, exceeds the critical value, $F_{U(6,393)} \cong 2.10$ (taken from Table E.5), so the null hypothesis may be rejected. There is evidence of a significant difference among the means of the transformed data pertaining to weekly work hours for the various occupational groupings. The p value is .041. Using the results shown in Figure 14.10 and employing the Tukey-Kramer procedure, it is found that the significant difference in the natural logarithmic transformed weekly work hours data occurs between the administrative support group (A) and the production and repair group (R). There is evidence that employees in administrative support work significantly fewer hours per week than do those in production and repair.

In response to Bud Conley's second question, whether there are significant differences in median personal income based on employee participation in work-related decisions, another descriptive analysis will be undertaken so that the assumptions of the test procedures can be evaluated.

Figure 14.11 presents output illustrating the box-and-whisker plots obtained from accessing STATISTIX (Reference 13). These plots indicate a right-skewness in the sampled populations. Although not shown, stem-and-leaf displays and normal probability plots were also obtained and confirmed the right-skewness present in the sampled populations.

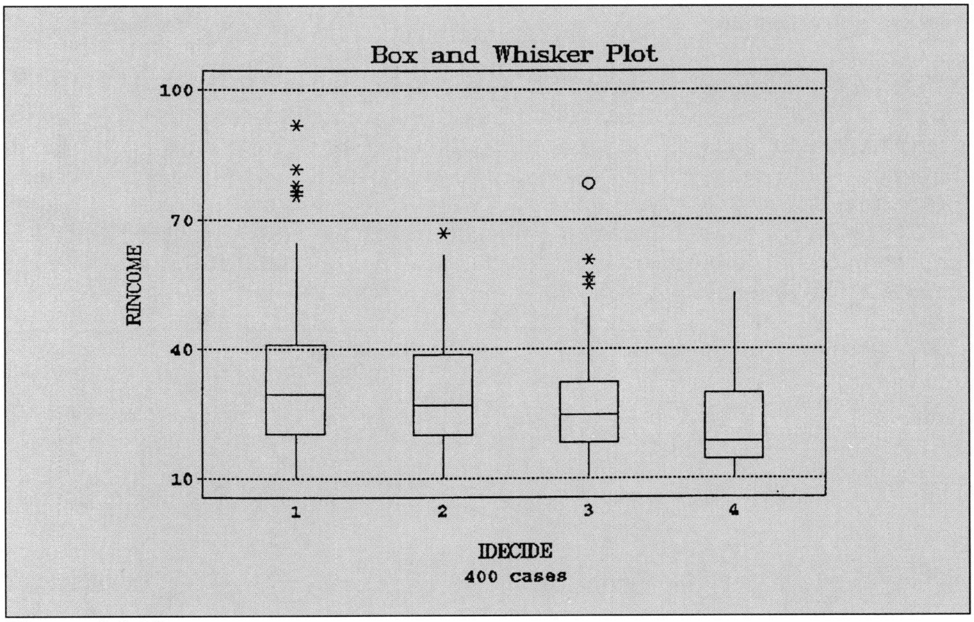

Figure 14.11
STATISTIX output displaying box-and-whisker plots of personal income based on level of perceived participation in work-related decisions.

To test for possible differences in the four population medians, SAS is accessed (Reference 12) and the Kruskal-Wallis rank test is used. Figure 14.12 on page 556 presents partial SAS output illustrating the results of the test. Using a .05 level of significance, to test the null hypothesis of equality of median personal income among the four levels of perceived decision-making participation

$$H_0: \quad M_1 = M_2 = M_3 = M_4$$

against the alternative

$$H_1: \quad \text{Not all the } M_j \text{ are equal (where } j = 1, \ldots, 4)$$

we observe that the H-test statistic, 21.172, exceeds 7.815, the upper-tailed critical value from the chi-square distribution with 3 degrees of freedom (see Table E.4). The p value is .0001. Therefore we reject the null hypothesis and we may conclude that there is evidence of a difference among the median employee personal incomes based on perceived level of participation in work-related decision making.

```
                              The SAS System

                     N P A R 1 W A Y   P R O C E D U R E

              Wilcoxon Scores (Rank Sums) for Variable RINCOME
                     Classified by Variable IDECIDE

                            Sum of       Expected        Std Dev          Mean
IDECIDE          N          Scores       Under H0       Under H0         Score

ALWAYS         138      30483.5000     27669.0000     1099.16981    220.894928
MUCH           157      32817.5000     31478.5000     1129.08741    209.028662
SOMETIMES       60      10789.0000     12030.0000      825.63802    179.816667
NEVER           45       6110.0000      9022.5000      730.62585    135.777778
                        Average Scores were used for Ties

          Kruskal-Wallis Test (Chi-Square Approximation)
          CHISQ=  21.172       DF=  3      Prob > CHISQ=       0.0001
```

Figure 14.12
Partial SAS output illustrating the Kruskal-Wallis rank test for differences in c medians.

From the output of Figure 14.12, we may follow up on the Kruskal-Wallis test by extracting the average ranks, obtaining the absolute differences in all pairwise combinations of groups, and computing the corresponding critical ranges [from Equation (14.9) on page 549] necessary to employ Dunn's multiple comparison procedure.

Using an overall .05 level of significance for these data, Z_U is +2.64 since the area contained in the upper tail of the standardized normal distribution is .05/12 or .0042. Thus we have

Absolute Differences	Critical Range	Decision
$\lvert\bar{R}_1 - \bar{R}_2\rvert = \lvert 220.9 - 209.0\rvert = 11.9$	$2.64\sqrt{\frac{(400)(401)}{12}\left(\frac{1}{138}+\frac{1}{157}\right)} = 35.6$	Not significant
$\lvert\bar{R}_1 - \bar{R}_3\rvert = \lvert 220.9 - 179.8\rvert = 41.1$	$2.64\sqrt{\frac{(400)(401)}{12}\left(\frac{1}{138}+\frac{1}{60}\right)} = 47.2$	Not significant
$\lvert\bar{R}_1 - \bar{R}_4\rvert = \lvert 220.9 - 135.8\rvert = 85.1$	$2.64\sqrt{\frac{(400)(401)}{12}\left(\frac{1}{138}+\frac{1}{45}\right)} = 52.4$	Significant
$\lvert\bar{R}_2 - \bar{R}_3\rvert = \lvert 209.0 - 179.8\rvert = 29.2$	$2.64\sqrt{\frac{(400)(401)}{12}\left(\frac{1}{157}+\frac{1}{60}\right)} = 46.3$	Not significant
$\lvert\bar{R}_2 - \bar{R}_4\rvert = \lvert 209.0 - 135.8\rvert = 73.2$	$2.64\sqrt{\frac{(400)(401)}{12}\left(\frac{1}{157}+\frac{1}{45}\right)} = 51.6$	Significant
$\lvert\bar{R}_3 - \bar{R}_4\rvert = \lvert 179.8 - 135.8\rvert = 44.0$	$2.64\sqrt{\frac{(400)(401)}{12}\left(\frac{1}{60}+\frac{1}{45}\right)} = 60.2$	Not significant

We should inform Bud Conley that the median personal incomes are significantly different with respect to perceived participation in work-related decisions. In particular, employees who state they always participate have a significantly higher median personal income than do workers who say they never participate and employees who say they participate much of the time also have a significantly higher median personal income than do workers who claim they never participate. All other differences between paired groups are due to chance.

Survey/Database Project for Section 14.6

The following problems refer to the sample data obtained from the questionnaire of Figure 2.6 on pages 28–29 and presented in Table 2.3 on pages 33–40. They should be solved with the aid of an available computer package.

Data file EMPSAT.DAT

Suppose you are hired as a research assistant to Bud Conley, the Vice-President of Human Resources at Kalosha Industries. He has given you a list of questions (see Problems 14.18 to 14.30) that he needs answered prior to his meeting with a representative from the B&L Corporation, the employee-benefits consulting firm that he has retained. A confirmatory statistical analysis based on the answers to these questions pertaining to the numerical variables in the Employee Satisfaction Survey will provide him with a better understanding of the composition of the Kalosha Industries full-time workforce and assist him in his deliberations with the B&L Corporation toward the development of an employee-benefits package.

From the responses to the questions dealing with numerical variables in the Employee Satisfaction Survey (see pages 33–40) in Problems 14.18 through 14.30 which follow,

(a) Obtain a set of descriptive statistics for each group.
(b) Develop the stem-and-leaf display for each group.
(c) Form the box-and-whisker plot for each group.
(d) Develop the normal probability plot for each group.
(e) Based on a descriptive analysis of the findings in (a) through (d) with respect to the assumptions of the one-way ANOVA F test and Kruskal-Wallis rank test, select an appropriate procedure and carry out the testing of hypotheses at the $\alpha = .05$ level of significance.
(f) ACTION Write a memo to Bud Conley discussing your test selection and your findings.

14.18 Is there evidence of a difference in the average weekly hours that full-time employees of Kalosha Industries work (see question 1) based on important job characteristics (see question 11)?

14.19 Is there evidence of a difference in the average weekly hours that full-time employees of Kalosha Industries work (see question 1) based on perceptions for getting ahead (see question 12)?

14.20 Is there evidence of a difference in the average weekly hours that full-time employees of Kalosha Industries work (see question 1) based on perceived participation in work-related decisions (see question 21)?

14.21 Is there evidence of a difference in the average weekly hours that full-time employees of Kalosha Industries work (see question 1) based on how proud they feel about being part of the organization (see question 23)?

14.22 Is there evidence of a difference in the average personal income of full-time employees at Kalosha Industries (see question 7) based on occupational grouping (see question 2)?

14.23 Is there evidence of a difference in the average personal income of full-time employees at Kalosha Industries (see question 7) based on important job characteristics (see question 11)?

14.24 Is there evidence of a difference in the average personal income of full-time employees at Kalosha Industries (see question 7) based on perceptions for getting ahead (see question 12)?

14.25 Is there evidence of a difference in the average personal income of full-time employees at Kalosha Industries (see question 7) based on advancement in the organization (see question 20)?

14.26 Is there evidence of a difference in the average personal income of full-time employees at Kalosha Industries (see question 7) based on perceived participation in work-related decisions (see question 21)?

14.27 Is there evidence of a difference in the average personal income of full-time employees at Kalosha Industries (see question 7) based on how proud they feel about being part of the organization (see question 23)?

14.28 Is there evidence of a difference in the average personal income of full-time employees at Kalosha Industries (see question 7) based on perception of importance of schooling (see question 27)?

14.29 Is there evidence of a difference in longevity (i.e., average time spent working on a full-time basis at Kalosha Industries—see question 16) based on job satisfaction (question 9)?

14.30 Is there evidence of a difference in employee longevity (i.e., average time spent working on a full-time basis at Kalosha Industries—see question 16) based on how proud they feel about being part of the organization (see question 23)?

14.7 The Randomized Block Model

In Section 14.4 we developed the one-way ANOVA *F* test to evaluate differences in the means of *c* groups and in Section 14.5 we considered the Kruskal-Wallis rank test to evaluate differences in the medians of the *c* groups. Such testing procedures as these would be employed in experimental situations in which *n* homogeneous items or individuals (i.e., *experimental units*) are randomly assigned to the *c* levels of a factor of interest (i.e., the *treatment* groups). As we may recall from Section 14.3, such designed one-factor experiments are referred to as *one-way* or *completely randomized design models.*

Alternatively, in Sections 13.8–13.10 we used the *t* test for the mean difference or the Wilcoxon signed-ranks test for the median difference in situations involving repeated measurements or matched samples in order to evaluate differences between two treatment conditions. Suppose, now, we wish to extend this to situations in which there are more than two treatment groups or levels of a factor of interest. In such cases, the heterogeneous sets of items or individuals that have been matched (or on whom repeated measurements have been taken) are called **blocks.** Numerical data may then be obtained as the response or outcome of each treatment group and block combination. Thus, in designing experiments of this type, there would be two things to consider: treatments and blocks. However, with respect to our tests of hypotheses, we focus on the differences among the *c* levels of the factor of interest (i.e., the treatment groups).

Experimental situations such as this are referred to as **randomized block design models**. The purpose of blocking is to remove as much variability as possible so that we may focus on differences among the c treatment conditions. Thus, when appropriate, the purpose of selecting a randomized block design model instead of a completely randomized design model is to provide a more efficient analysis by reducing the experimental error and thereby obtaining more precise results (See References 1, 5, 6, and 11).

14.8 Randomized-Block *F* Test for Differences in *c* Means

14.8.1 Introduction

As in Section 14.4, when the outcome measurements across the c groups are continuous and certain assumptions are met, a methodology known as *analysis of variance* (or *ANOVA*) may be used to compare the means of the groups. In this section we will extend our earlier discussion of ANOVA and develop the **randomized block *F* test** in order to evaluate differences among the means in c groups.

We recall from Figure 14.1 on page 528 that in the completely randomized or one-way analysis-of-variance model, the total variation in the outcome measurements (SST) is subdivided into that which is attributable to differences *among* the c groups (SSA) and that which is due to chance or attributable to inherent variation *within* the c groups (SSW). Within-group variation is considered experimental error, and among-group variation is attributable to treatment effects.

For the randomized block design model, in order to filter out the effects of the blocking, we need to further subdivide the within-group variation (SSW) into that which is attributable to differences among the blocks (SSBL) and that which is attributable to inherent random error (SSE). Therefore, as presented in Figure 14.13, in a randomized block design model the total variation in the outcome measurements is the summation of three components—among-group variation (SSA), among-block variation (SSBL), and inherent random error (SSE).

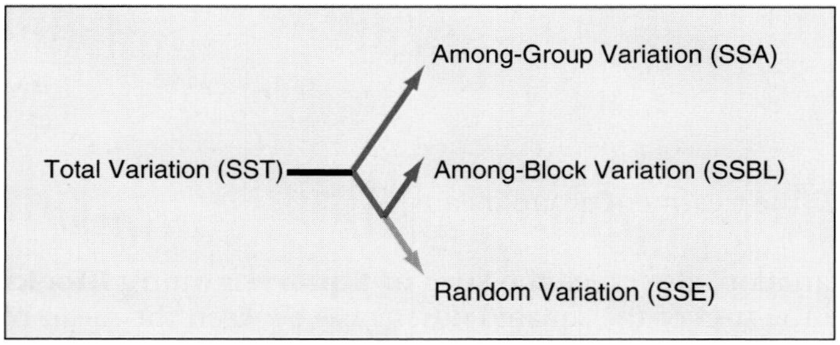

Figure 14.13 Partitioning the total variation in a randomized block design model.

14.8.2 Development

To develop the ANOVA procedure for the randomized block design model we need to define the terms at the top of page 560:

X_{ij} = the value in the ith block for the jth treatment group

$\bar{X}_{i.}$ = the mean of all the values in block i

$\bar{X}_{.j}$ = the mean of all the values for treatment group j

$\sum_{j=1}^{c}\sum_{i=1}^{r} X_{ij}$ = the summation of the values over all blocks and all groups—i.e., the grand total

r = the number of blocks

c = the number of groups

n = the total number of observations (where $n = rc$)

The total variation, also called **Sum of Squares Total (SST)**, is a measure of the variation among all the observations. It can be obtained by summing the squared differences between each individual observation and the overall or grand mean $\bar{\bar{X}}$ that is based on all n observations. SST would be computed as

$$\text{SST} = \sum_{j=1}^{c}\sum_{i=1}^{r}\left(X_{ij} - \bar{\bar{X}}\right)^2 \qquad \textbf{(14.10)}$$

where $\bar{\bar{X}} = \dfrac{\sum_{j=1}^{c}\sum_{i=1}^{r} X_{ij}}{rc}$ (i.e., the overall or grand mean)

The among-group variation, also called the **Sum of Squares Among** groups **(SSA)**, is measured by the sum of the squared differences between the sample mean of each group $\bar{X}_{.j}$ and the overall or grand mean $\bar{\bar{X}}$, weighted by the number of blocks r. The among-group variation is computed from

$$\text{SSA} = r\sum_{j=1}^{c}\left(\bar{X}_{.j} - \bar{\bar{X}}\right)^2 \qquad \textbf{(14.11)}$$

where $\bar{X}_{.j} = \dfrac{\sum_{i=1}^{r} X_{ij}}{r}$ (i.e., the treatment group means)

The among-block variation, also called the **Sum of Squares among Blocks (SSBL)**, is measured by the sum of the squared differences between the mean of each block $\bar{X}_{i.}$ and the overall or grand mean $\bar{\bar{X}}$, weighted by the number of groups c. The among-block variation is computed from

$$\text{SSBL} = c\sum_{i=1}^{r}\left(\bar{X}_{i.} - \bar{\bar{X}}\right)^2 \qquad \textbf{(14.12)}$$

$$\text{where } \bar{X}_{i.} = \frac{\sum_{j=1}^{c} X_{ij}}{c} \text{ (i.e., the block means)}$$

The **inherent random variation,** also called the **Sum of Squares Error (SSE),** is measured by the sum of the squared differences among all the observations after the effect of the particular treatments and blocks have been accounted for. SSE would be computed from

$$SSE = \sum_{j=1}^{c} \sum_{i=1}^{r} \left(X_{ij} - \bar{X}_{.j} - \bar{X}_{i.} + \bar{\bar{X}}\right)^2 \quad \textbf{(14.13)}$$

Since there are c treatment levels of the factor being compared, there are $c - 1$ degrees of freedom associated with the sum of squares among groups (SSA). Similarly, since there are r blocks, there are $r - 1$ degrees of freedom associated with the sum of squares among blocks (SSBL). Moreover, there are $n - 1$ degrees of freedom associated with the sum of squares total (SST) because each observation X_{ij} is being compared to the overall or grand mean $\bar{\bar{X}}$ based on all n observations. Therefore, since the degrees of freedom for each of the sources of variation must add to the degrees of freedom for the total variation, we may obtain the degrees of freedom for the sum of squares error (SSE) component by subtraction and algebraic manipulation.[4] The degrees of freedom is given by $(r - 1)(c - 1)$.

If each of the component sums of squares is divided by its associated degrees of freedom, we obtain the three *variances* or **mean square** terms (**MSA**, **MSBL**, and **MSE**) needed for ANOVA:

$$MSA = \frac{SSA}{c - 1} \quad \textbf{(14.14a)}$$

$$MSBL = \frac{SSBL}{r - 1} \quad \textbf{(14.14b)}$$

$$MSE = \frac{SSE}{(r - 1)(c - 1)} \quad \textbf{(14.14c)}$$

If the assumptions pertaining to the analysis of variance are met, the null hypothesis of no differences in the population means (i.e., no treatment effects)

$$H_0: \mu_{.1} = \mu_{.2} = \cdots = \mu_{.c}$$

may be tested against the alternative that not all the c population means are equal

$$H_1: \text{Not all } \mu_{.j} \text{ are equal (where } j = 1, 2, \ldots, c)$$

by computing the test statistic F as

$$F = \frac{\text{MSA}}{\text{MSE}} \tag{14.15}$$

The F-test statistic follows an F distribution with $c - 1$ and $(r - 1)(c - 1)$ degrees of freedom. For a given level of significance, α, we may reject the null hypothesis if the computed F-test statistic exceeds the upper-tailed critical value $F_{U[c-1,(r-1)(c-1)]}$ from the F distribution (see Table E.5). That is, we have the following decision rule:

Reject H_0 if $F > F_{U[c-1,(r-1)(c-1)]}$;

otherwise don't reject H_0.

To examine whether it is advantageous to block, some researchers suggest that the test of the null hypothesis of no block effects be performed. Thus we may test

$$H_0: \quad \mu_{1.} = \mu_{2.} = \cdots = \mu_{r.}$$

against the alternative

$$H_1: \quad \text{not all } \mu_{i.} \text{ are equal}$$

We form the F statistic

$$F = \frac{\text{MSBL}}{\text{MSE}} \tag{14.16}$$

and the null hypothesis would be rejected at the α level of significance if

$$F = \frac{\text{MSBL}}{\text{MSE}} > F_{U[r-1,(r-1)(c-1)]}$$

However, it may be argued that this is unnecessary, that the sole purpose of establishing the blocks was to provide a more efficient means of testing for treatment effects by reducing the experimental error.[5]

As in Section 14.4.2, the results of an analysis-of-variance procedure are usually displayed in an ANOVA summary table, the format for which is presented in Table 14.9.

14.8.3 Application

To illustrate the randomized block F test, suppose that a fast-food chain having four branches in a particular geographical area would like to evaluate the service at these restaurants. The research director for the chain hires 24 investigators (raters) with varied experiences in food-service evaluations. After preliminary consultations, the 24 investigators are stratified into six blocks of four—based on food-

Table 14.9 Analysis-of-variance table for the randomized block design.

Source	Degrees of Freedom	Sums of Squares	Mean Square (Variance)	F
Among treatments	$c - 1$	$SSA = r\sum_{j=1}^{c}(\bar{X}_{.j} - \bar{\bar{X}})^2$	$MSA = \frac{SSA}{c-1}$	$F = \frac{MSA}{MSE}$
Among blocks	$r - 1$	$SSBL = c\sum_{i=1}^{r}(\bar{X}_{i.} - \bar{\bar{X}})^2$	$MSBL = \frac{SSBL}{r-1}$	$F = \frac{MSBL}{MSE}$
Error	$(r-1)(c-1)$	$SSE = \sum_{j=1}^{c}\sum_{i=1}^{r}(X_{ij} - \bar{X}_{.j} - \bar{X}_{i.} + \bar{\bar{X}})^2$	$MSE = \frac{SSE}{(r-1)(c-1)}$	
Total	$rc - 1$	$SST = \sum_{j=1}^{c}\sum_{i=1}^{r}(X_{ij} - \bar{\bar{X}})^2$		

service evaluation experience—so that the four most experienced investigators are placed in block 1, the next four most experienced investigators are placed in block 2, and so on. Within each of the six homogeneous blocks, the four raters are then randomly assigned to evaluate the service at a particular restaurant using a rating scale from 0 (low) to 100 (high). The results are summarized in Table 14.10. The group totals, group means, block totals, block means, grand total, and grand mean are also presented in Table 14.10 and some of these statistics are highlighted along with the original data ratings in the scatter plot of Figure 14.14 (on page 564) to provide a visual impression of the results of the experiment.

Table 14.10 Restaurant ratings for four branches of a fast-food chain.

Blocks of Raters	Restaurants A	B	C	D	Totals	Means
1	70	61	82	74	287	71.75
2	77	75	88	76	316	79.00
3	76	67	90	80	313	78.25
4	80	63	96	76	315	78.75
5	84	66	92	84	326	81.50
6	78	68	98	86	330	82.50
Totals	465	400	546	476	1,887	
Means	77.50	66.67	91.00	79.33		78.625

In addition, from Table 14.10 we have

$$r = 6 \quad c = 4 \quad n = rc = 24$$

and, as stated

$$\bar{\bar{X}} = \frac{\sum_{j=1}^{c}\sum_{i=1}^{r} X_{ij}}{rc} = \frac{1,887}{24} = 78.625$$

Although it is strongly suggested that available statistical software packages (References 10, 12, and 13) be accessed to determine the results of a randomized

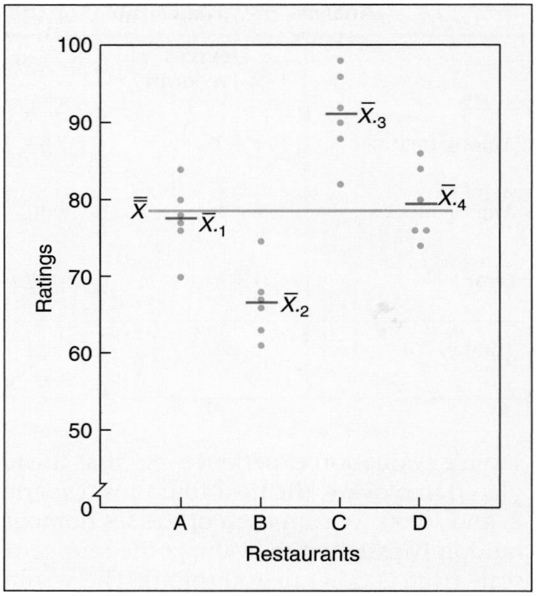

Figure 14.14
Scatter plot of ratings of services rendered at four restaurants.
Source: Table 14.10.

block design experiment, for illustrative purposes we set up the following computations:

Using Equation (14.10),

$$SST = \sum_{j=1}^{c}\sum_{i=1}^{r}(X_{ij}-\bar{\bar{X}})^2 = (70-78.625)^2 + (77-78.625)^2 + \cdots + (86-78.625)^2$$
$$= 2,295.63$$

Using Equation (14.11),

$$SSA = r\sum_{j=1}^{c}(\bar{X}_{.j}-\bar{\bar{X}})^2 = 6\left[(77.50-78.625)^2+(66.67-78.625)^2+\cdots+(79.33-78.625)^2\right]$$
$$= 1,787.46$$

Using Equation (14.12),

$$SSBL = c\sum_{i=1}^{r}(\bar{X}_{i.}-\bar{\bar{X}})^2 = 4\left[(71.75-78.625)^2+(79.00-78.625)^2+\cdots+(82.50-78.625)^2\right]$$
$$= 283.38$$

Using Equation (14.13),

$$\begin{aligned} SSE = \sum_{j=1}^{c}\sum_{i=1}^{r}(X_{ij}-\bar{X}_{.j}-\bar{X}_{i.}+\bar{\bar{X}})^2 &= (70-77.50-71.75+78.625)^2 \\ &+ (77-77.50-79.00+78.625)^2 \\ &+ \\ &\vdots \\ &+ (86-79.33-82.50+78.625)^2 \\ &= 224.79 \end{aligned}$$

Thus, using Equations (14.14a), (14.14b), and (14.14c),

$$MSA = \frac{SSA}{c - 1} = \frac{1,787.46}{3} = 595.820$$

$$MSBL = \frac{SSBL}{r - 1} = \frac{283.38}{5} = 56.676$$

$$MSE = \frac{SSE}{(r - 1)(c - 1)} = \frac{224.79}{15} = 14.986$$

In the fast-food chain study the computations just completed can be summarized in the analysis-of-variance table shown in Table 14.11.

Table 14.11 Analysis-of-variance table for the fast-food chain study.

Source	Degrees of Freedom	Sums of Squares	Mean Square (Variance)	F
Among groups (branches)	$4 - 1 = 3$	1,787.46	$MSA = \frac{1,787.46}{3} = 595.820$	$F = \frac{595.820}{14.986} = 39.758$
Among blocks (raters)	$6 - 1 = 5$	283.38	$MSBL = \frac{283.38}{5} = 56.676$	$F = \frac{56.676}{14.986} = 3.782$
Error	$(6 - 1)(4 - 1) = 15$	224.79	$MSE = \frac{224.79}{15} = 14.986$	
Total	$(6)(4) - 1 = 23$	2,295.63		

In addition to the entries in this table, the p value (i.e., the probability of obtaining an F statistic as large as or larger than the one obtained, given that the null hypothesis is true) is included in the ANOVA table of most statistical software packages. Since most individuals who need to evaluate data from randomized block design models would be performing their analysis by accessing statistical software, the inclusion of the p value enables us to make direct conclusions about the null hypothesis without referring to a table of critical values of the F distribution. If the p value is less than the specified level of significance α, the null hypothesis is rejected. Here, however, we will analyze the data that have been summarized in Table 14.11.

Substituting 1, 2, 3, 4 for A, B, C, D when testing for differences between the restaurants and using the .05 level of significance, the decision rule would be to reject the null hypothesis (H_0: $\mu_{.1} = \mu_{.2} = \mu_{.3} = \mu_{.4}$) if the calculated F value exceeds 3.29 (see Figure 14.15 on page 566). Since $F = 39.758 > F_{U(3,15)} = 3.29$, we may reject H_0 and conclude that there is evidence of a difference in the average rating between the different restaurants.

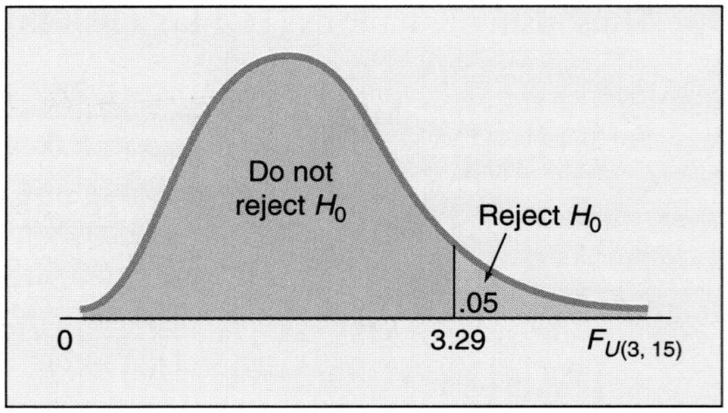

Figure 14.15
Regions of rejection and nonrejection for the fast food chain study at the .05 level of significance with 3 and 15 degrees of freedom.

As a check on the effectiveness of blocking, we may test for a difference among the groups of raters. The decision rule, using the .05 level of significance, would be to reject the null hypothesis (H_0: $\mu_{1.} = \mu_{2.} = \cdots = \mu_{6.}$) if the calculated F value exceeds 2.90 (see Figure 14.16). Since $F = 3.782 > F_{U(5,15)} = 2.90$, we may reject H_0 and conclude that there is evidence of a difference among the groups of raters. Thus we may conclude that the blocking has been advantageous in reducing the experimental error.

In addition to the assumptions of the one-way analysis of variance previously mentioned in Section 14.4.6, we need to assume that there is no *interacting effect* between the treatments and the blocks. That is, we need to assume that any differences between the treatments (the restaurants) are consistent across the entire set of blocks (the groups of raters). The concept of *interaction* will be discussed in Section 14.10.4.

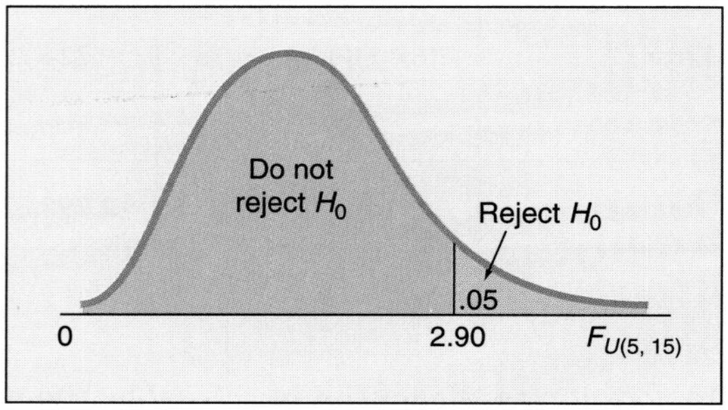

Figure 14.16
Regions of rejection and nonrejection for the fast food chain study at the .05 level of significance with 5 and 15 degrees of freedom.

14.8.4 Multiple Comparisons: The Tukey Procedure

As in the case of the one-way ANOVA model, once the null hypothesis of no differences between the treatment groups has been rejected, we need to determine which of these treatment groups are significantly different from each other. For the randomized block design model, since the sample sizes for each *treatment* group are equal, we use a procedure developed by John Tukey (see References 6 and 14). The critical range for the Tukey procedure is given in Equation (14.17):

$$\text{critical range} = Q_{U[c,(r-1)(c-1)]}\sqrt{\frac{\text{MSE}}{r}} \quad \textbf{(14.17)}$$

Each of the $c(c - 1)/2$ pairs of means is compared against the one critical range. A specific pair would be declared different if the absolute difference in the sample means $|\bar{X}_{.j} - \bar{X}_{.j'}|$ exceeds this critical range.

To apply the Tukey procedure, we return to our fast-food-chain study. Since there are four restaurants, there are $(4)(4 - 1)/2 = 6$ possible pairwise comparisons to be made. From Table 14.10 on page 563, the absolute mean differences are

1. $|\bar{X}_{.1} - \bar{X}_{.2}| = |77.50 - 66.67| = 10.83$
2. $|\bar{X}_{.1} - \bar{X}_{.3}| = |77.50 - 91.00| = 13.50$
3. $|\bar{X}_{.1} - \bar{X}_{.4}| = |77.50 - 79.33| = 1.83$
4. $|\bar{X}_{.2} - \bar{X}_{.3}| = |66.67 - 91.00| = 24.33$
5. $|\bar{X}_{.2} - \bar{X}_{.4}| = |66.67 - 79.33| = 12.66$
6. $|\bar{X}_{.3} - \bar{X}_{.4}| = |91.00 - 79.33| = 11.67$

To determine the critical range, we use Table 14.11 on page 565 to obtain MSE = 14.986 and $r = 6$. From Table E.12 [for $\alpha = .05$, $c = 4$, and $(r - 1)(c - 1) = (6 - 1)(4 - 1) = 15$], the upper-tailed critical value, $Q_{U[4,15]}$, is 4.08. Therefore, using Equation (14.17) we have

$$\text{critical range} = 4.08\sqrt{\frac{14.986}{6}} = 6.448$$

We note that all contrasts except $|\bar{X}_{.1} - \bar{X}_{.4}|$ are greater than the critical range. Therefore we may conclude that there is evidence of a significant difference in the average rating between all pairs of restaurant branches except for branches A and D. Moreover, branch C has the highest ratings (i.e., is most preferred) and branch B has the lowest (i.e., is least preferred).

14.8.5 Comparing the Randomized Block Design to the One-Way (Completely Randomized) Design

Now that we have developed the randomized block model and have used it in the fast-food-chain study, the question arises as to what effect the blocking had on the analysis. That is, did the blocking result in an increase in precision in comparing the different treatment groups?

The estimated **relative efficiency *(RE)*** of the randomized block design as compared with the completely randomized design may be computed as in Equation (14.18):

$$RE = \frac{(r - 1)\text{MSBL} + r(c - 1)\text{MSE}}{(rc - 1)\text{MSE}} \qquad \textbf{(14.18)}$$

Thus, from Table 14.11 on page 565 for the fast-food-chain study we have

$$RE = \frac{(5)(56.676) + (6)(3)(14.986)}{(23)(14.986)} = 1.60$$

This means that 1.6 times as many observations in each treatment group would be needed in a one-way ANOVA design to obtain the same precision for comparison of treatment group means as would be needed for our randomized block design.

Problems for Section 14.8

14.31 Explain the difference between the randomized block design model and the completely randomized design model. Discuss fully.

Data file COFFEE.DAT

14.32 A taste-testing experiment has been designed so that four brands of Colombian coffee are to be rated by nine experts. To avoid any *carryover* effects, the tasting sequence for the four brews is randomly determined for each of the nine expert tasters until a rating on a 7-point scale (1 = extremely unpleasing, 7 = extremely pleasing) is given for each of the following four characteristics: taste, aroma, richness, and acidity. The following table displays the *summated* ratings—accumulated over all four characteristics.

Summated ratings of four brands of Colombian coffee.

	Brand			
Expert	A	B	C	D
E.B	24	26	25	22
N.B.	27	27	26	24
M.D.	19	22	20	16
M.H.	24	27	25	23
B.J.	22	25	22	21
R.J.	26	27	24	24
B.K.	27	26	22	23
B.M.	25	27	24	21
J.S.	22	23	20	19

(a) Construct an appropriate graph, plot, or chart of the data and describe any trends or relationships that might be apparent among the treatment groups and among the blocks.

(b) At the .05 level of significance, completely analyze the data to determine whether there is evidence of a difference in the summated ratings of the four brands of Colombian coffee and, if so, which of the brands are rated highest (i.e., best).

(c) ACTION Based on your findings, write a draft of an article that you might send to the food editor in your local newspaper.

14.33 A medical researcher wishes to perform an experiment to determine if the choice of treatment substance affects the clotting time of plasma (in minutes). Five different clotting enhancement substances (i.e., treatments) are to be compared and seven female patients, all of whom are in their first term of pregnancy, are to be studied. For each patient, five vials of blood are drawn and one vial each is randomly assigned to one of the five treatments. The clotting time data are shown at the top of page 569:

Data file CLOTTING.DAT

Patient	Treatment Substance 1	2	3	4	5
1	8.4	8.1	8.5	8.6	8.5
2	10.3	10.0	9.9	10.6	10.2
3	12.4	11.8	12.3	12.5	12.2
4	9.7	9.8	9.9	10.4	10.4
5	8.6	8.4	9.7	9.9	9.5
6	9.3	9.6	10.3	10.5	10.2
7	11.1	10.6	11.6	10.9	11.4

(a) Construct an appropriate graph, plot, or chart of the data and describe any trends or relationships that might be apparent among the treatment groups and among the blocks.
(b) At the .05 level of significance, is there evidence of a difference in the average plasma clotting time among the five treatment substances?
(c) If appropriate, use the Tukey procedure to determine the treatment substances that differ in average clotting time.
(d) Determine the relative efficiency of the randomized block design as compared with the completely randomized design.
(e) ACTION Draft a report for the medical researcher based on the above findings.

14.34 A nutritionist wishes to compare three well-known dietary products. Using data on girth (i.e., a function of height and weight), age, and metabolism, he matches 18 of his male clients into six groups of three each and randomly assigns one member of each group to one of the three dietary treatments. The following data represent the amount of weight (in pounds) lost by the 18 clients after six weeks on treatment:

Client Groups	Dietary Treatment 1	2	3
1	10.4	12.1	9.0
2	9.8	14.5	9.6
3	7.3	10.0	9.8
4	7.5	9.9	10.7
5	8.6	14.2	11.1
6	10.7	10.5	10.5

Data file DIET.DAT

(a) Construct an appropriate graph, plot, or chart of the data and describe any trends or relationships that might be apparent among the treatment groups and among the blocks.
(b) At the .05 level of significance, is there evidence of a difference in the average amount of weight (in pounds) lost among the three dietary treatments?
(c) If appropriate, use the Tukey procedure to determine the dietary treatments that differ in average weight lost. (Use $\alpha = .05$.)
(d) Determine the relative efficiency of the randomized block design as compared with the completely randomized design.
(e) ACTION Draft a report that the nutritionist might use in future dietary recommendations to clients based on the findings of this study.

14.35 The dean of a well-known business school wanted to study the student-faculty evaluation process at his campus since it is used in reappointment, promotion, and tenure decisions. In particular, he was interested in determining the type of educational setting most conducive to higher faculty evaluations from students—MBA courses, advanced undergraduate courses, or required undergraduate courses. Since the faculty's semester work load at this institution is three courses, the dean took a random sample of ten faculty from his school who had been assigned one course in each of the three aforementioned types of educational settings and retrieved their end-of-semester evaluation forms. The results below are mean ratings on a 5-point scale (1 = very poor, 5 = outstanding) to the question: "Compared to other teachers you have had, how would you rate this individual's teaching ability?" and each of the ratings is from classes containing 25–30 students.

Data file RATING.DAT

	Type of Class		
Faculty Member	MBA Course	Advanced Undergrad	Required Undergrad
L.M.	4.12	4.06	3.38
N.R.	4.87	4.72	4.60
A.C.	3.46	3.49	2.39
J.K.	3.87	3.61	3.23
J.B.	4.04	3.83	3.55
D.B.	2.90	3.23	3.52
W.F.	4.16	4.07	3.68
R.S.	4.19	3.76	3.83
M.L.	4.75	4.39	4.22
V.P.	4.29	4.34	3.67

(a) Construct an appropriate graph, plot, or chart of the data and describe any trends or relationships that might be apparent among the treatment groups and among the blocks.
(b) Using a level of significance of $\alpha = .05$, is there evidence of a difference in the ratings based on type of class?
(c) If appropriate, use the Tukey procedure to determine which types of classes differ in the ratings? (Use $\alpha = .05$.)
(d) Determine the relative efficiency of the randomized block design as compared with the completely randomized design.
(e) ACTION Draft a memo the dean might send to his department heads regarding the findings from (b) and (c) that may assist them in decisions regarding reappointment, promotion, and/or tenure based on teaching ratings.

Interchapter Problems for Section 14.8

14.36 In Problem 13.62 on page 509 you used a t test to compare average wear based on two types of materials used on the soles of shoes.
(a) Use the randomized block F test on this data set. (Use $\alpha = .05$.)
(b) Square the value of t you computed in Problem 13.62; notice that it is the same (except for rounding error) as the F value. Express in your own words the relationship between t and F.

● 14.37 In Problem 13.63 on page 510 you used a t test to compare average gasoline mileage between regular and high-octane gas.
(a) Use the randomized block F test on this data set. (Use $\alpha = .05$.)
(b) Square the value of t you computed in Problem 13.63; notice that it is the same (except for rounding error) as the F value. Express in your own words the relationship between t and F.

14.9 Friedman Rank Test for Differences in c Medians

14.9.1 Introduction

It often happens that, although the randomized block design model is deemed appropriate for a particular experiment, we may prefer some distribution-free alternative to the classical randomized block F test for analyzing the data. If the data collected are only in rank form within each block or if normality cannot be assumed, a simple but fairly powerful distribution-free approach called the **Friedman rank test** can be utilized.

14.9.2 Development

The Friedman rank test is primarily used to test whether c sample groups have been drawn from populations having equal medians. That is, we may test

$$H_0\text{:}\quad M_1 = M_2 = \cdots = M_c$$

against the alternative

$$H_1\text{:}\quad \text{Not all } M_j \text{ are equal (where } j = 1, 2, \ldots, c)$$

To develop the test we first replace the data by their ranks on a block-to-block basis. That is, in each of the r independent blocks we replace the c observations by their corresponding ranks such that the rank 1 is given to the smallest observation in the block and the rank c to the largest. If any values in a block are tied, they are assigned the average of the ranks that they would otherwise have been given. Thus, R_{ij} is the rank (from 1 to c) associated with the jth group (where $j = 1, 2, \ldots, c$) in the ith block (where $i = 1, 2, \ldots, r$).

Under the null hypothesis of no differences in the c groups, each ranking within a block is equally likely. Thus there are $c!$ possible ways of ranking within a particular block and $(c!)^r$ possible arrangements of ranks over all r independent blocks. Moreover, if the null hypothesis is true, there would be no real differences among the average ranks for each group (taken over all r blocks).

From the above, the following test statistic F_R may be derived:

$$F_R = \frac{12}{rc(c+1)} \sum_{j=1}^{c} R_{.j}^2 - 3r(c+1) \qquad \textbf{(14.19)}$$

where $R_{.j}^2$ is the square of the rank total for group j ($j = 1, 2, \ldots, c$)
r is the number of independent blocks
c is the number of groups

As the number of blocks in the experiment gets large (greater than 5), the test statistic F_R may be approximated by the chi-square distribution with $c - 1$ degrees of freedom. Thus for any selected level of significance α, the decision rule would be to reject the null hypothesis if the computed value of F_R exceeds the upper-tail critical value for the chi-square distribution having $c - 1$ degrees of freedom as shown in Figure 14.17. That is,

$$\text{Reject } H_0 \text{ if } F_R > \chi^2_{U(c-1)};$$

$$\text{otherwise don't reject } H_0.$$

The critical χ^2 values are presented in Table E.4.

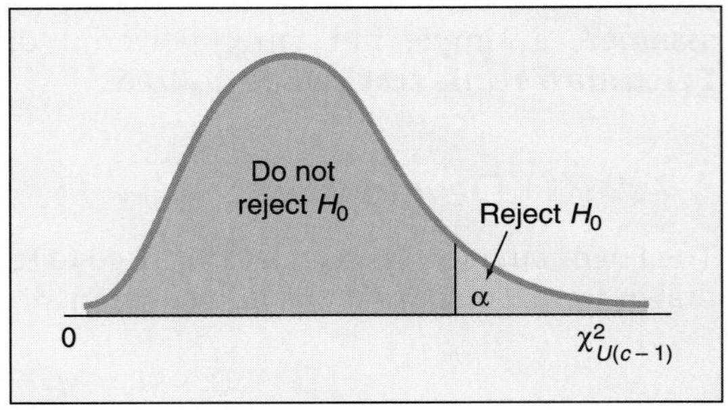

Figure 14.17
Determining the rejection region.

14.9.3 Application

To illustrate the Friedman rank test for differences in c medians, let us return to our fast-food-chain study of the previous section. We may recall that the research director for the chain designed a randomized block experiment in which 24 investigators were stratified into six blocks of four—based on food-service evaluation experience—and the four members of each block were randomly assigned to evaluate the service at one of the four restaurants owned by the chain. The results of the experiment were displayed in Table 14.10 on page 563 along with some summary computations and a scatter plot was presented in Figure 14.14 on page 564 so that a visual, exploratory evaluation of potential trends, relationships, and violations in assumptions of particular testing procedures could be made. If the research director did not want to make the assumption that the service ratings were normally distributed for each restaurant, the distribution-free Friedman rank test for differences in the four population medians could be used.

The null hypothesis to be tested is that the median service ratings for the four restaurants are equal; the alternative is that at least one of the restaurants differs from the others. Thus, substituting 1, 2, 3, 4 for A, B, C, D we have

$$H_0\text{: } M_{.1} = M_{.2} = M_{.3} = M_{.4}$$

$$H_1\text{: Not all the medians are equal}$$

Converting the 24 service ratings in Table 14.10 (page 563) to ranks within each block, we obtain Table 14.12. As shown in Table 14.12, the sum of the ranks for each group are obtained

$$\text{Rank sums: } R_{.1} = 14.5 \quad R_{.2} = 6.0 \quad R_{.3} = 24.0 \quad R_{.4} = 15.5$$

Table 14.12 **Converting data to ranks within blocks.**

Blocks of Raters	Restaurants			
	A	B	C	D
1	2.0	1.0	4.0	3.0
2	3.0	1.0	4.0	2.0
3	2.0	1.0	4.0	3.0
4	3.0	1.0	4.0	2.0
5	2.5	1.0	4.0	2.5
6	2.0	1.0	4.0	3.0
Rank totals:	14.5	6.0	24.0	15.5

Source: Data are taken from Table 14.10 on page 563.

As a check on the rankings we have

$$R_{.1} + R_{.2} + R_{.3} = \frac{rc(c+1)}{2} \tag{14.20}$$

For our data, using Equation (14.20):

$$14.5 + 6 + 24 + 15.5 = \frac{(6)(4)(5)}{2}$$

$$60 = 60$$

Using Equation (14.19), we obtain

$$\begin{aligned} F_R &= \frac{12}{rc(c+1)} \sum_{j=1}^{c} R_{.j}^2 - 3r(c+1) \\ &= \left\{\frac{12}{(6)(4)(5)}[14.5^2 + 6.0^2 + 24.0^2 + 15.5^2]\right\} - (3)(6)(5) \\ &= \left(\frac{12}{120}\right)(1{,}062.5) - 90 \\ &= 106.25 - 90 = 16.25 \end{aligned}$$

Since the computed F_R statistic exceeds 7.815, the upper-tail critical value under the chi-square distribution having $c - 1 = 3$ degrees of freedom (see Table E.4), the null hypothesis may be rejected at the $\alpha = .05$ level. We may conclude that there are significant differences (perceived by the raters) with respect to the service rendered at the four restaurants.

We may note that these are the same results that were obtained for these data using the randomized block F test in Section 14.8.

14.9.4 Multiple Comparisons: Nemenyi's Procedure

Since we rejected the null hypothesis and concluded that there was evidence of a significant difference among the restaurants with respect to their median service ratings, the next step would be a simultaneous comparison of all possible pairs of restaurants to determine which one or ones differ from the others. As a follow-up to the Friedman rank test, a *post hoc* or *a posteriori* multiple comparison procedure proposed by P. Nemenyi (see Reference 9) will be described.

In general, with c treatment groups or levels of a factor, there are $c(c - 1)/2$ possible pairwise comparisons to be made. The first step is to obtain the average rank $\bar{R}_{.j}$ for each of the j groups (where $j = 1, \ldots, c$). We may recall from Equation (14.19) that when computing the test-statistic F_R we obtained the rank total $R_{.j}$ in each group. Thus to compute the average rank $\bar{R}_{.j}$ for the jth group we have

$$\bar{R}_{.j} = \frac{R_{.j}}{n_j} \qquad (\text{where } j = 1, \ldots, c)$$

We then compute the differences $\bar{R}_{.j} - \bar{R}_{.j'}$ (where $j \neq j'$) among all $c(c - 1)/2$ pairs of average ranks. The critical range for Nemenyi's procedure is obtained from

$$\text{critical range} = Q_{U[c,\infty]}\sqrt{\frac{c(c+1)}{12r}} \qquad \textbf{(14.21)}$$

where, for a selected α overall level of significance, $Q_{U[c,\infty]}$ is the upper-tailed critical value from a **Studentized range distribution** (Table E.12) having c and ∞ degrees of freedom. We may recall that the Q statistic was also used in obtaining the critical ranges for multiple comparisons following the rejection of the null hypothesis of the equality of c means in both the classical one-way ANOVA F test and the randomized block F test.

The final step is to compare each of the $c(c - 1)/2$ pairs of average ranks against the one critical range obtained from Equation (14.21). A specific pair of groups would be declared significantly different if the absolute difference in their corresponding average ranks exceeds this critical range.

To apply the Nemenyi procedure in our fast-food-chain study, we first obtain the average ranks over the four groups from the corresponding rank sums given on page 572:

Rank sums:	$R_{.1} = 14.5$	$R_{.2} = 6.0$	$R_{.3} = 24.0$	$R_{.4} = 15.5$
Average ranks:	$\bar{R}_{.1} = 2.42$	$\bar{R}_{.2} = 1.00$	$\bar{R}_{.3} = 4.00$	$\bar{R}_{.4} = 2.58$

There are $(4)(4 - 1)/2 = 6$ pairwise comparisons to be made because there are four restaurants. The absolute differences in the average ranks are

1. $|\bar{R}_{.1} - \bar{R}_{.2}| = |2.42 - 1.00| = 1.42$
2. $|\bar{R}_{.1} - \bar{R}_{.3}| = |2.42 - 4.00| = 1.58$
3. $|\bar{R}_{.1} - \bar{R}_{.4}| = |2.42 - 2.58| = 0.16$

4. $|\bar{R}_{.2} - \bar{R}_{.3}| = |1.00 - 4.00| = 3.00$

5. $|\bar{R}_{.2} - \bar{R}_{.4}| = |1.00 - 2.58| = 1.58$

6. $|\bar{R}_{.3} - \bar{R}_{.4}| = |4.00 - 2.58| = 1.42$

Since the three groups each have the same sample size, there is only one critical range that would be used in all possible comparisons. Choosing an overall level of significance of .05, we determine the critical range from Equation (14.21) by first obtaining $Q_{U[4,\infty]} = 3.63$ from Table E.12 so that

$$\text{critical range} = Q_{U[4,\infty]}\sqrt{\frac{c(c+1)}{12r}}$$

$$= 3.63\sqrt{\frac{(4)(5)}{(12)(6)}}$$

$$= 1.91$$

We note that only the fourth pairwise comparison ($|\bar{R}_{.2} - \bar{R}_{.3}| = |1.00 - 4.00| = 3.00$) exceeds the critical range of 1.91, so the research director may conclude that the service at restaurant C is significantly better than that at restaurant B but there is no evidence of a significant difference in the service ratings between any other pairs of restaurants. We may recall that had the research director utilized the parametric Tukey procedure described in Section 14.8.4, he would have concluded that there was a significant difference in the average service ratings between all pairs of restaurants in the chain except for restaurants A and D.

Owing to our conflicting results, the Tukey procedure should be used after performing a randomized block *F* test if the assumption of normality in the four underlying populations is viable. If not, the Nemenyi procedure should be employed as a follow-up to the Friedman rank test.

Problems for Section 14.9

● 14.38 The President's Council on Physical Fitness and Sports asked a panel of medical experts to rank five diverse forms of exercise with respect to their special contributions to physical fitness and overall well-being. The rankings (1 = least beneficial, 5 = most beneficial) are displayed below for each of nine equally important characteristics of fitness and well-being.

Data file EXERCISE.DAT

	Exercise				
Characteristic	Bicycling	Calisthenics	Jogging	Swimming	Tennis
Balance	5.0	2.0	4.0	1.0	3.0
Digestion	2.5	1.0	4.5	4.5	2.5
Flexibility	1.5	5.0	1.5	4.0	3.0
Muscular definition	4.0	5.0	2.5	2.5	1.0
Muscular endurance	3.0	1.0	4.5	4.5	2.0
Muscular strength	3.5	3.5	5.0	1.5	1.5
Sleep	3.0	2.0	4.5	4.5	1.0
Stamina	3.0	1.0	4.5	4.5	2.0
Weight control	4.0	1.0	5.0	2.0	3.0

(a) Is there evidence of a difference in the "perceived benefit" ratings of the five forms of exercise? (Use $\alpha = .05$.) (*Hint:* Treat the nine characteristics as blocks.)
(b) Based on your results in (a), use the Nemenyi procedure to determine which exercises are most beneficial (Use $\alpha = .05$.)
(c) **ACTION** Write a letter to a friend describing your findings.

Data file BESTTEAM.DAT

14.39 A radio sports show host surveyed his audience to determine the most dominant teams in professional sports during the early 1990s. The teams selected by the fans were the Atlanta Braves in baseball, the Dallas Cowboys in football, the Chicago Bulls in basketball, and the Pittsburgh Penguins in hockey. He then assembled a panel of 10 experts and asked them to rank these teams from 1 (most dominant) to 4 (least dominant). The results are as follows:

	Team			
Expert	Baseball Braves	Football Cowboys	Basketball Bulls	Hockey Penguins
B.M.	3	2	1	4
L.D.	2	1	3	4
H.C.	1	3	4	2
F.C.	4	1	2	3
T.J.	2	3	1	4
T.A.	1	2	3	4
H.B.	3	1	2	4
H.D.	3	2	1	4
T.D.	3	1	2	4
S.D.	4	2	3	1

(a) At the .05 level of significance, completely analyze the data to determine if there is evidence that shows all four sports teams are not perceived as equally dominant and, if so, which ones are considered best.
(b) **ACTION** Write a letter to the sports show host indicating how you would interpret the findings in (a).

14.40 A psychological experiment is designed to determine if there are any differences in recall ability due to different levels of exposure to an object. Three levels of exposure (in msec) are considered. Eight sets of triplets are chosen as subjects. In each of the eight sets of triplets, the members are randomly assigned to be examined under an exposure level.

Data file RECALL.DAT

	Exposure Level		
Set of Triplets (Subjects)	Minimum	Moderate	High
I	55	68	67
II	78	83	84
III	34	53	54
IV	56	67	65
V	79	78	85
VI	20	29	30
VII	68	88	92
VIII	59	58	72

(a) At the .05 level of significance, is there evidence of a difference in the median recall ability between the exposure levels?

(b) If appropriate, use the Nemenyi procedure to determine the exposure levels that differ in average recall ability. (Use $\alpha = .05$.)

(c) ACTION Write a memo to your industrial psychology professor regarding the results and possible implications in advertising.

14.41 Use the Friedman rank test and, if appropriate, the Nemenyi procedure to answer part (b) of Problem 14.32 (coffee ratings) on page 568. (Use $\alpha = .05$.) Are there any differences in your present results from those previously obtained? Discuss.

14.42 Use the Friedman rank test and, if appropriate, the Nemenyi procedure to answer parts (b) and (c) of Problem 14.33 (clotting time) on page 568. (Use $\alpha = .05$.) Are there any differences in your present results from those previously obtained? Discuss.

14.43 Use the Friedman rank test and, if appropriate, the Nemenyi procedure to answer parts (b) and (c) of Problem 14.35 (student–faculty evaluation ratings) on page 570. (Use $\alpha = .05$.) Are there any differences in your present results from those previously obtained? Discuss.

14.10 The Factorial Design Model and Two-Way Analysis of Variance

14.10.1 Introduction

In Sections 14.3–14.6 we discussed the one-way analysis-of-variance or completely randomized design model and in Sections 14.7–14.9 the randomized block design model was introduced. In this section we shall extend our discussion to consider an experimental design model in which two factors are of interest. The two factors may differ with respect to the number of **levels** (or groups) they contain. However, we shall be concerned only with situations in which there are equal sample sizes n' for each combination of the levels of factor A with those of factor B. (See Reference 1 for a discussion of ANOVA models with unequal sample sizes.)

14.10.2 Development

In order to extend our discussion of the analysis of variance and develop the F-test procedures for the two-factor factorial design model with equal replication,[6] we need to define the following terms:

X_{ijk} = the value of the kth observation for level i of factor A and level j of factor B

$X_{ij.}$ = the sum of the values in cell ij (the observations at level i of factor A and level j of factor B)

$X_{i..}$ = the sum of the values for row i of factor A

$X_{.j.}$ = the sum of the values for column j of factor B

GT = the grand total of all values over all rows and columns

r = the number of levels of factor A

c = the number of levels of factor B

n' = the number of values (replications) for each cell

n = the total number of observations in the experiment (where $n = rcn'$)

These new terms are needed for using computational formulas in an analysis of data obtained from two-factor factorial designed experiments. For purposes of illustration, we present both conceptual and computational approaches for the decomposition of the total variation needed in the development of the *F*-test procedures. Nevertheless, owing to the numerical drudgery involved, particularly as the number of levels of each factor increases and the number of replications in each cell increases, we strongly suggest that one of the many available statistical software packages be used when analyzing data obtained from factorial design models.

From Figure 14.1 (page 528) we recall that in the completely randomized design model the sum of squares total (or SST) was subdivided into sum of squares among groups (or SSA) and sum of squares within groups (or SSW). Also, from Figure 14.13 (page 559) we note that in the randomized block design model the total variation (that is, SST) was subdivided into sum of squares among treatment groups (or SSA), sum of squares among blocks (or SSBL), and sum of squares error (or SSE). For the two-factor factorial design model with replication (that is, $n' > 1$ in each cell), we need to subdivide the total variation (SST) into sum of squares due to factor A (or SSFA), sum of squares due to factor B (or SSFB), sum of squares due to the interacting effect of A and B (or SSAB), and sum of squares due to inherent random error (or SSE). This decomposition of the total variation (SST) is displayed in Figure 14.18.

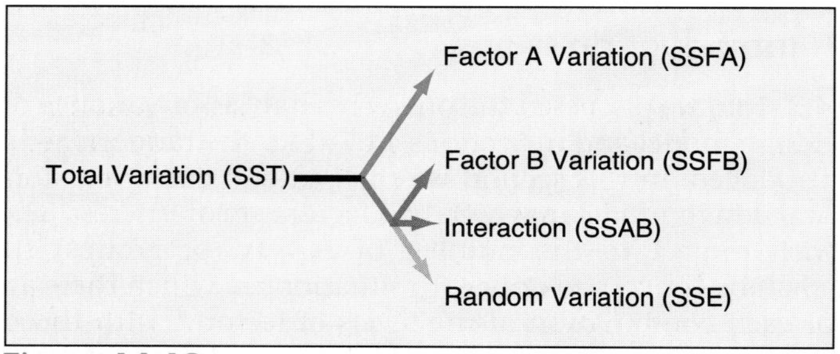

Figure 14.18
Partitioning the total variation in a two-factor factorial design model.

The **Sum of Squares Total** (or **SST**) represents the total variation among all the observations around the grand mean. SST would be computed as

$$\text{SST} = \sum_{i=1}^{r}\sum_{j=1}^{c}\sum_{k=1}^{n'}(X_{ijk} - \bar{\bar{X}})^2 = \sum_{i=1}^{r}\sum_{j=1}^{c}\sum_{k=1}^{n'}X_{ijk}^2 - \frac{(\text{GT})^2}{rcn'} \quad \textbf{(14.22)}$$

where $\bar{\bar{X}} = \dfrac{\sum_{i=1}^{r}\sum_{j=1}^{c}\sum_{k=1}^{n'}X_{ijk}}{rcn'}$ (i.e., the overall or grand mean)

The **Sum of Squares** due to **Factor A** (or **SSFA**) represents the differences among the various levels of factor A and the grand mean. SSFA would be computed as

$$\text{SSFA} = cn'\sum_{i=1}^{r}\left(\bar{X}_{i..} - \bar{\bar{X}}\right)^2 = \sum_{i=1}^{r}\frac{X_{i..}^2}{cn'} - \frac{(\text{GT})^2}{rcn'} \qquad \textbf{(14.23)}$$

where $\bar{X}_{i..} = \dfrac{\sum_{j=1}^{c}\sum_{k=1}^{n'} X_{ijk}}{cn'}$ (i.e., the mean of each level of factor A)

The **Sum of Squares** due to **Factor B** (or **SSFB**) represents the differences among the various levels of factor B and the grand mean. SSFB would be computed as

$$\text{SSFB} = rn'\sum_{j=1}^{c}(\bar{X}_{.j.} - \bar{\bar{X}})^2 = \sum_{j=1}^{c}\frac{X_{.j.}^2}{rn'} - \frac{(\text{GT})^2}{rcn'} \qquad \textbf{(14.24)}$$

where $\bar{X}_{.j.} = \dfrac{\sum_{i=1}^{r}\sum_{k=1}^{n'} X_{ijk}}{rn'}$ (i.e., the mean of each level of factor B)

The **Sum of Squares** due to the interacting effect of **A** and **B** (or **SSAB**) represents the effect of the combinations of factor A and factor B. SSAB would be computed as

$$\begin{aligned}\text{SSAB} &= n'\sum_{i=1}^{r}\sum_{j=1}^{c}(\bar{X}_{ij.} - \bar{X}_{i..} - \bar{X}_{.j.} + \bar{\bar{X}})^2 \\ &= \sum_{i=1}^{r}\sum_{j=1}^{c}\frac{X_{ij.}^2}{n'} - \sum_{i=1}^{r}\frac{X_{ij.}^2}{cn'} - \sum_{j=1}^{c}\frac{X_{.j.}^2}{rn'} + \frac{(\text{GT})^2}{rcn'}\end{aligned} \qquad \textbf{(14.25)}$$

where $\bar{X}_{ij.} = \sum_{k=1}^{n'}\dfrac{X_{ijk}}{n'}$ (i.e., the mean of each cell)

The **Sum of Squares Error** (or **SSE**) represents the differences among the observations within each cell and the corresponding cell mean. SSE would be computed as

$$\text{SSE} = \sum_{i=1}^{r}\sum_{j=1}^{c}\sum_{k=1}^{n'}(X_{ijk} - \bar{X}_{ij.})^2 = \sum_{i=1}^{r}\sum_{j=1}^{c}\sum_{k=1}^{n'}X_{ijk}^2 - \sum_{i=1}^{r}\sum_{j=1}^{c}\frac{X_{ij.}^2}{n'} \qquad \textbf{(14.26)}$$

Since there are c treatment levels of factor A, there are $c - 1$ degrees of freedom associated with SSA. Similarly, since there are r treatment levels of factor B, there are $r - 1$ degrees of freedom associated with SSB. Moreover, since there are n' replications in each of the rc cells, there are $rc(n' - 1)$ degrees of freedom associated with the inherent random error term. Carrying this further, there are $n - 1$ degrees of freedom associated with the sum of squares total (SST) because each observation X_{ijk} is being compared to the overall or grand mean $\bar{\bar{X}}$ based on all n observations. Therefore, since the degrees of freedom for each of the sources of variation must add to the degrees of freedom for the total variation (SST), we may obtain the degrees of freedom for the interaction component (SSAB) by subtraction and algebraic manipulation.[7] The degrees of freedom is given by $(r - 1)(c - 1)$.

If each of the sums of squares is divided by its associated degrees of freedom, we obtain the four *variances* or **mean square** terms (**MSFA**, **MSFB**, **MSAB**, and **MSE**) needed for ANOVA:

$$\text{MSFA} = \frac{\text{SSFA}}{r - 1} \quad \textbf{(14.27a)}$$

$$\text{MSFB} = \frac{\text{SSFB}}{c - 1} \quad \textbf{(14.27b)}$$

$$\text{MSAB} = \frac{\text{SSAB}}{(r - 1)(c - 1)} \quad \textbf{(14.27c)}$$

$$\text{MSE} = \frac{\text{SSE}}{rc(n' - 1)} \quad \textbf{(14.27d)}$$

In the two-factor ANOVA model there are three distinct tests that may be performed. If we assume that the levels of factor A and factor B have been specifically selected for analysis (rather than being randomly selected from a population of possible levels), then we would have the following three tests of hypotheses:

To test the hypothesis of no difference due to factor A

$$H_0: \ \mu_{1..} = \mu_{2..} = \cdots = \mu_{r..}$$

against the alternative

$$H_1: \ \text{Not all } \mu_{i..} \text{ are equal}$$

we form the F statistic

$$F = \frac{\text{MSFA}}{\text{MSE}} \quad \textbf{(14.28)}$$

and the null hypothesis would be rejected at the α level of significance if

$$F = \frac{\text{MSFA}}{\text{MSE}} > F_{U[(r-1), rc(n'-1)]}$$

To test the hypothesis of no difference due to factor B

$$H_0: \ \mu_{.1.} = \mu_{.2.} = \cdots = \mu_{.c.}$$

against the alternative

$$H_1: \ \text{Not all } \mu_{.j.} \text{ are equal}$$

we form the F statistic

$$F = \frac{\text{MSFB}}{\text{MSE}} \tag{14.29}$$

and the null hypothesis would be rejected at the α level of significance if

$$F = \frac{\text{MSFB}}{\text{MSE}} > F_{U[(c-1),rc(n'-1)]}$$

To test the hypothesis of no interaction of factors A and B

$$H_0: \ AB_{ij} = 0 \text{ (for all } i \text{ and } j)$$

against the alternative

$$H_1: \ AB_{ij} \neq 0$$

we form the F statistic

$$F = \frac{\text{MSAB}}{\text{MSE}} \tag{14.30}$$

and the null hypothesis would be rejected at the α level of significance if

$$F = \frac{\text{MSAB}}{\text{MSE}} > F_{U[(r-1)(c-1),rc(n'-1)]}$$

As in Sections 14.4.2 and 14.8.2, the entire set of steps may be summarized in an analysis-of-variance (ANOVA) table such as Table 14.13 on page 582.

14.10.3 Application

To illustrate the two-factor factorial design model, suppose that the marketing research director for a supermarket chain was interested in studying the effect of shelf location on sales of a product. Four different shelf locations were to be studied: normal location (A), additional location in store (B), new location only and

Table 14.13 Analysis-of-variance table for the two-factor model with replication.

Source	Degrees of Freedom	Sums of Squares	Mean Square (Variance)	F
A	$r-1$	$\sum_{i=1}^{r} \frac{X_{i..}^2}{cn'} - \frac{(\text{GT})^2}{rcn'}$	$\text{MSFA} = \frac{\text{SSFA}}{r-1}$	$F = \frac{\text{MSFA}}{\text{MSE}}$
B	$c-1$	$\sum_{j=1}^{c} \frac{X_{.j.}^2}{rn'} - \frac{(\text{GT})^2}{rcn'}$	$\text{MSFB} = \frac{\text{SSFB}}{c-1}$	$F = \frac{\text{MSFB}}{\text{MSE}}$
AB	$(r-1)(c-1)$	$\sum_{i=1}^{r}\sum_{j=1}^{c} \frac{X_{ij.}^2}{n'} - \sum_{i=1}^{r} \frac{X_{i..}^2}{cn'} - \sum_{j=1}^{c} \frac{X_{.j.}^2}{rn'} + \frac{(\text{GT})^2}{rcn'}$	$\text{MSAB} = \frac{\text{SSAB}}{(r-1)(c-1)}$	$F = \frac{\text{MSAB}}{\text{MSE}}$
Error	$rc(n'-1)$	$\sum_{i=1}^{r}\sum_{j=1}^{c}\sum_{k=1}^{n'} X_{ijk}^2 - \sum_{i=1}^{r}\sum_{j=1}^{c} \frac{X_{ij.}^2}{n'}$	$\text{MSE} = \frac{\text{SSE}}{rc(n'-1)}$	
Total	$rcn'-1$	$\sum_{i=1}^{r}\sum_{j=1}^{c}\sum_{k=1}^{n'} X_{ijk}^2 - \frac{(\text{GT})^2}{rcn'}$		

"shelf-talker" (C), and normal location and "ribboning" (D). Three different store sizes were to be considered: small, medium, and large. For each shelf location a random sample of two stores of each size was selected. The results in weekly sales are summarized in Table 14.14.

Table 14.14 Weekly sales by store size and shelf location.

	Shelf Location					
Store Size	A	B	C	D	Totals	Means
Small	45 50	56 63	65 71	48 53	451	56.375
Medium	57 65	69 78	73 80	60 57	539	67.375
Large	70 78	75 82	82 89	71 75	622	77.750
Totals	365	423	460	364	1,612	
Means	60.83	70.50	76.67	60.67		67.167

From this table we have

$r = 3,\quad c = 4,\quad n' = 2,\quad X_{1..} = 451,\quad X_{2..} = 539,\quad X_{3..} = 622$

$X_{.1.} = 365,\quad X_{.2.} = 423,\quad X_{.3.} = 460,\quad X_{.4.} = 364,\quad \text{GT} = 1,612$

$X_{11.} = 95,\quad X_{12.} = 119,\quad X_{13.} = 136,\quad X_{14.} = 101,\quad X_{21.} = 122,\quad X_{22.} = 147$

$X_{23.} = 153,\quad X_{24.} = 117,\quad X_{31.} = 148,\quad X_{32.} = 157,\quad X_{33.} = 171,\quad X_{34.} = 146$

$$\sum_{i=1}^{r}\sum_{j=1}^{c}\sum_{k=1}^{n'} X_{ijk}^2 = 45^2 + 50^2 + \cdots + 75^2 = 111,550$$

$$\sum_{i=1}^{r} \frac{X_{i..}^2}{cn'} = \frac{451^2 + 539^2 + 622^2}{(4)(2)} = 110,100.75$$

$$\sum_{j=1}^{c} \frac{X_{.j.}^2}{rn'} = \frac{365^2 + 423^2 + 460^2 + 364^2}{(3)(2)} = 109{,}375$$

$$\sum_{i=1}^{r}\sum_{j=1}^{c} \frac{X_{ij.}^2}{n'} = \frac{95^2 + 119^2 + \cdots + 146^2}{(2)} = 111{,}292$$

$$\frac{(\text{GT})^2}{rcn'} = \frac{1{,}612^2}{(3)(4)(2)} = 108{,}272.66$$

Using Equation (14.22),

$$\text{SST} = \sum_{i=1}^{r}\sum_{j=1}^{c}\sum_{k=1}^{n'} X_{ijk}^2 - \frac{(\text{GT})^2}{rcn'}$$

$$= 111{,}550 - 108{,}272.66 = 3{,}277.34$$

Using Equation (14.23),

$$\text{SSFA} = \sum_{i=1}^{r} \frac{X_{i..}^2}{cn'} - \frac{(\text{GT})^2}{rcn'}$$

$$= 110{,}100.75 - 108{,}272.66 = 1{,}828.09$$

Using Equation (14.24),

$$\text{SSFB} = \sum_{j=1}^{c} \frac{X_{.j.}^2}{rn'} - \frac{(\text{GT})^2}{rcn'}$$

$$= 109{,}375 - 108{,}272.66 = 1{,}102.34$$

Using Equation (14.25),

$$\text{SSAB} = \sum_{i=1}^{r}\sum_{j=1}^{c} \frac{X_{ij.}^2}{n'} - \sum_{i=1}^{r} \frac{X_{i..}^2}{cn'} - \sum_{j=1}^{c} \frac{X_{.j.}^2}{rn'} + \frac{(\text{GT})^2}{rcn'}$$

$$= 111{,}292 - 110{,}100.75 - 109{,}375 + 108{,}272.66$$

$$= 88.91$$

Using Equation (14.26),

$$\text{SSE} = \sum_{i=1}^{r}\sum_{j=1}^{c}\sum_{k=1}^{n'} X_{ijk}^2 - \sum_{i=1}^{r}\sum_{j=1}^{c} \frac{X_{ij.}^2}{n'}$$

$$= 111{,}550 - 111{,}292 = 258$$

To compute the variances we use Equations (14.27a) through (14.27d). From Equation (14.27a):

$$\text{MSFA} = \frac{\text{SSFA}}{r-1} = \frac{1{,}828.09}{3-1} = 914.045$$

From Equation (14.27b):

$$\text{MSFB} = \frac{\text{SSFB}}{c-1} = \frac{1{,}102.34}{4-1} = 367.447$$

From Equation (14.27c):

$$\text{MSAB} = \frac{\text{SSAB}}{(r-1)(c-1)} = \frac{88.91}{(3-1)(4-1)} = 14.818$$

From Equation (14.27d):

$$\text{MSE} = \frac{\text{SSE}}{rc(n'-1)} = \frac{258}{(3)(4)(2-1)} = 21.5$$

In the supermarket study the computations just completed can be summarized in Table 14.15.

Table 14.15 Analysis-of-variance table for the supermarket example.

Source	Degrees of Freedom	Sums of Squares	Mean Square (Variance)	F
A (Store Size)	3 − 1 = 2	110,100.75 − 108,272.66 = 1,828.09	$\text{MSFA} = \frac{1{,}828.09}{2} = 914.045$	$F = \frac{914.045}{21.5} = 42.51$
B (Shelf Location)	4 − 1 = 3	109,375 − 108,272.66 = 1,102.34	$\text{MSFB} = \frac{1{,}102.34}{3} = 367.447$	$F = \frac{367.447}{21.5} = 17.09$
AB (Store Size × Shelf Location)	(3 − 1)(4 − 1) = 6	111,292 − 110,100.75 − 109,375 +108,272.66 = 88.91	$\text{MSAB} = \frac{88.91}{6} = 14.818$	$F = \frac{14.818}{21.5} = .69$
Error	(3)(4)(2 − 1) = 12	111,550 − 111,292 = 258	$\text{MSE} = \frac{258}{12} = 21.5$	
Total	(3)(4)(2) − 1 = 23	111,550 − 108,272.66 = 3,277.34		

Using the .05 level of significance and testing for a difference between the store sizes, the decision rule would be to reject the null hypothesis (H_0: $\mu_{1..} = \mu_{2..} = \cdots = \mu_{r..}$) if the calculated F value exceeds 3.89 (see Figure 14.19). Since $F = 42.51 > F_{U(2,12)} = 3.89$, we may reject H_0 and conclude that there is evidence of a difference between the store sizes in terms of the average weekly sales.

Using the .05 level of significance and testing for a difference between the shelf locations, the decision rule would be to reject the null hypothesis (H_0: $\mu_{.1.} = \mu_{.2.} = \cdots = \mu_{.c.}$) if the calculated F value exceeds 3.49 (see Figure 14.20). Since $F = 17.09 > F_{U(3,12)} = 3.49$, we may reject H_0 and conclude that there is evidence of a difference between the shelf locations in terms of the average weekly sales.

Finally, we can test whether there is an interacting effect between factor A (store size) and factor B (shelf location). Using the .05 level of significance, the decision rule would be to reject the null hypothesis [$AB_{ij} = 0$ (for all i and j)] if the calculated F value exceeds 3.00 (see Figure 14.21). Since $F = .69 < F_{U(6,12)} = 3.00$, we do not reject H_0 and we conclude that there is no evidence of an interacting effect between store size and shelf location.

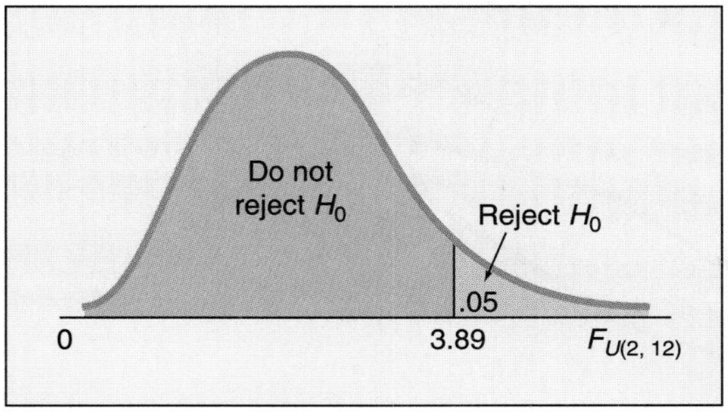

Figure 14.19
Regions of rejection and nonrejection at the .05 level of significance with 2 and 12 degrees of freedom.

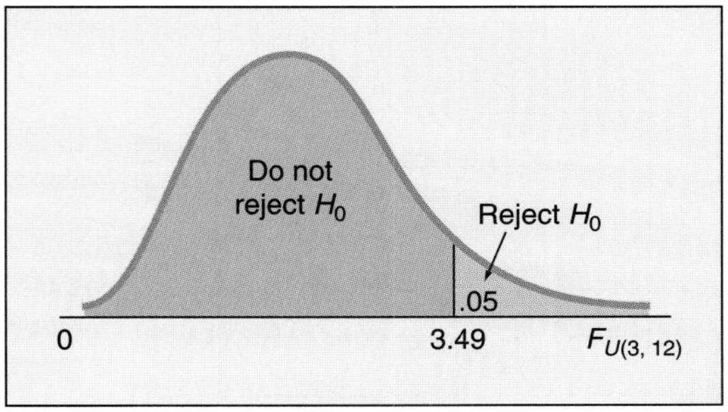

Figure 14.20
Regions of rejection and nonrejection at the .05 level of significance with 3 and 12 degrees of freedom.

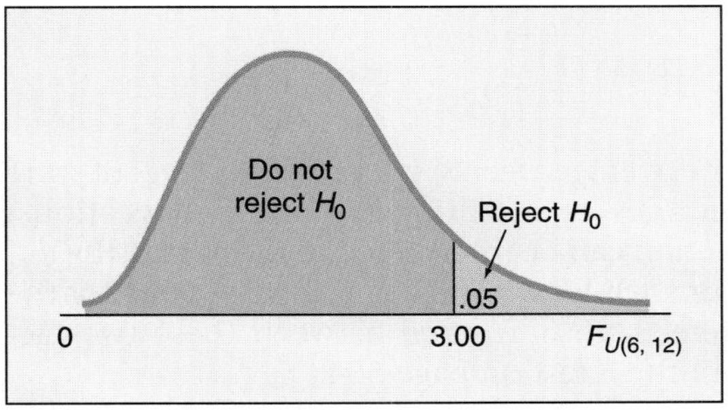

Figure 14.21
Regions of rejection and nonrejection at the .05 level of significance with 6 and 12 degrees of freedom.

14.10.4 Interpreting Interaction Effects

Now that the tests for the significance of factor A, factor B, and their interaction have been performed, we can get a better understanding of the interpretation of the concept of interaction by plotting the cell means as shown in Figure 14.22 on page 586. Since $\overline{X}_{ij.} = X_{ij.}/n'$, we have

$$\overline{X}_{11.} = \frac{95}{2} = 47.5, \quad \overline{X}_{12.} = \frac{119}{2} = 59.5, \quad \overline{X}_{13.} = \frac{136}{2} = 68.0, \quad \overline{X}_{14.} = \frac{101}{2} = 50.5$$

$$\overline{X}_{21.} = \frac{122}{2} = 61.0, \quad \overline{X}_{22.} = \frac{147}{2} = 73.5, \quad \overline{X}_{23.} = \frac{153}{2} = 76.5, \quad \overline{X}_{24.} = \frac{117}{2} = 58.5$$

$$\overline{X}_{31.} = \frac{148}{2} = 74.0, \quad \overline{X}_{32.} = \frac{157}{2} = 78.5, \quad \overline{X}_{33.} = \frac{171}{2} = 85.5, \quad \overline{X}_{34.} = \frac{146}{2} = 73.0$$

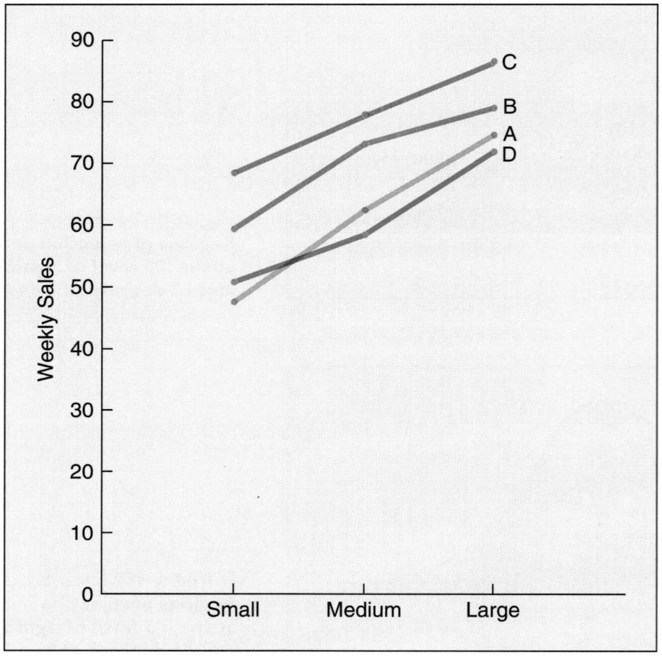

Figure 14.22
Average weekly sales based on store size for different shelf locations.

In Figure 14.22 we have plotted the average weekly sales for each level of store size and location. For our data the four lines (representing the four shelf locations) appear roughly parallel. This phenomenon can be interpreted to mean that the *difference* in weekly sales between the four shelf locations is virtually the same for the three store sizes. In other words, there is no *interaction* between these two factors—as was clearly substantiated from the F test on page 584.

What would be the interpretation if there were an *interacting effect?* In such a situation some levels of factor A would respond better with certain levels of factor B. For example, suppose that some shelf locations were better for large stores and other shelf locations were better for small stores. If this were true, the lines of Figure 14.22 would not be nearly as parallel and the interaction effect might be statistically significant. In such a situation, the differences between the various shelf locations would not be the same for all store sizes. Such an outcome would also serve to complicate the interpretation of the *main effects,* since differences in one factor (shelf locations) would not be consistent across the other factor (store sizes).

14.10.5 Multiple Comparisons: The Tukey Procedure

As in the case of the one-way and randomized block models, once the null hypothesis of no differences in the levels of a factor has been rejected, we need to determine the particular groups or levels that are significantly different from each other. A procedure developed by John Tukey (References 6 and 14) may be used for factor A as well as for factor B.

For factor A we have

$$\text{critical range} = Q_{U[r, rc(n'-1)]}\sqrt{\frac{\text{MSE}}{cn'}} \qquad \textbf{(14.31)}$$

and for factor B we have

$$\text{critical range} = Q_{U[c, rc(n'-1)]}\sqrt{\frac{\text{MSE}}{rn'}} \qquad \textbf{(14.32)}$$

As in Sections 14.4.5 and 14.8.4, each of the $c(c-1)/2$ or $r(r-1)/2$ pairs of means is compared against the appropriate critical range. A specific pair of means would be declared significantly different if the absolute difference in the sample means ($|\bar{X}_{i..} - \bar{X}_{i'..}|$ for factor A or $|\bar{X}_{.j.} - \bar{X}_{.j'.}|$ for factor B) exceeds its respective critical range.

To apply the Tukey procedure, we return to our supermarket example. With respect to factor A, since there are three groups, there are $(3)(3-1)/2 = 3$ possible paired comparisons to be made. From Table 14.14 on page 582, the absolute mean differences are

1. $|\bar{X}_{1..} - \bar{X}_{2..}| = |56.375 - 67.375| = 11.000$
2. $|\bar{X}_{1..} - \bar{X}_{3..}| = |56.375 - 77.750| = 21.375$
3. $|\bar{X}_{2..} - \bar{X}_{3..}| = |67.375 - 77.750| = 10.375$

To determine the critical range, from Table 14.15 on page 584 we have: MSE = 21.5, $r = 3$, and $c = 4$. From Table E.12, for $\alpha = .05$, $r = 3$, and $rc(n' - 1) = (3)(4)(2 - 1) = 12$, the upper-tailed critical value, $Q_{U(3,12)}$, is 3.77. From Equation (14.31), we have

$$\text{critical range} = 4.20\sqrt{\frac{21.5}{6}} = 7.95$$

We note that all the contrasts are greater than the critical range. Therefore we may conclude that small, medium, and large stores differ from each other in weekly sales.

With respect to factor B, since there are four groups, there are $(4)(4-1)/2 = 6$ possible pairwise comparisons to be made. From Table 14.14 on page 582, the absolute mean differences are

1. $|\bar{X}_{.1.} - \bar{X}_{.2.}| = |60.83 - 70.50| = 9.67$
2. $|\bar{X}_{.1.} - \bar{X}_{.3.}| = |60.83 - 76.67| = 15.84$
3. $|\bar{X}_{.1.} - \bar{X}_{.4.}| = |60.83 - 60.67| = .16$
4. $|\bar{X}_{.2.} - \bar{X}_{.3.}| = |70.50 - 76.67| = 6.17$
5. $|\bar{X}_{.2.} - \bar{X}_{.4.}| = |70.50 - 60.67| = 9.83$
6. $|\bar{X}_{.3.} - \bar{X}_{.4.}| = |76.67 - 60.67| = 16.00$

To determine the critical range, from Table 14.15 on page 584 we have: MSE = 21.5, $r = 3$, and $c = 4$. From Table E.12, for $\alpha = .05$, $c = 4$, and $rc(n' - 1) = (3)(4)(2 - 1) = 12$, the upper-tailed critical value, $Q_{U(4,12)}$, is 4.20. From Equation (14.32), we have

$$\text{critical range} = 4.20\sqrt{\frac{21.5}{6}} = 7.95$$

We note that $\bar{X}_{.1.}$ is different from $\bar{X}_{.2.}$ (9.67 > 7.95) and $\bar{X}_{.3.}$ (15.84 > 7.95), and $\bar{X}_{.4.}$ is different from $\bar{X}_{.2.}$ (9.83 > 7.95) and $\bar{X}_{.3.}$ (16 > 7.95). Thus we may conclude that shelf locations A (normal) and D (normal plus "ribboning") are each different from locations B (additional location in store) and C (new location only and "shelf-talker"). However, there is no evidence of a difference between locations A and D or between locations B and C.

14.10.6 Fixed, Random, and Mixed Models

In our discussion of analysis-of-variance models we have not focused on the manner in which the various levels of a factor have been selected. From this perspective, there are three alternative models:

1. Fixed-effects model (Model I)
2. Random-effects model (Model II)
3. Mixed-effects model (Model III)

The first model, the **fixed-effects model** (Model I), described thus far in this section, assumes that the levels of a factor have been *specifically* selected for analysis. This means that the levels of the factor have *not* been randomly selected from a population and that no inferences can be drawn about any other levels except the ones used in the study.

In contrast to the fixed-effects model, Model II, the **random-effects model,** contains factors in which the levels are *randomly selected* from a population. The objective for a random-effects model is not necessarily to examine differences among levels but, more important, to estimate the variability due to each factor (see Reference 5). For example, if we wished to study the effect on productivity of different workers and different machines, we might randomly select a sample of machines and assign a random sample of workers to each machine for a given number of days. Not only would we be able to measure whether workers and machines have significant effects on productivity, but we would also be able to estimate the variability due to different machines and the variability due to different workers.

The third model, the **mixed-effects model** (Model III), contains a mixture of fixed and random effects.

Although the random- and mixed-effect models can be discussed in much greater depth (see References 5 and 6), our focus involves the consequences of the various models on the F test. Since the components of the models differ in their assumptions, they also lead to different F tests in evaluating the significance of the main effects (factors A and B). Therefore, the appropriate F tests for each of the three models are summarized in Table 14.16.

Table 14.16 ***F* tests for two-factor ANOVA models with replication.**

Null Hypothesis	Fixed Effects (A & B Fixed)	Random Effects (A & B Random)	Mixed Effects (A Fixed, B Random)	Mixed Effects (A Random, B Fixed)
$\mu_{i..} = 0$	$F = \frac{\text{MSFA}}{\text{MSE}}$	$F = \frac{\text{MSFA}}{\text{MSAB}}$	$F = \frac{\text{MSFA}}{\text{MSAB}}$	$F = \frac{\text{MSFA}}{\text{MSE}}$
$\mu_{.j.} = 0$	$F = \frac{\text{MSFB}}{\text{MSE}}$	$F = \frac{\text{MSFB}}{\text{MSAB}}$	$F = \frac{\text{MSFB}}{\text{MSE}}$	$F = \frac{\text{MSFB}}{\text{MSAB}}$
$AB_{ij} = 0$	$F = \frac{\text{MSAB}}{\text{MSE}}$	$F = \frac{\text{MSAB}}{\text{MSE}}$	$F = \frac{\text{MSAB}}{\text{MSE}}$	$F = \frac{\text{MSAB}}{\text{MSE}}$

As we observe in Table 14.16, the tests for the *main effects* differ depending on the type of model selected. For the fixed-effects model, the *F* tests involve the ratio of MSFA or MSFB to MSE. For the random-effects model, the *F* tests (for the main effects) involve the ratio of MSFA or MSFB to MSAB. For the mixed model with factor A fixed and factor B random, the *F* test for factor A involves the ratio of MSFA to MSAB, and the test for factor B involves the ratio of MSFB to MSE. For the mixed model with factor A random and factor B fixed, the *F* test for factor A involves the ratio of MSFA to MSE, while the test for factor B involves the ratio of MSFB to MSAB.

Problems for Section 14.10

14.44 Explain the difference between
(a) The one-factor and two-factor ANOVA models
(b) The randomized block design model and the two-factor factorial design model

14.45 A videocassette recorder (VCR) repair service wished to study the effect of VCR brand and service center on the repair time measured in minutes. Three VCR brands (A, B, C) were specifically selected for analysis. Three service centers were also selected. Each service center was assigned to perform a particular repair on two VCRs of each brand. The results were as follows:

Data file
VCREPAIR.DAT

	VCR Brands		
Service Centers	A	B	C
1	52	48	59
	57	39	67
2	51	61	58
	43	52	64
3	37	44	65
	46	50	69

(a) At the .05 level of significance
(1) Is there an effect due to service centers?
(2) Is there an effect due to VCR brand?
(3) Is there an interaction due to service center and VCR brand?
(b) Plot a graph of average service time for each service center for each VCR brand.

(c) If appropriate, use the Tukey procedure to determine which service centers and which VCR brands differ in average service time. (Use $\alpha = .05$.)
(d) Based upon the results, what conclusions can you reach concerning average service time?
(e) **ACTION** Write a memo to the CEO of the repair service describing your findings.
(f) If the three service centers were randomly selected, how would your analysis in (a) be affected?

Data file
AMPLIFY.DAT

14.46 An experiment was designed to study the effect of two factors on the amplification of a stereo recording. The factors were type of amplifier (four brands) and type of receiver (two brands). For each combination of factor levels, three tests were performed in which decibel output was measured. A higher decibel output means a better result. The *coded* results were as follows:

	Amplifiers			
Receiver	A	B	C	D
	9	8	8	10
R_1	4	11	7	15
	12	16	1	9
	7	5	0	6
R_2	1	9	1	7
	4	6	7	5

(a) At the .01 level of significance
(1) Is there an effect due to receivers?
(2) Is there an effect due to amplifiers?
(3) Is there an interaction between receivers and amplifiers?
(b) Plot a graph of average decibel output for each receiver for each amplifier.
(c) If appropriate, use the Tukey procedure to determine which amplifiers differ in average decibel output. (Use $\alpha = .01$.)
(d) **ACTION** Based upon the results, what conclusions can you reach concerning average decibel output? Write a memo on this to your music professor.

Data file
KNEE.DAT

14.47 A hospital administrator wished to examine postsurgical hospitalization periods following knee surgery. A random sample of 30 patients was selected, five for each combination of age group and type of surgery. The results, in number of postsurgical hospitalization days, were as follows:

	Age Group		
Type of Knee Surgery	Under 30	30 to 50	Over 50
	1	4	3
	3	3	5
Arthroscopy	2	2	2
	6	3	3
	2	2	3
	3	4	4
	10	5	8
Arthrotomy	6	11	12
	7	5	10
	8	6	3

(a) At the .05 level of significance
   (1) Is there a difference between the types of surgery?
   (2) Is there a difference between the age groups?
   (3) Is there an interaction between type of knee surgery and age group?
(b) Plot a graph of the average number of days of postsurgical hospitalization for each type of surgery for each age group.
(c) If appropriate, use the Tukey procedure to determine the age groups that differ in average number of postsurgical hospitalization days. (Use $\alpha = .05$.)
(d) **ACTION** Based on the results, what conclusions can the hospital administrator draw? Write a letter to the administrator about this.

● 14.48 The quality control director for a clothing manufacturer wanted to study the effect of operators and machines on the breaking strength (in pounds) of wool serge material. A batch of the material was cut into square yard pieces and these were randomly assigned, three each, to all 12 combinations of four operators and three machines chosen specifically for the experiment. The results were as follows:

Data file BREAKING.DAT

	Machine		
Operator	I	II	III
	115	111	109
A	115	108	110
	119	114	107
	117	105	110
B	114	102	113
	114	106	114
	109	100	103
C	110	103	102
	106	101	105
	112	105	108
D	115	107	111
	111	107	110

(a) At the .05 level of significance
   (1) Is there an effect due to operator?
   (2) Is there an effect due to machine?
   (3) Is there an interaction due to operator and machine?
(b) Plot a graph of average breaking strength for each operator for each machine.
(c) If appropriate, use the Tukey procedure to determine which operators and which machines differ in average breaking strength. (Use $\alpha = .05$.)
(d) **ACTION** Prepare a report for the quality control director regarding your findings on average breaking strength.

14.49 The production manager for an appliance manufacturer wished to determine the optimal length of time for the washing cycle of a household clothes washer. An experiment was designed to measure the effect of detergent used and washing cycle time on the amount of dirt removed from standard household laundry loads. Four brands of detergent (A, B, C, D) and four levels of washing cycle (18, 20, 22, 24 minutes) were specifically selected for analysis. In order to run the experiment, 32 standard household laundry loads (having equal weight and dirt) were randomly assigned, two each, to the 16 detergent–washing cycle time combinations. The results (in pounds of dirt removed) were as shown at the top of page 592:

Data file LAUNDRY.DAT

Detergent Brand	Washing Cycle Time (in Minutes) 18	20	22	24
A	.11	.13	.17	.17
	.09	.13	.19	.18
B	.12	.14	.17	.19
	.10	.15	.18	.17
C	.08	.16	.18	.20
	.09	.13	.17	.16
D	.11	.12	.16	.15
	.13	.13	.17	.17

(a) At the .05 level of significance
  (1) Is there an effect due to detergent?
  (2) Is there an effect due to washing cycle time?
  (3) Is there an interaction due to detergent and washing cycle time?
(b) Plot a graph of average amount of dirt removed (in pounds) for each detergent for each washing cycle time.
(c) If appropriate, use the Tukey procedure to determine which detergents and which washing cycle times differ with respect to average amount of dirt removed. (Use $\alpha = .05$.)
(d) **ACTION** Prepare a report for the production manager regarding your findings on average amount of dirt removed. Be sure to offer a recommendation as to what the optimal washing cycle should be for this type of household clothes washer.
(e) If the four brands of detergents were randomly selected, how would your analysis in (a) be affected?

14.11 Potential Hypothesis-Testing Pitfalls and Ethical Issues

14.11.1 Potential Pitfalls

In this chapter we focused on several experimental design models and introduced various parametric and distribution-free procedures that may be employed when analyzing possible differences in the numerical outcomes or response measurements among the (treatment) levels of some factor of interest. We then extended this to consider the effects of two factors in designed experiments. Again, part of a good data analysis is to understand the assumptions underlying each of the hypothesis-testing procedures and, using this as well as other criteria, select the one most appropriate for a given set of conditions. Thus, even when our primary purpose is a confirmatory analysis of a particular experiment or set of data, we must always perform an exploratory, descriptive analysis first so that through observation we may better understand what the data are conveying and what potential trends, relations, and effects are being indicated. Failure to take this careful look at the data is not itself unethical. It usually results, however, in a less than optimal analysis.

14.11.2 Ethical Issues

Ethical considerations arise when a researcher is manipulative of the hypothesis-testing process in ways that enable personal gain. In coordinating and managing a project dealing with a long-term designed experiment or ongoing clinical trial in such industries as tobacco, pharmaceuticals, and health care, it is imperative that the principal investigator develop an operational plan or protocol dealing with the process of data collection, evaluation, and analysis. Particularly when many people are involved in the process, a system of checks and balances must be established to prevent fraud, plagiarism, faking data, or falsifying results. As previously discussed in Section 11.11.2 (pages 412 to 415), the following are some of the ethical issues that arise when designing experiments and analyzing the results:

- Data collection method—randomization
- Informed consent from human subjects being "treated"
- Type of test—two-tailed or one-tailed
- Choice of level of significance α
- Data snooping
- Cleansing and discarding of data
- Reporting of findings
- Meta-analysis

Lives can depend on the acceptance and application of faulty research. This becomes even more important when ethics are involved. You should go back and reread Section 11.11, so that it becomes ingrained in your thinking process. Again, when discussing ethical issues concerning the hypothesis-testing methodology, the key is *intent*. We must distinguish between poor confirmatory data analysis and unethical practice. Unethical behavior occurs when a researcher willfully causes a selection bias in data collection, manipulates the treatment of human subjects without informed consent, uses *data snooping* to select the type of test (two-tailed or one-tailed) and/or level of significance to his or her advantage, hides the facts by discarding observations that do not support a stated hypothesis, and fails to report pertinent findings.

14.12 Hypothesis Testing Based on *c* Samples of Numerical Data: A Review

As observed in the summary chart for this chapter (see page 594), we may distinguish among approaches for comparing *c* groups containing numerical data based on the experimental design model used for obtaining the measured responses or outcomes. Hypothesis-testing methodology was developed separately for analyzing data obtained from one-way or completely randomized design models, randomized block design models, and two-factor factorial design models. After focusing on an appropriate grouping of similar test procedures, we need to look carefully at the assumptions and other criteria prior to selecting a particular procedure. On page 526 of Section 14.1 you were given a list emphasizing the important points to be discussed in the chapter. Check over the list now to see whether you feel you have an understanding of these key points. To be sure, you should be able to answer the following conceptual questions (page 595):

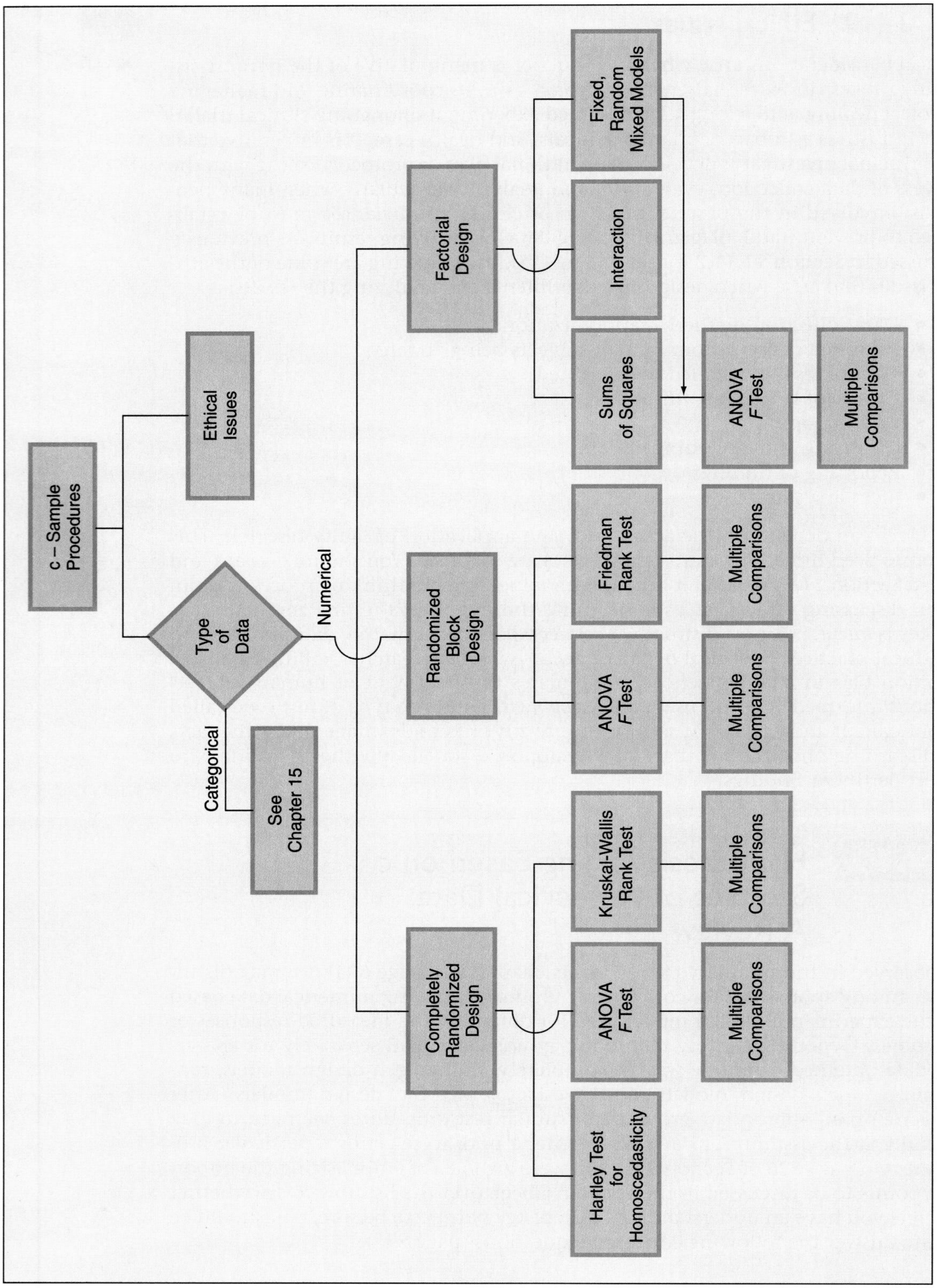

Chapter 14 summary chart.

1. What are the distinguishing features of the completely randomized design, randomized block design, and two-factor factorial design models?
2. What are the major assumptions of ANOVA?
3. Under what conditions should Hartley's F_{max} test be used?
4. Under what conditions should the one-way ANOVA *F* test to examine possible differences in the means of *c* independent populations be selected?
5. Under what conditions should the Kruskal-Wallis rank test to examine possible differences in the medians of *c* independent populations be selected?
6. When and how should multiple comparison procedures for evaluating pairwise combinations of the group means or medians be used?
7. Under what conditions should the randomized block *F* test to examine possible differences in the means of *c* related populations be selected?
8. Under what conditions should the Friedman rank test to examine possible differences in the medians of *c* related populations be selected?
9. Under what conditions should the two-way ANOVA *F* test to examine possible differences in the means of each factor in a factorial design be selected?
10. What do we mean by the concept of interaction in a factorial design?
11. How can we use the two-way ANOVA *F* test to examine possible interaction in the levels of the factors in a factorial design?

Check over the list of questions to see if indeed you know the answers and could (1) explain your answers to someone who did not read this chapter and (2) give reference to specific readings or examples that support your answer. Also, reread any of the sections that may have seemed fuzzy to see if they make sense now.

Getting It All Together

Key Terms

a posteriori *537*
among-block variation (SSBL) *560*
among-group variation (SSA) *530*
analysis of variance (ANOVA) *527*
ANOVA summary table *532*
blocks *558*
completely randomized design *527*
critical range *537*
Dunn procedure *549*
experimental error *528*
experimental units *558*
***F* distribution** *531*
factorial design *577*
fixed-effects model *588*
F_{max} distribution *540*
Friedman rank test *571*
Hartley's F_{max} test *540*
homogeneity of variance *539*
independence of errors *539*
inherent random error (SSE) *561*
interacting effects *586*
interaction *586*

Chapter Review Problems

14.50 ACTION Write a letter to a friend who has not taken a course in statistics and explain what this chapter has been about. To highlight this chapter's content, be sure to incorporate your answers to the 11 review questions on page 595.

14.51 The retailing manager of a food chain wishes to determine whether product location has any effect on the sale of pet toys. Three different aisle locations are to be considered: front, middle, and rear. A random sample of 18 stores was selected with 6 stores randomly assigned to each aisle location. The size of the display area and price of the product were constant for all stores. At the end of a one-week trial period, the sales volume (in $000) of the product in each store were as follows:

Data file LOCATE.DAT

Aisle Location		
Front	Middle	Rear
8.6	2.0	4.6
7.2	3.2	2.8
5.4	2.4	6.0
4.0	1.8	2.2
5.0	1.4	2.8
6.2	1.6	4.0

(a) At the .05 level of significance, is there evidence of a difference in the variances for the three aisle locations?
(b) At the .05 level of significance, is there evidence of a difference in average sales between the various aisle locations?

(c) If appropriate, use the Tukey-Kramer procedure to determine which aisle locations are different in average sales. (Use $\alpha = .05$.)

(d) What would the retailing manager conclude? Discuss fully the retailing manager's options with respect to aisle locations.

Suppose that, when setting up his experiment, the retailing manager is able to study the effects of shelf height in addition to aisle location. Thus, instead of the one-factor completely randomized design model above, he can set up a two-factor factorial design model, each factor having three levels with two replications for each aisle location–shelf height combination. That is, there are two factors to be studied: (1) location in aisle (front, middle, and rear) and (2) height of shelf (top, middle, and bottom). A sample of 18 stores are randomly assigned, two to each of the nine aisle location–shelf height level combinations. Again, the size of the display area and price of the product were constant for all stores. At the end of a 1-week trial period, the sales volume (in \$000) of the product in each store were as follows:

	Height of Shelf		
Aisle Location	Top	Middle	Bottom
Front	8.6	6.2	5.0
	7.2	5.4	4.0
Middle	3.2	2.0	1.8
	2.4	1.4	1.6
Rear	6.0	4.0	2.8
	4.6	2.8	2.2

Note that the sales volume (in \$000) are the same in both tables. Here, however, the six outcome measurements for each aisle location are listed in groups of two, corresponding to the three shelf height combinations.

(e) At the .05 level of significance
   (1) Is there an effect due to aisle location?
   (2) Is there an effect due to shelf height?
   (3) Is there an interaction between aisle location and shelf height?

(f) Plot a graph of average sales for each aisle location for each shelf height.

(g) If appropriate, use the Tukey procedure to determine which aisle locations and which shelf heights differ in sales. (Use $\alpha = .05$.)

(h) Based on the results, what conclusions can you reach concerning sales? Discuss.

(i) Compare and contrast your results here with those from the one-way experiment in parts (a)–(c). Discuss fully.

14.52 To examine effects of the work environment on attitude toward work, an industrial psychologist randomly assigned a group of 18 recently hired sales trainees to three "home rooms"—six trainees per room. Each room was identical except for wall color. One was light green, another was light blue, and the third was deep red.

During the week-long training program, the trainees stayed mainly in their respective home rooms. At the end of the program, an attitude scale was used to measure each trainee's attitude toward work (a low score indicates a poor attitude, a high score a good attitude). The following data were obtained (see page 598):

Data file
WORKATT.DAT

Room Color		
Light Green	Light Blue	Deep Red
46	59	34
51	54	29
48	47	43
42	55	40
58	49	45
50	44	34

Based on these data, the industrial psychologist wants to determine if there is evidence that work environment (i.e., color of room) has an effect on attitude toward work and, if so, which room color(s) significantly enhances attitude.

(a) Completely analyze the data. (Use $\alpha = .05$.)

(b) ACTION Based on this thorough analysis, draft a report discussing the implications of the findings for office design in large firms, knowing that the industrial psychologist might use this when meeting with the vice-president for human resources at the company.

Suppose that the vice-president wanted to know whether there was a gender effect present. If, in the above table, the first three observations for each level of the room color factor were for males and the last three observations for each level were for females:

(c) Completely reanalyze the data as a set of outcomes from a two-factor (fixed-effects) factorial design model where the gender factor has two levels, the room color factor has three levels, and there are three replications for each of the six gender–room color combinations. (Use $\alpha = .05$.)

(d) Compare and contrast your results here with those from the one-way experiment in part (a). Discuss fully.

14.53 A senior partner in a brokerage firm wishes to determine if there is really any difference between long-run performance of different categories of people hired as customers' representatives. The junior members of the firm are classified into four groups: professionals who have changed careers, recent business school graduates, former salesmen, and brokers hired from competing firms. A random sample of six individuals in each of these categories is selected and a detailed *performance score* is obtained.

Data file
BROKER.DAT

Customer Representative Backgrounds			
Professionals	Business School Grads	Salesmen	Brokers
88	65	61	83
85	73	67	87
95	54	74	90
96	72	65	84
91	81	68	92
88	69	77	94

(a) Is there evidence of a difference in the average performance score for the various categories? (Use $\alpha = .05$.)

(b) ACTION Write a memo to the senior partner explaining your findings.

Suppose that the senior partner wanted to know whether there was a gender effect present. If, in the prior table, the first three observations for each level of the customer representative background factor were for males and the last three observations for each level were for females:

(c) Completely reanalyze the data as a set of outcomes from a two-factor (fixed-effects) factorial design model where the gender factor has two levels, the background factor has four levels, and there are three replications for each of the eight gender–background combinations. (Use $\alpha = .05$.)

(d) Compare and contrast your results here with those from the one-way experiment in part (a). Discuss fully.

14.54 A recent wine tasting was held by the J. S. Wine Club in which eight wines were rated by club members. Information concerning the country of origin and the price were not known to the club members until after the tasting took place. The wines rated (and the prices paid for them) were

1. French white $8.59
2. Italian white $6.50
3. Italian red $6.50
4. French burgundy (red) $8.69
5. French burgundy (red) $9.75
6. California Beaujolais (red) $8.50
7. French white $7.75
8. California white $11.59

Data file WINE.DAT

The summated ratings over several characteristics for the 12 club members were as follows:

	Wine							
Respondent	1	2	3	4	5	6	7	8
A	10	17	15	9	12	6	15	9
B	9	14	11	5	16	2	15	7
C	10	18	10	5	18	5	10	10
D	9	11	13	10	17	11	14	9
E	10	16	12	8	18	8	10	10
F	6	16	3	8	4	2	2	5
G	9	12	14	9	9	6	6	5
H	7	12	11	8	15	9	12	8
I	10	18	12	12	16	10	10	16
J	16	9	10	13	18	11	15	14
K	14	16	13	12	15	15	17	11
L	15	17	10	13	15	16	16	13

(a) At the .01 level of significance, is there evidence of a difference in the average rating scores between the wines?

(b) What assumptions are necessary in order to do (a) of this problem? Comment on the validity of these assumptions.

(c) If appropriate, use the Tukey procedure to determine the wines that differ in average rating. (Use $\alpha = .01$.)

(d) Based upon your results in (c)
   (1) Do you think that country of origin has had an effect on the ratings?
   (2) Do you think that the type of wine (red versus white) has had an effect on the ratings?
   (3) Do you think that price has had an effect on the ratings? Discuss fully.

(e) Determine the relative efficiency of the randomized block design as compared with the completely randomized design.
(f) Ignore the blocking variable and "erroneously" reanalyze the data as a one-factor completely randomized design model where the one factor (brands of wines) has eight levels and each level contains a sample of 12 independent observations.
(g) Compare the SSBL and SSE terms in part (a) to the SSW term in part (f). Discuss.
(h) Using the results in parts (a), (f), and (g) as a basis, describe the problems that can arise when analyzing data if the wrong procedures are applied.

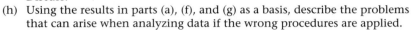

Collaborative Learning Mini-Case Projects

For each of the following, refer to the instructions on page 101.

Data file CEREAL.DAT

CL 14.1 Refer to CL 3.2 on page 101, CL 4.2 on page 165, and CL 5.2 on page 199. Your group, the ________ Corporation, has been hired by the food editor of a popular family magazine to study the cost and nutritional characteristics of ready-to-eat cereals. Armed with Special Data Set 2 of Appendix D on pages D6–D7, the ________ Corporation wishes to determine whether there is evidence of a difference in the average cost per serving of ready-to-eat cereals based on their classification as high fiber, moderate fiber, or low fiber.
(a) Thoroughly analyze the data.
(b) Write and submit an executive summary, clearly specifying all hypotheses, selected levels of significance, and the assumptions of the chosen test procedures.
(c) Prepare and deliver a five-minute oral presentation to the food editor of the magazine.

Data file FRAGRAN.DAT

CL 14.2 Refer to CL 3.3 on page 102, CL 4.3 on page 165, and CL 5.3 on page 199. Your group, the ________ Corporation, has been hired by the marketing director of a manufacturer of well-known men's and women's fragrances to study the characteristics of currently available fragrances. Armed with Special Data Set 3 of Appendix D on pages D8–D9, the ________ Corporation wishes to determine whether there is evidence of a difference in the average cost per ounce based on intensity (very strong, strong, medium, or mild).
(a) Thoroughly analyze the data.
(b) Write and submit an executive summary, clearly specifying all hypotheses, selected levels of significance, and the assumptions of the chosen test procedures.
(c) Prepare and deliver a five-minute oral presentation to the marketing director.

Data file CAMERA.DAT

CL 14.3 Refer to CL 3.4 on page 102, CL 4.4 on page 166, and CL 5.4 on page 200. Your group, the ________ Corporation, has been hired by the travel editor of a well-known newspaper who is preparing a feature article on compact 35-mm cameras. Armed with Special Data Set 4 of Appendix D on pages D10–D11, the ________ Corporation wishes to determine whether there is evidence of a difference in the average framing accuracy based on type of 35-mm camera.
(a) Thoroughly analyze the data.
(b) Write and submit an executive summary, clearly specifying all hypotheses, selected levels of significance, and the assumptions of the chosen test procedures.
(c) Prepare and deliver a five-minute oral presentation to the travel editor.

Case Study E Test-Marketing and Promoting a Ball-Point Pen

Data file
PEN.DAT

EPC Advertising has been hired by a well-established manufacturer of pens to develop a series of advertisements and national promotions for the upcoming holiday season. To prepare for this project, Nat Berry, research director at EPC Advertising, decided to initiate a study of the effect of advertising on product perception. After conferring with Kate Hansen, the chief of the statistics group, an experiment was designed in which five different advertisements were to be compared in the marketing of a ball-point pen. Advertisement A tended to greatly undersell the pen's characteristics. Advertisement B tended to slightly undersell the pen's characteristics. Advertisement C tended to slightly oversell the pen's characteristics. Advertisement D tended to greatly oversell the pen's characteristics. Advertisement E attempted to correctly state the pen's characteristics. A sample of 30 *adult* respondents, taken from a larger focus group, were randomly assigned to the five advertisements (so that there were six respondents to each). After reading the advertisement and developing a sense of "product expectation," all respondents unknowingly received the same pen to evaluate. The respondents were permitted to test their pen and the plausibility of the advertising copy. The respondents were then asked to rate the pen from 1 to 7 on three product characteristic scales as shown below:

	Extremely Poor			Neutral			Extremely Good
Appearance	1	2	3	4	5	6	7
Durability	1	2	3	4	5	6	7
Writing performance	1	2	3	4	5	6	7

The *combined* scores of three ratings (appearance, durability, and writing performance) are given below for the 30 respondents:

Product ratings for five advertisements

A	B	C	D	E
15	16	8	5	12
18	17	7	6	19
17	21	10	13	18
19	16	15	11	12
19	19	14	9	17
20	17	14	10	14

As a research assistant to Kate Hansen, you have been assigned to work on this project. You have an appointment to discuss this project with her at her office and when you arrive she states: "Hi, come on in and sit down. I'd offer you some coffee but I've just been called to a meeting so we'll make this brief today. You know, Nat Berry feels he is really on to something here—test-marketing the advertising copy—and I hope he's right. I had suggested that we set up a two-factor factorial design model with gender as one factor and with advertising copy as the second factor, but Nat said he would have trouble selling a sophisticated plan to our CEO. He reminded me about the KISS principle—Keep It Sound and Simple—and said that anything more than a one-way completely randomized design model just wouldn't fly with top management.

Anyway, Nat claimed he was primarily concerned with the content of the five possible ad copies, so the one-way model was what he got for this experiment. It is unfortunate, however, that we can't judge the gender factor in these potential holiday ads." Kate continued: "I want to be prepared for next week's meeting with the research team from our public relations group, so we don't take any heat for the design model that we've used. Please prepare an executive summary showing the advantages and disadvantages of the completely randomized design model and the two-factor factorial design model. That will allow us to discuss these alternatives and move forward with the client's project in a positive manner. Also, I'd like you to thoroughly analyze the data that were obtained from the completely randomized experiment. I'd like to have a detailed report on my desk in four days that summarizes your findings and includes, as an appendix, a discussion of the statistical analysis utilized. We'll then get together for lunch, go over the details, and prepare for the presentations to Nat and the CEO. Do you have any questions before you get started? No? Well, good luck—and don't hesitate to give me a call if something comes up."

————ONE WEEK LATER

Data file PENEAD.DAT

Following your presentation at the meeting with the research team from the agency's public relations group, their chief statistician John Mack suggested that an appropriate distribution-free method be considered for analyzing the data displayed in the table on page 601. "Many researchers would argue," John Mack observed, "that the 'product characteristic scales' used do not truly satisfy the criteria of *interval* or *ratio* scaling and, therefore, that distribution-free methods are more appropriate." "John, that might make for an interesting statistics argument," Nat Berry exclaimed, "we just have to be as sure as possible that we're practicing good data analysis" he continued. "Well Nat," Kate Hansen intervened. "That's just the point. I was, and still am, concerned about a potential gender effect that should be known if we really want to use effective advertising copy for our campaign. Fortunately, I'll have you know that my research assistant uncovered the fact that the first three ratings recorded in our table were provided by men and the last three by women in each of the five advertising copy samples. Now we can look at the experiment as a two-factor factorial design model and determine if there is a significant gender effect and study possible interaction as well." "Cool," said Nat Berry; "great," said John Mack. Kate Hansen continued: "We'd like to meet again next week and go over these findings. It'll give us all a chance to reflect on the advantages and disadvantages of the two-factor model versus the one-factor model that my assistant discussed twenty minutes ago." "Well, let's do it again next week, same time and place," ordered Nat Berry. John Mack followed: "Fine, but while you are preparing for next week's meeting, would you also take a look at the following data that I've collected in a similar manner to your experiment? In this new data set the sampled audience was composed only of high-school students who are part of a focus group, not adults. And the data are the ratings or responses only for students who were exposed to advertising copy E, which, as you may recall, attempted to correctly state the pen's characteristics. Please take a look at the corresponding group of adults in your study and analyze the differences in their responses."

The aforementioned combined ratings data for a sample of 8 *high-school student* respondents are:

14	13	15	9
11	13	12	16

You leave the room thinking about John Mack's last remarks. In addition to analyzing the original data as a two-factor factorial design model and preparing a discussion of the differences in the results obtained using that model and those obtained using the completely randomized design model that you just reported, you decide to answer John's requests by evaluating:

1. Whether there is evidence of a difference in the combined ratings of *adult* versus *high-school student* respondents subjected to advertisement E (which attempted to correctly state the pen's characteristics).
2. Whether there is evidence that the median combined ratings for the *high-school student* respondents subjected to advertisement E exceeds 12.

An answer to the former will identify possible differences in perceptions of adults and that of students with respect to the product. An answer to the latter will identify whether the product is "preferred" by the high school audience, since a median of 12 is the expected "neutral" combined characteristics rating—the sum of the appearance, durability, and writing performance ratings on three 7-point scales in which the value 4 is neutral. You will be preparing a detailed report on these matters for Kate Hansen.

Endnotes

1. Among-group variation is sometimes referred to as between-group variation. In these situations the sum of squares term is referred to as **Sum of Squares Between** groups or **SSB**.
2. In addition to this exploratory data analysis, a more confirmatory approach to examining the assumptions of a particular testing procedure would be taken before deciding if the procedure was viable for a given set of data. For the one-way ANOVA *F* test the major assumptions are that the sample data in each group are randomly and independently drawn from an underlying normal population and that these populations have equal variability (see Figures 14.2 and 14.3 on pages 528 and 529). To test for normality see Reference 2. To test for equality of population variances, a procedure developed by H. O. Hartley is presented in Section 14.4.7.
3. Since $c(c-1)/2$ simultaneous pairwise comparisons are being made, it is necessary to adjust the upper-tail area under the standardized normal distribution in order to obtain the appropriate critical value Z_U that enables the computed critical range to maintain an overall level of significance, α, throughout the experiment (see References 3 and 6).
4. The degrees of freedom associated with the sum of squares error (SSE) component are found by subtraction to be
$$n - 1 - (c - 1) - (r - 1)$$
and, since $n = rc$, through algebra we have
$$\begin{aligned} n - 1 - (c - 1) - (r - 1) &= rc - 1 - c + 1 - r + 1 \\ &= rc - c - r + 1 \\ &= (r - 1)(c - 1) \end{aligned}$$
5. In essence, in a randomized block design model the blocks are not given the same status of a factor. In Section 14.10 we shall see that when the blocks are considered important enough to be a second factor, the design is called a two-factor factorial model and tests for each factor effect would potentially be important.
6. We shall consider the general case in which there are n' observations for each combination of factor A and factor B (i.e., each cell). If there is only one observation per cell, the notation of the randomized block design model can be used with the blocks being considered a second factor of interest.
7. The degrees of freedom associated with the interaction (SSAB) component are found by subtraction to be
$$n - 1 - (c - 1) - (r - 1) - rc(n' - 1)$$
and, since $n = rcn'$, through algebra we have
$$\begin{aligned} n-1-(c-1)-(r-1)-rc(n'-1) &= rcn'-1-c+1-r+1-rcn'+rc \\ &= rc - c - r + 1 \\ &= (r - 1)(c - 1) \end{aligned}$$

References

1. Berenson, M. L., D. M. Levine, and M. Goldstein, *Intermediate Statistical Methods and Applications: A Computer Package Approach* (Englewood Cliffs, NJ: Prentice Hall, 1983).
2. Conover, W. J., *Practical Nonparametric Statistics*, 2d ed. (New York: Wiley, 1980).
3. Daniel, W. W., *Applied Nonparametric Statistics*, 2d ed. (Boston, MA: PWS Kent, 1990).
4. Dunn, O. J., "Multiple Comparisons Using Rank Sums," *Technometrics*, 1964, Vol. 6, pp. 241–252.
5. Hicks, C. R., *Fundamental Concepts in the Design of Experiments,* 3d ed. (New York: Holt, Rinehart and Winston, 1982).
6. Kirk, R. E., *Experimental Design,* 2d ed. (Belmont, CA: Brooks-Cole, 1982).
7. Kramer, C. Y., "Extension of Multiple Range Tests to Group Means with Unequal Numbers of Replications," *Biometrics*, 1956, Vol. 12, pp. 307–310.
8. *Microsoft EXCEL for Windows: Step by Step* (Redmond, WA: Microsoft Press, 1993).
9. Miller, R. G., *Simultaneous Statistical Inference*, 2d ed. (New York: Springer-Verlag, 1980).
10. *MINITAB Reference Manual Release 8* (State College, PA: Minitab, Inc., 1992).
11. Neter, J., W. Wasserman, and M. H. Kutner, *Applied Linear Statistical Models,* 3d ed. (Homewood, IL: Richard D. Irwin, 1990).
12. *SAS User's Guide Version 6* (Raleigh, NC: SAS Institute, 1988).
13. *STATISTIX Version 4.0* (Tallahassee, FL: Analytical Software, Inc., 1992).
14. Tukey, J. W., "Comparing Individual Means in the Analysis of Variance," *Biometrics*, 1949, Vol. 5, pp. 99–114.

chapter 15

Hypothesis Testing with Categorical Data

CHAPTER OBJECTIVE

To extend the basic principles of hypothesis-testing methodology to situations involving categorical variables.

15.1 Introduction

The analysis of **categorical data** for purposes of decision making is of vital importance in business, medical, and social science research. In conducting a survey, for example, questions are often written to solicit categorical rather than numerical responses. In the preceding four chapters we were concerned with hypothesis-testing procedures that are used when analyzing numerical data. In Chapters 11 and 12, a variety of one-sample tests were presented, in Chapter 13 several two-sample tests were described, and in Chapter 14 some *c*-sample tests were developed. In this chapter we shall extend our discussion of hypothesis-testing methodology to consider procedures that are used when analyzing categorical data. We will begin by focusing our attention on situations in which a single sample containing categorical data is drawn and interest centers on the testing of a hypothesis pertaining to a specified value of a population proportion. This will be followed by describing situations involving an analysis of differences in population proportions based on two independent samples, two related samples, and *c* independent samples. In addition, we shall extend our earlier discussions on the theory of probability in Sections 6.7 and 6.8 by presenting a more formal, confirmatory analysis of the hypothesis of independence in the joint responses to two categorical variables. Once again, in this chapter emphasis will be given to the assumptions behind the use of the various tests.

Upon completion of this chapter, you should be able to:

1. Know when and how to use the Z test for the population proportion p
2. Know when and how to use the Z test to examine possible differences in the proportions of two independent populations
3. Know when and how to use the χ^2 test to examine possible differences in the proportions of two independent populations
4. Understand the similarities and differences between the Z and χ^2 tests for differences in population proportions
5. Know when and how to use the χ^2 test to examine possible differences in the proportions of c independent populations
6. Know when and how to use the χ^2 test of independence in the joint responses to two categorical variables
7. Know when and how to use the McNemar test for a possible difference in proportions from two related populations

15.2 One-Sample Z Test for the Proportion

15.2.1 Introduction and Development

It is sometimes useful to test a hypothesis pertaining to a specified value of a population proportion p. For example, an individual may make a claim or statement about the value of the population proportion that corresponds to a categorical variable and then select a random sample from the population in order to perform a hypothesis test. The **sample proportion,** $p_s = X/n$, is computed and the value of this statistic must be compared to the hypothesized value of the parameter, p, so that a decision pertaining to the hypothesis can be made.

If certain assumptions can be met, we may recall from Section 9.3 that the sampling distribution of a proportion will follow a standardized normal distribution. Thus, to evaluate the magnitude of the difference between the sample proportion p_s and the hypothesized population proportion p, the test statistic Z is given in Equation (15.1):

$$Z \cong \frac{p_s - p}{\sqrt{\frac{p(1-p)}{n}}} \tag{15.1}$$

where

$$p_s = \frac{X}{n} = \frac{\text{number of successes in sample}}{\text{sample size}} = \text{observed proportion of successes}$$

p = proportion of successes from the null hypothesis

The test statistic Z is approximately normally distributed.

Alternatively, instead of examining the *proportion* of successes in a sample, as in Equation (15.1), we may wish to study the *number* of successes in a sample. The test statistic Z for determining the magnitude of the difference between the number of successes in a sample and the hypothesized or expected number of successes in the population is presented in Equation (15.2) as

$$Z \cong \frac{X - np}{\sqrt{np(1-p)}} \tag{15.2}$$

Again, this test statistic Z is approximately normally distributed. We may recall from Section 8.6 that although the random variable X (that is, the number of successes in the sample) follows a binomial distribution, if the sample size is large enough [that is, both $np \geq 5$ and $n(1 - p) \geq 5$], the normal distribution provides a good approximation to the binomial distribution.

Aside from possible rounding errors, the test statistic Z given by Equations (15.1) and (15.2) will provide exactly the same results. The two alternative forms of the test statistic are equivalent because the numerator of Equation (15.2) is n times the numerator of Equation (15.1) and the denominator of Equation (15.2) is also n times the denominator of Equation (15.1). The choice in a particular application is up to the user.

15.2.2 Application

To illustrate the use of the (one-sample) Z test for a hypothesized proportion, let us return to the cereal box packaging example discussed in Chapters 9–11. The production manager had also been concerned with the sealing process for filled boxes. Once the package inside the box is filled, it is supposed to be sealed so that it is airtight. Based on past experience, however, it is known that one out of ten packages (that is, 10% or .10) do not meet standards for sealing and must be "reworked" in order to pass inspection. To alter this situation, suppose that the production manager implements a newly developed packaging system on a trial basis. After a one-day "break-in" period, he takes a random sample of 200 boxes that represent daily output at the plant and, through inspection, finds that 11 need rework. The production manager would want to determine whether there is evidence that, under the new packaging system, the proportion of defective packages has improved (that is, has decreased below .10).

In terms of proportions (rather than percentages), the null and alternative hypotheses can be stated as follows:

$$H_0: \quad p \geq .10$$
$$H_1: \quad p < .10$$

Since the production manager is interested in whether or not there has been a significant reduction in the proportion of defective packages owing to the new process, the test is one-tailed. If a level of significance α of .05 is selected, the rejection and nonrejection regions would be set up as in Figure 15.1, and the decision rule would be

$$\text{Reject } H_0 \text{ if } Z < -1.645;$$
$$\text{otherwise do not reject } H_0.$$

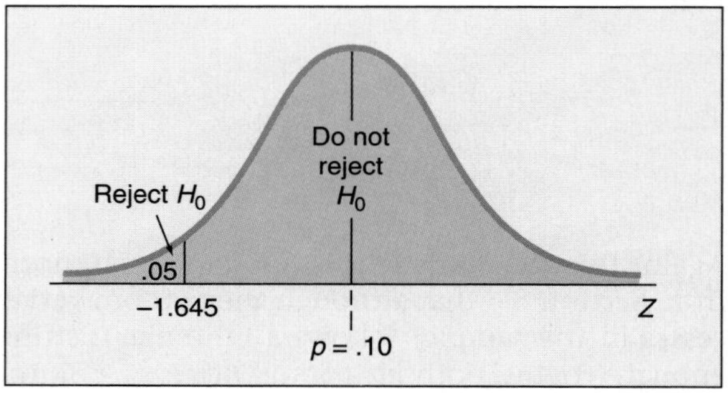

Figure 15.1
One-tailed test of hypothesis for a proportion at the .05 level of significance.

From our data,

$$p_s = \frac{11}{200} = .055$$

Using Equation (15.1) we have

$$Z \cong \frac{p_s - p}{\sqrt{\dfrac{p(1-p)}{n}}} = \frac{.055 - .10}{\sqrt{\dfrac{(.10)(.90)}{200}}} = \frac{-.045}{\sqrt{.00045}} = \frac{-.045}{.0212} = -2.12$$

or using Equation (15.2), we have

$$Z \cong \frac{X - np}{\sqrt{np(1-p)}} = \frac{11 - 200(.10)}{\sqrt{(200)(.10)(.90)}} = \frac{11 - 20}{\sqrt{18}} = \frac{-9}{4.243} = -2.12$$

Since $-2.12 < -1.645$, we reject H_0. Thus, the manager may conclude that there is evidence that the proportion of defectives with the new system is less than .10.

● **Finding the *p* Value** As an alternative approach toward making a hypothesis-testing decision, we may also compute the *p* value for this situation (see Sections 11.5 and 11.8). Since a one-tailed test is involved in which the rejection region is located only in the lower tail (see Figure 15.2), we need to find the area below a *Z* value of -2.12. From Table E.2, this probability will be $.5000 - .4830 = .0170$. Since this value is less than $\alpha = .05$, the null hypothesis can be rejected.

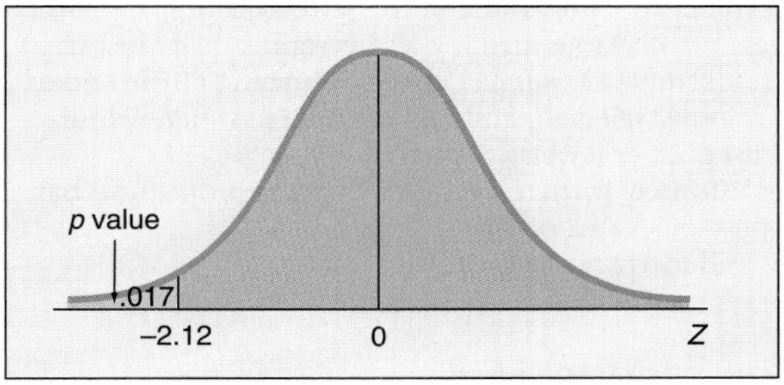

Figure 15.2
Determining the *p* value for a one-tailed test.

Problems for Section 15.2

15.1 Prove that the formula on the right-hand side of Equation (15.1) on page 607 is equivalent to the formula on the right-hand side of Equation (15.2).

15.2 A television manufacturer had claimed in its warranty that in the past not more than 10% of its television sets needed any repair during their first two years of operation. In order to test the validity of this claim a government testing agency selects a sample of 100 sets and finds that 14 sets required some repair within their first two years of operation. Using the .01 level of significance,
(a) Is the manufacturer's claim valid or is there evidence that the claim is not valid?
(b) Compute the *p* value and interpret its meaning.

15.3 The Giansante Company, provider of extermination services, claims that no more than 15% of its customers need repeated treatment after a 90-day warranty period. In order to determine the validity of this claim, a consumer organization selected a sample of 100 customers and found that 22 needed repeated treatment after the 90-day warranty period.
(a) Is there evidence at the .05 level of significance that the claim is not valid (i.e., that the proportion needing treatment is greater than .15)?
(b) Compute the *p* value and interpret its meaning.

● 15.4 The personnel director of a large insurance company is interested in reducing the turnover rate of data processing clerks in the first year of employment. Past records indicate that 25% of all new hires in this area are no longer employed at the end of one year. Extensive new training approaches are implemented for a sample of 150 new data processing clerks. At the end of a one-year period, of these 150 individuals, 29 are no longer employed.
(a) At the .01 level of significance, is there evidence that the proportion of data processing clerks who have gone through the new training and are no longer employed is less than .25?
(b) Compute the *p* value and interpret its meaning.

● 15.5 The marketing manager for an automobile manufacturer is interested in determining the proportion of new compact-car owners who would have purchased a driver's side inflatable auto bag if it had been available for an additional cost of \$300. The manager believes from previous information that the proportion is .30. Suppose that a survey of 200 new compact-car owners is selected and 79 indicate that they would have purchased the inflatable air bags.
(a) At the .10 level of significance, is there evidence that the population proportion is different from .30?
(b) Compute the p value and interpret its meaning.

15.6 The marketing branch of the Mexican Tourist Bureau would like to increase the proportion of tourists who purchase silver jewelry while vacationing in Mexico from its present estimated value of .40. Toward this end, promotional literature describing both the beauty and value of the jewelry is prepared and distributed to all passengers on airplanes arriving at a certain seaside resort during a one-week period. A sample of 500 passengers returning at the end of the one-week period is randomly selected, and 227 of these passengers indicate that they have purchased silver jewelry.
(a) At the .05 level of significance, is there evidence that the proportion has increased above the previous value of .40?
(b) Compute the p value and interpret its meaning.

Interchapter Problems for Section 15.2

15.7 Refer to Problem 5.3 on page 175.
(a) Suppose that a recently taken random sample of 250 individuals indicates that 158 are using Microsoft products as their primary Windows business applications software. At the .05 level of significance, is there evidence that the proportion has changed from the previous 1992 market share?
(b) Compute the p value and interpret its meaning.
(c) **Class Project** Consider your class to be a sample of all students at your school. Determine the proportion of students in your class who use Microsoft products as their primary Windows business applications software. At the .05 level of significance, is there evidence that this proportion is different from the 1992 market share?

15.8 Refer to Problem 5.8 on page 178.
(a) Suppose that a recently taken random sample of 200 individuals indicates that 78 preferred Kellogg products to those from all other companies producing ready-to-eat cereals. At the .05 level of significance, is there evidence that the proportion has changed from the 1992 market share?
(b) Compute the p value and interpret its meaning.
(c) **Class Project** Consider your class to be a sample of all students at your school. Determine the proportion of students in your class who prefer Kellogg products to those from all other companies producing ready-to-eat cereals. At the .05 level of significance, is there evidence that this proportion is different from the 1992 market share?

15.9 Refer to Problem 5.47 on page 194.
(a) Suppose that a random sample of 175 beer drinkers in New York State indicates that 93 prefer Anheuser-Busch beer to all other brands. At the .10 level of significance, is there evidence that the proportion has changed from the stated market share in a recent year?
(b) Compute the p value and interpret its meaning.

15.10 Refer to Problem 5.50 on page 196.
(a) Suppose that a random sample of 200 individuals, taken this year, indicates that 106 preferred caffeinated cola to all other carbonated soft drinks. At the .01 level of significance, is there evidence that the propor-

tion has changed from the recorded market share based on supermarket sales in a recent year?

(b) Compute the p value and interpret its meaning.

(c) **Class Project** Consider your class to be a sample of all students at your school. Determine the proportion of students in your class who prefer caffeinated cola to all other carbonated soft drinks. At the .01 level of significance, is there evidence that this proportion is different from the recorded market share based on supermarket sales in a recent year?

15.3 Z Test for Difference in Two Proportions (Independent Samples)

15.3.1 Introduction

A researcher is often concerned with making comparisons and analyzing differences between two populations in terms of some categorical characteristic. A test for the difference between two proportions based on independent samples can be performed using two different methods. In this section we present a procedure whose test statistic Z is approximated by a standard normal distribution. In Section 15.4 we will develop a procedure whose test statistic χ^2 is approximated by a chi-square distribution with one degree of freedom. The results will be equivalent.

15.3.2 Development

When evaluating differences between two proportions based on independent samples, a Z test can be employed. The test statistic Z used to determine the difference between the two population proportions is based on the difference between the two sample proportions $(p_{s_1} - p_{s_2})$. Because of the central limit theorem discussed in Section 9.2, this test statistic may be approximated by a standard normal distribution for large enough sample sizes. As shown in Equation (15.3), the Z-test statistic is

$$Z \cong \frac{(p_{s_1} - p_{s_2}) - (p_1 - p_2)}{\sqrt{\bar{p}(1-\bar{p})\left(\frac{1}{n_1} + \frac{1}{n_2}\right)}} \quad \textbf{(15.3)}$$

with

$$\bar{p} = \frac{X_1 + X_2}{n_1 + n_2} \qquad p_{s_1} = \frac{X_1}{n_1} \qquad p_{s_2} = \frac{X_2}{n_2}$$

where p_{s_1} = sample proportion obtained from population 1
X_1 = number of successes in sample 1
n_1 = size of the sample taken from population 1

p_1 = proportion of successes in population 1
p_{s_2} = sample proportion obtained from population 2
X_2 = number of successes in sample 2
n_2 = size of the sample taken from population 2
p_2 = proportion of successes in population 2
$\bar{p}$ = pooled estimate of the population proportion

Under the null hypothesis, it is assumed that the two population proportions are equal. We should note that $\bar{p}$, the pooled estimate for the population proportion, is based on the null hypothesis. Therefore, when testing for equality in the two population proportions we obtain an overall estimate of the common population proportion by combining or pooling together the two sample proportions. This estimate, $\bar{p}$, is simply the number of successes in the two samples combined $(X_1 + X_2)$ divided by the total sample size from the two sample groups $(n_1 + n_2)$.

A distinguishing feature of employing this Z test for evaluating differences in population proportions based on two independent samples is that we may be interested in either determining whether there is *any difference* in the proportion of successes in the two groups (two-tailed test) or whether one group had a *higher* proportion of successes than the other group (one-tailed test).[1]

Two-Tailed Test	One-Tailed Test	One-Tailed Test
H_0: $p_1 = p_2$	H_0: $p_1 \geq p_2$	H_0: $p_1 \leq p_2$
H_1: $p_1 \neq p_2$	H_1: $p_1 < p_2$	H_1: $p_1 > p_2$

where p_1 = proportion of successes in population 1
p_2 = proportion of successes in population 2

To test the null hypothesis of no difference in the proportions of two independent populations

$$H_0: \; p_1 = p_2$$

against the alternative that the two population proportions are not the same

$$H_1: \; p_1 \neq p_2$$

we may use the test statistic Z, given by Equation (15.3), and, for a given level of significance, α, we would reject the null hypothesis if the computed Z-test statistic exceeds the upper-tailed critical value from the standard normal distribution or if the computed test statistic falls below the lower-tailed critical value from the standard normal distribution.

15.3.3 Application

To illustrate the use of the Z test for homogeneity of two proportions, suppose that a personnel director was investigating employee perception of the fairness of two different methods of performance evaluation. To test for differences between the two methods, 160 employees were each randomly assigned to be evaluated by one of the two methods. A total of 78 employees were assigned to be evaluated under method 1, which enabled individuals to provide feedback to supervisory queries as part of their evaluation process. An additional 82 employees were assigned to method 2, which enabled individuals to provide self-assessment ratings of their work performance. Following the evaluations, the employees were asked whether they considered the performance evaluation process to have been fair or unfair. From the first sample, 63 employees felt that method 1 was fair. From the second sample, 49 employees believed that method 2 was fair. These results are displayed in Table 15.1.

Table 15.1 A comparison of employee perception of fairness based on two methods of performance evaluation.

	Evaluation Method	
	1	2
Sample sizes:	$n_1 = 78$	$n_2 = 82$
No. of employees perceiving method as fair:	$X_1 = 63$	$X_2 = 49$

The null and alternative hypotheses are

$$H_0\!: \quad p_1 = p_2 \text{ or the difference in proportions } p_1 - p_2 = 0$$

$$H_1\!: \quad p_1 \neq p_2 \text{ or the difference in proportions } p_1 - p_2 \neq 0$$

If the test were to be carried out at the .01 level of significance, the critical values would be −2.58 and +2.58 (see Figure 15.3) and our decision rule is

$$\text{Reject } H_0 \text{ if } Z > +2.58$$
$$\text{or if } Z < -2.58;$$
$$\text{otherwise do not reject } H_0.$$

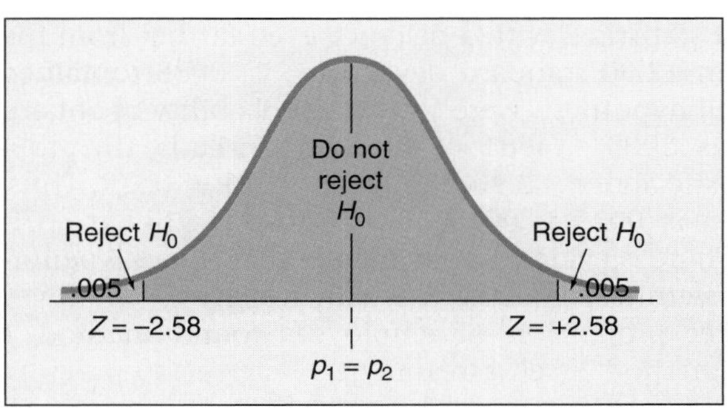

Figure 15.3 Testing a hypothesis about the difference between two proportions at the .01 level of significance.

For our data we have

$$Z \cong \frac{(p_{s_1} - p_{s_2}) - (p_1 - p_2)}{\sqrt{\bar{p}(1 - \bar{p})\left(\dfrac{1}{n_1} + \dfrac{1}{n_2}\right)}}$$

where

$$p_{s_1} = \frac{X_1}{n_1} = \frac{63}{78} = .808 \qquad p_{s_2} = \frac{X_2}{n_2} = \frac{49}{82} = .598$$

and

$$\bar{p} = \frac{X_1 + X_2}{n_1 + n_2} = \frac{63 + 49}{78 + 82} = \frac{112}{160} = .70$$

so that

$$\begin{aligned} Z &\cong \frac{.808 - .598}{\sqrt{(.70)(.30)\left(\frac{1}{78} + \frac{1}{82}\right)}} \\ &= \frac{.210}{\sqrt{(.2100)\left(\frac{160}{6{,}396}\right)}} \\ &= \frac{.210}{\sqrt{.005253}} \\ &= \frac{.210}{.0725} = +2.90 \end{aligned}$$

Using a .01 level of significance, the null hypothesis (H_0) is rejected because $Z = +2.90 > +2.58$. If the null hypothesis were true, there would be an $\alpha = .01$ probability of obtaining a Z-test statistic either larger than $+2.58$ standard deviations from the center of the Z distribution or smaller than -2.58 standard deviations from the center of the Z distribution. The p value, or probability of obtaining a difference between the two sample proportions even larger than the .210 observed here, which translates into a test statistic Z with a distance even farther from the center of the Z distribution than ± 2.90 standard deviations, is .0038 (obtained from Table E.2). That is, if the null hypothesis were true, the probability of obtaining a Z-test statistic below -2.90 is $.5000 - .4981 = .0019$ and, similarly, the probability of obtaining a Z-test statistic above $+2.90$ is $.5000 - .4981 = .0019$. Thus, for this two-tailed test, the p value is $.0019 + .0019 = .0038$. Since $.0038 < \alpha = .01$, the null hypothesis is rejected. There is evidence to conclude that the two performance evaluation methods are significantly different with respect to employee perception of fairness. A greater proportion of employees found method 1 (employee feedback) fairer than method 2 (self-assessment).

Problems for Section 15.3

15.11 We wish to determine if there is any difference in the popularity of football between college-educated males and non-college-educated males. A sample of 100 college-educated males revealed 55 who considered themselves football fans. A sample of 200 non-college-educated males revealed 125 who considered themselves football fans.

(a) Is there any evidence of a difference in football popularity between college-educated and non-college-educated males at the .01 level of significance?

(b) Compute the p value in (a) and interpret its meaning.

15.12 A marketing study conducted in a large city showed that of 100 married women who worked full time, 43 ate dinner at a restaurant at least one night during a typical workweek, and a sample of 100 married women who did not work full time indicated that 27 ate dinner at a restaurant at least one night during the typical workweek.

(a) Using a .05 level of significance, is there evidence of a difference between the two groups of married women in the proportion who eat dinner in a restaurant at least once during the typical workweek?

(b) Compute the p value in (a) and interpret its meaning.

15.13 A professor of accountancy was studying the readability of the annual reports of two major companies. A random sample of 100 certified public accountants was selected. Fifty were randomly assigned to read the annual report of Company A, and the other 50 were to read the annual report of Company B. Based upon a standard measure of readability, 17 found Company A's annual report "understandable" and 23 found Company B's annual report "understandable."

(a) At the .10 level of significance, is there any evidence of a difference between the two companies in the proportion of CPAs who would find the annual reports understandable?

(b) Compute the p value in (a) and interpret its meaning.

15.14 The director of marketing for a company manufacturing laundry detergent conducted an experiment to compare customer satisfaction with the laundry detergent product based on level of temperature used for the wash. A random sample of 500 individuals who agreed to participate in the experiment were asked to use the product on a standard-sized load under a low temperature setting. A second random sample of 500 participants were asked to use the product on a standard-sized load under a high temperature setting. Of the 500 participants who used a low temperature setting, 280 were happy with the cleansing outcome. Of the 500 participants who used a high temperature setting, 320 were happy with the results.

(a) At the .05 level of significance, is there evidence that the brand of laundry detergent is preferred when used with a high temperature setting than with a low temperature setting?

(b) Compute the p value in (a) and interpret its meaning.

15.15 The manager of a campus bookstore conducted a survey to investigate if there were any differences between males and females with respect to the consideration of the purchase of educational videotapes. Depending on the answer, her goal was to develop pertinent promotional material that would lead to increased sales in educational videotapes over the coming semester. Of the 40 males in the survey, 13 stated they would consider the purchase of educational videotapes. Of the 30 females in the survey, 15 said they would consider the purchase of educational videotapes.

(a) At the .01 level of significance, is there evidence of a difference in the proportion of males and females who would consider the purchase of educational videotapes?

(b) Compute the p value in (a) and interpret its meaning.

(c) Given the results in (a) and (b), what should the bookstore manager do with respect to a future promotional campaign? Discuss your recommendations.

Interchapter Problems for Section 15.3

15.16 Refer to Problem 6.5 on page 209.

(a) At the .05 level of significance, is there evidence of a difference in the proportion of students who have an entertainment credit card based on whether or not they have a bank credit card?

(b) Compute the p value in (a) and interpret its meaning.

(c) Suppose that instead of determining whether there is a difference between the two groups as in (a), we wish to know if there is evidence that students who have a bank credit card are more likely to have an entertainment credit card than students who do not have a bank credit card. At the .05 level of significance, what conclusion do you reach?

● 15.17 Refer to Problem 6.7 on page 210.

(a) Is there evidence of a difference in the proportion of males and females who enjoy shopping for clothing? (Use $\alpha = .05$.)

(b) Compute the p value in (a) and interpret its meaning.

(c) Suppose that instead of determining whether there is a difference between the two groups as in (a), we wish to know if there is evidence that the proportion of females who enjoy shopping for clothing is higher than the proportion of males. At the .05 level of significance, what conclusion do you reach?

15.18 Refer to Problem 6.8 on page 210.

(a) At the .10 level of significance, is there evidence of a difference in the proportion of males and females who use the health club facilities?

(b) Compute the p value in (a) and interpret its meaning.

15.19 Refer to Problem 6.65 on page 235.

(a) At the .05 level of significance, is there evidence that a greater proportion of non-college graduates (as opposed to college graduates) held their current jobs for more than five years?

(b) Compute the p value in (a) and interpret its meaning.

15.4 χ^2 Test for Difference in Two Proportions (Independent Samples)

15.4.1 Introduction

In the previous section we described the Z test for the difference between two proportions based on independent samples. Rather than directly comparing proportions of success, in this section we will view the data in terms of the frequency of success in two groups. We will develop a procedure whose test statistic χ^2 is approximated by a chi-square distribution with one degree of freedom. The results obtained by employing the χ^2 test are, except for possible rounding error, equivalent to those obtained by using the Z test of Section 15.3.

15.4.2 Development

If it is of interest to compare the tallies or counts of categorical responses between two independent groups, a two-way **table of cross-classifications** can be developed (see Section 5.5) to display the frequency of occurrence of successes and failures for each group. Such a table is also called a **contingency table**, which, as we may recall, was used in Chapter 6 to define and study probability, particularly from an objective empirical approach. In this section, however, we develop methodology for a more confirmatory analysis of data presented in such contingency tables.

To motivate this discussion, let us return to the personnel director's study of employee perception of fairness of two different methods of performance evaluation that was presented in the previous section. Table 15.2 is a schematic layout of a cross-classification table resulting from the study and Table 15.3 is the contingency table displaying the actual data from the study.

Table 15.2 Layout of a 2 × 2 contingency table for comparing employee perception of fairness based on two methods of performance evaluation.

Employee Perception	Evaluation Method 1	Evaluation Method 2	Totals
Fair	X_1	X_2	X
Unfair	$n_1 - X_1$	$n_2 - X_2$	$n - X$
Totals	n_1	n_2	n

where X_1 = number of employees who believe method 1 is fair

X_2 = number of employees who believe method 2 is fair

$n_1 - X_1$ = number of employees who believe method 1 is unfair

$n_2 - X_2$ = number of employees who believe method 2 is unfair

$X = X_1 + X_2$ = the number of fair ratings

$n - X = (n_1 - X_1) + (n_2 - X_2)$ = the number of unfair ratings

n_1 = number of employees in the sample evaluated under method 1

n_2 = number of employees in the sample evaluated under method 2

$n = n_1 + n_2$ = total number of employees evaluated in the study

Table 15.3 2 × 2 contingency table for comparing *observed* employee perception of fairness based on two methods of performance evaluation.

Employee Perception	Evaluation Method 1	Evaluation Method 2	Totals
Fair	63	49	112
Unfair	15	33	48
Totals	78	82	160

The contingency table displayed in Table 15.3 has two rows, indicating whether the employees perceived their evaluations to be fair (that is, success) or unfair (that is, failure), and two columns, one for each method of performance evaluation. Such a table is called a 2 × 2 table. The cells in the table indicate the frequency for each row and column combination. The row totals indicate the number of fair and unfair ratings; the column totals are the sample sizes or number of people evaluated by each method. The proportion of employees who perceive their performance evaluations to be fair may be obtained by dividing the number of fair ratings for a particular method by the number of employees evaluated by that method. A methodology known as the **χ^2 test for homogeneity of proportions** may then be employed to compare the proportions for the two methods.

To test the null hypothesis of no differences in the two population proportions

$$H_0: \quad p_1 = p_2$$

against the alternative that the two population proportions are different

$$H_1: \quad p_1 \neq p_2$$

we obtain the χ^2-test statistic given by

$$\chi^2 = \sum_{\text{all cells}} \frac{(f_o - f_e)^2}{f_e} \tag{15.4}$$

where f_o is the *observed frequency* or actual tally in a particular cell of a 2 × 2 contingency table

and

f_e is the frequency we would expect to find in a particular cell if the null hypothesis is true.

To compute the *expected frequency* (f_e) in any cell requires an understanding of its conceptual foundation. If the null hypothesis is true and the proportion of fair ratings is equal for each population, then the sample proportions computed from the two groups should differ from each other only by chance, since they would each be providing an estimate of the common population parameter p. In such a situation, a statistic that would pool or combine these two separate estimates into one overall or average estimate of the population parameter p would provide more information than any one of the two separate estimates. This statistic, given by the symbol $\bar{p}$, therefore represents the overall or average proportion of fair ratings for the two groups combined (that is, the total number of fair ratings divided by the total number of employees rated). Using the notation for proportions as in Section 15.3, this can be stated as

$$\bar{p} = \frac{(X_1 + X_2)}{(n_1 + n_2)} = \frac{X}{n} \tag{15.5}$$

and its complement, $1 - \bar{p}$, represents the overall or average proportion of unfair ratings over the two evaluation methods.

To obtain the expected frequency (f_e) for each cell pertaining to fair ratings (that is, the first row in the contingency table), we multiply the sample size (or column total) for an employee evaluation method by $\bar{p}$. To obtain the expected frequency (f_e) for each cell pertaining to unfair ratings (that is, the second row in the contingency table), we multiply the sample size (or column total) for an employee evaluation method by $(1 - \bar{p})$.

The test statistic shown in Equation (15.4) approximately follows a chi-square distribution with degrees of freedom equal to the number of rows in the contingency table minus one times the number of columns in the table minus one:

$$\text{degrees of freedom} = (r - 1)(c - 1)$$

where r = number of rows in the table
c = number of columns in the table

For our 2×2 contingency table there is one degree of freedom; that is,

$$\text{degrees of freedom} = (2 - 1)(2 - 1) = 1$$

Using a level of significance α, the null hypothesis may be rejected in favor of the alternative if the computed χ^2-test statistic exceeds $\chi^2_{U(1)}$, the upper-tailed critical value from the **chi-square distribution** having one degree of freedom. That is, the decision rule is to reject H_0 if

$$\chi^2 > \chi^2_{U(1)}$$

as illustrated in Figure 15.4.

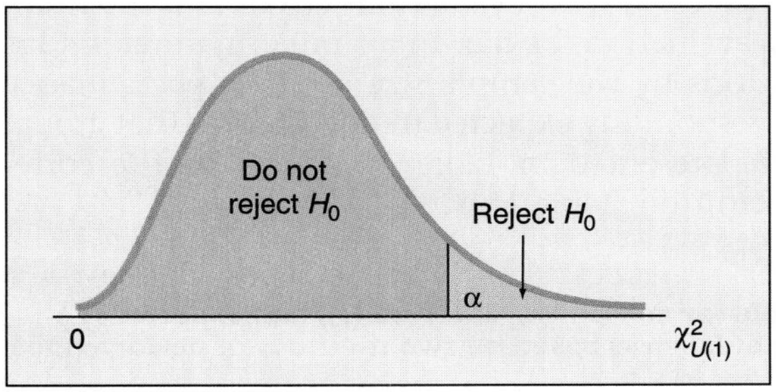

Figure 15.4
Testing a hypothesis for the difference between two proportions using the χ^2 test.

Referring to Equation (15.4), if the null hypothesis were true, the computed χ^2-test statistic should be close to 0 since the squared difference between what we actually observe in each cell (f_o) and what we theoretically expect (f_e) would be very small. On the other hand, if H_0 is false and there are real differences in the population proportions, we should expect the computed χ^2-test statistic to be large since the discrepancy between what we actually observe in each cell and what we theoretically expect will be magnified when we square the differences. However, what constitutes a large difference in a cell is relative. The same actual difference between f_o and f_e would contribute more to the χ^2-test statistic from a cell in which only a few observations are expected (f_e) than from a cell where many observations are expected. This is because a standardizing adjustment is made for the size of the cell—the squared difference between f_o and f_e is divided by the expected frequency

(f_e) in the cell. The χ^2-test statistic given in Equation (15.4) is then obtained by summing each standardized value $(f_o - f_e)^2/f_e$ over all the cells of the contingency table.

15.4.3 Application

To illustrate the use of the χ^2 test for homogeneity of two proportions, we again turn our attention to our personnel director's performance evaluation study. The results were displayed in Table 15.3 on page 617.

The null hypothesis (H_0: $p_1 = p_2$) states that, when comparing the two performance evaluation methods, there is no difference in the proportion of employees with respect to the perception of fairness. Using Equation (15.5) on page 618 we can estimate the common parameter p, the true proportion of employees who believe these performance evaluation methods to be fair. That is, $\bar{p}$, the overall or average proportion of employees who believe these performance evaluation methods are fair, is computed as

$$\bar{p} = \frac{(X_1 + X_2)}{(n_1 + n_2)} = \frac{X}{n}$$

$$= \frac{(63 + 49)}{(78 + 82)} = \frac{112}{160}$$

$$= .70$$

The estimated proportion of employees who do not feel that these evaluation methods are fair is the complement, $(1 - \bar{p})$, or .30. Multiplying these two proportions by the sample size used for performance evaluation method 1 gives the number of employees expected to perceive their evaluations as fair and the number not expected to perceive their evaluations as fair. In a similar manner, multiplying the two respective proportions by the sample size used for performance evaluation method 2 yields the corresponding expected frequencies for that group. All these expected frequencies are presented in Table 15.4, next to the corresponding observed frequencies taken from Table 15.3.

Table 15.4 **2 × 2 contingency table for comparing *observed* (f_o) and *expected* (f_e) employee perception of fairness based on two methods of performance evaluation.**

	Evaluation Method				
	1		2		
Employee Perception	Observed	Expected	Observed	Expected	Totals
Fair	63	54.6	49	57.4	112
Unfair	15	23.4	33	24.6	48
Totals	78	78	82	82	160

To test the null hypothesis of homogeneity of proportions

$$H_0: \ p_1 = p_2$$

against the alternative that the true population proportions are not equal

$$H_1: \ p_1 \neq p_2$$

we use the actual and expected data from Table 15.4 to compute the χ^2-test statistic given by Equation (15.4). The calculations are presented in Table 15.5.

Table 15.5 Computation of the χ^2-test statistic for the performance evaluation data.

f_o	f_e	$(f_o - f_e)$	$(f_o - f_e)^2$	$(f_o - f_e)^2/f_e$
63	54.6	+8.4	70.56	1.293
49	57.4	−8.4	70.56	1.229
15	23.4	−8.4	70.56	3.015
33	24.6	+8.4	70.56	2.868
				8.405

If a .01 level of significance is chosen, the critical value of the χ^2-test statistic could be obtained from Table E.4, a replica of which is presented as Table 15.6. The chi-square distribution is a skewed distribution whose shape depends solely on the number of degrees of freedom. As the number of degrees of freedom increases, the chi-square distribution becomes more symmetrical.

Table 15.6 Obtaining the χ^2 critical value from the chi-square distribution with 1 degree of freedom using a .01 level of significance.

Degrees of Freedom	Upper-Tail Areas (α)								
	.995	.99	.975	.95	. . .	.05	.025	.01	.005
1			0.001	0.004	. . .	3.841	5.024	6.635	7.879
2	0.010	0.020	0.051	0.103	. . .	5.991	7.378	9.210	10.597
3	0.072	0.115	0.216	0.352	. . .	7.815	9.348	11.345	12.838
4	0.207	0.297	0.484	0.711	. . .	9.488	11.143	13.277	14.860
5	0.412	0.554	0.831	1.145	. . .	11.071	12.833	15.086	16.750

Source: Extracted from Table E.4.

The values in the body of this table refer to selected upper-tailed areas of the chi-square distribution. Since a χ^2-test statistic for a 2 × 2 table has one degree of freedom, and we are testing at the $\alpha = .01$ level of significance, the critical value of the χ^2-test statistic is 6.635 (see Figure 15.5 on page 622). Since our computed χ^2-test statistic of 8.405 exceeds this critical value, the null hypothesis may be rejected. There is evidence to conclude that the two performance evaluation methods are significantly different with respect to employee perception of fairness. An examination of Table 15.1 on page 613 indicates that a greater proportion of employees found method 1 (employee feedback) fairer than method 2 (self-assessment).

As we shall observe in Section 15.7, computer output obtained by using various statistical software packages for contingency table analysis typically contains the cross-classification table with both the observed and expected frequencies, the computed χ^2-test statistic, and the probability (that is, p value) of obtaining a test

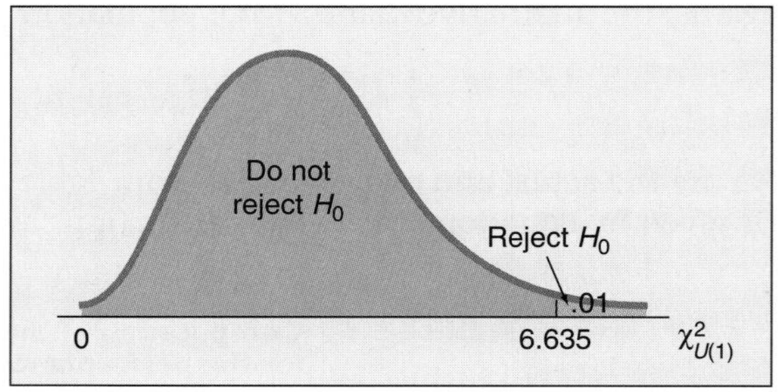

Figure 15.5
Finding the χ^2 critical value with 1 degree of freedom at the .01 level of significance.

statistic this large or more extreme if the null hypothesis were true. Whenever we have such a p value displayed in the computer output we will not need the critical value of the test statistic to make our decision. We can simply compare the obtained p value to our selected level of significance, α. If the p value is less than α, the null hypothesis is rejected; if the p value is greater than or equal to α, then H_0 is not rejected. In our performance evaluation study, since the p value = .0038 (obtained by accessing STATISTIX), which is less than $\alpha = .01$, the null hypothesis is rejected. There is evidence of a difference in the two proportions. The employees found method 1 (employee feedback) fairer than method 2 (self-assessment).

Caution For the test to give accurate results, the χ^2 test for 2×2 tables assumes that each expected frequency is at least five. If this assumption is not satisfied, other procedures, such as *Fisher's exact test* (see Reference 2), can be used.

15.4.4 Testing for the Equality of Two Proportions by Z and by χ^2: A Comparison of Results

We have seen in our personnel director's performance evaluation study that both the Z test based on the standard normal distribution and the χ^2 test based on the chi-square distribution having one degree of freedom have led to the same conclusion. This can be explained by the interrelationship between the standard normal distribution and a chi-square distribution having one degree of freedom. The χ^2-test statistic will always be the square of the Z-test statistic.[2] For instance, in the performance evaluation study the computed Z-test statistic was +2.90 and the computed χ^2-test statistic was 8.405. Except for rounding error, we note that this latter value is the square of +2.90 [that is, $(+2.90)^2 \cong 8.405$]. Also, if we compare the critical values of the test statistics from the two distributions, we can see that at the .01 level of significance the $\chi^2_{U(1)}$ value of 6.635 is the square of the Z values of ±2.58 (that is, $\chi^2_{U(1)} = Z^2$).

Advantage of the *Z* Test Over the χ^2 Test From this discussion it should be clear that when testing the null hypothesis of homogeneity of proportions

$$H_0:\ p_1 = p_2$$

against the alternative that the true population proportions are not equal

$$H_1:\ p_1 \neq p_2$$

the Z test and the χ^2 test are equivalent methods. However, if we are interested specifically in determining whether there is evidence of a *directional difference*, such as $p_1 > p_2$, then the Z test must be used with the entire rejection region located in one tail of the standard normal distribution.

Advantage of the χ^2 Test Over the Z Test On the other hand, if it is of interest to make comparisons and evaluate differences in the proportions among c groups or levels of some factor, we will be able to extend the χ^2 test for such purposes. The Z test, however, cannot be employed if there are more than two groups.

Problems for Section 15.4

15.20 Why shouldn't the χ^2 test for differences in proportions be used when the expected frequencies in some of the cells are too small?

15.21 Do Problem 15.11 on page 614 using the χ^2 test.

15.22 Do Problem 15.12 on page 615 using the χ^2 test.

15.23 Do Problem 15.13 on page 615 using the χ^2 test.

15.24 Do Problem 15.15 on page 615 using the χ^2 test.

15.25 In an effort to compare the efficacy of two medical approaches to removing plaque that clogs arteries, Dr. Eric J. Topol conducted a study in which he randomly assigned 1,012 heart patients to have either directional coronary atherectomy or balloon angioplasty (See E. Topol et al., "A Comparison of Directional Atherectomy with Coronary Angioplasty in Patients with Coronary Artery Disease," *The New England Journal of Medicine*, July 22, 1993, Vol. 329, pp. 221–227). Of the 512 patients given the atherectomy, 44 died or suffered heart attacks within six months of treatment. Of the 500 patients given angioplasty, 23 died or suffered heart attacks within six months of treatment.

(a) At the .01 level of significance, is there evidence of a difference in the two medical approaches with respect to the proportion of deaths or heart attacks within six months of treatment?

(b) Compute the p value in (a) and interpret its meaning.

(c) Given the results in (a) and/or (b), what should the physicians conclude with respect to the two approaches?

(d) How should a congressional policymaker react to the results in (a) and/or (b)?

(e) How should a hospital CEO react to the results in (a) and/or (b) if the medical facility is given a fixed sum of money per treatment?

(f) How should an insurance company CEO react to the results in (a) and/or (b) if the company has to pay for each treatment?

Interchapter Problems for Section 15.4

15.26 Refer to Problem 6.5 on page 209, and using the χ^2 test,

(a) At the .05 level of significance, is there evidence of a difference in the proportion of students who have an entertainment credit card based on whether or not they have a bank credit card?

(b) Compute the p value in (a) and interpret its meaning.

(c) Compare the results of (a) with those of Problem 15.16(a) on page 616. Discuss.

(d) Explain why the χ^2 test cannot be used to do Problem 15.16(c) on page 616.

- 15.27 Refer to Problem 6.7 on page 210, and using the χ^2 test,
  - (a) At the .05 level of significance, is there evidence of a difference in the proportion of males and females who enjoy shopping for clothing?
  - (b) Compute the p value in (a) and interpret its meaning.
  - (c) Compare the results of (a) with those of Problem 15.17(a) on page 616. Discuss.
  - (d) Explain why the χ^2 test cannot be used to do Problem 15.17(c) on page 616.

15.28 Refer to Problem 6.8 on page 210, and using the χ^2 test,
  - (a) At the .10 level of significance, is there evidence of a difference in the proportion of males and females who use the health club facilities?
  - (b) Compute the p value in (a) and interpret its meaning.
  - (c) Compare the results of (a) with those of Problem 15.18(a) on page 616. Discuss.

15.5 χ^2 Test for Differences in c Proportions (Independent Samples)

15.5.1 Introduction and Development

The χ^2 test can be extended to the general case in which there are c independent populations to be compared. Thus, if there is interest in evaluating differences in the proportions among c groups or levels of some factor, the χ^2 test can be used for such purposes. The contingency table would have two rows and c columns. To test the null hypothesis of no differences in the proportions among the c populations

$$H_0\colon\ p_1 = p_2 = \cdots = p_c$$

against the alternative that not all the c population proportions are equal

$$H_1\colon\ \text{Not all } p_j \text{ are equal (where } j = 1, 2, \ldots, c)$$

we use Equation (15.4) and compute the test statistic

$$\chi^2 = \sum_{\text{all cells}} \frac{(f_o - f_e)^2}{f_e}$$

where f_o is the observed frequency or actual tally in a particular cell of a $2 \times c$ contingency table and f_e is the theoretical tally or expected frequency in a particular cell if the null hypothesis were true.

To compute the expected frequency (f_e) in any cell, we must realize that if the null hypothesis were true and the proportions were equal across all c populations, then the c sample proportions should be differing from each other only by chance since they would each be providing estimates of the common population parameter p. In such a situation, a statistic that would pool or combine these c separate estimates into one overall or average estimate of the population parameter p would provide more information than any one of the c separate estimates alone. Expanding on Equation (15.5) on page 618, the statistic $\bar{p}$ represents the overall or average proportion over all c groups combined:

$$\bar{p} = \frac{(X_1 + X_2 + \cdots + X_c)}{(n_1 + n_2 + \cdots + n_c)} = \frac{X}{n} \quad \textbf{(15.5a)}$$

To obtain the expected frequency (f_e) for each cell in the first row in the contingency table, we multiply each respective sample size (or column total) by $\bar{p}$. To obtain the expected frequency (f_e) for each cell in the second row in the contingency table, we multiply each respective sample size (or column total) by $(1 - \bar{p})$. The test statistic shown in Equation (15.4) approximately follows a chi-square distribution with degrees of freedom equal to the number of rows in the contingency table minus one times the number of columns in the table minus one. For a $2 \times c$ contingency table there are $c - 1$ degrees of freedom; that is,

$$\text{degrees of freedom} = (2 - 1)(c - 1) = c - 1$$

Using a level of significance α, the null hypothesis may be rejected in favor of the alternative if the computed χ^2-test statistic exceeds the upper-tailed critical value from a χ^2 distribution having $c - 1$ degrees of freedom. That is, the decision rule is

$$\text{Reject } H_0 \text{ if } \chi^2 > \chi^2_{U(c-1)};$$
$$\text{otherwise don't reject } H_0.$$

This is depicted in Figure 15.6.

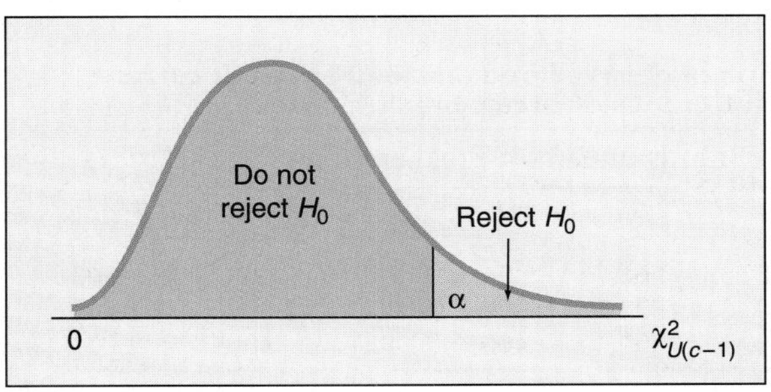

Figure 15.6
Testing for differences among c proportions using the χ^2 test.

For the χ^2 test to give accurate results when dealing with $2 \times c$ contingency tables, all expected frequencies must be large. For such situations, there has been much debate among statisticians as to the definition of "large." Some statistical researchers (see Reference 4) have found that the test gives accurate results as long as all expected frequencies equal or exceed 0.5. Other statisticians, more conservative in their approach, would require that no more than 20% of the cells contain expected frequencies less than 5 and no cells have expected frequencies less than 1 (see Reference 3). We suggest that a reasonable compromise between these points of view is to make sure that all expected frequencies are at least 1. To accomplish this, it may be necessary to collapse together two or more low frequency categories in our contingency table prior to performing the test. Such merging of categories usually results in expected frequencies sufficiently large to conduct the χ^2 test accurately. If the combining or pooling of categories is undesirable, alternative procedures are available (see References 1 and 6).

15.5.2 Application

To illustrate the χ^2 test for equality or homogeneity of proportions when there are more than two groups, let us suppose that a real estate developer has just received approval from the city council to develop a parcel of land that is to contain 40,000 apartment dwellings. Among the numerous items needed for each apartment that must be contracted is a circuit-breaker box to be installed in a kitchen cabinet. Several manufacturers make such circuit-breaker boxes and the real estate developer wishes to contract with only one supplier. From the architectural and engineering design specifications approved for this development, it is necessary that the circuit-breaker box be able to tolerate a stipulated current level without malfunctioning in order to be considered for use. Five vendors of circuit-breaker boxes who have placed a bid to obtain the contract have claimed that their products will meet the stipulated current requirement and have passed the first phase of the contract competition. However, since the circuit-breaker box is a relatively inexpensive item and the prices offered by the five vendors in bidding for the contract are very similar, the real estate project development director decides to design an experiment to evaluate the capability of each of the competing boxes. Random samples of 400 boxes are obtained from each vendor and subjected to a peak current test (that is, one in excess of the stipulated level of current). Table 15.7 presents, for each of the five vendors' products, the number of boxes that malfunction (i.e., at least one of the current breakers in the box fails to flip appropriately) during the test and the number of boxes that continue to work properly under the peak current condition.

Table 15.7 **Cross-classification of *observed* frequencies from peak current experiment on five brands of circuit-breaker boxes.**

Result of Peak Current Experiment	Brands of Circuit-Breaker Boxes					
	1	2	3	4	5	Totals
Malfunctioning boxes	92	66	94	144	104	500
Properly working boxes	308	334	306	256	296	1,500
Totals	400	400	400	400	400	2,000

Under the null hypothesis of no differences among the five vendors' products with respect to the proportion of malfunctioning or nonconforming boxes, we can use Equation (15.5a) to calculate an estimate of the common parameter p, the population proportion of malfunctioning circuit-breaker boxes. That is, $\bar{p}$, the overall or average proportion of malfunctioning boxes taken over all five competing vendors, is computed as

$$\bar{p} = \frac{(X_1 + X_2 + \cdots + X_c)}{(n_1 + n_2 + \cdots + n_c)} = \frac{X}{n}$$

$$= \frac{(92 + 66 + 94 + 144 + 104)}{(400 + 400 + 400 + 400 + 400)} = \frac{500}{2{,}000}$$

$$= .25$$

The estimated proportion of properly working circuit-breaker boxes in the population is the complement, $(1 - \bar{p})$, or .75. Multiplying these two proportions by the sample size used for each vendor's product results in the expected frequencies of malfunctioning and properly working boxes. These are presented in Table 15.8.

Table 15.8 **Cross-classification of *expected* frequencies from peak current experiment on five brands of circuit-breaker boxes.**

	Brands of Circuit-Breaker Boxes					
Result of Peak Current Experiment	1	2	3	4	5	Totals
Malfunctioning boxes	100	100	100	100	100	500
Properly working boxes	300	300	300	300	300	1,500
Totals	400	400	400	400	400	2,000

To test the null hypothesis of homogeneity or equality of proportions

$$H_0: \quad p_1 = p_2 = p_3 = p_4 = p_5$$

against the alternative that not all the five proportions are equal

$$H_1: \quad \text{Not all } p_j \text{ are equal (where } j = 1, 2, \ldots, 5)$$

we use the observed and expected data from Tables 15.7 and 15.8 to compute the χ^2-test statistic given by Equation (15.4). The calculations are presented in Table 15.9.

Table 15.9 **Computation of the χ^2-test statistic for peak current experiment on five brands of circuit-breaker boxes.**

f_o	f_e	$(f_o - f_e)$	$(f_o - f_e)^2$	$(f_o - f_e)^2/f_e$
92	100	−8	64	0.640
66	100	−34	1,156	11.560
94	100	−6	36	0.360
144	100	44	1,936	19.360
104	100	4	16	0.160
308	300	8	64	0.213
334	300	34	1,156	3.853
306	300	6	36	0.120
256	300	−44	1,936	6.453
296	300	−4	16	0.053
				42.772

If a .01 level of significance is chosen, the critical value of the χ^2-test statistic could be obtained from Table E.4. In our peak current experiment, since five vendor products are being evaluated, there are $(2 - 1)(5 - 1) = 4$ degrees of freedom. The critical value of χ^2 with four degrees of freedom at the $\alpha = .01$ level of significance is 13.277. Since our computed test statistic $\chi^2 = 42.772$ exceeds this critical value, the null hypothesis may be rejected (see Figure 15.7 on page 628). It may be stated that, at the .01 level of significance, there is sufficient evidence to conclude

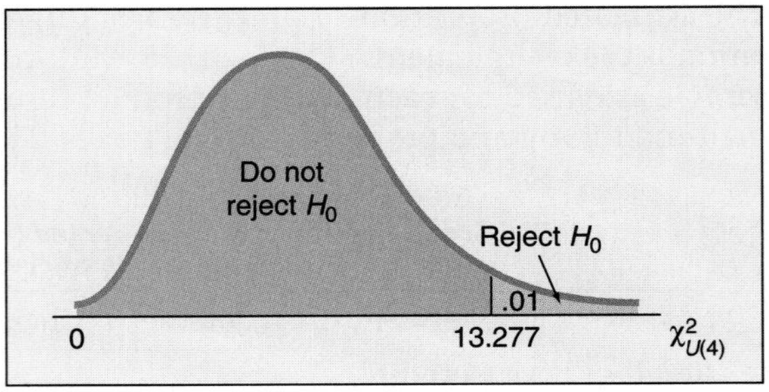

Figure 15.7
Testing for the equality of five proportions at the .01 level of significance with 4 degrees of freedom.

that the five brands of circuit-breaker boxes are different with respect to the proportion which malfunction during a peak current test.

Had we employed a statistical software package to perform the χ^2 test on our peak current experiment data, we could have read the p value for the computed χ^2-test statistic from the computer output and would not have needed the critical value of the test statistic to make our decision. Since the obtained p value is less than the selected level of significance, $\alpha = .01$, we reach the same conclusion—the null hypothesis can be rejected.

15.5.3 Multiple Comparisons: The Marascuilo Procedure

Rejecting the null hypothesis in a χ^2 test of equality of proportions in a $2 \times c$ table only allows us to conclude that not all the brands of circuit-breaker boxes are equal with respect to the proportion that malfunction during a peak current test. Interest would then focus on which brand or brands are different from the others with respect to performance. Since the result of the χ^2 test for homogeneity of proportions does not specifically answer these questions, other approaches are needed. One such confirmatory approach that may be used following rejection of the null hypothesis of equal proportions is the Marascuilo procedure (see References 5 and 6). This method is an example of a *post hoc* (or *a posteriori*) comparison procedure, since the hypotheses of interest are formulated *after* the data have been inspected.

The Marascuilo procedure enables us to examine simultaneously comparisons between all pairs of groups. The first step involved is to compute the differences $p_{s_j} - p_{s_{j'}}$ (where $j \neq j'$) among all $c(c - 1)/2$ pairs of proportions. The corresponding critical ranges for the Marascuilo procedure are then obtained from the quantity given in Equation (15.6):

$$\text{critical range} = \sqrt{\chi^2_{U(c-1)}}\sqrt{\frac{p_{s_j}(1 - p_{s_j})}{n_j} + \frac{p_{s_{j'}}(1 - p_{s_{j'}})}{n_{j'}}} \qquad \textbf{(15.6)}$$

where, for an overall α level of significance, $\sqrt{\chi^2_{U(c-1)}}$ is the square root of the upper-tailed critical value from a chi-square distribution having $c - 1$ degrees of freedom. A distinct critical range must be obtained for each pairwise comparison of sample proportions. The final step is to compare each of the $c(c - 1)/2$ pairs of proportions against its corresponding critical range. A specific pair would be declared significantly different if the absolute difference in the sample proportions $|p_{s_j} - p_{s_{j'}}|$ exceeds its critical range.

To apply the Marascuilo procedure we return to the peak current experiment. Using the χ^2 test we concluded that there was evidence of a significant difference among the population proportions. Since there are five brands, there are (5)(5 − 1)/2 = 10 possible pairwise comparisons to be made and ten critical ranges to compute. From Table 15.7 on page 626 the five sample proportions are:

$$p_{s_1} = \frac{X_1}{n_1} = \frac{92}{400} = .230$$

$$p_{s_2} = \frac{X_2}{n_2} = \frac{66}{400} = .167$$

$$p_{s_3} = \frac{X_3}{n_3} = \frac{94}{400} = .235$$

$$p_{s_4} = \frac{X_4}{n_4} = \frac{144}{400} = .360$$

$$p_{s_5} = \frac{X_5}{n_5} = \frac{104}{400} = .260$$

These five sample proportions are plotted along with the pooled estimate ($\bar{p}$) in Figure 15.8.

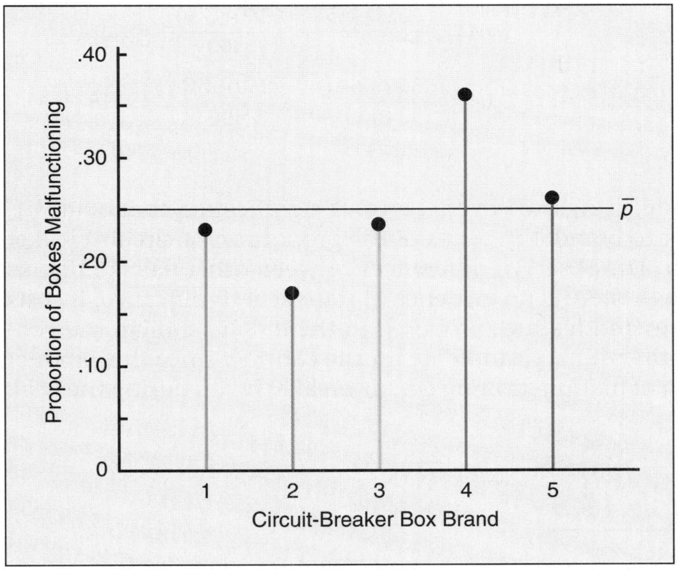

Figure 15.8
A comparison of the proportion of malfunctioning circuit-breaker boxes from five vendors.

For an overall level of significance of .01, the upper-tailed critical value of the χ^2-test statistic for a chi-square distribution having (c − 1) or four degrees of freedom is obtained from Table E.4 as 13.277. Thus, $\sqrt{\chi^2_{U(c-1)}} = \sqrt{13.277} = 3.644$.

From this we obtain the ten pairs of absolute differences in proportions and their corresponding critical ranges:

$\lvert p_{s_j} - p_{s_{j'}} \rvert$	critical range $= 3.644\sqrt{\dfrac{p_{s_j}(1-p_{s_j})}{n_j} + \dfrac{p_{s_{j'}}(1-p_{s_{j'}})}{n_{j'}}}$
$\lvert p_{s_1} - p_{s_2} \rvert = \lvert .230 - .167 \rvert = .063$	$3.644\sqrt{\dfrac{(.230)(.770)}{400} + \dfrac{(.167)(.833)}{400}} = .102$
$\lvert p_{s_1} - p_{s_3} \rvert = \lvert .230 - .235 \rvert = .005$	$3.644\sqrt{\dfrac{(.230)(.770)}{400} + \dfrac{(.235)(.765)}{400}} = .109$
$\lvert p_{s_1} - p_{s_4} \rvert = \lvert .230 - .360 \rvert = .130$	$3.644\sqrt{\dfrac{(.230)(.770)}{400} + \dfrac{(.360)(.640)}{400}} = .116$
$\lvert p_{s_1} - p_{s_5} \rvert = \lvert .230 - .260 \rvert = .030$	$3.644\sqrt{\dfrac{(.230)(.770)}{400} + \dfrac{(.260)(.740)}{400}} = .111$
$\lvert p_{s_2} - p_{s_3} \rvert = \lvert .167 - .235 \rvert = .068$	$3.644\sqrt{\dfrac{(.167)(.833)}{400} + \dfrac{(.235)(.765)}{400}} = .103$
$\lvert p_{s_2} - p_{s_4} \rvert = \lvert .167 - .360 \rvert = .193$	$3.644\sqrt{\dfrac{(.167)(.833)}{400} + \dfrac{(.360)(.640)}{400}} = .111$
$\lvert p_{s_2} - p_{s_5} \rvert = \lvert .167 - .260 \rvert = .093$	$3.644\sqrt{\dfrac{(.167)(.833)}{400} + \dfrac{(.260)(.740)}{400}} = .105$
$\lvert p_{s_3} - p_{s_4} \rvert = \lvert .235 - .360 \rvert = .125$	$3.644\sqrt{\dfrac{(.235)(.765)}{400} + \dfrac{(.360)(.640)}{400}} = .117$
$\lvert p_{s_3} - p_{s_5} \rvert = \lvert .235 - .260 \rvert = .025$	$3.644\sqrt{\dfrac{(.235)(.765)}{400} + \dfrac{(.260)(.740)}{400}} = .111$
$\lvert p_{s_4} - p_{s_5} \rvert = \lvert .360 - .260 \rvert = .100$	$3.644\sqrt{\dfrac{(.360)(.640)}{400} + \dfrac{(.260)(.740)}{400}} = .118$

From this we may conclude, using a .01 overall level of significance, that brands 1, 2, and 3 are each superior to brand 4 in terms of the proportion of circuit-breaker boxes which malfunction. The observed differences between other pairs of brands are due to chance. Although there is no evidence of statistical significance, if costs and other important issues and features proposed in the bid are similar, the vendor for brand 2 would be the primary candidate for the contract since this product had the lowest proportion of malfunctioning circuit-breaker boxes during the peak current experiment.

Problems for Section 15.5

15.29 Why can't the normal approximation be used for determining differences in the proportion of successes in more than two groups?

15.30 The faculty council of a large university would like to determine the opinion of various groups on a proposed trimester academic calendar. A random sample of 100 undergraduate students, 50 graduate students, and 50 faculty members is selected with the following results at the top of page 631:

Opinion	Undergraduate	Graduate	Faculty
Favor trimester	63	27	30
Oppose trimester	37	23	20
Totals	100	50	50

(a) At the .01 level of significance, is there evidence of a difference in attitude toward the trimester between the various groups? Completely analyze the results.

(b) ACTION What should the faculty council report to the president of the university concerning the attitude toward the trimester academic calendar?

15.31 Dr. Lawrence K. Altman reported the results of a clinical trial (*The New York Times*, May 1, 1993, p. 7) comparing the effectiveness of four drug regimens randomly assigned for treatment of patients following the onset of a heart attack. A total of 40,845 patients were studied. Each was given one of the four drug regimens. The outcome measure compared was the proportion of severe adverse events (i.e., deaths or disabling stroke) reported within 30 days of treatment. The data are presented below:

	Drug Regimen				
Result	A	B	C	D	Totals
Severe	714	785	754	820	3,073
Not severe	9,630	9,543	9,042	9,557	37,772
Totals	10,344	10,328	9,796	10,377	40,845

Note: A = accelerated TPA with intravenous heparin.
B = combined TPA and streptokinase, with intravenous heparin.
C = streptokinase with subcutaneous heparin.
D = streptokinase with intravenous heparin.

(a) Describe the methodology you are using to completely analyze the data (at $\alpha = .05$) to determine if there is evidence of a significant difference among the four drug regimens with respect to the proportion of patients suffering severe adverse events (i.e., death or disabling stroke) within 30 days following treatment for heart attack.

(b) State your findings.

(c) Discuss the impact that your findings may have on the community of health care administrators and policymakers if a dose of TPA costs $2,400 per patient while a dose of streptokinase costs $240 per patient.

15.32 The quality control manager of an automobile parts factory would like to know if there is a difference in the proportion of defective parts produced on different days of the work week. Random samples of 100 parts produced on each day of the week were selected with the following results:

Result	Mon.	Tues.	Wed.	Thurs.	Fri.
Number of defective parts	12	7	7	10	14
Number of acceptable parts	88	93	93	90	86
Totals	100	100	100	100	100

(a) At the .05 level of significance, is there evidence of a difference in the proportion of defective parts produced on the various days of the week? Completely analyze the results.

(b) **ACTION** Write an executive summary for the quality control manager.

15.33 A manufacturer of automobile batteries wishes to determine whether there are any differences in three media (magazine, TV, radio) in terms of recall of an ad. The results of an advertising study were as follows:

	Media			
Recall Ability	Magazine	TV	Radio	Totals
Number of persons remembering ad	25	10	7	42
Number of persons not remembering ad	73	93	108	274
Totals	98	103	115	316

(a) At the .10 level of significance, determine whether there is evidence of a media effect with respect to the proportion of individuals who can recall the ad. Completely analyze the results.

(b) **ACTION** Write an executive summary for the manufacturer.

● 15.34 The marketing director of a cable television company is interested in determining whether there is a difference in the proportion of households that adopt a cable television service based upon the type of residence (single-family dwelling, two- to four-family dwelling, and apartment house). A random sample of 400 households revealed the following:

	Type of Residence			
Adopt Cable TV?	Single-Family	Two-to Four-Family	Apartment House	Totals
Yes	94	39	77	210
No	56	36	98	190
Totals	150	75	175	400

(a) At the .01 level of significance, is there evidence of a difference among the types of residence with respect to the proportion of households that adopt the cable TV service? Completely analyze the results.

(b) **ACTION** Write an executive summary for the marketing director.

15.6 χ^2 Test of Independence

15.6.1 Introduction

We have just seen how the χ^2 test can be used to evaluate potential differences among the proportion of successes in any number of populations. For a contingency table that has *r* rows and *c* columns, the χ^2 test can be generalized as a test of independence. In these situations, we shall be able to extend our earlier discussions on the rules of probability in Sections 6.7 and 6.8 by presenting a more formal, confirmatory analysis based on a hypothesis of independence in the joint responses to two categorical variables.

15.6.2 Development

As a test of independence, the null and alternative hypotheses would be

H_0: The two categorical variables are independent (i.e., there is no relationship between them).

H_1: The two categorical variables are related (i.e., dependent).

And we once more use Equation (15.4) and compute the test statistic

$$\chi^2 = \sum_{\text{all cells}} \frac{(f_o - f_e)^2}{f_e}$$

The decision rule is to reject the null hypothesis at an α level of significance if the computed value of the test statistic exceeds the upper-tailed critical value from a chi-square distribution having $(r - 1)(c - 1)$ degrees of freedom (see Table E.4). That is,

$$\text{Reject } H_0 \text{ if } \chi^2 > \chi^2_{U(r-1)(c-1)};$$
$$\text{otherwise, don't reject } H_0.$$

Many researchers consider the χ^2 test for independence as an alternative approach to viewing the χ^2 test for equality of proportions. The test statistics are the same and the decision rules are the same. The stated hypotheses and the conclusion to be drawn are different. Thus, for example, in the performance evaluation study of Section 15.4, we concluded that there was evidence of a difference in the two methods with respect to the proportion of employees who perceived the evaluations as fair. From a different viewpoint we could conclude that there is a significant relationship between evaluation method used and perception of fairness. Similarly, in the peak current experiment of Section 15.5, we concluded that there was evidence of a difference in the brands with respect to the proportion which malfunction under a peak current test. Taking a different perspective, we could conclude that there is a significant relationship between brand of circuit-breaker box and performance under a specified peak current.

Nevertheless, there is a fundamental distinction between the two types of tests. The major difference is in the sampling scheme used.

In a test for equality of proportions, we have one factor of interest with two or more levels. These levels represent samples drawn from independent populations. The categorical responses in each sample group or level are usually dichotomized into two levels—*success* and *failure*. The objective is to make comparisons and evaluate differences in the proportions of success among the various levels.

On the other hand, in a test for independence, we have two factors of interest, each with two or more levels. One sample is drawn and the joint responses to the two categorical variables are tallied into the cells of the contingency table that represent particular levels of each variable.

15.6.3 Application

To illustrate the χ^2 test for independence, let us suppose that a survey has been taken by a central Nassau County (New York) branch of a large nationwide chain of real estate brokerage offices in order to profile single-family homes in some neighboring communities. One question of interest to the branch manager in profiling the central Nassau County houses is to determine whether there is a relationship between architectural style (cape, expanded ranch, colonial, ranch, and split-level) and geographic location (East Meadow, Farmingdale, and Levittown). Using the files from a United States Census Bureau *Current Housing Survey*, a random sample of $n = 233$ single-family homes is selected and a two-way tally is obtained for each combination of architectural style and geographic location. The resulting 5×3 contingency table is presented as Table 15.10.

Table 15.10 ***Observed*** **responses to a survey cross-classifying architectural style and geographical location for 233 single-family houses.**

	Geographical Location			
Style	East Meadow	Farmingdale	Levittown	Totals
Cape	31	14	52	97
Expanded ranch	2	1	12	15
Colonial	6	8	9	23
Ranch	16	20	24	60
Split-level	19	17	2	38
Totals	74	60	99	233

From the totals in the margins of Table 15.10 we observe that with respect to architectural style, 97 of the sampled homes are considered capes, 15 are expanded ranches, 23 are colonials, 60 are ranches, and 38 are split-levels. With respect to the neighboring Nassau County communities, 74 of the sampled homes are located in East Meadow, 60 are in Farmingdale, and 99 are in Levittown. The observed frequencies in the cells of the 5×3 contingency table represent the joint tallies from the 233 sampled single-family homes with respect to architectural style and geographic location.

The null and alternative hypotheses are

H_0: There is no relationship between architectural style and geographic location.

H_1: There is a relationship between architectural style and geographic location.

To test this null hypothesis of independence against the alternative that there is a relationship between the two categorical variables, we use Equation (15.4) and compute the test statistic

$$\chi^2 = \sum_{\text{all cells}} \frac{(f_o - f_e)^2}{f_e}$$

where f_o is the *observed frequency* or actual tally in a particular cell of the $r \times c$ contingency table and f_e is the theoretical frequency we would *expect* to find in a particular cell if the null hypothesis of independence was true.

To compute the *expected frequency* (f_e) in any cell, we may utilize probability rules developed in Section 6.8. That is, if the null hypothesis of independence were true, then we could use the multiplication rule for independent events discussed on page 223 [see Equation (6.8)] to determine the joint probability or proportion of responses expected for any cell combination. For example, under the null hypothesis of independence, the probability or proportion of responses expected in the upper-left corner cell representing cape houses in East Meadow would be the product of the two separate probabilities, that is,

$$P(\text{Cape } \textit{and} \text{ East Meadow}) = P(\text{Cape}) \times P(\text{East Meadow}).$$

Here, the proportion of cape houses, P(Cape), is 97/233 or 0.416, while the proportion of houses located in East Meadow, P(East Meadow), is 74/233 or 0.318. If the null hypothesis were true and architectural style and geographic location were independent, the expected proportion or probability P(Cape *and* East Meadow) would equal the product of the separate probabilities, 0.416×0.318, or 0.132. The expected frequency (f_e) for that particular cell combination would then be the product of the survey sample size n and this probability, that is, 233×0.132, or 30.8.

As a second example, to compute the expected frequency (f_e) for the lower-right corner cell representing split-level houses in Levittown under the null hypothesis of independence we have

$$P(\text{Split-level } \textit{and} \text{ Levittown}) = P(\text{Split-level}) \times P(\text{Levittown}).$$

Here, the proportion of split-level houses, P(Split-level), is 38/233 or 0.163, while the proportion of houses located in Levittown, P(Levittown), is 99/233 or 0.425. If the null hypothesis were true and architectural style and geographic location were independent, the expected proportion or probability P(Split-level *and* Levittown) would equal the product of the separate probabilities, 0.163×0.425, or 0.069. The expected frequency (f_e) for that particular cell combination would then be the product of the survey sample size n and this probability, that is, 233×0.069, or 16.1.

The f_e values for the remainder of the 5×3 contingency table would be obtained in a similar manner (see Table 15.11).

Table 15.11 ***Expected*** **frequency of responses to a survey cross-classifying architectural style and geographical location for 233 single-family houses.**

	Geographical Location			
Style	East Meadow	Farmingdale	Levittown	Totals
Cape	30.8	25.0	41.2	97
Expanded ranch	4.8	3.9	6.4	15
Colonial	7.3	5.9	9.8	23
Ranch	19.1	15.5	25.5	60
Split-level	12.1	9.8	16.1	38
Totals	74	60	99	233

An easier way to compute expected frequencies which does not require calculation of probabilities is

$$f_e = \frac{\text{row sum} \times \text{column sum}}{n} \tag{15.7}$$

where row sum is the sum of all the frequencies in the row
column sum is the sum of all the frequencies in the column
n is the survey sample size

For example, using Equation (15.7) for the upper-left corner cell representing cape houses in East Meadow we have

$$f_e = \frac{\text{row sum} \times \text{column sum}}{n} = \frac{(97)(74)}{233} = 30.8$$

while for the lower-right corner cell representing split-level houses in Levittown we have

$$f_e = \frac{\text{row sum} \times \text{column sum}}{n} = \frac{(38)(99)}{233} = 16.1$$

All other f_e values could be obtained in a similar manner (see Table 15.11 on page 635).

The test statistic shown in Equation (15.4) approximately follows a chi-square distribution with degrees of freedom equal to the number of rows in the contingency table minus one times the number of columns in the table minus one. For an $r \times c$ contingency table there are $(r - 1)(c - 1)$ degrees of freedom; that is,

$$\text{degrees of freedom} = (r - 1)(c - 1)$$

Using a level of significance α, the null hypothesis of independence may be rejected in favor of the alternative of a relationship between the two categorical variables if the computed χ^2-test statistic exceeds the upper-tailed critical value from a chi-square distribution having $(r - 1)(c - 1)$ degrees of freedom. That is, the decision rule is to

$$\text{Reject } H_0 \text{ if } \chi^2 > \chi^2_{U(r-1)(c-1)};$$

$$\text{otherwise don't reject } H_0.$$

This is depicted in Figure 15.9.

The χ^2-test statistic for these data would then be computed as indicated in Table 15.12. Using a level of significance, α, of .05, since $\chi^2 = 42.9607$ exceeds 15.507, the upper-tail critical value from the χ^2 distribution (see Table E.4) having $(5 - 1)(3 - 1) = 8$ degrees of freedom, the null hypothesis of independence is rejected (see Figure 15.10). There is evidence of a relationship between architectural style and geographical location. Examination of the table of observed frequencies (see Table 15.10 on page 634) shows that capes are found in greater abundance than expected in Levittown but are underrepresented in Farmingdale.

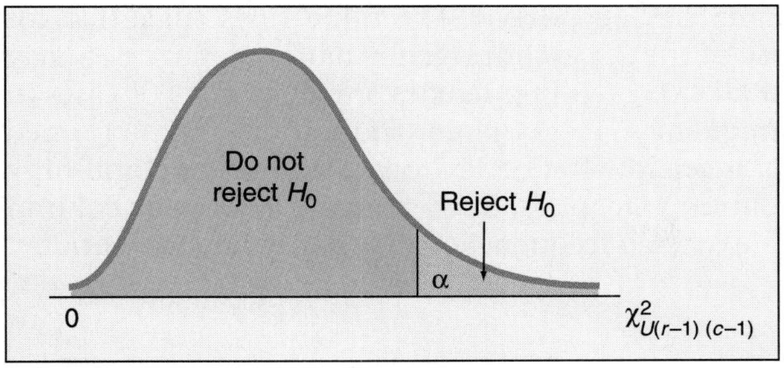

Figure 15.9
Testing for independence in an $r \times c$ contingency table using the χ^2 test.

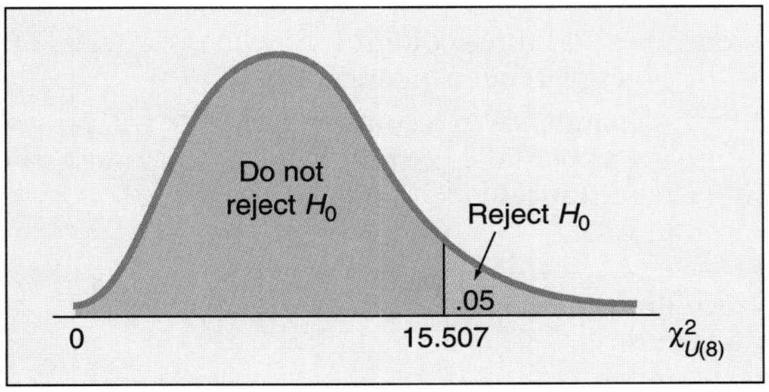

Figure 15.10
Testing for independence between style of house and geographical location at the .05 level of significance with 8 degrees of freedom.

Table 15.12 Computation of the chi-square test statistic for the style of house–geographical location contingency table.

f_o	f_e	$(f_o - f_e)$	$(f_o - f_e)^2$	$(f_o - f_e)^2/f_e$
31	30.8	+0.2	0.04	0.0013
14	25.0	−11.0	121.00	4.8400
52	41.2	+10.8	116.64	2.8311
2	4.8	−2.8	7.84	1.6333
1	3.9	−2.9	8.41	2.1564
12	6.4	+5.6	31.36	4.9000
6	7.3	−1.3	1.69	2.3151
8	5.9	+2.1	4.41	0.7475
9	9.8	−0.8	0.64	0.0653
16	19.1	−3.1	9.61	0.5031
20	15.5	+4.5	20.25	1.3065
24	25.5	−1.5	2.25	0.0882
19	12.1	+6.9	47.61	3.9347
17	9.8	+7.2	51.84	5.2898
2	16.1	−14.1	198.81	12.3484
				42.9607

Expanded ranches are overrepresented in Levittown. Split-levels are overrepresented in East Meadow and in Farmingdale but are vastly underrepresented in Levittown. The p value or probability of obtaining a χ^2 value of 42.9607 or one even more extreme if the null hypothesis of independence were true is less than .005.

● **Caution** As in the case of the $2 \times c$ contingency tables, in order to assure accurate results, use of the χ^2 test when dealing with $r \times c$ contingency tables requires that all expected frequencies be "large." The same rules suggested for employing the χ^2 test in the case of the $2 \times c$ contingency tables on page 625 may be used. Again, we suggest that all expected frequencies are at least 1. For cases in which one or more expected frequency is less than 1, the test may be performed after collapsing together two or more low-frequency row categories or combining two or more low-frequency column categories. Such merging of row or column categories will usually result in expected frequencies sufficiently large to conduct the χ^2 test accurately.

Problems for Section 15.6

15.35 If a contingency table had four rows and three columns, how many degrees of freedom would there be for the χ^2 test for independence?

15.36 When performing a χ^2 test for independence in a contingency table with r rows and c columns, determine the upper-tailed critical value of the χ^2-test statistic in each of the following circumstances:
(a) $\alpha = .05$, $r = 4$ rows, $c = 5$ columns.
(b) $\alpha = .01$, $r = 4$ rows, $c = 5$ columns.
(c) $\alpha = .01$, $r = 4$ rows, $c = 6$ columns.
(d) $\alpha = .01$, $r = 3$ rows, $c = 6$ columns.
(e) $\alpha = .01$, $r = 6$ rows, $c = 3$ columns.

15.37 A nationwide market research study was undertaken to determine the preferences of various age groups of males for different sports. A random sample of 1,000 men was selected, and each individual was asked to indicate his favorite sport. The results were as follows:

	Sport				
Age Group	Baseball	Football	Basketball	Hockey	Totals
Under 20	26	47	41	36	150
20–29	38	84	80	48	250
30–39	72	68	38	22	200
40–49	96	48	30	26	200
50 and over	134	44	18	4	200
Totals	366	291	207	136	1,000

(a) At the .01 level of significance, is there evidence of a relationship between age of men and preference for sports?
(b) Compute the p value in (a) and interpret its meaning.

● 15.38 Suppose that a survey has been taken to determine if there is a relationship between place of residence and automobile preference. A random sample of 200 car owners from large cities, 150 from suburbs, and 150 from rural areas were selected with the following results:

	Automobile Preference					
Residence	GM	Ford	Chrysler	European	Asian	Totals
Large City	64	40	26	8	62	200
Suburb	53	35	24	6	32	150
Rural	53	45	30	6	16	150
Totals	170	120	80	20	110	500

(a) At the .05 level of significance, is there evidence of a relationship between place of residence and automobile preference?
(b) Compute the p value in (a) and interpret its meaning.

15.39 During the Vietnam War a lottery system was instituted to choose males to be drafted into the military. Numbers representing days of the year were "randomly" selected; men born on days of the year with low numbers were drafted first, while those with high numbers were not drafted. The following shows how many low (1–122), medium (123–244), and high (245–366) numbers were drawn for birthdates in each quarter of the year:

	Quarter of Year				
Number Set	Jan.–Mar.	Apr.–Jun.	Jul.–Sep.	Oct.–Dec.	Totals
Low	21	28	35	38	122
Medium	34	22	29	37	122
High	36	41	28	17	122
Totals	91	91	92	92	366

(a) Is there evidence that the numbers drawn were related to the time of year? (Use $\alpha = .05$.)
(b) Would you conclude that the lottery drawing appears to have been random?
(c) Compute the p value in (a) and interpret its meaning.

15.40 A large corporation was interested in determining whether an association exists between commuting time of their employees and the level of stress-related problems observed on the job. A study of 116 assembly-line workers revealed the following:

	Stress			
Commuting Time	High	Moderate	Low	Totals
Under 15 min.	9	5	18	32
15 min. to 45 min.	17	8	28	53
Over 45 min.	18	6	7	31
Totals	44	19	53	116

(a) At the .01 level of significance, is there evidence of a relationship between commuting time and stress?
(b) Compute the p value in (a) and interpret its meaning.

15.7 Using the Computer for Hypothesis Testing with Categorical Data: The Kalosha Industries Employee Satisfaction Survey

15.7.1 Introduction

To this point in the chapter, several methods that involved cross-classifications of categorical variables have been considered. When dealing with large batches of either experimental or survey data of this type, we may recall from Section 5.6 that it was helpful to use available statistical software to assist in an exploratory, descriptive analysis. In this section we shall focus on using various statistical software packages (References 8–11) to assist in an inferential analysis of our data. To do this, let us return to the Kalosha Industries Employee Satisfaction Survey described in Chapter 2.

15.7.2 Kalosha Industries Employee Satisfaction Survey

Data file EMPSAT.DAT

Bud Conley, the Vice-President of Human Resources at Kalosha Industries, is preparing for another meeting with a representative from the B & L Corporation to discuss the contents of an employee-benefits package that is being developed. Prior to this meeting, answers to the following two questions (dealing with categorical variables) would be of particular concern in a confirmatory analysis of the survey data (Table 2.3 on pages 33–40):

1. Is there evidence of a significant relationship between gender and chosen occupation among the full-time employees of Kalosha Industries (see questions 5 and 2 in the Survey)?
2. Is there evidence of a significant relationship between job characteristic importance and job satisfaction (see questions 11 and 9 in the Survey)?

These and other questions raised by Bud Conley (see Survey/Database Project at the end of the section) require a detailed descriptive statistical analysis of the 400 responses to the Survey along with a confirmatory analysis.

15.7.3 Using Statistical Software for Categorical Data

In response to Bud Conley's first question, Figure 15.11 represents partial SAS output using PROC FREQ for these data. We observe from this 2×7 contingency table that, in general, there appears to be a relationship between gender and occupational grouping at Kalosha Industries. The percentage of males who work in some of the occupational settings differs substantially from that of their female counterparts. In particular, we note from the cells of the table that certain combinations of the two categorical variables stand out. When compared to males, we observe that females are employed in an administrative support capacity with far greater frequency than could be expected if there was no gender–occupation relationship. In addition, when compared to females, we note that males work either in production or as laborers with far greater frequency than could be expected if the two categorical variables were not related. This visual impression is confirmed from the χ^2 test for independence obtained by accessing PROC FREQ of SAS. Note the magnitude of the difference between the observed (f_o) and expected (f_e) frequency counts in these two cells. Moreover, we note that in addition to displaying the contingency table with observed and expected frequency counts in each cell, the FREQ procedure lists the computed χ^2-test statistic as well as provides the p value. Also, a warning message is printed by SAS whenever at least 5% of the cells have theoretical frequencies below 5.

For the gender–occupational group data, the χ^2-test statistic is 73.467 and the p value or probability of obtaining a value of χ^2 greater than 73.467 is .0000. Thus, the null hypothesis is rejected. Using the .05 level of significance, there is evidence

The SAS System

TABLE OF SEX BY OCCUP

SEX / OCCUP Frequency Expected	MGL	PROF	TEC/SAL	ADMSPT	SERV	PROD	LABOR	Total
MALES	36 37.863	33 38.445	34 33.203	14 37.863	18 16.893	51 31.455	47 37.28	233
FEMALES	29 27.138	33 27.555	23 23.798	51 27.138	11 12.108	3 22.545	17 26.72	167
Total	65	66	57	65	29	54	64	400

STATISTICS FOR TABLE OF SEX BY OCCUP

Statistic	DF	Value	Prob
Chi-Square	6	73.467	0.000
Likelihood Ratio Chi-Square	6	82.353	0.000
Mantel-Haenszel Chi-Square	1	17.250	0.000
Phi Coefficient		0.429	
Contingency Coefficient		0.394	
Cramer's V		0.429	

Sample Size = 400

Figure 15.11
Contingency table of gender and occupation from SAS output.
Note: SAS provides the user with numerous options when developing a contingency table. (See Figure 5.8 on page 187 for a contingency table used in a descriptive statistical analysis of the gender-occupation data.)

of a relationship between gender and occupational grouping.

Using another viewpoint, that of homogeneity of proportions, we may also conclude that there is evidence of a difference in the proportion of men employed across various occupational groupings. The Marascuilo procedure, however, is not currently available as part of any of the popular statistical software packages. A comparison of the $c(c - 1)/2 = (7)(6)/2 = 21$ possible pairs of proportions using a hand-held calculator indicates a significantly lower proportion of administrative support jobs held by males than there are in all other occupational groupings. In addition, a significantly greater proportion of production or repair work jobs than of managerial, professional, or technical/sales jobs are held by males. All other pairwise differences in the proportions of jobs held by males in the various occupational groupings are due to chance.

In response to Bud Conley's second question, Figure 15.12 presents SPSS output for these data. We observe from this 5×4 contingency table that, in general, there does not appear to be a relationship between job characteristic importance and job satisfaction at Kalosha Industries. The observed and expected frequencies in each cell are fairly similar. On the other hand, as flagged by the note on the SPSS output, the confirmatory analysis is in jeopardy here because of the sparseness of expected frequencies in several cells. The next step then is to combine adjacent column categories so that no cell has an expected frequency below 1. Figure 15.13 displays the revised contingency table along with the appropriate χ^2-test statistic. Note how drastically the small expected frequencies influenced the test statistic in the two contingency tables. Testing the null hypothesis of independence between job characteristic importance and job satisfaction at the .05 level of significance, from Figure 15.13 we observe that 2.224, the computed value of the χ^2-test statistic, is less than 15.507, the upper-tailed critical value from a chi-square distribution having $(5 - 1)(3 - 1) = 8$ degrees of freedom (see Table E.4). Therefore we cannot reject the null hypothesis. There is no evidence of a relationship between

JOBCHAR by SATJOB

SATJOB — Page 1 of 1

JOBCHAR (Count / Exp Val)	VS	MS	LD	VD	Row Total
Hi Income	46 47.6	43 44.0	10 7.0	4 4.4	103 25.8%
Not Fired	11 11.1	11 10.3	1 1.6	1 1.0	24 6.0%
Flexible HRS	7 8.8	10 8.1	1 1.3	1 .8	19 4.8%
Advancement	23 23.6	22 21.8	6 3.4	0 2.2	51 12.8%
Enjoying	98 93.9	85 86.8	9 13.7	11 8.6	203 50.8%
Column Total	185 46.3%	171 42.8%	27 6.8%	17 4.3%	400 100.0%

Chi-Square	Value	DF	Significance
Pearson	9.21346	12	.68460
Likelihood Ratio	11.07923	12	.52214

Minimum Expected Frequency - .808
Cells with Expected Frequency < 5 - 7 OF 20 (35.0%)

Number of Missing Observations: 0

Figure 15.12
Contingency table of job characteristic importance and job satisfaction from SPSS output.

JOBCHAR by SATJOBR

SATJOBR — Page 1 of 1

JOBCHAR (Count / Exp Val)	VS	MS	DIS	Row Total
Hi Income	46 47.6	43 44.0	14 11.3	103 25.8%
Not Fired	11 11.1	11 10.3	2 2.6	24 6.0%
Flexible HRS	7 8.8	10 8.1	2 2.1	19 4.8%
Advancement	23 23.6	22 21.8	6 5.6	51 12.8%
Enjoying	98 93.9	85 86.8	20 22.3	203 50.8%
Column Total	185 46.3%	171 42.8%	44 11.0%	400 100.0%

Chi-Square	Value	DF	Significance
Pearson	2.22399	8	.97336
Likelihood Ratio	2.19726	8	.97436

Minimum Expected Frequency - 2.090
Cells with Expected Frequency < 5 - 2 OF 15 (13.3%)

Number of Missing Observations: 0

Figure 15.13
Revised contingency table of job characteristic importance and job satisfaction from SPSS output.

job characteristic importance and job satisfaction. The differences between the observed and expected frequencies are due to chance. The p value given as .97336 exceeds the specified α of .05.

Survey/Database Project for Section 15.7

Data file EMPSAT.DAT

The following problems refer to the sample data obtained from the questionnaire of Figure 2.6 on pages 28–29 and presented in Table 2.3 on pages 33–40. They should be solved with the aid of an available computer package.

Suppose you are hired as a research assistant to Bud Conley, the Vice-President of Human Resources at Kalosha Industries. He has given you a list of questions (see Problems 15.41 to 15.54) that he needs answered prior to his meeting with the representative from the B & L Corporation, the employee-benefits consulting firm that he has retained.

From the responses to the questions dealing with categorical variables in the Employee Satisfaction Survey (see pages 33–40), in Problems 15.41 through 15.54 which follow,

(a) Form a contingency table and analyze the data using an $\alpha = .05$ level of significance. (*Note*: Combine adjacent row or column categories as necessary to perform the statistical test.)

(b) ACTION Write a memo to Bud Conley discussing your findings.

15.41 Is there evidence of a difference in the proportion of union members (question 14) among the various occupational groupings (question 2)?

15.42 Is there evidence of a difference in the proportion of individuals participating in budgetary decisions (question 22) among the various occupational groupings (question 2)?

15.43 Is there evidence of a relationship between gender (question 5) and job satisfaction (question 9)?

15.44 Is there evidence of a relationship between job characteristic importance (question 11) and job satisfaction (question 9)?

15.45 Is there evidence of a relationship between belief on getting ahead (question 12) and job satisfaction (question 9)?

15.46 Is there evidence of a relationship between union membership (question 14) and job satisfaction (question 9)?

15.47 Is there evidence of a relationship between likelihood of future promotion (question 18) and job satisfaction (question 9)?

15.48 Is there evidence of a relationship between perceived gender-based promotional opportunities (question 19) and job satisfaction (question 9)?

15.49 Is there evidence of a relationship between perceived advancement (question 20) and job satisfaction (question 9)?

15.50 Is there evidence of a relationship between perceived participation in decision making (question 21) and job satisfaction (question 9)?

15.51 Is there evidence of a relationship between budgetary decision making (question 22) and job satisfaction (question 9)?

15.52 Is there evidence of a relationship between attitude toward Kalosha Industries (question 23) and job satisfaction (question 9)?

15.53 Is there evidence of a relationship between perception of employee-management relations (question 25) and job satisfaction (question 9)?

15.54 Is there evidence of a relationship between perception of co-worker relations (question 26) and job satisfaction (question 9)?

15.8 McNemar Test for Differences in Two Related Proportions

15.8.1 Introduction

In Sections 15.3 and 15.4 we were concerned with situations involving an analysis of differences in population proportions based on two independent samples. However, as in Sections 13.9 (the t test for the mean difference) and 13.10 (the Wilcoxon signed-ranks test for the median difference) when dealing with numerical data, it is often the case that we wish to evaluate differences in population proportions based on related samples. Many such applications involving categorical data and proportions exist in public relations, advertising, food processing, pharmaceutical research, the social sciences, and medical research:

- Comparing a new product against a standard product
- Measuring the worth of advertising copy
- Studying brand switching and brand loyalty patterns
- Evaluating taste-testing experiments
- Investigating the efficacy of a drug
- Examining the results of a political debate

In some situations we may design an experiment that consists of matched pairs of individuals. For example, we might wish to determine whether there is evidence of a difference between two groups that have been matched according to some control characteristic. In other situations, however, it may be more appropriate to design an experiment that deals with the repeated responses from the same individuals. For example, we might wish to determine whether there has been a change in perception, attitude, belief, or behavior from one time period to another. To analyze differences between two proportions in situations such as those just described, a test developed by McNemar (References 2 and 7) may be employed.

15.8.2 Development

The McNemar test can be used either for determining whether there is evidence of a difference between the two related proportions (i.e., a two-tailed test) or for determining whether there is evidence of a significant directional change so that one group has a higher proportion than the other (i.e., a one-tailed test).

Two-Tailed Test	One-Tailed Test	One-Tailed Test
H_0: $p_1 = p_2$	H_0: $p_1 \geq p_2$	H_0: $p_1 \leq p_2$
H_1: $p_1 \neq p_2$	H_1: $p_1 < p_2$	H_1: $p_1 > p_2$

where p_1 = proportion of successes in population 1
p_2 = proportion of successes in population 2

To develop the McNemar test let us examine the following 2×2 schematic contingency table (Table 15.13).

Table 15.13 **2 x 2 schematic contingency table for the McNemar test.**

	Condition (Group 2)		
Condition (Group 1)	Yes	No	Totals
Yes	A	B	$A + B$
No	C	D	$C + D$
Totals	$A + C$	$B + D$	n

where A = number of respondents that answer yes to condition 1 and yes to condition 2

B = number of respondents that answer yes to condition 1 and no to condition 2

C = number of respondents that answer no to condition 1 and yes to condition 2

D = number of respondents that answer no to condition 1 and no to condition 2

n = number of respondents in the sample (i.e., the sample size)

The sample proportions of interest are

$p_{s_1} = \frac{A+B}{n}$ = proportion of respondents in the sample who answer yes to condition 1

$p_{s_2} = \frac{A+C}{n}$ = proportion of respondents in the sample who answer yes to condition 2

The test statistic for the McNemar test is given by

$$Z \cong \frac{B - C}{\sqrt{B + C}} \tag{15.8}$$

where the test statistic Z is approximately normally distributed. Thus, for example, using a level of significance α to test the null hypothesis of no differences in the related population proportions (H_0: $p_1 = p_2$) against the two-tailed alternative that there is evidence of a difference (H_1: $p_1 \neq p_2$), our decision rule would be to reject H_0 if the computed value of the test statistic Z exceeds the upper-tailed critical value from the standard normal distribution (Table E.2) or if the test statistic is less than the lower-tailed critical value from the standard normal distribution.

15.8.3 Application

To illustrate the McNemar test, let us again refer to the real estate survey, discussed in Section 15.6.3, that was taken by a central Nassau County (New York) branch of a large nationwide chain of real estate brokerage offices in order to profile single-family homes in some neighboring communities during a recent year. Suppose that the branch manager had wanted to know whether the proportion of homeowners in the Nassau County communities who stated that they intended to put their house up for sale in the next year differs from the actual proportion who placed their home up for sale during that year. If, during the survey, homeowners in the Nassau County communities were asked whether they intended to put their homes up for sale in the next year and each homeowner was then contacted one year later to determine if the house was actually put up for sale, we could form a contingency table (Table 15.14) to summarize the following results:

Table 15.14 Contingency table showing intentions and actual happenings with respect to putting houses up for sale in three Nassau County communities.

	Actually Put House up for Sale		
Intended to Put House up for Sale	Yes	No	Total
Yes	23	3	26
No	11	196	207
Total	34	199	233

The McNemar test is appropriate here because there are two categorical responses for each homeowner in the sample and this forms the basis for a repeated responses experiment as described in Section 15.8.1.

Since the branch manager wanted to determine whether the proportion of houses that were intended to be put up for sale is different from the proportion that were actually put up for sale in the three Nassau County communities (i.e., East Meadow, Farmingdale, and Levittown), the null and alternative hypotheses would be

$$H_0: \; p_1 = p_2$$
$$H_1: \; p_1 \neq p_2$$

If the test were carried out at the .05 level of significance, the critical values would be -1.96 and $+1.96$ (see Figure 15.14) and the decision rule would be

$$\text{Reject } H_0 \text{ if } Z < -1.96 \text{ or if } Z > +1.96;$$
$$\text{otherwise do not reject } H_0.$$

For our data,

$$A = 23 \quad B = 3 \quad C = 11 \quad D = 196$$

so that

$$p_{s_1} = \frac{A + B}{n} = \frac{26}{233} = .112 \qquad p_{s_2} = \frac{A + C}{n} = \frac{34}{233} = .146$$

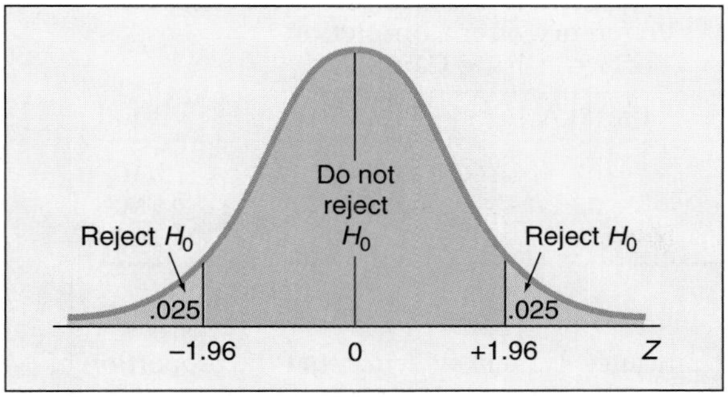

Figure 15.14
Two-tailed McNemar test at the .05 level of significance.

From Equation (15.8)

$$Z \cong \frac{B - C}{\sqrt{B + C}} = \frac{3 - 11}{\sqrt{3 + 11}} = \frac{-8}{\sqrt{14}} = -2.14$$

Since $Z = -2.14 < -1.96$, the null hypothesis can be rejected. Using the *p*-value approach, the probability of obtaining a *Z*-test statistic below -2.14 is $.5000 - .4838 = .0162$. Since a two-tailed test is being used, this value must be doubled to account for the area in the two tails. Since $.0324 < .05$, the null hypothesis can be rejected. The branch manager can conclude that there is evidence that the proportion of houses that are intended for sale is different from the proportion of houses that are actually put up for sale within one year. More houses were actually put up for sale than was intended.

● **Caution** It is essential to a good data analysis that we apply the appropriate statistical technique to a specific situation. For example, when comparing two population proportions based on independent samples, either the *Z* test or χ^2 test (see Sections 15.3 and 15.4) should be employed, depending on whether the alternative hypothesis of interest is, respectively, one-tailed or two-tailed. However, when comparing two population proportions based on related samples, the McNemar test should be used. It is interesting to note that if the branch manager had inadvertently treated the data as two independent samples and incorrectly applied the methods of either Section 15.3 or 15.4, erroneous conclusions would have been drawn because the null hypothesis of no differences in the proportions would not have been rejected. Statistical power would have been lost by using an improper procedure which fails to capture the design model of the experiment.

Problems for Section 15.8

15.55 What is the distinguishing feature between the McNemar test and the χ^2 test for the difference between proportions?

15.56 A market researcher wanted to study the effect of an advertising campaign for Brand A coffee to determine whether the proportion of coffee drinkers who preferred Brand A would increase as a result of the campaign. A random sample of 200 coffee drinkers was selected and they indicated their preference for either Brand A or Brand B prior to the beginning of the advertising campaign and at its completion. The results were as follows (see page 648):

Preference Prior to Advertising Campaign	Preference after Completion of Advertising Campaign: Brand A	Brand B	Total
Brand A	101	9	110
Brand B	22	68	90
Total	123	77	200

(a) At the .05 level of significance, is there evidence that the proportion of coffee drinkers who prefer Brand A is higher at the end of the advertising campaign than at the beginning?
(b) Compute the p value in (a) and interpret its meaning.

15.57 A political pollster wanted to evaluate the effect on voter preference of a televised debate between two candidates who were running for mayor of a city. A random sample of 500 registered voters was selected and each respondent was asked to indicate his or her preference prior to and after the debate. The results were as follows:

Preference Prior to Debate	Preference after Debate: Candidate A	Candidate B	Total
Candidate A	269	21	290
Candidate B	36	174	210
Total	305	195	500

(a) At the .01 level of significance, is there evidence of a difference in the proportion of voters who favor Candidate A prior to and after the debate?
(b) Compute the p value in (a) and interpret its meaning.

15.58 The coordinator of the Introduction to Computers course in a school of business would like to determine whether there is any difference in the proportion of students who intend to major in computers at the beginning of the course and after its completion. Each student enrolled in the course is asked to indicate on the first day of class and again after completion of the final exam whether he or she intends to major in computers. The results were as follows:

Computer Major Prior to Taking Computer Course	Computer Major after Completion of Computer Course: Yes	No	Total
Yes	52	32	84
No	13	230	243
Total	65	262	327

(a) At the .05 level of significance, is there evidence of a difference between the proportions of students who intend to major in computers prior to taking the computer course and after completion of the course?
(b) Compute the p value in (a) and interpret its meaning.

15.59 The personnel director of a large manufacturing company would like to reduce excessive absenteeism among assembly-line workers. She has decided to institute an experimental incentive plan that will provide financial rewards to employees who are absent less than 5 days in a given calendar year. A sample of 100 workers is selected at the end of the one-year trial period. For each of the two years, information is obtained for each employee selected that indicates whether the worker was absent less than 5 days in that year. The results were as follows:

	Year 2		
Year 1	< 5 Days Absent	≥ 5 Days Absent	Total
< 5 days absent	32	4	36
≥ 5 days absent	25	39	64
Total	57	43	100

(a) At the .01 level of significance, is there evidence that the proportion of employees absent less than 5 days is lower in year 1 than in year 2?
(b) Compute the p value in (a) and interpret its meaning.
(c) What conclusion should the personnel director reach regarding the effect of the incentive plan?

Interchapter Problem for Section 15.8

15.60 Refer to Problem 6.6 on page 210.
(a) Is there evidence of a difference in the proportion of individuals who planned to buy and actually purchased a new television? (Use $\alpha = .05$.)
(b) Compute the p value in (a) and interpret its meaning.

15.9 Potential Hypothesis-Testing Pitfalls and Exploring Ethical Issues

15.9.1 Potential Pitfalls

In this chapter several tests were developed to assist in a confirmatory analysis of categorical variables. Part of a good data analysis is to understand the assumptions underlying the various hypothesis-testing procedures and, using this as well as other criteria, select the one most appropriate for a given set of conditions.

For the test to give accurate results, the χ^2 test for 2×2 tables assumes that each expected frequency is at least 5. If this assumption is not satisfied, other procedures, such as *Fisher's exact test* (Reference 2), can be used. When dealing with $2 \times c$ contingency tables or with $r \times c$ contingency tables, in order to assure accurate results the χ^2 test requires that all expected frequencies be "large." We suggest that all expected frequencies are at least 1. For cases in which one or more expected frequency is less than 1, the test may be performed after collapsing together two or more low-frequency row categories or combining two or more low-frequency column categories. Such merging of row or column categories usually results in expected frequencies sufficiently large to conduct the χ^2 test accurately.

In addition, it is essential to a good data analysis that we apply the appropriate statistical technique to a specific situation. For example, when comparing two population proportions based on independent samples, either the Z test or χ^2 test (see Sections 15.3 and 15.4) should be employed, depending on whether the alternative hypothesis of interest is, respectively, one-tailed or two-tailed. However, when comparing two population proportions based on related samples, the McNemar-test should be used. Failure to choose the correct procedure that represents the design model of the experiment will result in a reduction in statistical power and may lead to erroneous conclusions.

15.9.2 Ethical Issues

As stated in Section 14.11.2, ethical considerations arise when a researcher is manipulative of the hypothesis-testing procedure in ways that enable personal gain. In coordinating and managing a project dealing with a long-term designed experiment or large-scale survey, it is imperative that the principal investigator develop an operational plan or protocol dealing with the process of data collection, evaluation, and analysis. Particularly when many people are involved in the process, a system of checks and balances must be established to prevent fraud, plagiarism, faking data, or falsifying results.

15.10 Hypothesis Testing Based on Categorical Data: A Review

As observed in the summary chart for this chapter (see page 651), we may distinguish among approaches to categorical data analysis. Hypothesis-testing methodology was developed separately for analyzing categorical response data obtained from one sample, two related samples, two independent samples, and c independent samples. In addition, we extended our earlier discussions on the rules of probability in Sections 6.7 and 6.8 by presenting a more formal, confirmatory analysis of the hypothesis of independence in the joint responses to two categorical variables. Once again, emphasis was given to the assumptions and conditions behind the use of the various tests. On page 606 of Section 15.1 you were given a list emphasizing the important points to be discussed in the chapter. Check over the list now to see whether you feel you have an understanding of these key points. To be sure, you should be able to answer the following conceptual questions:

1. Under what conditions should the Z test for the population proportion p be used?
2. Under what conditions should the Z test to examine possible differences in the proportions of two independent populations be used?
3. Under what conditions should the χ^2 test to examine possible differences in the proportions of two independent populations be used?
4. What are the similarities and distinctions between the Z and χ^2 tests for differences in population proportions?
5. Under what conditions should the χ^2 test to examine possible differences in the proportions of c independent populations be used?
6. Under what conditions should the χ^2 test of independence be used?
7. Under what conditions should the McNemar test be used?

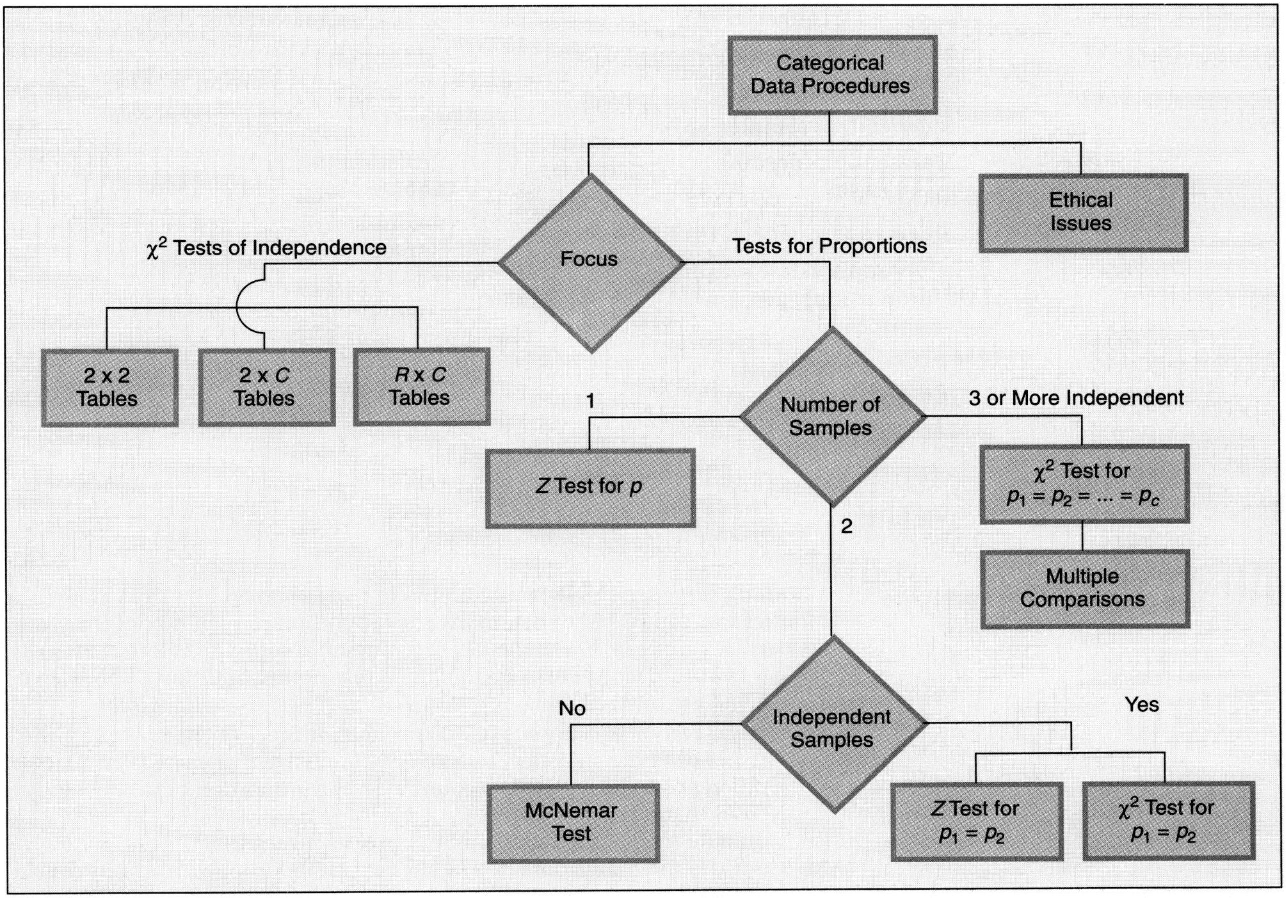

Chapter 15 summary chart.

Check over the list of questions to see if indeed you know the answers and could (1) explain your answers to someone who did not read this chapter and (2) give reference to specific readings or examples that support your answer. Also, reread any of the sections that may have seemed fuzzy to see if they make sense now.

Getting It All Together

Key Terms

2 × 2 table *617*
2 × *c* table *624*
χ^2 test for differences in *c* proportions *624*
χ^2 test for difference in two proportions *616*
χ^2 test for independence *632*
categorical data *606*
chi-square distribution *619*
contingency table *616*
cross-classification table *616*
degrees of freedom *619*
equality of proportions *622*

expected frequencies (f_e) *618*
homogeneity of proportions *618*
independence *633*
independent samples *624*
Marascuilo procedure *628*
McNemar test *644*
observed frequencies (f_o) *618*
one-sample Z test for the proportion *606*
pooled estimate ($\bar{p}$) of common population proportion *624–625*
population proportion p *607*
$r \times c$ table *632*
related samples *644*
sample proportion p_S *606*
theoretical or expected frequencies (f_e) *618*
Z test for difference in two proportions *611*

Chapter Review Problems

15.61 A housing survey of single-family homes in two suburban New York State counties was conducted to determine the proportion of such homes that have gas heat. A sample of 300 single-family homes in County A indicated that 185 had gas heat, and a sample of 200 single-family homes in County B indicated that 75 had gas heat.

Note: Use a level of significance of .01 throughout the problem.

(a) Use *two* different statistical tests to determine whether there is evidence of a difference between the two counties in the proportion of single-family houses that have gas heat.
(b) Compute the p value in (a) and interpret its meaning.
(c) Compare the results obtained by the two methods in (a). Are your conclusions the same?
(d) If you wanted to know whether there was evidence that County A had a higher proportion of single-family houses with gas heat, what method would you use to perform the statistical test?

● 15.62 A "physician's health study" of the effectiveness of aspirin in the reduction of heart attacks was begun in 1982 and completed in 1987 (see C. Hennekens et al., "Findings from the Aspirin Component of the Ongoing Physician's Health Study," *The New England Journal of Medicine,* January 28, 1988, Vol. 318, pp. 262–264). Of 11,037 male medical doctors in the United States who took one 325-mg buffered aspirin tablet every other day, 104 suffered heart attacks during the five-year period of the study. Of 11,034 male medical doctors in the United States who took a placebo (that is, a pill that, unknown to the participants in the study, contained no active ingredients), 189 suffered heart attacks during the five-year period of the study.

(a) At the .01 level of significance, is there evidence that the proportion having heart attacks is lower for the male medical doctors in the United States who received the buffered aspirin every other day than for those who received the placebo?
(b) Compute the p value in (a). Does this lead you to believe that taking one buffered aspirin pill every other day was effective in reducing the incidence of heart attacks? Explain.
(c) Why is it not appropriate to use the χ^2 test in (a)?

15.63 A statistician wished to study the distribution of three types of cars (subcompact, compact, and full size) sold in the four geographical regions of the

United States (Northeast, South, Midwest, West). A random sample of 200 cars was selected with the following results:

Of 60 cars sold in the Northeast, 25 were subcompacts, 20 were compacts, and 15 were full size.

Of 40 cars sold in the South, 10 were subcompacts, 10 were compacts, and 20 were full size.

Of 50 cars sold in the Midwest, 15 were subcompacts, 15 were compacts, and 20 were full size.

Of 50 cars sold in the West, 20 were subcompacts, 15 were compacts, and 15 were full size.

(a) At the .05 level of significance, is there evidence of a relationship between type of car and geographical region?
(b) Compute the p value in (a) and interpret its meaning.
(c) Estimate with 95% confidence the true proportion of full size cars sold in the Northeast.
(d) From the given data, can you obtain a reasonable 95% confidence interval estimate of the proportion of full-size cars sold throughout the United States? Discuss.

15.64 A market researcher was interested in the effect of advertisements on buyer intention for new car buyers. Suppose that prospective purchasers of new cars were first asked whether they preferred Toyota or GM and then were subjected to video advertisements of comparable brands of each car. After viewing the ads, the prospective customers again indicated their preferences. The results were as follows:

Preference before Ads	Preference after Ads: Toyota	GM	Total
Toyota	97	3	100
GM	11	89	100
Total	108	92	200

(a) Is there evidence of a difference in the proportion of respondents who prefer Toyota before and after viewing the ads? (Use $\alpha = .05$.)
(b) Compute the p value in (a) and interpret its meaning.

Suppose the following table was derived from the table in (a):

Preference	Toyota	GM	Total
Before ad	100	100	200
After ad	108	92	200
Total	208	192	400

(c) Explain how this table is obtained from the table in (a).
(d) Using the table in (c), is there evidence of a difference in preference for Toyota before and after viewing the ads? (Use $\alpha = .05$.)
(e) Compute the p value in (d) and explain its meaning.
(f) Explain the difference in the results in (a) and (d) of this problem. Which method of analyzing the data do you think is correct and which is incorrect? Why?

Suppose that the market researcher also wanted to determine if education level was related to automobile preference after viewing the advertisements. The following data were also obtained from the same prospective purchasers:

Education	Preference		Total
	Toyota	GM	
No college	26	49	75
Some college	34	16	50
College graduate	48	27	75
Total	108	92	200

(g) Is there evidence that education level is related to automobile preference after viewing the advertisements? (Use $\alpha = .05$.)

15.65 A market researcher was interested in studying the preference for Coca-Cola and Pepsi Cola before a taste test and after a taste test. A sample of 200 households was selected. The results were as follows:

Preference before Test	Preference after Test		Total
	Coca-Cola	Pepsi Cola	
Coca-Cola	104	6	110
Pepsi Cola	14	76	90
Total	118	82	200

(a) Is there evidence of a difference in the proportion of respondents who prefer Coca-Cola before and after the taste test? (Use $\alpha = .01$.)
(b) Compute the p value in (a) and interpret its meaning.

Suppose the following table was derived from the table in (a):

Preference	Coca-Cola	Pepsi Cola	Total
Before test	110	90	200
After test	118	82	200
Total	228	172	400

(c) Explain how this table is obtained from the table in (a).
(d) Using the table in (c), is there evidence of a difference in preference for Coca-Cola before and after the taste tests? (Use $\alpha = .01$.)
(e) Compute the p value in (d) and explain its meaning.
(f) Explain the difference in the results in (a) and (d) of this problem. Which method of analyzing the data do you think is correct and which is incorrect? Why?

Collaborative Learning Mini Case Project

For the following, refer to the instructions on page 101.

Data file
FRAGRAN.DAT

CL 15.1 Refer to CL 3.3 on page 102 and CL 5.3 on page 199. Your group, the ________ Corporation, has been hired by the marketing director of a manufacturer of well-known men's and women's fragrances to study the characteristics of currently available fragrances. Armed with Special Data Set 3 of Appendix D on pages D8–D9 displaying useful information on 83 such fragrances, the ________ Corporation wishes to determine whether there is evidence of a relationship between type of fragrance (perfume, cologne, or "other") and intensity of fragrance (very strong, strong, medium, or mild).

(a) Thoroughly analyze the data.

(b) Write and submit an executive summary, clearly specifying the hypotheses, selected level of significance, and the assumptions of the chosen test procedure.

(c) Prepare and deliver a five-minute oral presentation to the marketing director.

Case Study F—Airline Satisfaction Survey

As Chief Research Associate you are attending this week's meeting of the board of directors at the invitation of Mike Drucker, the Senior Vice-President for Marketing and Promotion at Ber Lev Airlines. Mike is at the podium. "Ladies and gentlemen, I'd like to inform you that the data from our quarterly survey have now been edited and entered into our computer system. This data set constitutes a random sample of 1,600 adult passengers who flew with us on mainland routes over the two-week period ending last Friday. It is imperative that we continue to monitor the airline services we provide through these quarterly surveys so that by keeping tabs on our customers and taking the market pulse, we can continue to make those improvements that will guarantee that our passengers remain loyal to Ber Lev and that through their satisfaction they will encourage their friends and families to fly with us." "Thank you Mike," interrupted Lorena Martinez-Moreno, the retired United States Air Force General who recently took over as CEO of Ber Lev Airlines, "will you kindly remind us of the central theme in this particular quarterly survey and give us your time table for presentation and discussion." "Certainly, General Martinez-Moreno. The theme, ma'am, for this survey deals with passenger satisfaction and its potential relationships with reasons for travel and disposition of luggage. Gender differences will also be explored. As for a time table, Dr. Elvin Axelrod, the Director of Central Processing, assures me that the initial printouts will be ready by noon today. Since it takes two working days for data cleansing and preliminary analyses, I will be ready to make our presentations at next week's board meeting. I'll need fifteen minutes for the presentation and I request an additional fifteen minutes for questions, answers, and board discussion." "Very good, Mike, it sounds like your Marketing and Promotion Department has got a handle on this one. On behalf of the board I would like to grant your requests and state that we look forward to your presentation next week. Muchas gracias!

(*Many thanks!*) Please keep me informed if you need anything to expedite the analysis." "Thank you, ma'am," Mike replied, and he took his seat.

Two Days Later in Mike Drucker's Office You are sitting in the office of the Senior Vice-President for Marketing and Promotion awaiting Mike Drucker's entrance. He is on the telephone in the vestibule discussing the computer printouts that he is holding. Suddenly, the conversation ends and Mike enters his office smiling. "Well, here it is," Mike Drucker exclaims triumphantly as he takes his seat at his desk. "Dr. Axelrod says that the data appear to be clean, all error checks worked. You did a great job!" You nod and acknowledge the praise, aware that this is just the beginning. Mike continues, "Let me turn this over to you again. Take a good look at the important five theme questions, along with the responses and the cross-tabulations. I'd like some confirmatory answers to the following:

1. Our well-known jingle says *'Ber Lev is the airline of first choice for one out of two, so if you're not the one, it's time you pick us too!'* I really hate that jingle—but people like it, it's catchy. But is it accurate? Do we have to get rid of it? Is there evidence that the proportion of passengers who state that Ber Lev is the airline of choice differs from .50?
2. As an indicator of increased customer satisfaction, a significantly greater proportion of passengers should be pleased with Ber Lev and claim it is the airline of choice *after* their recent flight than *before* it. Is there evidence of a significant increase in passenger satisfaction as a result of the recent flight?
3. For advertising and promotional purposes, it is important to know if a gender effect is present with respect to customer satisfaction. Is there evidence of a difference between males and females in terms of the proportion of these passengers who claim Ber Lev is the airline of choice ?
4. Many of us have argued over whether the primary reason for a trip affects customer satisfaction. Is there evidence of a difference among the various primary reasons for flying with respect to the proportion of the passengers who claim Ber Lev is the airline of choice?
5. For advertising and promotional purposes, it is important to study potential associations between such factors as primary reasons for flying and luggage disposition. If handling the luggage is perceived as an important service feature, we will have to make sure Ber Lev provides better and quicker service for those who check their luggage in as well as more room on the aircraft for those who carry their luggage on board. Is there evidence of a relationship between primary reasons for flying and luggage disposition?

I know we're under the gun on this one, but I'd like you to have this analysis on my desk first thing Monday morning. Please prepare an executive summary and attach all tables and charts, list all hypotheses that you are testing, the levels of significance you chose for the tests, and the conclusions to be drawn. Also, please prepare to give me and my staff an informal ten-minute presentation on your findings. Do you have any questions?" Taking a deep breath, you say "no, not at this time. I'm ready to work." "Thanks," Mike continues, as he ushers you out of his office, "I'll see you first thing Monday, but if any questions come up don't hesitate to call me."

Responses to Theme Portion of Ber Lev Airlines Quarterly Satisfaction Survey

1. What is your gender?
   Male....960 Female....640
2. *Before* this trip, would you have considered Ber Lev to be your airline of choice?
   Yes....816 No....784
3. Now that you've taken this trip, do you consider Ber Lev as your airline of choice?
   Yes....832 No....768
4. What was the primary reason for taking this trip?
   Business....880
   Emergency.....64
   Moving/Transit.....96
   Pleasure....560
5. What did you do with your baggage for this trip?
   Carried it all on board....768
   Checked it all in....592
   Carried some and checked some....192
   Had no luggage.....48

Cross-tabulations:

Theme Question 2

Choice *Before*	Choice *After* Yes	No	Totals
Yes	806	10	816
No	26	758	784
Totals	832	768	1,600

Theme Question 3

Choice *After*	Gender Males	Females	Totals
Yes	512	320	832
No	448	320	768
Totals	960	640	1,600

Theme Question 4

Choice *After*	Primary Reason: Business	Emergency	Moving/ In Transit	Pleasure	Totals
Yes	455	20	42	315	832
No	425	44	54	245	768
Totals	880	64	96	560	1,600

Theme Question 5

Primary Reason	Baggage Disposition: Carry All	Check All	Do Both	No Luggage	Totals
Business	653	83	103	41	880
Emergency	47	14	1	2	64
Moving/In Transit	6	78	9	3	96
Pleasure	62	417	79	2	560
Totals	768	592	192	48	1,600

Endnotes

1. If the hypothesized difference is 0 (that is, $p_1 - p_2 = 0$ or $p_1 = p_2$), the numerator in Equation (15.3) becomes $p_{S_1} = p_{S_2}$.

2 Examine Figures 15.3 and 15.5. Observe that in a two-tailed test, the two critical values, $+Z$ and $-Z$, denote rejection regions in the tails of the standard normal distribution. Each contains an area of $\alpha/2$. Also observe that the one critical value $\chi^2_{U(1)}$ denotes a rejection region in the upper tail of a chi-square distribution having one degree of freedom. The upper tail of this chi-square distribution contains an area of α. Since the Z-test statistic from the standard normal distribution ranges from $-\infty$ to $+\infty$ and the χ^2-test statistic from the chi-square distribution ranges from 0 to $+\infty$, we see that by squaring the Z value we obtain the $\chi^2_{U(1)}$ value.

References

1. Cohen, J., "An Alternative to Marascuilo's 'Large-Sample Multiple Comparisons' for Proportions," *Psychological Bulletin*, 1967, Vol. 67, pp. 199–201.
2. Daniel, W. W., *Applied Nonparametric Statistics*, 2d ed. (Boston, MA: PWS Kent, 1990).
3. Dixon, W. J., and F. J. Massey, Jr., *Introduction to Statistical Analysis*, 4th ed. (New York: McGraw-Hill, 1983).
4. Lewontin, R. C., and J. Felsenstein, "Robustness of Homogeneity Tests in 2 × *n* Tables," *Biometrics*, March 1965, Vol. 21, pp. 19–33.
5. Marascuilo, L. A., "Large-Sample Multiple Comparisons," *Psychological Bulletin*, 1966, Vol. 65, pp. 280–290.
6. Marascuilo, L. A., and M. McSweeney, *Nonparametric and Distribution-Free Methods for the Social Sciences* (Monterey, CA: Brooks/Cole, 1977).
7. McNemar, Q., "Note on the Sampling Error of the Difference Between Correlated Proportions or Percentages," *Psychometrika*, 1947, Vol. 12, pp. 153–157.
8.. *MINITAB Reference Manual Release 8* (State College, PA: Minitab, Inc., 1992).
9. Norusis, M., *SPSS Guide to Data Analysis for SPSS-X with Additional Instructions for SPSS/PC+* (Chicago, IL: SPSS Inc., 1986).
10. *SAS User's Guide Version 6* (Raleigh, NC: SAS Institute, 1988).
11. *STATISTIX Version 4.0* (Tallahassee, FL: Analytical Software, Inc., 1992).

chapter 16

Statistical Applications in Quality and Productivity Management

CHAPTER OBJECTIVES

To provide an introduction to the history of quality and Deming's 14 points of management, to demonstrate the use of a variety of control charts, and to show the interrelationship of management and statistical tools.

16.1 Introduction

In this chapter we will focus on statistical applications in quality and productivity management. The pioneer of this methodology, W. A. Shewhart, said more than a half-century ago (see Reference 21) that

> The long range contribution of statistics depends not so much upon getting a lot of highly trained statisticians into industry as it does in creating a statistically minded generation of physicists, chemists, engineers and others who will in any way have a hand in developing and directing the production processes of tomorrow.

In this chapter we will begin with a historical perspective on quality and productivity and discuss the evolution of management styles. We will then develop the underlying theory behind the topic of control charts. We will also consider two management planning tools useful for process improvement, the process flow diagram and the fishbone diagram. The subsequent discussion of Deming's 14 points of management sets the stage for the development of a variety of control charts used for different types of data. In addition, an intriguing experiment known as "the parable of the red beads" will be developed to highlight the different types of variation inherent in a set of data and to reinforce the importance of management's responsibility to improve systems.
Upon completion of this chapter, you should be able to:

1. Understand the differences between the four generations of management
2. Distinguish between special cause and common causes of variation
3. Develop process flow and fishbone diagrams
4. Appreciate the essential elements of Deming's 14 points of *management by process* and be able to indicate how this approach is different from the *management by control* approach
5. Develop control charts for both categorical and numerical variables.
6. Understand the circumstances in which each control chart should be used

16.2 Quality and Productivity: A Historical Perspective

By the mid-1980s it had become clear that a global economy had developed in which companies located in an individual country had to compete not only with local and national competitors but also with competitors from all parts of the world. This global economy developed because of many factors, including the rapid expansion in worldwide communications and the exponential increase in the availability and power of computer systems. In such an environment, it is vitally important that business organizations be able to respond rapidly to changes in market conditions by incorporating the most effective managerial approaches available.

The development of this global economy has also led to a reemergence of an interest in the area of quality improvement in the United States. Evidence of the renewed interest can be seen in the increasing importance being placed on the competition for the Malcolm Baldrige Award (see Reference 10), given annually to companies making the greatest strides in improving quality and customer satisfaction with their products and services. Among the companies who have won this award are Motorola, Xerox, Federal Express, Cadillac Motor Company, Ritz-Carlton Hotels, AT&T Universal Card Services, and Eastman Chemical Company.

We may understand the background to this reemergence of interest in quality and productivity by briefly examining the historical development of management in four phases (see References 4, 12, and 13). We may think of first-generation management as *management by doing,* the type of management practiced by primitive hunter-gatherer societies in which individuals produced something for themselves or their tribal unit whenever the product was required.

Beginning in the Middle Ages, the rise of craft guilds in Europe led to a second generation of management, *management by directing.* Craft guilds managed the training of apprentices and journeymen and determined standards of quality and workmanship for the products made by the guild.

The development of Watt's steam engine, Whitney's system of interchangeable parts, and many other inventions led to the Industrial Revolution of the nineteenth century (see Reference 6). The creation of the assembly line, pioneered by Henry Ford for the production of automobiles, brought about a third generation of management, *management by control,* in which workers were divided into those who actually did the work (i.e., the laborers) and those who planned and supervised the work (i.e., the managers). This approach took responsibility for quality out of the hands of the individual worker and put it under the inspector, foreman, and other managers. The American engineer Frederick W. Taylor attempted to develop methods for overcoming this division by developing a scientific management approach that required a detailed study of each job. The management by control style also contained a hierarchical structure that emphasized individual accountability to a set of predetermined goals. This approach has been commonly practiced in the United States since the development of the factory environment at the beginning of the twentieth century.

The last decade has seen the competitive advantage enjoyed by the United States in the post–World War II period gradually eliminated because of several factors. First, the United States emerged from World War II with its industrial base intact, unlike most European countries and Japan. Thus, the United States was in a monopolistic position and the rest of the world eagerly awaited whatever consumer products it was able to produce. In such a supplier's economy, both labor and management had little incentive to examine critically the ways in which they were operating with the goal of making production more efficient.

Second, the redevelopment of Japanese industry, beginning in 1950, with the assistance of individuals such as W. Edwards Deming, Joseph Juran, Kaoru Ishikawa, and others was based on an emphasis on quality and continuous improvement of products and services. The approach pioneered by these individuals has led to a fourth generation of management that has been called *management by process.* It is often referred to as **total quality management** or **TQM.** One of the principal features of this approach is a focus on the continuous improvement of processes. This managerial style is characterized by an emphasis on teamwork, a focus on the customer (broadened to include everyone who is a part of a process), and a fast reaction to change. Management by process has a strong statistical foundation based on a thorough knowledge of variability, a systems perspective, and belief in continuous improvement. Such statistical tools as Pareto diagrams, histograms, and control charts, and management planning tools such as process flow and fishbone diagrams, are an integral part of this approach.

Control charts, as a tool for studying the variability of a system, are useful for helping managers determine how to improve a process. A general introduction to the theory underlying control charts will be the subject of our next section.

16.3 The Theory of Control Charts

If we are examining data that have been collected sequentially over a period of time, it is imperative that a graph of the variable of interest be plotted at successive time periods. One such graph, originally developed by Shewhart (see References 21, 22, and 23) is the control chart.

The **control chart** is a means of monitoring variation in the characteristic of a product or service by (1) focusing on the time dimension in which the system produces products or services and (2) studying the nature of the variability in the system. The control chart may be used to study past performance and/or to evaluate present conditions. Data collected from a control chart may form the basis for process improvement. Control charts may be used for different types of variables—for categorical variables such as the proportion of airplane flights of a particular company that are more than 15 minutes late on a given day, for discrete variables such as the count of the number of paint blemishes in a panel of a car door, and for continuous variables such as the amount of apple juice contained in one-liter bottles.

In addition to providing a visual display of data representing a process, the principal focus of the control chart is the attempt to separate special or assignable causes of variation from chance or common causes of variation.

> **Special or assignable causes** of variation represent large fluctuations or patterns in the data that are not inherent to a process. These fluctuations are often caused by changes in a system that represent either problems to be fixed or opportunities to exploit.
>
> **Chance or common causes** of variation represent the inherent variability that exists in a system. These consist of the numerous small causes of variability that operate randomly or by chance.

The distinction between the two causes of variation is crucial because special causes of variation are considered to be those that are not part of a process and are correctable or exploitable without changing the system, whereas common causes of variation may be reduced only by changing the system. Such systemic changes are the responsibility of management.

Control charts allow us to monitor the process and determine the presence of special causes. There are two types of errors that control charts help prevent. The first type of error involves the belief that an observed value represents special cause variation when in fact it is due to the common cause variation of the system. Treating such a common cause of variation as special cause variation can result in *tampering* with or overadjustment of a process with an accompanying increase in variation. The second type of error involves treating special cause variation as if it is common cause variation and thus not taking immediate corrective action when it is necessary. Although these errors can still occur when a control chart is used, they are far less likely.

The most typical form of control chart will set control limits that are within ±3 standard deviations[1] of the statistical measure of interest (be it the average, the proportion, the range, etc.). In general, this may be stated as

$$\text{process average} \pm 3 \text{ standard deviations} \quad \textbf{(16.1)}$$

so that

$$\text{upper control limit} = \text{process average} + 3 \text{ standard deviations}$$
$$\text{lower control limit} = \text{process average} - 3 \text{ standard deviations}$$

Once these control limits are set, the control chart is evaluated from the perspective of (1) discerning any pattern that might exist in the values over time and (2) determining whether any points fall outside the control limits. Figure 16.1 illustrates three different situations.

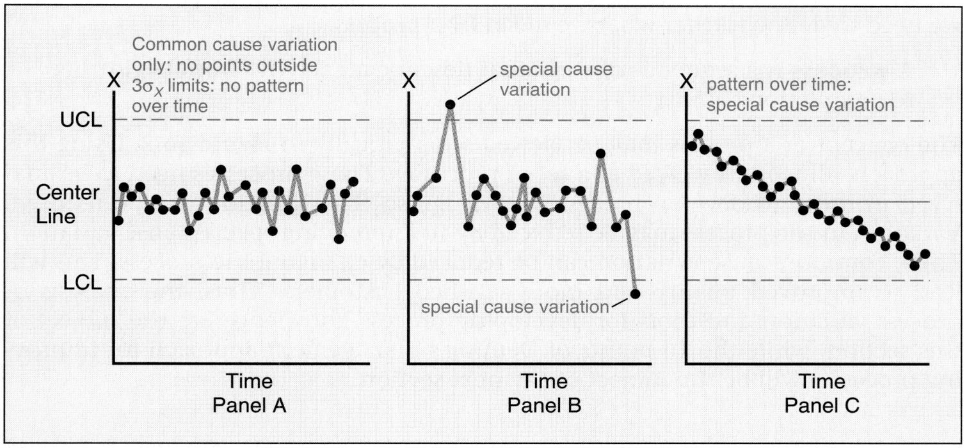

Figure 16.1
Three control chart patterns.

In panel A of Figure 16.1, we observe a process that is stable and contains only common cause variation, one in which there does not appear to be any pattern in the ordering of values over time, and one in which there are no points that fall outside the three standard deviation control limits. Panel B, on the contrary, contains two points that fall outside the three standard deviation control limits. Each of these points would have to be investigated to determine the special causes that led to their occurrence. Although panel C does not have any points outside the control limits, it has a series of consecutive points above the average value (the center line) as well as a series of consecutive points below the average value. In addition, a long-term overall downward trend in the value of the variable is clearly visible. Such a situation would call for corrective action to determine what might account for this pattern prior to initiating any changes in the system.

The detection of a trend is not always so obvious. Two easy rules[2] for indicating the presence of a trend are (1) having eight consecutive points above the center line (or eight consecutive points below the center line) or (2) having eight consecutive points that are increasing (or eight consecutive points that are decreasing).

Once all special causes of variation have been explained and eliminated, the process involved can be examined on a continuing basis until there are no patterns over time or points outside the three standard deviation limits. When the process contains only common cause variation, its performance is predictable (at least in the near future).

In order to reduce common cause variation it is necessary to alter the system that is producing the product or service. In our next section, we will discuss some management planning tools that are extremely valuable in helping to understand a process so that a system can be improved.

16.4 Some Tools for Studying a Process: Fishbone (Ishikawa) and Process Flow Diagrams

16.4.1 Introduction

Before we can determine the appropriate control charts to be used for a set of data, we need to define further what we mean by a process.

> A **process** is a sequence of steps that describe an activity from beginning to completion.

The concept of a process may be viewed schematically in Figure 16.2. Using this approach, all work is viewed as a set of processes. These processes need to be analyzed in order to develop process knowledge so that variation can be reduced. Variation in the process may be reduced by first removing special cause variation. Then, common cause variation can be reduced by changing the process. This will lead to improved quality and more satisfied customers. Thus, the analysis of process variation and tools for developing process knowledge are the subject of this section, while the 14 points of Deming's management approach for improving processes will be the subject of the next section.

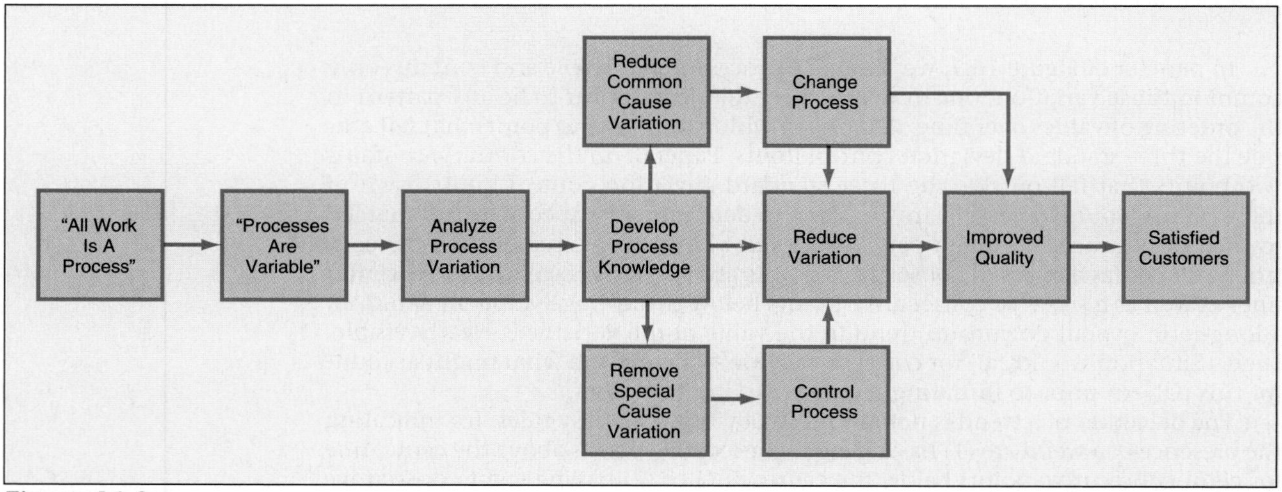

Figure 16.2
The concept of a process.
Source: Reprinted from R. Snee, "Statistical Thinking and Its Contributions to Total Quality," *American Statistician*, 1990, Vol. 44, pp. 116–121.

In order to understand any process of interest, it is useful to become familiar with two management planning tools, the *fishbone* (or *Ishikawa) diagram* and the *process flow diagram.*

16.4.2 The Fishbone (or Ishikawa) Diagram

The **fishbone diagram** was originally developed by Kaoru Ishikawa (see References 13, 16, and 27) to represent the relationship between some effect that could be measured and the set of possible causes that produce the effect. For this reason it is also known as a *cause-and-effect diagram.*

The name fishbone diagram (see Figure 16.3) comes from the way various causes are arranged on the diagram. Typically the effect or problem is shown on the right side and the major causes are listed on the left side of the diagram. These causes are often subdivided into four categories (typically manpower, methods, material, and machinery in a manufacturing environment and equipment, policies, procedures, and people in a service environment). Within each major category specific causes are listed as branches and subbranches of the major category tree. Thus the impression of the cause-and-effect diagram is that of a "fishbone."

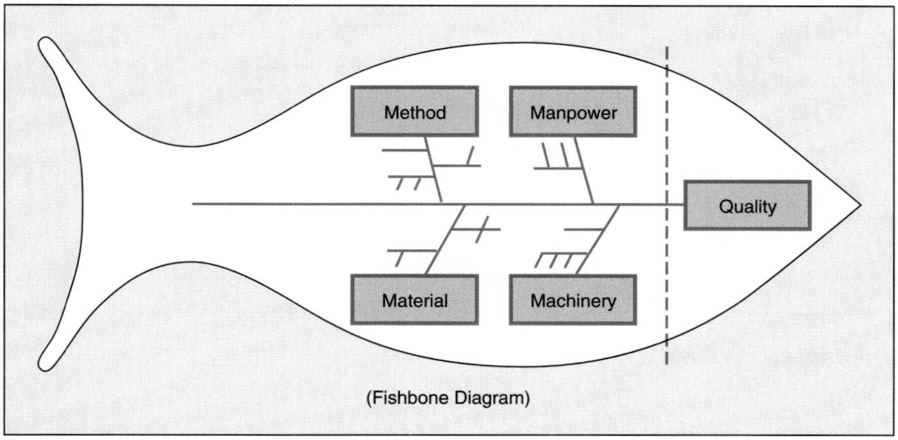

Figure 16.3
The fishbone diagram.
Source: Reprinted from *The Memory Jogger*, p. 24, Fig. 19.4. © copyright 1989 GOAL QPC, 13 Branch Street, Methuen, MA 01844, Tel. 508-685-3900. Used with permission.

The fishbone diagram can be useful in understanding processes in a variety of applications. Figure 16.4 on page 666 represents a fishbone diagram for a problem many have encountered—being late for work. Before a fishbone diagram can be constructed, a group of individuals should collectively "brainstorm" a list of causes for the effect of interest. In Figure 16.4 (which relates to Marilyn Levine, an elementary school teacher and the spouse of one of this text's authors) we observe that each of the four major categories has been subdivided into several causes, with subcauses appearing on each branch. For example, under materials, five causes are listed: gas, information, student lesson assignments, breakfast, and other meals. For student lesson assignments, such subcauses as accessing existing file, revising existing file, and printing copies are shown. Similar sets of causes are illustrated for methods, manpower, and machines.

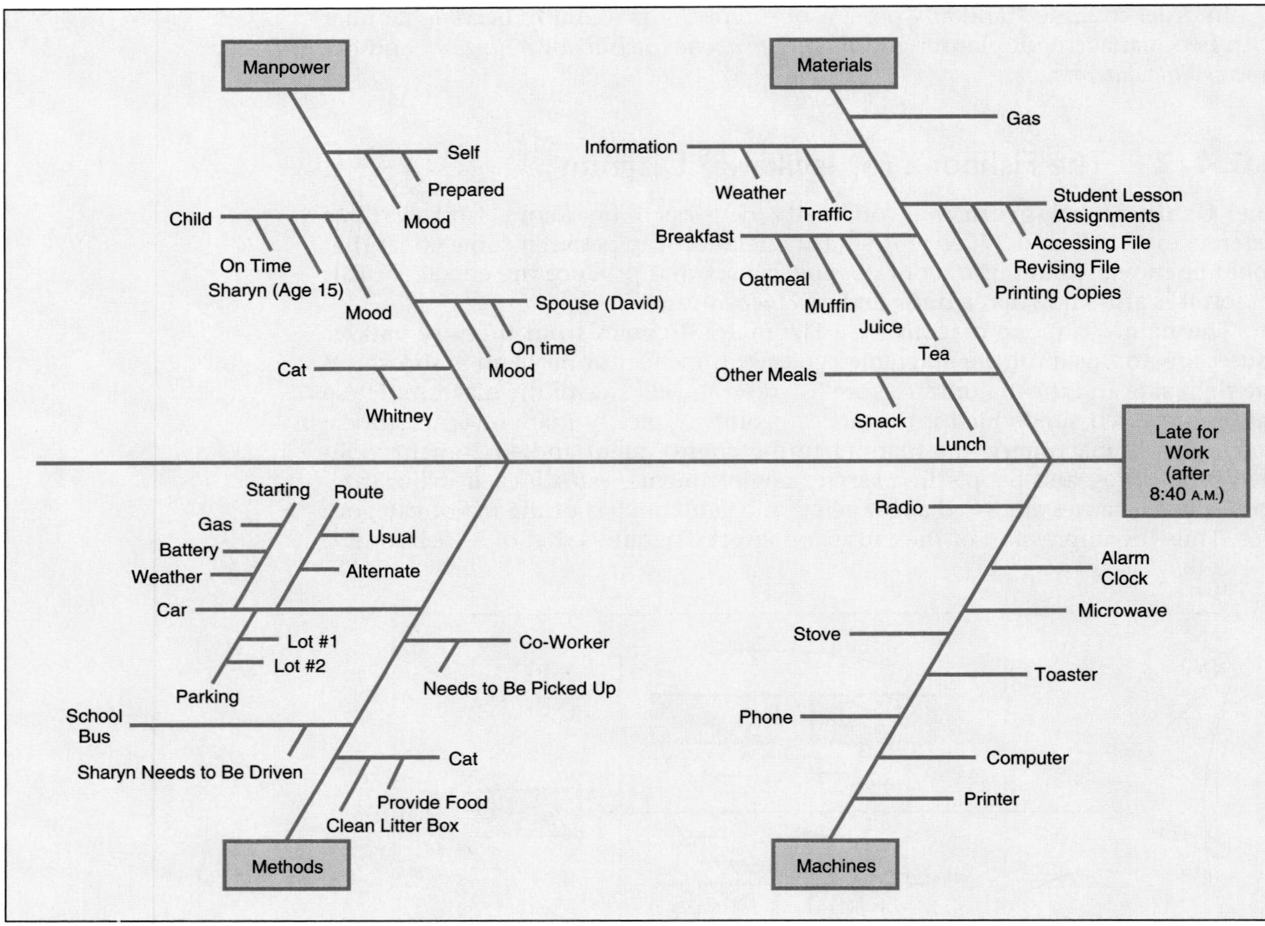

Figure 16.4
Fishbone diagram for Marilyn Levine's getting-to-work process.

As a second example, Figure 16.5 represents a fishbone diagram that might be constructed to represent the process of studying the utilization (or actually the underutilization) of hospital operating rooms on Saturday. Under the category of people, among the possible causes are registered nurse (RN) staff, medical doctor staff, and housekeeping staff, each with subbranches. Under the category of policies, among the causes are informed consent, personnel, supplies, and marketing plans. Similar detailed sets of causes are provided for the procedures and equipment categories.

16.4.3 Process Flow Diagrams

A second useful management planning tool for understanding a process is a **process flow diagram.** This diagram allows us to see a flow of steps in a process from its beginning to its termination. Such a diagram is invaluable in understanding a process. Three commonly used process flow symbols are depicted in Figure 16.6. The oval termination symbol is used at the beginning and end of a process as a start/stop symbol. The rectangular process symbol is used to indicate that a

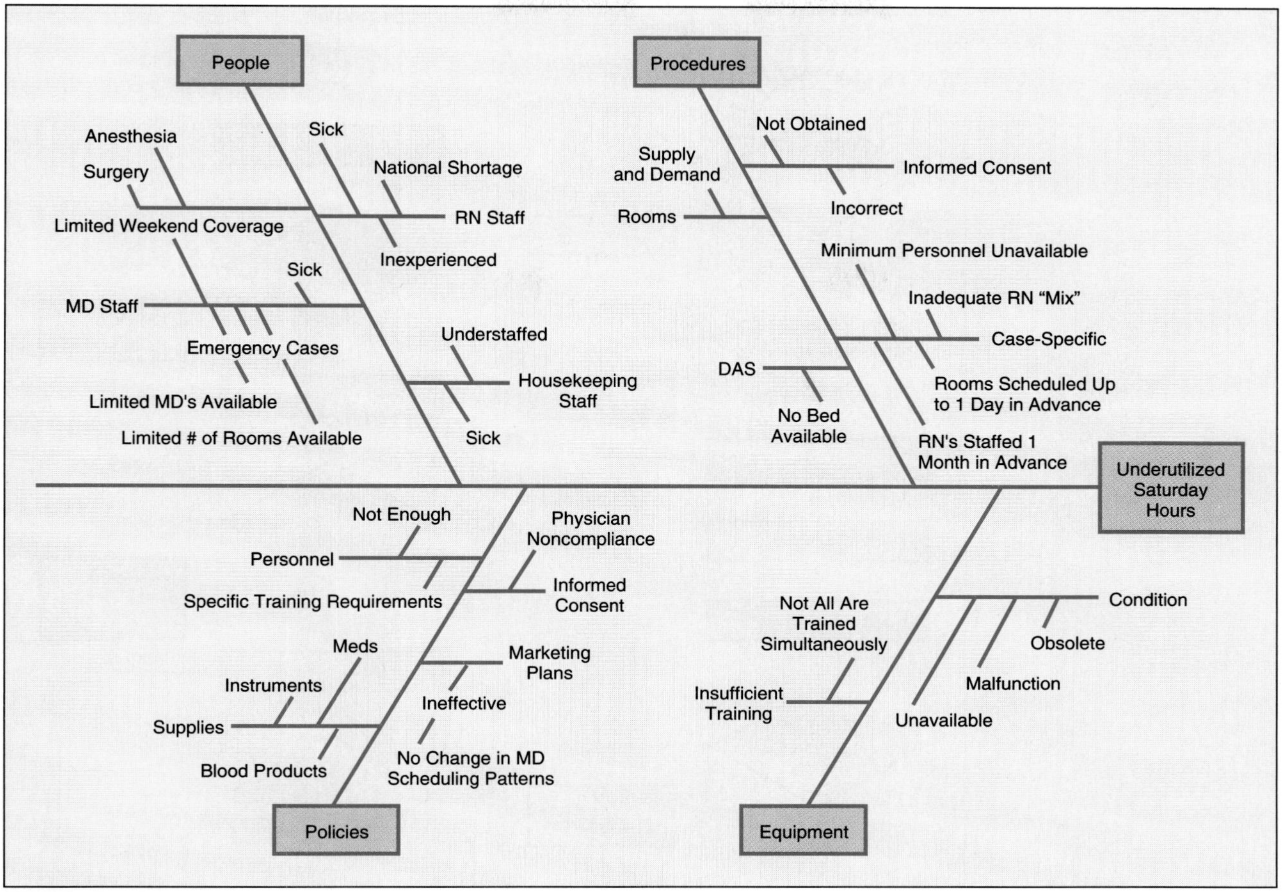

Figure 16.5
Fishbone diagram for the utilization of hospital operating rooms on Saturday.

step of the process is to be performed. The decision diamond symbol provides one way in, but at least two ways out, so that alternative paths in the process can be described.

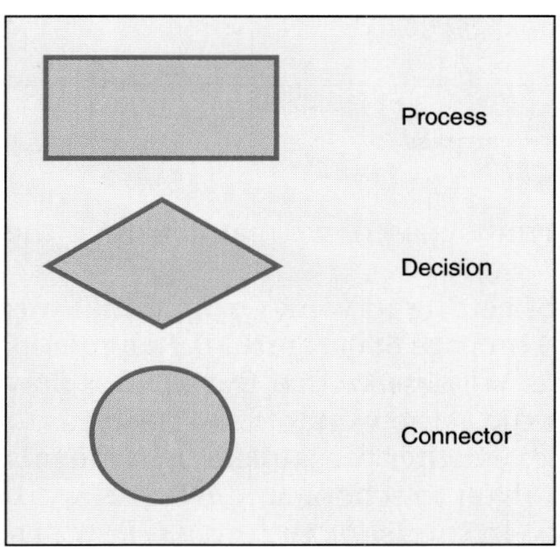

Figure 16.6
Process flow diagram symbols.

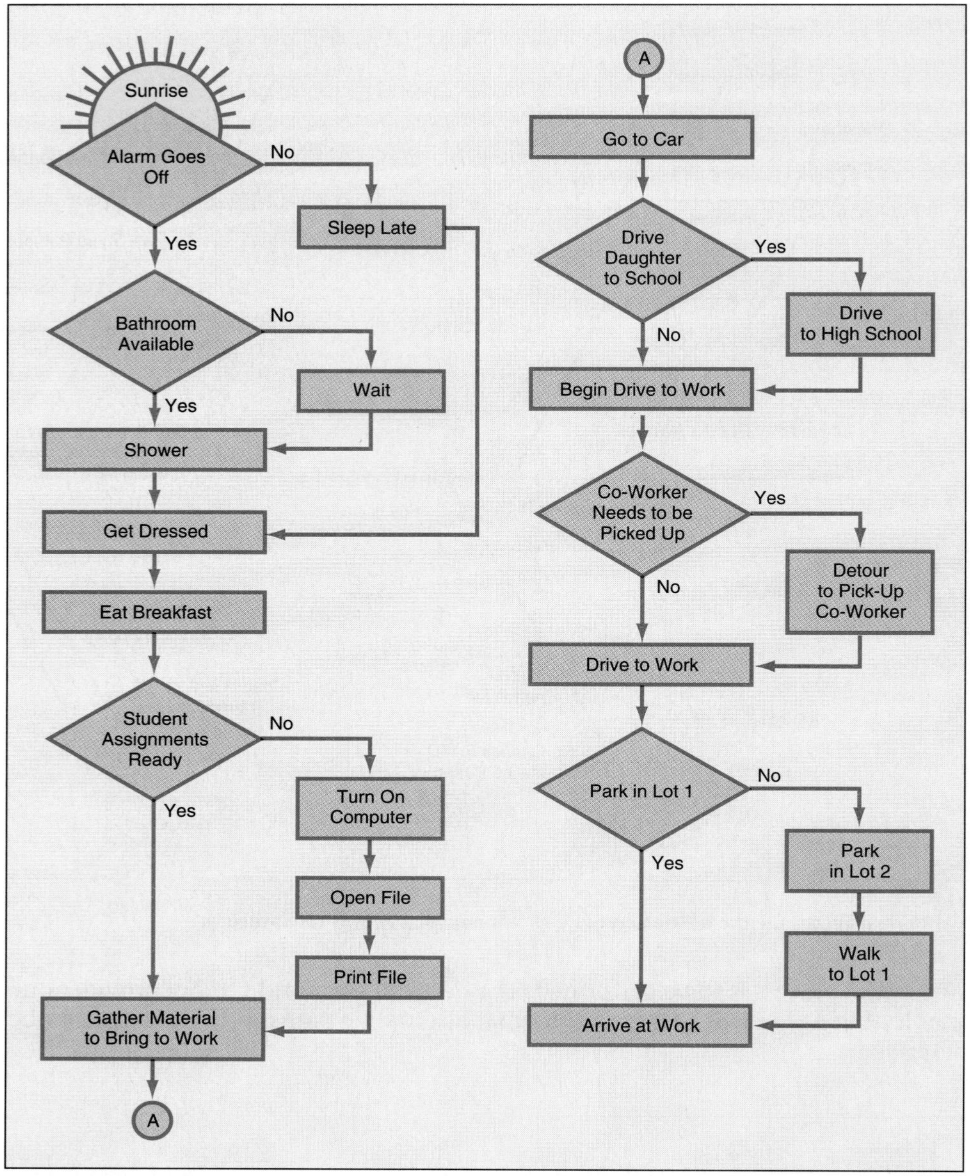

Figure 16.7
Process flow diagram for Marilyn Levine's getting-to-work process.

Now that we have defined some of the symbols used, we may illustrate the process flow diagram with two examples.

The first situation refers to the problem of getting to work on time that was depicted in the fishbone diagram of Figure 16.4 on page 666. Figure 16.7 represents the process flow diagram for this circumstance. We observe that this process flow diagram begins with a concern about whether the alarm went off and proceeds to whether the bathroom was currently available, whether the student assignments were ready, whether her daughter had to be driven to school, and whether a co-worker had to be picked up on the way to work. In summary, the process flow dia-

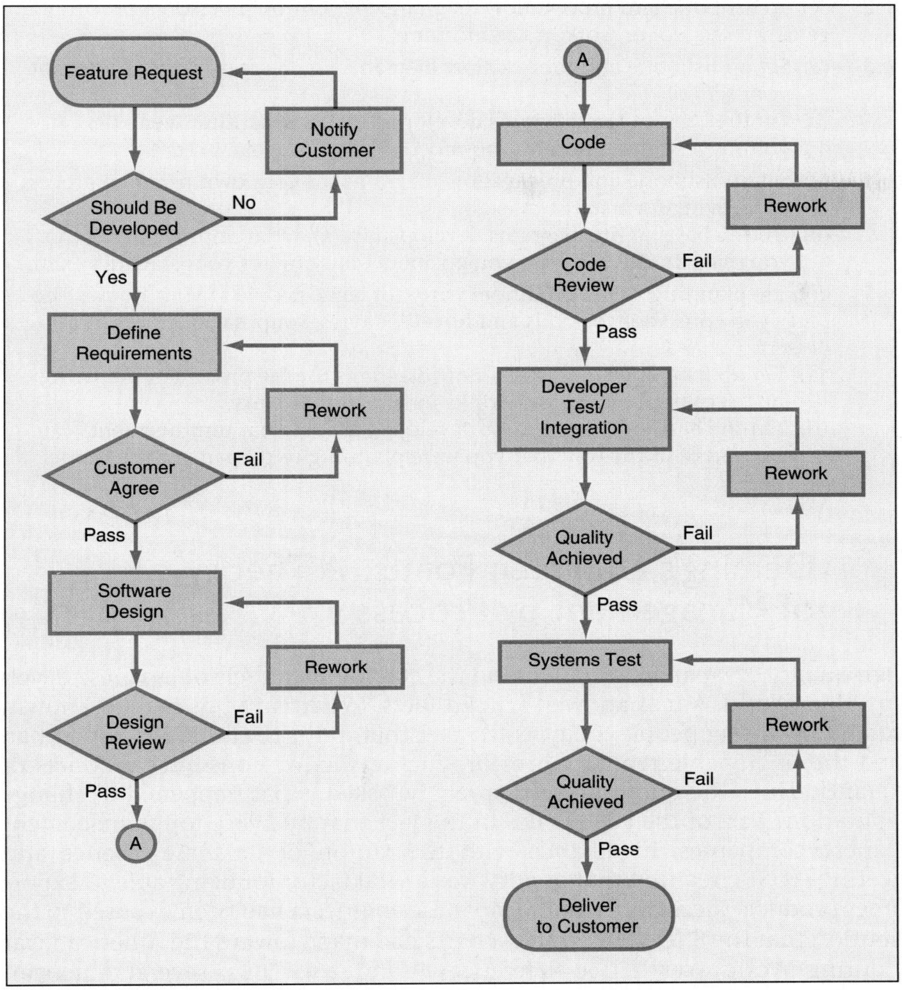

Figure 16.8
Process flow diagram for software development.

gram, by forcing us to document the process involved, provides us with a more thorough understanding of the various aspects of the process and gives insights into where delays may be likely to develop.

A second example of a process flow diagram, involving the development of software, is depicted in Figure 16.8. In this instance, the software to be developed passes through a series of steps, each of which contains an inspection aspect (with rework if necessary) before being delivered to a customer.

Now that we have discussed these management planning tools that are helpful in understanding a process, in the next section we will consider a managerial approach that may be used when processes need to be changed.

Problems for Section 16.4

16.1 Compare and contrast the fishbone diagram and the process flow diagram.

16.2 Set up fishbone and process flow diagrams for your own personal process of getting to school or work in the morning.

16.3 (a) Set up fishbone and process flow diagrams for the registration process at your school.
(b) On the basis of the diagrams developed in (a), what improvements can you suggest in the registration process?

16.4 (a) Set up fishbone and process flow diagrams for your own personal process of studying for a statistics exam.
(b) On the basis of the diagrams developed in (a), what improvements can you make in the way that you go about studying for your statistics exam?

16.5 You are planning to have a dinner party for eight people at your house. The party will consist of cocktails and hors d'oeuvres, soup, salad, entree, and dessert.
(a) Set up fishbone and process flow diagrams for the process of preparing and serving the food and drinks for the dinner party.
(b) On the basis of the diagrams developed in (a), what improvements can you make in the way that you were planning to prepare for the dinner party?

16.5 Deming's Fourteen Points: A Theory of Management by Process

The high quality of Japanese products and the economic miracle of Japanese development after World War II are well-known facts. What is not as readily known, particularly by young people today, is the fact that, prior to the 1950s, Japan had acquired the unenviable reputation of producing shoddy consumer products of poor quality. Thus, the question must surely be asked, what happened to change this reputation? Part of the answer lies in the fact that by 1950, top management of Japanese companies, in alliance with the Union of Japanese Science and Engineering (JUSE), realized that quality was a vital factor for being able to export consumer products successfully. Some Japanese engineers had been exposed to the contribution that the Shewhart control charts had made toward the American war effort during World War II (see References 9 and 26). Thus, several American experts, including W. Edwards Deming, were invited to Japan during the early 1950s. The rigorous application of a total quality management approach led to improved productivity in Japanese industry. Owing primarily to his experiences in Japan, Deming developed his approach to management based on the following 14 points:

1. Create constancy of purpose for improvement of product and service.
2. Adopt the new philosophy.
3. Cease dependence on inspection to achieve quality.
4. End the practice of awarding business on the basis of price tag alone. Instead, minimize total cost by working with a single supplier.
5. Improve constantly and forever every process for planning, production, and service.
6. Institute training on the job.
7. Adopt and institute leadership.
8. Drive out fear.
9. Break down barriers between staff areas.
10. Eliminate slogans, exhortations, and targets for the workforce.

11. Eliminate numerical quotas for the workforce and numerical goals for management.
12. Remove barriers that rob people of pride of workmanship. Eliminate the annual rating or merit system.
13. Institute a vigorous program of education and self-improvement for everyone.
14. Put everyone in the company to work to accomplish the transformation.

Point 1, create constancy of purpose, refers to how an organization deals with problems that arise both at present and in the future. The focus is on the constant improvement of a product or service. This improvement process is illustrated by the **Shewhart cycle** of Figure 16.9. Unlike the traditional manufacturing approach of "design it, make it, try to sell it," the Shewhart cycle represents a continuous cycle of "plan, do, study, and act." The first step, *planning,* represents the initial design phase for planning a change in a manufacturing or service process. The second step, *doing,* involves carrying out the change, preferably on a small scale. In doing this, planned experiments (see Chapter 14) can be a particularly valuable approach. The third step, *studying,* involves an analysis of the results using statistical tools to determine what was learned. The fourth step, *acting,* involves the acceptance of the change, its abandonment, or further study of the change under different conditions. With this approach, the process starts with the customer as the most important element of the production or service process.

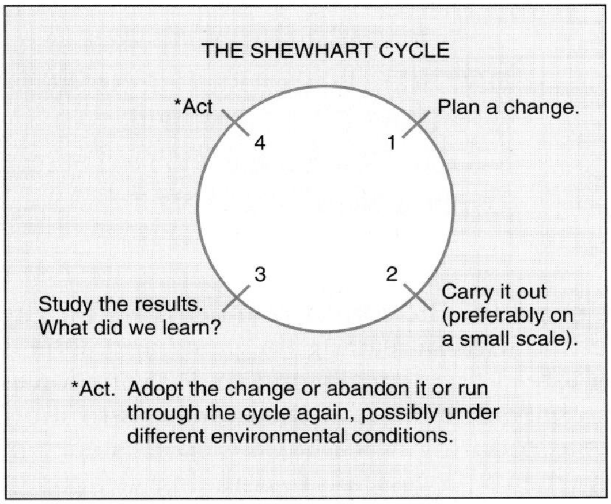

Figure 16.9
The Shewhart cycle.
Source: Adapted from *Out of the Crisis* by W. Edwards Deming by permission of MIT and W. Edwards Deming. Published by MIT, Center for Advanced Engineering Study, Cambridge, MA 02139. Copyright 1986 by W. Edwards Deming. Adapted from Figure 5, p. 88.

The key aspect of this approach is the dominance of the concern for future problems. Improvement of processes must go hand in hand with innovation and the development of new products. The importance of innovation can be illustrated by an analogy to the "Fosbury Flop" technique[3] of high jumping. Prior to the development of this technique of high jumping, the accepted approach was to use a technique called the "Western Roll." If someone had just worked at improving his ability using the Western Roll, he still would not have been able to compete with others who had adopted what was at that time the newer and better Fosbury Flop technique. Thus, innovation must go hand in hand with process improvement.

Point 2, adopt the new philosophy, refers to the urgency with which American companies need to realize that we are in a new economic age that differs drastically from the post–World War II period of American dominance (see Reference 9). It is a commonly accepted fact that, as part of human nature, people will not act until a crisis is at hand, because they prefer to continue doing things in ways that they believe have produced successful results in the past. However, in this new economic age, American management is often afflicted with what Deming called a set of "deadly diseases"—including lack of constancy of purpose, emphasis on short-term profits, fear of unfriendly takeover, evaluation of performance and merit rating systems, and excessive mobility of management. Finally, management philosophy needs to accept the notion that higher quality costs less, *not* more. However, it does require an up-front investment to get quality. This investment pays off with tremendous dividends.

Point 3, cease dependence on mass inspection to achieve quality, implies that any inspection whose purpose is to improve quality is too late because the quality is already built into the product. It would be better to focus on making it right the first time. Among the difficulties involved in mass inspection (besides high costs) are the failure of inspectors to agree on nonconforming items and the problem of separating good and bad items. Such difficulties can be illustrated with an example taken from Scherkenbach (see Reference 20) and depicted in Figure 16.10. Suppose that your job here is to read the sentence displayed in Figure 16.10. The process involves proofreading the sentence with the objective of counting the number of occurrences of the letter "F." Read the sentence and note the number of occurrences of the letter F that you discover.

Figure 16.10
An example of the proofreading process.
Source: W. W. Scherkenbach, *The Deming Route to Quality and Productivity: Road Maps and Roadblocks* (Washington, D.C.: CEEPress, 1987).

FINISHED FILES ARE THE RESULT
OF YEARS OF SCIENTIFIC
STUDY COMBINED WITH THE
EXPERIENCE OF MANY YEARS

People usually see either three Fs or six Fs. The correct number is six Fs. The number seen is dependent on the method used in reading the paragraph. One is likely to find three Fs if the paragraph is read phonetically and six Fs if one forces oneself to count the number of Fs carefully. The point of the exercise is to show that, if we have such a simple process as counting Fs leading to inconsistency of "inspector's" results, what will happen when a process fails to contain a clear operational definition of nonconforming? Certainly, in such situations much more variability from inspector to inspector will occur.

Point 4, ending the practice of awarding business on the basis of price tag alone, represents the antithesis of lowest-bidder awards. It focuses on the fact that there can be no real long-term meaning to price without a knowledge of the quality of the product. A lowest-bidder approach ignores the advantages in reduced variation of a single supplier and fails to consider the advantages of the development of a long-term relationship between purchaser and supplier. Such a relationship would allow the supplier to be innovative and would tend to make the supplier and purchaser partners in achieving success.

Point 5, improve constantly and forever the system of production and services, reinforces the importance of the continuous focus of the Shewhart cycle and the belief that quality needs to be built in at the design stage. Attaining quality is viewed as a never-ending process in which smaller variation translates into a reduction in the economic losses that occur in the manufacturing of a product whose characteristics are variable. Such an approach stands in contrast to one whose only concern is in meeting specifications. This latter all-or-none approach does not associate any economic loss with products whose characteristics are within specification limits.

Point 6, institute training, reflects the needs of all employees including production workers, engineers, and managers. It is critically important for management to understand the differences between special causes and common causes of variation, so that proper action can be taken in each circumstance. In particular, training needs to focus on developing standards for acceptable work that do not change on a daily basis. Also, management needs to realize that people learn in different ways; some learn better with written instructions, some learn better with verbal instructions. In addition, management must decide who should be trained in what. Often corporate training in areas such as statistics is wasted because of (1) limited screening of who gets trained, (2) no emphasis on how to manage what has been learned, and (3) little insistence on the use of what has been learned.

Point 7, adopt and institute leadership, relates to the distinction between leadership and supervision. The aim of leadership should be to improve the system and achieve greater consistency of performance.

Points 8 through 12 [drive out fear, break down barriers between staff areas, eliminate slogans, eliminate numerical quotas for the workforce and numerical goals for management, and remove barriers to pride of workmanship including the annual rating and merit system (this may be the most controversial point)] are all related to how the performance of an employee is to be evaluated.

The quota system for the production worker is viewed as detrimental for several reasons. First, it has a negative effect on the quality of the product, since supervisors are more inclined to pass inferior products through the system when they need to meet work quotas. Such flexible standards of work reduce the pride of workmanship of the individual and perpetuate a system in which peer pressure holds the upper half of the workers to no more than the quota rate.

In addition, the emphasis on targets and exhortations may place an improper burden on the worker, since it is management's job to improve the system, not to expect workers to produce beyond the system (this will be clearly illustrated in Section 16.7).

Third, the annual performance rating system can rob the manager of his or her pride of workmanship because this system of evaluation more often than not fails to provide a meaningful measure of performance. For many managers, the only customer is his or her supervisor. In too many cases (see Reference 24), efforts are focused on either distorting figures or distorting the system to produce the desired set of results, rather than on efforts to improve the system. Such an approach stifles teamwork, because there is often a reduced tangible reward for working together across functional areas. Finally, it rewards people who work successfully within the system, rather than people who work to improve the system.

Point 13, encourage education and self-improvement for everyone, reflects the notion that the most important resource of any organization is its people. Efforts that improve the knowledge of people in the organization also serve to increase the assets of the organization.

Point 14, take action to accomplish the transformation, again reflects the approach of management as a process in which one continuously strives toward improvement in a never-ending cycle.

Now that we have provided a brief introduction to the Deming philosophy and linked *management by process* to fundamental control chart ideas, in the following sections we shall develop several control charts that are used in industry.

16.6 Control Charts for the Proportion and Number of Nonconforming Items—The *p* and *np* Charts

16.6.1 Introduction

Let us turn our attention to various types of control charts that are used to monitor processes and determine whether special cause variation and common cause variation are present in a process. One type of commonly used control chart are the **attribute charts,** which are used when items that are sampled are classified according to whether they conform or do not conform to operationally defined requirements. The *p* and *np* charts that will be discussed in this section are based on either the proportion of nonconforming items (*p chart*) or the number of nonconforming items (*np* chart) in a sample. The *c* chart (to be discussed in Section 16.8) is based on a count of the number of nonconforming items per unit.

16.6.2 The *p* Chart

You may recall that we studied proportions in Chapters 5 through 9. We discussed the binomial distribution in Section 7.5 and the normal approximation to the binomial distribution in Section 8.6.2. Moreover, in Equation (9.7) of Section 9.3 we defined the proportion as X/n and in Equation (9.8) we defined the standard deviation of the proportions as

$$\sigma_{p_s} = \sqrt{\frac{p(1-p)}{n}}$$

Using Equation (16.1) on page 663, we may establish control limits for the proportion of nonconforming[4] items from the sample or subgroup data as

$$\bar{p} \pm 3\sqrt{\frac{\bar{p}(1-\bar{p})}{\bar{n}}} \qquad \textbf{(16.2)}$$

so that

$$LCL = \bar{p} - 3\sqrt{\frac{\bar{p}(1-\bar{p})}{\bar{n}}} \qquad \textbf{(16.3a)}$$

$$UCL = \bar{p} + 3\sqrt{\frac{\bar{p}(1-\bar{p})}{\bar{n}}} \qquad \textbf{(16.3b)}$$

where

X_i = number of nonconforming items in subgroup i
n_i = sample or subgroup size for subgroup i
$p_{S_i} = X_i / n_i$
k = number of subgroups taken
$\bar{n}$ = average subgroup size
$\bar{p}$ = average proportion of nonconforming items

For equal n_i,

$$\bar{n} = n_i \quad \text{and} \quad \bar{p} = \frac{\sum_{i=1}^{k} p_{S_i}}{k}$$

or in general,

$$\bar{n} = \frac{\sum_{i=1}^{k} n_i}{k} \quad \text{and} \quad \bar{p} = \frac{\sum_{i=1}^{k} X_i}{\sum_{i=1}^{k} n_i}$$

Any negative value for the lower control limit will mean that the lower control limit does not exist.

We may observe an application of the p chart by referring to the plan of a large hotel located in a resort city to improve the quality of its services. One aspect of its services to hotel guests is represented by the readiness of a hotel room when the guest first enters his or her assigned room. From the viewpoint of initial impression, it is particularly important that all amenities that are supposed to be available (soap, towels, complimentary guest basket, and so on) are actually available in the room and equally important that all appliances such as the radio, television, and telephone are in proper working order. Hotel management has decided that it would study this process for a four-week period by taking a daily sample of 200 rooms for which guests are holding reservations. Thus, it would be determined before their arrival as to whether the room contained any nonconformances in terms of the availability of amenities and the working order of all appliances. Table 16.1 on page 676 represents the number and proportion of rooms that were considered to be nonconforming for each day in the four-week period.

For these data, $k = 28, \sum_{i=1}^{k} p_{S_i} = 2.315$, and $n_i = 200$

Thus,

$$\bar{p} = \frac{2.315}{28} = .0827$$

so that using Equation (16.2) we have

$$.0827 \pm 3\sqrt{\frac{(.0827)(.9173)}{200}}$$

$$.0827 \pm .0584$$

Table 16.1
Rooms considered nonconforming at check-in over a four-week period.

Day	Rooms Studied	Rooms Not Ready	Proportion	Day	Rooms Studied	Rooms Not Ready	Proportion
1	200	16	.080	15	200	18	.090
2	200	7	.035	16	200	13	.065
3	200	21	.105	17	200	15	.075
4	200	17	.085	18	200	10	.050
5	200	25	.125	19	200	14	.070
6	200	19	.095	20	200	25	.125
7	200	16	.080	21	200	19	.095
8	200	15	.075	22	200	12	.060
9	200	11	.055	23	200	6	.030
10	200	12	.060	24	200	12	.060
11	200	22	.110	25	200	18	.090
12	200	20	.100	26	200	15	.075
13	200	17	.085	27	200	20	.100
14	200	26	.130	28	200	22	.110

Thus,

$$UCL = .0827 + .0584 = .1411$$

and

$$LCL = .0827 - .0584 = .0243$$

The control chart for the data of Table 16.1 is displayed in Figure 16.11. An examination of Figure 16.11 seems to indicate a process in a state of statistical control, with the individual points distributed around $\bar{p}$ without any pattern. Thus, any improvement in this system of making rooms ready for guests must come from the reduction of common cause variation. As we have stated previously, such system alterations are the responsibility of management.

Now that we have examined a situation in which the sample or subgroup size was equal, we need to turn to a more general situation in which the subgroup size may vary over time. As a rule, as long as none of the subgroup sizes n_i differs from the average subgroup size $\bar{n}$ by more than ±25% of $\bar{n}$ (see Reference 8), Equation (16.2) may be used to obtain the control limits for the p chart.

We may observe the application of a p chart when the subgroup sizes are unequal by studying a process related to the production of gauze sponges at a factory. The number of nonconforming sponges and the number of sponges produced daily for a period of 32 days are displayed in Table 16.2.

For these data, $k = 32$, $\sum_{i=1}^{k} n_i = 19{,}926$ and $\sum_{i=1}^{k} X_i = 665$.

Thus,

$$\bar{n} = \frac{19{,}926}{32} = 622.69 \quad \text{and} \quad \bar{p} = \frac{665}{19{,}926} = .033$$

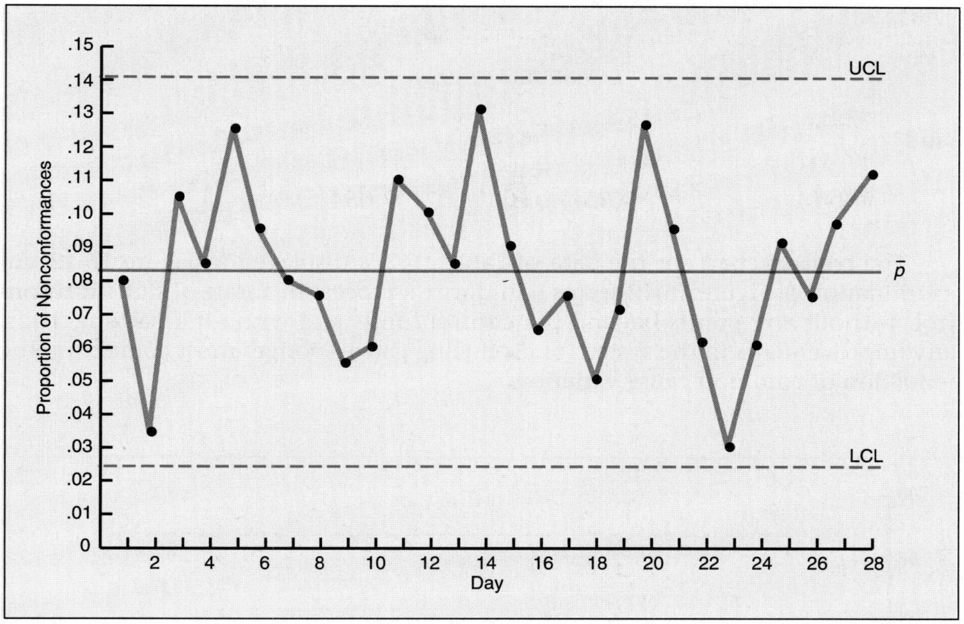

Figure 16.11
***p* chart for the proportion of nonconforming rooms upon guest arrival.**
Source: Table 16.1.

so that we have

$$.033 \pm 3\sqrt{\frac{(.033)(1-.033)}{622.69}}$$

$$.033 \pm .021$$

Table 16.2
Nonconforming sponges produced daily for a 32-day period.

Day	Number Produced	Number Nonconforming	Proportion	Day	Number Produced	Number Nonconforming	Proportion
1	690	21	.030	17	575	20	.035
2	580	22	.038	18	610	16	.026
3	685	20	.029	19	596	15	.025
4	595	21	.035	20	630	24	.038
5	665	23	.035	21	625	25	.040
6	596	19	.032	22	615	21	.034
7	600	18	.030	23	575	23	.040
8	620	24	.039	24	572	20	.035
9	610	20	.033	25	645	24	.037
10	595	22	.037	26	651	25	.038
11	645	19	.029	27	660	21	.032
12	675	23	.034	28	685	19	.028
13	670	22	.033	29	671	17	.025
14	590	26	.044	30	660	22	.033
15	585	17	.029	31	595	24	.040
16	560	16	.029	32	600	16	.027

Thus,

$$LCL = .033 - .021 = .012$$

and

$$UCL = .033 + .021 = .054$$

The control chart for the data of Table 16.2 is displayed in Figure 16.12. An examination of Figure 16.12 seems to indicate a process in a state of statistical control, without any points beyond the control limits and without a pattern. Thus, any improvements in the system of producing gauze sponges must come from the reduction of common cause variation.

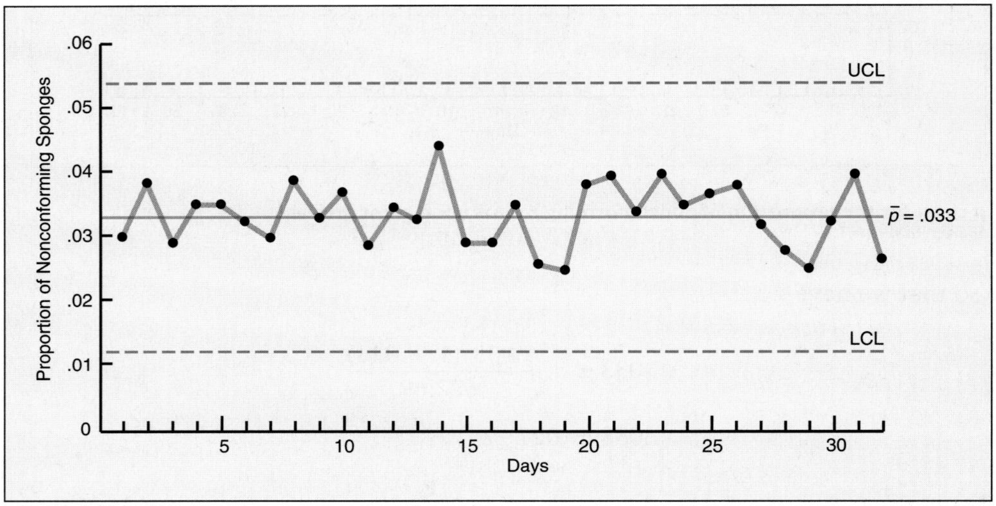

Figure 16.12
***p* chart for proportion of nonconforming sponges.**

16.6.3 The *np* Chart

When the subgroups are all of the same size, an alternative to the p chart that is desirable is the np chart. You may recall from the normal approximation to the binomial distribution in Section 8.6.2 that we defined the standard error of the number of "successes" or nonconforming items as

$$\sigma_x = \sqrt{np(1 - p)}$$

Thus, using the normal approximation to the binomial and Equation (16.1), we may set up control limits for the number of nonconforming items as follows:

$$\bar{X} \pm 3\sqrt{\bar{X}(1 - \bar{p})} \quad \textbf{(16.4)}$$

so that

$$LCL = \bar{X} - 3\sqrt{\bar{X}(1 - \bar{p})} \quad \textbf{(16.5a)}$$

$$UCL = \bar{X} + 3\sqrt{\bar{X}(1 - \bar{p})} \quad \textbf{(16.5b)}$$

where

$$\bar{X} = \frac{\sum_{i=1}^{k} X_i}{k}$$

$$\bar{p} = \frac{\sum_{i=1}^{k} X_i}{nk}$$

n = the subgroup size

k = number of subgroups

To illustrate the *np* chart, we return to the data of Table 16.1 on page 676 that were used previously for the *p* chart.

For these data, $k = 28$, $n = 200$, and $\sum_{i=1}^{k} X_i = 463$

Thus,

$$\bar{X} = \frac{463}{28} = 16.536 \text{ and } \bar{p} = \frac{463}{(200)28} = .0827$$

so that we have

$$16.536 \pm 3\sqrt{16.536(1 - .0827)}$$

$$16.536 \pm 11.684$$

Thus,

$$UCL = 16.536 + 11.684 = 28.22$$

and

$$LCL = 16.536 - 11.684 = 4.852$$

The *np* chart for these data is displayed in Figure 16.13 on page 680. We note that this figure provides precisely the same results as the *p* chart of Figure 16.11 on page 677, but indicates the number of nonconforming items rather than the proportion. The choice between the two approaches (available only for equal subgroups) is a matter of personal preference.

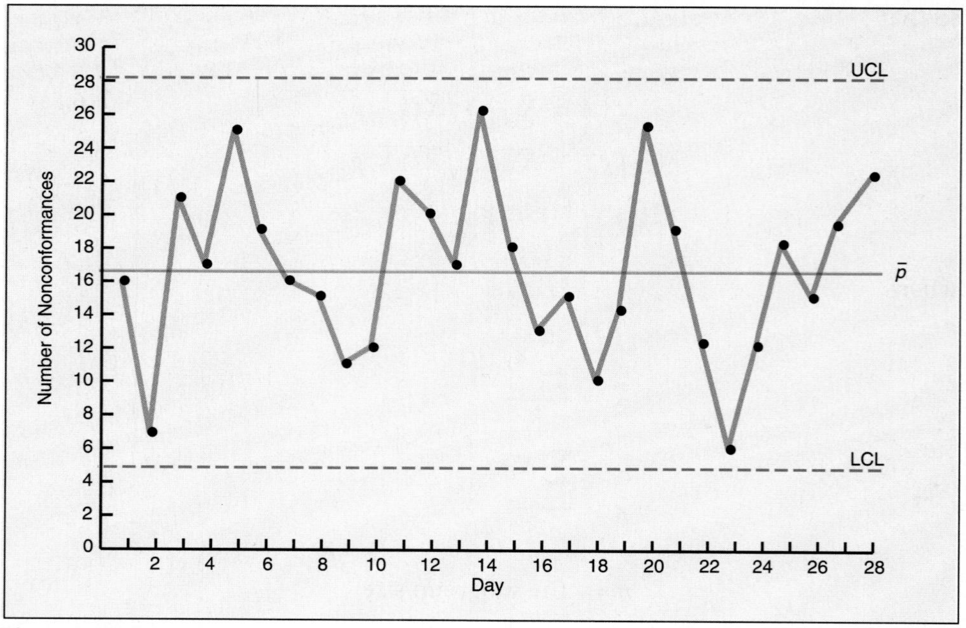

Figure 16.13
***np* chart for the number of nonconforming rooms upon guest arrival.**
Source: Data are taken from Table 16.1 on page 676.

Problems for Section 16.6

16.6 The Commuters Watchdog Council of a railroad that serves a large metropolitan area wishes to monitor the on-time performance of the railroad during the morning rush hour. Suppose that a train is defined as being late if it arrives more than five minutes after the scheduled arrival time. A total of 235 trains are scheduled during the rush hour each morning. The results for a four-week period (based on a five-day work week) were as follows:

Data file RRSPC.DAT

Day	Number of Late Arrivals	Day	Number of Late Arrivals
1	17	11	21
2	25	12	23
3	22	13	67
4	27	14	24
5	32	15	35
6	23	16	18
7	16	17	23
8	24	18	24
9	20	19	26
10	36	20	35

(a) Set up a *p* chart for the proportion of late arrivals and indicate whether the arrival process is in statistical control during this period.

(b) Set up an *np* chart for the number of late arrivals and indicate whether the arrival process is in statistical control during this period.
(c) Compare the results of the *p* chart obtained in (a) with those of the *np* chart obtained in (b).
(d) What effect would it have on your conclusion in (a) or (b) if you knew that there had been a four-inch snowstorm on the morning of day 13?

16.7 A professional basketball player has embarked on a program to study his ability to shoot foul shots. On each day in which a game is not scheduled, he intends to shoot 100 foul shots. He maintains records over a period of 40 days of practice, with the following results:

Data file
FOULSPC.DAT

Day	Number of Foul Shots Made	Day	Number of Foul Shots Made
1	73	21	64
2	75	22	67
3	69	23	72
4	72	24	70
5	77	25	74
6	71	26	76
7	68	27	75
8	70	28	78
9	67	29	76
10	74	30	80
11	75	31	78
12	72	32	83
13	70	33	84
14	74	34	81
15	73	35	86
16	76	36	85
17	69	37	86
18	68	38	87
19	72	39	85
20	70	40	85

(a) Set up a *p* chart for the proportion of successful foul shots. Do you think that the player's foul-shooting process is in statistical control? If not, why not?
(b) Set up an *np* chart for the number of foul shots made and indicate whether the foul-shooting process is in statistical control during this period.
(c) What if you were told that after the first 20 days, the player changed his method of shooting foul shots? How might this information change the conclusions you drew in (a) or (b)?
(d) If you knew this information prior to doing (a) or (b), how might you have done the *p* or *np* chart differently?

16.8 A private mail delivery service has a policy of guaranteeing delivery by 10:30 A.M. of the morning after a package is picked up. Suppose that management of the service wishes to study delivery performance in a particular geographic area over a four-week time period based on a five-day work week. The total number of packages delivered daily and the number of packages that were not delivered by 10:30 A.M. were recorded with the result displayed at the top of page 682:

Day	Number of Packages Delivered	Number of Packages Not Arriving before 10:30 A.M.	Day	Number of Packages Delivered	Number of Packages Not Arriving before 10:30 A.M.
1	136	4	11	157	6
2	153	6	12	150	9
3	127	2	13	142	8
4	157	7	14	137	10
5	144	5	15	147	8
6	122	5	16	132	7
7	154	6	17	136	6
8	132	3	18	137	7
9	160	8	19	153	11
10	142	7	20	141	7

Data file
MAILSPC.DAT

(a) Set up a p chart for the proportion of packages that are not delivered before 10:30 A.M.
(b) Does the process give an out-of-control signal?

16.9 The superintendent of a school district was interested in studying student absenteeism at a particular elementary school during December and January. The school had 537 students registered during this time period. The results were recorded as follows:

Data file
ABSSPC.DAT

Day	Number of Students Absent	Day	Number of Students Absent
1	39	19	54
2	46	20	52
3	38	21	46
4	46	22	45
5	53	23	42
6	52	24	44
7	56	25	49
8	61	26	39
9	51	27	72
10	55	28	55
11	52	29	50
12	49	30	42
13	44	31	48
14	39	32	46
15	53	33	45
16	68	34	49
17	101	35	41
18	70	36	47

Note: The first 17 days were from December and the last 19 days were from January.

(a) Set up a p chart for the proportion of students who were absent during December and January. Does the process give an out-of-control signal?
(b) Set up an np chart for the number of students absent and indicate whether the absenteeism process is in statistical control during this period.
(c) Compare the results of the p chart obtained in (a) with those of the np chart obtained in (b).
(d) If the superintendent wants to develop a process to reduce absenteeism, how should she proceed?

16.10 A bottling company of Sweet Suzy's Sugarless Cola maintains daily records of the occurrences of unacceptable cans flowing from the filling and sealing machine. Nonconformities such as improper filling amount, dented cans, and cans that are improperly sealed are noted. Data for one month's production (based on a five-day work week) were as follows:

Day	Number of Cans Filled	Number of Unacceptable Cans	Day	Number of Cans Filled	Number of Unacceptable Cans
1	5,043	47	12	5,314	70
2	4,852	51	13	5,097	64
3	4,908	43	14	4,932	59
4	4,756	37	15	5,023	75
5	4,901	78	16	5,117	71
6	4,892	66	17	5,099	68
7	5,354	51	18	5,345	78
8	5,321	66	19	5,456	88
9	5,045	61	20	5,554	83
10	5,113	72	21	5,421	82
11	5,247	63	22	5,555	87

Data file COLASPC.DAT

(a) Set up a p chart for the proportion of unacceptable cans for the month. Does the process give an out-of-control signal?

(b) If management wants to develop a process for reducing the proportion of unacceptable cans, how should it proceed?

16.11 The manager of the accounting office of a large hospital was interested in studying the problem of errors in the entry of account numbers into the computer system. A group of 200 account numbers was selected from each day's output and each was inspected to determine whether it was a conforming item. The results for a period of 39 days were as follows:

Data file ERRORSPC.DAT

Day	Number of Nonconforming Items	Day	Number of Nonconforming Items
1	3	21	13
2	5	22	5
3	2	23	2
4	11	24	0
5	6	25	14
6	15	26	10
7	8	27	9
8	1	28	7
9	25	29	6
10	4	30	1
11	0	31	21
12	6	32	2
13	9	33	4
14	2	34	2
15	8	35	8
16	28	36	30
17	16	37	0
18	5	38	0
19	10	39	1
20	30		

(a) Set up a p chart for the proportion of nonconforming items. Does the process give an out-of-control signal?
(b) Set up an np chart for the number of nonconforming items and indicate whether the process is in statistical control during this period.
(c) Compare the results of the p chart obtained in (a) with those of the np chart obtained in (b).
(d) On the basis of the results of (a) or (b), what would you now do as a manager to improve the process of account number entry?

16.12 A manager of a regional office of a local telephone company has as one of her job responsibilities the task of processing requests for additions, changes, or deletions of telephone service. A service improvement team that was formed decided to look at the corrections in terms of central office equipment and facilities required to process the orders that were issued to service requests. Data collected over a period of 30 days revealed the following:

Data file TELESPC.DAT

Day	Number of Orders	Number of Corrections	Day	Number of Orders	Number of Corrections
1	690	80	16	831	91
2	676	88	17	816	80
3	896	74	18	701	96
4	707	94	19	761	78
5	694	70	20	851	85
6	765	95	21	678	65
7	788	73	22	915	74
8	794	103	23	698	68
9	694	100	24	821	72
10	784	103	25	750	101
11	812	70	26	600	91
12	759	83	27	744	64
13	781	64	28	698	67
14	682	64	29	820	105
15	802	72	30	732	112

(a) Set up a p chart for the proportion of corrections. Does the process give an out-of-control signal?
(b) What would you now do as a manager to improve the processing of requests for changes in telephone service?

16.7 The Red Bead Experiment: Understanding Process Variability

We began this chapter with a review of the history of quality and productivity and then developed the concept of common cause variation and special cause variation that led us to a discussion of Deming's 14 points. We discussed the important management planning tools of fishbone and process flow diagrams for understanding a process and thus far we have covered the p and np control chart procedures. In this section, in order to enhance our understanding of these two types of variation, we will discuss what has become a famous parable, the **red bead experiment.**

The experiment involves the selection of beads from a box that typically contains 4,000 beads.[5] Several different scenarios can be used for conducting the experiment. The one that we will use here begins with the following:

A facilitator (who will play the role of company foreman) asks the audience to volunteer for the jobs of workers (at least four are needed), inspectors (two are needed), chief inspector (one is needed), and recorder (one is needed). A worker's job consists of using a paddle that has five rows of ten bead-size holes to select 50 beads from the box of beads.

Once the participants have been selected, the foreman explains the jobs to the particular participants. The job of the workers is to produce white beads, since red beads are unacceptable to the customers. Strict procedures are to be followed. Work standards call for the production of 50 beads by each worker (a strict quota system), no more and no less than 50. Management has established a standard that no more than two red beads per worker are to be produced on any given day. The paddle is dipped into the box of beads so that when it is removed each of the 50 holes contains a bead. Once this is done, the paddle is carried to each of the two inspectors, who independently record the count of red beads. The chief inspector compares their counts and announces the results to the audience. The recorder writes down the number of red beads next to the name of the worker.

Once all the people know their jobs, "production" can begin. Suppose that on the first "day," the number of red beads "produced" by the four workers (call them Alyson, David, Peter, and Sharyn) was 9, 12, 13, and 7, respectively. How should management react to the day's production when the standard says that no more than two red beads per worker should be produced? Should all the workers be reprimanded, or should only David and Peter be given a stern warning that they will be fired if they don't improve?

Suppose that production continues for an additional two days and the results for all three days are summarized in Table 16.3.

Table 16.3 Red bead experiment results for four workers over three days.

	Day			
Name	1	2	3	All 3 Days
Alyson	9	11	6	26
David	12	12	8	32
Peter	13	6	12	31
Sharyn	7	9	8	24
All four workers	41	38	34	113
Average ($\bar{X}$)	10.25	9.5	8.5	9.42

From Table 16.3 we may observe several phenomena. On each day, some of the workers were above the average and some below the average. On day 1 Sharyn did best, but on day 2 Peter (who had the worst record on day 1) was best, and on day 3 Alyson was best.

How then can we explain all this variation? An answer can be provided by using Equation (16.4) to develop an *np* chart. For these data we have

$$k = 4 \text{ workers} \times 3 \text{ days} = 12, \quad n = 50 \quad \text{and} \quad \sum_{i=1}^{k} X_i = 113$$

Thus,

$$\overline{X} = \frac{113}{12} = 9.42 \text{ and } \bar{p} = \frac{113}{(50)(12)} = .1883$$

so that we have

$$\overline{X} \pm 3\sqrt{\overline{X}(1 - \bar{p})}$$

$$9.42 \pm 3\sqrt{(9.42)(1 - .1883)}$$

$$9.42 \pm 8.30$$

Thus,

$$UCL = 9.42 + 8.30 = 17.72$$

and

$$LCL = 9.42 - 8.30 = 1.12$$

Figure 16.14 represents the *np* control chart for the data of Table 16.2. We observe from Figure 16.14 that all of the points are within the control limits and there are no patterns in the results. The differences between the workers *merely represent common cause variation inherent in a stable system.*

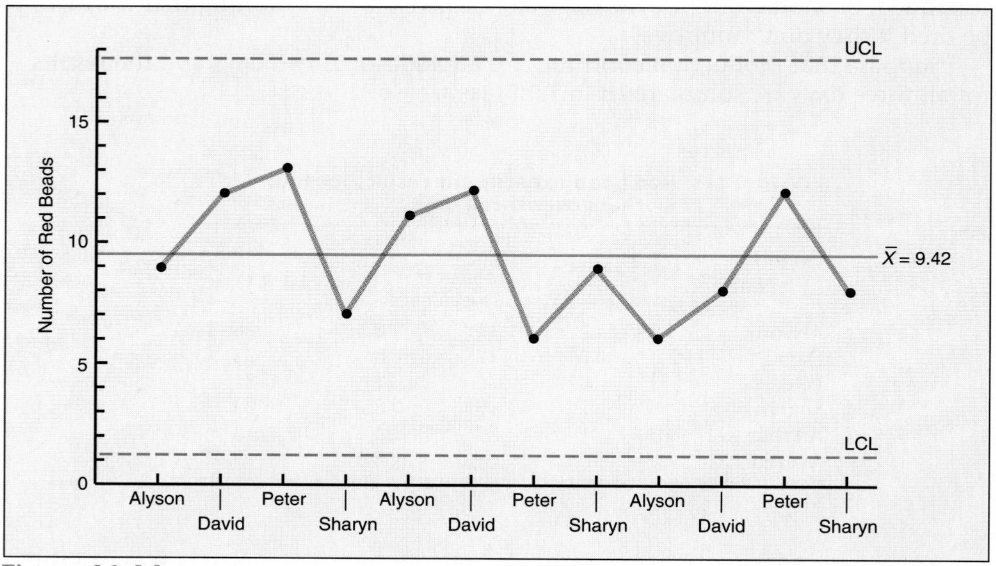

Figure 16.14
***np* chart for the red bead experiment.**

In conclusion, there are four morals to the parable of the red beads:

1. Variation is an inherent part of any process.
2. Workers work within a system over which they have little control. It is the system that primarily determines their performance.
3. Only management can change the system.

4. Some workers will always be above the average and some workers will always be below the average.

Problems for Section 16.7

16.13 How do you think many managers would have reacted after day 1? Day 2? Day 3?

16.14 **(Class Project)** Obtain a version of the red bead experiment for your class.
(a) Conduct the experiment in the same way as described in Section 16.7.
(b) Remove 400 red beads from the bead box before beginning the experiment. How do your results differ from those obtained in (a)? What does this tell you about the effect of the "system" on the workers?

16.8 The *c* Chart: A Control Chart for the Number of Occurrences per Unit

In Section 16.6 we considered the *p* chart for the proportion of nonconforming items and the *np* chart for the number of nonconforming items. In other instances we may be interested in determining the number of nonconformities (or occurrences) in a unit (often called an **area of opportunity**) where the subgroup size is very large and the probability of a nonconformity occurring in any part of the unit is very small. This differs from the *p* and *np* chart in that we are not classifying each unit as conforming or nonconforming, but are counting the number of occurrences in the unit.

We may recall from Section 7.6 that such a situation fits the assumptions of a Poisson distribution. Among the phenomena that could be described by this process would be the number of flaws in a square foot of carpet, the number of typographical errors on a printed page, the number of breakdowns per day in an academic computer center, and the number of "turnovers" per game by a basketball team.

In Section 7.6, for the Poisson distribution, we defined the standard deviation of the number of occurrences as being equal to the square root of the average number of occurrences (λ). Assuming that the size of each subgroup unit remains constant,[6] we may set up control limits for the number of occurrences per unit using the normal approximation to the Poisson distribution. Using Equation (16.1), the control limits for the average number of occurrences would be

$$\bar{c} \pm 3\sqrt{\bar{c}} \qquad \textbf{(16.6)}$$

so that

$$LCL = \bar{c} - 3\sqrt{\bar{c}} \qquad \textbf{(16.7a)}$$

$$UCL = \bar{c} + 3\sqrt{\bar{c}} \qquad \textbf{(16.7b)}$$

where

$$\bar{c} = \frac{\sum_{i=1}^{k} c_i}{k}$$

where $\bar{c}$ = average number of occurrences
k = number of units sampled
c_i = number of occurrences in unit i

As an application of the c chart, suppose that the production manager of a large baking factory that makes Marilyn's pumpkin chocolate chip cupcakes for the Halloween holiday season needed to study the baking process to determine the number of chocolate chips that were contained in the cupcakes being baked. A subgroup of 50 cupcakes was selected from the production line. The results, listed in the order of selection, are summarized in Table 16.4.

Table 16.4 Number of chocolate chips in a subgroup of 50 cupcakes.

Cupcake	Number of Chocolate Chips	Cupcake	Number of Chocolate Chips
1	8	26	7
2	10	27	5
3	6	28	8
4	7	29	6
5	5	30	7
6	7	31	5
7	9	32	5
8	8	33	4
9	7	34	4
10	9	35	3
11	10	36	5
12	7	37	2
13	8	38	4
14	11	39	3
15	10	40	3
16	9	41	4
17	8	42	2
18	7	43	4
19	10	44	5
20	11	45	5
21	8	46	3
22	7	47	2
23	8	48	5
24	6	49	4
25	7	50	4

For these data

$$k = 50 \text{ and } \sum_{i=1}^{k} c_i = 312$$

Thus,

$$\bar{c} = \frac{312}{50} = 6.24$$

so that using Equations (16.6) and (16.7), we have

$$6.24 \pm 3\sqrt{6.24}$$

$$6.24 \pm 7.494$$

Thus,

$$UCL = 6.24 + 7.494 = 13.734$$

and

$$LCL = 6.24 - 7.494 \text{ so that } LCL \text{ does not exist.}$$

The control chart for the data of Table 16.4 is displayed in Figure 16.15. An examination of Figure 16.15 does not indicate any points outside the control limits. However, there is a clear pattern to the number of chocolate chips per cupcake over time, with cupcakes baked in the first half of the sequence almost always having more than the average number of chocolate chips and cupcakes in the latter half of the sequence having fewer than the average number of chocolate chips. Thus, the production manager should immediately investigate the process to determine the special causes that have produced this pattern of variation. The best place to start would be to ask the workers on the production line.

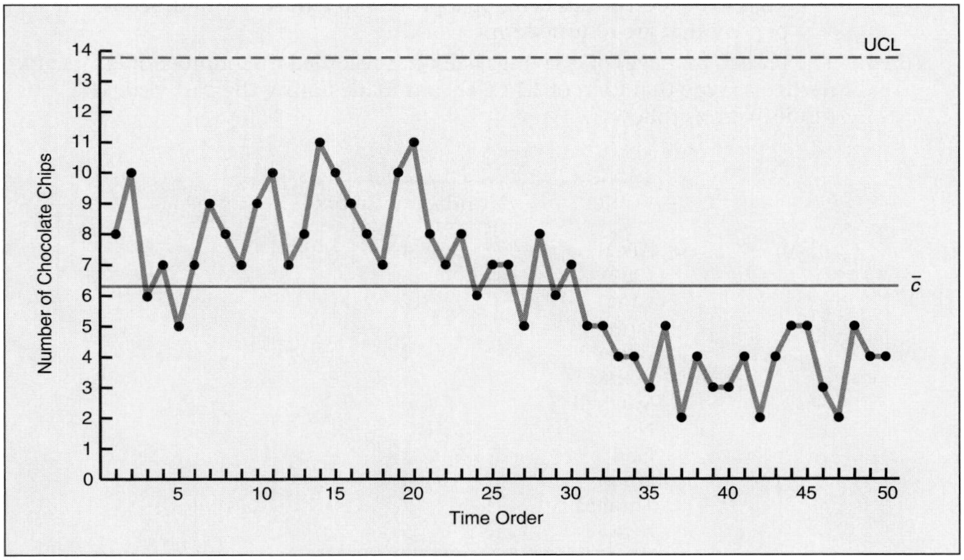

Figure 16.15
c chart for number of chocolate chips per cupcake.

Problems for Section 16.8

Data file DRYCLEAN.DAT

● 16.15 The owner of a dry cleaning business, in an effort to measure the quality of the services provided, would like to study the number of dry-cleaned items that are returned for rework per day. Records were kept for a four-week period (the store is open Monday–Saturday) with the results indicated as follows:

Day	Items Returned for Rework	Day	Items Returned for Rework
1	4	13	5
2	6	14	8
3	3	15	3
4	7	16	4
5	6	17	10
6	8	18	9
7	6	19	6
8	4	20	5
9	8	21	8
10	6	22	6
11	5	23	7
12	12	24	9

(a) Set up a *c* chart for the number of items per day that are returned for rework. Do you think that the process is in a state of statistical control?
(b) Should the owner of the dry cleaning store take action to investigate why 12 items were returned for rework on day 12? Explain. Would your answer be the same if 20 items were returned for rework on day 12?
(c) On the basis of the results in (a), how should the owner of the dry cleaning store proceed in setting up a process to reduce the number of items per day that are returned for rework?

16.16 The branch manager of a savings bank has recorded the number of errors of a particular type that each of 12 tellers has made during the past year. The results were as follows:

Data file TELLER.DAT

Teller	Number of Errors
Alice	4
Carl	7
Gina	12
Jane	6
Linda	2
Marla	5
Mitchell	6
Nora	3
Paul	5
Susan	4
Thomas	7
Vera	5

(a) Do you think the bank manager will single out Gina for any disciplinary action regarding her performance in the last year?
(b) Set up a *c* chart for the number of errors committed by the 12 tellers. Is the number of errors in a state of statistical control?
(c) Based on the *c* chart developed in (b), do you now think that Gina should be singled out for disciplinary action regarding her performance? Does your conclusion now agree with what you expected the manager to do?

(d) On the basis of the results in (b), how should the branch manager go about setting up a program to reduce this particular type of error?

16.17 Falls are one source of preventable hospital injury. Although most patients who fall are not hurt, a risk of serious injury is involved. The following data represent the number of patient falls per month over a 28-month period in a 19-bed AIDS unit at a major metropolitan hospital.

Data file PTFALLS.DAT

Month	Number of Patient Falls	Month	Number of Patient Falls
1	2	15	6
2	4	16	5
3	2	17	3
4	4	18	8
5	3	19	6
6	3	20	3
7	1	21	9
8	4	22	4
9	5	23	5
10	11	24	0
11	8	25	2
12	7	26	6
13	9	27	5
14	10	28	7

(a) Set up a *c* chart for the number of patient falls per month. Is the process of patient falls per month in a state of statistical control?
(b) What effect would it have on your conclusions if you knew that the unit was started only one month prior to the beginning of data collection?
(c) What other factors might contribute to special cause variation in this problem?

16.18 The director of operations for an airline was interested in studying the number of pieces of baggage that are lost (temporarily or permanently) at a large airport. Records indicating the number of lost baggage claims filed per day over a one-month period were as follows:

Data file BAGGAGE.DAT

Day	Number of Claims	Day	Number of Claims
1	14	16	28
2	23	17	20
3	17	18	13
4	25	19	26
5	27	20	42
6	42	21	38
7	35	22	23
8	29	23	28
9	30	24	19
10	23	25	26
11	15	26	14
12	27	27	30
13	41	28	37
14	50	29	17
15	23	30	24

(a) Set up a control chart for the number of claims per day. Is the process in a state of statistical control? Explain.

(b) Suppose that the total number of pieces of baggage per day was available for the 30-day period. Explain how you might proceed with a different control chart than the one you used in (a). Indicate what the advantages could be of using this alternative control chart compared with the control chart that was used in (a).

16.19 The University of Southwest North Carolina has recently completed its basketball season. Its basketball coach, the legendary Raving Rick Rawng, has maintained records of the number of turnovers (times the ball was lost without taking a shot) per game. The results are as follows:

Data file TURNOVER.DAT

Game	Number of Turnovers	Game	Number of Turnovers
1	16	14	18
2	12	15	26
3	25	16	14
4	17	17	12
5	11	18	16
6	19	19	29
7	17	20	11
8	23	21	7
9	12	22	15
10	9	23	12
11	13	24	17
12	16	25	22
13	21	26	14

(a) Set up a c chart for the number of turnovers per game. Is the process in a state of statistical control?

(b) On the basis of the results of (a), how should the coach proceed in setting up a process to reduce the number of turnovers in the future?

16.9 Control Charts for the Mean ($\overline{X}$) and the Range (R)

16.9.1 Introduction

Whenever a characteristic of interest is measured on an interval or ratio scale, **variables control charts** can be used to monitor a process. Because measures from these more powerful scales provide more information than the proportion or number of nonconforming items, these charts are more sensitive in detecting special cause variation than the p, np, or c charts. Variables charts are typically used in pairs. One chart monitors the variation in a process, while the other monitors the process average. The chart that monitors variability must be examined first because if it indicates the presence of out-of-control conditions, the interpretation of the chart for the average will be misleading. Although several alternative pairs of charts can be considered (see References 8, 13, 17, and 19) in this text we will study the control chart for the range and the control chart for the average.

16.9.2 The *R* Chart: A Control Chart for Dispersion

Before obtaining control limits for the mean, we need to develop a control chart for the range. This will enable us to determine whether the variability in a process is in control or whether shifts are occurring over time. If the process range is in control, then it can be used to develop the control limits for the average.

From Equation (16.1), we observe that to obtain control limits for the range we need to obtain an estimate of the average range and the standard deviation of the range. As seen in Equation (16.8), these control limits are a function not only of the d_2 factor, which represents the relationship between the standard deviation and the range for varying sample sizes, but also the d_3 factor, which represents the relationship between the standard deviation and the standard deviation of the range for varying sample sizes. Values for these factors are presented in Table E.13. Thus, we may set up the following control limits for the range over k consecutive sequences or periods of time.

$$\bar{R} \pm 3\bar{R}\frac{d_3}{d_2} \tag{16.8}$$

so that

$$LCL = \bar{R} - 3\bar{R}\frac{d_3}{d_2} \tag{16.9a}$$

and

$$UCL = \bar{R} + 3\bar{R}\frac{d_3}{d_2} \tag{16.9b}$$

where

$$\bar{R} = \frac{\sum_{i=1}^{k} R_i}{k}$$

Referring to Equations (16.9a) and (16.9b), we may simplify the calculations by utilizing the D_3 factor, equal to $1 - 3(d_3/d_2)$, and the D_4 factor, equal to $1 + 3(d_3/d_2)$, to obtain the control limits as shown in Equations (16.10a) and (16.10b).

$$LCL = D_3 \bar{R} \tag{16.10a}$$

$$UCL = D_4 \bar{R} \tag{16.10b}$$

To illustrate the application of the *R* chart, let us refer to the following example. Suppose that the management of the hotel discussed in Section 16.6 also wanted to analyze the check-in process. In particular, they wanted to study the amount of time it took for the delivery of luggage (as measured from the time the guest completed check-in procedures to the time the luggage arrived in the guest's room). Data were recorded over a four-week (Sunday–Saturday) period and subgroups of 5 deliveries were selected (on a certain shift) on each day for analysis. The summary results (in minutes) are recorded in Table 16.5.

Table 16.5 Subgroup average and range for luggage delivery times over a four-week period.

Day	Subgroup Average $\bar{X}_i$ (in minutes)	Subgroup Range R_i (in minutes)	Day	Subgroup Average $\bar{X}_i$ (in minutes)	Subgroup Range R_i (in minutes)
1	5.32	3.85	15	5.21	3.26
2	6.59	4.27	16	4.68	2.92
3	4.88	3.28	17	5.32	3.37
4	5.70	2.99	18	4.90	3.55
5	4.07	3.61	19	4.44	3.73
6	7.34	5.04	20	5.80	3.86
7	6.79	4.22	21	5.61	3.65
8	4.93	3.69	22	4.77	3.38
9	5.01	3.33	23	4.37	3.02
10	3.92	2.96	24	4.79	3.80
11	5.66	3.77	25	5.03	4.11
12	4.98	3.09	26	5.11	3.75
13	6.83	5.21	27	6.94	4.57
14	5.27	3.84	28	5.71	4.29

For these data

$$k = 28 \text{ and } \sum_{i=1}^{k} R_i = 104.41$$

Thus,

$$\bar{R} = \frac{104.41}{28} = 3.729$$

From Table E.13 for $n = 5$, we obtain $d_2 = 2.326$ and $d_3 = .864$. Using Equations (16.8) and (16.9), we have

$$3.729 \pm 3\frac{(.864)(3.729)}{2.326}$$

$$3.729 \pm 4.155$$

so that

$$UCL = 3.729 + 4.155 = 7.884$$

and

$$LCL = 3.729 - 4.155 \quad \text{so that } LCL \text{ does not exist.}$$

Alternatively, using Equation (16.10), from Table E.13, $D_3 = 0$ and $D_4 = 2.114$. Thus,

$$UCL = (2.114)(3.729) = 7.883$$

and

$$LCL \text{ does not exist.}$$

We note that the lower control limit for R does not exist since a negative range is impossible to attain. The R chart is displayed in Figure 16.16. An examination of Figure 16.16 does not indicate any individual ranges outside the control limits.

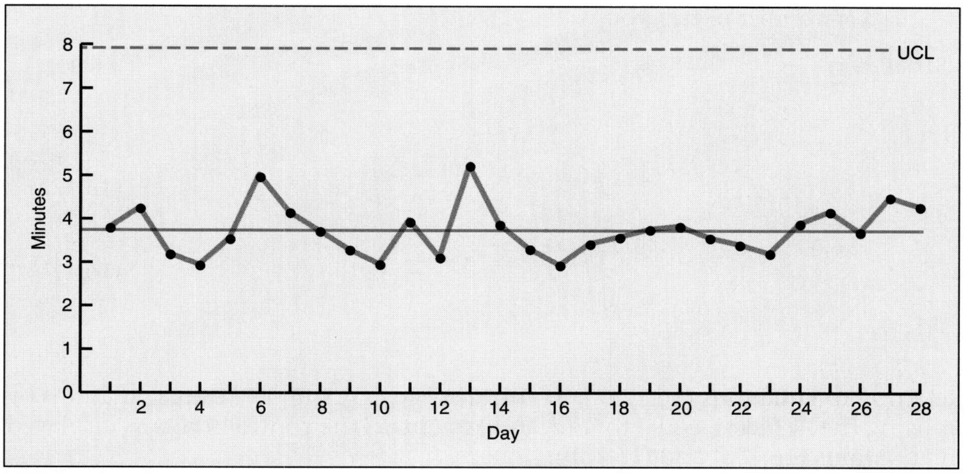

Figure 16.16
The *R* chart for luggage delivery times.
Source: Table 16.5.

16.9.3 The $\overline{X}$ Chart

Now that we have determined that the control chart for the range is in control, we may continue by examining the control chart for the process average.

The control chart for $\bar{X}$ uses subgroups of size n that are obtained over k consecutive sequences or periods of time. From Equation (16.1), we observe that to obtain control limits for the average we need to obtain an estimate of the average of the subgroup averages (which we shall call $\bar{\bar{X}}$) and the standard deviation of the average (which we called the standard error of the mean $\sigma_{\bar{X}}$ in Chapter 9). These control limits are a function of the d_2 factor, which represents the relationship between the standard deviation and the range for varying sample sizes. The range may be used to estimate the standard deviation as long as the subgroup size is no more than 10 (see References 13, 17, 19). Thus, we may set up the following control limits:

$$\bar{\bar{X}} \pm 3\frac{\bar{R}}{d_2\sqrt{n}} \tag{16.11}$$

where

$$\bar{\bar{X}} = \frac{\sum_{i=1}^{k} \bar{X}_i}{k} \quad \text{and} \quad \bar{R} = \frac{\sum_{i=1}^{k} R_i}{k}$$

where $\bar{X}_i$ = the sample mean of n observations at time i
R_i = the range of n observations at time i
k = number of subgroups

so that

$$LCL = \bar{\bar{X}} - 3\frac{\bar{R}}{d_2\sqrt{n}} \tag{16.12a}$$

$$UCL = \bar{\bar{X}} + 3\frac{\bar{R}}{d_2\sqrt{n}} \tag{16.12b}$$

Referring to Equations (16.12a) and (16.12b), we may simplify the calculations by utilizing the A_2 factor, equal to $3/(d_2\sqrt{n})$, to obtain the control limits as displayed in Equations (16.13a) and (16.13b).

$$LCL = \bar{\bar{X}} - A_2\bar{R} \tag{16.13a}$$

$$UCL = \bar{\bar{X}} + A_2\bar{R} \tag{16.13b}$$

Thus, returning to our example concerning luggage delivery times, from Table 16.5 we may compute

$$k = 28 \qquad \sum_{i=1}^{k} \overline{X}_i = 149.97 \qquad \sum_{i=1}^{k} R_i = 104.41$$

so that

$$\overline{\overline{X}} = \frac{149.97}{28} = 5.356 \text{ and } \overline{R} = \frac{104.41}{28} = 3.729$$

From Table E.13 for $n = 5$, we obtain $d_2 = 2.326$. Thus, using Equation (16.12) we have

$$5.356 \pm 3\frac{3.729}{(2.326)\sqrt{5}}$$

$$5.356 \pm 2.151$$

Therefore

$$LCL = 5.356 - 2.151 = 3.205$$

and

$$UCL = 5.356 + 2.151 = 7.507$$

Alternatively, using Equation (16.13), from Table E.13 $A_2 = .577$ and

$$LCL = 5.356 - (.577)(3.729) = 5.356 - 2.152 = 3.204$$

$$UCL = 5.356 + (.577)(3.729) = 5.356 + 2.152 = 7.508$$

These results are the same, except for rounding error.

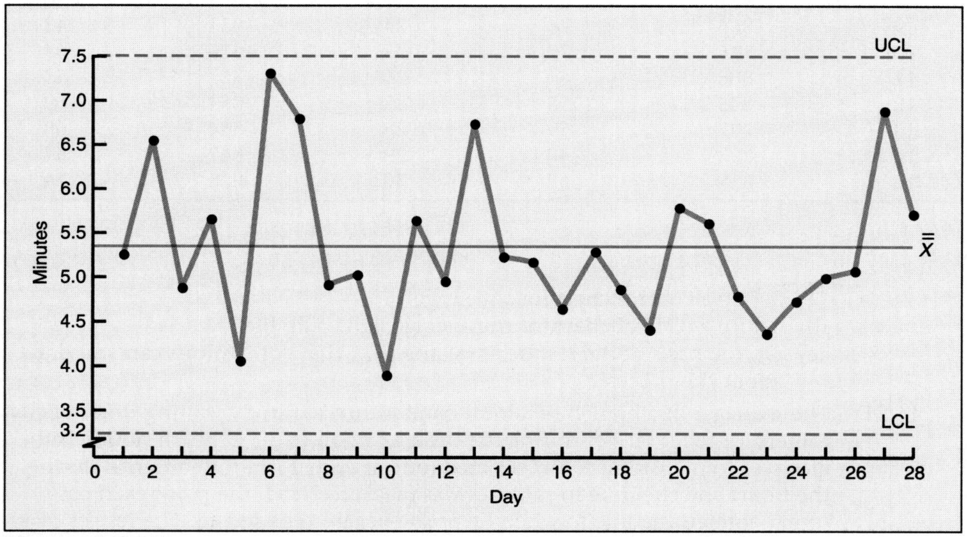

Figure 16.17
$\overline{X}$ chart for average luggage delivery time. *Source:* Table 16.5.

The control chart for the data of Table 16.5 is displayed in Figure 16.17 at the bottom of page 697. An examination of Figure 16.17 does not reveal any points outside the control limits, although there is a large amount of variability among the 28 subgroup means. However, a closer evaluation does seem to indicate a series of six consecutive points and a series of five consecutive points that are below the overall average. Since these occurred on days 14–19 and 22–26, which correspond to Saturday–Thursday and Sunday–Thursday, it would seem that further study might be necessary to determine whether a different system of delivery exists during the midweek and weekend periods. For example, the proportion of filled rooms may vary or the number of available workers may vary during these times. After conclusion of such a study, any further improvements in the delivery times would have to be brought about by changes in the management of the delivery service.

Problems for Section 16.9

16.20 The following data pertaining to incandescent light bulbs represent the average life and range for 30 subgroups of 5 light bulbs each.

Data file
BULBLIFE.DAT

Subgroup Number	Subgroup Mean $\bar{X}_i$	Subgroup Range R_i	Subgroup Number	Subgroup Mean $\bar{X}_i$	Subgroup Range R_i
1	790	52	16	845	42
2	845	56	17	891	38
3	857	116	18	859	65
4	846	89	19	826	70
5	843	65	20	828	37
6	877	73	21	854	52
7	861	38	22	847	49
8	891	84	23	868	40
9	866	76	24	851	43
10	816	72	25	870	64
11	806	61	26	857	53
12	835	55	27	851	59
13	797	59	28	834	68
14	803	47	29	842	57
15	818	69	30	825	74

(a) Set up a control chart for the range.
(b) Set up a control chart for the average light bulb life.
(c) On the basis of the results of (a) and (b), what conclusions can you draw about the process?

16.21 The manager of a branch of a local bank wanted to study waiting times of customers for teller service during the peak 12 noon to 1 P.M. lunch hour. A subgroup of four customers was selected (one at each 15-minute interval during the hour) and the time in minutes was measured from the point each customer entered the line to when he or she began to be served. The results over a four-week period are presented at the top of page 699:

Data file
BANKTIME.DAT

Day	Time in Minutes			
1	7.2	8.4	7.9	4.9
2	5.6	8.7	3.3	4.2
3	5.5	7.3	3.2	6.0
4	4.4	8.0	5.4	7.4
5	9.7	4.6	4.8	5.8
6	8.3	8.9	9.1	6.2
7	4.7	6.6	5.3	5.8
8	8.8	5.5	8.4	6.9
9	5.7	4.7	4.1	4.6
10	1.7	4.0	3.0	5.2
11	2.6	3.9	5.2	4.8
12	4.6	2.7	6.3	3.4
13	4.9	6.2	7.8	8.7
14	7.1	6.3	8.2	5.5
15	7.1	5.8	6.9	7.0
16	6.7	6.9	7.0	9.4
17	5.5	6.3	3.2	4.9
18	4.9	5.1	3.2	7.6
19	7.2	8.0	4.1	5.9
20	6.1	3.4	7.2	5.9

(a) Set up control charts for the arithmetic mean and the range.
(b) On the basis of the results in (a), indicate whether the process is in control in terms of these charts.

● 16.22 The manager of a warehouse for a local telephone company was involved in an important process to receive expensive circuit boards and return them to central stock so that they may be used at a later date when a circuit or new telephone service is needed. The timely return and processing of these units are critical in providing good service to field customers and reducing capital expenditures of the corporation. The following data represent the number of units handled by each of a subgroup of five employees over a 30-day period.

Data file
WAREHSE.DAT

	Employee				
Day	1	2	3	4	5
1	114	499	106	342	55
2	219	319	162	44	87
3	64	302	38	83	93
4	258	110	98	78	154
5	127	140	298	518	275
6	151	176	188	268	77
7	24	183	202	81	104
8	41	249	342	338	69
9	93	189	209	444	151
10	111	207	143	318	129
11	205	281	250	468	79
12	121	261	183	606	287
13	225	83	198	223	180
14	235	439	102	330	190
15	91	32	190	70	150

(Continued on page 700)

(Continued)

Day	Employee 1	2	3	4	5
16	181	191	182	444	124
17	52	190	310	245	156
18	90	538	277	308	171
19	78	587	147	172	299
20	45	265	126	137	151
21	410	227	179	298	342
22	68	375	195	67	72
23	140	266	157	92	140
24	145	170	231	60	191
25	129	74	148	119	139
26	143	384	263	147	131
27	86	229	474	181	40
28	164	313	295	297	280
29	257	310	217	152	351
30	106	134	175	153	69

(a) Set up control charts for the arithmetic mean and the range.
(b) On the basis of the results in (a), indicate whether the process is in control in terms of these charts.

16.23 The service manager of a large automobile dealership wanted to study the length of time required for a particular type of repair in his shop. A subgroup of 10 cars needing this repair was selected on each day for a period of four weeks. The results (service time in hours) were recorded in a table as follows:

Data file AUTOREP.DAT

Day	Subgroup Average $\bar{X}_i$	Subgroup Range R_i	Day	Subgroup Average $\bar{X}_i$	Subgroup Range R_i
1	3.73	5.23	11	3.64	5.37
2	3.16	4.82	12	3.27	4.42
3	3.56	4.98	13	3.16	4.85
4	3.01	4.28	14	3.39	4.44
5	3.87	5.74	15	3.85	5.06
6	3.90	5.42	16	3.90	4.99
7	3.54	4.08	17	3.72	4.67
8	3.32	4.55	18	3.51	4.37
9	3.29	4.48	19	3.34	4.53
10	3.83	5.09	20	3.99	5.28

(a) Set up all appropriate control charts and determine whether the service time process is in a state of statistical control.
(b) If the service manager wanted to develop a process to reduce service time, how should he proceed?

16.24 The manager of a private swimming pool facility monitors the pH (alkalinity-acidity) level of the swimming pool by taking hourly readings from 8 A.M. to 6 P.M. daily. The results for a three-week period summarized on a daily basis are presented at the top of page 701:

Data file PHLEVEL.DAT

Day	Average $\bar{X}_i$	Range R_i	Day	Average $\bar{X}_i$	Range R_i
1	7.34	0.16	12	7.39	0.16
2	7.41	0.12	13	7.40	0.18
3	7.30	0.11	14	7.35	0.17
4	7.28	0.19	15	7.39	0.22
5	7.23	0.17	16	7.42	0.20
6	7.30	0.20	17	7.40	0.18
7	7.35	0.15	18	7.37	0.18
8	7.38	0.19	19	7.41	0.22
9	7.32	0.14	20	7.36	0.15
10	7.38	0.19	21	7.40	0.12
11	7.43	0.23			

(a) Set up a control chart for the range.
(b) Set up a control chart for the average daily pH level.
(c) On the basis of the results of (a) and (b), what conclusions can you draw about the process?

16.10 Control Charts for Individual Values (X Charts)

In some circumstances, it is not feasible to obtain data in samples or subgroups of more than one item. Situations such as these require an individual value or *X* chart. When only a single individual value is available, we can consider each item as its own subgroup. However, unlike the case of the $\bar{X}$ chart, we cannot obtain an estimate of the variation within the subgroup since each subgroup is of size $n = 1$. In such a situation, we may estimate the standard deviation by using the moving range.[7]

The **moving range** is defined as the difference between the largest and smallest observations in a subset of *n* observations.

Since it is most common to use subsets of 2 observations, the *i*th moving range (MR_i) can be defined as in Equation (16.14)

$$MR_i = \left| X_{i-1} - X_i \right| \quad (16.14)$$

where n = number of observations in each subset
k = number of observations

This produces $(k - 1)$ moving ranges from which the average moving range ($\overline{MR}$) is computed in Equation (16.15)

$$\overline{MR} = \frac{\sum_{i=1}^{k-1} MR_i}{k - 1} \quad (16.15)$$

From Equation (16.1), we observe that to obtain control limits for the individual value we need to obtain an estimate of the average and the standard deviation. Since the relationship of the range to the standard deviation varies with sample size, when the moving range is used as an estimate of the process standard deviation, a constant factor called d_2 that reflects this relationship is used to develop the control limits. The d_2 factor is obtained from Table E.13 on page E32. The control limits for the individual value are

$$\bar{X} \pm 3\frac{\overline{MR}}{d_2} \tag{16.16}$$

so that

$$LCL = \bar{X} - 3\frac{\overline{MR}}{d_2} \tag{16.17a}$$

$$UCL = \bar{X} + 3\frac{\overline{MR}}{d_2} \tag{16.17b}$$

From Table E.13, for subgroups of size $n = 2$, the d_2 factor is 1.128.

We may simplify the calculations by utilizing the E_2 factor, equal to $3/d_2$, to obtain the control limits as displayed in Equation (16.18).

$$\bar{X} \pm E_2\overline{MR} \tag{16.18}$$

In order to illustrate this individual value control chart, we turn to an example involving the accounts receivable of a hardware distributor over a 30-day period. These data are summarized in Table 16.6.

In order to develop the control limits in accordance with Equations (16.14)–(16.17), we first need to obtain the moving range. These computations for the data of Table 16.6 are summarized in Table 16.7.

Thus,

$$\bar{X} = \frac{650.3}{30} = 21.677$$

and

$$\overline{MR} = \frac{220.5}{29} = 7.603$$

Table 16.6 Accounts receivable balance of a hardware distributor over a 30-day period.

Day	Receivable ($000)	Day	Receivable ($000)
1	33.6	16	15.5
2	18.4	17	26.5
3	10.2	18	19.1
4	16.9	19	20.3
5	35.1	20	22.2
6	25.1	21	16.3
7	16.9	22	26.1
8	13.4	23	35.1
9	23.5	24	19.2
10	29.0	25	33.3
11	20.5	26	25.7
12	25.2	27	16.5
13	18.9	28	18.8
14	19.3	29	22.0
15	12.6	30	15.1

Table 16.7 Computation of the moving range.

Receivable ($000)	Sample	High	Low	Moving Range
33.6	1	33.6	18.4	15.2
18.4	2	18.4	10.2	8.2
10.2	3	16.9	10.2	6.7
16.9	4	35.1	16.9	18.2
35.1	5	35.1	25.1	10.0
25.1	6	25.1	16.9	8.2
16.9	7	16.9	13.4	3.5
13.4	8	23.5	13.4	10.1
23.5	9	29.0	23.5	5.5
29.0	10	29.0	20.5	8.5
20.5	11	25.2	20.5	4.7
25.2	12	25.2	18.9	6.3
18.9	13	19.3	18.9	0.4
19.3	14	19.3	12.6	6.7
12.6	15	15.5	12.6	2.9
15.5	16	26.5	15.5	11.0
26.5	17	26.5	19.1	7.4
19.1	18	20.3	19.1	1.2
20.3	19	22.2	20.3	1.9
22.2	20	22.2	16.3	5.9
16.3	21	26.1	16.3	9.8
26.1	22	35.1	26.1	9.0
35.1	23	35.1	19.2	15.9
19.2	24	33.3	19.2	14.1
33.3	25	33.3	25.7	7.6
25.7	26	25.7	16.5	9.2
16.5	27	18.8	16.5	2.3
18.8	28	22.0	18.8	3.2
22.0	29	22.0	15.1	6.9
15.1				
$\sum_{i=1}^{k} X_i = 650.3$				$\sum_{i=1}^{k-1} MR_i = 220.5$

Using Equations (16.16) and (16.17), we may set up control limits as

$$21.677 \pm 3\left(\frac{7.603}{1.128}\right)$$

$$21.677 \pm 20.221$$

Thus,

$$LCL = 21.677 - 20.221 = 1.456$$

and

$$UCL = 21.677 + 20.221 = 41.898$$

Alternatively, using Equation (16.18), with $E_2 = 2.66$ for $n = 2$, we would have

$$21.677 \pm (2.66)(7.603)$$

$$21.677 \pm 20.22$$

so that

$$LCL = 21.677 - 20.22 = 1.457$$

and

$$UCL = 21.677 + 20.22 = 41.897$$

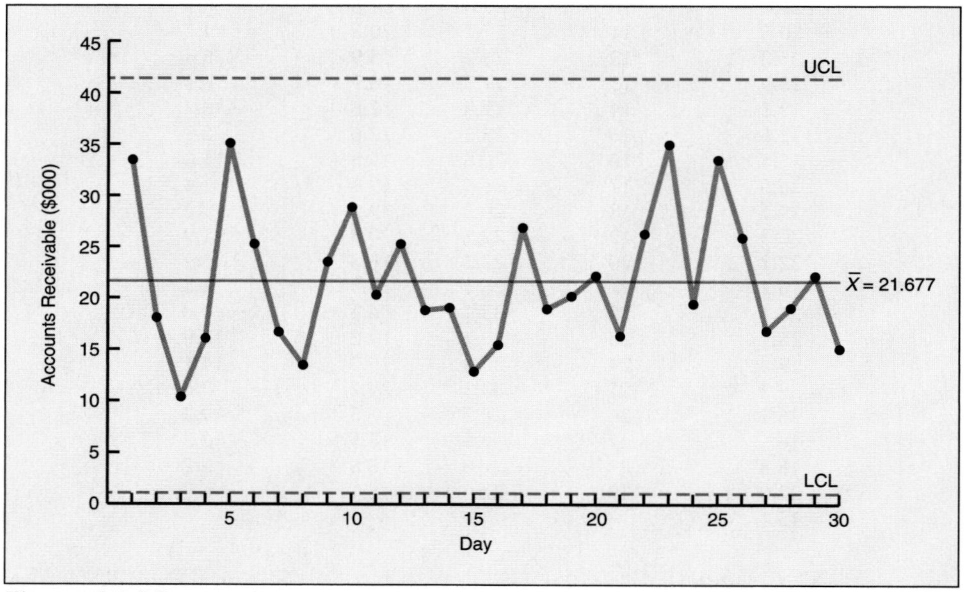

Figure 16.18
Individual value control chart for the amount of accounts receivable.

The results are the same, except for rounding error.

Figure 16.18 at the bottom of page 704 represents the control chart for the data of Table 16.6. An examination of Figure 16.18 does not reveal any points either above the upper control limit (*UCL*) or below the lower control limit (*LCL*), nor does it reveal any patterns over time, since there is no indication of a series of consecutive points either above or below the center line, or a series of consecutive points that are increasing or decreasing. However, although the process can be considered stable since there is no evidence of special cause variation, the existence of a large amount of common cause variation means that management's work in improving the accounts receivable process is just beginning. Process flow and fishbone diagrams should be developed to help in the understanding of the processes involved in order to facilitate continuous reduction in the variation of the accounts receivable. Deming's 14 points may be applied to improve the management of this accounts receivable function.

Problems for Section 16.10

● 16.25 A subgroup of 25 consecutive balls selected from a machine manufacturing softballs is selected during the production process. The circumferences (inches) are recorded below in row sequence (from left to right).

Data file
SOFTBALL.DAT

11.965	11.983	12.058	12.080	12.080
11.985	11.981	11.927	11.969	12.017
11.955	12.012	12.019	12.035	11.983
11.956	12.031	11.969	11.998	11.996
12.008	11.975	11.972	11.989	12.052

(a) Set up a control chart for the circumference of the softballs.
(b) Are the circumferences of the softballs in control?

16.26 A sample of 50 consecutive packages of table salt is selected from the output of a filling machine. The weights (in grams) of these packages are recorded below in row sequence (from left to right).

Data file
SALT.DAT

739	745	741	749	746	754	748	745	746	740
738	735	733	734	729	725	726	721	726	732
734	733	736	740	742	741	745	748	749	751
750	748	745	746	741	740	739	737	736	732
729	730	725	720	730	732	735	738	740	744

(a) Set up a control chart of the weights of the packages.
(b) Are the weights of the packages in control?

16.27 A corporation that manufactures jewelry products wanted to study the developmental engineering process for new jewelry products. In particular, it wanted to examine the development time, defined as the amount of time that passes between the cost approval day (the day in which the item is officially on the sales plan) to the production standard approval day. Thirty products (all developed during the same quarter of a particular year) were analyzed. The results (with the products listed in sequential order of their cost approval day) were as follows (see top of page 706):

Data file
JEWELRY.DAT

Product	Development Time (days)	Product	Development Time (days)
1	74	16	87
2	147	17	126
3	99	18	113
4	41	19	173
5	130	20	170
6	41	21	130
7	191	22	120
8	144	23	118
9	131	24	68
10	137	25	102
11	96	26	144
12	122	27	202
13	102	28	41
14	144	29	104
15	85	30	117

(a) Set up a control chart for the amount of development time.
(b) Is there evidence that the amount of development time is out of control?
(c) Assuming that all sources of special cause variation had been eliminated, what would you then do as a manager to reduce the amount of variation in the development time?

16.28 The following data represent the daily usage of water (1 unit = 748 gallons) at a small independent public water utility in the San Francisco area for a period of eight weeks (Monday to Friday only) during September and October in a recent year.

Data file
WATERUSE.DAT

Day	Water Usage	Day	Water Usage
1	2,503	21	2,610
2	2,668	22	2,638
3	2,725	23	2,915
4	2,638	24	2,100
5	4,453	25	3,175
6	2,739	26	2,393
7	3,307	27	2,306
8	2,984	28	2,227
9	2,759	29	2,549
10	2,633	30	2,635
11	2,468	31	2,578
12	2,592	32	2,492
13	3,700	33	2,428
14	3,152	34	2,389
15	2,305	35	3,224
16	2,302	36	2,330
17	2,504	37	2,269
18	3,310	38	2,302
19	2,483	39	2,286
20	2,224	40	2,200

(a) Set up a control chart for the amount of daily water usage.
(b) Is there any evidence that the daily usage is out of control?

(c) If the manager of the water district wanted to study ways of reducing both the average daily usage and the variation in the daily usage, what would you recommend?

Interchapter Problems

16.29 Referring to the data on the length of jeans in Problem 3.41 on page 82

Data file JEANS.DAT

(a) Set up a control chart for the length of jeans.
(b) Is there evidence that the length of jeans is out of control?

16.30 Referring to the data of Problem 3.42 on page 82

Data file QMILE.DAT

(a) Set up a control chart for Victor Sternberg's time trials.
(b) Do you think his time trials are in control?
(c) Compare the control chart obtained in (a) with the digidot plot of Problem 3.42. Discuss.

16.31 Referring to the data of Problem 3.43 on page 82

Data file DRESS.DAT

(a) Set up a control chart of gross daily sales receipts.
(b) What patterns, if any, can you observe from the control chart?

16.11 Summary and Overview

As we observe in the summary chart on page 708, in this chapter we have introduced the topic of quality and productivity by discussing the Deming approach to management and by developing several different types of control charts. Readers interested in the Deming approach are encouraged to examine References 1, 4, 5, 7, 12, 13, 14, 15, 18, 24, 27, and 28. Readers interested in additional control chart procedures should access References 8, 13, 17 and 19. On page 660 of Section 16.1 you were given a list emphasizing the important points to be discussed in the chapter. Check over the list now to see whether you feel you have an understanding of these key points. To be sure, you should be able to answer the following conceptual questions:

1. What are the differences in approach between management by control and management by process?
2. What is the difference between common causes of variation and special causes of variation?
3. What should be done to improve a process when special causes of variation are present?
4. What should be done to improve a process when only common causes of variation are present?
5. How can process flow and fishbone diagrams be used to improve processes?
6. Under what circumstances can the *np* chart be used?
7. What is the difference between attribute control charts and variables control charts?
8. What is the difference in the circumstances in which *p* charts can be used and *c* charts can be used?
9. Why are the $\overline{X}$ and range charts used together?
10. What principles did you learn from the red bead experiment?

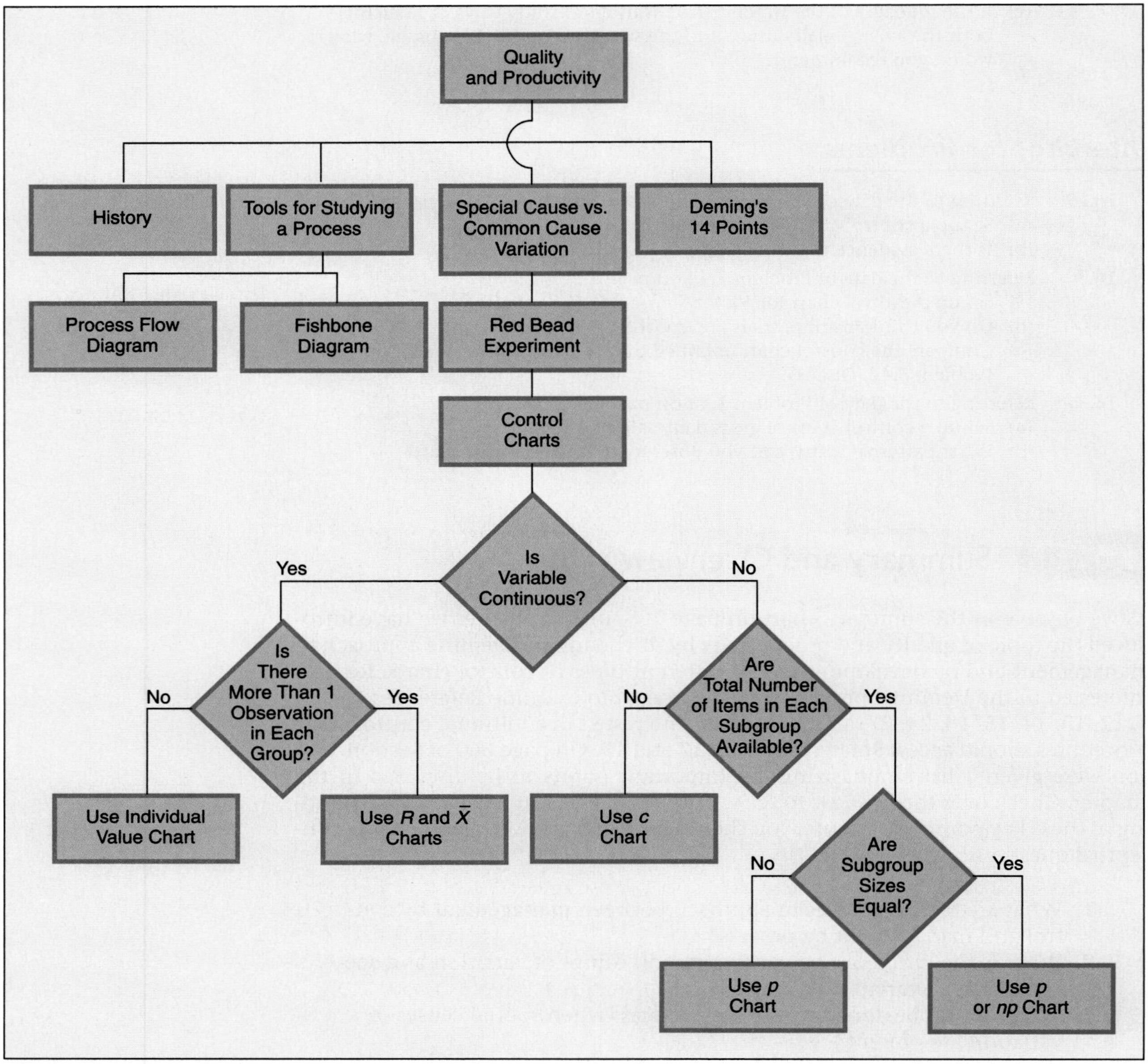

Chapter 16 summary chart.

Getting It All Together

Key Terms

A_2 factor *696*
area of opportunity *687*
attribute charts *674*
c chart *687*
common causes of variation *662*
control charts *662*
d_2 factor *693*
d_3 factor *693*

Chapter Review Problems

16.32 (a) On each morning for a period of four weeks, record your pulse rate (in beats per minute) just after you get out of bed. Set up a control chart for the pulse rate and determine whether it is in a state of statistical control. Explain.
(b) Record your pulse rate (in beats per minute) just before you go to sleep at night. Set up $\overline{X}$ and range charts for pulse rate and determine whether it is in a state of statistical control. Explain.
(c) Why might you prefer using the $\overline{X}$ and range charts in (b) as compared to the chart that was used in (a)? Why might you prefer using the chart that was used in (a) as compared to the $\overline{X}$ and range charts in (b)?

16.33 On each day for a period of four weeks, record the time (in minutes) that it takes you to get from where you reside to your first class or job—whichever you go to. Set up a control chart for this commuting time. Do you think that your commuting time is stable or is it out of statistical control? Explain.

16.34 **(Class Project)** The table of random numbers (Table E.1) can be used to simulate the selecting of different colored balls from an urn as follows:

1. Start in the row corresponding to the day of the month you were born plus the year in which you were born. For example, if you were born Oct. 15, 1971, you would start in row 15+71 = 86. If your total exceeds 100, subtract 100 from the total.
2. Two-digit random numbers are to be selected.
3. If the random number between 00 and 94 is selected, consider the ball to be white; if the random number is 95–99, consider the ball to be red.

Each student is to select 100 such two-digit random numbers and report the number of "red balls" in the sample. A control chart is to be set up of the number of (or the proportion of) red balls. What conclusions can you draw about the system of selecting red balls? Are all the students part of the system? Is anyone outside the system? If so, what explanation can you give for someone who has too many red balls? If a bonus was paid to the top 10% of the students (those 10% with the fewest red balls), what effect would that have on the rest of the students? Discuss.

Case Study G—Applying TQM in a Community Hospital

As Chief Operating Officer of a local community hospital, you have just returned from a three-day seminar on quality and productivity. It is your intention to implement at your own hospital many of the ideas that you learned at the seminar. You have decided to maintain control charts for the upcoming month for the following variables: number of daily admissions, proportion of rework in the laboratory (based on 1,000 daily samples), and time (in hours) between receipt of a specimen at the laboratory and completion of the work (based on a subgroup of 10 specimens per day). The data collected are summarized in Table 16.8.

Table 16.8 Hospital summary data.

		Processing Time		
Day	Number of Admissions	$\overline{X}_i$	R_i	Proportion of Rework in Laboratory
1	27	1.72	3.57	0.048
2	36	2.03	3.98	0.052
3	23	2.18	3.54	0.047
4	28	1.90	3.49	0.046
5	19	2.53	3.99	0.039
6	22	2.26	3.34	0.086
7	18	2.11	3.36	0.051
8	30	2.35	3.52	0.043
9	33	2.06	3.39	0.046
10	35	2.01	3.24	0.040
11	29	2.13	3.62	0.045
12	28	2.18	3.37	0.036
13	22	2.31	3.97	0.048
14	26	2.37	4.06	0.057
15	32	2.78	4.27	0.052
16	30	2.12	3.21	0.046
17	28	2.27	3.48	0.041
18	27	2.49	3.62	0.032
19	27	2.32	3.19	0.042
20	18	2.43	3.67	0.053
21	19	2.25	3.10	0.041
22	25	2.31	3.58	0.037
23	23	2.07	3.26	0.039
24	28	2.33	3.40	0.050
25	34	2.36	3.52	0.048
26	25	2.47	3.82	0.054
27	21	2.28	3.97	0.046
28	20	2.17	3.60	0.035
29	40	2.54	3.92	0.075
30	31	2.63	3.86	0.046

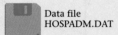

Data file
HOSPADM.DAT

You are to make a presentation to the Chief Executive Officer of the hospital and the Board of Directors. You need to prepare a report that summarizes the conclusions obtained from analyzing control charts for these variables. In addition, it is expected that you will recommend additional variables for which control charts are to be maintained. Finally, it is your intention to explain how the Deming philosophy of management by process can be implemented in the context of your hospital's environment.

Endnotes

1. We recall from Section 8.3 that in the normal distribution $\mu_x \pm 3\sigma_x$ includes almost all (99.73%) of the observations in the population. Although the calculations used in control charts are based on the normal distribution, we should stress that in analytical studies the concept of population has no applicability. The subject of interest is a process, not a population from which a sample is taken.
2. One rule relies on the concept of runs and is based on the Wald-Wolfowitz procedure discussed in Section 12.6. For a more detailed discussion of additional rules see Reference 8.
3. This example was related to the authors by Brian Joiner of Joiner Associates, who credits it to Ed Pindy of Philadelphia Electric Company.
4. In this chapter we use the terminology *nonconforming items,* while in Chapters 6–10 when we discussed proportions we used the terminology *success*.
5. Unknown to the participants in the experiment, 3,200 of the beads are white and 800 are red.
6. If the size of the sample unit varies appreciably, the *u* chart may be used instead of the *c* chart (see References 8, 13, 17, and 19).
7. Cryer and Ryan (see Reference 3) argue that the standard deviation is a better measure than the moving range.

References

1. Aguayo, R., *Dr. Deming The American Who Taught the Japanese about Quality* (New York: Lyle Stuart, 1990).
2. Brassard, M., *The Memory Jogger Plus* (Methuen, MA: GOAL/QPC, 1989).
3. Cryer, J. D., and T. P. Ryan, "The estimation of sigma for an *X* chart: MR/d_2 or S/c_4?" *Journal of Quality Technology*, 1990, Vol. 22, pp. 187–192.
4. Deming, W. E., *Out of the Crisis* (Cambridge, MA: MIT Center for Advanced Engineering Study, 1986).
5. Deming, W. E., *The New Economics for Business, Industry, and Government* (Cambridge, MA: MIT Center for Advanced Engineering Study, 1993).
6. Dobson, J. M., *A History of American Enterprise* (Englewood Cliffs, NJ: Prentice-Hall, 1988).
7. Gabor, A., *The Man Who Discovered Quality* (New York: Time Books, 1990).
8. Gitlow, H., A. Oppenheim, and R. Oppenheim, *Tools and Methods for the Improvement of Quality*, 2nd ed. (Homewood, Ill.: Richard D. Irwin, 1994).
9. Halberstam, D., *The Reckoning* (New York: William Morrow, 1986).
10. Holusha, J., "The Baldrige badge of courage—and quality," *New York Times,* October 21, 1990, p. F12.
11. Joiner, B. J., "The key role of statisticians in the transformation of North American industry," *American Statistician,* 1985, Vol. 39, pp. 224–234.
12. Joiner, B. L. *Fourth Generation Management* (New York: McGraw-Hill, 1994).
13. Levine, D. M., P. P. Ramsey, and M. L. Berenson, *Business Statistics for Quality and Productivity* (Englewood Cliffs, N J: Prentice-Hall, 1995).
14. Main, J., "The curmudgeon who talks tough on quality," *Fortune,* June 25, 1984, pp. 118–122.
15. Mann, N. R., *The Keys to Excellence: The Story of the Deming Philosophy* (Los Angeles: Prestwick Books, 1987).
16. *The Memory Jogger II: A Pocket Guide of Tools for Continuous Improvement and Effective Planning* (Methuen, MA: GOAL/QPC, 1994).
17. Montgomery, D. C. *Introduction to Statistical Quality Control,* 2nd ed.(New York: John Wiley, 1991).
18. Port, O., "The push for quality," *Business Week,* June 8, 1987, pp. 130–135.
19. Ryan, T. P., *Statistical Methods for Quality Improvement.* (New York: John Wiley, 1989).
20. Scherkenbach, W. W., *The Deming Route to Quality and Productivity: Road Maps and Roadblocks* (Washington, D.C.: CEEP Press, 1986).
21. Shewhart, W. A., "The applications of statistics as an aid in maintaining quality of manufactured products," *Journal of the American Statistical Association,* 1925, Vol. 20, pp. 546–548.
22. Shewhart, W. A., *Economic Control of Quality of Manufactured Products* (New York: Van Nostrand-Reinhard, 1931, reprinted by the American Society for Quality Control, Milwaukee, 1980).
23. Shewhart, W. A., and W. E. Deming, *Statistical Methods from the Viewpoint of Quality Control* (Washington, D.C.: Graduate School, Dept. of Agriculture, 1939, Dover Press, 1986).
24. Sholtes, P. R., *An Elaboration on Deming's Teaching on Performance Appraisal* (Madison, WI: Joiner Associates, 1987).
25. Skrebec, Q. R., "Ancient process control and its modern implications," *Quality Progress,* 1990, Vol. 23, pp. 49–52.
26. Wallis, W. A., "The statistical research group 1942–1945," *Journal of the American Statistical Association,* 1980, Vol. 75, pp. 320–335.
27. Walton, M., *The Deming Management Method* (New York: Perigee Books, Putnam Publishing Group, 1986).
28. Walton, M., *Deming Management at Work* (New York: G. P. Putnam, 1990).

chapter 17

Simple Linear Regression and Correlation

CHAPTER OBJECTIVE

To develop both descriptively and inferentially the simple linear regression and correlation models as a means of using one variable to predict another variable and to measure the strength of the association between two variables.

17.1 Introduction

In previous chapters we have focused primarily on a single numerical response variable such as personal income. We studied various measures of statistical description (see Chapter 4) and applied different techniques of statistical inference to make estimates and draw conclusions about our numerical response variable (see Chapters 10 to 14). In this and the following chapter we will concern ourselves with problems involving two or more numerical variables as a means of viewing the relationships that exist between them. Two techniques will be discussed: regression and correlation.

Regression analysis is used primarily for the purpose of prediction. Our goal in regression analysis is the development of a statistical model that can be used to predict the values of a **dependent** or **response variable** based upon the values of at least one **explanatory** or **independent variable.** In this chapter we shall focus on a simple regression model—one that would utilize a single numerical independent variable X to predict the numerical dependent variable Y. In Chapter 18 we shall develop a multiple regression model—one that could utilize several explanatory variables ($X_1, X_2, \ldots, X_p$) to predict a numerical dependent variable Y.[1]

Referring to our Employee Satisfaction Survey, for example, suppose that Bud Conley would like to develop a statistical model that can assist in predicting the personal income of full-time employees of Kalosha Industries. Although in practice several variables would actually be considered, it would seem that the number of years of full-time employment in the workforce might be a useful predictor of personal income. For this model, the dependent or response variable Y (the one to be predicted) would be personal income and the explanatory or independent variable X used to obtain the prediction is the number of years of full-time employment in the workforce.

Correlation analysis, in contrast to regression, is used to measure the strength of the association between numerical variables. For example, in Section 17.7 we will determine the correlation between the price of a six-pack of soft drink and the price of chicken in different international cities. In this instance, the objective is not to use one variable to predict another, but rather to measure the strength of the association or covariation that exists between two numerical variables. Upon completion of this chapter, you should be able to:

1. Interpret the regression coefficients obtained by using the least squares method of regression
2. Interpret the coefficients of determination and correlation
3. Be able to make the distinction between different measures of variation in regression analysis
4. Be familiar with the assumptions of regression analysis
5. Use residual analysis to determine whether the appropriate model has been fit to the data
6. Use influence analysis to determine whether any observations are unduly influencing the regression model
7. Make inferences about the regression coefficients
8. Make inferences about the predicted value of a response variable

MAN→ ACCTG→ ECON→

17.2 The Scatter Diagram

Methods of regression and correlation analysis will be applied to two problems in this chapter. In the first, suppose that the management of a chain of package delivery stores would like to develop a model for predicting the weekly sales (in thousands of dollars) for individual stores. A random sample of 20 stores was selected from among all the stores in the chain. In developing such a model many explanatory variables would be considered. For pedagogical purposes, we shall begin our discussion with a simple regression model in which only one numerical explanatory variable is used to predict the values of a dependent variable. Thus we will develop a model to predict weekly sales (the dependent variable Y) based upon the number of customers (the explanatory or independent variable X). The results for a sample of 20 stores are summarized in Table 17.1. Such data, however, can be presented in a form that is more visually interpretable.

Table 17.1 Number of customers and weekly sales for a sample of 20 package delivery stores.

Stores	Customers	Sales ($000)
1	907	11.20
2	926	11.05
3	506	6.84
4	741	9.21
5	789	9.42
6	889	10.08
7	874	9.45
8	510	6.73
9	529	7.24
10	420	6.12
11	679	7.63
12	872	9.43
13	924	9.46
14	607	7.64
15	452	6.92
16	729	8.95
17	794	9.33
18	844	10.23
19	1,010	11.77
20	621	7.41

Data file PACKAGE.DAT

In Chapter 3, when information concerning the tuition rates for out-of-state students in Texas was studied, various graphs (such as histograms, polygons, and ogives) were developed for data presentation. In a regression analysis involving one independent and one dependent variable the individual values are plotted on a two-dimensional graph called a **scatter diagram.** Each value is plotted at its particular X and Y coordinates. The scatter diagram for the data in Table 17.1 is shown in Figure 17.1 on page 716.

An examination of Figure 17.1 indicates a clearly increasing relationship between number of customers (X) and weekly sales (Y). As the number of customers increases, weekly sales increase. The exact mathematical form of the model

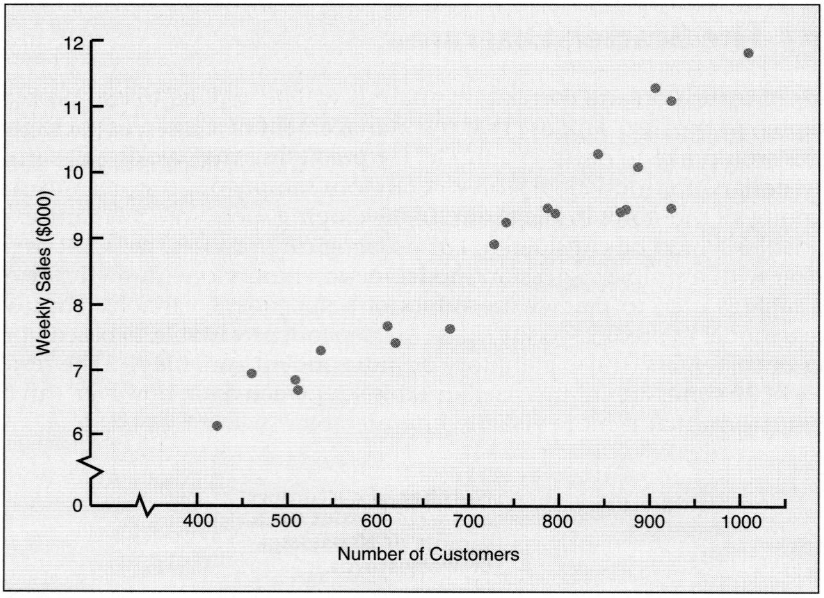

Figure 17.1
Scatter diagram of weekly sales and number of customers.
Source: Data are taken from Table 17.1 on page 715.

expressing the relationship as well as methods for estimating the weekly sales for a given number of customers will be examined in subsequent sections of this chapter.

Problems for Section 17.2

● 17.1 The marketing manager of a large supermarket chain would like to determine the effect of shelf space on the sales of pet food. A random sample of 12 equal-sized stores was selected and the results are presented below:

Data file PETFOOD.DAT

Pet food sales problem

Store	Shelf Space, *X* (feet)	Weekly Sales, *Y* (hundreds of dollars)
1	5	1.6
2	5	2.2
3	5	1.4
4	10	1.9
5	10	2.4
6	10	2.6
7	15	2.3
8	15	2.7
9	15	2.8
10	20	2.6
11	20	2.9
12	20	3.1

Set up a scatter diagram.

17.2 Over the past 25 years, a chain of discount ladies clothing stores has increased market share by increasing the number of locations in the chain. A systematic approach to site selection was never utilized. Site selection was primarily based on what was considered to be a great location or a great lease. This year, with a strategic plan for opening several new stores, the director of special projects and planning was asked to develop an approach to forecasting annual sales for any new store that was opened. The following represents the square footage and the annual sales (in thousands of dollars) for a sample of 14 stores in the chain.

Site selection problem.

Store	Square Feet	Annual Sales ($000)
1	1,726	3,681
2	1,642	3,895
3	2,816	6,653
4	5,555	9,543
5	1,292	3,418
6	2,208	5,563
7	1,313	3,660
8	1,102	2,694
9	3,151	5,468
10	1,516	2,898
11	5,161	10,674
12	4,567	7,585
13	5,841	11,760
14	3,008	4,085

Data file SITE.DAT

Set up a scatter diagram.

17.3 A company manufacturing parts would like to develop a model to estimate the number of worker-hours required for production runs of varying lot size. A random sample of 14 production runs (2 each for lot sizes 20, 30, 40, 50, 60, 70, and 80) was selected with the result shown below:

Production worker-hours problem.

Lot Size	Worker-Hours
20	50
20	55
30	73
30	67
40	87
40	95
50	108
50	112
60	128
60	135
70	148
70	160
80	170
80	162

Data file WORKHRS.DAT

Set up a scatter diagram.

17.4 An agronomist would like to determine the effect of a natural organic fertilizer on the yield of tomatoes. Five differing amounts of fertilizer are to be used on 10 equivalent plots of land: 0, 10, 20, 30, and 40 pounds per 100 square feet. The levels of fertilizer are randomly assigned to the plots of land with the results given below:

Data file
TOMYIELD.DAT

Tomato yield problem.

Plot	Amount of Fertilizer, X (in pounds per 100 square feet)	Yield, Y (in pounds)
1	0	6
2	0	8
3	10	11
4	10	14
5	20	18
6	20	23
7	30	25
8	30	28
9	40	30
10	40	34

Set up a scatter diagram.

17.5 The operations manager of a market research company would like to develop a model to predict the number of interviews conducted by interviewers on a given day. She believes that interviewer experience (as measured in weeks worked) is the primary determinant of the number of interviews that can be completed. A sample of 10 interviewers is selected and the number of interviews completed along with the corresponding weeks of experience are recorded with the following results:

Data file
INTERVW.DAT

Interviewer productivity problem.

Weeks of Experience	Number of Interviews Completed
15	4
41	9
58	12
18	6
37	8
52	10
28	6
24	5
45	10
33	7

Set up a scatter diagram.

17.6 A limousine service operating from a suburban county wanted to determine the length of time it would take to transport passengers from various locations to a major metropolitan airport during nonpeak times. A sample of 12 trips on a particular day during nonpeak times indicated the following:

Airport travel problem.

Distance (miles)	Time (minutes)
10.3	19.71
11.6	18.15
12.1	21.88
14.3	24.21
15.7	27.08
16.1	22.96
18.4	29.38
20.2	37.24
21.8	36.84
24.3	40.59
25.4	41.21
26.7	38.19

Data file
LIMO.DAT

Set up a scatter diagram.

17.3 Types of Regression Models

In the scatter diagram plotted in Figure 17.1 a rough idea of the type of relationship that exists between the variables can be observed. The nature of the relationship can take many forms, ranging from simple ones to extremely complicated mathematical functions. The simplest relationship consists of a straight-line or **linear relationship.** An example of this relationship is shown in Figure 17.2.

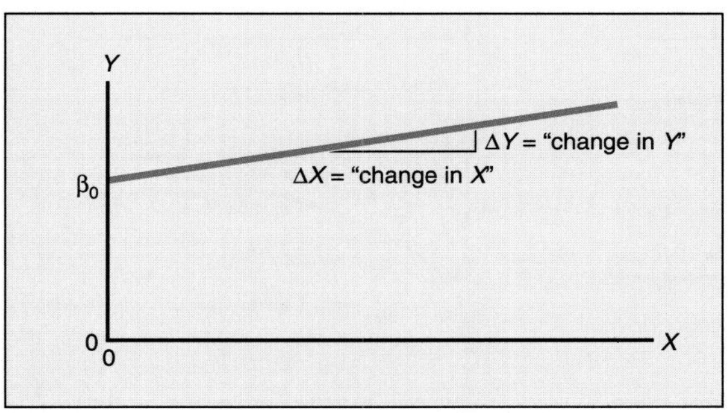

Figure 17.2
A positive straight-line relationship.

The straight-line (linear) model can be represented as

$$Y_i = \beta_0 + \beta_1 X_i + \epsilon_i \qquad (17.1)$$

where β_0 = Y intercept for the population
β_1 = slope for the population
ϵ_i = random error in Y for observation i

In this model, the slope of the line β_1 represents the expected change in Y per unit change in X; that is, it represents the amount that Y changes (either positively or negatively) for a particular unit change in X. On the other hand, the Y intercept β_0 represents the average value of Y when X equals zero. Moreover, the last component of the model, ϵ_i, represents the random error in Y for each observation i that occurs.

The proper mathematical model to be selected is influenced by the distribution of the X and Y values on the scatter diagram. This can be seen readily from an examination of panels A through F in Figure 17.3. Clearly, from panel A we note that the values of Y are generally increasing linearly as X increases. This panel is similar to Figure 17.1, which illustrates the relationship between number of customers and sales. Panel B is an example of a negative linear relationship. As X increases, we note that the values of Y are decreasing. An example of this type of relationship might be the price of a particular product and the amount of sales. Panel C shows a set of data in which there is very little or no relationship between X and Y. High and low values of Y appear at each value of X. The data in panel D show a positive curvilinear relationship between X and Y. The values of Y are increasing as X increases, but this increase tapers off beyond certain values of X. An example of this positive curvilinear relationship might be the age and maintenance cost of a machine. As a machine gets older, the maintenance cost may rise

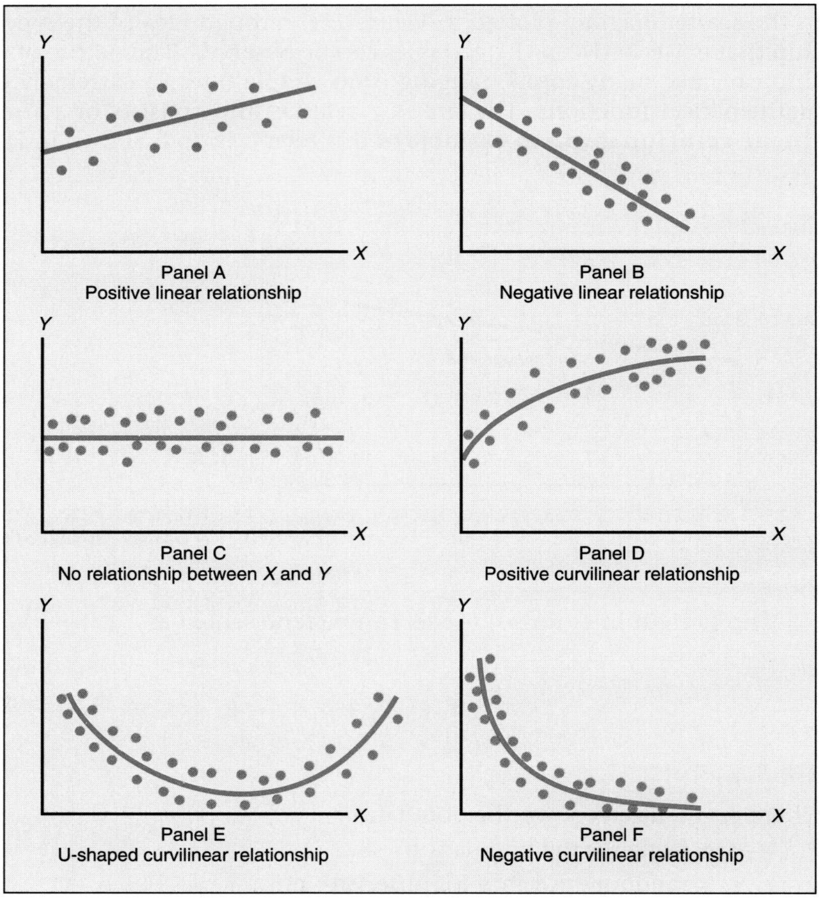

Figure 17.3
Examples of types of relationships found in scatter diagrams.

rapidly at first, but then level off beyond a certain number of years. Panel E shows a parabolic or U-shaped relationship between X and Y. As X increases, at first Y decreases; but as X continues to increase, Y not only stops decreasing but actually increases above its minimum value. An example of this type of relationship could be the number of errors per hour at a task and the number of hours worked. The number of errors per hour would decrease as the individual becomes more proficient at the task, but then would increase beyond a certain point because of factors such as fatigue and boredom. Finally, panel F indicates an exponential or negative curvilinear relationship between X and Y. In this case Y decreases very rapidly as X first increases, but then decreases much less rapidly as X increases further. An example of this exponential relationship could be the resale value of a particular type of automobile and its age. In the first year the resale value drops drastically from its original price; however, the resale value then decreases much less rapidly in subsequent years.

In this section we have briefly examined a variety of different models that could be used to represent the relationship between two variables. Although scatter diagrams can be extremely helpful in determining the mathematical form of the relationship, more sophisticated statistical procedures are available to determine the most appropriate model for a set of variables. In subsequent sections of this chapter we shall primarily focus on building statistical models for fitting *linear* relationships between variables.

17.4 Determining the Simple Linear Regression Equation

If we refer to the scatter diagram in Figure 17.1 on page 716, we notice that sales appear to increase linearly as a function of the number of customers. The question that must be addressed in regression analysis involves the determination of the particular straight-line model that is the best fit to these data.

17.4.1 The Least Squares Method

In the preceding section we hypothesized a statistical model to represent the relationship between two variables in a population. However, as noted in Table 17.1 on page 715, we have obtained data from only a random sample of the population. If certain assumptions are valid (see Section 17.8), the sample Y intercept (b_0) and the sample slope (b_1) can be used as estimates of the respective population parameters (β_0 and β_1). Thus the sample regression equation representing the straight-line regression model would be

$$\hat{Y}_i = b_0 + b_1 X_i \qquad \textbf{(17.1a)}$$

where $\hat{Y}_i$ is the predicted value of Y for observation i and X_i is the value of X for observation i.

This equation requires the determination of two coefficients—b_0 (the Y intercept) and b_1 (the slope) in order to predict values of Y. Once b_0 and b_1 are obtained, the straight line is known and can be plotted on the scatter diagram. Then we could make a visual comparison of how well our particular statistical model (a straight line) fits the original data. That is, we can see whether the original data lie close to the fitted line or deviate greatly from the fitted line.

Simple linear regression analysis is concerned with finding the straight line that fits the data best. The best fit means that we wish to find the straight line for which the differences between the actual values (Y_i) and the values that would be predicted from the fitted line of regression ($\hat{Y}_i$) are as small as possible. Because these differences will be both positive and negative for different observations, mathematically we minimize

$$\sum_{i=1}^{n}(Y_i - \hat{Y}_i)^2$$

where Y_i = actual value of Y for observation i

$\hat{Y}_i$ = predicted value of Y for observation i

Since $\hat{Y}_i = b_0 + b_1X_i$, we are minimizing

$$\sum_{i=1}^{n}[Y_i - (b_0 + b_1X_i)]^2$$

which has two unknowns, b_0 and b_1.

A mathematical technique that determines the values of b_0 and b_1 that best fit the observed data is known as the **least squares method.** Any values for b_0 and b_1 other than those determined by the least squares method would result in a greater sum of squared differences between the actual value of Y and the predicted value of Y.

In using the least squares method, we obtain the following two equations, called the normal equations:

$$\text{I.} \quad \sum_{i=1}^{n} Y_i = nb_0 + b_1 \sum_{i=1}^{n} X_i \qquad \textbf{(17.2a)}$$

$$\text{II.} \quad \sum_{i=1}^{n} X_iY_i = b_0 \sum_{i=1}^{n} X_i + b_1 \sum_{i=1}^{n} X_i^2 \qquad \textbf{(17.2b)}$$

From these two equations, we must solve for b_1 and b_0. However, in this text, we shall take the position that in solving regression equations, statistical software will be accessed to perform the (often tedious) calculations. Such packages are discussed in Section 17.15. Nevertheless, in order to understand how the results displayed in the output of these packages have been developed for the case of simple linear regression, we will directly illustrate many of the computations involved.

Referring back to Equations (17.2a) and (17.2b), since there are two equations with two unknowns, we can solve these equations simultaneously for b_1 and b_0 as follows:

$$b_1 = \frac{\sum_{i=1}^{n} X_iY_i - n\overline{X}\overline{Y}}{\sum_{i=1}^{n} X_i^2 - n\overline{X}^2} \qquad \textbf{(17.3)}$$

and

$$b_0 = \overline{Y} - b_1\overline{X} \qquad \textbf{(17.4)}$$

where

$$\overline{Y} = \frac{\sum_{i=1}^{n} Y_i}{n} \quad \text{and} \quad \overline{X} = \frac{\sum_{i=1}^{n} X_i}{n}$$

Examining Equations (17.3) and (17.4), we see that there are five quantities that must be calculated in order to determine b_0 and b_1. These are n, the sample size; $\sum_{i=1}^{n} X_i$, the sum of the X values; $\sum_{i=1}^{n} Y_i$, the sum of the Y values; $\sum_{i=1}^{n} X_i^2$, the sum of the squared X values, and $\sum_{i=1}^{n} X_iY_i$ the sum of the cross product of X and Y. For our data in Table 17.1, the number of customers was used to predict the weekly sales in a store. The computation of the various sums needed (including $\sum_{i=1}^{n} Y_i^2$, the sum of the squared Y values that will be used in Section 17.5) are presented in Table 17.2.

Table 17.2 Computations for the package delivery problem.

Store	Customers X	Sales Y	X^2	Y^2	XY
1	907	11.20	822,649	125.4400	10,158.40
2	926	11.05	857,476	122.1025	10,232.30
3	506	6.84	256,036	46.7856	3,461.04
4	741	9.21	549,081	84.8241	6,824.61
5	789	9.42	622,521	88.7364	7,432.38
6	889	10.08	790,321	101.6064	8,961.12
7	874	9.45	763,876	89.3025	8,259.30
8	510	6.73	260,100	45.2929	3,432.30
9	529	7.24	279,841	52.4176	3,829.96
10	420	6.12	176,400	37.4544	2,570.40
11	679	7.63	461,041	58.2169	5,180.77
12	872	9.43	760,384	88.9249	8,222.96
13	924	9.46	853,776	89.4916	8,741.04
14	607	7.64	368,449	58.3696	4,637.48
15	452	6.92	204,304	47.8864	3,127.84
16	729	8.95	531,441	80.1025	6,524.55
17	794	9.33	630,436	87.0489	7,408.02
18	844	10.23	712,336	104.6529	8,634.12
19	1,010	11.77	1,020,100	138.5329	11,887.70
20	621	7.41	385,641	54.9081	4,601.61
Totals	14,623	176.11	11,306,209	1,602.0971	134,127.90

Using Equations (17.3) and (17.4), we can compute the values of b_0 and b_1:

$$b_1 = \frac{\sum_{i=1}^{n} X_i Y_i - n\overline{X}\overline{Y}}{\sum_{i=1}^{n} X_i^2 - n\overline{X}^2}$$

where

$$\overline{Y} = \frac{\sum_{i=1}^{n} Y_i}{n} = \frac{176.11}{20} = 8.8055$$

$$\overline{X} = \frac{\sum_{i=1}^{n} X_i}{n} = \frac{14{,}623}{20} = 731.15$$

so that

$$b_1 = \frac{134{,}127.90 - (20)(731.15)(8.8055)}{11{,}306{,}209 - 20(731.15)^2}$$

$$= \frac{5{,}365.08}{614{,}603} = +.00873$$

and

$$b_0 = \overline{Y} - b_1\overline{X}$$

$$b_0 = 8.8055 - (.00873)(731.15) = +2.423$$

Thus the equation for the *best* straight line for these data is

$$\hat{Y}_i = 2.423 + .00873X_i$$

The slope b_1 was computed as +.00873. This means that for each increase of one unit in X, the value of Y is estimated to increase by an average of .00873 unit. That is, for each increase of one customer, the fitted model predicts that the expected weekly sales are estimated to increase by .00873 thousands of dollars or \$8.73 (or we can say that for each increase of 100 customers, weekly sales are expected to increase by \$873). Hence the slope can be viewed as representing the portion of the weekly sales that are estimated to vary according to the number of customers.

The Y intercept b_0 was computed to be +2.423 (thousands of dollars). The Y intercept represents the average value of Y when X equals zero. Since the number of customers is unlikely to be zero, this Y intercept can be viewed as expressing the portion of the weekly sales that varies with factors other than the number of customers.

The regression model that has been fit to the data can now be used to predict the weekly sales. For example, let us say that we would like to use the fitted model to predict the average weekly sales for a store with 600 customers.

We can determine the predicted value by substituting $X = 600$ into our regression equation,

$$\hat{Y}_i = 2.423 + .00873(600) = 7.661$$

Thus the predicted average weekly sales for a store with 600 customers is 7.661 thousands of dollars or $7,661.

17.4.2 Predictions in Regression Analysis: Interpolation versus Extrapolation

When using a regression model for prediction purposes, it is important that we consider only the relevant range of the independent variable in making our predictions. This **relevant range** encompasses all values from the smallest to the largest X used in developing the regression model. Hence, when predicting Y for a given value of X, we may *interpolate* within this relevant range of the X values, but we may not *extrapolate* beyond the range of X values. For example, when we use the number of customers to predict weekly sales we note from Table 17.1 that the number of customers varies from 420 to 1,010. Therefore, predictions of weekly sales should be made only for stores that have between 420 and 1,010 customers. Any prediction of weekly sales outside this range of number of customers presumes that the fitted relationship holds outside the 420 to 1,010 range.

Problems for Section 17.4

● 17.7 Referring to Problem 17.1, the pet food sales problem, on page 716, Data file PETFOOD.DAT

(a) Assuming a linear relationship, use the least squares method to compute the regression coefficients b_0 and b_1.
(b) Interpret the meaning of the slope b_1 in this problem.
(c) Predict the average weekly sales (in hundreds of dollars) of pet food for stores with 8 feet of shelf space for pet food.

17.8 Referring to Problem 17.2, the site selection problem, on page 716, Data file SITE.DAT

(a) Assuming a linear relationship, use the least squares method to find the regression coefficients b_0 and b_1.
(b) Interpret the meaning of the slope b_1 in this problem.
(c) Predict the average annual sales for a store that contains 4,000 square feet.

● 17.9 Referring to Problem 17.3, the production worker-hours problem, on page 717, Data file WORKHRS.DAT

(a) Assuming a linear relationship, use the least squares method to find the regression coefficients b_0 and b_1.
(b) Interpret the meaning of the Y intercept b_0 and the slope b_1 in this problem.
(c) Predict the average number of worker-hours required for a production run with a lot size of 45.
(d) Why would it not be appropriate to predict the average number of worker-hours required for a production run with lot size of 100? Explain.

17.10 Referring to Problem 17.4, the tomato yield problem, on page 718, Data file TOMYIELD.DAT

(a) Assuming a linear relationship, use the least squares method to find the regression coefficients b_0 and b_1.
(b) Interpret the meaning of the Y intercept b_0 and the slope b_1 in this problem.
(c) Predict the average yield of tomatoes for a plot that has been given 15 pounds per 100 square feet of natural organic fertilizer.
(d) Why would it not be appropriate to predict the average yield for a plot that has been fertilized with 100 pounds per 100 square feet? Explain.

Data file INTERVW.DAT

17.11 Referring to Problem 17.5, the interviewer productivity problem, on page 718,
(a) Assuming a linear relationship, use the least squares method to find the regression coefficients b_0 and b_1.
(b) Interpret the meaning of the Y intercept b_0 and the slope b_1 in this problem.
(c) Use the regression model developed in (a) to predict the number of interviews completed for an interviewer who has 30 weeks of experience.

Data file LIMO.DAT

17.12 Referring to Problem 17.6, the airport travel problem, on page 718,
(a) Assuming a linear relationship, use the least squares method to find the regression coefficients b_0 and b_1.
(b) Interpret the meaning of the Y intercept b_0 and the slope b_1 in this problem.
(c) Use the regression model developed in (a) to predict the number of minutes to transport someone from a location that is 21 miles from the airport.

17.5 Standard Error of the Estimate

In the preceding section we used the least squares method to develop an equation to predict the weekly sales based on the number of customers. Although the least squares method results in the line that fits the data with the minimum amount of variation, the regression equation is not a perfect predictor, unless all the observed data points fall on the regression line. Just as we cannot expect all data values to be located exactly at their arithmetic mean, in the same way we cannot expect all data points to fall exactly on the regression line. The regression line serves only as an approximate predictor of a Y value for a given value of X. Therefore, we need to develop a statistic that measures the variability of the actual Y values, from the predicted Y values, in the same way that we developed (see Chapter 4) a measure of the variability of each observation around its mean. The measure of variability around the line of regression (its standard deviation) is called the **standard error of the estimate.**

The variability around the line of regression is illustrated in Figure 17.4 for the package delivery sales problem. We can see from Figure 17.4 that, although the predicted line of regression falls near many of the actual values of Y, there are several values above the line of regression as well as below the line of regression, so that

$$\sum_{i=1}^{n}(Y_i - \hat{Y}_i) = 0$$

The standard error of the estimate, given by the symbol S_{YX}, is defined as

$$S_{YX} = \sqrt{\frac{\sum_{i=1}^{n}(Y_i - \hat{Y}_i)^2}{n-2}} \qquad (17.5)$$

where Y_i = actual value of Y for a given X_i
$\hat{Y}_i$ = predicted value of Y for a given X_i

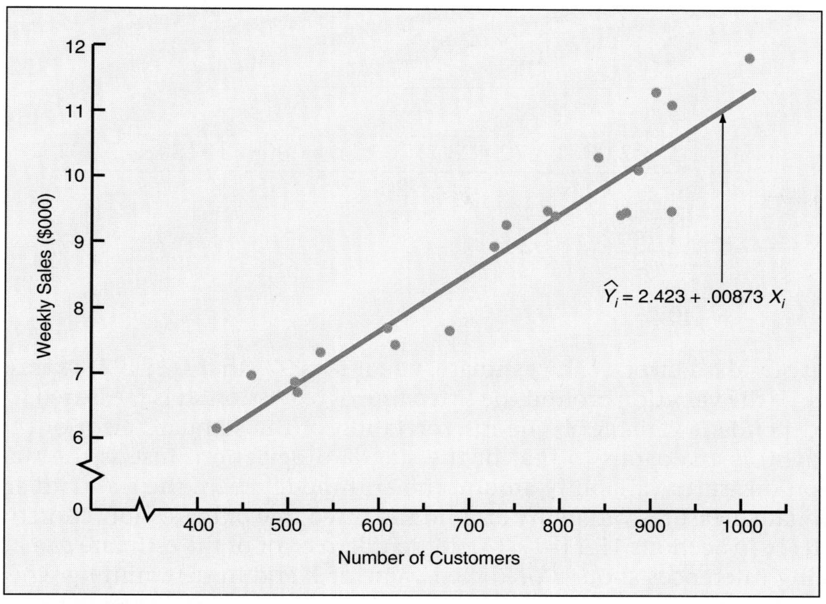

Figure 17.4
Scatter diagram and line of regression for the package delivery problem.

The computation of the standard error of the estimate using Equation (17.5) would first require determining the predicted value of Y for each X value in the sample. The computation can be simplified because of the following identity:

$$\sum_{i=1}^{n}(Y_i - \hat{Y}_i)^2 = \sum_{i=1}^{n} Y_i^2 - b_0\sum_{i=1}^{n} Y_i - b_1\sum_{i=1}^{n} X_iY_i$$

The standard error of the estimate S_{YX} can thus be obtained using the following computational formula:

$$S_{YX} = \sqrt{\frac{\sum_{i=1}^{n} Y_i^2 - b_0\sum_{i=1}^{n} Y_i - b_1\sum_{i=1}^{n} X_iY_i}{n-2}} \tag{17.6}$$

For the package delivery sales problem, from Table 17.2 on page 723 we have determined that

$$\sum_{i=1}^{n} Y_i^2 = 1{,}602.0971 \qquad \sum_{i=1}^{n} Y_i = 176.11 \qquad \sum_{i=1}^{n} X_iY_i = 134{,}127.90$$

$$b_0 = 2.423 \qquad b_1 = +.00873$$

Therefore, using Equation (17.6), the standard error of the estimate S_{YX} can be computed as

$$S_{YX} = \sqrt{\frac{\sum_{i=1}^{n} Y_i^2 - b_0 \sum_{i=1}^{n} Y_i - b_1 \sum_{i=1}^{n} X_i Y_i}{n-2}}$$

$$= \sqrt{\frac{1{,}602.0971 - (2.423)(176.11) - (.00873)(134{,}127.90)}{20 - 2}}$$

$$= \sqrt{\frac{4.446}{18}} = \sqrt{.247}$$

$$= .497$$

This standard error of the estimate, equal to .497 (that is, \$497) represents a measure of the variation around the fitted line of regression. It is measured in units of the dependent variable *Y*. The interpretation of the standard error of the estimate, then, is analogous to that of the standard deviation. Just as the standard deviation measures variability around the arithmetic mean, the standard error of the estimate measures variability around the fitted line of regression. Moreover, as we shall see in Sections 17.11–17.13, the standard error of the estimate can be used in making inferences about a predicted value of *Y* and in determining whether a statistically significant relationship exists between the two variables.

Problems for Section 17.5

Data file PETFOOD.DAT

● 17.13 Referring to the pet food sales problem (pages 716 and 725), compute the standard error of the estimate.

Data file SITE.DAT

17.14 Referring to the site selection problem (pages 716 and 725), compute the standard error of the estimate.

Data file WORKHRS.DAT

● 17.15 Referring to the production worker-hours problem (pages 717 and 725), compute the standard error of the estimate.

Data file TOMYIELD.DAT

17.16 Referring to the tomato yield problem (pages 718 and 725), compute the standard error of the estimate.

Data file INTERVW.DAT

17.17 Referring to the interviewer productivity problem (pages 718 and 726), compute the standard error of the estimate.

Data file LIMO.DAT

17.18 Referring to the airport travel problem (pages 718 and 726), compute the standard error of the estimate.

17.6 Measures of Variation in Regression and Correlation

In order to examine how well the independent variable predicts the dependent variable in our statistical model, we need to develop several measures of variation. The first measure, the **total sum of squares** (SST), is a measure of variation of the Y_i values around their mean $\bar{Y}$. In a regression analysis the total sum of squares can be subdivided into **explained variation** or **sum of squares** due to **regression** (SSR), that which is attributable to the relationship between *X* and *Y*, and **unexplained variation** or **error sum of squares** (SSE), that which is attributable to factors other than the relationship between *X* and *Y*. These different measures of variation can be seen in Figure 17.5.

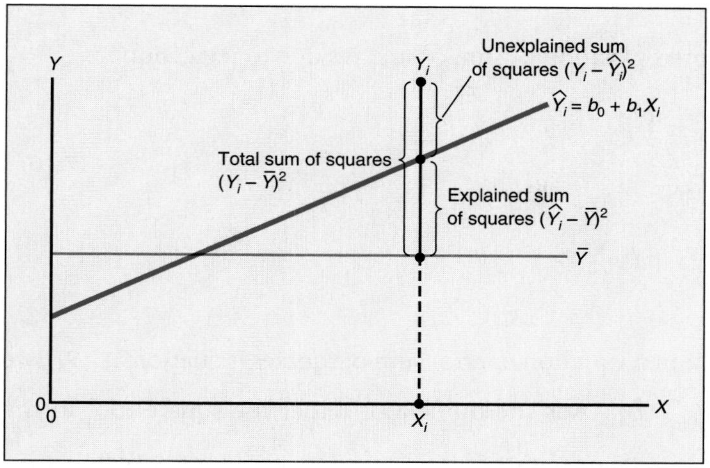

Figure 17.5
Measures of variation in regression.

The sum of squares due to regression (SSR) represents the difference between $\bar{Y}$ (the average value of Y) and $\hat{Y}_i$ (the value of Y that would be predicted from the regression relationship). The error sum of squares (SSE) represents that part of the variation in Y that is not explained by the regression. It is based upon the difference between Y_i and $\hat{Y}_i$.

These measures of variation can be represented as follows:

$$\text{total sum of squares} = \text{sum of squares due to regression} + \text{error sum of squares}$$
$$\text{SST} = \text{SSR} + \text{SSE} \tag{17.7}$$

where

$$\text{SST} = \text{total sum of squares} = \sum_{i=1}^{n}(Y_i - \bar{Y})^2 = \sum_{i=1}^{n}Y_i^2 - n\bar{Y}^2 \tag{17.8}$$

$$\begin{aligned}\text{SSE} &= \text{unexplained variation or error sum of squares} \\ &= \sum_{i=1}^{n}(Y_i - \hat{Y}_i)^2 \\ &= \sum_{i=1}^{n}Y_i^2 - b_0\sum_{i=1}^{n}Y_i - b_1\sum_{i=1}^{n}X_iY_i\end{aligned} \tag{17.9}$$

$$
\begin{aligned}
\text{SSR} &= \text{explained variation or sum of squares due to regression} \\
&= \sum_{i=1}^{n} (\hat{Y}_i - \overline{Y})^2 \\
&= \text{SST} - \text{SSE} \\
&= b_0 \sum_{i=1}^{n} Y_i + b_1 \sum_{i=1}^{n} X_i Y_i - n\overline{Y}^2
\end{aligned}
\tag{17.10}
$$

Examining the unexplained variation or error sum of squares [Equation (17.9)], we may recall that $\sum_{i=1}^{n} (Y_i - \hat{Y}_i)^2$ was the numerator under the square root in the computation of the standard error of the estimate [see Equation (17.5)]. Therefore, in the process of computing the standard error of the estimate, we have already computed the following error sum of squares:

$$
\begin{aligned}
\text{SSE} &= \sum_{i=1}^{n} Y_i^2 - b_0 \sum_{i=1}^{n} Y_i - b_1 \sum_{i=1}^{n} X_i Y_i \\
&= 1{,}602.0971 - (2.423)(176.11) - (.00873)(134{,}127.90) \\
&= 4.446
\end{aligned}
$$

Moreover,

$$
\begin{aligned}
\text{SST} &= \text{total sum of squares} \\
&= \sum_{i=1}^{n} Y_i^2 - n\overline{Y}^2 \\
&= 1{,}602.0971 - 20(8.8055)^2 \\
&= 1{,}602.0971 - 1{,}550.7366 \\
&= 51.3605
\end{aligned}
$$

and

$$
\begin{aligned}
\text{SSR} &= \text{explained variation or sum of squares due to regression} \\
&= b_0 \sum_{i=1}^{n} Y_i + b_1 \sum_{i=1}^{n} X_i Y_i - n\overline{Y}^2 \\
&= (2.423)(176.11) + (.00873)(134{,}127.90) - 20(8.8055)^2 \\
&= 46.9145
\end{aligned}
$$

We note also, from Equation (17.7), that

$$
\begin{aligned}
\text{SST} &= \text{SSR} + \text{SSE} \\
51.3605 &= 46.9145 + 4.4460
\end{aligned}
$$

Now the **coefficient of determination** r^2 can be defined as

$$r^2 = \frac{\text{sum of squares due to regression}}{\text{total sum of squares}} = \frac{\text{SSR}}{\text{SST}} \qquad \textbf{(17.11a)}$$

That is, the coefficient of determination measures the proportion of variation that is explained by the independent variable in the regression model. For the package delivery problem,

$$r^2 = \frac{46.9145}{51.3605} = .913$$

Therefore, 91.3% of the variation in weekly sales can be explained by the variability in the number of customers from store to store. This is an example where there is a strong linear relationship between two variables, since the use of a regression model has reduced the variability in predicting weekly sales by 91.3%. Only 8.7% of the sample variability in weekly sales can be explained by factors other than what is accounted for by the linear regression model.

To interpret the coefficient of determination—particularly when dealing with multiple regression models—some researchers suggest than an *adjusted* r^2 be computed to reflect both the number of explanatory variables in the model and the sample size. In simple linear regression, however, we denote the **adjusted r^2** as

$$r^2_{\text{adj}} = 1 - \left[(1 - r^2)\frac{n-1}{n-2}\right] \qquad \textbf{(17.11b)}$$

Thus for our package delivery data, since $r^2 = .913$ and $n = 20$,

$$\begin{aligned} r^2_{\text{adj}} &= 1 - \left[(1 - r^2)\frac{20-1}{20-2}\right] \\ &= 1 - \left[(1 - .913)\frac{19}{18}\right] \\ &= 1 - .092 \\ &= .908 \end{aligned}$$

This result is similar to the one obtained without adjustment for degrees of freedom.

Problems for Section 17.6

Data file PETFOOD.DAT

● 17.19 Referring to Problem 17.7 (pet food sales) on page 725,
(a) Compute the coefficient of determination r^2 and interpret its meaning.
(b) Compute the adjusted r^2.

Data file SITE.DAT

17.20 Referring to Problem 17.8 (site selection) on page 725,
(a) Compute the coefficient of determination r^2 and interpret its meaning.
(b) Compute the adjusted r^2.

Data file WORKHRS.DAT

17.21 Referring to Problem 17.9 (production worker-hours) on page 725,
(a) Compute the coefficient of determination r^2 and interpret its meaning.
(b) Compute the adjusted r^2.

Data file TOMYIELD.DAT

17.22 Referring to Problem 17.10 (tomato yield) on page 725,
(a) Compute the coefficient of determination r^2 and interpret its meaning.
(b) Compute the adjusted r^2.

Data file INTERVW.DAT

17.23 Referring to Problem 17.11 (interviewer productivity) on page 726,
(a) Compute the coefficient of determination r^2 and interpret its meaning.
(b) Compute the adjusted r^2.

Data file LIMO.DAT

17.24 Referring to Problem 17.12 (airport travel) on page 726,
(a) Compute the coefficient of determination r^2 and interpret its meaning.
(b) Compute the adjusted r^2.

17.25 When will the unexplained variation or error sum of squares be equal to 0?

17.26 When will the explained variation or sum of squares due to regression be equal to 0?

17.7 Correlation—Measuring the Strength of the Association

In our discussion of the relationship between two variables thus far in this chapter we have been concerned with the prediction of the dependent variable Y based on the independent variable X. In contrast to a regression analysis, in a correlation analysis we are only interested in measuring the degree of association between two variables.

The strength of a relationship between two variables in a population is usually measured by the **coefficient of correlation** ρ, whose values range from -1 for perfect negative correlation up to $+1$ for perfect positive correlation. Figure 17.6 illustrates these three different types of association between variables. In Panel A of Figure 17.6 there is a perfect negative linear relationship between X and Y so that Y will decrease in a perfectly predictable manner as X increases. Panel B of Figure 17.6 is an example in which there is no relationship between X and Y. As X increases, there is no change in Y, so that there is no association between the val-

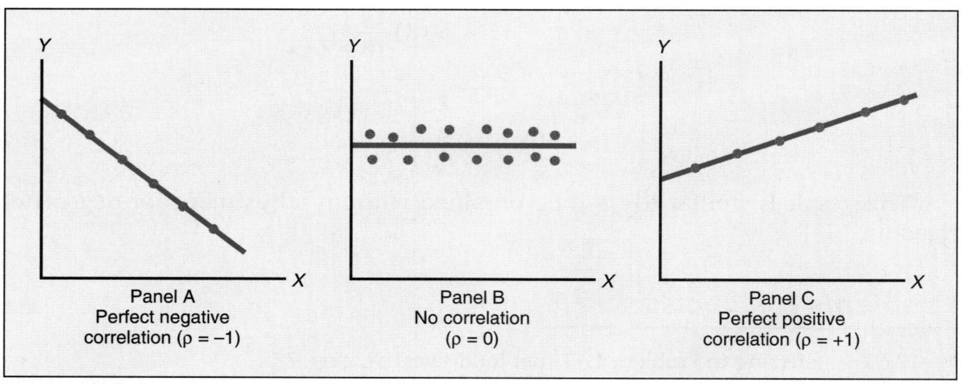

Figure 17.6
Types of association between variables.

ues of X and the values of Y. On the other hand, panel C of Figure 17.6 depicts a perfect positive correlation between X and Y. In this case Y increases in a perfectly predictable manner as X increases.

For regression-oriented problems, the sample coefficient of correlation (r) may be obtained from Equation (17.11a) as follows:

$$r^2 = \frac{\text{sum of squares due to regression}}{\text{total sum of squares}} = \frac{\text{SSR}}{\text{SST}}$$

so that

$$r = \sqrt{r^2} \qquad \textbf{(17.12)}$$

In simple linear regression r takes the sign of b_1. If b_1 is positive, r is positive. If b_1 is negative, r is negative. If b_1 is zero, r is zero.

In the package delivery problem, since $r^2 = .913$ and the slope b_1 is positive, the coefficient of correlation is computed as $+.956$. The closeness of the correlation coefficient to $+1.0$ implies a strong association between number of customers and weekly sales.

We have now computed and interpreted the correlation coefficient in terms of its regression viewpoint. As we mentioned at the beginning of this chapter, however, regression and correlation are two separate techniques, with regression being concerned with prediction and correlation with association. In many applications we are concerned only with measuring association between variables, not with using one variable to predict another.

If only a correlation analysis is being performed on a set of data, the sample correlation coefficient r can be computed directly using the following formula:

$$r = \frac{\sum_{i=1}^{n}(X_i - \overline{X})(Y_i - \overline{Y})}{\sqrt{\sum_{i=1}^{n}(X_i - \overline{X})^2}\sqrt{\sum_{i=1}^{n}(Y_i - \overline{Y})^2}} \qquad \textbf{(17.13a)}$$

or, alternatively, the "calculator" formula:

$$r = \frac{\sum_{i=1}^{n} X_iY_i - n\overline{X}\overline{Y}}{\sqrt{\sum_{i=1}^{n} X_i^2 - n\overline{X}^2}\sqrt{\sum_{i=1}^{n} Y_i^2 - n\overline{Y}^2}} \qquad \textbf{(17.13b)}$$

To illustrate such an example, suppose that we wanted to measure the strength of the association in the price of two different commodities in various cities throughout the world. The price of a six-pack of a brand name cola soft drink and of one pound of chicken was determined at a supermarket located in a sample of nine different cities. The results are summarized in Table 17.3.

Table 17.3 **Price (in $) of a six-pack of a brand name cola soft drink and of one pound of chicken in a sample of nine cities.**

City	Six-Pack of Brand Name Cola (X)	One Pound of Chicken (Y)
Frankfurt	3.27	3.06
Hong Kong	2.22	2.34
London	2.28	2.27
Manila	3.04	1.51
Mexico City	2.33	1.87
New York	2.69	1.65
Paris	4.07	3.09
Sydney	2.78	2.36
Tokyo	5.97	4.85

For the data of Table 17.3, we compute the following values:

$$\sum_{i=1}^{n} X_i = 28.65 \qquad \sum_{i=1}^{n} X_i^2 = 102.66 \qquad \sum_{i=1}^{n} Y_i = 23.00$$

$$n = 9 \qquad \sum_{i=1}^{n} Y_i^2 = 67.132 \qquad \sum_{i=1}^{n} X_iY_i = 81.854$$

From this we obtain

$$\overline{X} = \frac{28.65}{9} = 3.183$$

$$\overline{Y} = \frac{23.00}{9} = 2.5556$$

so that from Equation (17.13b)

$$r = \frac{\sum_{i=1}^{n} X_iY_i - n\overline{X}\overline{Y}}{\sqrt{\sum_{i=1}^{n} X_i^2 - n\overline{X}^2}\sqrt{\sum_{i=1}^{n} Y_i^2 - n\overline{Y}^2}}$$

$$= \frac{81.854 - 9(3.183)(2.5556)}{\sqrt{102.66 - 9(3.183)^2}\sqrt{67.132 - 9(2.5556)^2}}$$

$$= \frac{81.8540 - 73.2172}{\sqrt{11.4594}\sqrt{8.3522}}$$

$$r = +.883$$

The coefficient of correlation, $r = +.883$, between the price of a six-pack of a brand name cola soft drink and of one pound of chicken indicates a very strong association. A higher price of a six-pack of cola is strongly associated with a higher price of a pound of chicken. In Section 17.13, we will use these sample results to determine whether there is any evidence of a significant association between these variables in the population.

Problems for Section 17.7

17.27 Under what circumstances will the coefficient of correlation be negative?

17.28 Referring to Problem 17.19 (pet food sales) on page 731, compute the coefficient of correlation. Data file PETFOOD.DAT

17.29 Referring to Problem 17.20 (site selection) on page 732, compute the coefficient of correlation. Data file SITE.DAT

17.30 Referring to Problem 17.21 (production worker-hours) on page 732, compute the coefficient of correlation. Data file WORKHRS.DAT

17.31 Referring to Problem 17.22 (tomato yield) on page 732, compute the coefficient of correlation. Data file TOMYIELD.DAT

17.32 Referring to Problem 17.23 (interviewer productivity problem) on page 732, compute the coefficient of correlation. Data file INTERVW.DAT

17.33 Referring to Problem 17.24 (airport travel) on page 732, compute the coefficient of correlation. Data file LIMO.DAT

17.34 Suppose that we also wanted to measure the strength of the association in the price (in $) of a six-pack of a brand name cola soft drink and 100 tablets of a brand name pain reliever in supermarkets in a sample of nine different international cities. The results are as follows:

City	Six-Pack of Brand Name Cola	100 Tablets of Pain Reliever
Frankfurt	3.27	17.22
Hong Kong	2.22	6.21
London	2.28	9.17
Manila	3.04	14.61
Mexico City	2.33	4.85
New York	2.69	6.09
Paris	4.07	13.08
Sydney	2.78	8.04
Tokyo	5.97	8.39

Data file SUPERMKT.DAT

(a) Compute the coefficient of correlation r between the price of a six-pack of a brand name cola soft drink and 100 tablets of a brand name pain reliever.

(b) Is the price of a six-pack of a brand name cola soft drink more correlated with the price of chicken or with that of 100 tablets of a brand name pain reliever? Explain.

17.35 Suppose that we also wanted to measure the strength of the association in the price (in $) of a women's hairdresser and a men's brand name dress shirt in a sample of nine different international cities. The results are as follows (see the top of page 736): Data file CITY.DAT

City	Women's Hairdresser	Men's Dress Shirt
Frankfurt	29.85	49.41
Hong Kong	22.56	29.32
London	33.79	42.12
Manila	12.04	35.22
Mexico City	15.49	25.04
New York	34.87	37.85
Paris	27.73	55.28
Sydney	25.64	38.58
Tokyo	27.45	38.69

Compute the coefficient of correlation r between the price of a women's hairdresser and a men's brand name dress shirt.

17.8 Assumptions of Regression and Correlation

In our investigations into hypothesis testing and the analysis of variance we have noted that the appropriate application of a particular statistical procedure is dependent on how well a set of assumptions for that procedure are met. The assumptions necessary for regression and correlation analysis are analogous to those of the analysis of variance, since they fall under the general heading of linear models (References 5 and 12). Although there are some differences in the assumptions made by the regression model and by correlation (see References 5 and 12), this topic is beyond the scope of this text and we will consider only the former.

The four major assumptions of regression are

1. Normality
2. Homoscedasticity
3. Independence of error
4. Linearity

The first assumption, **normality,** requires that the values of Y be normally distributed at each value of X (see Figure 17.7). Like the t test and analysis-of-variance F test, regression analysis is fairly robust against departures from the normality assumption. As long as the distribution of Y_i values around each level of X is

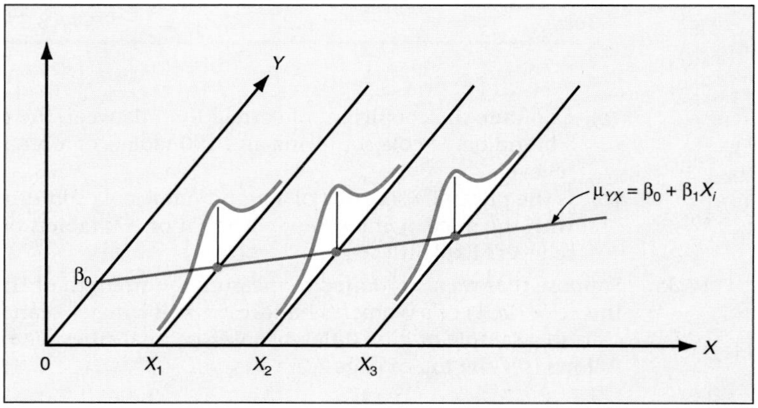

Figure 17.7
Assumptions of regression.

not extremely different from a normal distribution, inferences about the line of regression and the regression coefficients will not be seriously affected.

The second assumption, **homoscedasticity,** requires that the variation around the line of regression be constant for all values of X. This means that Y varies the same amount when X is a low value as when X is a high value (see Figure 17.7). The homoscedasticity assumption is important for using the least squares method of determining the regression coefficients. If there are serious departures from this assumption, either data transformations or weighted least squares methods (References 5 and 12) can be applied.

The third assumption, **independence of error,** requires that the error (residual difference between an observed and predicted value of Y) should be independent for each value of X. This assumption often refers to data that are collected over a period of time. When data are collected in this manner the residuals for a particular time period are often correlated with those of the previous time period.

The fourth assumption, **linearity,** states that the relationship among variables is linear. Two variables could be perfectly related in a nonlinear fashion and the linear correlation coefficient would be zero, indicating no relationship. Such nonlinear models will be discussed in Sections 18.11 and 18.13.

17.9 Regression Diagnostics: Residual Analysis

17.9.1 Introduction

In our discussion of our package delivery data throughout this chapter we have relied upon a simple regression model in which the dependent variable was predicted based upon a straight-line relationship with a single independent variable. In this section we shall use a graphical approach called **residual analysis** to evaluate the appropriateness of the regression model that has been fitted to the data. In addition, this approach will also allow us to study potential violations in the assumptions of our regression model (see Section 17.8).

17.9.2 Evaluating the Aptness of the Fitted Model

The residual or estimated error values (e_i) are defined as the difference between the observed (Y_i) and predicted ($\hat{Y}_i$) values of the dependent variable for given values X_i. Thus

$$e_i = Y_i - \hat{Y}_i \quad \textbf{(17.14)}$$

We may evaluate the aptness of the fitted regression model by plotting the residuals on the vertical axis against the corresponding X_i values of the independent variable on the horizontal axis. If the fitted model is appropriate for the data, there will be no apparent pattern in this plot of the residuals versus X_i. However, if the fitted model is not appropriate, there will be a relationship between the X_i values and the residuals e_i. Such a pattern can be observed in Figure 17.8 on page 738. Figure 17.8 (a) depicts a situation in which there is a significant simple linear relationship between X and Y. However, a curvilinear model between the

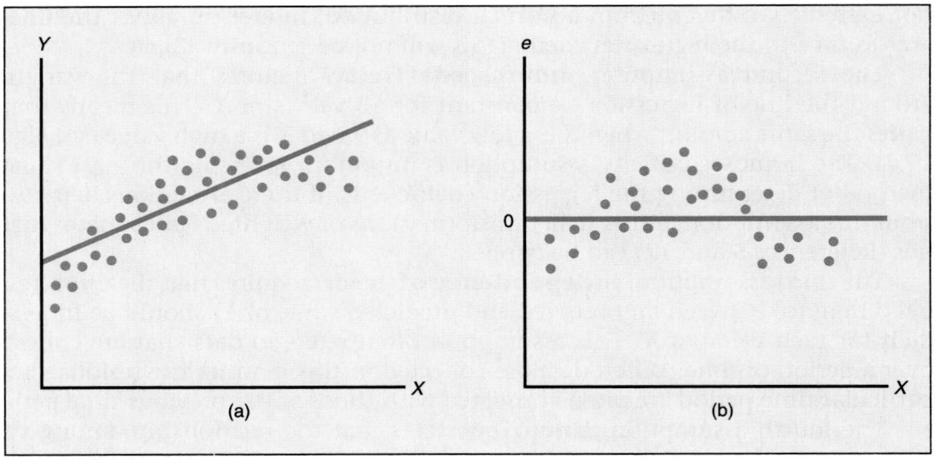

Figure 17.8
Studying the appropriateness of the simple linear regression model.

two variables seems more appropriate. This effect is highlighted in Figure 17.8 (b), the residual plot of e_i versus X_i. In (b) there is an obvious curvilinear effect between X_i and e_i. By plotting the residuals we have essentially filtered out or removed the *linear* trend of X with Y, thereby exposing the lack of fit in the simple linear model. Thus, from (a) and (b) we may conclude that the curvilinear model may be a better fit and should be evaluated in place of the simple linear model (see Section 18.11 for further discussion of fitting curvilinear models).

Having considered Figure 17.8, let us return to the evaluation of the package delivery data. Table 17.4 represents the observed, predicted, and residual values of

Table 17.4 Observed, predicted, and residual values for the package delivery data.

Observation	Number of Customers X_i	Weekly Sales Observed	Predicted	Residual	Standardized Residual, SR_i
1	907	11.200	10.341	0.859	1.81
2	926	11.050	10.506	0.544	1.15
3	506	6.840	6.840	−0.000	−0.00
4	741	9.210	8.891	0.319	0.65
5	789	9.420	9.310	0.110	0.22
6	889	10.080	10.183	−0.103	−0.22
7	874	9.450	10.052	−0.602	−1.25
8	510	6.730	6.875	−0.145	−0.31
9	529	7.240	7.041	0.199	0.42
10	420	6.120	6.089	0.031	0.07
11	679	7.630	8.350	−0.720	−1.48
12	872	9.430	10.035	−0.605	−1.26
13	924	9.460	10.489	−1.029	−2.18
14	607	7.640	7.722	−0.082	−0.17
15	452	6.920	6.369	0.551	1.21
16	729	8.950	8.787	0.163	0.33
17	794	9.330	9.354	−0.024	−0.05
18	844	10.230	9.791	0.439	0.91
19	1,010	11.770	11.240	0.530	1.17
20	621	7.410	7.844	−0.434	−0.90

the response variable (weekly sales) in the simple linear model we have fitted. We have also computed the standardized residuals. These represent each residual divided by its standard error. The standardized residual is expressed as Equation (17.15).

Standardized Residual

$$SR_i = \frac{e_i}{S_{YX}\sqrt{1 - h_i}} \qquad \textbf{(17.15)}$$

where

$$h_i = \frac{1}{n} + \frac{(X_i - \overline{X})^2}{\sum_{i=1}^{n} X_i^2 - n\overline{X}^2}$$

These standardized values allow us to consider the magnitude of the residuals in units that reflect the standardized variation around the line of regression. The standardized residuals have been plotted against the independent variable (number of customers) in Figure 17.9. From this we may observe that although there is widespread scatter in the residual plot, there is no apparent pattern or relationship between the standardized residuals and X_i. The residuals appear to be evenly spread above and below 0 for differing values of X. Thus we may conclude for the package delivery data that the fitted model appears to be appropriate.

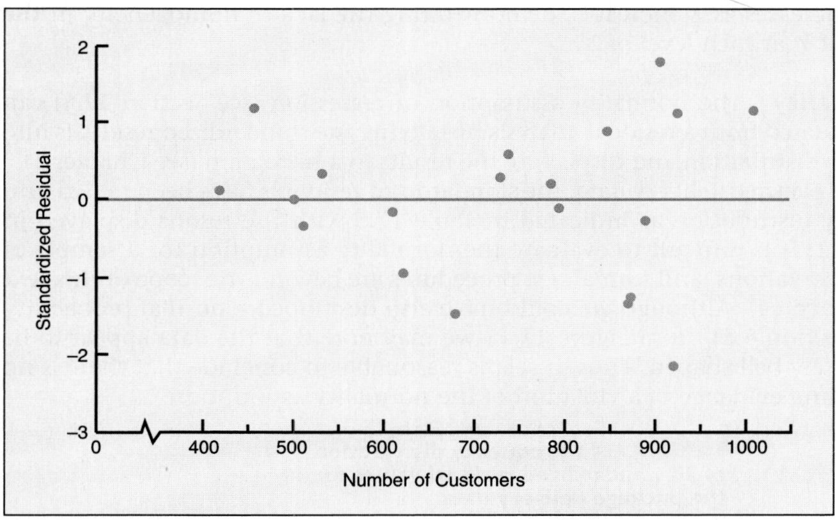

Figure 17.9
Plotting the standardized residuals versus the number of customers.

17.9.3 Evaluating the Assumptions

● **Homoscedasticity** The assumption of homoscedasticity (see Section 17.8) can also be evaluated from a plot of SR_i with X_i. For the package delivery data, there do not appear to be major differences in the variability of SR_i for different X_i

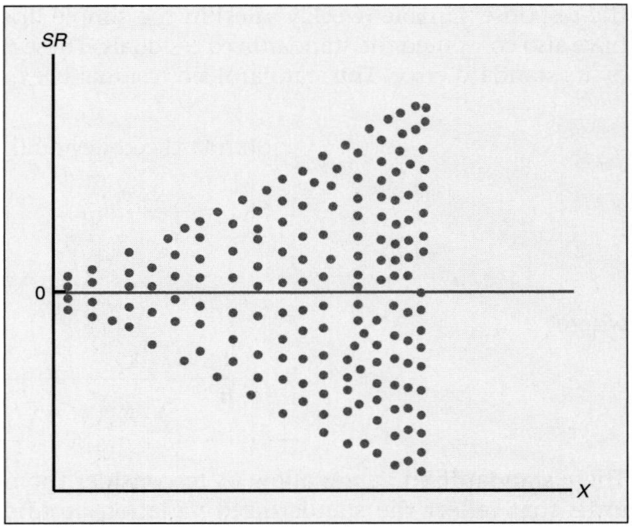

Figure 17.10
Violations in homoscedasticity.

values (see Figure 17.9 on page 739). Thus we may conclude that for our fitted model there is no apparent violation in the assumption of equal variance at each level of X.

If we wish to observe a case in which the homoscedasticity assumption is violated, we should examine the *hypothetical* plot of SR_i with X_i in Figure 17.10. In this hypothetical plot there appears to be a *fanning effect* in which the variability of the residuals increases as X increases, demonstrating the lack of homogeneity in the variances of Y_i at each level of X.

● **Normality** The normality assumption of regression (see Section 17.8) can also be evaluated from a residual analysis by tallying the standardized residuals into a frequency distribution and displaying the results in a histogram (see Chapter 3).

For the package delivery data, the standardized residuals have been tallied into a frequency distribution as indicated in Table 17.5 with the results displayed in Figure 17.11. It is difficult to evaluate the normality assumption for a sample of only 20 observations, and formal test procedures are beyond the scope of this text (see Reference 14). Although we could have also developed a normal probability plot (see Section 8.5), from Figure 17.11 we may note that the data appear to be approximately bell shaped. Thus, it seems reasonable to conclude that there is no overwhelming evidence of a violation of the normality assumption.

Table 17.5 Frequency distribution of 20 standardized residual values for the package delivery data.

Standardized Residuals	No.
−2.8 but less than −2.0	1
−2.0 but less than −1.2	3
−1.2 but less than −0.4	2
−0.4 but less than +0.4	8
+0.4 but less than +1.2	4
+1.2 but less than +2.0	2
+2.0 but less than +2.8	0
Totals	20

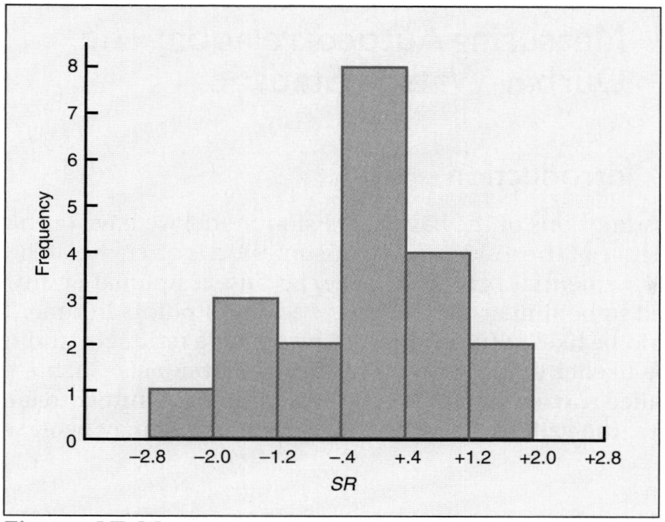

Figure 17.11
Plotting the standardized residuals for the package delivery data.

● **Independence** The independence assumption discussed in Section 17.8 can be evaluated by plotting the residuals in the order or sequence in which the observed data were obtained. Data collected over periods of time often exhibit an *autocorrelation* effect among successive observations. That is, there exists a correlation between a particular observation and the values that precede and succeed it. Such patterns, which violate the assumption of independence, are readily apparent in the plot of the residuals versus the time in which they were collected. This effect may be measured by the Durbin-Watson statistic, which will be the subject of Section 17.10.

Problems for Section 17.9

● 17.36 Referring to the pet food sales problem (pages 716, 725, and 728), perform a residual analysis on your results and determine the adequacy of the fit of the model. Data file PETFOOD.DAT

17.37 Referring to the site selection problem (pages 716, 725, and 728), perform a residual analysis on your results and determine the adequacy of the fit of the model. Data file SITE.DAT

● 17.38 Referring to the production worker-hours problem (pages 717, 725, and 728), perform a residual analysis on your results and determine the adequacy of the fit of the model. Data file WORKHRS.DAT

17.39 Referring to the tomato yield problem (pages 718, 725, and 728), perform a residual analysis on your results and determine the adequacy of the fit of the model. Data file TOMYIELD.DAT

17.40 Referring to the interviewer productivity problem (pages 718, 726, and 728), perform a residual analysis on your results and determine the adequacy of the fit of the model. Data file INTERVW.DAT

17.41 Referring to the airport travel problem (pages 718, 726, and 728), perform a residual analysis on your results and determine the adequacy of the fit of the model. Data file LIMO.DAT

17.10 Measuring Autocorrelation: The Durbin-Watson Statistic

17.10.1 Introduction

One of the assumptions of the basic regression model we have been considering is the independence of the residuals. This assumption is often violated when data are collected over sequential periods of time, because a residual at any one point in time may tend to be similar to residuals at adjacent points in time. Thus, positive residuals would be likelier to be followed by positive residuals, and negative residuals would be likelier to be followed by negative residuals. Such a pattern in the residuals is called **autocorrelation.** When substantial autocorrelation is present in a set of data, the validity of a fitted regression model may be in serious doubt.

17.10.2 Residual Plots to Detect Autocorrelation

As mentioned in Section 17.9, the easiest way to detect autocorrelation in a set of data is to plot the residuals or standardized residuals in time order. If a positive autocorrelation effect is present, clusters of residuals with the same sign will be present, and an apparent pattern will be readily detected. To illustrate the autocorrelation effect, we shall consider the following example.

Recall that in Sections 17.2–17.4 we developed a regression model to predict weekly sales based on the number of customers for a sample of 20 package delivery stores. Suppose that the manager of the seventeenth package delivery store listed in Table 17.1 on page 715 wanted to predict weekly sales based on the number of customers for a period of 15 weeks. In this situation, since data are collected over a period of 15 *consecutive* weeks at the same store, we would need to be concerned with the autocorrelation effect of the residuals. The data for this store are summarized in Table 17.6 on page 743.

Figure 17.12 represents partial MINITAB output.

```
The regression equation is
Sales = - 16.0 + 0.0308 Customer

Predictor        Coef        Stdev     t-ratio        p
Constant      -16.032        5.310       -3.02    0.010
Customer     0.030760     0.006158        5.00    0.000

s = 0.9360        R-sq = 65.7%       R-sq(adj) = 63.1%


Durbin-Watson statistic = 0.88
```

Figure 17.12
MINITAB output for the data of Table 17.6.

Table 17.6 Customers and sales for a period of 15 consecutive weeks.

Week	Customers	Sales ($000)
1	794	9.33
2	799	8.26
3	837	7.48
4	855	9.08
5	845	9.83
6	844	10.09
7	863	11.01
8	875	11.49
9	880	12.07
10	905	12.55
11	886	11.92
12	843	10.27
13	904	11.80
14	950	12.15
15	841	9.64

We note from Figure 17.12 that r^2 is .657, indicating that 65.7% of the variation in sales can be explained by variation in the number of customers. In addition, the Y intercept, b_0, is -16.032, while the slope, b_1, is .03076. However, before we can accept the validity of this model, we must undertake proper analyses of the residuals. Since the data have been collected over a consecutive period of 15 weeks, the residuals should be plotted over time to see if a pattern exists. Figure 17.13 represents such a plot for the data of Table 17.6.

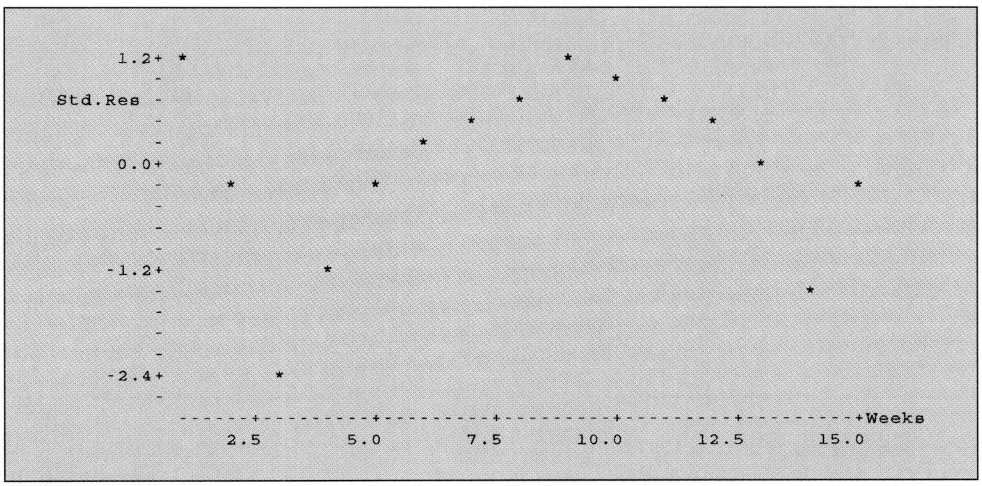

Figure 17.13
MINITAB plot of standardized residuals over time for the data of Table 17.6.

From Figure 17.13, we observe that the points tend to fluctuate up and down in a cyclical pattern. This cyclical pattern would give us strong cause for concern about the autocorrelation of the residuals and, hence, a violation in the assumption of independence of the residuals.

17.10.3 The Durbin-Watson Procedure

In addition to residual plots, autocorrelation can also be detected and measured by using the **Durbin-Watson statistic.** This statistic measures the correlation of each residual and the residual for the time period immediately preceding the one of interest. The Durbin-Watson statistic (D) is defined as:

$$D = \frac{\sum_{i=2}^{n}(e_i - e_{i-1})^2}{\sum_{i=1}^{n} e_i^2} \qquad (17.16)$$

where e_i = the residual at time period i.

Although the computation of the Durbin-Watson statistic is readily available from the output of most statistical software packages (see Figure 17.12 on page 742), for illustrative purposes, the computations for the data of Table 17.6 are summarized in Table 17.7.

Table 17.7 Computation of the Durbin-Watson statistic for the regression analysis of the single package delivery store.

Week	Sales (Y_i)	$\hat{Y}_i$	$e_i = Y_i - \hat{Y}_i$	e_{i-1}	$(e_i - e_{i-1})$	$(e_i - e_{i-1})^2$	e_i^2
1	9.33	8.3914	0.93857	*	*	*	0.88092
2	8.26	8.5452	−0.28523	0.93857	−1.22380	1.49769	0.08136
3	7.48	9.7141	−2.23412	−0.28523	−1.94889	3.79817	4.99128
4	9.08	10.2678	−1.18780	−2.23412	1.04632	1.09478	1.41087
5	9.83	9.9602	−0.13020	−1.18780	1.05760	1.11852	0.01695
6	10.09	9.9294	0.16056	−0.13020	0.29076	0.08454	0.02578
7	11.01	10.5139	0.49612	0.16056	0.33556	0.11260	0.24613
8	11.49	10.8830	0.60699	0.49612	0.11088	0.01229	0.36844
9	12.07	11.0368	1.03319	0.60699	0.42620	0.18165	1.06749
10	12.55	11.8058	0.74419	1.03319	−0.28901	0.08352	0.55381
11	11.92	11.2214	0.69863	0.74419	−0.04556	0.00208	0.48809
12	10.27	9.8987	0.37132	0.69863	−0.32731	0.10713	0.13788
13	11.80	11.7751	0.02495	0.37132	−0.34637	0.11997	0.00062
14	12.15	13.1900	−1.04002	0.02495	−1.06497	1.13416	1.08165
15	9.64	9.8372	−0.19716	−1.04002	0.84286	0.71042	0.03887
						$\sum_{i=2}^{n}(e_i - e_{i-1})^2 = 10.058$	$\sum_{i=1}^{n} e_i^2 = 11.39$

To have a better understanding of what the Durbin-Watson statistic is measuring, we need to examine the composition of the D statistic presented in Equation (17.16). The numerator $\sum_{i=2}^{n}(e_i - e_{i-1})$ represents the squared difference in two successive residuals, summed from the second observation to the nth observation. The denominator represents the sum of the squared residuals, $\sum_{i=1}^{n} e_i^2$. When successive residuals are positively autocorrelated, the value of D will approach zero. If the residuals are not correlated, the value of D will be close to 2.

(If there is negative autocorrelation, which rarely happens, D will be greater than 2 and could even approach its maximum value of 4.)

For our data in Table 17.7, we use Equation (17.16) and obtain

$$D = \frac{10.058}{11.39} = .883$$

The crux of the issue in using the Durbin-Watson statistic is the determination of when the autocorrelation is large enough to make the D statistic fall sufficiently below 2 to cause concern about the validity of the model. The answer to this question is dependent on the number of observations being analyzed and the number of independent variables in the model (in simple linear regression, $p = 1$). Table 17.8 has been extracted from Appendix E, Table E.14, the table of the Durbin-Watson statistic.

Table 17.8 Finding critical values of the Durbin-Watson statistic.

	$\alpha = .05$									
	$p = 1$		$p = 2$		$p = 3$		$p = 4$		$p = 5$	
n	d_L	d_U	d_L	d_U	d_L	d_U	d_L	d_U	d_L	d_U
15	1.08	1.36	.95	1.54	.82	1.75	.69	1.97	.56	2.21
16	1.10	1.37	.98	1.54	.86	1.73	.74	1.93	.62	2.15
17	1.13	1.38	1.02	1.54	.90	1.71	.78	1.90	.67	2.10
18	1.16	1.39	1.05	1.53	.93	1.69	.82	1.87	.71	2.06
19	1.18	1.40	1.08	1.53	.97	1.68	.86	1.85	.75	2.02
⋮	⋮	⋮	⋮	⋮	⋮	⋮	⋮	⋮	⋮	⋮
90	1.63	1.68	1.61	1.70	1.59	1.73	1.57	1.75	1.54	1.78
95	1.64	1.69	1.62	1.71	1.60	1.73	1.58	1.75	1.56	1.78
100	1.65	1.69	1.63	1.72	1.61	1.74	1.59	1.76	1.57	1.78

	$\alpha = .01$									
	$p = 1$		$p = 2$		$p = 3$		$p = 4$		$p = 5$	
n	d_L	d_U	d_L	d_U	d_L	d_U	d_L	d_U	d_L	d_U
15	.81	1.07	.70	1.25	.59	1.46	.49	1.70	.39	1.96
16	.84	1.09	.74	1.25	.63	1.44	.53	1.66	.44	1.90
17	.87	1.10	.77	1.25	.67	1.43	.57	1.63	.48	1.85
18	.90	1.12	.80	1.26	.71	1.42	.61	1.60	.52	1.80
19	.93	1.13	.83	1.26	.74	1.41	.65	1.58	.56	1.77
⋮	⋮	⋮	⋮	⋮	⋮	⋮	⋮	⋮	⋮	⋮
90	1.50	1.54	1.47	1.56	1.45	1.59	1.43	1.61	1.41	1.64
95	1.51	1.55	1.49	1.57	1.47	1.60	1.45	1.62	1.42	1.64
100	1.52	1.56	1.50	1.58	1.48	1.60	1.46	1.63	1.44	1.65

Note: n = number of observations; p = number of independent variables.
Source: Table E.14.

From Table 17.8 we observe that two values are shown in the table for each combination of level of significance (α), n (sample size), and p (the number of independent variables in the model). The first value, d_L, represents the lower critical value when there is no autocorrelation in the data. If D is below d_L we may conclude that there is evidence of autocorrelation among the residuals. Under such a circumstance, the least squares methods that we have considered in this chapter are inappropriate and alternative methods need to be used (see References 5 and 12). The second value, d_U, represents the upper critical value of D above which we would conclude that there is no evidence of autocorrelation among the residuals. If D is between d_L and d_U we are unable to make a definite conclusion.

Thus, as illustrated in Table 17.8, for our data concerning the single package delivery store, with one independent variable ($p = 1$), and 15 observations ($n = 15$), $d_L = 1.08$ and $d_U = 1.36$. Since $D = 0.883 < 1.08$, we may conclude that there is autocorrelation among the residuals. Our analysis of the data of Figure 17.12, which were obtained under the assumption that the least squares method was appropriate, *must not* be continued due to the presence of serious autocorrelation among the residuals. We need to consider alternative approaches discussed in References 5 and 12.

Problems for Section 17.10

17.42 Under what circumstances would it be important to compute the Durbin-Watson statistic? Explain.

Data file PETFOOD.DAT

● 17.43 Referring to Problem 17.1 (the pet food sales problem), is it necessary to compute the Durbin-Watson statistic? Explain.

17.44 Referring to Problem 17.1 (the pet food sales problem), under what circumstances would it be necessary to obtain the Durbin-Watson statistic before proceeding with the least squares method of regression analysis?

17.45 Suppose that the residuals for a set of data collected over ten consecutive time periods were as follows:

Data file RESID1.DAT

Time Period	Residual
1	−5
2	−4
3	−3
4	−2
5	−1
6	+1
7	+2
8	+3
9	+4
10	+5

(a) Plot the residuals over time. What conclusions can you reach about the pattern of the residuals over time?
(b) Compute the Durbin-Watson statistic.
(c) Based on (a) and (b), what conclusion can you reach about the autocorrelation of the residuals?

17.46 Suppose that the residuals for a set of data collected over 15 consecutive time periods were as follows:

Data file RESID2.DAT

Time Period	Residual
1	+4
2	−6
3	−1
4	−5
5	+2
6	+5
7	−2
8	+7
9	+6
10	−3
11	+1
12	+3
13	0
14	−4
15	−7

(a) Plot the residuals over time. What conclusions can you reach about the pattern of the residuals over time?

(b) Compute the Durbin-Watson statistic. At the .05 level of significance, is there evidence of positive autocorrelation among the residuals?
(c) Based on (a) and (b), what conclusion can you reach about the autocorrelation of the residuals?

17.11 Confidence-Interval Estimate for Predicting μ_{YX}

In Sections 17.1–17.7 we were concerned with the use of regression and correlation solely for the purpose of description. The least squares method has been utilized to determine the regression coefficients and to predict the value of Y from a given value of X. In addition, the standard error of the estimate has been discussed along with the coefficients of correlation and determination.

Now that we have used residual analysis in Section 17.9 to assure ourselves that the assumptions of the least squares regression model have not been violated and that the straight-line model is appropriate, we may concern ourselves with making inferences about the relationship between the variables in a population based on our sample results. In this section, we will discuss methods of making predictive inferences about the mean of Y, and in the following section we will predict an individual response value Y_I.

We may recall that in Section 17.4 the fitted regression equation was used to make predictions about the value of Y for a given X. In the package delivery problem, for example, we predicted that the average weekly sales for stores with 600 customers would be 7.661 (in thousands of dollars). This estimate, however, is merely a point estimate of the true average value. In Chapter 10 we developed the concept of the confidence interval as an estimate of the population average. In a similar fashion, a confidence-interval estimate can now be developed to make inferences about the average predicted value of Y:

$$\hat{Y}_i \pm t_{n-2} S_{YX} \sqrt{h_i} \qquad \textbf{(17.17)}$$

where[2]

$$h_i = \frac{1}{n} + \frac{(X_i - \overline{X})^2}{\sum_{i=1}^{n} X_i^2 - n\overline{X}^2}$$

$\hat{Y}_i$ = predicted value of Y; $\hat{Y}_i = b_0 + b_1 X_i$

S_{YX} = standard error of the estimate

n = sample size

X_i = given value of X

An examination of Equation (17.17) indicates that the width of the confidence interval is dependent on several factors. For a given level of confidence, increased variation around the line of regression, as measured by the standard error of the estimate, results in a wider interval. However, as would be expected, increased sample size reduces the width of the interval. Moreover, the width of the interval also varies at different values of X. When predicting Y for values of X close to $\overline{X}$, the interval is much narrower than for predictions for X values more distant from the mean. This effect can be seen from the square-root portion of Equation (17.17) and from Figure 17.14.

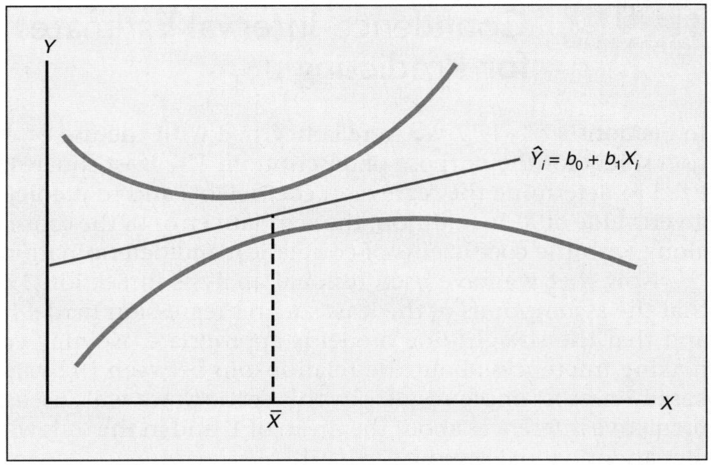

Figure 17.14
Interval estimates of μ_{YX} for different values of X.

As indicated in Figure 17.14, the interval estimate of the true mean of Y varies *hyperbolically* as a function of the closeness of the given X to $\overline{X}$. When predictions are to be made for X values that are distant from the average value of X, the much wider interval is the tradeoff for predicting at such values of X. Thus, as depicted in Figure 17.14, we observe a *confidence-band effect* for the predictions.

Let us now use Equation (17.17) for our package delivery problem. Suppose that we desire a 95% confidence interval estimate of the true average weekly sales for all stores with 600 customers. We compute the following:

$$\hat{Y}_i = 2.423 + .00873X_i$$

and for $X_i = 600$ we obtain $\hat{Y}_i = 7.661$.

Also,

$$\overline{X} = 731.15 \qquad S_{YX} = .497$$

$$\sum_{i=1}^{n} X_i^2 = 11{,}306{,}209$$

From Table E.3, $t_{18} = 2.1009$. Thus

$$\hat{Y}_i \pm t_{n-2} S_{YX} \sqrt{h_i}$$

where

$$h_i = \frac{1}{n} + \frac{(X_i - \overline{X})^2}{\sum_{i=1}^{n} X_i^2 - n\overline{X}^2}$$

so that we have

$$\hat{Y}_i \pm t_{n-2} S_{YX} \sqrt{\frac{1}{n} + \frac{(X_i - \overline{X})^2}{\sum_{i=1}^{n} X_i^2 - n\overline{X}^2}}$$

and

$$7.661 \pm (2.1009)(.497)\sqrt{\frac{1}{20} + \frac{(600 - 731.15)^2}{11{,}306{,}209 - 20(731.15)^2}}$$

$$= 7.661 \pm (1.044)\sqrt{\frac{1}{20} + \frac{(-131.15)^2}{11{,}306{,}209 - 10{,}691{,}606}}$$

$$= 7.661 \pm (1.044)\sqrt{.078}$$

$$= 7.661 \pm .292$$

so

$$7.369 \leq \mu_{YX} \leq 7.953$$

Therefore, our estimate is that the average weekly sales is between 7.369 (that is, $7,369) and 7.953 (that is, $7,953) for stores with 600 customers.

Problems for Section 17.11

● 17.47 Referring to the pet food sales problem (pages 716, 725, and 728), set up a 90% confidence interval estimate of the average weekly sales for all stores that have 8 feet of shelf space for pet food. Data file PETFOOD.DAT

17.48 Referring to the site selection problem (pages 716, 725, and 728), set up a 95% confidence interval estimate of the average sales for stores with 4,000 square feet. Data file SITE.DAT

● 17.49 Referring to the production worker-hours problem (pages 717, 725, and 728), set up a 90% confidence interval estimate of the average worker hours for all production runs with a lot size of 45. Data file WORKHRS.DAT

17.50 Referring to the tomato yield problem (pages 718, 725, and 728), set up a 90% confidence interval estimate of the average yield for all tomatoes that have been fertilized with 15 pounds per 100 square feet of natural organic fertilizer. Data file TOMYIELD.DAT

17.51 Referring to the interviewer productivity problem (pages 718, 726, and 728), set up a 95% confidence interval estimate of the average number of interviews completed for all interviewers with 20 weeks of experience. Data file INTERVW.DAT

17.52 Referring to the airport travel problem (pages 718, 726, and 728), set up a 95% confidence interval estimate of the average travel time for all distances of 21 miles. Data file LIMO.DAT

17.12 Prediction Interval for an Individual Response Y_I

In addition to the need to obtain a confidence-interval estimate for the average value, it is often important to be able to predict the response that would be obtained for an individual value. Although the form of the prediction interval is similar to the confidence-interval estimate of Equation (17.17), the prediction interval is estimating an individual value, not a parameter. Thus the prediction interval for an individual response Y_I at a particular value X_i is provided in Equation (17.18).

$$\hat{Y}_i \pm t_{n-2} S_{YX} \sqrt{1 + h_i} \tag{17.18}$$

where h_i, $\hat{Y}_i$, S_{YX}, n, and X_i are defined as in Equation (17.17) on page 747.

Suppose we desire a 95% prediction interval estimate of the weekly sales for an individual store with 600 customers. We compute the following:

$$\hat{Y}_i = 2.423 + .00873X_i$$

and for $X_i = 600$, $\hat{Y}_i = 7.661$.

Also

$$\bar{X} = 731.15 \qquad S_{YX} = .497$$

$$\sum_{i=1}^{n} X_i = 14{,}623 \qquad \sum_{i=1}^{n} X_i^2 = 11{,}306{,}209$$

From Table E.3, $t_{18} = 2.1009$. Thus

$$\hat{Y}_i \pm t_{n-2} S_{YX} \sqrt{1 + h_i}$$

so that

$$\hat{Y}_i \pm t_{n-2} S_{YX} \sqrt{1 + \frac{1}{n} + \frac{(X_i - \bar{X})^2}{\sum_{i=1}^{n} X_i^2 - n\bar{X}^2}}$$

and

$$7.661 \pm (2.1009)(.497)\sqrt{1 + \frac{1}{20} + \frac{(600 - 731.15)^2}{11{,}306{,}209 - 20(731.15)^2}}$$

$$= 7.661 \pm (1.044)\sqrt{1 + \frac{1}{20} + \frac{(-131.15)^2}{11{,}306{,}209 - 10{,}691{,}606}}$$

$$= 7.661 \pm (1.044)\sqrt{1.078}$$

$$= 7.661 \pm 1.084$$

so

$$6.577 \le Y_I \le 8.745$$

Therefore with 95% confidence, our estimate is that the weekly sales for an individual store that has 600 customers is between 6.577 (that is, \$6,577) and 8.745

(that is, \$8,745). We note that this prediction interval is much wider than the confidence interval estimate obtained in Section 17.11 for the average value.

Problems for Section 17.12

● 17.53 Referring to Problem 17.47 on page 749 (the pet food sales problem) — Data file PETFOOD.DAT
(a) Set up a 90% prediction interval of the weekly sales of an individual store that has 8 feet of shelf space for pet food.
(b) Explain the difference in the results obtained in (a) and those obtained in Problem 17.47.

17.54 Referring to Problem 17.48 on page 749 (the site selection problem) — Data file SITE.DAT
(a) Set up a 95% prediction interval of the sales of an individual store that has 4,000 square feet.
(b) Explain the difference in the results obtained in (a) and those obtained in Problem 17.48.

● 17.55 Referring to Problem 17.49 on page 749 (the production worker-hours problem), set up a 90% prediction interval of the number of worker-hours for a single lot size of 45. — Data file WRKHRS.DAT

17.56 Referring to Problem 17.50 on page 749 (the tomato yield problem), set up a 90% prediction interval of the yield of tomatoes for an individual plot that has been fertilized with 25 pounds per 100 square feet of natural organic fertilizer. — Data file TOMYIELD.DAT

17.57 Referring to Problem 17.51 on page 749 (the interviewer productivity problem), set up a 95% prediction interval of the number of interviews completed for an interviewer with 20 weeks of experience. — Data file INTERVW.DAT

17.58 Referring to Problem 17.52 on page 749 (the airport travel problem), set up a 95% prediction interval of the travel time for an individual trip of 21 miles. — Data file LIMO.DAT

17.13 Inferences about the Population Parameters in Regression and Correlation

In the preceding two sections we used statistical inference to develop a confidence interval estimate for μ_{YX}, the true mean value of Y, and a prediction interval for Y_I, an individual observation. In this section, statistical inference will be used to draw conclusions about the population slope β_1 and the population correlation coefficient ρ.

We can determine whether a significant relationship between the variables X and Y exists by testing whether β_1 (the true slope) is equal to zero. If this hypothesis is rejected, one could conclude that there is evidence of a linear relationship. The null and alternative hypotheses could be stated as follows:

H_0: $\beta_1 = 0$ (There is no relationship.)

H_1: $\beta_1 \neq 0$ (There is a relationship.)

and the test statistic for this is given by

$$t = \frac{b_1 - \beta_1}{S_{b_1}} \quad (17.19)$$

where

$$S_{b_1} = \frac{S_{YX}}{\sqrt{\sum_{i=1}^{n} X_i^2 - n\overline{X}^2}}$$

and the test statistic t follows a t distribution having $n - 2$ degrees of freedom.

Turning to our package delivery problem, let us now test whether the sample results enable us to conclude that a significant relationship between the number of customers and the weekly sales exists at the .05 level of significance. The results from Sections 17.4 and 17.5 gave the following information:

$$b_1 = +.00873 \qquad n = 20 \qquad S_{YX} = .497$$

$$\overline{X} = 731.15 \qquad \sum_{i=1}^{n} X_i^2 = 11,306,209$$

Therefore, to test the existence of a relationship at the .05 level of significance we have (see Figure 17.15)

$$S_{b_1} = \frac{S_{YX}}{\sqrt{\sum_{i=1}^{n} X_i^2 - n\overline{X}^2}}$$

$$= \frac{.497}{\sqrt{11,306,209 - 20(731.15)^2}} = \frac{.497}{\sqrt{614,603}} = .000634$$

and, under the null hypothesis, $\beta_1 = 0$ so that

$$t = \frac{b_1}{S_{b_1}}$$

$$= \frac{.00873}{.000634} = 13.77$$

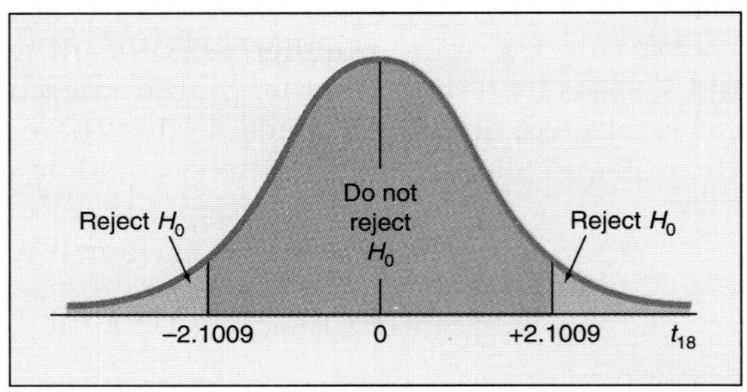

Figure 17.15
Testing a hypothesis about the population slope at the .05 level of significance with 18 degrees of freedom.

Since $t = 13.77 > t_{18} = 2.1009$, we reject H_0. Hence, we can conclude that there is a significant linear relationship between average weekly sales and number of customers.

A second, equivalent method for testing the existence of a linear relationship between the variables is to set up a confidence-interval estimate of β_1 and to determine whether the hypothesized value ($\beta_1 = 0$) is included in the interval. The confidence-interval estimate of β_1 would be obtained by using the following formula:

$$b_1 \pm t_{n-2} S_{b_1} \qquad (17.20)$$

If there were a 95% confidence interval estimate desired here, we would have $b_1 = +.00873$, $t_{18} = 2.1009$, and $S_{b_1} = .000634$. Thus

$$b_1 \pm t_{n-2} S_{b_1} = +.00873 \pm (2.1009)(.000634)$$
$$= +.00873 \pm .00133$$
$$+.0074 \leq \beta_1 \leq +.01006$$

From Equation (17.20) the true slope is estimated with 95% confidence to be between +.0074 and +.01006 (that is, \$7.40 to \$10.06). Since these values are above zero, we can conclude that there is a significant linear relationship between weekly sales and number of customers.

On the other hand, had the interval included zero, no relationship would have been determined.

A third method for examining the existence of a linear relationship between two variables involves the sample correlation coefficient r. The existence of a relationship between X and Y, which was tested using Equation (17.19), could be tested in terms of the correlation coefficient with equivalent results. Testing for the existence of a linear relationship between two variables is the same as determining whether there is any significant correlation between them. The population correlation coefficient ρ is hypothesized as equal to zero. Thus the null and alternative hypotheses would be

H_0: $\rho = 0$ (There is no correlation.)

H_1: $\rho \neq 0$ (There is correlation.)

The test statistic for determining the existence of a significant correlation is given by

$$t = \frac{r - \rho}{\sqrt{\dfrac{1 - r^2}{n - 2}}} \qquad (17.21)$$

where the test statistic t follows a t distribution having $n - 2$ degrees of freedom.

In order to demonstrate that this statistic produces the same result as the test for the existence of a slope [Equation (17.19)], we will use the package delivery data. For these data $r = +.956$, $r^2 = .913$, and $n = 20$ so testing the null hypothesis we have

$$t = \frac{r}{\sqrt{\dfrac{1 - r^2}{n - 2}}}$$

$$t = \frac{.956}{\sqrt{\dfrac{1 - .913}{20 - 2}}} = 13.75$$

We may note that this t value is, except for possible rounding error, the same as that obtained by using Equation (17.19). Therefore, in a linear regression analysis Equations (17.19) and (17.21) give equivalent alternative ways of determining the existence of a relationship between two variables. However, if the sole purpose of a particular study is to determine the existence of correlation, then Equation (17.21) is more appropriate. For instance, in Section 17.7 we studied the association of the price of a six-pack of a brand name cola soft drink and the price of chicken. Had we wanted to determine the significance of the correlation between these two variables, we could have used Equation (17.21) as follows:

H_0: $\rho = 0$ (There is no correlation.)

H_1: $\rho \neq 0$ (There is correlation.)

If a level of significance of .05 was selected, we would have (see Figure 17.16)

$$t = \frac{r}{\sqrt{\dfrac{1 - r^2}{n - 2}}}$$

$$= \frac{.883}{\sqrt{\dfrac{1 - (.883)^2}{9 - 2}}} = \frac{.883}{.1774} = +4.98$$

Since $t = 4.98 > t_7 = 2.3646$, we reject H_0.

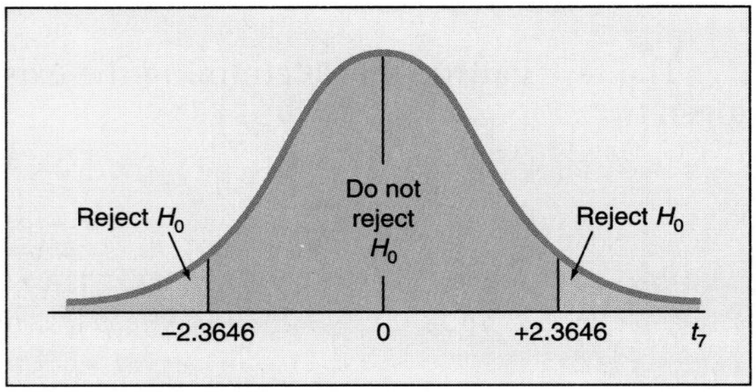

Figure 17.16
Testing for the existence of correlation at the .05 level of significance with 7 degrees of freedom.

Since the null hypothesis has been rejected, we would conclude that there is evidence of an association between the price of a six-pack of a brand name cola soft drink and the price of chicken.

When inferences concerning the population slope were discussed, confidence intervals and tests of hypothesis were used interchangeably. However, when examining the correlation coefficient, the development of a confidence interval becomes more complicated because the shape of the sampling distribution of the statistic r varies for different values of the true correlation coefficient. Methods for developing a confidence interval estimate for the correlation coefficient are presented in References 5 and 12.

Problems for Section 17.13

● 17.59 Referring to the pet food sales problem (pages 716, 725, and 728), at the .10 level of significance, is there evidence of a linear relationship between shelf space and sales? Data file PETFOOD.DAT

17.60 Referring to the site selection problem (pages 716, 725, and 728), at the .05 level of significance, is there evidence of a linear relationship between annual sales and square footage? Data file SITE.DAT

● 17.61 Referring to the production worker-hours problem (pages 717, 725, and 728), at the .10 level of significance, is there evidence of a linear relationship between lot size and worker-hours? Data file WORKHRS.DAT

17.62 Referring to the tomato yield problem (pages 718, 725, and 728), at the .10 level of significance, is there evidence of a linear relationship between the amount of fertilizer used and the yield of tomatoes? Data file TOMYIELD.DAT

17.63 Referring to the interviewer productivity problem (pages 718, 726, and 728), at the .05 level of significance, is there evidence of a relationship between weeks of experience and number of interviews completed? Data file INTERVW.DAT

17.64 Referring to the airport travel problem (pages 718, 726, and 728), at the .05 level of significance, is there evidence of a relationship between distance and travel time? Data file LIMO.DAT

17.65 Referring to Problem 17.34 on page 735, at the .01 level of significance, is there evidence of a linear relationship between the price of a six-pack of brand name cola soft drink and 100 tablets of a pain reliever? Data file SUPERMKT.DAT

17.66 Referring to Problem 17.35 on page 735, at the .10 level of significance, is there evidence of a linear relationship between the cost of a women's hairdresser and a men's dress shirt? Data file CITY.DAT

17.14 Regression Diagnostics: Influence Analysis

17.14.1 Introduction

Regression diagnostics deals with both the evaluation of the aptness of a particular model and the potential effect or *influence* of each particular point on that fitted model. In Section 17.19 we have utilized methods of residual analysis to study the aptness of our fitted model. In this section we will consider several methods that measure the influence of particular data points. Among a variety of recently developed criteria (see References 1, 4, 6, 9, 20) we shall consider the following:

1. The hat matrix elements, h_i
2. The Studentized deleted residuals, t_i^*
3. Cook's distance statistic, D_i

Table 17.9 on page 756 represents the values of these statistics for the package delivery data of Table 17.1, which have been obtained from MINITAB. We note from Table 17.9 that certain data points have been highlighted for further analysis.

Table 17.9 Influence statistics for the package delivery data.

Observation	Number of Customers X_i	Weekly Sales Y_i	Residual	h_i	Studentized Deleted Residual, t_i^*	Cook's D_i
1	907	11.20	0.859	0.100314	1.94065	0.181993
2	926	11.05	0.544	0.111774	1.16119	0.083228
3	506	6.84	−0.000	0.132480	−0.00019	0.000000
4	741	9.21	0.319	0.050158	0.64093	0.011213
5	789	9.42	0.110	0.055445	0.21866	0.001482
6	889	10.08	−0.103	0.090541	−0.21044	0.002328
7	874	9.45	−0.602	0.083202	−1.27646	0.071436
8	510	6.73	−0.145	0.129576	−0.30200	0.007149
9	529	7.24	0.199	0.116490	0.41260	0.011765
10	420	6.12	0.031	0.207523	0.06669	0.000616
11	679	7.63	−0.720	0.054425	−1.53115	0.062781
12	872	9.43	−0.605	0.082279	−1.28165	0.071097
13	924	9.46	−1.029	0.110512	−2.46260	0.294008
14	607	7.64	−0.082	0.075078	−0.16486	0.001166
15	452	6.92	0.551	0.176789	1.22864	0.157629
16	729	8.95	0.163	0.050008	0.32562	0.002937
17	794	9.33	−0.024	0.056427	−0.04816	0.000073
18	844	10.23	0.439	0.070721	0.90428	0.031434
19	1,010	11.77	0.530	0.176516	1.17779	0.145544
20	621	7.41	−0.434	0.069741	−0.89209	0.030173

17.14.2 The Hat Matrix Elements, h_i

We may recall from Section 17.11 that when we developed a confidence-interval estimate μ_{YX} we defined the **hat matrix diagonal elements, h_i,** as

$$h_i = \frac{1}{n} + \frac{(X_i - \overline{X})^2}{\sum_{i=1}^{n} X_i^2 - n\overline{X}^2} \qquad (17.22)$$

Each h_i reflects the influence of each X_i on the fitted regression model. If such influential points are present, we may need to reevaluate the necessity for keeping them in the model. In simple linear regression[3] Hoaglin and Welsch (see Reference 9) suggest the following decision rule:

> If $h_i > 4/n$, then X_i is an influential point and may be considered a candidate for removal from the model.

For our package delivery data, since $n = 20$, our criterion would be to "flag" any h_i value greater than 4/20 = .200. Referring to Table 17.9, we note that the tenth value of h_i (X_{10}) is .2075. This tenth X observation then is a potential candidate for removal from our package delivery model. However, other criteria for measuring influence must be considered prior to making such a decision.

17.14.3 The Studentized Deleted Residuals, t_i^*

In our discussion of residual analysis in Section 17.9, we defined the standardized residuals in Equation (17.15) as

$$SR_i = \frac{e_i}{S_{YX}\sqrt{1 - h_i}}$$

In an effort to better measure the adverse impact of each individual case on the model, Hoaglin and Welsch (see Reference 9) also developed the Studentized deleted residual t_i^* given in Equation (17.23):

$$t_i^* = \frac{e_{(i)}}{S_{(i)}\sqrt{1 - h_i}} \qquad (17.23)$$

where $e_{(i)}$ = the difference between the observed Y_i and $\hat{Y}_i$ based on a model that includes all observations except observation i.

$S_{(i)}$ = the standard error of the estimate for a model that includes all observations except observation i.

Thus this Studentized deleted residual measures the difference of each observation Y_i from that predicted by a model that includes all other observations. For example, t_1^* represents a measure of the difference between the actual weekly sales for the first store ($Y_1 = 11.20$) and the weekly sales that would be predicted for this store based on a model that included only the second through the twentieth stores. In simple linear regression, Hoaglin and Welsch suggest that if

$$|t_i^*| > t_{.10, n-3}$$

then this would mean that the observed and predicted Y values are so different that X_i is an influential point that adversely affects the model and may be considered a candidate for removal.

For our package delivery data, since $n = 20$, our criterion would be to flag any t_i^* value greater than 1.7396 (see Table E.3). Referring to Table 17.9, we note that $t_1^* = 1.941$ and $t_{13}^* = -2.463$. Thus, the first and thirteenth stores may each have an adverse effect on the model. We note that the tenth observation was flagged according to the h_i criterion but the first and thirteenth were not. Hence, with this lack of consistency we should consider another criterion, Cook's D_i, which is based on both the h_i and the standardized residual statistics.

17.14.4 Cook's Distance Statistic, D_i

The use of h_i and t_i^* in the search for potential troublesome data points is complementary. Neither criterion is sufficient by itself. When h_i is small, t_i^* may be large (see observations 1 and 13). On the other hand, when h_i is large, t_i^* may be moderate or small because the observed Y_i is consistent with the model and the rest of the data. To decide whether a point that has been flagged by either the h_i or t_i^* criterion is unduly affecting the model, Cook and Weisberg (see Reference 6) suggest the use of the D_i statistic. In the simple linear regression model[4] D_i is shown in Equation (17.24):

$$D_i = \frac{SR_i^2 h_i}{2(1 - h_i)} \tag{17.24}$$

where SR_i is the standardized residual of Equation (17.15).

In simple linear regression, Cook and Weisberg suggest that if

$$D_i > F_{.50,2,n-2}$$

this would mean that the observation may have an impact on the results of fitting the linear regression model.

For our package delivery data, since $n = 20$, our criterion would be to flag any $D_i > F_{.50,2,18} = .720$. (See Table E.5a.) Referring to Table 17.9 on page 756, we note that there are no D_i values that meet this criterion. Since these results are not consistent with those obtained from the h_i and t_i^* criteria, there is no clear basis for removing any of the observations from the fitted regression model.

17.14.5 Summary

In this section we have discussed several criteria for evaluating the influence of each observation on the regression model. As we have noted, the various statistics often do not yield consistent results. Under such circumstances, most statisticians would conclude that there is insufficient evidence for the removal of such observations from the model.

In addition to the three criteria presented here, other measures of influence have been developed (see References 1 and 10). While different researchers seem to prefer particular measures, currently there is no consensus as to the "best" measures. Hence, only when there is consistency in a selected set of measures is it appropriate to consider the removal of particular observations.

In conclusion, we should also realize that because of the computations involved in both residual analysis and influence analysis, diagnostic evaluation is not practical without the aid of a computer package. However, as Tukey (see Reference 18) has noted, the actual decision concerning the deletion of any observation is best left in the hands of the user rather than delegating such a decision to the computer package itself.

Problems for Section 17.14

17.67 What is the difference between residual analysis and influence analysis?

17.68 Explain the difference between the h_i measure and t_i^*.

For the data of Problems 17.69–17.74 perform an influence analysis and determine whether any observations may be deleted from the model. If necessary, reanalyze the regression model after deleting these observations and compare your results with the original model.

● 17.69 Refer to the pet food sales problem (pages 716, 725, and 728). Data file PETFOOD.DAT

17.70 Refer to the site selection problem (pages 716, 725, and 728). Data file SITE.DAT

● 17.71 Refer to the production worker-hours problem (pages 717, 725, and 728). Data file WORKHRS.DAT

17.72 Refer to the tomato yield problem (pages 718, 725, and 728). Data file TOMYIELD.DAT

17.73 Refer to the interviewer productivity problem (pages 718, 726, and 728). Data file INTERVW.DAT

17.74 Refer to the airport travel problem (pages 718, 726, and 728). Data file LIMO.DAT

17.15 Regression, Computers, and the Employee Satisfaction Survey

17.15.1 Introduction

When we discussed descriptive statistics and hypothesis testing, we used the Employee Satisfaction Survey to illustrate the role of the computer as an aid in data analysis. The role of computer software becomes even more important when applied to regression and correlation analysis and, in particular, to problems in multiple regression that will be discussed in Chapter 18. It is reasonable to state that with the development of residual analysis and influence analysis techniques, the role of the computer has become crucial even when only a simple linear regression model is being considered.

17.15.2 Using SAS, STATISTIX, and MINITAB for Regression Analysis

Data file EMPSAT.DAT

To first demonstrate the role of the computer in regression and correlation analysis, let us refer to the complete data set of Table 17.1 on page 715. If the SAS package was being accessed (see Reference 16), procedures such as PLOT and REG could be utilized. Figure 17.17 represents partial output from PROC REG for the package delivery data. We may note that in addition to the various regression statistics such

```
DEP VARIABLE: SALES
ANALYSIS OF VARIANCE
                              SUM OF              MEAN
SOURCE        DF             SQUARES            SQUARE    F VALUE      PROB>F
MODEL          1       SSR 46.83354090       46.83354090   186.219      0.0001
ERROR         18       SSE  4.52695410        0.25149745
C TOTAL       19       SST 51.36049500
     ROOT MSE                0.5014952        R-SQUARE      0.9119
     DEP MEAN                   8.8055        ADJ R-SQ      0.9070
     C.V.                      5.69525
PARAMETER ESTIMATES
                             PARAMETER             STANDARD          T FOR H0:
VARIABLE        DF            ESTIMATE                ERROR        PARAMETER=0
INTERCEP         1     b0   2.42304440      0.48096461 Sb0               5.038
CUSTOMER         1     b1  0.008729338     0.000639690 Sb1              13.646
VARIABLE           PROB>|T|                 TYPE I SS               TYPE II SS
INTERCEP             0.0001                1550.73660               6.38307706
CUSTOMER             0.0001               46.83354090              46.83354090
```

Figure 17.17
SAS output for the package delivery problem.

as the regression coefficients, standard errors, and r^2, we may obtain several statistics that relate to residual analysis and influence analysis.

In a similar manner, the STATISTIX package (see Reference 17) may also be accessed for regression and correlation analysis. Figure 17.18 represents partial output for the package delivery data. We may note that in addition to the regression coefficients, standard errors, and r^2 that are displayed, we may also obtain several statistics that relate to residual analysis and influence analysis.

```
STATISTIX 4.0

UNWEIGHTED LEAST SQUARES LINEAR REGRESSION OF SALES

PREDICTOR
VARIABLES        COEFFICIENT       STD ERROR       STUDENT'S T        P

CONSTANT         b0  2.42304        0.48096          5.04          0.0001
CUSTOMER         b1  0.00872       6.397E-04        13.65          0.0000

R-SQUARED              r2  0.9119     RESID. MEAN SQUARE (MSE)   0.25149
ADJUSTED R-SQUARED         0.9070     STANDARD DEVIATION         0.50149 Syx

SOURCE             DF            SS              MS           F          P

REGRESSION          1     SSR 46.8335         46.8335      186.22     0.0000
RESIDUAL           18     SSE 4.52695         0.25149
TOTAL              19         51.3604
```

Figure 17.18 **STATISTIX output for the package delivery problem.**

If the MINITAB package was being accessed (see References 11 and 15) for a regression analysis, Figure 17.19 represents partial output of the package delivery data. We observe that the regression coefficients, standard errors, and r^2 are pro-

```
The regression equation is
Sales = 2.42 + 0.00873 Customer

Predictor         Coef          Stdev      t-ratio        p
Constant  b0    2.4230         0.4810 Sb0     5.04      0.000
Customer  b1 0.0087293      0.0006397 Sb1    13.65      0.000

s = 0.5015      R-sq = 91.2%      R-sq(adj) = 90.7%

Analysis of Variance

SOURCE         DF          SS            MS          F        p
Regression      1   SSR 46.834       46.834     186.22    0.000
Error          18   SSE  4.527        0.251
Total          19   SST 51.360
```

Figure 17.19
MINITAB output for the package delivery problem.

vided here. The various statistics related to residual analysis and influence analysis were already displayed in Table 17.9 on page 756.

We note from the highlighted regression statistics displayed in Figures 17.17–17.19 some discrepancies with the results presented earlier in the chapter. These differences are due to rounding errors. The results from the computer packages are more precise.

17.15.3 Computers and the Employee Satisfaction Survey

Now that we have illustrated how packages such as SAS, STATISTIX, and MINITAB can be used in regression analysis, we return to the Employee Satisfaction Survey. Suppose that Bud Conley, the Vice-President for Human Resources, would like to develop a statistical model to forecast the personal income of Kalosha Industries employees. Although realistically one may need to include several explanatory variables in an analysis, he has decided to begin by using only a single independent variable for prediction—the number of years the employee has worked at a full-time job. He has also decided to do the analysis for separate groups of employees based on their occupation. The analysis that we will discuss here concerns only the 57 employees whose occupation is classified as technical/sales.

We begin our analysis by examining Figure 17.20, a scatter diagram of these two variables provided by the MINITAB package. We observe that there appears to be an increasing relationship between number of years in the workforce and income. Although there is some variability in the plotted data, it seems reasonable to start our regression analysis by assuming a linear relationship between the two variables. Using the MINITAB package, we obtain the output presented in Figure 17.21 on page 762.

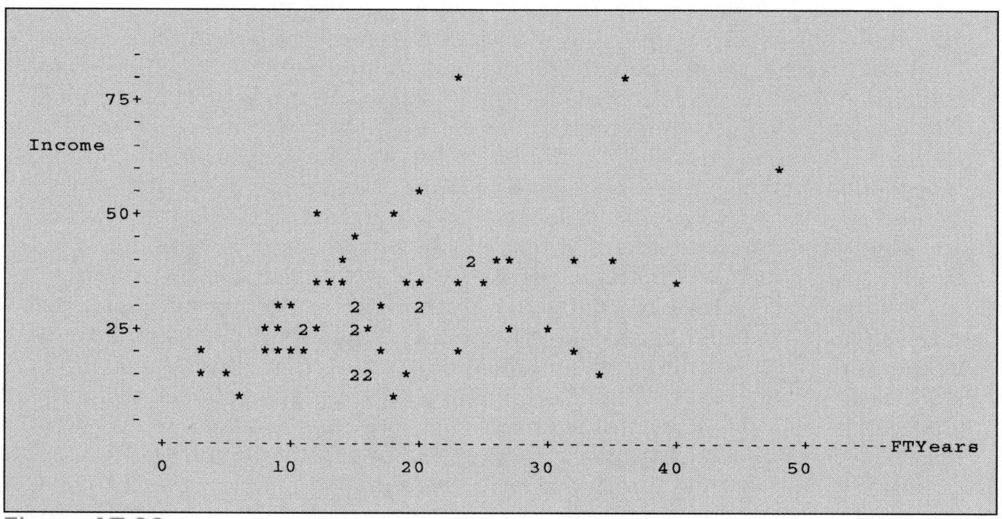

Figure 17.20
Scatter diagram obtained from MINITAB for the 57 employees whose occupation is classified as technical/sales.

We first note that the regression equation is

$$\hat{Y}_i = 17.195 + 0.73X_i$$

```
The regression equation is
Income = 17.2 + 0.730 FTYears

Predictor        Coef        Stdev     t-ratio        p
Constant       17.195        3.757        4.58    0.000
FTYears        0.7303       0.1834        3.98    0.000

s = 12.90       R-sq = 22.4%       R-sq(adj) = 21.0%

Analysis of Variance

SOURCE        DF        SS          MS         F        p
Regression     1     2639.2      2639.2     15.86    0.000
Error         55     9151.7       166.4
Total         56    11791.0
```

Figure 17.21
MINITAB output for the 57 employees whose occupation is classified as technical/sales.

where Y = yearly income in thousands of dollars
X = number of years in the workforce (full-time).

We may interpret the slope 0.73 to mean that for each additional one year in the workforce, the predicted average yearly income increases by .73 thousand dollars or \$730. We note that, for these 57 technical/sales employees, X varies from 3 to 48 years, so that predictions within this range of full-time workforce experience may be undertaken. The Y intercept 17.195 represents the predicted average annual income for an employee without any full-time experience in the workforce. Although this interpretation is appropriate in general, for our data we cannot make any predictions for employees without experience, since the smallest amount of experience of any of the technical/sales employees was three years.

Now that the regression model has been fit and the regression coefficients have been interpreted, we use residual analysis to determine the aptness of the model. First we may examine several plots of the standardized residuals (see Section 17.9). Figure 17.22 represents a MINITAB plot of the standardized residuals versus the independent variable, number of years in the workforce. We observe from Figure 17.22 that there appears to be little or no pattern in this plot; there are both high and low standardized residuals at many different levels of X. In addition, we observe little evidence of heterogeneity of variance at different levels of X.

We may also evaluate the normality assumption by plotting the standardized residuals as in Figure 17.23. From Figure 17.23 we note some departure from normality, with right skewness of the residuals apparent from the histogram with two large positive values of 3.09 and 3.45. This would give us cause for concern about whether the simple straight-line model was sufficiently useful for predicting income or whether a data transformation should be considered (see Section 18.13).

If we wish to continue with our analysis, we may evaluate whether there is a statistically significant linear relationship between these two variables. From Figure 17.21 we observe that r^2 is .224 or 22.4%. Thus 22.4% of the variation in income can be explained by variation in the number of years in the workforce. We note that the t statistic for the significance of the slope is 3.98, which, with $57 - 2 = 55$ degrees of freedom, is clearly significant even at the .01 level ($t = 3.98 > t_{55} = 2.6682$). The p value is .000. Thus there is reason to believe that there is evidence of a linear relationship between the two variables.

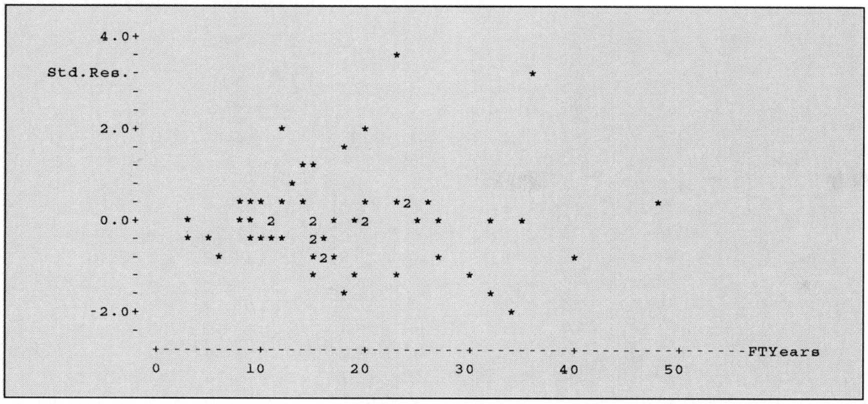

Figure 17.22
Residual plot obtained from MINITAB for the regression model of Figure 17.21.

```
Histogram of Std.Res.    N = 57

Midpoint    Count
   -2.0         1  *
   -1.5         2  **
   -1.0        11  ***********
   -0.5         7  *******
    0.0        19  *******************
    0.5        10  **********
    1.0         1  *
    1.5         2  **
    2.0         2  **
    2.5         0
    3.0         1  *
    3.5         1  *
```

Figure 17.23
Histogram of standardized residuals obtained from MINITAB for the data of Figure 17.21.

Practical Significance versus Statistical Significance Therefore, we have what seems to be an anomaly, a relatively low r^2 that is highly significant. This represents a difference between what may be *practically significant* and useful to a manager such as Bud Conley and what is *statistically significant.* The signifi cance of r^2 as tested using Equation (17.19) or (17.21) depends on the value of r and also the sample size. It is conceivable that if we take a large enough sample size, even an r^2 value of .01 could be highly statistically significant, but of little practical importance.

Although the relationship in our model was statistically significant, since r^2 was relatively small, more than 75% of the variation in income is explained by factors other than number of years in the workforce. This might lead one to consider using additional explanatory variables in a multiple regression model or a curvilinear regression model (see Chapter 18).

Once our preliminary regression analysis has been completed, we may wish to study the influence of individual observations on the model.

Although we have evidence that the model is not a strong fitting one, we may continue with our diagnostic analysis through the use of influence measures. This approach will enable us to determine whether any observations are unduly affecting the model. Figure 17.24 represents additional MINITAB output obtained for our model. Included are the standardized residual, the predicted Y value, h_i value, Studentized deleted residual, and Cook's D_i statistic for each observation.

ROW	Income	FTYears	Std.Res.	Yhat	h	tresids	cookd
1	20.2	3	0.06529	19.3854	0.064513	0.06469	0.000147
2	35.7	40	-0.88137	46.4065	0.113178	-0.87956	0.049570
3	33.3	12	0.57654	25.9581	0.025426	0.57301	0.004336
4	32.0	20	0.01561	31.8005	0.018166	0.01546	0.000002
5	35.7	20	0.30508	31.8005	0.018166	0.30255	0.000861
6	33.8	14	0.50002	27.4187	0.021186	0.49658	0.002706
7	30.3	15	0.16841	28.1490	0.019673	0.16692	0.000285
8	20.4	11	-0.37965	25.2278	0.028153	-0.37668	0.002088
9	18.4	23	-1.22228	33.9914	0.022112	-1.22791	0.016891
10	11.8	18	-1.45005	30.3399	0.017556	-1.46509	0.018787
11	40.2	14	1.00151	27.4187	0.021186	1.00154	0.010855
12	22.0	8	-0.08199	23.0369	0.038757	-0.08125	0.000136
13	78.0	23	3.45003	33.9914	0.022112	3.86184	0.134570
14	13.7	15	-1.13131	28.1490	0.019673	-1.13426	0.012842
15	40.8	35	-0.15752	42.7550	0.074269	-0.15612	0.000995
16	23.0	15	-0.40315	28.1490	0.019673	-0.40006	0.001631
17	10.3	6	-0.89587	21.5763	0.047847	-0.89424	0.020165
18	22.7	15	-0.42664	28.1490	0.019673	-0.42345	0.001826
19	81.7	36	3.09072	43.4853	0.081243	3.36901	0.422352
20	55.3	20	1.83852	31.8005	0.018166	1.88043	0.031270
21	16.0	19	-1.17874	31.0702	0.017659	-1.18302	0.012488
22	33.1	19	0.15876	31.0702	0.017659	0.15735	0.000227
23	25.1	30	-1.11115	39.1035	0.045464	-1.11357	0.029403
24	16.1	3	-0.26333	19.3854	0.064513	-0.26109	0.002391
25	25.2	9	0.11306	23.7672	0.034818	0.11204	0.000231
26	30.6	15	0.19190	28.1490	0.019673	0.19021	0.000370
27	41.5	26	0.41850	36.1823	0.029695	0.41534	0.002680
28	38.1	32	-0.19659	40.5641	0.055774	-0.19486	0.001141
29	24.7	27	-0.96279	36.9126	0.033031	-0.96214	0.015832
30	13.6	5	-0.57724	20.8460	0.052998	-0.57371	0.009324
31	45.7	15	1.37419	28.1490	0.019673	1.38563	0.018947
32	39.0	27	0.16456	36.9126	0.033031	0.16310	0.000463
33	26.9	11	0.13150	25.2278	0.028153	0.13032	0.000250
34	17.3	16	-0.90611	28.8793	0.018563	-0.90461	0.007765
35	38.3	24	0.28082	34.7217	0.024235	0.27846	0.000979
36	34.4	13	0.60485	26.6884	0.023104	0.60133	0.004326
37	23.1	12	-0.22444	25.9581	0.025426	-0.22249	0.000657
38	26.0	8	0.23429	23.0369	0.038757	0.23227	0.001107
39	23.4	11	-0.14374	25.2278	0.028153	-0.14245	0.000299
40	36.6	25	0.09021	35.4520	0.026763	0.08939	0.000112
41	25.0	16	-0.30357	28.8793	0.018563	-0.30105	0.000871
42	27.5	17	-0.16502	29.6096	0.017857	-0.16356	0.000248
43	58.0	48	0.49736	52.2489	0.196449	0.49393	0.030238
44	20.8	9	-0.23414	23.7672	0.034818	-0.23212	0.000989
45	51.6	12	2.01360	25.9581	0.025426	2.07308	0.052892
46	31.2	20	-0.04698	31.8005	0.018166	-0.04655	0.000020
47	31.3	9	0.59440	23.7672	0.034818	0.59087	0.006373
48	17.6	10	-0.54328	24.4975	0.031283	-0.53977	0.004766
49	36.9	23	0.22802	33.9914	0.022112	0.22604	0.000588
50	17.4	34	-1.97708	42.0247	0.067700	-2.03258	0.141922
51	17.8	17	-0.92380	29.6096	0.017857	-0.92255	0.007758
52	38.4	24	0.28867	34.7217	0.024235	0.28625	0.001035
53	16.5	15	-0.91208	28.1490	0.019673	-0.91067	0.008347
54	48.4	18	1.41252	30.3399	0.017556	1.42572	0.017827
55	21.9	32	-1.48902	40.5641	0.055774	-1.50609	0.065482
56	16.4	16	-0.97654	28.8793	0.018563	-0.97612	0.009018
57	28.9	10	0.34676	24.4975	0.031283	0.34397	0.001942

Figure 17.24 **MINITAB output for influence analysis.**

We may begin our influence analysis by examining the h_i statistic. Using the Hoaglin-Welsch criterion ($h_i > 4/n$), with $n = 57$, we would consider an observation influential if $h_i > 4/57 = .07$. From Figure 17.24 we observe that observations 2, 15, 19, and 43 (whose h_i values equal .113, .074, .081, and .196, respectively) exceed this criterion. Thus, based on the h_i criterion, observations 2, 15, 19, and 43 are possible candidates for removal from the model. However, other criteria for measuring influence must be considered prior to making such a decision.

The second criterion for measuring influence is the t_i^* statistic that involves the Studentized deleted residuals. Using the Hoaglin-Welsch criterion, ($|t_i^*| > t_{.10,n-3}$), for $n = 57$ we would consider an observation influential if $|t_i^*| > 1.6736$. From Figure 17.24 we observe that t_i^* equals 3.86 for observation 13, 3.37 for observation 19, 1.88 for observation 20, 2.07 for observation 45, and −2.03 for observation 50. We also observe that t_i^* is −.88 for observation 2, −.15 for observation 15, and .49 for observation 43. Thus observations 2, 15, and 43, which had high h_i values, did not adversely affect the model. However, observation 19 has been flagged according to both criteria, leading us to believe that this point may be unduly influencing the model. This provides even more evidence that other variables need to be evaluated for inclusion in a possible multiple regression model, so that the large residual apparent in this observation might be reduced.

In order to complete our influence analysis we need to look at Cook's D_i statistic, which measures the combined effect of h_i and SR_i. Using Cook and Weisberg's criterion for linear regression ($D_i > F_{.50,2,n-2}$), for $n = 57$ we should consider an observation influential if $D_i > .701$. Referring to Figure 17.24, we note that there are no D_i values that meet this criterion. However, the largest value is for observation 19 (.422). These results are not consistent with those obtained from the h_i and t_i^* criteria. Hence we would have no clear basis for removing any of the observations from the fitted regression model. Regardless of whether any observations are to be deleted from this model, it is clear that, at best, this model is only a marginally useful predictor of income and that other independent variables and/or a nonlinear relationship should be investigated for possible inclusion in a model. This subject will be the topic of our next chapter.

17.16 Pitfalls in Regression and Ethical Issues

17.16.1 Introduction

Regression and correlation analysis are perhaps the most widely used and, unfortunately, the most widely abused statistical techniques that are applied to business and economics. The difficulties that arise frequently come from the following sources:

1. Lacking an awareness of the assumptions of least squares regression.
2. Knowing how to evaluate the assumptions of least squares regression.
3. Knowing what are the alternatives to least squares regression if a particular assumption is violated.
4. Thinking that correlation implies causation.
5. Using a regression model without knowledge of the subject matter.

17.16.2 The Pitfalls of Regression

The widespread availability of spreadsheet and statistical software has removed the computational block that prevented many users from applying regression analysis to situations that required forecasting. With this positive development of availability comes the realization that, for many users, the access to powerful techniques has not been accompanied by an understanding of how to use regression analysis properly. How can a user be expected to know what the alternatives to least squares regression are if a particular assumption is violated, when he or she in many instances is not even aware of the assumptions of regression, let alone how the assumptions can be evaluated?

The necessity of going beyond the basic number crunching—of computing the Y intercept, the slope, and r^2—can be illustrated by referring to Table 17.10, a classical pedagogical piece of statistical literature that deals with the importance of observation through scatter plots and residual analysis.

Table 17.10 Four sets of artificial data.

Set A		Set B		Set C		Set D	
X_i	Y_i	X_i	Y_i	X_i	Y_i	X_i	Y_i
10	8.04	10	9.14	10	7.46	8	6.58
14	9.96	14	8.10	14	8.84	8	5.76
5	5.68	5	4.74	5	5.73	8	7.71
8	6.95	8	8.14	8	6.77	8	8.84
9	8.81	9	8.77	9	7.11	8	8.47
12	10.84	12	9.13	12	8.15	8	7.04
4	4.26	4	3.10	4	5.39	8	5.25
7	4.82	7	7.26	7	6.42	19	12.50
11	8.33	11	9.26	11	7.81	8	5.56
13	7.58	13	8.74	13	12.74	8	7.91
6	7.24	6	6.13	6	6.08	8	6.89

Source: F. J. Anscombe, "Graphs in Statistical Analysis," *American Statistician,* Vol. 27 (1973), pp. 17–21.

Anscombe (Reference 2) showed that for the four data sets given in Table 17.10 the following results would be obtained:

$$\hat{Y}_i = 3.0 + .5X_i$$

$$S_{YX} = 1.236$$

$$S_{b_1} = .118$$

$$r^2 = .667$$

$$\text{SSR} = \text{explained variation} = \sum_{i=1}^{n} (\hat{Y}_i - \overline{Y}_i)^2 = 27.50$$

$$\text{SSE} = \text{unexplained variation} = \sum_{i=1}^{n} (Y_i - \hat{Y}_i)^2 = 13.75$$

$$\text{SST} = \text{total variation} = \sum_{i=1}^{n} (Y_i - \overline{Y})^2 = 41.25$$

Thus with respect to the pertinent statistics associated with a simple linear regression, the four data sets are identical. Had we stopped our analysis at this point, valuable information in the data would be lost.

Table 17.11 gives the standardized residuals e_i/S_{YX} for each of the data sets.

Table 17.11 Standardized residuals.

	Data Set A	Data Set B	Data Set C	Data Set D	
X_i	e_i/S_{YX}	e_i/S_{YX}	e_i/S_{YX}	X_i	e_i/S_{YX}
4	−.599	−1.536	.314	8	−.340
5	.145	−.614	.185	8	−1.003
6	1.002	.105	.064	8	.574
7	−1.359	.614	−.065	8	1.489
8	−.041	.922	−.186	8	1.189
9	1.059	1.027	−.315	8	.032
10	.032	.922	−.437	8	−1.416
11	−.138	.614	−.558	19	.000
12	1.487	.105	−.687	8	−1.165
13	−1.554	−.614	2.622	8	.736
14	−.033	−1.536	−.937	8	−.089

Source: F. J. Anscombe, "Graphs in Statistical Analysis," *American Statistician*, Vol. 27 (1973), pp. 17–21.

When the standardized residuals are plotted against $\hat{Y}$,[5] we see how different the data sets are. Panels A, B, C, and D of Figure 17.25 on page 768 graphically depict, for each data set, a plot of the standardized residuals against the fitted values $\hat{Y}$. While the plot for data set A does not show any obvious anomalies, this is not the case for data sets B, C, and D. The parabolic form of the residual plot for B probably indicates that the basic simple linear regression model should be augmented to include a curvilinear term, as will be developed in Section 18.10. The plot for data set C clearly depicts what may very well be an *outlying* observation. If this is the case, we may deem it appropriate to remove the outlier and reestimate the basic model. The result of this exercise would probably be a relationship much different from what was originally uncovered. Similarly, the plot for data set D would be evaluated cautiously because the fitted model is so dependent on the outcome of a single response ($X_8 = 19$ and $Y_8 = 12.50$).

In summary, residual plots are of vital importance to a complete regression analysis. The information they impart is so basic to a credible analysis that such plots should *always* be included as part of a regression analysis.

Thus, a strategy that might be employed to avoid the first three pitfalls of regression listed would involve the following approach:

1. Always start with a scatter plot to observe the possible relationship between X and Y.
2. Check the assumptions of regression after the regression model has been fit, *before* moving on to using the results of the model.
3. Plot the residuals (or standardized residuals) versus the independent variable. This will enable you to determine whether the model fit to the data is an appropriate one and will allow you to check visually for violations of the homoscedasticity assumption.

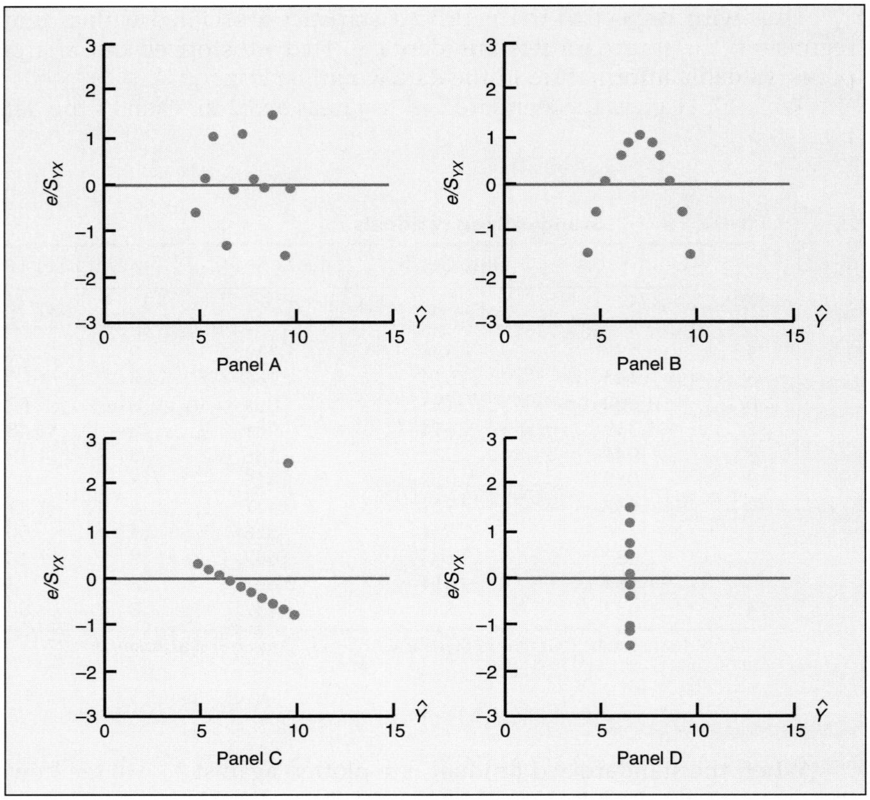

Figure 17.25
Plot of $\hat{Y}_i$ versus standardized residuals.
Source: F. J. Anscombe, "Graphs in statistical analysis," *American Statistician,* Vol. 27 (1973), pp. 17–21.

4. Use a histogram, stem-and-leaf display, box-and-whisker plot, or normal probability plot of the residuals to evaluate graphically whether the normality assumption has been seriously violated.
5. If the data have been collected in sequential order, plot the residuals in time order and compute the Durbin-Watson statistic.
6. Use influence analysis to determine whether any observations are outliers or are unduly influencing the model.
7. If the evaluation done in 3–6 indicates violations in the assumptions, use alternative methods to least squares regression or alternative least squares models (curvilinear or multiple regression), depending on what the evaluation has indicated.
8. If the evaluation done in 3–6 does not indicate violations in the assumptions, then the inferential aspects of the regression analysis can be undertaken. Confidence and prediction intervals can be developed and tests for the significance of the regression coefficients can be done.

Cautions In addition to the first three pitfalls considered above, two other pitfalls need to be mentioned. One involves the mistaken belief that correlation implies causation. In many instances the covariation between variables is spurious in that the relationship is actually caused by a third factor that has not been or cannot be measured.

Another pitfall involves the fact that a good-fitting model does not necessarily mean that the model can be used for prediction. An individual with knowledge of the subject matter would have to be convinced that the process that produced the data will remain stable in the future in order to use the model for predictive purposes.

17.16.3 Ethical Considerations

Ethical considerations arise when a user wishing to develop forecasts is manipulative of the process of developing the regression model. The key here is intent. Unethical behavior occurs when someone uses regression analysis to:

1. Forecast a response variable of interest with the willful intent of possibly excluding certain variables from consideration in the model.
2. Delete observations from the model to obtain a better model without giving reasons for deleting these observations.
3. Make forecasts without providing an evaluation of the assumptions when he or she knows that the assumptions of least squares regression have been violated.

All of these situations should make us realize even more the importance of following the steps given in Section 17.16.2 and knowing the assumptions of regression, how to evaluate them, and what to do if any of them have been violated.

17.17 Summary and Overview

As seen in the summary chart for this chapter (page 770), we developed the simple linear regression model, discussed the assumptions of the model, and how these assumptions could be evaluated. On page 714 of Section 17.1, you were given a list emphasizing the important points to be discussed in the chapter. Check over the list now to see whether you feel that you have an understanding of these key points. To be sure, you should be able to answer the following conceptual questions:

1. What is the interpretation of the Y intercept and the slope in a regression model?
2. What is the interpretation of the coefficient of determination?
3. Why should a residual analysis always be done as part of the development of a regression model?
4. What are the assumptions of regression analysis and how can they be evaluated?
5. What is the Durbin-Watson statistic and when and how should it be used in regression analysis?
6. What is the difference between a confidence interval estimate of the mean response μ_{YX} and a prediction interval estimate of Y_I?
7. What is the difference between residual analysis and influence analysis?
8. Under what circumstances can observations be considered for removal from the regression model?

In Chapter 18 we will continue our discussion of regression analysis by considering a variety of multiple regression models.

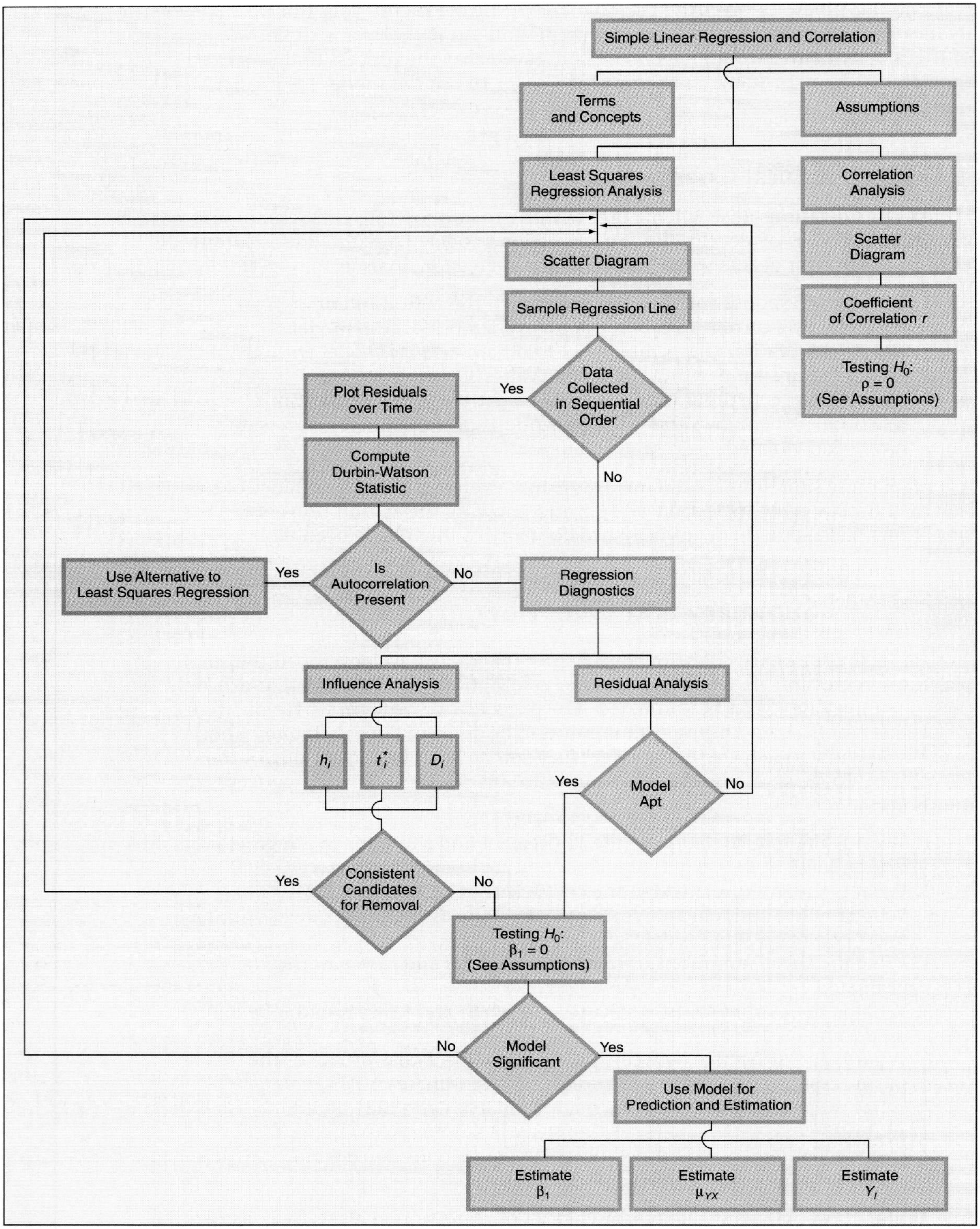

Chapter 17 summary chart.

Getting It All Together

Key Terms

adjusted r^2 *731*
assumptions of regression *736*
autocorrelation *742*
coefficient of correlation *732*
coefficient of determination *731*
confidence interval estimate for the mean response *747*
Cook's *D* statistic *757*
dependent variable *714*
Durbin-Watson statistic *744*
error sum of squares (SSE) *728*
explanatory variable *714*
hat matrix elements h_i *756*
homoscedasticity *737*
independence of error *737*
independent variable *714*
influence analysis *755*
least squares method *722*
linearity *737*
linear relationship *719*
normality *736*
prediction interval for an individual response *749*
regression coefficient *721*
regression sum of squares (SSR) *728*
relevant range *725*
residuals *737*
residual analysis *737*
response variable *714*
scatter diagram *715*
simple linear regression *721*
slope *721*
standard error of the estimate *726*
standardized residuals *739*
Studentized deleted residuals *757*
testing for correlation *753*
testing for the slope *751*
total sum of squares (SST) *728*
Y intercept *721*

Chapter Review Problems

● 17.75 A statistician for an American automobile manufacturer would like to develop a statistical model for predicting delivery time (the days between the ordering of the car and the actual delivery of the car) of custom-ordered new automobiles. The statistician believes that there is a linear relationship between the number of options ordered on the car and delivery time. A random sample of 16 cars is selected with the results given as follows:

Data file DELIVERY.DAT

Relating delivery time with options ordered (Problem 17.75).

Car	Number of Options Ordered, X	Delivery Time, Y (in days)	Car	Number of Options Ordered, X	Delivery Time, Y (in days)
1	3	25	9	12	44
2	4	32	10	12	51
3	4	26	11	14	53
4	7	38	12	16	58
5	7	34	13	17	61
6	8	41	14	20	64
7	9	39	15	23	66
8	11	46	16	25	70

(a) Set up a scatter diagram.
(b) Use the least squares method to find the regression coefficients b_0 and b_1.
(c) Interpret the meaning of the Y intercept b_0 and the slope b_1 in this problem.
(d) If a car was ordered that had 16 options, how many days would you predict it would take to be delivered?

(e) Compute the standard error of the estimate.
(f) Compute the coefficient of determination r^2 and interpret its meaning in this problem.
(g) Compute the adjusted r^2 and compare it with the coefficient of determination r^2.
(h) Compute the coefficient of correlation r.
(i) Set up a 95% confidence interval estimate of the average delivery time for all cars ordered with 16 options.
(j) Set up a 95% prediction interval estimate of the delivery time for an individual car that was ordered with 16 options.
(k) At the .05 level of significance, is there evidence of a linear relationship between number of options and delivery time?
(l) Set up a 95% confidence-interval estimate of the true slope.
(m) Perform a residual analysis on your results and determine the adequacy of the fit of the model.
(n) Perform an influence analysis and determine whether any observations should be deleted from the model. If necessary, reanalyze the regression model after deleting these observations and compare your results to the original model.
(o) What assumptions about the relationship between the number of options and delivery time would the statistician need to make in order to use this regression model for predictive purposes in the future?

17.76 An official of a local racetrack would like to develop a model to forecast the amount of money bet (in millions of dollars) based on attendance. A random sample of 15 days is selected with the results given in the table below.

Relating betting with attendance (Problem 17.76).

Day	Attendance (thousands)	Amount Bet (millions of dollars)	Day	Attendance (thousands)	Amount Bet (millions of dollars)
1	14.5	0.70	9	16.3	0.71
2	21.2	0.83	10	32.1	1.04
3	11.6	0.62	11	27.6	0.97
4	31.7	1.10	12	34.8	1.13
5	46.8	1.27	13	29.3	0.91
6	31.4	1.02	14	19.2	0.68
7	40.0	1.15	15	16.3	0.63
8	21.0	0.80			

Data file
BET.DAT

Hint: Determine which are the independent and dependent variables.
(a) Set up a scatter diagram.
(b) Assuming a linear relationship, use the least squares method to find the regression coefficients b_0 and b_1.
(c) Interpret the meaning of the slope b_1 in this problem.
(d) Predict the amount bet for a day on which attendance is 20,000.
(e) Compute the standard error of the estimate.
(f) Compute the coefficient of determination r^2 and interpret its meaning in this problem.
(g) Compute the coefficient of correlation r.
(h) Compute the Durbin-Watson statistic and, at the .05 level of significance, determine whether there is any autocorrelation in the residuals.
(i) Based on the results of (h), what conclusions can you reach concerning the validity of the model fit in (b)?
(j) Set up a 95% confidence-interval estimate of the average amount of money bet when attendance is 20,000.
(k) Set up a 95% prediction interval for the amount of money bet on a day in which attendance is 20,000.

(l) At the .05 level of significance, is there evidence of a linear relationship between the amount of money bet and attendance?
(m) Set up a 95% confidence-interval estimate of the true slope.
(n) Discuss why you should not predict the amount bet on a day in which the attendance exceeded 46,800 or was below 11,600.
(o) Perform a residual analysis on your results and determine the adequacy of the fit of the model.
(p) Perform an influence analysis and determine whether any observations should be deleted from the model. If necessary, reanalyze the regression model after deleting these observations and compare your results with the original model.

17.77 The owner of a large chain of ice-cream stores would like to study the effect of atmospheric temperature on sales during the summer season. A random sample of 21 days is selected with the results given as follows:

Relating sales to temperature (Problem 17.77).

Day	Daily High Temperature (°F)	Sales per Store (in $000)	Day	Daily High Temperature (°F)	Sales per Store (in $000)
1	63	1.52	12	75	1.92
2	70	1.68	13	98	3.40
3	73	1.80	14	100	3.28
4	75	2.05	15	92	3.17
5	80	2.36	16	87	2.83
6	82	2.25	17	84	2.58
7	85	2.68	18	88	2.86
8	88	2.90	19	80	2.26
9	90	3.14	20	82	2.14
10	91	3.06	21	76	1.98
11	92	3.24			

Data file
ICECREAM.DAT

(a) Set up a scatter diagram.
(b) Assuming a linear relationship, use the least squares method to compute the regression coefficients b_0 and b_1.
(c) Interpret the meaning of the slope b_1 in this problem.
(d) Predict the sales per store for a day in which the temperature is 83°F.
(e) Compute the standard error of the estimate.
(f) Compute the coefficient of determination r^2 and interpret its meaning in this problem.
(g) Compute the coefficient of correlation r.
(h) Compute the adjusted r^2 and compare it with the coefficient of determination r^2.
(i) Compute the Durbin-Watson statistic and, at the .05 level of significance, determine whether there is any autocorrelation in the residuals.
(j) Based on the results of (i), what conclusions can you reach concerning the validity of the model fit in (b)?
(k) Set up a 95% confidence-interval estimate of the average sales per store for all days in which the temperature is 83°F.
(l) Set up a 95% prediction interval for the sales per store on a day in which the temperature is 83°F.
(m) At the .05 level of significance, is there evidence of a linear relationship between temperature and sales?
(n) Set up a 95% confidence-interval estimate of the true slope.

(k) Set up a 90% confidence-interval estimate of the population slope.
(l) Perform a residual analysis on your results and determine the adequacy of the fit of the model.
(m) Perform an influence analysis and determine whether any observations may be deleted from the model. If necessary, reanalyze the regression model after deleting these observations and compare your results with the original model.

17.80 The Director of Graduate Studies at a large college of business would like to be able to predict the grade-point index (GPI) of students in an MBA program based upon Graduate Management Aptitude Test (GMAT) score. A sample of 20 students who had completed two years in the program was selected; the results are as follows:

Data file GPIGMAT.DAT

Relating GPI to GMAT score (Problem 17.80).

Observation	GMAT Score	GPI	Observation	GMAT Score	GPI
1	688	3.72	11	567	3.07
2	647	3.44	12	542	2.86
3	652	3.21	13	551	2.91
4	608	3.29	14	573	2.79
5	680	3.91	15	536	3.00
6	617	3.28	16	639	3.55
7	557	3.02	17	619	3.47
8	599	3.13	18	694	3.60
9	616	3.45	19	718	3.88
10	594	3.33	20	759	3.76

Hint: First determine which is the independent variable and which is the dependent variable.
(a) Assuming a linear relationship, use the least squares method to find the regression coefficients b_0 and b_1.
(b) Interpret the meaning of the Y intercept b_0 and the slope b_1 in this problem.
(c) Use the regression model developed in (a) to predict the grade-point index for a student with a GMAT score of 600.
(d) Compute the standard error of the estimate.
(e) Compute the coefficient of determination r^2 and interpret its meaning in this problem.
(f) Compute the coefficient of correlation r.
(g) Compute the adjusted r^2 and compare it with the coefficient of determination r^2.
(h) At the .05 level of significance, is there evidence of a linear relationship between GMAT score and grade-point index?
(i) Set up a 95% confidence-interval estimate for the average grade-point index of students with a GMAT score of 600.
(j) Set up a 95% prediction interval estimate of the grade-point index for a particular student with a GMAT score of 600.
(k) Set up a 95% confidence-interval estimate of the population slope.
(l) Perform a residual analysis on your results and determine the adequacy of the fit of the model.
(m) Perform an influence analysis and determine whether any observations may be deleted from the model. If necessary, reanalyze the regression model after deleting the observations and compare your results with the original model.

17.81 The manager of the purchasing department of a large banking organization would like to develop a model to predict the amount of time it would take to process invoices. Data were collected from a sample of 30 days with the following results shown as follows:

Relating time to invoices processed (Problem 17.81).

Data file INVOICE.DAT

Day	Number of Invoices Processed	Amount of Time (hours)	Day	Number of Invoices Processed	Amount of Time (hours)
1	149	2.1	16	169	2.5
2	60	1.8	17	190	2.9
3	188	2.3	18	233	3.4
4	19	0.3	19	289	4.1
5	201	2.7	20	45	1.2
6	58	1.0	21	193	2.5
7	77	1.7	22	70	1.8
8	222	3.1	23	241	3.8
9	181	2.8	24	103	1.5
10	30	1.0	25	163	2.8
11	110	1.5	26	120	2.5
12	83	1.2	27	201	3.3
13	60	0.8	28	135	2.0
14	25	0.4	29	80	1.7
15	173	2.0	30	29	0.5

(a) Set up a scatter diagram.
(b) Assuming a linear relationship, use the least squares method to find the regression coefficients b_0 and b_1.
(c) Interpret the meaning of the Y intercept b_0 and the slope b_1 in this problem.
(d) Use the regression model developed in (b) to predict the amount of time it would take to process 150 invoices.
(e) Compute the standard error of the estimate.
(f) Compute the coefficient of determination r^2 and interpret its meaning.
(g) Compute the coefficient of correlation.
(h) Compute the Durbin-Watson statistic and, at the .05 level of significance, determine whether there is any autocorrelation in the residuals.
(i) Based on the results of (h), what conclusions can you reach concerning the validity of the model fit in (b)?
(j) At the .05 level of significance, is there evidence of a relationship between the amount of time and the number of invoices processed?
(k) Set up a 95% confidence interval estimate of the average amount of time taken to process 150 invoices.
(l) Set up a 95% prediction interval of the amount of time it would take to process 150 invoices on a particular day.
(m) Perform a residual analysis on your results and determine the adequacy of the fit of the model.
(n) Perform an influence analysis and determine whether any observations may be deleted from the model. If necessary, reanalyze the regression model after deleting the observations and compare your results with the original model.

Data file
BB93.DAT

17.82 Crazy Dave, a well-known baseball analyst, would like to study various team statistics for a recent baseball season to determine the variables that might be useful in predicting the number of wins achieved by teams during the season. He has decided to begin by using the team earned run average (E.R.A.) to predict the number of wins. The data for the 28 major league teams were as follows:

Relating wins to E.R.A. (Problem 17.82).

American League			National League		
Team	Wins	E.R.A.	Team	Wins	E.R.A.
Boston	80	3.77	Florida	64	4.13
Cleveland	76	4.58	Cincinnati	73	4.51
Kansas City	84	4.04	Chicago Cubs	84	4.18
Minnesota	71	4.71	San Francisco	103	3.61
Toronto	95	4.21	Los Angeles	81	3.50
California	71	4.34	Pittsburgh	75	4.77
Seattle	82	4.20	San Diego	71	4.23
Texas	86	4.28	New York Mets	59	4.05
Detroit	85	4.65	St. Louis	87	4.09
Chicago White Sox	94	3.70	Philadelphia	97	3.95
Milwaukee	69	4.45	Atlanta	104	3.14
Oakland	68	4.90	Montreal	94	3.55
Baltimore	85	4.31	Houston	85	3.49
New York Yankees	88	4.13	Colorado	67	5.41

(a) Set up a scatter diagram.
(b) Assuming a linear relationship, use the least squares method to find the regression coefficients b_0 and b_1.
(c) Interpret the meaning of the Y intercept b_0 and the slope b_1 in this problem.
(d) Use the regression model developed in (b) to predict the number of wins for a team with an E.R.A. of 4.00.
(e) Compute the standard error of the estimate.
(f) Compute the coefficient of determination r^2 and interpret its meaning.
(g) Compute the coefficient of correlation.
(h) At the .05 level of significance, is there evidence of a relationship between the number of wins and the E.R.A.?
(i) Set up a 95% confidence interval estimate of the average number of wins for a team with an E.R.A. of 4.00.
(j) Set up a 95% prediction interval of the number of wins for an individual team that has an E.R.A. of 4.00.
(k) Set up a 95% confidence-interval estimate of the slope.
(l) Perform a residual analysis on your results and determine the adequacy of the fit of the model.
(m) Perform an influence analysis and determine whether any observations may be deleted from the model. If necessary, reanalyze the regression model after deleting the observations and compare your results with the original model.
(n) The 28 teams constitute a population. In order to use statistical inference [as in (h)–(k)], the data must be assumed to represent a random sample. What "population" would this sample be drawing conclusions about?

Survey Database Project for Chapter 17

Data file
EMPSAT.DAT

The following problem refers to the sample data obtained from the questionnaire of Figure 2.6 on pages 28–29 and presented in Table 2.3 on pages 33–40. It should be solved with the aid of an available computer package.

17.83 In Section 17.15.3 we used regression analysis to develop a model to forecast the personal income of Kalosha Industries employees whose occupational grouping was technical/sales based on the number of years the employee has worked at a full-time job. Suppose that we would like to do similar analyses for each of the other six occupational groups. Develop each of these models and write an executive summary to Bud Conley discussing your findings.

Case Study H—Predicting Sunday Newspaper Circulation

Data file
NEWSCIRC.DAT

You are employed in the marketing department of a large nationwide newspaper chain. The parent company is interested in investigating the feasibility of beginning a Sunday edition for some of its newspapers. However, before proceeding with a final decision, it needs to estimate the amount of Sunday circulation that would be expected. In particular, it wishes to predict the Sunday circulation that would be obtained by newspapers (in three different cities) that have daily circulations of 200,000, 400,000, and 600,000, respectively.

Toward this end, data (summarized in the table below) have been collected from a sample of 35 newspapers.

You have been asked to develop a model that would enable you to make a prediction of the expected Sunday circulation and to write a report that presents your results and summarizes your findings.

	Circulation (in 000)	
Paper	Sunday	Daily
Des Moines Register	344.522	206.204
Philadelphia Inquirer	982.663	515.523
Tampa Tribune	408.343	321.626
New York Times	1,762.015	1,209.225
New York News	983.240	781.796
Sacramento Bee	338.355	273.844
Los Angeles Times	1,531.527	1,164.388
Boston Globe	798.298	516.981
Cincinnati Enquirer	348.744	198.832
Orange Co. Register	407.760	354.843
Miami Herald	553.479	444.581
Chicago Tribune	1,133.249	733.775
Detroit News	1,215.149	481.766
Houston Chronicle	620.752	449.755
Kansas City Star	423.305	288.571
Omaha World Herald	284.611	223.748
Denver Post	417.779	252.624
St. Louis Post-Dispatch	585.681	391.286
Portland Oregonian	440.923	337.672
Washington Post	1,165.567	838.902
Long Island Newsday	960.308	825.512
San Francisco Chronicle	704.322	570.364

(continued on page 780)

Paper	Circulation (in 000) Sunday	Daily
Chicago Sun Times	559.093	537.780
Minneapolis Star Tribune	685.975	412.871
Baltimore Sun	488.506	391.952
Pittsburgh Press	557.000	220.465
Rocky Mountain News	432.502	374.009
Boston Herald	235.084	355.628
New Orleans Times-Picayune	324.241	272.280
Charlotte Observer	299.451	238.555
Hartford Courant	323.084	231.177
Rochester Democrat and Chronicle	262.048	133.239
St. Paul Pioneer Press	267.781	201.860
Providence Journal-Bulletin	268.060	197.120
L.A. Daily News	202.614	185.736

Source: From *Gale Directory of Publications:* 1994, 126th edition. Edited by Donald P. Boyden and John Krol. Gale Research, 1994 Copyright © 1994 by Gale Research, Inc. Reprinted by permission of the publisher.

Endnotes

1. In Section 18.12 we shall investigate multiple regression models in which at least one of the independent variables is categorical (see dummy variable models), while in Section 18.17 we will develop a model to predict a categorical response variable using logistic regression.
2. h_i are the "hat matrix diagonal elements," which reflect the influence (see Section 17.14) of each X_i in the simple linear regression model.
3. The more general criteria for multiple regression will be discussed in Section 18.16.
4. See note 3 above.
5. It is interesting and instructive to note that had we constructed the residual plots by using the independent variable as the X axis (instead of the estimated values $\hat{Y}$) the same conclusions would prevail.

References

1. Andrews, D. F., and D. Pregibon, "Finding the Outliers that Matter," *Journal of the Royal Statistical Society,* Ser. B., 1978, Vol. 40, pp. 85–93.
2. Anscombe, F. J., "Graphs in Statistical Analysis," *American Statistician,* 1973, Vol. 27, pp. 17–21.
3. Atkinson, A. C., "Robust and Diagnostic Regression Analysis," *Communications in Statistics,* 1982, Vol. 11, pp. 2559–2572.
4. Belsley, D. A., E. Kuh, and R. Welsch, *Regression Diagnostics: Identifying Influential Data and Sources of Collinearity* (New York: John Wiley, 1980).
5. Berenson, M. L., D. M. Levine, and M. Goldstein, *Intermediate Statistical Methods and Applications: A Computer Package Approach* (Englewood Cliffs, NJ: Prentice-Hall, 1983).
6. Cook, R. D., and S. Weisberg, *Residuals and Influence in Regression* (New York: Chapman and Hall, 1982).
7. Conover, W. J., *Practical Nonparametric Statistics,* 2d ed. (New York: John Wiley, 1980).
8. Draper, N. R., and H. Smith, *Applied Regression Analysis,* 2d ed. (New York: John Wiley, 1981).
9. Hoaglin, D. C., and R. Welsch, "The Hat Matrix in Regression and ANOVA," *The American Statistician,* 1978, Vol. 32, pp. 17–22.
10. Hocking, R. R., "Developments in Linear Regression Methodology: 1959–1982," *Technometrics,* 1983, Vol. 25, pp. 219–250.
11. *MINITAB Reference Manual Release 8* (State College, PA.: MINITAB, Inc., 1992).
12. Neter, J., W. Wasserman, and M. H. Kutner, *Applied Linear Statistical Models,* 3d ed. (Homewood, IL: Richard D. Irwin, 1990).
13. Pregibon, D. "Logistic Regression Diagnostics," *Annals of Statistics,* 1981, Vol. 9, pp. 705–724.
14. Ramsey, P. P., and P. H. Ramsey, "Simple Tests of Normality in Small Samples," *Journal of Quality Technology,* 1990, Vol. 22, pp. 299–309.
15. Ryan, B. F., and B. L. Joiner, *Minitab Student Handbook,* 3rd ed. (North Scituate, MA: Duxbury Press, 1994).
16. *SAS Language and Procedures Usage,* Version 6 (Cary, NC: SAS Institute, 1988).
17. *STATISTIX Version 4.0* (Tallahassee, FL: Analytical Software, 1992).
18. Tukey, J. W., "Data Analysis, Computation and Mathematics," *Quarterly Journal of Applied Mathematics,* 1972, Vol. 30, pp. 51–65.
19. Velleman, P. F., and R. Welsch, "Efficient Computing of Regression Diagnostics," *The American Statistician,* 1981, Vol. 35, pp. 234–242.
20. Weisberg, S., *Applied Linear Regression* (New York: John Wiley, 1980).

chapter 18

Multiple Regression Models

CHAPTER OBJECTIVES

To develop the multiple regression model as an extension of the simple linear regression model and to evaluate the contribution of each independent variable to the regression model. In addition, to extend the inferential procedures to predict the average value of *Y*; to measure the coefficient of partial determination; to develop and test the curvilinear regression model; to introduce dummy variables in regression analysis; to illustrate the model-building process; and to introduce the logistic regression model.

18.1 Introduction

In our discussion of the simple regression model in the preceding chapter, we focused on a model in which one independent or explanatory variable X was used to predict the value of a dependent or response variable Y. We may recall that a simple regression model was developed in order to predict the weekly sales based on the number of customers in a chain of package delivery stores. It is often the case that a better-fitting model can be developed if more than one explanatory variable is considered. Thus, in this chapter we shall extend our discussion to consider **multiple regression** models in which several explanatory variables can be used to predict the value of a dependent variable.

Upon completion of this chapter, you should be able to:

1. Interpret the regression coefficients
2. Use the multiple regression model for predicting the response variable
3. Determine whether there is a relationship between the response variable and the independent variables included in the model
4. Determine which independent variables make a significant contribution to the regression model
5. Interpret the coefficient of multiple determination
6. Interpret the coefficients of partial determination
7. Consider the inclusion of curvilinear terms in the regression model
8. Understand how categorical independent variables can be included in the regression model
9. Understand regression models that include interaction terms and models that involve transformed variables
10. Understand the problem of multicollinearity and how it can be measured
11. Use residual and influence analysis in multiple regression
12. Use stepwise and best subset approaches to build a multiple regression model
13. Use logistic regression to predict a categorical response variable

18.2 Developing the Multiple Regression Model

Suppose that we wanted to develop a regression model in order to predict the consumption of heating oil by single-family homes during the month of January. A sample of 15 similar homes built by a particular housing developer throughout the United States was selected for analysis. Although many variables could be considered, for simplicity only two explanatory variables are to be evaluated here—the average daily atmospheric temperature, as measured in degrees Fahrenheit, outside the house during that month (X_1) and the amount of insulation, as measured in inches, in the attic of the house (X_2). The results are presented in Table 18.1.

With two explanatory variables in the multiple regression model, a scatter diagram of the points can be plotted on a three-dimensional graph as shown in Figure 18.1.

For a particular investigation, when there are several explanatory variables present, the simple linear regression model of the preceding chapter [Equation

Table 18.1 Consumption of heating oil, atmospheric temperature, and amount of attic insulation for a random sample of 15 single-family homes.

Observation	Monthly Consumption of Heating Oil, (Gallons)	Average Daily Atmospheric Temperature, (°F)	Amount of Attic Insulation, (Inches)
1	275.3	40	3
2	363.8	27	3
3	164.3	40	10
4	40.8	73	6
5	94.3	64	6
6	230.9	34	6
7	366.7	9	6
8	300.6	8	10
9	237.8	23	10
10	121.4	63	3
11	31.4	65	10
12	203.5	41	6
13	441.1	21	3
14	323.0	38	3
15	52.5	58	10

Data file HTNGOIL.DAT

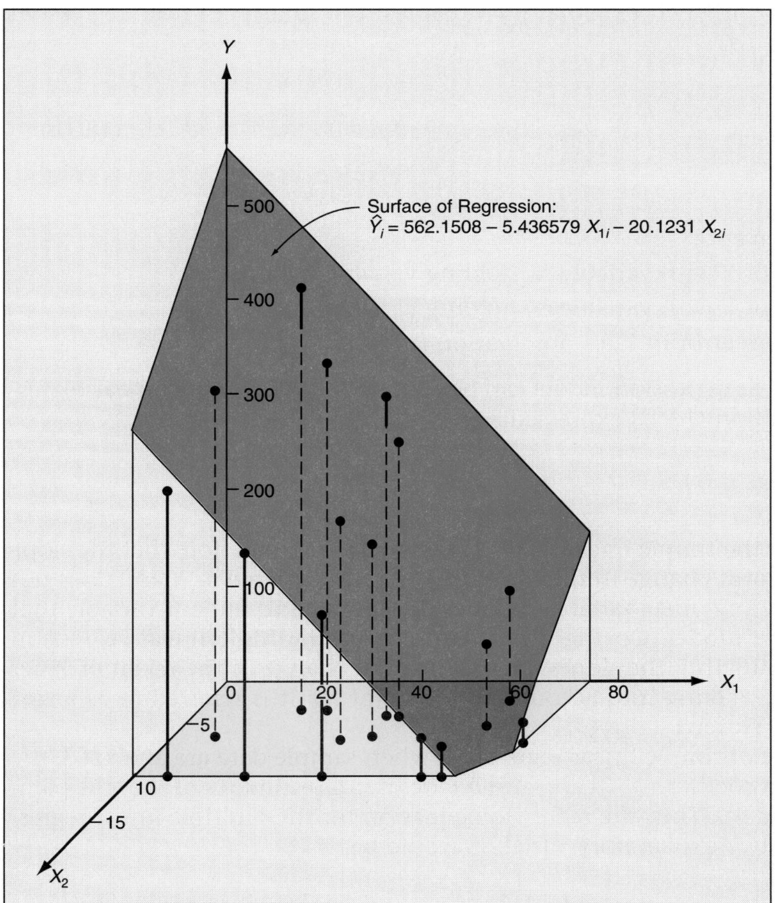

Figure 18.1
Scatter diagram of average daily atmospheric temperature X_1, amount of attic insulation X_2, and monthly consumption of heating oil Y with indicated regression plane fitted by least squares method.

(17.1)] can merely be extended by assuming a linear relationship between each explanatory variable and the dependent variable. For example, with P explanatory variables the multiple linear regression model is expressed as

$$Y_i = \beta_0 + \beta_1 X_{1i} + \beta_2 X_{2i} + \beta_3 X_{3i} + \cdots + \beta_P X_{Pi} + \epsilon_i \qquad \textbf{(18.1a)}$$

where β_0 = Y intercept

β_1 = slope of Y with variable X_1 holding variables $X_2, X_3, \ldots, X_P$ constant

β_2 = slope of Y with variable X_2 holding variables $X_1, X_3, \ldots, X_P$ constant

β_3 = slope of Y with variable X_3 holding variables $X_1, X_2, X_4, \ldots, X_P$ constant

.

.

.

β_P = slope of Y with variable X_P holding variables $X_1, X_2, X_3, \ldots, X_{P-1}$ constant

ϵ_i = random error in Y for observation i

For our data with two explanatory variables, the multiple linear regression model is expressed as

$$Y_i = \beta_0 + \beta_1 X_{1i} + \beta_2 X_{2i} + \epsilon_i \qquad \textbf{(18.1b)}$$

where β_0 = Y intercept

β_1 = slope of Y with variable X_1 holding variable X_2 constant

β_2 = slope of Y with variable X_2 holding variable X_1 constant

ϵ_i = random error in Y for observation i

This multiple linear regression model can be compared to the simple linear regression model [Equation (17.1)] expressed as

$$Y_i = \beta_0 + \beta_1 X_i + \epsilon_i$$

In the case of the simple linear regression model we should note that the slope β_1 represents the unit change in the mean of Y per unit change in X and does not take into account any other variables besides the single independent variable that is included in the model. On the other hand, in the multiple linear regression model [Equation 18.1b)], the slope β_1 represents the change in the mean of Y per unit change in X_1, taking into account the effect of X_2. It is referred to as a **net regression coefficient.**

As in the case of simple linear regression, when sample data are analyzed, the sample regression coefficients (b_0, b_1, and b_2) are used as estimates of the true parameters (β_0, β_1, and β_2). Thus the regression equation for the multiple linear regression model with two explanatory variables would be

$$\hat{Y}_i = b_0 + b_1X_{1i} + b_2X_{2i} \tag{18.2}$$

Using the least squares method, the values of the three sample regression coefficients may be obtained by accessing an appropriate computer package (see References 12, 14, 15, 17). Figure 18.2 presents partial output from the SAS REG procedure for the data of Table 18.1. Figure 18.24 (see page 844) presents output from the STATISTIX package, and Figure 18.25 (see page 845) presents output from MINITAB.

```
DEP VARIABLE: OIL                    (VARIANCE)
                         SUM OF           MEAN
SOURCE      DF          SQUARES         SQUARE          F VALUE    PROB > F

MODEL        2           228015         114007          168.471      0.0001
ERROR       12         8120.603        676.717
C TOTAL     14           236135
    ROOT MSE          26.013783     R-SQUARE             0.9656
    DEP MEAN            216.493     ADJ R-SQ             0.9599
    C.V.               12.01597

                      PARAMETER      STANDARD       T FOR HO:
VARIABLE    DF         ESTIMATE         ERROR     PARAMETER=0  PROB > :T:

INTERCEP     1  b0      562.151   21.093104 Sb0        26.651      0.0001
TEMPF        1  b1    -5.436581    0.336216 Sb1       -16.170      0.0001
INSU         1  b2   -20.012321    2.342505 Sb2        -8.543      0.0001

VARIABLE    DF        TYPE I SS     TYPE II SS

INTERCEP     1           703040         480653
TEMPF        1           178624         176938
INSU         1        49390.202      49390.202

                    PREDICT   STD ERR   LOWER95%   UPPER95%
  OBS    ACTUAL       VALUE   PREDICT       MEAN       MEAN   RESIDUAL

    1   275.300     284.651    10.300    262.210    307.092     -9.351
    2   363.800     355.326    11.196    330.932    379.721      8.474
    3   164.300     144.565    10.905    120.805    168.324     19.735
    4    40.800      45.207    12.923     17.050     73.363     -4.407
    5    94.300      94.136    10.465     71.335    116.936   0.164072
    6   230.900     257.233     7.081    241.806    272.661    -26.333
    7   366.700     393.148    12.493    365.927    420.369    -26.448
    8   300.600     318.535    15.435    284.905    352.165    -17.935
    9   237.800     236.986    12.389    209.994    263.979   0.813551
   10   121.400     159.609    12.867    131.574    187.645    -38.209
   11    31.400       8.650    13.666    -21.126     38.426     22.750
   12   203.500     219.177     6.767    204.434    233.921    -15.677
   13   441.100     387.946    12.130    361.516    414.375     53.154
   14   323.000     295.524    10.323    273.033    318.015     27.476
   15    52.500      46.706    12.390     19.710     73.703      5.794
```

Figure 18.2
Partial output for the data of Table 18.1 obtained from SAS PROC REG.

From Figure 18.2, we observe that the computed values of the regression coefficients in this problem are

$$b_0 = 562.151 \qquad b_1 = -5.43658 \qquad b_2 = -20.0123$$

Therefore, the multiple regression equation can be expressed as

$$\hat{Y}_i = 562.151 - 5.43658X_{1i} - 20.0123X_{2i}$$

where $\hat{Y}_i$ = predicted average amount of home heating oil consumed (gallons) during January for observation i

X_{1i} = average daily atmospheric temperature (°F) during January for observation i

X_{2i} = amount of attic insulation (inches) for observation i

The interpretation of the regression coefficients is analogous to that of the simple linear regression model. The Y intercept b_0, computed as 562.151, estimates the expected number of gallons of home heating oil that would be consumed in January when the average daily atmospheric temperature was 0° for a home that was not insulated (that is, a house with 0 inches of attic insulation). The slope of average daily atmospheric temperature with consumption of heating oil (b_1, computed as −5.43658) can be interpreted to mean that for a home with a *given* number of inches of attic insulation, the expected consumption of heating oil is estimated to *decrease* by 5.43658 gallons per month for each 1°F increase in average daily atmospheric temperature. Furthermore, the slope of amount of attic insulation with consumption of heating oil (b_2, computed as −20.0123) can be interpreted to mean that for a month with a *given* average daily atmospheric temperature, the expected consumption of heating oil is estimated to *decrease* by 20.0123 gallons for each additional inch of attic insulation.

Problems for Section 18.2

18.1 Explain the difference in the interpretation of the regression coefficients in simple linear regression and multiple linear regression.

● 18.2 A marketing analyst for a major shoe manufacturer is considering the development of a new brand of running shoes. In particular, the marketing analyst wishes to determine the variables that can be used in predicting durability (or the effect of long-term impact). The following two independent variables are to be considered:

X_1 (FOREIMP), a measurement of the forefoot shock-absorbing capability

X_2 (MIDSOLE), a measurement of the change in impact properties over time

along with the dependent variable Y (LTIMP), which is a measure of the long-term ability to absorb shock after a repeated impact test. A random sample of 15 types of currently manufactured running shoes was selected for testing. Using SAS, the following (partial) output is provided:

Shoe durability problem.

```
  DEP VARIABLE              LTIMP          (VARIANCE)
                           SUM OF               MEAN
SOURCE DF                 SQUARES             SQUARE              F VALUE          PROB > F
MODEL            2       12.61020            6.30510                97.69            0.0001
ERROR           12        0.77453            0.06454
C TOTAL         14       13.38473

                        PARAMETER           STANDARD             T FOR HO:
VARIABLE        DF       ESTIMATE              ERROR         PARAMETER = 0        PROB > |T|
INTERCEP         1      - 0.02686             .06905                - 0.39
FOREIMP          1        0.79116             .06295                 12.57             .0000
MIDSOLE          1        0.60484             .07174                  8.43             .0000

VARIABLE        DF      TYPE I SS         TYPE II SS
FOREIMP          1        8.02166           10.19682
MIDSOLE          1        4.58854            4.58854
```

(a) Assuming that each independent variable is linearly related to long-term impact, state the multiple regression equation.
(b) Interpret the meaning of the slopes in this problem.

18.3 A mail-order catalog business selling personal computer supplies, software, and hardware maintains a centralized warehouse for the distribution of products ordered. Management is currently examining the process of distribution from the warehouse and is interested in studying the factors that affect warehouse distribution costs. Currently, a small handling fee is added to the order, regardless of the amount of the order. Data have been collected over the past 24 months indicating the warehouse distribution costs, the sales, and the number of orders received. The results were as follows:

Distribution cost problem.

Month	Distribution Cost ($000)	Sales ($000)	Number of Orders	Month	Distribution Cost ($000)	Sales ($000)	Number of Orders
1	52.95	386	4,015	13	62.98	372	3,977
2	71.66	446	3,806	14	72.30	328	4,428
3	85.58	512	5,309	15	58.99	408	3,964
4	63.69	401	4,262	16	79.38	491	4,582
5	72.81	457	4,296	17	94.44	527	5,582
6	68.44	458	4,097	18	59.74	444	3,450
7	52.46	301	3,213	19	90.50	623	5,079
8	70.77	484	4,809	20	93.24	596	5,735
9	82.03	517	5,237	21	69.33	463	4,269
10	74.39	503	4,732	22	53.71	389	3,708
11	70.84	535	4,413	23	89.18	547	5,387
12	54.08	353	2,921	24	66.80	415	4,161

Access a computer package and perform a multiple linear regression analysis. Based on the results obtained

Data file WARECOST.DAT

(a) State the multiple regression equation.
(b) Interpret the meaning of the slopes in this problem.

Data file
ADRADTV.DAT

18.4 Suppose that a large consumer products company wanted to measure the effectiveness of different types of advertising media in the promotion of its products. Specifically, two types of advertising media were to be considered: radio and television advertising and newspaper advertising (including the cost of discount coupons). A sample of 22 cities with approximately equal populations was selected for study during a test period of one month. Each city was allocated a specific expenditure level for both radio and television advertising as well as newspaper advertising. The sales of the product (in thousands of dollars) during the test month was recorded along with the levels of media expenditure with the following results:

Advertising media problem.

City	Sales ($000)	Radio and Television Advertising ($000)	Newspaper Advertising ($000)	City	Sales ($000)	Radio and Television Advertising ($000)	Newspaper Advertising ($000)
1	973	0	40	12	1,577	45	45
2	1,119	0	40	13	1,044	50	0
3	875	25	25	14	914	50	0
4	625	25	25	15	1,329	55	25
5	910	30	30	16	1,330	55	25
6	971	30	30	17	1,405	60	30
7	931	35	35	18	1,436	60	30
8	1,177	35	35	19	1,521	65	35
9	882	40	25	20	1,741	65	35
10	982	40	25	21	1,866	70	40
11	1,628	45	45	22	1,717	70	40

Access a computer package and perform a multiple linear regression analysis. Based on the results obtained

(a) State the multiple regression equation.
(b) Interpret the meaning of the slopes in this problem.

18.5 The personnel department of a large industrial corporation would like to develop a model to predict the weekly salary based upon the length of employment and age of its managerial employees. A random sample of 16 managerial employees is selected with the results displayed below:

Data file
SALARY2.DAT

Employee salary problem.

Employee	Weekly Salary	Length of Employment (Months)	Age (Years)	Employee	Weekly Salary	Length of Employment (Months)	Age (Years)
1	$839	330	46	9	752	352	55
2	946	569	65	10	729	256	61
3	870	375	57	11	656	87	28
4	718	113	47	12	874	337	51
5	802	215	41	13	606	42	28
6	812	343	59	14	729	129	37
7	748	252	45	15	728	216	46
8	791	348	57	16	792	327	56

Access a computer package and perform a multiple linear regression analysis. Based on the results obtained
(a) State the multiple regression equation.
(b) Interpret the meaning of the slopes in this problem.

18.6 The director of broadcasting operations for a television station wanted to study the issue of "standby hours," hours in which unionized graphic artists at the station are paid but are not actually involved in any activity. The variables to be considered are

Standby hours (Y)—the total number of standby hours per week.

Total staff present (X_1)—the weekly total of people-days worked over a seven-day week.

Remote hours (X_2)—the total number of hours worked by employees at locations away from the central plant.

The results for a period of 26 weeks were as follows:

Data file STANDBY.DAT

Standby hours problem.

Week	Standby Hours	Total Staff Present	Remote Hours	Week	Standby Hours	Total Staff Present	Remote Hours
1	245	338	414	14	161	307	402
2	177	333	598	15	274	322	151
3	271	358	656	16	245	335	228
4	211	372	631	17	201	350	271
5	196	339	528	18	183	339	440
6	135	289	409	19	237	327	475
7	195	334	382	20	175	328	347
8	118	293	399	21	152	319	449
9	116	325	343	22	188	325	336
10	147	311	338	23	188	322	267
11	154	304	353	24	197	317	235
12	146	312	289	25	261	315	164
13	115	283	388	26	232	331	270

Access a computer package and develop a multiple regression model to predict standby hours based on total staff present and the number of remote hours. On the basis of the results obtained
(a) State the multiple regression model.
(b) Interpret the meaning of the slopes in this problem.

18.3 Prediction of the Dependent Variable Y for Given Values of the Explanatory Variables

Now that the multiple regression model has been fitted to these data, various procedures, analogous to those discussed for simple linear regression, could be developed. In this section we shall use the multiple regression model to predict the monthly consumption of heating oil.

Suppose that we wanted to predict the number of gallons of heating oil consumed in a house that had 6 inches of attic insulation during a month in which the average daily atmospheric temperature was 30°F. Using our multiple regression equation

$$\hat{Y}_i = 562.151 - 5.43658X_{1i} - 20.0123X_{2i}$$

with $X_{1i} = 30$ and $X_{2i} = 6$, we have

$$\hat{Y}_i = 562.151 - (5.43658)(30) - (20.0123)(6)$$

and thus

$$\hat{Y}_i = 278.9798$$

Therefore, we would estimate that an average of 278.98 gallons of heating oil would be used in houses with 6 inches of insulation when the average temperature was 30°F.

Problems for Section 18.3

Data file WARECOST.DAT

18.7 Referring to Problem 18.3 (the distribution cost problem) on page 787, predict the monthly warehouse distribution costs when sales are $400,000 and the number of orders is 4,500.

Data file ADRADTV.DAT

● 18.8 Referring to Problem 18.4 (the advertising media problem) on page 788, predict the sales for a city in which radio and television advertising is $20,000. and newspaper advertising is $20,000.

Data file SALARY2.DAT

18.9 Referring to Problem 18.5 (the employee salary problem) on page 788, predict the weekly salary for a managerial employee who has been employed for 15 years and is 47 years old.

Data file STANDBY.DAT

18.10 Referring to Problem 18.6 (the standby hours problem) on page 789, predict the standby hours for a week in which the total staff present is 310 people-days and the remote hours are 400.

18.4 Measuring Association in the Multiple Regression Model

We may recall from Section 17.6 that once a regression model has been developed, the coefficient of determination r^2 could be computed. In multiple regression, since there are at least two explanatory variables, the **coefficient of multiple determination** represents the proportion of the variation in Y that is explained by the set of explanatory variables selected. In our example containing two explanatory variables, the coefficient of multiple determination ($r^2_{Y.12}$) is given by

$$r^2_{Y.12} = \frac{\text{SSR}}{\text{SST}} \quad (18.3)$$

where

$$\text{SSR} = b_0\sum_{i=1}^{n} Y_i + b_1\sum_{i=1}^{n} X_{1i}Y_i + b_2\sum_{i=1}^{n} X_{2i}Y_i - n\overline{Y}^2$$

$$\text{SST} = \sum_{i=1}^{n} Y_i^2 - n\overline{Y}^2$$

In the home heating oil consumption problem we have already computed SSR = 228,015 and SST = 236,135 (rounded). Thus, as is displayed in the SAS output of Figure 18.2 on page 785,

$$r^2_{Y.12} = \frac{\text{SSR}}{\text{SST}} = \frac{228{,}015}{236{,}135} = .9656$$

This coefficient of multiple determination, computed as .9656, can be interpreted to mean that from the sample, 96.56% of the variation in the consumption of home heating oil can be explained by the variation in the average daily atmospheric temperature and the variation in the amount of attic insulation.

However, we may recall from Section 17.6 that when dealing with multiple regression models some researchers suggest that an *adjusted* r^2 be computed to reflect both the number of explanatory variables in the model and the sample size. This is especially necessary when we are comparing two or more regression models that predict the same dependent variable but have different numbers of explanatory or predictor variables. Thus, in multiple regression, we may denote the adjusted r^2 as

$$r^2_{\text{adj}} = 1 - \left[(1 - r^2_{Y.12\ldots P})\frac{n-1}{n-P-1}\right] \quad \textbf{(18.4)}$$

where P is the number of explanatory variables in the regression equation.

Thus, for our heating oil data, since $r^2_{Y.12} = .9656$, $n = 15$, and $P = 2$,

$$\begin{aligned} r^2_{\text{adj}} &= 1 - \left[(1 - r^2_{Y.12})\frac{(15-1)}{(15-2-1)}\right] \\ &= 1 - \left[(1 - .9656)\frac{14}{12}\right] \\ &= 1 - .04 \\ &= .96 \end{aligned}$$

Hence 96% of the variation in home heating oil usage can be explained by our multiple regression model—adjusted for number of predictors and sample size.

In order to further study the relationship among the variables, it is often useful to examine the correlation between each pair of variables included in the model. Such a *correlation matrix* that indicates the coefficient of correlation between each pair of variables is displayed in Table 18.2 at the top of page 792.

Table 18.2 Correlation matrix for the heating oil consumption problem.

	Y (Heating Oil)	X_1 (Temperature)	X_2 (Attic Insulation)
Y (Heating Oil)	$r_{YY} = 1.0$	$r_{Y1} = -.86974$	$r_{Y2} = -.46508$
X_1 (Temperature)	$r_{Y1} = -.86974$	$r_{11} = 1.0$	$r_{12} = .00892$
X_2 (Attic Insulation)	$r_{Y2} = -.46508$	$r_{12} = .00892$	$r_{22} = 1.0$

From Table 18.2, we observe that the correlation between the amount of heating oil consumed and temperature is −.86974, indicating a strong negative association between the variables. We may also observe that the correlation between the amount of heating oil consumed and attic insulation is −.46508, indicating a moderate negative correlation between these variables. Furthermore, we also note that there is virtually no correlation (.00892) between the two explanatory variables, temperature and attic insulation. Finally, we may note that the correlation coefficients along the main diagonal of the matrix (r_{YY}, r_{11}, r_{22}) are each 1.0, since there will be perfect correlation between a variable and itself.

Problems for Section 18.4

For Problems 18.11–18.15

(a) Compute the coefficient of multiple determination $r^2_{Y.12}$ and interpret its meaning.

(b) Compute the adjusted r^2.

- ● 18.11 Refer to Problem 18.2 (the shoe durability problem) on page 786.
- 18.12 Refer to Problem 18.3 (the distribution cost problem) on page 787.
- ● 18.13 Refer to Problem 18.4 (the advertising media problem) on page 788.
- 18.14 Refer to Problem 18.5 (the employee salary problem) on page 788.
- ● 18.15 Refer to Problem 18.6 (the standby hours problem) on page 789.

18.5 Residual Analysis in Multiple Regression

In Section 17.9 we utilized residual analysis to evaluate whether the simple linear regression model was appropriate for the set of data being studied. When examining a multiple linear regression model with two explanatory variables, the following residual plots are of particular interest:

1. Standardized residuals versus $\hat{Y}_i$
2. Standardized residuals versus X_{1i}
3. Standardized residuals versus X_{2i}
4. Standardized residuals versus time

The first residual plot examines the pattern of residuals for the predicted values of Y. If the standardized residuals appear to vary for different levels of the predicted Y value, it provides evidence of a possible curvilinear effect in at least one explanatory variable and/or the need to transform the dependent variable. The second and third residual plots involve the explanatory variables. Patterns in the plot of the standardized residuals versus an explanatory variable may indicate the existence of a curvilinear effect and, therefore, lead to the possible transformation of that explanatory variable. The fourth type of plot is used to investigate patterns in the residuals when the data have been collected in time order. Associated with the residuals plot versus time, as in Section 17.10, the Durbin-Watson statistic can be computed and the existence of positive autocorrelation among the residuals can be determined.

The residual plots are available as part of the output of virtually all computer packages. Figure 18.3 consists of the residual plots obtained from MINITAB for the

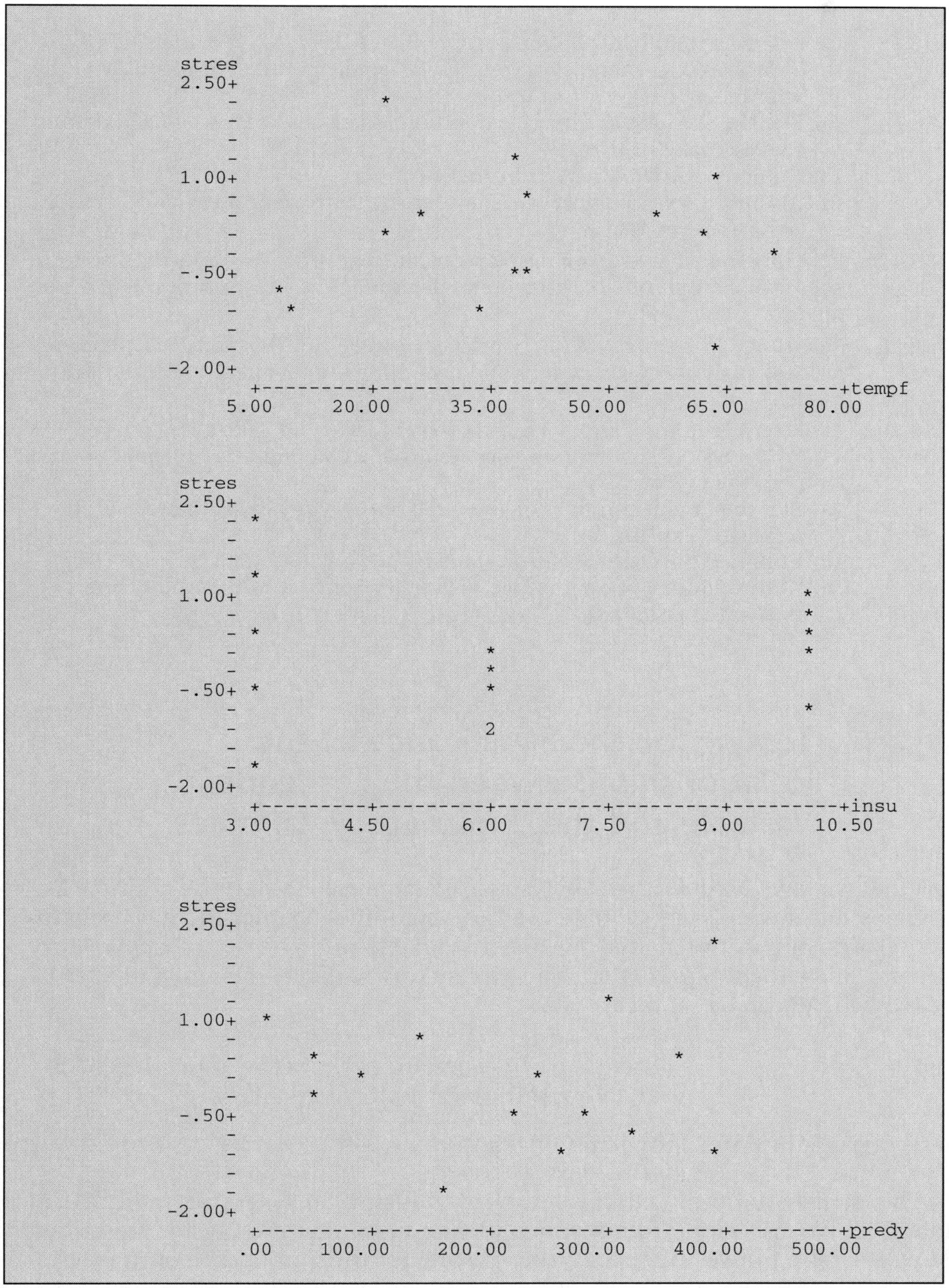

Figure 18.3
Residual plots for the heating oil usage model obtained from MINITAB.

heating oil problem. We can observe from Figure 18.3 that there appears to be very little or no pattern in the relationship between the standardized residuals and either the predicted value of Y, the value of X_1 (temperature), or the value of X_2 (attic insulation). Thus we may conclude that the multiple linear regression model is appropriate for predicting heating oil usage.

Problems for Section 18.5

18.16 (a) Referring to the distribution cost problem (pages 787, 790, and 792), perform a residual analysis on your results and determine the adequacy of the fit of the model.
(b) Plot the residuals against the months. Is there any evidence of a pattern in the residuals? Explain.
(c) Compute the Durbin-Watson statistic.
(d) At the .05 level of significance, is there evidence of positive autocorrelation in the residuals?

● 18.17 Referring to the advertising media problem (pages 788, 790, and 792), perform a residual analysis on your results and determine the adequacy of the fit of the model.

18.18 Referring to the employee salary problem (pages 788, 790, and 792), perform a residual analysis on your results and determine the adequacy of the fit of the model.

18.19 (a) Referring to the standby hours problem (pages 789, 790, and 792), perform a residual analysis on your results and determine the adequacy of the fit of the model.
(b) Plot the residuals against the weeks. Is there evidence of a pattern in the residuals? Explain.
(c) Compute the Durbin-Watson statistic.
(d) At the .05 level of significance, is there evidence of positive autocorrelation in the residuals?

18.6 Testing for the Significance of the Relationship Between the Dependent Variable and the Explanatory Variables

Now that we have used residual analysis to assure ourselves that the multiple linear regression model is appropriate, we can determine whether there is a significant relationship between the dependent variable and the set of explanatory variables. Since there is more than one explanatory variable, the null and alternative hypotheses can be set up as follows:

H_0: $\beta_1 = \beta_2 = 0$ (There is no linear relationship between the dependent variable and the explanatory variables.)

H_1: At least one $\beta_j \neq 0$ (At least one regression coefficient is not equal to zero.)

This null hypothesis may be tested by utilizing an F test, as indicated in Table 18.3. We may recall from Sections 13.6 and 14.4 that the F test is used when testing the ratio of two variances. When testing for the significance of the regres-

Table 18.3 **Analysis-of-variance table for testing the significance of a set of regression coefficients in a multiple regression model containing $P = 2$ explanatory variables.**

Source	df	Sums of Squares	Mean Square (Variance)	F
Regression	P	$SSR = b_0\sum_{i=1}^{n} Y_i + b_1\sum_{i=1}^{n} X_{1i}Y_i + b_2\sum_{i=1}^{n} X_{2i}Y_i - n\bar{Y}^2$	$MSR = \frac{SSR}{P}$	$F = \frac{MSR}{MSE}$
Error	$n - P - 1$	$SSE = \sum_{i=1}^{n} Y_i^2 - b_0\sum_{i=1}^{n} Y_i - b_1\sum_{i=1}^{n} X_{1i}Y_i - b_2\sum_{i=1}^{n} X_{2i}Y_i$	$MSE = \frac{SSE}{n - P - 1}$	
Total	$n - 1$	$SST = \sum_{i=1}^{n} Y_i^2 - n\bar{Y}^2$		

sion coefficients, the measure of random error is called the *error variance,* so that the F test is the ratio of the variance due to the regression divided by the error variance as shown in Equation (18.5):

$$F = \frac{MSR}{MSE} \quad \textbf{(18.5)}$$

where P is the number of explanatory variables in the regression model and F follows an F distribution with P and $n - P - 1$ degrees of freedom.

The decision rule is:

Reject H_0 at the α level of significance if $F > F_{U(P,n-P-1)}$; otherwise don't reject H_0.

For the data of the heating oil consumption problem, the ANOVA table (also displayed as part of Figure 18.2 on page 785) is presented in Table 18.4.

Table 18.4 **Analysis-of-variance table for testing the significance of a set of regression coefficients for the heating oil consumption problem.**

Source	df	Sums of Squares	Mean Square (Variance)	F
Regression	2	$(562.151)(3{,}247.4) + (-5.43658)(98{,}060.1) + (-20.0123)(18{,}057) - 15(216.493)^2 = 228{,}014.6263$	$\frac{228{,}014.6263}{2} = 114{,}007.31315$	$\frac{114{,}007.31315}{676.71692} = 168.47$
Error	$15 - 2 - 1 = 12$	$939{,}175.68 - (562.151)(3{,}247.4) - (-5.43658)(98{,}060.1) - (-20.0123)(18{,}057) = 8{,}120.6030$	$\frac{8{,}120.6030}{12} = 676.71692$	
Total	$15 - 1 = 14$	$939{,}175.68 - 15(216.443)^2 = 236{,}135.2293$		

Source: Format of Table 18.3.

If a level of significance of .05 is chosen, from Table E.5 we determine that the critical value on the F distribution (with 2 and 12 degrees of freedom) is 3.89, as depicted in Figure 18.4. From Equation (18.5), since $F = 168.47 > F_{U(2,\ 12)} = 3.89$, we can reject H_0 and conclude that *at least* one of the explanatory variables (temperature and/or insulation) is related to heating oil consumption.

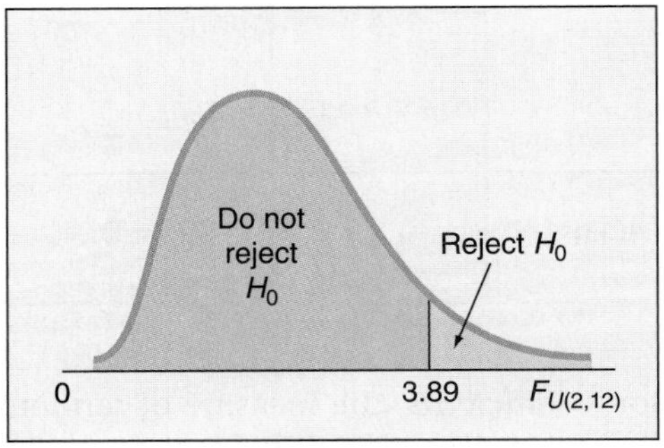

Figure 18.4
Testing of the significance of a set of regression coefficients at the .05 level of significance with 2 and 12 degrees of freedom.

Problems for Section 18.6

● 18.20 Referring to Problem 18.2 (the shoe durability problem) on page 786:
(a) Determine whether there is a significant relationship between long-term impact and the two explanatory variables at the .05 level of significance.
(b) Compute the p value and interpret its meaning.

Data file WARECOST.DAT

18.21 Referring to Problem 18.3 (the distribution cost problem) on page 787:
(a) Determine whether there is a significant relationship between distribution cost and the two explanatory variables (sales and number of orders) at the .05 level of significance.
(b) Compute the p value and interpret its meaning.

Data file ADRADTV.DAT

● 18.22 Referring to Problem 18.4 (the advertising media problem) on page 788:
(a) Determine whether there is a significant relationship between sales and the two explanatory variables (radio and television advertising and newspaper advertising) at the .05 level of significance.
(b) Compute the p value and interpret its meaning.

Data file SALARY2.DAT

18.23 Referring to Problem 18.5 (the employee salary problem) on page 788:
(a) Determine whether there is a significant relationship between weekly salary and the two explanatory variables (length of employment and age) at the .05 level of significance.
(b) Compute the p value and interpret its meaning.

Data file STANDBY.DAT

18.24 Referring to Problem 18.6 (the standby hours problem) on page 789:
(a) Determine whether there is a significant relationship between standby hours and the two explanatory variables (total staff present and remote hours) at the .05 level of significance.
(b) Compute the p value and interpret its meaning.

18.7 Testing Portions of a Multiple Regression Model

In developing a multiple regression model, the objective is to utilize only those explanatory variables that are useful in predicting the value of a dependent variable. If an explanatory variable is not helpful in making this prediction, it could be deleted from the multiple regression model and a model with fewer explanatory variables could be utilized in its place.

One method for determining the contribution of an explanatory variable is called **the partial *F*-test criterion** (see Reference 4). It involves determining the contribution to the regression sum of squares made by each explanatory variable after all the other explanatory variables have been included in a model. The new explanatory variable would only be included if it significantly improved the model. To apply the partial *F*-test criterion in our heating oil consumption problem containing two explanatory variables, we need to evaluate the contribution of the variable attic insulation (X_2) once average daily atmospheric temperature (X_1) has been included in the model and, conversely, we also must evaluate the contribution of the variable average daily atmospheric temperature (X_1) once attic insulation (X_2) has been included in the model.

The contribution of each explanatory variable to be included in the model can be determined by taking into account the regression sum of squares of a model that includes all explanatory variables except the one of interest, SSR (*all variables except k*). Thus, in general, to determine the contribution of variable *k* given that all other variables are already included, we would have

$$\begin{aligned} &\text{SSR}(X_k \mid \textit{all variables except } k) \\ &= \text{SSR}(\textit{all variables including } k) - \text{SSR}(\textit{all variables except } k) \end{aligned} \quad \textbf{(18.6a)}$$

If, as in the heating oil consumption problem, there are two explanatory variables, the contribution of each can be determined from Equations (18.6b) and (18.6c).

Contribution of Variable X_1 Given X_2 Has Been Included

$$\text{SSR}(X_1 \mid X_2) = \text{SSR}(X_1 \textit{ and } X_2) - \text{SSR}(X_2) \quad \textbf{(18.6b)}$$

Contribution of Variable X_2 Given X_1 Has Been Included

$$\text{SSR}(X_2 \mid X_1) = \text{SSR}(X_1 \textit{ and } X_2) - \text{SSR}(X_1) \quad \textbf{(18.6c)}$$

The term SSR(X_2) represents the sum of squares due to regression for a model that includes *only* the explanatory variable X_2 (amount of attic insulation); the term SSR(X_1) represents the sum of squares due to regression for a model that includes *only* the explanatory variable X_1 (average daily atmospheric temperature). Computer output obtained from the SAS REG procedure for these two models is presented in Figures 18.5 and 18.6 on page 798.

We can observe from Figure 18.5 that

$$\text{SSR}(X_2) = 51{,}076 \text{ (rounded)}$$

and, therefore, from Equation (18.6b),

$$\text{SSR}(X_1 \mid X_2) = \text{SSR}(X_1 \textit{ and } X_2) - \text{SSR}(X_2)$$

```
DEP VARIABLE: OIL                    (VARIANCE)
                        SUM OF           MEAN
SOURCE      DF        SQUARES          SQUARE        F VALUE     PROB > F
MODEL        1          51076           51076           3.59        .0807
ERROR       13      185058.76           14235
C TOTAL     14         236135
    ROOT MSE        119.31051        R-SQUARE         0.2163
    DEP MEAN          216.493        ADJ R-SQ         0.1560

                    PARAMETER        STANDARD      T FOR HO:
VARIABLE    DF       ESTIMATE           ERROR    PARAMETER=0  PROB > :T:

INTERCEP     1        345.378       74.690659           4.62       0.0005
INSU         1        -20.351       10.743429          -1.89       0.0807

VARIABLE    DF      TYPE I SS     TYPE II SS

INSU         1      51076.465      51076.465
```

Figure 18.5
Partial output of simple linear regression model of amount of heating oil consumed and amount of attic insulation (obtained from SAS PROC REG).

```
DEP VARIABLE: OIL                    (VARIANCE)
                        SUM OF           MEAN
SOURCE      DF        SQUARES          SQUARE        F VALUE     PROB > F
MODEL        1         178624          178624          40.38       0.0001
ERROR       13      57510.805            4424
C TOTAL     14         236135
    ROOT MSE           66.513        R-SQUARE         0.7565
    DEP MEAN          216.493        ADJ R-SQ         0.7378

                    PARAMETER        STANDARD      T FOR HO:
VARIABLE    DF       ESTIMATE           ERROR    PARAMETER=0  PROB > :T:

INTERCEP     1        436.438       38.639709          11.30       0.0001
TEMPF        1      -5.462208        0.859609          -6.35       0.0001

VARIABLE    DF      TYPE I SS     TYPE II SS

TEMPF        1         178624         178624
```

Figure 18.6
Partial output of simple linear regression model of amount of heating oil consumed and average daily atmospheric temperature (obtained from SAS PROC REG).

we have

$$\text{SSR}(X_1|X_2) = 228{,}015 - 51{,}076 = 176{,}939$$

We should note that this value, except for rounding, is also shown as the *Type II sum of squares (SS)* obtained from the SAS REG procedure for the regression model with two explanatory variables (see Figure 18.2 on page 785).

In order to determine whether X_1 significantly improves the model after X_2 has been included, we can now subdivide the regression sum of squares into two component parts as shown in Table 18.5.

The null and alternative hypotheses to test for the contribution of X_1 to the model would be

H_0: Variable X_1 does not significantly improve the model once variable X_2 has been included.

Table 18.5 **Analysis-of-variance table dividing the regression sum of squares into components to determine the contribution of variable X_1.**

Source	df	Sums of Squares	Mean Square (Variance)	F
Regression	2	228,015	114,007.5	
$\begin{Bmatrix} X_2 \\ X_1 \mid X_2 \end{Bmatrix}$	$\begin{Bmatrix} 1 \\ 1 \end{Bmatrix}$	$\begin{Bmatrix} 51{,}076 \\ 176{,}939 \end{Bmatrix}$	51,076 176,939	261.47
Error	12	8,120	MSE = 676.717	
Total	14	236,135		

H_1: Variable X_1 significantly improves the model once variable X_2 has been included.

The partial F-test criterion is expressed by

$$F = \frac{SSR(X_k \mid \text{all variables except } k)}{\text{MSE}} \tag{18.7}$$

where P is the number of explanatory variables in the regression model and F follows an F distribution with 1 and $n - P - 1$ degrees of freedom.

Thus, from Table 18.5 we have

$$F = \frac{176{,}939}{676.717} = 261.47$$

Since there are 1 and 12 degrees of freedom, respectively, if a level of significance of .05 is selected, from Table E.5 we can observe that the critical value is 4.75 (see Figure 18.7). Since the computed F value exceeds this critical F value ($261.47 > 4.75$), our decision would be to reject H_0 and conclude that the addition of variable X_1 (average daily atmospheric temperature) significantly improves a multiple regression model that already contains variable X_2 (attic insulation).

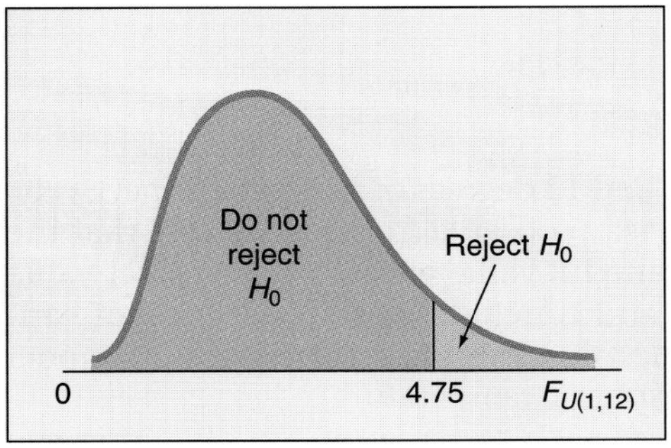

Figure 18.7
Testing for the contribution of a regression coefficient to a multiple regression model at the .05 level of significance with 1 and 12 degrees of freedom.

In order to evaluate the contribution of variable X_2 (attic insulation) to a model in which variable X_1 has been included we need to use Equation (18.6c):

$$SSR(X_2|X_1) = SSR(X_1\ and\ X_2) - SSR(X_1)$$

From Figures 18.2 and 18.6 we determine that

$$SSR(X_1) = 178{,}624$$

Therefore,

$$SSR(X_2|X_1) = 228{,}015 - 178{,}624 = 49{,}391 \text{ (rounded)}$$

Thus, in order to determine whether X_2 significantly improves a model after X_1 has been included, the regression sum of squares can be subdivided into two component parts as shown in Table 18.6.

Table 18.6 **Analysis-of-variance table dividing the regression sum of squares into components to determine the contribution of variable X_2.**

Source	df	Sums of Squares	Mean Square (Variance)	F
Regression	2	228,015	114,007.5	
X_1	1	178,624	178,624	
$X_2 \mid X_1$	1	49,391	49,391	72.99
Error	12	8,120	MSE = 676.717	
Total	14	236,135		

The null and alternative hypotheses to test for the contribution of X_2 to the model would be

H_0: Variable X_2 does not significantly improve the model once variable X_1 has been included.

H_1: Variable X_2 significantly improves the model once variable X_1 has been included.

Using Equation (18.7), we obtain

$$F = \frac{49{,}391}{676.717} = 72.99$$

as indicated in Table 18.6. Since there are 1 and 12 degrees of freedom, respectively, if a .05 level of significance is selected, we again observe from Figure 18.7 that the critical value of F is 4.75. Since the computed F value exceeds this critical value (72.99 > 4.75), our decision is to reject H_0 and conclude that the addition of variable X_2 (attic insulation) significantly improves the multiple regression model already containing X_1 (average daily atmospheric temperature).

Thus, by testing for the contribution of each explanatory variable after the other had been included in the model, we determine that each of the two explanatory variables contributed by significantly improving the model. Therefore, our multiple regression model should include both average daily atmospheric temperature X_1 and the amount of attic insulation X_2 in predicting the consumption of home heating oil.

Problems for Section 18.7

In Problems 18.25–18.29 at the .05 level of significance

(a) Determine whether each explanatory variable makes a significant contribution to the regression model. On the basis of these results, indicate the regression model that should be utilized in the problem.

(b) Compute the p values and interpret their meaning.

● 18.25 Refer to Problem 18.2 (the shoe durability problem) on page 786.

18.26 Refer to Problem 18.3 (the distribution cost problem) on page 787. Data file WARECOST.DAT

● 18.27 Refer to Problem 18.4 (the advertising media problem) on page 788. Data file ADRADTV.DAT

18.28 Refer to Problem 18.5 (the employee salary problem) on page 788. Data file SALARY2.DAT

18.29 Refer to Problem 18.6 (the standby hours problem) on page 789. Data file STANDBY.DAT

18.8 Inferences Concerning the Population Regression Coefficients

In Section 17.13 we may recall that tests of hypotheses were performed on the regression coefficients in a simple linear regression model in order to determine the significance of the relationship between X and Y. In addition, confidence intervals were used to estimate the population values of these regression coefficients. In this section, these procedures will be extended to situations involving multiple regression.

18.8.1 Tests of Hypothesis

To test the hypothesis that the population slope β_1 was zero, we used Equation (17.19):

$$t = \frac{b_1}{S_{b_1}}$$

However, this equation can be generalized for multiple regression as follows:

$$t = \frac{b_k}{S_{b_k}} \tag{18.8}$$

where P = number of explanatory variables in the regression equation

S_{b_k} = standard error of the regression coefficient b_k

and t follows a t distribution with $n - P - 1$ degrees of freedom.

Since the formulas for the standard errors of the regression coefficients are unwieldy with a large number of variables, it is fortunate that the results are provided as part of the output obtained from a computer package (see Figures 18.2, 18.24, and 18.25 on pages 785, 844, and 845, respectively).

Thus, if we wish to determine whether variable X_2 (amount of attic insulation) has a significant effect on the consumption of home heating oil, taking into account the average daily atmospheric temperature, the null and alternative hypotheses would be

$$H_0: \beta_2 = 0$$
$$H_1: \beta_2 \neq 0$$

From Equation (18.8) we have

$$t = \frac{b_2}{S_{b_2}}$$

and from the data of this problem,

$$b_2 = -20.0123 \quad \text{and} \quad S_{b_2} = 2.3425$$

so that

$$t = \frac{-20.0123}{2.3425} = -8.5431$$

If a level of significance of .05 is selected, from Table E.3 we can observe that for 12 degrees of freedom the critical values of t are -2.1788 and $+2.1788$ (see Figure 18.8).

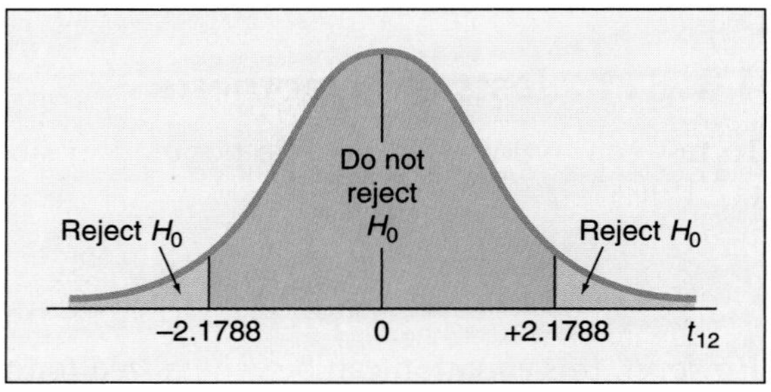

Figure 18.8
Testing for the significance of a regression coefficient at the .05 level of significance with 12 degrees of freedom.

Since we have $t = -8.5431 < t_{12} = -2.1788$, we reject H_0 and conclude that there is a significant relationship between variable X_2 (amount of attic insulation) and the consumption of heating oil, taking into account the average daily atmospheric temperature X_1.

In order to focus on the interpretation of this conclusion, we should note that there is a relationship between the value of the t-test statistic obtained from Equation (18.8) and the partial F-test statistic [Equation (18.7)] used to determine the contribution of X_2 to the multiple regression model. The t value was computed to be -8.5431, and the corresponding computed value of F was 72.99—which is the square of -8.5431. This points up the following relationship between t and F:[1]

$$t_a^2 = F_{1,a} \tag{18.9}$$

where a is the number of degrees of freedom.

Thus the test of significance for a particular regression coefficient (in this case b_2) is actually a test for the significance of adding a particular variable into a regression model given that the other variables have been included. Therefore, the t test for the regression coefficient is equivalent to testing for the contribution of each explanatory variable as discussed in Section 18.7.

18.8.2 Confidence Interval Estimation

Rather than attempting to determine the significance of a regression coefficient, we may be more concerned with estimating the population value of a regression coefficient. In multiple regression analysis a confidence interval estimate can be obtained from

$$b_k \pm t_{n-P-1} S_{b_k} \tag{18.10}$$

For example, if we wish to obtain a 95% confidence-interval estimate of the population slope β_1 (that is, the effect of average daily temperature X_1 on consumption of heating oil Y, holding constant the effect of attic insulation X_2), we would have, from Equation (18.10) and Figure 18.2,

$$b_1 \pm t_{12} S_{b_1}$$

Since the critical value of t at the 95% confidence level with 12 degrees of freedom is 2.1788 (see Table E.3), we have

$$-5.43658 \pm (2.1788)(.33622)$$

$$-5.43658 \pm .732556$$

$$-6.169136 \le \beta_1 \le -4.704024$$

Thus, taking into account the effect of attic insulation, we estimate that the effect of average daily atmospheric temperature is to reduce average consumption of heating oil by between approximately 4.7 and 6.17 gallons for each 1°F increase in temperature. Furthermore, we have 95% confidence that this interval correctly estimates the true relationship between these variables. Of course, from a hypothesis-testing viewpoint, since this confidence interval does not include zero, the regression coefficient β_1 would be considered to have a significant effect.

Problems for Section 18.8

● 18.30 Referring to Problem 18.2 (the shoe durability problem) on page 786, set up a 95% confidence interval estimate of the population slope between long-term impact and forefoot impact.

Data file WARECOST.DAT

18.31 Referring to Problem 18.3 (the distribution cost problem) on page 787, set up a 95% confidence interval estimate of the population slope between distribution cost and sales.

Data file ADRADTV.DAT

● 18.32 Referring to Problem 18.4 (the advertising media problem) on page 788, set up a 95% confidence interval estimate of the population slope between sales and radio and television advertising.

Data file SALARY2.DAT

18.33 Referring to Problem 18.5 (the employee salary problem) on page 788, set up a 95% confidence interval estimate of the population slope between weekly salary and length of employment.

Data file STANDYBY.DAT

18.34 Referring to Problem 18.6 (the standby hours problem) on page 789, set up a 95% confidence interval estimate of the population slope between standby hours and total staff present.

18.9 Confidence Interval Estimates for Predicting μ_{YX} and Y_I

In Section 18.3, we used the multiple regression equation to obtain a prediction of the average consumption of heating oil for a house that has 6 inches of attic insulation in a month in which the average daily temperature was 30°F. A confidence interval estimate of μ_{YX}, the true mean value of Y, and a prediction interval estimate of an individual value Y_I can be obtained by extending the procedures discussed in Sections 17.11 and 17.12 to the multiple regression model. However, as indicated in the previous section in our discussion of the standard error of the regression coefficients, the formulas used for predicting μ_{YX} and Y_I also become unwieldy when there are several explanatory variables included in a multiple regression model and, therefore, they usually are expressed in terms of matrix notation (see Reference 4). However, these interval estimates are available as optional procedures in most software packages. Figure 18.9 illustrates the confidence interval estimate for each observation in the sample obtained from the SAS REG procedure.

OBS	ACTUAL	PREDICT VALUE	STD ERR PREDICT	LOWER95% MEAN	UPPER95% MEAN	RESIDUAL
1	275.300	284.651	10.300	262.210	307.092	-9.351
2	363.800	355.326	11.196	330.932	379.721	8.474
3	164.300	144.565	10.905	120.805	168.324	19.735
4	40.800	45.207	12.923	17.050	73.363	-4.407
5	94.300	94.136	10.465	71.335	116.936	0.164072
6	230.900	257.233	7.081	241.806	272.661	-26.333
7	366.700	393.148	12.493	365.927	420.369	-26.448
8	300.600	318.535	15.435	284.905	352.165	-17.935
9	237.800	236.986	12.389	209.994	263.979	0.813551
10	121.400	159.609	12.867	131.574	187.645	-38.209
11	31.400	8.650	13.666	-21.126	38.426	22.750
12	203.500	219.177	6.767	204.434	233.921	-15.677
13	441.100	387.946	12.130	361.516	414.375	53.154
14	323.000	295.524	10.323	273.033	318.015	27.476
15	52.500	46.706	12.390	19.710	73.703	5.794

Figure 18.9
Confidence intervals obtained from SAS PROC REG for the heating oil usage model.

18.10 Coefficient of Partial Determination

In Section 18.4 we discussed the coefficient of multiple determination ($r^2_{Y.12}$), which measured the proportion of the variation in Y that was explained by variation in the two explanatory variables. Now that we have examined ways in which the contribution of each explanatory variable to the multiple regression model can be evaluated, we can also compute the **coefficients of partial determination** ($r^2_{Y1.2}$ and $r^2_{Y2.1}$). The coefficients measure the proportion of the variation in the dependent variable that is explained by each explanatory variable while controlling for, or holding constant, the other explanatory variable(s). Thus in a multiple regression model with two explanatory variables we have

$$r^2_{Y1.2} = \frac{\text{SSR}(X_1|X_2)}{\text{SST} - \text{SSR}(X_1 \text{ } and \text{ } X_2) + \text{SSR}(X_1|X_2)} \quad \textbf{(18.11a)}$$

and also

$$r^2_{Y2.1} = \frac{\text{SSR}(X_2|X_1)}{\text{SST} - \text{SSR}(X_1 \text{ } and \text{ } X_2) + \text{SSR}(X_2|X_1)} \quad \textbf{(18.11b)}$$

where SSR($X_1|X_2$) = sum of squares of the contribution of variable X_1 to the regression model given that variable X_2 has been included in the model

SST = total sum of squares for Y

SSR(X_1 *and* X_2) = regression sum of squares when variables X_1 and X_2 are both included in the multiple regression model

SSR($X_2|X_1$) = sum of squares of the contribution of variable X_2 to the regression model given that variable X_1 has been included in the model

while in a multiple regression model containing several (P) explanatory variables, we have

$$r^2_{Yk.(\textit{all variables except } k)} = \frac{\text{SSR}(X_k|\textit{all variables except } k)}{\text{SST} - \text{SSR}(\textit{all variables including } k) + \text{SSR}(X_k|\textit{all variables except } k)} \quad \textbf{(18.12)}$$

For our heating oil consumption problem we can compute

$$r^2_{Y1.2} = \frac{176{,}939}{236{,}135 - 228{,}015 + 176{,}939}$$
$$= 0.9561$$

and

$$r^2_{Y2.1} = \frac{49{,}391}{236{,}135 - 228{,}015 + 49{,}391}$$
$$= 0.8588$$

The coefficient of partial determination of variable Y with X_1 while holding X_2 constant ($r^2_{Y1.2}$) can be interpreted to mean that for a fixed (constant) amount of attic insulation, 95.61% of the variation in the consumption of heating oil in January can be explained by the variation in the average daily atmospheric temperature in that month. Moreover, the coefficient of partial determination of variable Y with X_2 while holding X_1 constant ($r^2_{Y2.1}$) can be interpreted to mean that for a given (constant) average daily atmospheric temperature, 85.88% of the variation in the consumption of heating oil in January can be explained by variation in the amount of attic insulation.

Problems for Section 18.10

● 18.35 Referring to Problem 18.2 (the shoe durability problem) on page 786, compute the coefficients of partial determination $r^2_{Y1.2}$ and $r^2_{Y2.1}$ and interpret their meaning.

Data file WARECOST.DAT

18.36 Referring to Problem 18.3 (the distribution cost problem) on page 787, compute the coefficients of partial determination $r^2_{Y1.2}$ and $r^2_{Y2.1}$ and interpret their meaning.

Data file ADRADTV.DAT

● 18.37 Referring to Problem 18.4 (the advertising media problem) on page 788, compute the coefficients of partial determination $r^2_{Y1.2}$ and $r^2_{Y2.1}$ and interpret their meaning.

Data file SALARY2.DAT

18.38 Referring to Problem 18.5 (the employee salary problem) on page 788, compute the coefficients of partial determination $r^2_{Y1.2}$ and $r^2_{Y2.1}$ and interpret their meaning.

Data file STANDBY.DAT

18.39 Referring to Problem 18.6 (the standby hours problem) on page 789, compute the coefficients of partial determination $r^2_{Y1.2}$ and $r^2_{Y2.1}$ and interpret their meaning.

18.11 The Curvilinear Regression Model

In our discussion of simple regression in Chapter 17 and multiple regression in this chapter, we have thus far assumed that the relationship between Y and each explanatory variable is linear. However, we may recall that several different types of relationships between variables were introduced in Section 17.3. One of the more common nonlinear relationships that was illustrated was a curvilinear polynomial relationship between two variables (see Figure 17.3 on page 720, panels D to F) in which Y increases (or decreases) at a *changing rate* for various values of X. This model of a polynomial relationship between X and Y can be expressed as

$$Y_i = \beta_0 + \beta_1 X_{1i} + \beta_{11} X_{1i}^2 + \epsilon_i \quad \textbf{(18.13)}$$

where β_0 = Y intercept

β_1 = *linear* effect on Y

β_{11} = *curvilinear* effect on Y

ϵ_i = random error in Y for observation i

This regression model is similar to the multiple regression model with two explanatory variables [see Equation (18.1a) on page 784] except that the second explanatory variable in this instance is merely the square of the first explanatory variable.

As in the case of multiple linear regression, when sample data are analyzed the sample regression coefficients (b_0, b_1, and b_{11}) are used as estimates of the population parameters (β_0, β_1, and β_{11}). Thus the regression equation for the curvilinear polynomial model having one explanatory variable (X_1) and a dependent variable (Y) is

$$\hat{Y}_i = b_0 + b_1 X_{1i} + b_{11} X_{1i}^2 \quad \textbf{(18.13a)}$$

An alternative approach to the curvilinear regression model expressed in Equation (18.13a) is to center the data by subtracting the mean of the explanatory variable from each value in the model. This centered regression model is expressed as Equation (18.14):

$$\hat{Y}_i = b_0' + b_1'(X_{1i} - \overline{X}_1) + b_{11}(X_{1i} - \overline{X}_1)^2 \quad \textbf{(18.14)}$$

Centering such a model may be done both for numerical and statistical reasons. First, from a computational perspective, more accuracy may be achieved if the mean is subtracted from each value before a regression equation is solved numerically. Second, and perhaps more important, the variance of the explanatory variable may be greatly inflated because X_1 and X_1^2 are positively correlated. Since X_1 and X_1^2 carry essentially the same information, it is sometimes difficult to determine whether the X_1 term is truly statistically significant. It is also possible that the slope of the X_1 term will have a sign opposite of the trend indicated by a scatter diagram. To avoid such problems, some researchers (see Reference 10) recommend centering the X_1 variable in a curvilinear regression model.

Mathematically, Equation (18.13a) and Equation (18.14) are equivalent. They give the same values for $\hat{Y}_i$ and b_{11} and they explain the same amount of the total variation. The difference between the two models occurs in the *intercept* (b_0 versus b_0') and *linear effect* (b_1 versus b_1') terms.

18.11.1 Finding the Regression Coefficients and Predicting Y

In order to illustrate the curvilinear regression model, let us suppose that the marketing department of a large supermarket chain wanted to study the price elasticity of packages of disposable razors. A sample of 15 stores with equal store traffic and product placement (i.e., at the checkout counter) was selected. Five stores were randomly assigned to each of three price levels (79 cents, 99 cents, and \$1.19) for the package of razors. The number of packages sold and the price at each store are presented in Table 18.7.

Data file DISPRAZ.DAT

Table 18.7 **Sales and price of packages of disposable razors for a sample of 15 stores.**

Sales	Price (cents)	Sales	Price (cents)
142	79	115	99
151	79	126	99
163	79	77	119
168	79	86	119
176	79	95	119
91	99	100	119
100	99	106	119
107	99		

In order to investigate the selection of the proper model expressing the relationship between price and sales, a scatter diagram is plotted as in Figure 18.10. An examination of Figure 18.10 indicates that the decrease in sales levels off for an

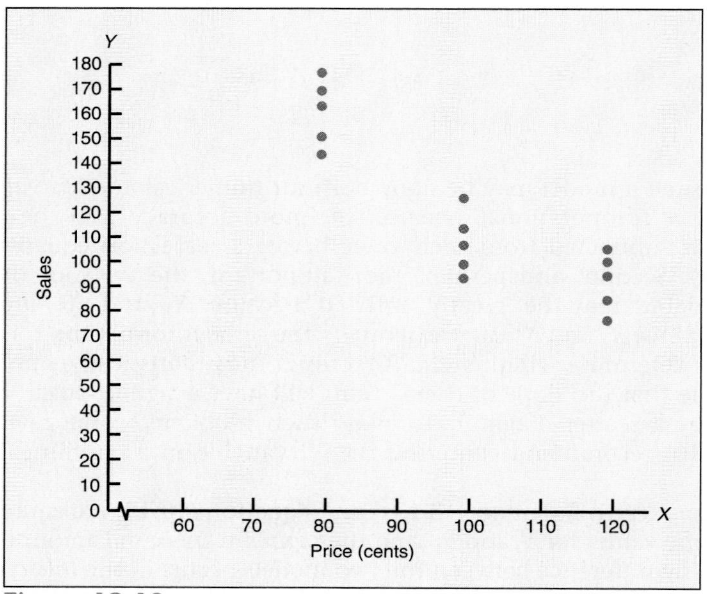

Figure 18.10
Scatter diagram of price (X) and sales (Y).

increasing price. Therefore, it appears that it may be more appropriate to use a curvilinear model to estimate sales based on price rather than a linear model.

As in the case of multiple regression, the values of the three sample regression coefficients (b_0', b_1', and b_{11}) can be obtained most easily by accessing a computer package (see References 12, 14, 15, 17).

Figure 18.11 presents partial output from MINITAB for the data of Table 18.7 using the centered model [Equation (18.14)]. From Figure 18.11 we observe that

$$b_0' = 107.8 \qquad b_1' = -1.68 \qquad b_{11} = .0465$$

Therefore the centered curvilinear model can be expressed as

$$\hat{Y}_i = 107.8 - 1.68(X_{1i} - \overline{X}_1) + .0465(X_{1i} - \overline{X}_1)^2$$

where $\hat{Y}_i$ = predicted average sales for store i
X_{1i} = price of disposable razors in store i

```
The regression equation is
sales = 108-1.68 pricecen+0.0465 prcensq

Predictor          Coef            Stdev          t-ratio
Constant        107.800            5.756            18.73
pricecen        -1.6800           0.2035            -8.26
prcensq         0.04650          0.01762             2.64

s = 12.87          R-sq = 86.2%          R-sq(adj) = 83.9%

Analysis of Variance

SOURCE           DF          SS            MS
Regression        2     12442.8        6221.4
Error            12      1987.6         165.6
Total            14     14430.4

SOURCE           DF      SEQ SS
pricecen          1     11289.6
prcensq           1      1153.2
```

Figure 18.11
Partial output for the data of Table 18.7 using MINITAB.

As depicted in Figure 18.12 on page 810, this curvilinear regression equation is plotted on the scatter diagram to indicate how well the selected regression model fits the original data. From our curvilinear regression equation and Figure 18.12, the Y intercept (b_0', computed as 107.80) can be interpreted to mean that the predicted sales for $X_{1i} = \overline{X}_1 = 99$ is 107.8 packages. To interpret the coefficients b_1' and b_{11}, we see from Figure 18.12 that the sales decrease with increasing price; nevertheless, we also observe that these decreases in sales level off or become reduced with increasing price. This can be seen by predicting average sales for packages priced at 79 cents, 99 cents, and 119 cents ($1.19). Using our curvilinear regression equation

$$\hat{Y}_i = 107.8 - 1.68(X_{1i} - \overline{X}_1) + .0465(X_{1i} - \overline{X}_1)^2$$

where $\overline{X}_1 = 99$,

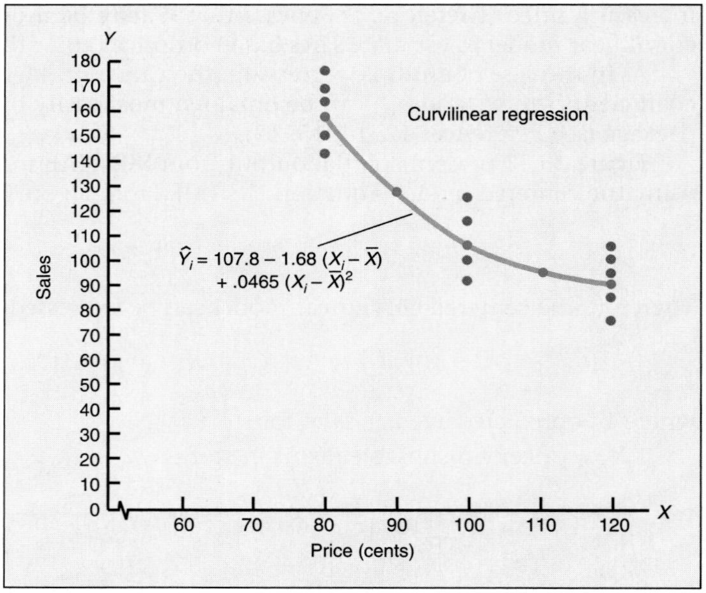

Figure 18.12
Scatter diagram expressing the curvilinear relationship between price (*X*) and sales (*Y*).

for $X_{1i} = 79$ we have

$$\hat{Y}_i = 107.8 - (1.68)(79 - 99) + (.0465)(79 - 99)^2 = 160$$

for $X_{1i} = 99$ we have

$$\hat{Y}_i = 107.8 - (1.68)(99 - 99) + (.0465)(99 - 99)^2 = 107.8$$

for $X_{1i} = 119$ we have

$$\hat{Y}_i = 107.8 - (1.68)(119 - 99) + (.0465)(119 - 99)^2 = 92.8$$

Thus we observe that a store selling the razors for 79 cents is expected to sell 52.2 more packages than a store selling them for 99 cents, but a store selling them for 99 cents is expected to sell only 15 more packages than a store selling them for \$1.19.

18.11.2 Testing for the Significance of the Curvilinear Model

Now that the curvilinear model has been fitted to the data, we can determine whether there is a significant curvilinear relationship between sales, *Y*, and price, *X*. In a manner similar to multiple regression (see Section 18.6), the null and alternative hypotheses can be set up as follows:

H_0: $\beta_1 = \beta_{11} = 0$ (There is no relationship between X_1 and Y.)

H_1: β_1 and/or $\beta_{11} \neq 0$ (At least one regression coefficient is not equal to zero.)

The null hypothesis can be tested by utilizing an *F* test [Equation (18.5)] as indicated in Table 18.8.

Table 18.8 Analysis-of-variance table testing the significance of a curvilinear polynomial relationship.

Source	df	Sums of Squares	Mean Square (Variance)	F
Regression	2	$SSR = b_0' \sum_{i=1}^{n} Y_i + b_1' \sum_{i=1}^{n} (X_{1i} - \bar{X}_1)Y_i + b_{11} \sum_{i=1}^{n} (X_{1i} - \bar{X}_1)^2 Y_i - n\bar{Y}^2$	$MSR = \frac{SSR}{2}$	$\frac{MSR}{MSE}$
Error	$n - 3$	$SSE = \sum_{i=1}^{n} Y_i^2 - b_0' \sum_{i=1}^{n} Y_i - b_1' \sum_{i=1}^{n} (X_{1i} - \bar{X}_1)Y_i - b_{11} \sum_{i=1}^{n} (X_{1i} - \bar{X}_1)^2 Y_i$	$MSE = \frac{SSE}{n - 3}$	
Total	$n - 1$	$SST = \sum_{i=1}^{n} Y_i^2 - n\bar{Y}^2$		

For the data of Table 18.7 the ANOVA table is displayed as part of the computer output in Figure 18.11 on page 809.

If a level of significance of .05 is chosen, from Table E.5 we find that for 2 and 12 degrees of freedom the critical value on the F distribution is 3.89 (see Figure 18.13). Utilizing Equation (18.5), since

$$F = \frac{MSR}{MSE} = \frac{6{,}221.4}{165.6} = 37.57 > F_{U(2,12)} = 3.89$$

we can reject the null hypothesis (H_0) and conclude that there is a significant curvilinear relationship between sales and price of razors.

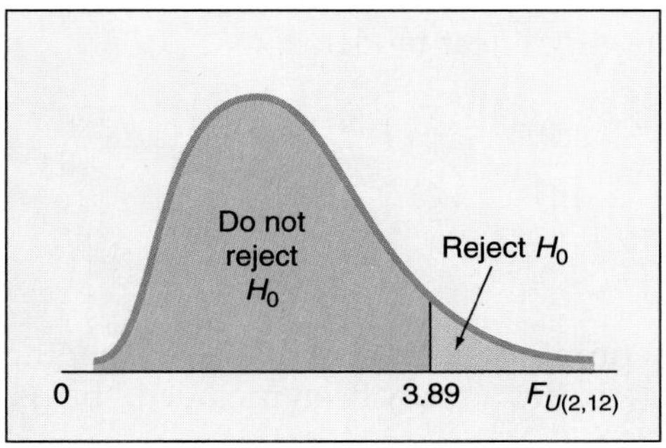

Figure 18.13
Testing for the existence of a curvilinear relationship at the .05 level of significance with 2 and 12 degrees of freedom.

In the multiple regression model we computed the coefficient of multiple determination $r^2_{Y.12}$ (see Section 18.4) to represent the proportion of variation in Y that is explained by variation in the explanatory variables. In curvilinear regression analysis, this coefficient can be computed from Equation (18.3):

$$r^2_{Y.12} = \frac{SSR}{SST}$$

From Figure 18.11,

$$SSR = 12{,}442.8 \qquad SST = 14{,}430.4$$

Thus, as displayed in Figure 18.11,

$$r^2_{Y.12} = \frac{\text{SSR}}{\text{SST}} = \frac{12{,}442.8}{14{,}430.4} = .862$$

This coefficient of multiple determination, computed as .862, can be interpreted to mean that 86.2% of the variation in sales can be explained by the curvilinear relationship between sales (Y) and price (X). We may also recall from Section 18.4 that we computed an *adjusted* $r^2_{Y.12}$ to take into account the number of explanatory variables and the number of degrees of freedom. In our curvilinear regression model, $P = 2$, since we have two explanatory variables, X_1 and its square (X_1^2). Thus, using Equation (18.4) for the razor sales data, we have

$$\begin{aligned} r^2_{\text{adj}} &= 1 - \left[\left(1 - r^2_{Y.12}\right)\frac{(15-1)}{(15-2-1)}\right] \\ &= 1 - \left[(1 - .862)\frac{14}{12}\right] \\ &= 1 - .161 \\ &= .839 \end{aligned}$$

18.11.3 Testing the Curvilinear Effect

In using a regression model to examine a relationship between two variables, we would like to fit not only the most accurate model but also the simplest model expressing that relationship. Therefore, it becomes important to examine whether there is a significant difference between the curvilinear model

$$Y_i = \beta_0' + \beta_1'(X_{1i} - \overline{X}_1) + \beta_{11}(X_{1i} - \overline{X}_1)^2 + \epsilon_i$$

and the linear model

$$Y_i = \beta_0 + \beta_1 X_i + \epsilon_i$$

These two models may be compared by determining the regression effect of adding the curvilinear term, given that the linear term has already been included, that is, $\text{SSR}(X_1^2|X_1)$.

We may recall that in Section 18.8.1 we used the t test for the regression coefficient to determine whether each particular variable made a significant contribution to the regression model. From Figure 18.11 on page 809, we observe that the standard error of each regression coefficient and its corresponding t statistic is available as part of the MINITAB output. Thus we may test the significance of the contribution of the curvilinear effect with the following null and alternative hypotheses:

H_0: Including the curvilinear effect does not significantly improve the model ($\beta_{11} = 0$).

H_1: Including the curvilinear effect significantly improves the model ($\beta_{11} \neq 0$).

For our data

$$t = \frac{b_{11}}{S_{b_{11}}}$$

so that

$$t = \frac{.0465}{.01762} = 2.64$$

If a level of significance of .05 is selected, from Table E.3, we find with 12 degrees of freedom that the critical values are −2.1788 and +2.1788 (see Figure 18.14). Since $t = 2.64 > t_{12} = 2.1788$, our decision would be to reject H_0 and conclude that the curvilinear model is significantly better than the linear model in representing the relationship between sales and price.

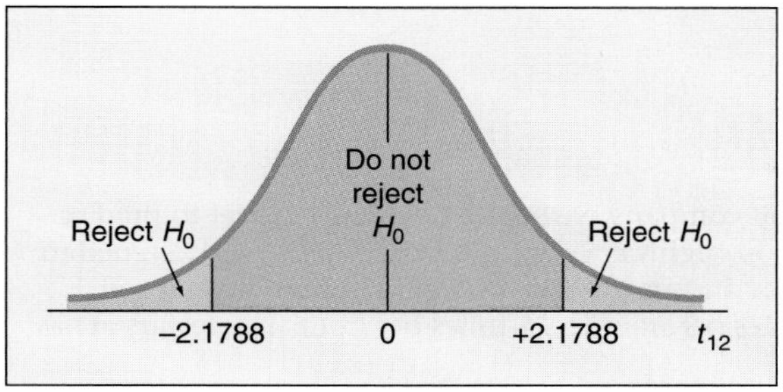

Figure 18.14
Testing for the contribution of the curvilinear effect to a regression model at the .05 level of significance with 12 degrees of freedom.

18.11.4 Testing the Linear Effect

Now that we have tested for the curvilinear effect, we should also determine whether there is a significant difference between the curvilinear model

$$Y_i = \beta'_0 + \beta'_1(X_{1i} - \overline{X}_1) + \beta_{11}(X_{1i} - \overline{X}_1)^2 + \epsilon_i$$

and the model that includes *only* the curvilinear effect

$$Y_i = \beta'_0 + \beta_{11}(X_{1i} - \overline{X}_1)^2 + \epsilon_i$$

As in the case of the curvilinear effect, we may use the *t* test to determine the contribution of the linear effect given that the curvilinear effect is already in the model.

For our data,

$$t = \frac{b'_1}{S_{b'_1}}$$

so that

$$t = \frac{-1.68}{.2035} = -8.26$$

The null and alternative hypotheses to test for the contribution of the linear effect to the regression model are

H_0: $\beta_1' = 0$ (Including the linear effect does not improve the curvilinear effect model.)

H_1: $\beta_1' \neq 0$ (Including the linear effect does improve the curvilinear effect model.)

If a level of significance of .05 is selected, from Table E.3, we find with 12 degrees of freedom that the critical values are −2.1788 and +2.1788 (see Figure 18.14 on page 813). Since $t = -8.26 < t_{12} = -2.1788$, our decision would be to reject H_0 and conclude that the curvilinear model that includes the linear effect is significantly better than the model that includes only the curvilinear effect.

Problems for Section 18.11

Data file SPEED.DAT

● 18.40 A researcher for a major oil company wished to develop a model to predict miles per gallon based upon highway speed. An experiment was designed in which a test car was driven during two trial periods at a particular speed, which ranged from 10 miles per hour to 75 miles per hour. The results are as follows:

Observation	Miles per Gallon	Speed (Miles per Hour)	Observation	Miles per Gallon	Speed (Miles per Hour)
1	4.8	10	15	21.3	45
2	5.7	10	16	22.0	45
3	8.6	15	17	20.5	50
4	7.3	15	18	19.7	50
5	9.8	20	19	18.6	55
6	11.2	20	20	19.3	55
7	13.7	25	21	14.4	60
8	12.4	25	22	13.7	60
9	18.2	30	23	12.1	65
10	16.8	30	24	13.0	65
11	19.9	35	25	10.1	70
12	19.0	35	26	9.4	70
13	22.4	40	27	8.4	75
14	23.5	40	28	7.6	75

Assuming a curvilinear polynomial relationship between speed and mileage, access a computer package to perform a regression analysis. On the basis of the results obtained

(a) Set up a scatter diagram between speed and miles per gallon.
(b) State the equation for the curvilinear model.
(c) Predict the mileage obtained when the car is driven at 55 miles per hour.

(d) Determine whether there is a significant curvilinear relationship between mileage and speed at the .05 level of significance.
(e) Interpret the meaning of the coefficient of multiple determination $r^2_{Y.12}$.
(f) Compute the adjusted r^2.
(g) Perform a residual analysis on your results and determine the adequacy of the fit of the model.
(h) At the .05 level of significance, determine whether the curvilinear model is a better fit than the linear regression model.

18.41 An industrial psychologist would like to develop a model to predict the number of typing errors based upon the amount of alcoholic consumption. A random sample of 15 typists was selected with the following results:

Typist	X Alcoholic Consumption (ounces)	Y Number of Errors
1	0	2
2	0	6
3	0	3
4	1	7
5	1	5
6	1	9
7	2	12
8	2	7
9	2	9
10	3	13
11	3	18
12	3	16
13	4	24
14	4	30
15	4	22

Data file ALCOHOL.DAT

Assuming a curvilinear relationship between alcoholic consumption and the number of errors, access a computer package to perform the regression analysis. On the basis of the results obtained
(a) Set up a scatter diagram between alcoholic consumption X and the number of errors Y.
(b) State the equation for the curvilinear model.
(c) Predict the number of errors made by a typist who has consumed 2.5 ounces of alcohol.
(d) Determine whether there is a significant curvilinear relationship between alcoholic consumption and the number of errors made at the .05 level of significance.
(e) Interpret the meaning of the coefficient of multiple determination $r^2_{Y.12}$.
(f) Compute the adjusted r^2.
(g) Perform a residual analysis on your results and determine the adequacy of the fit of the model.
(h) At the .05 level of significance, determine whether the curvilinear model is a better fit than the linear regression model.

18.42 Suppose that an agronomist wanted to design a study in which a wide range of fertilizer levels (pounds per hundred square feet) were to be used in order to determine whether the relationship between the yield of tomatoes and amount of fertilizer would be fit by a curvilinear model. Six fertilizer levels were to be utilized: 0, 20, 40, 60, 80, and 100 pounds per hundred square feet.

Data file TOMYLD2.DAT

These six levels were randomly assigned to plots of land with the following results:

Plot	Amount of Fertilizer (pounds per 100 square feet)	Yield (pounds)
1	0	6
2	0	9
3	20	19
4	20	24
5	40	32
6	40	38
7	60	46
8	60	50
9	80	48
10	80	54
11	100	52
12	100	58

Assuming a curvilinear relationship between the amount of fertilizer used and tomato yield, access a computer package to perform the regression analysis.

(a) Set up a scatter diagram between amount of fertilizer and yield.
(b) State the regression equation for the curvilinear model.
(c) Predict the yield of tomatoes (in pounds) for a plot that has been fertilized with 70 pounds per hundred square feet of natural organic fertilizer.
(d) Determine whether there is a significant relationship between the amount of fertilizer used and tomato yield at the .05 level of significance.
(e) Compute the *p* value in (d) and interpret its meaning.
(f) Compute the coefficient of multiple determination $r^2_{Y.12}$ and interpret its meaning.
(g) Compute the adjusted r^2.
(h) At the .05 level of significance, determine whether the curvilinear model is superior to the linear regression model.
(i) Compute the *p* value in (h) and interpret its meaning.
(j) Perform a residual analysis on your results and determine the adequacy of the fit of the model.

Data file HINGOIL.DAT

18.43 Referring to the data of Table 18.1 on page 783
(a) Fit a multiple regression model that includes a linear relationship between oil consumption and temperature and a curvilinear relationship between oil consumption and the amount of attic insulation.
(b) Perform a residual analysis of your results and determine the adequacy of the fit of the model.
(c) At the .05 level of significance, determine whether the model that includes this curvilinear term is superior to the multiple linear regression model.

18.12 Dummy-Variable Models

In our discussion of multiple regression models thus far we have assumed that each explanatory (or independent) variable is numerical. However, there are many occasions in which categorical variables need to be considered as part of the model development process. For example, referring to the Employee Satisfaction Survey, we recall that in Section 17.15 we used the number of years in the workforce to develop a model to predict income. In addition, we may also wish to include the effect of such factors as gender, whether the individuals participate in budgetary decisions, whether they take part in decisions that affect their work, and whether they are proud to be working for the organization.

The use of **dummy variables** is the vehicle that permits us to consider categorical explanatory variables as part of the regression model. If a given categorical explanatory variable has two categories, then only one dummy variable will be needed to represent the two categories. The particular dummy variable (X_d) would be defined as

$$X_d = \begin{cases} 0 & \text{if the observation was in category 1} \\ 1 & \text{if the observation was in category 2} \end{cases}$$

In order to illustrate the application of dummy variables in regression, let us examine a model for predicting the income of employees based on the number of years in the workforce (X_1) and whether or not the individual participates in budgetary decisions. Thus a dummy variable for participates in budgetary decisions (X_2) could be defined as

$$X_2 = \begin{cases} 0 & \text{if the individual does not participate in budgetary decisions} \\ 1 & \text{if the individual participates in budgetary decisions} \end{cases}$$

Assuming that the slope between income and the number of years in the workforce is the same for both groups,[2] the regression model could be stated as

$$Y_i = \beta_0 + \beta_1 X_{1i} + \beta_2 X_{2i} + \epsilon_i \qquad (18.15)$$

where Y_i = income for employee i

β_0 = Y intercept

β_1 = slope of income with number of years in the workforce, holding constant whether the individual participates in budgetary decisions

β_2 = incremental effect of the individual participating in budgetary decisions, holding number of years in the workforce constant

ϵ_i = random error in Y for employee i

Using the sample of 57 employees whose occupation is classified as technical/sales, the model stated in Equation (18.15) was fitted. The resulting sample regression coefficients (b_0, b_1, and b_2), standard errors, and t values are summarized in Table 18.9.

Table 18.9 Summary of results for dummy variable model.

Variable Name	Regression Coefficient	Standard Error	t
Constant	13.936	3.850	3.62
Years	0.7314	0.1759	4.16
Participation in budget decisions	8.027	3.341	2.40

Note the following:

1. Holding constant the effect of whether the individual participates in budgetary decisions, each additional year in the workforce is estimated to be worth an average of $731.40 toward the employee's income.
2. b_2 measures the effect on income of having participated in budgetary decisions ($X_2 = 1$) as compared with not having participated in budgetary decisions ($X_2 = 0$). Thus, holding the number of years in the workforce constant, we estimate that an employee who participates in budgetary decisions will have, on average, an income $8,027 above that of someone who does not participate in budgetary decisions.

Using the results of Table 18.9, the model for these data may be stated as

$$\hat{Y}_i = 13.936 + 0.7314X_{1i} + 8.027X_{2i}$$

For employees who do not participate in budgetary decisions the model reduces to

$$\hat{Y}_i = 13.936 + 0.7314X_{1i}$$

since $X_2 = 0$.
For employees who do participate in budgetary decisions the model reduces to

$$\hat{Y}_i = 21.963 + 0.7314X_{1i}$$

since $X_2 = 1$.

The fitted models for the two types of employees are displayed in Figure 18.15.

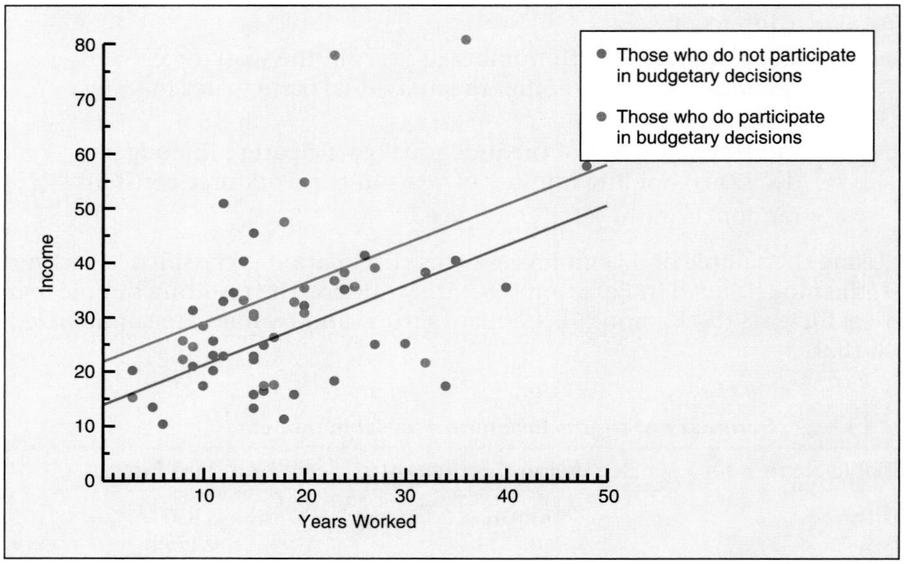

Figure 18.15
Regression models for employees who do and do not participate in budgetary decisions.

Problems for Section 18.12

18.44 Under what circumstances would we want to include a dummy variable in a regression model?

18.45 What assumption concerning the slope between the response variable Y and the explanatory variable X must be made when a dummy variable is included in a regression model?

18.46 A bank would like to develop a model to predict the total sum of money that customers withdraw from automatic teller machines (ATMs) on a weekend based on the median value of homes in the neighborhood the ATM is located and the ATM's location (0 = not a shopping center; 1 = shopping center). A random sample of 15 ATMs is selected with the following results:

Data file ATM2.DAT

ATM Number	Amount Withdrawn ($000)	Median Value of Homes ($000)	Location of ATM
1	120	225	1
2	99	170	0
3	91	153	1
4	82	132	0
5	124	237	1
6	104	187	1
7	127	245	1
8	80	125	1
9	115	215	1
10	97	170	0
11	117	223	0
12	86	147	0
13	109	197	1
14	94	167	0
15	112	210	0

Access an available computer package to perform a multiple linear regression analysis. On the basis of the results obtained

(a) State the multiple regression equation.
(b) Interpret the meaning of the slopes in this problem.
(c) Predict the average amount of money withdrawn for a neighborhood in which the median value of homes is $200,000 for an ATM that is located in a shopping center.
(d) Determine whether there is a significant relationship between the amount of money withdrawn and the two explanatory variables (median value of houses and the dummy variable of ATM location) at the .05 level of significance.
(e) Interpret the meaning of the coefficient of multiple determination $r^2_{Y.12}$.
(f) Compute the adjusted r^2.
(g) Perform a residual analysis of your results and determine the adequacy of the fit of the model.
(h) At the .05 level of significance, determine whether each explanatory variable makes a contribution to the regression model. On the basis of these results, indicate the regression model that should be used in this problem.
(i) Set up 95% confidence interval estimates of the population slope for the relationship between the amount of money withdrawn and median value of homes, and for the amount of money withdrawn and ATM location.
(j) Compute the coefficients of partial determination and interpret them.
(k) What assumption about the slope of amount of money withdrawn with median value of homes must be made in this problem?

Interchapter Problems for Section 18.12

Data file PETFOOD.DAT

● 18.47 Referring to Problem 17.1 on page 716, suppose that in addition to studying the effect of shelf space on the sales of pet food, the marketing manager also wanted to study the effect of product placement on sales. Suppose that in stores 2, 6, 9, and 12 the pet food was placed at the front of the aisle while in the other stores it was placed at the back of the aisle. Access an available computer package to perform a multiple linear regression analysis. On the basis of the results obtained

(a) State the multiple regression equation.
(b) Interpret the meaning of the slopes in this problem.
(c) Predict the average weekly sales of pet food for a store with 8 square feet of shelf space that is situated at the back of the aisle.
(d) Determine whether there is a significant relationship between sales and the two explanatory variables (shelf space and the dummy variable aisle position) at the .05 level of significance.
(e) Interpret the meaning of the coefficient of multiple determination $r^2_{Y.12}$.
(f) Compute the adjusted r^2.
(g) Compare $r^2_{Y.12}$ with the r^2 value computed in Problem 17.19(a) on page 731 and the adjusted r^2 of (f) with the adjusted r^2 computed in Problem 17.19(b). Explain the results.
(h) At the .05 level of significance, determine whether each explanatory variable makes a contribution to the regression model. On the basis of these results, indicate the regression model that should be used in this problem.
(i) Set up 95% confidence-interval estimates of the population slope for the relationship between sales and shelf space and between sales and aisle location.
(j) Compare the slope obtained in (b) with the slope for the simple linear regression model of Problem 17.7 on page 725. Explain the difference in the results.
(k) Compute the coefficients of partial determination and interpret their meaning.
(l) What assumption about the slope of shelf space with sales must be made in this problem?
(m) Perform a residual analysis on your results and determine the adequacy of the fit of the model.

Data file BB93.DAT

18.48 Referring to Problem 17.82 on page 778, suppose that in addition to using earned run average to predict the number of wins, we wanted to include the league (American versus National) as an explanatory variable.

(a) State the multiple regression equation.
(b) Interpret the meaning of the slopes in this problem.
(c) Predict the average number of wins for a team with an E.R.A. of 4.00 in the American League.
(d) Determine whether there is a significant relationship between wins and the two explanatory variables (E.R.A. and league) at the .05 level of significance.
(e) Interpret the meaning of the coefficient of multiple determination $r^2_{Y.12}$.
(f) Compute the adjusted r^2.
(g) Compare $r^2_{Y.12}$ with the value computed in Problem 17.82(f) on page 778. Explain the results.
(h) Perform a residual analysis on your results and determine the adequacy of the fit of the model.
(i) At the .05 level of significance, determine whether each explanatory variable makes a contribution to the regression model. On the basis of these results, indicate the regression model that should be used in this problem.
(j) Set up 95% confidence interval estimates of the population slope for the relationship between wins and E.R.A. and between wins and league.

(k) Compare the slope obtained in (b) with the slope for the simple linear regression model of Problem 17.83 on page 778. Explain the difference in the results.
(l) Compute the coefficients of partial determination and interpret their meaning.
(m) What assumption about the slope of wins with E.R.A. must be made in this problem?

18.13 Other Types of Regression Models

In our discussion of multiple regression models we have thus far examined the multiple linear model [Equation (18.1a)], the curvilinear polynomial model [Equations (18.13) and (18.14)], and the dummy-variable model [Equation (18.15)]. See pages 784, 807, and 817.

18.13.1 Interaction Terms in Regression Models

In the multiple linear regression model [Equation (18.1b)] we have included only terms that express a relationship between the explanatory variables and a dependent variable Y. However, in some situations the relationship between X_1 and Y changes for differing values of X_2. In such a case, an *interaction* term involving the product of explanatory variables may be included. With two explanatory variables, such an interaction model may be stated as

$$Y_i = \beta_0 + \beta_1 X_{1i} + \beta_2 X_{2i} + \beta_3 X_{1i} X_{2i} + \epsilon_i \qquad \textbf{(18.16)}$$

As an example of an interaction model we may refer back to the dummy-variable model discussed in Section 18.12. We may recall that Equation (18.15) postulated a dummy-variable model in which the slope of X_1 was constant for each category of the dummy variable (X_2). If in fact the slope of income with number of years in the workforce was different for employees who did or did not make budgetary decisions, an interaction term consisting of the product of the two explanatory variables should be included. For that example the model would be stated as

$$Y_i = \beta_0 + \beta_1 X_{1i} + \beta_2 X_{2i} + \beta_3 X_{1i} X_{2i} + \epsilon_i$$

where Y_i = income
β_0 = Y intercept
β_1 = slope of income with number of years in the workforce, holding participation in budgetary decisions constant
β_2 = incremental effect of participation in budgetary decisions, holding number of years in the workforce constant
β_3 = slope representing interaction of number of years in the workforce and participation in budgetary decisions
ϵ_i = random error in Y for employee i

18.13.2 Regression Models Using Transformations

In Section 17.8 we discussed the assumptions of normality, homoscedasticity, and independence of error that are involved in the regression model. In many circumstances the effect of violations of these assumptions can be overcome by transforming the dependent variable, the explanatory variables, or both.

By using the transformed variables we are often able to obtain a simpler model than had we maintained the original variables. By reexpressing X and/or Y we may simplify the relationship to one that is linear in its transformation. Unfortunately, the choice of an appropriate transformation is often not an easy one to make. Among the transformations along a "ladder of powers" discussed by Tukey (see References 4 and 19) are the square-root transformation, the logarithmic transformation, and the reciprocal transformation. If a **square-root transformation** were applied to the values of each of two explanatory variables, the multiple regression model would be

$$Y_i = \beta_0 + \beta_1 \sqrt{X_{1i}} + \beta_2 \sqrt{X_{2i}} + \epsilon_i \qquad (18.17)$$

If a **logarithmic transformation** had been applied, the model would be

$$Y_i = \beta_0 + \beta_1 \ln X_{1i} + \beta_2 \ln X_{2i} + \epsilon_i \qquad (18.18)$$

If a **reciprocal transformation** were applied, the model would be

$$Y_i = \beta_0 + \beta_1 \frac{1}{X_{1i}} + \beta_2 \frac{1}{X_{2i}} + \epsilon_i \qquad (18.19)$$

Interestingly, in some situations the use of a transformation can change what appears to be a nonlinear model into a linear model. For example, the multiplicative model

$$Y_i = \beta_0 X_{1i}^{\beta_1} X_{2i}^{\beta_2} \epsilon_i \qquad (18.20)$$

can be transformed (by taking natural logarithms of both the dependent and explanatory variables) to the model

$$\ln Y_i = \ln \beta_0 + \beta_1 \ln X_{1i} + \beta_2 \ln X_{2i} + \ln \epsilon_i \qquad (18.21)$$

Hence Equation (18.21) is linear in the natural logarithms. In a similar fashion the **exponential** model

$$Y_i = e^{\beta_0 + \beta_1 X_{1i} + \beta_2 X_{2i}} \epsilon_i \quad (18.22)$$

can also be transformed to linear form (by taking natural logarithms of both the dependent and explanatory variables). The resulting model is

$$\ln Y_i = \beta_0 + \beta_1 X_{1i} + \beta_2 X_{2i} + \ln \epsilon_i \quad (18.23)$$

Problems for Section 18.13

18.49 Referring to Problem 18.46 on page 819, suppose that we wish to include a term in the multiple regression model that represents the interaction of median value of homes and ATM location. Reanalyze the data by accessing a computer package for this model. On the basis of the results obtained

(a) State the multiple regression equation.

Data file ATM2.DAT

(b) At the .05 level of significance, determine whether the inclusion of an interaction term made a significant contribution to the model that already included the median value of homes and the ATM location. On the basis of these results indicate the regression model that should be used in this problem.

18.50 Referring to the data of Problem 18.40 on page 814 use a square-root transformation of the explanatory variable (speed) as in Equation (18.17) and access a computer package to reanalyze the data using this model. On the basis of your results

Data file SPEED.DAT

(a) State the regression equation.

(b) Predict the average mileage obtained when the car is driven at 55 miles per hour.

(c) Perform a residual analysis on your results and determine the adequacy of the fit of the model.

(d) At the .05 level of significance, is there a significant relationship between mileage and the square root of speed?

(e) Interpret the meaning of the coefficient of determination r^2 in this problem.

(f) Compute the adjusted r^2.

(g) Compare your results with those obtained in Problem 18.40. Which model would you choose? Why?

18.51 Referring to the data of Problem 18.40 on page 814, use a logarithmic transformation of the explanatory variable (speed) as in Equation (18.18) and access a computer package to reanalyze the data using this model. On the basis of your results

(a) State the regression equation.

Data file SPEED.DAT

(b) Predict the average mileage obtained when the car is driven at 55 miles per hour.

(c) Perform a residual analysis on your results and determine the adequacy of the fit of the model.

(d) At the .05 level of significance, is there a significant relationship between mileage and the natural logarithm of speed?

(e) Interpret the meaning of the coefficient of determination r^2 in this problem.

(f) Compute the adjusted r^2.

(g) Compare your results with those obtained in Problems 18.40 and 18.50. Which model would you choose? Why?

Interchapter Problems for Section 18.13

Data file PETFOOD.DAT

● 18.52 Referring to Problem 18.47 on page 820, suppose that we wish to include a term in the multiple regression model that represents the interaction of shelf space and aisle location. Reanalyze the data by accessing a computer package for this model. On the basis of the results obtained
(a) State the multiple regression equation.
(b) At the .05 level of significance, determine whether the inclusion of the interaction term made a significant contribution to the model that already included shelf space and aisle location. On the basis of these results indicate the regression model that should be used in this problem.

Data file BB93.DAT

18.53 Referring to Problem 18.48 on page 820, suppose that we wish to include a term in the multiple regression model that represents the interaction of E.R.A. and league. Reanalyze the data by accessing a computer package for this model. On the basis of the results obtained
(a) State the multiple regression equation.
(b) At the .05 level of significance, determine whether the inclusion of an interaction term made a significant contribution to the model that already included E.R.A. and league. On the basis of these results indicate the regression model that should be used in this problem.

18.14 Multicollinearity

One important problem in the application of multiple regression analysis involves the possible **multicollinearity** of the explanatory variables. This condition refers to situations in which some of the explanatory variables are highly correlated with each other. In such situations collinear variables do not provide new information, and it becomes difficult to separate the effect of such variables on the dependent or response variable. In such cases, the values of the regression coefficients for the correlated variables may fluctuate drastically, depending on which variables are included in the model.

One method of measuring collinearity uses the **variance inflationary factor (VIF)** for each explanatory variable. This VIF is defined as in Equation (18.24):

$$\text{VIF}_j = \frac{1}{1 - R_j^2} \qquad \textbf{(18.24)}$$

where R_j^2 represents the coefficient of multiple determination of explanatory variable X_j with all other X variables.

If there are only two explanatory variables, R_j^2 is merely the coefficient of determination between X_1 and X_2. If, for example, there were three explanatory variables, then R_1^2 would be the coefficient of multiple determination of X_1 with X_2 and X_3.

If a set of explanatory variables are uncorrelated, then VIF_j will be equal to 1. If the set were highly intercorrelated, then VIF_j might even exceed 10. Marquardt (see Reference 10) suggests that if VIF_j is greater than 10, there is too much correlation between variable X_j and the other explanatory variables. However, other researchers (see Reference 16) suggest a more conservative criterion that would employ alternatives to least squares regression if the maximum VIF_j were to exceed 5.

If we examine our heating oil data, we note from Table 18.2 on page 792 that the correlation between the two explanatory variables, temperature and attic insulation, is only .00892. Therefore, since there are only two explanatory variables in the model, we may compute the VIF_j from Equation (18.24):

$$\text{VIF}_1 = \text{VIF}_2 = \frac{1}{1 - (.00892)^2}$$

$$\text{VIF}_1 = \text{VIF}_2 \cong 1.00$$

Thus we may conclude that there is no reason to suspect any multicollinearity for the heating oil data.

We shall return to this subject of multicollinearity in Section 18.16 when we discuss model building.

Problems for Section 18.14

18.54 Referring to Problem 18.3 on page 787, the distribution cost problem, determine the VIF for each explanatory variable in the model. Is there reason to suspect the existence of multicollinearity?

18.55 Referring to Problem 18.4 on page 788, the advertising media problem, determine the VIF for each explanatory variable in the model. Is there reason to suspect the existence of multicollinearity?

18.56 Referring to Problem 18.5 on page 788, the employee salary problem, determine the VIF for each explanatory variable in the model. Is there reason to suspect the existence of multicollinearity?

18.57 Referring to Problem 18.6 on page 789, the standby hours problem, determine the VIF for each explanatory variable in the model. Is there reason to suspect the existence of multicollinearity?

18.15 Influence Analysis in Multiple Regression

18.15.1 Introduction

Now that we have considered the issue of whether multicollinearity exists between the explanatory variables and have evaluated the aptness of the fitted model through the use of residual analysis, we are ready to utilize the influence-analysis techniques discussed in Section 17.14 to determine whether any individual observations have undue influence on the fitted model.

We may recall that in Section 17.14 we considered three measures:

1. The hat matrix elements h_i
2. The Studentized deleted residuals t_i^*
3. Cook's distance statistic D_i

Figure 18.16 presents the values of these statistics for the heating oil data of Table 18.1, which have been obtained from the MINITAB computer package. We note from Figure 18.16 that certain data points have been highlighted for further analysis.

ROW	htngoil	predy	stress	hi	stdelres	cookd
1	275.3	284.651	-0.39144	0.156757	-0.37720	0.009495
2	363.8	355.326	0.36087	0.185246	0.34740	0.009870
3	164.3	144.565	0.83561	0.175717	0.82438	0.049616
4	40.8	45.207	-0.19519	0.246777	-0.18717	0.004161
5	94.3	94.136	0.00689	0.161823	0.00660	0.000003
6	230.9	257.233	-1.05200	0.074084	-1.05714	0.029517
7	366.7	393.148	-1.15911	0.230654	-1.17765	0.134267
8	300.6	318.535	-0.85651	0.352057	-0.84632	0.132868
9	237.8	236.986	0.03557	0.226801	0.03405	0.000124
10	121.4	159.609	-1.69004	0.244667	-1.85367	0.308398
11	31.4	8.650	1.02779	0.275988	1.03043	0.134224
12	203.5	219.177	-0.62414	0.067663	-0.60751	0.009424
13	441.1	387.946	2.30980	0.217438	2.96740	0.494131
14	323.0	295.524	1.15068	0.157465	1.16802	0.082488
15	52.5	46.706	0.25330	0.226864	0.24317	0.006276

Figure 18.16
Influence statistics obtained from MINITAB for the heating oil data.

18.15.2 Using the Hat Matrix Elements h_i

We may recall from Section 17.14 that each h_i reflects the influence of each X_i value on the fitted regression model. In a multiple regression model containing P explanatory variables, Hoaglin and Welsch (see Reference 7) suggest the following decision rule:

$$\text{If } h_i > 2(P+1)/n$$

then X_i is an influential point and may be considered a candidate for removal from the model.

For our heating oil data, since $n = 15$ and $P = 2$, our criterion would be to flag any h_i value greater than .40. Referring to Figure 18.16, we observe that none of the h_i values exceed .36; therefore, based on this criterion, there do not appear to be any observations that can be considered for removal from the model.

18.15.3 Using the Studentized Deleted Residuals t_i^*

We may recall from Section 17.14 that the Studentized deleted residual measures the difference between each observed value Y_i and predicted value $\hat{Y}_i$ obtained from a model that includes all observations other than i. In the multiple regression model, Hoaglin and Welsch suggested that

$$\text{if } \left|t_i^*\right| > t_{.10,n-P-2}$$

then the observed and predicted values are so different that observation i is an influential point that adversely affects the model and may be considered a candidate for removal.

For our heating oil data, since $n = 15$ and $P = 2$, our criterion would be to flag any t_i^* value greater than $|1.7959|$ (see Table E.3). Referring to Figure 18.16, we note that $t_{10}^* = -1.854$ and $t_{13}^* = 2.967$. Thus the tenth and thirteenth observations may each have an adverse effect on the model. We should note that these points were not previously flagged according to the h_i criterion. Hence, we shall consider Cook's D_i statistic that is based on both h_i and the standardized residual.

18.15.4 Using Cook's Distance Statistic D_i

Now that the h_i and t_i^* statistics have been considered, we turn to the D_i statistic that was discussed in Section 17.14.4. In the multiple regression model, Cook and Weisberg (see Reference 5) suggest that

$$\text{if } D_i > F_{.50,P+1,n-P-1}$$

then the observation may have an impact on the results of fitting a multiple regression model.

For our heating oil data, since $n = 15$ and $P = 2$, our criterion would be to flag any $D_i > F_{.50,3,12} = .835$. Referring to Figure 18.16, we note that none of the D_i values exceed .495, so that according to this criterion there are no values that may be deleted (although we should note that the largest D_i values are for observations 13 and 10, respectively). Hence, we would have no clear basis for removing any of the observations from the multiple regression model.

18.15.5 Summary

In this section we have discussed several criteria for evaluating the influence of each observation on the multiple regression model. The various statistics did not lead to a consistent set of conclusions. According to both the h_i and D_i criteria, none of the observations are candidates for removal from the model. However, according to the t_i^* criterion, observations 13 and 10 may be adversely affecting the fit of the model. Although some statisticians might argue for their removal, it seems reasonable to keep them in the model both because of the inconsistency of the influence statistics and because the model fits extremely well ($r^2_{Y.12} = .96$) regardless of whether or not these observations are included.

The use of regression diagnostics (such as residual analysis and influence analysis) has provided us with the opportunity to closely evaluate each point in the data. Perhaps we might be able to explain the large residuals in observations 13 and 10 as being due to other factors beside the atmospheric temperature and amount of attic insulation. For example, it is quite possible that the large positive residual for observation 13 might be explained by the fact that the thermostatic control was turned to an especially high level in a month in which the average monthly temperature was only 21 degrees Fahrenheit. On the other hand, the large negative residual of observation 10 might be explained by the fact that in a month in which the average atmospheric temperature was 63 degrees Fahrenheit, the thermostat was turned on less than what would have been expected for such situations.

Problems

For Problems 18.58–18.66, perform an influence analysis and determine whether any observations should be deleted from the model. If necessary, reanalyze the regression model after deleting these observations and compare your results to the original model.

Data file WARECOST.DAT
18.58 Refer to the distribution cost problem (see pages 787, 790, and 792).

Data file ADRADTV.DAT
● 18.59 Refer to the advertising media problem (see pages 788, 790, and 792).

Data file SALARY2.DAT
18.60 Refer to the employee salary problem (see pages 788, 790, and 792).

Data file STANDBY.DAT
18.61 Refer to the standby hours problem (pages 789, 790, and 792).

Data file SPEED.DAT
● 18.62 Refer to Problem 18.40 on page 814.

Data file ALCOHOL.DAT
18.63 Refer to Problem 18.41 on page 815.

Data file TOMYIELD.DAT
18.64 Refer to Problem 18.42 on page 815.

Data file HTNGOIL.DAT
18.65 Refer to Problem 18.43 on page 816.

Data file ATM2DAT
18.66 Refer to Problem 18.46 on page 819.

Data file SPEED.DAT
18.67 Referring to Problem 18.50 on page 823
(a) Perform an influence analysis and determine whether any observations should be deleted from the model. If necessary, reanalyze the regression model after deleting these observations and compare your results with the original model.
(b) Compare the results in (a) with those obtained in Problem 18.62.

Data file SPEED.DAT
18.68 Referring to Problem 18.51 on page 823
(a) Perform an influence analysis and determine whether any observations should be deleted from the model. If necessary, reanalyze the regression model after deleting these observations and compare your results with the original model.
(b) Compare the results in (a) with those obtained in Problems 18.62 and 18.67.

Interchapter Problems for Section 18.15

Data file PETFOOD.DAT
● 18.69 Referring to Problem 18.47 on page 820.
(a) Perform an influence analysis and determine whether any observations should be deleted from the model. If necessary, reanalyze the regression model after deleting these observations and compare your results with the original model.
(b) Compare the results in (a) with those obtained in Problem 17.69 on page 759.

Data file BB93.DAT
18.70 Referring to Problem 18.48 on page 820.
(a) Perform an influence analysis and determine whether any observations should be deleted from the model. If necessary, reanalyze the regression model after deleting these observations and compare your results with the original model.
(b) Compare the results in (a) with those obtained in Problem 17.82 on page 778.

18.16 An Example of Model Building

18.16.1 Introduction

In this chapter we have developed the multiple linear regression model and subsequently discussed the curvilinear polynomial model, models involving dummy variables, and models involving transformations of variables. In this section we shall conclude our discussion of regression by developing a model that includes a set of several categorical and numerical explanatory variables.

We may recall that in Chapter 17, only one numerical variable (number of years in the workforce) was used in the development of a regression model for predicting income in our Employee Satisfaction Survey. Let us now reevaluate this regression model by also considering such other numerical explanatory variables as years of education, years worked for the company, and number of promotions received. In addition, let us consider such categorical explanatory variables as gender (0 = female, 1 = male), participation in budgetary decisions, participation in decisions that affect work (recoded as 0 for sometimes or never and 1 for always or much of the time), importance of formal schooling, and importance of on-the-job training (each recoded as 0 for somewhat important or not-at-all important and 1 for very important or important), and proud to be working for this organization (recoded as 0 for indifferent or not at all proud and 1 for very proud or somewhat proud).

Before we begin to develop a model to predict income, we should keep in mind that a widely used criterion of model building is parsimony. This means that we wish to develop a regression model that includes the fewest number of explanatory variables that permits an adequate interpretation of the dependent variable of interest. Regression models with fewer explanatory variables are inherently easier to interpret, particularly since they are less likely to be affected by the problem of multicollinearity (see Section 18.14).

In addition, we should realize that the selection of an appropriate model when 10 explanatory variables are to be considered involves complexities that are not present for a model that contains only two explanatory variables. First, the evaluation of all possible regression models becomes more computationally complex. Second, although competing models can be quantitatively evaluated, there may not exist a uniquely *best* model but rather several equally appropriate models.

We shall begin our analysis of the Employee Satisfaction Survey data by first measuring the amount of collinearity that exists between the explanatory variables through the use of the variance inflationary factor [see Equation (18.24)]. Figure 18.17 on page 830 represents partial MINITAB output for a multiple linear regression model in which income is predicted from the 10 explanatory variables. We may observe that most of the VIF values are relatively small, ranging from a high of 2.0 for the years worked for the company to a low of 1.2 for the importance of formal schooling for the job. Thus, based on the criteria developed by Marquardt (see References 10 and 11), there is little evidence of multicollinearity among the set of explanatory variables. We also note that the coefficient of multiple determination is .410 and the adjusted r^2 is .281.

18.16.2 The Stepwise Regression Approach to Model Building

We may now continue our analysis of these data by attempting to determine the explanatory variables that might be deleted from the complete model. We shall first utilize a widely used search procedure called **stepwise regression,** which attempts to find the "best" regression model without examining all possible regressions. Once a *best* model has been found, residual analysis is utilized to evaluate the aptness of the model and influence measures are computed to determine whether any observations may be deleted.

```
The regression equation is

INCOME = 0.9 + 4.33 ORGMONEY + 0.765 WRKYEARS + 1.01 EDUC
         + 3.25 SEX - 0.205 EMPYEARS + 0.31 NUMPROMO + 5.38 IDECIDE
         - 1.75 PROUDORG + 2.13 SCHOOLNG - 6.78 TRAINING

Predictor       Coef      Stdev    t-ratio        p        VIF
Constant        0.88      14.00       0.06    0.950
ORGMONEY       4.329      3.724       1.16    0.251        1.3
WRKYEARS      0.7651     0.2274       3.37    0.002        1.7
EDUC          1.0097     0.8611       1.17    0.247        1.3
SEX            3.246      4.025       0.81    0.424        1.5
EMPYEARS     -0.2051     0.3310      -0.62    0.538        2.0
NUMPROMO       0.307      1.713       0.18    0.859        1.4
IDECIDE        5.382      4.765       1.13    0.265        1.2
PROUDORG      -1.748      6.009      -0.29    0.772        1.3
SCHOOLNG       2.133      3.932       0.54    0.590        1.2
TRAINING      -6.782      4.856      -1.40    0.169        1.3

s = 12.30      R-sq = 41.0%     R-sq(adj) = 28.1%
```

Figure 18.17
MINITAB output for full regression model with ten explanatory variables.

We may recall that in Section 18.7 the partial *F*-test criterion was used to evaluate portions of a multiple regression model. Stepwise regression extends this partial *F*-test criterion to a model with any number of explanatory variables. An important feature of this stepwise process is that an explanatory variable that has entered into the model at an early stage may subsequently be removed once other explanatory variables are considered. That is, in stepwise regression, variables are either added to or deleted from the regression model at each step of the model-building process. The stepwise procedure terminates with the selection of a *best-fitting* model when no variables can be added to or deleted from the last model fitted.

We may now observe this stepwise process for our data. Figure 18.18 represents a partial output obtained from MINITAB. For this example a significance level of .05 was utilized either to enter a variable into the model or to delete a variable from the model. The first variable entered into the model is WRKYEARS (number of years in the workforce). Since the *t* value of 3.98 is greater than the critical value for $\alpha = .05$ (that is, $t_{.05,55} = \pm 2.004$), WRKYEARS is included in the regression model.

```
STEPWISE REGRESSION OF INCOME ON 10 PREDICTORS, WITH N =  57

     STEP          1        2
 CONSTANT      17.19    13.94

 WRKYEARS       0.73     0.73
 T-RATIO        3.98     4.16

 ORGMONEY                 8.0
 T-RATIO                 2.40

 S              12.9     12.4
 R-SQ          22.38    29.88
```

Figure 18.18
Partial output for the model predicting income using stepwise regression obtained from MINITAB.

The next step involves the evaluation of the second variable to be included in this model. The variable to be chosen is that which will make the largest contribution to the model, given that the first explanatory variable has already been selected. For this model, the second variable is ORGMONEY (participation in budgetary decisions). Since the t value of 2.40 for ORGMONEY is greater than the critical value for $\alpha = .05$ (that is, $t_{.05,\ 54} = 2.0049$), ORGMONEY is included in the regression model.

Now that ORGMONEY has been entered into the model, we may determine whether WRKYEARS is still an important contributing variable or whether it may be eliminated from the model. Since the t value of 4.16 for WRKYEARS is also greater than the critical value for $\alpha = .05$ (that is, $t_{.05,54} = \pm 2.0049$), WRKYEARS should remain in the regression model.

The next step involves the determination of whether any of the remaining variables should be added to the model. Since none of the other variables meet the .05 criterion for entry into the model, the stepwise procedure terminates with a model that includes the number of years in the workforce and whether the employee participates in budgetary decisions.

Before we use residual analysis to test the aptness of this model, it would be appropriate to determine whether or not a model containing an interaction term between WRKYEARS and ORGMONEY is justified (see Section 18.13). Figure 18.19 presents partial output from MINITAB for a model that includes the interaction between the number of years in the workforce and whether the employee participates in budgetary decisions. From this model we observe that the t value for the contribution of WRKYEARS*ORGMONEY | WRKYEARS, ORGMONEY is 1.90. Since $1.90 < 2.0057$ (that is, when $\alpha = .05$, $t_{.05,\ 53} = \pm 2.0057$), we may conclude that the interaction term should not be included as part of the model. Thus, the model selected to predict income includes only the number of years in the workforce and whether the employee participates in budgetary decisions.

```
The regression equation is
INCOME = 17.6 + 0.531 WRKYEARS - 5.28 ORGMONEY + 0.730 WRK*ORG

Predictor        Coef       Stdev    t-ratio       p
Constant       17.595       4.227       4.16   0.000
WRKYEARS       0.5310      0.2017       2.63   0.011
ORGMONEY       -5.282       7.741      -0.68   0.498
WRK*ORG        0.7301      0.3851       1.90   0.063
```

Figure 18.19
Partial output from the interaction model obtained from MINITAB.

Now that the explanatory variables to be included in the model have been selected, a residual analysis should be undertaken to evaluate the aptness of the fitted model. Figure 18.20 on pages 832–833 presents partial output obtained from MINITAB for these purposes. We may observe from Figure 18.20 that the plots of the standardized residuals versus the number of years in the workforce and employee participation in budgetary decisions reveal no apparent pattern. In addition, a histogram of the standardized residuals indicates only moderate departure from normality. Since the residual analysis appeared to confirm the aptness of the fitted model, we may now utilize various influence measures to determine whether any of the observations have unduly influenced the fitted model. Figure 18.21 on page 834 represents the values of the h_i, t_i^*, and Cook's D_i statistics for our fitted

```
The regression equation is
INCOME = 13.9 + 0.731 WRKYEARS + 8.03 ORGMONEY

Predictor       Coef       Stdev    t-ratio       p      VIF
Constant      13.936       3.850       3.62   0.001
WRKYEARS      0.7314      0.1759       4.16   0.000      1.0
ORGMONEY       8.027       3.341       2.40   0.020      1.0

s = 12.37       R-sq = 29.9%     R-sq(adj) = 27.3%

Analysis of Variance

SOURCE        DF        SS         MS          F        P
Regression     2    3523.3     1761.7      11.51    0.000
Error         54    8267.7      153.1
Total         56   11791.0

SOURCE        DF    SEQ SS
WRKYEARS       1    2639.2
ORGMONEY       1     884.1



Histogram of STRES N = 57

Midpoint    Count
    -2.0        2  **
    -1.5        3  ***
    -1.0        5  *****
    -0.5       12  ************
     0.0       19  *******************
     0.5        5  *****
     1.0        5  *****
     1.5        3  ***
     2.0        0
     2.5        1  *
     3.0        2  **
```

Figure 18.20
MINITAB output for a model that includes work years and participation in budgetary decisions.

model. We note from Figure 18.21 that certain data points have been highlighted for further analysis.

For our fitted model, since $n = 57$ and $P = 2$, using the decision rule suggested by Hoaglin and Welsch (see Section 18.15.2), our criterion would be to flag any h_i value greater than $2(2 + 1)/57 = .1053$. Referring to Figure 18.21, we note that observations 2 ($h_2 = .1249$), 19 ($h_{19} = .1074$), and 43 ($h_{43} = .2081$) have h_i values that exceed .1053 and therefore are considered to be possible candidates for deletion from the model.

Turning to the Studentized deleted residual measure t_i^*, for our model, since $P = 2$ and $n = 57$ and using the decision rule suggested by Hoaglin and Welsch (see Section 18.15.3), our criterion would be to flag any $|t_i^*|$ value greater than 1.6741 (see Table E.3). Referring to Figure 18.21, we note that $t_{13}^* = 3.58772$, $t_{19}^* = 3.07302$, $t_{31}^* = 1.73938$, $t_{45}^* = 2.49158$, $t_{50}^* = -1.84229$, and $t_{55}^* = -2.03612$. Thus these observations may have an adverse effect on the model. We note that observation 19 was also flagged according to the h_i criterion, but observations 13, 31, 45, 50, and 55 were not.

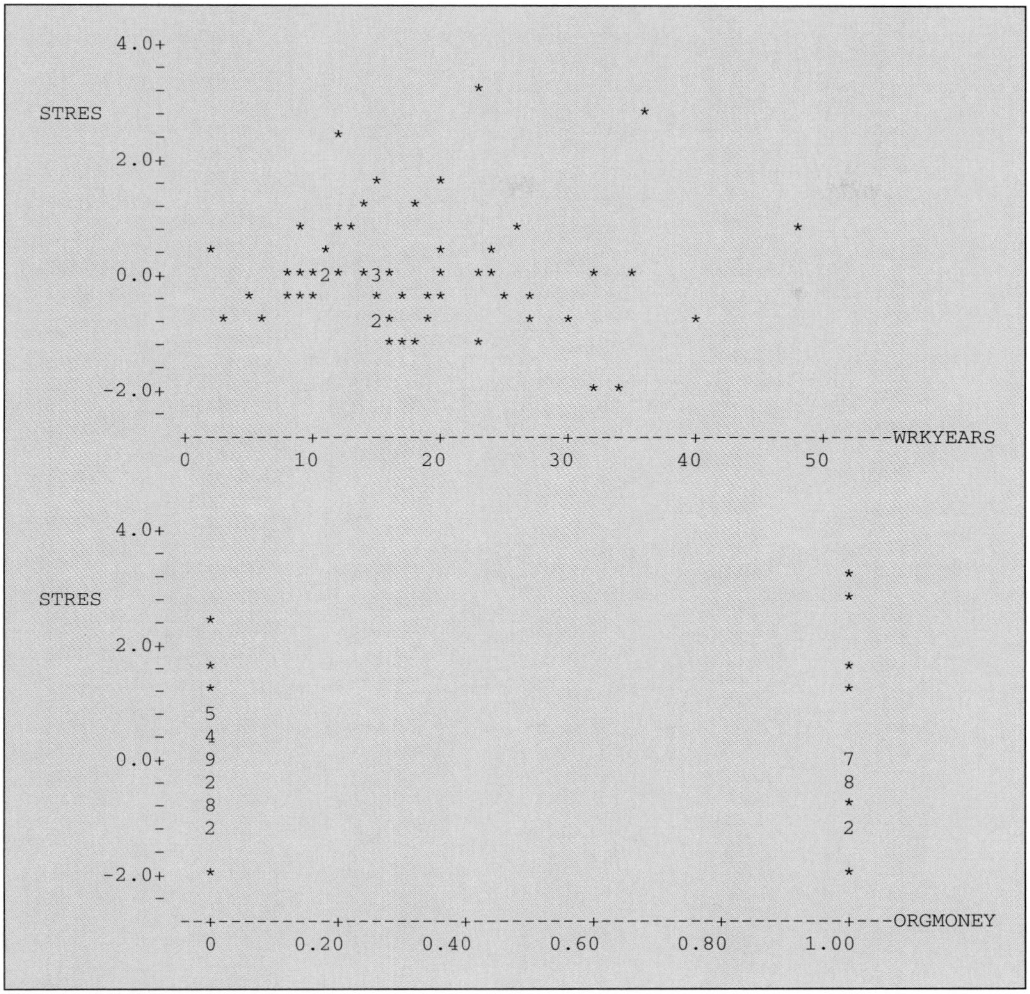

Figure 18.20 ***(continued)***

Thus, because of the lack of consistency between h_i and t_i^*, we should consider a third criterion, Cook's D_i statistic, which is based on both h_i and the standardized residual. For our model, in which $P = 2$ and $n = 57$, using the decision rule suggested by Cook and Weisberg (see Section 18.15.4), our criterion would be to flag any $D_i > F_{.50,(3,54)} = .800$. Referring to Figure 18.21, we note that none of the D_i values exceed .327, so according to this criterion there are no values that should be deleted. Hence, we would have no clear basis for removing any observations from the multiple regression model.

From Figure 18.20 we note that the VIF values are 1.0, so there is no multicollinearity between the two explanatory variables. The coefficient of multiple determination is .299 and the adjusted r^2 is .273. This compares favorably with the adjusted r^2 of .281 for the 10-explanatory-variable model. Thus our fitted model can be expressed as

$$\hat{Y}_i = 13.936 + 0.7314X_{1i} + 8.027X_{2i}$$

ROW	INCOME	YHAT	stres	hi	tresids	cookd
1	20.2	16.1302	0.34226	0.076499	0.33944	0.003234
2	35.7	43.1903	-0.64710	0.124879	-0.64358	0.019918
3	33.3	22.7124	0.87210	0.037343	0.87014	0.009834
4	32.0	28.5632	0.28202	0.030020	0.27960	0.000821
5	35.7	36.5907	-0.07362	0.044120	-0.07294	0.000083
6	33.8	32.2026	0.13225	0.047073	0.13104	0.000288
7	30.3	32.9339	-0.21789	0.045570	-0.21596	0.000756
8	20.4	21.9811	-0.13042	0.040077	-0.12922	0.000237
9	18.4	30.7573	-1.01608	0.033943	-1.01639	0.012092
10	11.8	27.1005	-1.25516	0.029426	-1.26203	0.015921
11	40.2	24.1751	1.31706	0.033087	1.32629	0.019786
12	22.0	27.8144	-0.48586	0.064574	-0.48239	0.005432
13	78.0	38.7847	3.24837	0.048101	3.58772	0.177733
14	13.7	24.9065	-0.92032	0.031566	-0.91900	0.009202
15	40.8	39.5335	0.10706	0.086008	0.10608	0.000360
16	23.0	24.9065	-0.15657	0.031566	-0.15515	0.000266
17	10.3	18.3243	-0.66881	0.059809	-0.66535	0.009485
18	22.7	24.9065	-0.18120	0.031566	-0.17957	0.000357
19	81.7	48.2923	2.85772	0.107380	3.07302	0.327472
20	55.3	36.5907	1.54654	0.044120	1.56726	0.036799
21	16.0	27.8319	-0.97066	0.029521	-0.97013	0.009553
22	33.1	35.8593	-0.22803	0.043602	-0.22602	0.000790
23	25.1	35.8768	-0.89700	0.057242	-0.89535	0.016285
24	16.1	24.1577	-0.68275	0.090274	-0.67934	0.015419
25	25.2	28.5458	-0.27899	0.060647	-0.27660	0.001675
26	30.6	32.9339	-0.19307	0.045570	-0.19134	0.000593
27	41.5	32.9513	0.70568	0.041503	0.70236	0.007188
28	38.1	37.3395	0.06365	0.067536	0.06306	0.000098
29	24.7	33.6827	-0.74280	0.044832	-0.73968	0.008632
30	13.6	17.5929	-0.33372	0.064968	0.33096	0.002579
31	45.7	24.9065	1.70765	0.031566	1.73938	0.031683
32	39.0	41.7101	-0.22580	0.059066	-0.22380	0.001067
33	26.9	21.9811	0.40575	0.040077	0.40259	0.002291
34	17.3	33.6653	-1.35303	0.044472	-1.36376	0.028401
35	38.3	39.5161	-0.10085	0.050236	-0.09992	0.000179
36	34.4	23.4438	0.90138	0.035013	0.89979	0.009826
37	23.1	22.7124	0.03193	0.037343	0.03163	0.000013
38	26.0	27.8144	-0.15162	0.064574	-0.15024	0.000529
39	23.4	21.9811	0.11704	0.040077	0.11597	0.000191
40	36.6	40.2474	-0.30288	0.052775	-0.30031	0.001704
41	25.0	25.6378	-0.05235	0.030448	-0.05186	0.000029
42	27.5	34.3966	-0.56998	0.043778	-0.56639	0.004958
43	58.0	49.0411	0.81362	0.208089	0.81104	0.057982
44	20.8	20.5184	0.02331	0.046757	0.02310	0.000009
45	51.6	22.7124	2.37947	0.037343	2.49158	0.073211
46	31.2	36.5907	-0.44560	0.044120	-0.44227	0.003055
47	31.3	20.5184	0.89246	0.046757	0.89075	0.013023
48	17.6	21.2497	-0.30155	0.043215	-0.29899	0.001369
49	36.9	38.7847	-0.15612	0.048101	-0.15470	0.000411
50	17.4	38.8022	-1.80276	0.079447	-1.84229	0.093494
51	17.8	34.3966	-1.37166	0.043778	-1.38321	0.028712
52	38.4	31.4886	0.56891	0.036059	0.56532	0.004036
53	16.5	24.9065	-0.69037	0.031566	-0.68699	0.005178
54	48.4	35.1280	1.09673	0.043488	1.09883	0.018229
55	21.9	45.3669	-1.97928	0.081865	-2.03612	0.116437
56	16.4	25.6378	-0.75821	0.030448	-0.75519	0.006016
57	28.9	29.2772	-0.03139	0.057124	-0.03110	0.000020

Figure 18.21
Influence statistics obtained from MINITAB for the model of Figure 18.20.

From this model we may conclude that holding constant the effect of whether the individual participates in budgetary decisions, each additional year in the workforce is estimated to be worth an average of $731.40 toward income. Holding the number of years in the workforce constant, an employee who participates in budgetary decisions is estimated to have an income $8,027 above that of someone who does not participate in budgetary decisions.

Compared with the simple linear regression model, the inclusion of the second independent variable, participating in budgetary decisions, has improved the adjusted r^2 to .2728 from .2097. However, we should realize that with more than 70% of the variation in income still unexplained, this model may be of limited practical value as a predictor of income, even though each of the two independent variables makes a statistically significant contribution to the regression model.

18.16.3 The Best Subset Approach to Model Building

Although stepwise regression has been used extensively for model building, in recent years with the increase of computer power available in statistical software packages has come the ability to examine either all possible regression models for a given set of independent variables or at least the best subset of models for a given number of independent variables. Figure 18.22 on page 836 represents partial output obtained from the STATISTIX package in which the best three regression models for a given number of parameters were provided according to two widely used criteria. The first criterion that is often used is the adjusted r^2, which adjusts the r^2 of each model to account for the number of variables in the model (see Section 18.4). Since models with different numbers of independent variables are to be compared, the adjusted r^2 is the appropriate criterion here rather than r^2.

Referring to Figure 18.22, we observe that the adjusted r^2 reaches a maximum value of .3283 when there are five independent variables (WRKYEARS, ORGMONEY, EDUC, IDECIDE, and TRAINING) plus the intercept term (for a total of six terms). Other models with similar adjusted r^2 are .3138 for the four-variable model consisting of WRKYEARS, ORGMONEY, EDUC, and TRAINING; .3268 for the six-variable model consisting of WRKYEARS, ORGMONEY, EDUC, SEX, IDECIDE, and TRAINING; and .3199 for the seven-variable model consisting of WRKYEARS, ORGMONEY, EDUC, SEX, EMPYEARS (years employed by Kalosha Industries), IDECIDE, and TRAINING. Thus, the best subset approach, unlike stepwise regression, has provided us with several alternative models to evaluate in greater depth using other criteria such as parsimony, interpretability, departure from model assumptions (as evaluated by residual analysis), and the influence of individual observations.

A second criterion often used in the evaluation of competing models is based on the statistic developed by Mallows (see References 4 and 17). This statistic, which is called C_{p*}, measures the differences of a fitted regression model from a true model, along with random error. The **C_{p*} statistic** is defined as

$$C_{p*} = \frac{\left(1 - R_{p*}^2\right)(n - T)}{1 - R_T^2} - \left(n - 2p^*\right) \quad (18.25)$$

BEST SUBSET REGRESSION MODELS FOR INCOME

UNFORCED INDEPENDENT VARIABLES: (A)WRKYEARS (B)ORGMONEY (C)EDUC (D)SEX (E)EMPYEARS (F)NUMPROMO (G)IDECIDE (H)PROUDORG (I)SCHOOLNG (J)TRAINING
3 "BEST" MODELS FROM EACH SUBSET SIZE LISTED.

P	CP	ADJUSTED R SQUARE	R SQUARE	RESID SS	MODEL VARIABLES
1	22.9	0.0000	0.0000	11790.9	INTERCEPT ONLY
2	7.5	0.2097	0.2238	9151.71	A
2	17.4	0.0805	0.0970	10647.6	D
2	17.9	0.0740	0.0905	10723.7	J
3	3.6	0.2728	0.2988	8267.65	A B
3	4.0	0.2676	0.2937	8327.46	A D
3	5.1	0.2534	0.2801	8488.35	A C
4	2.8	0.2979	0.3355	7834.57	A B C
4	3.2	0.2919	0.3299	7901.70	A B J
4	3.3	0.2903	0.3283	7919.75	A B D
5	2.7	0.3138	0.3628	7513.25	A B C J
5	2.8	0.3119	0.3611	7533.53	A B D J
5	2.9	0.3106	0.3598	7548.03	A B C G
6	2.7	0.3283	0.3882	7213.33	A B C G J
6	3.2	0.3207	0.3814	7294.40	A B C D J
6	3.4	0.3180	0.3789	7323.81	A C D G J
7	3.8	0.3268	0.3990	7086.73	A B C D G J
7	4.3	0.3205	0.3933	7153.96	A B C E G J
7	4.3	0.3198	0.3927	7161.06	A B C G I J
8	5.4	0.3199	0.4049	7016.77	A B C D E G J
8	5.5	0.3183	0.4035	7032.86	A B C D G I J
8	5.8	0.3140	0.3998	7077.40	A B C D F G J
9	7.1	0.3095	0.4082	6978.28	A B C D E G I J
9	7.3	0.3067	0.4057	7007.30	A B C D E G H J
9	7.4	0.3059	0.4051	7014.91	A B C D E F G J
10	9.0	0.2961	0.4092	6965.72	A B C D E G H I J
10	9.1	0.2953	0.4086	6973.68	A B C D E F G I J
10	9.3	0.2921	0.4059	7005.38	A B C D E F G H J
11	11.0	0.2813	0.4096	6960.87	A B C D E F G H I J

Figure 18.22
Best subset regression output obtained from STATISTIX.

where

$p^* = P + 1$, number of parameters included in a regression model with P independent variables

T = total number of parameters to be considered for inclusion in the regression model

$R^2_{p^*}$ = coefficient of multiple determination for a regression model that has p^* parameters

R^2_T = coefficient of multiple determination for a regression model that contains all T parameters

Using Equation (18.25) to compute C_{p^*} for the model containing five independent variables (WRKYEARS, ORGMONEY, EDUC, IDECIDE, and TRAINING) we would have

$$n = 57 \qquad p^* = 6 \qquad T = 10 + 1 = 11 \qquad R^2_{p^*} = .3882$$

so that

$$C_{p^*} = \frac{(1 - .3882)(57 - 11)}{1 - .4096} - [57 - 2(6)]$$

$$C_{p^*} = 2.667$$

When a regression model with P independent variables contains only random differences from a true model, the average value of C_{p^*} is p^*, the number of parameters. Thus, in evaluating many alternative regression models our goal is to find models whose C_{p^*} is close to or below p^*.

From Figure 18.22, we observe that numerous models contain C_{p^*} values that are below p^*. As was the case with the adjusted r^2 criterion, C_{p^*} has provided several alternative models for us to evaluate in greater depth using other criteria such as parsimony, interpretability, departure from model assumptions (as evaluated by residual analysis), and the influence of individual observations.

Problem for Section 18.16

18.71 Refer to Figure 18.22, and select three alternative models for further analysis. Compare the results obtained for these models to those of the model discussed in Section 18.16.

Data file EMPSAT.DAT

18.17 Logistic Regression

18.17.1 Development of the Logistic Regression Model

In our discussion of the simple linear regression model in Chapter 17 and multiple regression models in Sections 18.1–18.16, we have limited ourselves to considering response variables that are numerical. However, in many instances, the response variable is categorical and takes on one of only two values. The use of simple or multiple least squares regression for this type of response variable would often lead to predicted values that are less than zero or greater than one, values that cannot possibly occur.

An alternative approach, **logistic regression,** originally applied to survival data in the health sciences (see Reference 9), has been developed to enable us to use regression models to predict the probability of a particular categorical response for a given set of explanatory variables (which could be numerical or categorical). This logistic regression model is based on the **odds ratio,** which represents the probability of a success compared to the probability of failure. The odds ratio may be expressed as

$$\text{Odds ratio} = \frac{\text{probability of success}}{1 - \text{probability of success}} \qquad \textbf{(18.26)}$$

Using Equation (18.26), if the probability of success for an event was .50, the odds ratio would be

$$\text{Odds ratio} = \frac{.50}{1 - .50} = 1.0 \text{ or } 1 \text{ to } 1$$

and if the probability of success for an event was .75, the odds ratio would be

$$\text{Odds ratio} = \frac{.75}{1 - .75} = 3.0 \text{ or } 3 \text{ to } 1$$

The logistic regression model is based on the natural logarithm of this odds ratio.[3] A mathematical method called *maximum likelihood estimation* is usually used to develop a regression model to predict the natural logarithm of this odds ratio. This model may be expressed as

$$\ln(\text{Odds ratio}_i) = \beta_0 + \beta_1 X_{1i} + \beta_2 X_{2i} + \cdots + \beta_K X_{Ki} + \epsilon_i \qquad (18.27)$$

where K is the number of independent variables in the model and ϵ_i is random error for observation i. For sample data we would have

$$\ln(\text{Estimated odds ratio}_i) = b_0 + b_1 X_{1i} + b_2 X_{2i} + \cdots + b_K X_{Ki} \qquad (18.28)$$

Once the logistic regression model has been fit to a set of data, the estimated odds ratio may be obtained by raising the mathematical constant e to the power equal to the natural logarithm of the estimated odds ratio. This may be expressed as

$$\text{Estimated odds ratio} = e^{\ln(\text{Estimated odds ratio})} \qquad (18.29)$$

Once the estimated odds ratio has been obtained, we may find the estimated probability of success from

$$\text{Estimated probability of success} = \frac{\text{Estimated odds ratio}}{1 + \text{Estimated odds ratio}} \qquad (18.30)$$

18.17.2 Application

To illustrate the logistic regression model, let us suppose that the marketing department for a travel and entertainment credit card company was about to embark on a periodic campaign to convince existing holders of the company's standard credit card to upgrade to one of the company's premium cards for a nominal annual fee. The major decision facing the marketing department concerns which of the current standard credit card holders should be targeted for the campaign. Data available from a sample of 30 credit card holders who were contacted during last year's campaign indicate the following: whether the credit card holder upgraded from the standard to a premium card (0 = no, 1 = yes), the total amount of credit card purchases (in thousands of dollars) using the company's credit card in the one year prior to the campaign (X_1), and whether the credit card holder possessed additional credit cards (which required an additional charge) for other members of the household (X_2: 0 = no, 1 = yes). The data are presented in Table 18.10.

Data file
LOGPURCH.DAT

Table 18.10 Purchase behavior, annual credit card spending, and possession of additional credit cards for a sample of 30 credit card holders.

Observation	Purchase Behavior	Annual Spending	Possession of Additional Credit Card	Observation	Purchase Behavior	Annual Spending	Possession of Additional Credit Card
1	0	32.1007	0	16	0	23.7609	0
2	1	34.3706	1	17	0	35.0388	1
3	0	4.8749	0	18	1	49.7388	1
4	0	8.1263	0	19	0	24.7372	0
5	0	12.9783	0	20	1	26.1315	1
6	0	16.0471	0	21	0	31.3220	1
7	0	20.6648	0	22	1	40.1967	1
8	1	42.0483	1	23	0	35.3899	0
9	0	42.2264	1	24	0	30.2280	0
10	1	37.9900	1	25	1	50.3778	0
11	1	53.6063	1	26	0	52.7713	0
12	0	38.7936	0	27	0	27.3728	0
13	0	27.9999	0	28	1	59.2146	1
14	1	42.1694	0	29	1	50.0686	1
15	1	56.1997	1	30	1	35.4234	1

Figure 18.23 represents partial output for the logistic regression model obtained from STATISTIX.

```
UNWEIGHTED LOGISTIC REGRESSION OF BUY

PREDICTOR                                   Wald Statistic
VARIABLES      COEFFICIENT     STD ERROR       COEF/SE          P

CONSTANT       b0 -6.92923       2.83241        -2.45        0.0144
SPENDING       b1  0.13925       0.06594         2.11        0.0347
EXTRA          b2  2.77118       1.16550         2.38        0.0174

DEVIANCE                      20.08
P-VALUE                      0.8275
DEGREES OF FREEDOM               27

CASES INCLUDED 30   MISSING CASES 0
```

Figure 18.23
Partial logistic regression output for the data of Table 18.10 obtained from STATISTIX.

From Figure 18.23, we note that the response variable has been named BUY, X_1 is named SPENDING, and X_2 is named EXTRA. The regression coefficients b_0, b_1, and b_2 may be interpreted as follows:

1. The regression constant b_0 is equal to -6.92923.

2. The regression coefficient b_1 is equal to 0.13925. This can be interpreted to mean that holding constant the effect of whether the credit card holder has additional cards for members of the household, for each increase of \$1,000 in annual credit card spending using the company's card, we estimate that the *natural logarithm of the odds ratio of purchasing the premium card* will increase by 0.13925.
3. The regression coefficient b_2 is equal to 2.77118. This can be interpreted to mean that holding constant the annual credit card spending, we estimate that the *natural logarithm of the odds ratio of purchasing the premium card* will increase by 2.77118 for a credit card holder who has additional cards for members of the household compared to one who does not have additional cards.

As was the case with least squares regression models, one of the main purposes of undertaking logistic regression analysis is to provide predictions of a response variable. Suppose that we wanted to predict the probability that a credit card holder who had charged \$36,000 on the company's card last year would purchase the premium card during the marketing campaign. If we were predicting for a credit card holder who has obtained extra cards for members of the household, we would have $X_1 = 36$ and $X_2 = 1$ and, from Equation (18.28), the results for the regression model fitted in Figure 18.23 would be

$$\begin{aligned}\ln(\text{Estimated odds of purchasing versus not purchasing}) &= -6.92923 + (0.13925)(36) \\ &\quad + (2.77118)(1) \\ &= 0.85495\end{aligned}$$

Using Equation (18.29), we would have

$$\text{Estimated odds ratio} = e^{.85495} = 2.3513$$

This can be interpreted to mean that the odds that a credit card holder who spent \$36,000 last year and has additional cards would purchase the premium card during the campaign rather than not purchasing are 2.3513 to 1. This can be converted to a probability by using Equation (18.30), so that

$$\begin{aligned}\text{Estimated probability of purchasing premium card} &= \frac{2.3513}{1 + 2.3513} \\ &= .7016\end{aligned}$$

Thus we would estimate that the probability is .7016 that a credit card holder who spent \$36,000 last year and has additional cards would purchase the premium card during the campaign. In other words, 70.16% of such individuals could be expected to purchase the premium card.

Now that we have used the logistic regression model for prediction, we shall consider two other aspects of the model-fitting process: whether the model fit is a good-fitting model, and whether each of the independent variables included in the model makes a significant contribution to the model. One statistic that is sometimes used to evaluate the question of whether the model fit is a good-fitting one is the **deviance statistic.** This statistic measures the fit of the current model compared to a model that has as many parameters as there are data points (what is called a *saturated model*). The deviance statistic follows a chi-square distribution with $n - K - 1$ degrees of freedom. The null and alternative hypotheses for this statistic are

H_0: The model is a good-fitting model.

H_1: The model is not a good-fitting model.

Using a level of significance α, the decision rule is:

$$\text{Reject } H_0 \text{ if deviance} > \chi^2_{U(n-K-1)};$$
$$\text{otherwise, don't reject } H_0.$$

From Figure 18.23 on page 839, and using a .05 level of significance, we observe that

$$\text{deviance} = 20.08 < \chi^2_{U(27)} = 40.113.$$

Thus H_0 would not be rejected. The p value of .8275 exceeds .05. We would conclude that the model is a good-fitting one.

Now that we have concluded that the model is a good-fitting one, we need to evaluate whether each of the independent variables makes a significant contribution to the model. As was the case with linear regression in Sections 17.13 and 18.8, the test statistic is based on the ratio of the regression coefficient to the standard error of the regression coefficient. In logistic regression, this ratio is called the **Wald statistic** and follows the normal distribution. From Figure 18.23, we observe that the Wald statistic is 2.11 for X_1 and 2.38 for X_2. Each of these is greater than the critical value of 1.96 for the normal distribution at the .05 level of significance (the p values are .0347 and .0174). Thus, we may conclude that each of the two explanatory variables makes a contribution to the model and should be included.

Problems for Section 18.17

18.72 Referring to the data of Figure 18.23 on page 839: (Data file LOGPURCH.DAT)

(a) Predict the probability that a credit card holder who had charged \$36,000 on the company's card last year and did not have any additional credit cards for members of the household would purchase the premium card during the marketing campaign.

(b) Compare the results obtained in (a) with those on page 840.

(c) Predict the probability that a credit card holder who had charged \$18,000 on the company's card last year and did not have any additional credit cards for members of the household would purchase the premium card during the marketing campaign.

(d) Compare the results of (a) and (c) and indicate what implications these results might have for the strategy for the marketing campaign.

18.73 The director of graduate studies at a well-known college of business would like to predict the success of students in an MBA program. Two explanatory variables, undergraduate grade point average and GMAT score, were available for a random sample of 30 students, 20 of whom had successfully completed the program (coded as 1) and 10 of whom had not successfully completed the program in the required amount of time (coded as 0). The results were as displayed on page 842:

Data file MBA.DAT

Success in MBA Program	Undergraduate Grade Point Average	GMAT Score
0	2.93	617
0	3.05	557
0	3.11	599
0	3.24	616
0	3.36	594
0	3.41	567
0	3.45	542
0	3.60	551
0	3.64	573
0	3.57	536
1	2.75	688
1	2.81	647
1	3.03	652
1	3.10	608
1	3.06	680
1	3.17	639
1	3.24	632
1	3.41	639
1	3.37	619
1	3.46	665
1	3.57	694
1	3.62	641
1	3.66	594
1	3.69	678
1	3.70	624
1	3.78	654
1	3.84	718
1	3.77	692
1	3.79	632
1	3.97	784

(a) Fit a logistic regression model to predict the probability of successful completion of the MBA program based on undergraduate grade point average and GMAT score.
(b) Explain the meaning of the regression coefficients for the model fit in (a).
(c) Predict the probability of successful completion of the program for a student with an undergraduate grade point average of 3.25 and a GMAT score of 600.
(d) At the .05 level of significance, is there evidence that a logistic regression model that uses undergraduate grade point average and GMAT score to predict probability of success in the MBA program is a good-fitting model?
(e) At the .05 level of significance, is there evidence that undergraduate grade point average and GMAT score each make a significant contribution to the logistic regression model?
(f) Fit a logistic regression model that includes only undergraduate grade point average to predict probability of success in the MBA program.
(g) Fit a logistic regression model that includes only GMAT score to predict probability of success in the MBA program.
(h) Compare the models fit in (f) and (g) to the model fit in (a). How might you evaluate whether there is a difference between the models?

18.74 The marketing manager for a large nationally franchised lawn service company would like to study the characteristics that differentiate homeowners who do and do not have a lawn service. A random sample of 30 homeowners located in a suburban area of a large city was selected: 15 who did not have a lawn service (code 0) and 15 who had a lawn service (code 1). Information for these 30 homeowners was also available that indicated family income (in thousands of dollars), lawn size (in thousands of square feet), attitude toward outdoor recreational activities (0 = unfavorable, 1 = favorable), number of teenagers in the household, and age of the head of the household. The results were as follows:

Lawn Service	Income	Lawn Size	Attitude	Teenagers	Age
0	24.3	3.0	0	2	38
0	25.6	4.3	1	1	45
0	61.7	1.9	1	2	47
0	34.9	4.5	1	0	37
0	37.2	1.7	0	1	39
0	27.5	3.2	0	2	37
0	40.0	4.6	1	1	45
0	33.1	7.9	1	1	46
0	35.3	5.6	1	3	37
0	44.8	6.0	1	2	39
0	27.9	4.5	1	2	47
0	54.6	9.1	1	3	36
0	32.3	4.2	1	1	38
0	40.6	9.4	1	2	44
0	48.9	2.3	0	0	32
1	57.3	6.9	1	1	43
1	74.1	8.3	1	1	39
1	44.6	10.8	0	2	40
1	70.1	10.1	0	1	55
1	71.4	10.3	1	1	49
1	63.1	6.8	1	2	53
1	84.1	7.2	0	0	51
1	44.7	3.3	1	3	48
1	36.2	4.7	0	2	41
1	52.9	5.7	1	0	45
1	39.5	10.9	0	2	43
1	84.6	8.3	0	0	62
1	67.4	7.8	0	3	52
1	51.6	6.3	0	0	34
1	56.4	7.2	1	1	45

Data file
LAWN.DAT

(a) Fit a logistic regression model to predict the probability of using a lawn service based on family income (in thousands of dollars), lawn size (in thousands of square feet), attitude toward outdoor recreational activities (0 = unfavorable, 1 = favorable), number of teenagers in the household, and age of the head of the household.

(b) Explain the meaning of the regression coefficients for the model fit in (a).

(c) Predict the probability of purchasing a lawn service for a 48-year-old homeowner with a family income of $50,000, a lawn size of 5,000 square feet, a negative attitude toward outdoor recreation, and one teenager in the household.

(d) At the .05 level of significance, is there evidence that a logistic regression model that uses family income, lawn size, attitude toward outdoor recreation, number of teenagers in the household, and age of the head of the household is a good-fitting model?

1. Willfully fails to: (1) remove variables from consideration that exhibit a high multicollinearity with other independent variables or (2) use methods other than least squares regression when the assumptions necessary for least squares regression have been violated.
2. Uses a single stepwise regression approach without taking the opportunity to consider other alternative models.

18.20 Summary and Overview

In this chapter we developed the multiple regression model, including dummy variables, multicollinearity, transformations, model building, and logistic regression. On page 782 of Section 18.1, you were given a list emphasizing the important points to be discussed in the chapter. Check over the list now to see whether you feel that you have an understanding of these key points. To be sure, you should be able to answer the following conceptual questions:

1. How does the interpretation of the regression coefficients differ in multiple regression as compared to simple regression?
2. How does testing the significance of the entire regression model differ from testing the contribution of each independent variable in the multiple regression model?
3. How do the coefficients of partial determination differ from the coefficient of multiple determination?
4. Why and how are dummy variables used?
5. How can we evaluate whether the slope of an independent variable with the response variable is the same for each level of the dummy variable?
6. What is the purpose of using transformations in multiple regression analysis?
7. How do we evaluate whether independent variables are intercorrelated?
8. Why is best subset regression useful in the selection of a regression model?
9. Under what circumstances is logistic regression appropriate?
10. How is the interpretation of the regression coefficients different in logistic regression as compared to simple linear regression?

In Chapter 19 we will continue our discussion of forecasting by considering a variety of time series forecasting models.

Getting It All Together

Key Terms

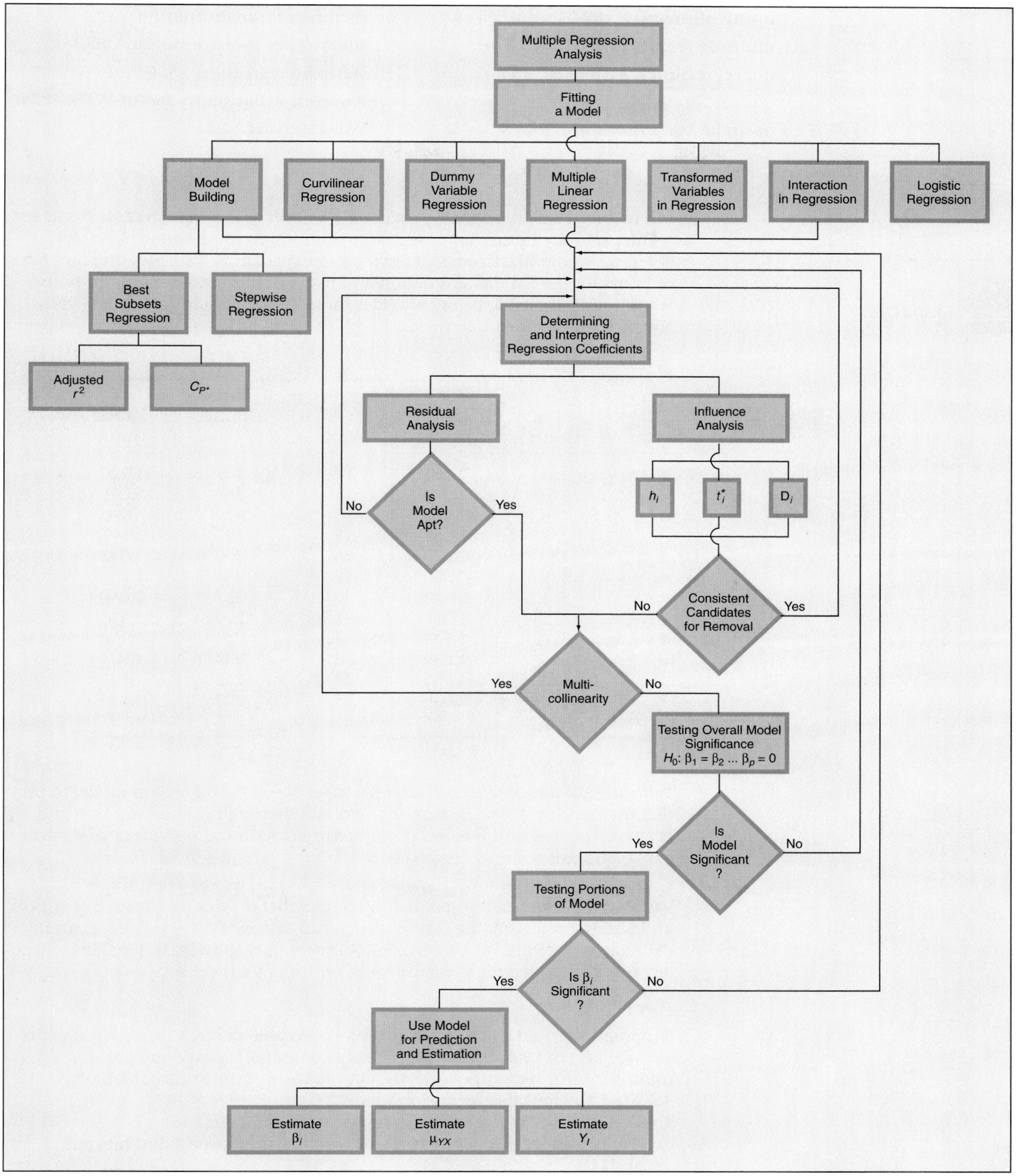

Chapter 18 summary chart.

multicollinearity *824*
multiple regression *782*
net regression coefficient *784*
odds ratio *837*
partial *F*-test criterion *797*
reciprocal transformation *822*
square root transformation *822*
stepwise regression *829*
variance inflationary factor (VIF) *824*
Wald statistic *841*

Chapter Review Problems

Data file
STANDBY.DAT

18.75 In Problem 18.6 on page 789, we used total staff present and remote hours to predict standby hours. Suppose that in addition to these two explanatory variables, we would like to consider two other explanatory variables: the number of hours for the Dubner animation and text machine and the total in-house labor hours. The table below provides information pertaining to these variables for the sample of 26 weeks.

Week	Dubner Hours	Total Labor Hours	Week	Total Dubner Hours	Labor Hours
1	323	2,001	14	207	1,720
2	340	2,030	15	287	2,056
3	340	2,226	16	290	1,890,
4	352	2,154	17	355	2,187,
5	380	2,078	18	300	2,032
6	339	2,080	19	284	1,856
7	331	2,073	20	337	2,068
8	311	1,758	21	279	1,813
9	328	1,624	22	244	1,808
10	353	1,889	23	253	1,834
11	518	1,988	24	272	1,973
12	440	2,049	25	223	1,839
13	276	1,796	26	272	1,935

With the assistance of a computer package, develop a regression model to predict the standby hours. Be sure to perform a thorough residual analysis and evaluate the various measures of influence. In addition, provide a detailed explanation of your results.

18.76 Crazy Dave, a well-known baseball analyst, would like to determine which variables are important in predicting the number of wins by a team in a season, and also in predicting the team's earned run average (E.R.A.). The following data related to wins, runs scored, E.R.A., saves, hits allowed, walks allowed, and errors was collected for a recent season with the results displayed on page 849.

Part I

Suppose we would like to develop a model to predict E.R.A. based on the hits allowed and the walks allowed. Access a computer package and perform a multiple linear regression analysis. On the basis of the results obtained

(a) State the multiple regression model.
(b) Interpret the meaning of the slopes in this problem.
(c) Predict the average E.R.A. for a team that has allowed 1,500 hits and 500 walks.
(d) Determine whether there is a significant relationship between the E.R.A. and the two explanatory variables (hits allowed and walks allowed) at the .05 level of significance.

Team	Wins	Runs	ERA	Saves	Hits Allowed	Walks Allowed	Errors
1	89	705	3.79	48	1,419	538	93
2	73	599	3.58	39	1,403	535	139
3	72	579	3.84	42	1,449	532	134
4	86	738	3.82	52	1,400	550	129
5	76	674	4.11	46	1,507	566	141
6	75	791	4.60	36	1,534	564	116
7	72	610	3.81	44	1,426	512	122
8	92	740	3.43	39	1,344	435	89
9	90	747	3.70	50	1,391	479	95
10	76	733	4.21	44	1,453	612	114
11	96	745	3.73	58	1,396	601	125
12	64	679	4.55	30	1,466	661	112
13	77	682	4.09	42	1,471	598	154
14	96	780	3.91	49	1,346	541	93
15	98	682	3.14	41	1,321	489	109
16	78	593	3.39	37	1,337	575	114
17	90	660	3.46	55	1,362	470	96
18	81	608	3.72	45	1,386	539	114
19	63	548	3.41	29	1,401	553	174
20	87	648	3.25	49	1,296	525	124
21	72	599	3.66	34	1,404	482	116
22	70	686	4.11	34	1,387	549	131
23	96	693	3.35	43	1,410	455	101
24	83	631	3.38	47	1,405	400	94
25	82	617	3.56	46	1,444	439	115
26	72	574	3.61	30	1,385	502	113

Data file
BB92.DAT

(e) Compute the p value in (d) and interpret its meaning.
(f) Interpret the meaning of the coefficient of multiple determination $r^2_{Y.12}$.
(g) Compute the adjusted r^2.
(h) At the .05 level of significance, determine whether each explanatory variable makes a significant contribution to the regression model. On the basis of these results, indicate the regression model that should be utilized in this problem.
(i) Compute the p values in (h) and interpret their meaning.
(j) Set up a 95% confidence interval estimate of the population slope between the E.R.A. and the number of hits allowed.
(k) Compute the coefficients of partial determination $r^2_{Y1.2}$ and $r^2_{Y2.1}$ and interpret their meaning.
(l) Determine the VIF for each explanatory variable in the model. Is there reason to suspect the existence of multicollinearity?
(m) Perform a residual analysis on your results and determine the adequacy of the fit of the model.
(n) Perform an influence analysis and determine whether any observations should be deleted from the model. If necessary, reanalyze the regression model after deleting these observations and compare your results with the original model.

Part II

Suppose we would like to develop a model to predict the number of wins. Evaluate the other six variables provided (runs, E.R.A., saves, hits allowed, walks allowed, and errors) as possible explanatory variables to be included in the model. Be sure to perform a thorough residual analysis and evaluate the measures of influence. In addition, provide a detailed explanation of your results.

18.77 A headline on page 1 of *The New York Times* of March 4, 1990, read "Wine Equation Puts Some Noses Out of Joint." The article proceeded to explain that Professor Orley Ashenfelter, a Princeton University economist, had developed a multiple regression model to predict the quality of French Bordeaux based on the amount of winter rain, the average temperature during the growing season, and the harvest rain. The equation developed was

$$Q = -12.145 + .00117\ WR + .6164\ TMP - .00386\ HR$$

where

Q = logarithmic index of quality where 1961 equals 100
WR = winter rain (October through March) in millimeters
TMP = average temperature during the growing season (April through September), in degrees Centigrade
HR = harvest rain (August to September) in millimeters

You are at a cocktail party, sipping a glass of wine, when one of your friends mentions to you that she has read the article. She asks you to explain the meaning of the coefficients in the equation and also asks you what analyses that might have been done have not been included in the article. You respond . . . .

Interchapter Problems for Chapter 18

Data file DELIVERY.DAT

● 18.78 Referring to Problem 17.75 on page 771, the statistician for the automobile manufacturer believes that a second explanatory variable, shipping mileage, may be related to delivery time. The shipping mileage (in hundreds of miles) for the 16 cars in the sample is presented below.

Car	Shipping Mileage (hundreds of miles)
1	7.5
2	13.3
3	4.7
4	14.6
5	8.4
6	12.6
7	6.2
8	16.4
9	9.7
10	17.2
11	10.6
12	11.3
13	9.0
14	12.3
15	8.2
16	11.5

Access a computer package and perform a multiple linear regression analysis. On the basis of the results obtained
(a) State the multiple regression equation.
(b) Interpret the meaning of the slopes in this problem.

(c) If a car was ordered with 10 options and had to be shipped 800 miles, what would you predict the average delivery time to be?
(d) Determine whether there is a significant relationship between delivery time and the two explanatory variables (number of options and shipping mileage) at the .05 level of significance.
(e) Indicate the p values in (d) and interpret their meaning.
(f) Interpret the meaning of the coefficient of multiple determination $r^2_{Y.12}$ in this problem.
(g) Compute the adjusted r^2.
(h) At the .05 level of significance, determine whether each explanatory variable makes a significant contribution to the regression model. Based upon these results, indicate the regression model that should be utilized in this problem.
(i) Indicate the p values in (h) and interpret their meaning.
(j) Compute the coefficients of partial determination $r^2_{Y1.2}$ and $r^2_{Y2.1}$ and interpret their meaning.
(k) Determine the VIF for each explanatory variable in the model. Is there reason to suspect the existence of multicollinearity?
(l) Perform a residual analysis on your results and determine the adequacy of the fit of the model.
(m) Perform an influence analysis and determine whether any observations should be deleted from the model. If necessary, reanalyze the regression model after deleting these observations and compare your results with the original model.

18.79 Referring to Problem 17.78 on page 774, suppose that we also wish to include the time period in which the house was sold in the model. The following table represents the time period (in months) in which each of the 30 houses was sold.

Data file HOUSE1.DAT

Time period for the sample of 30 houses.

House	Time Period (months)	House	Time Period (months)
1	10	16	12
2	10	17	5
3	11	18	14
4	2	19	1
5	5	20	3
6	4	21	14
7	17	22	12
8	13	23	11
9	6	24	12
10	5	25	2
11	7	26	6
12	4	27	12
13	11	28	4
14	10	29	9
15	17	30	12

Access a computer package and perform a multiple regression analysis. Based on the results obtained
(a) State the multiple regression equation.
(b) Interpret the meaning of the slopes in this equation.
(c) Predict the average selling price for a house that has an assessed value of $70,000 and was sold in time period 12.

(d) Determine whether there is a significant relationship between selling price and the two explanatory variables (assessed value and time period) at the .05 level of significance.
(e) Compute the p value in (d) and interpret its meaning.
(f) Interpret the meaning of the coefficient of multiple determination $r^2_{Y.12}$ in this problem.
(g) Compute the adjusted r^2.
(h) At the .05 level of significance, determine whether each explanatory variable makes a significant contribution to the regression model. On the basis of these results, indicate the regression model that should be used in this problem.
(i) Compute the p values in (h) and interpret their meaning.
(j) Set up a 95% confidence-interval estimate of the true population slope between selling price and assessed value. How does the interpretation of the slope here differ from Problem 17.78 (k)?
(k) Compute the coefficients of partial determination $r^2_{Y1.2}$ and $r^2_{Y2.1}$ and interpret their meaning.
(l) Determine the VIF for each explanatory variable in the model. Is there reason to suspect the existence of multicollinearity?
(m) Perform a residual analysis on your results and determine the adequacy of the fit of the model.
(n) Perform an influence analysis and determine whether any observations should be deleted from the model. If necessary, reanalyze the regression model after deleting these observations and compare your results with the original model.

Data file
BET.DAT

18.80 Referring to the data of Problem 17.76 on page 772, suppose that we wish to fit a curvilinear model to predict the amount bet based upon attendance. Access a computer package to perform the regression analysis.
(a) State the regression equation.
(b) Predict the average amount bet for a day in which attendance is 30,000.
(c) Determine whether there is a significant relationship between attendance and amount bet at the .05 level of significance.
(d) Compute the p value in (c) and interpret its meaning.
(e) Compute the coefficient of multiple determination $r^2_{Y.12}$ and interpret its meaning.
(f) Compute the adjusted r^2.
(g) At the .05 level of significance, determine whether the curvilinear model is superior to the linear regression model.
(h) Compute the p value in (g) and interpret its meaning.
(i) Perform a residual analysis on your results and determine the adequacy of the fit of the model.
(j) Perform an influence analysis and determine whether any observations should be deleted from the model. If necessary, reanalyze the regression model after deleting these observations and compare your results with the original model.

Data file
HOUSE1.DAT

18.81 Referring to Problem 17.78 on page 774, suppose that in addition to using assessed value to predict sales price, we also wanted to use information concerning whether the house was brand new. Houses 1, 2, 10, 14, 18, 20, 22, 24, 25, 26, 28, and 30 are brand new. Access a computer package and perform a multiple regression analysis. On the basis of the results obtained
(a) State the multiple regression equation.
(b) Interpret the meaning of the slopes in this problem.
(c) Predict the average selling price for a brand new house with an assessed value of $75,000.
(d) Determine whether there is a significant relationship between selling price and the two explanatory variables (assessed value and whether the house is brand new) at the .05 level of significance.

(e) Compute the p value in (d) and interpret its meaning.
(f) Interpret the meaning of the coefficient of multiple determination $r^2_{Y.12}$.
(g) Compute the adjusted r^2.
(h) At the .05 level of significance, determine whether each explanatory variable makes a significant contribution to the regression model. On the basis of these results, indicate the regression model that should be used in this problem.
(i) Compute the p values in (h) and interpret their meaning.
(j) Set up a 95% confidence-interval estimate of the population slope between selling price and assessed value.
(k) Compute the coefficients of partial determination $r^2_{Y1.2}$ and $r^2_{Y2.1}$ and interpret their meaning.
(l) Determine the VIF for each explanatory variable in the model. Is there reason to suspect the existence of multicollinearity?
(m) What assumption about the slope of selling price and assessed value must be made in this problem?
(n) At the .05 level of significance, determine whether the inclusion of an interaction term makes a significant contribution to the model that already contains assessed value and whether the house is brand new. On the basis of these results, indicate the regression model that should be used in this problem.
(o) Compute the p value in (n) and interpret its meaning.
(p) Perform a residual analysis on your results and determine the adequacy of the fit of the model.
(q) Perform an influence analysis and determine whether any observations should be deleted from the model. If necessary, reanalyze the regression model after deleting these observations and compare your results with the original model.

18.82 In Problem 17.78 on page 774 we used assessed value to predict the selling price of houses; in Problem 18.79 on page 851 we also considered the time period in which the house was purchased, and in Problem 18.81 we considered whether or not the house was brand new. Suppose that we wanted to consider all three of these explanatory variables—assessed value, time period, and whether or not the house was brand new. With the assistance of a computer package, develop a regression model to predict the selling price of houses. Be sure to perform a thorough residual analysis and evaluate the various measures of influence. In addition, provide a detailed explanation of your results.

Survey/Database Project for Chapter 18

The following problems refer to the sample data obtained from the questionnaire of Figure 2.6 *on pages* 28–29 *and presented in Table* 2.3 *on pages* 33–40. *They should be solved with the aid of an available computer package.*

Data file EMPSAT.DAT

18.83 In Section 18.16 we used multiple regression analysis to develop a model to forecast the income of Kalosha Industries employees whose occupation was classified as technical/sales based on the possible consideration of ten explanatory variables. Suppose that we would like to do a similar analysis for each of the other six occupational groups in addition to the employees who are in technical/sales occupations. Develop each of these models and write an executive summary to Bud Conley discussing your findings.

18.84 We would like to develop a regression model to be able to determine which factors are important in job satisfaction. In order to develop such a model, we will reclassify job satisfaction (SATJOB) into two categories, very satisfied (code 1) and not very satisfied (recoded value of 0, original codes of 2, 3, and 4). We shall consider six explanatory variables for inclusion in the model, sex (SEX, recoded as 0 = female, 1 = male), personal income (RINCOME), number of years worked for Kalosha Industries (EMPYEARS), number of promotions received (NUMPROMO), whether the job allows participation in making decisions that affect work (IDECIDE, recoded as 1 = always and 0 = other than always), and whether there is participation in budgetary decisions (ORGMONEY, recoded as 0 = no, 1 = yes). Develop several models using these explanatory variables to predict job satisfaction. Compare and contrast the models fitted. Write an executive summary to Bud Conley discussing your findings.

Case Study I—The Mountain States Potato Company

The Mountain States Potato Company is a potato-processing firm in eastern Idaho. A by-product of the process, called a filter cake, has been sold to area feedlots as cattle feed. Recently, one of the feedlot owners complained that the cattle were not gaining weight and believed that the problem was the filter cake they purchased from the Mountain States Potato Company.

Initially, all that was known of the filter cake system was that historical records showed that the solids had been running in the neighborhood of 11.5% in years past. Presently, the solids were running in the 8 to 9% range. Several additions had been made to the plant in the intervening years which had significantly increased the water and solids volume and the clarifier temperature. What was actually affecting the solids was a mystery, but since the plant needed to get rid of its solid waste if it was going to run, something had to be done quickly. The only practical solution was to determine some way to get the solids content back up to the previous levels. Individuals involved in the process were asked to identify variables that might be manipulated that could in turn affect the solids content. This review turned up six variables that would affect solids. The variables are:

SOLIDS	Percent solids in filter cake.
PH	Acidity. This indicates bacterial action in the clarifier. As bacterial action progresses, organic acids are produced that can be measured using pH. This is controlled by the downtime of the system.
LOWER	Pressure of vacuum line below fluid line on rotating drum.
UPPER	Pressure of vacuum line above fluid line on rotating drum.
THICK	Cake thickness measured on drum.
VARIDRIV	Setting used to control drum speed. May differ from DRUMSPD due to mechanical inefficiencies.
DRUMSPD	Speed at which the drum was rotated when collecting filter cake. Measured with a stopwatch.

Data obtained by monitoring the process several times daily for 20 days are reported on page 855.

Develop a regression model to predict the percent of solids. Write an executive summary of your findings to the president of the Mountain Potato Company.

Obs.	SOLIDS	PH	LOWER	UPPER	THICK	VARIDRIV	DRUMSPD
1	9.7	3.7	13	14	0.250	6	33.00
2	9.4	3.8	17	18	0.875	6	30.43
3	10.5	3.8	14	15	0.500	6	34.00
4	10.9	3.9	14	14	0.500	6	34.00
5	11.6	4.3	17	18	0.375	6	36.24
6	10.9	4.2	16	17	0.500	6	31.76
7	11.0	4.3	16	19	0.375	6	34.00
8	10.7	3.9	15	16	0.375	6	32.13
9	11.8	3.6	8	8	0.375	6	37.00
10	9.7	4.0	18	18	0.500	6	36.00
11	11.6	4.0	12	13	0.313	5	45.00
12	10.9	3.9	15	15	0.500	5	50.00
13	10.0	3.8	17	18	0.625	5	46.91
14	10.3	3.8	13	14	0.500	4	57.50
15	10.1	3.6	17	17	0.625	4	60.40
16	9.9	3.8	17	18	0.500	4	53.14
17	9.5	3.5	17	18	0.625	6	34.40
18	10.5	3.8	15	17	0.500	6	33.96
19	10.8	3.9	15	17	0.750	6	35.00
20	10.4	3.9	14	15	0.500	6	35.00
21	10.9	4.0	15	16	0.500	6	34.00
22	11.2	4.4	17	19	0.375	6	34.00
23	9.5	3.8	17	17	0.500	6	33.49
24	10.7	3.9	15	17	0.500	6	33.38
25	10.1	3.8	15	17	0.500	6	41.00
26	10.5	3.8	17	17	0.500	6	36.00
27	10.9	4.0	15	17	0.250	6	34.00
28	15.5	4.3	13	15	0.625	6	41.00
29	13.1	4.0	17	17	0.500	6	35.00
30	11.0	4.0	14	15	0.375	6	36.00
31	12.5	4.2	15	17	0.313	6	37.72
32	11.7	4.2	14	14	0.250	6	36.00
33	11.9	4.4	15	16	0.375	6	36.52
34	11.7	3.4	8	10	0.313	6	38.08
35	17.8	4.3	12	12	0.313	6	38.00
36	11.8	4.5	14	15	0.250	6	33.00
37	10.0	3.7	12	13	0.250	5	48.00
38	10.3	3.7	15	15	0.500	5	48.00
39	9.8	3.8	14	15	0.500	5	47.24
40	10.0	3.7	13	14	0.500	6	37.00
41	10.6	4.1	14	15	0.500	6	33.70
42	11.2	3.9	13	14	0.375	6	38.26
43	10.9	3.7	13	14	0.313	6	38.00
44	11.0	4.1	13	14	0.375	6	37.00
45	11.0	4.1	14	15	0.375	6	38.00
46	11.7	4.5	14	14	0.250	6	36.26
47	11.8	4.4	13	14	0.250	6	37.45
48	12.0	4.2	13	13	0.375	6	38.00
49	11.8	4.6	14	14	0.375	6	36.90
50	11.1	4.0	14	15	0.500	6	37.00
51	11.6	3.9	14	14	0.500	6	37.50
52	11.0	4.0	14	15	0.500	6	36.00
53	11.2	3.9	15	15	0.313	6	35.00
54	11.0	4.2	14	14	0.375	6	37.00

Source: Midwest Society for Case Research, 1994.

Data file POTATO.DAT

Endnotes

1. The relationship between t and F indicated in Equation (18.9) holds when t is a two-tailed test.
2. If the two groups have different slopes, an *interaction* term needs to be included in the model (see Section 18.13 and Reference 4).
3. The natural logarithm, usually abbreviated as ln, is the logarithm to the base e, the mathematical constant that is equal to 2.71828.

References

1. Andrews, D. F., and D. Pregibon, "Finding the Outliers That Matter," *Journal of the Royal Statistical Society,* Ser. B., 1978, Vol. 40, pp. 85–93.
2. Atkinson, A. C., "Robust and Diagnostic Regression Analysis," *Communications in Statistics,* 1982, Vol. 11, pp. 2559–2572.
3. Belsley, D. A., E. Kuh, and R. Welsch, *Regression Diagnostics: Identifying Influential Data and Sources of Collinearity* (New York: John Wiley, 1980).
4. Berenson, M. L., D. M. Levine, and M. Goldstein, *Intermediate Statistical Methods and Applications: A Computer Package Approach* (Englewood Cliffs, NJ: Prentice-Hall, 1983).
5. Cook, R. D., and S. Weisberg, *Residuals and Influence in Regression* (New York: Chapman and Hall, 1982).
6. Dillon, W. R., and M. Goldstein, *Multivariate Analysis: Methods and Applications,* 2d ed. (New York: John Wiley, 1988).
7. Hoaglin, D. C., and R. Welsch, "The Hat Matrix in Regression and ANOVA," *The American Statistician,* 1978, Vol. 32, pp. 17–22.
8. Hocking, R. R., "Developments in Linear Regression Methodology: 1959–1982," *Technometrics,* 1983, Vol. 25, 219–250.
9. Hosmer, D., and S. Lemeshow, *Applied Logistic Regression* (New York: John Wiley, 1989).
10. Marquardt, D. W., "You Should Standardize the Predictor Variables in Your Regression Models," discussion of "A Critique of Some Ridge Regression Methods," by G. Smith and F. Campbell, *Journal of the American Statistical Association,* 1980, Vol. 75, pp. 87–91.
11. Marquardt, D. W., and R. D. Snee, "Ridge Regression in Practice," *The American Statistician,* 1975, Vol. 29, pp. 3–19.
12. Norusis, M. J. *SPSS for Windows Base System User's Guide Release 5.0* (Chicago, IL: SPSS, Inc., 1992).
13. Pregibon, D., "Logistic Regression Diagnostics," *Annals of Statistics,* 1981, Vol. 9, pp. 705–724.
14. Ryan, B. F., and B. L. Joiner, *Minitab Student Handbook,* 3rd ed. (North Scituate, MA: Duxbury Press, 1994).
15. *SAS Language and Procedures Usage,* Version 6 (Cary, NC: SAS Institute, 1988).
16. Snee, R. D., "Some Aspects of Nonorthogonal Data Analysis, Part I. Developing Prediction Equations," *Journal of Quality Technology,* 1973, Vol. 5, pp. 67–79.
17. *STATISTIX User's Guide* (Tallahassee, FL: Analytical Software, 1992).
18. Tukey, J. W. "Data Analysis, Computation and Mathematics," *Quarterly Journal of Applied Mathematics,* 1972, Vol. 30, pp. 51–65.
19. Tukey, J. W., *Exploratory Data Analysis* (Reading, MA: Addison-Wesley, 1977).
20. Velleman, P. F., and R. Welsch, "Efficient Computing of Regression Diagnostics," *The American Statistician,* 1981, Vol. 35, pp. 234–242.
21. Weisberg, S., *Applied Linear Regression* (New York: John Wiley, 1980).

chapter 19

Time-Series Forecasting

CHAPTER OBJECTIVE To introduce a variety of time-series models for forecasting purposes.

19.1 Introduction

In the preceding two chapters we discussed the topic of regression analysis as a tool for model building and prediction. In these respects, regression analysis provides a useful guide to managerial decision making. In this chapter we shall develop other business forecasting methods. Upon completion of this chapter, you should be able to:

1. Understand the components of the classical time-series model
2. Forecast the future value of a time series using least squares methods
3. Use exponential smoothing and moving average methods
4. Use the Holt-Winters and autoregressive forecasting models
5. Use the mean absolute deviation (MAD) to measure forecasting error
6. Use the least squares trend projection and the seasonal index for forecasting purposes with monthly data

19.2 The Importance of Business Forecasting

19.2.1 Introduction to Forecasting

Since economic and business conditions vary over time, business leaders must find ways to keep abreast of the effects that such changes will have on their operations. One technique that business leaders may use as an aid in planning for the level of future operational needs is **forecasting.** Although numerous forecasting methods have been devised, they all have one common goal, to make predictions of future events so that these projections can then be incorporated into the decision-making process. As examples, the government must be able to forecast such things as unemployment, inflation, industrial production, and expected revenues from personal and corporate income taxes in order to formulate its policies, and the marketing department of a large retailing corporation must be able to forecast product demand, sales revenues, consumer preferences, inventory, etc., in order to make timely decisions regarding its advertising strategies.

19.2.2 Types of Forecasting Methods

There are basically two approaches to forecasting: *qualitative* and *quantitative.* Qualitative forecasting methods are especially important when historical data are unavailable, as would be the case, for example, if the marketing department wanted to predict the sales of a new product. Qualitative forecasting methods are considered to be highly subjective and judgmental. These include the *factor listing method, expert opinion,* and the *Delphi technique* (see Reference 4). On the other hand, quantitative forecasting methods make use of historical data. The goal is to study past happenings in order to better understand the underlying structure of the data and thereby provide the means necessary for predicting future occurrences.

Quantitative forecasting methods can be subdivided into two types: *time series* and *causal.* Causal forecasting methods involve the determination of factors that relate to the variable to be predicted. These include multiple regression analysis with lagged variables, econometric modeling, leading indicator analysis, and diffusion indexes and other economic barometers (see References 5 and 6). On the other hand, time-series forecasting methods involve the projection of future values of a variable based entirely on the past and present observations of that variable. It is these methods that we shall be concerned with here.

19.2.3 Introduction to Time-Series Analysis

A **time series** is a set of numerical data that is obtained at regular periods over time.

For example, the *daily* closing prices of a particular stock on the New York Stock Exchange constitutes a time series. Other examples of economic or business time series are the *monthly* publication of the Consumer Price Index; the *quarterly* statements of gross national product (GNP); as well as the *annually* recorded total sales revenues of a particular firm. Time series, however, are not restricted to economic or business data. As an example, the Dean of Students at your college may wish to investigate whether there is an indication of persistent grade inflation during the past decade. To accomplish this, on an annual basis either the percentage of freshmen and sophomore students on the Dean's List may be examined or the percentage of seniors graduating with honors may be studied.

19.2.4 Objectives of Time-Series Analysis

The basic assumption underlying time-series analysis is that the factors that have influenced patterns of activity in the past and present will continue to do so in more or less the same manner in the future. Thus the major goals of time-series analysis are to identify and isolate these influencing factors for predictive (forecasting) purposes as well as for managerial planning and control.

19.3 Component Factors of the Classical Multiplicative Time-Series Model

19.3.1 Introduction

To achieve these goals, many mathematical models have been devised for exploring the fluctuations among the component factors of a time series. Perhaps the most fundamental is the **classical multiplicative model** for data recorded annually, quarterly, or monthly. It is this model that will be considered in this text.

To demonstrate the classical multiplicative time-series model, Figure 19.1 on page 860 presents the net sales for Eastman Kodak Company from 1970 to 1992. If we may characterize these time-series data, it is clear that net sales has shown a tendency to increase over this 23-year period. This overall long-term tendency or impression (of upward or downward movements) is known as a **trend.**

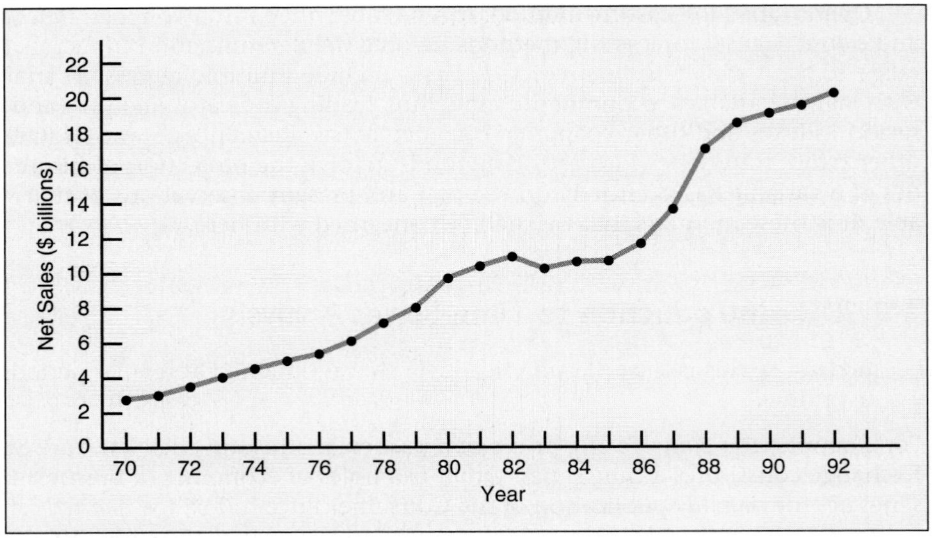

Figure 19.1
Net sales (in billions of dollars) for Eastman Kodak Company (1970–1992).
Source: Moody's Handbook of Common Stocks, 1980, 1989, 1993.

However, trend is not the only component factor influencing either these particular data or other annual time series. Two other factors, the *cyclical* component and the *irregular* component, are also present in the data. The **cyclical component** depicts the up and down swings or movements through the series. Cyclical movements vary in length, usually lasting from 2 to 10 years; differ in intensity or amplitude; and are often correlated with a business cycle. In some years the values will be higher than what would be predicted by a simple trend line (i.e., they are at or near the *peak* of a cycle), whereas in other years the values will be lower than what would be predicted by a trend line (i.e., they are at or near the bottom or *trough* of a cycle). Any observed data that do not follow the trend curve modified by the cyclical component are indicative of the **irregular or random component.** When data are recorded monthly or quarterly, in addition to the trend, cyclical, and irregular components, an additional component called the *seasonal factor* will be considered (see Section 19.9).

19.3.2 The Classical Multiplicative Time-Series Model

We have thus far mentioned that there are three or four component factors, respectively, which influence an economic or business time series. These are summarized in Table 19.1 on page 861. The classical multiplicative time-series model states that any observed value in a time series is the *product* of these influencing factors; that is, when the data are obtained annually, an observation Y_i recorded in the year i may be expressed as

$$Y_i = T_i \cdot C_i \cdot I_i \tag{19.1}$$

Table 19.1 Factors influencing time-series data.

Component	Classification of Component	Definition	Reason for Influence	Duration
Trend	Systematic	Overall or persistent, long-term upward or downward pattern of movement	Changes in technology, population, wealth, value	Several years
Seasonal	Systematic	Fairly regular periodic fluctuations that occur within each 12-month period year after year	Weather conditions, social customs, religious customs	Within 12 months (or monthly or quarterly data)
Cyclical	Systematic	Repeating up-and-down swings or movements through four phases: from peak (prosperity) to contraction (recession) to trough (depression) to expansion (recovery or growth)	Interactions of numerous combinations of factors influencing the economy	Usually 2–10 years with differing intensity for a complete cycle
Irregular	Unsystematic	The erratic or "residual" fluctuations in a time series that exist after taking into account the systematic effects—trend, seasonal, and cyclical	Random variations in data or due to unforeseen events such as strikes, hurricanes, floods, political assassinations, etc.	Short duration and nonrepeating

where in the year i,

T_i = value of the trend component
C_i = value of the cyclical component
I_i = value of the irregular component

On the other hand, when the data are obtained either quarterly or monthly, an observation Y_i recorded in time period i may be given as

$$Y_i = T_i \cdot S_i \cdot C_i \cdot I_i \qquad \textbf{(19.2)}$$

where, in the time period i, T_i, C_i, and I_i are the values of the trend, cyclical, and irregular components, respectively, and S_i is the value of the seasonal component.

The first step in a time-series analysis is to plot the data and observe their tendencies over time. We must first determine whether there appears to be a long-term upward or downward movement in the series (i.e., a trend) or whether the series seems to oscillate about a horizontal line over time. If the latter is the case (that is, there is no long-term upward or downward trend), then the method of moving averages or the method of exponential smoothing may be employed to smooth the series and provide us with an overall long-term impression (see

Section 19.4). On the other hand, if a trend is actually present, a variety of time-series forecasting methods can be considered (see Sections 19.5–19.8) when dealing with annual data. For monthly time-series data, forecasting will be developed in Section 19.9.

19.4 Smoothing the Annual Time Series: Moving Averages and Exponential Smoothing

Table 19.2 presents the annual worldwide factory sales (in millions of units) of cars, trucks, and buses manufactured by General Motors Corporation over the 23-year period from 1970 to 1992 and Figure 19.2 is a time-series plot of these data. When we examine annual data such as these, our visual impression of the overall long-term tendencies or trend movements in the series is obscured by the amount of variation from year to year. It then becomes difficult to judge whether any long-term upward or downward trend effect really exists in the series.

In situations such as these, the method of *moving averages* or the method of *exponential smoothing* may be used to smooth a series and thereby provide us with an overall impression of the pattern of movement in the data over time.

Table 19.2 Factory sales (in millions of units*) for General Motors Corp. (1970–1992).

Year	Factory Sales	Year	Factory Sales
1970	5.3	1982	6.2
1971	7.8	1983	7.8
1972	7.8	1984	8.3
1973	8.7	1985	9.3
1974	6.7	1986	8.6
1975	6.6	1987	7.8
1976	8.6	1988	8.1
1977	9.1	1989	7.9
1978	9.5	1990	7.5
1979	9.0	1991	7.0
1980	7.1	1992	7.2
1981	6.8		

Source: Moody's Handbook of Common Stocks, 1980, 1989, 1993.
*From all sources including passenger cars, trucks and buses, and overseas plants.

19.4.1 Moving Averages

The method of moving-averages for smoothing a time series is highly subjective and dependent upon the length of the period selected for constructing the averages. To eliminate the cyclical fluctuations, the period chosen should be an integer value that corresponds to (or is a multiple of) the estimated average length of a cycle in the series.

But what are moving averages and how are they computed?

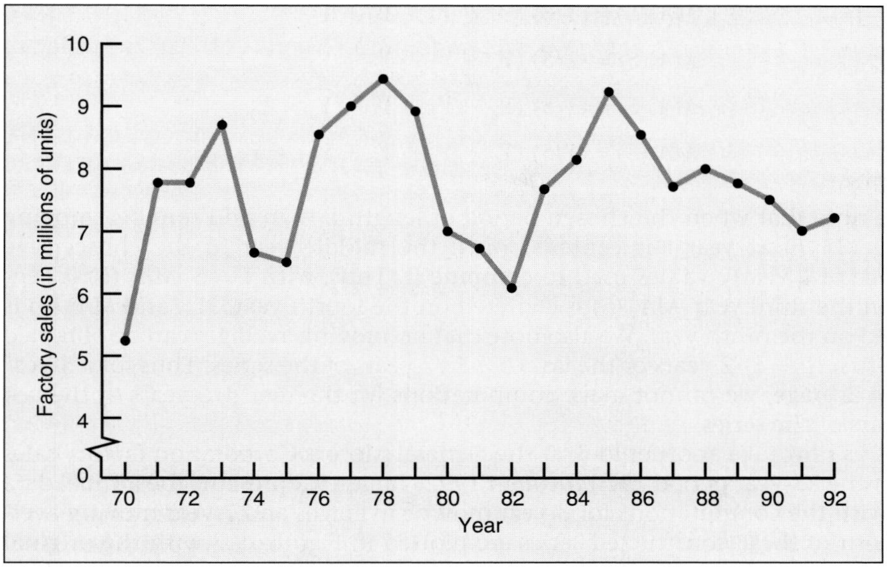

Figure 19.2
Factory sales (in millions of units) for General Motors Corp. (1970–1992).
Source: Data are taken from Table 19.2.

Moving averages for a chosen period of length L consists of a series of arithmetic means computed over time such that each mean is calculated for a sequence of observed values having that particular length L.

For example, 5-year moving averages consist of a series of means obtained over time by averaging out consecutive sequences containing five observed values. In general, for any series composed of n years, a moving average of length L (given by the symbol $MA_i(L)$ can be computed at year i as follows:

$$MA_i(L) = \frac{1}{L} \sum_{t=(1-L)/2}^{(L-1)/2} Y_{i+t} \tag{19.3}$$

where L = an *odd* number of years

and

$$i = \left(\frac{L-1}{2}\right) + 1, \left(\frac{L-1}{2}\right) + 2, \ldots, n - \left(\frac{L-1}{2}\right)$$

To illustrate the use of Equation (19.3), suppose we desire to compute 5-year moving averages from a series containing $n = 11$ years. Since $L = 5$, then $i = 3, 4, 5, 6, 7, 8, 9$. Therefore we have

$$MA_3(5) = (1/5)\,(Y_1 + Y_2 + Y_3 + Y_4 + Y_5)$$
$$MA_4(5) = (1/5)\,(Y_2 + Y_3 + Y_4 + Y_5 + Y_6)$$
$$MA_5(5) = (1/5)\,(Y_3 + Y_4 + Y_5 + Y_6 + Y_7)$$

$$MA_6(5) = (1/5)\ (Y_4 + Y_5 + Y_6 + Y_7 + Y_8)$$

$$MA_7(5) = (1/5)\ (Y_5 + Y_6 + Y_7 + Y_8 + Y_9)$$

$$MA_8(5) = (1/5)\ (Y_6 + Y_7 + Y_8 + Y_9 + Y_{10})$$

$$MA_9(5) = (1/5)\ (Y_7 + Y_8 + Y_9 + Y_{10} + Y_{11})$$

We note that when the chosen period of length L is an odd number, moving average $MA_i(L)$ at year i is centered on i, the middle year in the consecutive sequence of L yearly values used to compute it. Thus, with $L = 5$, MA_3 (5) is centered on the third year, MA_4 (5) is centered on the fourth year, . . ., and MA_9 (5) is centered on the ninth year. We also note that no moving averages can be obtained for the first $(L - 1)/2$ years or the last $(L - 1)/2$ years of the series. Thus for a 5-year moving average, we cannot make computations for the first two years or the last two years of the series.

Let us now take another look at the General Motors Corporation factory sales data for the 23-year period 1970 through 1992. Table 19.3 presents the annual data along with the computations for 3-year moving averages and 7-year moving averages. Both of these constructed series are plotted in Figure 19.3 with the original data (see page 865).

In practice, to compute 3-year moving averages, we first obtain a series of 3-year moving totals as indicated in column (3) of Table 19.3 and then divide each of these totals by 3. The results are given in column (4). For example, since our observed time series was first recorded in 1970, the first 3-year moving total consists

Table 19.3 3-year and 7-year moving averages of factory sales at General Motors Corp. (1970–1992).

(1) Year	(2) Factory Sales (in millions)	(3) 3-Year Moving Total	(4) 3-Year Moving Average	(5) 7-Year Moving Total	(6) 7-Year Moving Average
1970	5.3	—	—	—	—
1971	7.8	20.9	7.0	—	—
1972	7.8	24.3	8.1	—	—
1973	8.7	23.2	7.7	51.5	7.4
1974	6.7	22.0	7.3	55.3	7.9
1975	6.6	21.9	7.3	57.0	8.1
1976	8.6	24.3	8.1	58.2	8.3
1977	9.1	27.2	9.1	56.6	8.1
1978	9.5	27.6	9.2	56.7	8.1
1979	9.0	25.6	8.5	56.3	8.0
1980	7.1	22.9	7.6	55.5	7.9
1981	6.8	20.1	6.7	54.7	7.8
1982	6.2	20.8	6.9	54.5	7.8
1983	7.8	22.3	7.4	54.1	7.7
1984	8.3	25.4	8.5	54.8	7.8
1985	9.3	26.2	8.7	56.1	8.0
1986	8.6	25.7	8.6	57.8	8.3
1987	7.8	24.5	8.2	57.5	8.2
1988	8.1	23.8	7.9	56.2	8.0
1989	7.9	23.5	7.8	54.1	7.7
1990	7.5	22.4	7.5	—	—
1991	7.0	21.7	7.2	—	—
1992	7.2	—	—	—	—

Source: Data are taken from Table 19.2.

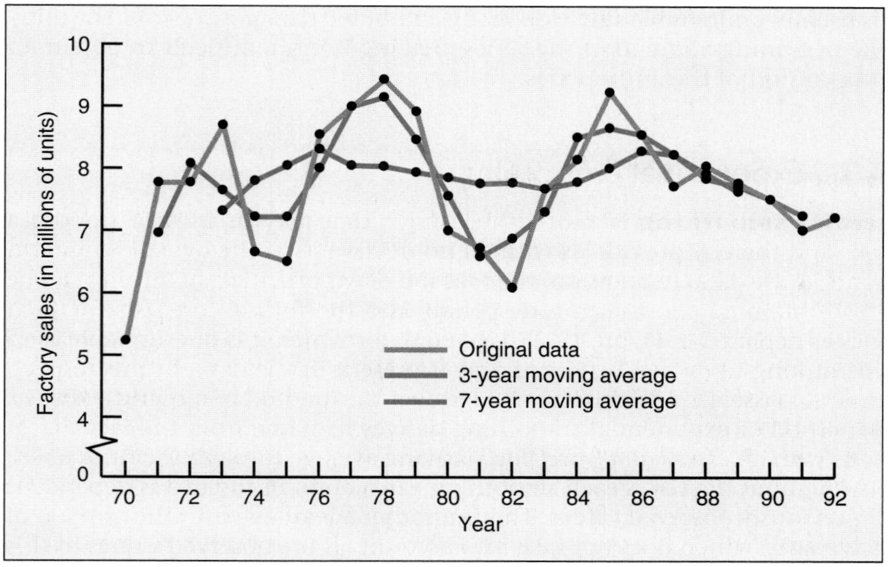

Figure 19.3
Plotting the 3-year and 7-year moving averages.
Source: Data are taken from Table 19.3.

of the sum of the first three annually recorded values—5.3, 7.8, and 7.8. This moving total, 20.9, is then centered so that the recording is made against the year 1971. To obtain the moving total for the year 1972—which consists of the observed annual sales data for the years 1971, 1972, and 1973—we add the next observed value in the time series (year 1973) to the previous moving total and then subtract the first (oldest) value in the series. This process continues so that the 3-year moving total for any particular year i in the series represents the sum of the observed value for year i along with the observed values for the year preceding it and the year following it. On the other hand, with 7-year moving totals, the result computed and recorded for the year i consists of the observed value in the time series for year i plus the three observed values that precede it and the three observed values that follow it. To "move" the 7-year total from one year to the next, we add on to the previous total the next observed value in the time series and remove the oldest value that had appeared in the previous total. This process continues through the series. The 7-year moving averages are then obtained by dividing the series of moving totals by 7.

We note from columns (3) and (4) of Table 19.3 that, in obtaining the 3-year moving averages, no result can be computed for the first or last observed value in the time series. Moreover, as seen in columns (5) and (6), when computing 7-year moving averages there are no results for the first three observed values or the last three values. This occurs because the first 7-year moving total for the data at hand consists of factory sales during the years 1970 through 1976, which is centered at 1973, and the last moving total consists of factory sales recorded in 1986 through 1992, which is centered at 1989.

From Figure 19.3 we can see that the 7-year moving averages smooth the series a great deal more than do the 3-year moving averages, since the period is of longer duration. Unfortunately, however, as we previously had noted, the longer the period, the fewer the number of moving average values that can be computed and plotted. Therefore, selecting moving averages with periods of length greater than

7 years is usually undesirable since too many computed data points would be missing at the beginning and end of the series, making it more difficult to obtain an overall impression of the entire series.

19.4.2 Exponential Smoothing

Exponential smoothing is another technique that may be used to smooth a time series and thereby provide us with an impression as to the overall long-term movements in the data. In addition, the method of exponential smoothing can be utilized for obtaining short-term (one period into the future) forecasts for such time series as depicted in Figure 19.2 (page 863), for which it is questionable as to what type of long-term trend effect, if any, is present in the data. In this respect, the technique possesses a distinct advantage over the method of moving averages.

The method of exponential smoothing derives its name from the fact that it provides us with an *exponentially weighted* moving average through the time series; that is, throughout the series each smoothing calculation or forecast is dependent upon all previously observed values. This is another advantage over the method of moving averages, which does not take into account all the observed values in this manner. With exponential smoothing, the weights assigned to the observed values decrease over time so that when a calculation is made, the most recently observed value receives the highest weight, the previously observed value receives the second highest weight, and so on, with the initially observed value receiving the lowest weight.

Although the magnitude of work involved from this description may seem formidable, we should realize that exponential smoothing as well as moving average methods are usually available among the procedures provided in many statistical software packages (see, for example, References 8, 9, and 10).

If we may focus on the smoothing aspects of the technique (rather than the forecasting aspects), the formulas developed for exponentially smoothing a series in any time period i are based only on three terms—the presently observed value in the time series Y_i, the previously computed exponentially smoothed value E_{i-1}, and some subjectively assigned weight or smoothing coefficient W. Thus to smooth a series at any time period i we have the following expression:[1]

$$E_i = WY_i + (1 - W)E_{i-1} \qquad \textbf{(19.4)}$$

where E_i = value of the exponentially smoothed series being computed in time period i

E_{i-1} = value of the exponentially smoothed series already computed in time period $i - 1$

Y_i = observed value of the time series in period i

W = subjectively assigned weight or smoothing coefficient (where $0 < W < 1$)

The choice of a smoothing coefficient or weight that we should assign to our time series is quite important since it will affect our results. Unfortunately, this selection is rather subjective. However, in regard to smoothing ability, we may observe from Figures 19.3 (page 865) and 19.4 (page 868) that a series of L term moving averages is related to an exponentially smoothed series having weight W as follows:

$$W = \frac{2}{L+1} \tag{19.5}$$

or

$$L = \frac{2}{W} - 1 \tag{19.6}$$

From Equations (19.5) and (19.6) we note that with respect to smoothing ability, similarities are found between the 3-year series of moving averages (Figure 19.3) and the exponentially smoothed series having weight $W = .50$ (see Figure 19.4). In addition, we see that the series of 7-year moving averages (Figure 19.3) corresponds to the exponentially smoothed series having weight $W = .25$ (see Figure 19.4). By examining how our two smoothing series (one with $W = .25$ and the other with $W = .50$) fit the observed data in Figure 19.4, we may realize that the choice of a particular smoothing coefficient W is dependent upon the purpose of the user. If we desire only to smooth a series by eliminating unwanted cyclical and irregular variations, we should select a small value for W (closer to zero). On the other hand, if our goal is forecasting, we should choose a larger value for W (closer to 1). In the former case, the overall long-term tendencies of the series will be apparent; in the latter case, future short-term directions may be more adequately predicted.

Smoothing Table 19.4 presents the exponentially smoothed values (using smoothing coefficients of $W = .50$ and $W = .25$) for annual factory sales of the General Motors Corporation over the 23-year period 1970 through 1992. As previously indicated, the two smoothed series are plotted in Figure 19.4 on page 868 along with the original time-series data.

To demonstrate the computations for the exponentially smoothed values as shown in Table 19.4 on page 868, let us consider the series having a smoothing coefficient of $W = .25$. As a starting point we may use the initial observed value $Y_{1970} = 5.3$ as our first smoothed value ($E_{1970} = 5.3$). Now using the observed value of the time series for the year 1971 ($Y_{1971} = 7.8$), we may smooth the series for the year 1971 by computing

$$\begin{aligned} E_{1971} &= WY_{1971} + (1 - W)E_{1970} \\ &= (.25)(7.8) + (.75)(5.3) = 5.9 \text{ million} \end{aligned}$$

To smooth the series for the year 1972, we have

$$\begin{aligned} E_{1972} &= WY_{1972} + (1 - W)E_{1971} \\ &= (.25)(7.8) + (.75)(5.9) = 6.4 \text{ million} \end{aligned}$$

Table 19.4 Exponentially smoothed series of factory sales of General Motors Corp. (1970–1992).

Year	Factory Sales (in millions)	$W = .50$	$W = .25$
1970	5.3	5.3	5.3
1971	7.8	6.6	5.9
1972	7.8	7.2	6.4
1973	8.7	7.9	7.0
1974	6.7	7.3	6.9
1975	6.6	7.0	6.8
1976	8.6	7.8	7.3
1977	9.1	8.4	7.7
1978	9.5	9.0	8.2
1979	9.0	9.0	8.4
1980	7.1	8.0	8.1
1981	6.8	7.4	7.7
1982	6.2	6.8	7.4
1983	7.8	7.3	7.5
1984	8.3	7.8	7.7
1985	9.3	8.6	8.1
1986	8.6	8.6	8.2
1987	7.8	8.2	8.1
1988	8.1	8.1	8.1
1989	7.9	8.0	8.1
1990	7.5	7.8	8.0
1991	7.0	7.4	7.8
1992	7.2	7.3	7.7

Source: Data are taken from Table 19.2.

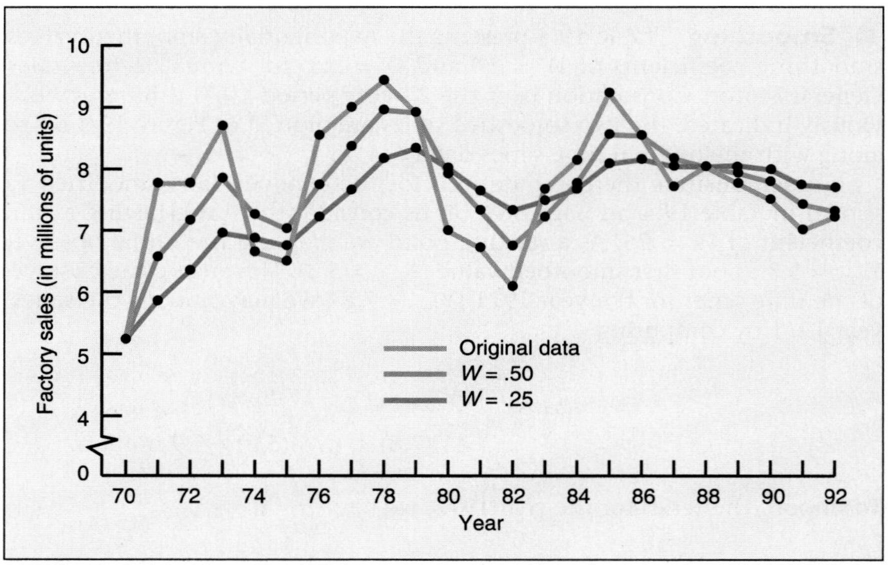

Figure 19.4
Plotting the exponentially smoothed series (W = .50 and W = .25).
Source: Data are taken from Table 19.4.

To smooth the series for the year 1973, we have

$$E_{1973} = WY_{1973} + (1 - W)E_{1972}$$
$$= (.25)(8.7) + (.75)(6.4) = 7.0 \text{ million}$$

This process continues until exponentially smoothed values have been obtained for all 23 years in the series as shown in Table 19.4 and Figure 19.4.

Forecasting To use the exponentially weighted moving average for purposes of forecasting rather than for smoothing, we take the smoothed value in our current period of time (say time period i) as our projected estimate of the observed value of the time series in the following time period, $i + 1$—that is,

$$\hat{Y}_{i+1} = E_i \qquad (19.7)$$

For example, to forecast the number of units sold from all General Motors Corporation plants during the year 1993, we would use the smoothed value for the year 1992 as its estimate. From Table 19.4, for a smoothing coefficient of $W = .50$, that projection is 7.3 million units.

Once the observed data for the year 1993 become available, we can use Equation (19.4) to make a forecast for the year 1994 by obtaining the smoothed value for 1993 as follows:

$$E_{1993} = WY_{1993} + (1 - W)E_{1992}$$
$$\text{current smoothed value} = (W)(\text{current observed value}) + (1 - W)(\text{previous smoothed value})$$

or, in terms of forecasting,

$$\hat{Y}_{1994} = WY_{1993} + (1 - W)\hat{Y}_{1993}$$
$$\text{new forecast} = (W)(\text{current observed value}) + (1 - W)(\text{current forecast})$$

Problems for Section 19.4

19.1 The data at the top of page 870 represent the median income of families in the United States (in 1990 dollars) for all races, for whites, and for blacks, for the 14-year period from 1977 to 1990. For each of the three sets of data (all races, whites, and blacks):
(a) Plot the data on a chart.
(b) Fit a 3-year moving average to your data and plot the results on your chart.
(c) Using a smoothing coefficient of .50, exponentially smooth the series and plot your results on your chart.
(d) What is your exponentially smoothed forecast for the trend in 1991?

(e) ACTION Go to your library and record the actual 1991 value from the table available from the U.S. Department of Commerce. Compare your results to the forecast you made in (d). Discuss.

(f) ACTION Write a letter to one of your United States Senators explaining the trend in median family income for each of the two groups and all races combined for the period from 1977 to 1990.

Data file
MEDINCUS.DAT

Median family income in the United States (1977–1990).

Year	All Races	Whites	Blacks
1977	34,528	36,104	20,625
1978	35,361	36,821	21,808
1979	35,262	36,796	20,836
1980	33,346	34,743	20,103
1981	32,190	33,814	19,074
1982	31,738	33,322	18,417
1983	32,378	33,905	19,108
1984	33,251	34,827	19,411
1985	33,689	35,410	20,390
1986	35,129	36,740	20,993
1987	35,632	37,260	21,177
1988	35,565	37,470	21,355
1989	36,062	37,919	21,301
1990	35,353	36,915	21,423

Source: U.S. Department of Commerce, Bureau of the Census, Table B-28.

19.2 The data in the accompanying table represent the annual earnings per share of TRW Inc. over the 23-year period 1970 through 1992.

Data file
TRW.DAT

Earnings per share at TRW Inc. (1970–1992).

Year	Earnings per Share	Year	Earnings per Share
1970	2.39	1982	5.49
1971	1.85	1983	5.53
1972	2.22	1984	6.66
1973	2.95	1985	3.79
1974	2.76	1986	7.25
1975	3.08	1987	4.01
1976	4.02	1988	4.23
1977	4.77	1989	4.31
1978	5.42	1990	3.39
1979	5.86	1991	2.30
1980	6.15	1992	3.09
1981	6.60		

Source: Moody's Handbook of Common Stocks, 1980, 1989, 1993.

(a) Plot the data on a chart.
(b) Fit a 3-year moving average to the data and plot the results on your chart.
(c) Using a smoothing coefficient of .50, exponentially smooth the series and plot the results on your chart.

(d) What is your exponentially smoothed forecast for the trend in 1993?

19.3 The data given in the accompanying table represent the annual number of employees (in thousands) in an oil supply company for the years 1974 through 1993.

Data file OILSUPP.DAT

Number of employees (in thousands).

Year	Number	Year	Number	Year	Number
1974	1.45	1982	2.06	1990	1.88
1975	1.55	1983	1.80	1991	2.00
1976	1.61	1984	1.73	1992	2.08
1977	1.60	1985	1.77	1993	1.88
1978	1.74	1986	1.90		
1979	1.92	1987	1.82		
1980	1.95	1988	1.65		
1981	2.04	1989	1.73		

(a) Plot the data on a chart.
(b) Fit a 3-year moving average to the data and plot the results on your chart.
(c) Using a smoothing coefficient of .50, exponentially smooth the series and plot the results on your chart.
(d) What is your exponentially smoothed forecast for the trend in 1994?

19.4 The data given in the accompanying table represent the annual sales dollars (in millions) for a food-processing company for the years 1968 through 1993.

Data file FOODTIME.DAT

Annual sales dollars (millions).

Year	Sales Dollars	Year	Sales Dollars	Year	Sales Dollars
1968	41.6	1977	53.2	1986	36.4
1969	48.0	1978	53.3	1987	38.4
1970	51.7	1979	51.6	1988	42.6
1971	55.9	1980	49.0	1989	34.8
1972	51.8	1981	38.6	1990	28.4
1973	57.0	1982	37.3	1991	23.9
1974	64.4	1983	43.8	1992	27.8
1975	60.8	1984	41.7	1993	42.1
1976	56.3	1985	38.3		

(a) Plot the data on a chart.
(b) Fit a 7-year moving average to the data and plot the results on your chart.
(c) Using a smoothing coefficient of .25, exponentially smooth the series and plot the results on your chart.
(d) What is your exponentially smoothed forecast for the trend in 1994?

19.5 Time-Series Analysis of Annual Data: Least Squares Trend Fitting and Forecasting

The component factor of a time series most often studied is trend. Primarily, we study trend for predictive purposes; that is, we may wish to study trend directly as an aid in making intermediate and long-range forecasting projections. Secondly, we may wish to study trend in order to isolate and then eliminate its influencing effects on the time-series model as a guide to short-run (1 year or less) forecasting of general business cycle conditions. As depicted in Figure 19.1 on page 860, to obtain some visual impression or feeling of the overall long-term movements in a time series we construct a chart in which the observed data (dependent variable) are plotted on the vertical axis and the time periods (independent variable) are plotted on the horizontal axis. If it appears that a straight-line trend could be adequately fitted to the data, the two most widely used methods of trend fitting are the method of least squares (see Section 17.4) and the method of *double exponential smoothing* (References 1, 2, and 4). If the time-series data indicate some long-run downward or upward curvilinear movement, the two most widely used trend-fitting methods are the method of least squares (see Section 18.11) and the method of *triple exponential smoothing* (References 1, 2, and 4). In this section we shall focus on least squares methods for fitting linear and curvilinear trends as guides to forecasting. In Sections 19.6 and 19.7 other, more elaborate, forecasting approaches will be described.

19.5.1 The Linear Model

We recall from Section 17.4 that the least squares method permits us to fit a straight line of the form

$$\hat{Y}_i = b_0 + b_1 X_i \tag{19.8}$$

such that the values we calculate for the two coefficients—the intercept b_0 and the slope b_1—result in the sum of squared differences between each observed value Y_i in the data and each predicted value $\hat{Y}_i$ along the trend line being minimized; that is

$$\sum_{i=1}^{n} (Y_i - \hat{Y}_i)^2 = \text{minimum}$$

To obtain such a line, we recall that in linear regression analysis we compute the slope from

$$b_1 = \frac{\sum_{i=1}^{n} X_i Y_i - n\overline{X}\overline{Y}}{\sum_{i=1}^{n} X_i^2 - n\overline{X}^2} \tag{19.9}$$

and the intercept from

$$b_0 = \overline{Y} - b_1 \overline{X} \tag{19.10}$$

Once this is accomplished and the line $\hat{Y}_i = b_0 + b_1X_i$ is obtained, we may substitute values for X into Equation (19.8) to predict various values for Y.

When using the method of least squares for fitting trends in time series, our computational efforts can be simplified if we properly code the X values. The first observation in our time series is selected as the origin and assigned a code value of $X = 0$. All successive observations are then assigned consecutively increasing integer codes: 1, 2, 3, . . . , so that the nth and last observation in the series has code $n - 1$. Thus, for example, for time-series data recorded annually over 23 years, the first year will be assigned a coded value of 0, the second year will be coded as 1, the third year will be coded as 2, . . . , and the final (twenty-third) year will be coded as 22.

The annual time series presented in Table 19.5 and plotted in Figure 19.1 on page 860 represents the net sales (in billions of dollars) for the Eastman Kodak Company over the 23-year period 1970 through 1992. Coding the consecutive X values 0 through 22 and then using Equations (19.9) and (19.10) or, as demonstrated in Figure 19.5, employing a statistical software package such as MINITAB (see Reference 8), we determine that

$$\hat{Y}_i = 1.2011 + 0.8003X_i$$

where the origin is 1970 and X units = 1 year.

Table 19.5 Net sales (in billions of dollars) for Eastman Kodak Company (1970–1992).

Year	Net Sales (billions of dollars)	Year	Net Sales (billions of dollars)
1970	2.8	1982	10.8
1971	3.0	1983	10.2
1972	3.5	1984	10.6
1973	4.0	1985	10.6
1974	4.6	1986	11.5
1975	5.0	1987	13.3
1976	5.4	1988	17.0
1977	6.0	1989	18.4
1978	7.0	1990	18.9
1979	8.0	1991	19.4
1980	9.7	1992	20.1
1981	10.3		

Source: Moody's Handbook of Common Stocks, 1980, 1989, 1993.

```
The regression equation is
sales = 1.20 + 0.800 years

Predictor        Coef      Stdev    t-ratio       p
Constant  b0  1.2011     0.5510       2.18    0.041
years     b1  0.80030   0.04290      18.66    0.000

s = 1.365       R-sq = 94.3%     R-sq(adj) = 94.0%
```

Figure 19.5
Partial MINITAB output for fitting linear regression model to forecast annual net sales at Eastman Kodak Company.

The intercept $b_0 = 1.2011$ is the fitted trend value reflecting the net sales (in billions of dollars) at Eastman Kodak during the origin or base year, 1970. The slope $b_1 = 0.8003$ indicates that net sales are increasing at a rate of 0.8003 billions of dollars per year.

To project the trend in the net sales to the year 1993, we substitute $X = 23$, the code for 1993, into the equation and our forecast is

$$1993 \quad \hat{Y}_{24} = 1.2011 + (0.8003)(23) = 19.6 \text{ billions of dollars}$$

The fitted trend line projected to 1993 is plotted in Figure 19.6 along with the original time series. A careful examination of Figure 19.6 reveals that a marked increase has occurred in the more recent years of the series. Perhaps, then, a curvilinear trend model would better fit the series? Two such models—a *quadratic* trend model and an *exponential* trend model—are presented in Sections 19.5.2 and 19.5.3, respectively.

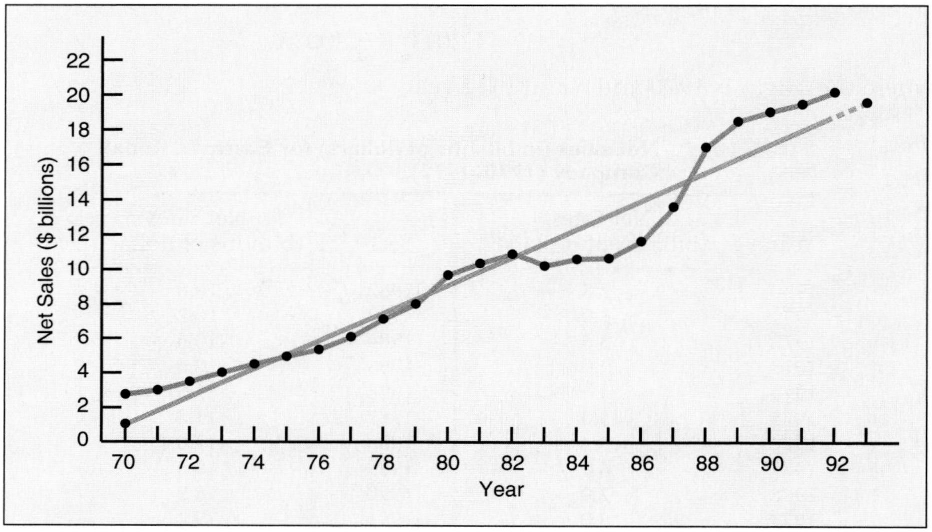

Figure 19.6
Fitting the least squares trend line.

19.5.2 The Quadratic Model

A quadratic model or *second-degree polynomial* is the simplest of the curvilinear models. Using the least squares method of Section 18.11, we may fit a quadratic trend equation of the form

$$\hat{Y}_i = b_0 + b_1 X_i + b_{11} X_i^2 \tag{19.11}$$

where b_0 = estimated Y intercept
b_1 = estimated *linear* effect on Y
b_{11} = estimated *curvilinear* effect on Y

```
The regression equation is
sales = 2.92 + 0.309 years + 0.0223 yearsq

Predictor         Coef       Stdev    t-ratio        p
Constant  b0    2.9217      0.5949       4.91    0.000
years     b1    0.3087      0.1253       2.46    0.023
yearsq    b11 0.022346    0.005499       4.06    0.001

s = 1.035       R-sq = 96.9%     R-sq(adj) = 96.6%
```

Figure 19.7
Partial MINITAB output for fitting a quadratic regression model to forecast annual net sales at Eastman Kodak Company.

Once again, we may access a statistical software package to perform the computations necessary to obtain the least squares fit. Figure 19.7 provides MINITAB output for the quadratic model representing annual net sales at Eastman Kodak. From this we determine that

$$\hat{Y}_i = 2.9217 + 0.3087X_i + 0.0223X_i^2$$

where the origin is 1970 and X units = 1 year.

To use the quadratic trend equation for forecasting purposes we substitute the appropriate coded X values into this equation. For example, to predict the trend in net sales for the year 1993 (that is, $X = 23$), we have

$$1993 \quad \hat{Y}_{24} = 2.9217 + (0.3087)(23) + (0.0223)(23^2)$$
$$= 21.82 \text{ billions of dollars}$$

The fitted quadratic trend equation projected to 1993 is plotted in Figure 19.8 together with the original time series.

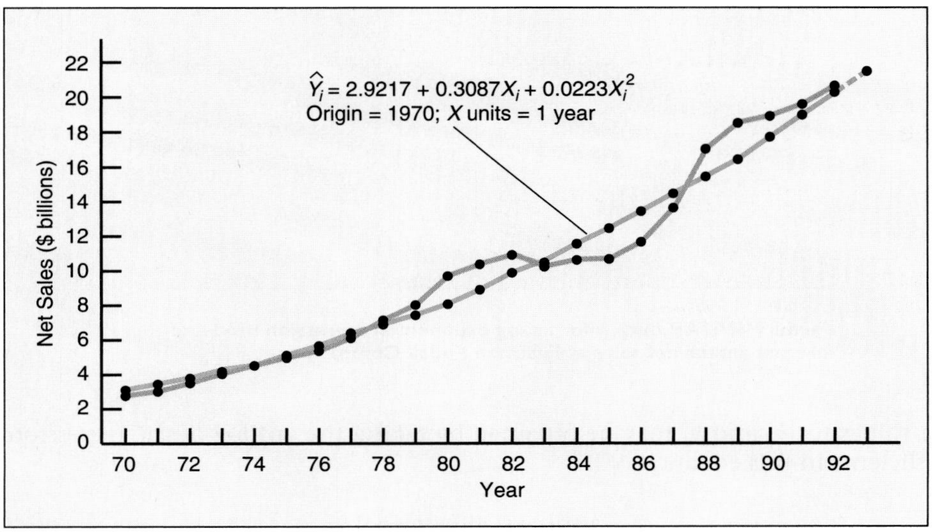

Figure 19.8
Fitting the quadratic trend equation.

19.5.3 The Exponential Model

When a series appears to be *increasing at an increasing rate* such that the *percent difference* from observation to observation is *constant,* we may fit an exponential trend equation of the form

$$\hat{Y}_i = b_0 b_1^{X_i} \qquad \textbf{(19.12)}$$

where b_0 = estimated Y intercept

$(b_1 - 1) \times 100\%$ = estimated annual *compound growth rate* (in percent)

If we take the logarithm (base 10) of both sides of Equation (19.12), we have

$$\log \hat{Y}_i = \log b_0 + X_i \log b_1 \qquad \textbf{(19.13)}$$

Since Equation (19.13) is linear in form, we may use the method of least squares by working with the log Y_i values instead of the Y_i values and obtain the slope (log b_1) and intercept (log b_0). Once again, we may access a statistical software package to accomplish the necessary calculations.

Figure 19.9 represents MINITAB output for an exponential model of annual net sales at Eastman Kodak. From this we determine that

$$\log \hat{Y}_i = 0.49949 + 0.0389X_i$$

where the origin is 1970 and X units = 1 year.

```
The regression equation is
logsales = 0.499 + 0.0389 years

Predictor        Coef       Stdev    t-ratio        p
Constant      0.49949     0.01918      26.04    0.000
years        0.038902    0.001493      26.05    0.000

s = 0.04751       R-sq = 97.0%     R-sq(adj) = 96.9%
```

Figure 19.9
Partial MINITAB output for fitting exponential regression model to forecast annual net sales at Eastman Kodak Company.

The values for b_0 and b_1 may be obtained by taking the antilog of the regression coefficients in this equation:

$$b_0 = \text{antilog } 0.49949 = 3.155$$

$$b_1 = \text{antilog } 0.0389 = 1.0937$$

Thus the fitted exponential trend equation can be expressed as

$$\hat{Y}_i = (3.155)(1.0937)^{X_i}$$

where the origin is 1970 and X units = 1 year.

The intercept $b_0 = 3.155$ is the fitted trend value representing net sales in the base year 1970. The value $(b_1 - 1) \times 100\% = 9.37\%$ is the annual compound growth rate in net sales at Eastman Kodak.

For forecasting purposes, we may substitute the appropriate coded X values into either of the two equations. For example, to predict the trend in net sales for the year 1993 (that is, $X = 23$), we have

$$1993 \qquad \log \hat{Y}_{24} = 0.49949 + (0.0389)(23) = 1.3937$$
$$\hat{Y}_{24} = \text{antilog } 1.3937 = 24.76 \text{ billions of dollars}$$

or

$$1993 \qquad \hat{Y}_{24} = (3.155)(1.0937)^{23} = 24.76 \text{ billions of dollars}$$

The fitted exponential trend equation projected to 1993 is plotted in Figure 19.10 together with the original time series.

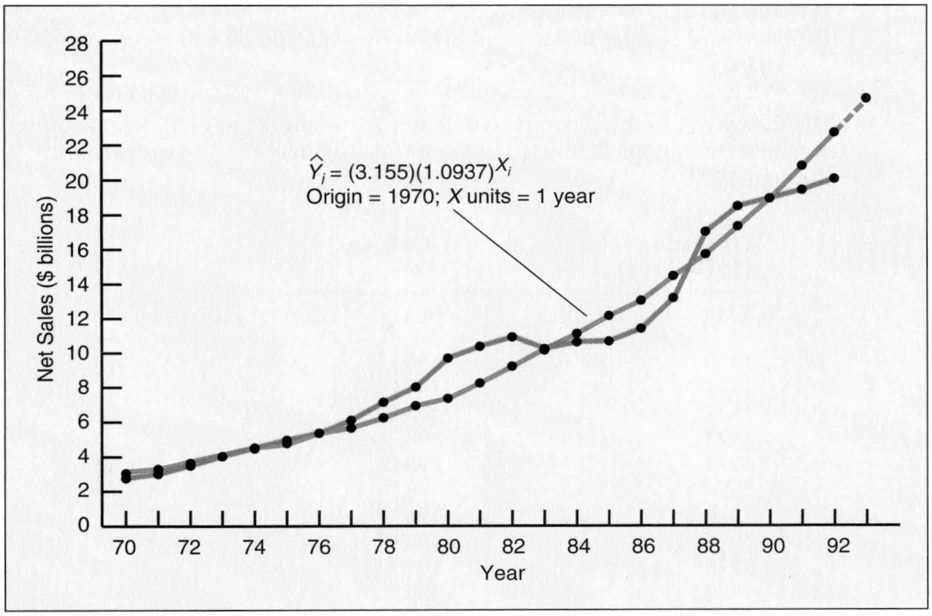

Figure 19.10
Fitting the exponential trend equation.

We have now seen the annual net sales data at Eastman Kodak fitted by three different models: linear, quadratic, and exponential. In Section 19.8 we will compare the results of these and other forecasting models to determine, *a posteriori,* the best fit. In Problems 19.51 and 19.52 on pages 914 and 915 the student will have the opportunity to use *a priori* methods to determine the appropriate model for a given series of data.

Problems for Section 19.5

● 19.5 The data given in the accompanying table represent the annual net sales (in billions of dollars) of Upjohn Co. over the 23-year period 1970 through 1992.

Data file
UPJOHN.DAT

Net sales at Upjohn Co. (1970–1992).

Year	Sales	Year	Sales
1970	0.4	1982	1.8
1971	0.4	1983	1.7
1972	0.5	1984	1.9
1973	0.7	1985	2.0
1974	0.8	1986	2.3
1975	0.9	1987	2.5
1976	1.0	1988	2.7
1977	1.1	1989	2.9
1978	1.3	1990	3.0
1979	1.5	1991	3.4
1980	1.8	1992	3.6
1981	1.9		

Source: Moody's Handbook of Common Stocks, 1980, 1989, 1993.

(a) Plot the data on a chart.
(b) Fit a least squares linear trend line to the data and plot the line on your chart.
(c) What are your trend forecasts for the years 1993, 1994, 1995, and 1996?

19.6 The data given in the accompanying table represent the annual net operating revenues (in billions of dollars) of Coca-Cola Co. over the 23-year period 1970 through 1992.

Data file
COKE.DAT

Operating revenues at Coca-Cola Co. (1970–1992).

Year	Revenues	Year	Revenues
1970	1.6	1982	5.9
1971	1.7	1983	6.6
1972	1.9	1984	7.2
1973	2.1	1985	7.9
1974	2.5	1986	7.0
1975	2.9	1987	7.7
1976	3.1	1988	8.3
1977	3.6	1989	9.0
1978	4.3	1990	10.2
1979	4.5	1991	11.6
1980	5.3	1992	13.0
1981	5.5		

Source: Moody's Handbook of Common Stocks, 1980, 1989, 1993.

(a) Plot the data on a chart.
(b) Fit a least squares linear trend line to the data and plot the line on your chart.
(c) What are your trend forecasts for the years 1993, 1994, 1995, and 1996?

19.7 The data given below represent the annual net sales (in billions of dollars) of Gillette Company, Inc., over the 23-year period 1970 through 1992.

Data file GILLETTE.DAT

Net sales at Gillette Company, Inc. (1970–1992).

Year	Sales	Year	Sales
1970	0.7	1982	2.2
1971	0.7	1983	2.2
1972	0.8	1984	2.3
1973	1.1	1985	2.4
1974	1.2	1986	2.8
1975	1.4	1987	3.2
1976	1.5	1988	3.6
1977	1.6	1989	3.8
1978	1.7	1990	4.3
1979	2.0	1991	4.7
1980	2.3	1992	5.2
1981	2.3		

Source: Moody's Handbook of Common Stocks, 1980, 1989, 1993.

(a) Plot the data on a chart.
(b) Fit a least squares linear trend line to the data and plot the line on your chart.
(c) What are your trend forecasts for the years 1993, 1994, 1995, and 1996?

19.8 The data given in the accompanying table represent the annual net sales (in billions of dollars) of the Georgia-Pacific Corp. over the 23-year period 1970 through 1992.

Data file GAPAC.DAT

Net sales at Georgia-Pacific Corp. (1970–1992).

Year	Sales	Year	Sales
1970	1.1	1982	5.4
1971	1.3	1983	6.5
1972	1.8	1984	6.7
1973	2.2	1985	6.7
1974	2.4	1986	7.2
1975	2.4	1987	8.6
1976	3.0	1988	9.5
1977	3.7	1989	10.1
1978	4.4	1990	12.7
1979	5.2	1991	11.5
1980	5.0	1992	11.8
1981	5.4		

Source: Moody's Handbook of Common Stocks, 1980, 1989, 1993.

(a) Plot the data on a chart.
(b) Fit a least squares linear trend line to the data and plot the line on your chart.
(c) What are your trend forecasts for the years 1993, 1994, 1995, and 1996?

19.9 The data given in the accompanying table represent the annual total income (in billions of dollars) of Boeing Co. over the 23-year period 1970 through 1992.

Data file
BOEING.DAT

Total income of Boeing Co. (1970–1992).

Year	Income	Year	Income
1970	3.7	1982	9.2
1971	3.0	1983	11.3
1972	2.4	1984	10.6
1973	3.4	1985	14.0
1974	3.8	1986	16.8
1975	3.8	1987	15.8
1976	4.0	1988	17.3
1977	4.1	1989	20.6
1978	5.6	1990	27.5
1979	8.5	1991	29.3
1980	9.8	1992	30.2
1981	10.1		

Source: Moody's Handbook of Common Stocks, 1980, 1989, 1993.

(a) Plot the data on a chart.
(b) Fit a quadratic trend equation to the data and plot the curve on your chart.
(c) What are your trend forecasts for the years 1993, 1994, 1995, and 1996?

19.10 The following data represent average SAT scores (verbal and math) for males and females, along with the average total SAT score for the 20-year period from 1972 to 1991.

Data file
SAT.DAT

SAT scores (1972–1991).

	Male		Female		
Year	Verbal	Math	Verbal	Math	Total
1972	454	505	452	461	937
1973	446	502	443	460	926
1974	447	501	442	459	924
1975	437	495	431	449	906
1976	433	497	430	446	903
1977	431	497	427	445	899
1978	433	494	425	444	897
1979	431	493	423	443	894
1980	428	491	420	443	890
1981	430	492	418	443	890
1982	431	493	421	443	893
1983	430	493	420	445	893
1984	433	495	420	449	897
1985	437	499	425	452	906
1986	437	501	426	451	906
1987	435	500	425	453	906

(Continued)

SAT scores 1972–1991 (continued).

Year	Male		Female		Total
	Verbal	Math	Verbal	Math	
1988	435	498	422	455	904
1989	434	500	421	454	903
1990	429	499	419	455	900
1991	426	497	418	453	896

Source: New York Times, August 27, 1991, p. A20 and August 28, 1990.

For each of the five variables (average SAT score in verbal and math for males and females, along with the average total score):
(a) Plot the data on a chart.
(b) Fit a quadratic trend equation to the data.
(c) What are your trend forecasts for the years 1992, 1993, and 1994?

19.11 The following data represent the receipts and expenditures for state and local governments for the 22-year period from 1970 to 1991.

State and local government expenditures (1970–1991).

Data file STATE.DAT

Year	Receipts	Expenditures	Surplus or Deficit
1970	129.0	127.2	1.80
1971	145.3	142.8	2.50
1972	169.7	156.3	13.40
1973	185.3	171.9	13.40
1974	200.6	193.5	7.10
1975	225.6	221.0	4.60
1976	253.9	239.3	14.60
1977	281.9	256.3	25.60
1978	309.3	278.2	31.10
1979	330.6	305.4	25.20
1980	361.4	336.6	24.80
1981	390.8	362.3	28.50
1982	409.0	382.1	26.90
1983	443.4	403.2	40.20
1984	492.2	434.1	58.10
1985	528.7	472.6	56.10
1986	571.2	517.0	54.20
1987	594.3	554.2	40.10
1988	631.3	593.0	38.30
1989	677.0	635.9	41.10
1990	724.5	698.8	25.70
1991	770.6	741.1	29.50

Source: U.S. Department of Commerce, Bureau of Economic Analysis, Table B-77.

For each of the three variables (receipts, expenditures, and surplus or deficit):
(a) Plot the data on a chart.
(b) Fit a linear trend equation to the data.
(c) Fit a quadratic trend equation to the data.
(d) Using the models fit in (b) and (c), make annual forecasts for 1992, 1993, and 1994.

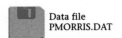

(e) **ACTION** Go to your library and record the actual 1992, 1993, and 1994 values from the table available from the Department of Commerce. Compare your results to that in (d). Discuss.

Data file PMORRIS.DAT

19.12 The data given in the accompanying table represent the annual operating revenues (in billions of dollars) of Philip Morris Companies, Inc., over the 23-year period 1970 through 1992.

Operating revenues of Philip Morris Companies, Inc. (1970–1992).

Year	Revenue	Year	Revenue
1970	1.5	1982	11.6
1971	1.9	1983	13.0
1972	2.1	1984	13.8
1973	2.6	1985	16.0
1974	3.0	1986	25.9
1975	3.6	1987	28.2
1976	4.3	1988	31.7
1977	5.2	1989	44.8
1978	6.6	1990	51.3
1979	8.1	1991	56.5
1980	9.6	1992	59.1
1981	10.7		

Source: Moody's Handbook of Common Stocks, 1980, 1989, 1993.

(a) Plot the data on a chart.
(b) Fit an exponential trend equation to the data and plot the curve on your chart.
(c) What are your trend forecasts for the years 1993, 1994, 1995, and 1996?

● 19.13 The data given below represent the annual net sales (in billions of dollars) of Black & Decker Corp. over the 23-year period 1970 through 1992.

Data file BDECKER.DAT

Net sales of Black & Decker Corp. (1970–1992).

Year	Sales	Year	Sales
1970	0.3	1982	1.2
1971	0.3	1983	1.2
1972	0.3	1984	1.5
1973	0.4	1985	1.7
1974	0.6	1986	1.8
1975	0.7	1987	1.9
1976	0.7	1988	2.3
1977	0.8	1989	3.2
1978	1.0	1990	4.8
1979	1.2	1991	4.7
1980	1.2	1992	4.8
1981	1.2		

Source: Moody's Handbook of Common Stocks, 1980, 1989, 1993.

(a) Plot the data on a chart.
(b) Fit an exponential trend equation to the data and plot the curve on your chart.
(c) What are your trend forecasts for the years 1993, 1994, 1995, and 1996?

19.14 The following data represent the number of employees (in thousands) on nonagricultural payrolls for the 42-year period from 1950 to 1991. The data are divided into goods-producing industries, nongovernment service producing industries, federal government, and state and local government:

Data file EMPLOYEE.DAT

Total employment on nonagricultural payrolls (1950–1991).

Year	Goods Producing	Nongovernment Services	Federal Government	State and Local Government
1950	45,197	20,665	1,928	4,098
1951	47,819	21,471	2,302	4,087
1952	48,793	21,987	2,420	4,188
1953	50,202	22,483	2,305	4,340
1954	48,990	22,488	2,188	4,563
1955	50,641	23,214	2,187	4,727
1956	52,369	23,988	2,209	5,069
1957	52,853	24,273	2,217	5,399
1958	51,324	23,972	2,191	5,648
1959	53,268	24,774	2,233	5,850
1960	54,189	25,402	2,270	6,083
1961	53,999	25,548	2,279	6,315
1962	55,549	26,208	2,340	6,550
1963	56,653	26,787	2,358	6,868
1964	58,283	27,682	2,348	7,248
1965	60,765	28,765	2,378	7,696
1966	63,901	29,959	2,564	8,220
1967	65,803	31,104	2,719	8,672
1968	67,897	32,321	2,737	9,102
1969	70,384	33,828	2,758	9,437
1970	70,880	34,748	2,731	9,823
1971	71,214	35,397	2,696	10,185
1972	73,675	36,674	2,684	10,649
1973	76,790	38,166	2,663	11,068
1974	78,265	39,301	2,724	11,446
1975	76,945	39,660	2,748	11,937
1976	79,382	41,159	2,733	12,138
1977	82,471	42,999	2,727	12,399
1978	86,697	45,441	2,753	12,919
1979	89,823	47,416	2,773	13,174
1980	90,406	48,507	2,866	13,375
1981	91,156	49,628	2,772	13,259
1982	89,566	49,916	2,739	13,098
1983	90,200	50,996	2,774	13,096
1984	94,496	53,746	2,807	13,216
1985	97,519	56,266	2,875	13,519
1986	99,525	58,274	2,899	13,794
1987	102,200	60,482	2,943	14,067
1988	105,536	62,977	2,971	14,415
1989	108,329	65,228	2,988	14,791
1990	109,971	66,692	3,085	15,237
1991	108,975	66,720	2,965	15,469

Source: U.S. Department of Labor, Bureau of Labor Statistics, Table B-41.

For each of the four variables (employees in goods-producing industries, non-government service producing industries, federal government, and state and local government):

(a) Plot the data on a chart.
(b) Fit a linear trend equation to the data.
(c) Fit an exponential trend equation to the data.
(d) Using the models fit in (b) and (c), make annual forecasts for 1992, 1993, and 1994.
(e) ACTION Go to your library and record the actual 1992, 1993, and 1994 values from the table available from the Department of Labor. Compare your results to that in (d). Discuss.

19.6 The Holt-Winters Method for Trend Fitting and Forecasting

The **Holt-Winters method** (Reference 7) is a sophisticated extension of the exponential smoothing approach described in Section 19.4.2. Whereas the exponential smoothing procedure provides an impression of the overall, long-term movements in the data and permits short-term forecasting, the more elaborate Holt-Winters technique also allows for the study of trend through intermediate and/or long-term forecasting into the future. The differences between the two procedures are highlighted in Figure 19.11.

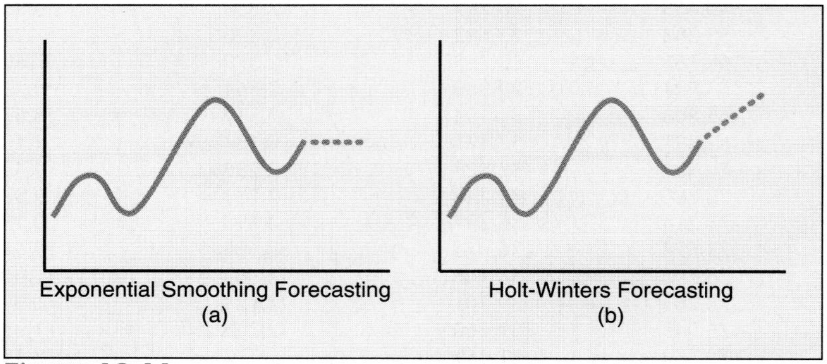

Figure 19.11
Exponential smoothing and the Holt-Winters method.

From (a) we observe that exponential smoothing can be used most effectively for short-term forecasting (one period into the future). We can, of course, extend this forecast numerous time periods into the future. This would be meaningful if there were no overall upward or downward trend in the series. However, if any upward or downward movement does exist, such a horizontal projection will entirely miss it. On the other hand, the Holt-Winters forecasting method of (b) is designed to detect such phenomena. Hence, the Holt-Winters technique concurrently provides for study of overall movement level and future trend in a series.

To use the Holt-Winters method at any time period i we must continuously estimate the level of the series (that is, the smoothed value E_i) and the value of trend (T_i). This is achieved through the solution to the following equations:

Level	$E_i = U(E_{i-1} + T_{i-1}) + (1 - U)Y_i$	**(19.14a)**
Trend	$T_i = VT_{i-1} + (1 - V)(E_i - E_{i-1})$	**(19.14b)**

where E_i = level of the smoothed series being computed in time period i
E_{i-1} = level of the smoothed series already computed in time period $i - 1$
T_i = value of the trend component being computed in time period i
T_{i-1} = value of the trend component already computed in time period $i - 1$
Y_i = observed value of the time series in period i
U = subjectively assigned smoothing constant (where $0 < U < 1$)
V = subjectively assigned smoothing constant (where $0 < V < 1$)

To begin computations, we set $E_2 = Y_2$ and $T_2 = Y_2 - Y_1$ and choose smoothing constants for U and V. We then compute E_i and T_i for all i years, $i = 3, 4, \ldots, n$.

To illustrate the Holt-Winters method, let us return to the time series presented in Table 19.5 (on page 873) and plotted in Figure 19.1 (on page 860), which represents the net sales (in billions of dollars) for the Eastman Kodak Company over the 23-year period 1970 through 1992. The computations are shown in Table 19.6 with selected constants $U = .3$ and $V = .3$.

Table 19.6 **Using the Holt-Winters method on annual net sales (in billions of dollars) for Eastman Kodak Company (1970–1992).**

Year	i	Net Sales Y_i	$(U)(E_{i-1} + T_{i-1}) + (1 - U)(Y_i) = E_i$	$(V)(T_{i-1}) + (1 - V)(E_i - E_{i-1}) = T_i$
1970	1	2.8	****	****
1971	2	3.0	3.0	0.2
1972	3	3.5	(.3)(3.0 + 0.2) + (.7)(3.5) = 3.4	(.3)(0.2) + (.7)(3.4 − 3.0) = 0.3
1973	4	4.0	(.3)(3.4 + 0.3) + (.7)(4.0) = 3.9	(.3)(0.3) + (.7)(3.9 − 3.4) = 0.4
1974	5	4.6	(.3)(3.9 + 0.4) + (.7)(4.6) = 4.5	(.3)(0.4) + (.7)(4.5 − 3.9) = 1.6
1975	6	5.0	(.3)(4.5 + 1.6) + (.7)(5.0) = 5.3	(.3)(1.6) + (.7)(5.3 − 4.5) = 1.0
1976	7	5.4	(.3)(5.3 + 1.0) + (.7)(5.4) = 5.7	(.3)(1.0) + (.7)(5.7 − 5.3) = 0.6
1977	8	6.0	(.3)(5.7 + 0.6) + (.7)(6.0) = 6.1	(.3)(0.6) + (.7)(6.1 − 5.7) = 0.5
1978	9	7.0	(.3)(6.1 + 0.5) + (.7)(7.0) = 6.9	(.3)(0.5) + (.7)(6.9 − 6.1) = 0.7
1979	10	8.0	(.3)(6.9 + 0.7) + (.7)(8.0) = 7.9	(.3)(0.7) + (.7)(7.9 − 6.9) = 0.9
1980	11	9.7	(.3)(7.9 + 0.9) + (.7)(9.7) = 9.4	(.3)(0.9) + (.7)(9.4 − 7.9) = 1.3
1981	12	10.3	(.3)(9.4 + 1.3) + (.7)(10.3) = 10.4	(.3)(1.3) + (.7)(10.4 − 9.4) = 1.1
1982	13	10.8	(.3)(10.4 + 1.1) + (.7)(10.8) = 11.0	(.3)(1.1) + (.7)(11.0 − 10.4) = 0.8
1983	14	10.2	(.3)(11.0 + 0.8) + (.7)(10.2) = 10.7	(.3)(0.8) + (.7)(10.7 − 11.0) = 0.0
1984	15	10.6	(.3)(10.7 + 0.0) + (.7)(10.6) = 10.6	(.3)(0.0) + (.7)(10.6 − 10.7) = −0.1
1985	16	10.6	(.3)(10.6 − 0.1) + (.7)(10.6) = 10.6	(.3)(−0.1) + (.7)(10.6 − 10.6) = 0.0
1986	17	11.5	(.3)(10.6 + 0.0) + (.7)(11.5) = 11.2	(.3)(0.0) + (.7)(11.2 − 10.6) = 0.4
1987	18	13.3	(.3)(11.2 + 0.4) + (.7)(13.3) = 12.8	(.3)(0.4) + (.7)(12.8 − 11.2) = 1.2
1988	19	17.0	(.3)(12.8 + 1.2) + (.7)(17.0) = 16.1	(.3)(1.2) + (.7)(16.1 − 12.8) = 2.7
1989	20	18.4	(.3)(16.1 + 2.7) + (.7)(18.4) = 18.5	(.3)(2.7) + (.7)(18.5 − 16.1) = 2.5
1990	21	18.9	(.3)(18.5 + 2.5) + (.7)(18.9) = 19.5	(.3)(2.5) + (.7)(19.5 − 18.5) = 1.5
1991	22	19.4	(.3)(19.5 + 1.5) + (.7)(19.4) = 19.9	(.3)(1.5) + (.7)(19.9 − 19.5) = 0.7
1992	23	20.1	(.3)(19.9 + 0.7) + (.7)(20.1) = 20.2	(.3)(0.7) + (.7)(20.2 − 19.9) = 0.5

Source: Data are taken from Table 19.5.

To begin, we set

$$E_2 = Y_2 = 3.0$$

and

$$T_2 = Y_2 - Y_1 = 3.0 - 2.8 = 0.2$$

Choosing smoothing constants $U = .3$ and $V = .3$, Equations (19.14a) and (19.14b) become

$$E_i = (.3)(E_{i-1} + T_{i-1}) + (.7)(Y_i)$$

and

$$T_i = (.3)(T_{i-1}) + (.7)(E_i - E_{i-1})$$

As an example, for 1972, the third year, $i = 3$ and we have

$$E_3 = (.3)(3.0 + 0.2) + (.7)(3.5) = 3.4$$

and

$$T_3 = (.3)(0.2) + (.7)(3.4 - 3.0) = 0.3$$

Continuing, these values would then be used in Equations (19.14a) and (19.14b) to obtain E_4 and T_4, and so on, yielding the results shown in Table 19.6 on page 885.

To use the Holt-Winters method for forecasting we assume that all future trend movements will continue from the most recent smoothed level E_n. Hence, to forecast j years into the future we have

$$\hat{Y}_{n+j} = E_n + j(T_n) \qquad \textbf{(19.15)}$$

where $\hat{Y}_{n+j}$ = forecasted value j years into the future
E_n = level of the smoothed series computed in the most recent time period n
T_n = value of the trend component computed in the most recent time period n
j = number of years into the future

Using E_{23} and T_{23}, the latest estimates of current level and trend, respectively, our forecasts of net sales for the years 1993 through 1996 are obtained from Equation (19.15) as follows:

$$\hat{Y}_{n+j} = E_n + j(T_n)$$

1993: 1 year ahead $\hat{Y}_{24} = E_{23} + (1)(T_{23}) = 20.2 + (1)(0.5) = 20.7$ billions of dollars

1994: 2 years ahead $\hat{Y}_{25} = E_{23} + (2)(T_{23}) = 20.2 + (2)(0.5) = 21.2$ billions of dollars

1995: 3 years ahead $\hat{Y}_{26} = E_{23} + (3)(T_{23}) = 20.2 + (3)(0.5) = 21.7$ billions of dollars

1996: 4 years ahead $\hat{Y}_{27} = E_{23} + (4)(T_{23}) = 20.2 + (4)(0.5) = 22.2$ billions of dollars

The data, the fit, and the forecasts are plotted in Figure 19.12.

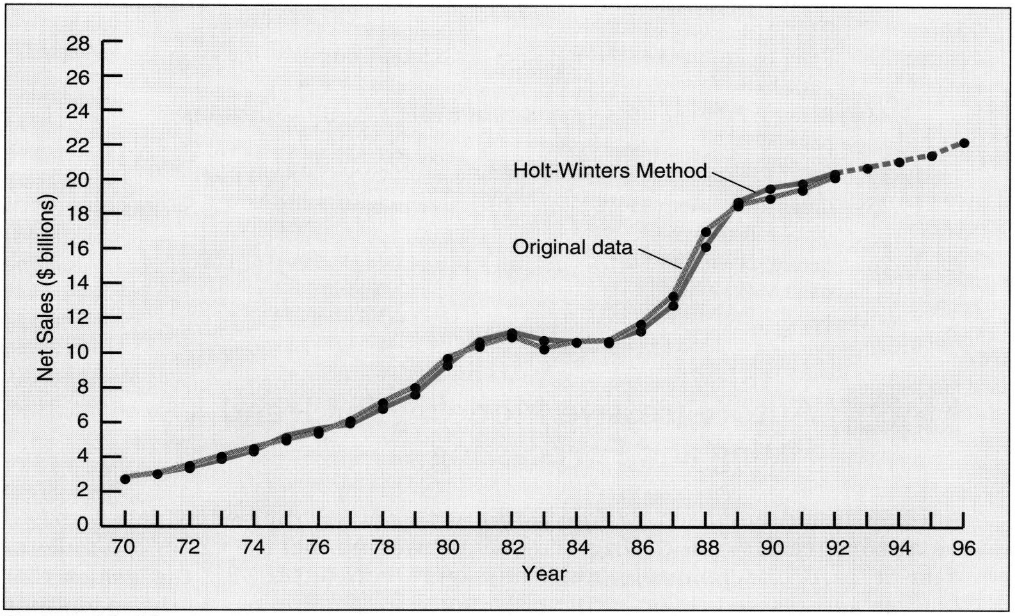

Figure 19.12
Using the Holt-Winters method on Eastman Kodak Company data.
Source: Data are taken from Tables 19.5 and 19.6.

Problems for Section 19.6

● 19.15 Given an annual time series with 20 consecutive observations, if the smoothed level for the most recent value is 34.2 and the corresponding trend level is computed to be 5.6,
(a) What is your forecast for the coming year?
(b) What is your forecast five years from now?

19.16 Given the following series from $n = 15$ consecutive time periods:

3	5	6	8	10	10	12	15	16	13	16	17	22	19	24

Use the Holt-Winters method (with $U = .30$ and $V = .30$) to forecast the series for the sixteenth through twentieth periods.

19.17 Given the following series from $n = 10$ consecutive time periods:

137	125	116	110	103	96	86	79	72	66

Use the Holt-Winters method (with $U = .20$ and $V = .20$) to forecast the series for the eleventh through fourteenth periods.

19.18 The Holt-Winters method was described as a sophisticated extension of the exponential smoothing approach presented in Section 19.4.2. Under what conditions would it still be preferable to employ the exponential smoothing procedure? Discuss.

In Problems 19.19–19.26 use the Holt-Winters method (with $U = .30$ and $V = .30$) to provide annual forecasts from 1993 through 1996.

19.19 Refer to Problem 19.2—earnings per share at TRW Inc.—on page 870.

19.20 Refer to Problem 19.5—net sales at Upjohn Co.—on page 878.

19.21 Refer to Problem 19.6—net operating revenues of Coca-Cola Co.—on page 878.

19.22 Refer to Problem 19.7—net sales at Gillette Company, Inc.—on page 879.

19.23 Refer to Problem 19.8—net sales at Georgia-Pacific Corp.—on page 879.

19.24 Refer to Problem 19.9—total income at Boeing Co.—on page 880.

19.25 Refer to Problem 19.12—operating revenues at Philip Morris Companies, Inc.—on page 882.

19.26 Refer to Problem 19.13—net sales at Black & Decker Corp.—on page 882.

19.7 Autoregressive Modeling for Trend Fitting and Forecasting

Another useful approach to forecasting with annual time-series data is based on **autoregressive modeling.**[2] Frequently, we find that the values of a series of data at particular points in time are highly correlated with the values that precede and succeed them. A first-order autocorrelation refers to the magnitude of the association between consecutive values in a time series. A second-order autocorrelation refers to the magnitude of the relationship between values two periods apart. Moreover, a pth-order autocorrelation refers to the size of the correlation between values in a time series that are p periods apart. To obtain a better historical fit of our data and, at the same time, be able to make useful forecasts of their future behavior, we may take advantage of the potential autocorrelation features inherent in such data by considering autoregressive modeling methods.

A set of autoregressive models are expressed by Equations (19.16), (19.17), and (19.18).

First-Order Autoregressive Model

$$Y_i = \omega + \psi_1 Y_{i-1} + \delta_i \quad (19.16)$$

Second-Order Autoregressive Model

$$Y_i = \omega + \psi_1 Y_{i-1} + \psi_2 Y_{i-2} + \delta_i \quad (19.17)$$

***p*th-Order Autoregressive Model**

$$Y_i = \omega + \psi_1 Y_{i-1} + \psi_2 Y_{i-2} + \cdots + \psi_p Y_{i-p} + \delta_i \quad (19.18)$$

where Y_i = the observed value of the series at time i

Y_{i-1} = the observed value of the series at time $i - 1$

Y_{i-2} = the observed value of the series at time $i - 2$

Y_{i-p} = the observed value of the series at time $i - p$

ω = fixed parameter to be estimated from least squares regression analysis

$\psi_1, \psi_2, \ldots, \psi_p$ = autoregression parameters to be estimated from least squares regression analysis

δ_i = a nonautocorrelated random (error) component (with 0 mean and constant variance)

We note that the first-order autoregressive model [Equation (19.16)] is similar in form to the simple linear regression model [Equation (17.1) on page 719] and the *p*th-order autoregressive model [Equation (19.18)] is similar in form to the multiple linear regression model [Equation 18.1a) on page 784]. In the regression models the regression parameters were given by the symbols $\beta_0, \beta_1, \ldots, \beta_p$, with corresponding statistics denoted by $b_0, b_1, \ldots, b_p$. In the autoregressive models the analogous parameters are given by the symbols $\omega, \psi_1, \ldots, \psi_p$ with corresponding estimates denoted by $\hat{\omega}, \hat{\psi}_1, \ldots, \hat{\psi}_p$.

A first-order autoregressive model [Equation (19.16)] is concerned with only the correlation between consecutive values in a series. A second-order autoregressive model [Equation (19.17)] considers the effects of the relationship between consecutive values in a series as well as the correlation between values two periods apart. A *p*th-order autoregressive model [Equation (19.18)] deals with the effects of relationships between consecutive values, values two periods apart, and so on—up to values p periods apart. The selection of an appropriate autoregressive model, then, is no easy task. We must weigh the advantages due to parsimony with the concern of failing to take into account important autocorrelation behavior inherent in the data. On the other hand, we must be equally concerned with selecting a high-order model requiring the estimation of numerous, unnecessary parameters—especially if n, the number of observations in the series, is not too large. The reason for this is that p out of n data values will be lost in obtaining an estimate of ψ_p when comparing each data value Y_i with its "fairly near neighbor" Y_{i-p}, which is p periods apart (that is, the comparisons are Y_{1+p} versus Y_1, Y_{2+p} versus $Y_2, \ldots,$ and Y_n versus Y_{n-p}). To illustrate this, suppose we have the following series of $n = 7$ consecutive values:

31 34 37 35 36 43 40

Comparison schema for autoregressive models of order one and order two are established in the accompanying table:

i	First-Order Autoregressive Model (Y_i versus Y_{i-1})	Second-Order Autoregressive Model (Y_i versus Y_{i-1} and Y_i versus Y_{i-2})
1	$31 \leftrightarrow \cdots$	$31 \leftrightarrow \cdots$ and $31 \leftrightarrow \cdots$
2	$34 \leftrightarrow 31$	$34 \leftrightarrow 31$ and $34 \leftrightarrow \cdots$
3	$37 \leftrightarrow 34$	$37 \leftrightarrow 34$ and $37 \leftrightarrow 31$
4	$35 \leftrightarrow 37$	$35 \leftrightarrow 37$ and $35 \leftrightarrow 34$
5	$36 \leftrightarrow 35$	$36 \leftrightarrow 35$ and $36 \leftrightarrow 37$
6	$43 \leftrightarrow 36$	$43 \leftrightarrow 36$ and $43 \leftrightarrow 35$
7	$40 \leftrightarrow 43$	$40 \leftrightarrow 43$ and $40 \leftrightarrow 36$
	(1 comparison is lost for regression analysis)	(2 comparisons are lost for regression analysis)

Once a model is selected and least squares regression methods are used to obtain estimates of the parameters, the next step would be to determine the appropriateness of this model. Either we can select a given pth-order autoregressive model based on previous experiences with similar data, or else we may select, as a starting point, a model with several parameters and then eliminate those that do not contribute significantly. In this latter approach, Newbold (Reference 7) suggests the following test for the significance of the highest-order autoregressive parameter in the fitted model:

$$H_0: \quad \psi_p = 0 \text{ (The highest-order parameter is 0)}$$

against the two-sided alternative

$$H_1: \quad \psi_p \neq 0 \text{ (The parameter } \psi_p \text{ is significantly meaningful)}$$

The test statistic, readily obtainable from the output of various multiple regression programs (which provide estimates of regression coefficients and standard errors), is approximated by

$$Z \cong \frac{\hat{\psi}_p}{S_{\hat{\psi}_p}} \tag{19.19}$$

where $\hat{\psi}_p$ = the estimate of the highest-order parameter ψ_p in the autoregressive model

$S_{\hat{\psi}_p}$ = the standard deviation of $\hat{\psi}_p$

Using an α level of significance, the decision rule is to reject H_0 if $Z > +Z_{\alpha/2}$ (the upper-tail critical value from a standardized normal distribution) or if $Z < -Z_{\alpha/2}$ (the lower-tail critical value from a standardized normal distribution), and not to reject H_0 if $-Z_{\alpha/2} \leq Z \leq +Z_{\alpha/2}$.

If the null hypothesis that $\psi_p = 0$ is not rejected, we may conclude that the selected model contains too many estimated parameters. The highest-order term would then be discarded and an autoregressive model of order $p - 1$ would be obtained through least squares regression. A test of the hypothesis that the new highest-order term is 0 would then be repeated.

This testing and modeling procedure continues until we reject H_0. When this occurs, we know that our highest-order parameter is significant and we are ready to use the particular model for forecasting purposes.

The fitted pth-order autoregressive model has the following form:

$$\hat{Y}_i = \hat{\omega} + \hat{\psi}_1 Y_{i-1} + \hat{\psi}_2 Y_{i-2} + \cdots + \hat{\psi}_p Y_{i-p} \tag{19.20}$$

where
$\hat{Y}_i$ = the fitted value of the series at time i
Y_{i-1} = the observed value of the series at time $i - 1$
Y_{i-2} = the observed value of the series at time $i - 2$
Y_{i-p} = the observed value of the series at time $i - p$
$\hat{\omega}, \hat{\psi}_1, \hat{\psi}_2, \ldots, \hat{\psi}_p$ = regression estimates of the parameters $\omega, \psi_1, \psi_2, \ldots, \psi_p$

To forecast j years into the future from the current nth time period we have

$$\hat{Y}_{n+j} = \hat{\omega} + \hat{\psi}_1 \hat{Y}_{n+j-1} + \hat{\psi}_2 \hat{Y}_{n+j-2} + \cdots + \hat{\psi}_p \hat{Y}_{n+j-p} \tag{19.21}$$

where $\hat{\omega}, \hat{\psi}_1, \hat{\psi}_2, \ldots, \hat{\psi}_p$ are the regression estimates of the parameters $\omega, \psi_1, \psi_2, \ldots, \psi_p$; where j is the number of years into the future; and where, for $k > 0$, $\hat{Y}_{n+k}$ is the forecast of Y_{n+k} from the current time period, while for $k \le 0$, $\hat{Y}_{n+k}$ is the observed value Y_{n+k}.

Thus, to make forecasts j years into the future from, say, a p = third-order autoregressive model, we need only the most recent $p = 3$ observed data values Y_n, Y_{n-1}, and Y_{n-2} and the estimates of the parameters ω, ψ_1, ψ_2, and ψ_3 from a multiple regression program. To forecast one year ahead, Equation (19.21) becomes

$$\hat{Y}_{n+1} = \hat{\omega} + \hat{\psi}_1 Y_n + \hat{\psi}_2 Y_{n-1} + \hat{\psi}_3 Y_{n-2}$$

To forecast two years ahead, Equation (19.21) becomes

$$\hat{Y}_{n+2} = \hat{\omega} + \hat{\psi}_1 \hat{Y}_{n+1} + \hat{\psi}_2 Y_n + \hat{\psi}_3 Y_{n-1}$$

To forecast three years ahead, Equation (19.21) becomes

$$\hat{Y}_{n+3} = \hat{\omega} + \hat{\psi}_1 \hat{Y}_{n+2} + \hat{\psi}_2 \hat{Y}_{n+1} + \hat{\psi}_3 Y_n$$

To forecast four years ahead, Equation (19.21) becomes

$$\hat{Y}_{n+4} = \hat{\omega} + \hat{\psi}_1 \hat{Y}_{n+3} + \hat{\psi}_2 \hat{Y}_{n+2} + \hat{\psi}_3 \hat{Y}_{n+1}$$

and so on.

To demonstrate the autoregressive modeling technique let us return once more to the time series presented in Table 19.5 (on page 873) and plotted in Figure

19.1 (on page 860), which represents the net sales (in billions of dollars) for the Eastman Kodak Company over the 23-year period 1970 through 1992. Table 19.7 displays the setup for first-order, second-order, and third-order autoregressive models. All the columns in this table are needed for fitting third-order autoregressive models. The last column would be omitted when fitting second-order autoregressive models and the last two columns would be eliminated when fitting first-order autoregressive models. Thus we note that either $p = 1$, 2, or 3 observations out of $n = 23$ are lost in the comparisons needed for developing these first-order, second-order, and third-order autoregressive models.

Table 19.7 Developing first-order, second-order, and third-order autoregressive models on net sales for Eastman Kodak Company (1970–1992).

			Predictor Variables		
Year	i	"Dependent" Variable Y_i	Y_{i-1}	Y_{i-2}	Y_{i-3}
1970	1	2.8	*	*	*
1971	2	3.0	2.8	*	*
1972	3	3.5	3.0	2.8	*
1973	4	4.0	3.5	3.0	2.8
1974	5	4.6	4.0	3.5	3.0
1975	6	5.0	4.6	4.0	3.5
1976	7	5.4	5.0	4.6	4.0
1977	8	6.0	5.4	5.0	4.6
1978	9	7.0	6.0	5.4	5.0
1979	10	8.0	7.0	6.0	5.4
1980	11	9.7	8.0	7.0	6.0
1981	12	10.3	9.7	8.0	7.0
1982	13	10.8	10.3	9.7	8.0
1983	14	10.2	10.8	10.3	9.7
1984	15	10.6	10.2	10.8	10.3
1985	16	10.6	10.6	10.2	10.8
1986	17	11.5	10.6	10.6	10.2
1987	18	13.3	11.5	10.6	10.6
1988	19	17.0	13.3	11.5	10.6
1989	20	18.4	17.0	13.3	11.5
1990	21	18.9	18.4	17.0	13.3
1991	22	19.4	18.9	18.4	17.0
1992	23	20.1	19.4	18.9	18.4

Using MINITAB (Reference 8), the following third-order autoregressive model is fitted to the Eastman Kodak net sales data (see Figure 19.13):

$$\hat{Y}_i = 0.446 + 1.534Y_{i-1} - 0.739Y_{i-2} + 0.218Y_{i-3}$$

where the origin is 1973 and "Y" units = 1 year.

Next, one might test for the significance of the highest-order parameter. On the other hand, if one's experiences with similar data permit him or her to hypothesize that a third-order autoregressive model is appropriate for this time series, our fitted model can be used directly for forecasting purposes without the need for testing for parameter significance. Therefore, to demonstrate the forecasting procedure for our third-order autoregressive model, we use the estimates

```
The regression equation is
sales = 0.446 + 1.53 lag1year - 0.739 lag2year + 0.218 lag3year

20 cases used 3 cases contain missing values

Predictor          Coef        Stdev    t-ratio        p
Constant   ω̂    0.4459       0.4301       1.04    0.315
lag1year   ψ̂1   1.5343       0.2450       6.26    0.000
lag2year   ψ̂2  -0.7394       0.4226      -1.75    0.099
lag3year   ψ̂3   0.2179       0.2688       0.81    0.429

s = 0.8251        R-sq = 97.9%      R-sq(adj) = 97.5%
```

Figure 19.13
Partial MINITAB output for third-order autoregressive model.

$$\hat{\omega} = 0.446, \qquad \hat{\psi}_1 = 1.534, \qquad \hat{\psi}_2 = -0.739, \qquad \hat{\psi}_3 = 0.218$$

as well as the three most current data values

$$Y_{21} = 18.9, \qquad Y_{22} = 19.4, \qquad Y_{23} = 20.1$$

Our forecasts of net sales at Eastman Kodak for the years 1993 to 1996 are obtained from Equation (19.21) as follows:

$$\hat{Y}_{n+j} = 0.446 + 1.534\hat{Y}_{n+j-1} - 0.739\hat{Y}_{n+j-2} + 0.218\hat{Y}_{n+j-3}$$

1993: 1 year ahead $\hat{Y}_{24} = 0.446 + (1.534)(20.1) - (0.739)(19.4) + (0.218)(18.9)$
$= 21.0$ billions of dollars

1994: 2 years ahead $\hat{Y}_{25} = 0.446 + (1.534)(21.0) - (0.739)(20.1) + (0.218)(19.4)$
$= 21.9$ billions of dollars

1995: 3 years ahead $\hat{Y}_{26} = 0.446 + (1.534)(21.9) - (0.739)(21.0) + (0.218)(20.1)$
$= 22.9$ billions of dollars

1996: 4 years ahead $\hat{Y}_{27} = 0.446 + (1.534)(22.9) - (0.739)(21.9) + (0.218)(21.0)$
$= 23.8$ billions of dollars

Prior to forecasting, however, most researchers would have preferred to test for the significance of the parameters of a fitted model. Using the output from MINITAB (Figure 19.13), the highest-order parameter estimate $\hat{\psi}_3$ for the fitted third-order autoregressive model is 0.218 (rounded) with a standard deviation $S_{\hat{\psi}_3}$ of 0.269 (rounded).

To test

$$H_0: \quad \psi_3 = 0$$

against

$$H_1: \quad \psi_3 \neq 0$$

we have, from Equation (19.19),

$$Z \cong \frac{\hat{\psi}_3}{S_{\hat{\psi}_3}} = \frac{0.218}{0.269} = 0.81$$

Using a .05 level of significance, the two-tailed test has critical Z values of ±1.96. Since $Z = +0.81 < +1.96$, the upper-tail critical value under the standardized normal distribution (Table E.2), we may not reject H_0, and we would conclude that the third-order parameter of the autoregressive model is not significantly important and can be deleted.

Using MINITAB once again, a second-order autoregressive model is obtained and partial output is displayed in Figure 19.14.

```
The regression equation is
sales = 0.473 + 1.45 lag1year - 0.455 lag2year

21 cases used 2 cases contain missing values

Predictor          Coef       Stdev     t-ratio        p
Constant    ω̂    0.4728      0.3825        1.24    0.232
lag1year    ψ̂1   1.4530      0.2132        6.82    0.000
lag2year    ψ̂2  -0.4548      0.2249       -2.02    0.058

s = 0.7939     R-sq = 98.0%     R-sq(adj) = 97.8%
```

Figure 19.14
Partial MINITAB output for second-order autoregressive model.

The second-order autoregressive model is

$$\hat{Y}_i = 0.473 + 1.453\, Y_{i-1} - 0.455\, Y_{i-2}$$

where the origin is 1972 and "Y" units = 1 year.

From the MINITAB output the highest-order parameter estimate $\hat{\psi}_2$ is -0.455 (rounded) with a standard deviation $S_{\hat{\psi}_2} = 0.225$ (rounded).

To test

$$H_0\colon \quad \psi_2 = 0$$

against

$$H_1\colon \quad \psi_2 \neq 0$$

we have, from Equation (19.19),

$$Z \cong \frac{\hat{\psi}_2}{S_{\hat{\psi}_2}} = \frac{-0.455}{0.225} = -2.02$$

Testing again at the .05 level of significance, since $Z = -2.02 < -1.96$, we may reject H_0, and we would conclude that the second-order parameter of the autoregressive model is significantly important and it should be included in the model.

Our model-building approach has led to the selection of the second-order autoregressive model as the most appropriate for the given data. Using the estimates $\hat{\omega} = 0.473$, $\hat{\psi}_1 = 1.453$, and $\hat{\psi}_2 = -0.455$, as well as the two most recent data values $Y_{22} = 19.4$ and $Y_{23} = 20.1$, our forecasts of net sales at Eastman Kodak for the years 1993 to 1996 are obtained from Equation (19.21) as follows:

$$\hat{Y}_{n+j} = 0.473 + 1.453\hat{Y}_{n+j-1} - 0.455\,\hat{Y}_{n+j-2}$$

1993: 1 year ahead
$$\hat{Y}_{24} = 0.473 + (1.453)(20.1) - 0.455(19.4) = 20.8 \text{ billions of dollars}$$

1994: 2 years ahead
$$\hat{Y}_{25} = 0.473 + (1.453)(20.8) - 0.455(20.1) = 21.4 \text{ billions of dollars}$$

1995: 3 years ahead
$$\hat{Y}_{26} = 0.473 + (1.453)(21.4) - 0.455(20.8) = 22.1 \text{ billions of dollars}$$

1996: 4 years ahead
$$\hat{Y}_{27} = 0.473 + (1.453)(22.1) - 0.455(21.4) = 22.7 \text{ billions of dollars}$$

The data and the forecasts are plotted in Figure 19.15.

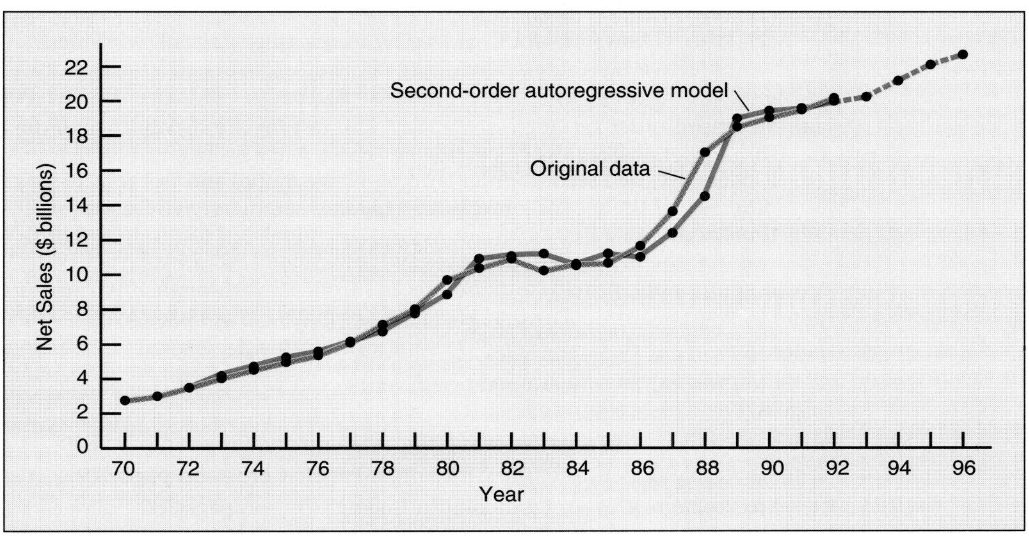

Figure 19.15
Using a second-order autoregressive model on annual net sales at Eastman Kodak Company.

Problems for Section 19.7

● 19.27 Given an annual time series with 40 consecutive observations, if you were to fit a fifth-order autoregressive model:

(a) How many comparisons would be lost in the development of the autoregressive model?
(b) How many parameters would you need to estimate?
(c) Which of the original 40 values would you need for forecasting?
(d) Express the model.
(e) Write a general equation to indicate how you would forecast j years into the future.

19.28 An annual time series with 17 consecutive values is obtained. A third-order autoregressive model is fitted to the data and has the following estimated parameters and standard deviations:

$$\hat{\omega} = 4.50, \quad \hat{\psi}_1 = 1.80, \quad \hat{\psi}_2 = .70, \quad \hat{\psi}_3 = .20$$

$$S_{\hat{\psi}_1} = .50, \quad S_{\hat{\psi}_2} = .30, \quad S_{\hat{\psi}_3} = .10$$

At the .05 level of significance, test the appropriateness of the fitted model.

19.29 Refer to Problem 19.28. The three most recent observations are

$$Y_{15} = 23, \quad Y_{16} = 28, \quad Y_{17} = 34$$

(a) Forecast the series for the next year and the following year.
(b) Suppose, when testing for the appropriateness of the fitted model in Problem 19.28, the standard deviations were

$$S_{\hat{\psi}_1} = .45, \quad S_{\hat{\psi}_2} = .35, \quad S_{\hat{\psi}_2} = .15$$

(1) What would you conclude?
(2) Discuss how you would proceed if forecasting were still your main objective.

For Problems 19.30–19.37:

(a) Fit a third-order autoregressive model and test for the significance of the third-order autoregressive parameter. (Use $\alpha = .05$.)
(b) If necessary, fit a second-order autoregressive model and test for the significance of the second-order autoregressive parameter. (Use $\alpha = .05$.)
(c) If necessary, fit a first-order autoregressive model and test for the significance of the first-order autoregressive parameter. (Use $\alpha = .05$.)
(d) If appropriate, provide annual forecasts from 1993 through 1996.

19.30 Refer to Problem 19.2—earnings per share at TRW Inc.—on page 870.

● 19.31 Refer to Problem 19.5—net sales at Upjohn Co.—on page 878.

19.32 Refer to Problem 19.6—net operating revenues of Coca-Cola Co.—on page 878.

19.33 Refer to Problem 19.7—net sales at Gillette Company, Inc.—on page 879.

19.34 Refer to Problem 19.8—net sales at Georgia-Pacific Corp.—on page 879.

19.35 Refer to Problem 19.9—total income at Boeing Co.—on page 880.

19.36 Refer to Problem 19.12—operating revenues at Philip Morris Companies, Inc.—on page 882.

● 19.37 Refer to Problem 19.13—net sales at Black & Decker Corp.—on page 882.

19.8 Choosing an Appropriate Forecasting Model

In Sections 19.5–19.7 seven alternative time-series forecasting methods were developed. In Section 19.5 we studied three commonly used models that are based on the method of least squares—the linear model, the quadratic model, and the exponential model. In Section 19.6 we described the Holt-Winters method and in Section 19.7 we covered three autoregressive methods—the first-order, second-order, and third-order models.

A major question must be answered at this time: Is there a *best* model? That is, among such models as these, which *one* should we select if we are interested in time-series forecasting? Three approaches are offered as guidelines for model selection:

1. Perform a residual analysis.
2. Measure the magnitude of the residual error.
3. Use the principle of parsimony.

The most widely used methods of determining the adequacy of a particular forecasting model are based on a judgment of how well it has fit a given set of time-series data. These methods, of course, assume that future movements in the series can be projected by a study of past behavior patterns. One such method is to perform a residual analysis; a second is to measure the magnitude of the residual error; a third approach is to select the simplest, least cumbersome model that fits the data well (that is, the *principle of parsimony*).

19.8.1 Residual Analysis

We may recall from our study of regression analysis in Sections 17.9 and 18.5 that differences between the observed and fitted data are known as *residuals*. Thus, for the *i*th year in an annual time series of n years, the residual is defined as

$$e_i = (Y_i - \hat{Y}_i) \qquad \textbf{(19.22)}$$

where Y_i is the observed value in year i
and $\hat{Y}_i$ is the fitted value in year i

Once a particular model has been fitted to a given time series we may plot the residuals over the n time periods. As depicted in Figure 19.16(a) on page 898, if the particular model fits adequately, the residuals represent the irregular component of the time series and, therefore, they should be randomly distributed throughout the series. On the other hand, as illustrated in the three remaining panels of Figure 19.16, if the particular model does not fit adequately, the residuals may demonstrate some systematic pattern such as a failure to account for trend (b), a failure to account for cyclical variation (c), or, with monthly data, a failure to account for seasonal variation (d).

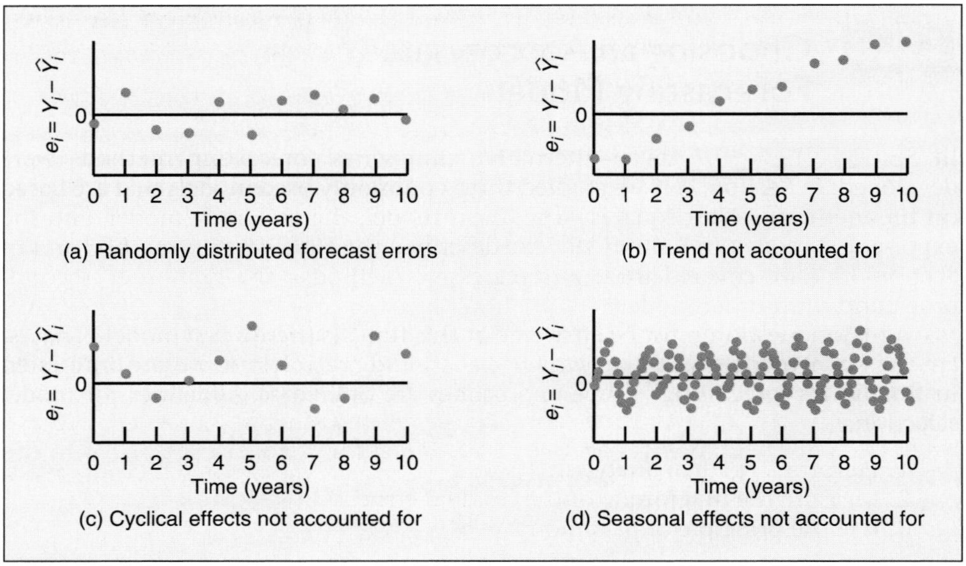

Figure 19.16
A residual analysis to study error patterns.

19.8.2 Measuring the Magnitude of the Residual Error

If, after performing a residual analysis, we still believe that two or more models appear to fit the data adequately, then a second method used for model selection is based on some measure of the magnitude of the residual error. Numerous measures have been proposed (see References 1, 2, 7, and 11) and unfortunately, there is no consensus among researchers as to which particular measure is best for determining the most appropriate forecasting model.

Based on the principle of least squares, one measure that we have already used in regression analysis (see Sections 17.5 and 17.6) is the *unexplained variation:*

$$\text{SSE} = \text{unexplained variation} = \sum_{i=1}^{n} (Y_i - \hat{Y}_i)^2 \qquad \textbf{(19.23)}$$

For a particular model, this measure is based on the sum of squared differences between the actual and fitted values in a given time series. If a model were to fit the past time-series data *perfectly,* then unexplained variation would be zero. On the other hand, if a model were to fit the past time-series data *poorly,* the unexplained variation would be large. Thus when comparing the adequacy of two or more forecasting models, the one with the *minimum* unexplained variation can be selected as most appropriate based on past fits of the given time series.

Nevertheless, a major drawback to using the unexplained variation measure when comparing forecasting models is that it penalizes a model too much for large, individual forecasting errors. That is, whenever there is a large discrepancy between Y_i and $\hat{Y}_i$ the computation for unexplained variation becomes magnified through the squaring process. For this reason, a measure that most researchers seem to prefer for assessing the appropriateness of various forecasting models is the **mean absolute deviation (MAD):**

$$\text{MAD} = \frac{\sum_{i=1}^{n} |Y_i - \hat{Y}_i|}{n} \qquad \textbf{(19.24)}$$

For a particular model, the MAD is a measure of the average of the absolute discrepancies between the actual and fitted values in a given time series. If a model were to fit the past time-series data *perfectly,* the MAD would be zero, whereas if a model were to fit the past time-series data *poorly,* the MAD would be large. Hence when comparing the merits of two or more forecasting models, the one with the *minimum* MAD can be selected as most appropriate on the basis of past fits of the given time series.

19.8.3 Principle of Parsimony

If, after performing a residual analysis and comparing the obtained MAD measures we still believe that two or more models appear to adequately fit the data, then a third method for model selection is based on the **principle of parsimony.** That is, we should select the *simplest* model that gets the job done adequately.

Among the seven forecasting models studied in this chapter, the least squares linear and quadratic models and the first-order autoregressive model would be regarded by most researchers as the simplest. Their ranking would likely be in the order given. The least squares exponential model and the Holt-Winters method would qualify as the most complex of the techniques presented.

19.8.4 A Comparison of Five Forecasting Methods

To illustrate the model selection process, we again consider the annual time-series data on net sales at Eastman Kodak Company over the 23-year period 1970–1992. Five of the forecasting methods described in Sections 19.5–19.7 are to be compared: the linear model, the quadratic model, the exponential model, the Holt-Winters model, and the second-order autoregressive model. (There is no need to further study the third-order autoregressive model for this time series since we may recall from Section 19.7 that this model did not significantly improve the fit over the simpler second-order autoregressive model.)

Panels (a) through (e) of Figure 19.17 on pages 900 and 901 display the residual plots for the linear, quadratic, exponential, Holt-Winters, and second-order autoregressive models on the Eastman Kodak net sales data. When drawing conclusions from such residual plots caution must be used since only 23 data points have been observed.

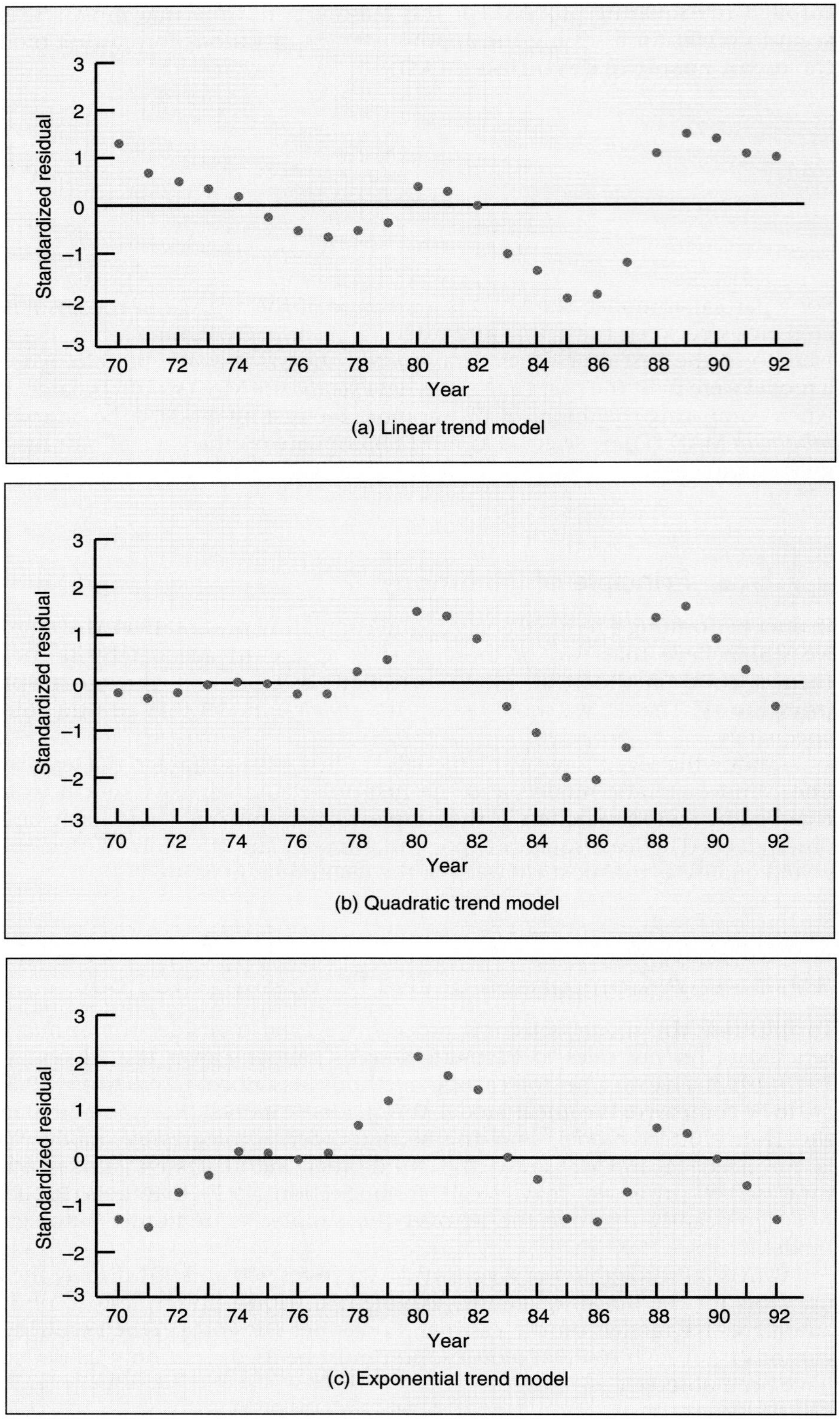

Figure 19.17
Residual plots for five forecasting methods.
Source: Data are taken from Table 19.8 on page 902.

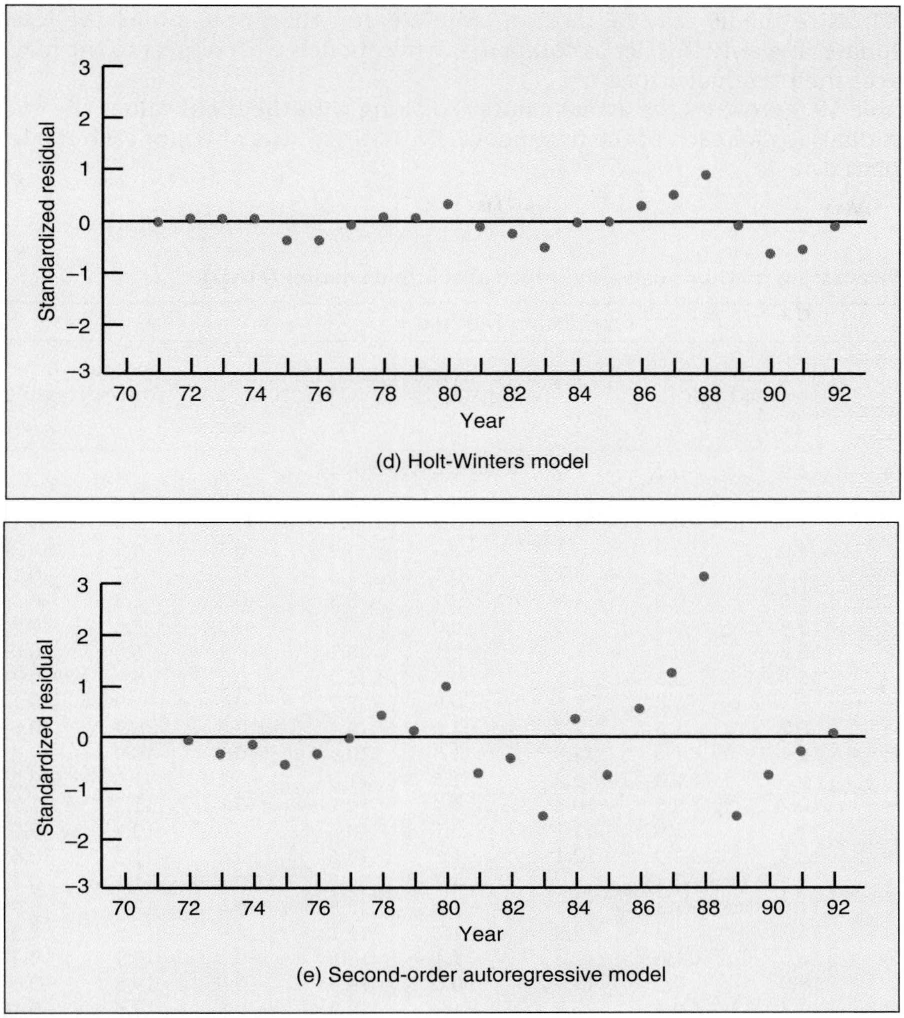

Figure 19.17 ***(Continued)***

From panels (a), (b), and (c) we note that the cyclical effects were unaccounted for in each of the least squares models. However, the residual plots for the quadratic and exponential models seem to suggest that these models provide a better fit to the series than does the linear model because panels (b) and (c) display more randomness (that is, less systematic pattern) in the residuals over the first eight years of the series. On the other hand, the increasing (wider) amplitude observed in the later years of all five residual plots may be suggesting that none of the models examined here performed outstandingly with respect to capturing the large net sales movements that have occurred in these recent years. Nevertheless, from panels (d) and (e) we observe that the Holt-Winters method seems to provide the closest fit but the second-order autoregressive model exhibits the least amount of systematic structure.

To summarize, on the basis of the residual analyses of all five forecasting models, it would appear that the Holt-Winters model and the second-order autoregressive model may be most appropriate and the linear model the least appropriate. To verify this, let us compare the five models with respect to the magnitude of their residual errors.

Table 19.8 provides the actual values (Y_i) along with the fitted values ($\hat{Y}_i$) and the residuals (e_i) for each of the five models. In addition, the MAD for each model is displayed.

Table 19.8 Comparison of five forecasting methods using the mean absolute deviation (MAD).

		Forecasting Method									
	Net Sales	Linear		Quadratic		Exponential		Holt-Winters		Second-Order Autoregressive	
Year	Y_i	$\hat{Y}_i$	e_i	$\hat{Y}_i$	e_1	$\hat{Y}_i$	e_i	$\hat{Y}_i$	e_i	$\hat{Y}_i$	e_i
1970	2.8	1.2	1.6	2.9	−0.1	3.2	−0.4	—	—	—	—
1971	3.0	2.0	1.0	3.3	−0.3	3.5	−0.5	3.0	0.0	—	—
1972	3.5	2.8	0.7	3.6	−0.1	3.8	−0.3	3.4	0.1	3.6	−0.1
1973	4.0	3.6	0.4	4.0	0.0	4.1	−0.1	3.9	0.1	4.2	−0.2
1974	4.6	4.4	0.2	4.5	0.1	4.5	0.1	4.5	0.1	4.7	−0.1
1975	5.0	5.2	−0.2	5.0	0.0	4.9	0.1	5.3	−0.3	5.3	−0.3
1976	5.4	6.0	−0.6	5.6	−0.2	5.4	0.0	5.7	−0.3	5.6	−0.2
1977	6.0	6.8	−0.8	6.2	−0.2	5.9	0.1	6.1	−0.1	6.0	0.0
1978	7.0	7.6	−0.6	6.8	0.2	6.5	0.5	6.9	0.1	6.7	0.3
1979	8.0	8.4	−0.4	7.5	0.5	7.1	0.9	7.9	0.1	7.9	0.1
1980	9.7	9.2	0.5	8.2	1.5	7.7	2.0	9.4	0.3	8.9	0.8
1981	10.3	10.0	0.3	9.0	1.3	8.5	1.8	10.4	−0.1	10.9	−0.6
1982	10.8	10.8	0.0	9.8	1.0	9.3	1.5	11.0	−0.2	11.0	−0.2
1983	10.2	11.6	−1.4	10.7	−0.5	10.1	0.1	10.7	−0.5	11.5	−1.3
1984	10.6	12.4	−1.8	11.6	−1.0	11.1	0.5	10.6	0.0	10.4	0.2
1985	10.6	13.2	−2.6	12.6	−2.0	12.1	−1.5	10.6	0.0	11.2	−0.6
1986	11.5	14.0	−2.5	13.6	−2.1	13.2	1.7	11.2	0.3	11.1	0.4
1987	13.3	14.8	−1.5	14.6	−1.3	14.5	−1.2	12.8	0.5	12.4	0.9
1988	17.0	15.6	1.4	15.7	1.3	15.8	1.2	16.1	0.9	14.6	2.4
1989	18.4	16.4	2.0	16.8	1.6	17.3	1.1	18.5	−0.1	19.1	−0.7
1990	18.9	17.2	1.7	18.0	0.9	18.9	0.0	19.5	−0.6	19.5	−0.6
1991	19.4	18.0	1.4	19.3	0.1	20.7	−1.3	19.9	−0.5	19.6	−0.2
1992	20.1	18.8	1.3	20.5	−0.4	22.7	−2.6	20.2	−0.1	20.1	0.0
Absolute Sum			24.9		16.7		19.5		5.3		10.2
MAD		$\frac{24.9}{23} = 1.08$		$\frac{16.7}{23} = 0.73$		$\frac{19.5}{23} = 0.85$		$\frac{5.3}{22} = 0.24$		$\frac{10.2}{21} = 0.49$	

A comparison of the MAD for each of the models clearly indicates that the simplest model, the linear model, is, for this time series, the one with the poorest fit. Moreover, the other two least squares models (quadratic and exponential) do not show sufficient improvement over the linear model. As anticipated from the residual analysis (Figure 19.17 on pages 900–901), the models with the smallest MAD are the Holt-Winters model and the second-order autoregressive model. Although the Holt-Winters model may be slightly superior, on the basis of the principle of parsimony, the second-order autoregressive model is the one selected for purposes of forecasting net sales of the Eastman Kodak Company.

19.8.5 Model Selection: A Warning

Once a particular forecasting model is selected, it becomes imperative that we appropriately monitor the chosen model. After all, the objective in selecting the model is to be able to project or forecast future movements in a set of time-series data. Unfortunately, such forecasting models are generally poor at detecting changes in the underlying structure of the time series. It is important then that such projections be examined together with those obtained through other types of forecasting methods (such as the use of leading indicators). As soon as a *new* data value (Y_t) is observed in time period t it must be compared with its projection ($\hat{Y}_t$). If the difference is too large, the forecasting model should be revised. Such *adaptive-control procedures* are described in Reference 2.

Problems for Section 19.8

For Problems 19.38–19.45:

(a) Perform a residual analysis for each fitted model.

(b) Compute the MAD for each fitted model.

(c) On the basis of (a), (b), and parsimony, which model would you select for purposes of forecasting? Discuss.

19.38 Refer to Problems 19.2 on page 870, 19.19 (page 888), and 19.30 (page 896)—earnings per share at TRW Inc.

● 19.39 Refer to Problems 19.5 (page 878), 19.20 (page 888), and 19.31 (page 896)—net sales at Upjohn Co.

19.40 Refer to Problems 19.6 (page 878), 19.21 (page 888), and 19.32 (page 896)—net operating revenues at Coca-Cola Co.

19.41 Refer to Problems 19.7 (page 879), 19.22 (page 888), and 19.33 (page 896)—net sales at Gillette Company, Inc.

19.42 Refer to Problems 19.8 (page 879), 19.23 (page 888), and 19.34 (page 896)—net sales at Georgia-Pacific Corp.

19.43 Refer to Problems 19.9 (page 880), 19.24 (page 888), and 19.35 (page 896)—total income at Boeing Co.

19.44 Refer to Problems 19.12 (page 882), 19.25 (page 888), and 19.36 (page 896)—operating revenues at Philip Morris Companies, Inc.

● 19.45 Refer to Problems 19.13 (page 882), 19.26 (page 888), and 19.37 (page 896)—net sales at Black & Decker Corp.

19.9 Time-Series Forecasting of Monthly Data

Table 19.9 on page 904 presents the monthly private residential construction expenditures (in millions of dollars) in a small city in the United States from January 1988 through December 1993. This time series is displayed in Figure 19.18 (page 904). For such monthly time series as these the classical multiplicative time-series model includes the **seasonal component** in addition to the trend, cyclical, and irregular components. The model is expressed by Equation (19.2) on page 861 as

$$Y_i = T_i \cdot S_i \cdot C_i \cdot I_i$$

Table 19.9 **Monthly private residential construction expenditures (in millions of dollars) in a small city in the United States (January 1988–December 1993).**

	Year					
Month	1988	1989	1990	1991	1992	1993
January	10.2	11.2	12.5	12.6	13.2	13.0
February	9.7	11.0	12.0	12.0	12.5	12.7
March	11.3	12.7	13.9	14.2	14.4	14.8
April	12.4	14.3	15.4	15.6	15.8	15.9
May	13.6	16.2	17.0	17.1	17.1	17.1
June	14.5	17.7	18.2	18.3	18.1	17.7
July	14.8	18.4	18.6	18.9	18.7	17.9
August	15.3	18.6	18.8	19.3	18.9	18.0
September	15.0	18.1	18.4	18.7	18.1	16.8
October	15.0	18.0	18.2	18.7	17.8	16.3
November	14.2	16.7	17.1	17.7	16.7	14.7
December	12.4	14.2	14.5	15.0	14.0	12.2

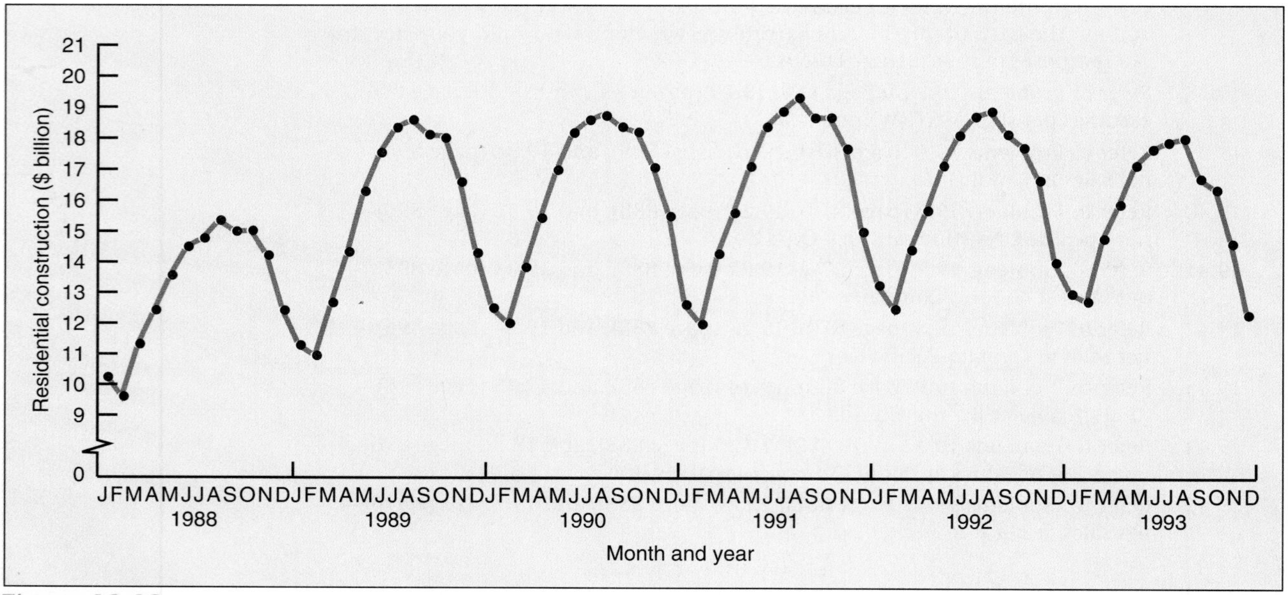

Figure 19.18
Private residential construction (in millions of dollars) in a small city in the United States from January 1988 to December 1993.
Source: Data are taken from Table 19.9.

19.9.1 Least Squares Trend Fitting and Forecasting

To fit a least squares trend line to the six-year monthly series we code the consecutive X values 0 through 71 and employ a statistical software package such as MINITAB. As observed from the MINITAB output displayed in Figure 19.19, the linear model is given by

$$\hat{Y}_i = 14.033 + 0.043X_i$$

where the origin[3] is January 15, 1988 and X units = 1 month.

```
The regression equation is
rescon = 14.0 + 0.0431 months

Predictor          Coef          Stdev      t-ratio          p
Constant     b0  14.0330        0.5599        25.07      0.000
months       b1  0.04312       0.01361         3.17      0.002

s = 2.400       R-sq = 12.5%    R-sq(adj) = 11.3%
```

Figure 19.19
Partial MINITAB output for fitting linear regression model to monthly time-series data on private residential construction expenditures.

The intercept $b_0 =$ 14.033 is the fitted trend value reflecting the private residential construction expenditures (in millions of dollars) during the origin or base month, January 1988. The slope $b_1 = 0.043$ indicates that private residential construction expenditures are increasing at a rate of 0.043 millions of dollars (that is, 43 thousands of dollars) per month over this six-year period. This is depicted in Figure 19.20, where the slope of the fitted monthly trend line exhibits a slight tendency to increase over time. This equation may be used to project future monthly trend values in private residential construction expenditures. However such monthly time series are influenced by seasonal factors, and we must develop a **seasonal index** that accounts for the month-to-month fluctuations.

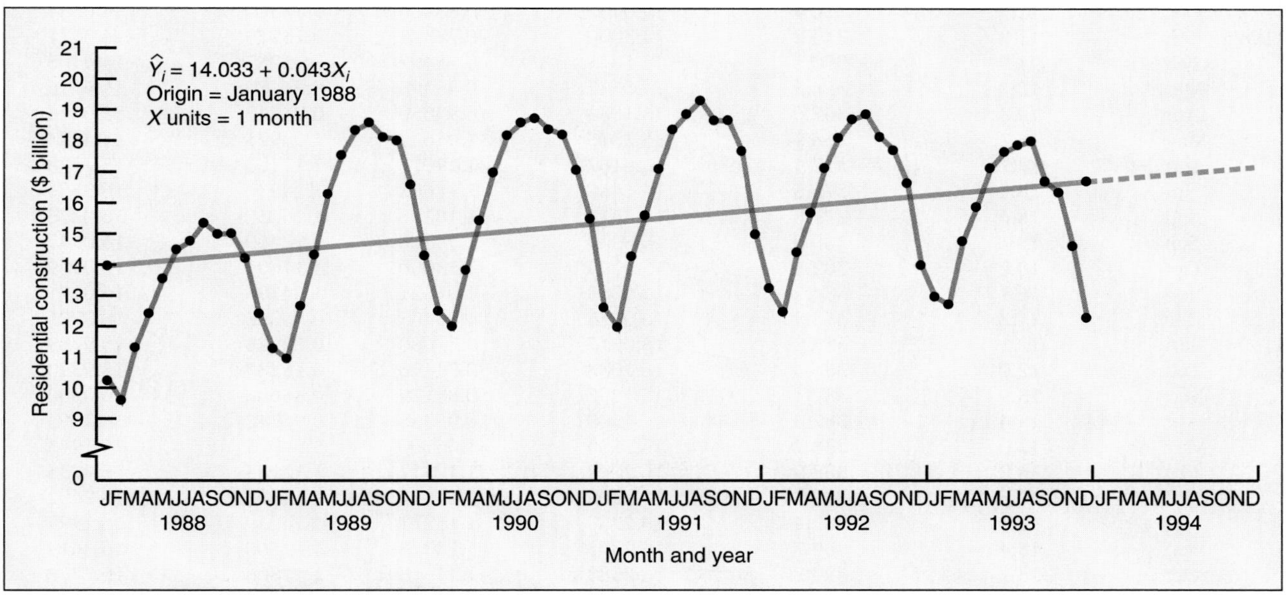

Figure 19.20 **Fitting the least squares trend line.**
Source: Data are taken from Table 19.8 and fitted trend line from Figure 19.19.

19.9.2 Computing the Seasonal Index

It is important to isolate and study the seasonal movements in a monthly time series for two reasons. First, by knowing the value of the seasonal component for any particular month, we can easily adjust and improve upon trend projections for forecasting purposes. Second, by knowing the value of the seasonal component we can decompose the time series by eliminating its influences—along with those pertaining to trend and irregular fluctuations—and thereby concentrate on the cyclical movements of the series. If, as is often assumed, the seasonal movements are fairly constant over time, the construction of a seasonal index may be illustrated from Tables 19.10 and 19.11 on pages 906–908.

Table 19.10 Developing the seasonal index.

(1) Year and Month		(2) Private Residential Construction Expenditures ($ millions)	(3) 13-Month Weighted Moving Totals	(4) Weighted (13-Month) Moving Averages	(5) Ratios to Moving Average	(6) Seasonal Index	(7) Deseasonalized Data
1988	Jan	10.2	*	*	*	0.78228	13.0388
	Feb	9.7	*	*	*	0.75287	12.8841
	Mar	11.3	*	*	*	0.86680	13.0364
	Apr	12.4	*	*	*	0.95242	13.0194
	May	13.6	*	*	*	1.04975	12.9555
	Jun	14.5	*	*	*	1.12183	12.9253
	Jul	14.8	317.8	13.2417	1.11768	1.14530	12.9223
	Aug	15.3	320.1	13.3375	1.14714	1.16050	13.1840
	Sep	15.0	322.8	13.4500	1.11524	1.12460	13.3381
	Oct	15.0	326.1	13.5875	1.10396	1.11809	13.4157
	Nov	14.2	330.6	13.7750	1.03085	1.04196	13.6282
	Dec	12.4	336.4	14.0167	0.88466	0.88360	14.0334
1989	Jan	11.2	343.2	14.3000	0.78322	0.78228	14.3171
	Feb	11.0	350.1	14.5875	0.75407	0.75287	14.6108
	Mar	12.7	356.5	14.8542	0.85498	0.86680	14.6516
	Apr	14.3	362.6	15.1083	0.94650	0.95242	15.0144
	May	16.2	368.1	15.3375	1.05623	1.04975	15.4323
	Jun	17.7	372.4	15.5167	1.14071	1.12183	15.7778
	Jul	18.4	375.5	15.6458	1.17603	1.14530	16.0656
	Aug	18.6	377.8	15.7417	1.18158	1.16050	16.0276
	Sep	18.1	380.0	15.8333	1.14316	1.12460	16.0947
	Oct	18.0	382.3	15.9292	1.13000	1.11809	16.0989
	Nov	16.7	384.2	16.0083	1.04321	1.04196	16.0275
	Dec	14.2	385.5	16.0625	0.88405	0.88360	16.0705
1990	Jan	12.5	386.2	16.0917	0.77680	0.78228	15.9789
	Feb	12.0	386.6	16.1083	0.74496	0.75287	15.9390
	Mar	13.9	387.1	16.1292	0.86179	0.86680	16.0360
	Apr	15.4	387.6	16.1500	0.95356	0.95242	16.1693
	May	17.0	388.2	16.1750	1.05100	1.04975	16.1943
	Jun	18.2	388.9	16.2042	1.12317	1.12183	16.2235
	Jul	18.6	389.3	16.2208	1.14667	1.14530	16.2402
	Aug	18.8	389.4	16.2250	1.15871	1.16050	16.1999
	Sep	18.4	389.7	16.2375	1.13318	1.12460	16.3614
	Oct	18.2	390.2	16.2583	1.11943	1.11809	16.2778
	Nov	17.1	390.5	16.2708	1.05096	1.04196	16.4114
	Dec	14.5	390.7	16.2792	0.89071	0.88360	16.4101
1991	Jan	12.6	391.1	16.2958	0.77320	0.78228	16.1067
	Feb	12.0	391.9	16.3292	0.73488	0.75287	15.9390
	Mar	14.2	392.7	16.3625	0.86784	0.86680	16.3821
	Apr	15.6	393.5	16.3958	0.95146	0.95242	16.3793
	May	17.1	394.6	16.4417	1.04004	1.04975	16.2896
	Jun	18.3	395.7	16.4875	1.10993	1.12183	16.3127

Table 19.10 (continued)

(1) Year and Month		(2) Private Residential Construction Expenditures ($ millions)	(3) 13-Month Weighted Moving Totals	(4) Weighted (13-Month) Moving Averages	(5) Ratios to Moving Average	(6) Seasonal Index	(7) Deseasonalized Data
	Jul	18.9	396.8	16.5333	1.14315	1.14530	16.5022
	Aug	19.3	397.9	16.5792	1.16411	1.16050	16.6308
	Sep	18.7	398.6	16.6083	1.12594	1.12460	16.6282
	Oct	18.7	399.0	16.6250	1.12481	1.11809	16.7250
	Nov	17.7	399.2	16.6333	1.06413	1.04196	16.9872
	Dec	15.0	399.0	16.6250	0.90226	0.88360	16.9759
1992	Jan	13.2	398.6	16.6083	0.79478	0.78228	16.8737
	Feb	12.5	398.0	16.5833	0.75377	0.75287	16.6032
	Mar	14.4	397.0	16.5417	0.87053	0.86680	16.6128
	Apr	15.8	395.5	16.4792	0.95879	0.95242	16.5893
	May	17.1	393.6	16.4000	1.04268	1.04975	16.2896
	Jun	18.1	391.6	16.3167	1.10930	1.12183	16.1344
	Jul	18.7	390.4	16.2667	1.14959	1.14530	16.3275
	Aug	18.9	390.4	16.2667	1.16189	1.16050	16.2861
	Sep	18.1	391.0	16.2917	1.11100	1.12460	16.0947
	Oct	17.8	391.5	16.3125	1.09119	1.11809	15.9200
	Nov	16.7	391.6	16.3167	1.02349	1.04196	16.0275
	Dec	14.0	391.2	16.3000	0.85890	0.88360	15.8442
1993	Jan	13.0	390.0	16.2500	0.80000	0.78228	16.6181
	Feb	12.7	388.3	16.1792	0.78496	0.75287	16.8688
	Mar	14.8	386.1	16.0875	0.91997	0.86680	17.0743
	Apr	15.9	383.3	15.9708	0.99556	0.95242	16.6943
	May	17.1	379.8	15.8250	1.08057	1.04975	16.2896
	Jun	17.7	376.0	15.6667	1.12979	1.12183	15.7778
	Jul	17.9	*	*	*	1.14530	15.6290
	Aug	18.0	*	*	*	1.16050	15.5106
	Sep	16.8	*	*	*	1.12460	14.9387
	Oct	16.3	*	*	*	1.11809	14.5784
	Nov	14.7	*	*	*	1.04196	14.1080
	Dec	12.2	*	*	*	0.88360	13.8071

Note: A MINITAB macro was written (Reference 8) to produce columns (3) through (7).
Source: Data are taken from Table 19.9.

To start, a series of 13-month weighted moving totals is obtained. To compute a 13-month weighted moving total the first and last months receive a weight of 1 and the middle 11 months receive a weight of 2. Thus, for example, the first 13-month weighted moving total is obtained by adding the January 1988 and January 1989 private residential construction expenditure values to twice the expenditure values given to the middle 11 months (February 1988 through December 1988). That is

$$(1)(\text{Jan } 88) + (2)(\text{Feb } 88) + (2)(\text{Mar } 88) + \cdots + (2)(\text{Dec } 88) + (1)(\text{Jan } 89)$$

The resulting moving total, 317.8, is recorded in the middle month—July 1988. The second 13-month weighted moving total is obtained by adding the February 1988 and February 1989 private residential construction expenditure values to twice the expenditure values given to the middle 11 months (March 1988 through January 1989). That is,

$$(1)(\text{Feb } 88) + (2)(\text{Mar } 88) + (2)(\text{Apr } 88) + \cdots + (2)(\text{Jan } 89) + (1)(\text{Feb } 89)$$

The resulting moving total, 320.1, is recorded in the middle month—August 1988.

This process continues by always adding the extremes, the values representing the first and last months of the moving total, to twice the 11 middle month values. As observed in column (3) of Table 19.10, when recording these values the results are displayed in the middle month comprising each respective moving total.

By dividing these moving totals in column (3) by 24, the *weighted moving averages* are obtained as shown in column (4). These **weighted moving averages** are said to consist of the *trend and cyclical components* of the series. The original data [column (2)] are then divided by the respective weighted moving averages [column (4)], yielding the *ratios to moving averages* depicted in column (5). Essentially, these **ratios to moving averages** represent the *seasonal and irregular fluctuations* in the series, since the division of the observed data [column (2)] by the weighted moving averages [column (4)] effectively eliminates trend and cyclical influences, as demonstrated in Equation (19.25):

$$\frac{Y_i}{\text{weighted moving average}_i} = \frac{T_i \cdot S_i \cdot C_i \cdot I_i}{T_i \cdot C_i} = S_i \cdot I_i \qquad (19.25)$$

To form the seasonal index, the ratios to moving averages data from Table 19.10 are rearranged according to monthly values, as depicted in Table 19.11.

Table 19.11 Computing the seasonal index from the median of monthly ratios to moving averages.

	Year							
Month	1988	1989	1990	1991	1992	1993	Median	Seasonal Index
January	*	0.78322	0.77680	0.77320	0.79478	0.80000	0.78322	0.78228
February	*	0.75407	0.74496	0.73488	0.75377	0.78496	0.75377	0.75287
March	*	0.85498	0.86179	0.86784	0.87053	0.91997	0.86784	0.86680
April	*	0.94650	0.95356	0.95146	0.95879	0.99556	0.95356	0.95242
May	*	1.05623	1.05100	1.04004	1.04268	1.08057	1.05100	1.04975
June	*	1.14071	1.12317	1.10993	1.10930	1.12979	1.12317	1.12183
July	1.11768	1.17603	1.14667	1.14315	1.14959	*	1.14667	1.14530
August	1.14714	1.18158	1.15871	1.16411	1.16189	*	1.16189	1.16050
September	1.11524	1.14316	1.13318	1.12594	1.11100	*	1.12594	1.12460
October	1.10396	1.13000	1.11943	1.12481	1.09119	*	1.11943	1.11809
November	1.03085	1.04321	1.05096	1.06413	1.02349	*	1.04321	1.04196
December	0.88466	0.88405	0.89071	0.90226	0.85890	*	0.88466	0.88360
							12.01436	12.00000

$$\text{seasonal index} = \frac{(12.0)(\text{median})}{12.01436}$$

Source: Data are taken from Table 19.10.

From Table 19.11 it is seen that for each month the irregular variations can be eliminated if the median of the various obtained ratios to moving averages is used as an indicator of seasonal activity over time. As shown in Table 19.11, these median values are then adjusted so that the total value of the seasonal indexes over the year is 12.0 and the average value of each (monthly) seasonal index is 1.0. Thus we note that a seasonal index of 0.782 for the month of January indicates that the value of private residential construction expenditures in January is only 78.2% of the monthly average. A seasonal index of 1.145 for the month of July indicates that the value of expenditures in July is 14.5% higher than average.

19.9.3 Using the Seasonal Index in Forecasting

To use the seasonal index to adjust a trend projection for forecasting purposes, we merely multiply the projected trend value for a particular month by the corresponding seasonal index for that month. For example, using our model, the projected monthly trend values in private residential construction expenditures over the year 1996 are listed in column (1) of Table 19.12. The respective monthly seasonal indexes are displayed in column (2). Adjusting for seasonal fluctuations, the product of the various projected monthly trend values with their respective seasonal indexes yields the set of monthly forecasts shown in column (3).

Table 19.12 Adjusting least squares trend projections by seasonal indexes for forecasting purposes.

Month	(1) Monthly Trend Projection for Year 1996	(2) Seasonal Index	(3) Forecast
January	18.173	0.78228	14.216
February	18.216	0.75287	13.714
March	18.259	0.86680	15.827
April	18.302	0.95242	17.431
May	18.345	1.04975	19.258
June	18.389	1.12183	20.629
July	18.432	1.14530	21.110
August	18.475	1.16050	21.440
September	18.518	1.12460	20.825
October	18.561	1.11809	20.753
November	18.604	1.04196	19.385
December	18.647	0.88360	16.477

Source: Data are taken from Table 19.11 and the annual trend model.

Problems for Section 19.9

● 19.46 The data given in the table at the top of page 910 represent the monthly outlays (in thousands of dollars) by a municipality to its sanitation department from January 1985 through December 1994.

Data file OUTLAYS.DAT

Monthly outlays.

Month	Year									
	1985	1986	1987	1988	1989	1990	1991	1992	1993	1994
January	262	259	271	251	298	260	275	315	354	417
February	295	276	241	231	283	291	321	342	365	408
March	333	310	301	252	315	307	352	370	389	416
April	252	238	265	293	287	293	322	316	198	398
May	274	270	255	278	301	279	309	361	366	397
June	245	292	301	447	185	287	314	320	389	452
July	377	289	278	216	368	344	299	324	341	423
August	291	289	262	247	310	359	355	320	413	456
September	273	273	246	267	313	250	324	344	387	356
October	266	271	249	281	312	368	310	300	384	479
November	286	272	246	297	325	359	339	350	415	425
December	285	284	221	288	326	345	320	333	328	499

(a) Plot the data on a chart.
(b) Compute the seasonal index.
(c) Fit a least squares linear-trend line to the monthly time series.
(d) Use the monthly trend equation and the seasonal index to forecast the monthly outlays for all twelve months of 1995 and 1996.

For Problems 19.47–19.49:
(a) Plot the data on a chart.
(b) Compute the seasonal index.
(c) Fit a least squares trend line to the monthly time series.
(d) Use the monthly trend equation and the seasonal index to make a forecast for all twelve months of 1994.
(e) ACTION Go to your library and, using appropriate sources, record the actual data in 1994. Compare the actual values with the forecasted values. Discuss.
(f) Make a forecast for December 1996.

19.47 The data in the accompanying table represent the total monthly retail sales in the United States from January 1989 to December 1993.

Data file RETSALES.DAT

Monthly retail sales.

Month	Year				
	1989	1990	1991	1992	1993
January	124.2	133.3	134.5	141.9	148.4
February	120.5	128.0	131.6	142.8	145.0
March	141.9	149.2	152.7	154.5	164.6
April	140.4	145.8	151.7	158.8	170.3
May	151.0	155.0	163.5	165.7	176.1
June	149.8	154.4	157.5	164.2	175.7
July	145.3	149.7	158.3	165.4	177.7
August	153.8	158.2	163.4	165.9	177.1
September	144.8	146.3	149.8	160.2	171.1
October	143.1	151.5	155.5	168.7	176.4
November	149.6	156.1	159.1	167.0	180.9
December	177.4	179.7	185.2	204.0	218.3

Source: U.S. Department of Commerce, 1994.

19.48 The data in the accompanying table represent the monthly total sales of mutual funds (in millions of dollars) from January 1990 through December 1993.

Data file MUTFUNDS.DAT

Monthly total market value of mutual funds.

	Year			
Month	1990	1991	1992	1993
January	13,719	13,409	32,589	36,877
February	11,818	13,915	26,230	35,419
March	13,843	15,810	30,411	42,212
April	14,052	20,429	29,906	40,936
May	12,517	18,255	26,984	36,901
June	13,027	16,964	28,287	40,692
July	12,502	19,398	32,016	43,629
August	13,221	19,993	29,777	45,575
September	9,974	20,892	29,360	43,199
October	10,564	24,618	28,049	45,526
November	10,444	22,677	28,740	50,650
December	15,379	28,167	38,796	55,830

Source: New York Stock Exchange, 1994.

19.49 The data in the accompanying table represent average monthly retail gasoline prices (in cents per gallon) in the United States from January 1989 through December 1993.

Data file GASPRICE.DAT

Average monthly retail gasoline prices.

	Year				
Month	1989	1990	1991	1992	1993
January	91.8	104.2	124.7	107.3	111.7
February	92.6	103.7	114.3	105.4	110.8
March	94.0	102.3	108.2	105.8	109.8
April	106.5	104.4	110.4	107.9	111.2
May	111.9	106.1	115.6	113.6	112.9
June	111.4	108.8	116.0	117.9	113.0
July	109.2	108.4	112.7	117.5	110.9
August	105.7	119.0	114.0	115.8	109.7
September	102.9	129.4	114.3	115.8	108.5
October	102.7	137.8	112.2	115.4	112.7
November	99.9	137.7	113.4	115.9	111.3
December	98.0	135.4	112.3	113.6	107.0

Source: U.S. Energy Information Administration, 1994.

19.10 Pitfalls Concerning Time-Series Analysis

The value of such forecasting methodology as time-series analysis, which utilizes past and present information as a guide to the future, was recognized and most eloquently expressed more than two centuries ago by the American statesman Patrick Henry, who said:

> I have but one lamp by which my feet are guided, and that is the lamp of experience. I know no way of judging the future but by the past. *[Speech at Virginia Convention (Richmond) March 23, 1775]*

If it were true (as time-series analysis assumes) that the factors that have affected particular patterns of economic activity in the past and present will continue to do so in a similar manner in the future, time-series analysis, by itself, would certainly be a most appropriate and effective forecasting tool as well as an aid in the managerial control of present activities.

On the other hand, critics of classical time-series methods have argued that these techniques are overly naïve and mechanical; that is, a mathematical model based on the past should not be utilized for mechanically extrapolating trends into the future without considering personal judgments, business experiences, or changing technologies, habits, and needs. (See Problem 19.50 on page 914.) Thus in recent years, econometricians have been concerned with including such factors in developing highly sophisticated computerized models of economic activity for forecasting purposes. Such forecasting methods, however, are beyond the scope of this text (References 1–5, 7, and 11).

Nevertheless, as we have seen from the preceding sections of this chapter, time-series methods provide useful guides to business leaders for projecting future trends (on a long-term and short-term basis). If used properly—in conjunction with other forecasting methods as well as business judgment and experience—time-series methods will continue to be an excellent managerial tool for decision making.

19.11 Summary and Overview

As we observe in the summary chart for this chapter, we have developed numerous approaches for time-series forecasting, including moving averages, exponential smoothing, the linear, quadratic, and exponential trend models, the Holt-Winters approach, the autoregressive model, and we have described the contribution and use of the seasonal index. On page 858 of Section 19.1, you were given a list emphasizing the important points to be discussed in the chapter. Check over the list now to see whether you feel that you have an understanding of these key points. To be sure, you should be able to answer the following conceptual questions:

1. How do the time-series forecasting models developed in this chapter differ from the regression and multiple regression models considered in Chapters 17 and 18?
2. What is the difference between moving averages and exponential smoothing?
3. Under what circumstances would the exponential trend model be most likely to be appropriate?
4. What is the difference between exponential smoothing and the Holt-Winters method?
5. How do autoregressive modeling approaches differ from the other approaches to forecasting?

6. What are the alternative approaches to choosing an appropriate forecasting model?
7. How does forecasting for monthly or quarterly data differ from forecasting for annual data?

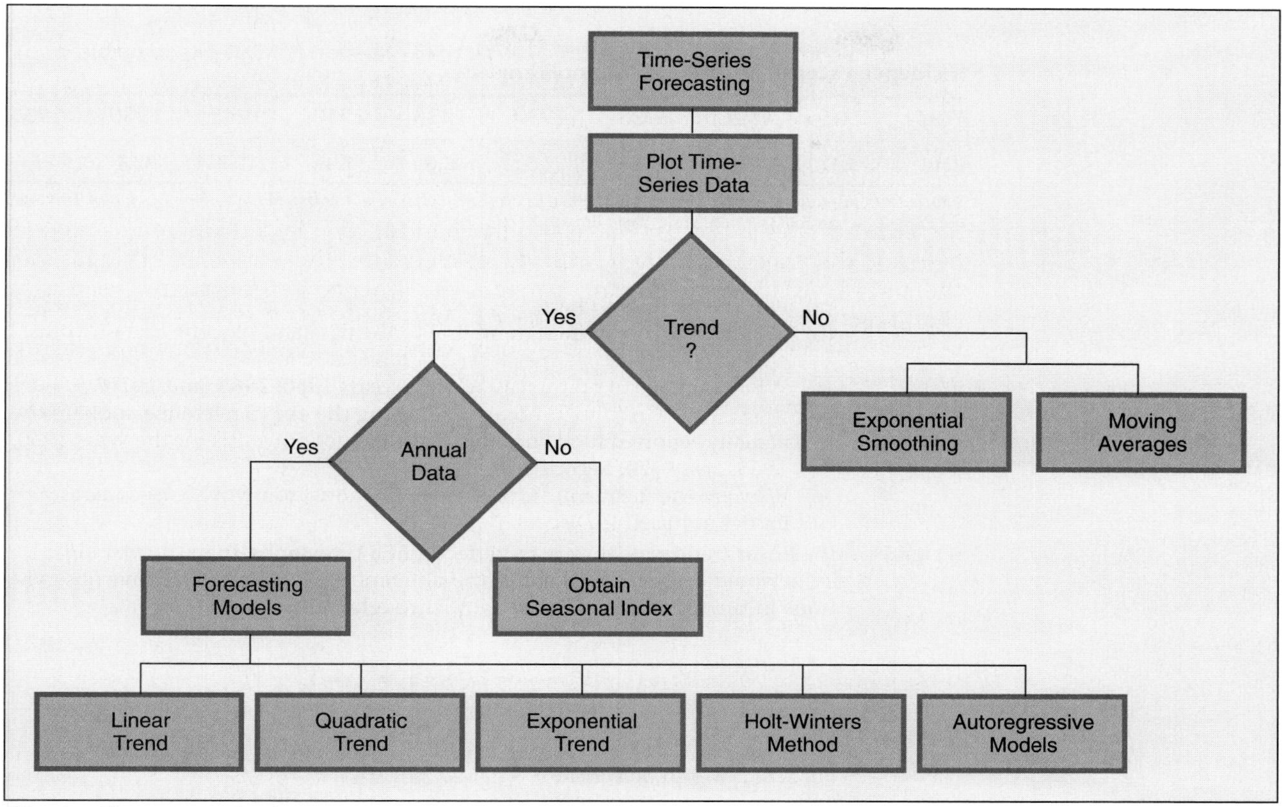

Chapter 19 summary chart.

Getting It All Together

Key Terms

autoregressive modeling *888*
classical multiplicative model *859*
cyclical component *860*
exponential smoothing *866*
exponential trend model *876*
forecasting *858*
Holt-Winters method *884*
irregular component *860*
linear trend model *872*
mean absolute deviation (MAD) *899*
moving averages *863*
principle of parsimony *899*
quadratic trend model *874*
ratio to moving averages *908*
seasonal component *903*
seasonal index *905*
time series *859*
trend *859*
weighted moving averages *908*

Chapter Review Problems

19.50 The data given below represent the annual incidence rates (per 100,000 persons) of reported acute poliomyelitis recorded over 5-year periods from 1915 to 1955.

Incidence rates of reported acute poliomyelitis.

Year	1915	1920	1925	1930	1935	1940	1945	1950	1955
Rate	3.1	2.2	5.3	7.5	8.5	7.4	10.3	22.1	17.6

Source: Data are taken from B. Wattenberg, ed., *The Statistical History of the United States: From Colonial Times to the Present* (Series B303), (New York: Basic Books, 1976).

(a) Plot the data on a chart.
(b) Fit a least squares linear trend line to the data and plot the line on your chart.
(c) What are your trend forecasts for the years 1960, 1965, and 1970?
(d) ACTION Go to your library and, using the above reference, look up the actually reported incidence rates of acute poliomyelitis for the years 1960, 1965, and 1970. Record your results.
(e) Why are the mechanical trend extrapolations from your least squares model not useful? Discuss.

19.51 If a linear trend model were to perfectly fit a time series, then the *first differences* would be constant. That is, the differences between consecutive observations in the series would be the same throughout:

$$Y_2 - Y_1 = Y_3 - Y_2 = \cdots = Y_{i+1} - Y_i = \cdots = Y_n - Y_{n-1}$$

If a quadratic trend model were to perfectly fit a time series, then the *second differences* would be constant. That is,

$$\begin{aligned}[(Y_3 - Y_2) - (Y_2 - Y_1)] &= [(Y_4 - Y_3) - (Y_3 - Y_2)] \\ &= \cdots = [(Y_{i+2} - Y_{i+1}) - (Y_{i+1} - Y_i)] \\ &= \cdots = [(Y_n - Y_{n-1}) - (Y_{n-1} - Y_{n-2})]\end{aligned}$$

If an exponential trend model were to perfectly fit a time series, then the *percentage differences* between consecutive observations would be constant. That is,

$$\begin{aligned}\left(\frac{Y_2 - Y_1}{Y_1}\right) \times 100\% &= \left(\frac{Y_3 - Y_2}{Y_2}\right) \times 100\% \\ &= \left(\frac{Y_{i+1} - Y_i}{Y_i}\right) \times 100\% \\ &= \cdots \\ &= \left(\frac{Y_n - Y_{n-1}}{Y_{n-1}}\right) \times 100\%\end{aligned}$$

Although we should not expect a perfectly fitting model for any particular set of time-series data, we nevertheless can evaluate the first differences, second differences, and percentage differences for a given series as a guide for determining an appropriate model to choose.

For each of the time-series data sets presented below:

(a) Determine the most appropriate model to fit.
(b) Develop this trend equation.
(c) Forecast the trend value for the year 1999.

	Year									
	1985	1986	1987	1988	1989	1990	1991	1992	1993	1994
Time series I	10.0	15.1	24.0	36.7	53.8	74.8	100.0	129.2	162.4	199.0
Time series II	30.0	33.1	36.4	39.9	43.9	48.2	53.2	58.2	64.5	70.7
Time series III	60.0	67.9	76.1	84.0	92.2	100.0	108.0	115.8	124.1	132.0

19.52 A time-series plot often aids the forecaster in determining an appropriate model to use. For each of the time-series data sets presented below:

(a) Plot the observed data (Y) over time (X) and plot the logarithm of the observed data ($\log Y$) over time (X) to determine whether a linear trend model or an exponential trend model is more appropriate. *Hint:* Recall from Section 19.5.3 that if the plot of $\log Y$ versus X appears to be linear, an exponential trend model provides an appropriate fit.
(b) Develop this trend equation.
(c) Forecast the trend value for the year 1999.

	Year									
	1985	1986	1987	1988	1989	1990	1991	1992	1993	1994
Time series I	100.0	115.2	130.1	144.9	160.0	175.0	189.8	204.9	219.8	235.0
Time series II	100.0	115.2	131.7	150.8	174.1	200.0	230.8	266.1	305.5	351.8

19.53 The data given in the accompanying table represent the annual gross revenues (in millions of dollars) obtained by a utility company for the years 1981 through 1994.

Data file AGREV.DAT

Annual gross revenues.

Year	1981	1982	1983	1984	1985	1986	1987	1988	1989	1990	1991	1992	1993	1994
Gross revenues	13.0	14.1	15.7	17.0	18.4	20.9	23.5	26.2	29.0	32.8	36.5	41.0	45.4	50.8

(a) Compare the first differences, second differences, and percent differences (see Problem 19.51) to determine the most appropriate model to fit.
(b) Develop this trend equation.
(c) What has been the annual growth in gross revenues over the 14 years?
(d) Forecast the trend value for the year 1999.

19.54 The data given in the accompanying table represent the annual revenues (in millions of dollars) of an advertising agency for the years 1975 through 1994.

Data file
ADTIME.DAT

Annual revenues (millions of dollars).

Year	Revenues	Year	Revenues	Year	Revenues
1975	51.0	1982	93.0	1989	100.9
1976	54.1	1983	102.8	1990	110.9
1977	56.4	1984	98.0	1991	133.3
1978	58.1	1985	83.6	1992	192.8
1979	69.5	1986	81.0	1993	234.0
1980	79.2	1987	87.0	1994	238.9
1981	89.2	1988	102.9		

(a) Plot the data over time and plot the logarithm of the data over time to determine whether a linear trend model or an exponential trend model is more appropriate (see Problem 19.52).
(b) Develop this trend equation.
(c) What has been the annual growth in advertising revenues over the 20 years?
(d) Forecast the trend value for the year 1998.

For Problems 19.55–19.58:
(a) Plot the data on a chart.
(b) Fit a least squares linear trend line to the data.
(c) Fit a quadratic trend equation to the data.
(d) Fit an exponential trend equation to the data.
(e) Use the Holt-Winters method (with $U = .30$ and $V = .30$) to fit the time series.
(f) Fit a third-order autoregressive model and test for the significance of the third-order autoregressive parameter. (Use $\alpha = .05$.)
(g) If necessary, fit a second-order autoregressive model and test for the significance of the second-order autoregressive parameter. (Use $\alpha = .05$.)
(h) If necessary, fit a first-order autoregressive model and test for the significance of the first-order autoregressive parameter. (Use $\alpha = .05$.)
(i) Perform a residual analysis for each of the fitted models in (b)–(e) and the most appropriate autoregressive model in (f)–(h).
(j) Compute the MAD for each corresponding model in (i).
(k) On the basis of (i), (j), and parsimony, which model would you select for purposes of forecasting? Discuss.
(l) Using the selected model in (k), make an annual forecast from 1993 through 1996.

19.55 The data in the accompanying table represent the annual total sales (in billions of dollars) of International Business Machines Corp. (IBM) over the 23-year period 1970 through 1992.

Data file
IBM.DAT

Total sales at IBM (1970–1992).

Year	Sales	Year	Sales	Year	Sales
1970	7.5	1978	21.1	1986	52.2
1971	8.3	1979	22.9	1987	55.3
1972	9.5	1980	26.2	1988	59.7
1973	11.0	1981	29.1	1989	62.7
1974	12.7	1982	34.4	1990	43.9
1975	14.4	1983	40.2	1991	37.0
1976	16.3	1984	45.9	1992	33.8
1977	18.1	1985	50.1		

Source: Moody's Handbook of Common Stocks, 1980, 1989, 1993.

19.56 The data in the accompanying table represent the annual total revenues (in billions of dollars) of McDonald's Corp. over the 22-year-period 1971 through 1992.

Data file
MCDONALD.DAT

Total revenues at McDonald's Corp. (1971–1992).

Year	Revenues	Year	Revenues	Year	Revenues
1971	0.3	1979	1.9	1987	4.9
1972	0.4	1980	2.2	1988	5.6
1973	0.6	1981	2.5	1989	6.1
1974	0.7	1982	2.8	1990	5.0
1975	1.0	1983	3.1	1991	4.9
1976	1.2	1984	3.4	1992	5.1
1977	1.4	1985	3.8		
1978	1.7	1986	4.2		

Source: Moody's Handbook of Common Stocks, 1980, 1989, 1993.

19.57 The data in the accompanying table represent the annual total revenues (in billions of dollars) of Sears, Roebuck & Co. over the 23-year period 1970 through 1992.

Data file
SEARS.DAT

Total revenues at Sears, Roebuck & Co. (1970–1992).

Year	Revenues	Year	Revenues	Year	Revenues
1970	8.9	1978	22.9	1986	42.3
1971	9.3	1979	24.5	1987	45.9
1972	10.0	1980	25.2	1988	50.3
1973	11.0	1981	27.4	1989	53.8
1974	12.3	1982	30.0	1990	50.3
1975	13.1	1983	35.9	1991	50.9
1976	17.7	1984	38.8	1992	52.3
1977	19.6	1985	40.7		

Source: Moody's Handbook of Common Stocks, 1980, 1989, 1993.

19.58 The data in the accompanying table at the top of page 918 represent the annual net sales (in billions of dollars) of Xerox Corporation over the 23-year period 1970 through 1992.

Data file
XEROX.DAT

Year	Revenues	Year	Revenues	Year	Revenues
1970	1.7	1978	5.9	1986	13.3
1971	2.0	1979	6.9	1987	15.1
1972	2.4	1980	8.0	1988	16.4
1973	3.0	1981	8.5	1989	17.6
1974	3.5	1982	8.5	1990	13.6
1975	4.1	1983	8.3	1991	13.8
1976	4.4	1984	8.6	1992	14.7
1977	5.1	1985	9.0		

Case Study J—Currency Trading

As a member of a financial firm that has been hired by a group of investors to undertake trading in various currencies, you have been assigned the task of studying the long-term trends in the exchange rates of the Canadian dollar, the French franc, the German mark, the Japanese yen, and the English pound in terms of the U.S. dollar. The data in the following table have been collected for the 26-year period from 1967 to 1992:

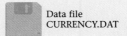

Data file CURRENCY.DAT

Exchange Rates of Five Currencies in Terms of the U.S. Dollar.

Year	Canadian Dollar	French Franc	German Mark	Japanese Yen	English Pound
1967	1.0789	4.9206	3.9865	362.13	275.04
1968	1.0776	4.9529	3.9920	360.55	239.35
1969	1.0769	5.1999	3.9251	358.36	239.01
1970	1.0444	5.5288	3.6465	358.16	239.15
1971	1.0099	5.5100	3.4830	347.79	244.42
1972	0.9907	5.0444	3.1886	303.13	250.34
1973	1.0002	4.4535	2.6715	271.31	245.25
1974	0.9780	4.8107	2.5868	291.84	234.03
1975	1.0175	4.2877	2.4614	296.78	222.17
1976	0.9863	4.7825	2.5185	296.45	180.48
1977	1.0633	4.9161	2.3236	268.62	174.49
1978	1.1405	4.5091	2.0097	210.39	191.84
1979	1.1713	4.2567	1.8343	219.02	212.24
1980	1.1693	4.2251	1.8175	226.63	232.46
1981	1.1990	5.4397	2.2632	220.63	202.43
1982	1.2344	6.5794	2.4281	249.06	174.80
1983	1.2325	7.6204	2.5539	237.55	151.59
1984	1.2952	8.7356	2.8455	237.46	133.68
1985	1.3659	8.9800	2.9420	238.47	129.74
1986	1.3896	6.9257	2.1705	168.35	146.77
1987	1.3259	6.0122	1.7981	144.60	163.98
1988	1.2306	5.9595	1.7570	128.17	178.13
1989	1.1842	6.3802	1.8808	138.07	163.82
1990	1.1668	5.4467	1.6166	145.00	178.41
1991	1.1460	5.6468	1.6610	134.59	176.74
1992	1.2085	5.2935	1.5618	126.78	176.63

Source: Board of Governors of the Federal Reserve System, Table B-107.

Develop forecasting models for the exchange rate of each of these five currencies based on these data. Be sure to indicate which forecasting model you have chosen for each currency and what limitations there are to the model. Provide forecasts for the years 1995, 1996, and 1997 for each currency. Write an executive summary for a presentation that is scheduled to be made to the investor's group next week.

Endnotes

1. That all i observed values in the time series are included in the computation of the exponentially smoothed value in time period i can be seen by noting that the present smoothed value is calculated using the smoothed value of the previous period, and that value, in turn, was calculated using the smoothed value from its previous period, and so on. Algebraically, this can be stated as follows:
   In time period 1,
   $$E_1 = Y_1$$
   In time period 2,
   $$E_2 = WY_2 + (1 - W)E_1 = WY_2 + (1 - W)Y_1$$
   In time period 3,
   $$E_3 = WY_3 + (1 - W)E_2 = WY_3 + (1 - W)[WY_2 + (1 - W)Y_1]$$
   $$= WY_3 + W(1 - W)Y_2 + (1 - W)^2Y_1$$
   In general, in time period i,
   $$E_i = WY_i + (1 - W)E_{i-1} = WY_i + W(1 - W)Y_{i-1} + W(1 - W)^2 Y_{i-2} + \cdots + (1 - W)^{(i-1)}Y_1$$
   Thus we see that over time, as the integer value i gets large, the weights assigned to the earlier (older) values in the time series may become negligible.
2. It should be noted that the exponential smoothing model of Section 19.4.2, the Holt-Winters model of Section 19.6, and the autoregressive models of Section 19.7 are all special cases of *autoregressive integrated moving average (ARIMA)* models developed by Box and Jenkins (Reference 3). The Box-Jenkins approach, however, is beyond the scope of this text.
3. Monthly data are usually recorded and plotted in the middle of the month. Therefore the origin here is presented as January 15, 1988.

References

1. Bails, D. G., and L. C. Peppers, *Business Fluctuations: Forecasting Techniques and Applications* (Englewood Cliffs, NJ: Prentice-Hall, 1982).
2. Bowerman, B. L., and R. T. O'Connell, *Forecasting and Time-Series,* 3rd ed. (North Scituate, MA: Duxbury Press, 1990).
3. Box, G. E. P., and G. M. Jenkins, *Time Series Analysis: Forecasting and Control,* 2d ed. (San Francisco, CA: Holden-Day, 1977).
4. Brown, R. G., *Smoothing, Forecasting, and Prediction* (Englewood Cliffs, NJ: Prentice-Hall, 1963).
5. Chambers, J. C., S. K. Mullick, and D. D. Smith, "How to Choose the Right Forecasting Technique," *Harvard Business Review,* Vol. 49, No. 4, July-August 1971, pp. 45–74.
6. Mahmoud, E., "Accuracy in Forecasting: A Survey," *Journal of Forecasting,* Vol. 3, 1984, pp. 139–159.
7. Newbold, P., *Statistics for Business and Economics,* 4th ed. (Englewood Cliffs, NJ: Prentice-Hall, 1994).
8. Ryan, B. F. and B. L. Joiner, *MINITAB Student Handbook,* 3d ed. (North Scituate, MA: Duxbury Press, 1994).
9. *SAS-ETS User's Guide* (Cary, NC: SAS Institute, 1988).
10. *STATISTIX 4.0* (Tallahassee, FL: Analytical Software, 1992).
11. Wilson, J. H., and B. Keating, *Business Forecasting* (Homewood, IL: Richard D. Irwin, 1990).

Answers to Selected Problems (•)

Chapter 2

2.4 (a) discrete numerical, ratio
(b) categorical, nominal
(c) discrete numerical, ratio
(d) continuous numerical, ratio
(e) categorical, nominal
(f) continuous numerical, ratio
(g) categorical, nominal
(h) discrete numerical, ratio
(i) continuous numerical, ratio
(j) categorical, nominal
(k) categorical, nominal

2.32 (a) continuous numerical, ratio
(b) discrete numerical, ratio
(c) categorical, nominal
(d) continuous numerical, ratio
(e) categorical, nominal
(f) discrete numerical, ratio
(g) categorical, nominal

2.33 (a) categorical, nominal
(b) categorical, nominal
(c) continuous numerical, ratio
(d) continuous numerical, ratio
(e) categorical, nominal
(f) categorical, nominal
(g) discrete numerical, ratio
(h) discrete numerical, ratio
(i) continuous numerical, ratio
(j) continuous numerical, ratio
(k) categorical, nominal
(l) discrete numerical, ratio
(m) continuous numerical, ratio
(n) continuous numerical, ratio
(o) categorical, nominal
(p) continuous numerical, ratio

2.37 $N = 93$ $n = 15$ Sample *without* replacement
Row 29: 12 47 83 76 22 65 93 10 61 36 89 58 86 92 71

2.42 line 401—EDUC 41
line 402—RICHWORK 4

line 403—AGE 10
line 404—SCHOOLNG 5
line 405—SEX 3

Chapter 3

3.1 (a)

9	147
10	02238
11	135566777
12	223489
13	02

(c) Stem-and-leaf display shows how data distribute and cluster.

3.5 (b) Stem-and-leaf display: Book values

0L	4
0H	55666677777788888888999999
1L	000000001122334
1H	555668
2L	3

3.12 (b) Frequencies are 4 7 9 13 9 5 3

3.14 Classes are 0 < 5, 5 < 10, etc.
Frequencies are 1 27 15 6 1

3.19 The percentages are
8 14 18 26 18 10 6

3.21 The percentages are
2 54 30 12 2

3.34 (a) The cumulative frequencies are
0 4 11 20 33 42 47 50

(b) The cumulative percentages are
0 8 22 40 66 84 94 100

3.36 (a) The cumulative frequencies are
0 1 28 43 49 50

(b) The cumulative percentages are
0 2 56 86 98 100

3.72 (a) The classes are:
0 < 10; 10 < 20, etc.
The frequencies from the ASE are:
16 6 1 1 1 0 0 0 0
The frequencies from the NYSE are:
13 8 15 7 2 3 1 0 1

Chapter 4

4.2 (a)

	Batch 1	Batch 2
mean	4	14
median	3	13
mode	2	12
midrange	6	16
midhinge	3.5	13.5

(b) The observations in Batch 1 are each 10 units less than the observations in Batch 2.

4.10 (a) Revised stem-and-leaf display

2	8
3	458
4	1
5	01157
6	28
7	16
8	5
9	
10	
11	9
12	
13	12
14	19
15	9

(b) mean = 7.78; median = 6.20; mode = 5.10
midrange = 9.35; midhinge = 8.53

(c) The midrange and midhinge are largest.

4.15 (a) mean = 147.1; median = 148.5; bimodal;
midrange = 147.5; midhinge = 147.5

4.17 (a)

	Batch 1	Batch 2
range	8	8
IQR	3	3
variance	8.33	8.33
S	2.89	2.89
CV	72.2%	20.6%

4.22 (a) range = 13.10; interquartile range = 7.95;
variance = 17.95; $S = 4.24$; CV = 54.5%

(b) The majority of the data fall within ±4.24 of the mean.

4.27 (a) range = 131; interquartile range = 41;
$S = 31.7$; CV = 21.5%

4.28 (a) and (b) For each batch the data are positive or right-skewed since mean > median.

4.31 The data are positive or right-skewed since mean > median.

4.36 The data are approximately symmetric.

4.39 (a) Five-number summary (MINITAB):
2.80 4.55 6.20 12.50 15.90

(b) and (c) The data are right-skewed.

4.44 (a) Five-number summary:
82 127 148.5 168 213

(b) and (c) The data are approximately symmetric.

4.45 (a) mean = 6.0; median = 6.5; mode = 8.0;
midrange = 6.0; midhinge = 5.5

(b) range = 10.0; interquartile range = 5.0;
variance = 9.40; $\sigma_x = 3.07$; $CV_{pop} = 51.1\%$

(c) No. Data are approximately symmetrical.

4.50 (d) (1) mean ≅ 7.7; median ≅ 4.2; mode ≅ 5.0;
midrange ≅ 9.0; midhinge ≅ 5.8

(d) (2) range ≅ 14.0; interquartile
range ≅ 6.8; $S \cong 4.3$; CV ≅ 55.8%

(d) (3) The data are right-skewed.

4.53 (a) (1) mean ≅ 148.8; median ≅ 148; mode ≅ 150; midrange ≅ 150; midhinge ≅ 147
(a) (2) range ≅ 140; interquartile range ≅ 48; $S \cong 29.7$; CV ≅ 20.0%
(a) (3) The data are approximately symmetrical.

4.55 (b) (1) mean = 9.8; median = 9.0; mode = 8.0; midrange = 13.5; midhinge = 9.0
(b) (2) range = 19.0; interquartile range = 4.0; $S = 3.7$; CV = 37.8%
(b) (3) The data are right-skewed.

4.76 (a) mean = \$41.78; median = \$42.00; midrange = \$58.50; midhinge = \$39.88; range = \$97.00; IQR = \$29.75; S = \$21.30; CV = 51.0%.
Five-number summary (MINITAB):
\$10.00 \$25.00 \$42.00 \$54.75 \$107.00

Chapter 5

5.15 (a)

Financial Condition	Education Level H.S. Degree or Lower	Some College	Degree or Higher	Totals
Worse off now than before	60.7	30.0	18.1	43.4
No difference	24.2	45.6	19.5	27.2
Better off now than before	15.1	24.4	62.4	29.4
Totals	100.0	100.0	100.0	100.0

(b) More people felt that they were worse off than better off.

Chapter 6

6.5 (a) Having a bank credit card, since there is only one criterion satisfied.
(b) Having a bank credit card *and* having a travel and entertainment credit card, since two criteria are involved.
(c) Not having a bank credit card is the complement of having a bank credit card, since it involves all events other than having a bank credit card.
(d) It satisfies two criteria, having a bank credit card and having a travel and entertainment credit card.

6.7 (b) Enjoying shopping for clothing is a simple event since it satisfies one criterion.
(c) A male who enjoys shopping for clothing is a joint event since it satisfies two criteria.
(d) Not enjoying shopping for clothing is the complement.

6.10 (a) $P(B) = 120/200$
(b) $P(B') = 80/200$
(c) $P(T) = 75/200$
(d) $P(T') = 125/200$

6.12 (a) 240/500
(b) 360/500
(c) 260/500
(d) 140/500

6.15 (a) $P(B \text{ and } T) = 60/200$
(b) $P(B' \text{ and } T) = 15/200$
(c) $P(B' \text{ and } T') = 65/200$

6.17 (a) 224/500
(b) 104/500
(c) 136/500

6.23 (a) 135/200
(b) 140/200
(c) 200/200 = 1.0

6.25 (a) 396/500
(b) 276/500
(c) 500/500 = 1.0

6.28 (a) 60/120
(b) 60/125
(c) Since $P(T|B) = 60/120 \neq P(T) = 75/200$; not statistically independent.

6.30 (a) 36/260
(b) 136/360
(c) $P(\text{Enjoys}|\text{female}) = 224/260 \neq P(\text{Enjoys}) = 360/500$; not statistically independent.

6.33 $(120/200)(75/200) \neq 60/200$; not statistically independent.

6.35 $(360/500)(240/500) \neq (136/500)$; not statistically independent.

6.44 (a) 2/3
(b) .36

6.46 (a) .625
(b) .56
(c) .325

6.49 (a) 27,000
(b) 1/27,000 = .000037
(c) "Dial combinations" follow counting rule 1 in which *K* different mutually exclusive and collectively exhaustive events can occur on each of *n* trials.

6.55 720

6.58 35

6.65 (a) (1) 80/200
(2) 55/200
(3) 125/200
(b) 55/80
(c) $P(< 5|\text{College grad}) = 55/80 \neq P(< 5) = 130/200$; not statistically independent).

Chapter 7

7.1 (a) A: 1.00; B: 3.00
(b) A: 1.22; B: 1.22
(c) A: Right-skewed; B: Left-skewed

7.6 (a) 7
(b) $\sigma_x^2 = 5.8333$
$\sigma_x = 2.42$
(d) $-.056$
(e) Lose 5.6¢ per bet.
(f) Gain 5.6¢ per bet.

7.10 $E(500) = \$500$; $E(1{,}000) = \$800$; $E(2{,}000) = \$600$
Purchase 1,000 pounds.

7.19 (a) .0778
(b) .6826
(c) $P(X = 0) = .0102$ $P(X = 1) = .0768$ $P(X = 2) = .2304$.
Distribution is slightly left skewed.

7.23 (a) .2851
(b) .1606
(c) .7149
(d) .2945

7.35 (a) (1) .6496
(2) .1503
(3) .1493
(c) 7; $P(X = 7) = .2668$
(d) 1.449
$7 \pm 2(1.449)$; $P(4 < X < 10) = .9244$

7.36 (a) (1) .8171
(2) .1667
(3) .0162
(b) (1) .8187
(2) .1637
(3) .0176
(c) (1) .3679
(2) .3679
(3) .2642

Chapter 8

8.3 (a) (1) .3599
(2) .6401
(3) .0832
(4) .9168
(5) .8599
(6) .5832
(7) .4431
(8) .5569
(b) (1) .1401
(2) .4168
(3) .3918
(4) .8349
(5) .1151
(c) 0
(d) -1.00
(e) $+1.00$

8.7 (a) .4082
(b) .0669
(c) 25.08%
(d) 749.2 trucks
(e) $Z = -0.84$ so that $X = 39.92$ thousand miles.

8.14 (a) .1587
(b) .0466
(c) .7865
(d) 46.4 hours
(e) 40.0 hours
(f) 6.7 hours

8.20 Area under normal curve covered: .1429 .2857 .4286 .5714 .7143 .8571
Standardized normal quantile value: −1.07 −0.57 −0.18 +0.18 +0.57 +1.07

8.21 (a) $\overline{X} = \$106.80$
$S = \$38.16$
(b) *Five Number Summary*
\$40 \$80 \$100 \$135 \$200
(c) Slightly right skewed

8.33 (a) (1) .2051
(2) .8282
(3) .3770
(4) .1719
(5) .6231
(6) .7735
(b) (1) .2034
(2) .8289
(3) .3745
(4) .1711
(5) .6255
(6) .7718

8.37 (a) (1) .0821
(2) .5438
(b) (1) .1841
(2) .7286

8.40 (a) (1) .3413
(2) .3413
(3) .6826
(4) .8413
(5) .1587
(6) .1160
(b) 10.0 million dollars
(c) 6.8 million dollars
(d) 3.4 million dollars

8.44 Height: $P(X > 67) = .2119$
Weight: $P(X > 135) = .1587$
Weight is slightly the more unusual characteristic

8.47 (a) Exact (binomial) using Table E.7.
(1) .5325
(2) .9961

(b) Approximate (normal)
 (1) .2776
 (2) .99926

8.50 (a) .0287
(b) .0108

Chapter 9

9.6 $\mu_x = 1.30$; $\sigma_x = 0.04$
(a) .1915
(b) .1747
(c) 1.2664 to 1.3336
(d) (1) $\mu_{\overline{X}} = 1.30$; $\sigma_{\overline{X}} = 0.01$
 (2) normal
 (3) .4772
 (4) .15735
 (5) 1.2916 to 1.3084
(e) and (f) Since samples of size 16 are being taken, rather than individual values (samples of $n = 1$), because $\sigma_{\overline{X}} = \sigma_x / \sqrt{n}$, more values lie closer to the mean with the increased sample size and fewer values lie far away from the mean.
(g) They are equally likely to occur (probability = .1587) since, as n increases, more sample means will be closer to the population mean.

9.9 (a) .2486
(b) .0918
(c) .1293 and .2514
(d) A percent defective above 10.5% is more likely to occur since it is only .33 standard deviation above the population value of 10%.

9.14 .14833

9.16 .2549 and .0823

Chapter 10

10.5 (a) $.9877 \le \mu_x \le 1.0023$
(b) Since the value of 1.0 is included in the interval, there is no reason to believe that the average is below 1.0.
(c) No, since σ_x is known and $n = 50$, from the central limit theorem we may assume that $\overline{X}$ is normally distributed.
(d) An individual value of .98 is only .75 standard deviation below the sample mean of .995. The confidence interval represents the estimate of the average of a sample of 50, not an individual value.

10.9 $\$1,067.40 \le \mu_x \le \$1,332.60$

10.13 (a) $87.769 \le \mu_x \le 109.964$

10.29 (a) $\$653.37 < X_f < \$1,746.73$

10.32 (a) $34.477 < X_f < 143.255$

10.39 $.2246 \le p \le .3754$

10.41 $.342 \le p \le .478$

10.48 $n = 97$
10.50 $n = 167$
10.56 $n = 323$
10.57 $n = 271$
10.61 (a) $322.62 \le \mu_x \le 377.38$
(b) $n = 93$
10.63 (a) $.2284 \le p \le .3716$
(b) $n = 214$
10.85 (a) $14.085 \le \mu_x \le 16.515$
(b) $.530 \le p \le .820$
(c) $7.52 < X_f < 23.08$
(e) $n = 25$
(f) $n = 784$

Chapter 11

11.12 $-1.96 < Z = -0.80 < +1.96$. Don't reject H_0. There is no evidence that the average amount dispensed is different from 8 ounces.
11.16 (a) H_0: $\mu_x = 375$
H_1: $\mu_x \neq 375$
(b) $-1.96 < Z = -1.768 < +1.96$. Don't reject H_0. There is no evidence that the mean is different from 375 hours ($\alpha = .05$).
11.20 p value $= .4238$
11.24 p value $= .0768$
11.29 (a) H_0: $\mu_x \ge 2.8$; H_1: $\mu_x < 2.8$
(b) $Z = -1.75 < -1.645$. Reject H_0. The average is significantly less than 2.8, and, therefore, we can conclude that there is evidence that the process is not working properly.
11.36 $Z = -1.75$; the p value $= .5000 - .4599 = .0401$, which is less than $\alpha = .05$. There is evidence the process is not working properly.
11.42 (a) Power = .6387; $\beta = .3613$
(b) Power = .9908; $\beta = .0092$
11.43 (a) Power = .3707; $\beta = .6293$
(b) Power = .9525; $\beta = .0475$
(c) The decrease in α has increased β and decreased power.
11.44 (a) Power = .8037; $\beta = .1963$
(b) Power = .9996; $\beta = .0004$
(c) The increase in sample size has increased power.
11.49 $n = 64$
11.52 $n = 17$
11.55 (a) $n = 19$ households
(b) Power = .3707
(c) Power = .99988
(d) By almost doubling the sample size (from 19 to 36), the power increased from .975 to .99988.
(e) Power = .6387; $\beta = .3613$
(f) Power = 1.0; $\beta = 0$
(g) The increase in α has reduced β and increased power.
(h) Power = .6331; $\beta = .3669$
(i) Power = 1.0; $\beta = 0$

(j) The increase in sample size has increased power.
(k) $\$14.37 \le \mu_x \le \$16.95$
(l) $Z = 3.32 > 2.33$. Reject H_0.
(m) Institute the bagel and breakfast delivery service. There is sufficient evidence that the average order will exceed $14.

Chapter 12

12.1 (a) $t = 3.30 > t_{35} = 2.0301$. Reject H_0. There is evidence that the EER is different from 9.0.
(b) The data are approximately normally distributed.
(c) Using SAS, p value is .0022.

12.3 (a) $t = 3.552 > t_{99} = 1.9842$. Reject H_0. There is evidence that the average balance is different from $75.
(b) p value $< .005$

12.10 (a) $t = 1.714 < t_{14} = 1.7613$. Don't reject H_0. There is no evidence that the average waiting time exceeds 90 days.
(b) The data are measured on either an interval or ratio scale and the underlying population is approximately normally distributed.
(c) $.05 < p$ value $< .10$

12.12 $-2.58 < Z = -0.35 < 2.58$. Don't reject H_0. There is no evidence that the median tar content of this new brand is different from 17 milligrams.

12.14 (a) $W = 471$ so $Z = 2.96 > 1.96$. Reject H_0. There is evidence that the median EER of the air conditioners is different from 9.0.
(b) The data are measured on either an interval or ratio scale and the underlying population is approximately symmetrical.
(c) The results are the same. The p value here is .0030.

12.19 (a) $W = 87 < W_U = 95$. Don't reject H_0. There is no evidence that the median waiting time exceeds 90 days.
(b) The data are measured on either an interval or ratio scale and the underlying population is approximately symmetrical.
(c) The results are the same.

12.21 (a) $\chi^2 = 88.81 > \chi^2_{U(29)} = 42.557$. Reject H_0. There is evidence that the population standard deviation has increased above 1.2°.
(b) The data are measured on either an interval or ratio scale and the underlying population is approximately normally distributed.
(c) p value $< .005$ in upper tail.

12.27 (a) $\chi^2_{L(19)} = 8.907 < \chi^2 = 22.29 < \chi^2_{U(19)} = 32.852$. Don't reject H_0.
(b) The data are measured on either an interval or ratio scale and the underlying population is approximately normally distributed.
(c) $.10 < p$ value $< .25$ in upper tail.

12.28 (a) $\chi^2 = 24.8004 > \chi^2_{U(9)} = 21.666$. Reject H_0. There is evidence that the process standard deviation has increased.
(b) The data are measured on either an interval or ratio scale and the underlying population is approximately normally distributed.
(c) p value $< .005$.

12.31 Median = 4.15. $U = 4 < U_L = 11$, so reject H_0. There is evidence of a trend.

12.34 $U = 8 < U_L = 10$, so reject H_0. There is evidence the process is out of control.

12.40 (a) Yes, from a normal probability plot.
(b) $t = -2.98 < t_{14} = -1.7613$. Reject H_0. There is evidence that the mean time is less than 30 seconds.
(c) $W = 16 < W_L = 25$. Reject H_0. There is evidence that the median time is less than 30 seconds.
(d) Since the two tests give the same results, the shape of the underlying population does not affect the conclusions.
(e) If it is cost efficient, the new method should be implemented.
(f) $\chi^2 = 28.88 > \chi^2_{U(14)} = 26.119$. Reject H_0. There is evidence that the standard deviation has increased.

Chapter 13

13.1 (a) $Z = +0.39 < +1.96$. Don't reject H_0. There is no evidence of a difference in the average life of bulbs produced by the two machines.
(b) p value = .6966

13.3 (a) $t = +1.91 > t_{198} = +1.645$. Reject H_0. There is evidence of a difference in output between the day and evening shift.
(b) $.05 < p$ value $< .10$ (or .0562 estimated from a normal distribution.)

13.8 (a) $t = -2.19 < t_{48} = -2.0106$. Reject H_0. There is evidence of a difference in the average talking time prior to recharge. The newly developed battery lasts longer.
(b) Normality in each population and equality of variances.
(c) $.02 < p$ value $< .05$

13.11 (a) $t = +2.948 > t_{28} = +2.7633$. Reject H_0. There is evidence that the average tuition is higher at prep schools in the Northeast than in the Midwest.
(b) p value $< .01$

13.13 Yes. $t' = +4.18 > +1.9905$. Reject H_0. Appraised values are higher in Farmingdale.

13.18 (a) $t' = -2.19 < t'_{47} = -2.0117$. Reject H_0. There is evidence of a difference in the average talking time prior to recharge. The p value is between .02 and .05.
(b) Normality in each population.
(c) The results are very similar.

13.23 (a) Yes. $t' = +2.948 > +2.6245$. Reject H_0.
(b) The results of Problems 13.11 and 13.23 (a) are very similar.

13.25 No. Since $78 < T_1 = 84 < 132$, don't reject H_0

13.29 (a) Yes. Let the nickel-cadmium sample be group 1. Thus $T_1 = 502.5$; since $Z = -2.62 < -1.96$, reject H_0.
(b) Equal variability in the two populations.
(c) The results are all similar.

13.34 (a) Yes, reject H_0. $T_1 = 292.5$ so that $Z = +2.49 > +2.33$. The p value is .0064.
(b) The results are all similar.

13.36 (a) $F_{L(9,9)} = 0.248 < F = 0.811 < F_{U(9,9)} = 4.03$. Don't reject H_0.
(b) One is the reciprocal of the other if the sample sizes are equal.
(c) Reject H_0 if $F > F_{U(9,9)} = 3.18$.
(d) Reject H_0 if $F < F_{L(9,9)} = 0.314$.

13.41 (a) No. We don't reject H_0 since $F_{L(24,24)} = 0.441 < F = 0.867 < F_{U(24,24)} = 2.27$. Note that the p value > .05.
(b) Depending on the assumption of underlying normality in the populations, either the pooled-variance t test or the Wilcoxon rank sum test is more appropriate.

13.46 (a) No. We don't reject H_0 since $F_{L(14,14)} = 0.232 < F = 0.478 < F_{U(14,14)} = 4.31$.
(b) p value > .05

13.63 (a) Since $t = -0.46$ falls between $t_9 = \pm 2.2622$, don't reject H_0. There is no evidence of a difference in the average gasoline mileage between regular and high octane gasoline.
(b) p value > .50

13.69 (a) No. Since $W_L = 5 < W = 18.5 < W_U = 40$, don't reject H_0.
(b) These results are the same as those obtained previously.

13.72 (a) [$\$104.64 \leq \mu_x \leq \$115.36$]
(b) $t = -10.0 < t_{24} = -2.7969$. Reject H_0. The average monthly balance of Plan A accounts is not equal to $105.
(c) $F_{L(24,49)} = 0.37 < F = 1.125 < F_{U(24,49)} = 2.40$. Don't reject H_0. There is no evidence of a difference in the variances between Plan A and Plan B.
(d) $t = -9.903 < t_{73} = -2.6449$. Reject H_0. There is evidence of a difference in the average monthly balance between Plan A and Plan B. Plan A's balance is lower.
(e) The respective p values are < .01, > .05, and < .01.

Chapter 14

14.5 (d) $F_{max} = 1.184 < F_{max(4,7)} = 8.44$. Don't reject H_0.
(e) Yes, we can proceed.
(f) $F = 4.22 > F_{U(3,28)} = 2.95$. Reject H_0. There is evidence of a difference.
(g) Program A is superior to B and C.

14.8 (a) $F_{max} = 7.22 < F_{max(5,3)} = 50.7$. Don't reject H_0. $F = 10.30 > F_{U(4,17)} = 2.96$. Reject H_0. Alloy 2 is weakest..

14.12 $H = 0.635 < \chi^2_{U(2)} = 9.210$. Don't reject H_0.

14.15 $H = 9.51 > \chi^2_{U(3)} = 7.815$. Reject H_0. Using the Dunn procedure, the critical range is 12.38. Program A is superior to C.

14.33 (b) $F = 7.02 > F_{U(4,24)} = 2.78$. Reject H_0. There is evidence of a difference.
(c) Critical range = 0.472. Treatment substance 2 results in a significantly faster clotting time than does substance 3, 4, or 5. Treatment substance 1 is also significantly faster than 4. Other pairwise differences are not significant.
(d) $RE = 15.9$.

14.37 (a) $F = 0.21 < F_{U(1,9)} = 5.12$. Don't reject H_0.
(b) $F_{U(1,df)} = t^2_{df}$

14.38 (a) $F_R = 8.244 < \chi^2_{U(4)} = 9.488$. Don't reject H_0.
(b) Nemenyi's procedure is not used because H_0 was not rejected in (a).

14.42 $F_R = 14.71 > \chi^2_{U(4)} = 9.488$. Reject H_0. Using Nemenyi's procedure, the critical range is 2.31. Substance 2 results in a significantly faster clotting time than substance 4.

14.45 (a) (1) $F = 0.51 < F_{U(2,9)} = 4.26$. Don't reject H_0. There is no effect due to service center.
(2) $F = 17.58 > F_{U(2,9)} = 4.26$. Reject H_0. There is evidence of a brand effect.
(3) $F = 3.36 < F_{U(4,9)} = 3.63$. Don't reject H_0. Interaction is not significant.
(c) Critical range = 8.36. Brand C requires a significantly greater amount of time for repair than either of the others.
(f) The test for equality of brands would have been

$$F = \frac{472.9}{90.3} = 5.24 < F_{U(2,4)} = 6.94,$$ and the Tukey procedure for pairwise comparisons would not have been made.

14.48 (a) (1) $F = 26.57 > F_{U(3,24)} = 3.01$. Reject H_0. There is evidence of an operator effect.
(2) $F = 43.60 > F_{U(2,24)} = 3.40$. Reject H_0. There is evidence of a machine effect.
(3) $F = 3.81 > F_{U(6,24)} = 2.51$. Reject H_0. There is evidence of a significant interaction between operator and machine.
(c) The Tukey procedure for pairwise comparisons is not used. The significant interaction makes the study of the main effects difficult.

Chapter 15

15.4 (a) $Z = -1.60 > -2.33$. Don't reject H_0. There is no evidence that the proportion is less than .25.
(b) p value = .0548.

15.5 (a) $Z = +2.93 > +1.645$. Reject H_0. There is evidence that the proportion is different from .30.
(b) p value = .0034.

15.12 (a) $Z = +2.37 > +1.96$. Reject H_0. There is evidence of a difference in the proportion of women in the two groups who eat dinner in a restaurant during the work week.
(b) p value = .0178.

15.14 (a) $Z = +2.58 > +1.645$. Reject H_0. There is evidence that a high temperature setting is preferred.
(b) p value = .005.

15.17 (a) $Z = +7.34 > +1.96$. Reject H_0.
(b) p value = .0000.
(c) $Z = +7.34 > +1.645$. Reject H_0.

15.22 $\chi^2 = 5.617 > \chi^2_{U(1)} = 3.841$. Reject H_0.

15.27 (a) $\chi^2 = 53.826 > \chi^2_{U(1)} = 3.841$. Reject H_0.
(b) p value = .0000.
(d) The χ^2 test can be used only to test for a *difference* between two proportions.

15.30 (a) $\chi^2 = 1.125 < \chi^2_{U(2)} = 9.210$. Don't reject H_0. There is no evidence of a difference in attitude toward the trimester among the various groups.

15.34 (a) $\chi^2 = 11.2949 > \chi^2_{U(2)} = 9.210$. Reject H_0. There is evidence of a difference. Using the Marascuilo procedure, residents of single-family dwellings adopt a cable TV service significantly more than do residents of apartment houses.

15.38 (a) $\chi^2 = 22.780 > \chi^2_{U(8)} = 15.507$. Reject H_0. There is a relationship between type of area of residence and manufacturer preference in an automobile purchase.

(b) p value $< .005$.

15.40 (a) $\chi^2 = 9.82 < \chi^2_{U(4)} = 13.277$. Don't reject H_0. There is no evidence of a relationship between commuting time and stress.

(b) $.025 < p$ value $< .05$.

15.59 (a) $Z = -3.90 < -2.33$. Reject H_0. There is evidence that the proportion of employees with fewer than 5 days absent is lower in year 1.

(b) p value $= .00005$.

(c) That the incentive plan significantly reduced absenteeism.

15.60 (a) $Z = -4.08 < -1.96$. Reject H_0. There is evidence of a difference.

(b) p value $= .0000$.

15.62 (a) $Z = -5.00 < -2.33$. Reject H_0. There is evidence that the proportion of male medical doctors having heart attacks is lower for those who had the aspirin than for those who did not have the aspirin.

(b) p value is less than .0001. There is very little likelihood that these results could have occurred if the aspirin did not reduce the incidence of heart attacks.

(c) The χ^2 test is not appropriate because we have a directional alternative hypothesis.

Chapter 16

16.6 (a) $\bar{p} = .1145$; $LCL = .0522$; $UCL = .1768$. The proportion of late arrivals on day 13 is substantially out of control. Possible special causes of this value should be investigated. In addition, the next four highest points all occur on Friday.

(b) $\bar{X} = 26.9$; $UCL = 41.54$; and $LCL = 12.26$.

(c) The results are exactly the same. The p chart expresses the results in terms of the proportion and the np chart expresses the results in terms of the number of successes.

(d) The snowstorm would explain why the proportion of late arrivals was so high on day 13.

16.10 (a) $\bar{p} = .01288$; $UCL = .01753$; $LCL = .00823$. Although none of the points are outside the control limits, there is evidence of a pattern over time, since the last eight points are all above the mean and most of the earlier points are below the mean. Thus, the special causes that might be contributing to this pattern should be investigated before any change in the system of operation is contemplated.

(b) Once special causes have been eliminated and the process is stable, process flow and fishbone diagrams of the process should be developed to increase process knowledge. Then Deming's 14 points can be applied to improve the system.

16.15 (a) $\bar{c} = 6.458$; $UCL = 14.082$. The process appears to be in control since there are no points outside the upper control limit and there is no pattern in the results over time.

(b) The value of 12 is within the control limits, so that it should be identified as a source of common cause variation. Thus, no action should be taken concerning this value. If the value was 20 instead of 12, $\bar{c}$ would be 6.792 and *UCL* would be 14.61. In this situation, a value of 20 would be substantially above the *UCL*, and action should be taken to explain this special cause of variation.

(c) Since the process is in control, process flow and fishbone diagrams of the process should be developed to increase process knowledge.

16.22 (a) $\bar{R} = 271.57$; $UCL = 574.20$. *LCL* does not exist. There are no points outside the control limits and no evidence of a pattern in the range chart. $\bar{\bar{X}} = 198.67$; $UCL = 355.31$; and $LCL = 42.03$. There are no points outside the control limits and no evidence of a pattern in the $\bar{X}$ chart.

16.23 (a) $\bar{R} = 4.8325$, $\bar{\bar{X}} = 3.549$, For *R* chart $LCL = 1.0786$, and $UCL = 8.5864$. For $\bar{X}$ chart: $LCL = 2.06$, and $UCL = 5.038$. The process appears to be in control since there are no points outside the lower and upper control limits and there is no pattern in the results over time.

(b) Since the process is in control, it is up to management to reduce the common cause variation by application of the 14 points of the Deming theory of management by process. In addition, process knowledge would be enhanced by the development of process flow and fishbone diagrams.

16.25 (a) $\bar{X} = 11.9998$; $\overline{MR} = .03648$; $LCL = 11.903$; $UCL = 12.097$.

(b) There is no evidence of special cause variation.

Chapter 17

17.7 (a) $b_0 = 1.45$; $b_1 = .074$.

(b) For each increase of 1 foot of shelf space, sales will increase by $7.40 per week.

(c) $\hat{Y}_i = 2.042$ or $204.20.

17.9 (a) $b_0 = 12.6786$; $b_1 = 1.9607$.

(b) The *Y* intercept b_0 (equal to 12.6786) represents the portion of the worker hours that is not affected by variation in lot size. The slope b_1 (equal to 1.96) means that for each increase in lot size of one unit, worker hours are predicted to increase by 1.96.

(c) $\hat{Y}_i = 100.91$.

(d) Lot size varied from 20 to 80, so that predicting for a lot size of 100 would be extrapolating beyond the range of the *X* variable.

17.13 $S_{YX} = 0.308$.

17.15 $S_{YX} = 4.71$.

17.19 (a) $r^2 = .684$; 68.4% of the variation in sales can be explained by variation in shelf space.
(b) $r^2_{adj} = .652$.

17.21 (a) $r^2 = .9878$. 98.78% of the variation in worker hours can be explained by variation in lot size.
(b) $r^2_{adj} = .987$.

17.28 $r = +.827$.

17.30 $r = +.9939$.

17.36 Based on a residual analysis the model appears to be adequate.

17.38 Based on a residual analysis the model appears to be adequate.

17.43 The data have been collected for a single time period from a set of stores. There is no sequential nature about the set of stores. Therefore it is not necessary to compute the Durbin-Watson statistic.

17.47 $1.835 \le \mu_{YX} \le 2.249$.

17.49 $98.597 \le \mu_{YX} \le 103.223$.

17.53 (a) $1.447 \le Y_I \le 2.637$.
(b) This is an estimate for an individual response rather than an average predicted value.

17.55 (a) $92.20 \le Y_I \le 109.62$.
(b) This is an estimate for an individual response rather than an average predicted value.

17.59 $t = 4.653 > t_{10} = 1.8125$. Reject H_0. There is evidence of a linear relationship.

17.61 $t = 31.15 > t_{12} = 1.7823$. Reject H_0. There is evidence of a significant relationship between lot size and worker hours.

17.69 Max $h_i = .2333 < .3333$; max $|t_i^*| = 1.49 < 1.8331$; max $D_i = .369 < .743$. Thus there is no evidence that any observations should be removed from the model.

17.71 Max $h_i = .232 < .286$; observations 12 and 14 have large Studentized deleted residuals ($|t_i^*| = 2.967$ and $|t_i^*| = 2.057 > 1.7823$); Cook's D_i for these observations are .445 and .504 < .735. However, these are the largest D_i values. Thus, one might wish to consider these observations as being influential and delete them from the model; however, the model is an extremely good-fitting one with or without them.

17.75 (b) $b_0 = 21.9256$; $b_1 = +2.0687$.
(c) If the cars have no options, delivery time averages approximately 22 days; for each option ordered delivery time increases by 2.0687 days.
(d) 55.0248 days.
(e) $S_{YX} = 3.0448$.
(f) $r_2 = .9575$; 95.75% of the variation in delivery time can be explained by variation in the number of options ordered.
(g) $r = +.9785$.
(h) $r^2_{adj} = .955$.
(i) $53.1115 \le \mu_{YX} \le 56.9381$.
(j) $48.22 \le Y_I \le 61.83$.
(k) $t = 17.769 > t_{14} = 2.1448$. Reject H_0. There is evidence of a linear relationship.

(l) $+1.8187 \le \beta_1 \le +2.3187$.
(m) Based on a residual analysis the model appears to be adequate.
(n) h_i for observation $16 = .3096 > .25$. Therefore this observation is an influential point. However, $|t_i^*|$ for observation $16 = 1.50 < 1.76$, so that this observation does not adversely affect the model. The largest $|t_i^*| = 1.57 < 1.76$. The largest Cook's D_i is for observation $16 = .465 < .73$. Thus, we may conclude that there is insufficient reason to delete this observation from the model.

17.80 (a) $b_0 = +0.30$; $b_1 = +.00487$.
(b) The slope b_1 can be interpreted to mean that for each increase of one point in GMAT score, grade point index is predicted to increase by .00487 points (or for each increase of 100 points in GMAT score, grade point index is predicted to increase by 0.487 points). The Y intercept b_0 represents the portion of the grade point index that varies with factors other than GMAT score.
(c) $\hat{Y} = 3.222$.
(d) $S_{YX} = .158$.
(e) $r^2 = .793$; 79.3% of the variation in grade point index can be explained by variation in GMAT score.
(f) $r = +.891$.
(g) $r^2_{\text{adj}} = .781$.
(h) $t = 8.31 > t_{18} = 2.1009$. Reject H_0. There is evidence of a linear relationship between GMAT score and grade point index.
(i) $3.143 \le \mu_{YX} \le 3.301$.
(j) $2.881 \le Y_I \le 3.563$.
(k) $+.00364 \le \beta_1 \le +.00610$.
(l) There is widespread scatter in the residual plot, but there is no pattern in the relationship between the residuals and X_i.
(m) $h_{20} = .305 > .20$, so observation 20 is an influential point. However, the $|t^*|$ values for observations 3, 5, 14, and 20, equal to 1.879, 2.216, 2.228, and 1.961, are each > 1.7396. We note that Cook's D for observation $20 = .729 > .720$. Thus, we need to explore an alternative model that does not include observation 20. For this model $r^2 = .819$ and the fitted model is $\hat{Y}_i = -.0799 + .0055081\, X_i$. The removal of the twentieth observation has changed the regression coefficients somewhat and has produced a slightly better fitting model. However, the range for the predictions of GPI that can be made has been narrowed from GMAT scores of 536 to 759 to GMAT scores of 536 to 718.

Chapter 18

18.2 (a) $\hat{Y}_i = -.02686 + .79116\, X_{1_i} + .60484\, X_{2_i}$.
(b) For a given midsole impact, each increase of one unit in forefoot impact absorbing capability results in an increase in the long-term ability to absorb shock by .79116 units. For a given forefoot impact absorbing capability, each increase in one unit in midsole impact results in an increase in the long-term ability to absorb shock by .60484 units.

18.4 (a) $\hat{Y}_i = 156.4 + 13.081 X_{1_i} + 16.795_{2_i}$ where X_1 = radio and television advertising in thousands of dollars and X_2 = newspaper advertising in thousands of dollars.
(b) Holding the amount of newspaper advertising constant, for each increase of \$1,000 in radio and television advertising, sales are predicted to increase by \$13,081. Holding the amount of radio and television advertising constant, for each increase of \$1,000 in newspaper advertising, sales are predicted to increase by \$16,795.

18.8 $\hat{Y}_i = \$753.95$

18.11 (a) $r^2_{Y.12} = .9421$; 94.21% of the variation in the long-term ability to absorb shock can be explained by variation in forefoot impact-absorbing capability and variation in midsole impact.
(b) $r^2_{\text{adj}} = .9263$.

18.13 (a) $r^2_{Y.12} = .809$; 80.9% of the variation in sales can be explained by variation in radio and television advertising and in newspaper advertising.
(b) $r^2_{\text{adj}} = .789$.

18.15 (a) $r^2_{Y.12} = .490$; 49.0% of the variation in standby hours can be explained by variation in the total staff and remote hours.
(b) $r^2_{\text{adj}} = .446$.

18.17 There appears to be a curvilinear relationship in the plot of the residuals against both radio and television advertising and against newspaper advertising. Thus, curvilinear terms for each of these explanatory variables should be considered for inclusion in the model.

18.20 $F = 97.69 > F_{U(2,12)} = 3.89$. Reject H_0. At least one of the independent variables is related to the dependent variable Y.

18.22 $F = 40.16 > F_{U(2,19)} = 3.52$. Reject H_0. There is a significant relationship between sales and radio and television advertising and newspaper advertising.

18.25 $F = 157.98 > F_{U(1,12)} = 4.75$ and $F = 71.09 > F_{U(1,12)} = 4.75$ or $t = 12.57 > t_{12} = 2.1788$ and $t = 8.43 > t_{12} = 2.1788$; each independent variable makes a significant contribution in the presence of the other variable, and both variables should be included in the model.

18.27 $t = 7.43 > t_{19} = 2.093$ and $t = 5.67 > t_{19} = 2.093$. Each explanatory variable makes a significant contribution and should be included in the model.

18.30 $.654 \le \beta_1 \le .928$.

18.32 $9.399 \le \beta_1 \le 16.763$.

18.35 $r^2_{Y1.2} = .9294$; holding the effect of midsole impact constant, 92.94% of the variation in the long-term ability to absorb shock can be explained by variation in forefoot impact-absorbing capability. $r^2_{Y2.1} = .8556$; holding the effect of forefoot impact-absorbing capability constant, 85.56% of the variation in the long-term ability to absorb shock can be explained by variation in midsole impact.

18.37 $r^2_{Y1.2} = .7442$. For a given amount of newspaper advertising, 74.42% of the variation in sales can be explained by variation in radio and television advertising. $r^2_{Y2.1} = .6283$. For a given amount of radio and television advertising, 62.83% of the variation in sales can be explained by variation in newspaper advertising.

18.40 (a) $\hat{Y}_i = 20.2983 + .03908(X_i - \bar{X}) - .0145(X_i - \bar{X})^2$.

(c) $\hat{Y}_i = 18.52$.

(d) $F = 141.46 > F_{U(2,25)} = 3.39$. Reject H_0. There is evidence of a curvilinear relationship between speed and miles per gallon.

(e) $r^2_{Y.12} = .9188$; 91.88% of the variation in miles per gallon can be explained by the curvilinear relationship with speed.

(f) $r^2_{adj} = .912$

(g) The model indicates only positive residuals for intermediate values of $(X_i - \bar{X})$

(h) $t = -16.63 < -2.0595$. The curvilinear effect makes a significant contribution to the model.

18.47 (a) $\hat{Y}_i = 1.30 + .074X_{1_i} + .45X_{2_i}$, where X_1 = shelf space, $X_2 = 0$ for back of aisle and 1 for front of aisle.

(b) Holding constant the effect of aisle location, for each additional foot of shelf space, predicted sales increase by .074 hundreds of dollars ($7.40). For a given amount of shelf space, a front-of-aisle location increases average sales by .45 hundreds of dollars ($45).

(c) $\hat{Y}_i = 1.892$ or $189.20.

(d) $F = 28.562 > F_{U(2,9)} = 4.26$. Reject H_0. There is evidence of a relationship between sales and the two independent variables.

(e) $r^2_{Y.12} = .864$; 86.4% of the variation in sales can be explained by variation in shelf space and variation in aisle location.

(f) $r^2_{adj} = .834$.

(g) $r^2_{Y.12} = .864$ while $r^2 = .684$; $r^2_{adj} = .834$ as compared to .652. The inclusion of the aisle-location variable has resulted in an increase in the r^2.

(h) $t = 6.72 > t_9 = 2.2622$ and $t = 3.45 > t_9 = 2.2622$. Therefore each explanatory variable makes a significant contribution and should be included in the model.

(i) $[.049 \leq \beta_1 \leq .099]$.
$[.155 \leq \beta_2 \leq .745]$.

(j) The slope $b_1 = .074$ is unchanged in this case because shelf space and aisle location are not correlated.

(k) $r^2_{Y1.2} = .834$. Holding constant the effect of aisle location, 83.4% of the variation in sales can be explained by variation in shelf space; $r^2_{Y2.1} = .569$; for a given amount of shelf space 56.9% of the variation in sales can be explained by variation in aisle location.

(l) That the slope of shelf space and sales is the same regardless of whether the aisle location is front or back.

(m) Based on a residual analysis, the model appears adequate.

18.52 (a) $\hat{Y}_i = 1.20 + .082X_{1_i} + .75X_{2_i} - .024X_{1_i}X_{2_i}$, where X_1 = shelf space, $X_2 = 0$ for back of aisle and 1 for front of aisle.

(b) $t = -1.03 > t_8 = -2.306$. Do not reject H_0. No evidence that the interaction term makes a contribution to the model. Therefore we should use the $\hat{Y}_i = b_0 + b_1X_{1_i}, + b_2X_{2_i}$, model.

18.59 Observations 1, 2, 13, 14 are influential (h_1 and $h_2 = .2924$, and h_{13} and $h_{14} = .3564 > .2727$). Observations 2, 4, 7, and 13 had an effect on the model ($t^*_2 = 2.44$, $t^*_4 = 1.98$, $t^*_7 = 1.87$, $t^*_{13} = 1.96 >$ $|1.7341|$. The largest values for Cook's D_i are .652 for observation 2 and .619 for observation 13, which were less than .82. However, since these values were substantially above the D_i values for the other observations and were also found to have had an effect on the model and to be influential, a model was studied with observations 2 and 13 deleted. In this model, r^2 was .88, while $b_0 = -24.7$, $b_1 = 14.932$, and $b_2 = 19.107$.

18.62 None of the observations exceeds $h_i = .214$. However, observation 14 has $|t^*_i| = 2.30 > 1.7081$. Since the largest Cook's D_i is for observation 27 ($D_{27} = .238$) and the second largest is for observation 14 ($D_{14} = .131$), each < .81, there is insufficient evidence that any observations should be removed from the model.

18.69 (a) Max $h_i = .40 < .50$; max $|t^*_i| = 2.16 > 1.8331$; max $D_i = .28 <$ .845. Observation 5, which has $|t^*_5| = 2.16$, has $D_5 = .183$. Thus there is insufficient evidence that this observation should be deleted from the model.

(b) The model in Problem 17.69 did not have any observations to be deleted; it did not have any significant $|t^*_i|$ observations.

18.78 (a) $\hat{Y}_i = 16.19567 + 2.03779\, X_{1_i} + 0.56262\, X_{2_i}$.

(b) For a given shipping mileage, each additional option ordered increases delivery time by 2.03779 days. For a given number of options, each 100-mile increase in shipping mileage increases delivery time by 0.56262 days.

(c) 41.07 days.

(d) $F = 270.58 > F_{U(2,13)} = 3.81$, so we may reject H_0. There is a significant relationship between delivery time and number of options ordered and shipping mileage.

(e) $r^2_{Y.12} = .9765$; 97.65% of the variation in delivery time can be explained by variation in the number of options ordered and the shipping mileage.

(f) $r^2_{\text{adj}} = .973$.

(g) $F = 509.16 > F_{U(1,13)} = 4.67$ and $F = 10.53 > F_{U(1,13)} = 4.67$. Each independent variable makes a significant contribution and should be included in the model.

(h) $.18794 \le \beta_2 \le .93730$.

(j) $r^2_{Y1.2} = .9751$. For a given shipping mileage, 97.51% of the variation in delivery time can be explained by variation in the number of options.

$r^2_{Y2.1} = .4474$. For a given number of options, 44.74% of the variation in delivery time can be explained by variation in shipping mileage.

(k) $VIF_1 = 1.0$. $VIF_2 = 1.0$. There is no reason to suspect the existence of multicollinearity.

(l) On the basis of a residual analysis the model appears adequate.

(m) Max $h_i = .333 < .375$; no reason to indicate influential observations. $|t^*_i|$ for observations 13 (2.91) and 16 (2.07) are each

> 1.7709. $D_{13} = .253$ and $D_{16} = .511 < .826$, so that there appears to be insufficient evidence to delete any observations. However, since these D_i values are far in excess of all others, the counterargument can be made that there should be further investigation into whether these observations should be deleted from the model.

Chapter 19

19.3 (b) and (c)

Pd.	Year	Y_i	3-Year Moving Total	3-Year Moving Avg.	(W = .50) ϵ_i
1	1972	1.45	**	**	1.45
2	1973	1.55	4.61	1.54	1.50
3	1974	1.61	4.76	1.59	1.55
4	1975	1.60	4.95	1.65	1.58
5	1976	1.74	5.26	1.75	1.66
6	1977	1.92	5.61	1.87	1.79
7	1978	1.95	5.91	1.97	1.87
8	1979	2.04	6.05	2.02	1.95
9	1980	2.06	5.90	1.97	2.01
10	1981	1.80	5.59	1.86	1.90
11	1982	1.73	5.30	1.77	1.82
12	1983	1.77	5.40	1.80	1.79
13	1984	1.90	5.49	1.83	1.85
14	1985	1.82	5.37	1.79	1.83
15	1986	1.65	5.20	1.73	1.74
16	1987	1.73	5.26	1.75	1.74
17	1988	1.88	5.61	1.87	1.81
18	1989	2.00	5.96	1.99	1.90
19	1990	2.08	5.96	1.99	1.99
20	1991	1.88	**	**	1.94

(d) $\hat{Y}_{1994} = \epsilon_{1993} = 1.94$.

19.5 (b) $\hat{Y}_i = 0.216 + .139\, X_i$, where origin = 1970 and X units = 1 year.

(c) 1993: 3.413
1994: 3.552
1995: 3.691
1996: 3.830

19.13 (b) $\log \hat{Y}_i = -.51371 + .0532\, X_i$ or $\hat{Y}_i = (.3062)\,(1.1303)^{X_i}$ where origin = 1970 and X units = 1 year.

(c) 1993: 5.12
1994: 5.79
1995: 6.55
1996: 7.40

19.15 (a) 39.8
(b) 62.2

19.20 $\hat{Y}_{n+j} = 3.597 + (j)(0.259)$

19.26 $\hat{Y}_{n+j} = 5.076 + (j)(0.289)$

19.27 (a) 5.
(b) 6.
(c) The most recent five observed values—Y_{36}, Y_{37}, Y_{38}, Y_{39}, and Y_{40}.
(d) $\hat{Y}_i = \hat{\omega} + \hat{\psi}_1 Y_{i-1} + \hat{\psi}_2 Y_{i-2} + \ldots + \hat{\psi}_5 Y_{i-5}$.
(e) $\hat{Y}_{n+j} = \hat{\omega} + \hat{\psi}_1 \hat{Y}_{n+j-1} + \hat{\psi}_2 \hat{Y}_{n+j-2} + \ldots + \hat{\psi}_5 \hat{Y}_{n+j-5}$.

19.31 First-order model is chosen.

19.37 Third-order model is chosen.

19.39 $MAD_{LT} = 0.09$; $MAD_{A1} = 0.11$

19.45 $MAD_{ET} = 0.25$; $MAD_{A3} = 0.21$

19.46 (c) $\hat{Y}_i = 244.909 + 1.192X_i$, where origin = Jan. 15, 1985, and X units = 1 month.

Year	Month	S_i	$\hat{Y}_i$	Forecast
1995	Jan.	0.941	387.97	364.884
	Feb.	0.985	389.16	383.165
	Mar.	1.072	390.35	418.389
	Apr.	0.964	391.54	377.389
	May	0.964	392.73	378.620
	Jun.	1.045	393.93	411.640
	Jul.	1.033	395.12	407.969
	Aug.	1.027	396.31	407.136
	Sep.	0.985	397.50	391.649
	Oct.	0.964	398.70	384.482
	Nov.	1.024	399.89	409.323
	Dec.	0.997	401.08	399.923
1996	Jan.	0.941	402.27	378.339
	Feb.	0.985	403.46	397.250
	Mar.	1.072	404.66	433.722
	Apr.	0.964	405.85	391.177
	May	0.964	407.04	392.412
	Jun.	1.045	408.23	426.589
	Jul.	1.033	409.42	422.740
	Aug.	1.027	410.62	421.832
	Sep.	0.985	411.81	405.744
	Oct.	0.964	413.00	398.278
	Nov.	1.024	414.19	423.967
	Dec.	0.997	415.39	414.187

(e) Obtaining the cyclical relatives (C_i) for 1993 and 1994.

Year	Month	Y_i	S_i	$T_iC_iI_i$	$\hat{Y}_i$	C_iI_i	Weighted Moving Total	C_i
1993	Jan.	354.00	0.941	376.39	359.35	1.047	4.055	1.014
	Feb.	365.00	0.985	370.71	360.55	1.028	4.107	1.027
	Mar.	389.00	1.072	362.93	361.74	1.003	3.601	0.900
	Apr.	198.00	0.964	205.43	362.93	0.566	3.178	0.794
	May	366.00	0.964	379.64	364.12	1.043	3.670	0.918
	Jun.	389.00	1.045	372.26	365.32	1.019	3.982	0.995
	Jul.	341.00	1.033	330.26	366.51	0.901	3.915	0.979
	Aug.	413.00	1.027	402.02	367.70	1.093	4.153	1.038
	Sep.	387.00	0.985	392.78	368.89	1.065	4.299	1.075
	Oct.	384.00	0.964	398.19	370.08	1.076	4.309	1.077
	Nov.	415.00	1.024	405.43	371.28	1.092	4.143	1.036
	Dec.	328.00	0.997	328.95	372.47	0.883	4.045	1.011
1994	Jan.	417.00	0.941	443.38	373.66	1.187	4.362	1.090
	Feb.	408.00	0.985	414.38	374.85	1.105	4.430	1.107
	Mar.	416.00	1.072	388.12	376.04	1.032	4.264	1.066
	Apr.	398.00	0.964	412.93	377.24	1.095	4.310	1.077
	May	397.00	0.964	411.80	378.43	1.088	4.410	1.103
	Jun.	452.00	1.045	432.55	379.62	1.139	4.443	1.111
	Jul.	423.00	1.033	409.68	380.81	1.076	4.453	1.113
	Aug.	456.00	1.027	443.88	382.01	1.162	4.343	1.086
	Sep.	356.00	0.985	361.32	383.20	0.943	4.340	1.085
	Oct.	479.00	0.964	496.71	384.39	1.292	4.604	1.151
	Nov.	425.00	1.024	415.20	385.58	1.077	4.740	1.185
	Dec.	499.00	0.997	500.44	386.77	1.294	**	**

APPENDIX A

Review of Arithmetic and Algebra

A.1 Rules for Arithmetic Operations

The following is a summary of various rules for arithmetic operations with each rule illustrated by a numerical example.

Rule	Example
1. $a + b = c$ and $b + a = c$	$2 + 1 = 3$ and $1 + 2 = 3$
2. $a + (b + c) = (a + b) + c$	$5 + (7 + 4) = (5 + 7) + 4 = 16$
3. $a - b = c$ but $b - a \neq c$	$9 - 7 = 2$ but $7 - 9 = -2$
4. $a \times b = b \times a$	$7 \times 6 = 6 \times 7 = 42$
5. $a \times (b + c) = (a \times b) + (a \times c)$	$2 \times (3 + 5) = (2 \times 3) + (2 \times 5) = 16$
6. $a \div b \neq b \div a$	$12 \div 3 \neq 3 \div 12$
7. $\frac{a + b}{c} = \frac{a}{c} + \frac{b}{c}$	$\frac{7 + 3}{2} = \frac{7}{2} + \frac{3}{2} = 5$
8. $\frac{a}{b + c} \neq \frac{a}{b} + \frac{a}{c}$	$\frac{3}{4 + 5} \neq \frac{3}{4} + \frac{3}{5}$
9. $\frac{1}{a} + \frac{1}{b} = \frac{b + a}{ab}$	$\frac{1}{3} + \frac{1}{5} = \frac{5 + 3}{(3)(5)} = \frac{8}{15}$
10. $\frac{a}{b} \times \frac{c}{d} = \frac{a \times c}{b \times d}$	$\frac{2}{3} \times \frac{6}{7} = \frac{2 \times 6}{3 \times 7} = \frac{12}{21}$
11. $\frac{a}{b} \div \frac{c}{d} = \frac{a \times d}{b \times c}$	$\frac{5}{8} \div \frac{3}{7} = \frac{5 \times 7}{8 \times 3} = \frac{35}{24}$

A.2 Rules for Algebra: Exponents and Square Roots

The following is a summary of various rules for arithmetic operations with each rule illustrated by a numerical example:

Rule	Example
1. $X^a \cdot X^b = X^{a+b}$	$4^2 \cdot 4^3 = 4^5$
2. $(X^a)^b = X^{ab}$	$(2^2)^3 = 2^6$
3. $(X^a/X^b) = X^{a-b}$	$\frac{3^5}{3^3} = 3^2$
4. $\frac{X^a}{X^a} = X^0 = 1$	$\frac{3^4}{3^4} = 3^0 = 1$
5. $\sqrt{XY} = \sqrt{X}\sqrt{Y}$	$\sqrt{(25)(4)} = \sqrt{25}\sqrt{4} = 10$
6. $\sqrt{\frac{X}{Y}} = \frac{\sqrt{X}}{\sqrt{Y}}$	$\sqrt{\frac{16}{100}} = \frac{\sqrt{16}}{\sqrt{100}} = .40$

APPENDIX B

Summation Notation

Since the operation of addition occurs so frequently in statistics, the special symbol Σ (sigma) is used to denote "taking the sum of." Suppose, for example, that we have a set of n values for some variable X. The expression $\sum_{i=1}^{n} X_i$ means that these n values are to be added together. Thus

$$\sum_{i=1}^{n} X_i = X_1 + X_2 + X_3 + \cdots + X_n$$

The use of the summation notation can be illustrated in the following problem. Suppose that we have five observations of a variable X: $X_1 = 2$, $X_2 = 0$, $X_3 = -1$, $X_4 = 5$, and $X_5 = 7$. Thus

$$\sum_{i=1}^{5} X_i = X_1 + X_2 + X_3 + X_4 + X_5 = 2 + 0 + (-1) + 5 + 7 = 13$$

In statistics we are also frequently involved with summing the squared values of a variable. Thus

$$\sum_{i=1}^{n} X_i^2 = X_1^2 + X_2^2 + X_3^2 + \cdots + X_n^2$$

and, in our example, we have

$$\begin{aligned}\sum_{i=1}^{5} X_i^2 &= X_1^2 + X_2^2 + X_3^2 + X_4^2 + X_5^2 \\ &= 2^2 + 0^2 + (-1)^2 + 5^2 + 7^2 \\ &= 4 + 0 + 1 + 25 + 49 \\ &= 79\end{aligned}$$

We should realize here that $\sum_{i=1}^{n} X_i^2$, the summation of the squares, is not the same as $\left(\sum_{i=1}^{n} X_i\right)^2$, the square of the sum, that is,

$$\sum_{i=1}^{n} X_i^2 \neq \left(\sum_{i=1}^{n} X_i\right)^2$$

In our example the summation of squares is equal to 79. This is not equal to the square of the sum, which is $13^2 = 169$.

Another frequently used operation involves the summation of the product. That is, suppose that we have two variables, X and Y, each having n observations. Then,

$$\sum_{i=1}^{n} X_iY_i = X_1Y_1 + X_2Y_2 + X_3Y_3 + \cdots + X_nY_n$$

Continuing with our previous example, suppose that there is also a second variable Y whose five values are $Y_1 = 1$, $Y_2 = 3$, $Y_3 = -2$, $Y_4 = 4$, and $Y_5 = 3$. Then,

$$\begin{aligned}\sum_{i=1}^{5} X_iY_i &= X_1Y_1 + X_2Y_2 + X_3Y_3 + X_4Y_4 + X_5Y_5\\ &= (2)(1) + (0)(3) + (-1)(-2) + (5)(4) + (7)(3)\\ &= 2 + 0 + 2 + 20 + 21\\ &= 45\end{aligned}$$

In computing $\sum_{i=1}^{n} X_iY_i$ we must realize that the first value of X is multiplied by the first value of Y, the second value of X is multiplied by the second value of Y, and so on. These cross products are then summed in order to obtain the desired result. However, we should note here that the summation of cross products is not equal to the product of the individual sums, that is,

$$\sum_{i=1}^{n} X_iY_i \neq \left(\sum_{i=1}^{n} X_i\right)\left(\sum_{i=1}^{n} Y_i\right)$$

In our example, $\sum_{i=1}^{5} X_i = 13$ and $\sum_{i=1}^{5} Y_i = 1 + 3 + (-2) + 4 + 3 = 9$ so that $\left(\sum_{i=1}^{5} X_i\right)\left(\sum_{i=1}^{5} Y_i\right) = (13)(9) = 117$. This is not the same as $\sum_{i=1}^{n} X_iY_i$, which equals 45.

Before studying the four basic rules of performing operations with summation notation, it would be helpful to present the values for each of the five observations of X and Y in a tabular format.

Observation	X_i	Y_i
1	2	1
2	0	3
3	−1	−2
4	5	4
5	7	3
	$\sum_{i=1}^{5} X_i = 13$	$\sum_{i=1}^{5} Y_i = 9$

Rule 1: The summation of the values of two variables is equal to the sum of the values of each summed variable.

$$\sum_{i=1}^{n}(X_i + Y_i) = \sum_{i=1}^{n} X_i + \sum_{i=1}^{n} Y_i$$

Thus, in our example,

$$\sum_{i=1}^{5}(X_i + Y_i) = (2+1) + (0+3) + (-1+(-2)) + (5+4) + (7+3)$$

$$= 3 + 3 + (-3) + 9 + 10$$

$$= 22 = \sum_{i=1}^{5} X_i + \sum_{i=1}^{5} Y_i = 13 + 9 = 22$$

Rule 2: The summation of a difference between the values of two variables is equal to the difference between the summed values of the variables.

$$\sum_{i=1}^{n}(X_i - Y_i) = \sum_{i=1}^{n} X_i - \sum_{i=1}^{n} Y_i$$

Thus, in our example,

$$\sum_{i=1}^{5}(X_i - Y_i) = (2-1) + (0-3) + (-1-(-2)) + (5-4) + (7-3)$$

$$= 1 + (-3) + 1 + 1 + 4$$

$$= 4 = \sum_{i=1}^{5} X_i - \sum_{i=1}^{5} Y_i = 13 - 9 = 4$$

Rule 3: The summation of a constant times a variable is equal to that constant times the summation of the values of the variable.

$$\sum_{i=1}^{n} cX_i = c\sum_{i=1}^{n} X_i$$

where c is a constant.

Thus, in our example, if $c = 2$,

$$\sum_{i=1}^{5} cX_i = \sum_{i=1}^{5} 2X_i = (2)(2) + (2)(0) + (2)(-1) + (2)(5) + (2)(7)$$

$$= 4 + 0 + (-2) + 10 + 14$$

$$= 26 = 2\sum_{i=1}^{5} X = (2)(13) = 26$$

Rule 4: A constant summed n times will be equal to n times the value of the constant.

$$\sum_{i=1}^{n} c = nc$$

where c is a constant. Thus, if the constant $c = 2$ is summed five times, we would have

$$\sum_{i=1}^{5} c = 2 + 2 + 2 + 2 + 2$$

$$= 10 = (5)(2) = 10$$

To illustrate how these summation rules are used, we may demonstrate one of the mathematical properties pertaining to the average or arithmetic mean (see Section 4.5.3), that is,

$$\sum_{i=1}^{n} (X_i - \overline{X}) = 0$$

This property states that the summation of the differences between each observation and the arithmetic mean is zero. This can be proven mathematically in the following manner.

1. From Equation (4.1),

$$\overline{X} = \frac{\sum_{i=1}^{n} X_i}{n}$$

   Thus, using summation rule 2, we have

$$\sum_{i=1}^{n} (X_i - \overline{X}) = \sum_{i=1}^{n} X_i - \sum_{i=1}^{n} \overline{X}$$

2. Since, for any fixed set of data, $\overline{X}$ can be considered a constant, from summation rule 4 we have

$$\sum_{i=1}^{n} \overline{X} = n\overline{X}$$

Therefore,

$$\sum_{i=1}^{n}(X_i - \bar{X}) = \sum_{i=1}^{n} X_i - n\bar{X}$$

3. However, from Equation (4.1), since

$$\bar{X} = \frac{\sum_{i=1}^{n} X_i}{n} \quad \text{then} \quad n\bar{X} = \sum_{i=1}^{n} X_i$$

Therefore,

$$\sum_{i=1}^{n}(X_i - \bar{X}) = \sum_{i=1}^{n} X_i - \sum_{i=1}^{n} X_i$$

Thus we have shown that

$$\sum_{i=1}^{n}(X_i - \bar{X}) = 0$$

Problem

Suppose that there are six observations for the variables X and Y such that $X_1 = 2$, $X_2 = 1$, $X_3 = 5$, $X_4 = -3$, $X_5 = 1$, $X_6 = -2$, and $Y_1 = 4$, $Y_2 = 0$, $Y_3 = -1$, $Y_4 = 2$, $Y_5 = 7$, and $Y_6 = -3$. Compute each of the following:

(a) $\sum_{i=1}^{6} X_i$

(b) $\sum_{i=1}^{6} Y_i$

(c) $\sum_{i=1}^{6} X_i^2$

(d) $\sum_{i=1}^{6} Y_i^2$

(e) $\sum_{i=1}^{6} X_i Y_i$

(f) $\sum_{i=1}^{6} (X_i + Y_i)$

(g) $\sum_{i=1}^{6} (X_i - Y_i)$

(h) $\sum_{i=1}^{6} \left(X_i - 3Y_i + 2X_i^2\right)$

(i) $\sum_{i=1}^{6} (cX_i)$, where $c = -1$

(j) $\sum_{i=1}^{6} (X_i - 3Y_i + c)$, where $c = +3$

References

1. Bashaw, W. L., *Mathematics for Statistics* (New York: Wiley, 1969).
2. Lanzer, P. *Video Review of Arithmetic* (Roslyn Heights, NY: Video Aided Instruction, 1990).
3. Levine, D. *Video Review of Statistics* (Roslyn Heights, NY: Video Aided Instruction, 1989).
4. Shane, H., *Video Review of Elementary Algebra* (Roslyn Heights, NY: Video Aided Instruction, 1990).

APPENDIX C

Statistical Symbols and Greek Alphabet

C.1 Statistical Symbols

$+$ add

$-$ subtract

$=$ equals

$\cong$ approximately equal to

$>$ greater than

$\geq$ or $\geqslant$ greater than or equal to

$\times$ multiply

$\div$ divide

$\neq$ not equal

$<$ less than

$\leq$ or $\leqslant$ less than or equal to

C.2 Greek Alphabet

Greek Letter	Greek Name	English Equivalent	Greek Letter	Greek Name	English Equivalent
Α α	Alpha	a	Ν ν	Nu	n
Β β	Beta	b	Ξ ξ	Xi	x
Γ γ	Gamma	g	Ο ο	Omicron	ŏ
Δ δ	Delta	d	Π π	Pi	p
Ε ϵ	Epsilon	ĕ	Ρ ρ	Rho	r
Ζ ζ	Zeta	z	Σ σ	Sigma	s
Η η	Eta	ē	Τ τ	Tau	t
Θ θ	Theta	th	Υ υ	Upsilon	u
Ι ι	Iota	i	Φ ϕ	Phi	ph
Κ κ	Kappa	k	Χ χ	Chi	ch
Λ λ	Lambda	l	Ψ ψ	Psi	ps
Μ μ	Mu	m	Ω ω	Omega	ō

APPENDIX D

Special Data Sets (for Collaborative Learning Mini-Case Projects)

D.1 Special Data Set 1

Data file C&U.DAT

This is a file containing data for the states of Texas, North Carolina, and Pennsylvania regarding out-of-state tuition payments. There are 60 schools in Texas, 45 in North Carolina, and 90 in Pennsylvania. To use the file, note the following codes for the data:

- Out-of-state tuition charges (in $000)
- Type of institution: 1 = private; 2 = public
- Location of institution: 1 = rural; 2 = suburb; 3 = urban
- Academic calendar: 1 = qtr.; 2 = sem.; 3 = tri.; 4 = 414; 5 = other
- Classification of institution: 1 = NLA; 2 = NU; 3 = RLA; 4 = RU; 5 = SS
- State: 1 = Texas; 2 = North Carolina; 3 = Pennsylvania

Moreover, for academic calendar: Qtr. = quarter, Sem. = semester; Tri. = Trimester; 414 = 4–1–4. For institutional classification: NLA = national liberal arts school; NU = national university; RLA = regional liberal arts school; RU = regional university; SS = school with special focus.

School	Tuition (in $000)	Type	Setting	Calendar	Class
Texas					
Abilene Christian Univ.	7.2	Private	Suburb	Sem.	RU
Angelo State Univ.	4.9	Public	Urban	Sem.	RU
Austin College	10.7	Private	Urban	414	NLA
Baylor Univ.	10.4	Private	Urban	Sem.	NU
Concordia Lutheran College	6.4	Private	Urban	Sem.	RLA
Dallas Baptist Univ.	4.8	Private	Urban	414	RLA
East Texas Baptist Univ.	4.7	Private	Urban	414	RLA
East Texas State Univ.	4.6	Private	Urban	Qtr.	RU
Hardin Simmons Univ.	6.0	Private	Urban	Sem.	RU
Houston Baptist Univ.	5.4	Private	Urban	Qtr.	RU
Howard Payne Univ.	4.8	Private	Rural	Sem.	RLA
Huston-Tillotson College	4.7	Private	Urban	Sem.	RLA

School	Tuition (in $000)	Type	Setting	Calendar	Class
Texas (Continued)					
Incarnate Word College	8.3	Private	Urban	Sem.	RLA
Jarvis Christian College	3.8	Private	Rural	Sem.	RLA
Lamar Univ.	4.8	Public	Urban	Sem.	RU
LeTourneau Univ.	8.3	Private	Urban	Sem.	RLA
Lubbock Christian Univ.	6.4	Private	Urban	Sem.	RLA
McMurry Univ.	6.6	Private	Urban	414	RLA
Midwestern State Univ.	4.5	Public	Urban	Sem.	RU
Our Lady of the Lake	8.0	Private	Urban	Sem.	RU
Paul Quinn College	3.6	Private	Urban	Sem.	RLA
Prairie View A&M Univ.	2.4	Public	Rural	Sem.	RU
Rice Univ.	8.5	Private	Urban	Sem.	NU
St. Edward's Univ.	8.8	Private	Urban	Sem.	RU
St. Mary's Univ.	7.7	Private	Urban	Sem.	RU
Sam Houston State Univ.	4.9	Public	Rural	Sem.	RU
Schreiner College	8.6	Private	Rural	414	RLA
Southern Methodist Univ.	12.0	Private	Suburb	Sem.	NU
Southwest Texas State Univ.	4.9	Public	Urban	Sem.	RU
Southwestern Adventist	7.0	Private	Rural	Sem.	RLA
Southwestern Univ.	11.0	Private	Suburb	Sem.	RLA
Stephen F. Austin State U.	4.9	Public	Rural	Sem.	RU
Sul Ross State Univ.	3.9	Public	Rural	Sem.	RU
Tarleton State Univ.	4.9	Public	Rural	Sem.	RU
Texas A&I Univ.	4.4	Public	Rural	Sem.	RU
Texas A&M Univ.	4.9	Public	Urban	Sem.	NU
Texas A&M at Galveston	4.9	Public	Urban	Sem.	RU
Texas Christian Univ.	8.0	Private	Urban	Sem.	NC
Texas College	3.6	Private	Urban	Sem.	RLA
Texas Lutheran College	7.4	Private	Urban	Sem.	RLA
Texas Southern Univ.	7.9	Public	Urban	Sem.	RU
Texas Tech Univ.	4.9	Public	Urban	Sem.	RU
Texas Wesleyan Univ.	5.8	Private	Urban	Sem.	RU
Texas Woman's Univ.	3.9	Public	Urban	Sem.	RU
Trinity Univ.	11.6	Private	Urban	Sem.	RU
U. of Dallas	10.3	Private	Suburb	Sem.	NLA
U. of Houston	3.4	Public	Urban	Sem.	NU
U. of Houston-Downtown	3.9	Public	Urban	Sem.	RU
U. of Mary Hardin-Baylor	5.0	Private	Suburb	Sem.	RLA
U. of North Texas	3.9	Public	Urban	Sem.	RU
U. of St. Thomas	8.0	Private	Urban	Sem.	RU
U. of Texas at Arlington	3.5	Public	Suburb	Sem.	NU
U. of Texas at Austin	4.9	Public	Urban	Sem.	NU
U. of Texas at Dallas	5.8	Public	Suburb	Sem.	RU
U. of Texas at El Paso	4.1	Public	Urban	Sem.	RU
U. of Texas-Pan American	3.5	Public	Urban	Sem.	RU
U. of Texas, San Antonio	3.9	Public	Urban	Sem.	RU
Wayland Baptist Univ.	4.8	Private	Urban	414	RU
West Texas State Univ.	5.9	Public	Rural	Sem.	RU
Wiley College	3.6	Private	Urban	Sem.	RLA

School	Tuition (in $000)	Type	Location	Calendar	Class
North Carolina					
Appalachian State Univ.	6.5	Public	Rural	Sem.	RU
Barber Scotia College	4.0	Private	Urban	Sem.	RLA
Barton College	7.1	Private	Urban	Sem.	RLA
Belmont Abbey College	8.3	Private	Suburb	Sem.	RLA
Bennett College	5.4	Private	Urban	Sem.	RLA
Campbell Univ.	7.6	Private	Rural	Sem.	RU
Catawba College	9.0	Private	Suburb	Sem.	RLA
Davidson College	15.7	Private	Suburb	Sem.	NLA
Duke Univ.	16.7	Private	Urban	Sem.	NU
East Carolina Univ.	6.4	Public	Urban	Sem.	RU
Elizabeth City State Univ.	5.0	Public	Rural	Sem.	RU
Elon College	8.5	Private	Suburb	414	RU
Fayetteville State Univ.	5.7	Public	Urban	Sem.	RU
Gardner Webb Univ.	7.7	Private	Rural	Sem.	RU
Greensboro College	7.2	Private	Urban	Sem.	RLA
Guilford College	12.4	Private	Suburb	Sem.	NLA
High Point Univ.	7.1	Private	Urban	Sem.	RU
Johnson C. Smith Univ.	5.5	Private	Urban	Sem.	RLA
Lenoir-Rhyne College	9.7	Private	Suburb	Sem.	RLA
Livingston College	4.4	Private	Urban	Sem.	RLA
Mars Hill College	7.0	Private	Rural	Sem.	RLA
Meredith College	6.3	Private	Urban	Sem.	RU
Methodist College	8.3	Private	Urban	Sem.	RLA
N.C. A&T Univ.	6.9	Public	Urban	Sem.	RU
N.C. Central Univ.	5.7	Public	Urban	Sem.	RU
N.C. School of the Arts	7.6	Public	Urban	Tri.	SS
N.C. State Univ.	7.9	Public	Urban	Sem.	NU
N.C. Wesleyan College	7.9	Private	Suburb	Sem.	RLA
Pembroke State Univ.	6.0	Public	Rural	Sem.	RU
Pfeiffer College	8.2	Private	Rural	Sem.	RLA
Queens College	10.4	Private	Suburb	Sem.	RLA
St. Andrews Presbyterian Coll.	9.9	Private	Rural	414	RLA
St. Augustine's College	3.9	Private	Urban	Sem.	RU
Salem College	9.8	Private	Urban	414	RU
Shaw Univ.	8.2	Private	Urban	Sem.	RU
U. of N.C. at Asheville	5.6	Public	Urban	Sem.	RU
U. of N.C. at Chapel Hill	7.9	Public	Urban	Sem.	NU
U. of N.C. at Charlotte	6.4	Public	Urban	Sem.	RU
U. of N.C. at Greensboro	7.4	Public	Urban	Sem.	NU
U. of N.C. at Wilmington	7.0	Public	Suburb	Sem.	RU
Wake Forest Univ.	13.0	Private	Suburb	Sem.	RU
Warren Wilson College	8.7	Private	Rural	Sem.	RLA
Western Carolina Univ.	6.4	Public	Rural	Sem.	RU
Wingate College	6.7	Private	Suburb	Sem.	RU
Winston–Salem State Univ.	7.4	Private	Urban	Sem.	RU

School	Tuition (in $000)	Type	Location	Calendar	Class
Pennsylvania					
Albright College	14.9	Private	Suburb	414	NLA
Allegheny College	16.4	Private	Rural	Sem.	NLA
Allentown College St. Francis	9.3	Private	Rural	Sem.	RLA
Alvernia College	8.4	Private	Urban	Sem.	RLA
Beaver College	12.3	Private	Suburb	Sem.	RU
Bloomsburg Univ. of Pa.	4.9	Public	Rural	Sem.	RU
Bryn Mawr College	17.1	Private	Suburb	Sem.	NLA
Bucknell Univ.	17.7	Private	Rural	Sem.	NLA
Cabrini College	9.7	Private	Suburb	Sem.	RLA
California Univ. of Pa.	6.3	Public	Rural	Sem.	RU
Carlow College	9.4	Private	Urban	Sem.	RLA
Carnegie Mellon Univ.	17.0	Private	Urban	Sem.	NU
Cedar Crest College	13.7	Private	Suburb	Sem.	RLA
Chatham College	12.6	Private	Urban	414	NLA
Chestnut Hill College	9.5	Private	Suburb	Sem.	NLA
Cheyney Univ.	2.7	Public	Rural	Sem.	RU
Clarion Univ. of Pa.	4.4	Public	Rural	Sem.	RU
College Misericordia	10.0	Private	Suburb	Sem.	RLA
Delaware Valley College	11.6	Private	Suburb	Sem.	RU
Dickinson College	17.7	Private	Urban	Sem.	NLA
Drexel Univ.	11.7	Private	Urban	Qtr.	NU
Duquesne Univ.	10.6	Private	Urban	Sem.	NU
East Stroudsburg Univ.	4.9	Public	Rural	Sem.	RU
Eastern College	10.6	Private	Suburb	Sem.	RLA
Edinboro Univ.	6.1	Public	Rural	Sem.	RU
Elizabethtown College	13.2	Private	Suburb	Sem.	RU
Franklin & Marshall College	22.3	Private	Urban	Sem.	NLA
Gannon Univ.	9.1	Private	Urban	Sem.	RU
Geneva College	8.9	Private	Suburb	Sem.	RLA
Gettysburg College	18.9	Private	Rural	Sem.	NLA
Grove City College	5.0	Private	Rural	Sem.	RU
Gwynedd Mercy College	10.2	Private	Suburb	Sem.	RU
Hahnemann Univ.	8.3	Private	Urban	Sem.	NU
Haverford College	17.9	Private	Suburb	Sem.	NLA
Holy Family College	8.3	Private	Urban	Sem.	RLA
Immaculata College	9.4	Private	Suburb	Sem.	RLA
Indiana Univ. of Pa.	4.9	Public	Rural	Sem.	RU
Juniata College	14.2	Private	Rural	Sem.	NLA
King's College	10.2	Private	Urban	Sem.	RU
Kutztown Univ.	8.3	Public	Rural	Sem.	RU
LaRoche College	8.4	Private	Suburb	Sem.	RU
LaSalle Univ.	11.5	Private	Urban	Sem.	RU
Lafayette College	17.9	Private	Urban	Sem.	NLA
Lebanon Valley College	13.3	Private	Rural	Sem.	NLA
Lehigh Univ.	17.8	Private	Urban	Sem.	NU
Lincoln Univ.	4.0	Public	Rural	Tri.	RLA
Lock Haven Univ. of Pa.	6.1	Public	Rural	Sem.	RU

School	Tuition (in $000)	Type	Location	Calendar	Class
Pennsylvania (Continued)					
Lycoming College	13.0	Private	Rural	414	RLA
Mansfield Univ.	5.5	Public	Rural	Sem.	RU
Marywood College	9.6	Private	Suburb	Sem.	RU
Mercyhurst College	9.3	Private	Suburb	Other	RU
Messiah College	9.7	Private	Suburb	414	RU
Millersville Univ. of Pa.	7.7	Public	Suburb	414	RU
Moore College Art and Design	13.5	Private	Urban	Sem.	SS
Moravian College	14.3	Private	Urban	Sem.	RU
Muhlenberg College	16.4	Private	Suburb	Sem.	NLA
Neumann College	10.0	Private	Suburb	Sem.	NLA
Penn State U. at Erie	10.1	Public	Urban	Sem.	RU
Penn State U. at College Park	9.6	Public	Urban	Sem.	NU
Philadelphia College Tex. & Sci.	11.7	Private	Suburb	Sem.	RU
Point Park College	9.3	Private	Urban	Sem.	RU
Robert Morris College	6.0	Private	Urban	Sem.	SS
Rosemont College	10.7	Private	Suburb	Sem.	RLA
St. Francis College	10.3	Private	Rural	Sem.	RU
St. Joseph's Univ.	11.9	Private	Urban	Sem.	RU
St. Vincent College	10.2	Private	Rural	Sem.	RLA
Seton Hill College	10.2	Private	Suburb	Sem.	RLA
Shippensburg Univ.	6.1	Public	Rural	Sem.	RU
Slippery Rock Univ.	7.7	Public	Rural	Sem.	RU
Susquehanna Univ.	15.6	Private	Rural	Sem.	RU
Swarthmore College	18.3	Private	Suburb	Sem.	NLA
Temple Univ.	9.1	Public	Urban	Sem.	NU
Thiel College	10.4	Private	Rural	Sem.	RLA
Univ. of the Arts	11.2	Private	Urban	Sem.	SS
Univ. of Pennsylvania	16.1	Private	Urban	Sem.	NU
Univ. of Pittsburgh	10.3	Public	Urban	Sem.	NU
U. of Pittsburgh at Bradford	9.7	Public	Rural	Sem.	RLA
U. of Pittsburgh at Greensburg	10.3	Public	Suburb	Sem.	RLA
U. of Pittsburgh at Johnstown	9.7	Public	Suburb	Sem.	RU
U. of Scranton	10.7	Private	Urban	414	RU
Ursinus College	14.1	Private	Suburb	Sem.	NLA
Villanova Univ.	15.2	Private	Suburb	Sem.	RU
Washington and Jefferson College	15.4	Private	Suburb	414	NLA
Waynesburg College	8.4	Private	Rural	Sem.	RU
West Chester Univ.	6.1	Public	Suburb	Sem.	RU
Westminster College	11.4	Private	Rural	414	RLA
Widener Univ.	11.7	Private	Suburb	Sem.	RU
Wilkes Univ.	9.5	Private	Suburb	Sem.	RU
Wilson College	11.4	Private	Rural	Tri.	RLA
York College of Pa.	4.8	Private	Suburb	Sem.	RU

Source: "America's Best Colleges, 1994 College Guide," *U.S. News & World Report,* extracted from College Counsel 1993 of Natick, Mass. Reprinted by special permission, *U.S. News & World Report,* © 1993 by *U.S. News & World Report* and by College Counsel.

Data file
CEREAL.DAT

D.2 Special Data Set 2

This is a file containing data for a sample of $n = 84$ ready-to-eat cereals. To use the file, note the following:

- Type product (H for high fiber, M for moderate fiber, L for low fiber)
- Cost per serving (in cents)
- Weight per serving (in ounces)
- Calories per serving
- Sugar per serving (in grams)

Ready-to-Eat Cereal	Type Product	Cost	Weight	Calories	Sugar
All-Bran with Extra Fiber	H	38	2.0	100	0
Fiber One	H	34	2.0	120	0
Bran Buds	H	21	1.5	110	12
100% Bran	H	23	1.5	110	9
All-Bran Original	H	23	1.5	110	8
100% Organic Raisin Bran Flakes	H	51	2.0	180	17
Uncle Sam	H	28	2.0	220	0
Bran Flakes	H	23	1.5	140	8
Bran Flakes	H	21	1.5	140	8
Crunchy Corn Bran	H	28	1.5	140	9
Fiberwise	H	43	1.5	140	8
Raisin Bran	H	30	2.0	180	21
Multi Bran Chex	H	25	1.5	140	9
Shredded Wheat 'N Bran	H	28	1.5	140	0
Fruit & Fibre Peaches, Raisins Almonds & Oat Clusters	H	38	2.0	180	12
Fruitful Bran	H	43	2.0	180	18
Raisin Bran	H	29	1.75	160	16
Shredded Wheat	H	29	1.67	160	0
Cracklin' Oat Bran	H	44	2.0	220	14
Skinner's Raisin Bran	H	27	2.0	220	12
Frosted Wheat Squares	H	40	2.0	200	12
Grape-Nuts	H	29	2.0	220	6
Shredded Wheat Spoon Size	M	28	1.5	140	0
Common Sense Oat Bran	M	27	1.33	130	8
Frosted Mini-Wheats	M	28	1.25	130	8
Grape-Nuts Flakes	M	19	1.25	110	6
Whole Grain Total	M	23	1.0	100	3
Whole Grain Wheat Chex	M	24	1.5	150	5
Whole Grain Wheaties	M	16	1.0	100	3
Total Raisin Bran	M	30	1.5	140	14
Raisin Nut Bran	M	37	2.0	220	16
Raisin Squares	M	38	2.0	180	12
Oatios with Extra Oat Bran	M	27	1.0	110	2
Nutri-Grain Almond Raisin	M	47	2.0	210	11
Crispy Wheats 'N Raisins	M	24	1.33	130	13
Life	M	26	1.5	150	9
Multi Grain Cheerios	M	24	1.0	100	6
Oat Squares	M	34	2.0	200	12
Mueslix Crispy Blend	M	50	2.25	240	19
Cheerios	M	15	.8	90	0
Cinnamon Oat Squares	M	35	2.0	220	14

Ready to Eat Cereal	Type Product	Cost	Weight	Calories	Sugar
Clusters	M	44	2.0	220	25
100% Natural Whole Grain with Raisins (Low Fat)	M	39	2.0	220	14
Honey Bunches of Oats with Almonds	M	29	1.5	180	9
Low-Fat Granola with Raisins	M	36	1.67	180	14
Basic 4	M	38	1.75	170	11
Just Right with Fruit & Nuts	M	42	1.75	190	12
Apple Cinnamon Cheerios	M	25	1.33	150	13
Honey Nut Cheerios	M	26	1.33	150	13
Oatmeal Raisin Crisp	M	47	2.5	260	20
Nut & Honey Crunch	M	29	1.5	170	12
Puffed Rice	L	13	.5	50	0
Puffed Wheat	L	15	.5	50	0
Kix	L	16	.67	70	2
Honey-Comb	L	17	.75	80	8
Corn Flakes	L	10	1.0	100	2
Product 19	L	23	1.0	100	3
Rice Chex	L	17	1.0	100	2
Apple Jacks	L	23	1.0	110	14
Cocoa Puffs	L	21	1.0	110	13
Cookie-Crisp Chocolate Chip	L	25	1.0	110	13
Corn Chex	L	19	1.0	110	3
Corn Pops	L	22	1.0	110	12
Crispix	L	21	1.0	110	3
Froot Loops	L	21	1.0	110	13
Lucky Charms	L	22	1.0	110	12
Marshmallow Alpha-Bits	L	23	1.0	110	14
Rice Krispies	L	18	1.0	110	3
Special K	L	22	1.0	110	3
Teenage Mutant Ninja Turtles	L	23	1.0	110	11
Total Corn Flakes	L	28	1.0	110	3
Trix	L	24	1.0	110	12
Fruity Pebbles	L	24	1.25	130	14
Wheaties Honey Gold	L	20	1.33	130	13
Cocoa Krispies	L	28	1.33	150	15
Cocoa Pebbles	L	28	1.33	150	17
Frosted Flakes	L	20	1.33	150	15
Golden Grahams	L	27	1.33	150	12
Kenmei Rice Bran	L	22	1.33	150	5
Smacks	L	24	1.33	150	20
Triples	L	22	1.33	150	4
Cap'n Crunch	L	22	1.33	160	16
Cap'n Crunch's Crunch Berries	L	23	1.33	160	16
Cinnamon Toast Crunch	L	26	1.33	160	12

Source: Copyright 1992 by Consumers Union of United States, Inc., Yonkers, NY 10703. Adapted by permission from *Consumer Reports,* November 1992, pp. 693–695.

Data file
FRAGRAN.DAT

D.3 Special Data Set 3

This is a file containing data for a sample of $n = 83$ fragrances. To use the file, note the following:

- Gender (W = women and M = men with respective coded values of 1 and 2)
- Type of fragrance (P = perfume, C = cologne, O = other; the respective coded values are 1, 2, 3)
- Cost in dollars per ounce
- Intensity or strength of the fragrance
  (VS = very strong, S = strong, Me = medium, and Mi = mild;
  the respective coded values are 1, 2, 3, 4)

Fragrance	Gender	Type	Cost	Intensity
Gio	W	P	300	S
Cabochard	W	P	175	S
"Delicious"	W	P	190	S
Chanel No. 5	W	C	19	S
Obsession	W	P	180	S
Sublime	W	P	250	VS
Vivid	W	P	230	S
Chanel No. 5	W	O	32	S
Dune	W	P	185	S
Ninja	W	C	8	S
Safari	W	C	19	Mi
Soft Musk	W	C	7	Me
360	W	P	200	S
Tresor	W	P	220	Mi
Venzia	W	P	320	VS
Coco	W	P	215	S
Opium	W	O	30	VS
Oscar de la Renta	W	P	200	Mi
Volupte	W	P	220	Me
Wild Heart	W	C	9	S
"An Impression of Chanel # 5"	W	C	4	Me
Chanel No. 5	W	P	215	S
Chloe	W	P	170	Mi
Mesmerize	W	C	9	Me
Obsession	W	C	21	Mi
Aliage Sport Fragrance	W	C	12	Me
Chanel No. 5	W	O	19	Mi
Passion	W	P	185	S
Charlie	W	C	8	S
Realm Women	W	O	29	VS
Shalimar	W	P	205	VS

Fragrance	Gender	Type	Cost	Intensity
White Diamonds	W	P	225	VS
Wind Song	W	C	10	Me
Incognito	W	C	13	S
L'Air du Temps	W	P	260	Me
Ma Griffe	W	P	152	VS
Navy	W	C	11	S
White Linen	W	P	170	Mi
Samsara	W	P	210	Me
Tuscany per Donna	W	P	190	VS
Versus	W	O	22	S
Halston	W	P	272	S
Red	W	C	23	S
Catalyst	W	P	250	S
Escape	W	C	23	S
Liz Claiborne	W	P	135	VS
Chantilly	W	C	2	S
Amarige	W	P	280	S
Feminite du Bois	W	P	240	S
Opium	W	P	205	Me
Giorgio	W	C	23	VS
Primo	W	C	7	S
Anais Anais	W	P	210	Me
Chloe Narcisse	W	P	225	VS
Donna Karan New York	W	P	350	VS
Tribu	W	P	220	VS
Our Version of White Diamonds	W	C	8	VS
Calyx	W	C	26	Me
Jean Naté	W	C	7	S
White Diamonds	W	O	23	VS
Emeraude	W	C	6	Mi
Poison	W	C	19	VS
Joy	W	P	300	VS
Angelfire	W	C	9	Mi
Caliente	W	C	11	Mi
Dewberry	W	O	19	S
Drakkar Noir	M	O	15	VS
Lancer	M	C	4	S
Eternity for Men	M	C	12	VS
Realm for Men	M	C	29	S
Preferred Stock	M	C	9	S
Gravity	M	C	10	S
Obsession for Men	M	C	10	S
Escape for Men	M	O	21	S
Old Spice	M	C	2	S
Tribute	M	C	7	S
Egoiste	M	C	16	S
Stetson	M	C	7	Me
English Leather	M	C	3	Me
Safari for Men	M	O	14	VS
Aramis	M	C	9	S
Brut	M	C	4	VS
Polo	M	C	10	VS

Source: Copyright 1993 by Consumers Union of United States, Inc., Yonkers, NY 10703. Adapted by permission from *Consumer Reports,* December 1993, pp. 772–773.

Data file
CAMERA.DAT

D.4 Special Data Set 4

This is a file containing data for a sample of $n = 59$ compact 35mm cameras. To use the file, note the following:

- Type of camera (ML = multiple long focal length, MM = multiple medium focal length, MS = multiple short focal length, SA = automatic single focal length, and SF = fixed single focal length; the respective values are coded 1, 2, 3, 4, 5)
- Average price in dollars
- Weight in ounces
- Close-up ability or "smallest field"
- Range or longest distance to shoot a flash picture in feet
- Framing accuracy or percentage of image area in the print that can be seen in the viewfinder
- Number of 24-exposure rolls per battery or "battery life"

Camera Brand and Model	Average Price ($)	Weight (oz.)	Close-up Ability	Range (ft.)	Framing Accuracy (%)	Number of Rolls
Ricoh Mirai Zoom 3	265	20	6	18	85	31
Nikon Zoom-Touch 800	359	19	7	22	83	33
Olympus Infinity SuperZoom 3000	277	13	4	17	78	45
Pentax IQ Zoom 105-R	271	19	14	16	92	46
Konica AIBORG	339	21	7	18	80	36
Canon Sure Shot Megazoom 105	224	17	5	15	76	46
Yashica Zoomtec 105 Super	292	17	5	15	76	46
Samsung AF Zoom 1050	230	16	10	18	73	52
Nikon Zoom-Touch 400	171	14	10	15	74	38
Canon Sure Shot Megazoom 76	175	15	8	14	79	40
Nikon Zoom-Touch 500S	185	15	9	14	76	34
Sigma Zoom Super 70	200	18	13	19	71	41
Olympus Infinity Zoom 210	186	14	8	11	77	37
Yashica Zoomtec 70	199	13	16	14	79	30
Canon Sure Shot Tele Max	113	9	10	14	75	33
Chinon Auto 4001	150	14	16	14	81	46
Konica Big Mini BM-311Z	159	12	16	16	76	33
Samsung AF Zoom 777i	158	15	16	20	72	35
Ricoh Shotmaster Zoom Super	219	14	7	12	80	38
Leica C2-Zoom	430	13	11	17	89	36
Minolta Freedom 70c	184	11	11	14	68	23
Olympus Infinity Tele	133	11	13	18	82	44
Rokinon 35AFZ	150	13	16	13	85	10
Fuji Discovery Mini Dual QD Plus	185	8	11	11	82	30
Olympus Infinity Zoom 220 Panorama	217	14	13	16	74	36
Vivitar 320Z Series 1	121	12	20	10	81	21
Yashica Twintec	100	11	33	14	79	25
Minolta Freedom Dual C	115	11	19	10	84	48
Yashica T4	183	7	11	10	74	51
Olympus Infinity Stylus	130	7	11	12	78	33

Camera Brand and Model	Average Price ($)	Weight (oz.)	Close-up Ability	Range (ft.)	Framing Accuracy (%)	Number of Rolls
Contax T2	718	12	22	12	90	45
Minolta Freedom Escort	132	7	21	12	89	25
Konica Big Mini BM-201	171	8	11	10	84	49
Leica Mini	289	7	22	11	88	33
Nikon One-Touch 200	105	9	22	11	79	29
Canon Sure Shot Max	86	9	20	11	78	41
Ricoh Shotmaster AF-P	89	9	36	11	92	50
Canon Snappy LX	57	10	55	9	80	46
Konica MT-100	100	9	45	9	81	24
Ansco Mini AF	99	8	45	8	86	13
Kodak Star 835 AF	65	11	46	12	82	37
Minolta Freedom AF 35	92	10	34	8	81	35
Fuji Discovery 80	83	10	47	11	77	30
Chinon Splash AF-2	119	15	47	13	81	17
Kalimar Spirit AF	70	10	40	7	82	60
Kalimar Spirit 2	45	10	55	6	97	63
Samsung AF-200	79	11	55	6	74	53
Vivitar EZ200	57	10	40	7	56	33
Yashica J Mini	65	9	48	10	91	37
Chinon Auto GL-AF	85	10	47	11	75	45
Yashica Sensation	67	10	47	13	92	41
Konica TOP's	50	9	57	8	76	41
Kodak Star 935	62	10	43	9	79	39
Fuji DL-25	52	10	55	11	88	34
Ricoh Shotmaster FF	58	10	36	9	78	62
Chinon Auto GL-II	50	11	47	10	79	26
Olympus Trip Junior	52	9	38	11	67	37
Minolta Freedom 50N	58	9	55	10	83	32
Olympus Trip AF Super	76	10	47	10	64	29

Source: Copyright 1992 by Consumers Union of United States, Inc., Yonkers, NY 10703. Adapted by permission from *Consumer Reports,* December 1992, pp. 762–765.

APPENDIX E

Tables

TABLE E.1 Table of Random Numbers

Row	Column							
	00000 12345	00001 67890	11111 12345	11112 67890	22222 12345	22223 67890	33333 12345	33334 67890
01	49280	88924	35779	00283	81163	07275	89863	02348
02	61870	41657	07468	08612	98083	97349	20775	45091
03	43898	65923	25078	86129	78496	97653	91550	08078
04	62993	93912	30454	84598	56095	20664	12872	64647
05	33850	58555	51438	85507	71865	79488	76783	31708
06	97340	03364	88472	04334	63919	36394	11095	92470
07	70543	29776	10087	10072	55980	64688	68239	20461
08	89382	93809	00796	95945	34101	81277	66090	88872
09	37818	72142	67140	50785	22380	16703	53362	44940
10	60430	22834	14130	96593	23298	56203	92671	15925
11	82975	66158	84731	19436	55790	69229	28661	13675
12	39087	71938	40355	54324	08401	26299	49420	59208
13	55700	24586	93247	32596	11865	63397	44251	43189
14	14756	23997	78643	75912	83832	32768	18928	57070
15	32166	53251	70654	92827	63491	04233	33825	69662
16	23236	73751	31888	81718	06546	83246	47651	04877
17	45794	26926	15130	82455	78305	55058	52551	47182
18	09893	20505	14225	68514	46427	56788	96297	78822
19	54382	74598	91499	14523	68479	27686	46162	83554
20	94750	89923	37089	20048	80336	94598	26940	36858
21	70297	34135	53140	33340	42050	82341	44104	82949
22	85157	47954	32979	26575	57600	40881	12250	73742
23	11100	02340	12860	74697	96644	89439	28707	25815
24	36871	50775	30592	57143	17381	68856	25853	35041
25	23913	48357	63308	16090	51690	54607	72407	55538
26	79348	36085	27973	65157	07456	22255	25626	57054
27	92074	54641	53673	54421	18130	60103	69593	49464
28	06873	21440	75593	41373	49502	17972	82578	16364
29	12478	37622	99659	31065	83613	69889	58869	29571
30	57175	55564	65411	42547	70457	03426	72937	83792
31	91616	11075	80103	07831	59309	13276	26710	73000
32	78025	73539	14621	39044	47450	03197	12787	47709
33	27587	67228	80145	10175	12822	86687	65530	49325
34	16690	20427	04251	64477	73709	73945	92396	68263
35	70183	58065	65489	31833	82093	16747	10386	59293
36	90730	35385	15679	99742	50866	78028	75573	67257
37	10934	93242	13431	24590	02770	48582	00906	58595
38	82462	30166	79613	47416	13389	80268	05085	96666
39	27463	10433	07606	16285	93699	60912	94532	95632
40	02979	52997	09079	92709	90110	47506	53693	49892
41	46888	69929	75233	52507	32097	37594	10067	67327
42	53638	83161	08289	12639	08141	12640	28437	09268
43	82433	61427	17239	89160	19666	08814	37841	12847
44	35766	31672	50082	22795	66948	65581	84393	15890
45	10853	42581	08792	13257	61973	24450	52351	16602
46	20341	27398	72906	63955	17276	10646	74692	48438
47	54458	90542	77563	51839	52901	53355	83281	19177
48	26337	66530	16687	35179	46560	00123	44546	79896
49	34314	23729	85264	05575	96855	23820	11091	79821
50	28603	10708	68933	34189	92166	15181	66628	58599

TABLE E.1 *(Continued)*

	Column							
Row	00000 12345	00001 67890	11111 12345	11112 67890	22222 12345	22223 67890	33333 12345	33334 67890
51	66194	28926	99547	16625	45515	67953	12108	57846
52	78240	43195	24837	32511	70880	22070	52622	61881
53	00833	88000	67299	68215	11274	55624	32991	17436
54	12111	86683	61270	58036	64192	90611	15145	01748
55	47189	99951	05755	03834	43782	90599	40282	51417
56	76396	72486	62423	27618	84184	78922	73561	52818
57	46409	17469	32483	09083	76175	19985	26309	91536
58	74626	22111	87286	46772	42243	68046	44250	42439
59	34450	81974	93723	49023	58432	67083	36876	93391
60	36327	72135	33005	28701	34710	49359	50693	89311
61	74185	77536	84825	09934	99103	09325	67389	45869
62	12296	41623	62873	37943	25584	09609	63360	47270
63	90822	60280	88925	99610	42772	60561	76873	04117
64	72121	79152	96591	90305	10189	79778	68016	13747
65	95268	41377	25684	08151	61816	58555	54305	86189
66	92603	09091	75884	93424	72586	88903	30061	14457
67	18813	90291	05275	01223	79607	95426	34900	09778
68	38840	26903	28624	67157	51986	42865	14508	49315
69	05959	33836	53758	16562	41081	38012	41230	20528
70	85141	21155	99212	32685	51403	31926	69813	58781
71	75047	59643	31074	38172	03718	32119	69506	67143
72	30752	95260	68032	62871	58781	34143	68790	69766
73	22986	82575	42187	62295	84295	30634	66562	31442
74	99439	86692	90348	66036	48399	73451	26698	39437
75	20389	93029	11881	71685	65452	89047	63669	02656
76	39249	05173	68256	36359	20250	68686	05947	09335
77	96777	33605	29481	20063	09398	01843	35139	61344
78	04860	32918	10798	50492	52655	33359	94713	28393
79	41613	42375	00403	03656	77580	87772	86877	57085
80	17930	00794	53836	53692	67135	98102	61912	11246
81	24649	31845	25736	75231	83808	98917	93829	99430
82	79899	34061	54308	59358	56462	58166	97302	86828
83	76801	49594	81002	30397	52728	15101	72070	33706
84	36239	63636	38140	65731	39788	06872	38971	53363
85	07392	64449	17886	63632	53995	17574	22247	62607
86	67133	04181	33874	98835	67453	59734	76381	63455
87	77759	31504	32832	70861	15152	29733	75371	39174
88	85992	72268	42920	20810	29361	51423	90306	73574
89	79553	75952	54116	65553	47139	60579	09165	85490
90	41101	17336	48951	53674	17880	45260	08575	49321
91	36191	17095	32123	91576	84221	78902	82010	30847
92	62329	63898	23268	74283	26091	68409	69704	82267
93	14751	13151	93115	01437	56945	89661	67680	79790
94	48462	59278	44185	29616	76537	19589	83139	28454
95	29435	88105	59651	44391	74588	55114	80834	85686
96	28340	29285	12965	14821	80425	16602	44653	70467
97	02167	58940	27149	80242	10587	79786	34959	75339
98	17864	00991	39557	54981	23588	81914	37609	13128
99	79675	80605	60059	35862	00254	36546	21545	78179
00	72335	82037	92003	34100	29879	46613	89720	13274

Source: Partially extracted from The Rand Corporation, *A Million Random Digits with 100,000 Normal Deviates* (Glencoe, Ill.: The Free Press, 1955).

TABLE E.2 The Standardized Normal Distribution

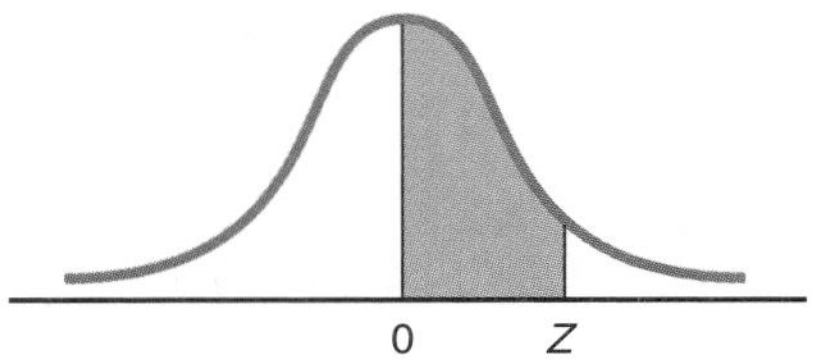

Entry represents area under the standardized normal distribution from the mean to Z

Z	.00	.01	.02	.03	.04	.05	.06	.07	.08	.09
0.0	.0000	.0040	.0080	.0120	.0160	.0199	.0239	.0279	.0319	.0359
0.1	.0398	.0438	.0478	.0517	.0557	.0596	.0636	.0675	.0714	.0753
0.2	.0793	.0832	.0871	.0910	.0948	.0987	.1026	.1064	.1103	.1141
0.3	.1179	.1217	.1255	.1293	.1331	.1368	.1406	.1443	.1480	.1517
0.4	.1554	.1591	.1628	.1664	.1700	.1736	.1772	.1808	.1844	.1879
0.5	.1915	.1950	.1985	.2019	.2054	.2088	.2123	.2157	.2190	.2224
0.6	.2257	.2291	.2324	.2357	.2389	.2422	.2454	.2486	.2518	.2549
0.7	.2580	.2612	.2642	.2673	.2704	.2734	.2764	.2794	.2823	.2852
0.8	.2881	.2910	.2939	.2967	.2995	.3023	.3051	.3078	.3106	.3133
0.9	.3159	.3186	.3212	.3238	.3264	.3289	.3315	.3340	.3365	.3389
1.0	.3413	.3438	.3461	.3485	.3508	.3531	.3554	.3577	.3599	.3621
1.1	.3643	.3665	.3686	.3708	.3729	.3749	.3770	.3790	.3810	.3830
1.2	.3849	.3869	.3888	.3907	.3925	.3944	.3962	.3980	.3997	.4015
1.3	.4032	.4049	.4066	.4082	.4099	.4115	.4131	.4147	.4162	.4177
1.4	.4192	.4207	.4222	.4236	.4251	.4265	.4279	.4292	.4306	.4319
1.5	.4332	.4345	.4357	.4370	.4382	.4394	.4406	.4418	.4429	.4441
1.6	.4452	.4463	.4474	.4484	.4495	.4505	.4515	.4525	.4535	.4545
1.7	.4554	.4564	.4573	.4582	.4591	.4599	.4608	.4616	.4625	.4633
1.8	.4641	.4649	.4656	.4664	.4671	.4678	.4686	.4693	.4699	.4706
1.9	.4713	.4719	.4726	.4732	.4738	.4744	.4750	.4756	.4761	.4767
2.0	.4772	.4778	.4783	.4788	.4793	.4798	.4803	.4808	.4812	.4817
2.1	.4821	.4826	.4830	.4834	.4838	.4842	.4846	.4850	.4854	.4857
2.2	.4861	.4864	.4868	.4871	.4875	.4878	.4881	.4884	.4887	.4890
2.3	.4893	.4896	.4898	.4901	.4904	.4906	.4909	.4911	.4913	.4916
2.4	.4918	.4920	.4922	.4925	.4927	.4929	.4931	.4932	.4934	.4936
2.5	.4938	.4940	.4941	.4943	.4945	.4946	.4948	.4949	.4951	.4952
2.6	.4953	.4955	.4956	.4957	.4959	.4960	.4961	.4962	.4963	.4964
2.7	.4965	.4966	.4967	.4968	.4969	.4970	.4971	.4972	.4973	.4974
2.8	.4974	.4975	.4976	.4977	.4977	.4978	.4979	.4979	.4980	.4981
2.9	.4981	.4982	.4982	.4983	.4984	.4984	.4985	.4985	.4986	.4986
3.0	.49865	.49869	.49874	.49878	.49882	.49886	.49889	.49893	.49897	.49900
3.1	.49903	.49906	.49910	.49913	.49916	.49918	.49921	.49924	.49926	.49929
3.2	.49931	.49934	.49936	.49938	.49940	.49942	.49944	.49946	.49948	.49950
3.3	.49952	.49953	.49955	.49957	.49958	.49960	.49961	.49962	.49964	.49965
3.4	.49966	.49968	.49969	.49970	.49971	.49972	.49973	.49974	.49975	.49976
3.5	.49977	.49978	.49978	.49979	.49980	.49981	.49981	.49982	.49983	.49983
3.6	.49984	.49985	.49985	.49986	.49986	.49987	.49987	.49988	.49988	.49989
3.7	.49989	.49990	.49990	.49990	.49991	.49991	.49992	.49992	.49992	.49992
3.8	.49993	.49993	.49993	.49994	.49994	.49994	.49994	.49995	.49995	.49995
3.9	.49995	.49995	.49996	.49996	.49996	.49996	.49996	.49996	.49997	.49997

TABLE E.3 Critical Values of t

For a particular number of degrees of freedom, entry represents the critical value of t corresponding to a specified upper tail area (α)

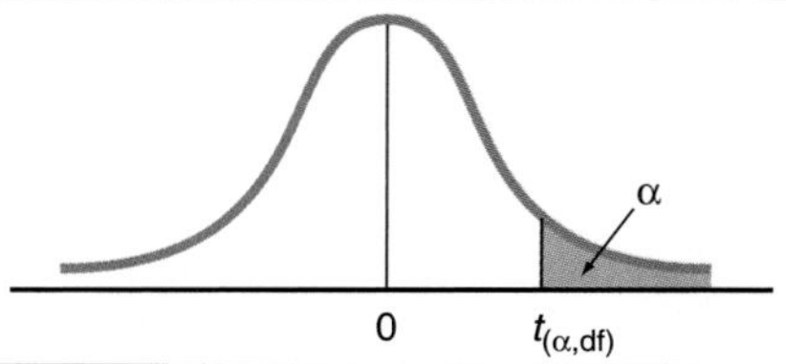

Degrees of Freedom	Upper Tail Areas (α) .25	.10	.05	.025	.01	.005
1	1.0000	3.0777	6.3138	12.7062	31.8207	63.6574
2	0.8165	1.8856	2.9200	4.3027	6.9646	9.9248
3	0.7649	1.6377	2.3534	3.1824	4.5407	5.8409
4	0.7407	1.5332	2.1318	2.7764	3.7469	4.6041
5	0.7267	1.4759	2.0150	2.5706	3.3649	4.0322
6	0.7176	1.4398	1.9432	2.4469	3.1427	3.7074
7	0.7111	1.4149	1.8946	2.3646	2.9980	3.4995
8	0.7064	1.3968	1.8595	2.3060	2.8965	3.3554
9	0.7027	1.3830	1.8331	2.2622	2.8214	3.2498
10	0.6998	1.3722	1.8125	2.2281	2.7638	3.1693
11	0.6974	1.3634	1.7959	2.2010	2.7181	3.1058
12	0.6955	1.3562	1.7823	2.1788	2.6810	3.0545
13	0.6938	1.3502	1.7709	2.1604	2.6503	3.0123
14	0.6924	1.3450	1.7613	2.1448	2.6245	2.9768
15	0.6912	1.3406	1.7531	2.1315	2.6025	2.9467
16	0.6901	1.3368	1.7459	2.1199	2.5835	2.9208
17	0.6892	1.3334	1.7396	2.1098	2.5669	2.8982
18	0.6884	1.3304	1.7341	2.1009	2.5524	2.8784
19	0.6876	1.3277	1.7291	2.0930	2.5395	2.8609
20	0.6870	1.3253	1.7247	2.0860	2.5280	2.8453
21	0.6864	1.3232	1.7207	2.0796	2.5177	2.8314
22	0.6858	1.3212	1.7171	2.0739	2.5083	2.8188
23	0.6853	1.3195	1.7139	2.0687	2.4999	2.8073
24	0.6848	1.3178	1.7109	2.0639	2.4922	2.7969
25	0.6844	1.3163	1.7081	2.0595	2.4851	2.7874
26	0.6840	1.3150	1.7056	2.0555	2.4786	2.7787
27	0.6837	1.3137	1.7033	2.0518	2.4727	2.7707
28	0.6834	1.3125	1.7011	2.0484	2.4671	2.7633
29	0.6830	1.3114	1.6991	2.0452	2.4620	2.7564
30	0.6828	1.3104	1.6973	2.0423	2.4573	2.7500
31	0.6825	1.3095	1.6955	2.0395	2.4528	2.7440
32	0.6822	1.3086	1.6939	2.0369	2.4487	2.7385
33	0.6820	1.3077	1.6924	2.0345	2.4448	2.7333
34	0.6818	1.3070	1.6909	2.0322	2.4411	2.7284
35	0.6816	1.3062	1.6896	2.0301	2.4377	2.7238
36	0.6814	1.3055	1.6883	2.0281	2.4345	2.7195
37	0.6812	1.3049	1.6871	2.0262	2.4314	2.7154
38	0.6810	1.3042	1.6860	2.0244	2.4286	2.7116
39	0.6808	1.3036	1.6849	2.0227	2.4258	2.7079
40	0.6807	1.3031	1.6839	2.0211	2.4233	2.7045
41	0.6805	1.3025	1.6829	2.0195	2.4208	2.7012
42	0.6804	1.3020	1.6820	2.0181	2.4185	2.6981
43	0.6802	1.3016	1.6811	2.0167	2.4163	2.6951
44	0.6801	1.3011	1.6802	2.0154	2.4141	2.6923
45	0.6800	1.3006	1.6794	2.0141	2.4121	2.6896
46	0.6799	1.3002	1.6787	2.0129	2.4102	2.6870
47	0.6797	1.2998	1.6779	2.0117	2.4083	2.6846
48	0.6796	1.2994	1.6772	2.0106	2.4066	2.6822
49	0.6795	1.2991	1.6766	2.0096	2.4049	2.6800
50	0.6794	1.2987	1.6759	2.0086	2.4033	2.6778

TABLE E.3 *(Continued)*

Degrees of Freedom	*Upper Tail Areas* (α)					
	.25	*.10*	*.05*	*.025*	*.01*	*.005*
51	0.6793	1.2984	1.6753	2.0076	2.4017	2.6757
52	0.6792	1.2980	1.6747	2.0066	2.4002	2.6737
53	0.6791	1.2977	1.6741	2.0057	2.3988	2.6718
54	0.6791	1.2974	1.6736	2.0049	2.3974	2.6700
55	0.6790	1.2971	1.6730	2.0040	2.3961	2.6682
56	0.6789	1.2969	1.6725	2.0032	2.3948	2.6665
57	0.6788	1.2966	1.6720	2.0025	2.3936	2.6649
58	0.6787	1.2963	1.6716	2.0017	2.3924	2.6633
59	0.6787	1.2961	1.6711	2.0010	2.3912	2.6618
60	0.6786	1.2958	1.6706	2.0003	2.3901	2.6603
61	0.6785	1.2956	1.6702	1.9996	2.3890	2.6589
62	0.6785	1.2954	1.6698	1.9990	2.3880	2.6575
63	0.6784	1.2951	1.6694	1.9983	2.3870	2.6561
64	0.6783	1.2949	1.6690	1.9977	2.3860	2.6549
65	0.6783	1.2947	1.6686	1.9971	2.3851	2.6536
66	0.6782	1.2945	1.6683	1.9966	2.3842	2.6524
67	0.6782	1.2943	1.6679	1.9960	2.3833	2.6512
68	0.6781	1.2941	1.6676	1.9955	2.3824	2.6501
69	0.6781	1.2939	1.6672	1.9949	2.3816	2.6490
70	0.6780	1.2938	1.6669	1.9944	2.3808	2.6479
71	0.6780	1.2936	1.6666	1.9939	2.3800	2.6469
72	0.6779	1.2934	1.6663	1.9935	2.3793	2.6459
73	0.6779	1.2933	1.6660	1.9930	2.3785	2.6449
74	0.6778	1.4931	1.6657	1.9925	2.3778	2.6439
75	0.6778	1.2929	1.6654	1.9921	2.3771	2.6430
76	0.6777	1.2928	1.6652	1.9917	2.3764	2.6421
77	0.6777	1.2926	1.6649	1.9913	2.3758	2.6412
78	0.6776	1.2925	1.6646	1.9908	2.3751	2.6403
79	0.6776	1.2924	1.6644	1.9905	2.3745	2.6395
80	0.6776	1.2922	1.6641	1.9901	2.3739	2.6387
81	0.6775	1.2921	1.6639	1.9897	2.3733	2.6379
82	0.6775	1.2920	1.6636	1.9893	2.3727	2.6371
83	0.6775	1.2918	1.6634	1.9890	2.3721	2.6364
84	0.6774	1.2917	1.6632	1.9886	2.3716	2.6356
85	0.6774	1.2916	1.6630	1.9883	2.3710	2.6349
86	0.6774	1.2915	1.6628	1.9879	2.3705	2.6342
87	0.6773	1.2914	1.6626	1.9876	2.3700	2.6335
88	0.6773	1.2912	1.6624	1.9873	2.3695	2.6329
89	0.6773	1.2911	1.6622	1.9870	2.3690	2.6322
90	0.6772	1.2910	1.6620	1.9867	2.3685	2.6316
91	0.6772	1.2909	1.6618	1.9864	2.3680	2.6309
92	0.6772	1.2908	1.6616	1.9861	2.3676	2.6303
93	0.6771	1.2907	1.6614	1.9858	2.3671	2.6297
94	0.6771	1.2906	1.6612	1.9855	2.3667	2.6291
95	0.6771	1.2905	1.6611	1.9853	2.3662	2.6286
96	0.6771	1.2904	1.6609	1.9850	2.3658	2.6280
97	0.6770	1.2903	1.6607	1.9847	2.3654	2.6275
98	0.6770	1.2902	1.6606	1.9845	2.3650	2.6269
99	0.6770	1.2902	1.6604	1.9842	2.3646	2.6264
100	0.6770	1.2901	1.6602	1.9840	2.3642	2.6259
110	0.6767	1.2893	1.6588	1.9818	2.3607	2.6213
120	0.6765	1.2886	1.6577	1.9799	2.3578	2.6174
∞	0.6745	1.2816	1.6449	1.9600	2.3263	2.5758

TABLE E.4 Critical Values of χ^2

For a particular number of degrees of freedom, entry represents the critical value of χ^2 corresponding to a specified upper tail area (α)

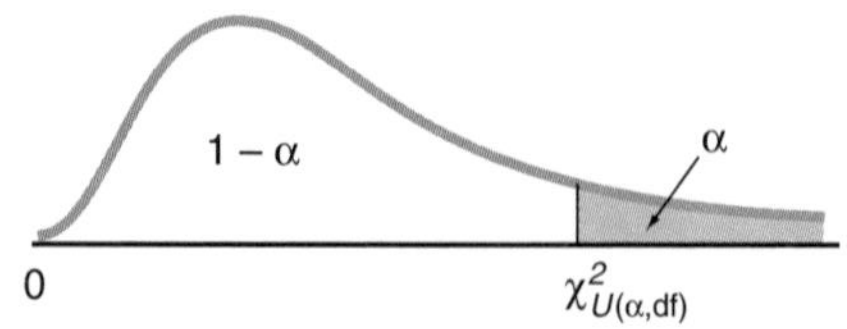

Degrees of Freedom	Upper Tail Areas (α)											
	.995	.99	.975	.95	.90	.75	.25	.10	.05	.025	.01	.005
1			0.001	0.004	0.016	0.102	1.323	2.706	3.841	5.024	6.635	7.879
2	0.010	0.020	0.051	0.103	0.211	0.575	2.773	4.605	5.991	7.378	9.210	10.597
3	0.072	0.115	0.216	0.352	0.584	1.213	4.108	6.251	7.815	9.348	11.345	12.838
4	0.207	0.297	0.484	0.711	1.064	1.923	5.385	7.779	9.488	11.143	13.277	14.860
5	0.412	0.554	0.831	1.145	1.610	2.675	6.626	9.236	11.071	12.833	15.086	16.750
6	0.676	0.872	1.237	1.635	2.204	3.455	7.841	10.645	12.592	14.449	16.812	18.548
7	0.989	1.239	1.690	2.167	2.833	4.255	9.037	12.017	14.067	16.013	18.475	20.278
8	1.344	1.646	2.180	2.733	3.490	5.071	10.219	13.362	15.507	17.535	20.090	21.955
9	1.735	2.088	2.700	3.325	4.168	5.899	11.389	14.684	16.919	19.023	21.666	23.589
10	2.156	2.558	3.247	3.940	4.865	6.737	12.549	15.987	18.307	20.483	23.209	25.188
11	2.603	3.053	3.816	4.575	5.578	7.584	13.701	17.275	19.675	21.920	24.725	26.757
12	3.074	3.571	4.404	5.226	6.304	8.438	14.845	18.549	21.026	23.337	26.217	28.299
13	3.565	4.107	5.009	5.892	7.042	9.299	15.984	19.812	22.362	24.736	27.688	29.819
14	4.075	4.660	5.629	6.571	7.790	10.165	17.117	21.064	23.685	26.119	29.141	31.319
15	4.601	5.229	6.262	7.261	8.547	11.037	18.245	22.307	24.996	27.488	30.578	32.801
16	5.142	5.812	6.908	7.962	9.312	11.912	19.369	23.542	26.296	28.845	32.000	34.267
17	5.697	6.408	7.564	8.672	10.085	12.792	20.489	24.769	27.587	30.191	33.409	35.718
18	6.265	7.015	8.231	9.390	10.865	13.675	21.605	25.989	28.869	31.526	34.805	37.156
19	6.844	7.633	8.907	10.117	11.651	14.562	22.718	27.204	30.144	32.852	36.191	38.582
20	7.434	8.260	9.591	10.851	12.443	15.452	23.828	28.412	31.410	34.170	37.566	39.997
21	8.034	8.897	10.283	11.591	13.240	16.344	24.935	29.615	32.671	35.479	38.932	41.401
22	8.643	9.542	10.982	12.338	14.042	17.240	26.039	30.813	33.924	36.781	40.289	42.796
23	9.260	10.196	11.689	13.091	14.848	18.137	27.141	32.007	35.172	38.076	41.638	44.181
24	9.886	10.856	12.401	13.848	15.659	19.037	28.241	33.196	36.415	39.364	42.980	45.559
25	10.520	11.524	13.120	14.611	16.473	19.939	29.339	34.382	37.652	40.646	44.314	46.928
26	11.160	12.198	13.844	15.379	17.292	20.843	30.435	35.563	38.885	41.923	45.642	48.290
27	11.808	12.879	14.573	16.151	18.114	21.749	31.528	36.741	40.113	43.194	46.963	49.645
28	12.461	13.565	15.308	16.928	18.939	22.657	32.620	37.916	41.337	44.461	48.278	50.993
29	13.121	14.257	16.047	17.708	19.768	23.567	33.711	39.087	42.557	45.722	49.588	52.336
30	13.787	14.954	16.791	18.493	20.599	24.478	34.800	40.256	43.773	46.979	50.892	53.672

For larger values of degrees of freedom (df) the expression $Z = \sqrt{2\chi^2} - \sqrt{2(df) - 1}$ may be used and the resulting upper tail area can be obtained from the table of the standardized normal distribution (Table E.2).

TABLE E.5 Critical Values of F

For a particular combination of numerator and denominator degrees of freedom, entry represents the critical values of F corresponding to a specified upper tail area (α).

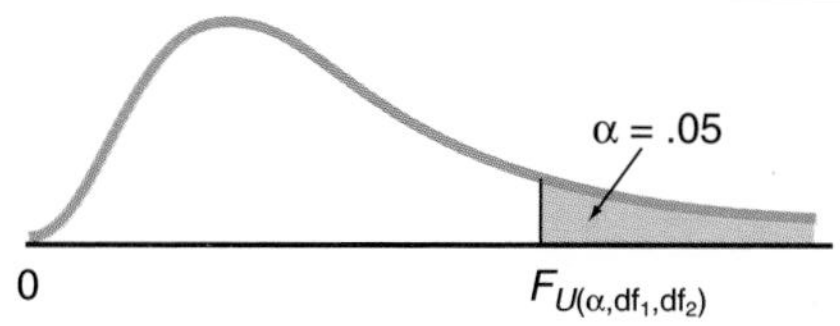

Denominator df_2	Numerator df_1: 1	2	3	4	5	6	7	8	9	10	12	15	20	24	30	40	60	120	∞
1	161.4	199.5	215.7	224.6	230.2	234.0	236.8	238.9	240.5	241.9	243.9	245.9	248.0	249.1	250.1	251.1	252.2	253.3	254.3
2	18.51	19.00	19.16	19.25	19.30	19.33	19.35	19.37	19.38	19.40	19.41	19.43	19.45	19.45	19.46	19.47	19.48	19.49	19.50
3	10.13	9.55	9.28	9.12	9.01	8.94	8.89	8.85	8.81	8.79	8.74	8.70	8.66	8.64	8.62	8.59	8.57	8.55	8.53
4	7.71	6.94	6.59	6.39	6.26	6.16	6.09	6.04	6.00	5.96	5.91	5.86	5.80	5.77	5.75	5.72	5.69	5.66	5.63
5	6.61	5.79	5.41	5.19	5.05	4.95	4.88	4.82	4.77	4.74	4.68	4.62	4.56	4.53	4.50	4.46	4.43	4.40	4.36
6	5.99	5.14	4.76	4.53	4.39	4.28	4.21	4.15	4.10	4.06	4.00	3.94	3.87	3.84	3.81	3.77	3.74	3.70	3.67
7	5.59	4.74	4.35	4.12	3.97	3.87	3.79	3.73	3.68	3.64	3.57	3.51	3.44	3.41	3.38	3.34	3.30	3.27	3.23
8	5.32	4.46	4.07	3.84	3.69	3.58	3.50	3.44	3.39	3.35	3.28	3.22	3.15	3.12	3.08	3.04	3.01	2.97	2.93
9	5.12	4.26	3.86	3.63	3.48	3.37	3.29	3.23	3.18	3.14	3.07	3.01	2.94	2.90	2.86	2.83	2.79	2.75	2.71
10	4.96	4.10	3.71	3.48	3.33	3.22	3.14	3.07	3.02	2.98	2.91	2.85	2.77	2.74	2.70	2.66	2.62	2.58	2.54
11	4.84	3.98	3.59	3.36	3.20	3.09	3.01	2.95	2.90	2.85	2.79	2.72	2.65	2.61	2.57	2.53	2.49	2.45	2.40
12	4.75	3.89	3.49	3.26	3.11	3.00	2.91	2.85	2.80	2.75	2.69	2.62	2.54	2.51	2.47	2.43	2.38	2.34	2.30
13	4.67	3.81	3.41	3.18	3.03	2.92	2.83	2.77	2.71	2.67	2.60	2.53	2.46	2.42	2.38	2.34	2.30	2.25	2.21
14	4.60	3.74	3.34	3.11	2.96	2.85	2.76	2.70	2.65	2.60	2.53	2.46	2.39	2.35	2.31	2.27	2.22	2.18	2.13
15	4.54	3.68	3.29	3.06	2.90	2.79	2.71	2.64	2.59	2.54	2.48	2.40	2.33	2.29	2.25	2.20	2.16	2.11	2.07
16	4.49	3.63	3.24	3.01	2.85	2.74	2.66	2.59	2.54	2.49	2.42	2.35	2.28	2.24	2.19	2.15	2.11	2.06	2.01
17	4.45	3.59	3.20	2.96	2.81	2.70	2.61	2.55	2.49	2.45	2.38	2.31	2.23	2.19	2.15	2.10	2.06	2.01	1.96
18	4.41	3.55	3.16	2.93	2.77	2.66	2.58	2.51	2.46	2.41	2.34	2.27	2.19	2.15	2.11	2.06	2.02	1.97	1.92
19	4.38	3.52	3.13	2.90	2.74	2.63	2.54	2.48	2.42	2.38	2.31	2.23	2.16	2.11	2.07	2.03	1.98	1.93	1.88
20	4.35	3.49	3.10	2.87	2.71	2.60	2.51	2.45	2.39	2.35	2.28	2.20	2.12	2.08	2.04	1.99	1.95	1.90	1.84
21	4.32	3.47	3.07	2.84	2.68	2.57	2.49	2.42	2.37	2.32	2.25	2.18	2.10	2.05	2.01	1.96	1.92	1.87	1.81
22	4.30	3.44	3.05	2.82	2.66	2.55	2.46	2.40	2.34	2.30	2.23	2.15	2.07	2.03	1.98	1.94	1.89	1.84	1.78
23	4.28	3.42	3.03	2.80	2.64	2.53	2.44	2.37	2.32	2.27	2.20	2.13	2.05	2.01	1.96	1.91	1.86	1.81	1.76
24	4.26	3.40	3.01	2.78	2.62	2.51	2.42	2.36	2.30	2.25	2.18	2.11	2.03	1.98	1.94	1.89	1.84	1.79	1.73
25	4.24	3.39	2.99	2.76	2.60	2.49	2.40	2.34	2.28	2.24	2.16	2.09	2.01	1.96	1.92	1.87	1.82	1.77	1.71
26	4.23	3.37	2.98	2.74	2.59	2.47	2.39	2.32	2.27	2.22	2.15	2.07	1.99	1.95	1.90	1.85	1.80	1.75	1.69
27	4.21	3.35	2.96	2.73	2.57	2.46	2.37	2.31	2.25	2.20	2.13	2.06	1.97	1.93	1.88	1.84	1.79	1.73	1.67
28	4.20	3.34	2.95	2.71	2.56	2.45	2.36	2.29	2.24	2.19	2.12	2.04	1.96	1.91	1.87	1.82	1.77	1.71	1.65
29	4.18	3.33	2.93	2.70	2.55	2.43	2.35	2.28	2.22	2.18	2.10	2.03	1.94	1.90	1.85	1.81	1.75	1.70	1.64
30	4.17	3.32	2.92	2.69	2.53	2.42	2.33	2.27	2.21	2.16	2.09	2.01	1.93	1.89	1.84	1.79	1.74	1.68	1.62
40	4.08	3.23	2.84	2.61	2.45	2.34	2.25	2.18	2.12	2.08	2.00	1.92	1.84	1.79	1.74	1.69	1.64	1.58	1.51
60	4.00	3.15	2.76	2.53	2.37	2.25	2.17	2.10	2.04	1.99	1.92	1.84	1.75	1.70	1.65	1.59	1.53	1.47	1.39
120	3.92	3.07	2.68	2.45	2.29	2.17	2.09	2.02	1.96	1.91	1.83	1.75	1.66	1.61	1.55	1.50	1.43	1.35	1.25
∞	3.84	3.00	2.60	2.37	2.21	2.10	2.01	1.94	1.88	1.83	1.75	1.67	1.57	1.52	1.46	1.39	1.32	1.22	1.00

TABLE E.5 (Continued)

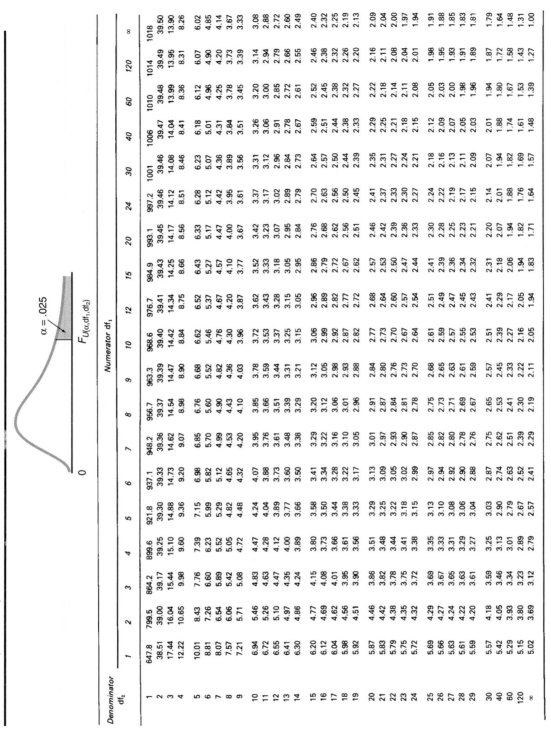

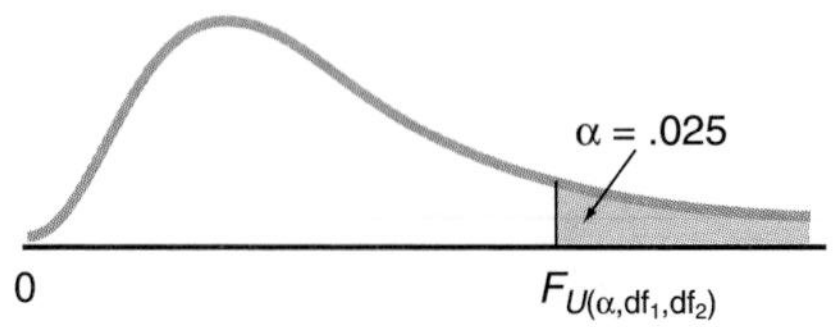

Denominator df_2	Numerator df_1																		
	1	2	3	4	5	6	7	8	9	10	12	15	20	24	30	40	60	120	∞
1	647.8	799.5	864.2	899.6	921.8	937.1	948.2	956.7	963.3	968.6	976.7	984.9	993.1	997.2	1001	1006	1010	1014	1018
2	38.51	39.00	39.17	39.25	39.30	39.33	39.36	39.37	39.39	39.40	39.41	39.43	39.45	39.46	39.46	39.47	39.48	39.49	39.50
3	17.44	16.04	15.44	15.10	14.88	14.73	14.62	14.54	14.47	14.42	14.34	14.25	14.17	14.12	14.08	14.04	13.99	13.95	13.90
4	12.22	10.65	9.98	9.60	9.36	9.20	9.07	8.98	8.90	8.84	8.75	8.66	8.56	8.51	8.46	8.41	8.36	8.31	8.26
5	10.01	8.43	7.76	7.39	7.15	6.98	6.85	6.76	6.68	6.62	6.52	6.43	6.33	6.28	6.23	6.18	6.12	6.07	6.02
6	8.81	7.26	6.60	6.23	5.99	5.82	5.70	5.60	5.52	5.46	5.37	5.27	5.17	5.12	5.07	5.01	4.96	4.90	4.85
7	8.07	6.54	5.89	5.52	5.29	5.12	4.99	4.90	4.82	4.76	4.67	4.57	4.47	4.42	4.36	4.31	4.25	4.20	4.14
8	7.57	6.06	5.42	5.05	4.82	4.65	4.53	4.43	4.36	4.30	4.20	4.10	4.00	3.95	3.89	3.84	3.78	3.73	3.67
9	7.21	5.71	5.08	4.72	4.48	4.32	4.20	4.10	4.03	3.96	3.87	3.77	3.67	3.61	3.56	3.51	3.45	3.39	3.33
10	6.94	5.46	4.83	4.47	4.24	4.07	3.95	3.85	3.78	3.72	3.62	3.52	3.42	3.37	3.31	3.26	3.20	3.14	3.08
11	6.72	5.26	4.63	4.28	4.04	3.88	3.76	3.66	3.59	3.53	3.43	3.33	3.23	3.17	3.12	3.06	3.00	2.94	2.88
12	6.55	5.10	4.47	4.12	3.89	3.73	3.61	3.51	3.44	3.37	3.28	3.18	3.07	3.02	2.96	2.91	2.85	2.79	2.72
13	6.41	4.97	4.35	4.00	3.77	3.60	3.48	3.39	3.31	3.25	3.15	3.05	2.95	2.89	2.84	2.78	2.72	2.66	2.60
14	6.30	4.86	4.24	3.89	3.66	3.50	3.38	3.29	3.21	3.15	3.05	2.95	2.84	2.79	2.73	2.67	2.61	2.55	2.49
15	6.20	4.77	4.15	3.80	3.58	3.41	3.29	3.20	3.12	3.06	2.96	2.86	2.76	2.70	2.64	2.59	2.52	2.46	2.40
16	6.12	4.69	4.08	3.73	3.50	3.34	3.22	3.12	3.05	2.99	2.89	2.79	2.68	2.63	2.57	2.51	2.45	2.38	2.32
17	6.04	4.62	4.01	3.66	3.44	3.28	3.16	3.06	2.98	2.92	2.82	2.72	2.62	2.56	2.50	2.44	2.38	2.32	2.25
18	5.98	4.56	3.95	3.61	3.38	3.22	3.10	3.01	2.93	2.87	2.77	2.67	2.56	2.50	2.44	2.38	2.32	2.26	2.19
19	5.92	4.51	3.90	3.56	3.33	3.17	3.05	2.96	2.88	2.82	2.72	2.62	2.51	2.45	2.39	2.33	2.27	2.20	2.13
20	5.87	4.46	3.86	3.51	3.29	3.13	3.01	2.91	2.84	2.77	2.68	2.57	2.46	2.41	2.35	2.29	2.22	2.16	2.09
21	5.83	4.42	3.82	3.48	3.25	3.09	2.97	2.87	2.80	2.73	2.64	2.53	2.42	2.37	2.31	2.25	2.18	2.11	2.04
22	5.79	4.38	3.78	3.44	3.22	3.05	2.93	2.84	2.76	2.70	2.60	2.50	2.39	2.33	2.27	2.21	2.14	2.08	2.00
23	5.75	4.35	3.75	3.41	3.18	3.02	2.90	2.81	2.73	2.67	2.57	2.47	2.36	2.30	2.24	2.18	2.11	2.04	1.97
24	5.72	4.32	3.72	3.38	3.15	2.99	2.87	2.78	2.70	2.64	2.54	2.44	2.33	2.27	2.21	2.15	2.08	2.01	1.94
25	5.69	4.29	3.69	3.35	3.13	2.97	2.85	2.75	2.68	2.61	2.51	2.41	2.30	2.24	2.18	2.12	2.05	1.98	1.91
26	5.66	4.27	3.67	3.33	3.10	2.94	2.82	2.73	2.65	2.59	2.49	2.39	2.28	2.22	2.16	2.09	2.03	1.95	1.88
27	5.63	4.24	3.65	3.31	3.08	2.92	2.80	2.71	2.63	2.57	2.47	2.36	2.25	2.19	2.13	2.07	2.00	1.93	1.85
28	5.61	4.22	3.63	3.29	3.06	2.90	2.78	2.69	2.61	2.55	2.45	2.34	2.23	2.17	2.11	2.05	1.98	1.91	1.83
29	5.59	4.20	3.61	3.27	3.04	2.88	2.76	2.67	2.59	2.53	2.43	2.32	2.21	2.15	2.09	2.03	1.96	1.89	1.81
30	5.57	4.18	3.59	3.25	3.03	2.87	2.75	2.65	2.57	2.51	2.41	2.31	2.20	2.14	2.07	2.01	1.94	1.87	1.79
40	5.42	4.05	3.46	3.13	2.90	2.74	2.62	2.53	2.45	2.39	2.29	2.18	2.07	2.01	1.94	1.88	1.80	1.72	1.64
60	5.29	3.93	3.34	3.01	2.79	2.63	2.51	2.41	2.33	2.27	2.17	2.06	1.94	1.88	1.82	1.74	1.67	1.58	1.48
120	5.15	3.80	3.23	2.89	2.67	2.52	2.39	2.30	2.22	2.16	2.05	1.94	1.82	1.76	1.69	1.61	1.53	1.43	1.31
∞	5.02	3.69	3.12	2.79	2.57	2.41	2.29	2.19	2.11	2.05	1.94	1.83	1.71	1.64	1.57	1.48	1.39	1.27	1.00

TABLE E.5 *(Continued)*

Denominator df_2	Numerator df_1: 1	2	3	4	5	6	7	8	9	10	12	15	20	24	30	40	60	120	∞
1	4052	4999.5	5403	5625	5764	5859	5928	5982	6022	6056	6106	6157	6209	6235	6261	6287	6313	6339	6366
2	98.50	99.00	99.17	99.25	99.30	99.33	99.36	99.37	99.39	99.40	99.42	99.43	99.45	99.46	99.47	99.47	99.48	99.49	99.50
3	34.12	30.82	29.46	28.71	28.24	27.91	27.67	27.49	27.35	27.23	27.05	26.87	26.69	26.60	26.50	26.41	26.32	26.22	26.13
4	21.20	18.00	16.69	15.98	15.52	15.21	14.98	14.80	14.66	14.55	14.37	14.20	14.02	13.93	13.84	13.75	13.65	13.56	13.46
5	16.26	13.27	12.06	11.39	10.97	10.67	10.46	10.29	10.16	10.05	9.89	9.72	9.55	9.47	9.38	9.29	9.20	9.11	9.02
6	13.75	10.92	9.78	9.15	8.75	8.47	8.26	8.10	7.98	7.87	7.72	7.56	7.40	7.31	7.23	7.14	7.06	6.97	6.88
7	12.25	9.55	8.45	7.85	7.46	7.19	6.99	6.84	6.72	6.62	6.47	6.31	6.16	6.07	5.99	5.91	5.82	5.74	5.65
8	11.26	8.65	7.59	7.01	6.63	6.37	6.18	6.03	5.91	5.81	5.67	5.52	5.36	5.28	5.20	5.12	5.03	4.95	4.86
9	10.56	8.02	6.99	6.42	6.06	5.80	5.61	5.47	5.35	5.26	5.11	4.96	4.81	4.73	4.65	4.57	4.48	4.40	4.31
10	10.04	7.56	6.55	5.99	5.64	5.39	5.20	5.06	4.94	4.85	4.71	4.56	4.41	4.33	4.25	4.17	4.08	4.00	3.91
11	9.65	7.21	6.22	5.67	5.32	5.07	4.89	4.74	4.63	4.54	4.40	4.25	4.10	4.02	3.94	3.86	3.78	3.69	3.60
12	9.33	6.93	5.95	5.41	5.06	4.82	4.64	4.50	4.39	4.30	4.16	4.01	3.86	3.78	3.70	3.62	3.54	3.45	3.36
13	9.07	6.70	5.74	5.21	4.86	4.62	4.44	4.30	4.19	4.10	3.96	3.82	3.66	3.59	3.51	3.43	3.34	3.25	3.17
14	8.86	6.51	5.56	5.04	4.69	4.46	4.28	4.14	4.03	3.94	3.80	3.66	3.51	3.43	3.35	3.27	3.18	3.09	3.00
15	8.68	6.36	5.42	4.89	4.56	4.32	4.14	4.00	3.89	3.80	3.67	3.52	3.37	3.29	3.21	3.13	3.05	2.96	2.87
16	8.53	6.23	5.29	4.77	4.44	4.20	4.03	3.89	3.78	3.69	3.55	3.41	3.26	3.18	3.10	3.02	2.93	2.84	2.75
17	8.40	6.11	5.18	4.67	4.34	4.10	3.93	3.79	3.68	3.59	3.46	3.31	3.16	3.08	3.00	2.92	2.83	2.75	2.65
18	8.29	6.01	5.09	4.58	4.25	4.01	3.84	3.71	3.60	3.51	3.37	3.23	3.08	3.00	2.92	2.84	2.75	2.66	2.57
19	8.18	5.93	5.01	4.50	4.17	3.94	3.77	3.63	3.52	3.43	3.30	3.15	3.00	2.92	2.84	2.76	2.67	2.58	2.49
20	8.10	5.85	4.94	4.43	4.10	3.87	3.70	3.56	3.46	3.37	3.23	3.09	2.94	2.86	2.78	2.69	2.61	2.52	2.42
21	8.02	5.78	4.87	4.37	4.04	3.81	3.64	3.51	3.40	3.31	3.17	3.03	2.88	2.80	2.72	2.64	2.55	2.46	2.36
22	7.95	5.72	4.82	4.31	3.99	3.76	3.59	3.45	3.35	3.26	3.12	2.98	2.83	2.75	2.67	2.58	2.50	2.40	2.31
23	7.88	5.66	4.76	4.26	3.94	3.71	3.54	3.41	3.30	3.21	3.07	2.93	2.78	2.70	2.62	2.54	2.45	2.35	2.26
24	7.82	5.61	4.72	4.22	3.90	3.67	3.50	3.36	3.26	3.17	3.03	2.89	2.74	2.66	2.58	2.49	2.40	2.31	2.21
25	7.77	5.57	4.68	4.18	3.85	3.63	3.46	3.32	3.22	3.13	2.99	2.85	2.70	2.62	2.54	2.45	2.36	2.27	2.17
26	7.72	5.53	4.64	4.14	3.82	3.59	3.42	3.29	3.18	3.09	2.96	2.81	2.66	2.58	2.50	2.42	2.33	2.23	2.13
27	7.68	5.49	4.60	4.11	3.78	3.56	3.39	3.26	3.15	3.06	2.93	2.78	2.63	2.55	2.47	2.38	2.29	2.20	2.10
28	7.64	5.45	4.57	4.07	3.75	3.53	3.36	3.23	3.12	3.03	2.90	2.75	2.60	2.52	2.44	2.35	2.26	2.17	2.06
29	7.60	5.42	4.54	4.04	3.73	3.50	3.33	3.20	3.09	3.00	2.87	2.73	2.57	2.49	2.41	2.33	2.23	2.14	2.03
30	7.56	5.39	4.51	4.02	3.70	3.47	3.30	3.17	3.07	2.98	2.84	2.70	2.55	2.47	2.39	2.30	2.21	2.11	2.01
40	7.31	5.18	4.31	3.83	3.51	3.29	3.12	2.99	2.89	2.80	2.66	2.52	2.37	2.29	2.20	2.11	2.02	1.92	1.80
60	7.08	4.98	4.13	3.65	3.34	3.12	2.95	2.82	2.72	2.63	2.50	2.35	2.20	2.12	2.03	1.94	1.84	1.73	1.60
120	6.85	4.79	3.95	3.48	3.17	2.96	2.79	2.66	2.56	2.47	2.34	2.19	2.03	1.95	1.86	1.76	1.66	1.53	1.38
∞	6.63	4.61	3.78	3.32	3.02	2.80	2.64	2.51	2.41	2.32	2.18	2.04	1.88	1.79	1.70	1.59	1.47	1.32	1.00

TABLE E.5 *(Continued)*

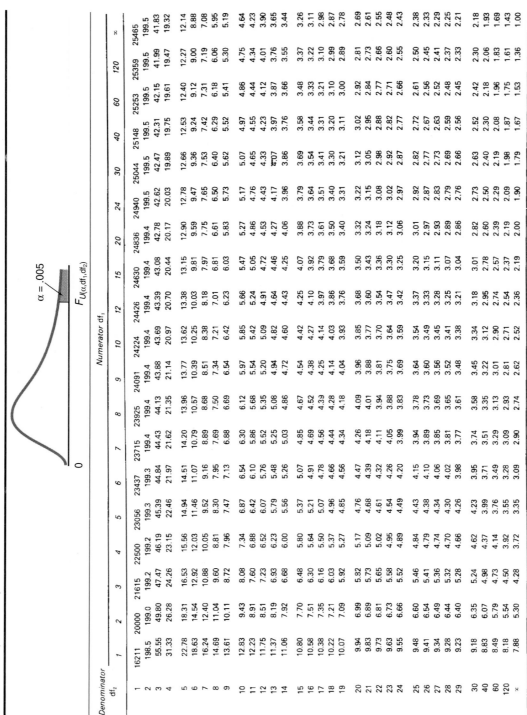

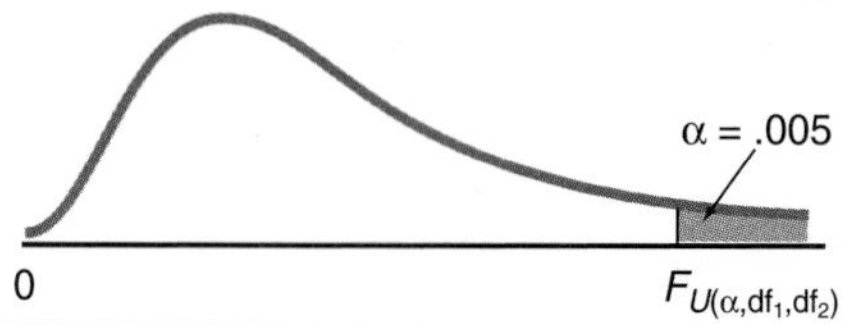

Denominator df_2	Numerator df_1 1	2	3	4	5	6	7	8	9	10	12	15	20	24	30	40	60	120	∞
1	16211	20000	21615	22500	23056	23437	23715	23925	24091	24224	24426	24630	24836	24940	25044	25148	25253	25359	25465
2	198.5	199.0	199.2	199.2	199.3	199.3	199.4	199.4	199.4	199.4	199.4	199.4	199.4	199.5	199.5	199.5	199.5	199.5	199.5
3	55.55	49.80	47.47	46.19	45.39	44.84	44.43	44.13	43.88	43.69	43.39	43.08	42.78	42.62	42.47	42.31	42.15	41.99	41.83
4	31.33	26.28	24.26	23.15	22.46	21.97	21.62	21.35	21.14	20.97	20.70	20.44	20.17	20.03	19.89	19.75	19.61	19.47	19.32
5	22.78	18.31	16.53	15.56	14.94	14.51	14.20	13.96	13.77	13.62	13.38	13.15	12.90	12.78	12.66	12.53	12.40	12.27	12.14
6	18.63	14.54	12.92	12.03	11.46	11.07	10.79	10.57	10.39	10.25	10.03	9.81	9.59	9.47	9.36	9.24	9.12	9.00	8.88
7	16.24	12.40	10.88	10.05	9.52	9.16	8.89	8.68	8.51	8.38	8.18	7.97	7.75	7.65	7.53	7.42	7.31	7.19	7.08
8	14.69	11.04	9.60	8.81	8.30	7.95	7.69	7.50	7.34	7.21	7.01	6.81	6.61	6.50	6.40	6.29	6.18	6.06	5.95
9	13.61	10.11	8.72	7.96	7.47	7.13	6.88	6.69	6.54	6.42	6.23	6.03	5.83	5.73	5.62	5.52	5.41	5.30	5.19
10	12.83	9.43	8.08	7.34	6.87	6.54	6.30	6.12	5.97	5.85	5.66	5.47	5.27	5.17	5.07	4.97	4.86	4.75	4.64
11	12.23	8.91	7.60	6.88	6.42	6.10	5.86	5.68	5.54	5.42	5.24	5.05	4.86	4.76	4.65	4.55	4.44	4.34	4.23
12	11.75	8.51	7.23	6.52	6.07	5.76	5.52	5.35	5.20	5.09	4.91	4.72	4.53	4.43	4.33	4.23	4.12	4.01	3.90
13	11.37	8.19	6.93	6.23	5.79	5.48	5.25	5.08	4.94	4.82	4.64	4.46	4.27	4.17	4.07	3.97	3.87	3.76	3.65
14	11.06	7.92	6.68	6.00	5.56	5.26	5.03	4.86	4.72	4.60	4.43	4.25	4.06	3.96	3.86	3.76	3.66	3.55	3.44
15	10.80	7.70	6.48	5.80	5.37	5.07	4.85	4.67	4.54	4.42	4.25	4.07	3.88	3.79	3.69	3.58	3.48	3.37	3.26
16	10.58	7.51	6.30	5.64	5.21	4.91	4.69	4.52	4.38	4.27	4.10	3.92	3.73	3.64	3.54	3.44	3.33	3.22	3.11
17	10.38	7.35	6.16	5.50	5.07	4.78	4.56	4.39	4.25	4.14	3.97	3.79	3.61	3.51	3.41	3.31	3.21	3.10	2.98
18	10.22	7.21	6.03	5.37	4.96	4.66	4.44	4.28	4.14	4.03	3.86	3.68	3.50	3.40	3.30	3.20	3.10	2.99	2.87
19	10.07	7.09	5.92	5.27	4.85	4.56	4.34	4.18	4.04	3.93	3.76	3.59	3.40	3.31	3.21	3.11	3.00	2.89	2.78
20	9.94	6.99	5.82	5.17	4.76	4.47	4.26	4.09	3.96	3.85	3.68	3.50	3.32	3.22	3.12	3.02	2.92	2.81	2.69
21	9.83	6.89	5.73	5.09	4.68	4.39	4.18	4.01	3.88	3.77	3.60	3.43	3.24	3.15	3.05	2.95	2.84	2.73	2.61
22	9.73	6.81	5.65	5.02	4.61	4.32	4.11	3.94	3.81	3.70	3.54	3.36	3.18	3.08	2.98	2.88	2.77	2.66	2.55
23	9.63	6.73	5.58	4.95	4.54	4.26	4.05	3.88	3.75	3.64	3.47	3.30	3.12	3.02	2.92	2.82	2.71	2.60	2.48
24	9.55	6.66	5.52	4.89	4.49	4.20	3.99	3.83	3.69	3.59	3.42	3.25	3.06	2.97	2.87	2.77	2.66	2.55	2.43
25	9.48	6.60	5.46	4.84	4.43	4.15	3.94	3.78	3.64	3.54	3.37	3.20	3.01	2.92	2.82	2.72	2.61	2.50	2.38
26	9.41	6.54	5.41	4.79	4.38	4.10	3.89	3.73	3.60	3.49	3.33	3.15	2.97	2.87	2.77	2.67	2.56	2.45	2.33
27	9.34	6.49	5.36	4.74	4.34	4.06	3.85	3.69	3.56	3.45	3.28	3.11	2.93	2.83	2.73	2.63	2.52	2.41	2.29
28	9.28	6.44	5.32	4.70	4.30	4.02	3.81	3.65	3.52	3.41	3.25	3.07	2.89	2.79	2.69	2.59	2.48	2.37	2.25
29	9.23	6.40	5.28	4.66	4.26	3.98	3.77	3.61	3.48	3.38	3.21	3.04	2.86	2.76	2.66	2.56	2.45	2.33	2.21
30	9.18	6.35	5.24	4.62	4.23	3.95	3.74	3.58	3.45	3.34	3.18	3.01	2.82	2.73	2.63	2.52	2.42	2.30	2.18
40	8.83	6.07	4.98	4.37	3.99	3.71	3.51	3.35	3.22	3.12	2.95	2.78	2.60	2.50	2.40	2.30	2.18	2.06	1.93
60	8.49	5.79	4.73	4.14	3.76	3.49	3.29	3.13	3.01	2.90	2.74	2.57	2.39	2.29	2.19	2.08	1.96	1.83	1.69
120	8.18	5.54	4.50	3.92	3.55	3.28	3.09	2.93	2.81	2.71	2.54	2.37	2.19	2.09	1.98	1.87	1.75	1.61	1.43
∞	7.88	5.30	4.28	3.72	3.35	3.09	2.90	2.74	2.62	2.52	2.36	2.19	2.00	1.90	1.79	1.67	1.53	1.36	1.00

Source: Reprinted from E. S. Pearson and H. O. Hartley, eds., *Biometrika Tables for Statisticians,* 3rd ed., 1966, by permission of the *Biometrika* Trustees.

TABLE E.5a Selected Critical Values of F for Cook's D_i Statistic

	$\alpha = .50$											
	Numerator df $= P + 1$											
Denominator df $= n - P - 1$	2	3	4	5	6	7	8	9	10	12	15	20
10	.743	.845	.899	.932	.954	.971	.983	.992	1.00	1.01	1.02	1.03
11	.739	.840	.893	.926	.948	.964	.977	.986	.994	1.01	1.02	1.03
12	.735	.835	.888	.921	.943	.959	.972	.981	.989	1.00	1.01	1.02
15	.726	.826	.878	.911	.933	.949	.960	.970	.977	.989	1.00	1.01
20	.718	.816	.868	.900	.922	.938	.950	.959	.966	.977	.989	1.00
24	.714	.812	.863	.895	.917	.932	.944	.953	.961	.972	.983	.994
30	.709	.807	.858	.890	.912	.927	.939	.948	.955	.966	.978	.989
40	.705	.802	.854	.885	.907	.922	.934	.943	.950	.961	.972	.983
60	.701	.798	.849	.880	.901	.917	.928	.937	.945	.956	.967	.978
120	.697	.793	.844	.875	.896	.912	.923	.932	.939	.950	.961	.972
∞	.693	.789	.839	.870	.891	.907	.918	.927	.934	.945	.956	.967

Source: Extracted from E. S. Pearson and H. O. Hartley, eds., *Biometrika Tables for Statisticians*, 3rd ed., 1966, by permission of the *Biometrika* Trustees.

TABLE E.6 Table of Poisson Probabilities

For a given value of λ, entry indicates the probability of obtaining a specified value of X

X	λ 0.1	0.2	0.3	0.4	0.5	0.6	0.7	0.8	0.9	1.0
0	.9048	.8187	.7408	.6703	.6065	.5488	.4966	.4493	.4066	.3679
1	.0905	.1637	.2222	.2681	.3033	.3293	.3476	.3595	.3659	.3679
2	.0045	.0164	.0333	.0536	.0758	.0988	.1217	.1438	.1647	.1839
3	.0002	.0011	.0033	.0072	.0126	.0198	.0284	.0383	.0494	.0613
4	.0000	.0001	.0003	.0007	.0016	.0030	.0050	.0077	.0111	.0153
5	.0000	.0000	.0000	.0001	.0002	.0004	.0007	.0012	.0020	.0031
6	.0000	.0000	.0000	.0000	.0000	.0000	.0001	.0002	.0003	.0005
7	.0000	.0000	.0000	.0000	.0000	.0000	.0000	.0000	.0000	.0001

X	λ 1.1	1.2	1.3	1.4	1.5	1.6	1.7	1.8	1.9	2.0
0	.3329	.3012	.2725	.2466	.2231	.2019	.1827	.1653	.1496	.1353
1	.3662	.3614	.3543	.3452	.3347	.3230	.3106	.2975	.2842	.2707
2	.2014	.2169	.2303	.2417	.2510	.2584	.2640	.2678	.2700	.2707
3	.0738	.0867	.0998	.1128	.1255	.1378	.1496	.1607	.1710	.1804
4	.0203	.0260	.0324	.0395	.0471	.0551	.0636	.0723	0812	.0902
5	.0045	.0062	.0084	.0111	.0141	.0176	.0216	.0260	.0309	.0361
6	.0008	.0012	.0018	.0026	.0035	.0047	.0061	.0078	.0098	.0120
7	.0001	.0002	.0003	.0005	.0008	.0011	.0015	.0020	.0027	.0034
8	.0000	.0000	.0001	.0001	.0001	.0002	.0003	.0005	.0006	.0009
9	.0000	.0000	.0000	.0000	.0000	.0000	.0001	.0001	.0001	.0002

X	λ 2.1	2.2	2.3	2.4	2.5	2.6	2.7	2.8	2.9	3.0
0	.1225	.1108	.1003	.0907	.0821	.0743	.0672	.0608	.0550	.0498
1	.2572	.2438	.2306	.2177	.2052	.1931	.1815	.1703	.1596	.1494
2	.2700	.2681	.2652	.2613	.2565	.2510	.2450	.2384	.2314	.2240
3	.1890	.1966	.2033	.2090	.2138	.2176	.2205	.2225	.2237	.2240
4	.0992	.1082	.1169	.1254	.1336	.1414	.1488	.1557	.1622	.1680
5	.0417	.0476	.0538	.0602	.0668	.0735	.0804	.0872	.0940	.1008
6	.0146	.0174	.0206	.0241	.0278	.0319	.0362	.0407	.0455	.0504
7	.0044	.0055	.0068	.0083	.0099	.0118	.0139	.0163	.0188	.0216
8	.0011	.0015	.0019	.0025	.0031	.0038	.0047	.0057	.0068	.0081
9	.0003	.0004	.0005	.0007	.0009	.0011	.0014	.0018	.0022	.0027
10	.0001	.0001	.0001	.0002	.0002	.0003	.0004	.0005	.0006	.0008
11	.0000	.0000	.0000	.0000	.0000	.0001	.0001	.0001	.0002	.0002
12	.0000	.0000	.0000	.0000	.0000	.0000	.0000	.0000	.0000	.0001

X	λ 3.1	3.2	3.3	3.4	3.5	3.6	3.7	3.8	3.9	4.0
0	.0450	.0408	.0369	.0334	.0302	.0273	.0247	.0224	.0202	.0183
1	.1397	.1304	.1217	.1135	.1057	.0984	.0915	.0850	.0789	.0733
2	.2165	.2087	.2008	.1929	.1850	.1771	.1692	.1615	.1539	.1465
3	.2237	.2226	.2209	.2186	.2158	.2125	.2087	.2046	.2001	.1954
4	.1734	.1781	.1823	.1858	.1888	.1912	.1931	.1944	.1951	.1954
5	.1075	.1140	.1203	.1264	.1322	.1377	.1429	.1477	.1522	.1563
6	.0555	.0608	.0662	.0716	.0771	.0826	.0881	.0936	.0989	.1042
7	.0246	.0278	.0312	.0348	.0385	.0425	.0466	.0508	.0551	.0595
8	.0095	.0111	.0129	.0148	.0169	.0191	.0215	.0241	.0269	.0298
9	.0033	.0040	.0047	.0056	.0066	.0076	.0089	.0102	.0116	.0132

TABLE E.6 *(Continued)*

X	λ = 3.1	3.2	3.3	3.4	3.5	3.6	3.7	3.8	3.9	4.0
10	.0010	.0013	.0016	.0019	.0023	.0028	.0033	.0039	.0045	.0053
11	.0003	.0004	.0005	.0006	.0007	.0009	.0011	.0013	.0016	.0019
12	.0001	.0001	.0001	.0002	.0002	.0003	.0003	.0004	.0005	.0006
13	.0000	.0000	.0000	.0000	.0001	.0001	.0001	.0001	.0002	.0002
14	.0000	.0000	.0000	.0000	.0000	.0000	.0000	.0000	.0000	.0001

X	λ = 4.1	4.2	4.3	4.4	4.5	4.6	4.7	4.8	4.9	5.0
0	.0166	.0150	.0136	.0123	.0111	.0101	.0091	.0082	.0074	.0067
1	.0679	.0630	.0583	.0540	.0500	.0462	.0427	.0395	.0365	.0337
2	.1393	.1323	.1254	.1188	.1125	.1063	.1005	.0948	.0894	.0842
3	.1904	.1852	.1798	.1743	.1687	.1631	.1574	.1517	.1460	.1404
4	.1951	.1944	.1933	.1917	.1898	.1875	.1849	.1820	.1789	.1755
5	.1600	.1633	.1662	.1687	.1708	.1725	.1738	.1747	.1753	.1755
6	.1093	.1143	.1191	.1237	.1281	.1323	.1362	.1398	.1432	.1462
7	.0640	.0686	.0732	.0778	.0824	.0869	.0914	.0959	.1002	.1044
8	.0328	.0360	.0393	.0428	.0463	.0500	.0537	.0575	.0614	.0653
9	.0150	.0168	.0188	.0209	.0232	.0255	.0280	.0307	.0334	.0363
10	.0061	.0071	.0081	.0092	.0104	.0118	.0132	.0147	.0164	.0181
11	.0023	.0027	.0032	.0037	.0043	.0049	.0056	.0064	.0073	.0082
12	.0008	.0009	.0011	.0014	.0016	.0019	.0022	.0026	.0030	.0034
13	.0002	.0003	.0004	.0005	.0006	.0007	.0008	.0009	.0011	.0013
14	.0001	.0001	.0001	.0001	.0002	.0002	.0003	.0003	.0004	.0005
15	.0000	.0000	.0000	.0000	.0001	.0001	.0001	.0001	.0001	.0002

X	λ = 5.1	5.2	5.3	5.4	5.5	5.6	5.7	5.8	5.9	6.0
0	.0061	.0055	.0050	.0045	.0041	.0037	.0033	.0030	.0027	.0025
1	.0311	.0287	.0265	.0244	.0225	.0207	.0191	.0176	.0162	.0149
2	.0793	.0746	.0701	.0659	.0618	.0580	.0544	.0509	.0477	.0446
3	.1348	.1293	.1239	.1185	.1133	.1082	.1033	.0985	.0938	.0892
4	.1719	.1681	.1641	.1600	.1558	.1515	.1472	.1428	.1383	.1339
5	.1753	.1748	.1740	.1728	.1714	.1697	.1678	.1656	.1632	.1606
6	.1490	.1515	.1537	.1555	.1571	.1584	.1594	.1601	.1605	.1606
7	.1086	.1125	.1163	.1200	.1234	.1267	.1298	.1326	.1353	.1377
8	.0692	.0731	.0771	.0810	.0849	.0887	.0925	.0962	.0998	.1033
9	.0392	.0423	.0454	.0486	.0519	.0552	.0586	.0620	.0654	.0688
10	.0200	.0220	.0241	.0262	.0285	.0309	.0334	.0359	.0386	.0413
11	.0093	.0104	.0116	.0129	.0143	.0157	.0173	.0190	.0207	.0225
12	.0039	.0045	.0051	.0058	.0065	.0073	.0082	.0092	.0102	.0113
13	.0015	.0018	.0021	.0024	.0028	.0032	.0036	.0041	.0046	.0052
14	.0006	.0007	.0008	.0009	.0011	.0013	.0015	.0017	.0019	.0022
15	.0002	.0002	.0003	.0003	.0004	.0005	.0006	.0007	.0008	.0009
16	.0001	.0001	.0001	.0001	.0001	.0002	.0002	.0002	.0003	.0003
17	.0000	.0000	.0000	.0000	.0000	.0000	.0001	.0001	.0001	.0001

X	λ = 6.1	6.2	6.3	6.4	6.5	6.6	6.7	6.8	6.9	7.0
0	.0022	.0020	.0018	.0017	.0015	.0014	.0012	.0011	.0010	.0009
1	.0137	.0126	.0116	.0106	.0098	.0090	.0082	.0076	.0070	.0064
2	.0417	.0390	.0364	.0340	.0318	.0296	.0276	.0258	.0240	.0223
3	.0848	.0806	.0765	.0726	.0688	.0652	.0617	.0584	.0552	.0521
4	.1294	.1249	.1205	.1162	.1118	.1076	.1034	.0992	.0952	.0912

TABLE E.6 *(Continued)*

					λ					
X	6.1	6.2	6.3	6.4	6.5	6.6	6.7	6.8	6.9	7.0
5	.1579	.1549	.1519	.1487	.1454	.1420	.1385	.1349	.1314	.1277
6	.1605	.1601	.1595	.1586	.1575	.1562	.1546	.1529	.1511	.1490
7	.1399	.1418	.1435	.1450	.1462	.1472	.1480	.1486	.1489	.1490
8	.1066	.1099	.1130	.1160	.1188	.1215	.1240	.1263	.1284	.1304
9	.0723	.0757	.0791	.0825	.0858	.0891	.0923	.0954	.0985	.1014
10	.0441	.0469	.0498	.0528	.0558	.0588	.0618	.0649	.0679	.0710
11	.0245	.0265	.0285	.0307	.0330	.0353	.0377	.0401	.0426	.0452
12	.0124	.0137	.0150	.0164	.0179	.0194	.0210	.0227	.0245	.0264
13	.0058	.0065	.0073	.0081	.0089	.0098	.0108	.0119	.0130	.0142
14	.0025	.0029	.0033	.0037	.0041	.0046	.0052	.0058	.0064	.0071
15	.0010	.0012	.0014	.0016	.0018	.0020	.0023	.0026	.0029	.0033
16	.0004	.0005	.0005	.0006	.0007	.0008	.0010	.0011	.0013	.0014
17	.0001	.0002	.0002	.0002	.0003	.0003	.0004	.0004	.0005	.0006
18	.0000	.0001	.0001	.0001	.0001	.0001	.0001	.0002	.0002	.0002
19	.0000	.0000	.0000	.0000	.0000	.0000	.0000	.0001	.0001	.0001

					λ					
X	7.1	7.2	7.3	7.4	7.5	7.6	7.7	7.8	7.9	8.0
0	.0008	.0007	.0007	.0006	.0006	.0005	.0005	.0004	.0004	.0003
1	.0059	.0054	.0049	.0045	.0041	.0038	.0035	.0032	.0029	.0027
2	.0208	.0194	.0180	.0167	.0156	.0145	.0134	.0125	.0116	.0107
3	.0492	.0464	.0438	.0413	.0389	.0366	.0345	.0324	.0305	.0286
4	.0874	.0836	.0799	.0764	.0729	.0696	.0663	.0632	.0602	.0573
5	.1241	.1204	.1167	.1130	.1094	.1057	.1021	.0986	.0951	.0916
6	.1468	.1445	.1420	.1394	.1367	.1339	.1311	.1282	.1252	.1221
7	.1489	.1486	.1481	.1474	.1465	.1454	.1442	.1428	.1413	.1396
8	.1321	.1337	.1351	.1363	.1373	.1382	.1388	.1392	.1395	.1396
9	.1042	.1070	.1096	.1121	.1144	.1167	.1187	.1207	.1224	.1241
10	.0740	.0770	.0800	.0829	.0858	.0887	.0914	.0941	.0967	.0993
11	.0478	.0504	.0531	.0558	.0585	.0613	.0640	.0667	.0695	.0722
12	.0283	.0303	.0323	.0344	.0366	.0388	.0411	.0434	.0457	.0481
13	.0154	.0168	.0181	.0196	.0211	.0227	.0243	.0260	.0278	.0296
14	.0078	.0086	.0095	.0104	.0113	.0123	.0134	.0145	.0157	.0169
15	.0037	.0041	.0046	.0051	.0057	.0062	.0069	.0075	.0083	.0090
16	.0016	.0019	.0021	.0024	.0026	.0030	.0033	.0037	.0041	.0045
17	.0007	.0008	.0009	.0010	.0012	.0013	.0015	.0017	.0019	.0021
18	.0003	.0003	.0004	.0004	.0005	.0006	.0006	.0007	.0008	.0009
19	.0001	.0001	.0001	.0002	.0002	.0002	.0003	.0003	.0003	.0004
20	.0000	.0000	.0001	.0001	.0001	.0001	.0001	.0001	.0001	.0002
21	.0000	.0000	.0000	.0000	.0000	.0000	.0000	.0000	.0001	.0001

					λ					
X	8.1	8.2	8.3	8.4	8.5	8.6	8.7	8.8	8.9	9.0
0	.0003	.0003	.0002	.0002	.0002	.0002	.0002	.0002	.0001	.0001
1	.0025	.0023	.0021	.0019	.0017	.0016	.0014	.0013	.0012	.0011
2	.0100	.0092	.0086	.0079	.0074	.0068	.0063	.0058	.0054	.0050
3	.0269	.0252	.0237	.0222	.0208	.0195	.0183	.0171	.0160	.0150
4	.0544	.0517	.0491	.0466	.0443	.0420	.0398	.0377	.0357	.0337
5	.0882	.0849	.0816	.0784	.0752	.0722	.0692	.0663	.0635	.0607
6	.1191	.1160	.1128	.1097	.1066	.1034	.1003	.0972	.0941	.0911
7	.1378	.1358	.1338	.1317	.1294	.1271	.1247	.1222	.1197	.1171
8	.1395	.1392	.1388	.1382	.1375	.1366	.1356	.1344	.1332	.1318
9	.1256	.1269	.1280	.1290	.1299	.1306	.1311	.1315	.1317	.1318

TABLE E.6 *(Continued)*

X	λ = 8.1	8.2	8.3	8.4	8.5	8.6	8.7	8.8	8.9	9.0
10	.1017	.1040	.1063	.1084	.1104	.1123	.1140	.1157	.1172	.1186
11	.0749	.0776	.0802	.0828	.0853	.0878	.0902	.0925	.0948	.0970
12	.0505	.0530	.0555	.0579	.0604	.0629	.0654	.0679	.0703	.0728
13	.0315	.0334	.0354	.0374	.0395	.0416	.0438	.0459	.0481	.0504
14	.0182	.0196	.0210	.0225	.0240	.0256	.0272	.0289	.0306	.0324
15	.0098	.0107	.0116	.0126	.0136	.0147	.0158	.0169	.0182	.0194
16	.0050	.0055	.0060	.0066	.0072	.0079	.0086	.0093	.0101	.0109
17	.0024	.0026	.0029	.0033	.0036	.0040	.0044	.0048	.0053	.0058
18	.0011	.0012	.0014	.0015	.0017	.0019	.0021	.0024	.0026	.0029
19	.0005	.0005	.0006	.0007	.0008	.0009	.0010	.0011	.0012	.0014
20	.0002	.0002	.0002	.0003	.0003	.0004	.0004	.0005	.0005	.0006
21	.0001	.0001	.0001	.0001	.0001	.0002	.0002	.0002	.0002	.0003
22	.0000	.0000	.0000	.0000	.0001	.0001	.0001	.0001	.0001	.0001

X	λ = 9.1	9.2	9.3	9.4	9.5	9.6	9.7	9.8	9.9	10
0	.0001	.0001	.0001	.0001	.0001	.0001	.0001	.0001	.0001	.0000
1	.0010	.0009	.0009	.0008	.0007	.0007	.0006	.0005	.0005	.0005
2	.0046	.0043	.0040	.0037	.0034	.0031	.0029	.0027	.0025	.0023
3	.0140	.0131	.0123	.0115	.0107	.0100	.0093	.0087	.0081	.0076
4	.0319	.0302	.0285	.0269	.0254	.0240	.0226	.0213	.0201	.0189
5	.0581	.0555	.0530	.0506	.0483	.0460	.0439	.0418	.0398	.0378
6	.0881	.0851	.0822	.0793	.0764	.0736	.0709	.0682	.0656	.0631
7	.1145	.1118	.1091	.1064	.1037	.1010	.0982	.0955	.0928	.0901
8	.1302	.1286	.1269	.1251	.1232	.1212	.1191	.1170	.1148	.1126
9	.1317	.1315	.1311	.1306	.1300	.1293	.1284	.1274	.1263	.1251
10	.1198	.1210	.1219	.1228	.1235	.1241	.1245	.1249	.1250	.1251
11	.0991	.1012	.1031	.1049	.1067	.1083	.1098	.1112	.1125	.1137
12	.0752	.0776	.0799	.0822	.0844	.0866	.0888	.0908	.0928	.0948
13	.0526	.0549	.0572	.0594	.0617	.0640	.0662	.0685	.0707	.0729
14	.0342	.0361	.0380	.0399	.0419	.0439	.0459	.0479	.0500	.0521
15	.0208	.0221	.0235	.0250	.0265	.0281	.0297	.0313	.0330	.0347
16	.0118	.0127	.0137	.0147	.0157	.0168	.0180	.0192	.0204	.0217
17	.0063	.0069	.0075	.0081	.0088	.0095	.0103	.0111	.0119	.0128
18	.0032	.0035	.0039	.0042	.0046	.0051	.0055	.0060	.0065	.0071
19	.0015	.0017	.0019	.0021	.0023	.0026	.0028	.0031	.0034	.0037
20	.0007	.0008	.0009	.0010	.0011	.0012	.0014	.0015	.0017	.0019
21	.0003	.0003	.0004	.0004	.0005	.0006	.0006	.0007	.0008	.0009
22	.0001	.0001	.0002	.0002	.0002	.0002	.0003	.0003	.0004	.0004
23	.0000	.0001	.0001	.0001	.0001	.0001	.0001	.0001	.0002	.0002
24	.0000	.0000	.0000	.0000	.0000	.0000	.0000	.0001	.0001	.0001

X	λ = 20	X	λ = 20
0	.0000	20	.0888
1	.0000	21	.0846
2	.0000	22	.0769
3	.0000	23	.0669
4	.0000	24	.0557
5	.0001	25	.0446
6	.0002	26	.0343
7	.0005	27	.0254
8	.0013	28	.0181
9	.0029	29	.0125
10	.0058	30	.0083
11	.0106	31	.0054
12	.0176	32	.0034
13	.0271	33	.0020
14	.0387	34	.0012
15	.0516	35	.0007
16	.0646	36	.0004
17	.0760	37	.0002
18	.0844	38	.0001
19	.0888	39	.0001

Source: Extracted from William H. Beyer, ed., *CRC Basic Statistical Tables* (Cleveland, Ohio: The Chemical Rubber Co., 1971). Reprinted with permission. © The Chemical Rubber Co., CRC Press, Inc.

TABLE E.7 Table of Binomial Probabilities

For a given combination of *n* and *p*, entry indicates the probability of obtaining a specified value of *X*. To locate entry: **when $p \le .50$,** read *p* across the top heading and both *n* and *X* down the left margin; **when $p \ge .50$,** read *p* across the bottom heading and both *n* and *X* up the right margin

		p																			
n	*X*	0.01	0.02	0.03	0.04	0.05	0.06	0.07	0.08	0.09	0.10	0.11	0.12	0.13	0.14	0.15	0.16	0.17	0.18	*X*	*n*
2	0	0.9801	0.9604	0.9409	0.9216	0.9025	0.8836	0.8649	0.8464	0.8281	0.8100	0.7921	0.7744	0.7569	0.7396	0.7225	0.7056	0.6889	0.6724	2	
	1	0.0198	0.0392	0.0582	0.0768	0.0950	0.1128	0.1302	0.1472	0.1638	0.1800	0.1958	0.2112	0.2262	0.2408	0.2550	0.2688	0.2822	0.2952	1	
	2	0.0001	0.0004	0.0009	0.0016	0.0025	0.0036	0.0049	0.0064	0.0081	0.0100	0.0121	0.0144	0.0169	0.0196	0.0225	0.0256	0.0289	0.0324	0	2
3	0	0.9703	0.9412	0.9127	0.8847	0.8574	0.8306	0.8044	0.7787	0.7536	0.7290	0.7050	0.6815	0.6585	0.6361	0.6141	0.5927	0.5718	0.5514	3	
	1	0.0294	0.0576	0.0847	0.1106	0.1354	0.1590	0.1816	0.2031	0.2236	0.2430	0.2614	0.2788	0.2952	0.3106	0.3251	0.3387	0.3513	0.3631	2	
	2	0.0003	0.0012	0.0026	0.0046	0.0071	0.0102	0.0137	0.0177	0.0221	0.0270	0.0323	0.0380	0.0441	0.0506	0.0574	0.0645	0.0720	0.0797	1	
	3	0.0000	0.0000	0.0000	0.0001	0.0001	0.0002	0.0003	0.0005	0.0007	0.0010	0.0013	0.0017	0.0022	0.0027	0.0034	0.0041	0.0049	0.0058	0	3
4	0	0.9606	0.9224	0.8853	0.8493	0.8145	0.7807	0.7481	0.7164	0.6857	0.6561	0.6274	0.5997	0.5729	0.5470	0.5220	0.4979	0.4746	0.4521	4	
	1	0.0388	0.0753	0.1095	0.1416	0.1715	0.1993	0.2252	0.2492	0.2713	0.2916	0.3102	0.3271	0.3424	0.3562	0.3685	0.3793	0.3888	0.3970	3	
	2	0.0006	0.0023	0.0051	0.0088	0.0135	0.0191	0.0254	0.0325	0.0402	0.0486	0.0575	0.0669	0.0767	0.0870	0.0975	0.1084	0.1195	0.1307	2	
	3	0.0000	0.0000	0.0001	0.0002	0.0005	0.0008	0.0013	0.0019	0.0027	0.0036	0.0047	0.0061	0.0076	0.0094	0.0115	0.0138	0.0163	0.0191	1	
	4	—	—	0.0000	0.0000	0.0000	0.0000	0.0000	0.0000	0.0001	0.0001	0.0001	0.0002	0.0003	0.0004	0.0005	0.0007	0.0008	0.0010	0	4
5	0	0.9510	0.9039	0.8587	0.8154	0.7738	0.7339	0.6957	0.6591	0.6240	0.5905	0.5584	0.5277	0.4984	0.4704	0.4437	0.4182	0.3939	0.3707	5	
	1	0.0480	0.0922	0.1328	0.1699	0.2036	0.2342	0.2618	0.2866	0.3086	0.3280	0.3451	0.3598	0.3724	0.3829	0.3915	0.3983	0.4034	0.4069	4	
	2	0.0010	0.0038	0.0082	0.0142	0.0214	0.0299	0.0394	0.0498	0.0610	0.0729	0.0853	0.0981	0.1113	0.1247	0.1382	0.1517	0.1652	0.1786	3	
	3	0.0000	0.0001	0.0003	0.0006	0.0011	0.0019	0.0030	0.0043	0.0060	0.0081	0.0105	0.0134	0.0166	0.0203	0.0244	0.0289	0.0338	0.0392	2	
	4	—	0.0000	0.0000	0.0000	0.0000	0.0001	0.0001	0.0002	0.0003	0.0004	0.0007	0.0009	0.0012	0.0017	0.0022	0.0028	0.0035	0.0043	1	
	5	—	—	—	—	—	0.0000	0.0000	0.0000	0.0000	0.0000	0.0000	0.0000	0.0000	0.0001	0.0001	0.0001	0.0001	0.0002	0	5
6	0	0.9415	0.8858	0.8330	0.7828	0.7351	0.6899	0.6470	0.6064	0.5679	0.5314	0.4970	0.4644	0.4336	0.4046	0.3771	0.3513	0.3269	0.3040	6	
	1	0.0571	0.1085	0.1546	0.1957	0.2321	0.2642	0.2922	0.3164	0.3370	0.3543	0.3685	0.3800	0.3888	0.3952	0.3993	0.4015	0.4018	0.4004	5	
	2	0.0014	0.0055	0.0120	0.0204	0.0305	0.0422	0.0550	0.0688	0.0833	0.0984	0.1139	0.1295	0.1452	0.1608	0.1762	0.1912	0.2057	0.2197	4	
	3	0.0000	0.0002	0.0005	0.0011	0.0021	0.0036	0.0055	0.0080	0.0110	0.0146	0.0188	0.0236	0.0289	0.0349	0.0415	0.0486	0.0562	0.0643	3	
	4	—	0.0000	0.0000	0.0000	0.0001	0.0002	0.0003	0.0005	0.0008	0.0012	0.0017	0.0024	0.0032	0.0043	0.0055	0.0069	0.0086	0.0106	2	
	5	—	—	—	—	0.0000	0.0000	0.0000	0.0000	0.0000	0.0001	0.0001	0.0001	0.0002	0.0003	0.0004	0.0005	0.0007	0.0009	1	
	6	—	—	—	—	—	—	—	—	—	0.0000	0.0000	0.0000	0.0000	0.0000	0.0000	0.0000	0.0000	0.0000	0	6

7	0	0.9321	0.8681	0.8080	0.7514	0.6983	0.6485	0.6017	0.5578	0.5168	0.4783	0.4423	0.4087	0.3773	0.3479	0.3206	0.2951	0.2714	0.2493	7	
	1	0.0659	0.1240	0.1749	0.2192	0.2573	0.2897	0.3170	0.3396	0.3578	0.3720	0.3827	0.3901	0.3946	0.3965	0.3960	0.3935	0.3891	0.3830	6	
	2	0.0020	0.0076	0.0162	0.0274	0.0406	0.0555	0.0716	0.0886	0.1061	0.1240	0.1419	0.1596	0.1769	0.1936	0.2097	0.2248	0.2391	0.2523	5	
	3	0.0000	0.0003	0.0008	0.0019	0.0036	0.0059	0.0090	0.0128	0.0175	0.0230	0.0292	0.0363	0.0441	0.0525	0.0617	0.0714	0.0816	0.0923	4	
	4	—	0.0000	0.0000	0.0001	0.0002	0.0004	0.0007	0.0011	0.0017	0.0026	0.0036	0.0049	0.0066	0.0086	0.0109	0.0136	0.0167	0.0203	3	
	5	—	—	—	0.0000	0.0000	0.0000	0.0000	0.0001	0.0001	0.0002	0.0003	0.0004	0.0006	0.0008	0.0012	0.0016	0.0021	0.0027	2	
	6	—	—	—	—	—	—	—	0.0000	0.0000	0.0000	0.0000	0.0000	0.0000	0.0000	0.0001	0.0001	0.0001	0.0002	1	
	7	—	—	—	—	—	—	—	—	—	—	—	—	—	—	0.0000	0.0000	0.0000	0.0000	0	7
8	0	0.9227	0.8508	0.7837	0.7214	0.6634	0.6096	0.5596	0.5132	0.4703	0.4305	0.3937	0.3596	0.3282	0.2992	0.2725	0.2479	0.2252	0.2044	8	
	1	0.0746	0.1389	0.1939	0.2405	0.2793	0.3113	0.3370	0.3570	0.3721	0.3826	0.3892	0.3923	0.3923	0.3897	0.3847	0.3777	0.3691	0.3590	7	
	2	0.0026	0.0099	0.0210	0.0351	0.0515	0.0695	0.0888	0.1087	0.1288	0.1488	0.1684	0.1872	0.2052	0.2220	0.2376	0.2518	0.2646	0.2758	6	
	3	0.0001	0.0004	0.0013	0.0029	0.0054	0.0089	0.0134	0.0189	0.0255	0.0331	0.0416	0.0511	0.0613	0.0723	0.0839	0.0959	0.1084	0.1211	5	
	4	0.0000	0.0000	0.0001	0.0002	0.0004	0.0007	0.0013	0.0021	0.0031	0.0046	0.0064	0.0087	0.0115	0.0147	0.0185	0.0228	0.0277	0.0332	4	
	5	—	—	0.0000	0.0000	0.0000	0.0000	0.0001	0.0001	0.0002	0.0004	0.0006	0.0009	0.0014	0.0019	0.0026	0.0035	0.0045	0.0058	3	
	6	—	—	—	—	—	—	0.0000	0.0000	0.0000	0.0000	0.0000	0.0001	0.0001	0.0002	0.0002	0.0003	0.0005	0.0006	2	
	7	—	—	—	—	—	—	—	—	—	—	—	0.0000	0.0000	0.0000	0.0000	0.0000	0.0000	0.0000	1	
	8	—	—	—	—	—	—	—	—	—	—	—	—	—	—	—	—	—	—	0	8
9	0	0.9135	0.8337	0.7602	0.6925	0.6302	0.5730	0.5204	0.4722	0.4279	0.3874	0.3504	0.3165	0.2855	0.2573	0.2316	0.2082	0.1869	0.1676	9	
	1	0.0830	0.1531	0.2116	0.2597	0.2985	0.3292	0.3525	0.3695	0.3809	0.3874	0.3897	0.3884	0.3840	0.3770	0.3679	0.3569	0.3446	0.3312	8	
	2	0.0034	0.0125	0.0262	0.0433	0.0629	0.0840	0.1061	0.1285	0.1507	0.1722	0.1927	0.2119	0.2295	0.2455	0.2597	0.2720	0.2823	0.2908	7	
	3	0.0001	0.0006	0.0019	0.0042	0.0077	0.0125	0.0186	0.0261	0.0348	0.0446	0.0556	0.0674	0.0800	0.0933	0.1069	0.1209	0.1349	0.1489	6	
	4	0.0000	0.0000	0.0001	0.0003	0.0006	0.0012	0.0021	0.0034	0.0052	0.0074	0.0103	0.0138	0.0179	0.0228	0.0283	0.0345	0.0415	0.0490	5	
	5	—	—	0.0000	0.0000	0.0000	0.0001	0.0002	0.0003	0.0005	0.0008	0.0013	0.0019	0.0027	0.0037	0.0050	0.0066	0.0085	0.0108	4	
	6	—	—	—	—	—	0.0000	0.0000	0.0000	0.0000	0.0001	0.0001	0.0002	0.0003	0.0004	0.0006	0.0008	0.0012	0.0016	3	
	7	—	—	—	—	—	—	—	—	—	0.0000	0.0000	0.0000	0.0000	0.0000	0.0000	0.0001	0.0001	0.0001	2	
	8	—	—	—	—	—	—	—	—	—	—	—	—	—	—	—	0.0000	0.0000	0.0000	1	
	9	—	—	—	—	—	—	—	—	—	—	—	—	—	—	—	—	—	—	0	9
10	0	0.9044	0.8171	0.7374	0.6648	0.5987	0.5386	0.4840	0.4344	0.3894	0.3487	0.3118	0.2785	0.2484	0.2213	0.1969	0.1749	0.1552	0.1374	10	
	1	0.0914	0.1667	0.2281	0.2770	0.3151	0.3438	0.3643	0.3777	0.3851	0.3874	0.3854	0.3798	0.3712	0.3603	0.3474	0.3331	0.3178	0.3017	9	
	2	0.0042	0.0153	0.0317	0.0519	0.0746	0.0988	0.1234	0.1478	0.1714	0.1937	0.2143	0.2330	0.2496	0.2639	0.2759	0.2856	0.2929	0.2980	8	
	3	0.0001	0.0008	0.0026	0.0058	0.0105	0.0168	0.0248	0.0343	0.0452	0.0574	0.0706	0.0847	0.0995	0.1146	0.1298	0.1450	0.1600	0.1745	7	
	4	0.0000	0.0000	0.0001	0.0004	0.0010	0.0019	0.0033	0.0052	0.0078	0.0112	0.0153	0.0202	0.0260	0.0326	0.0401	0.0483	0.0573	0.0670	6	
	5	—	—	0.0000	0.0000	0.0001	0.0001	0.0003	0.0005	0.0009	0.0015	0.0023	0.0033	0.0047	0.0064	0.0085	0.0111	0.0141	0.0177	5	
	6	—	—	—	—	0.0000	0.0000	0.0000	0.0000	0.0001	0.0001	0.0002	0.0004	0.0006	0.0009	0.0012	0.0018	0.0024	0.0032	4	
	7	—	—	—	—	—	—	—	—	0.0000	0.0000	0.0000	0.0000	0.0000	0.0001	0.0001	0.0002	0.0003	0.0004	3	
	8	—	—	—	—	—	—	—	—	—	—	—	—	—	0.0000	0.0000	0.0000	0.0000	0.0000	2	
	9	—	—	—	—	—	—	—	—	—	—	—	—	—	—	—	—	—	—	1	
	10	—	—	—	—	—	—	—	—	—	—	—	—	—	—	—	—	—	—	0	10
n	*X*	0.99	0.98	0.97	0.96	0.95	0.94	0.93	0.92	0.91	0.90	0.89	0.88	0.87	0.86	0.85	0.84	0.83	0.82	*X*	*n*
											P										

TABLE E.7 *(Continued)*

		P																			
n	X	0.19	0.20	0.21	0.22	0.23	0.24	0.25	0.26	0.27	0.28	0.29	0.30	0.31	0.32	0.33	0.34	0.35	0.36	X	n
2	0	0.6561	0.6400	0.6241	0.6084	0.5929	0.5776	0.5625	0.5476	0.5329	0.5184	0.5041	0.4900	0.4761	0.4624	0.4489	0.4356	0.4225	0.4096	2	
	1	0.3078	0.3200	0.3318	0.3432	0.3542	0.3648	0.3750	0.3848	0.3942	0.4032	0.4118	0.4200	0.4278	0.4352	0.4422	0.4488	0.4550	0.4608	1	
	2	0.0361	0.0400	0.0441	0.0484	0.0529	0.0576	0.0625	0.0676	0.0729	0.0784	0.0841	0.0900	0.0961	0.1024	0.1089	0.1156	0.1225	0.1296	0	2
3	0	0.5314	0.5120	0.4930	0.4746	0.4565	0.4390	0.4219	0.4052	0.3890	0.3732	0.3579	0.3430	0.3285	0.3144	0.3008	0.2875	0.2746	0.2621	3	
	1	0.3740	0.3840	0.3932	0.4015	0.4091	0.4159	0.4219	0.4271	0.4316	0.4355	0.4386	0.4410	0.4428	0.4439	0.4444	0.4443	0.4436	0.4424	2	
	2	0.0877	0.0960	0.1045	0.1133	0.1222	0.1313	0.1406	0.1501	0.1597	0.1693	0.1791	0.1890	0.1989	0.2089	0.2189	0.2289	0.2389	0.2488	1	
	3	0.0069	0.0080	0.0093	0.0106	0.0122	0.0138	0.0156	0.0176	0.0197	0.0220	0.0244	0.0270	0.0298	0.0328	0.0359	0.0393	0.0429	0.0467	0	3
4	0	0.4305	0.4096	0.3895	0.3702	0.3515	0.3336	0.3164	0.2999	0.2840	0.2687	0.2541	0.2401	0.2267	0.2138	0.2015	0.1897	0.1785	0.1678	4	
	1	0.4039	0.4096	0.4142	0.4176	0.4200	0.4214	0.4219	0.4214	0.4201	0.4180	0.4152	0.4116	0.4074	0.4025	0.3970	0.3910	0.3845	0.3775	3	
	2	0.1421	0.1536	0.1651	0.1767	0.1882	0.1996	0.2109	0.2221	0.2331	0.2439	0.2544	0.2646	0.2745	0.2841	0.2933	0.3021	0.3105	0.3185	2	
	3	0.0222	0.0256	0.0293	0.0332	0.0375	0.0420	0.0469	0.0520	0.0575	0.0632	0.0693	0.0756	0.0822	0.0891	0.0963	0.1038	0.1115	0.1194	1	
	4	0.0013	0.0016	0.0019	0.0023	0.0028	0.0033	0.0039	0.0046	0.0053	0.0061	0.0071	0.0081	0.0092	0.0105	0.0119	0.0134	0.0150	0.0168	0	4
5	0	0.3487	0.3277	0.3077	0.2887	0.2707	0.2536	0.2373	0.2219	0.2073	0.1935	0.1804	0.1681	0.1564	0.1454	0.1350	0.1252	0.1160	0.1074	5	
	1	0.4089	0.4096	0.4090	0.4072	0.4043	0.4003	0.3955	0.3898	0.3834	0.3762	0.3685	0.3601	0.3513	0.3421	0.3325	0.3226	0.3124	0.3020	4	
	2	0.1919	0.2048	0.2174	0.2297	0.2415	0.2529	0.2637	0.2739	0.2836	0.2926	0.3010	0.3087	0.3157	0.3220	0.3275	0.3323	0.3364	0.3397	3	
	3	0.0450	0.0512	0.0578	0.0648	0.0721	0.0798	0.0879	0.0962	0.1049	0.1138	0.1229	0.1323	0.1418	0.1515	0.1613	0.1712	0.1811	0.1911	2	
	4	0.0053	0.0064	0.0077	0.0091	0.0108	0.0126	0.0146	0.0169	0.0194	0.0221	0.0251	0.0283	0.0319	0.0357	0.0397	0.0441	0.0488	0.0537	1	
	5	0.0002	0.0003	0.0004	0.0005	0.0006	0.0008	0.0010	0.0012	0.0014	0.0017	0.0021	0.0024	0.0029	0.0034	0.0039	0.0045	0.0053	0.0060	0	5
6	0	0.2824	0.2621	0.2431	0.2252	0.2084	0.1927	0.1780	0.1642	0.1513	0.1393	0.1281	0.1176	0.1079	0.0989	0.0905	0.0827	0.0754	0.0687	6	
	1	0.3975	0.3932	0.3877	0.3811	0.3735	0.3651	0.3560	0.3462	0.3358	0.3251	0.3139	0.3025	0.2909	0.2792	0.2673	0.2555	0.2437	0.2319	5	
	2	0.2331	0.2458	0.2577	0.2687	0.2789	0.2882	0.2966	0.3041	0.3105	0.3160	0.3206	0.3241	0.3267	0.3284	0.3292	0.3290	0.3280	0.3261	4	
	3	0.0729	0.0819	0.0913	0.1011	0.1111	0.1214	0.1318	0.1424	0.1531	0.1639	0.1746	0.1852	0.1957	0.2061	0.2162	0.2260	0.2355	0.2446	3	
	4	0.0128	0.0154	0.0182	0.0214	0.0249	0.0287	0.0330	0.0375	0.0425	0.0478	0.0535	0.0595	0.0660	0.0727	0.0799	0.0873	0.0951	0.1032	2	
	5	0.0012	0.0015	0.0019	0.0024	0.0030	0.0036	0.0044	0.0053	0.0063	0.0074	0.0087	0.0102	0.0119	0.0137	0.0157	0.0180	0.0205	0.0232	1	
	6	0.0000	0.0001	0.0001	0.0001	0.0001	0.0002	0.0002	0.0003	0.0004	0.0005	0.0006	0.0007	0.0009	0.0011	0.0013	0.0015	0.0018	0.0022	0	6

n	X	0.81	0.80	0.79	0.78	0.77	0.76	0.75	0.74	0.73	0.72	0.71	0.70	0.69	0.68	0.67	0.66	0.65	0.64	X	n
7	0	0.2288	0.2097	0.1920	0.1757	0.1605	0.1465	0.1335	0.1215	0.1105	0.1003	0.0910	0.0824	0.0745	0.0672	0.0606	0.0546	0.0490	0.0440	7	
	1	0.3756	0.3670	0.3573	0.3468	0.3356	0.3237	0.3115	0.2989	0.2860	0.2731	0.2600	0.2471	0.2342	0.2215	0.2090	0.1967	0.1848	0.1732	6	
	2	0.2643	0.2753	0.2850	0.2935	0.3007	0.3067	0.3115	0.3150	0.3174	0.3186	0.3186	0.3177	0.3156	0.3127	0.3088	0.3040	0.2985	0.2922	5	
	3	0.1033	0.1147	0.1263	0.1379	0.1497	0.1614	0.1730	0.1845	0.1956	0.2065	0.2169	0.2269	0.2363	0.2452	0.2535	0.2610	0.2679	0.2740	4	
	4	0.0242	0.0287	0.0336	0.0389	0.0447	0.0510	0.0577	0.0648	0.0724	0.0803	0.0886	0.0972	0.1062	0.1154	0.1248	0.1345	0.1442	0.1541	3	
	5	0.0034	0.0043	0.0054	0.0066	0.0080	0.0097	0.0115	0.0137	0.0161	0.0187	0.0217	0.0250	0.0286	0.0326	0.0369	0.0416	0.0466	0.0520	2	
	6	0.0003	0.0004	0.0005	0.0006	0.0008	0.0010	0.0013	0.0016	0.0020	0.0024	0.0030	0.0036	0.0043	0.0051	0.0061	0.0071	0.0084	0.0098	1	
	7	0.0000	0.0000	0.0000	0.0000	0.0000	0.0000	0.0001	0.0001	0.0001	0.0001	0.0002	0.0002	0.0003	0.0003	0.0004	0.0005	0.0006	0.0008	0	7
8	0	0.1853	0.1678	0.1517	0.1370	0.1236	0.1113	0.1001	0.0899	0.0806	0.0722	0.0646	0.0576	0.0514	0.0457	0.0406	0.0360	0.0319	0.0281	8	
	1	0.3477	0.3355	0.3226	0.3092	0.2953	0.2812	0.2670	0.2527	0.2386	0.2247	0.2110	0.1977	0.1847	0.1721	0.1600	0.1484	0.1373	0.1267	7	
	2	0.2855	0.2936	0.3002	0.3052	0.3087	0.3108	0.3115	0.3108	0.3089	0.3058	0.3017	0.2965	0.2904	0.2835	0.2758	0.2675	0.2587	0.2494	6	
	3	0.1339	0.1468	0.1596	0.1722	0.1844	0.1963	0.2076	0.2184	0.2285	0.2379	0.2464	0.2541	0.2609	0.2668	0.2717	0.2756	0.2786	0.2805	5	
	4	0.0393	0.0459	0.0530	0.0607	0.0689	0.0775	0.0865	0.0959	0.1056	0.1156	0.1258	0.1361	0.1465	0.1569	0.1673	0.1775	0.1875	0.1973	4	
	5	0.0074	0.0092	0.0113	0.0137	0.0165	0.0196	0.0231	0.0270	0.0313	0.0360	0.0411	0.0467	0.0527	0.0591	0.0659	0.0732	0.0808	0.0888	3	
	6	0.0009	0.0011	0.0015	0.0019	0.0025	0.0031	0.0038	0.0047	0.0058	0.0070	0.0084	0.0100	0.0118	0.0139	0.0162	0.0188	0.0217	0.0250	2	
	7	0.0001	0.0001	0.0001	0.0002	0.0002	0.0003	0.0004	0.0005	0.0006	0.0008	0.0010	0.0012	0.0015	0.0019	0.0023	0.0028	0.0033	0.0040	1	
	8	0.0000	0.0000	0.0000	0.0000	0.0000	0.0000	0.0000	0.0000	0.0000	0.0000	0.0001	0.0001	0.0001	0.0001	0.0001	0.0002	0.0002	0.0003	0	8
9	0	0.1501	0.1342	0.1199	0.1069	0.0952	0.0846	0.0751	0.0665	0.0589	0.0520	0.0458	0.0404	0.0355	0.0311	0.0272	0.0238	0.0207	0.0180	9	
	1	0.3169	0.3020	0.2867	0.2713	0.2558	0.2404	0.2253	0.2104	0.1960	0.1820	0.1685	0.1556	0.1433	0.1317	0.1206	0.1102	0.1004	0.0912	8	
	2	0.2973	0.3020	0.3049	0.3061	0.3056	0.3037	0.3003	0.2957	0.2899	0.2831	0.2754	0.2668	0.2576	0.2478	0.2376	0.2270	0.2162	0.2052	7	
	3	0.1627	0.1762	0.1891	0.2014	0.2130	0.2238	0.2336	0.2424	0.2502	0.2569	0.2624	0.2668	0.2701	0.2721	0.2731	0.2729	0.2716	0.2693	6	
	4	0.0573	0.0661	0.0754	0.0852	0.0954	0.1060	0.1168	0.1278	0.1388	0.1499	0.1608	0.1715	0.1820	0.1921	0.2017	0.2109	0.2194	0.2272	5	
	5	0.0134	0.0165	0.0200	0.0240	0.0285	0.0335	0.0389	0.0449	0.0513	0.0583	0.0657	0.0735	0.0818	0.0904	0.0994	0.1086	0.1181	0.1278	4	
	6	0.0021	0.0028	0.0036	0.0045	0.0057	0.0070	0.0087	0.0105	0.0127	0.0151	0.0179	0.0210	0.0245	0.0284	0.0326	0.0373	0.0424	0.0479	3	
	7	0.0002	0.0003	0.0004	0.0005	0.0007	0.0010	0.0012	0.0016	0.0020	0.0025	0.0031	0.0039	0.0047	0.0057	0.0069	0.0082	0.0098	0.0116	2	
	8	0.0000	0.0000	0.0000	0.0000	0.0001	0.0001	0.0001	0.0001	0.0002	0.0002	0.0003	0.0004	0.0005	0.0007	0.0008	0.0011	0.0013	0.0016	1	
	9	—	—	—	—	0.0000	0.0000	0.0000	0.0000	0.0000	0.0000	0.0000	0.0000	0.0000	0.0000	0.0000	0.0001	0.0001	0.0001	0	9
10	0	0.1216	0.1074	0.0947	0.0834	0.0733	0.0643	0.0563	0.0492	0.0430	0.0374	0.0326	0.0282	0.0245	0.0211	0.0182	0.0157	0.0135	0.0115	10	
	1	0.2852	0.2684	0.2517	0.2351	0.2188	0.2030	0.1877	0.1730	0.1590	0.1456	0.1330	0.1211	0.1099	0.0995	0.0898	0.0808	0.0725	0.0649	9	
	2	0.3010	0.3020	0.3011	0.2984	0.2942	0.2885	0.2816	0.2735	0.2646	0.2548	0.2444	0.2335	0.2222	0.2107	0.1990	0.1873	0.1757	0.1642	8	
	3	0.1883	0.2013	0.2134	0.2244	0.2343	0.2429	0.2503	0.2563	0.2609	0.2642	0.2662	0.2668	0.2662	0.2644	0.2614	0.2573	0.2522	0.2462	7	
	4	0.0773	0.0881	0.0993	0.1108	0.1225	0.1343	0.1460	0.1576	0.1689	0.1798	0.1903	0.2001	0.2093	0.2177	0.2253	0.2320	0.2377	0.2424	6	
	5	0.0218	0.0264	0.0317	0.0375	0.0439	0.0509	0.0584	0.0664	0.0750	0.0839	0.0933	0.1029	0.1128	0.1229	0.1332	0.1434	0.1536	0.1636	5	
	6	0.0043	0.0055	0.0070	0.0088	0.0109	0.0134	0.0162	0.0195	0.0231	0.0272	0.0317	0.0368	0.0422	0.0482	0.0547	0.0616	0.0689	0.0767	4	
	7	0.0006	0.0008	0.0011	0.0014	0.0019	0.0024	0.0031	0.0039	0.0049	0.0060	0.0074	0.0090	0.0108	0.0130	0.0154	0.0181	0.0212	0.0247	3	
	8	0.0001	0.0001	0.0001	0.0002	0.0002	0.0003	0.0004	0.0005	0.0007	0.0009	0.0011	0.0014	0.0018	0.0023	0.0028	0.0035	0.0043	0.0052	2	
	9	0.0000	0.0000	0.0000	0.0000	0.0000	0.0000	0.0000	0.0000	0.0001	0.0001	0.0001	0.0001	0.0002	0.0002	0.0003	0.0004	0.0005	0.0006	1	
	10	—	—	—	—	—	—	—	—	0.0000	0.0000	0.0000	0.0000	0.0000	0.0000	0.0000	0.0000	0.0000	0.0000	0	10
												P									

TABLE E.7 *(Continued)*

		P															
n	X	0.37	0.38	0.39	0.40	0.41	0.42	0.43	0.44	0.45	0.46	0.47	0.48	0.49	0.50	X	n
2	0	0.3969	0.3844	0.3721	0.3600	0.3481	0.3364	0.3249	0.3136	0.3025	0.2916	0.2809	0.2704	0.2601	0.2500	2	
	1	0.4662	0.4712	0.4758	0.4800	0.4838	0.4872	0.4902	0.4928	0.4950	0.4968	0.4982	0.4992	0.4998	0.5000	1	
	2	0.1369	0.1444	0.1521	0.1600	0.1681	0.1764	0.1849	0.1936	0.2025	0.2116	0.2209	0.2304	0.2401	0.2500	0	2
3	0	0.2500	0.2383	0.2270	0.2160	0.2054	0.1951	0.1852	0.1756	0.1664	0.1575	0.1489	0.1406	0.1327	0.1250	3	
	1	0.4406	0.4382	0.4354	0.4320	0.4282	0.4239	0.4191	0.4140	0.4084	0.4024	0.3961	0.3894	0.3823	0.3750	2	
	2	0.2587	0.2686	0.2783	0.2880	0.2975	0.3069	0.3162	0.3252	0.3341	0.3428	0.3512	0.3594	0.3674	0.3750	1	
	3	0.0507	0.0549	0.0593	0.0640	0.0689	0.0741	0.0795	0.0852	0.0911	0.0973	0.1038	0.1106	0.1176	0.1250	0	3
4	0	0.1575	0.1478	0.1385	0.1296	0.1212	0.1132	0.1056	0.0983	0.0915	0.0850	0.0789	0.0731	0.0677	0.0625	4	
	1	0.3701	0.3623	0.3541	0.3456	0.3368	0.3278	0.3185	0.3091	0.2995	0.2897	0.2799	0.2700	0.2600	0.2500	3	
	2	0.3260	0.3330	0.3396	0.3456	0.3511	0.3560	0.3604	0.3643	0.3675	0.3702	0.3723	0.3738	0.3747	0.3750	2	
	3	0.1276	0.1361	0.1447	0.1536	0.1627	0.1719	0.1813	0.1908	0.2005	0.2102	0.2201	0.2300	0.2400	0.2500	1	
	4	0.0187	0.0209	0.0231	0.0256	0.0283	0.0311	0.0342	0.0375	0.0410	0.0448	0.0488	0.0531	0.0576	0.0625	0	4
5	0	0.0992	0.0916	0.0845	0.0778	0.0715	0.0656	0.0602	0.0551	0.0503	0.0459	0.0418	0.0380	0.0345	0.0312	5	
	1	0.2914	0.2808	0.2700	0.2592	0.2484	0.2376	0.2270	0.2164	0.2059	0.1956	0.1854	0.1755	0.1657	0.1562	4	
	2	0.3423	0.3441	0.3452	0.3456	0.3452	0.3442	0.3424	0.3400	0.3369	0.3332	0.3289	0.3240	0.3185	0.3125	3	
	3	0.2010	0.2109	0.2207	0.2304	0.2399	0.2492	0.2583	0.2671	0.2757	0.2838	0.2916	0.2990	0.3060	0.3125	2	
	4	0.0590	0.0646	0.0706	0.0768	0.0834	0.0902	0.0974	0.1049	0.1128	0.1209	0.1293	0.1380	0.1470	0.1562	1	
	5	0.0069	0.0079	0.0090	0.0102	0.0116	0.0131	0.0147	0.0165	0.0185	0.0206	0.0229	0.0255	0.0282	0.0312	0	5
6	0	0.0625	0.0568	0.0515	0.0467	0.0422	0.0381	0.0343	0.0308	0.0277	0.0248	0.0222	0.0198	0.0176	0.0156	6	
	1	0.2203	0.2089	0.1976	0.1866	0.1759	0.1654	0.1552	0.1454	0.1359	0.1267	0.1179	0.1095	0.1014	0.0937	5	
	2	0.3235	0.3201	0.3159	0.3110	0.3055	0.2994	0.2928	0.2856	0.2780	0.2699	0.2615	0.2527	0.2436	0.2344	4	
	3	0.2533	0.2616	0.2693	0.2765	0.2831	0.2891	0.2945	0.2992	0.3032	0.3065	0.3091	0.3110	0.3121	0.3125	3	
	4	0.1116	0.1202	0.1291	0.1382	0.1475	0.1570	0.1666	0.1763	0.1861	0.1958	0.2056	0.2153	0.2249	0.2344	2	
	5	0.0262	0.0295	0.0330	0.0369	0.0410	0.0455	0.0503	0.0554	0.0609	0.0667	0.0729	0.0795	0.0864	0.0937	1	
	6	0.0026	0.0030	0.0035	0.0041	0.0048	0.0055	0.0063	0.0073	0.0083	0.0095	0.0108	0.0122	0.0138	0.0156	0	6

n	X	0.63	0.62	0.61	0.60	0.59	0.58	0.57	0.56	0.55	0.54	0.53	0.52	0.51	0.50	X	n
7	0	0.0394	0.0352	0.0314	0.0280	0.0249	0.0221	0.0195	0.0173	0.0152	0.0134	0.0117	0.0103	0.0090	0.0078	7	
	1	0.1619	0.1511	0.1407	0.1306	0.1211	0.1119	0.1032	0.0950	0.0872	0.0798	0.0729	0.0664	0.0604	0.0547	6	
	2	0.2853	0.2778	0.2698	0.2613	0.2524	0.2431	0.2336	0.2239	0.2140	0.2040	0.1940	0.1840	0.1740	0.1641	5	
	3	0.2793	0.2838	0.2875	0.2903	0.2923	0.2934	0.2937	0.2932	0.2918	0.2897	0.2867	0.2830	0.2786	0.2734	4	
	4	0.1640	0.1739	0.1838	0.1935	0.2031	0.2125	0.2216	0.2304	0.2388	0.2468	0.2543	0.2612	0.2676	0.2734	3	
	5	0.0578	0.0640	0.0705	0.0774	0.0847	0.0923	0.1003	0.1086	0.1172	0.1261	0.1353	0.1447	0.1543	0.1641	2	
	6	0.0113	0.0131	0.0150	0.0172	0.0196	0.0223	0.0252	0.0284	0.0320	0.0358	0.0400	0.0445	0.0494	0.0547	1	
	7	0.0009	0.0011	0.0014	0.0016	0.0019	0.0023	0.0027	0.0032	0.0037	0.0044	0.0051	0.0059	0.0068	0.0078	0	7
8	0	0.0248	0.0218	0.0192	0.0168	0.0147	0.0128	0.0111	0.0097	0.0084	0.0072	0.0062	0.0053	0.0046	0.0039	8	
	1	0.1166	0.1071	0.0981	0.0896	0.0816	0.0742	0.0672	0.0608	0.0548	0.0493	0.0442	0.0395	0.0352	0.0312	7	
	2	0.2397	0.2297	0.2194	0.2090	0.1985	0.1880	0.1776	0.1672	0.1569	0.1469	0.1371	0.1275	0.1183	0.1094	6	
	3	0.2815	0.2815	0.2806	0.2787	0.2759	0.2723	0.2679	0.2627	0.2568	0.2503	0.2431	0.2355	0.2273	0.2187	5	
	4	0.2067	0.2157	0.2242	0.2322	0.2397	0.2465	0.2526	0.2580	0.2627	0.2665	0.2695	0.2717	0.2730	0.2734	4	
	5	0.0971	0.1058	0.1147	0.1239	0.1332	0.1428	0.1525	0.1622	0.1719	0.1816	0.1912	0.2006	0.2098	0.2187	3	
	6	0.0285	0.0324	0.0367	0.0413	0.0463	0.0517	0.0575	0.0637	0.0703	0.0774	0.0848	0.0926	0.1008	0.1094	2	
	7	0.0048	0.0057	0.0067	0.0079	0.0092	0.0107	0.0124	0.0143	0.0164	0.0188	0.0215	0.0244	0.0277	0.0312	1	
	8	0.0004	0.0004	0.0005	0.0007	0.0008	0.0010	0.0012	0.0014	0.0017	0.0020	0.0024	0.0028	0.0033	0.0039	0	8
9	0	0.0156	0.0135	0.0117	0.0101	0.0087	0.0074	0.0064	0.0054	0.0046	0.0039	0.0033	0.0028	0.0023	0.0020	9	
	1	0.0826	0.0747	0.0673	0.0605	0.0542	0.0484	0.0431	0.0383	0.0339	0.0299	0.0263	0.0231	0.0202	0.0176	8	
	2	0.1941	0.1831	0.1721	0.1612	0.1506	0.1402	0.1301	0.1204	0.1110	0.1020	0.0934	0.0853	0.0776	0.0703	7	
	3	0.2660	0.2618	0.2567	0.2508	0.2442	0.2369	0.2291	0.2207	0.2119	0.2027	0.1933	0.1837	0.1739	0.1641	6	
	4	0.2344	0.2407	0.2462	0.2508	0.2545	0.2573	0.2592	0.2601	0.2600	0.2590	0.2571	0.2543	0.2506	0.2461	5	
	5	0.1376	0.1475	0.1574	0.1672	0.1769	0.1863	0.1955	0.2044	0.2128	0.2207	0.2280	0.2347	0.2408	0.2461	4	
	6	0.0539	0.0603	0.0671	0.0743	0.0819	0.0900	0.0983	0.1070	0.1160	0.1253	0.1348	0.1445	0.1542	0.1641	3	
	7	0.0136	0.0158	0.0184	0.0212	0.0244	0.0279	0.0318	0.0360	0.0407	0.0458	0.0512	0.0571	0.0635	0.0703	2	
	8	0.0020	0.0024	0.0029	0.0035	0.0042	0.0051	0.0060	0.0071	0.0083	0.0097	0.0114	0.0132	0.0153	0.0176	1	
	9	0.0001	0.0002	0.0002	0.0003	0.0003	0.0004	0.0005	0.0006	0.0008	0.0009	0.0011	0.0014	0.0016	0.0020	0	9
10	0	0.0098	0.0084	0.0071	0.0060	0.0051	0.0043	0.0036	0.0030	0.0025	0.0021	0.0017	0.0014	0.0012	0.0010	10	
	1	0.0578	0.0514	0.0456	0.0403	0.0355	0.0312	0.0273	0.0238	0.0207	0.0180	0.0155	0.0133	0.0114	0.0098	9	
	2	0.1529	0.1419	0.1312	0.1209	0.1111	0.1017	0.0927	0.0843	0.0763	0.0688	0.0619	0.0554	0.0494	0.0439	8	
	3	0.2394	0.2319	0.2237	0.2150	0.2058	0.1963	0.1865	0.1765	0.1665	0.1564	0.1464	0.1364	0.1267	0.1172	7	
	4	0.2461	0.2487	0.2503	0.2508	0.2503	0.2488	0.2462	0.2427	0.2384	0.2331	0.2271	0.2204	0.2130	0.2051	6	
	5	0.1734	0.1829	0.1920	0.2007	0.2087	0.2162	0.2229	0.2289	0.2340	0.2383	0.2417	0.2441	0.2456	0.2461	5	
	6	0.0849	0.0934	0.1023	0.1115	0.1209	0.1304	0.1401	0.1499	0.1596	0.1692	0.1786	0.1878	0.1966	0.2051	4	
	7	0.0285	0.0327	0.0374	0.0425	0.0480	0.0540	0.0604	0.0673	0.0746	0.0824	0.0905	0.0991	0.1080	0.1172	3	
	8	0.0063	0.0075	0.0090	0.0106	0.0125	0.0147	0.0171	0.0198	0.0229	0.0263	0.0301	0.0343	0.0389	0.0439	2	
	9	0.0008	0.0010	0.0013	0.0016	0.0019	0.0024	0.0029	0.0035	0.0042	0.0050	0.0059	0.0070	0.0083	0.0098	1	
	10	0.0000	0.0001	0.0001	0.0001	0.0001	0.0002	0.0002	0.0003	0.0003	0.0004	0.0005	0.0006	0.0008	0.0010	0	10
n	X	0.63	0.62	0.61	0.60	0.59	0.58	0.57	0.56	0.55	0.54	0.53	0.52	0.51	0.50	X	n
									P								

TABLE E.7 *(Continued)*

		P																		
n	*X*	0.01	0.02	0.03	0.04	0.05	0.06	0.07	0.08	0.09	0.10	0.11	0.12	0.13	0.14	0.15	0.16	0.17	0.18	*X*	*n*
20	0	0.8179	0.6676	0.5438	0.4420	0.3585	0.2901	0.2342	0.1887	0.1516	0.1216	0.0972	0.0776	0.0617	0.0490	0.0388	0.0306	0.0241	0.0189	20	
	1	0.1652	0.2725	0.3364	0.3683	0.3774	0.3703	0.3526	0.3282	0.3000	0.2702	0.2403	0.2115	0.1844	0.1595	0.1368	0.1165	0.0986	0.0829	19	
	2	0.0159	0.0528	0.0988	0.1458	0.1887	0.2246	0.2521	0.2711	0.2818	0.2852	0.2822	0.2740	0.2618	0.2466	0.2293	0.2109	0.1919	0.1730	18	
	3	0.0010	0.0065	0.0183	0.0364	0.0596	0.0860	0.1139	0.1414	0.1672	0.1901	0.2093	0.2242	0.2347	0.2409	0.2428	0.2410	0.2358	0.2278	17	
	4	0.0000	0.0006	0.0024	0.0065	0.0133	0.0233	0.0364	0.0523	0.0703	0.0898	0.1099	0.1299	0.1491	0.1666	0.1821	0.1951	0.2053	0.2125	16	
	5	—	0.0000	0.0002	0.0009	0.0022	0.0048	0.0088	0.0145	0.0222	0.0319	0.0435	0.0567	0.0713	0.0868	0.1028	0.1189	0.1345	0.1493	15	
	6	—	—	0.0000	0.0001	0.0003	0.0008	0.0017	0.0032	0.0055	0.0089	0.0134	0.0193	0.0266	0.0353	0.0454	0.0566	0.0689	0.0819	14	
	7	—	—	—	0.0000	0.0000	0.0001	0.0002	0.0005	0.0011	0.0020	0.0033	0.0053	0.0080	0.0115	0.0160	0.0216	0.0282	0.0360	13	
	8	—	—	—	—	—	0.0000	0.0000	0.0001	0.0002	0.0004	0.0007	0.0012	0.0019	0.0030	0.0046	0.0067	0.0094	0.0128	12	
	9	—	—	—	—	—	—	—	0.0000	0.0000	0.0001	0.0001	0.0002	0.0004	0.0007	0.0011	0.0017	0.0026	0.0038	11	
	10	—	—	—	—	—	—	—	—	—	0.0000	0.0000	0.0000	0.0001	0.0001	0.0002	0.0004	0.0006	0.0009	10	
	11	—	—	—	—	—	—	—	—	—	—	—	—	0.0000	0.0000	0.0000	0.0001	0.0001	0.0002	9	
	12	—	—	—	—	—	—	—	—	—	—	—	—	—	—	—	0.0000	0.0000	0.0000	8	
	13	—	—	—	—	—	—	—	—	—	—	—	—	—	—	—	—	—	—	7	
	14	—	—	—	—	—	—	—	—	—	—	—	—	—	—	—	—	—	—	6	
	15	—	—	—	—	—	—	—	—	—	—	—	—	—	—	—	—	—	—	5	
	16	—	—	—	—	—	—	—	—	—	—	—	—	—	—	—	—	—	—	4	
	17	—	—	—	—	—	—	—	—	—	—	—	—	—	—	—	—	—	—	3	
	18	—	—	—	—	—	—	—	—	—	—	—	—	—	—	—	—	—	—	2	
	19	—	—	—	—	—	—	—	—	—	—	—	—	—	—	—	—	—	—	1	
	20	—	—	—	—	—	—	—	—	—	—	—	—	—	—	—	—	—	—	0	20
n	*X*	0.99	0.98	0.97	0.96	0.95	0.94	0.93	0.92	0.91	0.90	0.89	0.88	0.87	0.86	0.85	0.84	0.83	0.82	*X*	*n*
		P																			

n	X	0.19	0.20	0.21	0.22	0.23	0.24	0.25	0.26	0.27	0.28	0.29	0.30	0.31	0.32	0.33	0.34	0.35	0.36	X	n
		P																			
20	0	0.0148	0.0115	0.0090	0.0069	0.0054	0.0041	0.0032	0.0024	0.0018	0.0014	0.0011	0.0008	0.0006	0.0004	0.0003	0.0002	0.0002	0.0001	20	
	1	0.0693	0.0576	0.0477	0.0392	0.0321	0.0261	0.0211	0.0170	0.0137	0.0109	0.0087	0.0068	0.0054	0.0042	0.0033	0.0025	0.0020	0.0015	19	
	2	0.1545	0.1369	0.1204	0.1050	0.0910	0.0783	0.0669	0.0569	0.0480	0.0403	0.0336	0.0278	0.0229	0.0188	0.0153	0.0124	0.0100	0.0080	18	
	3	0.2175	0.2054	0.1920	0.1777	0.1631	0.1484	0.1339	0.1199	0.1065	0.0940	0.0823	0.0716	0.0619	0.0531	0.0453	0.0383	0.0323	0.0270	17	
	4	0.2168	0.2182	0.2169	0.2131	0.2070	0.1991	0.1897	0.1790	0.1675	0.1553	0.1429	0.1304	0.1181	0.1062	0.0947	0.0839	0.0738	0.0645	16	
	5	0.1627	0.1746	0.1845	0.1923	0.1979	0.2012	0.2023	0.2013	0.1982	0.1933	0.1868	0.1789	0.1698	0.1599	0.1493	0.1384	0.1272	0.1161	15	
	6	0.0954	0.1091	0.1226	0.1356	0.1478	0.1589	0.1686	0.1768	0.1833	0.1879	0.1907	0.1916	0.1907	0.1881	0.1839	0.1782	0.1712	0.1632	14	
	7	0.0448	0.0545	0.0652	0.0765	0.0883	0.1003	0.1124	0.1242	0.1356	0.1462	0.1558	0.1643	0.1714	0.1770	0.1811	0.1836	0.1844	0.1836	13	
	8	0.0171	0.0222	0.0282	0.0351	0.0429	0.0515	0.0609	0.0709	0.0815	0.0924	0.1034	0.1144	0.1251	0.1354	0.1450	0.1537	0.1614	0.1678	12	
	9	0.0053	0.0074	0.0100	0.0132	0.0171	0.0217	0.0271	0.0332	0.0402	0.0479	0.0563	0.0654	0.0750	0.0849	0.0952	0.1056	0.1158	0.1259	11	
	10	0.0014	0.0020	0.0029	0.0041	0.0056	0.0075	0.0099	0.0128	0.0163	0.0205	0.0253	0.0308	0.0370	0.0440	0.0516	0.0598	0.0686	0.0779	10	
	11	0.0003	0.0005	0.0007	0.0010	0.0015	0.0022	0.0030	0.0041	0.0055	0.0072	0.0094	0.0120	0.0151	0.0188	0.0231	0.0280	0.0336	0.0398	9	
	12	0.0001	0.0001	0.0001	0.0002	0.0003	0.0005	0.0008	0.0011	0.0015	0.0021	0.0029	0.0039	0.0051	0.0066	0.0085	0.0108	0.0136	0.0168	8	
	13	0.0000	0.0000	0.0000	0.0000	0.0001	0.0001	0.0002	0.0002	0.0003	0.0005	0.0007	0.0010	0.0014	0.0019	0.0026	0.0034	0.0045	0.0058	7	
	14	—	—	—	—	0.0000	0.0000	0.0000	0.0000	0.0001	0.0001	0.0001	0.0002	0.0003	0.0005	0.0006	0.0009	0.0012	0.0016	6	
	15	—	—	—	—	—	—	—	—	0.0000	0.0000	0.0000	0.0000	0.0001	0.0001	0.0001	0.0002	0.0003	0.0004	5	
	16	—	—	—	—	—	—	—	—	—	—	—	—	0.0000	0.0000	0.0000	0.0000	0.0000	0.0001	4	
	17	—	—	—	—	—	—	—	—	—	—	—	—	—	—	—	—	—	0.0000	3	
	18	—	—	—	—	—	—	—	—	—	—	—	—	—	—	—	—	—	—	2	
	19	—	—	—	—	—	—	—	—	—	—	—	—	—	—	—	—	—	—	1	
	20	—	—	—	—	—	—	—	—	—	—	—	—	—	—	—	—	—	—	0	20
n	X	0.81	0.80	0.79	0.78	0.77	0.76	0.75	0.74	0.73	0.72	0.71	0.70	0.69	0.68	0.67	0.66	0.65	0.64	X	n
		P																			

TABLE E.7 *(Continued)*

n	X	0.37	0.38	0.39	0.40	0.41	0.42	0.43	0.44	0.45	0.46	0.47	0.48	0.49	0.50	X	n
20	0	0.0001	0.0001	0.0001	0.0000	0.0000	0.0000	0.0000	0.0000	0.0000	0.0000	0.0000	—	—	—	20	
	1	0.0011	0.0009	0.0007	0.0005	0.0004	0.0003	0.0002	0.0001	0.0001	0.0001	0.0001	0.0000	0.0000	0.0000	19	
	2	0.0064	0.0050	0.0040	0.0031	0.0024	0.0018	0.0014	0.0011	0.0008	0.0006	0.0005	0.0003	0.0002	0.0002	18	
	3	0.0224	0.0185	0.0152	0.0123	0.0100	0.0080	0.0064	0.0051	0.0040	0.0031	0.0024	0.0019	0.0014	0.0011	17	
	4	0.0559	0.0482	0.0412	0.0350	0.0295	0.0247	0.0206	0.0170	0.0139	0.0113	0.0092	0.0074	0.0059	0.0046	16	
	5	0.1051	0.0945	0.0843	0.0746	0.0656	0.0573	0.0496	0.0427	0.0365	0.0309	0.0260	0.0217	0.0180	0.0148	15	
	6	0.1543	0.1447	0.1347	0.1244	0.1140	0.1037	0.0936	0.0839	0.0746	0.0658	0.0577	0.0501	0.0432	0.0370	14	
	7	0.1812	0.1774	0.1722	0.1659	0.1585	0.1502	0.1413	0.1318	0.1221	0.1122	0.1023	0.0925	0.0830	0.0739	13	
	8	0.1730	0.1767	0.1790	0.1797	0.1790	0.1768	0.1732	0.1683	0.1623	0.1553	0.1474	0.1388	0.1296	0.1201	12	
	9	0.1354	0.1444	0.1526	0.1597	0.1658	0.1707	0.1742	0.1763	0.1771	0.1763	0.1742	0.1708	0.1661	0.1602	11	
	10	0.0875	0.0974	0.1073	0.1171	0.1268	0.1359	0.1446	0.1524	0.1593	0.1652	0.1700	0.1734	0.1755	0.1762	10	
	11	0.0467	0.0542	0.0624	0.0710	0.0801	0.0895	0.0991	0.1089	0.1185	0.1280	0.1370	0.1455	0.1533	0.1602	9	
	12	0.0206	0.0249	0.0299	0.0355	0.0417	0.0486	0.0561	0.0642	0.0727	0.0818	0.0911	0.1007	0.1105	0.1201	8	
	13	0.0074	0.0094	0.0118	0.0146	0.0178	0.0217	0.0260	0.0310	0.0366	0.0429	0.0497	0.0572	0.0653	0.0739	7	
	14	0.0022	0.0029	0.0038	0.0049	0.0062	0.0078	0.0098	0.0122	0.0150	0.0183	0.0221	0.0264	0.0314	0.0370	6	
	15	0.0005	0.0007	0.0010	0.0013	0.0017	0.0023	0.0030	0.0038	0.0049	0.0062	0.0078	0.0098	0.0121	0.0148	5	
	16	0.0001	0.0001	0.0002	0.0003	0.0004	0.0005	0.0007	0.0009	0.0013	0.0017	0.0022	0.0028	0.0036	0.0046	4	
	17	0.0000	0.0000	0.0000	0.0000	0.0001	0.0001	0.0001	0.0002	0.0002	0.0003	0.0005	0.0006	0.0008	0.0011	3	
	18	—	—	—	—	0.0000	0.0000	0.0000	0.0000	0.0000	0.0000	0.0001	0.0001	0.0001	0.0002	2	
	19	—	—	—	—	—	—	—	—	—	—	0.0000	0.0000	0.0000	0.0000	1	
	20	—	—	—	—	—	—	—	—	—	—	—	—	—	—	0	20
n	X	0.63	0.62	0.61	0.60	0.59	0.58	0.57	0.56	0.55	0.54	0.53	0.52	0.51	0.50	X	n

P (top header spans 0.37–0.50; bottom header spans 0.63–0.50)

TABLE E.8 Critical Values of Hartley's F_{max} Test $\left(F_{max} = \frac{s^2_{largest}}{s^2_{smallest}} \sim F_{max_{1-\alpha(c,\nu)}}\right)$

Upper 5% points (α = .05)

ν \ c	2	3	4	5	6	7	8	9	10	11	12
2	39.0	87.5	142	202	266	333	403	475	550	626	704
3	15.4	27.8	39.2	50.7	62.0	72.9	83.5	93.9	104	114	124
4	9.60	15.5	20.6	25.2	29.5	33.6	37.5	41.1	44.6	48.0	51.4
5	7.15	10.8	13.7	16.3	18.7	20.8	22.9	24.7	26.5	28.2	29.9
6	5.82	8.38	10.4	12.1	13.7	15.0	16.3	17.5	18.6	19.7	20.7
7	4.99	6.94	8.44	9.70	10.8	11.8	12.7	13.5	14.3	15.1	15.8
8	4.43	6.00	7.18	8.12	9.03	9.78	10.5	11.1	11.7	12.2	12.7
9	4.03	5.34	6.31	7.11	7.80	8.41	8.95	9.45	9.91	10.3	10.7
10	3.72	4.85	5.67	6.34	6.92	7.42	7.87	8.28	8.66	9.01	9.34
12	3.28	4.16	4.79	5.30	5.72	6.09	6.42	6.72	7.00	7.25	7.48
15	2.86	3.54	4.01	4.37	4.68	4.95	5.19	5.40	5.59	5.77	5.93
20	2.46	2.95	3.29	3.54	3.76	3.94	4.10	4.24	4.37	4.49	4.59
30	2.07	2.40	2.61	2.78	2.91	3.02	3.12	3.21	3.29	3.36	3.39
60	1.67	1.85	1.96	2.04	2.11	2.17	2.22	2.26	2.30	2.33	2.36
∞	1.00	1.00	1.00	1.00	1.00	1.00	1.00	1.00	1.00	1.00	1.00

Upper 1% points (α = .01)

ν \ c	2	3	4	5	6	7	8	9	10	11	12
2	199	448	729	1036	1362	1705	2063	2432	2813	3204	3605
3	47.5	85	120	151	184	21(6)	24(9)	28(1)	31(0)	33(7)	36(1)
4	23.2	37	49	59	69	79	89	97	106	113	120
5	14.9	22	28	33	38	42	46	50	54	57	60
6	11.1	15.5	19.1	22	25	27	30	32	34	36	37
7	8.89	12.1	14.5	16.5	18.4	20	22	23	24	26	27
8	7.50	9.9	11.7	13.2	14.5	15.8	16.9	17.9	18.9	19.8	21
9	6.54	8.5	9.9	11.1	12.1	13.1	13.9	14.7	15.3	16.0	16.6
10	5.85	7.4	8.6	9.6	10.4	11.1	11.8	12.4	12.9	13.4	13.9
12	4.91	6.1	6.9	7.6	8.2	8.7	9.1	9.5	9.9	10.2	10.6
15	4.07	4.9	5.5	6.0	6.4	6.7	7.1	7.3	7.5	7.8	8.0
20	3.32	3.8	4.3	4.6	4.9	5.1	5.3	5.5	5.6	5.8	5.9
30	2.63	3.0	3.3	3.4	3.6	3.7	3.8	3.9	4.0	4.1	4.2
60	1.96	2.2	2.3	2.4	2.4	2.5	2.5	2.6	2.6	2.7	2.7
∞	1.00	1.0	1.0	1.0	1.0	1.0	1.0	1.0	1.0	1.0	1.0

$s^2_{largest}$ is the largest and $s^2_{smallest}$ the smallest in a set of c independent mean squares, each based on ν degrees of freedom.

Source: Reprinted from E. S. Pearson and H. O. Hartley eds., *Biometrika Tables for Statisticians* 3rd ed., 1966, by permission of the *Biometrika* Trustees.

TABLE E.9 Lower and Upper Critical Values *U* for the Runs Test for Randomness

Part 1. Lower Tail ($\alpha = .025$)

n_1 \ n_2	2	3	4	5	6	7	8	9	10	11	12	13	14	15	16	17	18	19	20
2											2	2	2	2	2	2	2	2	2
3					2	2	2	2	2	2	2	2	2	3	3	3	3	3	3
4				2	2	2	3	3	3	3	3	3	3	3	4	4	4	4	4
5			2	2	3	3	3	3	3	4	4	4	4	4	4	4	5	5	5
6		2	2	3	3	3	3	4	4	4	4	5	5	5	5	5	5	6	6
7		2	2	3	3	3	4	4	5	5	5	5	5	6	6	6	6	6	6
8		2	3	3	3	4	4	5	5	5	6	6	6	6	6	7	7	7	7
9		2	3	3	4	4	5	5	5	6	6	6	7	7	7	7	8	8	8
10		2	3	3	4	5	5	5	6	6	7	7	7	7	8	8	8	8	9
11		2	3	4	4	5	5	6	6	7	7	7	8	8	8	9	9	9	9
12	2	2	3	4	4	5	6	6	7	7	7	8	8	8	9	9	9	10	10
13	2	2	3	4	5	5	6	6	7	7	8	8	9	9	9	10	10	10	10
14	2	2	3	4	5	5	6	7	7	8	8	9	9	9	10	10	10	11	11
15	2	3	3	4	5	6	6	7	7	8	8	9	9	10	10	11	11	11	12
16	2	3	4	4	5	6	6	7	8	8	9	9	10	10	11	11	11	12	12
17	2	3	4	4	5	6	7	7	8	9	9	10	10	11	11	11	12	12	13
18	2	3	4	5	5	6	7	8	8	9	9	10	10	11	11	12	12	13	13
19	2	3	4	5	6	6	7	8	8	9	10	10	11	11	12	12	13	13	13
20	2	3	4	5	6	6	7	8	9	9	10	10	11	12	12	13	13	13	14

Part 2. Upper Tail ($\alpha = .025$)

n_1 \ n_2	2	3	4	5	6	7	8	9	10	11	12	13	14	15	16	17	18	19	20
2																			
3																			
4				9	9														
5			9	10	10	11	11												
6			9	10	11	12	12	13	13	13	13								
7				11	12	13	13	14	14	14	14	15	15	15					
8				11	12	13	14	14	15	15	16	16	16	16	17	17	17	17	17
9					13	14	14	15	16	16	16	17	17	18	18	18	18	18	18
10					13	14	15	16	16	17	17	18	18	18	19	19	19	20	20
11					13	14	15	16	17	17	18	19	19	19	20	20	20	21	21
12					13	14	16	16	17	18	19	19	20	20	21	21	21	22	22
13						15	16	17	18	19	19	20	20	21	21	22	22	23	23
14						15	16	17	18	19	20	20	21	22	22	23	23	23	24
15						15	16	18	18	19	20	21	22	22	23	23	24	24	25
16							17	18	19	20	21	21	22	23	23	24	25	25	25
17							17	18	19	20	21	22	23	23	24	25	25	26	26
18							17	18	19	20	21	22	23	24	25	25	26	26	27
19							17	18	20	21	22	23	23	24	25	26	26	27	27
20							17	18	20	21	22	23	24	25	25	26	27	27	28

Source: Adapted from F. S. Swed and C. Eisenhart, *Ann. Math. Statist.*, vol. 14, 1943, pp. 83–86.

TABLE E.10 **Lower and Upper Critical Values *W* of Wilcoxon Signed-Ranks Test**

n	*One-Tailed:* $\alpha = .05$ *Two-Tailed:* $\alpha = .10$	$\alpha = .025$ $\alpha = .05$	$\alpha = .01$ $\alpha = .02$	$\alpha = .005$ $\alpha = .01$
	(Lower, Upper)			
5	0,15	—,—	—,—	—,—
6	2,19	0,21	—,—	—,—
7	3,25	2,26	0,28	—,—
8	5,31	3,33	1,35	0,36
9	8,37	5,40	3,42	1,44
10	10,45	8,47	5,50	3,52
11	13,53	10,56	7,59	5,61
12	17,61	13,65	10,68	7,71
13	21,70	17,74	12,79	10,81
14	25,80	21,84	16,89	13,92
15	30,90	25,95	19,101	16,104
16	35,101	29,107	23,113	19,117
17	41,112	34,119	27,126	23,130
18	47,124	40,131	32,139	27,144
19	53,137	46,144	37,153	32,158
20	60,150	52,158	43,167	37,173

Source: Adapted from Table 2 of F. Wilcoxon and R. A. Wilcox, *Some Rapid Approximate Statistical Procedures* (Pearl River, N.Y.: Lederle Laboratories, 1964), with permission of the American Cyanamid Company.

TABLE E.11 Lower and Upper Critical Values T_1 of Wilcoxon Rank Sum Test

n_2	α One-Tailed	α Two-Tailed	n_1 = 4	5	6	7	8	9	10
4	.05	.10	11,25						
	.025	.05	10,26						
	.01	.02	—,—						
	.005	.01	—,—						
5	.05	.10	12,28	19,36					
	.025	.05	11,29	17,38					
	.01	.02	10,30	16,39					
	.005	.01	—,—	15,40					
6	.05	.10	13,31	20,40	28,50				
	.025	.05	12,32	18,42	26,52				
	.01	.02	11,33	17,43	24,54				
	.005	.01	10,34	16,44	23,55				
7	.05	.10	14,34	21,44	29,55	39,66			
	.025	.05	13,35	20,45	27,57	36,69			
	.01	.02	11,37	18,47	25,59	34,71			
	.005	.01	10,38	16,49	24,60	32,73			
8	.05	.10	15,37	23,47	31,59	41,71	51,85		
	.025	.05	14,38	21,49	29,61	38,74	49,87		
	.01	.02	12,40	19,51	27,63	35,77	45,91		
	.005	.01	11,41	17,53	25,65	34,78	43,93		
9	.05	.10	16,40	24,51	33,63	43,76	54,90	66,105	
	.025	.05	14,42	22,53	31,65	40,79	51,93	62,109	
	.01	.02	13,43	20,55	28,68	37,82	47,97	59,112	
	.005	.01	11,45	18,57	26,70	35,84	45,99	56,115	
10	.05	.10	17,43	26,54	35,67	45,81	56,96	69,111	82,128
	.025	.05	15,45	23,57	32,70	42,84	53,99	65,115	78,132
	.01	.02	13,47	21,59	29,73	39,87	49,103	61,119	74,136
	.005	.01	12,48	19,61	27,75	37,89	47,105	58,122	71,139

Source: Adapted from Table 1 of F. Wilcoxon and R. A. Wilcox, *Some Rapid Approximate Statistical Procedures* (Pearl River, N.Y., Lederle Laboratories, 1964), with permission of the American Cyanamid Company.

TABLE E.12 Critical Values[a] of the Studentized Range Q

Upper 5% points ($\alpha = .05$)

v \ η	2	3	4	5	6	7	8	9	10	11	12	13	14	15	16	17	18	19	20
1	18.0	27.0	32.8	37.1	40.4	43.1	45.4	47.4	49.1	50.6	52.0	53.2	54.3	55.4	56.3	57.2	58.0	58.8	59.6
2	6.09	8.3	9.8	10.9	11.7	12.4	13.0	13.5	14.0	14.4	14.7	15.1	15.4	15.7	15.9	16.1	16.4	16.6	16.8
3	4.50	5.91	6.82	7.50	8.04	8.48	8.85	9.18	9.46	9.72	9.95	10.15	10.35	10.52	10.69	10.84	10.98	11.11	11.24
4	3.93	5.04	5.76	6.29	6.71	7.05	7.35	7.60	7.83	8.03	8.21	8.37	8.52	8.66	8.79	8.91	9.03	9.13	9.23
5	3.64	4.60	5.22	5.67	6.03	6.33	6.58	6.80	6.99	7.17	7.32	7.47	7.60	7.72	7.83	7.93	8.03	8.12	8.21
6	3.46	4.34	4.90	5.31	5.63	5.89	6.12	6.32	6.49	6.65	6.79	6.92	7.03	7.14	7.24	7.34	7.43	7.51	7.59
7	3.34	4.16	4.68	5.06	5.36	5.61	5.82	6.00	6.16	6.30	6.43	6.55	6.66	6.76	6.85	6.94	7.02	7.09	7.17
8	3.26	4.04	4.53	4.89	5.17	5.40	5.60	5.77	5.92	6.05	6.18	6.29	6.39	6.48	6.57	6.65	6.73	6.80	6.87
9	3.20	3.95	4.42	4.76	5.02	5.24	5.43	5.60	5.74	5.87	5.98	6.09	6.19	6.28	6.36	6.44	6.51	6.58	6.64
10	3.15	3.88	4.33	4.65	4.91	5.12	5.30	5.46	5.60	5.72	5.83	5.93	6.03	6.11	6.20	6.27	6.34	6.40	6.47
11	3.11	3.82	4.26	4.57	4.82	5.03	5.20	5.35	5.49	5.61	5.71	5.81	5.90	5.99	6.06	6.14	6.20	6.26	6.33
12	3.08	3.77	4.20	4.51	4.75	4.95	5.12	5.27	5.40	5.51	5.62	5.71	5.80	5.88	5.95	6.03	6.09	6.15	6.21
13	3.06	3.73	4.15	4.45	4.69	4.88	5.05	5.19	5.32	5.43	5.53	5.63	5.71	5.79	5.86	5.93	6.00	6.05	6.11
14	3.03	3.70	4.11	4.41	4.64	4.83	4.99	5.13	5.25	5.36	5.46	5.55	5.64	5.72	5.79	5.85	5.92	5.97	6.03
15	3.01	3.67	4.08	4.37	4.60	4.78	4.94	5.08	5.20	5.31	5.40	5.49	5.58	5.65	5.72	5.79	5.85	5.90	5.96
16	3.00	3.65	4.05	4.33	4.56	4.74	4.90	5.03	5.15	5.26	5.35	5.44	5.52	5.59	5.66	5.72	5.79	5.84	5.90
17	2.98	3.63	4.02	4.30	4.52	4.71	4.86	4.99	5.11	5.21	5.31	5.39	5.47	5.55	5.61	5.68	5.74	5.79	5.84
18	2.97	3.61	4.00	4.28	4.49	4.67	4.82	4.96	5.07	5.17	5.27	5.35	5.43	5.50	5.57	5.63	5.69	5.74	5.79
19	2.96	3.59	3.98	4.25	4.47	4.65	4.79	4.92	5.04	5.14	5.23	5.32	5.39	5.46	5.53	5.59	5.65	5.70	5.75
20	2.95	3.58	3.96	4.23	4.45	4.62	4.77	4.90	5.01	5.11	5.20	5.28	5.36	5.43	5.49	5.55	5.61	5.66	5.71
24	2.92	3.53	3.90	4.17	4.37	4.54	4.68	4.81	4.92	5.01	5.10	5.18	5.25	5.32	5.38	5.44	5.50	5.54	5.59
30	2.89	3.49	3.84	4.10	4.30	4.46	4.60	4.72	4.83	4.92	5.00	5.08	5.15	5.21	5.27	5.33	5.38	5.43	5.48
40	2.86	3.44	3.79	4.04	4.23	4.39	4.52	4.63	4.74	4.82	4.91	4.98	5.05	5.11	5.16	5.22	5.27	5.31	5.36
60	2.83	3.40	3.74	3.98	4.16	4.31	4.44	4.55	4.65	4.73	4.81	4.88	4.94	5.00	5.06	5.11	5.16	5.20	5.24
120	2.80	3.36	3.69	3.92	4.10	4.24	4.36	4.48	4.56	4.64	4.72	4.78	4.84	4.90	4.95	5.00	5.05	5.09	5.13
∞	2.77	3.31	3.63	3.86	4.03	4.17	4.29	4.39	4.47	4.55	4.62	4.68	4.74	4.80	4.85	4.89	4.93	4.97	5.01

TABLE E.12 *(Continued)*

Upper 1% points ($\alpha = .01$)

v \ η	2	3	4	5	6	7	8	9	10	11	12	13	14	15	16	17	18	19	20
1	90.0	135	164	186	202	216	227	237	246	253	260	266	272	277	282	286	290	294	298
2	14.0	19.0	22.3	24.7	26.6	28.2	29.5	30.7	31.7	32.6	33.4	34.1	34.8	35.4	36.0	36.5	37.0	37.5	37.9
3	8.26	10.6	12.2	13.3	14.2	15.0	15.6	16.2	16.7	17.1	17.5	17.9	18.2	18.5	18.8	19.1	19.3	19.5	19.8
4	6.51	8.12	9.17	9.96	10.6	11.1	11.5	11.9	12.3	12.6	12.8	13.1	13.3	13.5	13.7	13.9	14.1	14.2	14.4
5	5.70	6.97	7.80	8.42	8.91	9.32	9.67	9.97	10.24	10.48	10.70	10.89	11.08	11.24	11.40	11.55	11.68	11.81	11.93
6	5.24	6.33	7.03	7.56	7.97	8.32	8.61	8.87	9.10	9.30	9.49	9.65	9.81	9.95	10.08	10.21	10.32	10.43	10.54
7	4.95	5.92	6.54	7.01	7.37	7.68	7.94	8.17	8.37	8.55	8.71	8.86	9.00	9.12	9.24	9.35	9.46	9.55	9.65
8	4.74	5.63	6.20	6.63	6.96	7.24	7.47	7.68	7.87	8.03	8.18	8.31	8.44	8.55	8.66	8.76	8.85	8.94	9.03
9	4.60	5.43	5.96	6.35	6.66	6.91	7.13	7.32	7.49	7.65	7.78	7.91	8.03	8.13	8.23	8.32	8.41	8.49	8.57
10	4.48	5.27	5.77	6.14	6.43	6.67	6.87	7.05	7.21	7.36	7.48	7.60	7.71	7.81	7.91	7.99	8.07	8.15	8.22
11	4.39	5.14	5.62	5.97	6.25	6.48	6.67	6.84	6.99	7.13	7.25	7.36	7.46	7.56	7.65	7.73	7.81	7.88	7.95
12	4.32	5.04	5.50	5.84	6.10	6.32	6.51	6.67	6.81	6.94	7.06	7.17	7.26	7.36	7.44	7.52	7.59	7.66	7.73
13	4.26	4.96	5.40	5.73	5.98	6.19	6.37	6.53	6.67	6.79	6.90	7.01	7.10	7.19	7.27	7.34	7.42	7.48	7.55
14	4.21	4.89	5.32	5.63	5.88	6.08	6.26	6.41	6.54	6.66	6.77	6.87	6.96	7.05	7.12	7.20	7.27	7.33	7.39
15	4.17	4.83	5.25	5.56	5.80	5.99	6.16	6.31	6.44	6.55	6.66	6.76	6.84	6.93	7.00	7.07	7.14	7.20	7.26
16	4.13	4.78	5.19	5.49	5.72	5.92	6.08	6.22	6.35	6.46	6.56	6.66	6.74	6.82	6.90	6.97	7.03	7.09	7.15
17	4.10	4.74	5.14	5.43	5.66	5.85	6.01	6.15	6.27	6.38	6.48	6.57	6.66	6.73	6.80	6.87	6.94	7.00	7.05
18	4.07	4.70	5.09	5.38	5.60	5.79	5.94	6.08	6.20	6.31	6.41	6.50	6.58	6.65	6.72	6.79	6.85	6.91	6.96
19	4.05	4.67	5.05	5.33	5.55	5.73	5.89	6.02	6.14	6.25	6.34	6.43	6.51	6.58	6.65	6.72	6.78	6.84	6.89
20	4.02	4.64	5.02	5.29	5.51	5.69	5.84	5.97	6.09	6.19	6.29	6.37	6.45	6.52	6.59	6.65	6.71	6.76	6.82
24	3.96	4.54	4.91	5.17	5.37	5.54	5.69	5.81	5.92	6.02	6.11	6.19	6.26	6.33	6.39	6.45	6.51	6.56	6.61
30	3.89	4.45	4.80	5.05	5.24	5.40	5.54	5.65	5.76	5.85	5.93	6.01	6.08	6.14	6.20	6.26	6.31	6.36	6.41
40	3.82	4.37	4.70	4.93	5.11	5.27	5.39	5.50	5.60	5.69	5.77	5.84	5.90	5.96	6.02	6.07	6.12	6.17	6.21
60	3.76	4.28	4.60	4.82	4.99	5.13	5.25	5.36	5.45	5.53	5.60	5.67	5.73	5.79	5.84	5.89	5.93	5.98	6.02
120	3.70	4.20	4.50	4.71	4.87	5.01	5.12	5.21	5.30	5.38	5.44	5.51	5.56	5.61	5.66	5.71	5.75	5.79	5.83
∞	3.64	4.12	4.40	4.60	4.76	4.88	4.99	5.08	5.16	5.23	5.29	5.35	5.40	5.45	5.49	5.54	5.57	5.61	5.65

[a]**Range/$S_Y \sim Q_{1-\alpha;\eta,v}$. η is the size of the sample from which the range is obtained, and v is the number of degrees of freedom of S_Y.**

Source: Reprinted from E. S. Pearson and H. O. Hartley, eds., Table 29 of *Biometrika Tables for Statisticians, Vol. 1,* 3rd ed., 1966, by permission of the *Biometrika* Trustees, London.

TABLE E.13 Control Chart Factors

Number of Observations in Sample	d_2	d_3	D_3	D_4	A_2	E_2
2	1.128	0.853	0	3.267	1.880	2.659
3	1.693	0.888	0	2.575	1.023	1.772
4	2.059	0.880	0	2.282	0.729	1.457
5	2.326	0.864	0	2.114	0.577	1.290
6	2.534	0.848	0	2.004	0.483	1.184
7	2.704	0.833	0.076	1.924	0.419	1.109
8	2.847	0.820	0.136	1.864	0.373	1.054
9	2.970	0.808	0.184	1.816	0.337	1.010
10	3.078	0.797	0.223	1.777	0.308	0.975
11	3.173	0.787	0.256	1.744	0.285	0.946
12	3.258	0.778	0.283	1.717	0.266	0.921
13	3.336	0.770	0.307	1.693	0.249	0.899
14	3.407	0.763	0.328	1.672	0.235	0.881
15	3.472	0.756	0.347	1.653	0.223	0.864
16	3.532	0.750	0.363	1.637	0.212	0.849
17	3.588	0.744	0.378	1.622	0.203	0.836
18	3.640	0.739	0.391	1.609	0.194	0.824
19	3.689	0.733	0.404	1.596	0.187	0.813
20	3.735	0.729	0.415	1.585	0.180	0.803
21	3.778	0.724	0.425	1.575	0.173	0.794
22	3.819	0.720	0.435	1.565	0.167	0.785
23	3.858	0.716	0.443	1.557	0.162	0.778
24	3.895	0.712	0.452	1.548	0.157	0.770
25	3.931	0.708	0.459	1.541	0.153	0.763

Source: Reprinted from ASTM-STP 15D by kind permission of the American Society for Testing and Materials.

TABLE E.14 Critical Values d_L and d_U of the Durbin-Watson Statistic D (Critical Values Are One-Sided)[a].

	α = .05										α = .01									
	$P = 1$		$P = 2$		$P = 3$		$P = 4$		$P = 5$		$P = 1$		$P = 2$		$P = 3$		$P = 4$		$P = 5$	
n	d_L	d_U	d_L	d_U	d_L	d_U	d_L	d_U	d_L	d_U	d_L	d_U	d_L	d_U	d_L	d_U	d_L	d_U	d_L	d_U
15	1.08	1.36	.95	1.54	.82	1.75	.69	1.97	.56	2.21	.81	1.07	.70	1.25	.59	1.46	.49	1.70	.39	1.96
16	1.10	1.37	.98	1.54	.86	1.73	.74	1.93	.62	2.15	.84	1.09	.74	1.25	.63	1.44	.53	1.66	.44	1.90
17	1.13	1.38	1.02	1.54	.90	1.71	.78	1.90	.67	2.10	.87	1.10	.77	1.25	.67	1.43	.57	1.63	.48	1.85
18	1.16	1.39	1.05	1.53	.93	1.69	.82	1.87	.71	2.06	.90	1.12	.80	1.26	.71	1.42	.61	1.60	.52	1.80
19	1.18	1.40	1.08	1.53	.97	1.68	.86	1.85	.75	2.02	.93	1.13	.83	1.26	.74	1.41	.65	1.58	.56	1.77
20	1.20	1.41	1.10	1.54	1.00	1.68	.90	1.83	.79	1.99	.95	1.15	.86	1.27	.77	1.41	.68	1.57	.60	1.74
21	1.22	1.42	1.13	1.54	1.03	1.67	.93	1.81	.83	1.96	.97	1.16	.89	1.27	.80	1.41	.72	1.55	.63	1.71
22	1.24	1.43	1.15	1.54	1.05	1.66	.96	1.80	.86	1.94	1.00	1.17	.91	1.28	.83	1.40	.75	1.54	.66	1.69
23	1.26	1.44	1.17	1.54	1.08	1.66	.99	1.79	.90	1.92	1.02	1.19	.94	1.29	.86	1.40	.77	1.53	.70	1.67
24	1.27	1.45	1.19	1.55	1.10	1.66	1.01	1.78	.93	1.90	1.04	1.20	.96	1.30	.88	1.41	.80	1.53	.72	1.66
25	1.29	1.45	1.21	1.55	1.12	1.66	1.04	1.77	.95	1.89	1.05	1.21	.98	1.30	.90	1.41	.83	1.52	.75	1.65
26	1.30	1.46	1.22	1.55	1.14	1.65	1.06	1.76	.98	1.88	1.07	1.22	1.00	1.31	.93	1.41	.85	1.52	.78	1.64
27	1.32	1.47	1.24	1.56	1.16	1.65	1.08	1.76	1.01	1.86	1.09	1.23	1.02	1.32	.95	1.41	.88	1.51	.81	1.63
28	1.33	1.48	1.26	1.56	1.18	1.65	1.10	1.75	1.03	1.85	1.10	1.24	1.04	1.32	.97	1.41	.90	1.51	.83	1.62
29	1.34	1.48	1.27	1.56	1.20	1.65	1.12	1.74	1.05	1.84	1.12	1.25	1.05	1.33	.99	1.42	.92	1.51	.85	1.61
30	1.35	1.49	1.28	1.57	1.21	1.65	1.14	1.74	1.07	1.83	1.13	1.26	1.07	1.34	1.01	1.42	.94	1.51	.88	1.61
31	1.36	1.50	1.30	1.57	1.23	1.65	1.16	1.74	1.09	1.83	1.15	1.27	1.08	1.34	1.02	1.42	.96	1.51	.90	1.60
32	1.37	1.50	1.31	1.57	1.24	1.65	1.18	1.73	1.11	1.82	1.16	1.28	1.10	1.35	1.04	1.43	.98	1.51	.92	1.60
33	1.38	1.51	1.32	1.58	1.26	1.65	1.19	1.73	1.13	1.81	1.17	1.29	1.11	1.36	1.05	1.43	1.00	1.51	.94	1.59
34	1.39	1.51	1.33	1.58	1.27	1.65	1.21	1.73	1.15	1.81	1.18	1.30	1.13	1.36	1.07	1.43	1.01	1.51	.95	1.59
35	1.40	1.52	1.34	1.58	1.28	1.65	1.22	1.73	1.16	1.80	1.19	1.31	1.14	1.37	1.08	1.44	1.03	1.51	.97	1.59
36	1.41	1.52	1.35	1.59	1.29	1.65	1.24	1.73	1.18	1.80	1.21	1.32	1.15	1.38	1.10	1.44	1.04	1.51	.99	1.59
37	1.42	1.53	1.36	1.59	1.31	1.66	1.25	1.72	1.19	1.80	1.22	1.32	1.16	1.38	1.11	1.45	1.06	1.51	1.00	1.59
38	1.43	1.54	1.37	1.59	1.32	1.66	1.26	1.72	1.21	1.79	1.23	1.33	1.18	1.39	1.12	1.45	1.07	1.52	1.02	1.58
39	1.43	1.54	1.38	1.60	1.33	1.66	1.27	1.72	1.22	1.79	1.24	1.34	1.19	1.39	1.14	1.45	1.09	1.52	1.03	1.58
40	1.44	1.54	1.39	1.60	1.34	1.66	1.29	1.72	1.23	1.79	1.25	1.34	1.20	1.40	1.15	1.46	1.10	1.52	1.05	1.58
45	1.48	1.57	1.43	1.62	1.38	1.67	1.34	1.72	1.29	1.78	1.29	1.38	1.24	1.42	1.20	1.48	1.16	1.53	1.11	1.58
50	1.50	1.59	1.46	1.63	1.42	1.67	1.38	1.72	1.34	1.77	1.32	1.40	1.28	1.45	1.24	1.49	1.20	1.54	1.16	1.59
55	1.53	1.60	1.49	1.64	1.45	1.68	1.41	1.72	1.38	1.77	1.36	1.43	1.32	1.47	1.28	1.51	1.25	1.55	1.21	1.59
60	1.55	1.62	1.51	1.65	1.48	1.69	1.44	1.73	1.41	1.77	1.38	1.45	1.35	1.48	1.32	1.52	1.28	1.56	1.25	1.60
65	1.57	1.63	1.54	1.66	1.50	1.70	1.47	1.73	1.44	1.77	1.41	1.47	1.38	1.50	1.35	1.53	1.31	1.57	1.28	1.61
70	1.58	1.64	1.55	1.67	1.52	1.70	1.49	1.74	1.46	1.77	1.43	1.49	1.40	1.52	1.37	1.55	1.34	1.58	1.31	1.61
75	1.60	1.65	1.57	1.68	1.54	1.71	1.51	1.74	1.49	1.77	1.45	1.50	1.42	1.53	1.39	1.56	1.37	1.59	1.34	1.62
80	1.61	1.66	1.59	1.69	1.56	1.72	1.53	1.74	1.51	1.77	1.47	1.52	1.44	1.54	1.42	1.57	1.39	1.60	1.36	1.62
85	1.62	1.67	1.60	1.70	1.57	1.72	1.55	1.75	1.52	1.77	1.48	1.53	1.46	1.55	1.43	1.58	1.41	1.60	1.39	1.63
90	1.63	1.68	1.61	1.70	1.59	1.73	1.57	1.75	1.54	1.78	1.50	1.54	1.47	1.56	1.45	1.59	1.43	1.61	1.41	1.64
95	1.64	1.69	1.62	1.71	1.60	1.73	1.58	1.75	1.56	1.78	1.51	1.55	1.49	1.57	1.47	1.60	1.45	1.62	1.42	1.64
100	1.65	1.69	1.63	1.72	1.61	1.74	1.59	1.76	1.57	1.78	1.52	1.56	1.50	1.58	1.48	1.60	1.46	1.63	1.44	1.65

[a] n = number of observations; P = number of independent variables.

Source: This table is reproduced from *Biometrika,* Vol. 41 (1951), pp. 173 and 175, with the permission of the *Biometrika* Trustees.

APPENDIX F

Documentation for Data Diskette for Basic Business Statistics Sixth Edition

This documentation is being provided to accompany the data diskette for the Sixth Edition of the Berenson-Levine *Basic Business Statistics* text. The 3½″ diskette is formatted for the IBM and close compatible personal computers. The diskette contains 189 files. Due to the number of files, the diskette has been divided into two directories named BL1 and BL2. All files from ABSSPC.DAT to KNEE.DAT are in directory BL1. All files from LAUNDRY.DAT to XEROX.DAT are in directory BL2. To access a file we need to indicate the disk drive in which the diskette is located followed by a colon (:), a slash (\), the name of the directory (either BL1 or BL2), another slash (\), and the particular file name. For example, to access the file EMPSAT.DAT we would enter

{drive}:\BL1\EMPSAT.DAT

while to access file TAX.DAT we would enter

{drive}:\BL2\TAX.DAT

Each raw data file has been formatted in an ASCII format. In addition, system files for the Employee Satisfaction Survey database (see pages 33-40 of the text) have been created in MINITAB (EMPSAT.MTW), SAS (EMPSAT.SD2), SPSS (EMPSAT.SYS), and STATISTIX (EMPSAT.SX). The following is an alphabetical listing and description of the 189 raw data files and system files contained on the diskette.

Name	Description
ABSSPC.DAT	Number of students absent and total number of students registered over 36 days from Problem 16.9 on page 682.
ACCRES.DAT	Computer processing time (in seconds) for 5 Accounting Department (=0) and 6 Research Department (=1) jobs from Problem 13.74 on page 521.
ADAMHA.DAT	Amount of funding (in millions of dollars) for 21 schools from Problem 4.10 on page 116.
	Problems 4.22, 4.31, 4.39, 4.50, and 8.31 also use the file.
ADRADTV.DAT	Sales (in thousands of dollars), radio/TV ads (in thousands of dollars), and newspaper ads (in thousands of dollars) for 22 cities from Problem 18.4 on page 788.
	Problems 18.8, 18.13, 18.17, 18.22, 18.27, 18.32, 18.37, 18.55, and 18.59 also use the file.
ADTIME.DAT	Data showing year and annual revenues (in millions of dollars) over the 20-year period 1975 through 1994 from Problem 19.54 on page 916.
AGREV.DAT	Data showing year and annual gross revenues (in millions of dollars) over the 14-year period 1981 through 1994 from Problem 19.53 on page 915.
ALCOHOL.DAT	Alcohol consumption (in ounces) and number of errors made by 15 typists from Problem 18.41 on page 815. Problem 18.63 also uses the file.
AMPHRS.DAT	Capacity (in ampere-hours) for 20 batteries from Problem 12.32 on page 448. The data appear on page 425.
AMPLIFY.DAT	Coded db ratings for three recordings based on two cartridges (C_1=1, C_2=2) and four amplifiers (A=1, B=2, C=3, D=4) from Problem 14.46 on page 590. The variables are cartridge, amplifier, and coded decibel output.
ATM1.DAT	Cash withdrawals by 25 customers at a local bank from Problem 8.21 on page 303.

Name	Description
ATM2.DAT	Amount withdrawn (in thousands of dollars), median value of homes (in thousands of dollars), and location of ATM (Not a shopping center=0, Shopping center=1) for 15 ATM locations from Problem 18.46 on page 819. Problems 18.49 and 18.66 also use the file.
AUTOBAT.DAT	Lengths of life (in months) for 28 automobile batteries from Problem 12.44 on page 459.
AUTOREP.DAT	Day, average service time, and range in service time (over a 20-day period) from Problem 16.22 on page 700.
BAGGAGE.DAT	30 days of baggage claims from Problem 16.17 on page 691.
BANKTIME.DAT	Waiting times of four bank customers per day for 20 days from Problem 16.20 on page 699.
BATFAIL.DAT	Times to failure (in hours) for groups of five batteries exposed to four pressure levels (low=1, normal=2, high=3, very high=4) from Problem 14.13 on page 551.
BB92.DAT	Wins, runs, E.R.A., saves, hits allowed, walks allowed, and errors committed for 26 teams from Problem 18.76 page 849.
BB93.DAT	Wins, E.R.A., league (American=0, National=1) fro 28 teams from Problem 17.82 on page 778 and Problem 18.48 on page 820. Problems 18.53 and 18.70 also use the file.
BDECKER.DAT	Data showing year and net sales (in billions of dollars) at Black & Decker Corp. over the 23-year period 1970 through 1992 from Problem 19.13 on page 882. Problems 19.26, 19.37, and 19.45 also use the file.
BESTTEAM. DAT	Rankings by 10 experts on each of four teams (Braves=1, Cowboys=2, Bulls=3, Penguins=4) in different sports from Problem 14.39 on page 576. The variables are expert, team, and ranking.
BET.DAT	Attendance (in thousands) and betting volume (in millions of dollars) for 15 days from Problem 17.76 on page 772. Problem 18.80 also uses the file.
BKPRICE.DAT	Prices of 12 books, sold on campus and sold off campus, from Problem 13.65 on page 511.
BLACKJ.DAT	Profits (in dollars) from five sessions in each of four strategies (dealer's=1, five count=2, basic ten count=3, advanced ten count=4) from Problem 14.9 on page 544. Problem 14.17 also uses the file.
BOEING.DAT	Data showing year and total income (in billions of dollars) at Boeing Co. over the 23-year period 1970 through 1992 from Problem 19.9 on page 880. Problems 19.24, 19.35, and 19.43 also use the file.
BREAKING.DAT	Breaking strength (in pounds) for three pieces of wool serge material based on four operators (A=1, B=2, C=3, D=4) and three machines (I=1, II=2, III=3) from Problem 14.48 on page 591. The variables are operator, machine, and breaking strength.
BROKER.DAT	Performance scores of three customer representatives based on gender (males=1, females=2) and four background (professionals=1, business school graduates=2, salesmen=3, brokers=4) combinations from Problem 14.53 on page 598. The variables are gender, background, and performance score.
BULBLIFE.DAT	30 days of data on the mean life, range, and subgroup number for subgroups of five bulbs from Problem 16.19 on page 698.
BULBS.DAT	Length of life of 40 light bulbs from Manufacturer A (=1) and 40 light bulbs from Manufacture B (=2) from Problem 3.18 on page 66. Problems 3.25, 3.32, 3.40, 8.29, and 10.97 also the file.

Name	Description
C&U.DAT	Tuition charges along with 4 categorical variables: type (private=1, public=2), setting (rural=1, suburb=2, urban=3), calendar (quarter=1, semester=2, trimester=3, 414=4), and classification (NLA=1, NU=2, RLA=3, RU=4, SS=5) for 195 schools in 3 states (Texas=1, North Carolina=2, Pennsylvania=3). This is Special Data Set 1 on pages D1–D5. See Problem 3.33 on page 77 and see CL3.1 on page 101, CL4.1 on page 165, and CL5.1 on page 199.
C&UNC.DAT	Tuition charges at 45 North Carolina schools along with 4 categorical variables: type (private=1, public=2), setting (rural=1, suburb=2, urban=3), calendar (quarter=1, semester=2, trimester=3, 414=4), and classification (NLA=1, NU=2, RLA=3, RU=4, SS=5). See Table 3.6 on page 69 and see Problem 9.5 on page 333.
C&UPENN.DAT	Tuition charges at 90 Pennsylvania schools along with 4 categorical variables: type (private=1, public=2), setting (rural=1, suburb=2, urban=3), calendar (quarter=1, semester=2, trimester=3, 414=4), and classification (NLA=1, NU=2, RLA=3, RU=4, SS=5). See CL3.1 on page 101 and CL8.1 on page 316.
C&UTEX.DAT	Tuition charges at 60 Texas schools along with 4 categorical variables: type (private=1, public=2), setting (rural=1, suburb=2, urban=3), calendar (quarter=1, semester=2, trimester=3, 414=4), and classification (NLA=1, NU=2, RLA=3, RU=4, SS=5). See Table 3.1 on page 55.
CAMERA.DAT	Data on 59 cameras with 7 variables: type (ML=1, MM=2, MS=3, SA=4, SF=5), price (in dollars), weight (in ounces), close-up ability (in inches), flash distance (in feet), percent image area, and number of 24-exposure rolls per battery. This is Special Data Set 4 on pages D10–D11. See CL3.4 on page 102, CL4.4 on page 166, CL5.4 on page 200, CL8.4 on page 317, CL10.3 on page 381, CL12.3 on page 460, CL13.3 on page 524, and CL14.3 on page 600.
CANCER.DAT	Cancer incidence rate in all 50 states for Problem 3.6 on page 60. Problems 3.15, 3.22, 3.29, 3.37, 4.48, 4.54, and 8.26 also use the file.
CEREAL.DAT	Data on 84 ready-to-eat cereals with 6 variables: type product (high fiber=1, moderate fiber=2, low fiber=3), cost (in cents per serving), weight (in ounces per serving), calories per serving, and sugar (in grams per serving). This is Special Data Set 2 on pages D6–D7. See CL3.2 on page 101, CL4.2 on page 165, CL5.2 on page 199, CL8.2 on page 316, CL10.1 on page 381, CL12.1 on page 459, CL13.1 on page 523, and CL14.1 on page 600.
CHARRED.DAT	Lengths of charred material (in centimeters) in 20 strips from Problem 12.43 on page 459.
CITY.DAT	Prices (in dollars) of hairdressers and shirts in 9 cities from Problem 17.35 on page 735. Problem 17.66 also uses the data.
CLAIM.DAT	Larceny claims (in dollars) filed by 18 insured individuals from Problem 12.13 on page 436.
CLOTTING.DAT	Plasma clotting times (in minutes) for seven patients whose blood is studied under each of five different treatment substances from Problem 14.33 on page 568. The variables are patient, treatment substance, and clotting time. Problem 14.42 also uses the file.
COFFEE.DAT	Ratings by nine experts on each of four brands of coffee (A=1, B=2, C=3 D=4) from Problem 14.32 on page 568. The variables are expert, brand, and rating. Problem 14.41 also uses the file.
COKE.DAT	Data showing year and operating revenues (in billions of dollars) at Coca-Cola Co. over the 23-year period 1970 through 1992 from Problem 19.6 on page 878. Problems 19.21, 19.32, and 19.40 also use the file.
COLASPC.DAT	Day, total number of cans filled, and number of unacceptable cans (over a 22-day period) from Problem 16.10 on page 683.

Name	Description
COMPFAIL.DAT	Failure time ranks for samples of eight stereo components under three different temperature levels (150=1, 200=2, 250=3) from Problem 14.14 on page 551.
COOKIES.DAT	Calories in 10 types of chocolate chip cookies from Problem 4.8 on page 116.
CORDPHON.DAT	Prices of 32 corded telephone models from Problem 4.76 on page 162. Problems 8.32, 10.21, 10.28, and 10.38 also use the file.
CURR.DAT	Test scores for three groups of 7,9, and 8 students assigned to study different sets of mathematical materials from Problem 14.7 on page 543. Problem 14.16 also uses the file.
CURRENCY.DAT	Data showing year and average annual exchange rates (against the U.S. dollar) for the Canadian dollar, French franc, German mark, Japanese yen, and English pound over the 26-year period 1967 through 1992 from Case Study J on page 918.
DELIVERY.DAT	Delivery time (in days), number of options ordered, and shipping mileage (in hundreds of miles) for 16 cars from Problems 17.75 on page 771 and 18.78 on page 850.
DENTAL.DAT	Annual family dental expenses for 10 employees from Problem 10.12 on page 354. Problem 10.31 also uses the file.
DIET.DAT	Weight losses (in pounds) for six client groups under three dietary treatments from Problem 14.34 on page 569. The variables are client group, dietary treatment, and amount of weight lost.
DISPRAZ.DAT	Price (in cents) and number of packages sold at 15 stores from Table 18.7 on page 808.
DRESS.DAT	Sales revenues (in thousands of dollars) over 28 consecutive days from Problem 3.43 on page 82. Problem 16.30 also uses the file.
DRYCLEAN.DAT	24 days of items returned for rework from Problem 16.14 on page 689.
ECOTESTS.DAT	Test scores for 11 students, on midterms and on final exams, from Problem 13.68 on page 517.
EER.DAT	Average energy efficiency ratings (EER) for 36 air conditioning units from Problem 12.1 on page 429. Problem 12.14 also uses the file.
EMPLOYEE.DAT	Data showing year, number of employees (in thousands) in goods-producing industries, in non-government service producing industries, in federal government, and in state and local government over the 42-year period 1950 through 1991 from Problem 19.14 on page 883.
EMPSAT.DAT	Responses to 28 questions by 400 employees in Kalosha Industries Employee Satisfaction Survey (see Figure 2.6 on pages 28-29 and Table 2.3 on pages 33-40) in ASCII format.
EMPSAT.MTW	Data for the Kalosha Industries Employee Satisfaction Survey (see Figure 2.6 on pages 28-29 and Table 2.3 on pages 33-40) in MINITAB format.
EMPSAT.SD2	Data for the Kalosha Industries Employee Satisfaction Survey (see Figure 2.6 on pages 28-29 and Table 2.3 on pages 33-40) in SAS format. To read the SAS system file, the SAS library must be defined as follows: LIBNAME IN 'A:\BL1'; Then specify the data set name as DATA = IN.EMPSAT for each procedure to be run. For example, PROC PRINT DATA = IN.EMPSAT; RUN; displays the data in the system file.

Name	Description
EMPSAT.SX	Data for the Kalosha Industries Employee Satisfaction Survey (See Figure 2.6 on pages 28-29 and Table 2.3 on page 33-40) in STATISTIX format.
EMPSAT.SYS	Data for the Kalosha Industries Employee Satisfaction Survey (See Figure 2.6 on pages 28-29 and Table 2.3 on page 33-40) in SPSS format.
ERRORSPC.DAT	Number of nonconforming items and number of accounts processed over 39 days from Problem 16.11 on page 683.
EXERCISE.DAT	Ratings of nine medical experts on five diverse forms of exercise (1=bicycling, 2=calisthenics, 3=jogging, 4=swimming, 5=tennis) from Problem 14.38 on page 575. The variables are expert, exercise, and rating.
FLASHBAT.DAT	Time to failure (in hours) for 13 flashlight batteries from Problem 4.5 on page 115. Problems 4.20, 4.29, 4.37, 10.19, 10.26, and 10.36 also use the file.
FOODTIME.DAT	Data showing year and annual sales (in millions of dollars) over the 26-year period 1968 through 1993 from Problem 19.4 on page 871.
FOULSPC.DAT	Number of foul shots made and number taken over 40 days from Problem 16.7 on page 681.
FRAGRAN.DAT	Data on 83 fragrances with 4 variables: gender (women=1, men=2), type of produce (perfume=1, cologne=2, "other"=3), cost (in dollars per ounce), intensity (very strong=1, strong=2, medium=3, mild=4). This is Special Data Set 3 on pages D8–D9. See CL3.3 on page 102, CL4.3 on page 165, CL5.3 on page 199, CL8.3 on page 316, CL10.2 on page 381, CL12.2 on page 460, CL13.2 on page 523, CL14.2 on page 600, and CL15.1 on page 655.
FUNRAISE.DAT	Amounts pledged (in thousands of dollars) in a fund raising campaign by 9 alumni from Problem 4.79 on page 163.
GAPAC.DAT	Data showing year and net sales (in billions of dollars) at Georgia-Pacific Corp. over the 23-year period 1970 through 1992 from Problem 19.8 on page 879. Problems 19.23, 19.34, and 19.42 also use the file.
GAS1.DAT	Amount of gasoline filled(in gallons) in 24 automobiles from Problem 8.24 on page 304.
GASERVE.DAT	Length of time (in days) to establish gasoline heating service in 15 houses from Problem 10.13 on page 354. Problems 10.22, 10.32, 12.10, and 12.19 also use the file.
GASMILE.DAT	Miles per gallon on 10 cars, each obtained with regular gas and with high octane gas from Problem 13.63 on page 510. Problems 13.69 and 14.37 also use the file.
GASPRICE.DAT	Data on average monthly retail gasoline prices (in cents per gallon) over the 60-month period from January 1989 through December 1993 from Problem 19.49 on page 911.
GAUGE.DAT	Prices of 30 pencil tire pressure gauges in Problem 3.70 on page 95.
GILLETTE.DAT	Data showing year and net sales (in billions of dollars) at Gillette Company, Inc. over the 23-year period 1970 through 1992 from Problem 19.7 on page 879. Problems 19.22, 19.33, and 19.41 also use the file.
GMAT.DAT	GMAT scores for 20 applicants to an MBA program from Problem 12.42 on page 458.
GPIGMAT.DAT	GMAT scores and GPI for 20 students from Problem 17.80 on page 776.
GRADES.DAT	Test scores for 33 students, on midterms and on final exams, from Problem 13.76 on page 523.
GROCERY.DAT	Grocery bills for 28 customers in a supermarket from Problem 8.22 on page 304.

Name	Description
HMO.DAT	Waiting times (in minutes) for 25 patients in an HMO from Problem 10.14 on page 354. Problems 10.23, 10.33, 12.11, and 12.20 also use the file.
HOSPADM.DAT	Day, number of admissions, average processing time (in hours), range of processing times, and proportion of laboratory rework (over a 30-day period) from Case Study G on page 710.
HOSPITAL.DAT	Occupancy rates (in percentages) at 16 urban hospitals (=0) and 16 suburban hospitals (=1) from Problem 13.20 on page 480. Problems 13.31 and 13.43 also use the file.
HOUSE1.DAT	Selling price (in thousands of dollars), assessed value (in thousands of dollars), type (new=0, old=1), and time period of sale for 30 houses from Problem 17.78 on page 774. Problems 18.79, 18.81, and 18.82 also the file.
HOUSE2.DAT	Assessed value (in thousands of dollars) and heating area (in thousands of square feet) for 15 houses from Problem 17.79 on page 775.
HTNGOIL.DAT	Temperature (in degrees Fahrenheit), attic insulation (in inches), and monthly consumption of heating oil (in gallons) from Table 18.1 on page 783 and used in Problems 18.43 on page 816 and 18.65 on page 828.
IBM.DAT	Data showing year and annual total sales (in billions of dollars) at IBM over the 23-year period 1970 through 1992 from Problem 19.55 on page 916.
ICECREAM.DAT	Daily temperature (in degrees Fahrenheit) and sales (in thousands of dollars) for 21 days from Problem 17.77 on page 773.
INDPSYCH.DAT	Reaction time rankings of assembly-line workers for three groups of 9,8, and 8 using different assembly methods from Problem 14.12 on page 550.
INSPGM.DAT	Test scores of eight employees in each of four groups (A=1, B=2, C=3, D=4) from Problem 14.5 on page 542. Problem 14.15 also uses the file.
INTERVW.DAT	Weeks of experience and number of interviews completed by 10 interviewers from Problem 17.5 on page 718. Problems 17.11, 17.17, 17.23, 17.32, 17.40, 17.51, 17.57, 17.63, and 17.73 also use the file.
INTRANK.DAT	Preference ranks for 10 MBA students (=0) and 12 MPH students (=1) from Problem 13.26 on page 487.
INVOICE.DAT	Number of invoices processed and amount of time (in hours) for 30 days from Problem 17.81 on page 777.
JEANS.DAT	Data on the lengths (in inches) of 30 consecutive pairs of manufactured jeans from Problem 3.41 on page 82. Problems 12.34 and 16.28 also use the file.
JEWELRY.DAT	Product assignment number and development time (in days) for 30 jewelry products from Problem 16.26 on page 705.
KNEE.DAT	Postsurgical recovery times (in days) for five patients in each combination of two types of surgery (arthroscopy=1, arthrotomy=2) and three age groups (under 30=1, 30 to 50=2, over 50=3) from Problem 14.47 on page 590. The variables are type surgery, age group, and length of stay.
LAUNDRY.DAT	Dirt (in pounds) removed from two laundry loads based on four detergent brands (A=1, B=2, C=3, D=4) and four cycle times (18 minutes, 20 minutes, 22 minutes, 24 minutes) from Problem 14.49 on page 591. The variables are brand, cycle time, and pounds of dirt removed.
LAWN.DAT	Lawn service (no=0, yes=1), family income (in thousands of dollars), lawn size (in thousands of square feet), attitude toward activities (unfavorable=0, favorable=1), number of teenagers, and age of household head at 30 houses from Problem 18.74 on page 843.

Name	Description
LIMO.DAT	Distance (in miles) and time (in minutes) for 12 trips from Problem 17.6 on page 718. Problems 17.12, 17.18, 17.24, 17.33, 17.41, 17.52, 17.58, 17.64, and 17.74 also use the file.
LIRE.DAT	Appraised values (in thousands of dollars) on 12 houses, each appraised by two real estate agents from Problem 13.61 on page 509.
LOCATE.DAT	Sales volume (in thousands of dollars) at two stores based on three aisle locations (front=1, middle=2, rear=3) and three shelf heights (top=1, middle=2, bottom=3) from Problem 14.51 on page 596. The variables are aisle location, height of shelf, and sales volume.
LOGPURCH.DAT	Annual spending (in thousands of dollars), purchasing behavior (no=0, yes=1), and possession of additional credit cards (no=0, yes=1) for 30 people from Table 18.10 on page 839. See Problem 18.72 on page 841.
MAILORD.DAT	Monthly billing records of 20 unpaid accounts in Problem 3.2 on page 58. Problems 4.13, 4.25, 4.34, 4.42, and 4.51 also use the file.
MAILSPC.DAT	Day, total number of packages, and late packages (over a 20-day period) from Problem 16.8 on page 681.
MBA.DAT	Success in program (no=0, yes=1), GPA, and GMAT score for 30 students from Problem 18.73 on page 842.
MCDONALD.DAT	Data showing year and annual total revenues (in billions of dollars) at McDonald's Corp. over the 22-year period 1971 through 1992 from Problem 19.56 on page 917.
MEDINCUS.DAT	Data showing year, median income for all races, median income for whites, and median income for blacks over the 14-year period 1977 through 1990 from Problem 19.1 on page 870.
MKTEST.DAT	Test scores for 19 students on a marketing exam from Problem 8.23 on page 304.
MOWER.DAT	Prices of 15 side-bagging lawn mowers from Problem 4.7 on page 115. Problems 4.21, 4.30, 4.38, 10.20, 10.27, 10.37, and 12.16 also use the file.
MUTFUNDS.DAT	Monthly total sales (in millions of dollars) of mutual funds over the 48-month period from January 1990 through December 1993 from Problem 19.48 on page 911.
NEWSCIRC.DAT	Sunday and Daily circulation (in thousands) for 34 newspapers from Case Study H on page 779.
NICKBAT.DAT	Talking times (in minutes) prior to recharging for 25 cadmium batteries (=0) and 25 metal hydride batteries (=1) from Problem 13.8 on page 470. Problems 13.18, 13.29, and 13.41 also use the file.
NYBANKS.DAT	Annual yields (in percent) for 10 commercial banks (=0) and 10 savings banks (=1) from Problem 13.9 on page 471. Problems 13.19, 13.30, and 13.42 also use the file.
NYSEAX.DAT	Stock prices for 25 AE (=1) and 50 NYSE (=2) companies from Problem 3.72 on page 97.
OILSUPP.DAT	Data showing year and number of employees (in thousands) over the 20-year period 1974 through 1993 from Problem 19.3 on page 871.
OILUSE.DAT	Data on annual amount of heating oil consumed (in gallons) in a sample of 35 single family houses from Table 10.2 on page 353.
OUTLAYS.DAT	Data on monthly outlays (in thousands of dollars) over the 120-month period January 1985 through December 1994 from Problem 19.46 on page 909.
PACKAGE.DAT	Number of customers and weekly sales (in thousands of dollars) at 20 stores from Table 17.1 on page 715.

Name	Description
PASCAL.DAT	Time (in minutes) for 9 students to write and run a PASCAL program from Problem 13.75 on page 522.
PEANUT.DAT	37 brands of peanut butter by type (0=creamy or 1=chunky), score, cost, and sodium from Problem 3.7 on page 60. Problems 3.16, 3.23, 3.30, 3.38, 8.27, and 10.94 also use the file.
PEN.DAT	Product ratings of three ball-point pens under gender (males=1, females=2) and five advertisement (A=1, B=2, C=3, D=4, E=5) combinations from Case Study E on page 601. The variables are gender, ad, and product rating.
PENEAD.DAT	Product ratings of ball-point pens by 6 adults (=0) and 8 HS students (=1) from Case Study E on page 602.
PETFOOD.DAT	Data on shelf space (in feet), weekly sales (in hundreds of dollars), and aisle location (back=0, front=1) from Problem 17.1 on page 716 and Problem 18.47 on page 820. Problems 17.7 17.13, 17.19, 17.28, 17.36, 17.43, 17.44, 17.47, 17.53, 17.59, 17.69, 18.52, and 18.69 also use the file.
PHLEVEL.DAT	Day, average pH level, and range in pH level (over 21 days) from Problem 16.23 on page 700.
PLASTIC.DAT	Hardness measurements (in Brinell units) for 50 plastic blocks from Problem 12.2 on page 429. Problem 12.15 also uses the file.
PMORRIS.DAT	Data showing year and operating revenues (in billions of dollars) at Philip Morris Companies, Inc. over the 23-year period 1970 through 1992 from Problems 19.12 on page 882. Problems 19.25, 19.36, and 19.44 also use the file.
POTATO.DAT	Percent solids content in filter cake, acidity (in pH), lower pressure, upper pressure, cake thickness, drum speed setting, and drum speed for 54 measurement from Case Study I on page 855.
PREP.DAT	Tuition charges at 15 NE (=0) and 15 MW (=1) prep schools from Problem 4.80 on page 163. Problems 10.98, 13.11, 13.23, 13.34, 13.46, and 14.10 also use the file.
PTFALLS.DAT	28 months of patient falls from Problem 16.16 on page 691.
QMILE.DAT	Quarter mile time trials (in seconds) over 27 consecutive days from Problem 3.42 on page 82. Problems 12.35 and 16.29 also use the file.
RAISINS.DAT.	Weights (in ounces) of 30 consecutively filled packages of raisins from Problem 12.39 on page 457. Problem 12.40 also uses the file.
RATING.DAT	Student ratings of ten faculty based on three classes they taught (MBA course=1, advanced undergraduate course=2, required undergraduate course=3) from Problem 14.35 on page 570. The variables are faculty member, course, and rating. Problem 14.43 also uses the file.
RECALL.DAT	Recall ability scores for eight sets of triplets examined under three stimuli (minimum=1, moderate=2, high=3) from Problem 14.40 on page 576. The variables are set of triplets, exposure level, and recall ability score.
RESID1.DAT	Data on 10 consecutive time periods and the corresponding residuals from Problem 17.45 on page 746.
RESID2.DAT	Data on 15 consecutive time periods and the corresponding residuals from Problem 17.46 on page 746.
RETSALES.DAT	Retail sales (in billions of dollars) over the 60-month period from January 1989 through December 1993 from Problem 19.47 on page 910.
ROAD.DAT	Acceleration time data for 22 German (=0) and 30 Japanese (=1) automobile models in Problem 3.8 on page 61. Problems 3.17, 3.24, 3.31, 3.39, 4.82, and 10.95 also use the file.
RRSPC.DAT	Number of late arrivals and total arrivals over 20 days from Problem 16.6 on page 680.

Name	Description
SALARY1.DAT	Starting salaries (in thousands of dollars) for 100 graduates in Problem 3.69 on page 95.
SALARY2.DAT	Weekly salary (in dollars), longevity (in months), and age (in years) for 16 employees from problem 18.5 on page 788. Problems 18.9, 18.14, 18.18, 18.23, 18.28, 18.33, 18.38, 18.56, and 18.60 also use the file.
SALESCMP.DAT	Sales (in thousands of dollars) at 13 matched pairs of stores, one with a sales campaign and one without a sales campaign from Problem 13.64 on page 510. Problem 13.70 also uses the file.
SALESVOL.DAT	Sales volume (in thousands of dollars) for 12 sales persons paid on an hourly rate (=0) and 12 paid on a commission basis (=1) from Problem 13.7 on page 470. Problems 13.17 and 13.28 also use the file.
SALT.DAT	Number of package and weight (in grams) for 50 packages of table salt from Problem 16.25 on page 705.
SAT.DAT	Data showing year, average verbal score and average math score for males and females, and average total score on the SAT, over the 20-year period 1972 through 1991 from Problem 19.10 on page 880.
SCALES.DAT	Prices of 39 bathroom scales from Problem 3.4 on page 59.
SEARS.DAT	Data showing year and annual total revenues (in billions of dollars) at Sears, Roebuck & Co. over the 23-year period 1970 through 1992 from Problem 19.57 on page 917.
SHAMPOO.DAT	Cost per ounce (in cents) for 31 normal (=1) and 29 fine (=2) hair products from Problem 3.9 on page 62. Problems 10.96, 13.12, 13.24, 13.35, 13.47, and 14.11 also use the file.
SHOESOLE.DAT	Sole wear "scores" for 10 pairs of shoes, one using new material and one using old material from Problem 13.62 on page 509. Problem 14.36 also uses the file.
SHOWER.DAT	Maximum flow rates (in gallons per minute) for 34 shower heads from Problem 3.3 on page 58. Problems 3.13, 3.20, 3.27, 3.35, 4.14, 4.26, 4.35, 4.43, 4.52, 8.25, 10.17, 10.24 and 10.34 also use the file.
SITE.DAT	Store number, square footage, and sales (in thousands of dollars) in 14 stores from Problem 17.2 on page 717. Problems 17.8, 17.14, 17.20, 17.29, 17.37, 17.48, 17.54, 17.60, and 17.70 also use the file.
SLEEP.DAT	Amount of sleep (in hours) for four groups of 7, 8, 6, and 7 subjects taking different sleep medications from Problem 14.6 on page 542.
SOFTBALL.DAT	Number of ball and circumference (in inches) for 25 softballs from Problem 16.24 on page 704.
SOUP.DAT	Data on 47 brands of canned soup with 7 variables: product (chicken noodle=1, vegetable=2, tomato=3), type (CC=1, CR=2, DC=3, DI=4) cost (in cents per 8-oz serving), calories (per 8-oz serving), fat (in grams per 8-oz serving), calories as a percentage of at (per 8-oz serving) from Case Study B on page 166.
SPEED.DAT	Miles per gallon, speed (in miles per hour), and speed difference for 28 road tests from Problem 18.40 on page 814. Problems 18.50, 18.51, 18.62, 18.65, and 18.68 also use the file.
STANDBY.DAT	Standby hours, staff, remote hours, Dubner hours, and labor hours for 26 weeks from Problem 18.6 on page 789 and Problem 18.75 on page 848. Problems 18.10, 18.15, 18.19, 18.24, 18.29, 18.34, 18.39, 18.57, and 18.61 also use the file.
STATE.DAT	Data showing year, receipts, expenditures, and surplus (all in billions of dollars) for state and local governments over the 22-year period 1970 through 1991 from Problem 19.11 on page 881.
STOCK1.DAT	Book values of 50 stocks for Problem 3.5 on page 60. Problems 3.14, 3.21, 3.28, 3.36, and 4.55 also use the file.

Name	Description
STUDIO.DAT	Monthly rents for 10 studio apartments in Manhattan (=0) and 10 studio apartments in Brooklyn Heights (=1) from Problem 4.77 on page 162. Problems 13.10, 13.22, 13.33, and 13.45 also use the file.
SUPERMKT.DAT	Prices (in dollars) of colas and of pain relievers in 9 cities from Problem 17.34 on page 735. Problem 17.65 also uses the file.
TAPES.DAT	Prices of 17 Type IV (metal) audiotapes from Problem 4.9 on page 116.
TAR.DAT	Tar content (in milligrams) of 24 cigarettes from Problem 12.12 on page 436.
TASKTIME.DAT	Times (in seconds) for 15 workers to complete a task on an assembly line from Problem 12.41 on page 458.
TAX.DAT	Quarterly sales tax receipts (in thousands of dollars) for all 50 business establishments from Problem 4.47 on page 140. Problem 4.49 also uses the file.
TAXRET.DAT	Amount of money owed by 11 taxpayers, as computed by a tax preparation company and by the individual, from Problem 13.66 on page 516.
TELESPC.DAT	Number of orders and number of corrections over 30 days from Problem 16.12 on page 684.
TELLER1.DAT	Length of time (in minutes) for a teller to service 24 consecutive customers from Figure 3.10 on page 79. See Problem 12.33 on page 448.
TELLER2.DAT	Errors by 12 bank tellers from Problem 16.15 on page 690.
TENSILE.DAT	Tensile strengths in five different alloys using respective samples of 5, 4, 5, 4, and 4 from Problem 14.8 on page 544.
TESTRANK.DAT	Rank scores for 10 students taught by a "traditional" method (T=0) and 10 students taught by an "experimental" method (E=1) from Problem 13.25 on page 487.
TOMYIELD.DAT	Yield (in pounds) and amount of fertilizer (in pounds per 100 square feet) on 10 plots of land from Problem 17.4 on page 718. Problems 17.10, 17.16, 17.22, 17.31, 17.39, 17.50, 17.56, 17.62, and 17.72 also use the file.
TOMYLD2.DAT	Amount of fertilizer (in pounds per 100 square feet) and yield (in pounds) for 12 plots of land from Problem 18.42 on page 815. Problem 18.64 also uses the file.
TOOTHPST.DAT	Cost and cleaning test score for 39 brands of tubed toothpaste in Problem 3.71 on page 96. Problem 8.30 also uses the file.
TRAIN1.DAT	Time (in minutes) early or late with respect to scheduled arrival for 10 trains from Problem 4.11 on page 117. Problems 4.23, 4.32, and 4.40 also use the file.
TRAIN2.DAT	Time (in minutes) early or late with respect to scheduled arrival for 10 LIRR trains (=0) and 12 NJT trains (=1) from Problem 13.27 on page 487.
TRAINING.DAT	Assembly times (in seconds) for 16 employees receiving computer-assisted, individual-based training (=0) and 16 employees receiving team-based training (=1) from Problem 13.21 on page 480. Problems 13.32 and 13.44 also use the file.
TRW.DAT	Data showing year and earnings per share at TRW Inc. over the 23-year period 1970 through 1992 from Problem 19.2 on page 870. Problems 19.19, 19.30, and 19.38 also use the file.
TURNOVER.DAT	26 days of game turnovers from Problem 16.18 on page 692.
UPJOHN.DAT	Data showing year and net sales (in billions of dollars) at Upjohn Co. over the 23-year period 1970 through 1992 from Problem 19.5 on page 878. Problems 19.20, 19.31, and 19.39 also use the file.

Name	Description
UTILITY.DAT	Utility charges for 50 three-bedroom apartments from Problem 3.12 on page 66. Problems 3.19, 3.26, 3.34, 4.15, 4.27, 4.36, 4.44, 4.53, 8.28, 10.18, 10.25, and 10.35 also use the file.
VCREPAIR.DAT	Repair times (in minutes) of two VCRs based on three service centers (1, 2, 3) and three VCR brands (A=1, B=2, C=3) from Problem 14.45 on page 589. The variables are center, brand, and repair time.
VET.DAT	Data for 34 years on percentage of total federal outlays for veterans' benefits and services from Problem 12.31 on page 448.
WARECOST.DAT	Distribution cost (in thousands of dollars), sales (in thousands of dollars), and number of orders for 24 months from Problem 18.3 on page 787. Problems 18.7, 18.12, 18.16, 18.21, 18.26, 18.31, 18.36, 18.54 and 18.58 also use the file.
WAREHSE.DAT	Number of units handled per day over 30 days by each of five employees from Problem 16.21 on page 699.
WATER.DAT	Annual water consumption (in thousands of gallons) for 15 families from Problem 4.12 on page 117. Problems 4.24, 4.33, and 4.41 also the file.
WATERUSE.DAT	40 days of water use (1 unit = 748 gallons) from Problem 16.27 on page 706.
WEATHER.DAT	Temperatures recorded (in degrees Fahrenheit) for 22 cities, over 2 days, from Problem 13.67 on page 517.
WINE.DAT	Summated ratings by 12 experts of eight wines from Problem 14.54 on page 599. The variables are expert, wine, and rating.
WORKATT.DAT	Attitude scale scores for three trainees based on gender (males=1, females=2) and three room color (green=1, blue=2, red=3) combinations from Problems 14.52 on page 597. The variables are gender, room color, and attitude scale score.
WORKHRS.DAT	Lotsizes and worker-hours for 14 production runs from Problem 17.3 on page 717. Problems 17.9, 17.15, 17.21, 17.30, 17.38, 17.49, 17.55, 17.61, and 17.71 also use the file.
XEROX.DAT	Data showing year and annual net sales (in billions of dollars) at Xerox Corp. over the 23-year period 1970 through 1992 from Problem 19.58 on page 917.

Index

U

V

W

X

Y

Z